Incropera's Principles of Heat and Mass Transfer

GLOBAL EDITION

THEODORE L. BERGMAN

Department of Mechanical Engineering
University of Kansas

ADRIENNE S. LAVINE

Mechanical and Aerospace Engineering
Department
University of California, Los Angeles

ISBN: 978-1-119-38291-1

Printed and bound by CPI Group (UK) Ltd, Croydon, CR0 4YY

C9781119382911_180924

Preface

In his *Forward to Preface*, Frank Incropera shared with readers the timeline for the multi-edition transition of authorship from Incropera and DeWitt to Bergman and Lavine. Throughout the 15 years of our involvement with the text, we have been inspired by, and mindful of, Frank's insistence that the quality of the expository material be of paramount importance. We have also attempted to demonstrate the relevance of heat transfer by providing a multitude of examples, ranging from traditional and non-traditional energy generation to potential climate change, where heat transfer plays a vital role.

Since 2007, unexpected developments have evolved in engineering education. For example, the escalating cost of higher education is now debated at all levels of political leadership. As classroom instructors and parents of college students, this concern is not lost on us. In response, we have taken steps to hold the cost of the text in check by reducing its page count and forgoing production of a new edition of the companion text, *Introduction to Heat Transfer*. On the pedagogical front, we have reduced the complexity of many example and end-of-chapter problems. In addition to introducing new end-of-chapter problems, we have modified a significant number of existing problems, often necessitating modified solution approaches.

Incropera's Principles of Heat and Mass Transfer provides a rigorous and systematic problem-solving methodology, and a broad range of fundamental as well as applications-motivated end-of-chapter problems that require students to hone and exercise the concepts of heat and mass transfer. We continue to strive to provide a text that will serve as a valuable resource for students and practicing engineers throughout their careers.

Approach and Organization

We adhere to four broad learning objectives:

1. The student should internalize the meaning of the terminology and physical principles associated with the subject.
2. The student should be able to delineate pertinent transport phenomena for any process or system involving heat or mass transfer.

3. The student should be able to use requisite inputs for computing heat or mass transfer rates and/or material temperatures or concentrations.

4. The student should be able to develop representative models of real processes and systems and draw conclusions concerning process/system design or performance from the attendant analysis.

The key concepts are reviewed and questions to test student understanding of the concepts are posed at the end of each chapter.

It is recommended that problems involving complex models and/or parameter sensitivity considerations be addressed using the *Interactive Heat Transfer* (*IHT*) software package that has been developed and refined in conjunction with the text. With its intuitive user interface, extensive built-in thermophysical property database, embedded convection correlations taken from the text, and other useful features, students can master the basic usage of *IHT* in about one hour. To facilitate use of *IHT*, selected example problems in the expository material are identified with an "IHT" icon as shown to the left. These problems are included as demonstrations in the *IHT* software, allowing students to observe how these problems can be solved easily and quickly. More information regarding *IHT* is available later in this preface. Due to the preponderance of readily available software packages capable of solving multi-dimensional conduction problems, the finite-element software package previously made available to students has been discontinued.

Some homework problems require a computer-based solution. Other problems include both a hand calculation and an extension that is computer based. The latter approach is time-tested and promotes the habit of checking computer-generated solutions with hand calculations. Once validated, the computer solution can be utilized to conduct parametric calculations. Problems involving both hand- and computer-generated solutions are identified by enclosing the exploratory part in a red rectangle, as, for example, (b), (c), or (d). This feature also allows instructors to focus their assignments on problems amenable to solution using hand calculations, and benefit from the richness of these problems without assigning the computer-based parts. Problems with a boxed number (for example, 1.25) require an entirely computer-based solution.

What's New to the Global Edition

Although the size of the text has been reduced, we have added some new problems and revised few end-of-chapter problems, with an emphasis on problems amenable to analytical solutions. Many of the revised problems require modified solution approaches. Within the text, the treatment of thermodynamics has been improved, with clarification of the various forms of energy and their relation to heat transfer. New material on micro- and nanoscale heat transfer and thermal boundary resistances has been added. Mixed convection is presented in a more rigorous manner.

Classroom Coverage

The content of the text has evolved over many years in response to the development of new, fundamental concepts of heat (and mass) transfer and novel ways that the principles of heat transfer are applied. A broad range of engineering disciplines and institutions, with varying missions, make use of this text. Moreover, it is used not only in introductory courses, but also in advanced courses at many colleges and universities. Mindful of this diversity, the authors' intent is *not* to assemble a text whose content is to be covered, in entirety, during a single semester- or quarter-long course. Rather, the text includes fundamental material that should be covered in any introductory heat transfer course, and optional material that can be covered, depending on the mission of the institution, the time available, or the interests of the instructor or practitioner.

Heat and Mass Transfer To assist instructors in preparing a syllabus for a *first course in heat and mass transfer*, we suggest the following (with suggestions for a *first course in heat transfer* further below).

Chapter 1 Introduction sets the stage for any discussion of heat transfer. It explains the science-based linkage between thermodynamics and heat transfer, and the relevance of heat transfer. It should be covered in its entirety. Much of the content of *Chapter 2 Introduction to Conduction* is critical in a first course, especially Section 2.1 The Conduction Rate Equation, Section 2.3 The Heat Diffusion Equation, and Section 2.4 Boundary and Initial Conditions. Section 2.2 The Thermal Properties of Matter need not be covered in depth in a first course.

Chapter 3 One-Dimensional, Steady-State Conduction includes some material that can be assigned depending on the instructor's interest. The optional material includes Section 3.1.5 Porous Media, and Section 3.7 Other Applications of One-Dimensional Steady-State Conduction. The content of *Chapter 4 Two-Dimensional, Steady-State Conduction* is important in that both fundamental concepts and approximate techniques are presented. We recommend that all of Chapter 4 be covered, although some instructors may elect to not include Section 4.4 Finite-Difference Equations and Section 4.5 Solving the Finite-Difference Equations if time is short. It is recommended that *Chapter 5 Transient Conduction* be covered in entirety, although some instructors may prefer to cover only some aspects of Sections 5.8 through 5.10.

The content of *Chapter 6 Introduction to Convection* is often difficult for students to absorb. However, Chapter 6 introduces fundamental concepts in a rigorous manner and sets the stage for Chapters 7 through 11. Chapter 6 should be covered in entirety in an introductory heat and mass transfer course.

Chapter 7 External Flow builds on Chapter 6, introduces several important concepts, and presents convection correlations that students will utilize throughout the remainder of the text and in subsequent professional practice. We recommend Sections 7.1 through 7.5 be included in any first course in heat and mass transfer. However, Sections 7.6 through 7.8 are optional. Likewise, *Chapter 8, Internal Flow* includes matter used in the remainder of the text and in professional practice. However, Sections 8.6 through 8.8 may be viewed as optional in a first course.

Buoyancy-induced flow is covered in *Chapter 9 Free Convection*. Most of Chapter 9 should be covered in a first course, although optional material includes Section 9.7 Free Convection Within Parallel Plate Channels. The content of *Chapter 10 Boiling and Condensation* that can be optional in a first course includes Section 10.5 Forced Convection Boiling, Section 10.9 Film Condensation on Radial Systems, and Section 10.10 Condensation in Horizontal Tubes. However, if time is short, Chapter 10 can be skipped without affecting students' ability to understand the remainder of the text. We recommend that *Chapter 11 Heat Exchangers* be covered in entirety, although Section 11.6 Additional Considerations may be de-emphasized in a first course.

A distinguishing feature of the text, from its inception, is the in-depth coverage of radiation heat transfer in *Chapter 12 Radiation: Processes and Properties*. The content of the chapter is perhaps more relevant today than ever. However, Section 12.9 can be covered in an advanced course. *Chapter 13 Radiation Exchange Between Surfaces* may be covered as time permits, or in an intermediate heat transfer course.

The material in *Chapter 14 Diffusion Mass Transfer* is relevant to many contemporary applications ranging from chemical processing to biotechnology, and should be covered in entirety in an introductory heat and mass transfer course. However, if problems involving stationary media are solely of interest, Section 14.2 may be omitted or covered in a follow-on course.

Heat Transfer   Usage of this text for a *first course in heat transfer* might be structured as follows.

The suggested coverage of *Chapters 1 through 5* is identical to that for a course in heat and mass transfer described above. Before beginning *Chapter 6 Introduction to Convection*, it is recommended that the definition of mass transfer, provided in the introductory remarks of *Chapter 14 Diffusion Mass Transfer*, be reviewed with students. With the definition of mass transfer firmly in hand, remaining content that focuses on, for example, Fick's law, Sherwood and Schmidt numbers, and evaporative cooling will be apparent and need not be covered. For example, within Chapter 6, Section 6.1.3 The Concentration Boundary Layer, Section 6.2.2 Mass Transfer, Section 6.7.1 The Heat and Mass Transfer Analogy, and Section 6.7.2. Evaporative Cooling may be skipped in entirety.

Chapter 7 External Flow coverage is the same as recommended for the first course in heat and mass transfer, above. Components of Chapter 7 that can be skipped, such as Example 7.3, will be evident. Section 8.9 Convection Mass Transfer may be skipped in *Chapter 8 Internal Flow* while Section 9.10 Convection Mass Transfer in *Chapter 9 Free Convection* need not be covered.

The recommended coverage in *Chapters 10 through 13* is the same as for a first course in heat and mass transfer, above. Except for its introductory remarks, *Chapter 14 Diffusion Mass Transfer* is not included in a heat transfer course.

End-of-chapter problems involving mass transfer and/or evaporative cooling that should not be assigned in a heat transfer course are clustered toward the end of problem sets, and are identified with appropriate headings.

Acknowledgements

We wish to thank our many colleagues and their students who have offered valuable suggestions through the years. For this edition, we thank Laurent Pilon of the University of California, Los Angeles for his suggestions that have enhanced the presentation of transient conduction in Chapter 5. We also express our appreciation to three practicing engineers, Haifan Liang, Umesh Mather, and Hilbert Li, for their advice that has improved the coverage of thermoelectric power generation and extended surfaces in Chapter 3, and gaseous radiation in Chapter 13.

Appreciation is extended to Matthew Jones of Brigham Young University for improving the table of blackbody radiation functions of Chapter 12. Finally, we are grateful to John Abraham of the University of St. Thomas for his many helpful suggestions regarding the content of Chapter 7.

In closing, we remain deeply grateful to our spouses, Tricia and Greg, for the love they have shared and the patience they have practiced over the past 15 years.

Theodore L. Bergman (tlbergman@ku.edu)
Lawrence, Kansas

Adrienne S. Lavine (lavine@seas.ucla.edu)
Los Angeles, California

Supplemental and Web Site Material

The companion web site for the text is www.wiley.com/college/bergman. By clicking on appropriate links, students may access **Answers to Selected Homework Problems** and the **Supplemental Material Handouts** of the text. Supplemental Sections are identified throughout the text with the icon shown in the margin to the left.

Material available *for instructors only* includes that which is available to students and the **Solutions Manual**, the **Lecture PowerPoint Slides**, and an **Image Gallery** that includes electronic versions of figures from the text for those wishing to prepare their own materials for classroom presentation. *The* Instructor Solutions Manual *is copyrighted material for use only by instructors who require the text for their course.*[1]

Interactive Heat Transfer 4.0 is available at the companion web site at no cost for both students and instructors. As described by the authors in the *Approach and Organization* section, this simple-to-use software tool provides modeling and computational features useful in solving many problems in the text, and it enables rapid what-if and exploratory analysis of many types of problems.

---

[1]Excerpts from the Solutions Manual may be reproduced by instructors for distribution on a not-for-profit basis for testing or instructional purposes only to students enrolled in courses for which the textbook has been adopted. *Any other reproduction or translation of the contents of the Solutions Manual beyond that permitted by Sections 107 or 108 of the 1976 United States Copyright Act without permission of the copyright owner is unlawful.*

Contents

CHAPTER **6** *Introduction to Convection*   **343**

CHAPTER **9** *Free Convection*   547

CHAPTER **10** *Boiling and Condensation*   603

Symbols

A	area, m^2	Fo	Fourier number
A_b	area of prime (unfinned) surface, m^2	Fr	Froude number
A_c	cross-sectional area, m^2	f	friction factor; similarity variable
A_p	fin profile area, m^2	G	irradiation, W/m^2; mass velocity, $kg/s \cdot m^2$
A_r	nozzle area ratio	Gr	Grashof number
a	acceleration, m/s^2; speed of sound, m/s	Gz	Graetz number
Bi	Biot number	g	gravitational acceleration, m/s^2
Bo	Bond number	H	nozzle height, m; Henry's constant, bar
C	molar concentration, $kmol/m^3$; heat capacity rate, W/K	h	convection heat transfer coefficient, $W/m^2 \cdot K$; Planck's constant, $J \cdot s$
C_D	drag coefficient	h_{fg}	latent heat of vaporization, J/kg
C_f	friction coefficient	h'_{fg}	modified heat of vaporization, J/kg
C_t	thermal capacitance, J/K	h_{sf}	latent heat of fusion, J/kg
Co	Confinement number	h_m	convection mass transfer coefficient, m/s
c	specific heat, $J/kg \cdot K$; speed of light, m/s	h_{rad}	radiation heat transfer coefficient, $W/m^2 \cdot K$
c_p	specific heat at constant pressure, $J/kg \cdot K$	I	electric current, A; radiation intensity, $W/m^2 \cdot sr$
c_v	specific heat at constant volume, $J/kg \cdot K$	i	electric current density, A/m^2; enthalpy per unit mass, J/kg
D	diameter, m		
D_{AB}	binary mass diffusivity, m^2/s	J	radiosity, W/m^2
D_b	bubble diameter, m	Ja	Jakob number
D_h	hydraulic diameter, m	J_i^*	diffusive molar flux of species i relative to the mixture molar average velocity, $kmol/s \cdot m^2$
d	diameter of gas molecule, nm		
E	thermal plus mechanical energy, J; electric potential, V; emissive power, W/m^2	j_i	diffusive mass flux of species i relative to the mixture mass average velocity, $kg/s \cdot m^2$
E^{tot}	total energy, J	j_H	Colburn j factor for heat transfer
Ec	Eckert number	j_m	Colburn j factor for mass transfer
$\dot{E}_g$	rate of energy generation, W	k	thermal conductivity, $W/m \cdot K$
$\dot{E}_{in}$	rate of energy transfer into a control volume, W	k_B	Boltzmann's constant, J/K
$\dot{E}_{out}$	rate of energy transfer out of control volume, W	k_0	zero-order, homogeneous reaction rate constant, $kmol/s \cdot m^3$
$\dot{E}_{st}$	rate of increase of energy stored within a control volume, W	k_1	first-order, homogeneous reaction rate constant, s^{-1}
e	thermal internal energy per unit mass, J/kg; surface roughness, m	k''_1	first-order, surface reaction rate constant, m/s
		L	length, m
F	force, N; fraction of blackbody radiation in a wavelength band; view factor	Le	Lewis number

$\dot{M}_i$	rate of transfer of mass for species i, kg/s	r_o	cylinder or sphere radius, m
$\dot{M}_{i,g}$	rate of increase of mass of species i due to chemical reactions, kg/s	r, ϕ, z	cylindrical coordinates
		r, θ, ϕ	spherical coordinates
$\dot{M}_{in}$	rate at which mass enters a control volume, kg/s	S	solubility, kmol/m³ · atm; shape factor for two-dimensional conduction, m; nozzle
$\dot{M}_{out}$	rate at which mass leaves a control volume, kg/s		pitch, m; plate spacing, m; Seebeck coefficient, V/K
$\dot{M}_{st}$	rate of increase of mass stored within a control volume, kg/s	S_c	solar constant, W/m²
$\mathcal{M}_i$	molecular weight of species i, kg/kmol	S_D, S_L, S_T	diagonal, longitudinal, and transverse pitch of a tube bank, m
Ma	Mach number	Sc	Schmidt number
m	mass, kg	Sh	Sherwood number
$\dot{m}$	mass flow rate, kg/s	St	Stanton number
m_i	mass fraction of species i, ρ_i/ρ	T	temperature, K
N	integer number	t	time, s
N_L, N_T	number of tubes in longitudinal and transverse directions	U	overall heat transfer coefficient, W/m² · K; internal energy, J
Nu	Nusselt number	u, v, w	mass average fluid velocity components, m/s
NTU	number of transfer units	$u*, v*, w*$	molar average velocity components, m/s
N_i	molar transfer rate of species i relative to fixed coordinates, kmol/s	V	volume, m³; fluid velocity, m/s
N_i''	molar flux of species i relative to fixed coordinates, kmol/s · m²	v	specific volume, m³/kg
		W	width of a slot nozzle, m
$\dot{N}_i$	molar rate of increase of species i per unit volume due to chemical reactions, kmol/s · m³	$\dot{W}$	rate at which work is performed, W
		We	Weber number
		X	vapor quality
$\dot{N}_i''$	surface reaction rate of species i, kmol/s · m²	X_{tt}	Martinelli parameter
		X, Y, Z	components of the body force per unit volume, N/m³
$\mathcal{N}$	Avogadro's number		
n_i''	mass flux of species i relative to fixed coordinates, kg/s · m²	x, y, z	rectangular coordinates, m
		x_c	critical location for transition to turbulence, m
$\dot{n}_i$	mass rate of increase of species i per unit volume due to chemical reactions, kg/s · m³	$x_{fd,c}$	concentration entry length, m
		$x_{fd,h}$	hydrodynamic entry length, m
		$x_{fd,t}$	thermal entry length, m
P	power, W; perimeter, m	x_i	mole fraction of species i, C_i/C
P_L, P_T	dimensionless longitudinal and transverse pitch of a tube bank	Z	thermoelectric material property, K⁻¹
Pe	Peclet number		
Pr	Prandtl number	Greek Letters	
p	pressure, N/m²	α	thermal diffusivity, m²/s; accommodation coefficient; absorptivity
Q	energy transfer, J		
q	heat transfer rate, W	β	volumetric thermal expansion coefficient, K⁻¹
$\dot{q}$	rate of energy generation per unit volume, W/m³	Γ	mass flow rate per unit width in film condensation, kg/s · m
q'	heat transfer rate per unit length, W/m	γ	ratio of specific heats
q''	heat flux, W/m²	δ	hydrodynamic boundary layer thickness, m
$q*$	dimensionless conduction heat rate	δ_c	concentration boundary layer thickness, m
R	cylinder radius, m; gas constant, J/kg · K	δ_p	thermal penetration depth, m
$\mathcal{R}$	universal gas constant, J/kmol · K	δ_t	thermal boundary layer thickness, m
Ra	Rayleigh number	ε	emissivity; porosity; heat exchanger effectiveness
Re	Reynolds number		
R_e	electric resistance, Ω	ε_f	fin effectiveness
R_f	fouling factor, m² · K/W	η	thermodynamic efficiency; similarity variable
R_m	mass transfer resistance, s/m³	η_f	fin efficiency
$R_{m,n}$	residual for the m, n nodal point	η_o	overall efficiency of fin array
R_t	thermal resistance, K/W	θ	zenith angle, rad; temperature difference, K
$R_{t,c}$	thermal contact resistance, K/W	κ	absorption coefficient, m⁻¹
$R_{t,f}$	fin thermal resistance, K/W	λ	wavelength, μm
$R_{t,o}$	thermal resistance of fin array, K/W	λ_{mfp}	mean free path length, nm
		μ	viscosity, kg/s · m

ν	kinematic viscosity, m²/s; frequency of radiation, s^{-1}	L	based on characteristic length
ρ	mass density, kg/m³; reflectivity	l	saturated liquid conditions
ρ_e	electric resistivity, Ω/m	lat	latent energy
σ	Stefan–Boltzmann constant, W/m² · K^4; electrical conductivity, $1/\Omega \cdot$ m; normal viscous stress, N/m²; surface tension, N/m	lm	log mean condition
		m	mean value over a tube cross section
		max	maximum
Φ	viscous dissipation function, s^{-2}	N	natural (free) convection
φ	volume fraction	o	center or midplane condition; tube outlet condition; outer
ϕ	azimuthal angle, rad		
ψ	stream function, m²/s	p	momentum
τ	shear stress, N/m²; transmissivity	ph	phonon
ω	solid angle, sr; perfusion rate, s^{-1}	R	reradiating surface
		$r,$ ref	reflected radiation
		rad	radiation

Subscripts

A, B	species in a binary mixture	S	solar conditions
abs	absorbed	s	surface conditions; solid properties; saturated solid conditions
am	arithmetic mean		
atm	atmospheric	sat	saturated conditions
b	base of an extended surface; blackbody	sens	sensible energy
C	Carnot	sky	sky conditions
c	cross-sectional; concentration; cold; critical	ss	steady state
		sur	surroundings
cr	critical insulation thickness	t	thermal
cond	conduction	tr	transmitted
conv	convection	υ	saturated vapor conditions
CF	counterflow	x	local conditions on a surface
D	diameter; drag	λ	spectral
dif	diffusion	∞	free stream conditions
e	excess; emission; electron		
evap	evaporation		
f	fluid properties; fin conditions; saturated liquid conditions; film		
F	forced convection		
fd	fully developed conditions		
g	saturated vapor conditions		
h	hydrodynamic; hot; helical		
i	general species designation; inner surface; initial condition; tube inlet condition; incident radiation		

Superscripts

*	molar average; dimensionless quantity
tot	total energy (all forms)

Overbar

‾	average conditions; time mean

Introduction

1

From the study of thermodynamics, you have learned that energy can be transferred by interactions of a system with its surroundings. These interactions are called work and heat. However, thermodynamics deals with the end states of the process during which an interaction occurs and provides no information concerning the nature of the interaction or the time rate at which it occurs. The objective of this text is to extend thermodynamic analysis through the study of the *modes* of heat transfer and through the development of relations to calculate heat transfer *rates*.

In this chapter we lay the foundation for much of the material treated in the text. We do so by raising several questions: *What is heat transfer? How is heat transferred? Why is it important?* One objective is to develop an appreciation for the fundamental concepts and principles that underlie heat transfer processes. A second objective is to illustrate the manner in which a knowledge of heat transfer may be used with the first law of thermodynamics (*conservation of energy*) to solve problems relevant to technology and society.

1.1   *What and How?*

A simple, yet general, definition provides sufficient response to the question: What is heat transfer?

> *Heat transfer (or heat) is thermal energy in transit due to a spatial temperature difference.*

Whenever a temperature difference exists in a medium or between media, heat transfer must occur.

As shown in Figure 1.1, we refer to different types of heat transfer processes as *modes*. When a temperature gradient exists in a stationary medium, which may be a solid or a fluid, we use the term *conduction* to refer to the heat transfer that will occur across the medium. In contrast, the term *convection* refers to heat transfer that will occur between a surface and a moving fluid when they are at different temperatures. The third mode of heat transfer is termed *thermal radiation*. All surfaces of finite temperature emit energy in the form of electromagnetic waves. Hence, in the absence of an intervening medium, there is net heat transfer by radiation between two surfaces at different temperatures.

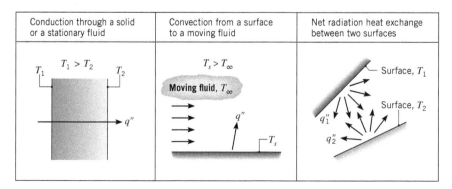

FIGURE 1.1   Conduction, convection, and radiation heat transfer modes.

1.2 *Physical Origins and Rate Equations*

As engineers, it is important that we understand the *physical mechanisms* which underlie the heat transfer modes and that we be able to use the rate equations that quantify the amount of energy being transferred per unit time.

1.2.1 Conduction

At mention of the word *conduction*, we should immediately conjure up concepts of *atomic* and *molecular activity* because processes at these levels sustain this mode of heat transfer. Conduction may be viewed as the transfer of energy from the more energetic to the less energetic particles of a substance due to interactions between the particles.

The physical mechanism of conduction is most easily explained by considering a gas and using ideas familiar from your thermodynamics background. Consider a gas in which a temperature gradient exists, and assume that there is *no bulk, or macroscopic, motion.* The gas may occupy the space between two surfaces that are maintained at different temperatures, as shown in Figure 1.2. We associate the temperature at any point with the energy of gas molecules in proximity to the point. This energy is related to the random translational motion, as well as to the internal rotational and vibrational motions, of the molecules.

Higher temperatures are associated with higher molecular energies. When neighboring molecules collide, as they are constantly doing, a transfer of energy from the more energetic to the less energetic molecules must occur. In the presence of a temperature gradient, energy transfer by conduction must then occur in the direction of decreasing temperature. This would be true even in the absence of collisions, as is evident from Figure 1.2. The hypothetical plane at x_o is constantly being crossed by molecules from above and below due to their *random* motion. However, molecules from above are associated with a higher temperature than those from below, in which case there must be a *net* transfer of energy in the positive x-direction. Collisions between molecules enhance this energy transfer. We may speak of the net transfer of energy by random molecular motion as a *diffusion* of energy.

The situation is much the same in liquids, although the molecules are more closely spaced and the molecular interactions are stronger and more frequent. Similarly, in a solid, conduction may be attributed to atomic activity in the form of lattice vibrations. The modern

FIGURE 1.2   Association of conduction heat transfer with diffusion of energy due to molecular activity.

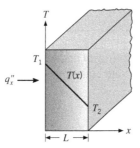

FIGURE 1.3   One-dimensional heat transfer by conduction (diffusion of energy).

view is to ascribe the energy transfer to *lattice waves* induced by atomic motion. In an electrical nonconductor, the energy transfer is exclusively via these lattice waves; in a conductor, it is also due to the translational motion of the free electrons. We treat the important properties associated with conduction phenomena in Chapter 2 and in Appendix A.

Examples of conduction heat transfer are legion. The exposed end of a metal spoon suddenly immersed in a cup of hot coffee is eventually warmed due to the conduction of energy through the spoon. On a winter day, there is significant energy loss from a heated room to the outside air. This loss is principally due to conduction heat transfer through the wall that separates the room air from the outside air.

Heat transfer processes can be quantified in terms of appropriate *rate equations*. These equations may be used to compute the amount of energy being transferred per unit time. For heat conduction, the rate equation is known as *Fourier's law*. For the one-dimensional plane wall shown in Figure 1.3, having a temperature distribution $T(x)$, the rate equation is expressed as

$$q_x'' = -k\frac{dT}{dx} \tag{1.1}$$

The *heat flux* $q_x''(\text{W/m}^2)$ is the heat transfer rate in the x-direction *per* unit area *perpendicular* to the direction of transfer, and it is proportional to the *temperature gradient, dT/dx*, in this direction. The parameter k is a *transport* property known as the *thermal conductivity* ($\text{W/m} \cdot \text{K}$) and is a characteristic of the wall material. The minus sign is a consequence of the fact that heat is transferred in the direction of decreasing temperature. Under the *steady-state conditions* shown in Figure 1.3, where the temperature distribution is *linear*, the temperature gradient may be expressed as

$$\frac{dT}{dx} = \frac{T_2 - T_1}{L}$$

and the heat flux is then

$$q_x'' = -k\frac{T_2 - T_1}{L}$$

or

$$q_x'' = k\frac{T_1 - T_2}{L} = k\frac{\Delta T}{L} \tag{1.2}$$

Note that this equation provides a *heat flux*, that is, the rate of heat transfer per *unit area*. The *heat rate* by conduction, $q_x(\text{W})$, through a plane wall of area A is then the product of the flux and the area, $q_x = q_x'' \cdot A$.

| EXAMPLE 1.1 | IHT * |

The wall of an industrial furnace is constructed from 0.15-m-thick fireclay brick having a thermal conductivity of 1.7 W/m · K. Measurements made during steady-state operation reveal temperatures of 1400 and 1150 K at the inner and outer surfaces, respectively. What is the rate of heat loss through a wall that is 0.5 m × 1.2 m on a side?

SOLUTION

Known: Steady-state conditions with prescribed wall thickness, area, thermal conductivity, and surface temperatures.

Find: Rate of heat loss through wall.

Schematic:

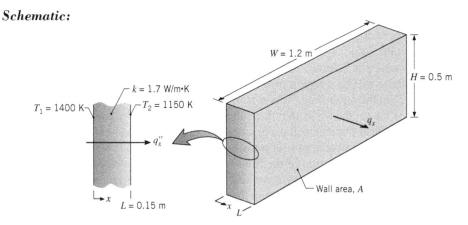

Assumptions:

1. Steady-state conditions.
2. One-dimensional conduction through the wall.
3. Constant thermal conductivity.

Analysis: Since heat transfer through the wall is by conduction, the heat flux may be determined from Fourier's law. Using Equation 1.2, we have

$$q''_x = k\frac{\Delta T}{L} = 1.7 \text{ W/m} \cdot \text{K} \times \frac{250 \text{ K}}{0.15 \text{ m}} = 2833 \text{ W/m}^2$$

The heat flux represents the rate of heat transfer through a section of unit area, and it is uniform (invariant) across the surface of the wall. The rate of heat loss through the wall of area $A = H \times W$ is then

$$q_x = (HW)q''_x = (0.5 \text{ m} \times 1.2 \text{ m})\,2833 \text{ W/m}^2 = 1700 \text{ W} \qquad \triangleleft$$

Comments: Note the direction of heat flow and the distinction between heat flux and heat rate.

*This icon identifies examples that are available in tutorial form in the *Interactive Heat Transfer (IHT)* software that accompanies the text. Each tutorial is brief and illustrates a basic function of the software. *IHT* can be used to solve simultaneous equations, perform parameter sensitivity studies, and graph the results. Use of *IHT* will reduce the time spent solving more complex end-of-chapter problems.

1.2.2 Convection

The convection heat transfer *mode* is comprised of *two mechanisms.* In addition to energy transfer due to *random molecular motion (diffusion)*, energy is also transferred by the *bulk,* or *macroscopic, motion* of the fluid. This fluid motion is associated with the fact that, at any instant, large numbers of molecules are moving collectively or as aggregates. Such motion, in the presence of a temperature gradient, contributes to heat transfer. Because the molecules in the aggregate retain their random motion, the total heat transfer is then due to a superposition of energy transport by the random motion of the molecules and by the bulk motion of the fluid. The term *convection* is customarily used when referring to this cumulative transport, and the term *advection* refers to transport due to bulk fluid motion alone.

We are especially interested in convection heat transfer between a fluid in motion and a bounding surface when the two are at different temperatures. Consider fluid flow over the hot surface of Figure 1.4. A consequence of the fluid–surface interaction is the development of a region in the fluid through which the velocity varies from zero at the surface to a finite value u_∞ associated with the flow. This region of the fluid is known as the *hydrodynamic*, or *velocity, boundary layer.* Moreover, if the surface and flow temperatures differ, there will be a region of the fluid through which the temperature varies from T_s at $y = 0$ to T_∞ in the outer flow. This region, called the *thermal boundary layer*, may be smaller, larger, or the same size as that through which the velocity varies. In any case, if $T_s > T_\infty$, convection heat transfer will occur from the surface to the outer flow.

The convection heat transfer mode is sustained both by random molecular motion and by the bulk motion of the fluid within the boundary layer. The contribution due to random molecular motion (diffusion) dominates near the surface where the fluid velocity is low. In fact, at the interface between the surface and the fluid ($y = 0$), the fluid velocity is zero, and heat is transferred by this mechanism only. The contribution due to bulk fluid motion originates from the fact that the boundary layer *grows* as the flow progresses in the x-direction. In effect, the heat that is conducted into this layer is swept downstream and is eventually transferred to the fluid outside the boundary layer. Appreciation of boundary layer phenomena is essential to understanding convection heat transfer. For this reason, the discipline of fluid mechanics will play a vital role in our later analysis of convection.

Convection heat transfer may be classified according to the nature of the flow. We speak of *forced convection* when the flow is caused by external means, such as by a fan, a pump, or atmospheric winds. As an example, consider the use of a fan to provide forced convection air cooling of hot electrical components on a stack of printed circuit boards (Figure 1.5*a*). In contrast, for *free* (or *natural) convection*, the flow is induced by buoyancy forces, which are due to density differences caused by temperature variations in the fluid. An example is the free convection heat transfer that occurs from hot components on

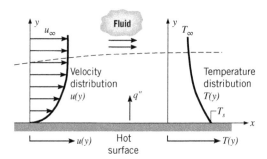

FIGURE 1.4 Boundary layer development in convection heat transfer.

a vertical array of circuit boards in air (Figure 1.5*b*). Air that makes contact with the components experiences an increase in temperature and hence a reduction in density. Since it is now lighter than the surrounding air, buoyancy forces induce a vertical motion for which warm air ascending from the boards is replaced by an inflow of cooler ambient air.

While we have presumed *pure* forced convection in Figure 1.5*a* and *pure* natural convection in Figure 1.5*b*, conditions corresponding to *mixed (combined) forced* and *natural convection* may exist. For example, if velocities associated with the flow of Figure 1.5*a* are small and/or buoyancy forces are large, a secondary flow that is comparable to the imposed forced flow could be induced. In this case, the buoyancy-induced flow would be normal to the forced flow and could have a significant effect on convection heat transfer from the components. In Figure 1.5*b*, mixed convection would result if a fan were used to force air upward between the circuit boards, thereby assisting the buoyancy flow, or downward, thereby opposing the buoyancy flow.

We have described the convection heat transfer mode as energy transfer occurring within a fluid due to the combined effects of conduction and bulk fluid motion. Typically, the energy that is being transferred is the *sensible*, or internal thermal, energy of the fluid. However, for some convection processes, there is, in addition, *latent* heat exchange. This latent heat exchange is generally associated with a phase change between the liquid and vapor states of the fluid. Two special cases of interest in this text are *boiling* and *condensation*. For example, convection heat transfer results from fluid motion induced by vapor bubbles generated at the bottom of a pan of boiling water (Figure 1.5*c*) or by the condensation of water vapor on the outer surface of a cold water pipe (Figure 1.5*d*).

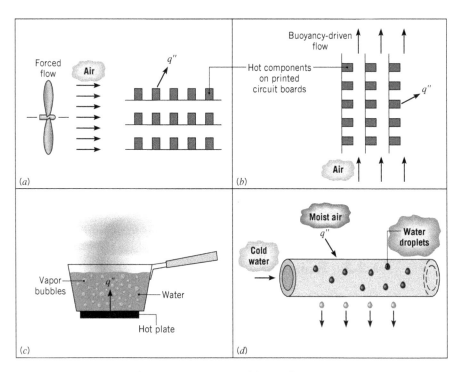

FIGURE 1.5 Convection heat transfer processes. (a) Forced convection. (b) Natural convection. (c) Boiling. (d) Condensation.

TABLE 1.1   Typical values of the convection heat transfer coefficient

Process	h (W/m^2 · K)
Free convection	
Gases	2–25
Liquids	50–1000
Forced convection	
Gases	25–250
Liquids	100–20,000
Convection with phase change	
Boiling or condensation	2500–100,000

Regardless of the nature of the convection heat transfer process, the appropriate rate equation is of the form

$$q'' = h(T_s - T_\infty) \tag{1.3a}$$

where q'', the convective *heat flux* (W/m^2), is proportional to the difference between the surface and fluid temperatures, T_s and T_∞, respectively. This expression is known as *Newton's law of cooling*, and the parameter h (W/m^2 · K) is termed the *convection heat transfer coefficient*. This coefficient depends on conditions in the boundary layer, which are influenced by surface geometry, the nature of the fluid motion, and an assortment of fluid thermodynamic and transport properties.

Any study of convection ultimately reduces to a study of the means by which h may be determined. Although consideration of these means is deferred to Chapter 6, convection heat transfer will frequently appear as a boundary condition in the solution of conduction problems (Chapters 2 through 5). In the solution of such problems we presume h to be known, using typical values given in Table 1.1.

When Equation 1.3a is used, the convection heat flux is presumed to be *positive* if heat is transferred *from* the surface ($T_s > T_\infty$) and *negative* if heat is transferred *to* the surface ($T_\infty > T_s$). However, nothing precludes us from expressing Newton's law of cooling as

$$q'' = h(T_\infty - T_s) \tag{1.3b}$$

in which case heat transfer is positive if it is to the surface.

1.2.3   Radiation

Thermal radiation is energy *emitted* by matter that is at a nonzero temperature. Although we will focus on radiation from solid surfaces, emission may also occur from liquids and gases. Regardless of the form of matter, the emission may be attributed to changes in the electron configurations of the constituent atoms or molecules. The energy of the radiation field is transported by electromagnetic waves (or alternatively, photons). While the transfer of energy by conduction or convection requires the presence of a material medium, radiation does not. In fact, radiation transfer occurs most efficiently in a vacuum.

Consider radiation transfer processes for the surface of Figure 1.6*a*. Radiation that is *emitted* by the surface originates from the thermal energy of matter bounded by the surface, and the rate at which energy is released per unit area (W/m^2) is termed the surface *emissive power, E*. There is an upper limit to the emissive power, which is prescribed by the *Stefan–Boltzmann law*

$$E_b = \sigma T_s^4 \tag{1.4}$$

where T_s is the *absolute temperature* (K) of the surface and σ is the *Stefan–Boltzmann constant* ($\sigma = 5.67 \times 10^{-8}$ W/m$^2 \cdot$ K^4). Such a surface is called an ideal radiator or *blackbody*.

The heat flux emitted by a real surface is less than that of a blackbody at the same temperature and is given by

$$E = \varepsilon \sigma T_s^4 \tag{1.5}$$

where ε is a radiative property of the surface termed the *emissivity*. With values in the range $0 \leq \varepsilon \leq 1$, this property provides a measure of how efficiently a surface emits energy relative to a blackbody. It depends strongly on the surface material and finish, and representative values are provided in Appendix A.

Radiation may also be *incident* on a surface from its surroundings. The radiation may originate from a special source, such as the sun, or from other surfaces to which the surface of interest is exposed. Irrespective of the source(s), we designate the rate at which all such radiation is incident on a unit area of the surface as the *irradiation G* (Figure 1.6*a*).

A portion, or all, of the irradiation may be *absorbed* by the surface, thereby increasing the thermal energy of the material. The rate at which radiant energy is absorbed per unit surface area may be evaluated from knowledge of a surface radiative property termed the *absorptivity α*. That is,

$$G_{abs} = \alpha G \tag{1.6}$$

where $0 \leq \alpha \leq 1$. If $\alpha < 1$ and the surface is *opaque*, portions of the irradiation are *reflected*. If the surface is *semitransparent*, portions of the irradiation may also be *transmitted*. However, whereas absorbed and emitted radiation increase and reduce, respectively, the thermal energy of matter, reflected and transmitted radiation have no effect on this

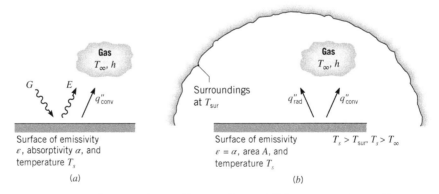

FIGURE 1.6   Radiation exchange: (a) at a surface and (b) between a surface and large surroundings.

energy. Note that the value of α depends on the nature of the irradiation, as well as on the surface itself. For example, the absorptivity of a surface to solar radiation may differ from its absorptivity to radiation emitted by the walls of a furnace.

In many engineering problems (a notable exception being problems involving solar radiation or radiation from other very high temperature sources), liquids can be considered opaque to radiation heat transfer, and gases can be considered transparent to it. Solids can be opaque (as is the case for metals) or *semitransparent* (as is the case for thin sheets of some polymers and some semiconducting materials).

A special case that occurs frequently involves radiation exchange between a small surface at T_s and a much larger, isothermal surface that completely surrounds the smaller one (Figure 1.6*b*). The *surroundings* could, for example, be the walls of a room or a furnace whose temperature T_{sur} differs from that of an enclosed surface ($T_{\text{sur}} \neq T_s$). We will show in Chapter 12 that, for such a condition, the irradiation may be approximated by emission from a blackbody at T_{sur}, in which case $G = \sigma T_{\text{sur}}^4$. If the surface is assumed to be one for which $\alpha = \varepsilon$ (a *gray surface*), the *net* rate of radiation heat transfer *from* the surface, expressed per unit area of the surface, is

$$q''_{\text{rad}} = \frac{q}{A} = \varepsilon E_b(T_s) - \alpha G = \varepsilon \sigma (T_s^4 - T_{\text{sur}}^4) \qquad (1.7)$$

This expression provides the difference between thermal energy that is released due to radiation emission and that gained due to radiation absorption.

For many applications, it is convenient to express the net radiation heat exchange in the form

$$q_{\text{rad}} = h_r A (T_s - T_{\text{sur}}) \qquad (1.8)$$

where, from Equation 1.7, the *radiation heat transfer coefficient* h_r is

$$h_r \equiv \varepsilon \sigma (T_s + T_{\text{sur}})(T_s^2 + T_{\text{sur}}^2) \qquad (1.9)$$

Here we have modeled the radiation mode in a manner similar to convection. In this sense we have *linearized* the radiation rate equation, making the heat rate proportional to a temperature difference rather than to the difference between two temperatures to the fourth power. Note, however, that h_r depends strongly on temperature, whereas the temperature dependence of the convection heat transfer coefficient h is generally weak.

The surfaces of Figure 1.6 may also simultaneously transfer heat by convection to an adjoining gas. For the conditions of Figure 1.6*b*, the total rate of heat transfer *from* the surface is then

$$q = q_{\text{conv}} + q_{\text{rad}} = hA(T_s - T_\infty) + \varepsilon A \sigma (T_s^4 - T_{\text{sur}}^4) \qquad (1.10)$$

EXAMPLE 1.2

An uninsulated steam pipe passes through a room in which the air and walls are at 25°C. The outside diameter of the pipe is 70 mm, and its surface temperature and emissivity are 200°C and 0.8, respectively. What are the surface emissive power and irradiation? If the coefficient associated with free convection heat transfer from the surface to the air is 15 W/m² · K, what is the rate of heat loss from the surface per unit length of pipe?

SOLUTION

Known: Uninsulated pipe of prescribed diameter, emissivity, and surface temperature in a room with fixed wall and air temperatures.

Find:

1. Surface emissive power and irradiation.
2. Rate of heat loss from pipe per unit length, q'.

Schematic:

Assumptions:

1. Steady-state conditions.
2. Radiation exchange between the pipe and the room is between a small surface and a much larger enclosure.
3. The surface emissivity and absorptivity are equal.

Analysis:

1. The surface emissive power may be evaluated from Equation 1.5, while the irradiation corresponds to $G = \sigma T_{sur}^4$. Hence

$$E = \varepsilon \sigma T_s^4 = 0.8(5.67 \times 10^{-8} \text{ W/m}^2 \cdot \text{K}^4)(473 \text{ K})^4 = 2270 \text{ W/m}^2 \qquad \triangleleft$$

$$G = \sigma T_{sur}^4 = 5.67 \times 10^{-8} \text{ W/m}^2 \cdot \text{K}^4(298 \text{ K})^4 = 447 \text{ W/m}^2 \qquad \triangleleft$$

2. Heat loss from the pipe is by convection to the room air and by radiation exchange with the walls. Hence, $q = q_{conv} + q_{rad}$ and from Equation 1.10, with $A = \pi DL$,

$$q = h(\pi DL)(T_s - T_\infty) + \varepsilon(\pi DL)\sigma(T_s^4 - T_{sur}^4)$$

The rate of heat loss per unit length of pipe is then

$$q' = \frac{q}{L} = 15 \text{ W/m}^2 \cdot \text{K} (\pi \times 0.07 \text{ m})(200 - 25)°\text{C}$$

$$+ 0.8(\pi \times 0.07 \text{ m}) \, 5.67 \times 10^{-8} \text{ W/m}^2 \cdot \text{K}^4(473^4 - 298^4) \text{K}^4$$

$$q' = 577 \text{ W/m} + 421 \text{ W/m} = 998 \text{ W/m} \qquad \triangleleft$$

Comments:

1. Note that temperature may be expressed in units of °C or K when evaluating the temperature difference for a convection (or conduction) heat transfer rate. However, temperature must be expressed in kelvins (K) when evaluating a radiation transfer rate.

2. The net rate of radiation heat transfer from the pipe may be expressed as

$$q'_{\text{rad}} = \pi D(E - \alpha G)$$

$$q'_{\text{rad}} = \pi \times 0.07 \text{ m} (2270 - 0.8 \times 447) \text{ W/m}^2 = 421 \text{ W/m}$$

3. In this situation, the radiation and convection heat transfer rates are comparable because T_s is large compared to T_{sur} and the coefficient associated with free convection is small. For more moderate values of T_s and the larger values of h associated with forced convection, the effect of radiation may often be neglected. The radiation heat transfer coefficient may be computed from Equation 1.9. For the conditions of this problem, its value is $h_r = 11 \text{ W/m}^2 \cdot \text{K}$.

1.2.4   The Thermal Resistance Concept

The three modes of heat transfer were introduced in the preceding sections. As is evident from Equations 1.2, 1.3, and 1.8, the heat transfer rate can be expressed in the form

$$q = q''A = \frac{\Delta T}{R_t} \tag{1.11}$$

where ΔT is a relevant temperature difference and A is the area normal to the direction of heat transfer. The quantity R_t is called a *thermal resistance* and takes different forms for the three different modes of heat transfer. For example, Equation 1.2 may be multiplied by the area A and rewritten as $q_x = \Delta T/R_{t,\text{cond}}$, where $R_{t,\text{cond}} = L/kA$ is a thermal resistance associated with conduction, having the units K/W. The thermal resistance concept will be considered in detail in Chapter 3 and will be seen to have great utility in solving complex heat transfer problems.

1.3   *Relationship to Thermodynamics*

If you have taken a thermodynamics course, you are aware that heat exchange plays a vital role in the first and second laws of thermodynamics because it is a primary mechanism for energy transfer between a system and its surroundings. While thermodynamics may be used to determine the *amount* of energy required in the form of heat for a system to pass from one state to another, it considers neither the mechanisms that provide for heat exchange nor the methods that exist for computing the *rate* of heat exchange. The discipline of heat transfer specifically seeks to quantify the rate at which heat is exchanged through the rate equations expressed, for example, by Equations 1.2, 1.3, and 1.7. Indeed, heat transfer principles enable the engineer to implement the concepts of thermodynamics. For example, the size of a power plant to be constructed cannot be determined from thermodynamics alone; the principles of heat transfer must also be invoked at the design stage.

This section considers the relationship of heat transfer to thermodynamics. Since the *first law* of thermodynamics (the *law of conservation of energy*) provides a useful, often essential, starting point for the solution of heat transfer problems, Section 1.3.1 will provide a development of the general formulations of the first law. The ideal (Carnot) efficiency of a *heat engine*, as determined by the *second law* of thermodynamics, will be reviewed in Section 1.3.2. It will be shown that a realistic description of the heat transfer between a heat engine and its surroundings limits the actual efficiency of a heat engine.

1.3.1   Relationship to the First Law of Thermodynamics (Conservation of Energy)

The first law of thermodynamics states that the total energy of a system is conserved, where *total energy* consists of mechanical energy (which is composed of kinetic and potential energy) and internal energy, as shown schematically in Figure 1.7. Internal energy can be subdivided into thermal energy (which will be defined more carefully later) and other forms of internal energy, such as chemical and nuclear energy. Since total energy is conserved, the only way that the amount of energy in a system can change is if energy crosses its boundaries. For a *closed system* (a region of fixed mass), there are only two ways energy can cross the system boundaries: heat transfer through the boundaries and work done on or by the system. This leads to the following statement of the first law for a closed system, which is familiar if you have taken a course in thermodynamics:

$$\Delta E_{\text{st}}^{\text{tot}} = Q - W \qquad (1.12a)$$

where $\Delta E_{\text{st}}^{\text{tot}}$ is the change in the total energy stored in the system, Q is the *net* heat transferred to the system, and W is the *net* work done by the system. This is schematically illustrated in Figure 1.8*a*.

The first law can also be applied to a *control volume* (or *open system*), a region of space bounded by a *control surface* through which mass may flow. Mass entering and leaving the control volume carries energy with it; this process, termed *energy advection*, adds a third way in which energy can cross the boundaries of a control volume. To summarize, the

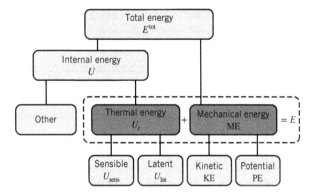

FIGURE 1.7   The components of total energy. The sum of thermal and mechanical energy, E, is of interest in the field of heat transfer.

FIGURE 1.8 Conservation of: (a) total energy for a closed system over a time interval and (b) thermal and mechanical energy for a control volume at an instant.

first law of thermodynamics can be simply stated as follows for both a control volume and a closed system:

1. Conservation of Total Energy: First Law of Thermodynamics over a Time Interval (Δt)

> *The increase in the amount of energy stored in a control volume must equal the amount of energy that enters the control volume, minus the amount of energy that leaves the control volume.*

As engineers, we often focus our attention on the thermal and mechanical forms of energy. We must recognize that the sum of thermal and mechanical energy is *not* conserved, because conversion can occur between other forms of energy and thermal or mechanical energy. For example, during combustion the amount of chemical energy in the system will decrease and the amount of thermal energy in the system will increase. If an electric motor operates within the system, it will cause conversion from electrical to mechanical energy. We can think of such energy conversions as resulting in *thermal or mechanical energy generation* (which can be either positive or negative). So a statement of the first law that is well suited for heat transfer analysis is:

2. Conservation of Thermal and Mechanical Energy over a Time Interval (Δt)

> *The increase in the amount of thermal and mechanical energy stored in a control volume must equal the amount of thermal and mechanical energy that enters the control volume, minus the amount of thermal and mechanical energy that leaves the control volume, plus the amount of thermal and mechanical energy that is generated within the control volume.*

This expression applies over a time interval Δt, and all the energy terms are measured in joules. Since the first law must be satisfied at every *instant* of time t, we can also formulate the law on a *rate basis*. That is, at any instant, there must be a balance between all *energy rates*, as measured in joules per second (W). In words, this is expressed as:

3. Conservation of Thermal and Mechanical Energy at an Instant (t)

> *The <u>rate</u> of increase of thermal and mechanical energy stored in a control volume must equal the <u>rate</u> at which thermal and mechanical energy enters the control volume, minus the <u>rate</u> at which thermal and mechanical energy leaves the control volume, plus the <u>rate</u> at which thermal and mechanical energy is generated within the control volume.*

If the inflow and generation of thermal and mechanical energy exceed the outflow, the amount of thermal and mechanical energy stored (accumulated) in the control volume must increase. If the outflow of thermal and mechanical energy exceeds the generation and inflow, the amount of stored thermal and mechanical energy in the control volume must decrease. If the inflow and generation equal the outflow, a *steady-state* condition must prevail such that there will be no change in the amount of thermal and mechanical energy stored in the control volume.

We will now rewrite the two boxed conservation of thermal and mechanical energy statements as equations. To do so, we let E stand for the sum of thermal and mechanical energy (in contrast to the symbol E^{tot} for total energy and as shown in Figure 1.7). The change in thermal and mechanical energy stored over the time interval Δt is then ΔE_{st}. The subscripts *in* and *out* refer to energy entering and leaving the control volume. Finally, thermal and mechanical energy generation is given the symbol E_g. Thus, boxed conservation statement 2 can be written as:

$$\Delta E_{st} = E_{in} - E_{out} + E_g \tag{1.12b}$$

Next, using a dot over a term to indicate a rate, boxed conservation statement 3 becomes:

$$\dot{E}_{st} \equiv \frac{dE_{st}}{dt} = \dot{E}_{in} - \dot{E}_{out} + \dot{E}_g \tag{1.12c}$$

This expression is illustrated schematically in Figure 1.8*b*.

Every application of the first law must begin with the identification of an appropriate control volume and its control surface, to which an analysis is subsequently applied. The first step is to indicate the control surface by drawing a dashed line. The second step is to decide whether to perform the analysis for a time interval Δt (Equation 1.12b) or on a rate basis (Equation 1.12c). This choice depends on the objective of the solution and on how information is given in the problem. The next step is to identify the energy terms that are relevant in the problem you are solving. To develop your confidence in taking this last step, the remainder of this section is devoted to clarifying the terms in the thermal and mechanical energy equations, Equations 1.12b and 1.12c.

Thermal and Mechanical Energy, E As noted earlier, *mechanical energy* is the sum of kinetic energy ($KE = \frac{1}{2}mV^2$, where m and V are mass and velocity, respectively) and potential energy ($PE = mgz$, where g is the gravitational acceleration and z is the vertical coordinate).

As shown in Figure 1.7, *thermal energy* consists of a *sensible component* U_{sens}, which accounts for translational, rotational, and/or vibrational motion of the atoms/molecules comprising the matter and a *latent component* U_{lat}, which relates to intermolecular forces influencing phase change between solid, liquid, and vapor states. The sensible energy is the portion associated mainly with changes in temperature (although it can also depend on pressure). The latent energy is the component associated with changes in phase. For example, if the material in the control volume increases in temperature, its sensible energy increases. If the material in the control volume changes from solid to liquid (*melting*) or from liquid to vapor (*vaporization, evaporation, boiling*), the latent energy increases. Conversely, if the phase change is from vapor to liquid (*condensation*) or from liquid to solid (*solidification, freezing*), the latent energy decreases.

Based on this discussion, the *change in stored thermal and mechanical energy* is given by $\Delta E_{st} = \Delta(KE + PE + U_t)$, where $U_t = U_{sens} + U_{lat}$. In many heat transfer problems, the changes in kinetic and potential energy are small relative to changes in the thermal energy, and can be neglected. In addition, if there is no phase change, then the only relevant term will be the change in sensible energy, that is, $\Delta E_{st} = \Delta U_{sens}$. For either an ideal gas or an *incompressible* substance (one whose density can be treated as constant), the change in sensible energy can be expressed as $\Delta U_{sens} = mc_v \Delta T$ or in terms of rates, $dU_{sens}/dt = mc_v dT/dt$.

Energy Generation, E_g To understand energy generation, it is helpful to remember that thermal energy is only a portion of the *internal energy*, the energy associated with the thermodynamic state of matter. In addition to the sensible and latent components, internal energy includes other components such as a *chemical component*, which accounts for energy stored in the chemical bonds between atoms; a *nuclear component*, which accounts for the binding forces in the nucleus; and components such as electrical or magnetic energy. *Energy generation, E_g*, is associated with conversion from some other form of internal energy (chemical, nuclear, electrical, or magnetic) to thermal or mechanical energy. It is a *volumetric phenomenon*. That is, it occurs within the control volume and is generally proportional to the size of this volume. We have already noted that energy generation occurs when chemical energy is converted to sensible energy by combustion. Another example is the conversion from electrical energy that occurs due to resistance heating when an electric current is passed through a conductor. That is, if an electric current I passes through a resistance R in the control volume, electrical energy is dissipated at a rate $I^2 R$, which corresponds to the rate at which thermal energy is generated (released) within the volume.

Inflow and Outflow, E_{in} and E_{out} Energy inflow and outflow are surface phenomena, that is, they are associated exclusively with processes occurring at the control surface. As discussed previously, the energy inflow and outflow terms include heat transfer and work interactions occurring at the system boundaries (e.g., due to displacement of a boundary, a rotating shaft, and/or electromagnetic effects). For cases in which mass crosses the control volume boundary (e.g., for situations involving fluid flow), the inflow and outflow terms also include energy (thermal and mechanical) that is advected (carried) by mass entering and leaving the control volume. For instance, if the mass flow rate entering through the boundary is $\dot{m}$, then the rate at which thermal and mechanical energy enters with the flow is $\dot{m}(u_t + \frac{1}{2}V^2 + gz)$, where u_t is the thermal energy per unit mass. The *mass flow rate* may be expressed as $\dot{m} = \rho V A_c$ where ρ is the fluid density and A_c is the cross-sectional area of the channel through which the fluid flows. The *volumetric flow rate* is simply $\dot{\forall} = V A_c = \dot{m}/\rho$.

When the first law is applied to a control volume with fluid crossing its boundary, it is customary to divide the work term into two contributions. The first contribution, termed *flow work*, is associated with the work done by pressure forces moving fluid through the boundary. For a *unit mass*, the amount of work is the product of the pressure and the specific volume of the fluid (pv). The symbol $\dot{W}$ is traditionally used for the rate at which the remaining work (not including flow work) is performed. If operation is under steady-state conditions ($dE_{st}/dt = 0$) and there is no thermal or mechanical energy generation, Equation 1.12c reduces to the following form of the steady-flow energy equation (see Figure 1.9), which will be familiar if you have taken a thermodynamics course:

$$\dot{m}(u_t + pv + \frac{1}{2}V^2 + gz)_{in} - \dot{m}(u_t + pv + \frac{1}{2}V^2 + gz)_{out} + q - \dot{W} = 0 \qquad (1.12d)$$

FIGURE 1.9 Conservation of energy for a steady-flow, open system.

Terms within the parentheses are expressed for a unit mass of fluid at the inflow and out-flow locations. When multiplied by the mass flow rate $\dot{m}$, they yield the rate at which the corresponding form of energy (thermal, flow work, kinetic, and potential) enters or leaves the control volume. The sum of thermal energy and flow work per unit mass may be replaced by the enthalpy per unit mass, $i = u_t + pv$.

In most open system applications of interest in this text, changes in latent energy between the inflow and outflow conditions of Equation 1.12d may be neglected, so the thermal energy reduces to only the sensible component. If the fluid is approximated as an *ideal gas* with *constant specific heats*, the difference in enthalpies (per unit mass) between the inlet and outlet flows may then be expressed as $(i_{in} - i_{out}) = c_p(T_{in} - T_{out})$, where c_p is the specific heat at constant pressure and T_{in} and T_{out} are the inlet and outlet temperatures, respectively. If the fluid is an *incompressible liquid*, its specific heats at constant pressure and volume are equal, $c_p = c_v \equiv c$, and for Equation 1.12d the change in sensible energy (per unit mass) reduces to $(u_{t,in} - u_{t,out}) = c(T_{in} - T_{out})$. Unless the pressure drop is extremely large, the difference in flow work terms, $(pv)_{in} - (pv)_{out}$, is negligible for a liquid. Thus for *either* ideal gases *or* incompressible liquids with no latent energy change, it is often appropriate to use the approximation $i_{in} - i_{out} = c_p(T_{in} - T_{out})$.

Simplified Steady-Flow Thermal Energy Equation Having already assumed steady-state conditions and no thermal or mechanical energy generation, we also adopt the approximation $i_{in} - i_{out} = c_p(T_{in} - T_{out})$, as discussed in the preceding paragraph. Further assuming negligible changes in kinetic and potential energy and negligible work, Equation 1.12d reduces to the *simplified steady-flow thermal energy equation*:

$$q = \dot{m}c_p(T_{out} - T_{in}) \tag{1.12e}$$

The right-hand side of Equation 1.12e represents the net rate of outflow of enthalpy (thermal energy plus flow work) for an ideal gas or of thermal energy for an incompressible liquid.

Since many engineering applications satisfy the assumptions of Equation 1.12e, it is commonly used for the analysis of heat transfer in moving fluids. It will be used in this text in the study of convection heat transfer in internal flow.

EXAMPLE 1.3

The blades of a wind turbine turn a large shaft at a relatively slow speed. The rotational speed is increased by a gearbox that has an efficiency of $\eta_{gb} = 0.93$. In turn, the gearbox output shaft drives an electric generator with an efficiency of $\eta_{gen} = 0.95$. The cylindrical *nacelle*, which houses the gearbox, generator, and associated equipment, is of length $L = 6$ m and diameter $D = 3$ m. If the turbine produces $P = 2.5$ MW of electrical power, and the air and surroundings temperatures are $T_\infty = 25°C$ and $T_{sur} = 20°C$, respectively, determine the exterior nacelle temperature. The emissivity of the nacelle is $\varepsilon = 0.83$, and the convective heat transfer coefficient is $h = 35$ W/m² · K. The surface of the nacelle that is adjacent to the blade hub can be considered to be adiabatic, and solar irradiation may be neglected.

SOLUTION

Known: Electrical power produced by a wind turbine. Gearbox and generator efficiencies, dimensions and emissivity of the nacelle, ambient and surrounding temperatures, and heat transfer coefficient.

Find: Exterior nacelle temperature.

Schematic:

Assumptions:

1. Steady-state conditions.
2. Large surroundings, $\alpha = \varepsilon$.
3. Surface of the nacelle that is adjacent to the hub is adiabatic.

Analysis: The first step is to perform an energy balance on the nacelle to determine the rate of heat transfer from the nacelle to the air and surroundings under steady-state conditions. This step can be accomplished using either conservation of *total* energy or conservation of *thermal and mechanical* energy; we will compare these two approaches.

Conservation of Total Energy The first of the three boxed statements of the first law in Section 1.3 can be converted to a rate basis and expressed in equation form as follows:

$$\frac{dE_{\mathrm{st}}^{\mathrm{tot}}}{dt} = \dot{E}_{\mathrm{in}}^{\mathrm{tot}} - \dot{E}_{\mathrm{out}}^{\mathrm{tot}} \tag{1}$$

Under steady-state conditions, this reduces to $\dot{E}_{\mathrm{in}}^{\mathrm{tot}} - \dot{E}_{\mathrm{out}}^{\mathrm{tot}} = 0$. The $\dot{E}_{\mathrm{in}}^{\mathrm{tot}}$ term corresponds to the mechanical work entering the nacelle $\dot{W}$, and the $\dot{E}_{\mathrm{out}}^{\mathrm{tot}}$ term includes the electrical power output P and the rate of heat transfer leaving the nacelle q. Thus

$$\dot{W} - P - q = 0 \tag{2}$$

Conservation of Thermal and Mechanical Energy Alternatively, we can express conservation of thermal and mechanical energy, starting with Equation 1.12c. Under steady-state conditions, this reduces to

$$\dot{E}_{\mathrm{in}} - \dot{E}_{\mathrm{out}} + \dot{E}_g = 0 \tag{3}$$

Here, $\dot{E}_{\mathrm{in}}$ once again corresponds to the mechanical work $\dot{W}$. However, $\dot{E}_{\mathrm{out}}$ now includes *only* the rate of heat transfer leaving the nacelle q. It does *not* include the electrical power, since E represents only the thermal and mechanical forms of energy. The electrical power appears in the generation term, because mechanical energy is converted to electrical energy in the generator, giving rise to a negative source of mechanical energy. That is, $\dot{E}_g = -P$. Thus, Equation (3) becomes

$$\dot{W} - q - P = 0 \tag{4}$$

which is equivalent to Equation (2), as it must be. Regardless of the manner in which the first law of thermodynamics is applied, the following expression for the rate of heat transfer evolves:

$$q = \dot{W} - P \tag{5}$$

The mechanical work and electrical power are related by the efficiencies of the gearbox and generator,

$$P = \dot{W}\eta_{\mathrm{gb}}\eta_{\mathrm{gen}} \tag{6}$$

Equation (5) can therefore be written as

$$q = P\left(\frac{1}{\eta_{\mathrm{gb}}\eta_{\mathrm{gen}}} - 1\right) = 2.5 \times 10^6\ \mathrm{W} \times \left(\frac{1}{0.93 \times 0.95} - 1\right) = 0.33 \times 10^6\ \mathrm{W} \tag{7}$$

Application of the Rate Equations   Heat transfer is due to convection and radiation from the exterior surface of the nacelle, governed by Equations 1.3a and 1.7, respectively. Thus

$$q = q_{rad} + q_{conv} = A[q''_{rad} + q''_{conv}]$$

$$= \left[\pi D L + \frac{\pi D^2}{4} \right] [\varepsilon \sigma (T_s^4 - T_{sur}^4) + h(T_s - T_\infty)] = 0.33 \times 10^6 \text{ W}$$

or

$$\left[\pi \times 3 \text{ m} \times 6 \text{ m} + \frac{\pi \times (3 \text{ m})^2}{4} \right]$$

$$\times [0.83 \times 5.67 \times 10^{-8} \text{ W/m}^2 \cdot \text{K}^4 (T_s^4 - (273 + 20)^4)\text{K}^4$$

$$+ 35 \text{ W/m}^2 \cdot \text{K}(T_s - (273 + 25)\text{K})] = 0.33 \times 10^6 \text{ W}$$

The preceding equation does not have a closed-form solution, but the nacelle surface temperature can be easily determined by trial and error or by using a software package such as the *Interactive Heat Transfer* (*IHT*) software accompanying your text. Doing so yields

$$T_s = 416 \text{ K} = 143°\text{C}$$

Comments:

1. The temperature inside the nacelle must be greater than the exterior surface temperature of the nacelle T_s, because the heat generated within the nacelle must be transferred from the interior of the nacelle to its surface, and from the surface to the air and surroundings.

2. The high temperature would preclude, for example, performance of routine maintenance by a worker, as illustrated in the problem statement. Thermal management approaches involving fans or blowers must be employed to reduce the temperature to an acceptable level.

3. Improvements in the efficiencies of either the gearbox or the generator would provide more electrical power and reduce the size and cost of the thermal management hardware. As such, improved efficiencies would increase revenue generated by the wind turbine and decrease both its capital and operating costs.

4. The heat transfer coefficient would not be a steady value but would vary periodically as the blades sweep past the nacelle. Therefore, the value of the heat transfer coefficient represents a *time-averaged* quantity.

EXAMPLE 1.4   IHT

A long conducting rod of diameter D and electrical resistance per unit length R'_e is initially in thermal equilibrium with the ambient air and its surroundings. This equilibrium is disturbed when an electrical current I is passed through the rod. Develop an equation that could be used to compute the variation of the rod temperature with time during the passage of the current.

SOLUTION

Known: Temperature of a rod of prescribed diameter and electrical resistance changes with time due to passage of an electrical current.

Find: Equation that governs temperature change with time for the rod.

Schematic:

Assumptions:

1. At any time t, the temperature of the rod is uniform.
2. Constant properties (ρ, c_v, $\varepsilon = \alpha$).
3. Radiation exchange between the outer surface of the rod and the surroundings is between a small surface and a large enclosure.

Analysis: In this problem, there is no mechanical energy component. So relevant energy terms include heat transfer by convection and radiation from the surface, thermal energy generation due to ohmic heating within the conductor, and a change in thermal energy storage. Since we wish to determine the rate of change of the temperature, the first law should be applied at an instant of time. Hence, applying Equation 1.12c to a control volume of length L about the rod, it follows that

$$\dot{E}_g - \dot{E}_{out} = \dot{E}_{st}$$

where thermal energy generation is due to the electric resistance heating,

$$\dot{E}_g = I^2 R_e' L$$

Heating occurs uniformly within the control volume and could also be expressed in terms of a volumetric heat generation rate $\dot{q}$(W/m^3). The generation rate for the entire control volume is then $\dot{E}_g = \dot{q}V$, where $\dot{q} = I^2 R_e'/(\pi D^2/4)$. Energy outflow is due to convection and net radiation from the surface, Equations 1.3a and 1.7, respectively,

$$\dot{E}_{out} = h(\pi DL)(T - T_\infty) + \varepsilon\sigma(\pi DL)(T^4 - T_{sur}^4)$$

and the change in energy storage is due only to sensible energy change,

$$\dot{E}_{st} = \frac{dU_{sens}}{dt} = mc_v\frac{dT}{dt} = \rho V c_v\frac{dT}{dt}$$

where ρ and c_v are the mass density and the specific heat, respectively, of the rod material, and V is the volume of the rod, $V = (\pi D^2/4)L$. Substituting the rate equations into the energy balance, it follows that

$$I^2 R'_e L - h(\pi DL)(T - T_\infty) - \varepsilon\sigma(\pi DL)(T^4 - T^4_{sur}) = \rho c_v \left(\frac{\pi D^2}{4} \right) L \frac{dT}{dt}$$

Hence

$$\frac{dT}{dt} = \frac{I^2 R'_e - \pi Dh(T - T_\infty) - \pi D\varepsilon\sigma(T^4 - T^4_{sur})}{\rho c_v (\pi D^2/4)} \qquad \lhd$$

Comments:

1. Solids and liquids can usually be treated as incompressible, in which case $c_v = c_p = c$. This assumption will be made in the remainder of this text.

2. The preceding equation could be solved for the time dependence of the rod temperature by integrating numerically. A steady-state condition would eventually be reached for which $dT/dt = 0$. The rod temperature is then determined by an algebraic equation of the form

$$\pi Dh(T - T_\infty) + \pi D\varepsilon\sigma(T^4 - T^4_{sur}) = I^2 R'_e$$

3. For fixed environmental conditions (h, T_∞, T_{sur}), as well as a rod of fixed geometry (D) and properties (ε, R'_e), the steady-state temperature depends on the rate of thermal energy generation and hence on the value of the electric current. Consider an uninsulated copper wire ($D = 1$ mm, $\varepsilon = 0.8$, $R'_e = 0.4$ Ω/m) in a large enclosure ($T_{sur} = 300$ K) through which cooling air is circulated ($h = 100$ W/m$^2 \cdot$ K, $T_\infty = 300$ K). Substituting these values into the foregoing equation, the rod temperature has been computed for operating currents in the range $0 \leq I \leq 10$ A, and the following results were obtained:

4. If a maximum operating temperature of $T = 60°C$ is prescribed for safety reasons, the current should not exceed 5.2 A. At this temperature, heat transfer by radiation (0.6 W/m) is much less than heat transfer by convection (10.4 W/m). Hence, if one wished to operate at a larger current while maintaining the rod temperature within the safety limit, the convection coefficient would have to be increased by increasing the velocity of the circulating air. For $h = 250$ W/m$^2 \cdot$ K, the maximum allowable current could be increased to 8.1 A.

5. The *IHT* software is especially useful for solving equations, such as the energy balance in Comment 1, and generating the graphical results of Comment 2.

EXAMPLE 1.5

Consider a mass m of ice at the fusion temperature ($T_f = 0°C$) that is enclosed in a cubical container of width W on a side. The container wall is of thickness L and thermal conductivity k. If the outer surface of the wall is heated to a temperature $T_1 > T_f$ to melt the ice, obtain an expression for the time needed to melt the entire mass of ice.

SOLUTION

Known: Mass and temperature of ice. Dimensions, thermal conductivity, and outer surface temperature of containing wall.

Find: Expression for time needed to melt the ice.

Schematic:

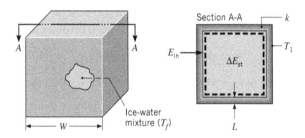

Assumptions:
1. Inner surface of wall is at T_f throughout the process.
2. Constant properties.
3. Steady-state, one-dimensional conduction through each wall.
4. Conduction area of one wall may be approximated as W^2 ($L \ll W$).

Analysis: Since we must determine the melting time t_m, the first law should be applied over the time interval $\Delta t = t_m$. Hence, applying Equation 1.12b to a control volume about the ice–water mixture, it follows that

$$E_{in} = \Delta E_{st} = \Delta U_{lat}$$

where the increase in energy stored within the control volume is due exclusively to the change in latent energy associated with conversion from the solid to liquid state. Heat is transferred to the ice by means of conduction through the container wall. Since the temperature difference across the wall is assumed to remain at ($T_1 - T_f$) throughout the melting process, the wall conduction rate is constant

$$q_{cond} = k(6W^2)\frac{T_1 - T_f}{L}$$

and the amount of energy inflow is

$$E_{in} = \left[k(6W^2)\frac{T_1 - T_f}{L} \right] t_m$$

The amount of energy required to effect such a phase change per unit mass of solid is termed the *latent heat of fusion* h_{sf}. Hence the increase in energy storage is

$$\Delta E_{st} = m h_{sf}$$

By substituting into the first law expression, it follows that

$$t_m = \frac{m h_{sf} L}{6 W^2 k (T_1 - T_f)} \qquad \triangleleft$$

Comments:

1. Several complications would arise if the ice were initially subcooled. The storage term would have to include the change in sensible (internal thermal) energy required to take the ice from the subcooled to the fusion temperature. During this process, temperature gradients would develop in the ice.

2. Consider a cavity of width $W = 100$ mm on a side, wall thickness $L = 5$ mm, and thermal conductivity $k = 0.05$ W/m · K. The mass of the ice in the cavity is

$$m = \rho_s (W - 2L)^3 = 920 \text{ kg/m}^3 \times (0.100 - 0.01)^3 \text{ m}^3 = 0.67 \text{ kg}$$

If the outer surface temperature is $T_1 = 30°C$, the time required to melt the ice is

$$t_m = \frac{0.67 \text{ kg} \times 334{,}000 \text{ J/kg} \times 0.005 \text{ m}}{6(0.100 \text{ m})^2 \times 0.05 \text{ W/m} \cdot \text{K}(30 - 0)°C} = 12{,}430 \text{ s} = 207 \text{ min}$$

The density and latent heat of fusion of the ice are $\rho_s = 920$ kg/m^3 and $h_{sf} = 334$ kJ/kg, respectively.

3. Note that the units of K and °C cancel each other in the foregoing expression for t_m. Such cancellation occurs frequently in heat transfer analysis and is due to both units appearing in the context of a *temperature difference*.

The Surface Energy Balance   We will frequently have occasion to apply the conservation of energy requirement at the surface of a medium. In this special case, the control surfaces are located on either side of the physical boundary and enclose no mass or volume (see Figure 1.10).

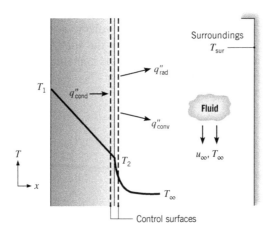

FIGURE 1.10   The energy balance for conservation of energy at the surface of a medium.

Accordingly, the generation and storage terms of the conservation expression, Equation 1.12c, are no longer relevant, and it is necessary to deal only with surface phenomena. For this case, the conservation requirement becomes

$$\dot{E}_{\text{in}} - \dot{E}_{\text{out}} = 0 \tag{1.13}$$

Even though energy generation may be occurring in the medium, the process would not affect the energy balance at the control surface. Moreover, this conservation requirement holds for both *steady-state* and *transient* conditions.

In Figure 1.10, three heat transfer terms are shown for the control surface. On a unit area basis, they are conduction from the medium *to* the control surface (q''_{cond}), convection *from* the surface to a fluid (q''_{conv}), and net radiation exchange from the surface to the surroundings (q''_{rad}). The energy balance then takes the form

$$q''_{\text{cond}} - q''_{\text{conv}} - q''_{\text{rad}} = 0 \tag{1.14}$$

and we can express each of the terms using the appropriate rate equations, Equations 1.2, 1.3a, and 1.7.

EXAMPLE 1.6

Humans are able to control their rates of heat production and heat loss to maintain a nearly constant core temperature of $T_c = 37°C$ under a wide range of environmental conditions. This process is called *thermoregulation*. From the perspective of calculating heat transfer between a human body and its surroundings, we focus on a layer of skin and fat, with its outer surface exposed to the environment and its inner surface at a temperature slightly less than the core temperature, $T_i = 35°C = 308$ K. Consider a person with a skin/fat layer of thickness $L = 3$ mm and effective thermal conductivity $k = 0.3$ W/m · K. The person has a surface area $A = 1.8$ m^2 and is dressed in a bathing suit. The emissivity of the skin is $\varepsilon = 0.95$.

1. When the person is in still air at $T_\infty = 297$ K, what is the skin surface temperature and rate of heat loss to the environment? Convection heat transfer to the air is characterized by a free convection coefficient of $h = 2$ W/m^2 · K.
2. When the person is in water at $T_\infty = 297$ K, what is the skin surface temperature and rate of heat loss? Heat transfer to the water is characterized by a convection coefficient of $h = 200$ W/m^2 · K.

SOLUTION

Known: Inner surface temperature of a skin/fat layer of known thickness, thermal conductivity, emissivity, and surface area. Ambient conditions.

Find: Skin surface temperature and rate of heat loss for the person in air and the person in water.

Schematic:

Assumptions:

1. Steady-state conditions.
2. One-dimensional heat transfer by conduction through the skin/fat layer.
3. Thermal conductivity is uniform.
4. In part 1, radiation exchange between the skin surface and the surroundings is between a small surface and a large enclosure at the air temperature.
5. Liquid water is opaque to thermal radiation.
6. Bathing suit has no effect on heat loss from body.
7. Solar radiation is negligible.
8. Body is completely immersed in water in part 2.

Analysis:

1. The skin surface temperature may be obtained by performing an energy balance at the skin surface. From Equation 1.13,

$$\dot{E}_{\text{in}} - \dot{E}_{\text{out}} = 0$$

It follows that, on a unit area basis,

$$q''_{\text{cond}} - q''_{\text{conv}} - q''_{\text{rad}} = 0$$

or, rearranging and substituting from Equations 1.2, 1.3a, and 1.7,

$$k\frac{T_i - T_s}{L} = h(T_s - T_\infty) + \varepsilon\sigma(T_s^4 - T_{\text{sur}}^4)$$

The only unknown is T_s, but we cannot solve for it explicitly because of the fourth-power dependence of the radiation term. Therefore, we must solve the equation iteratively, which can be done by hand or by using *IHT* or some other equation solver. To expedite a hand solution, we write the radiation heat flux in terms of the radiation heat transfer coefficient, using Equations 1.8 and 1.9:

$$k\frac{T_i - T_s}{L} = h(T_s - T_\infty) + h_r(T_s - T_{\text{sur}})$$

Solving for T_s, with $T_{sur} = T_\infty$, we have

$$T_s = \frac{\dfrac{kT_i}{L} + (h + h_r)T_\infty}{\dfrac{k}{L} + (h + h_r)}$$

We estimate h_r using Equation 1.9 with a guessed value of $T_s = 305$ K and $T_\infty = 297$ K, to yield $h_r = 5.9$ W/m$^2 \cdot$ K. Then, substituting numerical values into the preceding equation, we find

$$T_s = \frac{\dfrac{0.3 \text{ W/m} \cdot \text{K} \times 308 \text{ K}}{3 \times 10^{-3} \text{ m}} + (2 + 5.9) \text{ W/m}^2 \cdot \text{K} \times 297 \text{ K}}{\dfrac{0.3 \text{ W/m} \cdot \text{K}}{3 \times 10^{-3} \text{ m}} + (2 + 5.9) \text{ W/m}^2 \cdot \text{K}} = 307.2 \text{ K}$$

With this new value of T_s, we can recalculate h_r and T_s, which are unchanged. Thus the skin temperature is 307.2 K $\cong$ 34°C. ◁

The rate of heat loss can be found by evaluating the conduction through the skin/fat layer:

$$q_s = kA\frac{T_i - T_s}{L} = 0.3 \text{ W/m} \cdot \text{K} \times 1.8 \text{ m}^2 \times \frac{(308 - 307.2) \text{ K}}{3 \times 10^{-3} \text{ m}} = 146 \text{ W} \qquad ◁$$

2. Since liquid water is opaque to thermal radiation, heat loss from the skin surface is by convection only. Using the previous expression with $h_r = 0$, we find

$$T_s = \frac{\dfrac{0.3 \text{ W/m} \cdot \text{K} \times 308 \text{ K}}{3 \times 10^{-3} \text{ m}} + 200 \text{ W/m}^2 \cdot \text{K} \times 297 \text{ K}}{\dfrac{0.3 \text{ W/m} \cdot \text{K}}{3 \times 10^{-3} \text{ m}} + 200 \text{ W/m}^2 \cdot \text{K}} = 300.7 \text{ K} \qquad ◁$$

and

$$q_s = kA\frac{T_i - T_s}{L} = 0.3 \text{ W/m} \cdot \text{K} \times 1.8 \text{ m}^2 \times \frac{(308 - 300.7) \text{ K}}{3 \times 10^{-3} \text{ m}} = 1320 \text{ W} \qquad ◁$$

Comments:

1. When using energy balances involving radiation exchange, the temperatures appearing in the radiation terms must be expressed in kelvins, and it is good practice to use kelvins in all terms to avoid confusion.

2. In part 1, the rates of heat loss due to convection and radiation are 37 W and 109 W, respectively. Thus, it would not have been reasonable to neglect radiation. Care must be taken to include radiation when the heat transfer coefficient is small (as it often is for natural convection to a gas), even if the problem statement does not give any indication of its importance.

3. A typical rate of metabolic heat generation is 100 W. If the person stayed in the water too long, the core body temperature would begin to fall. The large rate of heat loss in water is due to the higher heat transfer coefficient, which in turn is due to the much larger thermal conductivity of water compared to air.

4. The skin temperature of 34°C in part 1 is comfortable, but the skin temperature of 28°C in part 2 is uncomfortably cold.

Application of the Conservation Laws: Methodology   In addition to being familiar with the transport rate equations described in Section 1.2, the heat transfer analyst must be able to work with the energy conservation requirements of Equations 1.12 and 1.13. The application of these balances is simplified if a few basic rules are followed.

1. The appropriate control volume must be defined, with the control surfaces represented by a dashed line or lines.

2. The appropriate time basis must be identified.

3. The relevant energy processes must be identified, and each process should be shown on the control volume by an appropriately labeled arrow.

4. The conservation equation must then be written, and appropriate rate expressions must be substituted for the relevant terms in the equation.

Note that the energy conservation requirement may be applied to a *finite* control volume or a *differential* (infinitesimal) control volume. In the first case, the resulting expression governs overall system behavior. In the second case, a differential equation is obtained that can be solved for conditions at each point in the system. Differential control volumes are introduced in Chapter 2, and both types of control volumes are used extensively throughout the text.

1.3.2   Relationship to the Second Law of Thermodynamics and the Efficiency of Heat Engines

In this section, we show how heat transfer plays a crucial role in managing and promoting the efficiency of a broad range of energy conversion devices. Recall that a heat engine is any device that operates continuously or cyclically and that converts heat to work. Examples include internal combustion engines, power plants, and thermoelectric devices (to be discussed in Section 3.7.2). Improving the efficiency of heat engines is a subject of great importance; for example, more efficient combustion engines consume less fuel to produce a given amount of work and reduce the corresponding emissions of pollutants and carbon dioxide. More efficient thermoelectric devices can generate more electricity from waste heat. Regardless of the energy conversion device, its size, weight, and cost can all be reduced through improvements in its energy conversion efficiency.

The second law of thermodynamics is often invoked when efficiency is of concern and can be expressed in a variety of different but equivalent ways. The *Kelvin–Planck statement* is particularly relevant to the operation of heat engines [1]. It states:

> *It is impossible for any system to operate in a thermodynamic cycle and deliver a net amount of work to its surroundings while receiving energy by heat transfer from a single thermal reservoir.*

Recall that a thermodynamic cycle is a process for which the initial and final states of the system are identical. Consequently, the energy stored in the system does not change between the initial and final states, and the first law of thermodynamics (Equation 1.12a) reduces to $W = Q$.

A consequence of the Kelvin–Planck statement is that a heat engine must exchange heat with two (or more) reservoirs, receiving thermal energy from the higher-temperature reservoir and rejecting thermal energy to the lower-temperature reservoir. Thus, converting all of the input heat to work is impossible, and $W = Q_{in} - Q_{out}$, where Q_{in} and Q_{out} are both defined to be positive. That is, Q_{in} is the heat transferred from the high temperature source to the heat engine, and Q_{out} is the heat transferred from the heat engine to the low temperature sink.

The efficiency of a heat engine is defined as the fraction of heat transferred into the system that is converted to work, namely

$$\eta \equiv \frac{W}{Q_{in}} = \frac{Q_{in} - Q_{out}}{Q_{in}} = 1 - \frac{Q_{out}}{Q_{in}} \tag{1.15}$$

The second law also tells us that, for a *reversible* process, the ratio Q_{out}/Q_{in} is equal to the ratio of the absolute temperatures of the respective reservoirs [1]. Thus, the efficiency of a heat engine undergoing a reversible process, called the *Carnot efficiency* η_C, is given by

$$\eta_C = 1 - \frac{T_c}{T_h} \tag{1.16}$$

where T_c and T_h are the absolute temperatures of the low- and high-temperature reservoirs, respectively. The Carnot efficiency is the maximum possible efficiency that any heat engine can achieve operating between those two temperatures. Any *real* heat engine, which will necessarily undergo an irreversible process, will have an efficiency lower than η_C.

From our knowledge of thermodynamics, we know that, for heat transfer to take place reversibly, it must occur through an infinitesimal temperature difference between the reservoir and heat engine. However, from our newly acquired knowledge of heat transfer mechanisms, as embodied, for example, in Equations 1.2, 1.3, and 1.7, we now realize that, for heat transfer to occur, there *must* be a nonzero temperature difference between the reservoir and the heat engine. This reality introduces irreversibility and reduces the efficiency.

With the concepts of the preceding paragraph in mind, we now consider a more realistic model of a heat engine [2–5] in which heat is transferred into the engine through a thermal resistance $R_{t,h}$, while heat is extracted from the engine through a second thermal resistance $R_{t,c}$ (Figure 1.11). The subscripts h and c refer to the hot and cold sides of the heat engine, respectively. As discussed in Section 1.2.4, these thermal resistances are associated with heat transfer between the heat engine and the reservoirs across a nonzero temperature difference, by way of the mechanisms of conduction, convection, and/or radiation. For example, the resistances could represent conduction through the walls separating the heat engine from the two reservoirs. Note that the reservoir temperatures are still T_h and T_c but that the temperatures seen by the heat engine are $T_{h,i} < T_h$ and $T_{c,i} > T_c$, as shown in the diagram. The heat engine is still assumed to be *internally* reversible, and its efficiency is still the Carnot efficiency. However, the Carnot efficiency is *now based on the internal*

FIGURE 1.11   Internally reversible heat engine exchanging heat with high- and low-temperature reservoirs through thermal resistances.

temperatures $T_{h,i}$ *and* $T_{c,i}$. Therefore, a modified efficiency that accounts for realistic (irreversible) heat transfer processes η_m is

$$\eta_m = 1 - \frac{Q_{\text{out}}}{Q_{\text{in}}} = 1 - \frac{q_{\text{out}}}{q_{\text{in}}} = 1 - \frac{T_{c,i}}{T_{h,i}} \qquad (1.17)$$

where the ratio of heat *flows* over a time interval, $Q_{\text{out}}/Q_{\text{in}}$, has been replaced by the corresponding ratio of heat *rates*, $q_{\text{out}}/q_{\text{in}}$. This replacement is based on applying energy conservation at an instant in time,[1] as discussed in Section 1.3.1. Utilizing the definition of a thermal resistance, the heat transfer rates into and out of the heat engine are given by

$$q_{\text{in}} = (T_h - T_{h,i})/R_{t,h} \qquad (1.18a)$$

$$q_{\text{out}} = (T_{c,i} - T_c)/R_{t,c} \qquad (1.18b)$$

Equations 1.18 can be solved for the internal temperatures, to yield

$$T_{h,i} = T_h - q_{\text{in}}R_{t,h} \qquad (1.19a)$$

$$T_{c,i} = T_c + q_{\text{out}}R_{t,c} = T_c + q_{\text{in}}(1 - \eta_m)R_{t,c} \qquad (1.19b)$$

In Equation 1.19b, q_{out} has been related to q_{in} and η_m, using Equation 1.17. The more realistic, modified efficiency can then be expressed as

$$\eta_m = 1 - \frac{T_{c,i}}{T_{h,i}} = 1 - \frac{T_c + q_{\text{in}}(1 - \eta_m)R_{t,c}}{T_h - q_{\text{in}}R_{t,h}} \qquad (1.20)$$

Solving for η_m results in

$$\eta_m = 1 - \frac{T_c}{T_h - q_{\text{in}}R_{\text{tot}}} \qquad (1.21)$$

---

[1]The heat engine is assumed to undergo a continuous, steady-flow process, so that all heat and work processes are occurring simultaneously, and the corresponding terms would be expressed in watts (W). For a heat engine undergoing a cyclic process with sequential heat and work processes occurring over different time intervals, we would need to introduce the time intervals for each process, and each term would be expressed in joules (J).

where $R_{tot} = R_{t,h} + R_{t,c}$. It is readily evident that $\eta_m = \eta_C$ only if the thermal resistances $R_{t,h}$ and $R_{t,c}$ could somehow be made infinitesimally small (or if $q_{in} = 0$). For realistic (nonzero) values of R_{tot}, $\eta_m < \eta_C$, and η_m further deteriorates as either R_{tot} or q_{in} increases. As an extreme case, note that $\eta_m = 0$ when $T_h = T_c + q_{in}R_{tot}$, meaning that no power could be produced even though the Carnot efficiency, as expressed in Equation 1.16, is nonzero.

In addition to the efficiency, another important parameter to consider is the power output of the heat engine, given by

$$\dot{W} = q_{in}\eta_m = q_{in}\left[1 - \frac{T_c}{T_h - q_{in}R_{tot}}\right] \tag{1.22}$$

It has already been noted in our discussion of Equation 1.21 that the efficiency is equal to the maximum Carnot efficiency ($\eta_m = \eta_C$) if $q_{in} = 0$. However, under these circumstances the power output $\dot{W}$ is zero according to Equation 1.22. To increase $\dot{W}$, q_{in} must be increased at the expense of decreased efficiency. In any real application, a balance must be struck between maximizing the efficiency and maximizing the power output. If provision of the heat input is inexpensive (for example, if waste heat is converted to power), a case could be made for sacrificing efficiency to maximize power output. In contrast, if fuel is expensive or emissions are detrimental (such as for a conventional fossil fuel power plant), the efficiency of the energy conversion may be as or more important than the power output. In any case, heat transfer and thermodynamic principles should be used to determine the actual efficiency and power output of a heat engine.

Although we have limited our discussion of the second law to heat engines, the preceding analysis shows how the principles of thermodynamics and heat transfer can be combined to address significant problems of contemporary interest.

EXAMPLE 1.7

In a large steam power plant, the combustion of coal provides a heat rate of $q_{in} = 2500$ MW at a flame temperature of $T_h = 1000$ K. Heat is rejected from the plant to a river flowing at $T_c = 300$ K. Heat is transferred from the combustion products to the exterior of large tubes in the boiler by way of radiation and convection, through the boiler tubes by conduction, and then from the interior tube surface to the working fluid (water) by convection. On the cold side, heat is extracted from the power plant by condensation of steam on the exterior condenser tube surfaces, through the condenser tube walls by conduction, and from the interior of the condenser tubes to the river water by convection. Hot and cold side thermal resistances account for the combined effects of conduction, convection, and radiation and, under *design conditions*, they are $R_{t,h} = 8 \times 10^{-8}$ K/W and $R_{t,c} = 2 \times 10^{-8}$ K/W, respectively.

1. Determine the efficiency and power output of the power plant, accounting for heat transfer effects to and from the cold and hot reservoirs. Treat the power plant as an internally reversible heat engine.

2. Over time, coal slag will accumulate on the combustion side of the boiler tubes. This *fouling process* increases the hot side resistance to $R_{t,h} = 9 \times 10^{-8}$ K/W. Concurrently, biological matter can accumulate on the river water side of the condenser tubes, increasing the cold side resistance to $R_{t,c} = 2.2 \times 10^{-8}$ K/W. Find the efficiency and power output of the plant under fouled conditions.

SOLUTION

Known: Source and sink temperatures and heat input rate for an internally reversible heat engine. Thermal resistances separating heat engine from source and sink under clean and fouled conditions.

Find:

1. Efficiency and power output for clean conditions.
2. Efficiency and power output under fouled conditions.

Schematic:

Assumptions:

1. Steady-state conditions.
2. Power plant behaves as an internally reversible heat engine, so its efficiency is the modified efficiency.

Analysis:

1. The modified efficiency of the internally reversible power plant, considering realistic heat transfer effects on the hot and cold sides of the power plant, is given by Equation 1.21:

$$\eta_m = 1 - \frac{T_c}{T_h - q_{in}R_{tot}}$$

where, for clean conditions

$$R_{tot} = R_{t,h} + R_{t,c} = 8 \times 10^{-8} \text{ K/W} + 2 \times 10^{-8} \text{ K/W} = 1.0 \times 10^{-7} \text{ K/W}$$

Thus

$$\eta_m = 1 - \frac{T_c}{T_h - q_{in}R_{tot}} = 1 - \frac{300 \text{ K}}{1000 \text{ K} - 2500 \times 10^6 \text{ W} \times 1.0 \times 10^{-7} \text{ K/W}} = 0.60 = 60\% \quad \triangleleft$$

The power output is given by

$$\dot{W} = q_{in}\eta_m = 2500 \text{ MW} \times 0.60 = 1500 \text{ MW} \quad \triangleleft$$

2. Under fouled conditions, the preceding calculations are repeated to find

$$\eta_m = 0.583 = 58.3\% \text{ and } \dot{W} = 1460 \text{ MW} \qquad \lhd$$

Comments:

1. The actual efficiency and power output of a power plant operating between these temperatures would be much less than the foregoing values, since there would be other irreversibilities internal to the power plant. Even if these irreversibilities were considered in a more comprehensive analysis, fouling effects would still reduce the plant efficiency and power output.

2. The Carnot efficiency is $\eta_C = 1 - T_c/T_h = 1 - 300 \text{ K}/1000 \text{ K} = 70\%$. The corresponding power output would be $\dot{W} = q_{in}\eta_C = 2500 \text{ MW} \times 0.70 = 1750 \text{ MW}$. Thus, if the effect of irreversible heat transfer from and to the hot and cold reservoirs, respectively, were neglected, the power output of the plant would be significantly overpredicted.

3. Fouling reduces the power output of the plant by $\Delta P = 40 \text{ MW}$. If the plant owner sells the electricity at a price of \$0.08/kW · h, the daily lost revenue associated with operating the fouled plant would be $C = 40{,}000 \text{ kW} \times \$0.08/\text{kW} \cdot \text{h} \times 24 \text{ h/day} = \$76{,}800/\text{day}$.

1.4 *Units and Dimensions*

The physical quantities of heat transfer are specified in terms of *dimensions*, which are measured in terms of units. Four *basic* dimensions are required for the development of heat transfer: length (L), mass (m), time (t), and temperature (T). All other physical quantities of interest may be related to these four basic dimensions.

In the United States, dimensions have been customarily measured in terms of the *English system of units*, for which the *base units* are:

Dimension		Unit
Length (L)	$\rightarrow$	foot (ft)
Mass (m)	$\rightarrow$	pound mass (lb_m)
Time (t)	$\rightarrow$	second (s)
Temperature (T)	$\rightarrow$	degree Fahrenheit (°F)

The units required to specify other physical quantities may then be inferred from this group.

In 1960, the SI (Système International d'Unités) system of units was defined by the Eleventh General Conference on Weights and Measures and recommended as a worldwide standard. In response to this trend, the American Society of Mechanical Engineers (ASME) has required the use of SI units in all of its publications since 1974. For this reason and because SI units are operationally more convenient than the English system, the SI system is used for calculations of this text. However, because for some time to come, engineers might also have to work with results expressed in the English system, you should be able to convert from one system to the other. For your convenience, conversion factors are provided after the Appendices.

TABLE 1.2   SI base and supplementary units

Quantity and Symbol	Unit and Symbol
Length (L)	meter (m)
Mass (m)	kilogram (kg)
Amount of substance	mole (mol)
Time (t)	second (s)
Electric current (I)	ampere (A)
Thermodynamic temperature (T)	kelvin (K)
Plane angle[a] (θ)	radian (rad)
Solid angle[a] (ω)	steradian (sr)

[a]Supplementary unit.

The SI *base* units required for this text are summarized in Table 1.2. With regard to these units, note that 1 mol is the amount of substance that has as many atoms or molecules as there are atoms in 12 g of carbon-12 (^{12}C); this is the gram-mole (mol). Although the mole has been recommended as the unit quantity of matter for the SI system, it is more consistent to work with the kilogram-mol (kmol, kg-mol). One kmol is simply the amount of substance that has as many atoms or molecules as there are atoms in 12 kg of ^{12}C. As long as the use is consistent within a given problem, no difficulties arise in using either mol or kmol. The molecular weight of a substance is the mass associated with a mole or kilogram-mole. For oxygen, as an example, the molecular weight $\mathcal{M}$ is 16 g/mol or 16 kg/kmol.

Although the SI unit of temperature is the kelvin, use of the Celsius temperature scale remains widespread. Zero on the Celsius scale (0°C) is equivalent to 273.15 K on the thermodynamic scale,[2] in which case

$$T(K) = T(°C) + 273.15$$

However, temperature *differences* are equivalent for the two scales and may be denoted as °C or K. Also, although the SI unit of time is the second, other units of time (minute, hour, and day) are so common that their use with the SI system is generally accepted.

The SI units comprise a coherent form of the metric system. That is, all remaining units may be derived from the base units using formulas that do not involve any numerical factors. *Derived* units for selected quantities are listed in Table 1.3. Note that force is measured in

TABLE 1.3   SI derived units for selected quantities

Quantity	Name and Symbol	Formula	Expression in SI Base Units
Force	newton (N)	$m \cdot kg/s^2$	$m \cdot kg/s^2$
Pressure and stress	pascal (Pa)	N/m^2	$kg/m \cdot s^2$
Energy	joule (J)	$N \cdot m$	$m^2 \cdot kg/s^2$
Power	watt (W)	J/s	$m^2 \cdot kg/s^3$

---

[2]The degree symbol is retained for designating the Celsius temperature (°C) to avoid confusion with the use of C for the unit of electrical charge (coulomb).

TABLE 1.4 Multiplying prefixes

Prefix	Abbreviation	Multiplier
femto	f	10^{-15}
pico	p	10^{-12}
nano	n	10^{-9}
micro	μ	10^{-6}
milli	m	10^{-3}
centi	c	10^{-2}
hecto	h	10^{2}
kilo	k	10^{3}
mega	M	10^{6}
giga	G	10^{9}
tera	T	10^{12}
peta	P	10^{15}
exa	E	10^{18}

newtons, where a 1-N force will accelerate a 1-kg mass at 1 m/s². Hence 1 N = 1 kg · m/s². The unit of pressure (N/m²) is often referred to as the pascal. In the SI system, there is one unit of energy (thermal, mechanical, or electrical) called the joule (J); 1 J = 1 N · m. The unit for energy rate, or power, is then J/s, where one joule per second is equivalent to one watt (1 J/s = 1 W). Since working with extremely large or small numbers is frequently necessary, a set of standard prefixes has been introduced to simplify matters (Table 1.4). For example, 1 megawatt (MW) = 10^6 W, and 1 micrometer (μm) = 10^{-6} m.

1.5 Analysis of Heat Transfer Problems: Methodology

A major objective of this text is to prepare you to solve engineering problems that involve heat transfer processes. To this end, numerous problems are provided at the end of each chapter. In working these problems you will gain a deeper appreciation for the fundamentals of the subject, and you will gain confidence in your ability to apply these fundamentals to the solution of engineering problems.

In solving problems, we advocate the use of a systematic procedure characterized by a prescribed format. We consistently employ this procedure in our examples, and we require our students to use it in their problem solutions. It consists of the following steps:

1. *Known:* After carefully reading the problem, state briefly and concisely what is known about the problem. Do not repeat the problem statement.

2. *Find:* State briefly and concisely what must be found.

3. *Schematic:* Draw a schematic of the physical system. If application of the conservation laws is anticipated, represent the required control surface or surfaces by dashed lines on the schematic. Identify relevant heat transfer processes by appropriately labeled arrows on the schematic.

4. *Assumptions:* List all pertinent simplifying assumptions.

5. *Properties:* Compile property values needed for subsequent calculations and identify the source from which they are obtained.

6. *Analysis:* Begin your analysis by applying appropriate conservation laws, and introduce rate equations as needed. Develop the analysis as completely as possible before substituting numerical values. Perform the calculations needed to obtain the desired results.

7. *Comments:* Discuss your results. Such a discussion may include a summary of key conclusions, a critique of the original assumptions, and an inference of trends obtained by performing additional *what-if* and *parameter sensitivity* calculations.

The importance of following steps 1 through 4 should not be underestimated. They provide a useful guide to thinking about a problem before effecting its solution. In step 7, we hope you will take the initiative to gain additional insights by performing calculations that may be computer based. The software accompanying this text provides a suitable tool for effecting such calculations.

EXAMPLE 1.8	IHT

The coating on a plate is cured by exposure to an infrared lamp providing a uniform irradiation of 2000 W/m². It absorbs 80% of the infrared irradiation and has an emissivity of 0.50. It is also exposed to an airflow and large surroundings for which temperatures are 20°C and 30°C, respectively.

1. If the convection coefficient between the plate and the ambient air is 15 W/m² · K, what is the cure temperature of the plate?

2. The final characteristics of the coating, including wear and durability, are known to depend on the temperature at which curing occurs. An airflow system is able to control the air velocity, and hence the convection coefficient, on the cured surface, but the process engineer needs to know how the temperature depends on the convection coefficient. Provide the desired information by computing and plotting the surface temperature as a function of h for $2 \leq h \leq 100$ W/m² · K. What value of h would provide a cure temperature of 50°C?

SOLUTION

Known: Coating with prescribed radiation properties is cured by irradiation from an infrared lamp. Heat transfer from the coating is by convection to ambient air and radiation exchange with the surroundings.

Find:

1. Cure temperature for $h = 15$ W/m² · K.

2. Effect of airflow on the cure temperature for $2 \leq h \leq 100$ W/m² · K. Value of h for which the cure temperature is 50°C.

Schematic:

Assumptions:

1. Steady-state conditions.

2. Negligible heat loss from back surface of plate.

3. Plate is small object in large surroundings, and coating has an absorptivity of $\alpha_{sur} = \varepsilon = 0.5$ with respect to irradiation from the surroundings.

Analysis:

1. Since the process corresponds to steady-state conditions and there is no heat transfer at the back surface, the plate must be isothermal ($T_s = T$). Hence the desired temperature may be determined by placing a control surface about the exposed surface and applying Equation 1.13 or by placing the control surface about the entire plate and applying Equation 1.12c. Adopting the latter approach and recognizing that there is no energy generation ($\dot{E}_g = 0$), Equation 1.12c reduces to

$$\dot{E}_{in} - \dot{E}_{out} = 0$$

where $\dot{E}_{st} = 0$ for steady-state conditions. With energy inflow due to absorption of the lamp irradiation by the coating and outflow due to convection and net radiation transfer to the surroundings, it follows that

$$(\alpha G)_{lamp} - q''_{conv} - q''_{rad} = 0$$

Substituting from Equations 1.3a and 1.7, we obtain

$$(\alpha G)_{lamp} - h(T - T_\infty) - \varepsilon\sigma(T^4 - T^4_{sur}) = 0$$

Substituting numerical values

$$0.8 \times 2000 \text{ W/m}^2 - 15 \text{ W/m}^2 \cdot \text{K}(T - 293) \text{ K}$$
$$- 0.5 \times 5.67 \times 10^{-8} \text{ W/m}^2 \cdot \text{K}^4 (T^4 - 303^4) \text{ K}^4 = 0$$

and solving by trial-and-error, we obtain

$$T = 377 \text{ K} = 104°C \qquad \triangleleft$$

2. Solving the foregoing energy balance for selected values of h in the prescribed range and plotting the results, we obtain

If a cure temperature of 50°C is desired, the airflow must provide a convection coefficient of

$$h(T = 50°C) = 51.0 \ W/m^2 \cdot K \qquad \lhd$$

Comments:

1. The coating (plate) temperature may be reduced by decreasing T_∞ and T_{sur}, as well as by increasing the air velocity and hence the convection coefficient.

2. The relative contributions of convection and radiation to heat transfer from the plate vary greatly with h. For $h = 2 \ W/m^2 \cdot K$, $T = 204°C$ and radiation dominates ($q''_{rad} \approx 1232 \ W/m^2$, $q''_{conv} \approx 368 \ W/m^2$). Conversely, for $h = 200 \ W/m^2 \cdot K$, $T = 28°C$ and convection dominates ($q''_{conv} \approx 1606 \ W/m^2$, $q''_{rad} \approx -6 \ W/m^2$). In fact, for this condition the plate temperature is slightly less than that of the surroundings and net radiation exchange is *to* the plate.

1.6 *Relevance of Heat Transfer*

We will devote much time to acquiring an understanding of heat transfer effects and to developing the skills needed to predict heat transfer rates and temperatures that evolve in certain situations. What is the value of this knowledge? To what problems may it be applied? A few examples will serve to illustrate the rich breadth of applications in which heat transfer plays a critical role.

The challenge of providing sufficient amounts of energy for humankind is well known. Adequate supplies of energy are needed not only to fuel industrial productivity, but also to supply safe drinking water and food for much of the world's population and to provide the sanitation necessary to control life-threatening diseases.

To appreciate the role heat transfer plays in the energy challenge, consider a flow chart that represents energy use in the United States, as shown in Figure 1.12*a*. Currently, about 61% of the 100 EJ of energy that is consumed annually in the United States is wasted in the form of heat. Nearly 70% of the energy used to generate electricity is lost in the form of heat. The transportation sector, which relies almost exclusively on petroleum-based fuels, utilizes only 21% of the energy it consumes; the remaining 79% is released in the form of

Figure 1.12   Flow charts for energy consumption and associated CO_2 emissions in the United States in 2012. (*a*) Energy production and consumption. (Based on data from U.S. Department of Energy and the Lawrence Livermore National Laboratory.) (*b*) Carbon dioxide by source of fossil fuel and end-use application [6]. Arrow widths represent relative magnitudes of the flow streams.

heat. Although the industrial and residential/commercial use of energy is relatively more efficient, opportunities for *energy conservation* abound. Creative thermal engineering, utilizing the tools of thermodynamics and heat transfer, can lead to new ways to (1) increase the efficiency by which energy is generated and converted, (2) reduce energy losses, and (3) harvest a large portion of the waste heat.

As evident in Figure 1.12*a*, *fossil fuels* (petroleum, natural gas, and coal) dominate the energy portfolio in many countries, such as the United States. The combustion of fossil fuels produces massive amounts of carbon dioxide; the amount of CO_2 released in the United States on an annual basis due to combustion is currently 5.07 Eg (5.07×10^{15} kg). As more CO_2 is pumped into the atmosphere, mechanisms of radiation heat transfer within the atmosphere are modified, resulting in potential changes in global temperatures. In a country like the United States, electricity generation and transportation are responsible for nearly 75% of the total CO_2 released into the atmosphere due to energy use (Figure 1.12*b*).

What are some of the ways engineers are applying the principles of heat transfer to address issues of energy and environmental *sustainability*?

The efficiency of a *gas turbine engine* can be significantly increased by increasing its operating temperature. Today, the temperatures of the combustion gases inside these engines far exceed the melting point of the exotic alloys used to manufacture the turbine blades and vanes. Safe operation is typically achieved by three means. First, relatively cool gases are injected through small holes at the leading edge of a turbine blade (Figure 1.13). These gases hug the blade as they are carried downstream and help insulate the blade from the hot combustion gases. Second, thin layers of a very low thermal conductivity, ceramic *thermal barrier coating* are applied to the blades and vanes to provide an extra layer of insulation. These coatings are produced by spraying molten ceramic powders onto the engine components using extremely high temperature sources such as plasma spray guns that can operate in excess of 10,000 kelvins. Third, the blades and vanes are designed with intricate, internal cooling passages, all carefully configured by the heat transfer engineer to allow the gas turbine engine to operate under such extreme conditions.

Alternative sources constitute a small fraction of the energy portfolio of many nations, as illustrated in the flow chart of Figure 1.12*a* for the United States. The intermittent nature of the power generated by sources such as the wind and solar irradiation limits their widespread utilization, and creative ways to store excess energy for use during low-power generation periods are urgently needed. For solar irradiation, the sun's rays can be concentrated and used to heat fluids to very high temperatures. The hot fluid can, in turn, be pumped to a power plant to drive a conventional steam turbine or other heat engine. If, however, some of the collected solar energy is stored during the day and retrieved at night, solar-generated electric power can be produced around the clock. Moreover, spreading the electric power generation over the entire day allows the same amount of electric energy to be produced each day, but at half the electric

(*a*)   (*b*)

FIGURE 1.13   Gas turbine blade. (*a*) External view showing holes for injection of cooling gases. (*b*) X ray view showing internal cooling passages. (Credit: Images courtesy of FarField Technology, Ltd., Christchurch, New Zealand.)

power rating. Hence, the size and cost of the power plant are reduced by a factor of approximately two with thermal energy storage. Transferring massive amounts of heat into and out of thermal energy storage devices requires novel heat transfer designs characterized by extremely low thermal resistances. Creation of such devices for thermal energy storage is a key factor to make solar electric power generation more efficient and affordable [7].

Due to the *information technology* revolution of the last several decades, strong industrial productivity growth has brought an improved quality of life worldwide. Many information technology breakthroughs have been enabled by advances in heat transfer engineering that have ensured the precise control of temperatures of systems ranging in size from nanoscale integrated circuits, to large data centers filled with heat-generating equipment. As electronic devices become faster and incorporate greater functionality, they generate more thermal energy. Simultaneously, the devices have become smaller. Inevitably, heat fluxes (W/m^2) and volumetric energy generation rates (W/m^3) keep increasing, but the operating temperatures of the devices must be held to reasonably low values to ensure their reliability. Further improvements in microprocessor technology are currently limited by our ability to cool these tiny devices [8]. Policy makers have voiced concern about limits to continually reduce the cost of computing and, in turn as a society, continue the growth in productivity that has marked the last 30 years, specifically citing the need to enhance heat transfer in electronics cooling [9]. How might your knowledge of heat transfer help ensure continued industrial productivity into the future?

Heat transfer is important not only in engineered systems but also in nature. Temperature regulates and triggers biological responses in all living systems and ultimately marks the boundary between sickness and health. Two common examples include *hypothermia*, which results from excessive cooling of the human body, and *heat stroke*, which is triggered in warm, humid environments. Both are deadly, and both are associated with core temperatures of the body exceeding physiological limits. Both are directly linked to the convection, radiation, and evaporation processes occurring at the surface of the body, the transport of heat within the body, and the metabolic energy generated volumetrically within the body.

Recent advances in *biomedical engineering*, such as laser surgery, have been enabled by successfully applying fundamental heat transfer principles [10, 11]. While increased temperatures resulting from contact with hot objects may cause thermal *burns*, beneficial *hyperthermal treatments* are used to purposely destroy, for example, cancerous lesions. In a similar manner, very low temperatures might induce *frostbite*, but purposeful localized freezing can selectively destroy diseased tissue during *cryosurgery*. Many medical therapies and devices therefore operate by destructively heating or cooling diseased tissue, while leaving the surrounding healthy tissue unaffected.

The ability to design many medical devices and to develop the appropriate protocol for their use hinges on the engineer's ability to predict and control the distribution of temperatures during thermal treatment and the distribution of chemical species in chemotherapies. The treatment of mammalian tissue is made complicated by its morphology, as shown in Figure 1.14. The flow of blood within the venular and capillary structure of a thermally treated area affects heat transfer through advection processes. Larger veins and arteries, which commonly exist in pairs throughout the body, carry blood at different temperatures and advect thermal energy at different rates. Therefore, the veins and arteries exist in a *counterflow heat exchange* arrangement with warm, arteriolar blood exchanging thermal energy with the cooler, venular blood through the intervening solid tissue. Networks of smaller capillaries can also affect local temperatures by *perfusing* blood through the treated area.

In subsequent chapters, example and homework problems will deal with the analysis of these and many other *thermal systems*.

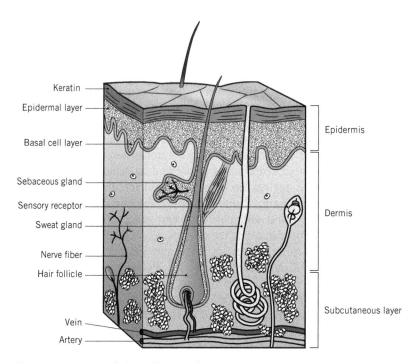

FIGURE 1.14   Morphology of human skin.

1.7 *Summary*

Although much of the material of this chapter will be discussed in greater detail, you should now have a reasonable overview of heat transfer. You should be aware of the several modes of transfer and their physical origins. You will be devoting much time to acquiring the tools needed to calculate heat transfer phenomena. However, before you can use these tools effectively, you must have the intuition to determine what is happening physically. Specifically, given a physical situation, you must be able to identify the relevant transport phenomena; the importance of developing this facility must not be underestimated. The example and problems at the end of this chapter will launch you on the road to developing this intuition.

You should also appreciate the significance of the rate equations and feel comfortable in using them to compute transport rates. These equations, summarized in Table 1.5, *should be committed to memory*. You must also recognize the importance of the conservation laws and the need to carefully identify control volumes. With the rate equations, the conservation laws may be used to solve numerous heat transfer problems.

Lastly, you should have begun to acquire an appreciation for the terminology and physical concepts that underpin the subject of heat transfer. Test your understanding of the important terms and concepts introduced in this chapter by addressing the following questions:

- What are the *physical mechanisms* associated with heat transfer by *conduction, convection,* and *radiation*?
- What is the driving potential for heat transfer? What are analogs to this potential and to heat transfer itself for the transport of electric charge?
- What is the difference between a heat *flux* and a heat *rate*? What are their units?

TABLE 1.5 Summary of heat transfer processes

Mode	Mechanism(s)	Rate Equation	Equation Number	Transport Property or Coefficient
Conduction	Diffusion of energy due to random molecular motion	$q''_x(\text{W/m}^2) = -k\dfrac{dT}{dx}$	(1.1)	k (W/m · K)
Convection	Diffusion of energy due to random molecular motion plus energy transfer due to bulk motion (advection)	$q''(\text{W/m}^2) = h(T_s - T_\infty)$	(1.3a)	h (W/m² · K)
Radiation	Energy transfer by electromagnetic waves	$q''(\text{W/m}^2) = \varepsilon\sigma(T_s^4 - T_{\text{sur}}^4)$ or $q(\text{W}) = h_r A(T_s - T_{\text{sur}})$	(1.7) (1.8)	ε h_r (W/m² · K)

- What is a *temperature gradient*? What are its units? What is the relationship of heat flow to a temperature gradient?
- What is the *thermal conductivity*? What are its units? What role does it play in heat transfer?
- What is *Fourier's law*? Can you write the equation from memory?
- If heat transfer by conduction through a medium occurs under *steady-state* conditions, will the temperature at a particular instant vary with location in the medium? Will the temperature at a particular location vary with time?
- What is the difference between *natural convection* and *forced convection*?
- What conditions are necessary for the development of a *hydrodynamic boundary layer*? A *thermal boundary layer*? What varies across a hydrodynamic boundary layer? Across a thermal boundary layer?
- If convection heat transfer for flow of a liquid or a vapor is not characterized by liquid/vapor phase change, what is the nature of the energy being transferred? What is it if there is such a phase change?
- What is *Newton's law of cooling*? Can you write the equation from memory?
- What role is played by the *convection heat transfer coefficient* in Newton's law of cooling? What are its units?
- What effect does convection heat transfer from or to a surface have on the solid bounded by the surface?
- What is predicted by the Stefan–Boltzmann law, and what unit of temperature must be used with the law? Can you write the equation from memory?
- What is the *emissivity*, and what role does it play in characterizing radiation transfer at a surface?
- What is *irradiation*? What are its units?
- What two outcomes characterize the response of an *opaque* surface to incident radiation? Which outcome affects the thermal energy of the medium bounded by the surface and how? What property characterizes this outcome?
- What conditions are associated with use of the *radiation heat transfer coefficient*?
- Can you write the equation used to express net radiation exchange between a small isothermal surface and a large isothermal enclosure?

- Consider the surface of a solid that is at an elevated temperature and exposed to cooler surroundings. By what mode(s) is heat transferred from the surface if (1) it is in intimate (perfect) contact with another solid, (2) it is exposed to the flow of a liquid, (3) it is exposed to the flow of a gas, and (4) it is in an evacuated chamber?
- What is the inherent difference between the application of conservation of energy over a *time interval* and at an *instant of time*?
- What is *thermal energy storage*? How does it differ from *thermal energy generation*? What role do the terms play in a surface energy balance?

EXAMPLE 1.9

A closed plastic thermos filled with hot coffee is in a room whose air and walls are at a fixed temperature. Identify all heat transfer processes that contribute to the cooling of the coffee. Comment on features that would contribute to a superior container design.

SOLUTION

Known: Hot coffee is separated from its cooler surroundings by a plastic flask, an air space, and a plastic cover.

Find: Relevant heat transfer processes.

Schematic:

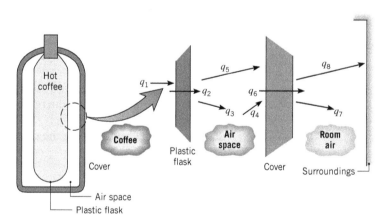

Pathways for energy transfer from the coffee are as follows:

q_1:   free convection from the coffee to the flask.

q_2:   conduction through the flask.

q_3:   free convection from the flask to the air.

q_4:   free convection from the air to the cover.

q_5:   net radiation exchange between the outer surface of the flask and the inner surface of the cover.

q_6: conduction through the cover.

q_7: free convection from the cover to the room air.

q_8: net radiation exchange between the outer surface of the cover and the surroundings.

Comments: Design improvements are associated with (1) use of aluminized (low-emissivity) surfaces for the flask and cover to reduce net radiation, and (2) evacuating the air space or using a filler material to retard free convection.

References

1. Moran, M. J., and H. N. Shapiro, *Fundamentals of Engineering Thermodynamics*, Wiley, Hoboken, NJ, 2004.

2. Curzon, F. L., and B. Ahlborn, *American J. Physics*, **43**, 22, 1975.

3. Novikov, I. I., *J. Nuclear Energy II*, **7**, 125, 1958.

4. Callen, H. B., *Thermodynamics and an Introduction to Thermostatistics*, Wiley, Hoboken, NJ, 1985.

5. Bejan, A., *American J. Physics*, **64**, 1054, 1996.

6. U.S. Environmental Protection Agency, *Inventory of U.S. Greenhouse Emissions and Sinks: 1990–2012*, Publication EPA 430-R-14-003, 2014.

7. Shabgard, H., T. L. Bergman, and A. Faghri, *Energy*, **60**, 474, 2013.

8. Bar-Cohen, A., and I. Madhusudan, *IEEE Trans. Components and Packaging Tech.*, **25**, 584, 2002.

9. Miller, R., *Business Week*, November 11, 2004.

10. Diller, K. R., and T. P. Ryan, *J. Heat Transfer*, **120**, 810, 1998.

11. Datta, A.K., *Biological and Bioenvironmental Heat and Mass Transfer*, Marcel Dekker, New York, 2002.

Problems

Conduction

1.1 The thermal conductivity of a sheet of rigid, extruded insulation is reported to be $k = 0.029$ W/m $\cdot$ K. The measured temperature difference across a 20-mm-thick sheet of the material is $T_1 - T_2 = 10°C$.

(a) What is the heat flux through a 2 m $\times$ 2 m sheet of the insulation?

(b) What is the rate of heat transfer through the sheet of insulation?

1.2 Consider the fireclay brick wall of Example 1.1 that is operating under different thermal conditions. The temperature distribution, at an instant in time, is $T(x) = a + bx$ where $a = 1400$ K and $b = -1000$ K/m. Determine the heat fluxes, q_x'', and heat rates, q_x, at $x = 0$ and $x = L$. Do steady-state conditions exist?

1.3 The heat flux that is applied to the left face of a plane wall is $q'' = 25$ W/m^2. The wall is of thickness $L = 12$ mm and of thermal conductivity $k = 12$ W/m $\cdot$ K. If the surface temperatures of the wall are measured to be 50°C on the left side and 30°C on the right side, do steady-state conditions exist?

1.4 A concrete wall, which has a surface area of 20 m^2 and is 0.30 m thick, separates conditioned room air from ambient air. The temperature of the inner surface of the wall is maintained at 25°C, and the thermal conductivity of the concrete is 1 W/m $\cdot$ K.

(a) Determine the rate of heat loss through the wall for outer surface temperatures ranging from −15°C to 38°C, which correspond to winter and summer extremes, respectively. Display your results graphically.

(b) On your graph, also plot the rate of heat loss as a function of the outer surface temperature for wall materials having thermal conductivities of 0.75 and 1.25 W/m $\cdot$ K. Explain the family of curves you have obtained.

1.5 The concrete slab of a basement is 11 m long, 8 m wide, and 0.20 m thick. During the winter, temperatures are nominally 17°C and 10°C at the top and bottom surfaces, respectively. If the concrete has a thermal conductivity of 1.4 W/m · K, what is the rate of heat loss through the slab? If the basement is heated by a gas furnace operating at an efficiency of $\eta_f = 0.90$ and natural gas is priced at $C_g = \$0.01/\text{MJ}$, what is the daily cost of the heat loss?

1.6 The heat flux through a wood slab 60 mm thick, whose inner and outer surface temperatures are 35 and 15°C, respectively, has been determined to be 35 W/m². What is the thermal conductivity of the wood?

1.7 The inner and outer surface temperatures of a glass window 5 mm thick are 15 and 5°C. What is the heat loss through a 1 m × 3 m window? The thermal conductivity of glass is 1.4 W/m · K.

1.8 The heat flux through a wood slab 50 mm thick, whose inner and outer surface temperatures are 40 and 20°C, respectively, has been determined to be 40 W/m². What is the thermal conductivity of the wood?

1.9 A thermodynamic analysis of a proposed Brayton cycle gas turbine yields $P = 5$ MW of net power production. The compressor, at an average temperature of $T_c = 400°C$, is driven by the turbine at an average temperature of $T_h = 1000°C$ by way of an $L = 1$-m-long, $d = 70$-mm-diameter shaft of thermal conductivity $k = 40$ W/m · K.

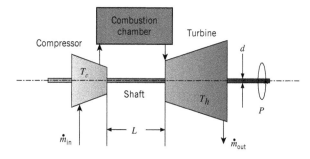

(a) Compare the steady-state conduction rate through the shaft connecting the hot turbine to the warm compressor to the net power predicted by the thermodynamics-based analysis.

(b) A research team proposes to scale down the gas turbine of part (a), keeping all dimensions in the same proportions. The team assumes that the same hot and cold temperatures exist as in part (a) and that the net power output of the gas turbine is proportional to the overall volume of the device. Plot the ratio of the conduction through the shaft to the net power output of the turbine over the range $0.005 \text{ m} \leq L \leq 1$ m. Is a scaled-down device with $L = 0.005$ m feasible?

1.10 The heat flux that is applied to one face of a plane wall is $q'' = 25$ W/m². The opposite face is exposed to air at temperature 30°C, with a convection heat transfer coefficient

of 25 W/m² · K. The surface temperature of the wall exposed to air is measured and found to be 50°C. Do steady-state conditions exist? If not, is the temperature of the wall increasing or decreasing with time?

1.11 What is the thickness required of a masonry wall having thermal conductivity 0.75 W/m · K if the heat rate is to be 80% of the heat rate through a composite structural wall having a thermal conductivity of 0.25 W/m · K and a thickness of 100 mm? Both walls are subjected to the same surface temperature difference.

1.12 The 5-mm-thick bottom of a 200-mm-diameter pan may be made from aluminum ($k = 240$ W/m · K) or copper ($k = 390$ W/m · K). When used to boil water, the surface of the bottom exposed to the water is nominally at 110°C. If heat is transferred from the stove to the pan at a rate of 600 W, what is the temperature of the surface in contact with the stove for each of the two materials?

1.13 A square silicon chip ($k = 150$ W/m · K) is of width $w = 5$ mm on a side and of thickness $t = 1$ mm. The chip is mounted in a substrate such that its side and back surfaces are insulated, while the front surface is exposed to a coolant. If 4 W are being dissipated in circuits mounted to the back surface of the chip, what is the steady-state temperature difference between back and front surfaces?

Convection

1.14 You've experienced convection cooling if you've ever extended your hand out the window of a moving vehicle or into a flowing water stream. With the surface of your hand at a temperature of 30°C, determine the convection heat flux for (a) a vehicle speed of 35 km/h in air at −5°C with a convection coefficient of 40 W/m² · K and (b) a velocity of 0.2 m/s in a water stream at 10°C with a convection coefficient of 900 W/m² · K. Which condition would *feel* colder? Contrast these results with a heat loss of approximately 30 W/m² under normal room conditions.

1.15 Air at 40°C flows over a long, 25-mm-diameter cylinder with an embedded electrical heater. In a series of tests, measurements were made of the power per unit length, P', required to maintain the cylinder surface temperature at 300°C for different free stream velocities V of the air. The results are as follows:

Air velocity, V (m/s)	1	2	4	8	12
Power, P' (W/m)	450	658	983	1507	1963

(a) Determine the convection coefficient for each velocity, and display your results graphically.

(b) Assuming the dependence of the convection coefficient on the velocity to be of the form $h = CV^n$, determine the parameters C and n from the results of part (a).

1.16 A wall has inner and outer surface temperatures of 22 and 10°C, respectively. The interior and exterior air temperatures are 26 and 10°C, respectively. The inner and outer convection heat transfer coefficients are 5 and 20 W/m² · K, respectively. Calculate the heat flux from the interior air to the wall, from the wall to the exterior air, and from the wall to the interior air. Is the wall under steady-state conditions?

1.17 The free convection heat transfer coefficient on a thin hot vertical plate suspended in still air can be determined from observations of the change in plate temperature with time as it cools. Assuming the plate is isothermal and radiation exchange with its surroundings is negligible, evaluate the convection coefficient at the instant of time when the plate temperature is 225°C and the change in plate temperature with time (dT/dt) is −0.022 K/s. The ambient air temperature is 25°C and the plate measures 0.3 × 0.3 m with a mass of 3.75 kg and a specific heat of 2770 J/kg · K.

1.18 An electric resistance heater is embedded in a long cylinder of diameter 30 mm. When water with a temperature of 25°C and velocity of 1 m/s flows crosswise over the cylinder, the power per unit length required to maintain the surface at a uniform temperature of 90°C is 28 kW/m. When air, also at 25°C, but with a velocity of 10 m/s is flowing, the power per unit length required to maintain the same surface temperature is 400 W/m. Calculate and compare the convection coefficients for the flows of water and air.

1.19 A transmission case measures $W = 0.30$ m on a side and receives a power input of $P_i = 150$ hp from the engine.

If the transmission efficiency is $\eta = 0.93$ and airflow over the case corresponds to $T_\infty = 30$°C and $h = 200$ W/m² · K, what is the surface temperature of the transmission? What is the thermal resistance associated with convection?

1.20 A cartridge electrical heater is shaped as a cylinder of length $L = 200$ mm and outer diameter $D = 20$ mm. Under normal operating conditions, the heater dissipates 2 kW while submerged in a water flow that is at 20°C and provides a convection heat transfer coefficient of $h = 5000$ W/m² · K. Neglecting heat transfer from the ends of the heater, determine its surface temperature T_s. If the water flow is inadvertently terminated while the heater continues to operate, the heater surface is exposed to air that is also at 20°C but for which $h = 50$ W/m² · K. What is the corresponding surface temperature? What are the consequences of such an event?

1.21 A common procedure for measuring the velocity of an airstream involves the insertion of an electrically heated wire (called a *hot-wire anemometer*) into the airflow, with the axis of the wire oriented perpendicular to the flow direction. The electrical energy dissipated in the wire is assumed to be transferred to the air by forced convection. Hence, for a prescribed electrical power, the temperature of the wire depends on the convection coefficient, which, in turn, depends on the velocity of the air. Consider a wire of length $L = 20$ mm and diameter $D = 0.5$ mm, for which a calibration of the form $V = 6.25 \times 10^{-5} h^2$ has been determined. The velocity V and the convection coefficient h have units of m/s and W/m² · K, respectively. In an application involving air at a temperature of $T_\infty = 20$°C, the surface temperature of the anemometer is maintained at $T_s = 70$°C with a voltage drop of 5 V and an electric current of 0.1 A. What is the velocity of the air?

1.22 A square isothermal chip is of width $w = 5$ mm on a side and is mounted in a substrate such that its side and back surfaces are well insulated; the front surface is exposed to the flow of a coolant at $T_\infty = 15$°C. From reliability considerations, the chip temperature must not exceed $T = 85$°C.

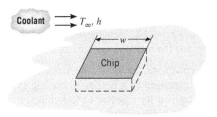

If the coolant is air and the corresponding convection coefficient is $h = 250$ W/m² · K, what is the maximum allowable chip power? If the coolant is a dielectric liquid for which $h = 3000$ W/m² · K, what is the maximum allowable power?

1.23 The temperature controller for a clothes dryer consists of a bimetallic switch mounted on an electrical heater attached to a wall-mounted insulation pad.

The switch is set to open at 70°C, the maximum dryer air temperature. To operate the dryer at a lower air temperature, sufficient power is supplied to the heater such that the switch reaches 70°C (T_{set}) when the air temperature T is less than T_{set}. If the convection heat transfer coefficient between the air and the exposed switch surface of 30 mm² is 25 W/m² · K, how much heater power P_e is required when the desired dryer air temperature is $T_\infty = 50°C$?

Radiation

1.24 An overhead 30-m-long, uninsulated industrial steam pipe of 120-mm diameter is routed through a building whose walls and air are at 30°C. Pressurized steam maintains a pipe surface temperature of 150°C, and the coefficient associated with natural convection is $h = 10$ W/m² · K. The surface emissivity is $\varepsilon = 0.8$.

(a) What is the rate of heat loss from the steam line?

(b) If the steam is generated in a gas-fired boiler operating at an efficiency of $\eta_f = 0.90$ and natural gas is priced at $C_g = \$0.02$ per MJ, what is the annual cost of heat loss from the line?

1.25 A one-dimensional plane wall is exposed to convective and radiative conditions at $x = 0$. The ambient and surrounding temperatures are $T_\infty = 20°C$ and $T_{sur} = 40°C$, respectively. The convection heat transfer coefficient is $h = 20$ W/m² · K, and the absorptivity of the exposed surface is $\alpha = 0.78$. Determine the convective and radiative heat fluxes to the wall at $x = 0$ if the wall surface temperature is $T_s = 24°C$. Assume the exposed wall surface is gray, and the surroundings are large.

1.26 Under conditions for which the same room temperature is maintained by a heating or cooling system, it is not uncommon for a person to feel chilled in the winter but comfortable in the summer. Provide a plausible explanation for this situation (with supporting calculations) by considering a room whose air temperature is maintained at 20°C throughout the year, while the walls of the room are nominally at 27°C and 14°C in the summer and winter, respectively. The exposed surface of a person in the room may be assumed to be at a temperature of 32°C throughout the year and to have an emissivity of 0.90. The coefficient associated with heat transfer by natural convection between the person and the room air is approximately 2 W/m² · K. What is the ratio of the thermal resistance due to convection to the thermal resistance due to radiation in the summer? What is the ratio of thermal resistances in the winter?

1.27 A spherical interplanetary probe of 0.6-m diameter contains electronics that dissipate 170 W. If the probe surface has an emissivity of 0.85 and the probe does not receive radiation from other surfaces, as, for example, from the sun, what is its surface temperature?

1.28 An instrumentation package has a spherical outer surface of diameter $D = 100$ mm and emissivity $\varepsilon = 0.25$. The package is placed in a large space simulation chamber whose walls are maintained at 77 K. If operation of the electronic components is restricted to the temperature range $40 \leq T \leq 85°C$, what is the range of acceptable power dissipation for the package? Display your results graphically, showing also the effect of variations in the emissivity by considering values of 0.20 and 0.30.

1.29 Consider the conditions of Problem 1.17. However, now the plate is in a vacuum with a surrounding temperature of 25°C. What is the emissivity of the plate? What is the rate at which radiation is emitted by the surface?

1.30 A vacuum system, as used in sputtering electrically conducting thin films on microcircuits, is comprised of a baseplate maintained by an electrical heater at 300 K and a shroud within the enclosure maintained at 77 K by a liquid-nitrogen coolant loop. The circular baseplate, insulated on the lower side, is 0.3 m in diameter and has an emissivity of 0.25.

(a) How much electrical power must be provided to the baseplate heater?

(b) At what rate must liquid nitrogen be supplied to the shroud if its heat of vaporization is 125 kJ/kg?

(c) To reduce the liquid nitrogen consumption, it is proposed to bond a thin sheet of aluminum foil ($\varepsilon = 0.09$) to the baseplate. Will this have the desired effect?

Relationship to Thermodynamics

1.31 Energy storage is necessary to, for example, allow solar-derived electricity to be generated around the clock. For a given mass of storage medium, show that the ratio of sensible thermal energy storage capacity to potential energy storage capacity may be expressed as

$$R = \frac{\Delta E_{st,t}}{\Delta E_{st,PE}} = \frac{c\Delta T}{gz}$$

where ΔT is the difference between the maximum and minimum temperatures associated with thermal energy storage, c is the specific heat, and z is the vertical coordinate for potential energy storage. Considering stone mix concrete as the storage medium, determine the value of R for $\Delta T = 100°C$ and $z = 100$ m. Which energy storage approach, thermal or potential, is more effective for the parameters of this problem?

1.32 An electrical resistor is connected to a battery, as shown schematically. After a brief transient, the resistor assumes a nearly uniform, steady-state temperature of 95°C, while the battery and lead wires remain at the ambient temperature of 25°C. Neglect the electrical resistance of the lead wires.

(a) Consider the resistor as a system about which a control surface is placed and Equation 1.12c is applied. Determine the corresponding values of $\dot{E}_{in}$ (W), $\dot{E}_g$ (W), $\dot{E}_{out}$ (W), and $\dot{E}_{st}$ (W). If a control surface is placed about the entire system, what are the values of $\dot{E}_{in}$, $\dot{E}_g$, $\dot{E}_{out}$, and $\dot{E}_{st}$?

(b) If electrical energy is dissipated uniformly within the resistor, which is a cylinder of diameter $D = 60$ mm and length $L = 250$ mm, what is the volumetric heat generation rate, $\dot{q}$ (W/m³)?

(c) Neglecting radiation from the resistor, what is the convection coefficient?

1.33 Pressurized water ($p_{in} = 10$ bar, $T_{in} = 110°C$) enters the bottom of an $L = 10$-m-long vertical tube of diameter $D = 100$ mm at a mass flow rate of $\dot{m} = 1.5$ kg/s. The tube is located inside a combustion chamber, resulting in heat transfer to the tube. Superheated steam exits the top of the tube at $p_{out} = 7$ bar, $T_{out} = 600°C$. Determine the change in the rate at which the following quantities enter and exit the tube: (a) the combined thermal and flow work, (b) the mechanical energy, and (c) the total energy of the water. Also, (d) determine the heat transfer rate, q. *Hint*: Relevant properties may be obtained from a thermodynamics text.

1.34 Consider the tube and inlet conditions of Problem 1.33. Heat transfer at a rate of $q = 3.89$ MW is delivered to the tube. For an exit pressure of $p = 8$ bar, determine (a) the temperature of the water at the outlet as well as the change in (b) combined thermal and flow work, (c) mechanical energy, and (d) total energy of the water from the inlet to the outlet of the tube. *Hint:* As a first estimate, neglect the change in mechanical energy in solving part (a). Relevant properties may be obtained from a thermodynamics text.

1.35 An internally reversible refrigerator has a modified coefficient of performance accounting for realistic heat transfer processes of

$$COP_m = \frac{q_{in}}{\dot{W}} = \frac{q_{in}}{q_{out} - q_{in}} = \frac{T_{c,i}}{T_{h,i} - T_{c,i}}$$

where q_{in} is the refrigerator cooling rate, q_{out} is the heat rejection rate, and $\dot{W}$ is the power input. Show that COP_m can be expressed in terms of the reservoir temperatures T_c and T_h, the cold and hot thermal resistances $R_{t,c}$ and $R_{t,h}$, and q_{in}, as

$$COP_m = \frac{T_c - q_{in}R_{tot}}{T_h - T_c + q_{in}R_{tot}}$$

where $R_{tot} = R_{t,c} + R_{t,h}$. Also, show that the power input may be expressed as

$$\dot{W} = q_{in}\frac{T_h - T_c + q_{in}R_{tot}}{T_c - q_{in}R_{tot}}$$

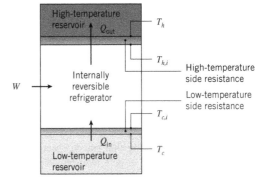

1.36 A household refrigerator operates with cold- and hot-temperature reservoirs of $T_c = 5°C$ and $T_h = 25°C$, respectively. When new, the cold and hot side resistances are $R_{c,n} = 0.05$ K/W and $R_{h,n} = 0.04$ K/W, respectively. Over time, dust accumulates on the refrigerator's condenser coil, which is located behind the refrigerator, increasing the hot side resistance to $R_{h,d} = 0.1$ K/W. It is desired to have a refrigerator cooling rate of $q_{in} = 750$ W. Using the results of Problem 1.35, determine the modified coefficient of performance and the required power input $\dot{W}$ under (a) clean and (b) dusty coil conditions.

Energy Balance and Multimode Effects

1.37 Chips of width $L = 15$ mm on a side are mounted to a substrate that is installed in an enclosure whose walls and air are maintained at a temperature of $T_{sur} = 25°C$. The chips have an emissivity of $\varepsilon = 0.60$ and a maximum allowable temperature of $T_s = 85°C$.

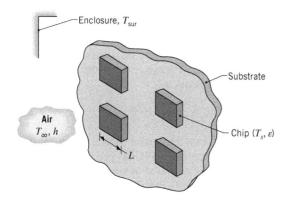

(a) If heat is rejected from the chips by radiation and natural convection, what is the maximum operating power of each chip? The convection coefficient depends on the chip-to-air temperature difference and may be approximated as $h = C(T_s - T_\infty)^{1/4}$, where $C = 4.2$ W/m$^2 \cdot$ K$^{5/4}$.

(b) If a fan is used to maintain airflow through the enclosure and heat transfer is by forced convection, with $h = 250$ W/m$^2 \cdot$ K, what is the maximum operating power?

1.38 Consider the transmission case of Problem 1.19, but now allow for radiation exchange with the ground/chassis, which may be approximated as large surroundings at $T_{sur} = 30°C$. If the emissivity of the case is $\varepsilon = 0.80$, what is the surface temperature? What are the values of the thermal resistances due to convection and radiation?

1.39 One method for growing thin silicon sheets for photovoltaic solar panels is to pass two thin strings of high melting temperature material upward through a bath of molten silicon. The silicon solidifies on the strings near the surface of the molten pool, and the solid silicon sheet is pulled slowly upward out of the pool. The silicon is replenished by supplying the molten pool with solid silicon powder. Consider a silicon sheet that is $W_{si} = 85$ mm wide and $t_{si} = 150$ μm thick that is pulled at a velocity of $V_{si} = 20$ mm/min. The silicon is melted by supplying electric power to the cylindrical growth chamber of height $H = 350$ mm and diameter $D = 300$ mm. The exposed surfaces of the growth chamber are at $T_s = 320$ K, the corresponding convection coefficient at the exposed surface is $h = 8$ W/m$^2 \cdot$ K, and the surface is characterized by an emissivity of $\varepsilon_s = 0.9$. The solid silicon powder is at $T_{si,i} = 298$ K, and the solid silicon sheet exits the chamber at $T_{si,o} = 420$ K. Both the surroundings and ambient temperatures are $T_\infty = T_{sur} = 298$ K.

(a) Determine the electric power, P_{elec}, needed to operate the system at steady state.

(b) If the photovoltaic panel absorbs a time-averaged solar flux of $q''_{sol} = 180$ W/m^2 and the panel has a conversion efficiency (the ratio of solar power absorbed to electric power produced) of $\eta = 0.20$, how long must the solar panel be operated to produce enough electric energy to offset the electric energy that was consumed in its manufacture?

1.40 Heat is transferred by radiation and convection between the inner surface of the nacelle of the wind turbine of Example 1.3 and the outer surfaces of the gearbox and generator. The convection heat flux associated with the gearbox and the generator may be described by $q''_{conv,gb} = h(T_{gb} - T_\infty)$ and $q''_{conv,gen} = h(T_{gen} - T_\infty)$, respectively, where the ambient temperature $T_\infty \approx T_s$ (which is the nacelle temperature) and $h = 40$ W/m$^2 \cdot$ K. The outer surfaces of both the gearbox and the generator are characterized by an emissivity of $\varepsilon = 0.9$. If the surface areas of the gearbox and generator are $A_{gb} = 6$ m^2 and $A_{gen} = 4$ m^2, respectively, determine their surface temperatures.

1.41 Radioactive wastes are packed in a long, thin-walled cylindrical container. The wastes generate thermal energy nonuniformly according to the relation $\dot{q} = \dot{q}_o[1 - (r/r_o)^2]$, where $\dot{q}$ is the local rate of energy generation per unit volume, $\dot{q}_o$ is a constant, and r_o is the radius of the container. Steady-state conditions are maintained by submerging the container in a liquid that is at T_∞ and provides a uniform convection coefficient h.

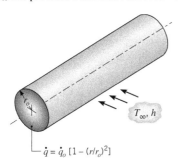

$$\dot{q} = \dot{q}_o [1 - (r/r_o)^2]$$

Obtain an expression for the total rate at which energy is generated in a unit length of the container. Use this result to obtain an expression for the temperature T_s of the container wall.

1.42 An aluminum plate 4 mm thick is mounted in a horizontal position, and its bottom surface is well insulated. A special, thin coating is applied to the top surface such that it absorbs 80% of any incident solar radiation, while having an emissivity of 0.25. The density ρ and specific heat c of aluminum are known to be 2700 kg/m³ and 900 J/kg · K, respectively.

(a) Consider conditions for which the plate is at a temperature of 35°C and its top surface is suddenly exposed to ambient air at $T_\infty = 30°C$ and to solar radiation that provides an incident flux of 900 W/m². The convection heat transfer coefficient between the surface and the air is $h = 20$ W/m² · K. What is the initial rate of change of the plate temperature?

(b) What will be the equilibrium temperature of the plate when steady-state conditions are reached?

1.43 A blood warmer is to be used during the transfusion of blood to a patient. This device is to heat blood taken from the blood bank at 10°C to 37°C at a flow rate of 200 ml/min. The blood passes through tubing of length 3 m, with a rectangular cross section 6.4 mm × 1.6 mm. At what rate must heat be added to the blood to accomplish the required temperature increase? If the fluid originates from a large tank with nearly zero velocity and flows vertically downward for its 2-m length, estimate the magnitudes of kinetic and potential energy changes. Assume the blood's properties are similar to those of water.

1.44 Consider a carton of milk that is refrigerated at a temperature of $T_m = 5°C$. The kitchen temperature on a hot summer day is $T_\infty = 30°C$. If the four sides of the carton are of height and width $L = 150$ mm and $w = 80$ mm, respectively, determine the heat transferred to the milk carton as it sits on the kitchen counter for durations of $t = 10\,s$, $60\,s$, and $300\,s$ before it is returned to the refrigerator. The convection coefficient associated with natural convection on the sides of the carton is $h = 10$ W/m² · K. The surface emissivity is 0.90. Assume the milk carton temperature remains at 5°C during the process. Your parents have taught you the importance of refrigerating certain foods from the food safety perspective. Comment on the importance of quickly returning the milk carton to the refrigerator from an energy conservation point of view.

1.45 The emissivity of galvanized steel sheet, a common roofing material, is $\varepsilon = 0.13$ at temperatures around 300 K, while its absorptivity for solar irradiation is $\alpha_S = 0.65$. Would the neighborhood cat be comfortable walking on a roof constructed of the material on a day when $G_S = 850$ W/m², $T_\infty = 18°C$, and $h = 7$ W/m² · K? Assume the bottom surface of the steel is insulated.

1.46 In one stage of an annealing process, 304 stainless steel sheet is taken from 300 K to 1250 K as it passes through an electrically heated oven at a speed of $V_s = 10$ mm/s. The sheet thickness and width are $t_s = 8$ mm and $W_s = 2$ m, respectively, while the height, width, and length of the oven are $H_o = 2$ m, $W_o = 2.4$ m, and $L_o = 25$ m, respectively. The top and four sides of the oven are exposed to ambient air and large surroundings, each at 300 K, and the corresponding surface temperature, convection coefficient, and emissivity are $T_s = 350$ K, $h = 10$ W/m² · K, and $\varepsilon_s = 0.8$. The bottom surface of the oven is also at 350 K and rests on a 0.5-m-thick concrete pad whose base is at 300 K. Estimate the required electric power input, P_{elec}, to the oven.

1.47 Annealing, an important step in semiconductor materials processing, can be accomplished by rapidly heating the silicon wafer to a high temperature for a short period of time. The schematic shows a method involving the use of a hot plate operating at an elevated temperature T_h. The wafer, initially at a temperature of $T_{w,i}$, is suddenly positioned at a gap separation distance L from the hot plate. The purpose of the analysis is to compare the heat fluxes by conduction through the gas within the gap and by radiation exchange between the hot plate and the cool wafer. The initial time rate of change in the temperature of the wafer, $(dT_w/dt)_i$, is also of interest. Approximating the surfaces of the hot plate and the wafer as blackbodies and assuming their diameter D to be much larger than the spacing L, the radiative heat flux may be expressed as $q''_{rad} = \sigma(T_h^4 - T_w^4)$. The silicon wafer has a thickness of $d = 0.78$ mm, a density of 2700 kg/m^3, and a specific heat of 875 J/kg · K. The thermal conductivity of the gas in the gap is 0.0436 W/m · K.

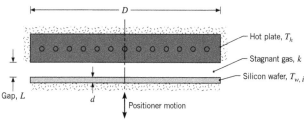

(a) For $T_h = 600°C$ and $T_{w,i} = 20°C$, calculate the radiative heat flux and the heat flux by conduction across a gap distance of $L = 0.2$ mm. Also determine the value of $(dT_w/dt)_i$, resulting from each of the heating modes.

(b) For gap distances of 0.2, 0.5, and 1.0 mm, determine the heat fluxes and temperature-time change as a function of the hot plate temperature for $300 \leq T_h \leq 1300°C$. Display your results graphically. Comment on the relative importance of the two heat transfer modes and the effect of the gap distance on the heating process. Under what conditions could a wafer be heated to 900°C in less than 10 s?

1.48 A furnace for processing semiconductor materials is formed by a silicon carbide chamber that is zone-heated on the top section and cooled on the lower section. With the elevator in the lowest position, a robot arm inserts the silicon wafer on the mounting pins. In a production operation, the wafer is rapidly moved toward the hot zone to achieve the temperature-time history required for the process recipe. In this position, the top and bottom surfaces of the wafer exchange radiation with the hot and cool zones, respectively, of the chamber. The zone temperatures are $T_h = 1500$ K and $T_c = 330$ K, and the emissivity and thickness of the wafer are $\varepsilon = 0.65$ and $d = 0.78$ mm, respectively. With the ambient gas at $T_\infty = 700$ K, convection coefficients at the upper and lower surfaces of the wafer are 8 and 4 W/m^2 · K, respectively. The silicon wafer has a density of 2700 kg/m^3 and a specific heat of 875 J/kg · K.

(a) For an initial condition corresponding to a wafer temperature of $T_{w,i} = 300$ K and the position of the wafer shown schematically, determine the corresponding time rate of change of the wafer temperature, $(dT_w/dt)_i$.

(b) Determine the steady-state temperature reached by the wafer if it remains in this position. How significant is convection heat transfer for this situation? Sketch how you would expect the wafer temperature to vary as a function of vertical distance.

1.49 Consider the conducting rod of Example 1.4 under steady-state conditions. As suggested in Comment 3, the temperature of the rod may be controlled by varying the speed of airflow over the rod, which, in turn, alters the convection heat transfer coefficient. To consider the effect of the convection coefficient, generate plots of T versus I for values of $h = 50$, 100, and 250 W/m^2 · K. Would variations in the surface emissivity have a significant effect on the rod temperature?

1.50 A small sphere of reference-grade iron with a specific heat of 447 J/kg · K and a mass of 0.60 kg is suddenly immersed in a water–ice mixture. Fine thermocouple wires suspend the sphere, and the temperature is observed to change from 15 to 14°C in 6.35 s. The experiment is repeated with a metallic sphere of the same diameter, but of unknown composition with a mass of 1.348 kg. If the same observed temperature change occurs in 4.59 s, what is the specific heat of the unknown material?

1.51 A 60 mm × 55 mm × 30 mm cell phone charger has a surface temperature of $T_s = 40°C$ when plugged into an electrical wall outlet but not in use. The surface of the charger is of emissivity $\varepsilon = 0.92$ and is subject to a free convection heat transfer coefficient of $h = 4.5$ W/m² · K. The room air and wall temperatures are $T_\infty = 22°C$ and $T_{sur} = 20°C$, respectively. If electricity costs $C = \$0.18/kW · h$, determine the daily cost of leaving the charger plugged in when not in use.

1.52 A spherical, stainless steel (AISI 302) canister is used to store reacting chemicals that provide for a uniform heat flux q_i'' to its inner surface. The canister is suddenly submerged in a liquid bath of temperature $T_\infty < T_i$, where T_i is the initial temperature of the canister wall.

$T_i = 500$ K
$\rho = 8055$ kg/m³
$c_p = 510$ J/kg·K

Canister — Reacting chemicals
$r_o = 0.6$ m
q_i''
$T_\infty = 300$ K
$h = 500$ W/m²·K
Bath
$r_i = 0.5$ m

(a) Assuming negligible temperature gradients in the canister wall and a constant heat flux q_i'', develop an equation that governs the variation of the wall temperature with time during the transient process. What is the initial rate of change of the wall temperature if $q_i'' = 10^5$ W/m²?

(b) What is the steady-state temperature of the wall?

(c) The convection coefficient depends on the velocity associated with fluid flow over the canister and whether the wall temperature is large enough to induce boiling in the liquid. Compute and plot the steady-state temperature as a function of h for the range $100 \le h \le 10,000$ W/m² · K. Is there a value of h below which operation would be unacceptable?

1.53 A freezer compartment is covered with a 2-mm-thick layer of frost at the time it malfunctions. If the compartment is in ambient air at 20°C and a coefficient of $h = 2$ W/m² · K characterizes heat transfer by natural convection from the exposed surface of the layer, estimate the time required to completely melt the frost. The frost may be assumed to have a mass density of 700 kg/m³ and a latent heat of fusion of 334 kJ/kg.

1.54 A vertical slab of Wood's metal is joined to a substrate on one surface and is melted as it is uniformly irradiated by a laser source on the opposite surface. The metal is initially at its fusion temperature of $T_f = 72°C$, and the melt runs off by gravity as soon as it is formed. The absorptivity of the metal to the laser radiation is $\alpha_l = 0.4$, and its latent heat of fusion is $h_{sf} = 33$ kJ/kg.

(a) Neglecting heat transfer from the irradiated surface by convection or radiation exchange with the surroundings, determine the instantaneous rate of melting in kg/s · m² if the laser irradiation is 5 kW/m². How much material is removed if irradiation is maintained for a period of 2 s?

(b) Allowing for convection to ambient air, with $T_\infty = 20°C$ and $h = 15$ W/m² · K, and radiation exchange with large surroundings ($\varepsilon = 0.4$, $T_{sur} = 20°C$), determine the instantaneous rate of melting during irradiation.

1.55 A computer consists of an array of five printed circuit boards (PCBs), each dissipating $P_b = 20$ W of power. Cooling of the electronic components on a board is provided by the forced flow of air, equally distributed in passages formed by adjoining boards, and the convection coefficient associated with heat transfer from the components to the air is approximately $h = 200$ W/m² · K. Air enters the computer console at a temperature of $T_i = 20°C$, and flow is driven by a fan whose power consumption is $P_f = 25$ W.

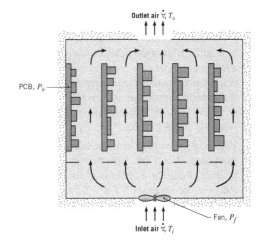

(a) If the temperature rise of the airflow, $(T_o = T_i)$, is not to exceed 15°C, what is the minimum allowable volumetric flow rate $\dot{\forall}$ of the air? The density and specific heat of the air may be approximated as $\rho = 1.161$ kg/m³ and $c_p = 1007$ J/kg · K, respectively.

(b) The component that is most susceptible to thermal failure dissipates 1 W/cm² of surface area. To minimize the potential for thermal failure, where should the component be installed on a PCB? What is its surface temperature at this location?

1.56 Consider the conditions of Problem 1.17, but the surroundings temperature is 25°C and radiation exchange with the surroundings is not negligible. If the convection coefficient is 6.4 W/m²·K and the emissivity of the plate is $\varepsilon = 0.42$, determine the time rate of change of the plate temperature, dT/dt, when the plate temperature is 225°C. Evaluate the heat loss by convection and the heat loss by radiation.

1.57 Most of the energy we consume as food is converted to thermal energy in the process of performing all our bodily functions and is ultimately lost as heat from our bodies. Consider a person who consumes 2100 kcal per day (note that what are commonly referred to as food calories are actually kilocalories), of which 2000 kcal is converted to thermal energy. (The remaining 100 kcal is used to do work on the environment.) The person has a surface area of 1.8 m² and is dressed in a bathing suit.

(a) The person is in a room at 20°C, with a convection heat transfer coefficient of 3 W/m²·K. At this air temperature, the person is not perspiring much. Estimate the person's average skin temperature.

(b) If the temperature of the environment were 33°C, what rate of perspiration would be needed to maintain a comfortable skin temperature of 33°C?

1.58 Consider Problem 1.1.

(a) If the exposed cold surface of the insulation is at $T_2 = 20$°C, what is the value of the convection heat transfer coefficient on the cold side of the insulation if the surroundings temperature is $T_{sur} = 320$ K, the ambient temperature is $T_\infty = 5$°C, and the emissivity is $\varepsilon = 0.95$? Express your results in units of W/m²·K and W/m²·°C.

(b) Using the convective heat transfer coefficient you calculated in part (a), determine the surface temperature, T_2, as the emissivity of the surface is varied over the range $0.05 \le \varepsilon \le 0.95$. The hot wall temperature of the insulation remains fixed at $T_1 = 30$°C. Display your results graphically.

1.59 The roof of a car in a parking lot absorbs a solar radiant flux of 800 W/m², and the underside is perfectly insulated. The convection coefficient between the roof and the ambient air is 12 W/m²·K.

(a) Neglecting radiation exchange with the surroundings, calculate the temperature of the roof under steady-state conditions if the ambient air temperature is 20°C.

(b) For the same ambient air temperature, calculate the temperature of the roof if its surface emissivity is 0.8.

(c) The convection coefficient depends on airflow conditions over the roof, increasing with increasing air speed. Compute and plot the roof temperature as a function of h for $2 \le h \le 200$ W/m²·K.

1.60 An experiment to determine the convection coefficient associated with airflow over the surface of a thick stainless steel casting involves the insertion of thermocouples into the casting at distances of 10 and 20 mm from the surface along a hypothetical line normal to the surface. The steel has a thermal conductivity of 15 W/m·K. If the thermocouples measure temperatures of 50 and 40°C in the steel when the air temperature is 100°C, what is the convection coefficient?

1.61 A rectangular forced air heating duct is suspended from the ceiling of a basement whose air and walls are at a temperature of $T_\infty = T_{sur} = 5$°C. The duct is 15 m long, and its cross section is 350 mm × 200 mm.

(a) For an uninsulated duct whose average surface temperature is 50°C, estimate the rate of heat loss from the duct. The surface emissivity and convection coefficient are approximately 0.5 and 4 W/m²·K, respectively.

(b) If heated air enters the duct at 58°C and a velocity of 4 m/s and the heat loss corresponds to the result of part (a), what is the outlet temperature? The density and specific heat of the air may be assumed to be $\rho = 1.10$ kg/m³ and $c_p = 1008$ J/kg·K, respectively.

1.62 Bus bars proposed for use in a power transmission station have a rectangular cross section of height $H = 600$ mm and width $W = 200$ mm. The electrical resistivity, $\rho_e(\mu\Omega \cdot m)$, of the bar material is a function of temperature, $\rho_e = \rho_{e,o}[1 + \alpha(T - T_o)]$, where $\rho_{e,o} = 0.0828 \; \mu\Omega \cdot m$, $T_o = 25$°C, and $\alpha = 0.0040$ K^{-1}. The emissivity of the bar's painted surface is 0.8, and the temperature of the surroundings is 30°C. The convection coefficient between the bar and the ambient air at 30°C is 10 W/m²·K.

(a) Assuming the bar has a uniform temperature T, calculate the steady-state temperature when a current of 60,000 A passes through the bar.

(b) Compute and plot the steady-state temperature of the bar as a function of the convection coefficient for $10 \le h \le 100$ W/m²·K. What minimum convection coefficient is required to maintain a safe-operating temperature below 120°C? Will increasing the emissivity significantly affect this result?

1.63 A solar flux of 800 W/m² is incident on a flat-plate solar collector used to heat water. The area of the collector is 4 m², and 90% of the solar radiation passes through the cover glass and is absorbed by the absorber plate. The remaining 10% is reflected away from the collector. Water flows through the tube passages on the back side of the absorber plate and is heated from an inlet temperature T_i to an outlet temperature T_o. The cover glass, operating at a temperature of 30°C, has an emissivity of 0.94 and experiences radiation exchange with the sky at −10°C. The convection coefficient between the cover glass and the ambient air at 25°C is 10 W/m²·K.

(a) Perform an overall energy balance on the collector to obtain an expression for the rate at which useful heat is collected per unit area of the collector, q_u''. Determine the value of q_u''.

(b) Calculate the temperature rise of the water, $T_o - T_i$, if the flow rate is 0.01 kg/s. Assume the specific heat of the water to be 4179 J/kg·K.

(c) The collector efficiency η is defined as the ratio of the useful heat collected to the rate at which solar energy is incident on the collector. What is the value of η?

Process Identification

1.64 In analyzing the performance of a thermal system, the engineer must be able to identify the relevant heat transfer processes. Only then can the system behavior be properly quantified. For the following systems, identify the pertinent processes, designating them by appropriately labeled arrows on a sketch of the system. Answer additional questions that appear in the problem statement. ·

(a) Identify the heat transfer processes that determine the temperature of an asphalt pavement on a summer day. Write an energy balance for the surface of the pavement.

(b) Microwave radiation is known to be transmitted by plastics, glass, and ceramics but to be absorbed by materials having polar molecules such as water. Water molecules exposed to microwave radiation align and reverse alignment with the microwave radiation at frequencies up to 10^9 s⁻¹, causing heat to be generated. Contrast cooking in a microwave oven with cooking in a conventional radiant or convection oven. In each case, what is the physical mechanism responsible for heating the food? Which oven has the greater energy utilization efficiency? Why? Microwave heating is being considered for drying clothes. How would the operation of a microwave clothes dryer differ from a conventional dryer? Which is likely to have the greater energy utilization efficiency? Why?

(c) Your grandmother is concerned about reducing her winter heating bills. Her strategy is to loosely fit rigid polystyrene sheets of insulation over her double-pane windows right after the first freezing weather arrives in the autumn. Identify the relevant heat transfer processes on a cold winter night when the foamed insulation sheet is placed (i) on the inner surface and (ii) on the outer surface of her window. To avoid condensation damage, which configuration is preferred? Condensation on the window pane does not occur when the foamed insulation is not in place.

Insulation on inner surface

Insulation on outer surface

(d) There is considerable interest in developing build-ing materials with improved insulating qualities. The development of such materials would do much to enhance energy conservation by reducing space heating requirements. It has been suggested that superior structural and insulating qualities could be obtained by using the composite shown. The mate-rial consists of a honeycomb, with cells of square cross section, sandwiched between solid slabs. The cells are filled with air, and the slabs, as well as the honeycomb matrix, are fabricated from plastics of low thermal conductivity. For heat transfer normal to the slabs, identify all heat transfer processes per-tinent to the performance of the composite. Suggest ways in which this performance could be enhanced.

Surface slabs

Cellular air spaces

(e) A thermocouple junction (bead) is used to measure the temperature of a hot gas stream flowing through a channel by inserting the junction into the mainstream of the gas. The surface of the channel is cooled such that its temperature is well below that of the gas. Identify the heat transfer processes associated with the junction surface. Will the junction sense a tem-perature that is less than, equal to, or greater than the gas temperature? A radiation shield is a small, open-ended tube that encloses the thermocouple junction, yet allows for passage of the gas through the tube. How does use of such a shield improve the accuracy of the temperature measurement?

Cool channel

Hot gases

Thermocouple bead

Shield

(f) A double-glazed, glass fire screen is inserted between a wood-burning fireplace and the interior

of a room. The screen consists of two vertical glass plates that are separated by a space through which room air may flow (the space is open at the top and bottom). Identify the heat transfer processes associ-ated with the fire screen.

Air channel

Glass plate

Air

(g) A thermocouple junction is used to measure the temperature of a solid material. The junction is inserted into a small circular hole and is held in place by epoxy. Identify the heat transfer processes associated with the junction. Will the junction sense a temperature less than, equal to, or greater than the solid temperature? How will the ther-mal conductivity of the epoxy affect the junction temperature?

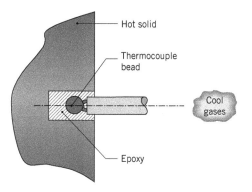

Hot solid

Thermocouple bead

Cool gases

Epoxy

1.65 In considering the following problems involving heat transfer in the natural environment (outdoors), recog-nize that solar radiation is comprised of long and short wavelength components. If this radiation is incident on a *semitransparent medium*, such as water or glass, two things will happen to the nonreflected portion of the radiation. The long wavelength component will be absorbed at the surface of the medium, whereas the short wavelength component will be transmitted by the surface.

(a) The number of panes in a window can strongly influence the heat loss from a heated room to the outside ambient air. Compare the single- and double-paned units shown by identifying relevant heat transfer processes for each case.

(b) In a typical flat-plate solar collector, energy is collected by a working fluid that is circulated through tubes that are in good contact with the back face of an absorber plate. The back face is insulated from the surroundings, and the absorber plate receives solar radiation on its front face, which is typically covered by one or more transparent plates. Identify the relevant heat transfer processes, first for the absorber plate with no cover plate and then for the absorber plate with a single cover plate.

(c) The solar energy collector design shown in the schematic has been used for agricultural applications. Air is blown through a long duct whose cross section is in the form of an equilateral triangle. One side of the triangle is comprised of a double-paned, semitransparent cover; the other two sides are constructed from aluminum sheets painted flat black on the inside and covered on the outside with a layer of styrofoam insulation. During sunny periods, air entering the system is heated for delivery to either a greenhouse, grain drying unit, or storage system.

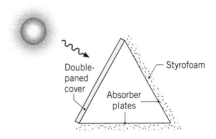

Identify all heat transfer processes associated with the cover plates, the absorber plate(s), and the air.

(d) Evacuated-tube solar collectors are capable of improved performance relative to flat-plate collectors. The design consists of an inner tube enclosed in an outer tube that is transparent to solar radiation. The annular space between the tubes is evacuated. The outer, opaque surface of the inner tube absorbs solar radiation, and a working fluid is passed through the tube to collect the solar energy. The collector design generally consists of a row of such tubes arranged in front of a reflecting panel. Identify all heat transfer processes relevant to the performance of this device.

Recall that *conduction* is the transport of energy in a medium due to a temperature gradient, and the physical mechanism is one of random atomic or molecular activity. In Chapter 1 we learned that conduction heat transfer is governed by *Fourier's law* and that use of the law to determine the heat flux depends on knowledge of the manner in which temperature varies within the medium (the *temperature distribution*). By way of introduction, we restricted our attention to simplified conditions (one-dimensional, steady-state conduction in a plane wall). However, Fourier's law is applicable to transient, multidimensional conduction in complex geometries.

The objectives of this chapter are twofold. First, we wish to develop a deeper understanding of Fourier's law. What are its origins? What form does it take for different geometries? How does its proportionality constant (the *thermal conductivity*) depend on the physical nature of the medium? Our second objective is to develop, from basic principles, the general equation, termed the *heat equation*, which governs the temperature distribution in a medium in which conduction is the only mode of heat transfer. The solution to this equation provides knowledge of the temperature distribution, which may then be used with Fourier's law to determine the heat flux.

2.1 *The Conduction Rate Equation*

Although the conduction rate equation, Fourier's law, was introduced in Section 1.2, it is now appropriate to consider its origin. Fourier's law is *phenomenological*; that is, it is developed from observed phenomena rather than being derived from first principles. Hence, we view the rate equation as a generalization based on much experimental evidence. For example, consider the steady-state conduction experiment of Figure 2.1. A cylindrical rod of known material is insulated on its lateral surface, while its end faces are maintained at different temperatures, with $T_1 > T_2$. The temperature difference causes conduction heat transfer in the positive x-direction. We are able to measure the heat transfer rate q_x, and we seek to determine how q_x depends on the following variables: ΔT, the temperature difference; Δx, the rod length; and A, the cross-sectional area.

We might imagine first holding ΔT and Δx constant and varying A. If we do so, we find that q_x is directly proportional to A. Similarly, holding ΔT and A constant, we observe that q_x varies inversely with Δx. Finally, holding A and Δx constant, we find that q_x is directly proportional to ΔT. The collective effect is then

$$q_x \propto A \frac{\Delta T}{\Delta x}$$

In changing the material (e.g., from a metal to a plastic), we would find that this proportionality remains valid. However, we would also find that, for equal values of A, Δx, and ΔT, the

FIGURE 2.1   Steady-state heat conduction experiment.

value of q_x would be smaller for the plastic than for the metal. This suggests that the proportionality may be converted to an equality by introducing a coefficient that is a measure of the material behavior. Hence, we write

$$q_x = kA\frac{\Delta T}{\Delta x}$$

where k, the *thermal conductivity* (W/m · K), is an important *property* of the material. Evaluating this expression in the limit as $\Delta x \to 0$, we obtain for the heat *rate*

$$q_x = -kA\frac{dT}{dx} \tag{2.1}$$

or for the heat *flux*

$$q_x'' = \frac{q_x}{A} = -k\frac{dT}{dx} \tag{2.2}$$

Recall that the minus sign is necessary because heat is always transferred in the direction of decreasing temperature.

Fourier's law, as written in Equation 2.2, implies that the heat flux is a directional quantity. In particular, the direction of q_x'' is *normal* to the cross-sectional area A. Or, more generally, the direction of heat flow will always be normal to a surface of constant temperature, called an *isothermal* surface. Figure 2.2 illustrates the direction of heat flow q_x'' in a plane wall for which the *temperature gradient dT/dx* is negative. From Equation 2.2, it follows that q_x'' is positive. Note that the isothermal surfaces are planes normal to the x-direction.

Recognizing that the heat flux is a vector quantity, we can write a more general statement of the conduction rate equation (*Fourier's law*) as follows:

$$q'' = -k\nabla T = -k\left(i\frac{\partial T}{\partial x} + j\frac{\partial T}{\partial y} + k\frac{\partial T}{\partial z}\right) \tag{2.3}$$

where ∇ is the three-dimensional del operator, i, j, and k are the unit vectors in the x, y, and z directions, and $T(x, y, z)$ is the scalar temperature field. It is implicit in Equation 2.3 that

FIGURE 2.2 The relationship between coordinate system, heat flow direction, and temperature gradient in one dimension.

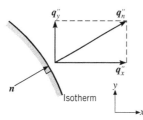

FIGURE 2.3   The heat flux vector normal to an isotherm in a two-dimensional coordinate system.

the heat flux vector is in a direction perpendicular to the isothermal surfaces. An alternative form of Fourier's law is therefore

$$q'' = q_n'' n = -k \frac{\partial T}{\partial n} n \tag{2.4}$$

where q_n'' is the heat flux in a direction n, which is normal to an *isotherm*, and n is the unit normal vector in that direction. This is illustrated for the two-dimensional case in Figure 2.3. The heat transfer is sustained by a temperature gradient along n. Note also that the heat flux vector can be resolved into components such that, in Cartesian coordinates, the general expression for q'' is

$$q'' = i q_x'' + j q_y'' + k q_z'' \tag{2.5}$$

where, from Equation 2.3, it follows that

$$q_x'' = -k \frac{\partial T}{\partial x} \qquad q_y'' = -k \frac{\partial t}{\partial y} \qquad q_z'' = -k \frac{\partial T}{\partial z} \tag{2.6}$$

Each of these expressions relates the heat flux *across a surface* to the temperature gradient in a direction perpendicular to the surface. It is also implicit in Equation 2.3 that the medium considered here is *isotropic*. For such a medium, the value of the thermal conductivity is independent of the coordinate direction.

Fourier's law is the cornerstone of conduction heat transfer, and its key features are summarized as follows. It is *not* an expression that may be derived from first principles; it is instead a generalization based on experimental evidence. It is an expression that *defines* an important material property, the thermal conductivity. In addition, Fourier's law is a vector expression indicating that the heat flux is normal to an isotherm and in the direction of decreasing temperature. Finally, note that Fourier's law applies for all matter, regardless of its state (solid, liquid, or gas).

2.2 *The Thermal Properties of Matter*

To use Fourier's law, the thermal conductivity of the material must be known. This property, which is referred to as a *transport property,* depends on the atomic and molecular structure of matter, and therefore, on the state of the matter. In this section we consider various forms of matter, identifying important aspects of their behavior and presenting typical property values.

2.2.1 Thermal Conductivity

From Fourier's law, Equation 2.6, the thermal conductivity associated with conduction in the x-direction is defined as

$$k_x \equiv -\frac{q_x''}{(\partial T/\partial x)}$$

Similar definitions are associated with thermal conductivities in the y- and z-directions (k_y, k_z), but for an isotropic material the thermal conductivity is independent of the direction of transfer, $k_x = k_y = k_z \equiv k$.

From the foregoing equation, it follows that, for a prescribed temperature gradient, the conduction heat flux increases with increasing thermal conductivity. In general, the thermal conductivity of a solid is larger than that of a liquid, which is larger than that of a gas. As illustrated in Figure 2.4, the thermal conductivity of a solid may be more than four orders of magnitude larger than that of a gas. This trend is due largely to differences in intermolecular spacing for the two states.

The Solid State In the modern view of materials, a solid may be comprised of free electrons and atoms bound in a periodic arrangement called the lattice. Accordingly, transport of thermal energy may be due to two effects: the migration of free electrons and lattice vibrational waves. When viewed as a particle-like phenomenon, the lattice vibration quanta are termed *phonons*. In pure metals, the electron contribution to conduction heat transfer dominates, whereas in nonconductors and semiconductors, the phonon contribution is dominant.

Kinetic theory yields the following expression for the thermal conductivity [1]:

$$k = \frac{1}{3}C\bar{c}\lambda_{\text{mfp}} \tag{2.7}$$

For conducting materials such as metals, $C \equiv C_e$ is the electron specific heat per unit volume, $\bar{c}$ is the mean electron velocity, and $\lambda_{\text{mfp}} \equiv \lambda_e$ is the electron mean free path, which is defined as the average distance traveled by an electron before it collides with either an imperfection in the material or with a phonon. In nonconducting solids, $C \equiv C_{\text{ph}}$ is the phonon specific heat, $\bar{c}$ is the average speed of sound, and $\lambda_{\text{mfp}} \equiv \lambda_{\text{ph}}$ is the phonon mean free

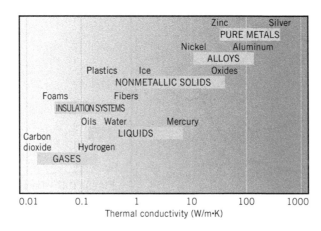

FIGURE 2.4 Range of thermal conductivity for various states of matter at normal temperatures and pressure.

path, which again is determined by collisions with imperfections or other phonons. In all cases, the thermal conductivity increases as the mean free path of the *energy carriers* (electrons or phonons) is increased.

When electrons and phonons carry thermal energy leading to conduction heat transfer in a solid, the thermal conductivity may be expressed as

$$k = k_e + k_{ph} \tag{2.8}$$

To a first approximation, k_e is inversely proportional to the electrical resistivity, ρ_e. For pure metals, which are of low ρ_e, k_e is much larger than k_{ph}. In contrast, for alloys, which are of substantially larger ρ_e, the contribution of k_{ph} to k is no longer negligible. For nonmetallic solids, k is determined primarily by k_{ph}, which increases as the frequency of interactions between the atoms and the lattice decreases. The regularity of the lattice arrangement has an important effect on k_{ph}, with crystalline (well-ordered) materials like quartz having a higher thermal conductivity than amorphous materials like glass. In fact, for crystalline, nonmetallic solids such as diamond and beryllium oxide, k_{ph} can be quite large, exceeding values of k associated with good conductors, such as aluminum.

The temperature dependence of k is shown in Figure 2.5 for representative metallic and nonmetallic solids. Values for selected materials of technical importance are also provided in Table A.1 (metallic solids) and Tables A.2 and A.3 (nonmetallic solids). More detailed treatments of thermal conductivity are available in the literature [2].

The Solid State: Micro- and Nanoscale Effects   In the preceding discussion, the *bulk* thermal conductivity is described, and the thermal conductivity values listed in Tables A.1

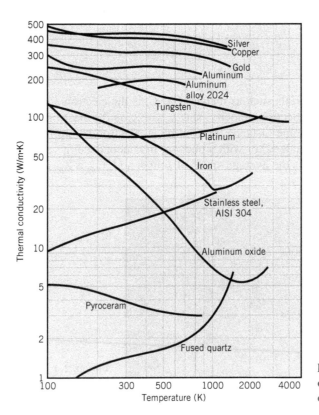

FIGURE 2.5   The temperature dependence of the thermal conductivity of selected solids.

through A.3 are appropriate for use when the physical dimensions of the material of interest are relatively large. This is the case in many commonplace engineering problems. However, in several areas of technology, such as microelectronics, the material's characteristic dimensions can be on the order of micrometers or nanometers, in which case care must be taken to account for the possible modifications of k that can occur as the physical dimensions become small.

Cross sections of *films* of the same material having thicknesses L_1 and L_2 are shown in Figure 2.6. Electrons or phonons that are associated with conduction of thermal energy are also shown qualitatively. Note that the physical boundaries of the film act to *scatter* the energy carriers and *redirect* their propagation. For large L/λ_{mfp}[1] (Figure 2.6a), the effect of the boundaries on reducing the *average* energy carrier path length is minor, and conduction heat transfer occurs as described for bulk materials. However, as the film becomes thin, the physical boundaries of the material can decrease the average *net* distance traveled by the energy carriers, as shown in Figure 2.6b. Moreover, electrons and phonons moving in the thin *x*-direction (representing conduction in the *x*-direction) are affected by the boundaries to a more significant degree than energy carriers moving in the *y*-direction. As such, for films characterized by small L/λ_{mfp}, we find that $k_x < k_y < k$, where k is the bulk thermal conductivity of the film material.

For $L/\lambda_{mfp} \geq 1$, the predicted values of k_x and k_y may be estimated to within 20% from the following expressions [1]:

$$\frac{k_x}{k} = 1 - \frac{\lambda_{mfp}}{3L} \tag{2.9a}$$

$$\frac{k_y}{k} = 1 - \frac{2\lambda_{mfp}}{3\pi L} \tag{2.9b}$$

Equations 2.9a,b reveal that the values of k_x and k_y are within approximately 5% of the bulk thermal conductivity if $L/\lambda_{mfp} > 7$ (for k_x) and $L/\lambda_{mfp} > 4.5$ (for k_y). Values of the mean free path as well as critical film thicknesses below which microscale effects must be considered, L_{crit}, are included in Table 2.1 for several materials at $T \approx 300$ K. For films with $\lambda_{mfp} < L < L_{crit}$, k_x

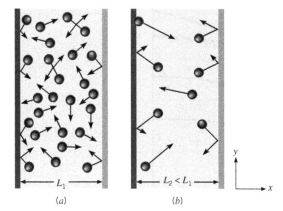

FIGURE 2.6 Electron or phonon trajectories in (a) a relatively thick film and (b) a relatively thin film with boundary effects.

---

[1]The quantity λ_{mfp}/L is a dimensionless parameter known as the Knudsen number. Large Knudsen numbers (small L/λ_{mfp}) suggest potentially significant nano- or microscale effects.

and k_y are reduced from the bulk value as indicated in Equations 2.9a,b. No general guidelines exist for predicting values of the thermal conductivities for $L/\lambda_{mfp} < 1$. Note that, in solids, the value of λ_{mfp} decreases as the temperature increases.

In addition to scattering from physical boundaries, as in the case of Figure 2.6*b*, energy carriers may be redirected by chemical *dopants* embedded within a material or by *grain boundaries* that separate individual clusters of material in otherwise homogeneous matter. An example of a *nanostructured material* is one that is chemically identical to its conventional counterpart, but is processed to provide very small grain sizes. This feature impacts heat transfer by increasing the scattering and reflection of energy carriers at grain boundaries.

Measured values of the thermal conductivity of a bulk, nanostructured yttria-stabilized zirconia material are shown in Figure 2.7. This particular ceramic is widely used for insulation purposes in high-temperature combustion devices. Conduction is dominated by phonon transfer, and the mean free path of the phonon energy carriers is, from Table 2.1, $\lambda_{mfp} = 25$ nm at 300 K. As the grain sizes are reduced to characteristic dimensions less than

TABLE 2.1   Mean free path and critical film thickness for various materials at $T \approx 300$ K [3,4]

Material	λ_{mfp} (nm)	$L_{crit,x}$ (nm)	$L_{crit,y}$ (nm)
Aluminum oxide	5.08	36	22
Diamond (IIa)	315	2200	1400
Gallium arsenide	23	160	100
Gold	31	220	140
Silicon	43	290	180
Silicon dioxide	0.6	4	3
Yttria-stabilized zirconia	25	170	110

FIGURE 2.7   Measured thermal conductivity of yttria-stabilized zirconia as a function of temperature and mean grain size, L [3].

25 nm (and more grain boundaries are introduced in the material per unit volume), significant reduction of the thermal conductivity occurs. Extrapolation of the results of Figure 2.7 to higher temperatures is not recommended, since the mean free path decreases with increasing temperature ($\lambda_{mfp} \approx 4$ nm at $T \approx 1525$ K) *and* grains of the material may coalesce, merge, and enlarge at elevated temperatures. Therefore, L/λ_{mfp} becomes larger at high temperatures, and reduction of k due to nanoscale effects becomes less pronounced. Research on heat transfer in nanostructured materials continues to reveal novel ways engineers can manipulate the nanostructure to reduce or increase thermal conductivity [5]. Potentially important consequences include applications such as gas turbine engine technology [6], microelectronics [7], and renewable energy [8].

The Fluid State The fluid state includes both liquids and gases. Because the intermolecular spacing is much larger and the motion of the molecules is more random for the fluid state than for the solid state, thermal energy transport is less effective. The thermal conductivity of gases and liquids is therefore generally smaller than that of solids.

The effect of temperature, pressure, and chemical species on the thermal conductivity of a gas may be explained in terms of the kinetic theory of gases [9]. From this theory it is known that the thermal conductivity is directly proportional to the density of the gas, the mean molecular speed $\bar{c}$, and the mean free path λ_{mfp}, which is the average distance traveled by an energy carrier (a molecule) before experiencing a collision.

$$k \approx \frac{1}{3} c_v \rho \bar{c} \lambda_{mfp} \tag{2.10}$$

For an ideal gas, the mean free path may be expressed as

$$\lambda_{mfp} = \frac{k_B T}{\sqrt{2} \pi d^2 p} \tag{2.11}$$

where k_B is Boltzmann's constant, $k_B = 1.381 \times 10^{-23}$ J/K, d is the diameter of the gas molecule, representative values of which are included in Figure 2.8, and p is the pressure. As expected, the mean free path is small for high pressure or low temperature, which causes densely packed molecules. The mean free path also depends on the diameter of the molecule, with larger molecules more likely to experience collisions than small molecules; in the limiting case of an infinitesimally small molecule, the molecules cannot collide, resulting in an infinite mean free path. The mean molecular speed, $\bar{c}$, can be determined from the kinetic theory of gases, and Equation 2.10 may ultimately be expressed as

$$k = \frac{9\gamma - 5}{4} \frac{c_v}{\pi d^2} \sqrt{\frac{\mathcal{M} k_B T}{\mathcal{N} \pi}} \tag{2.12}$$

where the parameter γ is the ratio of specific heats, $\gamma \equiv c_p/c_v$, and $\mathcal{N}$ is Avogadro's number, $\mathcal{N} = 6.022 \times 10^{23}$ molecules per mol. Equation 2.12 can be used to estimate the thermal conductivity of gas, although more accurate models have been developed [10].

FIGURE 2.8 The temperature dependence of the thermal conductivity of selected gases at normal pressures. Molecular diameters (d) are in nm [10]. Molecular weights ($\mathcal{M}$) of the gases are also shown.

It is important to note that the thermal conductivity is independent of pressure except in extreme cases as, for example, when conditions approach that of a perfect vacuum. Therefore, the assumption that k is independent of gas pressure for large volumes of gas is appropriate for the pressures of interest in this text. Accordingly, although the values of k presented in Table A.4 pertain to atmospheric pressure or the saturation pressure corresponding to the prescribed temperature, they may be used over a much wider pressure range.

Molecular conditions associated with the liquid state are more difficult to describe, and physical mechanisms for explaining the thermal conductivity are not well understood [11]. The thermal conductivity of nonmetallic liquids generally decreases with increasing temperature. As shown in Figure 2.9, water, glycerine, and engine oil are notable exceptions. The thermal conductivity of liquids is usually insensitive to pressure except near the thermodynamic critical point. Also, thermal conductivity generally decreases with increasing molecular weight. Values of the thermal conductivity are often tabulated as a function of temperature for the saturated state of the liquid. Tables A.5 and A.6 present such data for several common liquids.

Liquid metals are commonly used in high heat flux applications, such as occur in nuclear power plants. The thermal conductivity of such liquids is given in Table A.7. Note that the values are much larger than those of the nonmetallic liquids [12].

The Fluid State: Micro- and Nanoscale Effects   As for the solid state, the bulk thermal conductivity of a fluid may be modified when the characteristic dimension of the system becomes small, in particular for small values of L/λ_{mfp}. Similar to the situation of a thin solid film shown in Figure 2.6b, the molecular mean free path is restricted when a fluid is constrained by a small physical dimension, affecting conduction across a thin fluid layer.

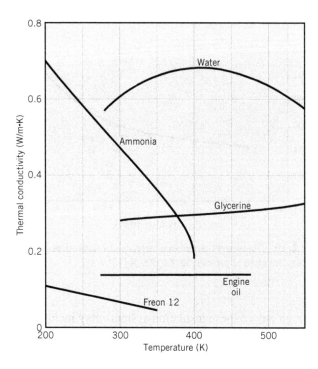

FIGURE 2.9 The temperature dependence of the thermal conductivity of selected nonmetallic liquids under saturated conditions.

Mixtures of fluids and solids can also be formulated to tailor the transport properties of the resulting *suspension*. For example, *nanofluids* are *base liquids* that are seeded with nanometer-sized solid particles. Their very small size allows the solid particles to remain suspended within the base liquid for a long time. From the heat transfer perspective, a nanofluid exploits the high thermal conductivity that is characteristic of most solids, as is evident in Figure 2.5, to boost the relatively low thermal conductivity of base liquids, typical values of which are shown in Figure 2.9. Common nanofluids involve liquid water seeded with nominally spherical nanoparticles of Al_2O_3 or CuO.

Insulation Systems Thermal insulations consist of low thermal conductivity materials combined to achieve an even lower system thermal conductivity. In conventional *fiber-*, *powder-*, and *flake*-type insulations, the solid material is finely dispersed throughout an air space. Such systems are characterized by an *effective thermal conductivity*, which depends on the thermal conductivity and surface radiative properties of the solid material, as well as the nature and volumetric fraction of the air or void space. A special parameter of the system is its bulk density (solid mass/total volume), which depends strongly on the manner in which the material is packed.

If small voids or hollow spaces are formed by bonding or fusing portions of the solid material, a rigid matrix is created. When these spaces are sealed from each other, the system is referred to as a *cellular* insulation. Examples of such rigid insulations are *foamed* systems, particularly those made from plastic and glass materials. *Reflective* insulations are composed of multilayered, parallel, thin sheets or foils of high reflectivity, which are spaced to reflect radiant energy back to its source. The spacing between the foils is designed to restrict the motion of air, and in high-performance insulations, the space is evacuated. In all types of insulation, evacuation of the air in the void space will reduce the effective thermal conductivity of the system.

FIGURE 2.10   Measured thermal conductivity of carbon-doped silica aerogel as a function of pressure at $T \approx 300$ K [15].

Heat transfer through any of these insulation systems may include several modes: conduction through the solid materials; conduction or convection through the air in the void spaces; and radiation exchange between the surfaces of the solid matrix. The effective thermal conductivity accounts for all of these processes, and values for selected insulation systems are summarized in Table A.3. Additional background information and data are available in the literature [13, 14].

As with thin films, micro- and nanoscale effects can influence the effective thermal conductivity of insulating materials. The value of k for a nanostructured silica aerogel material that is composed of approximately 5% by volume solid material and 95% by volume air that is trapped within pores of $L \approx 20$ nm is shown in Figure 2.10. Note that at $T \approx 300$ K, the mean free path for air at atmospheric pressure is approximately 80 nm. As the gas pressure is reduced, λ_{mfp} would increase for an unconfined gas, but the molecular motion of the trapped air is restricted by the walls of the small pores and k is reduced to extremely small values relative to the thermal conductivities of conventional matter reported in Figure 2.4.

2.2.2   Other Relevant Properties

In our analysis of heat transfer problems, it will be necessary to use several properties of matter. These properties are generally referred to as *thermophysical* properties and include two distinct categories, *transport* and *thermodynamic* properties. The transport properties include the diffusion rate coefficients such as k, the thermal conductivity (for heat transfer), and ν, the kinematic viscosity (for momentum transfer). Thermodynamic properties, on the other hand, pertain to the equilibrium state of a system. Density (ρ) and specific heat (c_p) are two such properties used extensively in thermodynamic analysis. The product ρc_p (J/m$^3 \cdot$ K), commonly termed the *volumetric heat capacity*, measures the ability of a material to store thermal energy. Because substances of large density are typically characterized by small specific heats, many solids and liquids, which are very good energy storage media, have comparable heat capacities ($\rho c_p > 1$ MJ/m$^3 \cdot$ K). Because of their very small densities, however, gases are poorly suited for thermal energy storage ($\rho c_p \approx 1$ kJ/m$^3 \cdot$ K).

Densities and specific heats are provided in the tables of Appendix A for a wide range of solids, liquids, and gases.

In heat transfer analysis, the ratio of the thermal conductivity to the heat capacity is an important property termed the *thermal diffusivity* α, which has units of m^2/s:

$$\alpha = \frac{k}{\rho c_p}$$

It measures the ability of a material to conduct thermal energy relative to its ability to store thermal energy. Materials of large α will respond quickly to changes in their thermal environment, whereas materials of small α will respond more sluggishly, taking longer to reach a new equilibrium condition.

The accuracy of engineering calculations depends on the accuracy with which the thermophysical properties are known [16–18]. Numerous examples could be cited of flaws in equipment and process design or failure to meet performance specifications that were attributable to misinformation associated with the selection of key property values. Selection of reliable property data is an integral part of any careful engineering analysis. Recommended data values for many thermophysical properties can be obtained from Reference 19.

EXAMPLE 2.1

The thermal diffusivity α is the controlling transport property for transient conduction. Using appropriate values of k, ρ, and c_p from Appendix A, calculate α for the following materials at the prescribed temperatures: pure aluminum, 300 and 700 K; silicon carbide, 1000 K; paraffin, 300 K.

SOLUTION

Known: Definition of the thermal diffusivity α.

Find: Numerical values of α for selected materials and temperatures.

Properties: Table A.1, pure aluminum (300 K):

$$\left.\begin{array}{l} \rho = 2702 \text{ kg/m}^3 \\ c_p = 903 \text{ J/kg} \cdot \text{K} \\ k = 237 \text{ W/m} \cdot \text{K} \end{array}\right\} \alpha = \frac{k}{\rho c_p} = \frac{237 \text{ W/m} \cdot \text{K}}{2702 \text{ kg/m}^3 \times 903 \text{ J/kg} \cdot \text{K}}$$

$$= 97.1 \times 10^{-6} \text{ m}^2/\text{s} \qquad \lhd$$

Table A.1, pure aluminum (700 K):

$$\begin{array}{ll} \rho = 2702 \text{ kg/m}^3 & \text{at 300 K} \\ c_p = 1090 \text{ J/kg} \cdot \text{K} & \text{at 700 K (by linear interpolation)} \\ k = 225 \text{ W/m} \cdot \text{K} & \text{at 700 K (by linear interpolation)} \end{array}$$

Hence

$$\alpha = \frac{k}{\rho c_p} = \frac{225 \text{ W/m} \cdot \text{K}}{2702 \text{ kg/m}^3 \times 1090 \text{ J/kg} \cdot \text{K}} = 76 \times 10^{-6} \text{ m}^2/\text{s} \qquad \lhd$$

Table A.2, silicon carbide (1000 K):

$$\left.\begin{array}{ll} \rho = 3160 \text{ kg/m}^3 & \text{at } 300 \text{ K} \\ c_p = 1195 \text{ J/kg} \cdot \text{K} & \text{at } 1000 \text{ K} \\ k = 87 \text{ W/m} \cdot \text{K} & \text{at } 1000 \text{ K} \end{array}\right\} \alpha = \frac{87 \text{ W/m} \cdot \text{K}}{3160 \text{ kg/m}^3 \times 1195 \text{ J/kg} \cdot \text{K}}$$

$$= 23 \times 10^{-6} \text{ m}^2/\text{s} \qquad \triangleleft$$

Table A.3, paraffin (300 K):

$$\left.\begin{array}{l} \rho = 900 \text{ kg/m}^3 \\ c_p = 2890 \text{ J/kg} \cdot \text{K} \\ k = 0.24 \text{ W/m} \cdot \text{K} \end{array}\right\} \alpha = \frac{k}{\rho c_p} = \frac{0.24 \text{ W/m} \cdot \text{K}}{900 \text{ kg/m}^3 \times 2890 \text{ J/kg} \cdot \text{K}}$$

$$= 9.2 \times 10^{-8} \text{ m}^2/\text{s} \qquad \triangleleft$$

Comments:

1. Note the temperature dependence of the thermophysical properties of aluminum and silicon carbide. For example, from Table A.2 for silicon carbide, $\alpha(1000 \text{ K}) \approx 0.1 \times \alpha(300 \text{ K})$; hence properties of this material have a strong temperature dependence.

2. The physical interpretation of α is that it provides a measure of heat transport (k) relative to energy storage (ρc_p). In general, metallic solids have higher α, whereas nonmetallics (e.g., paraffin) have lower values of α.

3. Linear interpolation of property values is generally acceptable for engineering calculations.

4. Use of the low-temperature (300 K) density at higher temperatures ignores thermal expansion effects but is also acceptable for engineering calculations.

5. The *IHT* software provides a library of thermophysical properties for selected solids, liquids, and gases that can be accessed from the toolbar button, *Properties*. See Example 2.1 in *IHT*.

EXAMPLE 2.2

The bulk thermal conductivity of a nanofluid containing uniformly dispersed, noncontacting spherical nanoparticles may be approximated by

$$k_{\text{nf}} = \left[\frac{k_p + 2k_{\text{bf}} + 2\varphi(k_p - k_{\text{bf}})}{k_p + 2k_{\text{bf}} - \varphi(k_p - k_{\text{bf}})}\right] k_{\text{bf}}$$

where φ is the volume fraction of the nanoparticles, and k_{bf}, k_p, and k_{nf} are the thermal conductivities of the base fluid, particle, and nanofluid, respectively. Likewise, the dynamic viscosity may be approximated as [20]

$$\mu_{\text{nf}} = \mu_{\text{bf}}(1 + 2.5\varphi)$$

Determine the values of k_{nf}, ρ_{nf}, $c_{p,\text{nf}}$, μ_{nf}, and α_{nf} for a mixture of water and Al_2O_3 nanoparticles at a temperature of $T = 300$ K and a particle volume fraction of $\varphi = 0.05$. The thermophysical properties of the particles are $k_p = 36.0$ W/m · K, $\rho_p = 3970$ kg/m^3, and $c_{p,p} = 0.765$ kJ/kg · K.

SOLUTION

Known: Expressions for the bulk thermal conductivity and viscosity of a nanofluid with spherical nanoparticles. Nanoparticle properties.

Find: Values of the nanofluid thermal conductivity, density, specific heat, dynamic viscosity, and thermal diffusivity.

Schematic:

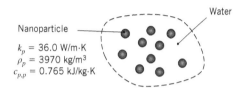

Nanoparticle
$k_p = 36.0$ W/m·K
$\rho_p = 3970$ kg/m³
$c_{p,p} = 0.765$ kJ/kg·K

Water

Assumptions:

1. Constant properties.
2. Density and specific heat are not affected by nanoscale phenomena.

Properties: Table A.6 ($T = 300$ K): Water; $k_{bf} = 0.613$ W/m · K, $\rho_{bf} = 997$ kg/m³, $c_{p,bf} = 4.179$ kJ/kg · K, $\mu_{bf} = 855 \times 10^{-6}$ N · s/m².

Analysis: From the problem statement,

$$k_{nf} = \left[\frac{k_p + 2k_{bf} + 2\varphi(k_p - k_{bf})}{k_p + 2k_{bf} - \varphi(k_p - k_{bf})} \right] k_{bf}$$

$$= \left[\frac{36.0 \text{ W/m} \cdot \text{K} + 2 \times 0.613 \text{ W/m} \cdot \text{K} + 2 \times 0.05(36.0 - 0.613) \text{ W/m} \cdot \text{K}}{36.0 \text{ W/m} \cdot \text{K} + 2 \times 0.613 \text{ W/m} \cdot \text{K} - 0.05(36.0 - 0.613) \text{ W/m} \cdot \text{K}} \right]$$

$$\times 0.613 \text{ W/m} \cdot \text{K}$$

$$= 0.705 \text{ W/m} \cdot \text{K} \qquad\qquad\qquad\qquad \triangleleft$$

Consider the control volume shown in the schematic to be of total volume V. Then the average density of the nanofluid, ρ_{nf}, is equal to the total mass within the volume divided by the volume:

$$\rho_{nf} = [\rho_{bf}V(1 - \varphi) + \rho_p V\varphi]/V$$

or,

$$\rho_{nf} = 997 \text{ kg/m}^3 \times (1 - 0.05) + 3970 \text{ kg/m}^3 \times 0.05 = 1146 \text{ kg/m}^3 \qquad \triangleleft$$

Similarly, the average specific heat can be defined as the total energy stored per unit temperature difference divided by the mass:

$$c_{p,nf} = [\rho_{bf}V(1 - \varphi)c_{p,bf} + \rho_p V\varphi c_{p,p}]/(\rho_{nf}V)$$

which yields

$$c_{p,\text{nf}} = \frac{997 \text{ kg/m}^3 \times 4.179 \text{ kJ/kg} \cdot \text{K} \times (1 - 0.05) + 3970 \text{ kg/m}^3 \times 0.765 \text{ kJ/kg} \cdot \text{K} \times (0.05)}{1146 \text{ kg/m}^3}$$

$$= 3.587 \text{ kJ/kg} \cdot \text{K} \qquad \triangleleft$$

From the problem statement, the dynamic viscosity of the nanofluid is

$$\mu_{\text{nf}} = 855 \times 10^{-6} \text{ N} \cdot \text{s/m}^2 \times (1 + 2.5 \times 0.05) = 962 \times 10^{-6} \text{ N} \cdot \text{s/m}^2 \qquad \triangleleft$$

The nanofluid's thermal diffusivity is

$$\alpha_{\text{nf}} = \frac{k_{\text{nf}}}{\rho_{\text{nf}} c_{p,\text{nf}}} = \frac{0.705 \text{ W/m} \cdot \text{K}}{1146 \text{ kg/m}^3 \times 3587 \text{ J/kg} \cdot \text{K}} = 171 \times 10^{-9} \text{ m}^2/\text{s} \qquad \triangleleft$$

Comments:

1. The ratio of the nanofluid thermal conductivity to the base fluid (water) thermal conductivity is

$$\frac{k_{\text{nf}}}{k_{\text{bf}}} = \frac{0.705 \text{ W/m} \cdot \text{K}}{0.613 \text{ W/m} \cdot \text{K}} = 1.150$$

Similarly, $\rho_{\text{nf}}/\rho_{\text{bf}} = 1.149$, $c_{p,\text{nf}}/c_{p,\text{bf}} = 0.858$, $\mu_{\text{nf}}/\mu_{\text{bf}} = 1.130$, and $\alpha_{\text{nf}}/\alpha_{\text{bf}} = 1.166$. The relatively large thermal conductivity and thermal diffusivity of the nanofluid enhance heat transfer rates in some applications. However, all of the thermophysical properties are affected by the addition of the nanoparticles, and, as will become evident in Chapters 6 through 9, properties such as the viscosity and specific heat are adversely affected. This condition can degrade thermal performance when the use of nanofluids involves convection heat transfer.

2. The expression for the nanofluid's thermal conductivity (and viscosity) is limited to dilute mixtures of noncontacting, spherical particles. In some cases, the particles do not remain separated but can *agglomerate* into long chains, providing effective paths for heat conduction through the fluid and larger bulk thermal conductivities. Hence, the expression for the thermal conductivity represents the *minimum* possible enhancement of the thermal conductivity by spherical nanoparticles. An expression for the maximum possible *isotropic* thermal conductivity of a nanofluid, corresponding to agglomeration of the spherical particles, is available [21], as are expressions for dilute suspensions of nonspherical particles [22]. Note that these expressions can also be applied to nano-structured *composite materials* consisting of a particulate phase interspersed within a host binding medium, as will be discussed in more detail in Chapter 3.

3. The nanofluid's density and specific heat are averaged values. As such, these properties do not depend on the manner in which the nanoparticles are dispersed within the base liquid.

2.3 *The Heat Diffusion Equation*

A major objective in a conduction analysis is to determine the *temperature field* in a medium resulting from conditions imposed on its boundaries. That is, we wish to know the *temperature distribution*, which represents how temperature varies with position in the medium. Once this distribution is known, the conduction heat flux at any point in the medium or on its surface may be computed from Fourier's law. Other important quantities of interest may also be determined. For a solid, knowledge of the temperature distribution could be used to ascertain structural integrity through determination of thermal stresses, expansions, and deflections. The temperature distribution could also be used to optimize the thickness of an insulating material or to determine the compatibility of special coatings or adhesives used with the material.

We now proceed to derive a differential equation whose solution provides the temperature distribution in the medium. Consider a homogeneous medium within which there is no bulk motion (advection) and the temperature distribution $T(x, y, z)$ is expressed in Cartesian coordinates. The medium is assumed to be incompressible, that is, its density can be treated as constant. Following the four-step methodology of applying conservation of energy (Section 1.3.1), we first define an infinitesimally small (differential) control volume, $dx \cdot dy \cdot dz$, as shown in Figure 2.11. Choosing to formulate the first law at an instant of time, the second step is to consider the energy processes that are relevant to this control volume. In the absence of motion, there are no changes in mechanical energy and no work being done on the system. Only thermal forms of energy need be considered. Specifically, if there are temperature gradients, conduction heat transfer will occur across each of the control surfaces. The conduction heat rates perpendicular to each of the control surfaces at the x-, y-, and z-coordinate locations are indicated by the terms q_x, q_y, and q_z, respectively. The conduction heat rates at the opposite surfaces can then be expressed as a Taylor series expansion where, neglecting higher-order terms,

$$q_{x+dx} = q_x + \frac{\partial q_x}{\partial x} dx \tag{2.13a}$$

$$q_{y+dy} = q_y + \frac{\partial q_y}{\partial y} dy \tag{2.13b}$$

$$q_{z+dz} = q_z + \frac{\partial q_z}{\partial z} dz \tag{2.13c}$$

In words, Equation 2.13a simply states that the x-component of the heat transfer rate at $x + dx$ is equal to the value of this component at x plus the amount by which it changes with respect to x times dx.

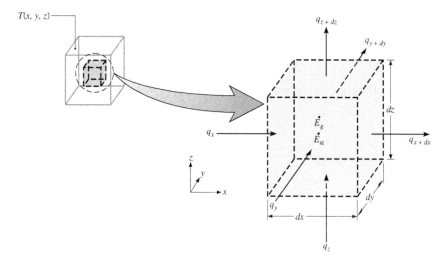

FIGURE 2.11 Differential control volume, $dx \, dy \, dz$, for conduction analysis in Cartesian coordinates.

Within the medium there may also be an *energy source* term associated with the rate of thermal energy generation. This term is represented as

$$\dot{E}_g = \dot{q}\, dx\, dy\, dz \qquad (2.14)$$

where $\dot{q}$ is the rate at which energy is generated per unit volume of the medium (W/m^3). In addition, changes may occur in the amount of the internal thermal energy stored by the material in the control volume. If the material is not experiencing a change in phase, latent energy effects are not pertinent, and the *energy storage* term reduces to the rate of change of sensible energy:

$$\dot{E}_{st} = \frac{\partial U_{sens}}{\partial t} = \rho c_v \frac{\partial T}{\partial t}\, dx\, dy\, dz = \rho c_p \frac{\partial T}{\partial t}\, dx\, dy\, dz \qquad (2.15)$$

Here, use has been made of the fact that $c_p = c_v$ for an incompressible substance.[2]

Once again it is important to note that the terms $\dot{E}_g$ and $\dot{E}_{st}$ represent different physical processes. The energy generation term $\dot{E}_g$ is a manifestation of some energy conversion process involving thermal energy on one hand and some other form of energy, such as chemical, electrical, or nuclear, on the other. The term is positive (a *source*) if thermal energy is being generated in the material at the expense of some other energy form; it is negative (a *sink*) if thermal energy is being consumed. In contrast, the energy storage term $\dot{E}_{st}$ refers to the rate of change of thermal energy stored by the matter.

The last step in the methodology outlined in Section 1.3.1 is to express conservation of energy using the foregoing rate equations. On a *rate* basis, the general form of the conservation of energy requirement is

$$\dot{E}_{in} + \dot{E}_g - \dot{E}_{out} = \dot{E}_{st} \qquad (1.12c)$$

Hence, recognizing that the conduction rates constitute the energy inflow $\dot{E}_{in}$ and outflow $\dot{E}_{out}$, and substituting Equations 2.14 and 2.15, we obtain

$$q_x + q_y + q_z + \dot{q}\, dx\, dy\, dz - q_{x+dx} - q_{y+dy} - q_{z+dz} = \rho c_p \frac{\partial T}{\partial t}\, dx\, dy\, dz \qquad (2.16)$$

Substituting from Equations 2.13, it follows that

$$-\frac{\partial q_x}{\partial x}\, dx - \frac{\partial q_y}{\partial y}\, dy - \frac{\partial q_z}{\partial z}\, dz + \dot{q}\, dx\, dy\, dz = \rho c_p \frac{\partial T}{\partial t}\, dx\, dy\, dz \qquad (2.17)$$

The conduction heat rates in an isotropic material may be evaluated from Fourier's law,

$$q_x = -k\, dy\, dz\, \frac{\partial T}{\partial x} \qquad (2.18a)$$

$$q_y = -k\, dx\, dz\, \frac{\partial T}{\partial y} \qquad (2.18b)$$

$$q_z = -k\, dx\, dy\, \frac{\partial T}{\partial z} \qquad (2.18c)$$

---

[2]Energy storage will be expressed using c_p in the remainder of the text. Use of c_p instead of c_v broadens applicability of Equation 2.15 to some scenarios beyond incompressible substances [23, 24], and is consistent with the simplified steady-flow thermal energy equation, Equation 1.12e.

where each heat flux component of Equation 2.6 has been multiplied by the appropriate control surface (differential) area to obtain the heat transfer rate. Substituting Equations 2.18 into Equation 2.17 and dividing out the dimensions of the control volume ($dx\ dy\ dz$), we obtain

$$\frac{\partial}{\partial x}\left(k\frac{\partial T}{\partial x}\right) + \frac{\partial}{\partial y}\left(k\frac{\partial T}{\partial y}\right) + \frac{\partial}{\partial z}\left(k\frac{\partial T}{\partial z}\right) + \dot{q} = \rho c_p \frac{\partial T}{\partial t} \qquad (2.19)$$

Equation 2.19 is the general form, in Cartesian coordinates, of the *heat diffusion equation*. This equation, often referred to as the *heat equation*, provides the basic tool for heat conduction analysis. From its solution, we can obtain the temperature distribution $T(x, y, z)$ as a function of time. The apparent complexity of this expression should not obscure the fact that it describes an important physical condition, that is, conservation of energy. You should have a clear understanding of the physical significance of each term appearing in the equation. For example, the term $\partial(k\ \partial T/\partial x)/\partial x$ is related to the *net* conduction heat flux *into* the control volume for the x-coordinate direction. That is, multiplying by dx,

$$\frac{\partial}{\partial x}\left(k\frac{\partial T}{\partial x}\right)dx = q''_x - q''_{x+dx} \qquad (2.20)$$

with similar expressions applying for the fluxes in the y- and z-directions. In words, the heat equation, Equation 2.19, therefore states that *at any point in the medium the net rate of energy transfer by conduction into a unit volume plus the volumetric rate of thermal energy generation must equal the rate of change of thermal energy stored within the volume.*

It is often possible to work with simplified versions of Equation 2.19. For example, if the thermal conductivity is constant, the heat equation is

$$\frac{\partial^2 T}{\partial x^2} + \frac{\partial^2 T}{\partial y^2} + \frac{\partial^2 T}{\partial z^2} + \frac{\dot{q}}{k} = \frac{1}{\alpha}\frac{\partial T}{\partial t} \qquad (2.21)$$

where $\alpha = k/\rho c_p$ is the *thermal diffusivity*. Additional simplifications of the general form of the heat equation are often possible. For example, under *steady-state* conditions, there can be no change in the amount of energy storage; hence Equation 2.19 reduces to

$$\frac{\partial}{\partial x}\left(k\frac{\partial T}{\partial x}\right) + \frac{\partial}{\partial y}\left(k\frac{\partial T}{\partial y}\right) + \frac{\partial}{\partial z}\left(k\frac{\partial T}{\partial z}\right) + \dot{q} = 0 \qquad (2.22)$$

Moreover, if the heat transfer is *one-dimensional* (e.g., in the x-direction) and there is *no energy generation*, Equation 2.22 reduces to

$$\frac{d}{dx}\left(k\frac{dT}{dx}\right) = 0 \qquad (2.23)$$

The important implication of this result is that, *under steady-state, one-dimensional conditions with no energy generation*, the heat flux is a constant in the direction of transfer ($dq''_x/dx = 0$).

The heat equation may also be expressed in cylindrical and spherical coordinates. The differential control volumes for these two coordinate systems are shown in Figures 2.12 and 2.13.

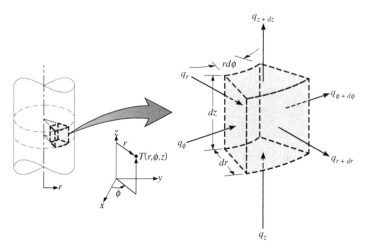

FIGURE 2.12   Differential control volume, $dr \cdot r\,d\phi \cdot dz$, for conduction analysis in cylindrical coordinates (r, ϕ, z).

Cylindrical Coordinates   When the del operator ∇ of Equation 2.3 is expressed in cylindrical coordinates, with i, j, and k representing the unit vectors in the r, ϕ, and z directions, the general form of the heat flux vector and hence of Fourier's law is

$$q'' = -k\nabla T = -k\left(i\frac{\partial T}{\partial r} + j\frac{1}{r}\frac{\partial T}{\partial \phi} + k\frac{\partial T}{\partial z} \right) \tag{2.24}$$

where

$$q_r'' = -k\frac{\partial T}{\partial r} \qquad q_\phi'' = -\frac{k}{r}\frac{\partial T}{\partial \phi} \qquad q_z'' = -k\frac{\partial T}{\partial z} \tag{2.25}$$

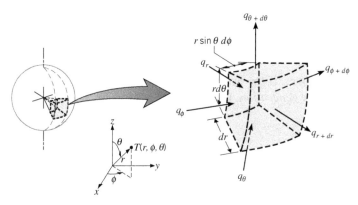

FIGURE 2.13   Differential control volume, $dr \cdot r\sin\theta\,d\phi \cdot r\,d\theta$, for conduction analysis in spherical coordinates (r, ϕ, θ).

are heat flux components in the radial, circumferential, and axial directions, respectively. Applying an energy balance to the differential control volume of Figure 2.12, the following general form of the heat equation is obtained:

$$\frac{1}{r}\frac{\partial}{\partial r}\left(kr\frac{\partial T}{\partial r}\right) + \frac{1}{r^2}\frac{\partial}{\partial \phi}\left(k\frac{\partial T}{\partial \phi}\right) + \frac{\partial}{\partial z}\left(k\frac{\partial T}{\partial z}\right) + \dot{q} = \rho c_p \frac{\partial T}{\partial t} \qquad (2.26)$$

Spherical Coordinates   In spherical coordinates, with i, j, and k representing the unit vectors in the r, θ, and ϕ directions, the general form of the heat flux vector and Fourier's law is

$$q'' = -k\nabla T = -k\left(i\frac{\partial T}{\partial r} + j\frac{1}{r}\frac{\partial T}{\partial \theta} + k\frac{1}{r\sin\theta}\frac{\partial T}{\partial \phi}\right) \qquad (2.27)$$

where

$$q''_r = -k\frac{\partial T}{\partial r} \qquad q''_\theta = -\frac{k}{r}\frac{\partial T}{\partial \theta} \qquad q''_\phi = -\frac{k}{r\sin\theta}\frac{\partial T}{\partial \phi} \qquad (2.28)$$

are heat flux components in the radial, polar, and azimuthal directions, respectively. Applying an energy balance to the differential control volume of Figure 2.13, the following general form of the heat equation is obtained:

$$\frac{1}{r^2}\frac{\partial}{\partial r}\left(kr^2\frac{\partial T}{\partial r}\right) + \frac{1}{r^2\sin^2\theta}\frac{\partial}{\partial \phi}\left(k\frac{\partial T}{\partial \phi}\right)$$

$$+ \frac{1}{r^2\sin\theta}\frac{\partial}{\partial \theta}\left(k\sin\theta\frac{\partial T}{\partial \theta}\right) + \dot{q} = \rho c_p \frac{\partial T}{\partial t} \qquad (2.29)$$

You should attempt to derive Equation 2.26 or 2.29 to gain experience in applying conservation principles to differential control volumes (see Problems 2.27 and 2.28). Note that the temperature gradient in Fourier's law must have units of K/m. Hence, when evaluating the gradient for an angular coordinate, it must be expressed in terms of the differential change in arc *length*. For example, the heat flux component in the circumferential direction of a cylindrical coordinate system is $q''_\phi = -(k/r)(\partial T/\partial \phi)$, *not* $q''_\phi = -k(\partial T/\partial \phi)$.

EXAMPLE 2.3

The temperature distribution across a wall 1 m thick at a certain instant of time is given as

$$T(x) = a + bx + cx^2$$

where T is in degrees Celsius and x is in meters, while $a = 900°C$, $b = -300°C/m$, and $c = -50°C/m^2$. A uniform heat generation, $\dot{q} = 1000$ W/m^3, is present in the wall of area 10 m^2 having the properties $\rho = 1600$ kg/m^3, $k = 40$ W/m · K, and $c_p = 4$ kJ/kg · K.

1. Determine the rate of heat transfer entering the wall ($x = 0$) and leaving the wall ($x = 1$ m).

2. Determine the rate of change of energy storage in the wall.

3. Determine the time rate of temperature change at $x = 0$, 0.25, and 0.5 m.

SOLUTION

Known: Temperature distribution $T(x)$ at an instant of time t in a one-dimensional wall with uniform heat generation.

Find:

1. Heat rates entering, q_{in} ($x = 0$), and leaving, q_{out} ($x = 1$ m), the wall.

2. Rate of change of energy storage in the wall, $\dot{E}_{st}$.

3. Time rate of temperature change at $x = 0$, 0.25, and 0.5 m.

Schematic:

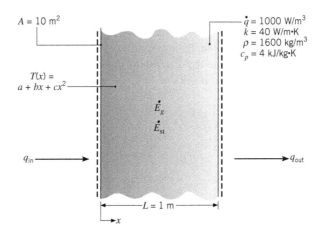

Assumptions:

1. One-dimensional conduction in the x-direction.

2. Incompressible, isotropic medium with constant properties.

3. Uniform internal heat generation, $\dot{q}$ (W/m^3).

Analysis:

1. Recall that once the temperature distribution is known for a medium, it is a simple matter to determine the conduction heat transfer rate at any point in the medium or at its surfaces by using Fourier's law. Hence the desired heat rates may be determined by using the prescribed temperature distribution with Equation 2.1. Accordingly,

$$q_{in} = q_x(0) = -kA\frac{\partial T}{\partial x}\bigg|_{x=0} = -kA(b + 2cx)_{x=0}$$

$$q_{in} = -bkA = 300°C/m \times 40 \text{ W/m} \cdot \text{K} \times 10 \text{ m}^2 = 120 \text{ kW} \quad \triangleleft$$

Similarly,

$$q_{out} = q_x(L) = -kA\frac{\partial T}{\partial x}\bigg|_{x=L} = -kA(b + 2cx)_{x=L}$$

$$q_{out} = -(b + 2cL)kA = -[-300°C/m$$

$$+ 2(-50°C/m^2) \times 1 \text{ m}] \times 40 \text{ W/m} \cdot \text{K} \times 10 \text{ m}^2 = 160 \text{ kW} \qquad \triangleleft$$

2. The rate of change of energy storage in the wall $\dot{E}_{st}$ may be determined by applying an overall energy balance to the wall. Using Equation 1.12c for a control volume about the wall,

$$\dot{E}_{in} + \dot{E}_g - \dot{E}_{out} = \dot{E}_{st}$$

where $\dot{E}_g = \dot{q}AL$, it follows that

$$\dot{E}_{st} = \dot{E}_{in} + \dot{E}_g - \dot{E}_{out} = q_{in} + \dot{q}AL - q_{out}$$

$$\dot{E}_{st} = 120 \text{ kW} + 1000 \text{ W/m}^3 \times 10 \text{ m}^2 \times 1 \text{ m} - 160 \text{ kW}$$

$$\dot{E}_{st} = -30 \text{ kW} \qquad \triangleleft$$

3. The time rate of change of the temperature at any point in the medium may be determined from the heat equation, Equation 2.21, rewritten as

$$\frac{\partial T}{\partial t} = \frac{k}{\rho c_p}\frac{\partial^2 T}{\partial x^2} + \frac{\dot{q}}{\rho c_p}$$

From the prescribed temperature distribution, it follows that

$$\frac{\partial^2 T}{\partial x^2} = \frac{\partial}{\partial x}\left(\frac{\partial T}{\partial x}\right)$$

$$= \frac{\partial}{\partial x}(b + 2cx) = 2c = 2(-50°C/m^2) = -100°C/m^2$$

Note that this derivative is independent of position in the medium. Hence the time rate of temperature change is also independent of position and is given by

$$\frac{\partial T}{\partial t} = \frac{40 \text{ W/m} \cdot \text{K}}{1600 \text{ kg/m}^3 \times 4 \text{ kJ/kg} \cdot \text{K}} \times (-100°C/m^2)$$

$$+ \frac{1000 \text{ W/m}^3}{1600 \text{ kg/m}^3 \times 4 \text{ kJ/kg} \cdot \text{K}}$$

$$\frac{\partial T}{\partial t} = -6.25 \times 10^{-4}°C/s + 1.56 \times 10^{-4}°C/s$$

$$= -4.69 \times 10^{-4}°C/s \qquad \triangleleft$$

Comments:

1. From this result, it is evident that the temperature at every point within the wall is decreasing with time.

2. Fourier's law can always be used to compute the conduction heat rate from knowledge of the temperature distribution, even for unsteady conditions with internal heat generation.

Microscale Effects   For most practical situations, the heat diffusion equations generated in this text may be used with confidence. However, these equations are based on Fourier's law, which does not account for the finite speed at which thermal information is propagated within the medium by the various energy carriers. The consequences of the finite propagation speed may be neglected if the heat transfer events of interest occur over a sufficiently long time scale, Δt, such that

$$\frac{\lambda_{\text{mfp}}}{\overline{c}\Delta t} \ll 1 \tag{2.30}$$

The heat diffusion equations of this text are likewise invalid for problems where boundary scattering must be explicitly considered. For example, the temperature distribution *within* the thin film of Figure 2.6*b* cannot be determined by applying the foregoing heat diffusion equations. Additional discussion of micro- and nanoscale heat transfer applications and analysis methods is available in the literature [1, 5, 10, 25].

2.4   *Boundary and Initial Conditions*

To determine the temperature distribution in a medium, it is necessary to solve the appropriate form of the heat equation. However, such a solution depends on the physical conditions existing at the *boundaries* of the medium and, if the situation is time dependent, on conditions existing in the medium at some *initial time*. With regard to the *boundary conditions,* there are several common possibilities that are simply expressed in mathematical form. Because the heat equation is second order in the spatial coordinates, two boundary conditions must be expressed for each coordinate needed to describe the system. Because the equation is first order in time, however, only one condition, termed the *initial condition*, must be specified.

Three kinds of boundary conditions commonly encountered in heat transfer are summarized in Table 2.2. The conditions are specified at the surface $x = 0$ for a one-dimensional system. Heat transfer is in the positive x-direction with the temperature distribution, which may be time dependent, designated as $T(x, t)$. The first condition corresponds to a situation for which the surface is maintained at a fixed temperature T_s. It is commonly termed a *Dirichlet condition*, or a boundary condition of the *first kind*. It is closely approximated, for example, when the surface is in contact with a melting solid or a boiling liquid. In both cases, there is heat transfer at the surface, while the surface remains at the temperature of the phase change process. The second condition corresponds to the existence of a fixed or constant heat flux q_s'' at the surface. This heat flux is related to the temperature gradient at the surface by Fourier's law, Equation 2.6, which may be expressed as

$$q_x''(0) = -k\left.\frac{\partial T}{\partial x}\right|_{x=0} = q_s''$$

It is termed a *Neumann condition*, or a boundary condition of the *second kind*, and may be realized by bonding a thin film electric heater to the surface. A special case of this condition

TABLE 2.2 Boundary conditions for the heat
diffusion equation at the surface ($x = 0$)

1. Constant surface temperature

   $$T(0,t) = T_s \qquad\qquad (2.31)$$

2. Constant surface heat flux
   (a) Finite heat flux

   $$-k\frac{\partial T}{\partial x}\bigg|_{x=0} = q_s'' \qquad\qquad (2.32)$$

   (b) Adiabatic or insulated surface

   $$\frac{\partial T}{\partial x}\bigg|_{x=0} = 0 \qquad\qquad (2.33)$$

3. Convection surface condition

   $$-k\frac{\partial T}{\partial x}\bigg|_{x=0} = h[T_\infty - T(0,t)] \qquad\qquad (2.34)$$

corresponds to the *perfectly insulated*, or *adiabatic*, surface for which $\partial T/\partial x\big|_{x=0} = 0$. The boundary condition of the *third kind* corresponds to the existence of convection heating (or cooling) at the surface and is obtained from the surface energy balance discussed in Section 1.3.1.

EXAMPLE 2.4

A long copper bar of rectangular cross section, whose width w is much greater than its thickness L, is maintained in contact with a *heat sink* (see Comment 1) at its lower surface, and the temperature throughout the bar is approximately equal to that of the sink, T_o. Suddenly, an electric current is passed through the bar and an airstream of temperature T_∞ is passed over the top surface, while the bottom surface continues to be maintained at T_o. Obtain the differential equation and the boundary and initial conditions that could be solved to determine the temperature as a function of position and time in the bar.

SOLUTION

Known: Copper bar initially in thermal equilibrium with a heat sink is suddenly heated by passage of an electric current.

Find: Differential equation and boundary and initial conditions needed to determine temperature as a function of position and time within the bar.

Schematic:

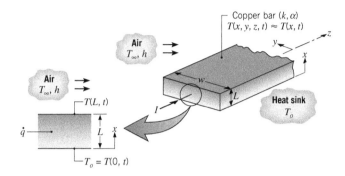

Assumptions:

1. Since the bar is long and $w \gg L$, end and side effects are negligible and heat transfer within the bar is primarily one dimensional in the x-direction.

2. Uniform volumetric heat generation, $\dot{q}$.

3. Constant properties.

Analysis: The temperature distribution is governed by the heat equation (Equation 2.19), which, for the one-dimensional and constant property conditions of the present problem, reduces to

$$\frac{\partial^2 T}{\partial x^2} + \frac{\dot{q}}{k} = \frac{1}{\alpha}\frac{\partial T}{\partial t} \qquad (1) \qquad \lhd$$

where the temperature is a function of position and time, $T(x, t)$. Since this differential equation is second order in the spatial coordinate x and first order in time t, there must be two boundary conditions for the x-direction and one condition, termed the initial condition, for time. The boundary condition at the bottom surface corresponds to case 1 of Table 2.2. In particular, since the temperature of this surface is maintained at a value, T_o, which is fixed with time, it follows that

$$T(0, t) = T_o \qquad (2) \qquad \lhd$$

The convection surface condition, case 3 of Table 2.2, is appropriate for the top surface. Hence

$$-k\frac{\partial T}{\partial x}\bigg|_{x=L} = h[T(L, t) - T_\infty] \qquad (3) \qquad \lhd$$

The initial condition is inferred from recognition that, before the change in conditions, the bar is at a uniform temperature T_o. Hence

$$T(x, 0) = T_o \qquad (4) \qquad \lhd$$

If T_o, T_∞, $\dot{q}$, and h are known, Equations 1 through 4 may be solved to obtain the time-varying temperature distribution $T(x, t)$ following imposition of the electric current.

Comments:

1. The heat sink at $x = 0$ could be maintained by exposing the surface to an ice bath or by attaching it to a *cold plate*. A cold plate contains coolant channels machined in a solid of large thermal conductivity (usually copper). By circulating a liquid (usually water) through the channels, the plate and hence the surface to which it is attached may be maintained at a nearly uniform temperature.

2. The temperature of the top surface $T(L, t)$ will change with time. This temperature is an unknown and may be obtained after finding $T(x, t)$.

3. We may use our physical intuition to sketch temperature distributions in the bar at selected times from the beginning to the end of the transient process. If we assume that $T_\infty > T_o$ and that the electric current is sufficiently large to heat the bar to temperatures in excess of T_∞, the following distributions would correspond to the initial condition ($t \le 0$), the final (steady-state) condition ($t \rightarrow \infty$), and two intermediate times.

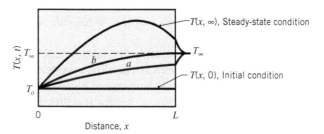

Note how the distributions comply with the initial and boundary conditions. What is a special feature of the distribution labeled (b)?

4. Our intuition may also be used to infer the manner in which the heat flux varies with time at the surfaces ($x = 0, L$) of the bar. On $q_x'' - t$ coordinates, the transient variations are as follows.

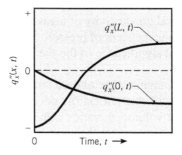

Convince yourself that the foregoing variations are consistent with the temperature distributions of Comment 3. For $t \rightarrow \infty$, how are $q_x''(0)$ and $q_x''(L)$ related to the volumetric rate of energy generation?

2.5 *Summary*

Despite the relative brevity of this chapter, its importance must not be underestimated. Understanding the conduction rate equation, Fourier's law, is essential. You must be cognizant of the importance of thermophysical properties; over time, you will develop a sense of the magnitudes of the properties of many real materials. Likewise, you must recognize that the heat equation is derived by applying the conservation of energy principle to a differential control volume and that it is used to determine temperature distributions within matter. From knowledge of the distribution, Fourier's law can be used to determine the corresponding conduction heat rates. A firm grasp of the various types of thermal boundary conditions that are used in conjunction with the heat equation is vital. Indeed, Chapter 2 is the foundation on which Chapters 3 through 5 are based, and you are encouraged to revisit this chapter often. You may test your understanding of various concepts by addressing the following questions.

- In the general formulation of *Fourier's law* (applicable to any geometry), what are the vector and scalar quantities? Why is there a minus sign on the right-hand side of the equation?
- What is an *isothermal surface*? What can be said about the heat flux at any location on this surface?
- What form does *Fourier's law* take for each of the orthogonal directions of Cartesian, cylindrical, and spherical coordinate systems? In each case, what are the units of the temperature gradient? Can you write each equation from memory?
- An important property of matter is defined by *Fourier's law*. What is it? What is its physical significance? What are its units?
- What is an *isotropic* material?
- Why is the thermal conductivity of a solid generally larger than that of a liquid? Why is the thermal conductivity of a liquid larger than that of a gas?
- Why is the thermal conductivity of an electrically conducting solid generally larger than that of a nonconductor? Why are materials such as beryllium oxide, diamond, and silicon carbide (see Table A.2) exceptions to this rule?
- Does the *effective thermal conductivity* of an insulation system account for heat transfer by conduction alone?
- Why does the thermal conductivity of a gas increase with increasing temperature? Why is it approximately independent of pressure?
- What is the physical significance of the *thermal diffusivity*? How is it defined and what are its units?
- What is the physical significance of each term appearing in the *heat equation*?
- Cite some examples of *thermal energy generation*. If the rate at which thermal energy is generated per unit volume, $\dot{q}$, varies with location in a medium of volume V, how can the rate of energy generation for the entire medium, $\dot{E}_g$, be determined from knowledge of $\dot{q}(x, y, z)$?
- For a chemically reacting medium, what kind of reaction provides a *source* of thermal energy ($\dot{q} > 0$)? What kind of reaction provides a *sink* for thermal energy ($\dot{q} < 0$)?
- To solve the *heat equation* for the temperature distribution in a medium, *boundary conditions* must be prescribed at the surfaces of the medium. What physical conditions are commonly suitable for this purpose?

References

1. Flik, M. I., B.-I. Choi, and K. E. Goodson, *J. Heat Transfer*, **114**, 666, 1992.

2. Klemens, P. G., "Theory of the Thermal Conductivity of Solids," in R. P. Tye, Ed., *Thermal Conductivity*, Vol. 1, Academic Press, London, 1969.

3. Yang, H.-S., G.-R. Bai, L. J. Thompson, and J. A. Eastman, *Acta Materialia*, **50**, 2309, 2002.

4. Chen, G., *J. Heat Transfer*, **118**, 539, 1996.

5. Carey, V. P., G. Chen, C. Grigoropoulos, M. Kaviany, and A. Majumdar, *Nano. and Micro. Thermophys. Engng.* **12**, 1, 2008.

6. Padture, N. P., M. Gell, and E. H. Jordan, *Science*, **296**, 280, 2002.

7. Schelling, P. K., L. Shi, and K. E. Goodson, *Mat. Today*, **8**, 30, 2005.

8. Baxter, J., Z. Bian, G. Chen, D. Danielson, M. S. Dresselhaus, A. G. Federov, T. S. Fisher, C. W. Jones, E. Maginn, W. Kortshagen, A. Manthiram, A. Nozik, D. R. Rolison, T. Sands, L. Shi, D. Sholl, and Y. Wu, *Energy and Environ. Sci.*, **2**, 559, 2009.

9. Vincenti, W. G., and C. H. Kruger Jr., *Introduction to Physical Gas Dynamics*, Wiley, New York, 1986.

10. Zhang, Z. M., *Nano/Microscale Heat Transfer*, McGraw-Hill, New York, 2007.

11. McLaughlin, E., "Theory of the Thermal Conductivity of Fluids," in R. P. Tye, Ed., *Thermal Conductivity*, Vol. 2, Academic Press, London, 1969.

12. Foust, O. J., Ed., "Sodium Chemistry and Physical Properties," in *Sodium-NaK Engineering Handbook*, Vol. 1, Gordon & Breach, New York, 1972.

13. Mallory, J. F., *Thermal Insulation*, Reinhold Book Corp., New York, 1969.

14. American Society of Heating, Refrigeration and Air Conditioning Engineers, *Handbook of Fundamentals*, Chapters 23–25 and 31, ASHRAE, New York, 2001.

15. Zeng, S. Q., A. Hunt, and R. Greif, *J. Heat Transfer*, **117**, 1055, 1995.

16. Sengers, J. V., and M. Klein, Eds., *The Technical Importance of Accurate Thermophysical Property Information*, National Bureau of Standards Technical Note No. 590, 1980.

17. Najjar, M. S., K. J. Bell, and R. N. Maddox, *Heat Transfer Eng.*, **2**, 27, 1981.

18. Hanley, H. J. M., and M. E. Baltatu, *Mech. Eng.*, **105**, 68, 1983.

19. Touloukian, Y. S., and C. Y. Ho, Eds., *Thermophysical Properties of Matter, The TPRC Data Series* (13 volumes on thermophysical properties: thermal conductivity, specific heat, thermal radiative, thermal diffusivity, and thermal linear expansion), Plenum Press, New York, 1970 through 1977.

20. Chow, T. S., *Phys. Rev. E*, **48**, 1977, 1993.

21. Keblinski, P., R. Prasher, and J. Eapen, *J. Nanopart. Res.*, **10**, 1089, 2008.

22. Hamilton, R. L., and O. K. Crosser, *I&EC Fundam.* **1**, 187, 1962.

23. Panton, R. L., *Incompressible Flow*, Wiley, Hoboken, 2013.

24. Burmeister, L. C., *Convective Heat Transfer*, Wiley-Interscience, New York, 1993.

25. Cahill, D. G., W. K. Ford, K. E. Goodson, G. D. Mahan, A. Majumdar, H. J. Maris, R. Merlin, and S. R. Phillpot, *App. Phys. Rev.*, **93**, 793, 2003.

Problems

Fourier's Law

2.1 Assume steady-state, one-dimensional heat conduction through the axisymmetric shape shown below.

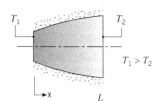

Assuming constant properties and no internal heat generation, sketch the temperature distribution on $T - x$ coordinates. Briefly explain the shape of your curve.

2.2 A hot water pipe with outside radius r_1 has a temperature T_1. A thick insulation, applied to reduce the heat loss, has an outer radius r_2 and temperature T_2. On $T - r$ coordinates, sketch the temperature distribution in the insulation for one-dimensional, steady-state heat transfer with constant properties. Give a brief explanation, justifying the shape of your curve.

2.3 Assume steady-state, one-dimensional heat conduction through the symmetric shape shown.

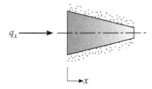

Assuming that there is no internal heat generation, derive an expression for the thermal conductivity $k(x)$ for these conditions: $A(x) = (1 - x)$, $T(x) = 300$ $(1 - 2x - x^3)$, and $q = 6000\,W$, where A is in square meters, T in kelvins, and x in meters.

2.4 To determine the effect of the temperature dependence of the thermal conductivity on the temperature distribution in a solid, consider a material for which this dependence may be represented as

$$k = k_o + aT$$

where k_o is a positive constant and a is a coefficient that may be positive or negative. Sketch the steady-state temperature distribution associated with heat transfer in a plane wall for three cases corresponding to $a > 0$, $a = 0$, and $a < 0$.

2.5 A young engineer is asked to design a thermal protection barrier for a sensitive electronic device that might be exposed to irradiation from a high-powered infrared laser. Having learned as a student that a low thermal conductivity material provides good insulating characteristics, the engineer specifies use of a nanostructured aerogel, characterized by a thermal conductivity of $k_a = 0.005\,W/m \cdot K$, for the protective barrier. The engineer's boss questions the wisdom of selecting the aerogel *because* it has a low thermal conductivity. Consider the sudden laser irradiation of (a) pure aluminum, (b) glass, and (c) aerogel. The laser provides irradiation of $G = 10 \times 10^6\,W/m^2$. The absorptivities of the materials are $\alpha = 0.2$, 0.9, and 0.8 for the aluminum, glass, and aerogel, respectively, and the initial temperature of the barrier is $T_i = 300\,K$. Explain why the boss is concerned. *Hint:* All materials experience thermal expansion (or contraction), and local stresses that develop within a material are, to a first approximation, proportional to the local temperature gradient.

2.6 A one-dimensional plane wall of thickness $2L = 80\,mm$ experiences uniform thermal energy generation of $\dot{q} = 1000\,W/m^3$ and is convectively cooled at $x = \pm 40$ mm by an ambient fluid characterized by $T_\infty = 30°C$. If the

steady-state temperature distribution within the wall is $T(x) = a(L^2 - x^2) + b$ where $a = 15°C/m^2$ and $b = 40°C$, what is the thermal conductivity of the wall? What is the value of the convection heat transfer coefficient, h?

2.7 Consider a plane wall 100 mm thick and of thermal conductivity $100\,W/m \cdot K$. Steady-state conditions are known to exist with $T_1 = 400\,K$ and $T_2 = 600\,K$. Determine the heat flux q_x'' and the temperature gradient dT/dx for the coordinate systems shown.

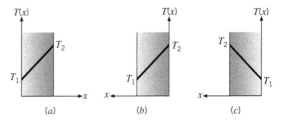

2.8 In the two-dimensional body illustrated, the gradient at surface A is found to be $\partial T/\partial y = 40$ K/m. What are $\partial T/\partial y$ and $\partial T/\partial x$ at surface B?

2.9 Steady-state, one-dimensional conduction occurs in a rod of constant thermal conductivity k and variable cross-sectional area $A_x(x) = A_o e^{ax}$, where A_o and a are constants. The lateral surface of the rod is well insulated.

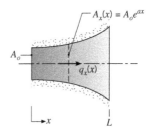

(a) Write an expression for the conduction heat rate, $q_x(x)$. Use this expression to determine the temperature distribution $T(x)$ and qualitatively sketch the distribution for $T(0) > T(L)$.

(b) Now consider conditions for which thermal energy is generated in the rod at a volumetric rate $\dot{q} = \dot{q}_o \exp(-ax)$ where $\dot{q}_o$ is a constant. Obtain an expression for $q_x(x)$ when the left face ($x = 0$) is well insulated.

Thermophysical Properties

2.10 An apparatus for measuring thermal conductivity employs an electrical heater sandwiched between two identical samples of diameter 30 mm and length 60 mm, which are pressed between plates maintained at a uniform temperature $T_o = 77°C$ by a circulating fluid. A conducting grease is placed between all the surfaces to ensure good thermal contact. Differential thermocouples are imbedded in the samples with a spacing of 15 mm. The lateral sides of the samples are insulated to ensure one-dimensional heat transfer through the samples.

(a) With two samples of SS316 in the apparatus, the heater draws 0.353 A at 100 V, and the differential thermocouples indicate $\Delta T_1 = \Delta T_2 = 25.0°C$. What is the thermal conductivity of the stainless steel sample material? What is the average temperature of the samples? Compare your result with the thermal conductivity value reported for this material in Table A.1.

(b) By mistake, an Armco iron sample is placed in the lower position of the apparatus with one of the SS316 samples from part (a) in the upper portion. For this situation, the heater draws 0.601 A at 100 V, and the differential thermocouples indicate $\Delta T_1 = \Delta T_2 = 15.0°C$. What are the thermal conductivity and average temperature of the Armco iron sample?

(c) What is the advantage in constructing the apparatus with two identical samples sandwiching the heater rather than with a single heater–sample combination? When would heat leakage out of the lateral surfaces of the samples become significant? Under what conditions would you expect $\Delta T_1 \neq \Delta T_2$?

2.11 An engineer desires to measure the thermal conductivity of an aerogel material. It is expected that

the aerogel will have an extremely small thermal conductivity.

(a) Explain why the apparatus of Problem 2.10 cannot be used to obtain an accurate measurement of the aerogel's thermal conductivity.

(b) The engineer designs a new apparatus for which an electric heater of diameter $D = 150$ mm is sandwiched between two thin plates of aluminum. The steady-state temperatures of the 5-mm-thick aluminum plates, T_1 and T_2, are measured with thermocouples. Aerogel sheets of thickness $t = 5$ mm are placed outside the aluminum plates, while a coolant with an inlet temperature of $T_{c,i} = 25°C$ maintains the exterior surfaces of the aerogel at a low temperature. The circular aerogel sheets are formed so that they encase the heater and aluminum sheets, providing insulation to minimize radial heat losses. At steady state, $T_1 = T_2 = 55°C$, and the heater draws 125 mA at 10 V. Determine the value of the aerogel thermal conductivity k_a.

(c) Calculate the temperature difference across the thickness of the 5-mm-thick aluminum plates. Comment on whether it is important to know the axial locations at which the temperatures of the aluminum plates are measured.

(d) If liquid water is used as the coolant with a total flow rate of $\dot{m} = 1$ kg/min (0.5 kg/min for each of the two streams), calculate the outlet temperature of the water, $T_{c,o}$.

2.12 Consider a 300 mm × 300 mm window in an aircraft. For a temperature difference of 80°C from the inner to the outer surface of the window, calculate the heat loss through $L = 10$-mm-thick polycarbonate, soda lime glass, and aerogel windows, respectively. The thermal conductivities of the aerogel and polycarbonate are $k_{ag} = 0.014$ W/m · K and $k_{pc} = 0.21$ W/m · K, respectively. Evaluate the thermal conductivity of the soda lime glass at 300 K. If the aircraft has 130 windows and the cost to heat the cabin air is $1/kW · h, compare the costs associated with the heat loss through the windows for an 8-hour intercontinental flight.

2.13 Use *IHT* to perform the following tasks.

(a) Graph the thermal conductivity of pure copper, 2024 aluminum, and AISI 302 stainless steel over the temperature range $300 \leq T \leq 600$ K. Include all data on a single graph, and comment on the trends you observe.

(b) Graph the thermal conductivity of helium and air over the temperature range $300 \leq T \leq 800$ K. Include the data on a single graph, and comment on the trends you observe.

(c) Graph the kinematic viscosity of engine oil, ethylene glycol, and liquid water over the temperature range $300 \leq T \leq 360$ K. Include all data on a single graph, and comment on the trends you observe.

(d) Graph the thermal conductivity of a water-Al_2O_3 nanofluid at $T = 300$ K over the volume fraction range $0 \leq \varphi \leq 0.08$. See Example 2.2.

2.14 Consider a small but known volume of metal that has a large thermal conductivity.

(a) Since the thermal conductivity is large, spatial temperature gradients that develop within the metal in response to mild heating are small. Neglecting spatial temperature gradients, derive a differential equation that could be solved for the temperature of the metal versus time $T(t)$ if the metal is subjected to a fixed surface heat rate q supplied by an electric heater.

(b) A student proposes to identify the unknown metal by comparing measured and predicted thermal responses. Once a match is made, relevant thermophysical properties might be determined, and, in turn, the metal may be identified by comparison to published property data. Will this approach work? Consider aluminum, gold, and silver as the candidate metals.

2.15 The thermal conductivity of helium at a certain temperature is 0.15 W/m · K. Calculate the helium temperature assuming ideal gas behavior, and compare it to the temperature found in Table A.4.

2.16 A method for determining the thermal conductivity k and the specific heat c_p of a material is illustrated in the sketch. Initially the two identical samples of diameter $D = 60$ mm and thickness $L = 10$ mm and the thin heater are at a uniform temperature of $T_i = 23.00°$C, while surrounded by an insulating powder. Suddenly the heater is energized to provide a uniform heat flux q''_o on each of the sample interfaces, and the heat flux is maintained constant for a period of time, Δt_o. A short time after sudden heating is initiated, the temperature at this interface T_o is related to the heat flux as

$$T_o(t) - T_i = 2q''_o \left(\frac{t}{\pi \rho c_p k} \right)^{1/2}$$

For a particular test run, the electrical heater dissipates 15.0 W for a period of $\Delta t_o = 120$ s, and the temperature at the interface is $T_o(30\text{ s}) = 24.57°$C after 30 s of heating. A long time after the heater is deenergized, $t \gg \Delta t_0$, the samples reach the uniform temperature of $T_o(\infty) = 33.50°$C. The density of the sample materials, determined by measurement of volume and mass, is $\rho = 3965$ kg/m³.

Determine the specific heat and thermal conductivity of the test material. By looking at values of the thermophysical properties in Table A.1 or A.2, identify the test sample material.

2.17 A cylindrical rod of stainless steel is insulated on its exterior surface except for the ends. The steady-state temperature distribution is $T(x) = a - bx/L$, where $a = 305$ K and $b = 10$ K. The diameter and length of the rod are $D = 20$ mm and $L = 100$ mm, respectively. Determine the heat flux along the rod, q''_x. *Hint:* The mass of the rod is $m = 0.248$ kg.

The Heat Equation

2.18 At a given instant of time, the temperature distribution within an infinite homogeneous body is given by the function

$$T(x, y, z) = x^2 - 2y^2 + z^2 - xy + 2yz$$

Assuming constant properties and no internal heat generation, determine the regions where the temperature changes with time.

2.19 Consider the temperature distributions associated with a dx differential control volume within the one-dimensional plane walls shown below.

(a) Steady-state conditions exist. Is thermal energy being generated within the differential control volume? If so, is the generation rate positive or negative?

(b) Steady-state conditions exist as in part (a). Is the volumetric generation rate positive or negative within the differential control volume?

(c) Steady-state conditions do not exist, and there is no volumetric thermal energy generation. Is the temperature of the material in the differential control volume increasing or decreasing with time?

(d) Transient conditions exist as in part (c). Is the temperature increasing or decreasing with time?

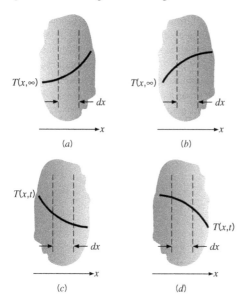

(a)　　　　(b)

(c)　　　　(d)

2.20 A pan is used to boil water by placing it on a stove, from which heat is transferred at a fixed rate q_o. There are two stages to the process. In Stage 1, the water is taken from its initial (room) temperature T_i to the boiling point, as heat is transferred from the pan by natural convection. During this stage, a constant value of the convection coefficient h may be assumed, while the bulk temperature of the water increases with time, $T_\infty = T_\infty(t)$. In Stage 2, the water has come to a boil, and its temperature remains at a fixed value, $T_\infty = T_b$, as heating continues. Consider a pan bottom of thickness L and diameter D, with a coordinate system

corresponding to $x = 0$ and $x = L$ for the surfaces in contact with the stove and water, respectively.

(a) Write the form of the heat equation and the boundary/initial conditions that determine the variation of temperature with position and time, $T(x, t)$, in the pan bottom during Stage 1. Express your result in terms of the parameters q_o, D, L, h, and T_∞, as well as appropriate properties of the pan material.

(b) During Stage 2, the surface of the pan in contact with the water is at a fixed temperature, $T(L, t) = T_L > T_b$. Write the form of the heat equation and boundary conditions that determine the temperature distribution $T(x)$ in the pan bottom. Express your result in terms of the parameters q_o, D, L, and T_L, as well as appropriate properties of the pan material.

2.21 Uniform internal heat generation at $\dot{q} = 5 \times 10^7$ W/m^3 is occurring in a cylindrical nuclear reactor fuel rod of 50-mm diameter, and under steady-state conditions the temperature distribution is of the form $T(r) = a + br^2$, where T is in degrees Celsius and r is in meters, while $a = 800°C$ and $b = -4.167 \times 10^5$ °C/m^2. The fuel rod properties are $k = 30$ W/m · K, $\rho = 1100$ kg/m^3, and $c_p = 800$ J/kg · K.

(a) What is the rate of heat transfer per unit length of the rod at $r = 0$ (the centerline) and at $r = 25$ mm (the surface)?

(b) If the reactor power level is suddenly increased to $\dot{q}_2 = 10^8$ W/m^3, what is the initial time rate of temperature change at $r = 0$ and $r = 25$ mm?

2.22 The temperature distribution across a wall 0.3 m thick at a certain instant of time is $T(x) = a + bx + cx^2$ where T is in degrees Celsius and x is in meters, $a = 200°C$, $b = -200°C/m$, and $c = 30°C/m^2$. The wall has a thermal conductivity of 1 W/m · K.

(a) On a unit surface area basis, determine the rate of heat transfer into and out of the wall and the rate of change of energy stored by the wall.

(b) If the cold surface is exposed to a fluid at 100°C, what is the convection coefficient?

2.23 A plane wall of thickness $2L = 40$ mm and thermal conductivity $k = 5$ W/m · K experiences uniform volumetric heat generation at a rate $\dot{q}$, while convection heat transfer occurs at both of its surfaces ($x = -L, +L$), each of which is exposed to a fluid of temperature $T_\infty = 20°C$. Under steady-state conditions, the temperature distribution in the wall is of the form $T(x) = a + bx + cx^2$ where $a = 82.0°C$, $b = -210°C/m$, $c = -2 \times 10^4 °C/m^2$, and x is in meters. The origin of the x-coordinate is at the midplane of the wall.

(a) Sketch the temperature distribution and identify significant physical features.

(b) What is the volumetric rate of heat generation $\dot{q}$ in the wall?

(c) Determine the surface heat fluxes, $q_x''(-L)$ and $q_x''(+L)$. How are these fluxes related to the heat generation rate?

(d) What are the convection coefficients for the surfaces at $x = -L$ and $x = +L$?

(e) Obtain an expression for the heat flux distribution $q_x''(x)$. Is the heat flux zero at any location? Explain any significant features of the distribution.

(f) If the source of the heat generation is suddenly deactivated ($\dot{q} = 0$), what is the rate of change of energy stored in the wall at this instant?

(g) What temperature will the wall eventually reach with $\dot{q} = 0$? How much energy must be removed by the fluid per unit area of the wall (J/m²) to reach this state? The density and specific heat of the wall material are 2600 kg/m³ and 800 J/kg · K, respectively.

2.24 Temperature distributions within a series of one-dimensional plane walls at an initial time, at steady state, and at several intermediate times are as shown.

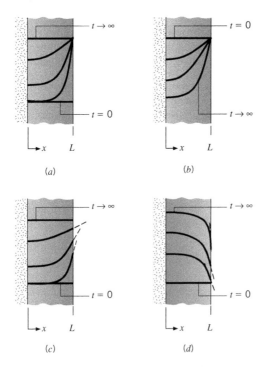

(a) *(b)*

(c) *(d)*

For each case, write the appropriate form of the heat diffusion equation. Also write the equations for the initial condition and the boundary conditions that are applied at

$x = 0$ and $x = L$. If volumetric generation occurs, it is uniform throughout the wall. The properties are constant.

2.25 A composite rod consists of two different materials, A and B, each of length 0.5L.

The thermal conductivity of Material A is half that of Material B, that is, $k_A/k_B = 0.5$. Sketch the steady-state temperature and heat flux distributions, $T(x)$ and q_x'', respectively. Assume constant properties and no internal heat generation in either material.

2.26 One-dimensional, steady-state conduction with uniform internal energy generation occurs in a plane wall with a thickness of 40 mm and a constant thermal conductivity of 5 W/m · K. For these conditions, the temperature distribution has the form $T(x) = a + bx + cx^2$. The surface at $x = 0$ has a temperature of $T(0) \equiv T_o = 130°C$ and experiences convection with a fluid for which $T_\infty = 30°C$ and $h = 500$ W/m² · K. The surface at $x = L$ is well insulated.

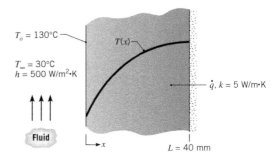

(a) Applying an overall energy balance to the wall, calculate the volumetric energy generation rate $\dot{q}$.

(b) Determine the coefficients a, b, and c by applying the boundary conditions to the prescribed temperature distribution. Use the results to calculate and plot the temperature distribution.

(c) Consider conditions for which the convection coefficient is halved, but the volumetric energy generation rate remains unchanged. Determine the new values of a, b, and c, and use the results to plot the temperature distribution. *Hint:* recognize that $T(0)$ is no longer 130°C.

(d) Under conditions for which the volumetric energy generation rate is doubled, and the convection coefficient remains unchanged ($h = 500$ W/m$^2 \cdot$ K), determine the new values of a, b, and c and plot the corresponding temperature distribution. Referring to the results of parts (b), (c), and (d) as Cases 1, 2, and 3, respectively, compare the temperature distributions for the three cases and discuss the effects of h and $\dot{q}$ on the distributions.

2.27 Derive the heat diffusion equation, Equation 2.26, for cylindrical coordinates beginning with the differential control volume shown in Figure 2.12.

2.28 Derive the heat diffusion equation, Equation 2.29, for spherical coordinates beginning with the differential control volume shown in Figure 2.13.

2.29 The steady-state temperature distribution in a semi-transparent material of thermal conductivity k and thickness L exposed to laser irradiation is of the form

$$T(x) = -\frac{A}{ka^2}e^{-ax} + Bx + C$$

where A, a, B, and C are known constants. For this situation, radiation absorption in the material is manifested by a distributed heat generation term, $\dot{q}(x)$.

(a) Obtain expressions for the conduction heat fluxes at the front and rear surfaces.

(b) Derive an expression for $\dot{q}(x)$.

(c) Derive an expression for the rate at which radiation is absorbed in the entire material, per unit surface area. Express your result in terms of the known constants for the temperature distribution, the thermal conductivity of the material, and its thickness.

2.30 One-dimensional, steady-state conduction with no energy generation is occurring in a cylindrical shell of inner radius r_1 and outer radius r_2. Under what condition is the linear temperature distribution shown possible?

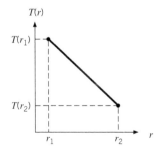

2.31 The steady-state temperature distribution in a one-dimensional wall of thermal conductivity k and thickness L is of the form $T = ax^2 + bx + c$. Derive expressions for the heat fluxes at the two wall faces ($x = 0, L$), and the energy generation rate in the wall per unit wall area.

2.32 One-dimensional, steady-state conduction with no energy generation is occurring in a plane wall of constant thermal conductivity.

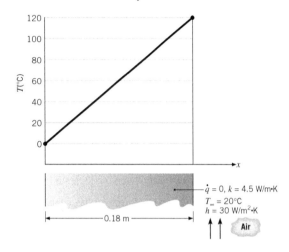

(a) Is the prescribed temperature distribution possible? Briefly explain your reasoning.

(b) With the temperature at $x = 0$ and the fluid temperature fixed at $T(0) = 0$°C and $T_\infty = 20$°C, respectively, compute and plot the temperature at $x = L$, $T(L)$, as a function of h for $10 \le h \le 100$ W/m$^2 \cdot$ K. Briefly explain your results.

2.33 A plane layer of coal of thickness $L = 1$ m experiences uniform volumetric generation at a rate of $\dot{q} = 10$ W/m^3 due to slow oxidation of the coal particles. Averaged over a daily period, the top surface of the layer transfers heat by convection to ambient air for which $h = 5$ W/m$^2 \cdot$ K and $T_\infty = 25$°C, while receiving solar irradiation in the amount $G_S = 400$ W/m^2. Irradiation from the atmosphere may be neglected. The solar absorptivity and emissivity of the surface are each $\alpha_S = \varepsilon = 0.95$.

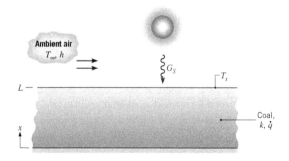

(a) Write the steady-state form of the heat diffusion equation for the layer of coal. Verify that this equation is satisfied by a temperature distribution of the form

$$T(x) = T_s + \frac{\dot{q}L^2}{2k}\left(1 - \frac{x^2}{L^2}\right)$$

From this distribution, what can you say about conditions at the bottom surface $(x = 0)$? Sketch the temperature distribution and label key features.

(b) Obtain an expression for the rate of heat transfer by conduction per unit area at $x = L$. Applying an energy balance to a control surface about the top surface of the layer, obtain an expression for T_s. Evaluate T_s and $T(0)$ for the prescribed conditions.

(c) Daily average values of G_S and h depend on a number of factors, such as time of year, cloud cover, and wind conditions. For $h = 5$ W/m² · K, compute and plot T_S and $T(0)$ as a function of G_S for $50 \leq G_S \leq 500$ W/m². For $G_S = 400$ W/m², compute and plot T_S and $T(0)$ as a function of h for $5 \leq h \leq 50$ W/m² · K.

2.34 The cylindrical system illustrated has negligible variation of temperature in the r- and z-directions. Assume that $\Delta r = r_o - r_i$ is small compared to r_i, and denote the length in the z-direction, normal to the page, as L.

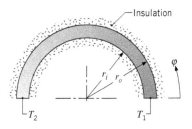

(a) Beginning with a properly defined control volume and considering energy generation and storage effects, derive the differential equation that prescribes the variation in temperature with the angular coordinate ϕ. Compare your result with Equation 2.26.

(b) For steady-state conditions with no internal heat generation and constant properties, determine the temperature distribution $T(\phi)$ in terms of the constants T_1, T_2, r_i, and r_o. Is this distribution linear in ϕ?

(c) For the conditions of part (b) write the expression for the heat rate q_ϕ.

2.35 Beginning with a differential control volume in the form of a cylindrical shell, derive the heat diffusion equation for a one-dimensional, cylindrical, radial coordinate system with internal heat generation. Compare your result with Equation 2.26.

2.36 Beginning with a differential control volume in the form of a spherical shell, derive the heat diffusion equation for a one-dimensional, spherical, radial coordinate system with internal heat generation. Compare your result with Equation 2.29.

2.37 A steam pipe is wrapped with insulation of inner and outer radii r_i and r_o, respectively. At a particular instant the temperature distribution in the insulation is known to be of the form

$$T(r) = C_1 \ln\left(\frac{r}{r_o}\right) + C_2$$

Are conditions steady-state or transient? How do the heat flux and heat rate vary with radius?

2.38 Passage of an electric current through a long conducting rod of radius r_i and thermal conductivity k_r results in uniform volumetric heating at a rate of $\dot{q}$. The conducting rod is wrapped in an electrically nonconducting cladding material of outer radius r_o and thermal conductivity k_c, and convection cooling is provided by an adjoining fluid.

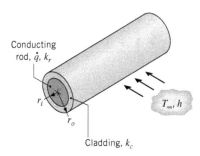

For steady-state conditions, write appropriate forms of the heat equations for the rod and cladding. Express appropriate boundary conditions for the solution of these equations.

2.39 Two-dimensional, steady-state conduction occurs in a hollow cylindrical solid of thermal conductivity $k = 16$ W/m · K, outer radius $r_o = 1$ m, and overall length $2z_o = 5$ m, where the origin of the coordinate system is located at the midpoint of the center line. The inner surface of the cylinder is insulated, and the temperature distribution within the cylinder has the form $T(r, z) = a + br^2 + c\ln r + dz^2$, where $a = -20°C$, $b = 150°C/m^2$, $c = -12°C$, $d = -300°C/m^2$ and r and z are in meters.

(a) Determine the inner radius r_i of the cylinder.

(b) Obtain an expression for the volumetric rate of heat generation, $\dot{q}$ (W/m³).

(c) Determine the axial distribution of the heat flux at the outer surface, $q_r''(r_o, z)$. What is the heat rate at the outer surface? Is it into or out of the cylinder?

(d) Determine the radial distribution of the heat flux at the end faces of the cylinder, $q_r''(r, +z_o)$ and $q_r''(r, -z_o)$. What are the corresponding heat rates? Are they into or out of the cylinder?

(e) Verify that your results are consistent with an overall energy balance on the cylinder.

2.40 An electric cable of radius r_1 and thermal conductivity k_c is enclosed by an insulating sleeve whose outer surface is of radius r_2 and experiences convection heat transfer and radiation exchange with the adjoining air and large surroundings, respectively. When electric current passes through the cable, thermal energy is generated within the cable at a volumetric rate $\dot{q}$.

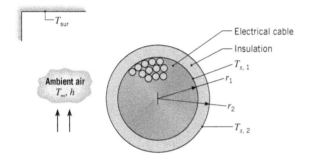

(a) Write the steady-state forms of the heat diffusion equation for the insulation and the cable. Verify that these equations are satisfied by the following temperature distributions:

Insulation: $T(r) = T_{s,2} + (T_{s,1} - T_{s,2}) \dfrac{\ln(r/r_2)}{\ln(r_1/r_2)}$

Cable: $T(r) = T_{s,1} + \dfrac{\dot{q}r_1^2}{4k_c}\left(1 - \dfrac{r^2}{r_1^2}\right)$

Sketch the temperature distribution, $T(r)$, in the cable and the sleeve, labeling key features.

(b) Applying Fourier's law, show that the rate of conduction heat transfer per unit length through the sleeve may be expressed as

$$q_r' = \frac{2\pi k_s(T_{s,1} - T_{s,2})}{\ln(r_2/r_1)}$$

Applying an energy balance to a control surface placed around the cable, obtain an alternative expression for q_r', expressing your result in terms of $\dot{q}$ and r_1.

(c) Applying an energy balance to a control surface placed around the outer surface of the sleeve, obtain an expression from which $T_{s,2}$ may be determined as a function of $\dot{q}, r_1, h, T_\infty, \varepsilon$, and T_{sur}.

(d) Consider conditions for which 250 A are passing through a cable having an electric resistance per unit length of $R_e' = 0.005\ \Omega/\text{m}$, a radius of $r_1 = 15$ mm, and a thermal conductivity of $k_c = 200$ W/m · K. For $k_s = 15$ W/m · K, $r_2 = 15.5$ mm, $h = 25$ W/m² · K, $\varepsilon = 0.9$, $T_\infty = 25°\text{C}$, and $T_{sur} = 35°\text{C}$, evaluate the surface temperatures, $T_{s,1}$ and $T_{s,2}$, as well as the temperature T_o at the centerline of the cable.

(e) With all other conditions remaining the same, compute and plot T_o, $T_{s,1}$, and $T_{s,2}$ as a function of r_2 for $15.5 \le r_2 \le 20$ mm.

2.41 A spherical shell of inner and outer radii r_i and r_o, respectively, contains heat-dissipating components, and at a particular instant the temperature distribution in the shell is known to be of the form

$$T(r) = \frac{C_1}{r} + C_2$$

Are conditions steady-state or transient? How do the heat flux and heat rate vary with radius?

Graphical Representations

2.42 The one-dimensional system of mass M with constant properties and no internal heat generation shown in the figure is initially at a uniform temperature T_i. The electrical heater is suddenly energized, providing a uniform heat flux q_o'' at the surface $x = 0$. The boundaries at $x = L$ and elsewhere are perfectly insulated.

(a) Write the differential equation, and identify the boundary and initial conditions that could be used to determine the temperature as a function of position and time in the system.

(b) On $T - x$ coordinates, sketch the temperature distributions for the initial condition ($t \le 0$) and for several times after the heater is energized. Will a steady-state temperature distribution ever be reached?

(c) On $q_x'' - t$ coordinates, sketch the heat flux $q_x''(x, t)$ at the planes $x = 0$, $x = L/2$, and $x = L$ as a function of time.

(d) After a period of time t_e has elapsed, the heater power is switched off. Assuming that the insulation is perfect, the system will eventually reach a final uniform temperature T_f. Derive an expression that can be used to determine T_f as a function of the parameters q''_o, t_e, T_i, and the system characteristics M, c_p, and A_s (the heater surface area).

2.43 A large plate of thickness $2L$ is at a uniform temperature of $T_i = 200°C$, when it is suddenly quenched by dipping it in a liquid bath of temperature $T_\infty = 20°C$. Heat transfer to the liquid is characterized by the convection coefficient h.

(a) If $x = 0$ corresponds to the midplane of the wall, on $T - x$ coordinates, sketch the temperature distributions for the following conditions: initial condition ($t \leq 0$), steady-state condition ($t \rightarrow \infty$), and two intermediate times.

(b) On $q''_x - t$ coordinates, sketch the variation with time of the heat flux at $x = L$.

(c) If $h = 100 \text{ W/m}^2 \cdot \text{K}$, what is the heat flux at $x = L$ and $t = 0$? If the wall has a thermal conductivity of $k = 50 \text{ W/m} \cdot \text{K}$ what is the corresponding temperature gradient at $x = L$?

(d) Consider a plate of thickness $2L = 20$ mm with a density of $\rho = 2770 \text{ kg/m}^3$ and a specific heat $c_p = 875 \text{ J/kg} \cdot \text{K}$. By performing an energy balance on the plate, determine the amount of energy per unit surface area of the plate (J/m^2) that is transferred to the bath over the time required to reach steady-state conditions.

(e) From other considerations, it is known that, during the quenching process, the heat flux at $x = +L$ and $x = -L$ decays exponentially with time according to the relation, $q''_x = A \exp(-Bt)$, where t is in seconds, $A = 1.80 \times 10^4 \text{ W/m}^2$, and $B = 4.126 \times 10^{-3} \text{ s}^{-1}$. Use this information to determine the energy per unit surface area of the plate that is transferred to the fluid during the quenching process.

2.44 The plane wall with constant properties and no internal heat generation shown in the figure is initially at a uniform temperature T_i. Suddenly the surface at $x = L$ is heated by a fluid at T_∞ having a convection heat transfer coefficient h. The boundary at $x = 0$ is perfectly insulated.

(a) Write the differential equation, and identify the boundary and initial conditions that could be used to determine the temperature as a function of position and time in the wall.

(b) On $T - x$ coordinates, sketch the temperature distributions for the following conditions: initial condition ($t \leq 0$), steady-state condition ($t \rightarrow \infty$), and two intermediate times.

(c) On $q''_x - t$ coordinates, sketch the heat flux at the locations $x = 0$, $x = L$. That is, show qualitatively how $q''_x(0, t)$ and $q''_x(L, t)$ vary with time.

(d) Write an expression for the total energy transferred to the wall per unit volume of the wall (J/m^3).

2.45 Consider the steady-state temperature distributions within a composite wall composed of Material A and Material B for the two cases shown. There is no internal generation, and the conduction process is one-dimensional.

Case 1   Case 2

Answer the following questions for each case. Which material has the higher thermal conductivity? Does the thermal conductivity vary significantly with temperature? If so, how? Describe the heat flux distribution $q''_x(x)$ through the composite wall. If the thickness and thermal conductivity of each material were both doubled and the boundary temperatures remained the same, what would be the effect on the heat flux distribution?

Case 1. Linear temperature distributions exist in both materials, as shown.
Case 2. Nonlinear temperature distributions exist in both materials, as shown.

2.46 A plane wall with constant properties is initially at a uniform temperature T_o. Suddenly, the surface at $x = L$ is exposed to a convection process with a fluid at $T_\infty (<T_o)$ having a convection coefficient h. Also, suddenly the wall experiences a uniform internal volumetric heating $\dot{q}$ that is sufficiently large to

induce a maximum steady-state temperature within the wall, which exceeds that of the fluid. The boundary at $x = 0$ remains at T_o.

(a) On $T - x$ coordinates, sketch the temperature distributions for the following conditions: initial condition ($t \leq 0$), steady-state condition ($t \to \infty$), and for two intermediate times. Identify key features of the distributions, especially the location of the maximum temperature and the temperature gradient at $x = L$.

(b) On $q''_x - t$ coordinates, sketch the heat flux for the locations $x = 0$ and $x = L$, that is, $q''_x(0, t)$ and $q''_x(L, t)$, respectively. Identify key features of the flux histories.

2.47 Consider the steady-state temperature distribution within a composite wall composed of Materials A and B.

The conduction process is one-dimensional. Within which material does uniform volumetric generation occur? What is the boundary condition at $x = -L_A$? How would the temperature distribution change if the thermal conductivity of Material A were doubled? How would the temperature distribution change if the thermal conductivity of Material B were doubled? Sketch the heat flux distribution $q''_x(x)$ through the composite wall.

2.48 A spherical particle of radius r_1 experiences uniform thermal generation at a rate of $\dot{q}$. The particle is encapsulated by a spherical shell of outside radius r_2 that is cooled by ambient air. The thermal conductivities of the particle and shell are k_1 and k_2, respectively, where $k_1 = 2k_2$.

(a) By applying the conservation of energy principle to spherical control volume A, which is placed at an arbitrary location within the sphere, determine a relationship between the temperature gradient dT/dr and the local radius r, for $0 \leq r \leq r_1$.

(b) By applying the conservation of energy principle to spherical control volume B, which is placed at an arbitrary location within the spherical shell, determine a relationship between the temperature gradient dT/dr and the local radius r, for $r_1 \leq r \leq r_2$.

(c) On $T - r$ coordinates, sketch the temperature distribution over the range $0 \leq r \leq r_2$.

2.49 A plane wall of thickness $L = 0.1$ m experiences uniform volumetric heating at a rate $\dot{q}$. One surface of the wall ($x = 0$) is insulated, and the other surface is exposed to a fluid at $T_\infty = 20°C$, with convection heat transfer characterized by $h = 1000$ W/m$^2 \cdot$ K. Initially, the temperature distribution in the wall is $T(x, 0) = a + bx^2$, where $a = 300°C$, $b = -1.0 \times 10^{4}°C/m^2$, and x is in meters. Suddenly, the volumetric heat generation is deactivated ($\dot{q} = 0$ for $t \geq 0$), while convection heat transfer continues to occur at $x = L$. The properties of the wall are $\rho = 7000$ kg/m^3, $c_p = 450$ J/kg $\cdot$ K, and $k = 90$ W/m $\cdot$ K.

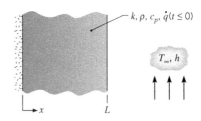

(a) Determine the magnitude of the volumetric energy generation rate $\dot{q}$ associated with the initial condition ($t < 0$).

(b) On $T - x$ coordinates, sketch the temperature distribution for the following conditions: initial condition ($t < 0$), steady-state condition ($t \to \infty$), and two intermediate conditions.

(c) On $q''_x - t$ coordinates, sketch the variation with time of the heat flux at the boundary exposed to the convection process, $q''_x(L, t)$. Calculate the corresponding value of the heat flux at $t = 0$, $q''_x(L, 0)$.

(d) Calculate the amount of energy removed from the wall per unit area (J/m^2) by the fluid stream as the wall cools from its initial to steady-state condition.

2.50 A plane wall that is insulated on one side ($x = 0$) is initially at a uniform temperature T_i, when its exposed surface at $x = L$ is suddenly raised to a temperature T_s.

(a) Verify that the following equation satisfies the heat equation and boundary conditions:

$$\frac{T(x,t) - T_s}{T_i - T_s} = C_1 \exp\left(-\frac{\pi^2}{4}\frac{\alpha t}{L^2}\right)\cos\left(\frac{\pi}{2}\frac{x}{L}\right)$$

where C_1 is a constant and α is the thermal diffusivity.

(b) Obtain expressions for the heat flux at $x = 0$ and $x = L$.

(c) Sketch the temperature distribution $T(x)$ at $t = 0$, at $t \to \infty$, and at an intermediate time. Sketch the variation with time of the heat flux at $x = L$, $q''_L(t)$.

(d) What effect does α have on the thermal response of the material to a change in surface temperature?

2.51 A composite one-dimensional plane wall is of overall thickness $2L$. Material A spans the domain $-L \le x < 0$ and experiences an *exothermic* chemical reaction leading to a uniform volumetric generation rate of $\dot{q}_A$. Material B spans the domain $0 \le x \le L$ and undergoes an *endothermic* chemical reaction corresponding to a uniform volumetric generation rate of $\dot{q}_B = -\dot{q}_A$. The surfaces at $x = \pm L$ are insulated. Sketch the steady-state temperature and heat flux distributions $T(x)$ and $q''_x(x)$, respectively, over the domain $-L \le x \le L$ for $k_A = k_B$, $k_A = 0.5k_B$, and $k_A = 2k_B$. Point out the important features of the distributions you have drawn. If $\dot{q}_B = -2\dot{q}_A$, can you sketch the steady-state temperature distribution?

2.52 Typically, air is heated in a hair dryer by blowing it across a coiled wire through which an electric current is passed. Thermal energy is generated by electric resistance heating within the wire and is transferred by convection from the surface of the wire to the air. Consider conditions for which the wire is initially at room temperature, T_i, and resistance heating is concurrently initiated with airflow at $t = 0$.

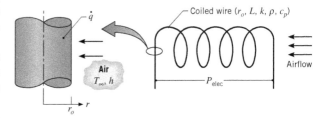

(a) For a wire radius r_o, an air temperature T_∞, and a convection coefficient h, write the form of the heat equation and the boundary/initial conditions that govern the transient thermal response, $T(r, t)$, of the wire.

(b) If the length and radius of the wire are 500 mm and 1 mm, respectively, what is the volumetric rate of thermal energy generation for a power consumption of $P_{elec} = 500$ W? What is the convection heat flux under steady-state conditions?

(c) On $T - r$ coordinates, sketch the temperature distributions for the following conditions: initial condition ($t \le 0$), steady-state condition ($t \to \infty$), and for two intermediate times.

(d) On $q''_r - t$ coordinates, sketch the variation of the heat flux with time for locations at $r = 0$ and $r = r_o$.

2.53 The steady-state temperature distribution in a composite plane wall of three different materials, each of constant thermal conductivity, is shown.

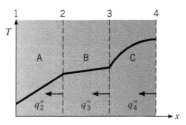

(a) Comment on the relative magnitudes of q''_2 and q''_3, and of q''_3 and q''_4.

(b) Comment on the relative magnitudes of k_A and k_B, and of k_B and k_C.

(c) Sketch the heat flux as a function of x.

One-Dimensional, Steady-State Conduction

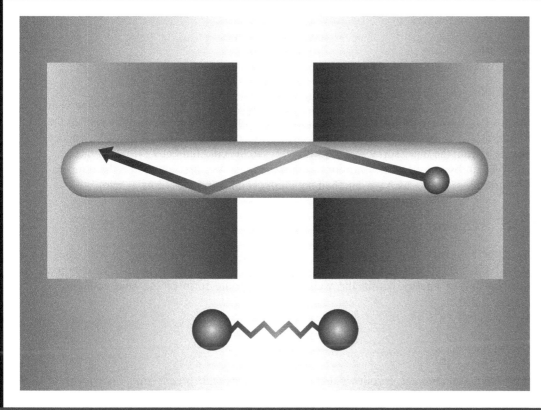

In this chapter we treat situations for which heat is transferred by diffusion under one-dimensional, steady-state conditions. The term *one-dimensional* refers to the fact that temperature gradients exist along only a single coordinate direction, and heat transfer occurs exclusively in that direction. The system is characterized by *steady-state* conditions if the temperature at each point is independent of time. Despite their inherent simplicity, one-dimensional, steady-state models may be used to accurately represent numerous engineering systems.

We begin our consideration of one-dimensional, steady-state conduction by discussing heat transfer with no internal generation of thermal energy (Sections 3.1 through 3.4). The objective is to determine expressions for the temperature distribution and heat transfer rate in common (planar, cylindrical, and spherical) geometries. For such geometries, an additional objective is to introduce the concept of *thermal resistance* and to show how *thermal circuits* may be used to model heat flow, much as electrical circuits are used for current flow. The effect of internal heat generation is treated in Section 3.5, and again our objective is to obtain expressions for temperature distributions and heat transfer rates. In Section 3.6, we consider the special case of one-dimensional, steady-state conduction for *extended surfaces*. In their most common form, these surfaces are termed *fins* and are used to *enhance* heat transfer by convection to an adjoining fluid. In addition to determining related temperature distributions and heat rates, our objective is to introduce *performance parameters* that may be used to determine their efficacy. Finally, in Section 3.7 we apply heat transfer and thermal resistance concepts to the human body, including the effects of *metabolic heat generation* and *perfusion*; to thermoelectric power generation; and to micro- and nanoscale conduction in *thin gas layers* and *thin solid films*.

3.1 *The Plane Wall*

For one-dimensional conduction in a plane wall, temperature is a function of the x-coordinate only and heat is transferred exclusively in this direction. In Figure 3.1a, a plane wall separates two fluids of different temperatures. Heat transfer occurs by convection from the hot fluid at $T_{\infty,1}$ to one surface of the wall at $T_{s,1}$, by conduction through the wall, and by convection from the other surface of the wall at $T_{s,2}$ to the cold fluid at $T_{\infty,2}$.

We begin by considering conditions *within* the wall. We first determine the temperature distribution, from which we can then obtain the conduction heat transfer rate.

3.1.1 Temperature Distribution

The temperature distribution in the wall can be determined by solving the heat equation with the proper boundary conditions. For steady-state conditions with no distributed source or sink of energy within the wall, the appropriate form of the heat equation is Equation 2.23

$$\frac{d}{dx}\left(k\frac{dT}{dx}\right) = 0 \tag{3.1}$$

Hence, from Equation 2.2, it follows that, for *one-dimensional, steady-state conduction in a plane wall with no heat generation, the heat flux is a constant, independent of x.* If the

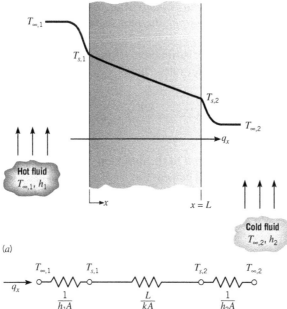

(a)

(b)

FIGURE 3.1 Heat transfer through a plane wall. (a) Temperature distribution. (b) Equivalent thermal circuit.

thermal conductivity of the wall material is assumed to be constant, the equation may be integrated twice to obtain the *general solution*

$$T(x) = C_1 x + C_2 \tag{3.2}$$

To obtain the constants of integration, C_1 and C_2, boundary conditions must be introduced. We choose to apply conditions of the first kind at $x = 0$ and $x = L$, in which case

$$T(0) = T_{s,1} \quad \text{and} \quad T(L) = T_{s,2}$$

Applying the condition at $x = 0$ to the general solution, it follows that

$$T_{s,1} = C_2$$

Similarly, at $x = L$,

$$T_{s,2} = C_1 L + C_2 = C_1 L + T_{s,1}$$

in which case

$$\frac{T_{s,2} - T_{s,1}}{L} = C_1$$

Substituting into the general solution, the temperature distribution is then

$$T(x) = (T_{s,2} - T_{s,1})\frac{x}{L} + T_{s,1} \tag{3.3}$$

From this result it is evident that, *for one-dimensional, steady-state conduction in a plane wall with no heat generation and constant thermal conductivity, the temperature varies linearly with x.*

Now that we have the temperature distribution, we may use Fourier's law, Equation 2.1, to determine the conduction heat transfer rate. That is,

$$q_x = -kA\frac{dT}{dx} = \frac{kA}{L}(T_{s,1} - T_{s,2}) \tag{3.4}$$

Note that A is the area of the wall *normal* to the direction of heat transfer and, for the plane wall, it is a constant independent of x. The heat flux is then

$$q_x'' = \frac{q_x}{A} = \frac{k}{L}(T_{s,1} - T_{s,2}) \tag{3.5}$$

Equations 3.4 and 3.5 indicate that both the heat rate q_x and heat flux q_x'' are constants, independent of x.

In the foregoing paragraphs we have used the *standard approach* to solving conduction problems. That is, the general solution for the temperature distribution is first obtained by solving the appropriate form of the heat equation. The boundary conditions are then applied to obtain the particular solution, which is used with Fourier's law to determine the heat transfer rate. Note that we have opted to prescribe surface temperatures at $x = 0$ and $x = L$ as boundary conditions, even though it is the fluid temperatures, not the surface temperatures, that are typically known. However, since adjoining fluid and surface temperatures are easily related through a surface energy balance (see Section 1.3.1), it is a simple matter to express Equations 3.3 through 3.5 in terms of fluid, rather than surface, temperatures. Alternatively, equivalent results could be obtained directly by using the surface energy balances as boundary conditions of the third kind in evaluating the constants of Equation 3.2 (see Problem 3.1).

3.1.2 Thermal Resistance

At this point we note that, for the special case of one-dimensional heat transfer with no internal energy generation and with constant properties, a very important concept is suggested by Equation 3.4. In particular, an analogy exists between the diffusion of heat and electrical charge. Just as an electrical resistance is associated with the conduction of electricity, a thermal resistance may be associated with the conduction of heat. Defining resistance as the ratio of a driving potential to the corresponding transfer rate, it follows from Equation 3.4 that the *thermal resistance for conduction* in a plane wall is

$$R_{t,\text{cond}} \equiv \frac{T_{s,1} - T_{s,2}}{q_x} = \frac{L}{kA} \tag{3.6}$$

Similarly, for electrical conduction in the same system, Ohm's law provides an electrical resistance of the form

$$R_e = \frac{E_{s,1} - E_{s,2}}{I} = \frac{L}{\sigma A} \tag{3.7}$$

The analogy between Equations 3.6 and 3.7 is obvious. A thermal resistance may also be associated with heat transfer by convection at a surface. From Newton's law of cooling,

$$q = hA(T_s - T_\infty) \tag{3.8}$$

The *thermal resistance for convection* is then

$$R_{t,\text{conv}} \equiv \frac{T_s - T_\infty}{q} = \frac{1}{hA} \tag{3.9}$$

Circuit representations provide a useful tool for both conceptualizing and quantifying heat transfer problems. The *equivalent thermal circuit* for the plane wall with convection surface conditions is shown in Figure 3.1*b*. The heat transfer rate may be determined from separate consideration of each element in the network. Since q_x is constant throughout the network, it follows that

$$q_x = \frac{T_{\infty,1} - T_{s,1}}{1/h_1 A} = \frac{T_{s,1} - T_{s,2}}{L/kA} = \frac{T_{s,2} - T_{\infty,2}}{1/h_2 A} \tag{3.10}$$

In terms of the *overall temperature difference*, $T_{\infty,1} - T_{\infty,2}$, and the *total thermal resistance*, R_{tot}, the heat transfer rate may also be expressed as

$$q_x = \frac{T_{\infty,1} - T_{\infty,2}}{R_{\text{tot}}} \tag{3.11}$$

Because the conduction and convection resistances are in series and may be summed, it follows that

$$R_{\text{tot}} = \frac{1}{h_1 A} + \frac{L}{kA} + \frac{1}{h_2 A} \tag{3.12}$$

Radiation exchange between the surface and surroundings may also be important if the convection heat transfer coefficient is small (as it often is for natural convection in a gas). A *thermal resistance for radiation* may be defined by reference to Equation 1.8:

$$R_{t,\text{rad}} = \frac{T_s - T_{\text{sur}}}{q_{\text{rad}}} = \frac{1}{h_r A} \tag{3.13}$$

For radiation between a surface and *large surroundings*, h_r is determined from Equation 1.9. Surface radiation and convection resistances act in parallel, and if $T_\infty = T_{\text{sur}}$, they may be combined to obtain a single, effective surface resistance.

3.1.3 The Composite Wall

Equivalent thermal circuits may also be used for more complex systems, such as *composite walls*. Such walls may involve any number of series and parallel thermal resistances due

FIGURE 3.2   Equivalent thermal circuit for a series composite wall.

to layers of different materials. Consider the series composite wall of Figure 3.2. The one-dimensional heat transfer rate for this system may be expressed as

$$q_x = \frac{T_{\infty,1} - T_{\infty,4}}{\Sigma R_t} \tag{3.14}$$

where $T_{\infty,1} - T_{\infty,4}$ is the *overall* temperature difference, and the summation includes all thermal resistances. Hence

$$q_x = \frac{T_{\infty,1} - T_{\infty,4}}{[(1/h_1 A) + (L_A/k_A A) + (L_B/k_B A) + (L_C/k_C A) + (1/h_4 A)]} \tag{3.15}$$

Alternatively, the heat transfer rate can be related to the temperature difference and resistance associated with each element. For example,

$$q_x = \frac{T_{\infty,1} - T_{s,1}}{(1/h_1 A)} = \frac{T_{s,1} - T_2}{(L_A/k_A A)} = \frac{T_2 - T_3}{(L_B/k_B A)} = \cdots \tag{3.16}$$

With composite systems, it is often convenient to work with an *overall heat transfer coefficient U*, which is defined by an expression analogous to Newton's law of cooling. Accordingly,

$$q_x \equiv UA \, \Delta T \tag{3.17}$$

where ΔT is the overall temperature difference. The overall heat transfer coefficient is related to the total thermal resistance, and from Equations 3.14 and 3.17 we see that $UA = 1/R_{\text{tot}}$. Hence, for the composite wall of Figure 3.2,

$$U = \frac{1}{R_{\text{tot}} A} = \frac{1}{[(1/h_1) + (L_A/k_A) + (L_B/k_B) + (L_C/k_C) + (1/h_4)]} \tag{3.18}$$

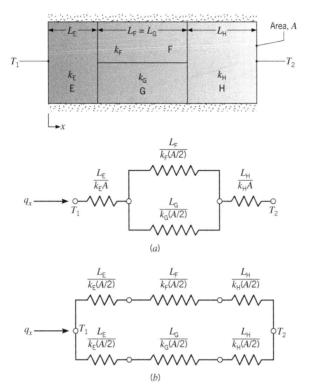

FIGURE 3.3 Equivalent thermal circuits for a series–parallel composite wall.

In general, we may write

$$R_{\text{tot}} = \Sigma R_t = \frac{\Delta T}{q} = \frac{1}{UA} \qquad (3.19)$$

Composite walls may also be characterized by series–parallel configurations, such as that shown in Figure 3.3. Although the heat flow is now multidimensional, it is often reasonable to assume one-dimensional conditions. Subject to this assumption, two different thermal circuits may be used. For case (*a*) it is presumed that surfaces normal to the *x*-direction are isothermal, whereas for case (*b*) it is assumed that surfaces parallel to the *x*-direction are adiabatic. Different results are obtained for R_{tot}, and the corresponding values of *q* bracket the actual heat transfer rate. These differences increase with increasing $|k_F - k_G|$, as multidimensional effects become more significant.

3.1.4 Contact Resistance

Although neglected until now, it is important to recognize that, in composite systems, the temperature drop across the interface between materials may be appreciable. This temperature change is attributed to what is known as the *thermal contact resistance*, $R_{t,c}$. The effect is shown in Figure 3.4, and for a unit area of the interface, the resistance is defined as

$$R''_{t,c} = \frac{T_A - T_B}{q''_x} \qquad (3.20)$$

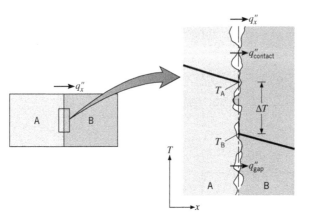

FIGURE 3.4 Temperature drop due to thermal contact resistance.

The existence of a finite contact resistance is due principally to surface roughness effects. Contact spots are interspersed with gaps that are, in most instances, air filled. Heat transfer is therefore due to conduction across the actual contact area and to conduction and/or radiation across the gaps. The contact resistance may be viewed as two parallel resistances: that due to the contact spots and that due to the gaps. The contact area is typically small, and, especially for rough surfaces, the major contribution to the resistance is made by the gaps.

For solids whose thermal conductivities exceed that of the interfacial fluid, the contact resistance may be reduced by increasing the area of the contact spots. Such an increase may be effected by increasing the contact pressure and/or by reducing the roughness of the mating surfaces. The contact resistance may also be reduced by selecting an interfacial fluid of large thermal conductivity. In this respect, no fluid (an evacuated interface) eliminates conduction across the gap, thereby increasing the contact resistance. Likewise, if the characteristic gap width L becomes small (as, for example, in the case of very smooth surfaces in contact), L/λ_{mfp} can approach values for which the thermal conductivity of the interfacial gas is reduced by microscale effects, as discussed in Section 2.2.

Although theories have been developed for predicting $R''_{t,c}$, the most reliable results are those that have been obtained experimentally. The effect of loading on metallic interfaces can be seen in Table 3.1*a*, which presents an approximate range of thermal resistances under vacuum conditions. The effect of interfacial fluid on the thermal resistance of an aluminum interface is shown in Table 3.1*b*.

TABLE 3.1 Thermal contact resistance for (*a*) metallic interfaces under vacuum conditions and (*b*) aluminum interface (10-μm surface roughness, 10^5 N/m^2) with different interfacial fluids [1]

Thermal Resistance, $R''_{t,c} \times 10^4$ (m$^2 \cdot$K/W)

(*a*) Vacuum Interface			(*b*) Interfacial Fluid	
Contact pressure	100 kN/m^2	10,000 kN/m^2	Air	2.75
Stainless steel	6–25	0.7–4.0	Helium	1.05
Copper	1–10	0.1–0.5	Hydrogen	0.720
Magnesium	1.5–3.5	0.2–0.4	Silicone oil	0.525
Aluminum	1.5–5.0	0.2–0.4	Glycerine	0.265

TABLE 3.2 Thermal resistance of representative solid/solid interfaces

Interface	$R''_{t,c} \times 10^4 \ (\text{m}^2 \cdot \text{K/W})$	Source
Silicon chip/lapped aluminum in air (27–500 kN/m^2)	0.3–0.6	[2]
Aluminum/aluminum with indium foil filler (~100 kN/m^2)	~0.07	[1, 3]
Stainless/stainless with indium foil filler (~3500 kN/m^2)	~0.04	[1, 3]
Aluminum/aluminum with metallic (Pb) coating	0.01–0.1	[4]
Aluminum/aluminum with Dow Corning 340 grease (~100 kN/m^2)	~0.07	[1, 3]
Stainless/stainless with Dow Corning 340 grease (~3500 kN/m^2)	~0.04	[1, 3]
Silicon chip/aluminum with 0.02-mm epoxy	0.2–0.9	[5]
Brass/brass with 15-μm tin solder	0.025–0.14	[6]

Contrary to the results of Table 3.1, many applications involve contact between dissimilar solids and/or a wide range of possible interstitial (filler) materials (Table 3.2). Any interstitial substance that fills the gap between contacting surfaces and whose thermal conductivity exceeds that of air will decrease the contact resistance. Two classes of materials that are well suited for this purpose are soft metals and thermal greases. The metals, which include indium, lead, tin, and silver, may be inserted as a thin foil or applied as a thin coating to one of the parent materials. Silicon-based thermal greases are attractive on the basis of their ability to completely fill the interstices with a material whose thermal conductivity is as much as 50 times that of air.

Unlike the foregoing interfaces, which are not permanent, many interfaces involve permanently bonded joints. The joint could be formed from an epoxy, a soft solder rich in lead, or a hard solder such as a gold/tin alloy. Due to interface resistances between the parent and bonding materials, the actual thermal resistance of the joint exceeds the theoretical value (L/k) computed from the thickness L and thermal conductivity k of the joint material. The thermal resistance of epoxied and soldered joints is also adversely affected by voids and cracks, which may form during manufacture or as a result of thermal cycling during normal operation.

Comprehensive reviews of thermal contact resistance results and models are provided by Snaith et al. [3], Madhusudana and Fletcher [7], and Yovanovich [8].

3.1.5 Porous Media

In many applications, heat transfer occurs within *porous media* that are combinations of a stationary solid and a fluid. When the fluid is *either* a gas *or* a liquid, the resulting porous medium is said to be *saturated*. In contrast, all three phases coexist in an *unsaturated* porous medium. Examples of porous media include beds of powder with a fluid occupying the interstitial regions between individual granules, as well as the insulation systems and nanofluids of Section 2.2.1. A saturated porous medium that consists of a stationary solid phase through which a fluid flows is referred to as a *packed bed* and is discussed in Section 7.8.

Consider a saturated porous medium that is subjected to surface temperatures T_1 at $x = 0$ and T_2 at $x = L$, as shown in Figure 3.5a. After steady-state conditions are reached and if $T_1 > T_2$, the heat rate may be expressed as

$$q_x = \frac{k_{\text{eff}} A}{L}(T_1 - T_2) \tag{3.21}$$

where k_{eff} is an effective thermal conductivity. Equation 3.21 is valid if fluid motion, as well as radiation heat transfer *within* the medium, are negligible. The effective thermal conductivity varies with the porosity or void fraction of the medium ε which is defined as the volume of fluid relative to the total volume (solid and fluid). In addition, k_{eff} depends on the thermal conductivities of each of the phases and, in this discussion, it is assumed that $k_s > k_f$. The detailed solid phase geometry, for example the size distribution and packing arrangement of individual powder particles, also affects the value of k_{eff}. Contact resistances that might evolve at interfaces between adjacent solid particles can impact the value of k_{eff}. As discussed in Section 2.2.1, nanoscale phenomena might also influence the effective thermal conductivity. Hence, prediction of k_{eff} can be difficult and, in general, requires detailed knowledge of parameters that might not be readily available.

Despite the complexity of the situation, the value of the effective thermal conductivity may be bracketed by considering the composite walls of Figures 3.5b and 3.5c. In Figure 3.5b, the medium is modeled as an equivalent, series composite wall consisting of a fluid region of length εL and a solid region of length $(1 - \varepsilon)L$. Applying Equations 3.17 and 3.18 to this model for which there is no convection ($h_1 = h_2 = 0$) and only two conduction terms, it follows that

$$q_x = \frac{A\,\Delta T}{(1 - \varepsilon)L/k_s + \varepsilon L/k_f} \tag{3.22}$$

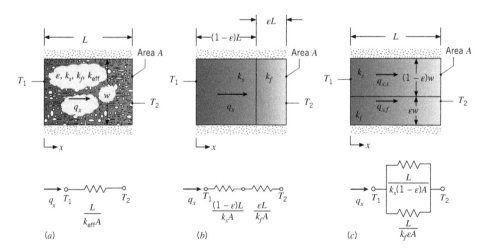

FIGURE 3.5 A porous medium. (*a*) The medium and its properties. (*b*) Series thermal resistance representation. (*c*) Parallel resistance representation.

Equating this result to Equation 3.21, we then obtain

$$k_{\text{eff,min}} = \frac{1}{(1-\varepsilon)/k_s + \varepsilon/k_f} \tag{3.23}$$

Alternatively, the medium of Figure 3.5*a* could be described by the equivalent, parallel composite wall consisting of a fluid region of width εw and a solid region of width $(1-\varepsilon)w$, as shown in Figure 3.5*c*. Combining Equation 3.21 with an expression for the equivalent resistance of two resistors in parallel gives

$$k_{\text{eff,max}} = \varepsilon k_f + (1-\varepsilon)k_s \tag{3.24}$$

While Equations 3.23 and 3.24 provide the minimum and maximum possible values of k_{eff}, more accurate expressions have been derived for specific composite systems within which nanoscale effects are negligible. Maxwell [9] derived an expression for the effective electrical conductivity of a solid matrix interspersed with uniformly distributed, noncontacting spherical inclusions. Noting the analogy between Equations 3.6 and 3.7, Maxwell's result may be used to determine the effective thermal conductivity of a saturated porous medium consisting of an interconnected solid phase within which a dilute distribution of spherical fluid regions exists, resulting in an expression of the form [10]

$$k_{\text{eff}} = \left[\frac{k_f + 2k_s - 2\varepsilon(k_s - k_f)}{k_f + 2k_s + \varepsilon(k_s - k_f)} \right] k_s \tag{3.25}$$

Equation 3.25 is valid for relatively small porosities ($\varepsilon \lesssim 0.25$) as shown schematically in Figure 3.5*a* [11]. It is equivalent to the expression introduced in Example 2.2 for a fluid that contains a dilute mixture of solid particles, but with reversal of the fluid and solid.

When analyzing conduction within porous media, it is important to consider the potential directional dependence of the effective thermal conductivity. For example, the media represented in Figure 3.5*b* or Figure 3.5*c* would *not* be characterized by isotropic properties, since the effective thermal conductivity in the *x*-direction is clearly different from values of k_{eff} in the vertical direction. Hence, although Equations 3.23 and 3.24 can be used to bracket the actual value of the effective thermal conductivity, they will generally overpredict the possible range of k_{eff} for isotropic media. For isotropic media, expressions have been developed to determine the minimum and maximum possible effective thermal conductivities based solely on knowledge of the porosity and the thermal conductivities of the solid and fluid. Specifically, the *maximum* possible value of k_{eff} in an isotropic porous medium is given by Equation 3.25, which corresponds to an interconnected, high thermal conductivity solid phase. The *minimum* possible value of k_{eff} for an isotropic medium corresponds to the case where the fluid phase forms long, randomly oriented fingers within the medium [12]. Additional information regarding conduction in saturated porous media is available [13].

EXAMPLE 3.1	IHT

In Example 1.6, we calculated the rate of heat loss from a human body in air and water environments. Now we consider the same conditions except that the surroundings (air or water) are at 10°C. To reduce the rate of heat loss, the person wears special sporting gear (snow suit or wet suit) made from silica aerogel insulation with an extremely low thermal conductivity of 0.014 W/m · K. The emissivity of the outer surface of the snow and wet suits is 0.95. What thickness of aerogel insulation is needed to reduce the rate of heat loss to 100 W (a typical metabolic heat generation rate) in air and water? What are the resulting skin temperatures?

SOLUTION

Known: Inner surface temperature of a skin/fat layer of known thickness, thermal conductivity, and surface area. Thermal conductivity and emissivity of snow and wet suits. Ambient conditions.

Find: Insulation thickness needed to reduce rate of heat loss to 100 W and corresponding skin temperature.

Schematic:

Assumptions:

1. Steady-state conditions.
2. One-dimensional heat transfer by conduction through the skin/fat and insulation layers.
3. Contact resistance is negligible.
4. Thermal conductivities are uniform.
5. Radiation exchange between the skin surface and the surroundings is between a small surface and a large enclosure at the air temperature.
6. Liquid water is opaque to thermal radiation.
7. Solar radiation is negligible.
8. Body is completely immersed in water in part 2.

Analysis: The thermal circuit can be constructed by recognizing that resistance to heat flow is associated with conduction through the skin/fat and insulation layers as well as convection and radiation at the outer surface. Accordingly, the circuit and the resistances are of the following form (with $h_r = 0$ for water):

The total thermal resistance needed to achieve the desired rate of heat loss is found from Equation 3.19,

$$R_{\text{tot}} = \frac{T_i - T_\infty}{q} = \frac{(35 - 10)\ \text{K}}{100\ \text{W}} = 0.25\ \text{K/W}$$

The total resistance between the inside of the skin/fat layer and the cold surroundings includes conduction resistances for the skin/fat and insulation layers as well as an effective resistance associated with convection and radiation, which act in parallel. Hence,

$$R_{\text{tot}} = \frac{L_{\text{sf}}}{k_{\text{sf}} A} + \frac{L_{\text{ins}}}{k_{\text{ins}} A} + \left(\frac{1}{1/hA} + \frac{1}{1/h_r A} \right)^{-1} = \frac{1}{A} \left(\frac{L_{\text{sf}}}{k_{\text{sf}}} + \frac{L_{\text{ins}}}{k_{\text{ins}}} + \frac{1}{h + h_r} \right)$$

This equation can be solved for the insulation thickness.

Air
The radiation heat transfer coefficient is approximated as having the same value as in Example 1.6: $h_r = 5.9\ \text{W/m}^2 \cdot \text{K}$.

$$L_{\text{ins}} = k_{\text{ins}} \left[A R_{\text{tot}} - \frac{L_{\text{sf}}}{k_{\text{sf}}} - \frac{1}{h + h_r} \right]$$

$$= 0.014\ \text{W/m} \cdot \text{K} \left[1.8\ \text{m}^2 \times 0.25\ \text{K/W} - \frac{3 \times 10^{-3}\ \text{m}}{0.3\ \text{W/m} \cdot \text{K}} - \frac{1}{(2 + 5.9)\ \text{W/m}^2 \cdot \text{K}} \right]$$

$$= 0.0044\ \text{m} = 4.4\ \text{mm} \qquad \qquad \triangleleft$$

Water

$$L_{\text{ins}} = k_{\text{ins}} \left[A R_{\text{tot}} - \frac{L_{\text{sf}}}{k_{\text{sf}}} - \frac{1}{h} \right]$$

$$= 0.014\ \text{W/m} \cdot \text{K} \left[1.8\ \text{m}^2 \times 0.25\ \text{K/W} - \frac{3 \times 10^{-3}\ \text{m}}{0.3\ \text{W/m} \cdot \text{K}} - \frac{1}{200\ \text{W/m}^2 \cdot \text{K}} \right]$$

$$= 0.0061\ \text{m} = 6.1\ \text{mm} \qquad \qquad \triangleleft$$

The skin temperature can be calculated by considering conduction through the skin/fat layer:

$$q = \frac{k_{sf} A (T_i - T_s)}{L_{sf}}$$

or solving for T_s,

$$T_s = T_i - \frac{q L_{sf}}{k_{sf} A} = 35°C - \frac{100 \text{ W} \times 3 \times 10^{-3} \text{ m}}{0.3 \text{ W/m} \cdot \text{K} \times 1.8 \text{ m}^2} = 34.4°C \qquad \lhd$$

The skin temperature is the same in both cases because the heat loss rate and skin/fat properties are the same.

Comments:

1. The silica aerogel is a porous material that is only about 5% solid. Its thermal conductivity is less than the thermal conductivity of the gas that fills its pores. As explained in Section 2.2, the reason for this seemingly impossible result is that the pore size is about 20 nm, which reduces the mean free path of the gas and hence decreases its thermal conductivity.

2. By reducing the rate of heat loss to 100 W, a person could remain in the cold environments indefinitely without becoming chilled. The skin temperature of 34.4°C would feel comfortable.

3. In the water case, the thermal resistance of the insulation dominates and all other resistances can be neglected.

4. The convection heat transfer coefficient associated with the air depends on the wind conditions, and it can vary over a broad range. As it changes, so will the outer surface temperature of the insulation layer. Since the radiation heat transfer coefficient depends on this temperature, it will also vary. We can perform a more complete analysis that takes this into account. The radiation heat transfer coefficient is given by Equation 1.9:

$$h_r = \varepsilon \sigma (T_{s,o} + T_{sur})(T_{s,o}^2 + T_{sur}^2) \qquad (1)$$

Here $T_{s,o}$ is the outer surface temperature of the insulation layer, which can be calculated from

$$T_{s,o} = T_i - q\left[\frac{L_{sf}}{k_{sf} A} + \frac{L_{ins}}{k_{ins} A}\right] \qquad (2)$$

where, from the problem solution:

$$L_{ins} = k_{ins}\left(AR_{tot} - \frac{L_{sf}}{k_{sf}} - \frac{1}{h + h_r}\right) \qquad (3)$$

Using all the values from above, these three equations have been solved for values of h in the range $0 \leq h \leq 100$ W/m$^2 \cdot$ K, and the required insulation thickness is represented graphically.

Increasing h reduces the corresponding convection resistance, which then requires additional insulation to maintain the heat transfer rate at 100 W. Once the heat transfer coefficient exceeds approximately 60 W/m² · K, the convection resistance is negligible and further increases in h have little effect on the required insulation thickness.

The outer surface temperature and radiation heat transfer coefficient can also be calculated. As h increases from 0 to 100 W/m² · K, $T_{s,o}$ decreases from 294 to 284 K, while h_r decreases from 5.2 to 4.9 W/m² · K. The initial estimate of $h_r = 5.9$ W/m² · K was not highly accurate. Using this more complete model of the radiation heat transfer, with $h = 2$ W/m² · K, the radiation heat transfer coefficient is 5.1 W/m² · K, and the required insulation thickness is 4.2 mm, close to the value calculated in the first part of the problem.

5. See Example 3.1 in *IHT*. This problem can also be solved using the thermal resistance network builder, *Models/Resistance Networks,* available in *IHT*.

EXAMPLE 3.2

A thin silicon chip and an 8-mm-thick aluminum substrate are separated by a 0.02-mm-thick epoxy joint. The chip and substrate are each 10 mm on a side, and their exposed surfaces are cooled by air, which is at a temperature of 25°C and provides a convection coefficient of 100 W/m² · K. If the chip dissipates 10^4 W/m² under normal conditions, will it operate below a maximum allowable temperature of 85°C?

SOLUTION

Known: Dimensions, heat dissipation, and maximum allowable temperature of a silicon chip. Thickness of aluminum substrate and epoxy joint. Convection conditions at exposed chip and substrate surfaces.

Find: Whether maximum allowable temperature is exceeded.

Schematic:

Assumptions:

1. Steady-state conditions.
2. One-dimensional conduction (negligible heat transfer from sides of composite).
3. Negligible chip thermal resistance (an isothermal chip).
4. Constant properties.
5. Negligible radiation exchange with surroundings.

Properties: Table A.1, pure aluminum ($T \sim 350$ K): $k = 239$ W/m · K.

Analysis: Heat dissipated in the chip is transferred to the air directly from the exposed surface and indirectly through the joint and substrate. Performing an energy balance on a control surface about the chip, it follows that, on the basis of a unit surface area,

$$q_c'' = q_1'' + q_2''$$

or

$$q_c'' = \frac{T_c - T_\infty}{(1/h)} + \frac{T_c - T_\infty}{R_{t,c}'' + (L/k) + (1/h)}$$

To conservatively estimate T_c, the maximum possible value of $R_{t,c}'' = 0.9 \times 10^{-4}$ m^2 · K/W is obtained from Table 3.2. Hence

$$T_c = T_\infty + q_c'' \left[h + \frac{1}{R_{t,c}'' + (L/k) + (1/h)} \right]^{-1}$$

or

$$T_c = 25°C + 10^4 \text{ W/m}^2$$

$$\times \left[100 + \frac{1}{(0.9 + 0.33 + 100) \times 10^{-4}} \right]^{-1} \text{m}^2 \cdot \text{K/W}$$

$$T_c = 25°C + 50.3°C = 75.3°C \qquad \triangleleft$$

Hence the chip will operate below its maximum allowable temperature.

Comments:

1. The joint and substrate thermal resistances are much less than the convection resistance. The joint resistance would have to increase to the unrealistically large value of $50 \times 10^{-4} \ \mathrm{m^2 \cdot K/W}$, before the maximum allowable chip temperature would be exceeded.

2. The allowable power dissipation may be increased by increasing the convection coefficients, either by increasing the air velocity and/or by replacing the air with a more effective heat transfer fluid. Exploring this option for $100 \leq h \leq 2000 \ \mathrm{W/m^2 \cdot K}$ with $T_c = 85°C$, the following results are obtained.

As $h \rightarrow \infty$, $q_2'' \rightarrow 0$ and virtually all of the chip power is transferred directly to the fluid stream.

3. As calculated, the *difference* between the air temperature ($T_\infty = 25°C$) and the chip temperature ($T_c = 75.3°C$) is 50.3 K. Keep in mind that this is a temperature *difference* and therefore is the same as 50.3°C.

4. Consider conditions for which airflow over the chip (upper) or substrate (lower) surface ceases due to a blockage in the air supply channel. If heat transfer from either surface is negligible, what are the resulting chip temperatures for $q_c'' = 10^4 \ \mathrm{W/m^2}$? [Answer, 126°C or 125°C]

EXAMPLE 3.3

A photovoltaic panel consists of (top to bottom) a 3-mm-thick ceria-doped glass ($k_g = 1.4$ W/m · K), a 0.1-mm-thick optical grade adhesive ($k_a = 145$ W/m · K), a *very thin* layer of silicon within which solar energy is converted to electrical energy, a 0.1-mm-thick solder layer, and a 2-mm-thick aluminum nitride substrate. The solar-to-electrical conversion efficiency within the silicon layer η decreases with increasing silicon temperature, T_{si}, and is described by the expression $\eta = a - bT_{si}$, where $a = 0.553$ and $b = 0.001 \ \mathrm{K^{-1}}$. The temperature T is expressed in kelvins over the range $300 \ \mathrm{K} \leq T_{si} \leq 525 \ \mathrm{K}$. Of the incident solar irradiation, $G_S = 700 \ \mathrm{W/m^2}$, 7% is reflected from the top surface of the glass, 10% is absorbed at the top surface of the glass, and 83% is transmitted to and absorbed within the silicon layer. Part of the solar irradiation absorbed in the silicon is converted to thermal energy, and the remainder is converted to electrical energy. The glass has an emissivity of $\varepsilon = 0.90$, and the bottom as well as the sides of the panel are insulated. Determine the electric power

P produced by an $L = 1$-m-long, $w = 0.1$-m-wide solar panel for conditions characterized by $h = 35$ W/m$^2 \cdot$ K and $T_\infty = T_{sur} = 20°$C.

SOLUTION

Known: Dimensions and materials of a photovoltaic solar panel. Material properties, solar irradiation, convection coefficient and ambient temperature, emissivity of top panel surface and surroundings temperature. Partitioning of the solar irradiation, and expression for the solar-to-electrical conversion efficiency.

Find: Electric power produced by the photovoltaic panel.

Schematic:

Assumptions:

1. Steady-state conditions.
2. One-dimensional heat transfer.
3. Constant properties.
4. Negligible thermal contact resistances.
5. Negligible temperature differences within the silicon layer.

Analysis: Recognize that there is no heat transfer to the bottom insulated surface of the solar panel. Hence, the solder layer and aluminum nitride substrate do not affect the solution, and all of the solar energy absorbed by the panel must ultimately leave the panel in the form of radiation and convection heat transfer from the top surface of the glass, and electric power to the grid, $P = \eta 0.83 G_S L w$. Performing an energy balance on the node associated with the silicon layer yields

$$0.83 G_S L w - \eta 0.83 G_S L w = \frac{T_{si} - T_{g,top}}{\dfrac{L_a}{k_a L w} + \dfrac{L_g}{k_g L w}}$$

Substituting the expression for the solar-to-electrical conversion efficiency and simplifying leads to

$$0.83 G_S (1 - a + b T_{si}) = \frac{T_{si} - T_{g,top}}{\dfrac{L_a}{k_a} + \dfrac{L_g}{k_g}} \tag{1}$$

Performing a second energy balance on the node associated with the top surface of the glass gives

$$0.83 G_S L w (1 - \eta) + 0.1 G_S L w = h L w (T_{g,top} - T_\infty) + \varepsilon \sigma L w (T_{g,top}^4 - T_{sur}^4)$$

Substituting the expression for the solar-to-electrical conversion efficiency into the preceding equation and simplifying provides

$$0.83 G_S (1 - a + b T_{si}) + 0.1 G_S = h (T_{g,top} - T_\infty) + \varepsilon \sigma (T_{g,top}^4 - T_{sur}^4) \tag{2}$$

Finally, substituting known values into Equations 1 and 2 and solving simultaneously yields $T_{si} = 307$ K $= 34°C$, providing a solar-to-electrical conversion efficiency of $\eta = 0.553 - 0.001$ K$^{-1} \times 307$ K $= 0.247$. Hence, the power produced by the photovoltaic panel is

$$P = \eta 0.83 G_S L w = 0.247 \times 0.83 \times 700 \ \text{W/m}^2 \times 1 \ \text{m} \times 0.1 \ \text{m} = 14.3 \ \text{W} \qquad \triangleleft$$

Comments:

1. The correct application of the conservation of energy requirement is crucial for determining the silicon temperature and the electric power. Note that solar energy is converted to *both* thermal and electrical energy, and the thermal circuit is used to quantify *only* the thermal energy transfer.

2. Because of the thermally insulated boundary condition, it is not necessary to include the solder or substrate layers in the analysis. This is because there is no conduction through these materials and, from Fourier's law, there can be no temperature gradients within these materials. At steady state, $T_{sdr} = T_{an} = T_{si}$.

3. As the convection coefficient increases, the temperature of the silicon decreases. This leads to a higher solar-to-electrical conversion efficiency and increased electric power output. For example, $P = 13.6$ W and 14.6 W for $h = 15$ W/m² · K and 55 W/m² · K, respectively.

4. The cost of a photovoltaic system can be reduced significantly by *concentrating* the solar energy onto the relatively expensive photovoltaic panel using inexpensive focusing mirrors or lenses. However, good thermal management then becomes even more important. For example, if the irradiation supplied to the panel were increased to $G_S = 7000$ W/m² through concentration, the conversion efficiency drops to $\eta = 0.160$ as the silicon temperature increases to $T_{si} = 119°C$, even for $h = 55$ W/m² · K. A key to reducing the cost of photovoltaic power generation is developing innovative cooling technologies for use in concentrating photovoltaic systems.

5. The simultaneous solution of Equations 1 and 2 may be achieved by using *IHT*, another commercial code, or a handheld calculator. A trial-and-error solution could also be obtained, but with considerable effort. Equations 1 and 2 could be combined to write a single transcendental expression for the silicon temperature, but the equation must still be solved numerically or by trial-and-error.

EXAMPLE 3.4

The thermal conductivity of a $D = 14$-nm-diameter carbon nanotube is measured with an instrument that is fabricated of a wafer of silicon nitride at a temperature of $T_\infty = 300$ K. The 20-μm-long nanotube rests on two 0.5-μm-thick, 10 μm × 10 μm square islands that are separated by a distance $s = 5$ μm. A thin layer of platinum is used as an electrical resistor on the *heated island* (at temperature T_h) to dissipate $q = 11.3$ μW of electrical power. On the *sensing island*, a similar layer of platinum is used to determine its temperature, T_s. The platinum's electrical resistance, $R(T_s) = E/I$, is found by measuring the voltage drop and electrical current across the platinum layer. The temperature of the sensing island, T_s, is then determined from the relationship of the platinum electrical resistance to its temperature. Each island is suspended by two $L_{sn} = 250$-μm-long silicon nitride beams that are $w_{sn} = 3$ μm wide and $t_{sn} = 0.5$ μm thick. A platinum line of width $w_{pt} = 1$ μm and thickness $t_{pt} = 0.2$ μm is deposited within each silicon nitride beam to power the heated island or to detect the voltage drop associated with the determination of T_s. The entire experiment is performed in a vacuum with $T_{sur} = 300$ K and at steady state, $T_s = 308.4$ K. Estimate the thermal conductivity of the carbon nanotube.

SOLUTION

Known: Dimensions, heat dissipated at the heated island, and temperatures of the sensing island and surrounding silicon nitride wafer.

Find: The thermal conductivity of the carbon nanotube.

Schematic:

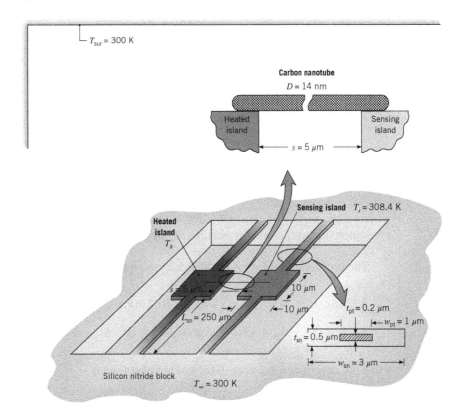

Assumptions:

1. Steady-state conditions.
2. One-dimensional heat transfer.
3. The heated and sensing islands are isothermal.
4. Radiation exchange between the surfaces and the surroundings is negligible.
5. Negligible convection losses.
6. Ohmic heating in the platinum signal lines is negligible.
7. Constant properties.
8. Negligible contact resistance between the nanotube and the islands.

Properties: Table A.1, platinum (325 K, assumed): $k_{pt} = 71.6$ W/m · K. Table A.2, silicon nitride (325 K, assumed): $k_{sn} = 15.5$ W/m · K.

Analysis: Thermal energy is conducted from the heated island through the two heated island support beams as well as through the carbon nanotube to the sensing island. Therefore, the thermal circuit is where each support beam provides a thermal resistance $R_{t,\text{sup}}$ that is composed of a resistance due to the silicon nitride (sn) in parallel with a resistance due to the platinum (pt) line.

The cross-sectional areas of the materials in the support beams are

$$A_{\text{pt}} = w_{\text{pt}}t_{\text{pt}} = (1 \times 10^{-6}\,\text{m}) \times (0.2 \times 10^{-6}\,\text{m}) = 2 \times 10^{-13}\,\text{m}^2$$

$$A_{\text{sn}} = w_{\text{sn}}t_{\text{sn}} - A_{\text{pt}} = (3 \times 10^{-6}\,\text{m}) \times (0.5 \times 10^{-6}\,\text{m}) - 2 \times 10^{-13}\,\text{m}^2 = 1.3 \times 10^{-12}\,\text{m}^2$$

while the cross-sectional area of the carbon nanotube is

$$A_{\text{cn}} = \pi D^2/4 = \pi(14 \times 10^{-9}\,\text{m})^2/4 = 1.54 \times 10^{-16}\,\text{m}^2$$

The thermal resistance of each support is

$$R_{t,\text{sup}} = \left[\frac{k_{\text{pt}}A_{\text{pt}}}{L_{\text{pt}}} + \frac{k_{\text{sn}}A_{\text{sn}}}{L_{\text{sn}}} \right]^{-1}$$

$$= \left[\frac{71.6\,\text{W/m} \cdot \text{K} \times 2 \times 10^{-13}\,\text{m}^2}{250 \times 10^{-6}\,\text{m}} + \frac{15.5\,\text{W/m} \cdot \text{K} \times 1.3 \times 10^{-12}\,\text{m}^2}{250 \times 10^{-6}\,\text{m}} \right]^{-1}$$

$$= 7.25 \times 10^6\,\text{K/W}$$

The combined rate of heat loss through both sensing island supports is

$$q_s = 2(T_s - T_\infty)/R_{t,\text{sup}} \tag{1}$$

Using Equation 1 and noting that $q_h = q - q_s$, consideration of the thermal circuit connecting T_h and T_∞ leads to the relation

$$T_h = T_\infty + \frac{1}{2}q_h R_{t,\text{sup}} = T_\infty + \frac{1}{2}q R_{t,\text{sup}} - (T_s - T_\infty) \tag{2}$$

From the portion of the circuit connecting T_h and T_s, $q_s = (T_h - T_s)/(s/(k_{cn}A_{cn}))$, which may be combined with Equations 1 and 2 to yield

$$k_{cn} = \frac{q_s s}{A_{cn}(T_h - T_s)} = \frac{2(T_s - T_\infty)s/R_{t,\text{sup}}}{A_{cn}\left(\frac{1}{2}qR_{t,\text{sup}} - 2(T_s - T_\infty)\right)}$$

$$= \frac{2 \times (308.4 \text{ K} - 300 \text{ K}) \times 5 \times 10^{-6} \text{ m}/7.25 \times 10^6 \text{ K/W}}{1.54 \times 10^{-16} \text{ m}^2 \times \left(\frac{1}{2} \times 11.3 \times 10^{-6} \text{ W} \times 7.25 \times 10^6 \text{ K/W} - 2 \times (308.4 \text{ K} - 300 \text{ K})\right)}$$

$$k_{cn} = 3113 \text{ W/m} \cdot \text{K} \qquad \triangleleft$$

Comments:

1. The measured thermal conductivity is extremely large, as evident by comparing its value to the thermal conductivities of pure metals shown in Figure 2.4.

2. Contact resistances between the carbon nanotube and the heated and sensing islands were neglected because little is known about such resistances at the nanoscale. However, *if* a contact resistance were included in the analysis, the measured thermal conductivity of the carbon nanotube would be even higher than the predicted value.

3. The significance of radiation heat transfer may be estimated by approximating the heated island as a blackbody radiating to T_{sur} from both its top and bottom surfaces. From the expression for T_h above, $T_h = 332.6$ K. Hence, $q_{\text{rad},b} \approx 5.67 \times 10^{-8}$ W/m$^2 \cdot$ K$^4 \times 2 \times (10 \times 10^{-6} \text{ m})^2 \times (332.6^4 - 300^4)K^4 = 4.7 \times 10^{-8}$ W $= 0.047$ μW, and radiation is negligible.

3.2 *An Alternative Conduction Analysis*

The conduction analysis of Section 3.1 was performed using the *standard approach.* That is, the heat equation was solved to obtain the temperature distribution, Equation 3.3, and Fourier's law was then applied to obtain the heat transfer rate, Equation 3.4. However, an alternative approach may be used for the conditions presently of interest. Considering conduction in the system of Figure 3.6, we recognize that, for *steady-state conditions* with *no heat generation* and *no heat loss from the sides*, the heat transfer rate q_x must be a constant independent of x. That is, for any differential element dx, $q_x = q_{x+dx}$. This condition is, of course, a consequence of the energy conservation requirement, and it must apply even if the area varies with position $A(x)$ and the thermal conductivity varies with temperature $k(T)$. Moreover, even though the temperature distribution may be two-dimensional, varying with x and y, it is often reasonable to neglect the y-variation and to assume a *one-dimensional* distribution in x.

For the above conditions it is possible to work exclusively with Fourier's law when performing a conduction analysis. In particular, since the conduction rate is a *constant,* the rate equation may be *integrated,* even though neither the rate nor the temperature distribution is known. Consider Fourier's law, Equation 2.1, which may be applied to

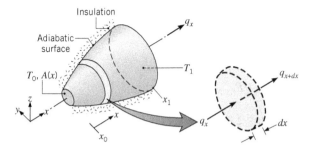

FIGURE 3.6   System with a constant conduction heat transfer rate.

the system of Figure 3.6. Although we may have no knowledge of the value of q_x or the form of $T(x)$, we do know that q_x is a constant. Hence we may express Fourier's law in the integral form

$$q_x \int_{x_0}^{x} \frac{dx}{A(x)} = -\int_{T_0}^{T} k(T)\, dT \qquad (3.26)$$

The cross-sectional area may be a known function of x, and the material thermal conductivity may vary with temperature in a known manner. If the integration is performed from a point x_0 at which the temperature T_0 is known, the resulting equation provides the functional form of $T(x)$. Moreover, if the temperature $T = T_1$ at some $x = x_1$ is also known, integration between x_0 and x_1 provides an expression from which q_x may be computed. Note that, if the area A is uniform and k is independent of temperature, Equation 3.26 reduces to

$$\frac{q_x \, \Delta x}{A} = -k\, \Delta T \qquad (3.27)$$

where $\Delta x = x_1 - x_0$ and $\Delta T = T_1 - T_0$.

We frequently elect to solve diffusion problems by working with integrated forms of the diffusion rate equations. However, the limiting conditions for which this may be done should be firmly fixed in our minds: *steady-state* and *one-dimensional* transfer with *no heat generation*.

EXAMPLE 3.5

The diagram shows a conical section fabricated from pyroceram. It is of circular cross section with the diameter $D = ax$, where $a = 0.25$. The small end is at $x_1 = 50$ mm and the large end at $x_2 = 250$ mm. The end temperatures are $T_1 = 400$ K and $T_2 = 600$ K, while the lateral surface is well insulated.

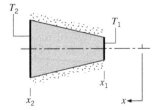

1. Derive an expression for the temperature distribution $T(x)$ in symbolic form, assuming one-dimensional conditions. Sketch the temperature distribution.
2. Calculate the heat rate q_x through the cone.

SOLUTION

Known: Conduction in a circular conical section having a diameter $D = ax$, where $a = 0.25$.

Find:

1. Temperature distribution $T(x)$.
2. Heat transfer rate q_x.

Schematic:

Assumptions:

1. Steady-state conditions.
2. One-dimensional conduction in the x-direction.
3. No internal heat generation.
4. Constant properties.

Properties: Table A.2, pyroceram (500 K): $k = 3.46$ W/m · K.

Analysis:

1. Since heat conduction occurs under steady-state, one-dimensional conditions with no internal heat generation, the heat transfer rate q_x is a constant independent of x. Accordingly, Fourier's law, Equation 2.1, may be used to determine the temperature distribution

$$q_x = -kA\frac{dT}{dx}$$

where $A = \pi D^2/4 = \pi a^2 x^2/4$. Separating variables,

$$\frac{4q_x dx}{\pi a^2 x^2} = -kdT$$

Integrating from x_1 to any x within the cone, and recalling that q_x and k are constants, it follows that

$$\frac{4q_x}{\pi a^2}\int_{x_1}^{x}\frac{dx}{x^2} = -k\int_{T_1}^{T}dT$$

Hence

$$\frac{4q_x}{\pi a^2}\left(-\frac{1}{x}+\frac{1}{x_1}\right) = -k(T-T_1)$$

or solving for T

$$T(x) = T_1 - \frac{4q_x}{\pi a^2 k}\left(\frac{1}{x_1}-\frac{1}{x}\right)$$

Although q_x is a constant, it is as yet an unknown. However, it may be determined by evaluating the above expression at $x = x_2$, where $T(x_2) = T_2$. Hence

$$T_2 = T_1 - \frac{4q_x}{\pi a^2 k}\left(\frac{1}{x_1}-\frac{1}{x_2}\right)$$

and solving for q_x

$$q_x = \frac{\pi a^2 k(T_1-T_2)}{4[(1/x_1)-(1/x_2)]}$$

Substituting for q_x into the expression for $T(x)$, the temperature distribution becomes

$$T(x) = T_1 + (T_1-T_2)\left[\frac{(1/x)-(1/x_1)}{(1/x_1)-(1/x_2)}\right] \qquad \triangleleft$$

From this result, temperature may be calculated as a function of x and the distribution is as shown.

Note that, since $dT/dx = -4q_x/k\pi a^2 x^2$ from Fourier's law, it follows that the temperature gradient and heat flux decrease with increasing x.

2. Substituting numerical values into the foregoing result for the heat transfer rate, it follows that

$$q_x = \frac{\pi(0.25)^2 \times 3.46 \text{ W/m} \cdot \text{K} (400-600) \text{ K}}{4(1/0.05 \text{ m} - 1/0.25 \text{ m})} = -2.12 \text{ W} \qquad \triangleleft$$

Comments: When the parameter a increases, the cross-sectional area changes more rapidly with distance, causing the one-dimensional assumption to become less appropriate.

3.3 *Radial Systems*

Cylindrical and spherical systems often experience temperature gradients in the radial direction only and may therefore be treated as one-dimensional. Moreover, under steady-state conditions with no heat generation, such systems may be analyzed by using the *standard* method, which begins with the appropriate form of the heat equation, or the *alternative* method, which begins with the appropriate form of Fourier's law. In this section, the cylindrical system is analyzed by means of the standard method and the spherical system by means of the alternative method.

3.3.1 The Cylinder

A common example is the hollow cylinder whose inner and outer surfaces are exposed to fluids at different temperatures (Figure 3.7). For steady-state conditions with no heat generation, the appropriate form of the heat equation, Equation 2.26, is

$$\frac{1}{r}\frac{d}{dr}\left(kr\frac{dT}{dr}\right) = 0 \tag{3.28}$$

where, for the moment, k is treated as a variable. The physical significance of this result becomes evident if we also consider the appropriate form of Fourier's law. The rate at which energy is conducted across any cylindrical surface in the solid may be expressed as

$$q_r = -kA\frac{dT}{dr} = -k(2\pi rL)\frac{dT}{dr} \tag{3.29}$$

where $A = 2\pi rL$ is the area normal to the direction of heat transfer. Since Equation 3.28 dictates that the quantity $kr(dT/dr)$ is independent of r, it follows from Equation 3.29 that the conduction *heat transfer rate* q_r (*not* the heat flux q_r'') is a *constant in the radial direction.*

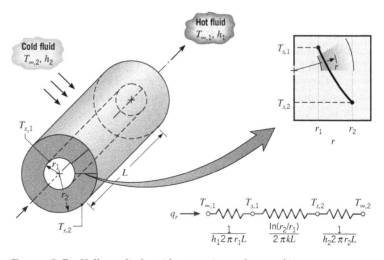

FIGURE 3.7 Hollow cylinder with convective surface conditions.

We may determine the temperature distribution in the cylinder by solving Equation 3.28 and applying appropriate boundary conditions. Assuming the value of k to be constant, Equation 3.28 may be integrated twice to obtain the general solution

$$T(r) = C_1 \ln r + C_2 \tag{3.30}$$

To obtain the constants of integration C_1 and C_2, we introduce the following boundary conditions:

$$T(r_1) = T_{s,1} \quad \text{and} \quad T(r_2) = T_{s,2}$$

Applying these conditions to the general solution, we then obtain

$$T_{s,1} = C_1 \ln r_1 + C_2 \quad \text{and} \quad T_{s,2} = C_1 \ln r_2 + C_2$$

Solving for C_1 and C_2 and substituting into the general solution, we then obtain

$$T(r) = \frac{T_{s,1} - T_{s,2}}{\ln(r_1/r_2)} \ln\left(\frac{r}{r_2}\right) + T_{s,2} \tag{3.31}$$

Note that the temperature distribution associated with radial conduction through a cylindrical wall is logarithmic, not linear, as it is for the plane wall under the same conditions. The logarithmic distribution is sketched in the inset of Figure 3.7.

If the temperature distribution, Equation 3.31, is now used with Fourier's law, Equation 3.29, we obtain the following expression for the heat transfer rate:

$$q_r = \frac{2\pi L k(T_{s,1} - T_{s,2})}{\ln(r_2/r_1)} \tag{3.32}$$

From this result it is evident that, for radial conduction in a cylindrical wall, the thermal resistance is of the form

$$R_{t,\text{cond}} = \frac{\ln(r_2/r_1)}{2\pi L k} \tag{3.33}$$

This resistance is shown in the series circuit of Figure 3.7. Note that since the value of q_r is independent of r, the foregoing result could have been obtained by using the alternative method, that is, by integrating Equation 3.29.

Consider now the composite system of Figure 3.8. Recalling how we treated the composite plane wall and neglecting the interfacial contact resistances, the heat transfer rate may be expressed as

$$q_r = \frac{T_{\infty,1} - T_{\infty,4}}{\dfrac{1}{2\pi r_1 L h_1} + \dfrac{\ln(r_2/r_1)}{2\pi k_A L} + \dfrac{\ln(r_3/r_2)}{2\pi k_B L} + \dfrac{\ln(r_4/r_3)}{2\pi k_C L} + \dfrac{1}{2\pi r_4 L h_4}} \tag{3.34}$$

The foregoing result may also be expressed in terms of an overall heat transfer coefficient. That is,

$$q_r = \frac{T_{\infty,1} - T_{\infty,4}}{R_{\text{tot}}} = UA(T_{\infty,1} - T_{\infty,4}) \tag{3.35}$$

FIGURE 3.8 Temperature distribution for a composite cylindrical wall.

If U is defined in terms of the inside area, $A_1 = 2\pi r_1 L$, Equations 3.34 and 3.35 may be equated to yield

$$U_1 = \cfrac{1}{\cfrac{1}{h_1} + \cfrac{r_1}{k_A}\ln\cfrac{r_2}{r_1} + \cfrac{r_1}{k_B}\ln\cfrac{r_3}{r_2} + \cfrac{r_1}{k_C}\ln\cfrac{r_4}{r_3} + \cfrac{r_1}{r_4}\cfrac{1}{h_4}} \qquad (3.36)$$

This definition is *arbitrary,* and the overall coefficient may also be defined in terms of A_4 or any of the intermediate areas. Note that

$$U_1 A_1 = U_2 A_2 = U_3 A_3 = U_4 A_4 = (\Sigma R_t)^{-1} \qquad (3.37)$$

and the specific forms of U_2, U_3, and U_4 may be inferred from Equations 3.34 and 3.35.

EXAMPLE 3.6

The possible existence of an optimum insulation thickness for radial systems is suggested by the presence of competing effects associated with an increase in this thickness. In particular, although the conduction resistance increases with the addition of insulation, the convection resistance decreases due to increasing outer surface area. Hence there may exist an insulation thickness that minimizes heat loss by maximizing the total resistance to heat transfer. Resolve this issue by considering the following system.

1. A thin-walled copper tube of radius r_i is used to transport a low-temperature refrigerant and is at a temperature T_i that is less than that of the ambient air at T_∞ around the tube. Is there an optimum thickness associated with application of insulation to the tube?

2. Confirm the above result by computing the total thermal resistance per unit length of tube for a 10-mm-diameter tube having the following insulation thicknesses: 0, 2, 5, 10, 20, and 40 mm. The insulation is composed of cellular glass, and the outer surface convection coefficient is 5 W/m^2 · K.

SOLUTION

Known: Radius r_i and temperature T_i of a thin-walled copper tube to be insulated from the ambient air.

Find:

1. Whether there exists an optimum insulation thickness that minimizes the heat transfer rate.

2. Thermal resistance associated with using cellular glass insulation of varying thickness.

Schematic:

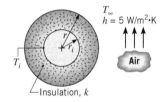

Assumptions:

1. Steady-state conditions.

2. One-dimensional heat transfer in the radial (cylindrical) direction.

3. Negligible tube wall thermal resistance.

4. Constant properties for insulation.

5. Negligible radiation exchange between insulation outer surface and surroundings.

Properties: Table A.3, cellular glass (285 K, assumed): $k = 0.055$ W/m · K.

Analysis:

1. The resistance to heat transfer between the refrigerant and the air is dominated by conduction in the insulation and convection in the air. The thermal circuit is therefore

$$q' \longleftarrow \quad T_i \,\,\text{\wwww}\,\, \text{\wwww}\,\, T_\infty$$
$$\frac{\ln(r/r_i)}{2\pi k} \qquad \frac{1}{2\pi rh}$$

where the conduction and convection resistances per unit length follow from Equations 3.33 and 3.9, respectively. The total thermal resistance per unit length of tube is then

$$R'_{\text{tot}} = \frac{\ln(r/r_i)}{2\pi k} + \frac{1}{2\pi rh}$$

where the rate of heat transfer per unit length of tube is

$$q' = \frac{T_\infty - T_i}{R'_{tot}}$$

An optimum insulation thickness would be associated with the value of r that minimized q' or maximized R'_{tot}. Such a value could be obtained from the requirement that

$$\frac{dR'_{tot}}{dr} = 0$$

Hence

$$\frac{1}{2\pi k r} - \frac{1}{2\pi r^2 h} = 0$$

or

$$r = \frac{k}{h}$$

To determine whether the foregoing result maximizes or minimizes the total resistance, the second derivative must be evaluated. Hence

$$\frac{d^2 R'_{tot}}{dr^2} = -\frac{1}{2\pi k r^2} + \frac{1}{\pi r^3 h}$$

or, at $r = k/h$,

$$\frac{d^2 R'_{tot}}{dr^2} = \frac{1}{\pi (k/h)^2} \left(\frac{1}{k} - \frac{1}{2k} \right) = \frac{1}{2\pi k^3/h^2} > 0$$

Since this result is always positive, it follows that $r = k/h$ is the insulation radius for which the total resistance is a minimum, not a maximum. Hence an *optimum* insulation thickness *does not exist*.

From the above result it makes more sense to think in terms of a *critical insulation radius*

$$r_{cr} \equiv \frac{k}{h}$$

which maximizes heat transfer, that is, below which q' increases with increasing r and above which q' decreases with increasing r.

2. With $h = 5$ W/m$^2 \cdot$ K and $k = 0.055$ W/m $\cdot$ K, the critical radius is

$$r_{cr} = \frac{0.055 \text{ W/m} \cdot \text{K}}{5 \text{ W/m}^2 \cdot \text{K}} = 0.011 \text{ m}$$

Hence $r_{cr} > r_i$ and heat transfer will increase with the addition of insulation up to a thickness of

$$r_{cr} - r_i = (0.011 - 0.005) \text{ m} = 0.006 \text{ m}$$

The thermal resistances corresponding to the prescribed insulation thicknesses may be calculated and are plotted as follows:

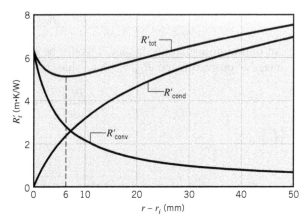

Comments:

1. The effect of the critical radius is revealed by the fact that, even for 20 mm of insulation, the total resistance is not as large as the value for no insulation.

2. If $r_i < r_{cr}$, as it is in this case, the total resistance decreases and the heat rate therefore increases with the addition of insulation. This trend continues until the outer radius of the insulation corresponds to the critical radius. The trend is desirable for electrical current flow through a wire, since the addition of electrical insulation would aid in transferring heat dissipated in the wire to the surroundings. Conversely, if $r_i > r_{cr}$, any addition of insulation would increase the total resistance and therefore decrease the heat loss. This behavior would be desirable for steam flow through a pipe, where insulation is added to reduce heat loss to the surroundings.

3. For radial systems, the problem of reducing the total resistance through the application of insulation exists only for small diameter wires or tubes and for small convection coefficients, such that $r_{cr} > r_i$. For a typical insulation ($k \approx 0.03$ W/m · K) and free convection in air ($h \approx 10$ W/m^2 · K), $r_{cr} = (k/h) \approx 0.003$ m. Such a small value tells us that, normally, $r_i > r_{cr}$ and we need not be concerned with the effects of a critical radius.

4. The existence of a critical radius requires that the heat transfer area change in the direction of transfer, as for radial conduction in a cylinder (or a sphere). In a plane wall the area perpendicular to the direction of heat flow is constant and there is no critical insulation thickness (the total resistance always increases with increasing insulation thickness).

3.3.2 The Sphere

Now consider applying the alternative method to analyzing conduction in the hollow sphere of Figure 3.9. For the differential control volume of the figure, energy conservation requires that $q_r = q_{r+dr}$ for steady-state, one-dimensional conditions with no heat generation. The appropriate form of Fourier's law is

$$q_r = -kA\frac{dT}{dr} = -k(4\pi r^2)\frac{dT}{dr} \tag{3.38}$$

where $A = 4\pi r^2$ is the area normal to the direction of heat transfer.

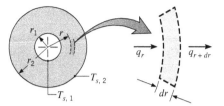

FIGURE 3.9 Conduction in a spherical shell.

Acknowledging that q_r is a constant, independent of r, Equation 3.38 may be expressed in the integral form

$$\frac{q_r}{4\pi} \int_{r_1}^{r_2} \frac{dr}{r^2} = -\int_{T_{s,1}}^{T_{s,2}} k(T) \, dT \tag{3.39}$$

Assuming constant k, we then obtain

$$q_r = \frac{4\pi k(T_{s,1} - T_{s,2})}{(1/r_1) - (1/r_2)} \tag{3.40}$$

Remembering that the thermal resistance is defined as the temperature difference divided by the heat transfer rate, we obtain

$$R_{t,\text{cond}} = \frac{1}{4\pi k}\left(\frac{1}{r_1} - \frac{1}{r_2}\right) \tag{3.41}$$

Note that the temperature distribution and Equations 3.40 and 3.41 could have been obtained by using the standard approach, which begins with the appropriate form of the heat equation.

Spherical composites may be treated in much the same way as composite walls and cylinders, where appropriate forms of the total resistance and overall heat transfer coefficient may be determined.

3.4 *Summary of One-Dimensional Conduction Results*

Many important problems are characterized by one-dimensional, steady-state conduction in plane, cylindrical, or spherical walls without thermal energy generation. Key results for these three geometries are summarized in Table 3.3, where ΔT refers to the temperature difference, $T_{s,1} - T_{s,2}$, between the inner and outer surfaces identified in Figures 3.1, 3.7, and 3.9. In each case, beginning with the heat equation, you should be able to derive the corresponding expressions for the temperature distribution, heat flux, heat rate, and thermal resistance.

3.5 *Conduction with Thermal Energy Generation*

In the preceding section we considered conduction problems for which the temperature distribution in a medium was determined solely by conditions at the boundaries of the medium. We now want to consider the additional effect on the temperature distribution of processes that may be occurring *within* the medium. In particular, we wish to consider situations for which thermal energy is being *generated* due to *conversion* from some other energy form.

TABLE 3.3   One-dimensional, steady-state solutions to the heat equation with no generation

	Plane Wall	**Cylindrical Wall**[a]	**Spherical Wall**[a]
Heat equation	$\dfrac{d^2T}{dx^2} = 0$	$\dfrac{1}{r}\dfrac{d}{dr}\left(r\dfrac{dT}{dr}\right) = 0$	$\dfrac{1}{r^2}\dfrac{d}{dr}\left(r^2\dfrac{dT}{dr}\right) = 0$
Temperature distribution	$T_{s,1} - \Delta T\dfrac{x}{L}$	$T_{s,2} + \Delta T\dfrac{\ln(r/r_2)}{\ln(r_1/r_2)}$	$T_{s,1} - \Delta T\left[\dfrac{1-(r_1/r)}{1-(r_1/r_2)}\right]$
Heat flux (q'')	$k\dfrac{\Delta T}{L}$	$\dfrac{k\,\Delta T}{r\ln(r_2/r_1)}$	$\dfrac{k\,\Delta T}{r^2[(1/r_1)-(1/r_2)]}$
Heat rate (q)	$kA\dfrac{\Delta T}{L}$	$\dfrac{2\pi L k\,\Delta T}{\ln(r_2/r_1)}$	$\dfrac{4\pi k\,\Delta T}{(1/r_1)-(1/r_2)}$
Thermal resistance ($R_{t,\text{cond}}$)	$\dfrac{L}{kA}$	$\dfrac{\ln(r_2/r_1)}{2\pi L k}$	$\dfrac{(1/r_1)-(1/r_2)}{4\pi k}$

[a]The critical radius of insulation is $r_{cr} = k/h$ for the cylinder and $r_{cr} = 2k/h$ for the sphere.

A common thermal energy generation process involves the conversion from *electrical to thermal energy* in a current-carrying medium (*Ohmic*, or *resistance*, or *Joule heating*). The rate at which energy is generated by passing a current I through a medium of electrical resistance R_e is

$$\dot{E}_g = I^2 R_e \tag{3.42}$$

If this power generation (W) occurs uniformly throughout the medium of volume V, the volumetric generation rate (W/m^3) is then

$$\dot{q} \equiv \frac{\dot{E}_g}{V} = \frac{I^2 R_e}{V} \tag{3.43}$$

Energy generation may also occur as a result of the deceleration and absorption of neutrons in the fuel element of a nuclear reactor or exothermic chemical reactions occurring within a medium. Endothermic reactions would, of course, have the inverse effect (a thermal energy sink) of converting thermal energy to chemical bonding energy. Finally, a conversion from electromagnetic to thermal energy may occur due to the absorption of radiation within the medium. The process occurs, for example, when gamma rays are absorbed in external nuclear reactor components (cladding, thermal shields, pressure vessels, etc.) or when visible radiation is absorbed in a semitransparent medium. Remember not to confuse energy generation with energy storage (Section 1.3.1).

3.5.1   The Plane Wall

Consider the plane wall of Figure 3.10*a*, in which there is *uniform* energy generation per unit volume ($\dot{q}$ is constant) and the surfaces are maintained at $T_{s,1}$ and $T_{s,2}$. For constant thermal conductivity k, the appropriate form of the heat equation, Equation 2.22, is

$$\frac{d^2T}{dx^2} + \frac{\dot{q}}{k} = 0 \tag{3.44}$$

The general solution is

$$T = -\frac{\dot{q}}{2k}x^2 + C_1 x + C_2 \tag{3.45}$$

where C_1 and C_2 are the constants of integration. For the prescribed boundary conditions,

$$T(-L) = T_{s,1} \quad \text{and} \quad T(L) = T_{s,2}$$

The constants may be evaluated and are of the form

$$C_1 = \frac{T_{s,2} - T_{s,1}}{2L} \quad \text{and} \quad C_2 = \frac{\dot{q}}{2k}L^2 + \frac{T_{s,1} + T_{s,2}}{2}$$

in which case the temperature distribution is

$$T(x) = \frac{\dot{q}L^2}{2k}\left(1 - \frac{x^2}{L^2}\right) + \frac{T_{s,2} - T_{s,1}}{2}\frac{x}{L} + \frac{T_{s,1} + T_{s,2}}{2} \tag{3.46}$$

The heat flux at any point in the wall may, of course, be determined by using Equation 3.46 with Fourier's law. Note, however, that *with generation the heat flux is no longer independent of x.*

The preceding result simplifies when both surfaces are maintained at a common temperature, $T_{s,1} = T_{s,2} \equiv T_s$. The temperature distribution is then *symmetrical* about the midplane, Figure 3.10b, and is given by

$$T(x) = \frac{\dot{q}L^2}{2k}\left(1 - \frac{x^2}{L^2}\right) + T_s \tag{3.47}$$

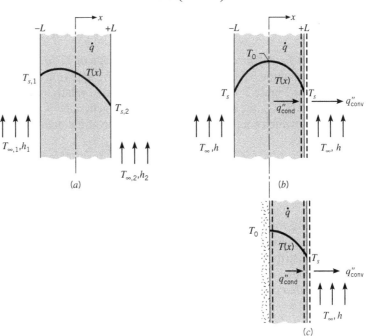

FIGURE 3.10   Conduction in a plane wall with uniform heat generation.
(a) Asymmetrical boundary conditions. (b) Symmetrical boundary conditions.
(c) Adiabatic surface at midplane.

The maximum temperature exists at the midplane

$$T(0) \equiv T_0 = \frac{\dot{q}L^2}{2k} + T_s \tag{3.48}$$

in which case the temperature distribution, Equation 3.47, may be expressed as

$$\frac{T(x) - T_0}{T_s - T_0} = \left(\frac{x}{L}\right)^2 \tag{3.49}$$

It is important to note that at the plane of symmetry in Figure 3.10b, the temperature gradient is zero, $(dT/dx)_{x=0} = 0$. Accordingly, there is no heat transfer across this plane, and it may be represented by the *adiabatic* surface shown in Figure 3.10c. One implication of this result is that Equation 3.47 also applies to plane walls that are perfectly insulated on one side ($x = 0$) and maintained at a fixed temperature T_s on the other side ($x = L$).

To use the foregoing results, the surface temperature(s) T_s must be known. However, a common situation is one for which it is the temperature of an adjoining fluid, T_∞, and not T_s, which is known. It then becomes necessary to relate T_s to T_∞. This relation may be developed by applying a surface energy balance. Consider the surface at $x = L$ for the symmetrical plane wall (Figure 3.10b) or the insulated plane wall (Figure 3.10c). Neglecting radiation and substituting the appropriate rate equations, the energy balance given by Equation 1.13 reduces to

$$-k \left.\frac{dT}{dx}\right|_{x=L} = h(T_s - T_\infty) \tag{3.50}$$

Substituting from Equation 3.47 to obtain the temperature gradient at $x = L$, it follows that

$$T_s = T_\infty + \frac{\dot{q}L}{h} \tag{3.51}$$

Hence T_s may be computed from knowledge of T_∞, $\dot{q}$, L, and h.

Equation 3.51 may also be obtained by applying an *overall* energy balance to the plane wall of Figure 3.10b or 3.10c. For example, relative to a control surface about the wall of Figure 3.10c, the rate at which energy is generated within the wall must be balanced by the rate at which energy leaves via convection at the boundary. Equation 1.12c reduces to

$$\dot{E}_g = \dot{E}_{out} \tag{3.52}$$

or, for a unit surface area,

$$\dot{q}L = h(T_s - T_\infty) \tag{3.53}$$

Solving for T_s, Equation 3.51 is obtained.

Equation 3.51 may be combined with Equation 3.47 to eliminate T_s from the temperature distribution, which is then expressed in terms of the known quantities $\dot{q}$, L, k, h, and T_∞. The same result may be obtained directly by using Equation 3.50 as a boundary condition to evaluate the constants of integration appearing in Equation 3.45.

EXAMPLE 3.7

A plane wall is a composite of two materials, A and B. The wall of material A has uniform heat generation $\dot{q} = 1.5 \times 10^6$ W/m³, $k_A = 75$ W/m · K, and thickness $L_A = 50$ mm. The wall material B has no generation with $k_B = 150$ W/m · K and thickness $L_B = 20$ mm. The inner surface of material A is well insulated, while the outer surface of material B is cooled by a water stream with $T_\infty = 30°C$ and $h = 1000$ W/m² · K.

1. Sketch the temperature distribution that exists in the composite under steady-state conditions.

2. Determine the temperature T_0 of the insulated surface and the temperature T_2 of the cooled surface.

SOLUTION

Known: Plane wall of material A with internal heat generation is insulated on one side and bounded by a second wall of material B, which is without heat generation and is subjected to convection cooling.

Find:

1. Sketch of steady-state temperature distribution in the composite.

2. Inner and outer surface temperatures of the composite.

Schematic:

Assumptions:

1. Steady-state conditions.

2. One-dimensional conduction in *x*-direction.

3. Negligible contact resistance between walls.

4. Inner surface of A adiabatic.

5. Constant properties for materials A and B.

Analysis:

1. From the prescribed physical conditions, the temperature distribution in the composite is known to have the following features, as shown:

   (a) Parabolic in material A.

   (b) Zero slope at insulated boundary.

   (c) Linear in material B.

(d) Slope change $= k_B/k_A = 2$ at interface.

The temperature distribution in the water is characterized by

(e) Large gradients near the surface.

2. The outer surface temperature T_2 may be obtained by performing an energy balance on a control volume about material B. Since there is no generation in this material, it follows that, for steady-state conditions and a unit surface area, the heat flux into the material at $x = L_A$ must equal the heat flux from the material due to convection at $x = L_A + L_B$. Hence

$$q'' = h(T_2 - T_\infty) \tag{1}$$

The heat flux q'' may be determined by performing a second energy balance on a control volume about material A. In particular, since the surface at $x = 0$ is adiabatic, there is no inflow and the rate at which energy is generated must equal the outflow. Accordingly, for a unit surface area,

$$\dot{q}L_A = q'' \tag{2}$$

Combining Equations 1 and 2, the outer surface temperature is

$$T_2 = T_\infty + \frac{\dot{q}L_A}{h}$$

$$T_2 = 30°C + \frac{1.5 \times 10^6 \ W/m^3 \times 0.05 \ m}{1000 \ W/m^2 \cdot K} = 105°C \qquad \lhd$$

From Equation 3.48 the temperature at the insulated surface is

$$T_0 = \frac{\dot{q}L_A^2}{2k_A} + T_1 \tag{3}$$

where T_1 may be obtained from the following thermal circuit:

$$q'' \longrightarrow \quad T_1 \underset{R''_{cond, B}}{-\!\!\!\wedge\!\!\wedge\!\!\wedge\!\!-} T_2 \underset{R''_{conv}}{-\!\!\!\wedge\!\!\wedge\!\!\wedge\!\!-} T_\infty$$

That is,

$$T_1 = T_\infty + (R''_{cond,B} + R''_{conv})q''$$

where the resistances for a unit surface area are

$$R''_{\text{cond,B}} = \frac{L_B}{k_B} \qquad R''_{\text{conv}} = \frac{1}{h}$$

Hence,

$$T_1 = 30°C + \left(\frac{0.02 \text{ m}}{150 \text{ W/m} \cdot \text{K}} + \frac{1}{1000 \text{ W/m}^2 \cdot \text{K}} \right)$$

$$\times 1.5 \times 10^6 \text{ W/m}^3 \times 0.05 \text{ m}$$

$$T_1 = 30°C + 85°C = 115°C$$

Substituting into Equation 3,

$$T_0 = \frac{1.5 \times 10^6 \text{ W/m}^3 (0.05 \text{ m})^2}{2 \times 75 \text{ W/m} \cdot \text{K}} + 115°C$$

$$T_0 = 25°C + 115°C = 140°C \qquad \triangleleft$$

Comments:

1. Material A, having heat generation, cannot be represented by a thermal circuit element.

2. Since the resistance to heat transfer by convection is significantly larger than that due to conduction in material B, $R''_{\text{conv}}/R''_{\text{cond}} = 7.5$, the surface-to-fluid temperature difference is much larger than the temperature drop across material B, $(T_2 - T_\infty)/(T_1 - T_2) = 7.5$. This result is consistent with the temperature distribution plotted in part 1.

3. The surface and interface temperatures (T_0, T_1, and T_2) depend on the generation rate $\dot{q}$, the thermal conductivities k_A and k_B, and the convection coefficient h. Each material will have a maximum allowable operating temperature, which must not be exceeded if thermal failure of the system is to be avoided. We explore the effect of one of these parameters by computing and plotting temperature distributions for values of $h = 200$ and $1000 \text{ W/m}^2 \cdot \text{K}$, which would be representative of air and liquid cooling, respectively.

Note that the shape of the two distributions is identical, with higher temperatures corresponding to the lower value of the convection heat transfer coefficient. Note the slight discontinuity in the temperature gradient, dT/dx, at $x = 50$ mm. What is the physical basis for this discontinuity? We have assumed negligible contact resistance at this location. What would be the effect of such a resistance on the temperature distribution throughout the system? Sketch a representative distribution. What would be the effect on the temperature distribution of an increase in $\dot{q}$, k_A, or k_B? Qualitatively sketch the effect of such changes on the temperature distribution.

4. This example is solved in the *Advanced* section of *IHT*.

3.5.2 Radial Systems

Heat generation may occur in a variety of radial geometries. Consider the long, solid cylinder of Figure 3.11, which could represent a current-carrying wire or a fuel element in a nuclear reactor. For steady-state conditions, the rate at which heat is generated within the cylinder must equal the rate at which heat is convected from the surface of the cylinder to a moving fluid. This condition allows the surface temperature to be maintained at a fixed value of T_s.

To determine the temperature distribution in the cylinder, we begin with the appropriate form of the heat equation. For constant thermal conductivity k, Equation 2.26 reduces to

$$\frac{1}{r}\frac{d}{dr}\left(r\frac{dT}{dr}\right) + \frac{\dot{q}}{k} = 0 \tag{3.54}$$

Separating variables and assuming uniform generation, this expression may be integrated to obtain

$$r\frac{dT}{dr} = -\frac{\dot{q}}{2k}r^2 + C_1 \tag{3.55}$$

Repeating the procedure, the general solution for the temperature distribution becomes

$$T(r) = -\frac{\dot{q}}{4k}r^2 + C_1 \ln r + C_2 \tag{3.56}$$

FIGURE 3.11   Conduction in a solid cylinder with uniform heat generation.

To obtain the constants of integration C_1 and C_2, we apply the boundary conditions

$$\left.\frac{dT}{dr}\right|_{r=0} = 0 \quad \text{and} \quad T(r_0) = T_s$$

The first condition results from the symmetry of the situation. That is, for the solid cylinder the centerline is a line of symmetry for the temperature distribution and the temperature gradient must be zero. Recall that similar conditions existed at the midplane of a wall having symmetrical boundary conditions (Figure 3.10b). From the symmetry condition at $r = 0$ and Equation 3.55, it is evident that $C_1 = 0$. Using the surface boundary condition at $r = r_o$ with Equation 3.56, we then obtain

$$C_2 = T_s + \frac{\dot{q}}{4k}r_o^2 \tag{3.57}$$

The temperature distribution is therefore

$$T(r) = \frac{\dot{q}r_o^2}{4k}\left(1 - \frac{r^2}{r_o^2}\right) + T_s \tag{3.58}$$

Evaluating Equation 3.58 at the centerline and dividing the result into Equation 3.58, we obtain the temperature distribution in nondimensional form,

$$\frac{T(r) - T_s}{T_o - T_s} = 1 - \left(\frac{r}{r_o}\right)^2 \tag{3.59}$$

where T_o is the centerline temperature. The heat rate at any radius in the cylinder may, of course, be evaluated by using Equation 3.58 with Fourier's law.

To relate the surface temperature, T_s, to the temperature of the cold fluid T_∞, either a surface energy balance or an overall energy balance may be used. Choosing the second approach, we obtain

$$\dot{q}(\pi r_o^2 L) = h(2\pi r_o L)(T_s - T_\infty)$$

or

$$T_s = T_\infty + \frac{\dot{q}r_o}{2h} \tag{3.60}$$

3.5.3 Tabulated Solutions

Appendix C provides a convenient and systematic procedure for treating the different combinations of surface conditions that may be applied to one-dimensional planar and radial (cylindrical and spherical) geometries with uniform thermal energy generation. From the tabulated results of this appendix, it is a simple matter to obtain distributions of the temperature, heat flux, and heat rate for boundary conditions of the *second kind* (a uniform surface heat flux) and the *third kind* (a surface heat flux that is proportional to a convection coefficient h or the overall heat transfer coefficient U). You are encouraged to become familiar with the contents of the appendix.

3.5.4 Application of Resistance Concepts

We conclude our discussion of heat generation effects with a word of caution. In particular, when such effects are present, the heat transfer rate is not a constant, independent of the spatial coordinate. Accordingly, it would be *incorrect* to use the conduction resistance concepts and the related heat rate equations developed in Sections 3.1 and 3.3.

EXAMPLE 3.8	IHT

Consider a long solid tube, insulated at the outer radius r_2 and cooled at the inner radius r_1, with uniform heat generation $\dot{q}(\text{W/m}^3)$ within the solid.

1. Obtain the general solution for the temperature distribution in the tube.

2. In a practical application a limit would be placed on the maximum temperature that is permissible at the insulated surface ($r = r_2$). Specifying this limit as $T_{s,2}$, identify appropriate boundary conditions that could be used to determine the arbitrary constants appearing in the general solution. Determine these constants and the corresponding form of the temperature distribution.

3. Determine the heat removal rate per unit length of tube.

4. If the coolant is available at a temperature T_∞, obtain an expression for the convection coefficient that would have to be maintained at the inner surface to allow for operation at prescribed values of $T_{s,2}$ and $\dot{q}$.

SOLUTION

Known: Solid tube with uniform heat generation is insulated at the outer surface and cooled at the inner surface.

Find:
1. General solution for the temperature distribution $T(r)$.
2. Appropriate boundary conditions and the corresponding form of the temperature distribution.
3. Heat removal rate for specified maximum temperature.
4. Corresponding required convection coefficient at the inner surface.

Schematic:

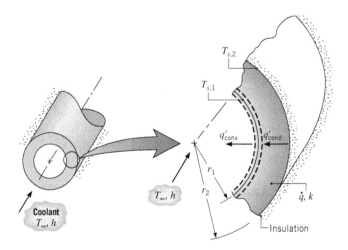

Assumptions:

1. Steady-state conditions.
2. One-dimensional radial conduction.
3. Constant properties.
4. Uniform volumetric heat generation.
5. Outer surface adiabatic.

Analysis:

1. To determine $T(r)$, the appropriate form of the heat equation, Equation 2.26, must be solved. For the prescribed conditions, this expression reduces to Equation 3.54, and the general solution is given by Equation 3.56. Hence, this solution applies in a cylindrical shell, as well as in a solid cylinder (Figure 3.11).

2. Two boundary conditions are needed to evaluate C_1 and C_2, and in this problem it is appropriate to specify both conditions at r_2. Invoking the prescribed temperature limit,

$$T(r_2) = T_{s,2} \tag{1}$$

and applying Fourier's law, Equation 3.29, at the adiabatic outer surface

$$\left. \frac{dT}{dr} \right|_{r_2} = 0 \tag{2}$$

Using Equations 3.56 and 1, it follows that

$$T_{s,2} = -\frac{\dot{q}}{4k} r_2^2 + C_1 \ln r_2 + C_2 \tag{3}$$

Similarly, from Equations 3.55 and 2

$$0 = -\frac{\dot{q}}{2k} r_2^2 + C_1 \tag{4}$$

Hence, from Equation 4,

$$C_1 = \frac{\dot{q}}{2k} r_2^2 \tag{5}$$

and from Equation 3

$$C_2 = T_{s,2} + \frac{\dot{q}}{4k} r_2^2 - \frac{\dot{q}}{2k} r_2^2 \ln r_2 \tag{6}$$

Substituting Equations 5 and 6 into the general solution, Equation 3.56, it follows that

$$T(r) = T_{s,2} + \frac{\dot{q}}{4k}(r_2^2 - r^2) - \frac{\dot{q}}{2k} r_2^2 \ln \frac{r_2}{r} \tag{7}$$

3. The heat removal rate may be determined by obtaining the conduction rate at r_1 or by evaluating the total generation rate for the tube. From Fourier's law

$$q_r' = -k 2\pi r \frac{dT}{dr}$$

Hence, substituting from Equation 7 and evaluating the result at r_1,

$$q_r'(r_1) = -k 2\pi r_1 \left(-\frac{\dot{q}}{2k} r_1 + \frac{\dot{q}}{2k} \frac{r_2^2}{r_1} \right) = -\pi \dot{q}(r_2^2 - r_1^2) \tag{8}$$

Alternatively, because the tube is insulated at r_2, the rate at which heat is generated in the tube must equal the rate of removal at r_1. That is, for a control volume about the tube, the energy conservation requirement, Equation 1.12c, reduces to $\dot{E}_g - \dot{E}_{out} = 0$, where $\dot{E}_g = \dot{q}\pi(r_2^2 - r_1^2)L$ and $\dot{E}_{out} = q_{cond}'L = -q_r'(r_1)L$. Hence

$$q_r'(r_1) = -\pi \dot{q}(r_2^2 - r_1^2) \tag{9}$$

4. Applying the energy conservation requirement, Equation 1.13, to the inner surface, it follows that

$$q_{cond}' = q_{conv}'$$

or

$$\pi \dot{q}(r_2^2 - r_1^2) = h 2\pi r_1(T_{s,1} - T_\infty)$$

Hence

$$h = \frac{\dot{q}(r_2^2 - r_1^2)}{2r_1(T_{s,1} - T_\infty)} \tag{10}$$

where $T_{s,1}$ may be obtained by evaluating Equation 7 at $r = r_1$.

Comments:

1. Note that, through application of Fourier's law in part 3, the sign on $q_r'(r_1)$ was found to be negative, Equation 8, implying that heat flow is in the negative r-direction. However, in applying the energy balance, we acknowledged that heat flow was *out* of the wall. Hence we expressed q_{cond}' as $-q_r'(r_1)$ and we expressed q_{conv}' in terms of $(T_{s,1} - T_\infty)$, rather than $(T_\infty - T_{s,1})$.

2. Results of the foregoing analysis may be used to determine the convection coefficient required to maintain the maximum tube temperature $T_{s,2}$ below a prescribed value. Consider a tube of thermal conductivity $k = 5$ W/m · K and inner and outer radii of $r_1 = 20$ mm and $r_2 = 25$ mm, respectively, with a maximum allowable temperature of $T_{s,2} = 350°C$. The tube experiences heat generation at a rate of $\dot{q} = 5 \times 10^6$ W/m³, and the coolant is at a temperature of $T_\infty = 80°C$. Obtaining $T(r_1) = T_{s,1} = 336.5°C$ from Equation 7 and substituting into Equation 10, the required convection coefficient is found to be $h = 110$ W/m² · K. Using the *IHT Workspace*, parametric calculations may be performed to determine the effects of the convection coefficient and the generation rate on the maximum tube temperature, and results are plotted as a function of h for three values of $\dot{q}$.

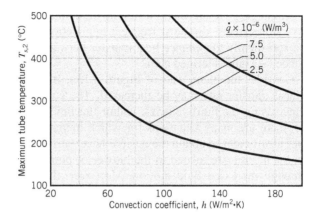

For each generation rate, the minimum value of h needed to maintain $T_{s,2} \leq 350°C$ may be determined from the figure.

3. The temperature distribution, Equation 7, may also be obtained by using the results of Appendix C. Applying a surface energy balance at $r = r_1$, with $q(r) = -\dot{q}\pi(r_2^2 - r_1^2)L$, $(T_{s,2} - T_{s,1})$ may be determined from Equation C.8 and the result substituted into Equation C.2 to eliminate $T_{s,1}$ and obtain the desired expression.

3.6 *Heat Transfer from Extended Surfaces*

The term *extended surface* is commonly used to depict an important special case involving heat transfer by conduction within a solid and heat transfer by convection (and/or radiation) from the boundaries of the solid. Until now, we have considered heat transfer from the boundaries of a solid to be in the same direction as heat transfer by conduction in the solid. In contrast, for an extended surface, the direction of heat transfer from the boundaries is perpendicular to the principal direction of heat transfer in the solid.

Consider a strut that connects two walls at different temperatures and across which there is fluid flow (Figure 3.12). With $T_1 > T_2$, temperature gradients in the x-direction sustain heat transfer by conduction in the strut. However, with $T_1 > T_2 > T_\infty$, there is concurrent heat transfer by convection to the fluid, causing q_x, and hence the magnitude of the temperature gradient, $|dT/dx|$, to decrease with increasing x.

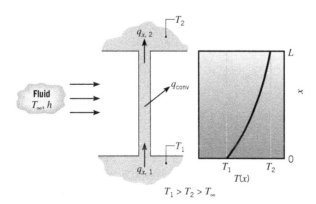

FIGURE 3.12 Combined conduction and convection in a structural element.

Although there are many different situations that involve such combined conduction–convection effects, the most frequent application is one in which an extended surface is used specifically to *enhance* heat transfer between a solid and an adjoining fluid. Such an extended surface is termed a *fin*.

Consider the plane wall of Figure 3.13*a*. If T_s is fixed, there are two ways in which the heat transfer rate may be increased. The convection coefficient h could be increased by increasing the fluid velocity, and/or the fluid temperature T_∞ could be reduced. However, there are many situations for which increasing h to the maximum possible value is either insufficient to obtain the desired heat transfer rate or the associated costs are prohibitive. Such costs are related to the blower or pump power requirements needed to increase h through increased fluid motion. Moreover, the second option of reducing T_∞ is often impractical. Examining Figure 3.13*b*, however, we see that there exists a third option. That is, the heat transfer rate may be increased by increasing the surface area across which the convection occurs. This may be done by employing *fins* that *extend* from the wall into the surrounding fluid. The thermal conductivity of the fin material can have a strong effect on the temperature distribution along the fin and therefore influences the degree to which the heat transfer rate is enhanced. Ideally, the fin material should have a large thermal conductivity to minimize temperature variations from its base to its tip. In the limit of infinite thermal conductivity, the entire fin would be at the temperature of the base surface, thereby providing the maximum possible heat transfer enhancement.

Examples of fin applications are easy to find. Consider the arrangement for cooling engine heads on motorcycles and lawn mowers or for cooling electric power transformers. Consider also the tubes with attached fins used to promote heat exchange between air and the working fluid of an air conditioner. Two common finned-tube arrangements are shown in Figure 3.14.

Different fin configurations are illustrated in Figure 3.15. A *straight fin* is any extended surface that is attached to a *plane wall*. It may be of uniform cross-sectional area, or its cross-sectional area may vary with the distance x from the wall. An *annular fin* is one that is circumferentially attached to a cylinder, and its cross section varies with radius from the wall of the cylinder. The foregoing fin types have rectangular cross sections, whose area may be expressed as a product of the fin thickness t and the width w for straight fins or the circumference $2\pi r$ for annular fins. In contrast a *pin fin*, or *spine,* is an extended surface of circular cross section. Pin fins may be of uniform or nonuniform cross section. In any application,

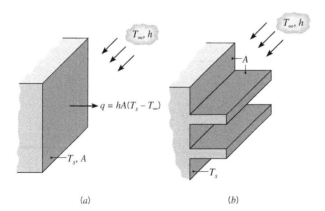

(*a*)   (*b*)

FIGURE 3.13   Use of fins to enhance heat transfer from a plane wall. (*a*) Bare surface. (*b*) Finned surface.

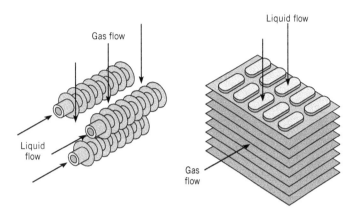

FIGURE 3.14 Schematic of typical finned-tube heat exchangers.

selection of a particular fin configuration may depend on space, weight, manufacturing, and cost considerations, as well as on the extent to which the fins reduce the surface convection coefficient and increase the pressure drop associated with flow over the fins.

3.6.1 A General Conduction Analysis

As engineers we are primarily interested in knowing the extent to which particular extended surfaces or fin arrangements could improve heat transfer from a surface to the surrounding fluid. To determine the heat transfer rate associated with a fin, we must first obtain the temperature distribution along the fin. As we have done for previous systems, we begin by performing an energy balance on an appropriate differential element. Consider the extended surface of Figure 3.16. The analysis is simplified if certain assumptions are made. We choose to assume one-dimensional conditions in the longitudinal (x-) direction, even though conduction within the fin is actually two-dimensional. The rate at which energy is convected to the fluid from any point on the fin surface must be balanced by the net rate at which energy reaches that point due to conduction in the transverse (y-, z-) direction. However, in practice the fin is thin, and temperature changes in the transverse direction within

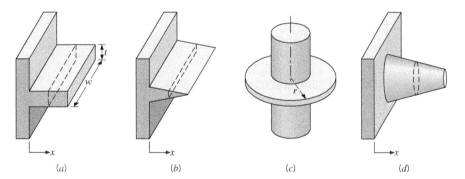

FIGURE 3.15 Fin configurations. (*a*) Straight fin of uniform cross section. (*b*) Straight fin of nonuniform cross section. (*c*) Annular fin. (*d*) Pin fin.

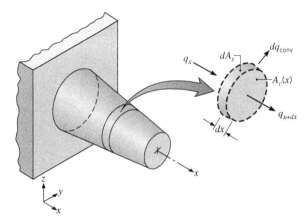

FIGURE 3.16 Energy balance for an extended surface.

the fin are small compared with the temperature difference between the fin and the environment. Hence, we may assume that the temperature is uniform across the fin thickness, that is, it is only a function of x. We will consider steady-state conditions and also assume that the thermal conductivity is constant, that radiation from the surface is negligible, that heat generation effects are absent, and that the convection heat transfer coefficient h is uniform over the surface.

Applying the conservation of energy requirement, Equation 1.12c, to the differential element of Figure 3.16, we obtain

$$q_x = q_{x+dx} + dq_{\text{conv}} \tag{3.61}$$

From Fourier's law we know that

$$q_x = -kA_c \frac{dT}{dx} \tag{3.62}$$

where A_c is the *cross-sectional* area, which may vary with x. Since the conduction heat rate at $x + dx$ may be expressed as

$$q_{x+dx} = q_x + \frac{dq_x}{dx} dx \tag{3.63}$$

it follows that

$$q_{x+dx} = -kA_c \frac{dT}{dx} - k \frac{d}{dx} \left(A_c \frac{dT}{dx} \right) dx \tag{3.64}$$

The convection heat transfer rate may be expressed as

$$dq_{\text{conv}} = hdA_s(T - T_\infty) = hPdx(T - T_\infty) \tag{3.65}$$

where dA_s is the *surface* area of the differential element and P is its perimeter. Substituting the foregoing rate equations into the energy balance, Equation 3.61, we obtain

$$\frac{d}{dx} \left(A_c \frac{dT}{dx} \right) - \frac{hP}{k} (T - T_\infty) = 0$$

or

$$\frac{d^2T}{dx^2} + \left(\frac{1}{A_c} \frac{dA_c}{dx} \right) \frac{dT}{dx} - \frac{hP}{kA_c}(T - T_\infty) = 0 \tag{3.66}$$

This result provides a general form of the energy equation for an extended surface. Its solution for appropriate boundary conditions provides the temperature distribution, which may be used with Equation 3.62 to calculate the conduction rate at any x.

3.6.2 Fins of Uniform Cross-Sectional Area

To solve Equation 3.66 it is necessary to be more specific about the geometry. We begin with the simplest case of straight rectangular and pin fins of uniform cross section (Figure 3.17). Each fin is attached to a base surface of temperature $T(0) = T_b$ and extends into a fluid of temperature T_∞.

For the prescribed fins, A_c and P are constant. Accordingly, with $dA_c/dx = 0$, Equation 3.66 reduces to

$$\frac{d^2T}{dx^2} - \frac{hP}{kA_c}(T - T_\infty) = 0 \tag{3.67}$$

To simplify the form of this equation, we transform the dependent variable by defining an *excess temperature* θ as

$$\theta(x) \equiv T(x) - T_\infty \tag{3.68}$$

where, since T_∞ is a constant, $d\theta/dx = dT/dx$. Substituting Equation 3.68 into Equation 3.67, we then obtain

$$\frac{d^2\theta}{dx^2} - m^2\theta = 0 \tag{3.69}$$

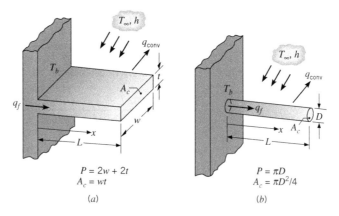

FIGURE 3.17 Straight fins of uniform cross section. (*a*) Rectangular fin. (*b*) Pin fin.

where

$$m^2 \equiv \frac{hP}{kA_c} \tag{3.70}$$

Equation 3.69 is a linear, homogeneous, second-order differential equation with constant coefficients. Its general solution is of the form[1]

$$\theta(x) = C_1 e^{mx} + C_2 e^{-mx} \tag{3.71}$$

To evaluate the constants C_1 and C_2 of Equation 3.71, it is necessary to specify appropriate boundary conditions. One such condition may be specified in terms of the temperature at the *base* of the fin ($x = 0$)

$$\theta(0) = T_b - T_\infty \equiv \theta_b \tag{3.72}$$

The second condition, specified at the fin tip ($x = L$), may correspond to one of four different physical situations.

The first condition, Case A, considers convection heat transfer from the fin tip. Applying an energy balance to a control surface about this tip (Figure 3.18), we obtain

$$hA_c[T(L) - T_\infty] = -kA_c \left.\frac{dT}{dx}\right|_{x=L}$$

or

$$h\theta(L) = -k \left.\frac{d\theta}{dx}\right|_{x=L} \tag{3.73}$$

That is, the rate at which energy is transferred to the fluid by convection from the tip must equal the rate at which energy reaches the tip by conduction through the fin. Substituting Equation 3.71 into Equations 3.72 and 3.73, we obtain, respectively,

$$\theta_b = C_1 + C_2 \tag{3.74}$$

and

$$h(C_1 e^{mL} + C_2 e^{-mL}) = km(C_2 e^{-mL} - C_1 e^{mL})$$

Solving for C_1 and C_2, it may be shown, after some manipulation, that

$$\frac{\theta}{\theta_b} = \frac{\cosh m(L - x) + (h/mk)\sinh m(L - x)}{\cosh mL + (h/mk)\sinh mL} \tag{3.75}$$

The form of this temperature distribution is shown schematically in Figure 3.18. Note that the magnitude of the temperature gradient decreases with increasing x. This trend is a consequence of the reduction in the conduction heat transfer rate $q_x(x)$ with increasing x due to continuous convection losses from the fin surface.

---

[1]By substitution it may readily be verified that Equation 3.71 is indeed a solution to Equation 3.69.

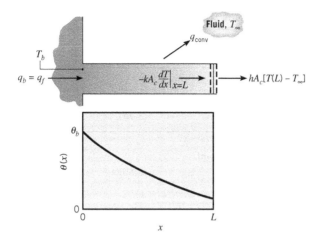

Figure 3.18 Conduction and convection in a fin of uniform cross section.

We are particularly interested in the rate of heat transferred from the entire fin. From Figure 3.18, it is evident that the fin heat transfer rate q_f may be evaluated in two alternative ways, both of which involve use of the temperature distribution. The simpler procedure, and the one that we will use, involves applying Fourier's law at the fin base. That is,

$$q_f = q_b = -kA_c \frac{dT}{dx}\bigg|_{x=0} = -kA_c \frac{d\theta}{dx}\bigg|_{x=0} \tag{3.76}$$

Hence, knowing the temperature distribution, $\theta(x)$, q_f may be evaluated, giving

$$q_f = \sqrt{hPkA_c}\,\theta_b \frac{\sinh mL + (h/mk)\cosh mL}{\cosh mL + (h/mk)\sinh mL} \tag{3.77}$$

Alternatively, conservation of energy dictates that the rate at which heat is transferred by convection from the fin must equal the rate at which it is conducted through the base of the fin. Accordingly, the alternative formulation for q_f is

$$q_f = \int_{A_f} h[T(x) - T_\infty]\, dA_s$$

$$q_f = \int_{A_f} h\theta(x)\, dA_s \tag{3.78}$$

where A_f is the *total,* including the tip, *fin surface area.* Substitution of Equation 3.75 into Equation 3.78 would yield Equation 3.77. Equations 3.75 and 3.77 are shown in Table 3.4 where $M = \sqrt{hPkA_c}\,\theta_b$.

The second tip condition, Case B, corresponds to the assumption that convective heat transfer at the fin tip is negligible, in which case the tip may be treated as adiabatic and

$$\frac{d\theta}{dx}\bigg|_{x=L} = 0 \tag{3.79}$$

Substituting from Equation 3.71 and dividing by m, we then obtain

$$C_1 e^{mL} - C_2 e^{-mL} = 0$$

TABLE 3.4   Temperature distributions and heat rates for fins of uniform cross section

Case	Tip Condition (x = L)	Temperature Distribution θ/θ_b	Fin Heat Transfer Rate q_f
A	Convection: $h\theta(L) = -kd\theta/dx\|_{x=L}$	$\dfrac{\cosh m(L-x) + (h/mk)\sinh m(L-x)}{\cosh mL + (h/mk)\sinh mL}$ (3.75)	$M\dfrac{\sinh mL + (h/mk)\cosh mL}{\cosh mL + (h/mk)\sinh mL}$ (3.77)
B	Adiabatic: $d\theta/dx\|_{x=L} = 0$	$\dfrac{\cosh m(L-x)}{\cosh mL}$ (3.80)	$M\tanh mL$ (3.81)
C	Prescribed temperature: $\theta(L) = \theta_L$	$\dfrac{(\theta_L/\theta_b)\sinh mx + \sinh m(L-x)}{\sinh mL}$ (3.82)	$M\dfrac{(\cosh mL - \theta_L/\theta_b)}{\sinh mL}$ (3.83)
D	Infinite fin ($L \to \infty$): $\theta(L) = 0$	e^{-mx} (3.84)	M (3.85)

$\theta \equiv T - T_\infty$ $m^2 \equiv hP/kA_c$

$\theta_b = \theta(0) = T_b - T_\infty$ $M \equiv \sqrt{hPkA_c}\,\theta_b$

A table of hyperbolic functions is given in Appendix B.1.

Using this expression with Equation 3.74 to solve for C_1 and C_2 and substituting the results into Equation 3.71, we obtain the fin temperature distribution, Equation 3.80 of Table 3.4. Using Equation 3.80 with Equation 3.76, the fin heat transfer rate is then given by Equation 3.81 of Table 3.4.

In the same manner, we can obtain the fin temperature distribution and fin heat transfer rate for Case C, where the temperature is prescribed at the fin tip. That is, the second boundary condition is $\theta(L) = \theta_L$, and the resulting expressions for the temperature distribution and heat rate are Equations 3.82 and 3.83, respectively, of Table 3.4. The very long fin, Case D, is an interesting extension of the preceding results. In particular, as, $L \to \infty$, $\theta_L \to 0$ and it is easily verified that the fin temperature distribution and heat rate are given by Equations 3.84 and 3.85, respectively, of Table 3.4.

EXAMPLE 3.9

A very long rod 5 mm in diameter has one end maintained at 100°C. The surface of the rod is exposed to ambient air at 25°C with a convection heat transfer coefficient of 100 W/m² · K.

1. Determine the temperature distributions along rods constructed from pure copper, 2024 aluminum alloy, and type AISI 316 stainless steel. What are the corresponding rates of heat loss from the rods?

2. Estimate how long the rods must be for the assumption of *infinite length* to yield an accurate estimate of the rate of heat loss.

SOLUTION

Known: A long circular rod exposed to ambient air.

Find:

1. Temperature distribution and rate of heat loss when rod is fabricated from copper, an aluminum alloy, or stainless steel.

2. How long rods must be to assume infinite length.

Schematic:

$T_b = 100°C$

Air

$T_\infty = 25°C$
$h = 100 \ W/m^2 \cdot K$

$k, L \rightarrow \infty, D = 5 \ mm$

Assumptions:

1. Steady-state conditions.

2. Temperature is uniform across the rod thickness.

3. Constant properties.

4. Negligible radiation exchange with surroundings.

5. Uniform heat transfer coefficient.

6. Infinitely long rod.

Properties: Table A.1, copper [$T = (T_b + T_\infty)/2 = 62.5°C \approx 335$ K]: $k = 398$ W/m · K. Table A.1, 2024 aluminum (335 K): $k = 180$ W/m · K. Table A.1, stainless steel, AISI 316 (335 K): $k = 14$ W/m · K.

Analysis:

1. Subject to the assumption of an infinitely long fin, the temperature distributions are determined from Equation 3.84, which may be expressed as

$$T = T_\infty + (T_b - T_\infty)e^{-mx}$$

where $m = (hP/kA_c)^{1/2} = (4h/kD)^{1/2}$. Substituting for h and D, as well as for the thermal conductivities of copper, the aluminum alloy, and the stainless steel, respectively, the values of m are 14.2, 21.2, and 75.6 m^{-1}. The temperature distributions may then be computed and plotted as follows:

From these distributions, it is evident that there is little additional heat transfer associated with extending the length of the rod much beyond 50, 200, and 300 mm, respectively, for the stainless steel, the aluminum alloy, and the copper.

From Equation 3.85, the rate of heat loss is

$$q_f = \sqrt{hPkA_c}\,\theta_b$$

Hence for copper,

$$
q_f = \Bigg[\, 100\ \text{W/m}^2 \cdot \text{K} \times \pi \times 0.005\ \text{m}
$$

$$
\times\, 398\ \text{W/m} \cdot \text{K} \times \frac{\pi}{4}(0.005\ \text{m})^2 \Bigg]^{1/2} (100-25)^\circ\text{C}
$$

$$
= 8.3\,\text{W} \qquad \qquad \qquad \qquad \triangleleft
$$

Similarly, for the aluminum alloy and stainless steel, respectively, the heat rates are $q_f = 5.6$ W and 1.6 W.

2. The validity of the infinite length assumption may be assessed by comparing Equations 3.81 and 3.85. To a satisfactory approximation, the expressions provide equivalent results if tanh $mL \geq 0.99$ or $mL \geq 2.65$. Hence a rod may be assumed to be infinitely long if

$$
L \geq L_\infty \equiv \frac{2.65}{m} = 2.65 \left(\frac{kA_c}{hP}\right)^{1/2}
$$

For copper,

$$
L_\infty = 2.65 \left[\frac{398\ \text{W/m} \cdot \text{K} \times (\pi/4)(0.005\ \text{m})^2}{100\ \text{W/m}^2 \cdot \text{K} \times \pi(0.005\ \text{m})}\right]^{1/2} = 0.19\ \text{m} \qquad \triangleleft
$$

Results for the aluminum alloy and stainless steel are $L_\infty = 0.13$ m and $L_\infty = 0.04$ m, respectively.

Comments:

1. The foregoing results suggest that the fin heat transfer rate may be accurately predicted from the infinite fin approximation if $mL \geq 2.65$. However, if the infinite fin approximation is to accurately predict the temperature distribution $T(x)$, a larger value of mL would be required. This value may be inferred from Equation 3.84 and the requirement that the tip temperature be close to the fluid temperature. For example, if we require that $\theta(L)/\theta_b = \exp(-mL) < 0.01$, it follows that $mL > 4.6$, in which case $L_\infty \approx 0.33$, 0.23, and 0.07 m for the copper, aluminum alloy, and stainless steel, respectively. These results are consistent with the distributions plotted in part 1.

2. This example is solved in the *Advanced* section of *IHT*.

3.6.3 Fin Performance Parameters

Recall that fins are used to enhance the heat transfer from a surface by increasing the effective surface area. However, the fin itself poses a conduction resistance. For this reason, there is no assurance that the heat transfer rate will increase through the use of fins.

Fin Effectiveness and Fin Resistance   One assessment of fin performance may be made by evaluating the fin effectiveness, ε_f. It is defined, for a surface of constant base temperature, as *the ratio of the fin heat transfer rate to the heat transfer rate that would exist without the fin.* Therefore

$$\varepsilon_f = \frac{q_f}{hA_{c,b}\theta_b} \tag{3.86}$$

where $A_{c,b}$ is the fin cross-sectional area at the base. In general, the use of fins may rarely be justified unless $\varepsilon_f \gtrsim 2$.

Subject to any one of the four tip conditions that have been considered in Section 3.6.2, the effectiveness for a fin of uniform cross section may be obtained by dividing the appropriate expression for q_f in Table 3.4 by $hA_{c,b}\theta_b$.[2] For example, for an infinitely long fin (Case D of Table 3.4) and assuming the convection coefficient of the finned surface is the same as that of the unfinned base, it follows that

$$\varepsilon_f = \left(\frac{kP}{hA_c} \right)^{1/2} \tag{3.87}$$

Several important trends can be inferred from this expression. Obviously, fin effectiveness is enhanced by the choice of a material of high thermal conductivity. Aluminum alloys and copper come to mind. However, although copper is superior from the standpoint of thermal conductivity, aluminum alloys are the more common choice because of additional benefits related to lower cost and weight. Fin effectiveness is also enhanced by increasing the ratio of the perimeter to the cross-sectional area. For this reason, the use of *thin,* but closely spaced fins, is preferred, with the proviso that the fin gap not be reduced to a value for which flow between the fins is severely impeded, thereby reducing the convection coefficient.

Equation 3.87 also suggests that the use of fins can be better justified under conditions for which the convection coefficient h is small. Hence from Table 1.1 it is evident that the need for fins is stronger when the fluid is a gas rather than a liquid. If fins are to be used on a surface separating a gas and a liquid, they are generally placed on the gas side. A common example is the tubing in an automobile radiator. Fins are applied to the outer tube surface, over which there is flow of ambient air (small h), and not to the inner surface, through which there is flow of water (large h). Note that, if $\varepsilon_f > 2$ is used as a criterion to justify the implementation of fins, Equation 3.87 yields the requirement that $(kP/hA_c) > 4$.

Equation 3.87 provides an upper limit to ε_f, which is reached as L approaches infinity. However, it is certainly not necessary to use very long fins to achieve near maximum

---

[2]Although the installation of fins will alter the surface convection coefficient, this effect is commonly neglected.

heat transfer enhancement. As seen in Example 3.9, 99% of the maximum possible fin heat transfer rate is achieved for $mL = 2.65$. Hence, it would make no sense to extend the fins beyond $L = 2.65/m$.

Fin performance may also be quantified in terms of a thermal resistance. Treating the difference between the base and fluid temperatures as the driving potential, a *fin resistance* may be defined as

$$R_{t,f} = \frac{\theta_b}{q_f} \tag{3.88}$$

This result is extremely useful, particularly when representing a finned surface with a thermal circuit. Note that, depending on the fin tip condition, an appropriate expression for q_f may be obtained from Table 3.4.

Combining the expression for the thermal resistance due to convection at the exposed base,

$$R_{t,b} = \frac{1}{hA_{c,b}} \tag{3.89}$$

with Equations 3.86 and 3.88 yields

$$\varepsilon_f = \frac{R_{t,b}}{R_{t,f}} \tag{3.90}$$

Hence the fin effectiveness may be interpreted as a ratio of thermal resistances, and to increase ε_f it is necessary to reduce the conduction/convection resistance of the fin. If the fin is to enhance heat transfer, its resistance must not exceed that of the exposed base.

Fin Efficiency   Another measure of fin performance is provided by the fin efficiency, η_f. A logical definition of the fin efficiency is the actual fin heat transfer rate, q_f, divided by the maximum possible heat transfer rate. The maximum heat transfer rate from the fin would occur *if* it were entirely at the base temperature, T_b. Therefore

$$n_f \equiv \frac{q_f}{q_{max}} = \frac{q_f}{hA_f(T_b - T_\infty)} = \frac{q_f}{hA_f\theta_b} \tag{3.91}$$

where A_f is the surface area of the fin. For a straight fin of uniform cross section and an adiabatic tip, Equations 3.81 and 3.91 yield

$$\eta_f = \frac{M \tanh mL}{hPL\theta_b} = \frac{\tanh mL}{mL} \tag{3.92}$$

Referring to Table B.1, this result tells us that η_f approaches its maximum and minimum values of 1 and 0, respectively, as L approaches 0 and ∞. Knowledge of the fin efficiency may be used to evaluate the fin resistance, where, from Equations 3.88 and 3.91, it follows that

$$R_{t,f} = \frac{1}{hA_f\eta_f} \tag{3.93}$$

Corrected Fin Lengths In lieu of the somewhat cumbersome expression for heat transfer from a straight rectangular fin with an active tip, Equation 3.77, it has been shown that approximate, yet accurate, predictions may be obtained by using the adiabatic tip result, Equation 3.81, with a corrected fin length of the form $L_c = L + (t/2)$ for a rectangular fin and $L_c = L + (D/4)$ for a pin fin [14]. The correction is based on assuming equivalence between heat transfer from the actual fin with tip convection and heat transfer from a longer, hypothetical fin with an adiabatic tip. Hence, with tip convection, the fin heat rate may be approximated as

$$q_f = M \tanh mL_c \tag{3.94}$$

and the corresponding efficiency as

$$\eta_f = \frac{\tanh mL_c}{mL_c} \tag{3.95}$$

Errors associated with the approximation are negligible if (ht/k) or $(hD/2k) \leqslant 0.0625$ [15].

If the width of a rectangular fin is much larger than its thickness, $w \gg t$, the perimeter may be approximated as $P = 2w$, and

$$mL_c = \left(\frac{hP}{kA_c}\right)^{1/2} L_c = \left(\frac{2h}{kt}\right)^{1/2} L_c$$

Multiplying numerator and denominator by $L_c^{1/2}$ and introducing a corrected fin profile area, $A_p = L_c t$, it follows that

$$mL_c = \left(\frac{2h}{kA_p}\right)^{1/2} L_c^{3/2} \tag{3.96}$$

Hence, as shown in Figure 3.19, the efficiency of a straight rectangular fin with tip convection may be represented as a function of $L_c^{3/2}(h/kA_p)^{1/2}$. The same functional dependence applies to all the fin types represented in Figures 3.19 and 3.20.

FIGURE 3.19 Efficiency of straight fins (rectangular, triangular, and parabolic profiles).

FIGURE 3.20   Efficiency of annular fins of rectangular profile.

3.6.4   Fins of Nonuniform Cross-Sectional Area

The second term of Equation 3.66 must be retained for fins of nonuniform cross-sectional area, and the solutions are no longer in the form of simple exponential or hyperbolic functions. As a special case, consider the annular fin shown in the inset of Figure 3.20. Although the fin thickness is uniform (t is independent of r), the cross-sectional area, $A_c = 2\pi r t$, varies with r. Replacing x by r in Equation 3.66 and expressing the surface area as $A_s = 2\pi(r^2 - r_1^2)$, the general form of the fin equation reduces to

$$\frac{d^2T}{dr^2} + \frac{1}{r}\frac{dT}{dr} - \frac{2h}{kt}(T - T_\infty) = 0$$

or, with $m^2 \equiv 2h/kt$ and $\theta \equiv T - T_\infty$,

$$\frac{d^2\theta}{dr^2} + \frac{1}{r}\frac{d\theta}{dr} - m^2\theta = 0$$

The foregoing expression is a *modified Bessel equation* of order zero, and its general solution is of the form

$$\theta(r) = C_1 I_0(mr) + C_2 K_0(mr)$$

where I_0 and K_0 are modified, zero-order Bessel functions of the first and second kinds, respectively. If the temperature at the base of the fin is prescribed, $\theta(r_1) = \theta_b$, and an adiabatic tip is presumed, $d\theta/dr|_{r_2} = 0$, C_1 and C_2 may be evaluated to yield a temperature distribution of the form

$$\frac{\theta}{\theta_b} = \frac{I_0(mr)K_1(mr_2) + K_0(mr)I_1(mr_2)}{I_0(mr_1)K_1(mr_2) + K_0(mr_1)I_1(mr_2)}$$

where $I_1(mr) = d[I_0(mr)]/d(mr)$ and $K_1(mr) = -d[K_0(mr)]/d(mr)$ are modified, first-order Bessel functions of the first and second kinds, respectively. The Bessel functions are tabulated in Appendix B.

With the fin heat transfer rate expressed as

$$q_f = -kA_{c,b} \frac{dT}{dr}\bigg|_{r=r_1} = -k(2\pi r_1 t)\frac{d\theta}{dr}\bigg|_{r=r_1}$$

it follows that

$$q_f = 2\pi k r_1 t \theta_b m \frac{K_1(mr_1)I_1(mr_2) - I_1(mr_1)K_1(mr_2)}{K_0(mr_1)I_1(mr_2) + I_0(mr_1)K_1(mr_2)}$$

from which the fin efficiency becomes

$$\eta_f = \frac{q_f}{h2\pi(r_2^2 - r_1^2)\theta_b} = \frac{2r_1}{m(r_2^2 - r_1^2)} \frac{K_1(mr_1)I_1(mr_2) - I_1(mr_1)K_1(mr_2)}{K_0(mr_1)I_1(mr_2) + I_0(mr_1)K_1(mr_2)} \quad (3.97)$$

This result may be applied for an active (convecting) tip, if the tip radius r_2 is replaced by a corrected radius of the form $r_{2c} = r_2 + (t/2)$. Results for the annular fin are represented graphically in Figure 3.20. Figure 3.19 includes results for triangular and parabolic straight fins.

Expressions for the efficiency and surface area of several common fin geometries are summarized in Table 3.5. Although results for the fins of uniform thickness or diameter were obtained by assuming an adiabatic tip, the effects of convection may be treated by using a corrected length (Equations 3.95 and 3.100) or radius (Equation 3.97). The

TABLE 3.5   Efficiency of common fin shapes

Straight Fins

Rectangular[a]
$A_f = 2wL_c$
$L_c = L + (t/2)$
$A_p = tL$

$$\eta_f = \frac{\tanh mL_c}{mL_c} \quad (3.95)$$

Triangular[a]
$A_f = 2w[L^2 + (t/2)^2]^{1/2}$
$A_p = (t/2)L$

$$\eta_f = \frac{1}{mL} \frac{I_1(2mL)}{I_0(2mL)} \quad (3.98)$$

Parabolic[a]
$A_f = w[C_1L + (L^2/t)\ln(t/L + C_1)]$
$C_1 = [1 + (t/L)^2]^{1/2}$
$A_p = (t/3)L$

$$\eta_f = \frac{2}{[4(mL)^2 + 1]^{1/2} + 1} \quad (3.99)$$

(continued)

TABLE 3.5 *Continued*

Annular Fin

Rectangular[a]

$A_f = 2\pi \, (r_{2c}^2 - r_1^2)$

$r_{2c} = r_2 + (t/2)$

$V = \pi \, (r_2^2 - r_1^2)t$

$$\eta_f = C_2 \, \frac{K_1(mr_1)I_1(mr_{2c}) - I_1(mr_1)K_1(mr_{2c})}{I_0(mr_1)K_1(mr_{2c}) + K_0(mr_1)I_1(mr_{2c})} \quad (3.97)$$

$$C_2 = \frac{(2r_1/m)}{(r_{2c}^2 - r_1^2)}$$

Pin Fins

Rectangular[b]

$A_f = \pi D L_c$

$L_c = L + (D/4)$

$V = (\pi D^2/4)L$

$$\eta_f = \frac{\tanh mL_c}{mL_c} \quad (3.100)$$

Triangular[b]

$A_f = \dfrac{\pi D}{2}[L^2 + (D/2)^2]^{1/2}$

$V = (\pi/12)D^2 L$

$$\eta_f = \frac{2}{mL} \frac{I_2(2mL)}{I_1(2mL)} \quad (3.101)$$

Parabolic[b]

$A_f = \dfrac{\pi L^3}{8D} \Big\{ C_3 C_4 - $

$\qquad \dfrac{L}{2D} \ln \left[(2DC_4/L) + C_3 \right] \Big\}$

$y = (D/2)(1 - x/L)^2$

$$\eta_f = \frac{2}{[4/9(mL)^2 + 1]^{1/2} + 1} \quad (3.102)$$

$C_3 = 1 + 2(D/L)^2$

$C_4 = [1 + (D/L)^2]^{1/2}$

$V = (\pi/20)D^2 L$

---

[a] $m = (2h/kt)^{1/2}$.

[b] $m = (4h/kD)^{1/2}$.

triangular and parabolic fins are of nonuniform thickness that reduces to zero at the fin tip. Expressions for the profile area, A_p, or the volume, V, of a fin are also provided in Table 3.5. The volume of a straight fin is simply the product of its width and profile area, $V = wA_p$.

Fin design is often motivated by a desire to minimize the fin material and/or related manufacturing costs required to achieve a prescribed cooling effectiveness. Hence, a straight *triangular* fin is attractive because, for equivalent heat transfer, it requires much less volume (fin material) than a rectangular profile. In this regard, heat dissipation per unit volume, $(q/V)_f$, is largest for a *parabolic* profile. However, since $(q/V)_f$ for the parabolic profile is only slightly larger than that for a triangular profile, its use can rarely be justified in view of its larger manufacturing costs. The *annular* fin of rectangular profile is commonly used to enhance heat transfer to or from circular tubes.

3.6.5 Overall Surface Efficiency

In contrast to the fin efficiency η_f, which characterizes the performance of a single fin, the *overall surface efficiency* η_o characterizes an *array* of fins and the base surface to which they are attached. Representative arrays are shown in Figure 3.21, where S designates the fin pitch. In each case the overall efficiency is defined as

$$\eta_o = \frac{q_t}{q_{max}} = \frac{q_t}{hA_t\theta_b} \tag{3.103}$$

where q_t is the total heat rate from the surface area A_t associated with both the fins and the exposed portion of the base (often termed the *prime* surface). If there are N fins in the array, each of surface area A_f, and the area of the prime surface is designated as A_b, the total surface area is

$$A_t = NA_f + A_b \tag{3.104}$$

The maximum possible heat rate would result if the entire fin surface, as well as the exposed base, were maintained at T_b.

The total rate of heat transfer by convection from the fins and the prime (unfinned) surface may be expressed as

$$q_t = N\eta_f hA_f\theta_b + hA_b\theta_b \tag{3.105}$$

where the convection coefficient h is assumed to be equivalent for the finned and prime surfaces and η_f is the efficiency of a single fin. Hence

$$q_t = h[N\eta_f A_f + (A_t - NA_f)]\theta_b = hA_t\left[1 - \frac{NA_f}{A_t}(1 - \eta_f)\right]\theta_b \tag{3.106}$$

Substituting Equation (3.106) into (3.103), it follows that

$$\eta_o = 1 - \frac{NA_f}{A_t}(1 - \eta_f) \tag{3.107}$$

From knowledge of η_o, Equation 3.103 may be used to calculate the total heat rate for a fin array.

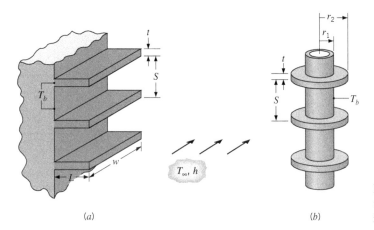

(a)   (b)

FIGURE 3.21 Representative fin arrays. (a) Rectangular fins. (b) Annular fins.

Recalling the definition of the fin thermal resistance, Equation 3.88, Equation 3.103 may be used to infer an expression for the thermal resistance of a fin array. That is,

$$R_{t,o} = \frac{\theta_b}{q_t} = \frac{1}{\eta_o h A_t} \tag{3.108}$$

where $R_{t,o}$ is an effective resistance that accounts for parallel heat flow paths by conduction/convection in the fins and by convection from the prime surface. Figure 3.22 illustrates the thermal circuits corresponding to the parallel paths and their representation in terms of an effective resistance.

If fins are machined as an integral part of the wall from which they extend (Figure 3.22*a*), there is no contact resistance at their base. However, more commonly, fins are manufactured separately and are attached to the wall by a metallurgical or adhesive joint. Alternatively, the attachment may involve a *press fit,* for which the fins are forced into slots machined on the wall material. In such cases (Figure 3.22*b*), there is a thermal contact resistance $R_{t,c}$, which may adversely influence overall thermal performance. An effective circuit resistance may again be obtained, where, with the contact resistance,

$$R_{t,o(c)} = \frac{\theta_b}{q_t} = \frac{1}{\eta_{o(c)} h A_t} \tag{3.109}$$

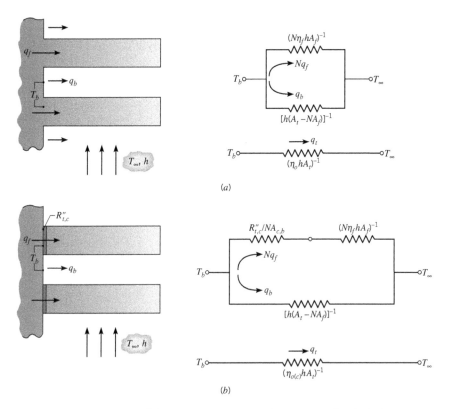

(a)

(b)

FIGURE 3.22   Fin array and thermal circuit. (*a*) Fins that are integral with the base. (*b*) Fins that are attached to the base.

It is readily shown that the corresponding overall surface efficiency is

$$\eta_{o(c)} = 1 - \frac{NA_f}{A_t}\left(1 - \frac{\eta_f}{C_1}\right) \tag{3.110a}$$

where

$$C_1 = 1 + \eta_f h A_f (R''_{t,c}/A_{c,b}) \tag{3.110b}$$

In manufacturing, care must be taken to render $R_{t,c} \ll R_{t,f}$.

EXAMPLE 3.10

The engine cylinder of a motorcycle is constructed of 2024-T6 aluminum alloy and is of height $H = 100$ mm and outer diameter $D = 2r_1 = 50$ mm. Under typical operating conditions, $q_t = 2$ kW of heat is transferred from the cylinder to ambient air at 300 K, with a convection coefficient of 75 W/m^2 · K. Annular fins are integrally cast with the cylinder to reduce the cylinder temperature. Consider ten equally-spaced fins, each of which are of thickness $t = 4$ mm and length $L = 20$ mm. What reduction in the cylinder temperature can be achieved by use of the fins?

SOLUTION

Known: Operating conditions of a finned motorcycle cylinder.

Find: Reduction in cylinder temperature associated with using fins.

Schematic:

Assumptions:

1. Steady-state conditions.

2. Temperature is uniform across the fin thickness.

3. Constant properties.

4. Negligible radiation exchange with surroundings.

5. Uniform convection coefficient over surface (with or without fins).

Properties: Table A.1, 2024-T6 aluminum ($T \approx 550$ K): $k = 186$ W/m · K.

Analysis: With the fins in place, Equation 3.106 can be rearranged to determine an expression for the cylinder temperature

$$T_b = T_\infty + \frac{q_t}{hA_t \left[1 - \dfrac{NA_f}{A_t}(1 - n_f) \right]}$$

where $A_f = 2\pi(r_{2c}^2 - r_1^2) = 2\pi[(0.047 \text{ m})^2 - (0.025 \text{ m})^2] = 0.00995 \text{ m}^2$ and, from Equation 3.104, $A_t = NA_f + 2\pi r_1(H - Nt) = 0.0995 \text{ m}^2 + 2\pi(0.025 \text{ m})[0.10 \text{ m} - 0.04 \text{ m}] = 0.109 \text{ m}^2$. With $r_{2c}/r_1 = 1.88$, $L_c = 0.022$ m, $A_p = 8.8 \times 10^{-5} \text{ m}^2$, we obtain $L_c^{3/2}(h/kA_p)^{1/2} = 0.221$. Hence, from Figure 3.20 (or Equation 3.96), the fin efficiency is $\eta_f \approx 0.96$. With the fins, the cylinder temperature is

$$T_b = 27°\text{C} + \frac{2000 \text{ W}}{75 \text{ W/m}^2 \cdot \text{K} \times 0.109 \text{ m}^2 \left[1 - \dfrac{10 \times 0.00995 \text{ m}^2}{0.109 \text{ m}^2}(1 - 0.96) \right]} = 282°\text{C}$$

Without the fins, the cylinder temperature would be

$$T_{b,\text{wo}} = T_\infty + \frac{q_t}{h(2\pi r_1 H)} = 27°\text{C} + \frac{2000 \text{ W}}{75 \text{ W/m}^2 \cdot \text{K}(2\pi \times 0.025 \text{ m} \times 0.10 \text{ m})} = 1725°\text{C}$$

Therefore, the reduction in cylinder temperature is

$$\Delta T_b = T_{b,\text{wo}} - T_b = 1725°\text{C} - 282°\text{C} = 1443°\text{C} \qquad \triangleleft$$

Comments:

1. From Table A.1, the melting temperature of 2024-T6 aluminum is 775 K = 502°C. The engine must be equipped with fins to avoid failure.

2. The fins are of high efficiency. Assuming isothermal fins ($\eta_f = 1$), the cylinder temperature would be $T_b = T_\infty + q_t/hA_t = 272°\text{C}$, which is only slightly lower than the predicted temperature.

3. Further reduction in the cylinder temperature could be achieved by adding more fins. Prescribing a fin clearance of 2 mm at each end of the array and a minimum fin gap of 4 mm, the maximum allowable number of fins is $N_{\max} = H/S = 0.10$ m/(0.004 + 0.004) m = 12.5, which we round down to $N_{\max} = 12$. Use of 12 fins reduces the cylinder temperature to $T_b = 245°\text{C}$.

4. The assumed temperature of 550 K (277°C) that was used to evaluate the thermal conductivity is reasonable.

5. The *Models/Extended Surfaces* option in the *Advanced* section of *IHT* provides ready-to-solve models for straight, pin, and annular fins, as well as for fin arrays. The models include the efficiency relations of Figures 3.19 and 3.20 and Table 3.5.

3.7 *Other Applications of One-Dimensional, Steady-State Conduction*

We conclude our discussion of one-dimensional, steady-state conduction by considering three special topics. Quantifying heat transfer within the human body is important because of the development of new medical treatments that involve extreme temperatures [16], and as we explore adverse thermal environments such as the Arctic, underwater, or in space. As will become evident, prediction of heat transfer in living tissue involves extensions of the material on conduction with thermal energy generation, Section 3.5, and heat transfer from extended surfaces, Section 3.6. Thermoelectric power generation pertains to solid-state conversion of heat to electric power, which can be utilized to reduce, perhaps significantly, the amount of wasted energy associated with power generation shown in Figure 1.12*a*. Because of the conversion from one form of energy (heat) to a second form (electric power), the analysis of thermoelectric energy conversion requires careful application of the first law of thermodynamics, as presented in Section 1.3. The analysis also utilizes the material related to conduction with thermal energy generation, presented in Section 3.5 and Appendix C. Finally, nanoscale effects, initially presented in Section 2.2.1 in the context of the thermophysical properties of matter, are further quantified in this section.

3.7.1 The Bioheat Equation

There are two main phenomena that make heat transfer in living tissues more complex than in conventional engineering materials: metabolic heat generation and the exchange of thermal energy between flowing blood and the surrounding tissue. Pennes [17] introduced a modification to the heat equation, now known as the Pennes or bioheat equation, to account for these effects. The bioheat equation is known to have limitations, but it continues to be a useful tool for understanding heat transfer in living tissues. In this section, we present a simplified version of the bioheat equation for the case of steady-state, one-dimensional heat transfer.

Both the metabolic heat generation and exchange of thermal energy with the blood can be viewed as effects of thermal energy generation. Therefore, we can rewrite Equation 3.44 to account for these two heat sources as

$$\frac{d^2T}{dx^2} + \frac{\dot{q}_m + \dot{q}_p}{k} = 0 \qquad (3.111)$$

where $\dot{q}_m$ and $\dot{q}_p$ are the *metabolic* and *perfusion* heat source terms, respectively. The perfusion term accounts for energy exchange between the blood and the tissue and is an energy source or sink according to whether heat transfer is from or to the blood, respectively. The thermal conductivity has been assumed constant in writing Equation 3.111.

Pennes proposed an expression for the perfusion term, $\dot{q}_p$, by assuming that within any small volume of tissue, the blood flowing in the small capillaries enters at an arterial temperature, T_a, and exits at the local tissue temperature, T. The rate at which heat is gained by the tissue is the rate at which heat is lost from the blood. If the perfusion rate is ω (m³/s of volumetric blood flow per m³ of tissue), the heat lost from the blood can be calculated from Equation 1.12e, or on a unit volume basis,

$$\dot{q}_p = \omega \rho_b c_b (T_a - T) \tag{3.112}$$

where ρ_b and c_b are the blood density and specific heat, respectively. Note that $\omega \rho_b$ is the blood mass flow rate per unit volume of tissue.

Substituting Equation 3.112 into Equation 3.111, we find

$$\frac{d^2 T}{dx^2} + \frac{\dot{q}_m + \omega \rho_b c_b (T_a - T)}{k} = 0 \tag{3.113}$$

Drawing on our experience with extended surfaces, it is convenient to define an excess temperature of the form $\theta \equiv T - T_a - \dot{q}_m / \omega \rho_b c_b$. Then, if we assume that T_a, $\dot{q}_m$, ω, and the blood properties are all constant, Equation 3.113 can be rewritten as

$$\frac{d^2 \theta}{dx^2} - \tilde{m}^2 \theta = 0 \tag{3.114}$$

where $\tilde{m}^2 = \omega \rho_b c_b / k$. This equation is identical in form to Equation 3.69. Depending on the form of the boundary conditions, it may therefore be possible to use the results of Table 3.4 to estimate the temperature distribution within the living tissue.

EXAMPLE 3.11

In Example 1.6, the temperature at the inner surface of the skin/fat layer was given as 35°C. In reality, this temperature depends on the existing heat transfer conditions, including phenomena occurring farther inside the body. Consider a region of muscle with a skin/fat layer over it. At a depth of $L_m = 30$ mm into the muscle, the temperature can be assumed to be at the core body temperature of $T_c = 37°C$. The muscle thermal conductivity is $k_m = 0.5$ W/m · K while the metabolic heat generation rate within the muscle is $\dot{q}_m = 700$ W/m³. The perfusion rate is $\omega = 0.0005$ s⁻¹; the blood density and specific heat are $\rho_b = 1000$ kg/m³ and $c_b = 3600$ J/kg · K, respectively, and the entering arterial blood temperature T_a is the same as the core body temperature. The thickness, emissivity, and thermal conductivity of the skin/fat layer are as given in Example 1.6; perfusion and metabolic heat generation within this layer can be neglected. We wish to predict the rate of heat loss from the body and the temperature at the inner surface of the skin/fat layer for the air and water environments of Example 1.6.

SOLUTION

Known: Dimensions and thermal conductivities of a muscle layer and a skin/fat layer. Skin emissivity and surface area. Metabolic heat generation rate and perfusion rate within the muscle layer. Core body and arterial temperatures. Blood density and specific heat. Ambient conditions.

Find: Rate of heat loss from body and temperature at inner surface of the skin/fat layer.

Schematic:

Assumptions:

1. Steady-state conditions.
2. One-dimensional heat transfer through the muscle and skin/fat layers.
3. Metabolic heat generation rate, perfusion rate, arterial temperature, blood properties, and thermal conductivities are all uniform.
4. Radiation heat transfer coefficient is known from Example 1.6.
5. Solar irradiation is negligible.

Analysis: We will combine an analysis of the muscle layer with a treatment of heat transfer through the skin/fat layer and into the environment. The rate of heat transfer through the skin/fat layer and into the environment can be expressed in terms of a total resistance, R_{tot}, as

$$q = \frac{T_i - T_\infty}{R_{tot}} \tag{1}$$

As in Example 3.1 and for exposure of the skin to the air, R_{tot} accounts for conduction through the skin/fat layer in series with heat transfer by convection and radiation, which act in parallel with each other. Thus,

$$R_{tot} = \frac{L_{sf}}{k_{sf}A} + \left(\frac{1}{1/hA} + \frac{1}{1/h_r A} \right)^{-1} = \frac{1}{A} \left(\frac{L_{sf}}{k_{sf}} + \frac{1}{h + h_r} \right)$$

Using the values from Example 1.6 for air,

$$R_{tot} = \frac{1}{1.8 \text{ m}^2} \left(\frac{0.003 \text{ m}}{0.3 \text{ W/m} \cdot \text{K}} + \frac{1}{(2 + 5.9) \text{ W/m}^2 \cdot \text{K}} \right) = 0.076 \text{ K/W}$$

For water, with $h_r = 0$ and $h = 200 \text{ W/m}^2 \cdot \text{K}$, $R_{tot} = 0.0083 \text{ W/m}^2 \cdot \text{K}$.

Heat transfer in the muscle layer is governed by Equation 3.114. The boundary conditions are specified in terms of the temperatures, T_c and T_i, where T_i is the temperature of the inner surface of the skin/fat layer and is, as yet, unknown. In terms of the excess temperature $\theta = T - T_a - \dot{q}_m/\omega\rho_b c_b$, the boundary conditions are then

$$\theta(0) = T_c - T_a - \frac{\dot{q}_m}{\omega\rho_b c_b} = \theta_c \quad \text{and} \quad \theta(L_m) = T_i - T_a - \frac{\dot{q}_m}{\omega\rho_b c_b} = \theta_i$$

Since we have two boundary conditions involving prescribed temperatures, the solution for θ is given by case C of Table 3.4,

$$\frac{\theta}{\theta_c} = \frac{(\theta_i/\theta_c)\sinh \tilde{m}x + \sinh \tilde{m}(L_m - x)}{\sinh \tilde{m}L_m}$$

The value of q_f given in Table 3.4 would correspond to the heat transfer rate at $x = 0$, but this is not of particular interest here. Rather, we seek the rate at which heat leaves the muscle and enters the skin/fat layer so that we can equate this quantity with the rate at which heat is transferred through the skin/fat layer and into the environment. Therefore, we calculate the heat transfer rate at $x = L_m$ as

$$q\bigg|_{x=L_m} = -k_m A \frac{dT}{dx}\bigg|_{x=L_m} = -k_m A \frac{d\theta}{dx}\bigg|_{x=L_m} = -k_m A\tilde{m}\theta_c \frac{(\theta_i/\theta_c)\cosh \tilde{m}L_m - 1}{\sinh \tilde{m}L_m} \quad (2)$$

Combining Equations 1 and 2 yields

$$-k_m A\tilde{m}\theta_c \frac{(\theta_i/\theta_c)\cosh \tilde{m}L_m - 1}{\sinh \tilde{m}L_m} = \frac{T_i - T_\infty}{R_{\text{tot}}}$$

This expression can be solved for T_i, recalling that T_i also appears in θ_i.

$$T_i = \frac{T_\infty \sinh \tilde{m}L_m + k_m A\tilde{m}R_{\text{tot}}\left[\theta_c + \left(T_a + \frac{\dot{q}_m}{\omega\rho_b c_b}\right)\cosh \tilde{m}L_m\right]}{\sinh \tilde{m}L_m + k_m A\tilde{m}R_{\text{tot}} \cosh \tilde{m}L_m}$$

where

$$\tilde{m} = \sqrt{\omega\rho_b c_b/k_m} = [0.0005 \text{ s}^{-1} \times 1000 \text{ kg/m}^3 \times 3600 \text{ J/kg} \cdot \text{K}/0.5 \text{ W/m} \cdot \text{K}]^{1/2}$$
$$= 60 \text{ m}^{-1}$$

$$\sinh(\tilde{m}L_m) = \sinh(60 \text{ m}^{-1} \times 0.03 \text{ m}) = 2.94$$
$$\cosh(\tilde{m}L_m) = \cosh(60 \text{ m}^{-1} \times 0.03 \text{ m}) = 3.11$$

and

$$\theta_c = T_c - T_a - \frac{\dot{q}_m}{\omega\rho_b c_b} = -\frac{\dot{q}_m}{\omega\rho_b c_b} = -\frac{700 \text{ W/m}^3}{0.0005 \text{ s}^{-1} \times 1000 \text{ kg/m}^3 \times 3600 \text{ J/kg} \cdot \text{K}}$$
$$= -0.389 \text{ K}$$

The excess temperature can be expressed in kelvins or degrees Celsius, since it is a temperature difference.
   Thus, for air:

$$T_i = \frac{\{24°C \times 2.94 + 0.5 \text{ W/m} \cdot \text{K} \times 1.8 \text{ m}^2 \times 60 \text{ m}^{-1} \times 0.076 \text{ K/W}[-0.389°C + (37°C + 0.389°C) \times 3.11]\}}{2.94 + 0.5 \text{ W/m} \cdot \text{K} \times 1.8 \text{ m}^2 \times 60 \text{ m}^{-1} \times 0.076 \text{ K/W} \times 3.11} = 34.8°C \quad \triangleleft$$

Next we can find the rate of heat loss:

$$q = \frac{T_i - T_\infty}{R_{tot}} = \frac{34.8°C - 24°C}{0.076 \ K/W} = 142 \ W \qquad \lhd$$

Repeating the calculation for water, we find

$$T_i = 28.2°C \qquad \lhd$$

$$q = 514 \ W \qquad \lhd$$

Comments:

1. The rate of heat loss in air agrees well with that calculated in Example 1.6. This is because the skin/fat inner surface temperature specified in Example 1.6 ($T_i = 35°C$) is close to the value calculated here ($T_i = 34.8°C$). In contrast, the rate of heat loss in water is in poor agreement with the value calculated in Example 1.6 because T_i determined here (28.2°C) differs substantially from the value specified in Example 1.6.

2. In reality, our bodies adjust in many ways to the thermal environment. For example, if we are too cold, we will shiver, which increases our metabolic heat generation rate. If we are too warm, the perfusion rate near the skin surface will increase, locally raising the skin temperature to increase heat loss to the environment.

3. The calculations can be repeated for a range of values of the perfusion rate, and the dependence of the heat loss rate on the perfusion rate is illustrated below. The effect is stronger for the case of the water environment, because the muscle temperature is lower and therefore the effect of perfusion by the warm arterial blood is more pronounced.

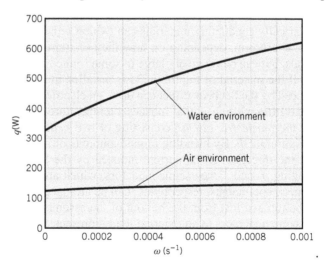

3.7.2 Thermoelectric Power Generation

As noted in Section 1.6, approximately 60% of the energy consumed globally is wasted in the form of low-grade heat. As such, an opportunity exists to harvest this energy stream and convert some of it to useful power. One approach involves *thermoelectric power generation,*

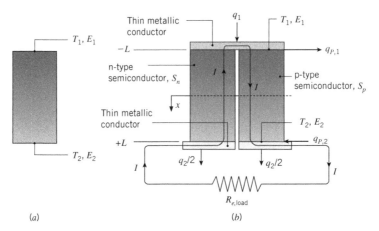

FIGURE 3.23   Thermoelectric phenomena. (*a*) The Seebeck effect.
(*b*) A simplified thermoelectric circuit consisting of one pair ($N = 1$) of
semiconducting pellets.

which operates on a fundamental principle termed the *Seebeck effect* that states when a temperature gradient is established within a material, a corresponding voltage gradient is induced. The *Seebeck coefficient S* is a material property representing the proportionality between voltage and temperature gradients and, accordingly, has units of volts/K. For a constant property material experiencing one-dimensional conduction, as illustrated in Figure 3.23*a*,

$$E_1 - E_2 = S(T_1 - T_2) \tag{3.115}$$

Electrically conducting materials can exhibit either positive or negative values of the Seebeck coefficient, depending on how they scatter electrons. The Seebeck effect is small in metals, but can be relatively large in some semiconducting materials.

If the material of Figure 3.23*a* is installed in an electric circuit, the voltage difference induced by the Seebeck effect can drive an electric current *I*, and electric power can be generated from waste heat that induces a temperature difference across the material. A simplified *thermoelectric circuit*, consisting of two *pellets* of semiconducting material, is shown in Figure 3.23*b*. By blending minute amounts of a secondary element into the pellet material, the direction of the current induced by the Seebeck effect can be manipulated. The resulting *p-* and *n-type* semiconductors, which are characterized by positive and negative Seebeck coefficients, respectively, can be arranged as shown in the figure. Heat is supplied to the top and lost from the bottom of the assembly, and thin metallic conductors connect the semiconductors to an external load represented by the electrical resistance, $R_{e,\text{load}}$. Ultimately, the amount of electric power that is produced is governed by the heat transfer rates to and from the pair of semiconducting pellets shown in Figure 3.23*b*.

In addition to inducing an electric current *I*, thermoelectric effects also induce the generation or absorption of heat *at the interface* between two dissimilar materials. This heat source or heat sink phenomenon is known as the *Peltier effect*, and the amount of heat absorbed q_P is related to the Seebeck coefficients of the adjoining materials by an equation of the form

$$q_P = I(S_p - S_n)T = IS_{p\text{-}n}T \tag{3.116}$$

where the individual Seebeck coefficients in the preceding expression, S_p and S_n, correspond to the p- and n-type semiconductors, and the differential Seebeck coefficient is $S_{p-n} \equiv S_p - S_n$. Temperature is expressed in kelvins in Equation 3.116. The heat absorption is positive (generation is negative) when the electric current flows from the n-type to the p-type semiconductor. Hence, in Figure 3.23*b*, Peltier heat absorption occurs at the warm interface between the semiconducting pellets and the upper, thin metallic conductor, while Peltier heat generation occurs at the cool interface between the pellets and the lower conductor.

When $T_1 > T_2$, the heat transfer rates to and from the device, q_1 and q_2, respectively, may be found by solving the appropriate form of the energy equation. For steady-state, one-dimensional conduction within the assembly of Figure 3.23*b* the analysis proceeds as follows.

Assuming the thin metallic connectors are of relatively high thermal and electrical conductivity, Ohmic dissipation occurs exclusively within the semiconducting pellets, each of which has a cross-sectional area $A_{c,s}$. The thermal resistances of the metallic conductors are assumed to be negligible, as is heat transfer within any gas trapped between the semiconducting pellets. Recognizing that the electrical resistance of each of the two pellets may be expressed as $R_{e,s} = \rho_{e,s}(2L)/A_{c,s}$ where $\rho_{e,s}$ is the electrical resistivity of the semiconducting material, Equation 3.43 may be used to find the uniform volumetric generation rate within each pellet

$$\dot{q} = \frac{I^2 \rho_{e,s}}{A_{c,s}^2} \tag{3.117}$$

Assuming negligible contact resistances and identical, as well as constant, thermophysical properties in each of the two pellets (with the exception being $S_p = -S_n$), Equation C.7 may be used to write expressions for the heat conduction out of and into the semiconducting material

$$q(x = L) = 2A_{c,s}\left[\frac{k_s}{2L}(T_1 - T_2) + \frac{I^2 \rho_{e,s} L}{A_{c,s}^2}\right] \tag{3.118a}$$

$$q(x = -L) = 2A_{c,s}\left[\frac{k_s}{2L}(T_1 - T_2) - \frac{I^2 \rho_{e,s} L}{A_{c,s}^2}\right] \tag{3.118b}$$

The factor of 2 outside the brackets accounts for heat transfer in *both* pellets and, as evident, $q(x = L) > q(x = -L)$.

Because of the Peltier effect, q_1 and q_2 are *not* equal to the heat transfer rates into and out of the pellets as expressed in Equations 3.118a,b. Incorporating Equation 3.116 in an energy balance for a control surface about the interface between the thin metallic conductor and the semiconductor material at $x = -L$ yields

$$q_1 = q(x = -L) + q_{P,1} = q(x = -L) + IS_{p-n}T_1 \tag{3.119}$$

Similarly at $x = L$,

$$q_2 = q(x = L) - IS_{n-p}T_2 = q(x = L) + IS_{p-n}T_2 \tag{3.120}$$

Combining Equations 3.118b and 3.119 yields

$$q_1 = \frac{A_{c,s}k_s}{L}(T_1 - T_2) + IS_{p\text{-}n}T_1 - 2\frac{I^2\rho_{e,s}L}{A_{c,s}} \tag{3.121}$$

Similarly, combining Equations 3.118a and 3.120 gives

$$q_2 = \frac{A_{c,s}k_s}{L}(T_1 - T_2) + IS_{p\text{-}n}T_2 + 2\frac{I^2\rho_{e,s}L}{A_{c,s}} \tag{3.122}$$

From an overall energy balance on the thermoelectric device, the electric power produced by the Seebeck effect is

$$P = q_1 - q_2 \tag{3.123}$$

Substituting Equations 3.121 and 3.122 into this expression yields

$$P = IS_{p\text{-}n}(T_1 - T_2) - 4\frac{I^2\rho_{e,s}L}{A_{c,s}} = IS_{p\text{-}n}(T_1 - T_2) - I^2R_{e,\text{tot}} \tag{3.124}$$

where $R_{e,\text{tot}} = 2R_{e,s}$.

The voltage difference induced by the Seebeck effect is relatively small for a single pair of semiconducting pellets. To amplify the voltage difference, thermoelectric *modules* are fabricated, as shown schematically in Figure 3.24*a* where $N \gg 1$ pairs of semiconducting pellets are wired in series. Thin layers of a dielectric material, usually a ceramic, sandwich the module to provide structural rigidity and electrical insulation from the surroundings. Assuming the thermal resistances of the thin ceramic layers are negligible, q_1, q_2, and the total module electric power, P_N, can be written by modifying Equations 3.121, 3.122, and 3.124 as

$$q_1 = \frac{1}{R_{t,\text{cond,mod}}}(T_1 - T_2) + IS_{p\text{-}n,\text{eff}}T_1 - \frac{I^2R_{e,\text{eff}}}{2} \tag{3.125}$$

$$q_2 = \frac{1}{R_{t,\text{cond,mod}}}(T_1 - T_2) + IS_{p\text{-}n,\text{eff}}T_2 + \frac{I^2R_{e,\text{eff}}}{2} \tag{3.126}$$

$$P_N = q_1 - q_2 = IS_{p\text{-}n,\text{eff}}(T_1 - T_2) - I^2R_{e,\text{eff}} \tag{3.127}$$

where $S_{p\text{-}n,\text{eff}} = NS_{p\text{-}n}$, and $R_{e,\text{eff}} = 2NR_{e,s}$ are the *effective* Seebeck coefficient and the total internal electrical resistance of the module while $R_{t,\text{cond,mod}} = L/NA_sk_s$ is the conduction resistance associated with the module's p-n semiconductor matrix. An equivalent thermal circuit for a convectively heated and cooled thermoelectric module is shown in Figure 3.24*b*. If heating or cooling were to be applied by radiation or conduction, the resistance network outside of the thermoelectric module portion of the circuit would be modified accordingly.

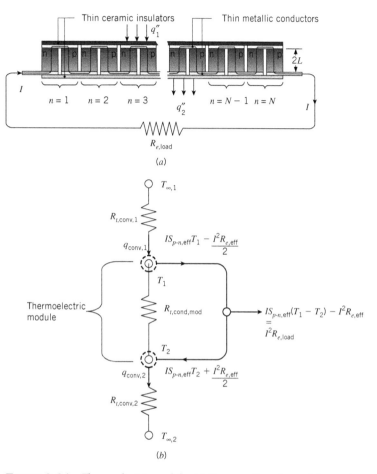

FIGURE 3.24 Thermoelectric module. (*a*) Cross-section of a module consisting of N semiconductor pairs. (*b*) Equivalent thermal circuit for a convectively heated and cooled module.

Returning to the single thermoelectric circuit of Figure 3.23*b*, the efficiency is defined as $\eta_{TE} \equiv P/q_1$. From Equations 3.121 and 3.124, it can be seen that efficiency depends on the electrical current in a complex manner. However, the efficiency can be maximized by adjusting the current through changes in the load resistance. The resulting maximum efficiency is given as [18]

$$\eta_{TE} = \left(1 - \frac{T_2}{T_1}\right) \frac{\sqrt{1 + Z\overline{T}} - 1}{\sqrt{1 + Z\overline{T}} + T_2/T_1} \tag{3.128}$$

where $\overline{T} = (T_1 + T_2)/2$, $S \equiv S_p = -S_n$, and

$$Z = \frac{S^2}{\rho_{e,s}k_s} \tag{3.129}$$

Since the efficiency increases with increasing $Z\overline{T}$, $Z\overline{T}$ may be seen as a dimensionless *figure of merit* associated with thermoelectric generation [19]. As $Z\overline{T} \to \infty$, $\eta_{TE} \to (1 - T_2/T_1) = (1 - T_c/T_h) \equiv \eta_C$ where η_C is the Carnot efficiency. As discussed in Section 1.3.2, the Carnot efficiency and, in turn, the thermoelectric efficiency cannot be determined until the appropriate hot and cold temperatures are calculated from a heat transfer analysis.

Because $Z\overline{T}$ is defined in terms of interrelated electrical and thermal conductivities, extensive research is being conducted to tailor the properties of the semiconducting pellets, primarily by manipulating the nanostructure of the material so as to independently control phonon and electron motion and, in turn, the thermal and electrical conductivities of the material. Currently, $Z\overline{T}$ values of approximately unity at room temperature are readily achieved. Finally, we note that thermoelectric modules can be operated in reverse; supplying electric power *to* the module allows one to control the heat transfer rates to or from the outer ceramic surfaces. Such *thermoelectric chillers* or *thermoelectric heaters* are used in a wide variety of applications. A comprehensive discussion of one-dimensional, steady-state heat transfer modeling associated with thermoelectric heating and cooling modules is available [20].

EXAMPLE 3.12

An array of $M = 48$ thermoelectric modules is installed on the exhaust of a sports car. Each module has an effective Seebeck coefficient of $S_{p\text{-}n,\text{eff}} = 0.1435$ V/K, and an internal electrical resistance of $R_{e,\text{eff}} = 8\ \Omega$. In addition, each module is of width and length $W = 54$ mm and contains $N = 100$ pairs of semiconducting pellets. Each pellet has an overall length of $2L = 5$ mm and cross-sectional area $A_{c,s} = 1.2 \times 10^{-5}$ m^2 and is characterized by a thermal conductivity of $k_s = 1.2$ W/m · K. The hot side of each module is exposed to exhaust gases at $T_{\infty,1} = 550°$C with $h_1 = 40$ W/m^2 · K, while the opposite side of each module is cooled by pressurized water at $T_{\infty,2} = 105°$C with $h_2 = 500$ W/m^2 · K. If the modules are wired in series, and the load resistance is $R_{e,\text{load}} = 400\ \Omega$, what is the electric power harvested from the hot exhaust gases?

SOLUTION

Known: Thermoelectric module properties and dimensions, number of semiconductor pairs in each module, and number of modules in the array. Temperature of exhaust gas and pressurized water, as well as convection coefficients at the hot and cold module surfaces. Modules are wired in series, and the electrical resistance of the load is known.

Find: Power produced by the module array.

Schematic:

Assumptions:

1. Steady-state conditions.
2. One-dimensional heat transfer.
3. Constant properties.
4. Negligible electrical and thermal contact resistances.
5. Negligible radiation exchange and negligible heat transfer within the gas inside the modules.
6. Negligible conduction resistance posed by the metallic contacts and ceramic insulators of the modules.

Analysis: We begin by analyzing a single module. The conduction resistance of each module's semiconductor array is

$$R_{t,\text{cond,mod}} = \frac{L}{N A_{c,s} k_s} = \frac{2.5 \times 10^{-3}\ \text{m}}{100 \times 1.2 \times 10^{-5}\ \text{m}^2 \times 1.2\ \text{W/m} \cdot \text{K}} = 1.736\ \text{K/W}$$

From Equation 3.125,

$$q_1 = \frac{1}{R_{t,\text{cond,mod}}}(T_1 - T_2) + I S_{p\text{-}n,\text{eff}} T_1 - \frac{I^2 R_{e,\text{eff}}}{2} = \frac{(T_1 - T_2)}{1.736\ \text{K/W}}$$

$$+\ I \times 0.1435\ \text{V/K} \times T_1 - I^2 \times 4\ \Omega \qquad (1)$$

while from Equation 3.126,

$$q_2 = \frac{1}{R_{t,\text{cond,mod}}}(T_1 - T_2) + I S_{p\text{-}n,\text{eff}} T_2 + \frac{I^2 R_{e,\text{eff}}}{2} = \frac{(T_1 - T_2)}{1.736\ \text{K/W}}$$

$$+\ I \times 0.1435\ \text{V/K} \times T_2 + I^2 \times 4\ \Omega \qquad (2)$$

At the hot surface, Newton's law of cooling may be written as

$$q_1 = h_1 W^2 (T_{\infty,1} - T_1) = 40 \text{ W/m}^2 \cdot \text{K} \times (0.054 \text{ m})^2 \times [(550 + 273)\text{K} - T_1] \qquad (3)$$

whereas at the cool surface,

$$q_2 = h_2 W^2 (T_2 - T_{\infty,2}) = 500 \text{ W/m}^2 \cdot \text{K} \times (0.054 \text{ m})^2 \times [T_2 - (105 + 273)\text{K}] \qquad (4)$$

Four equations have been written that include five unknowns, q_1, q_2, T_1, T_2, and I. An additional equation is obtained from the electrical circuit. With the modules wired in series, the total electric power produced by all $M = 48$ modules is equal to the electric power dissipated in the load resistance. Equation 3.127 yields

$$P_{tot} = MP_N = M[IS_{p\text{-}n,\text{eff}}(T_1 - T_2) - I^2 R_{e,\text{eff}}] = 48[I \times 0.1435 \text{ V/K} \times (T_1 - T_2) - I^2 \times 8 \text{ }\Omega] \qquad (5)$$

Since the electric power produced by the thermoelectric module is dissipated in the electrical load, it follows that

$$P_{tot} = I^2 R_{\text{load}} = I^2 \times 400 \text{ }\Omega \qquad (6)$$

Equations 1 through 6 may be solved simultaneously, yielding $P_{tot} = 46.9$ W.   ◁

Comments:

1. Equations 1 through 5 can be readily written by inspecting the equivalent thermal circuit of Figure 3.24*b*.

2. The module surface temperatures are $T_1 = 173°C$ and $T_2 = 134°C$, respectively. If these surface temperatures were specified in the problem statement, the electric power could be obtained directly from Equations 5 and 6. In any practical design of a thermoelectric generator, however, a heat transfer analysis must be conducted to determine the power generated.

3. Power generation is very sensitive to the convection heat transfer resistances. For $h_1 = h_2 \to \infty$, $P_{tot} = 5900$ W. To reduce the thermal resistance between the module and fluid streams, finned heat sinks are often used to increase the temperature difference across the modules and, in turn, increase their power output. Good thermal management and design are crucial to maximizing the power generation.

4. Harvesting the thermal energy contained in the exhaust with thermoelectrics can eliminate the need for an alternator, resulting in an increase in the net power produced by the engine, a reduction in the automobile's weight, and an increase in gas mileage of up to 10%.

5. Thermoelectric modules, operating in the heating mode, can be embedded in car seats and powered by thermoelectric exhaust harvesters, reducing energy costs associated with heating the entire passenger cabin. The seat modules can also be operated in the cooling mode, potentially eliminating the need for vapor compression air conditioning. Common refrigerants, such as R134a, are harmful greenhouse gases, and are emitted into the atmosphere by leakage through seals and connections, and by catastrophic leaks due to collisions. Replacing automobile vapor compression air conditioners with personalized thermoelectric seat coolers can eliminate the equivalent of 45 million metric tons of CO_2 released into the atmosphere every year in the United States alone.

3.7.3 Nanoscale Conduction

Lastly, we consider situations for which the pertinent physical dimensions are small, leading to potentially important nanoscale effects.

Conduction Through Thin Gas Layers   Figure 3.25 depicts instantaneous trajectories of gas molecules between two isothermal, solid surfaces separated by a distance L. As discussed in Section 1.2.1, even in the absence of *bulk* fluid motion individual molecules continually impinge on the two solid boundaries that are held at uniform surface temperatures $T_{s,1}$ and $T_{s,2}$, respectively. The molecules also collide with each other, exchanging energy *within* the gaseous medium. When the thickness of the gas layer is large, $L = L_1$ (Figure 3.25*a*), a particular gas molecule will collide more frequently with other gas molecules than with either of the solid boundaries. Alternatively, for a very thin gas layer, $L = L_2 \ll L_1$ (Figure 3.25*b*), the probability of a molecule striking either of the solid boundaries is high relative to the likelihood of it colliding with another molecule.

The energy content of a gas molecule is associated with its translational, rotational, and vibrational kinetic energies. It is this molecular-scale kinetic energy that ultimately defines the temperature of the gas, and collisions between individual molecules determine the value of the thermal conductivity, as discussed in Section 2.2.1. However, the manner in which a gas molecule is reflected or scattered from the solid walls also affects its level of kinetic energy and, in turn, its temperature. Hence, wall–molecule collisions can become important in determining the heat rate, q_x, as L/λ_{mfp} becomes small.

The collision with and subsequent scattering of an individual gas molecule from a solid wall can be described by a *thermal accommodation coefficient*, α_t,

$$\alpha_t = \frac{T_i - T_{sc}}{T_i - T_s} \tag{3.130}$$

where T_i is the effective molecule temperature just prior to striking the solid surface, T_{sc} is the temperature of the molecule immediately after it is scattered or reflected by the surface, and T_s is the surface temperature. When the temperature of the scattered molecule is identical to the wall temperature, $\alpha_t = 1$. Alternatively, if $T_{sc} = T_i$, the molecule's kinetic energy and temperature are unaffected by a collision with the wall and $\alpha_t = 0$.

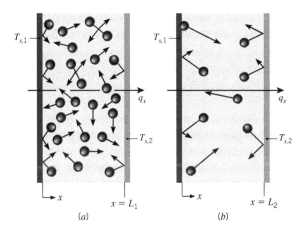

FIGURE 3.25 Molecule trajectories in (*a*) a relatively thick gas layer and (*b*) a relatively thin gas layer. Molecules collide with each other, and with the two solid walls.

For one-dimensional conduction within an ideal gas contained between two surfaces held at temperatures $T_{s,1}$ and $T_{s,2} < T_{s,1}$, the heat rate through the gas layer may be expressed as [21]

$$q = \frac{T_{s,1} - T_{s,2}}{(R_{t,m-m} + R_{t,m-s})} \tag{3.131}$$

where, at the molecular level, the thermal resistances are associated with molecule–molecule and molecule-surface collisions

$$R_{t,m-m} = \frac{L}{kA} \quad \text{and} \quad R_{t,m-s} = \frac{\lambda_{\text{mfp}}}{kA}\left[\frac{2 - \alpha_t}{\alpha_t}\right]\left[\frac{9\gamma - 5}{\gamma + 1}\right] \tag{3.132a,b}$$

In the preceding expression, $\gamma \equiv c_p/c_v$ is the specific heat ratio of the ideal gas. The two solids are assumed to be the same material with equal values of α_t, and the temperature difference is assumed to be small relative to the cold wall, $(T_{s,1} - T_{s,2})/T_{s,2} \ll 1$, where temperature is expressed in kelvins. Equations 3.132a,b may be combined to yield

$$\frac{R_{t,m-s}}{R_{t,m-m}} = \frac{\lambda_{\text{mfp}}}{L}\left[\frac{2 - \alpha_t}{\alpha_t}\right]\left[\frac{9\gamma - 5}{\gamma + 1}\right]$$

from which it is evident that $R_{t,m-s}$ may be neglected if L/λ_{mfp} is large and $\alpha_t \neq 0$. In this case, Equation 3.131 reduces to Equation 3.6. However, $R_{t,m-s}$ can be significant if L/λ_{mfp} is small. From Equation 2.11 the mean free path increases as the gas pressure is decreased. Hence, $R_{t,m-s}$ increases with decreasing gas pressure, and the heat rate can be pressure dependent when L/λ_{mfp} is small. Values of α_t for specific gas and surface combinations range from 0.87 to 0.97 for air–aluminum and air–steel, but can be less than 0.02 when helium interacts with clean metallic surfaces [21]. Equations 3.131, 3.132a,b may be applied to situations for which $L/\lambda_{\text{mfp}} \gtrsim 0.1$. For air at atmospheric pressure, this corresponds to $L \gtrsim 10$ nm.

Conduction Through Solid-Solid Interfaces   As shown in Figure 3.4, a thermal contact resistance is associated with the interface separating two materials. From Section 3.1.4 we know that for materials of conventional size, the value of the contact resistance depends on the roughness and topography of the mating surfaces, as well as the fluid that fills the gaps of the interface. However, a thermal resistance between two materials will exist even when the mating surfaces of the solids are perfectly smooth, with no intervening gap. This *thermal boundary resistance*, $R''_{t,b}$, is defined in the same manner as the thermal contact resistance (Equation 3.20)

$$R''_{t,b} = \frac{T_A - T_B}{q''_x} \tag{3.133}$$

and is ultimately due to the different atomic-level structures of the two materials. The step change in structure at the interface impedes phonon and electron transfer across the interface and, in turn, induces the thermal boundary resistance.

Measured values of the thermal boundary resistance between two perfectly smooth solids are shown in Table 3.6. A comparison of the thermal boundary resistances of Table 3.6 with the thermal contact resistances of Tables 3.1 and 3.2 shows that $R''_{t,b}$ is many orders of magnitude smaller than $R''_{t,c}$. Hence, thermal boundary resistances are usually negligible in applications

TABLE 3.6 Thermal boundary resistance, $R''_{t,b} \times 10^{10}$ (m² · K/W), for various solid-solid combinations at 300 K [22]

Solid A \ Solid B	Aluminum	Gold	Lead
Barium fluoride (BaF₂)	100	250	161
Aluminum oxide (Al₂O₃)	95	222	182
Diamond	217	250	323

involving materials of conventional size. However, for nanostructured materials or nanodevices, the thermal boundary resistance can be significant, as demonstrated in Example 3.13.

Conduction Through Thin Solid Films One-dimensional conduction across or along thin solid films was discussed in Section 2.2.1 in terms of the thermal conductivities k_x and k_y. The heat transfer rate across a thin solid film may be approximated by combining Equation 2.9a with Equation 3.5, yielding

$$q_x = \frac{k_x A}{L}(T_{s,1} - T_{s,2}) = \frac{k[1 - \lambda_{\mathrm{mfp}}/(3L)]A}{L}(T_{s,1} - T_{s,2}) \tag{3.134}$$

When L/λ_{mfp} is large, Equation (3.134) reduces to Equation 3.4. Many alternative expressions for k_x are available and are discussed in the literature [21].

EXAMPLE 3.13

Consider a room-temperature, nanocomposite plane wall that consists of $N = 10,000$ alternating layers of gold and aluminum oxide (sapphire), each of thickness δ. The wall is of total thickness $L = 1$ mm. Determine the thermal resistance of the nanocomposite wall, and compare its resistance to the thermal resistance posed by pure gold and sapphire plane walls, also of thickness $L = 1$ mm.

SOLUTION

Known: Composition, thickness, and number of alternating layers in a nanocomposite plane wall.

Find: Thermal resistance of nanocomposite, gold, and sapphire walls.

Schematic:

Assumptions:

1. Steady-state conditions.
2. One-dimensional conduction in the *x*-direction.
3. No internal heat generation.
4. Constant properties.

Properties: Table A.1, gold (300 K): $k_{au} = 317$ W/m · K. Table A.2, sapphire (300 K): $k_s = 46$ W/m · K.

Analysis: The thickness of each layer is $\delta_s = \delta_{au} = L/N = 1 \times 10^{-3}$ m/10,000 = 100 × 10^{-9} m = 100 nm. From Table 2.1, $\delta_s > L_{crit,x,s} = 36$ nm and the thermal conductivity of the sapphire is the same as its bulk value, $k_s = 46$ W/m · K. For the gold layers, $\delta_{au} < L_{crit,x,au} = 220$ nm. From Equation 2.9a and Table 2.1

$$k_{x,au} = k_{au}\left[1 - \frac{\lambda_{mfp,au}}{3\delta}\right] = 317 \text{ W/m} \cdot \text{K}\left[1 - \frac{31 \text{ nm}}{3 \times 100 \text{ nm}}\right] = 284 \text{ W/m} \cdot \text{K}$$

The nanocomposite wall is half gold by volume, and half sapphire. The conduction resistances of the gold layers, sapphire layers, and gold-sapphire interfaces are arranged in series. Therefore, the total thermal resistance of the nanocomposite wall per unit area is

$$R''_{tot} = \frac{L/2}{k_{x,au}} + \frac{L/2}{k_s} + NR''_{t,b}$$

From Table 3.6, gold-sapphire interface (300 K): $R''_{t,b} = 222 \times 10^{-10}$ m² · K/W. Hence,

$$R''_{tot} = \frac{(0.5 \times 10^{-3} \text{ m})}{284 \text{ W/m} \cdot \text{K}} + \frac{(0.5 \times 10^{-3} \text{ m})}{46 \text{ W/m} \cdot \text{K}} + 1 \times 10^4 \times 222 \times 10^{-10} \text{ m}^2 \cdot \text{K/W}$$

$$= 2.35 \times 10^{-4} \text{ m}^2 \cdot \text{K/W} \qquad \lhd$$

The thermal resistance of the gold wall is

$$R''_{tot,au} = \frac{L}{k_{au}} = \frac{1 \times 10^{-3} \text{ m}}{317 \text{ W/m} \cdot \text{K}} = 3.15 \times 10^{-6} \text{ m}^2 \cdot \text{K/W} \qquad \lhd$$

Similarly, for the sapphire wall

$$R''_{tot,s} = \frac{L}{k_s} = \frac{1 \times 10^{-3} \text{ m}}{46 \text{ W/m} \cdot \text{K}} = 21.7 \times 10^{-6} \text{ m}^2 \cdot \text{K/W} \qquad \lhd$$

Comments:

1. Building on the discussion of Section 2.2.1, this wall is another example of a *nano-structured material*. The thermal resistance of the nanocomposite is considerably larger than that of either the pure gold or pure sapphire walls.

2. Attempts have been made to *reduce* the thermal resistance of low thermal conductivity materials by blending the materials with solid nanoparticles of high thermal conductivity. For example, carbon nanotubes such as those of Example 3.4 could be blended with a plastic material. Such attempts involve a tradeoff between beneficial effects associated with the high thermal conductivity of the nanoparticles and adverse effects associated with boundary thermal resistances between the nanotubes and the surrounding

base material. The resulting nanocomposite may exhibit a thermal resistance that is either lower or higher than the base material.

3. If the wall were of thickness $L = 1$ m with $N = 10,000$, the layer thickness is $\delta = 100$ μm = 0.1 mm. In this case the thermal conductivities of both the gold and sapphire layers are their bulk values, and the total thermal resistances are 0.00315, 0.0127, and 0.0217 m$^2 \cdot$ K/W for the gold wall, the composite wall, and the sapphire wall, respectively. Now, the thermal resistance of the composite wall is bracketed by the thermal resistances of the gold and sapphire walls, as would be expected in applications that do not involve nanoscale heat transfer effects.

4. The conduction resistances of the 5000 layers of gold and 5000 layers of sapphire have been combined into single, equivalent resistances.

3.8 *Summary*

Despite its inherent mathematical simplicity, one-dimensional, steady-state heat transfer occurs in numerous engineering applications. Although one-dimensional, steady-state conditions may not apply exactly, the assumptions may often be made to obtain results of reasonable accuracy. You should therefore be thoroughly familiar with the means by which such problems are treated. In particular, you should be comfortable with the use of equivalent thermal circuits and with the expressions for the conduction resistances that pertain to each of the three common geometries. You should also be familiar with how the heat equation and Fourier's law may be used to obtain temperature distributions and the corresponding fluxes. The implications of an internally distributed source of energy should also be clearly understood. In addition, you should appreciate the important role that extended surfaces can play in the design of thermal systems and should have the facility to effect design and performance calculations for such surfaces. Finally, you should understand how the preceding concepts can be applied to analyze heat transfer in the human body, thermoelectric power generation, and nanoscale conduction.

You may test your understanding of this chapter's key concepts by addressing the following questions.

- Under what conditions may it be said that the *heat flux* is a constant, independent of the direction of heat flow? For each of these conditions, use physical considerations to convince yourself that the heat flux would not be independent of direction if the condition were not satisfied.
- For one-dimensional, steady-state conduction in a cylindrical or spherical shell without heat generation, is the radial heat flux independent of radius? Is the radial heat rate independent of radius?
- For one-dimensional, steady-state conduction without heat generation, what is the shape of the temperature distribution in a *plane wall*? In a *cylindrical shell*? In a *spherical shell*?
- What is the *thermal resistance*? How is it defined? What are its units?
- For conduction across a *plane wall*, can you write the expression for the thermal resistance from memory? Similarly, can you write expressions for the thermal resistance associated with conduction across *cylindrical* and *spherical* shells? From memory, can you express the thermal resistances associated with convection from a surface and net radiation exchange between the surface and large surroundings?

- What is the physical basis for existence of a *critical insulation radius*? How do the thermal conductivity and the convection coefficient affect its value?
- How is the conduction resistance of a solid affected by its thermal conductivity? How is the convection resistance at a surface affected by the convection coefficient? How is the radiation resistance affected by the surface emissivity?
- If heat is transferred from a surface by convection and radiation, how are the corresponding thermal resistances represented in a circuit?
- Consider steady-state conduction through a plane wall separating fluids of different temperatures, $T_{\infty,i}$ and $T_{\infty,o}$, adjoining the inner and outer surfaces, respectively. If the convection coefficient at the outer surface is five times larger than that at the inner surface, $h_o = 5h_i$, what can you say about relative proximity of the corresponding surface temperatures, $T_{s,o}$ and $T_{s,i}$, to their adjoining fluid temperatures?
- Can a thermal conduction resistance be applied to a *solid* cylinder or sphere?
- What is a *contact resistance*? How is it defined? What are its units for an interface of prescribed area? What are they for a unit area?
- How is the contact resistance affected by the roughness of adjoining surfaces?
- If the air in the contact region between two surfaces is replaced by helium, how is the thermal contact resistance affected? How is it affected if the region is evacuated?
- What is the *overall heat transfer coefficient*? How is it defined, and how is it related to the *total thermal resistance*? What are its units?
- In a solid circular cylinder experiencing uniform volumetric heating and convection heat transfer from its surface, how does the heat flux vary with radius? How does the heat rate vary with radius?
- In a solid sphere experiencing uniform volumetric heating and convection heat transfer from its surface, how does the heat flux vary with radius? How does the heat rate vary with radius?
- Is it possible to achieve steady-state conditions in a solid cylinder or sphere that is experiencing heat generation and whose surface is perfectly insulated? Explain.
- Can a material experiencing heat generation be represented by a thermal resistance and included in a circuit analysis? If so, why? If not, why not?
- What is the physical mechanism associated with cooking in a microwave oven? How do conditions differ from a conventional (convection or radiant) oven?
- If radiation is incident on the surface of a semitransparent medium and is absorbed as it propagates through the medium, will the corresponding volumetric rate of heat generation $\dot{q}$ be distributed uniformly in the medium? If not, how will $\dot{q}$ vary with distance from the surface?
- In what way is a plane wall that is of thickness $2L$ and experiences uniform volumetric heating and equivalent convection conditions at both surfaces similar to a plane wall that is of thickness L and experiences the same volumetric heating and convection conditions at one surface but whose opposite surface is well insulated?
- What purpose is served by attaching *fins* to a surface?
- In the derivation of the general form of the energy equation for an extended surface, why is the assumption of one-dimensional conduction an approximation? Under what conditions is it a good approximation?
- Consider a straight fin of uniform cross section (Figure 3.15*a*). For an *x*-location in the fin, sketch the temperature distribution in the transverse (*y*-) direction, placing the origin of the coordinate at the midplane of the fin ($-t/2 \le y \le t/2$). What is the form of a *surface* energy balance applied at the location ($x, t/2$)?

- What is the *fin effectiveness*? What is its range of possible values? Under what conditions are fins most effective?
- What is the *fin efficiency*? What is its range of possible values? Under what conditions will the efficiency be large?
- What is the *fin resistance*? What are its units?
- How are the effectiveness, efficiency, and thermal resistance of a fin affected if its thermal conductivity is increased? If the convection coefficient is increased? If the length of the fin is increased? If the thickness (or diameter) of the fin is increased?
- Heat is transferred from hot water flowing through a tube to air flowing over the tube. To enhance the rate of heat transfer, should fins be installed on the tube interior or exterior surface?
- A fin may be manufactured as an integral part of a surface by using a casting or extrusion process, or it may be separately brazed or adhered to the surface. From thermal considerations, which option is preferred?
- Describe the physical origins of the two heat source terms in the bioheat equation. Under what conditions is the perfusion term a heat sink?
- How do heat sinks increase the electric power generated by a thermoelectric device?
- Under what conditions do thermal resistances associated with molecule–wall interactions become important?

References

1. Fried, E., "Thermal Conduction Contribution to Heat Transfer at Contacts," in R. P. Tye, Ed., *Thermal Conductivity,* Vol. 2, Academic Press, London, 1969.

2. Eid, J. C., and V. W. Antonetti, "Small Scale Thermal Contact Resistance of Aluminum Against Silicon," in C. L. Tien, V. P. Carey, and J. K. Ferrel, Eds., *Heat Transfer—1986,* Vol. 2, Hemisphere, New York, 1986, pp. 659–664.

3. Snaith, B., P. W. O'Callaghan, and S. D. Probert, *Appl. Energy,* **16,** 175, 1984.

4. Yovanovich, M. M., "Theory and Application of Constriction and Spreading Resistance Concepts for Microelectronic Thermal Management," Presented at the International Symposium on Cooling Technology for Electronic Equipment, Honolulu, 1987.

5. Peterson, G. P., and L. S. Fletcher, "Thermal Contact Resistance of Silicon Chip Bonding Materials," Proceedings of the International Symposium on Cooling Technology for Electronic Equipment, Honolulu, 1987, pp. 438–448.

6. Yovanovich, M. M., and M. Tuarze, *AIAA J. Spacecraft Rockets,* **6,** 1013, 1969.

7. Madhusudana, C. V., and L. S. Fletcher, *AIAA J.,* **24,** 510, 1986.

8. Yovanovich, M. M., "Recent Developments in Thermal Contact, Gap and Joint Conductance Theories and Experiment," in C. L. Tien, V. P. Carey, and J. K. Ferrel, Eds., *Heat Transfer—1986,* Vol. 1, Hemisphere, New York, 1986, pp. 35–45.

9. Maxwell, J. C., *A Treatise on Electricity and Magnetism,* 3rd ed., Oxford University Press, Oxford, 1892.

10. Hamilton, R. L., and O. K. Crosser, *I&EC Fund.* **1,** 187, 1962.

11. Jeffrey, D. J., *Proc. Roy. Soc. A,* **335,** 355, 1973.

12. Hashin Z., and S. Shtrikman, *J. Appl. Phys.,* **33,** 3125, 1962.

13. Aichlmayr, H. T., and F. A. Kulacki, "The Effective Thermal Conductivity of Saturated Porous Media," in J. P. Hartnett, A. Bar-Cohen, and Y. I Cho, Eds., *Advances in Heat Transfer,* Vol. 39, Academic Press, London, 2006.

14. Harper, D. R., and W. B. Brown, "Mathematical Equations for Heat Conduction in the Fins of Air Cooled Engines," NACA Report No. 158, 1922.

15. Schneider, P. J., *Conduction Heat Transfer,* Addison-Wesley, Reading, MA, 1957.

16. Diller, K. R., and T. P. Ryan, *J. Heat Transfer,* **120,** 810, 1998.

17. Pennes, H. H., *J. Applied Physiology,* **85,** 5, 1998.

18. Goldsmid, H. J., "Conversion Efficiency and Figure-of-Merit," in D. M. Rowe, Ed., *CRC Handbook of Thermoelectrics,* Chap. 3, CRC Press, Boca Raton, 1995.

19. Majumdar, A., *Science,* **303,** 777, 2004.

20. Hodes, M., *IEEE Trans. Com. Pack. Tech.*, **28**, 218, 2005.

21. Zhang, Z. M., *Nano/Microscale Heat Transfer*, McGraw-Hill, New York, 2007.

22. Cahill, D. G., P. V. Braun, G. Chen, D. R. Clarke, S. Fan, K. E. Goodson, P. Keblinski, W. P. King, G. D. Mahan, A. Majumdar, J. M. Humphrey, S. R. Phillpot, E. Pop, and L. Shi, *App. Phys. Rev.*, **1**, 011305, 2014.

Problems

Plane and Composite Walls

3.1 Consider the plane wall of Figure 3.1, separating hot and cold fluids at temperatures $T_{\infty,1}$ and $T_{\infty,2}$, respectively. Using surface energy balances as boundary conditions at $x = 0$ and $x = L$ (see Equation 2.34), obtain the temperature distribution within the wall and the heat flux in terms of $T_{\infty,1}$, $T_{\infty,2}$, h_1, h_2, k, and L.

3.2 A new building to be located in a cold climate is being designed with a basement that has an $L = 250$-mm-thick wall. Inner and outer basement wall temperatures are $T_i = 20°C$ and $T_o = 0°C$, respectively. The architect can specify the wall material to be either aerated concrete block with $k_{ac} = 0.15$ W/m · K, or stone mix concrete. To reduce the conduction heat flux through the stone mix wall to a level equivalent to that of the aerated concrete wall, what thickness of extruded polystyrene sheet must be applied onto the inner surface of the stone mix concrete wall? Floor dimensions of the basement are 25 m × 40 m, and the expected rental rate is $50/m²/month. What is the yearly cost, in terms of lost rental income, if the stone mix concrete wall with polystyrene insulation is specified?

3.3 The rear window of an automobile is defogged by attaching a thin, transparent, film-type heating element to its inner surface. By electrically heating this element, a uniform heat flux may be established at the inner surface.

(a) For 4-mm-thick window glass, determine the electrical power required per unit window area to maintain an inner surface temperature of 15°C when the interior air temperature and convection coefficient are $T_{\infty,i} = 25°C$ and $h_i = 10$ W/m² · K, while the exterior (ambient) air temperature and convection coefficient are $T_{\infty,o} = -10°C$ and $h_o = 65$ W/m² · K.

(b) In practice $T_{\infty,o}$ and h_o vary according to weather conditions and car speed. For values of $h_o = 2$, 20, 65, and 100 W/m² · K, determine and plot the electrical power requirement as a function of $T_{\infty,o}$ for $-30 \leq T_{\infty,o} \leq 0°C$. From your results, what can you conclude about the need for heater operation at low values of h_o? How is this conclusion affected by the value of $T_{\infty,o}$? If $h \propto V^n$, where V is the vehicle speed and n is a positive exponent, how does the vehicle speed affect the need for heater operation?

3.4 A dormitory at a large university, built 50 years ago, has exterior walls constructed of $L_s = 25$-mm-thick sheathing with a thermal conductivity of $k_s = 0.1$ W/m · K. To reduce heat losses in the winter, the university decides to encapsulate the entire dormitory by applying an $L_i = 25$-mm-thick layer of extruded insulation characterized by $k_i = 0.029$ W/m · K to the exterior of the original sheathing. The extruded insulation is, in turn, covered with an $L_g = 5$-mm-thick architectural glass with $k_g = 1.4$ W/m · K. Determine the heat flux through the original and retrofitted walls when the interior and exterior air temperatures are $T_{\infty,i} = 22°C$ and $T_{\infty,o} = -20°C$, respectively. The inner and outer convection heat transfer coefficients are $h_i = 5$ W/m² · K and $h_o = 25$ W/m² · K, respectively.

3.5 In a manufacturing process, a transparent film is being bonded to a substrate as shown in the sketch. To cure the bond at a temperature T_0, a radiant source is used to provide a heat flux q_0'' (W/m²), all of which is absorbed at the bonded surface. The back of the substrate is maintained at T_1 while the free surface of the film is exposed to air at T_∞ and a convection heat transfer coefficient h.

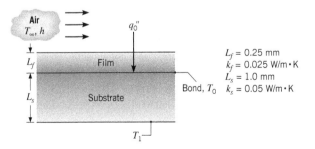

(a) Show the thermal circuit representing the steady-state heat transfer situation. Be sure to label *all* elements, nodes, and heat rates. Leave in symbolic form.

(b) Assume the following conditions: $T_\infty = 20°C$, $h = 50$ W/m² · K, and $T_1 = 30°C$. Calculate the heat flux q_0'' that is required to maintain the bonded surface at $T_0 = 60°C$.

(c) Compute and plot the required heat flux as a function of the film thickness for $0 \le L_f \le 1$ mm.

(d) If the film is not transparent and all of the radiant heat flux is absorbed at its upper surface, determine the heat flux required to achieve bonding. Plot your results as a function of L_f for $0 \le L_f \le 1$ mm.

3.6 The walls of a refrigerator are typically constructed by sandwiching a layer of insulation between sheet metal panels. Consider a wall made from fiberglass insulation of thermal conductivity $k_i = 0.046$ W/m · K and thickness $L_i = 50$ mm and steel panels, each of thermal conductivity $k_p = 60$ W/m · K and thickness $L_p = 3$ mm. If the wall separates refrigerated air at $T_{\infty,i} = 4°C$ from ambient air at $T_{\infty,o} = 25°C$, what is the heat gain per unit surface area? Coefficients associated with natural convection at the inner and outer surfaces may be approximated as $h_i = h_o = 5$ W/m² · K.

3.7 A composite wall is composed of an insulating material of thermal conductivity $k_{ins} = 1.5$ W/m · K sandwiched between two 1-mm-thick stainless steel sheets of thermal conductivity $k_{ss} = 15$ W/m · K. The wall separates two fluids of temperatures $T_{\infty,1} = 50°C$ and $T_{\infty,2} = 25°C$, respectively. Determine the thickness of insulation required to limit the heat flux through the composite wall to 60 W/m² for the cases shown in the table. The convection heat transfer coefficients are $h = 5$ W/m² · K for gas free convection, $h = 50$ W/m² · K for gas forced convection, $h = 500$ W/m² · K for liquid forced convection, and $h = 5000$ W/m² · K for either boiling or condensation.

Case	h_1 (W/m² · K)	h_2 (W/m² · K)
1	gas free convection	gas free convection
2	gas forced convection	gas free convection
3	gas forced convection	liquid forced convection
4	condensation	boiling

3.8 A $t = 12$-mm-thick horizontal layer of water has a top surface temperature of $T_c = -6°C$ and a bottom surface temperature of $T_h = 3°C$. Determine the location of the solid–liquid interface at steady state.

3.9 A technique for measuring convection heat transfer coefficients involves bonding one surface of a thin metallic foil to an insulating material and exposing the other surface to the fluid flow conditions of interest.

By passing an electric current through the foil, heat is dissipated uniformly within the foil and the corresponding flux, P_{elec}'', may be inferred from related voltage and current measurements. If the insulation thickness L and thermal conductivity k are known and the fluid, foil, and insulation temperatures (T_∞, T_s, T_b) are measured, the convection coefficient may be determined. Consider conditions for which $T_\infty = T_b = 25°C$, $P_{elec}'' = 2000$ W/m², $L = 10$ mm, and $k = 0.040$ W/m · K.

(a) With water flow over the surface, the foil temperature measurement yields $T_s = 27°C$. Determine the convection coefficient. What error would be incurred by assuming all of the dissipated power to be transferred to the water by convection?

(b) If, instead, air flows over the surface and the temperature measurement yields $T_s = 125°C$, what is the convection coefficient? The foil has an emissivity of 0.15 and is exposed to large surroundings at 25°C. What error would be incurred by assuming all of the dissipated power to be transferred to the air by convection?

(c) Typically, heat flux gages are operated at a fixed temperature (T_s), in which case the power dissipation provides a direct measure of the convection coefficient. For $T_s = 27°C$, plot P_{elec}'' as a function of h_o for $10 \le h_o \le 1000$ W/m² · K. What effect does h_o have on the error associated with neglecting conduction through the insulation?

3.10 The *wind chill*, which is experienced on a cold, windy day, is related to increased heat transfer from exposed human skin to the surrounding atmosphere. Consider a layer of fatty tissue that is 3 mm thick and whose interior surface is maintained at a temperature of 40°C. On a calm day the convection heat transfer coefficient at the outer surface is 25 W/m² · K, but with 30 km/h winds it reaches 65 W/m² · K. In both cases the ambient air temperature is −15°C.

(a) What is the ratio of the rate of heat loss per unit area from the skin for the calm day to that for the windy day?

(b) What will be the skin outer surface temperature for the calm day? For the windy day?

(c) What temperature would the air have to assume on the calm day to produce the same heat rate occurring with the air temperature at −15°C on the windy day?

3.11 A thermopane window consists of two pieces of glass 7 mm thick that enclose an air space 7 mm thick. The window separates room air at 20°C from outside ambient air at −10°C. The convection coefficient associated with the inner (room-side) surface is 10 W/m² · K.

   (a) If the convection coefficient associated with the outer (ambient) air is $h_o = 80$ W/m² · K, what is the heat loss through a window that is 0.8 m long by 0.5 m wide? Neglect radiation, and assume the air enclosed between the panes to be stagnant.

   (b) Compute and plot the effect of h_o on the heat loss for $10 \leq h_o \leq 100$ W/m² · K. Repeat this calculation for a triple-pane construction in which a third pane and a second air space of equivalent thickness are added.

3.12 A house has a composite wall of wood, fiberglass insulation, and plaster board, as indicated in the sketch. On a cold winter day, the convection heat transfer coefficients are $h_o = 60$ W/m² · K and $h_i = 30$ W/m² · K. The total wall surface area is 350 m².

(a) Determine a symbolic expression for the total thermal resistance of the wall, including inside and outside convection effects for the prescribed conditions.

(b) Determine the total rate of heat loss through the wall.

(c) If the wind were blowing violently, raising h_o to 300 W/m² · K, determine the percentage increase in the rate of heat loss.

(d) What is the controlling resistance that determines the amount of heat flow through the wall?

3.13 Consider the composite wall of Problem 3.12 under conditions for which the inside air is still characterized by $T_{\infty,i} = 20$°C and $h_i = 30$ W/m² · K. However, use the more realistic conditions for which the outside air is characterized by a diurnal (time) varying temperature of the form

$$T_{\infty,o}(\text{K}) = 273 + 5 \sin\left(\frac{2\pi}{24}t\right) \qquad 0 \leq t \leq 12 \text{ h}$$

$$T_{\infty,o}(\text{K}) = 273 + 11 \sin\left(\frac{2\pi}{24}t\right) \qquad 12 \leq t \leq 24 \text{ h}$$

with $h_o = 60$ W/m² · K. Assuming quasi-steady conditions for which changes in energy storage within the wall may be neglected, estimate the daily heat loss through the wall if its total surface area is 200 m².

3.14 The wall of a drying oven is constructed by sandwiching an insulation material of thermal conductivity $k = 0.05$ W/m · K between thin metal sheets. The oven air is at $T_{\infty,i} = 300$°C, and the corresponding convection coefficient is $h_i = 30$ W/m² · K. The inner wall surface absorbs a radiant flux of $q''_{\text{rad}} = 100$ W/m² from hotter objects within the oven. The room air is at $T_{\infty,o} = 25$°C, and the overall coefficient for convection and radiation from the outer surface is $h_o = 10$ W/m² · K.

(a) Draw the thermal circuit for the wall and label all temperatures, heat rates, and thermal resistances.

(b) What insulation thickness L is required to maintain the outer wall surface at a *safe-to-touch* temperature of $T_o = 40$°C?

3.15 Consider a composite wall that includes an 8-mm-thick hardwood siding, 40-mm by 130-mm hardwood studs on 0.65-m centers with glass fiber insulation (paper faced, 28 kg/m³), and a 12-mm layer of gypsum (vermiculite) wall board.

What is the thermal resistance associated with a wall that is 2.5 m high by 6.5 m wide (having 10 studs, each 2.5 m high)? Assume surfaces parallel to the *x*-direction are adiabatic.

3.16 Convection ovens operate on the principle of inducing forced convection inside the oven chamber with a fan. A small cake is to be baked in an oven when the convection feature is disabled. The walls of the oven consist of $L = 30$-mm-thick layers of insulation characterized by $k_{ins} = 0.03$ W/m · K that are sandwiched between two *thin* layers of sheet metal. The exterior surface of the oven is exposed to air at 23°C with $h_{ext} = 2$ W/m² · K. The interior oven air temperature is 180°C. Neglecting radiation heat transfer, determine the steady-state heat flux through the oven walls when the convection mode is disabled and the free convection coefficient at the inner oven surface is $h_{fr} = 3$ W/m² · K. Determine the heat flux through the oven walls when the convection mode is activated, in which case the forced convection coefficient at the inner oven surface is $h_{fo} = 27$ W/m² · K. Does operation of the oven in its convection mode result in significantly increased heat losses from the oven to the kitchen? Would your conclusion change if radiation were included in your analysis?

3.17 The composite wall of an oven consists of three materials, two of which are of known thermal conductivity, $k_A = 20$ Wm/· K and $k_C = 50$ W/m · K, and known thickness, $L_A = 0.30$ m and $L_C = 0.15$ m. The third material, B, which is sandwiched between materials A and C, is of known thickness, $L_B = 0.15$ m, but unknown thermal conductivity k_B.

Under steady-state operating conditions, measurements reveal an outer surface temperature of $T_{s,o} = 20$°C, an inner surface temperature of $T_{s,i} = 600$°C, and an oven air temperature of $T_\infty = 800$°C. The inside convection coefficient h is known to be 25 W/m² · K. What is the value of k_B?

3.18 The thermal characteristics of a small, dormitory refrigerator are determined by performing two separate experiments, each with the door closed and the refrigerator placed in ambient air at $T_\infty = 25$°C. In one case, an electric heater is suspended in the refrigerator cavity, while the refrigerator is unplugged. With the heater dissipating 20 W, a steady-state temperature of 90°C is recorded within the cavity. With the heater removed and the refrigerator now in operation, the second experiment involves maintaining a steady-state cavity temperature of 5°C for a fixed time interval and recording the electrical energy required to operate the refrigerator. In such an experiment for which steady operation is maintained over a 12-hour period, the input electrical energy is 125,000 J. Determine the refrigerator's coefficient of performance (COP).

3.19 A particular thermal system involves three objects of fixed shape with conduction resistances of $R_1 = 2$ K/W, $R_2 = 4$ K/W and $R_3 = 8$ K/W, respectively. An objective is to minimize the total thermal resistance R_{tot} associated with a combination of R_1, R_2, and R_3. The chief engineer is willing to invest limited funds to specify an alternative material for just one of the three objects; the alternative material will have a thermal conductivity that is twice its nominal value. Which object (1, 2, or 3) should be fabricated of the higher thermal conductivity material to most significantly decrease R_{tot}? *Hint*: Consider two cases, one for which the three thermal resistances are arranged in series, and the second for which the three resistances are arranged in parallel.

3.20 A firefighter's protective clothing, referred to as a *turn-out coat*, is typically constructed as an ensemble of three layers separated by air gaps, as shown schematically.

Representative dimensions and thermal conductivities for the layers are as follows.

Layer	*Thickness* (mm)	*k* (W/m · K)
Shell (s)	0.8	0.047
Moisture barrier (mb)	0.55	0.012
Thermal liner (tl)	3.5	0.038

The air gaps between the layers are 1 mm thick, and heat is transferred by conduction and radiation exchange through the stagnant air. The linearized radiation coefficient for a gap may be approximated as, $h_{rad} = \sigma(T_1 + T_2)(T_1^2 + T_2^2) \approx 4\sigma T_{avg}^3$, where T_{avg} represents the average temperature of the surfaces comprising the gap, and the radiation flux across the gap may be expressed as $q''_{rad} = h_{rad}(T_1 - T_2)$.

(a) Represent the turnout coat by a thermal circuit, labeling all the thermal resistances. Calculate and tabulate the thermal resistances per unit area ($m^2 \cdot K/W$) for each of the layers, as well as for the conduction and radiation processes in the gaps. Assume that a value of $T_{avg} = 470$ K may be used to approximate the radiation resistance of both gaps. Comment on the relative magnitudes of the resistances.

(b) For a *pre-flash-over* fire environment in which firefighters often work, the typical radiant heat flux on the fire-side of the turnout coat is 0.25 W/cm^2. What is the outer surface temperature of the turnout coat if the inner surface temperature is 66°C, a condition that would result in burn injury?

Contact Resistance

3.21 Consider the water of Example 1.5 to be initially liquid at its fusion temperature. The outer wall surface is suddenly reduced to $T_1 = -30$°C, resulting in solidification of the water. In reality, thermal contact resistances develop at the interface between the solid wall and the ice as the solidification progresses. Determine the interfacial thermal resistance, $R''_{t,c}$, shortly after the onset of freezing, given the dimensions in Comment 2. The initial solidification rate is 0.05 gram/s.

3.22 A composite wall separates combustion gases at 2600°C from a liquid coolant at 100°C, with gas- and liquid-side convection coefficients of 50 and 1000 $W/m^2 \cdot K$. The wall is composed of a 10-mm-thick layer of beryllium oxide on the gas side and a 20-mm-thick slab of stainless steel (AISI 304) on the liquid side. The contact resistance between the oxide and the steel is 0.05 $m^2 \cdot K/W$. What is the heat loss per unit surface area of the composite? Sketch the temperature distribution from the gas to the liquid.

3.23 Two stainless steel plates 15 mm thick are subjected to a contact pressure of 1 bar under vacuum conditions for which there is an overall temperature drop of 150°C across the plates. What is the heat flux through the plates? What is the temperature drop across the contact plane?

3.24 Consider a plane composite wall that is composed of two materials of thermal conductivities $k_A = 0.1$ W/m · K and $k_B = 0.04$ W/m · K and thicknesses $L_A = 10$ mm and $L_B = 20$ mm. The contact resistance at the interface between the two materials is known to be 0.30 $m^2 \cdot K/W$. Material A adjoins a fluid at 200°C for which $h = 10$ $W/m^2 \cdot K$, and material B adjoins a fluid at 40°C for which $h = 20$ $W/m^2 \cdot K$.

(a) What is the rate of heat transfer through a wall that is 2 m high by 2.5 m wide?

(b) Sketch the temperature distribution.

3.25 Approximately 10^6 discrete electrical components can be placed on a single integrated circuit (chip), with electrical heat dissipation as high as 30,000 W/m^2. The chip, which is very thin, is exposed to a dielectric liquid at its outer surface, with $h_o = 1000$ $W/m^2 \cdot K$ and $T_{\infty,o} = 20$°C, and is joined to a circuit board at its inner surface. The thermal contact resistance between the chip and the board is 10^{-4} $m^2 \cdot K/W$, and the board thickness and thermal conductivity are $L_b = 5$ mm and $k_b = 1$ W/m · K, respectively. The other surface of the board is exposed to ambient air for which $h_i = 40$ $W/m^2 \cdot K$ and $T_{\infty,i} = 20$°C.

(a) Sketch the equivalent thermal circuit corresponding to steady-state conditions. In variable form, label appropriate resistances, temperatures, and heat fluxes.

(b) Under steady-state conditions for which the chip heat dissipation is $q''_c = 30,000$ W/m^2, what is the chip temperature?

(c) The maximum allowable heat flux, $q''_{c,m}$, is determined by the constraint that the chip temperature must not exceed 85°C. Determine $q''_{c,m}$ for the foregoing conditions. If air is used in lieu of the dielectric liquid, the convection coefficient is reduced by approximately an order of magnitude. What is the value of $q''_{c,m}$ for $h_o = 100$ $W/m^2 \cdot K$? With air cooling, can significant improvements be realized by using an aluminum oxide circuit board and/or by using a conductive paste at the chip/board interface for which $R''_{t,c} = 10^{-5}$ $m^2 \cdot K/W$?

3.26 A commercial grade cubical freezer, 4 m on a side, has a composite wall consisting of an exterior sheet of 7.0-mm-thick plain carbon steel, an intermediate layer of 100-mm-thick cork insulation, and an inner sheet of 7.0-mm-thick aluminum alloy (2024). Adhesive interfaces between the insulation and the metallic strips are each characterized by a thermal contact resistance of $R''_{t,c} = 2.5 \times 10^{-4}$ $m^2 \cdot K/W$. What is the steady-state cooling load that must be maintained by the refrigerator under conditions for which the outer and inner surface temperatures are 25°C and −5°C, respectively?

3.27 Physicists have determined the theoretical value of the thermal conductivity of a carbon nanotube to be $k_{cn,T} = 5000 \text{ W/m} \cdot \text{K}$.

(a) Assuming the actual thermal conductivity of the carbon nanotube is the same as its theoretical value, find the thermal contact resistance, $R_{t,c}$, that exists between the carbon nanotube and the top surfaces of the heated and sensing islands in Example 3.4 .

(b) Using the value of the thermal contact resistance calculated in part (a), plot the fraction of the total resistance between the heated and sensing islands that is due to the thermal contact resistances for island separation distances of $5 \ \mu\text{m} \le s \le 20 \ \mu\text{m}$.

3.28 Consider a power transistor encapsulated in an aluminum case that is attached at its base to a square aluminum plate of thermal conductivity $k = 240 \text{ W/m} \cdot \text{K}$, thickness $L = 6 \text{ mm}$, and width $W = 20 \text{ mm}$. The case is joined to the plate by screws that maintain a contact pressure of 1 bar, and the back surface of the plate transfers heat by natural convection and radiation to ambient air and large surroundings at $T_\infty = T_{sur} = 25°\text{C}$. The surface has an emissivity of $\varepsilon = 0.9$, and the convection coefficient is $h = 4 \text{ W/m}^2 \cdot \text{K}$. The case is completely enclosed such that heat transfer may be assumed to occur exclusively through the base plate.

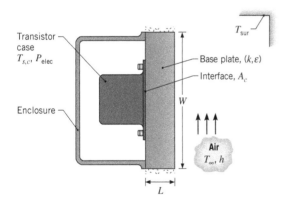

(a) If the air-filled aluminum-to-aluminum interface is characterized by an area of $A_c = 2 \times 10^{-4} \text{ m}^2$ and a roughness of $10 \ \mu\text{m}$, what is the maximum allowable power dissipation if the surface temperature of the case, $T_{s,c}$, is not to exceed $85°\text{C}$?

(b) The convection coefficient may be increased by subjecting the plate surface to a forced flow of air. Explore the effect of increasing the coefficient over the range $4 \le h \le 200 \text{ W/m}^2 \cdot \text{K}$.

Porous Media

3.29 *Ring-porous* woods, such as oak, are characterized by grains. The dark grains consist of very low-density material that forms early in the springtime. The surrounding lighter-colored wood is composed of high-density material that forms slowly throughout most of the growing season.

Assuming the low-density material is highly porous and the oak is dry, determine the fraction of the oak cross-section that appears as being grained. *Hint:* Assume the thermal conductivity parallel to the grains is the same as the radial conductivity of Table A.3.

3.30 A batt of glass fiber insulation is of density $\rho = 28 \text{ kg/m}^3$. Determine the maximum and minimum possible values of the effective thermal conductivity of the insulation at $T = 300 \text{ K}$, and compare with the value reported in Table A.3.

3.31 Determine the density, specific heat, and thermal conductivity of a lightweight aggregate concrete that is composed of 65% stone mix concrete and 35% air by volume. Evaluate properties at $T = 300 \text{ K}$.

3.32 A lightweight aggregate concrete slab of density $\rho = 1500 \text{ kg/m}^3$ consists of a solid, stone mix concrete matrix within which are small spherical pockets of air. Determine the effective thermal conductivity of the slab. Evaluate properties at $T = 300 \text{ K}$.

3.33 A one-dimensional plane wall of thickness L is constructed of a solid material with a linear, nonuniform porosity distribution described by $\varepsilon(x) = \varepsilon_{max}(x/L)$. Plot the steady-state temperature distribution, $T(x)$, for $k_s = 10 \text{ W/m} \cdot \text{K}$, $k_f = 0.1 \text{ W/m} \cdot \text{K}$, $L = 1 \text{ m}$, $\varepsilon_{max} = 0.25$, $T(x = 0) = 30°\text{C}$ and $q''_x = 100 \text{ W/m}^2$ using the expression for the minimum effective thermal conductivity of a porous medium, the expression for the maximum effective thermal conductivity of a porous medium, Maxwell's expression, and for the case where $k_{eff}(x) = k_s$.

Alternative Conduction Analysis

3.34 The diagram shows a conical section fabricated from pure aluminum. It is of circular cross section having diameter $D = ax^{1/2}$, where $a = 0.4$ m$^{1/2}$. The small end is located at $x_1 = 24$ mm and the large end at $x_2 = 124$ mm. The end temperatures are $T_1 = 600$ K and $T_2 = 400$ K, while the lateral surface is well insulated.

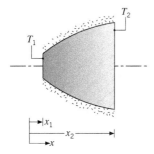

(a) Derive an expression for the temperature distribution $T(x)$ in symbolic form, assuming one-dimensional conditions. Sketch the temperature distribution.

(b) Calculate the heat rate q_x.

3.35 Use the alternative conduction analysis of Section 3.2 to derive an expression relating the radial heat rate, q_r, to the wall temperatures $T_{s,1}$ and $T_{s,2}$, for the hollow cylinder of Figure 3.7. Use your expression to calculate the heat transfer rate associated with a $L = 2$ m long cylinder of inner and outer radii of $r_1 = 50$ mm and $r_2 = 75$ mm, respectively. The thermal conductivity of the cylindrical wall is $k = 2.5$ W/m · K, and the inner and outer surface temperatures are $T_{s,1} = 100°C$ and $T_{s,2} = 67°C$, respectively.

3.36 A truncated solid cone is of circular cross section, and its diameter is related to the axial coordinate by an expression of the form $D = ax^{3/2}$, where $a = 1.0$ m$^{-1/2}$.

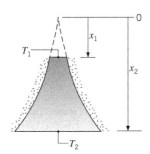

The sides are well insulated, while the top surface of the cone at x_1 is maintained at T_1 and the bottom surface at x_2 is maintained at T_2.

(a) Obtain an expression for the temperature distribution $T(x)$.

(b) What is the rate of heat transfer across the cone if it is constructed of pure aluminum with $x_1 = 0.075$ m, $T_1 = 100°C$, $x_2 = 0.225$ m, and $T_2 = 20°C$?

3.37 From Figure 2.5 it is evident that, over a wide temperature range, the temperature dependence of the thermal conductivity of many solids may be approximated by a linear expression of the form $k = k_o + aT$, where k_o is a positive constant and a is a coefficient that may be positive or negative. Obtain an expression for the heat flux across a plane wall whose inner and outer surfaces are maintained at T_0 and T_1, respectively. Sketch the forms of the temperature distribution corresponding to $a > 0$, $a = 0$, and $a < 0$.

3.38 Consider a tube wall of inner and outer radii r_i and r_o, whose temperatures are maintained at T_i and T_o, respectively. The thermal conductivity of the cylinder is temperature dependent and may be represented by an expression of the form $k = k_o(1 + aT)$, where k_o and a are constants. Obtain an expression for the heat transfer rate per unit length of the tube. What is the thermal resistance of the tube wall?

3.39 Measurements show that steady-state conduction through a plane wall without heat generation produced a convex temperature distribution such that the midpoint temperature was ΔT_o higher than expected for a linear temperature distribution.

Assuming that the thermal conductivity has a linear dependence on temperature, $k = k_o(1 + \alpha T)$, where α is a constant, develop a relationship to evaluate α in terms of ΔT_o, T_1, and T_2.

3.40 A device used to measure the surface temperature of an object to within a spatial resolution of approximately 50 nm is shown in the schematic. It consists of an extremely sharp-tipped stylus and an extremely small cantilever that is scanned across the surface. The probe tip is of circular cross section and is fabricated of polycrystalline silicon dioxide. The ambient temperature is measured at the pivoted end of the cantilever as $T_\infty = 25°C$, and the device is equipped with a sensor to measure the temperature at the upper end of the sharp tip, T_{sen}.

The thermal resistance between the sensing probe and the pivoted end is $R_t = 5 \times 10^6$ K/W.

(a) Determine the thermal resistance between the surface temperature and the sensing temperature.

(b) If the sensing temperature is $T_{\text{sen}} = 28.5°C$, determine the surface temperature.

Hint: Although nanoscale heat transfer effects may be important, assume that the conduction occurring in the air adjacent to the probe tip can be described by Fourier's law and the thermal conductivity found in Table A.4.

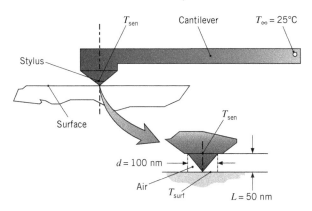

Cylindrical Wall

3.41 A steam pipe of 0.12-m outside diameter is insulated with a layer of calcium silicate.

(a) If the insulation is 20 mm thick and its inner and outer surfaces are maintained at $T_{s,1} = 800$ K and $T_{s,2} = 490$ K, respectively, what is the rate of heat loss per unit length (q') of the pipe?

(b) We wish to explore the effect of insulation thickness on the rate of heat loss q' and outer surface temperature $T_{s,2}$, with the inner surface temperature fixed at $T_{s,1} = 800$ K. The outer surface is exposed to an airflow ($T_\infty = 25°C$) that maintains a convection coefficient of $h = 25$ W/m$^2 \cdot$ K and to large surroundings for which $T_{\text{sur}} = T_\infty = 25°C$. The surface emissivity of calcium silicate is approximately 0.8. Compute and plot the temperature distribution in the insulation as a function of the dimensionless radial coordinate, $(r - r_1)/(r_2 - r_1)$, where $r_1 = 0.06$ m and r_2 is a variable ($0.06 < r_2 \leq 0.20$ m). Compute and plot the rate of heat loss as a function of the insulation thickness for $0 \leq (r_2 - r_1) \leq 0.14$ m.

3.42 The energy consumption associated with a home water heater has two components: (i) the energy that must be supplied to bring the temperature of groundwater to the heater storage temperature, as it is introduced to replace hot water that has been used; (ii) the energy needed to compensate for heat losses incurred while the water is stored at the prescribed temperature. We now wish to determine the energy needed to compensate for heat losses incurred while the water is stored at the prescribed temperature of 55°C. The cylindrical storage tank (with flat ends) has a capacity of 100 gal, and foamed urethane is used to insulate the side and end walls from ambient air at an annual average temperature of 20°C. The resistance to heat transfer is dominated by conduction in the insulation and by free convection in the air, for which $h \approx 2$ W/m$^2 \cdot$ K. If electric resistance heating is used to compensate for the losses and the cost of electric power is \$0.18/kWh, specify tank and insulation dimensions for which the annual cost associated with the heat losses is less than \$50.

3.43 A thin electrical heater is wrapped around the outer surface of a long cylindrical tube whose inner surface is maintained at a temperature of 5°C. The tube wall has inner and outer radii of 25 and 75 mm, respectively, and a thermal conductivity of 10 W/m $\cdot$ K. The thermal contact resistance between the heater and the outer surface of the tube (per unit length of the tube) is $R'_{t,c} = 0.01$ m $\cdot$ K/W. The outer surface of the heater is exposed to a fluid with $T_\infty = -10°C$ and a convection coefficient of $h = 100$ W/m$^2 \cdot$ K. Determine the heater power per unit length of tube required to maintain the heater at $T_o = 25°C$.

3.44 In Problem 3.43, the electrical power required to maintain the heater at $T_o = 25°C$ depends on the thermal conductivity of the wall material k, the thermal contact resistance $R'_{t,c}$ and the convection coefficient h. Compute and plot the separate effect of changes in k ($1 \leq k \leq 200$ W/m $\cdot$ K), $R'_{t,c}$ ($0 \leq R'_{t,c} \leq 0.1$ m $\cdot$ K/W), and h ($10 \leq h \leq 1000$ W/m$^2 \cdot$ K) on the total heater power requirement, as well as the rate of heat transfer to the inner surface of the tube and to the fluid.

3.45 Superheated steam at 575°C is routed from a boiler to the turbine of an electric power plant through steel tubes ($k = 35$ W/m $\cdot$ K) of 300-mm inner diameter and 30-mm wall thickness. To reduce heat loss to the surroundings and to maintain a *safe-to-touch* outer surface temperature, a layer of calcium silicate insulation ($k = 0.10$ W/m $\cdot$ K) is applied to the tubes, while degradation of the insulation is reduced by wrapping it in a thin sheet of aluminum having an emissivity of $\varepsilon = 0.20$. The air and wall temperatures of the power plant are 27°C.

(a) Assuming that the inner surface temperature of a steel tube corresponds to that of the steam and the convection coefficient outside the aluminum sheet is 6 W/m$^2 \cdot$ K, what is the minimum insulation thickness needed to ensure that the temperature of the aluminum does not exceed 50°C? What is the corresponding rate of heat loss per meter of tube length?

(b) Explore the effect of the insulation thickness on the temperature of the aluminum and the rate of heat loss per unit tube length.

3.46 A thin electrical heater is inserted between a long circular rod and a concentric tube with inner and outer radii of 15 and 30 mm. The rod (A) has a thermal conductivity of $k_A = 0.10$ W/m · K, while the tube (B) has a thermal conductivity of $k_B = 1.25$ W/m · K and its outer surface is subjected to convection with a fluid of temperature $T_\infty = -15°C$ and heat transfer coefficient 50 W/m² · K. The thermal contact resistance between the cylinder surfaces and the heater is negligible.

(a) Determine the electrical power per unit length of the cylinders (W/m) that is required to maintain the outer surface of cylinder B at 5°C.

(b) What is the temperature at the center of cylinder A?

3.47 A 2-mm-diameter electrical wire is insulated by a 2-mm-thick rubberized sheath ($k = 0.13$ W/m · K), and the wire/sheath interface is characterized by a thermal contact resistance of $R''_{t,c} = 3 \times 10^{-4}$ m² · K/W. The convection heat transfer coefficient at the outer surface of the sheath is 10 W/m² · K, and the temperature of the ambient air is 20°C. If the temperature of the insulation may not exceed 50°C, what is the maximum allowable electrical power that may be dissipated per unit length of the conductor? What is the critical radius of the insulation?

3.48 Electric current flows through a long rod generating thermal energy at a uniform volumetric rate of $\dot{q} = 2 \times 10^6$ W/m³. The rod is concentric with a hollow ceramic cylinder, creating an enclosure that is filled with air.

The thermal resistance per unit length due to radiation between the enclosure surfaces is $R'_{rad} = 0.30$ m · K/W, and the coefficient associated with free convection in the enclosure is $h = 20$ W/m² · K.

(a) Construct a thermal circuit that can be used to calculate the surface temperature of the rod, T_r. Label all temperatures, heat rates, and thermal resistances, and evaluate each thermal resistance.

(b) Calculate the surface temperature of the rod for the prescribed conditions.

3.49 The evaporator section of a refrigeration unit consists of thin-walled, 10-mm-diameter tubes through which refrigerant passes at a temperature of −18°C. Air is cooled as it flows over the tubes, maintaining a surface convection coefficient of 100 W/m² · K, and is subsequently routed to the refrigerator compartment.

(a) For the foregoing conditions and an air temperature of −3°C, what is the rate at which heat is extracted from the air per unit tube length?

(b) If the refrigerator's defrost unit malfunctions, frost will slowly accumulate on the outer tube surface. Assess the effect of frost formation on the cooling capacity of a tube for frost layer thicknesses in the range $0 \leq \delta \leq 4$ mm. Frost may be assumed to have a thermal conductivity of 0.4 W/m · K.

(c) The refrigerator is disconnected after the defrost unit malfunctions and a 2-mm-thick layer of frost has formed. If the tubes are in ambient air for which $T_\infty = 20°C$ and natural convection maintains a convection coefficient of 2 W/m² · K, how long will it take for the frost to melt? The frost may be assumed to have a mass density of 700 kg/m³ and a latent heat of fusion of 334 kJ/kg.

3.50 A composite cylindrical wall is composed of two materials of thermal conductivity k_A and k_B, which are separated by a very thin, electric resistance heater for which interfacial contact resistances are negligible.

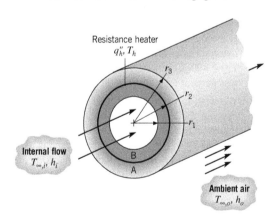

Liquid pumped through the tube is at a temperature $T_{\infty,i}$ and provides a convection coefficient h_i at the inner surface of the composite. The outer surface is exposed to ambient air, which is at $T_{\infty,o}$ and provides a convection coefficient of h_o. Under steady-state conditions, a uniform heat flux of q''_h is dissipated by the heater.

(a) Sketch the equivalent thermal circuit of the system and express all resistances in terms of relevant variables.

(b) Obtain an expression that may be used to determine the heater temperature, T_h.

(c) Obtain an expression for the ratio of heat flows to the outer and inner fluids, q'_o/q'_i. How might the variables of the problem be adjusted to minimize this ratio?

3.51 Steam flowing through a long, thin-walled pipe maintains the pipe wall at a uniform temperature of 600 K. The pipe is covered with an insulation blanket comprised of two different materials, A and B.

The interface between the two materials may be assumed to have an infinite contact resistance, and the entire outer surface is exposed to air for which $T_\infty = 320$ K and $h = 25$ W/m² · K.

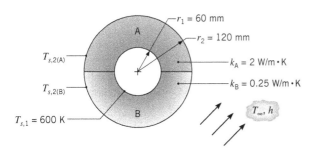

(a) Sketch the thermal circuit of the system. Label (using the preceding symbols) all pertinent nodes and resistances.

(b) For the prescribed conditions, what is the total heat loss from the pipe? What are the outer surface temperatures $T_{s,2(A)}$ and $T_{s,2(B)}$?

3.52 A bakelite coating is to be used with a 10-mm-diameter conducting rod, whose surface is maintained at 200°C by passage of an electrical current. The rod is in a fluid at 25°C, and the convection coefficient is 140 W/m² · K. What is the critical radius associated with the coating? What is the heat transfer rate per unit length for the bare rod and for the rod with a coating of bakelite that corresponds to the critical radius? How much bakelite should be added to reduce the heat transfer associated with the bare rod by 25%?

3.53 An uninsulated, thin-walled pipe of 100-mm diameter is used to transport water to equipment that operates outdoors and uses the water as a coolant. During particularly harsh winter conditions, the pipe wall achieves a temperature of −15°C and a cylindrical layer of ice forms on the inner surface of the wall. If the mean water temperature is 3°C and a convection coefficient of 2000 W/m² · K is maintained at the inner surface of the ice, which is at 0°C, what is the thickness of the ice layer?

Spherical Wall

3.54 A spherical Pyrex glass shell has inside and outside diameters of $D_1 = 0.1$ m and $D_2 = 0.2$ m, respectively. The inner surface is at $T_{s,1} = 100$°C while the outer surface is at $T_{s,2} = 45$°C.

(a) Determine the temperature at the midpoint of the shell thickness, $T(r_m = 0.075$ m).

(b) For the same surface temperatures and dimensions as in part (a), show how the midpoint temperature would change if the shell material were aluminum.

3.55 In Example 3.6, an expression was derived for the critical insulation radius of an insulated, cylindrical tube. Derive the expression that would be appropriate for an insulated sphere.

3.56 A hollow aluminum sphere, with an electrical heater in the center, is used in tests to determine the thermal conductivity of insulating materials. The inner and outer radii of the sphere are 0.15 and 0.18 m, respectively, and testing is done under steady-state conditions with the inner surface of the aluminum maintained at 250°C. In a particular test, a spherical shell of insulation is cast on the outer surface of the sphere to a thickness of 0.12 m. The system is in a room for which the air temperature is 20°C and the convection coefficient at the outer surface of the insulation is 30 W/m² · K. If 80 W are dissipated by the heater under steady-state conditions, what is the thermal conductivity of the insulation?

3.57 A spherical tank for storing liquid oxygen on the space shuttle is to be made from stainless steel of 0.80-m outer diameter and 5-mm wall thickness. The boiling point and latent heat of vaporization of liquid oxygen are 90 K and 213 kJ/kg, respectively. The tank is to be installed in a large compartment whose temperature is to be maintained at 240 K. Design a thermal insulation system that will maintain oxygen losses due to boiling below 1 kg/day.

3.58 A storage tank consists of a cylindrical section that has a length and inner diameter of $L = 2$ m and $D_i = 1$ m, respectively, and two hemispherical end sections. The tank is constructed from 20-mm-thick glass (Pyrex) and is exposed to ambient air for which the temperature is 300 K and the convection coefficient is 10 W/m² · K. The tank is used to store heated oil, which maintains the inner surface at a temperature of 400 K. Determine the electrical power that must be supplied to a heater submerged in the oil if the prescribed conditions are to be maintained. Radiation effects may be neglected, and the Pyrex may be assumed to have a thermal conductivity of 1.4 W/m · K.

3.59 The wall of a spherical tank of 1-m diameter contains an exothermic chemical reaction and is at 200°C when the ambient air temperature is 25°C. What thickness of urethane foam is required to reduce the exterior temperature to 40°C, assuming the convection coefficient is 20 W/m² · K for both situations? What is the percentage reduction in heat rate achieved by using the insulation?

3.60 A composite spherical shell of inner radius $r_1 = 0.30$ m is constructed from lead of outer radius $r_2 = 0.36$ m and AISI 302 stainless steel of outer radius $r_3 = 0.37$ m. The cavity is filled with radioactive wastes that generate heat at a rate of $\dot{q} = 5 \times 10^5$ W/m³. It is proposed to submerge the container in oceanic waters that are at a temperature of $T_\infty = 10$°C and provide a uniform convection coefficient of $h = 500$ W/m² · K at the outer surface of the container. Are there any problems associated with this proposal?

3.61 A spherical vessel used as a reactor for producing pharmaceuticals has a 10-mm-thick stainless steel wall ($k = 17$ W/m · K) and an inner diameter of 1 m. The exterior surface of the vessel is exposed to ambient air ($T_\infty = 25$°C) for which a convection coefficient of 6 W/m² · K may be assumed.

(a) During steady-state operation, an inner surface temperature of 50°C is maintained by energy generated within the reactor. What is the heat loss from the vessel?

(b) If a 20-mm-thick layer of fiberglass insulation ($k = 0.040$ W/m · K) is applied to the exterior of the vessel and the rate of thermal energy generation is unchanged, what is the inner surface temperature of the vessel?

3.62 A spherical shell of inner and outer radii r_i and r_o, respectively, is filled with a heat-generating material that provides for a uniform volumetric generation rate (W/m³) of $\dot{q}$. The outer surface of the shell is exposed to a fluid having a temperature T_∞ and a convection coefficient h. Obtain an expression for the steady-state temperature distribution $T(r)$ in the shell, expressing your result in terms of r_i, r_o, $\dot{q}$, h, T_∞, and the thermal conductivity k of the shell material.

3.63 A spherical tank of 3-m diameter contains a liquified-petroleum gas at −60°C. Insulation with a thermal conductivity of 0.06 W/m · K and thickness 250 mm is applied to the tank to reduce the heat gain.

(a) Determine the radial position in the insulation layer at which the temperature is 0°C when the ambient air temperature is 20°C and the convection coefficient on the outer surface is 6 W/m² · K.

(b) If the insulation is pervious to moisture from the atmospheric air, what conclusions can you reach

about the formation of ice in the insulation? What effect will ice formation have on heat gain to the LP gas? How could this situation be avoided?

3.64 A transistor, which may be approximated as a hemispherical heat source of radius $r_o = 0.1$ mm, is embedded in a large silicon substrate ($k = 125$ W/m · K) and dissipates heat at a rate q. All boundaries of the silicon are maintained at an ambient temperature of $T_\infty = 27$°C, except for the top surface, which is well insulated.

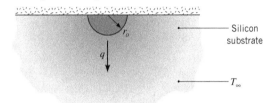

Obtain a general expression for the substrate temperature distribution and evaluate the surface temperature of the heat source for $q = 4$ W.

3.65 Liquid nitrogen ($T = 77$ K) is stored in a thin-walled, spherical container that is covered with a uniformly thick insulation layer of thermal conductivity $k = 0.15$ W/m · K. The outer surface temperature of the insulation is at $T_{s,2} = 20$°C. Due to space constraints, the outer radius of the insulation is fixed at $r_2 = 0.5$ m. Determine the radius of the thin-walled, spherical container that will yield the minimum heat transfer rate per unit nitrogen volume. Also calculate the minimum heat transfer rate per unit nitrogen volume.

Conduction with Thermal Energy Generation

3.66 An uncoated, solid cable of length $L = 1$ m and diameter $D = 40$ mm is exposed to convection conditions characterized by $h = 55$ W/m² · K and $T_\infty = 20$°C. Determine the maximum electric current that can be carried by the cable if it is pure copper, pure aluminum, or pure tin. Calculate the corresponding minimum wire temperatures. The electrical resistivity is $\rho_e = 10 \times 10^{-8}$ Ω · m for copper and aluminum at their melting points, while the electrical resistivity of tin is $\rho_e = 20 \times 10^{-8}$ Ω · m at its melting point.

3.67 The air *inside* a chamber at $T_{\infty,i} = 50$°C is heated convectively with $h_i = 20$ W/m² · K by a 200-mm-thick wall having a thermal conductivity of 4 W/m · K and a uniform heat generation of 1000 W/m³. To prevent any heat generated within the wall from being lost to the *outside* of the chamber at $T_{\infty,o} = 25$°C with $h_o = 5$ W/m² · K, a very thin electrical strip heater is placed on the outer wall to provide a uniform heat flux, q_o''.

(a) Sketch the temperature distribution in the wall on $T - x$ coordinates for the condition where no heat generated within the wall is lost to the *outside* of the chamber.

(b) What are the temperatures at the wall boundaries, $T(0)$ and $T(L)$, for the conditions of part (a)?

(c) Determine the value of q_o'' that must be supplied by the strip heater so that all heat generated within the wall is transferred to the *inside* of the chamber.

(d) If the heat generation in the wall were switched off while the heat flux to the strip heater remained constant, what would be the steady-state temperature, $T(0)$, of the outer wall surface?

3.68 A plane wall of thickness 0.1 m and thermal conductivity 25 W/m · K having uniform volumetric heat generation of 0.3 MW/m³ is insulated on one side, while the other side is exposed to a fluid at 92°C. The convection heat transfer coefficient between the wall and the fluid is 500 W/m² · K. Determine the maximum temperature in the wall.

3.69 Large, cylindrical bales of hay used to feed livestock in the winter months are $D = 2$ m in diameter and are stored end-to-end in long rows. Microbial energy generation occurs in the hay and can be excessive if the farmer bales the hay in a too-wet condition. Assuming the thermal conductivity of baled hay to be $k = 0.04$ W/m · K, determine the maximum steady-state hay temperature for dry hay ($\dot{q} = 1$ W/m³), moist hay ($\dot{q} = 10$ W/m³), and wet hay ($\dot{q} = 100$ W/m³). Ambient conditions are $T_\infty = 0°C$ and $h = 25$ W/m² · K.

3.70 Consider the cylindrical bales of hay in Problem 3.69. It is proposed to utilize the microbial energy generation associated with wet hay to heat water. Consider a 30-mm diameter, thin-walled tube inserted lengthwise through the middle of a cylindrical bale. The tube carries water at $T_{\infty,i} = 20°C$ with $h_i = 200$ W/m² · K.

(a) Determine the steady-state heat transfer rate to the water per unit length of tube.

(b) Plot the radial temperature distribution in the hay, $T(r)$.

(c) Plot the heat transfer rate to the water per unit length of tube for bale diameters of 0.2 m ≤ D ≤ 2 m.

3.71 Consider a plane composite wall that is composed of three materials (materials A, B, and C are arranged left to right) of thermal conductivities $k_A = 0.24$ W/m · K, $k_B = 0.13$ W/m · K, and $k_C = 0.50$ W/m · K. The thicknesses of the three sections of the wall are $L_A = 20$ mm, $L_B = 13$ mm, and $L_C = 20$ mm. A contact resistance of $R_{t,c}'' = 10^{-2}$ m² · K/W exists at the interface between materials A and B, as well as at the interface between materials B and C. The left face of the composite wall is insulated, while the right face is exposed to convective conditions characterized by $h = 10$ W/m² · K, $T_\infty = 20°C$. For Case 1, thermal energy is generated within material A at the rate $\dot{q}_A = 5000$ W/m³. For Case 2, thermal energy is generated within material C at the rate $\dot{q}_C = 5000$ W/m³.

(a) Determine the maximum temperature within the composite wall under steady-state conditions for Case 1.

(b) Sketch the steady-state temperature distribution on $T - x$ coordinates for Case 1.

(c) Sketch the steady-state temperature distribution for Case 2 on the same $T - x$ coordinates used for Case 1.

3.72 An air heater may be fabricated by coiling Nichrome wire and passing air in cross flow over the wire. Consider a heater fabricated from wire of diameter $D = 1$ mm, electrical resistivity $\rho_e = 10^{-6}$ Ω · m, thermal conductivity $k = 25$ W/m · K, and emissivity $\varepsilon = 0.20$. The heater is designed to deliver air at a temperature of $T_\infty = 50°C$ under flow conditions that provide a convection coefficient of $h = 250$ W/m² · K for the wire. The temperature of the housing that encloses the wire and through which the air flows is $T_{sur} = 50°C$.

If the maximum allowable temperature of the wire is $T_{max} = 1200°C$, what is the maximum allowable electric current I? If the maximum available voltage is $\Delta E = 110$ V, what is the corresponding length L of wire that may be used in the heater and the power rating of the heater? *Hint:* In your solution, assume negligible temperature variations within the wire, but after obtaining the desired results, assess the validity of this assumption.

3.73 Consider uniform thermal energy generation inside a one-dimensional plane wall of thickness L with one surface held at $T_{s,1}$ and the other surface insulated.

    (a) Find an expression for the conduction heat flux to the cold surface and the temperature of the hot surface $T_{s,2}$, expressing your results in terms of k, $\dot{q}$, L, and $T_{s,1}$.

    (b) Compare the heat flux found in part (a) with the heat flux associated with a plane wall without energy generation whose surface temperatures are $T_{s,1}$ and $T_{s,2}$.

3.74 Derive Equations C.2, C.5, and C.8 of Appendix C.

3.75 Derive Equations C.3, C.6, and C.9 of Appendix C.

3.76 A plane wall of thickness $2L$ and thermal conductivity k experiences a uniform volumetric generation rate $\dot{q}$. As shown in the sketch for Case 1, the surface at $x = -L$ is perfectly insulated, while the other surface is maintained at a uniform, constant temperature T_o. For Case 2, a very thin dielectric strip is inserted at the midpoint of the wall ($x = 0$) in order to electrically isolate the two sections, A and B. The thermal resistance of the strip is $R_t'' = 0.0005 \text{ m}^2 \cdot \text{K/W}$. The parameters associated with the wall are $k = 50 \text{ W/m} \cdot \text{K}$, $L = 20 \text{ mm}$, $\dot{q} = 5 \times 10^6 \text{ W/m}^3$, and $T_o = 50°\text{C}$.

Case 1   Case 2

    (a) Sketch the temperature distribution for Case 1 on $T - x$ coordinates. Describe the key features of this distribution. Identify the location of the maximum temperature in the wall and calculate this temperature.

    (b) Sketch the temperature distribution for Case 2 on the same $T - x$ coordinates. Describe the key features of this distribution.

    (c) What is the temperature difference between the two walls at $x = 0$ for Case 2?

    (d) What is the location of the maximum temperature in the composite wall of Case 2? Calculate this temperature.

3.77 A nuclear fuel element of thickness $2L$ is covered with a steel cladding of thickness b. Heat generated within the nuclear fuel at a rate $\dot{q}$ is removed by a fluid at T_∞, which adjoins one surface and is characterized by a convection coefficient h. The other surface is well insulated, and the fuel and steel have thermal conductivities of k_f and k_s, respectively.

    (a) Obtain an equation for the temperature distribution $T(x)$ in the nuclear fuel. Express your results in terms of $\dot{q}$, k_f, L, b, k_s, h, and T_∞.

    (b) Sketch the temperature distribution $T(x)$ for the entire system.

3.78 In Problem 3.67 the strip heater acts to *guard* against heat losses from the wall to the outside, and the required heat flux q_o'' depends on chamber operating conditions such as $\dot{q}$ and $T_{\infty,i}$. As a first step in designing a controller for the guard heater, compute and plot q_o'' and $T(0)$ as a function of $\dot{q}$ for $200 \le \dot{q} \le 2000 \text{ W/m}^3$ and $T_{\infty,i} = 30$, 50, and 70°C.

3.79 Steady-state conditions exist within the one-dimensional composite cylinder shown. The three materials have thermal conductivities of k_A, $k_B = 2k_A$, and $k_C = k_A$, respectively. Uniform volumetric energy generation, $\dot{q}$, occurs within Material B and contact resistances exist at r_A and r_B. The cylinder is placed within a vacuum chamber so that only radiation losses occur at $r = r_C$.

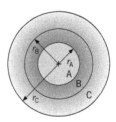

    (a) Write the appropriate form of the heat equation for each of the three materials.

    (b) Write an expression for the temperature $T(r_C)$ in terms of relevant quantities provided above, the temperature of the surroundings T_{sur}, and the emissivity of the exposed surface.

    (c) On $T - r$ coordinates, sketch the radial temperature distribution $T(r)$ for $0 \le r \le r_C$ for the case when $r_A = r_B - r_A = r_C - r_B$.

    (d) On $q'' - r$ coordinates, sketch the radial heat flux for $0 \le r \le r_C$.

3.80 The exposed surface ($x = 0$) of a plane wall of thermal conductivity k is subjected to microwave radiation that causes volumetric heating to vary as

$$\dot{q}(x) = \dot{q}_o \left(1 - \frac{x}{L}\right)$$

where $\dot{q}_o$ (W/m^3) is a constant. The boundary at $x = L$ is perfectly insulated, while the exposed surface is maintained at a constant temperature T_o. Determine the temperature distribution $T(x)$ in terms of x, L, k, $\dot{q}_o$, and T_o.

3.81 A quartz window of thickness L serves as a viewing port in a furnace used for annealing steel. The inner surface ($x = 0$) of the window is irradiated with a uniform heat flux q_o'' due to emission from hot gases in the furnace. A fraction, β, of this radiation may be assumed to be absorbed at the inner surface, while the remaining radiation is partially absorbed as it passes through the quartz. The volumetric heat generation due to this absorption may be described by an expression of the form

$$\dot{q}(x) = (1 - \beta)\, q_o'' \alpha e^{-\alpha x}$$

where α is the absorption coefficient of the quartz. Convection heat transfer occurs from the outer surface ($x = L$) of the window to ambient air at T_∞ and is characterized by the convection coefficient h. Convection and radiation emission from the inner surface may be neglected, along with radiation emission from the outer surface. Determine the temperature distribution in the quartz, expressing your result in terms of the foregoing parameters.

3.82 A cylindrical shell of inner and outer radii, r_i and r_o, respectively, is filled with a heat-generating material that provides a uniform volumetric generation rate (W/m^3) of $\dot{q}$. The inner surface is insulated, while the outer surface of the shell is exposed to a fluid at T_∞ and a convection coefficient h.

(a) Obtain an expression for the steady-state temperature distribution $T(r)$ in the shell, expressing your result in terms of r_i, r_o, $\dot{q}$, h, T_∞, and the thermal conductivity k of the shell material.

(b) Determine an expression for the heat rate, $q'(r_o)$, at the outer radius of the shell in terms of $\dot{q}$ and shell dimensions.

3.83 Consider the configuration of Example 3.8, where uniform volumetric heating within a stainless steel tube is induced by an electric current and heat is transferred by convection to air flowing through the tube. The tube wall has inner and outer radii of $r_1 = 25$ mm and $r_2 = 35$ mm, a thermal conductivity of $k = 15$ W/m · K, an electrical

resistivity of $\rho_e = 0.7 \times 10^{-6}$ Ω · m, and a maximum allowable operating temperature of 1400 K.

(a) Assuming the outer tube surface to be perfectly insulated and the airflow to be characterized by a temperature and convection coefficient of $T_{\infty,1} = 400$ K and $h_1 = 100$ W/m^2 · K, determine the maximum allowable electric current I.

(b) Compute and plot the radial temperature distribution in the tube wall for the electric current of part (a) and three values of h_1 (100, 500, and 1000 W/m^2 · K). For each value of h_1, determine the rate of heat transfer to the air per unit length of tube.

(c) In practice, even the best of insulating materials would be unable to maintain adiabatic conditions at the outer tube surface. Consider use of a refractory insulating material of thermal conductivity $k = 1.0$ W/m · K and neglect radiation exchange at its outer surface. For $h_1 = 100$ W/m^2 · K and the maximum allowable current determined in part (a), compute and plot the temperature distribution in the *composite* wall for two values of the insulation thickness ($\delta = 25$ and 50 mm). The outer surface of the insulation is exposed to room air for which $T_{\infty,2} = 300$ K and $h_2 = 25$ W/m^2 · K. For each insulation thickness, determine the rate of heat transfer per unit tube length to the inner airflow and the ambient air.

3.84 A nuclear reactor fuel element consists of a solid cylindrical pin of radius r_1 and thermal conductivity k_f. The fuel pin is in good contact with a cladding material of outer radius r_2 and thermal conductivity k_c. Consider steady-state conditions for which uniform heat generation occurs within the fuel at a volumetric rate $\dot{q}$ and the outer surface of the cladding is exposed to a coolant that is characterized by a temperature T_∞ and a convection coefficient h.

(a) Obtain equations for the temperature distributions $T_f(r)$ and $T_c(r)$ in the fuel and cladding, respectively. Express your results exclusively in terms of the foregoing variables.

(b) Consider a uranium oxide fuel pin for which $k_f = 2$ W/m · K and $r_1 = 6$ mm and cladding for which $k_c = 25$ W/m · K and $r_2 = 9$ mm. If $\dot{q} = 2 \times 10^8$ W/m^3, $h = 2000$ W/m^2 · K, and $T_\infty = 300$ K, what is the maximum temperature in the fuel element?

(c) Compute and plot the temperature distribution, $T(r)$, for values of $h = 2000$, 5000, and 10,000 W/m^2 · K. If the operator wishes to maintain the centerline temperature of the fuel element below 1000 K, can she do so by adjusting the coolant flow and hence the value of h?

3.85 A high-temperature, gas-cooled nuclear reactor consists of a composite cylindrical wall for which a thorium fuel element ($k \approx 57$ W/m · K) is encased in graphite ($k \approx 3$ W/m · K) and gaseous helium flows through an annular coolant channel. Consider conditions for which the helium temperature is $T_\infty = 600$ K and the convection coefficient at the outer surface of the graphite is $h = 2000$ W/m² · K.

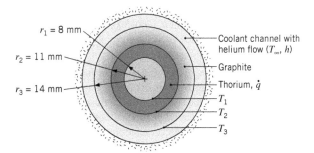

$r_1 = 8$ mm
$r_2 = 11$ mm
$r_3 = 14$ mm

Coolant channel with helium flow (T_∞, h)
Graphite
Thorium, $\dot{q}$
T_1
T_2
T_3

(a) If thermal energy is uniformly generated in the fuel element at a rate $\dot{q} = 10^8$ W/m³, what are the temperatures T_1 and T_2 at the inner and outer surfaces, respectively, of the fuel element?

(b) Compute and plot the temperature distribution in the composite wall for selected values of $\dot{q}$. What is the maximum allowable value of $\dot{q}$?

3.86 A long cylindrical rod of diameter 200 mm with thermal conductivity of 0.5 W/m · K experiences uniform volumetric heat generation of 24,000 W/m³. The rod is encapsulated by a circular sleeve having an outer diameter of 400 mm and a thermal conductivity of 4 W/m · K. The outer surface of the sleeve is exposed to cross flow of air at 27°C with a convection coefficient of 25 W/m² · K.

(a) Find the temperature at the interface between the rod and sleeve and on the outer surface.

(b) What is the temperature at the center of the rod?

3.87 A radioactive material of thermal conductivity k is cast as a solid sphere of radius r_o and placed in a liquid bath for which the temperature T_∞ and convection coefficient h are known. Heat is uniformly generated within the solid at a volumetric rate of $\dot{q}$. Obtain the steady-state radial temperature distribution in the solid, expressing your result in terms of r_o, $\dot{q}$, k, h, and T_∞.

3.88 Radioactive wastes ($k_{rw} = 20$ W/m · K) are stored in a spherical, stainless steel ($k_{ss} = 15$ W/m · K) container of inner and outer radii equal to $r_i = 0.5$ m and $r_o = 0.6$ m. Heat is generated volumetrically within the wastes at a uniform rate of $\dot{q} = 10^5$ W/m³, and the outer surface of

the container is exposed to a water flow for which $h = 1000$ W/m² · K and $T_\infty = 25$°C.

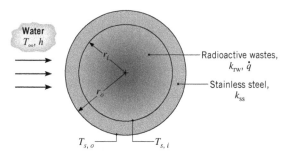

Water T_∞, h

Radioactive wastes, k_{rw}, $\dot{q}$
Stainless steel, k_{ss}

r_i
r_o
$T_{s,o}$
$T_{s,i}$

(a) Evaluate the steady-state outer surface temperature, $T_{s,o}$.

(b) Evaluate the steady-state inner surface temperature, $T_{s,i}$.

(c) Obtain an expression for the temperature distribution, $T(r)$, in the radioactive wastes. Express your result in terms of r_i, $T_{s,i}$, k_{rw}, and $\dot{q}$. Evaluate the temperature at $r = 0$.

(d) A proposed extension of the foregoing design involves storing waste materials having the same thermal conductivity but twice the heat generation ($\dot{q} = 2 \times 10^5$ W/m³) in a stainless steel container of equivalent inner radius ($r_i = 0.5$ m). Safety considerations dictate that the maximum system temperature not exceed 475°C and that the container wall thickness be no less than $t = 0.04$ m and preferably at or close to the original design ($t = 0.1$ m). Assess the effect of varying the outside convection coefficient to a maximum achievable value of $h = 5000$ W/m² · K (by increasing the water velocity) and the container wall thickness. Is the proposed extension feasible? If so, recommend suitable operating and design conditions for h and t, respectively.

3.89 Unique characteristics of biologically active materials such as fruits, vegetables, and other products require special care in handling. Following harvest and separation from producing plants, glucose is catabolized to produce carbon dioxide, water vapor, and heat, with attendant internal energy generation. Consider a carton of apples, each of 80-mm diameter, which is ventilated with air at 5°C and a velocity of 0.5 m/s. The corresponding value of the heat transfer coefficient is 7.5 W/m² · K. Within each apple thermal energy is uniformly generated at a total rate of 4000 J/kg · day. The density and thermal conductivity of the apple are 840 kg/m³ and 0.5 W/m · K, respectively.

Apple, 80 mm diameter

Air
$T_\infty = 5°C$

(a) Determine the apple center and surface temperatures.

(b) For the stacked arrangement of apples within the crate, the convection coefficient depends on the velocity as $h = C_1 V^{0.425}$, where $C_1 = 10.1$ W/m$^2 \cdot$ K $\cdot$ (m/s)$^{0.425}$. Compute and plot the center and surface temperatures as a function of the air velocity for $0.1 \leq V \leq 1$ m/s.

3.90 Consider the plane wall, long cylinder, and sphere shown schematically, each with the same characteristic length a, thermal conductivity k, and uniform volumetric energy generation rate $\dot{q}$.

Plane wall   Long cylinder   Sphere

(a) On the same graph, plot the steady-state dimensionless temperature, $[T(x \text{ or } r) - T(a)]/[(\dot{q}a^2)/2k]$, versus the dimensionless characteristic length, x/a or r/a, for each shape.

(b) Which shape has the smallest temperature difference between the center and the surface? Explain this behavior by comparing the ratio of the volume-to-surface area.

(c) Which shape would be preferred for use as a nuclear fuel element? Explain why.

Extended Surfaces and Simple Fins

3.91 The radiation heat gage shown in the diagram is made from constantan metal foil, which is coated black and is in the form of a circular disk of radius R and thickness t. The gage is located in an evacuated enclosure. The incident radiation flux absorbed by the foil, q_i'', diffuses toward the outer circumference and into the larger copper ring, which acts as a heat sink at the constant temperature $T(R)$. Two copper lead wires are attached to the center of the foil and

to the ring to complete a thermocouple circuit that allows for measurement of the temperature difference between the foil center and the foil edge, $\Delta T = T(0) - T(R)$.

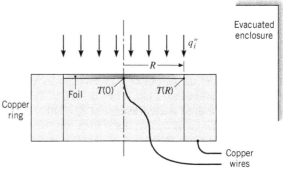

Obtain the differential equation that determines $T(r)$, the temperature distribution in the foil, under steady-state conditions. Solve this equation to obtain an expression relating ΔT to q_i''. You may neglect radiation exchange between the foil and its surroundings.

3.92 Copper tubing is joined to the absorber of a flat-plate solar collector as shown.

The aluminum alloy (2024-T6) absorber plate is 5 mm thick and well insulated on its bottom. The top surface of the plate is separated from a transparent cover plate by an evacuated space. The tubes are spaced a distance L of 0.018 m from each other, and water is circulated through the tubes to remove the collected energy. The water may be assumed to be at a uniform temperature of $T_w = 60°C$. Under steady-state operating conditions for which the *net* radiation heat flux to the surface is $q_{rad}'' = 775$ W/m^2, what is the maximum temperature on the plate and the heat transfer rate per unit length of tube? Note that q_{rad}'' represents the net effect of solar radiation absorption by the absorber plate and radiation exchange between the absorber and cover plates. You may assume the temperature of the absorber plate directly above a tube to be equal to that of the water.

3.93 One method that is used to grow nanowires (nanotubes with solid cores) is to initially deposit a small droplet of a liquid catalyst onto a flat surface. The surface and catalyst are heated and simultaneously exposed to a higher-temperature, low-pressure gas that contains a mixture of chemical species from which the nanowire is to be formed. The catalytic liquid *slowly* absorbs the species from the gas through its top surface and converts these to a solid material that is deposited onto the underlying liquid-solid interface, resulting in construction of the nanowire. The liquid catalyst remains suspended at the tip of the nanowire.

Consider the growth of a 15-nm-diameter silicon carbide nanowire onto a silicon carbide surface. The surface is maintained at a temperature of $T_s = 2400$ K, and the particular liquid catalyst that is used must be maintained in the range 2400 K $\leq T_c \leq$ 3000 K to perform its function. Determine the maximum length of a nanowire that may be grown for conditions characterized by $h = 10^5$ W/m$^2 \cdot$ K and $T_\infty = 8000$ K. Assume properties of the nanowire are the same as for bulk silicon carbide.

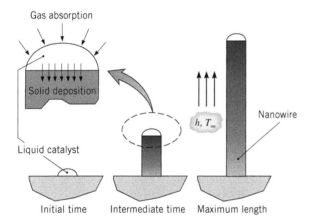

Gas absorption

Solid deposition

Liquid catalyst

h, T_∞

Nanowire

Initial time   Intermediate time   Maximum length

3.94 Consider the manufacture of photovoltaic silicon, as described in Problem 1.39. The thin sheet of silicon is pulled from the pool of molten material *very slowly* and is subjected to an ambient temperature of $T_\infty = 527°$C within the growth chamber. A convection coefficient of $h = 7.5$ W/m$^2 \cdot$ K is associated with the exposed surfaces of the silicon sheet when it is inside the growth chamber. Calculate the maximum allowable velocity of the silicon sheet V_{si}. The latent heat of fusion for silicon is $h_{sf} = 1.8 \times 10^6$ J/kg. It can be assumed that the thermal energy released due to solidification is removed by conduction along the sheet.

3.95 Copper tubing is joined to a solar collector plate of thickness t, and the working fluid maintains the temperature of the plate above the tubes at T_o. There is a uniform net radiation heat flux q''_{rad} to the top surface of the plate, while the bottom surface is well insulated. The top surface is also exposed to a fluid at T_∞ that provides for a uniform convection coefficient h.

(a) Derive the differential equation that governs the temperature distribution $T(x)$ in the plate.

(b) Obtain a solution to the differential equation for appropriate boundary conditions.

3.96 A thin flat plate of length L, thickness t, and width $W \gg L$ is thermally joined to two large heat sinks at different temperatures, $T(0) = T_o$ and $T(L) = T_L$, and with the bottom surface not insulated. Convection heat transfer is now allowed to occur between this surface and a fluid at T_∞, with a convection coefficient h.

(a) Derive the differential equation that determines the steady-state temperature distribution $T(x)$ in the plate.

(b) Solve the foregoing equation for the temperature distribution, and obtain an expression for the rate of heat transfer from the plate to the heat sinks.

(c) For $q''_o = 20,000$ W/m^2, $T_o = 100°$C, $T_L = 35°$C, $T_\infty = 25°$C, $k = 25$ W/m $\cdot$ K, $h = 50$ W/m$^2 \cdot$ K, $L = 100$ mm, $t = 5$ mm, and a plate width of $W = 30$ mm, plot the temperature distribution and determine the sink heat rates, $q_x(0)$ and $q_x(L)$. On the same graph, plot three additional temperature distributions corresponding to changes in the following parameters, with the remaining parameters unchanged: (i) $q''_o = 30,000$ W/m^2, (ii) $h = 200$ W/m$^2 \cdot$ K, and (iii) the value of q''_o for which $q_x(0) = 0$ when $h = 200$ W/m$^2 \cdot$ K.

3.97 A bonding operation utilizes a laser to provide a constant heat flux, q''_o, across the top surface of a thin adhesive-backed, plastic film to be affixed to a metal strip as shown in the sketch. The metal strip has a thickness $d = 1.25$ mm, and its width is large relative to that of the film. The thermophysical properties of the strip are $\rho = 7850$ kg/m^3, $c_p = 435$ J/kg $\cdot$ K, and $k = 60$ W/m $\cdot$ K. The thermal resistance of the plastic film of width $w_1 = 40$ mm is negligible. The upper and lower surfaces of the strip (including the plastic film) experience convection with air at 25°C and a convection coefficient of 10 W/m$^2 \cdot$ K. The strip and film are very long in the direction normal to the page. Assume the edges of the metal strip are at the air temperature (T_∞).

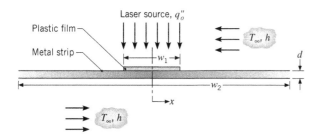

(a) Derive an expression for the temperature distribution in the portion of the steel strip with the plastic film ($-w_1/2 \le x \le +w_1/2$).

(b) If the heat flux provided by the laser is 10,000 W/m², determine the temperature of the plastic film at the center ($x = 0$) and its edges ($x = \pm w_1/2$).

(c) Plot the temperature distribution for the entire strip and point out its special features.

3.98 Consider a rod of diameter D, thermal conductivity k, and length $2L$ that is perfectly insulated over one portion of its length, $-L \le x \le 0$, and experiences convection with a fluid (T_∞, h) over the other portion, $0 \le x \le +L$. One end is maintained at T_1, while the other is separated from a heat sink at T_3 by an interfacial thermal contact resistance $R''_{t,c}$.

(a) Sketch the temperature distribution on $T - x$ coordinates and identify its key features. Assume that $T_1 > T_3 > T_\infty$.

(b) Derive an expression for the midpoint temperature T_2 in terms of the thermal and geometric parameters of the system.

(c) For $T_1 = 200°C$, $T_3 = 100°C$, and the conditions provided in the schematic, calculate T_2 and plot the temperature distribution. Describe key features of the distribution and compare it to your sketch of part (a).

3.99 A carbon nanotube is suspended across a trench of width $s = 5$ μm that separates two islands, each at $T_\infty = 300$ K. A focused laser beam irradiates the nanotube at a distance ξ from the left island, delivering $q = 10$ μW of energy to the nanotube. The nanotube temperature is measured at the midpoint of the trench using a point probe.

The measured nanotube temperature is $T_1 = 324.5$ K for $\xi_1 = 1.5$ μm and $T_2 = 326.4$ K for $\xi_2 = 3.5$ μm.

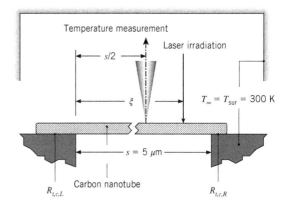

Determine the two contact resistances, $R_{t,c,L}$ and $R_{t,c,R}$ at the left and right ends of the nanotube, respectively. The experiment is performed in a vacuum with $T_{sur} = 300$ K. The nanotube thermal conductivity and diameter are $k_{cn} = 3100$ W/m · K and $D = 14$ nm, respectively.

3.100 A probe of overall length $L = 200$ mm and diameter $D = 12.5$ mm is inserted through a duct wall such that a portion of its length, referred to as the immersion length L_i, is in contact with the water stream whose temperature, $T_{\infty,i}$, is to be determined. The convection coefficients over the immersion and ambient-exposed lengths are $h_i = 1100$ W/m² · K and $h_o = 10$ W/m² · K, respectively. The probe has a thermal conductivity of 177 W/m · K and is in poor thermal contact with the duct wall.

(a) Derive an expression for evaluating the measurement error, $\Delta T_{err} = T_{tip} - T_{\infty,i}$, which is the difference between the tip temperature, T_{tip}, and the water temperature, $T_{\infty,i}$. *Hint:* Define a coordinate system with the origin at the duct wall and treat the probe as two fins extending inward and outward from the duct, but having the same base temperature. Use Case A results from Table 3.4.

(b) With the water and ambient air temperatures at 80 and 20°C, respectively, calculate the measurement error, ΔT_{err}, as a function of immersion length for the conditions $L_i/L = 0.225, 0.425$, and 0.625.

(c) Compute and plot the effects of probe thermal conductivity and water velocity (h_i) on the measurement error.

3.101 A metal rod of length $2L$, diameter D, and thermal conductivity k is inserted into a perfectly insulating wall, exposing one-half of its length to an airstream that is of temperature T_∞ and provides a convection coefficient h at the surface of the rod. An electromagnetic field induces volumetric energy generation at a uniform rate $\dot{q}$ within the *embedded* portion of the rod.

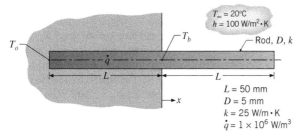

$$T_\infty = 20°C$$
$$h = 100 \ W/m^2 \cdot K$$
$$L = 50 \ mm$$
$$D = 5 \ mm$$
$$k = 25 \ W/m \cdot K$$
$$\dot{q} = 1 \times 10^6 \ W/m^3$$

(a) Derive an expression for the steady-state temperature T_b at the base of the exposed half of the rod. The exposed region may be approximated as a very long fin.

(b) Derive an expression for the steady-state temperature T_o at the end of the embedded half of the rod.

(c) Using numerical values provided in the schematic, plot the temperature distribution in the rod and describe key features of the distribution. Does the rod behave as a very long fin?

3.102 Turbine blades mounted to a rotating disc in a gas turbine engine are exposed to a gas stream that is at $T_\infty = 1200°C$ and maintains a convection coefficient of $h = 250 \ W/m^2 \cdot K$ over the blade.

The blades, which are fabricated from Inconel, $k \approx 20 \ W/m \cdot K$, have a length of $L = 50 \ mm$. The blade

profile has a uniform cross-sectional area of $A_c = 6 \times 10^{-4} \ m^2$ and a perimeter of $P = 110 \ mm$. A proposed blade-cooling scheme, which involves routing air through the supporting disc, is able to maintain the base of each blade at a temperature of $T_b = 300°C$.

(a) If the maximum allowable blade temperature is 1050°C and the blade tip may be assumed to be adiabatic, is the proposed cooling scheme satisfactory?

(b) For the proposed cooling scheme, what is the rate at which heat is transferred from each blade to the coolant?

3.103 In a test to determine the friction coefficient μ associated with a disk brake, one disk and its shaft are rotated at a constant angular velocity ω, while an equivalent disk/shaft assembly is stationary. Each disk has an outer radius of $r_2 = 180 \ mm$, a shaft radius of $r_1 = 20 \ mm$, a thickness of $t = 12 \ mm$, and a thermal conductivity of $k = 15 \ W/m \cdot K$. A known force F is applied to the system, and the corresponding torque τ required to maintain rotation is measured. The disk contact pressure may be assumed to be uniform (i.e., independent of location on the interface), and the disks may be assumed to be well insulated from the surroundings.

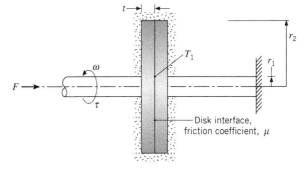

(a) Obtain an expression that may be used to evaluate μ from known quantities.

(b) For the region $r_1 \le r \le r_2$, determine the radial temperature distribution $T(r)$ in the disk, where $T(r_1) = T_1$ is presumed to be known.

(c) Consider test conditions for which $F = 200 \ N$, $\omega = 40 \ rad/s$, $\tau = 8 \ N \cdot m$, and $T_1 = 80°C$. Evaluate the friction coefficient and the maximum disk temperature.

3.104 Consider an extended surface of rectangular cross section with heat flow in the longitudinal direction.

In this problem we seek to determine conditions for which the transverse (*y*-direction) temperature difference within the extended surface is negligible compared to the temperature difference between the surface and the environment, such that the one-dimensional analysis of Section 3.6.1 is valid.

(a) Assume that the transverse temperature distribution is parabolic and of the form

$$\frac{T(y) - T_o(x)}{T_s(x) - T_o(x)} = \left(\frac{y}{t}\right)^2$$

where $T_s(x)$ is the surface temperature and $T_o(x)$ is the centerline temperature at any *x*-location. Using Fourier's law, write an expression for the conduction heat flux at the surface, $q_y''(t)$, in terms of T_s and T_o.

(b) Write an expression for the convection heat flux at the surface for the *x*-location. Equating the two expressions for the heat flux by conduction and convection, identify the parameter that determines the ratio $(T_o - T_s)/(T_s - T_\infty)$.

(c) From the foregoing analysis, develop a criterion for establishing the validity of the one-dimensional assumption used to model an extended surface.

3.105 A long, circular aluminum rod is attached at one end to a heated wall and transfers heat by convection to a cold fluid.

(a) If the diameter of the rod is doubled, by how much would the rate of heat removal change?

(b) If a copper rod of the same diameter is used in place of the aluminum, by how much would the rate of heat removal change?

3.106 A brass rod 100 mm long and 5 mm in diameter extends horizontally from a casting at 200°C. The rod is in an air environment with $T_\infty = 20°C$ and $h = 30$ W/m^2 · K. What is the temperature of the rod 25, 50, and 100 mm from the casting?

3.107 The extent to which the tip condition affects the thermal performance of a fin depends on the fin geometry and thermal conductivity, as well as the convection coefficient. Consider an alloyed aluminum ($k = 180$ W/m · K) rectangular fin of length $L = 10$ mm, thickness $t = 1$ mm, and width $w \gg t$. The base temperature of the fin is $T_b = 100°C$, and the fin is exposed to a fluid of temperature $T_\infty = 25°C$.

(a) Assuming a uniform convection coefficient of $h = 100$ W/m^2 · K over the entire fin surface, determine the fin heat transfer rate per unit width q_f', efficiency η_f, effectiveness ε_f, thermal resistance per unit width $R_{t,f}'$, and the tip temperature $T(L)$ for

Cases A and B of Table 3.4. Contrast your results with those based on an *infinite fin* approximation.

(b) Explore the effect of variations in the convection coefficient on the heat rate for $10 < h < 1000$ W/m^2 · K. Also consider the effect of such variations for a stainless steel fin ($k = 15$ W/m · K).

3.108 A pin fin of uniform, cross-sectional area is fabricated of an aluminum alloy ($k = 150$ W/m · K). The fin diameter is $D = 3$ mm, and the fin is exposed to convective conditions characterized by $h = 220$ W/m^2 · K. It is reported that the fin efficiency is $\eta_f = 0.65$. Determine the fin length L and the fin effectiveness ε_f. Account for tip convection.

3.109 The extent to which the tip condition affects the thermal performance of a fin depends on the fin geometry and thermal conductivity, as well as the convection coefficient. Consider an alloyed aluminum ($k = 180$ W/m · K) rectangular fin whose base temperature is $T_b = 100°C$. The fin is exposed to a fluid of temperature $T_\infty = 25°C$, and a uniform convection coefficient of $h = 100$ W/m^2 · K may be assumed for the fin surface.

(a) For a fin of length $L = 10$ mm, thickness $t = 1$ mm, and width $w \gg t$, determine the fin heat transfer rate per unit width q_f', efficiency η_f, effectiveness ε_f, thermal resistance per unit width $R_{t,f}'$, and tip temperature $T(L)$ for Cases A and B of Table 3.4. Contrast your results with those based on an *infinite fin* approximation.

(b) Explore the effect of variations in L on the heat rate for $3 < L < 50$ mm. Also consider the effect of such variations for a stainless steel fin ($k = 15$ W/m · K).

3.110 Two long copper rods of diameter $D = 10$ mm are soldered together end to end, with solder having a melting point of 650°C. The rods are in air at 25°C with a convection coefficient of 10 W/m^2 · K. What is the minimum power input needed to effect the soldering?

3.111 During the initial stages of the growth of the nanowire of Problem 3.93, a slight perturbation of the liquid catalyst droplet can cause it to be suspended on the top of the nanowire in an off-center position. The resulting nonuniform deposition of solid at the solid-liquid interface can be manipulated to form engineered shapes such as a *nanospring*, that is characterized by a spring radius r, spring pitch s, overall chord length L_c (length running along the spring), and end-to-end length L, as shown in the sketch. Consider a silicon carbide nanospring of diameter $D = 15$ nm, $r = 30$ nm, $s = 25$ nm, and $L_c = 425$ nm. From experiments, it is known that the average spring pitch $\bar{s}$ varies with average temperature $\bar{T}$ by the relation $d\bar{s}/d\bar{T} = 0.1$ nm/K.

Using this information, a student suggests that a *nanoactuator* can be constructed by connecting one end of the nanospring to a small heater and raising the temperature of that end of the nanospring above its initial value. Calculate the actuation distance ΔL for conditions where $h = 10^6$ W/m² · K, $T_\infty = T_i = 25°C$, with a base temperature of $T_b = 50°C$. If the base temperature can be controlled to within 1°C, calculate the accuracy to which the actuation distance can be controlled. *Hint:* Assume the spring radius does not change when the spring is heated. The overall spring length may be approximated by the formula,

$$L = \frac{\overline{s}}{2\pi} \frac{L_c}{\sqrt{r^2 + (\overline{s}/2\pi)^2}}$$

3.112 Consider two long, slender rods of the same diameter but different materials. One end of each rod is attached to a base surface maintained at 100°C, while the surfaces of the rods are exposed to ambient air at 20°C. By traversing the length of each rod with a thermocouple, it was observed that the temperatures of the rods were equal at the positions $x_A = 0.15$ m and $x_B = 0.075$ m, where x is measured from the base surface. If the thermal conductivity of rod A is known to be $k_A = 70$ W/m · K, determine the value of k_B for rod B.

3.113 An experimental arrangement for measuring the thermal conductivity of solid materials involves the use of two long rods that are equivalent in every respect, except that one is fabricated from a standard material of known thermal conductivity k_A while the other is fabricated from the material whose thermal conductivity k_B is desired. Both rods are attached at one end to a heat source of fixed temperature T_b, are exposed to a fluid of temperature T_∞, and are instrumented with thermocouples to measure the temperature at a fixed distance x_1 from the heat source. If the standard material is aluminum, with $k_A = 200$ W/m · K, and measurements reveal values of $T_A = 75°C$ and $T_B = 60°C$ at x_1 for $T_b = 100°C$ and $T_\infty = 25°C$, what is the thermal conductivity k_B of the test material?

Fin Systems and Arrays

3.114 Due to blockage of fluid flow, the convection heat transfer coefficient for the finned and prime surfaces of a fin array in general decreases as the number of fins increases. Consider the fin array shown in Figure 3.21a, with wall height of 20 mm, $w = 20$ mm, $t = 2$ mm, and $L = 20$ mm. The fins are aluminum and the base and environment temperatures are $T_b = 95°C$ and $T_\infty = 20°C$. As a first approximation, assume $h = h_{max} \times (1 - N/N_{max})$ where N_{max} is the maximum number of fins that can be placed on a surface, so that $N_{max} = 20$ mm/2 mm = 10. For $h_{max} = 50$ W/m² · K, determine the total heat rate for $N = 0, 3, 6,$ and 9 fins.

3.115 Finned passages are frequently formed between parallel plates to enhance convection heat transfer in compact heat exchanger cores. An important application is in electronic equipment cooling, where one or more air-cooled stacks are placed between heat-dissipating electrical components. Consider a single stack of rectangular fins of length L and thickness t, with convection conditions corresponding to h and T_∞.

(a) Obtain expressions for the fin heat transfer rates, $q_{f,o}$ and $q_{f,L}$, in terms of the base temperatures, T_o and T_L.

(b) In a specific application, a stack that is 200 mm wide and 100 mm deep contains 50 fins, each of length $L = 12$ mm. The entire stack is made from aluminum, which is everywhere 1.0 mm thick. If temperature limitations associated with electrical components joined to opposite plates dictate maximum allowable plate temperatures of $T_o = 400$ K and $T_L = 350$ K, what are the corresponding maximum power dissipations if $h = 150$ W/m² · K and $T_\infty = 300$ K?

3.116 The fin array of Problem 3.115 is commonly found in *compact heat exchangers,* whose function is to provide a large surface area per unit volume in transferring heat from one fluid to another. Consider conditions for which the second fluid maintains equivalent temperatures at the parallel plates,

$T_o = T_L$, thereby establishing symmetry about the midplane of the fin array. The heat exchanger is 1 m long in the direction of the flow of air (first fluid) and 1 m wide in a direction normal to both the airflow and the fin surfaces. The length of the fin passages between adjoining parallel plates is $L = 8$ mm, whereas the fin thermal conductivity and convection coefficient are $k = 200$ W/m · K (aluminum) and $h = 150$ W/m^2 · K, respectively.

(a) If the fin thickness and pitch are $t = 1$ mm and $S = 4$ mm, respectively, what is the value of the thermal resistance $R_{t,o}$ for a one-half section of the fin array?

(b) Subject to the constraints that the fin thickness and pitch may not be less than 0.5 and 3 mm, respectively, assess the effect of changes in t and S.

3.117 The engine cylinder of a motorcycle is constructed of 2024-T6 aluminum alloy and is of height $H = 0.15$ m and outside diameter $D = 50$ mm. Under typical operating conditions the outer surface of the cylinder is at a temperature of 500 K and is exposed to ambient air at 300 K, with a convection coefficient of 50 W/m^2 · K. Annular fins are integrally cast with the cylinder to increase heat transfer to the surroundings. Consider five such fins, which are of thickness $t = 6$ mm, length $L = 20$ mm, and equally spaced. What is the increase in heat transfer due to use of the fins?

3.118 In Example 1.5, we saw that to generate an electrical power of $P = 9$ W, the temperature of the PEM fuel cell had to be maintained at $T_c \approx 56.4°$C, which required removal of 11.25 W from the fuel cell and a cooling air velocity of $V = 9.4$ m/s for $T_\infty = 25°$C. To provide these convective conditions, the fuel cell is centered in a 50 mm × 26 mm rectangular duct, with 10-mm gaps between the exterior of the 50 mm × 50 mm × 6 mm fuel cell and the top and bottom of the well-insulated duct wall. A small fan, powered by the fuel cell, is used to circulate the cooling air. Inspection of a particular fan vendor's data sheets suggests that the ratio of the fan power consumption to the fan's volumetric flow rate is $P_f/\dot{V}_f = C = 1000$ W/(m^3/s) for the range $10^{-4} \le \dot{V}_f \le 10^{-2}$ m^3/s.

1. Determine the *net* electric power produced by the fuel cell–fan system, $P_{net} = P - P_f$.

2. Consider the effect of attaching an aluminum ($k = 200$ W/m · K) *finned heat sink*, of identical top and bottom sections, onto the fuel cell body. The contact joint has a thermal resistance of $R''_{t,c} = 10^{-3}$ m^2 · K/W, and the base of the heat sink is of thickness $t_b = 2$ mm. Each of the N rectangular fins is of length $L_f = 8$ mm and thickness $t_f = 1$ mm, and spans the entire length of the fuel cell, $L_c = 50$ mm. With the heat sink in place, radiation losses are negligible and the convective heat transfer coefficient may be related to the size and geometry of a typical air channel by an expression of the form $h = 1.78 k_{air} (L_f + a)/(L_f \cdot a)$, where a is the distance between fins. Draw an equivalent thermal circuit for part 2 and determine the total number of fins needed to reduce the fan power consumption to half of the value found in part 1.

3.119 As more and more components are placed on a single integrated circuit (chip), the amount of heat that is dissipated continues to increase. However, this increase is limited by the maximum allowable chip operating temperature, which is approximately 75°C. To maximize heat dissipation, it is proposed that a 4 × 4 array of copper pin fins be metallurgically joined to the outer surface of a square chip that is 12.7 mm on a side.

(a) Sketch the equivalent thermal circuit for the pin–chip–board assembly, assuming one-dimensional, steady-state conditions and negligible contact resistance between the pins and the chip. In variable form, label appropriate resistances, temperatures, and heat rates.

(b) For the conditions prescribed in Problem 3.25, what is the maximum rate at which heat can be dissipated in the chip when the pins are in place? That is, what is the value of q_c for $T_c = 75°$C? The pin diameter and length are $D_p = 1.5$ mm and $L_p = 15$ mm.

3.120 In Problem 3.119, the prescribed value of $h_o = 1000$ W/m^2 · K is large and characteristic of liquid cooling. In practice it would be far more preferable to use air cooling, for which a reasonable upper limit to the convection coefficient would be $h_o = 250$ W/m^2 · K. Assess the effect of changes in the pin fin geometry on the chip heat rate if the remaining conditions of Problem 3.119, including a maximum allowable chip temperature of 75°C, remain in effect. Parametric variations that may be considered include the total number of pins N in the square array, the pin diameter D_p, and the pin length L_p. However, the product $N^{1/2}D_p$ should not exceed 9 mm to ensure adequate airflow passage through the array. Recommend a design that enhances chip cooling.

3.121 Because of the large number of devices in today's PC chips, finned heat sinks are often used to maintain the chip at an acceptable operating temperature. Two fin designs are to be evaluated, both of which have base (unfinned) area dimensions of 53 mm × 57 mm. The fins are of square cross section and fabricated from an extruded aluminum alloy with a thermal conductivity of 175 W/m · K. Cooling air may be supplied at 25°C, and the maximum allowable chip temperature is 75°C. Other features of the design and operating conditions are tabulated.

Fin Dimensions

Design	Cross Section $w \times w$ (mm)	Length L (mm)	Number of Fins in Array	Convection Coefficient (W/m² · K)
A	3×3	30	6×9	125
B	1×1	7	14×17	375

54 pins, 9 × 6 array
(Design A)

Determine which fin arrangement is superior. In your analysis, calculate the heat rate, efficiency, and effectiveness of a single fin, as well as the total heat rate and overall efficiency of the array. Since real estate inside the computer enclosure is important, compare the total heat rate per unit volume for the two designs.

3.122 Consider design B of Problem 3.121. Over time, dust can collect in the fine grooves that separate the fins. Consider the buildup of a dust layer of thickness L_d, as shown in the sketch. Calculate and plot the total heat rate for design B for dust layers in the range $0 \leq L_d \leq 5$ mm.

The thermal conductivity of the dust can be taken as $k_d = 0.032$ W/m · K. Include the effects of convection from the fin tip.

3.123 A long rod of 20-mm diameter and a thermal conductivity of 1.5 W/m · K has a uniform internal volumetric thermal energy generation of 10^6 W/m³. The rod is covered with an electrically insulating sleeve of 2-mm thickness and thermal conductivity of 0.5 W/m · K. A spider with 12 ribs and dimensions as shown in the sketch has a thermal conductivity of 175 W/m · K, and is used to support the rod and to maintain concentricity with an 80-mm-diameter tube. Air at $T_\infty = 25$°C passes over the spider surface, and the convection coefficient is 20 W/m² · K. The outer surface of the tube is well insulated.

We wish to increase volumetric heating within the rod, while not allowing its centerline temperature to exceed 100°C. Determine the impact of the following changes, which may be effected independently or concurrently: (i) increasing the air speed and hence the convection coefficient; (ii) changing the number and/or thickness of the ribs; and (iii) using an electrically nonconducting sleeve material of larger thermal conductivity (e.g., amorphous carbon or quartz). Recommend a realistic configuration that yields a significant increase in $\dot{q}$.

3.124 Determine the percentage increase in heat transfer associated with attaching aluminum fins of rectangular profile to a plane wall. The fins are 50 mm long, 0.5 mm thick, and are equally spaced at a distance of 4 mm (250 fins/m). The convection coefficient associated with the bare wall is 40 W/m² · K, while that resulting from attachment of the fins is 30 W/m² · K.

3.125 Heat transfer from a transistor may be enhanced by inserting it in an aluminum sleeve ($k = 200$ W/m · K) having 12 integrally machined longitudinal fins on its outer surface. The transistor radius and height are $r_1 = 2.5$ mm and $H = 4$ mm, respectively, while the fins are of length $L = r_3 - r_2 = 8$ mm and uniform thickness $t = 0.8$ mm. The thickness of the sleeve base is $r_2 - r_1 = 1$ mm, and the contact resistance of the sleeve-transistor interface is $R''_{t,c} = 0.6 \times 10^{-3}$ m² · K/W. Air at $T_\infty = 20°C$ flows over the fin surface, providing an approximately uniform convection coeffficient of $h = 30$ W/m² · K.

(a) When the transistor case temperature is 80°C, what is the rate of heat transfer from the sleeve?

(b) Identify all of the measures that could be taken to improve design and/or operating conditions, such that heat dissipation may be increased while still maintaining a case temperature of 80°C. In words, assess the relative merits of each measure. Choose what you believe to be the three most promising measures, and numerically assess the effect of corresponding changes in design and/or operating conditions on thermal performance.

3.126 Aluminum fins of triangular profile are attached to a plane wall whose surface temperature is 250°C. The fin base thickness is 2 mm, and its length is 6 mm. The system is in ambient air at a temperature of 20°C, and the surface convection coefficient is 40 W/m² · K.

(a) What are the fin efficiency and effectiveness?

(b) What is the heat dissipated per unit width by a single fin?

3.127 Annular aluminum fins of rectangular profile are attached to a circular tube having an outside diameter of 50 mm and an outer surface temperature of 200°C. The fins are 4 mm thick and 15 mm long. The system is in ambient air at a temperature of 20°C, and the surface convection coefficient is 40 W/m² · K.

(a) What are the fin efficiency and effectiveness?

(b) If there are 125 such fins per meter of tube length, what is the rate of heat transfer per unit length of tube?

3.128 It is proposed to air-cool the cylinders of a combustion chamber by joining an aluminum casing with annular fins ($k = 240$ W/m · K) to the cylinder wall ($k = 50$ W/m · K).

The air is at 320 K and the corresponding convection coefficient is 100 W/m² · K. Although heating at the inner surface is periodic, it is reasonable to assume steady-state conditions with a time-averaged heat flux of $q''_i = 10^5$ W/m². Assuming negligible contact resistance between the wall and the casing, determine the wall inner temperature T_i, the interface temperature T_1, and the fin base temperature T_b. Determine these temperatures if the interface contact resistance is $R''_{t,c} = 10^{-4}$ m² · K/W.

3.129 Consider the air-cooled combustion cylinder of Problem 3.128, but instead of imposing a uniform heat flux at the inner surface, consider conditions for which the time-averaged temperature of the combustion gases is $T_g = 1100$ K and the corresponding convection coefficient is $h_g = 150$ W/m² · K. All other conditions, including the cylinder/casing contact resistance, remain the same. Determine the heat rate per unit length of cylinder (W/m), as well as the cylinder inner temperature T_i, the interface temperatures $T_{1,i}$ and $T_{1,o}$, and the fin base temperature T_b. Subject to the constraint that the fin gap is fixed at $\delta = 2$ mm, assess the effect of increasing the fin thickness at the expense of reducing the number of fins.

The Bioheat Equation

3.130 Consider the conditions of Example 3.11, except that the person is now exercising (in the air environment), which increases the metabolic heat generation rate by a factor of 8, to 5600 W/m^3. At what rate would the person have to perspire (in liters/s) to maintain the same skin temperature as in that example?

3.131 Consider the conditions of Example 3.11 for an air environment, except now the air and surroundings temperatures are both 15°C. Humans respond to cold by shivering, which increases the metabolic heat generation rate. What would the metabolic heat generation rate (per unit volume) have to be to maintain a comfortable skin temperature of 33°C under these conditions?

3.132 Consider heat transfer in a forearm, which can be approximated as a cylinder of muscle of radius 50 mm (neglecting the presence of bones), with an outer layer of skin and fat of thickness 3 mm. There is metabolic heat generation and perfusion within the muscle. The metabolic heat generation rate, perfusion rate, arterial temperature, and properties of blood, muscle, and skin/fat layer are identical to those in Example 3.11. The environment and surroundings are the same as for the air environment in Example 3.11.

(a) Write the bioheat transfer equation in radial coordinates. Write the boundary conditions that express symmetry at the centerline of the forearm and specified temperature at the outer surface of the muscle. Solve the differential equation and apply the boundary conditions to find an expression for the temperature distribution. Note that the derivatives of the modified Bessel functions are given in Section 3.6.4.

(b) Equate the heat flux at the outer surface of the muscle to the heat flux through the skin/fat layer and into the environment to determine the temperature at the outer surface of the muscle.

(c) Find the maximum forearm temperature.

Thermoelectric Power Generation

3.133 For one of the $M = 48$ modules of Example 3.12, determine a variety of different efficiency values concerning the conversion of waste heat to electrical energy.

(a) Determine the thermodynamic efficiency, $\eta_{\text{therm}} \equiv P_{M=1}/q_1$.

(b) Determine the figure of merit $Z\bar{T}$ for one module, and the thermoelectric efficiency, η_{TE} using Equation 3.128.

(c) Determine the Carnot efficiency, $\eta_{\text{Carnot}} = 1 - T_2/T_1$.

(d) Determine both the thermoelectric efficiency and the Carnot efficiency for the case where $h_1 = h_2 \to \infty$.

(e) The energy conversion efficiency of thermoelectric devices is commonly reported by evaluating Equation 3.128, but with $T_{\infty,1}$ and $T_{\infty,2}$ used instead of T_1 and T_2, respectively. Determine the value of η_{TE} based on the inappropriate use of $T_{\infty,1}$ and $T_{\infty,2}$, and compare with your answers for parts (b) and (d).

3.134 One of the thermoelectric modules of Example 3.12 is installed between a hot gas at $T_{\infty,1} = 450$°C and a cold gas at $T_{\infty,2} = 20$°C. The convection coefficient associated with the flowing gases is $h = h_1 = h_2 = 80$ W/m$^2 \cdot$K while the electrical resistance of the load is $R_{e,\text{load}} = 4$ Ω.

(a) Sketch the equivalent thermal circuit and determine the electric power generated by the module for the situation where the hot and cold gases provide convective heating and cooling directly to the module (no heat sinks).

(b) Two heat sinks ($k = 180$ W/m $\cdot$ K; see sketch), each with a base thickness of $L_b = 4$ mm and fin length $L_f = 20$ mm, are soldered to the upper and lower sides of the module. The fin spacing is 3 mm, while the solder joints each have a thermal resistance of $R''_{t,c} = 2.5 \times 10^{-6}$ m^2 · K/W. Each heat sink has $N = 11$ fins, so that $t = 2.182$ mm and $S = 5.182$ mm, as determined from the requirements that $W = (N-1)S + t$ and $S - t = 3$ mm. Sketch the equivalent thermal circuit and determine the electric power generated by the module. Compare the electric power generated to your answer for part (a). Assume adiabatic fin tips and convection coefficients that are the same as in part (a).

3.135 Thermoelectric modules have been used to generate electric power by tapping the heat generated by wood stoves. Consider the installation of the thermoelectric module of Example 3.12 on a vertical surface of a wood stove that has a surface temperature of $T_s = 375°$C. A thermal contact resistance of $R''_{t,c} = 5 \times 10^{-6}$ m^2 · K/W exists at the interface between the stove and the thermoelectric module, while the room air and walls are at $T_\infty = T_{sur} = 25°$C. The exposed surface of the thermoelectric module has an emissivity of $\varepsilon = 0.90$ and is subjected to a convection coefficient of $h = 15$ W/m^2 · K. Sketch the equivalent thermal circuit and determine the electric power generated by the module. The load electrical resistance is $R_{e,load} = 3$ Ω.

3.136 The electric power generator for an orbiting satellite is composed of a long, cylindrical uranium heat source that is housed within an enclosure of square cross section. The only way for heat that is generated by the uranium to leave the enclosure is through four rows of the thermoelectric modules of Example 3.12. The thermoelectric modules generate electric power and also radiate heat into deep space characterized by $T_{sur} = 4$ K. Consider the situation for which there are 20 modules in each row for a total of $M = 4 \times 20 = 80$ modules. The modules are wired in series with an electrical load of $R_{e,load} = 250$ Ω, and have an emissivity of $\varepsilon = 0.93$. Determine the electric power generated for $\dot{E}_g = 1, 10,$ and 100 kW. Also determine the surface temperatures of the modules for the three thermal energy generation rates.

3.137 Rows of the thermoelectric modules of Example 3.12 are attached to the flat absorber plate of Problem 3.92. The rows of modules are separated by $L_{sep} = 0.5$ m and the backs of the modules are cooled by water at a temperature of $T_w = 40°$C, with $h = 45$ W/m^2 · K.

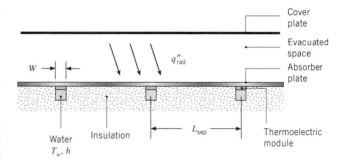

Determine the electric power produced by one row of thermoelectric modules connected in series electrically with a load resistance of 60 Ω. Calculate the heat transfer rate to the flowing water. Assume rows of 20 immediately adjacent modules, with the lengths of both the module rows and water tubing to be $L_{row} = 20$ W where $W = 54$ mm is the module dimension taken from Example 3.12. Neglect thermal contact resistances and the temperature drop across the tube wall, and assume that the high thermal conductivity tube wall creates a uniform temperature around the tube perimeter. Because of the thermal resistance provided by the thermoelectric modules, it is no longer appropriate to assume that the temperature of the absorber plate directly above a tube is equal to that of the water.

Nanoscale Conduction

3.138 Determine the conduction heat transfer rate through an air layer held between two 10 mm × 10 mm parallel aluminum plates. The plates are at temperatures $T_{s,1} = 305$ K and $T_{s,2} = 295$ K, respectively, and the air is at atmospheric pressure. Determine the conduction heat rate for plate spacings of $L = 1$ mm, $L = 1$ μm, and $L = 10$ nm. Assume a thermal accommodation coefficient of $\alpha_t = 0.92$.

3.139 Determine the parallel plate separation distance L, above which the thermal resistance associated with molecule-surface collisions $R_{t,m-s}$ is less than 1% of the resistance associated with molecule–molecule collisions, $R_{t,m-m}$ for (i) air between steel plates with $\alpha_t = 0.92$ and (ii) helium between clean aluminum plates with $\alpha_t = 0.02$. The gases are at atmospheric pressure, and the temperature is $T = 300$ K.

3.140 Determine the conduction heat flux through various plane layers that are subjected to boundary temperatures of $T_{s,1} = 301$ K and $T_{s,2} = 299$ K at atmospheric pressure. *Hint*: Do not account for nanoscale effects within the solid, and assume the thermal accommodation coefficient for an aluminum–air interface is $\alpha_t = 0.92$.

(a) Case A: The plane layer is aluminum. Determine the heat flux q''_x for $L_{tot} = 600$ μm and $L_{tot} = 600$ nm.

(b) Case B: Conduction occurs through an air layer. Determine the heat flux q''_x for $L_{tot} = 600$ μm and $L_{tot} = 600$ nm.

(c) Case C: The composite wall is composed of air held between two aluminum sheets. Determine the heat flux q''_x for $L_{tot} = 600$ μm (with aluminum sheet thicknesses of $\delta = 40$ μm) and $L_{tot} = 600$ nm (with aluminum sheet thicknesses of $\delta = 40$ nm).

(d) Case D: The composite wall is composed of 7 air layers interspersed between 8 aluminum sheets. Determine the heat flux q''_x for $L_{tot} = 600$ μm (with aluminum sheet and air layer thicknesses of $\delta = 40$ μm) and $L_{tot} = 600$ nm (with aluminum sheet and air layer thicknesses of $\delta = 40$ nm).

3.141 The Knudsen number, $K_n = \lambda_{mfp}/L$, is a dimensionless parameter used to describe potential micro- or nanoscale effects. Derive an expression for the ratio of the thermal resistance due to molecule–surface collisions to the thermal resistance associated with molecule–molecule collisions, $R_{t,m-s}/R_{t,m-m}$, in terms of the Knudsen number, the thermal accommodation coefficient α_t, and the ratio of specific heats γ, for an ideal gas. Plot the critical Knudsen number, Kn_{crit}, that is associated with $R_{t,m-s}/R_{t,m-m} = 0.01$ versus α_t, for $\gamma = 1.4$ and 1.67 (corresponding to air and helium, respectively).

3.142 A nanolaminated material is fabricated with an atomic layer deposition process, resulting in a series of stacked, alternating layers of tungsten and aluminum oxide (polycrystalline), each layer being $\delta = 0.5$ nm thick. Each tungsten–aluminum oxide interface is associated with a thermal resistance of $R''_{t,b} = 3.85 \times 10^{-9}$ m² · K/W. The theoretical values of the thermal conductivities of the *thin* aluminum oxide and tungsten layers are $k_A = 1.65$ W/m · K and $k_T = 6.10$ W/m · K, respectively. The properties are evaluated at $T = 300$ K.

(a) Determine the effective thermal conductivity of the nanolaminated material. Compare the value of the effective thermal conductivity to the bulk thermal conductivities of aluminum oxide (polycrystalline) and tungsten, given in Tables A.1 and A.2.

(b) Determine the effective thermal conductivity of the nanolaminated material assuming that the thermal conductivities of the tungsten and aluminum oxide layers are equal to their bulk values.

3.143 An aluminum tube of inner diameter $D_i = 1$ mm and wall thickness 0.1 mm is used to transport a warm biological fluid. The tube is coated with $N = 500$ alternating layers of aluminum and aluminum oxide (sapphire), each of which is 60 nm thick. Determine the thermal resistance in the radial direction for a unit tube length for the wall of the aluminum tube, as well as for the nanocomposite tube coating. Assume the aluminum mean free path is $\lambda_{mfp,Al} = 35$ nm. Comment on the effectiveness of the nanocomposite coating as a thermal insulator.

CHAPTER

Two-Dimensional, Steady-State Conduction

To this point, we have restricted our attention to conduction problems in which the temperature gradient is significant for only one coordinate direction. However, in many cases such problems are grossly oversimplified if a one-dimensional treatment is used, and it is necessary to account for multidimensional effects. In this chapter, we consider several techniques for treating two-dimensional systems under steady-state conditions.

We begin our consideration of two-dimensional, steady-state conduction by briefly reviewing alternative approaches to determining temperatures and heat rates (Section 4.1). The approaches range from *exact solutions*, which may be obtained for idealized conditions, to *approximate methods* of varying complexity and accuracy. In Section 4.2 we consider some of the mathematical issues associated with obtaining an exact solution. In Section 4.3, we present compilations of existing exact solutions for a variety of simple geometries. Our objective in Sections 4.4 and 4.5 is to show how *numerical methods* may be used to accurately predict temperatures and heat rates within the medium and at its boundaries.

4.1 *General Considerations and Solution Techniques*

Consider a long, prismatic solid in which there is two-dimensional heat conduction (Figure 4.1). With two surfaces insulated and the other surfaces maintained at different temperatures, $T_1 > T_2$, heat transfer by conduction occurs from surface 1 to 2. According to Fourier's law, Equation 2.3 or 2.4, the local heat flux in the solid is a vector that is everywhere perpendicular to lines of constant temperature (*isotherms*). The directions of the heat flux vector are represented by the *heat flow lines* of Figure 4.1, and the vector itself is the resultant of heat flux components in the *x*- and *y*-directions. These components are determined by Equation 2.6. Since the heat flow lines are, by definition, in the direction of heat flow, *no heat can be conducted across a heat flow line*, and they are therefore sometimes referred to as *adiabats*. Conversely, adiabatic surfaces (or symmetry lines) are heat flow lines.

Recall that, in any conduction analysis, there exist two major objectives. The first objective is to determine the temperature distribution in the medium, which, for the present problem, necessitates determining $T(x, y)$. This objective is achieved by solving the appropriate form of the heat equation. For two-dimensional, steady-state conditions with no generation and constant thermal conductivity, this form is, from Equation 2.22,

$$\frac{\partial^2 T}{\partial x^2} + \frac{\partial^2 T}{\partial y^2} = 0 \tag{4.1}$$

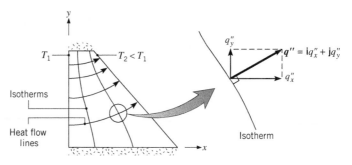

FIGURE 4.1 Two-dimensional conduction.

If Equation 4.1 can be solved for $T(x, y)$, it is then a simple matter to satisfy the second major objective, which is to determine the heat flux components q_x'' and q_y'' by applying the rate equations (2.6). Methods for solving Equation 4.1 include the use of *analytical*, *graphical*, and *numerical* (*finite-difference*, *finite-element*, or *boundary-element*) approaches.

The analytical method involves effecting an exact mathematical solution to Equation 4.1. The problem is more difficult than those considered in Chapter 3, since it now involves a partial, rather than an ordinary, differential equation. Although several techniques are available for solving such equations, the solutions typically involve complicated mathematical series and functions and may be obtained for only a restricted set of simple geometries and boundary conditions [1–5]. Nevertheless, the solutions are of value, since the dependent variable T is determined as a continuous function of the independent variables (x, y). Hence the solution could be used to compute the temperature at *any* point of interest in the medium. To illustrate the nature and importance of analytical techniques, an exact solution to Equation 4.1 is obtained in Section 4.2 by the method of *separation of variables*. Conduction shape factors and dimensionless conduction heat rates (Section 4.3) are compilations of existing solutions for geometries that are commonly encountered in engineering practice.

In contrast to the analytical methods, which provide *exact* results at *any* point, graphical and numerical methods can provide only *approximate* results at *discrete* points. Although superseded by computer solutions based on numerical procedures, the graphical, or flux-plotting, method may be used to obtain a quick estimate of the temperature distribution. Its use is restricted to two-dimensional problems involving adiabatic and isothermal boundaries. The method is based on the fact that isotherms must be perpendicular to heat flow lines, as noted in Figure 4.1. Unlike the analytical or graphical approaches, numerical methods (Sections 4.4 and 4.5) may be used to obtain accurate results for complex, two- or three-dimensional geometries involving a wide variety of boundary conditions.

4.2 *The Method of Separation of Variables*

To appreciate how the method of separation of variables may be used to solve two-dimensional conduction problems, we consider the system of Figure 4.2. Three sides of a thin rectangular plate or a long rectangular rod are maintained at a constant temperature T_1, while the fourth side is maintained at a constant temperature $T_2 \neq T_1$. Assuming negligible heat transfer from the surfaces of the plate or the ends of the rod, temperature gradients normal to the x–y plane may be neglected ($\partial^2 T/\partial z^2 \approx 0$) and conduction heat transfer is primarily in the x- and y-directions.

We are interested in the temperature distribution $T(x, y)$, but to simplify the solution we introduce the transformation

$$\theta \equiv \frac{T - T_1}{T_2 - T_1} \tag{4.2}$$

Substituting Equation 4.2 into Equation 4.1, the transformed differential equation is then

$$\frac{\partial^2 \theta}{\partial x^2} + \frac{\partial^2 \theta}{\partial y^2} = 0 \tag{4.3}$$

The graphical method is described, and its use is demonstrated, in Section 4S.1.

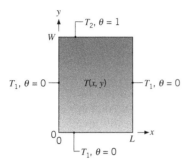

FIGURE 4.2 Two-dimensional conduction in a thin rectangular plate or a long rectangular rod.

Since the equation is second order in both x and y, two boundary conditions are needed for each of the coordinates. They are

$$\theta(0, y) = 0 \quad \text{and} \quad \theta(x, 0) = 0$$
$$\theta(L, y) = 0 \quad \text{and} \quad \theta(x, W) = 1$$

Note that, through the transformation of Equation 4.2, three of the four boundary conditions are now homogeneous and the value of θ is restricted to the range from 0 to 1.

We now apply the separation of variables technique by assuming that the desired solution can be expressed as the product of two functions, one of which depends only on x while the other depends only on y. That is, we assume the existence of a solution of the form

$$\theta(x, y) = X(x) \cdot Y(y) \tag{4.4}$$

Substituting into Equation 4.3 and dividing by XY, we obtain

$$-\frac{1}{X}\frac{d^2X}{dx^2} = \frac{1}{Y}\frac{d^2Y}{dy^2} \tag{4.5}$$

and it is evident that the differential equation is, in fact, separable. That is, the left-hand side of the equation depends only on x and the right-hand side depends only on y. Hence the equality can apply in general (for any x or y) only if both sides are equal to the same constant. Identifying this, as yet unknown, *separation constant* as λ^2, we then have

$$\frac{d^2X}{dx^2} + \lambda^2X = 0 \tag{4.6}$$

$$\frac{d^2Y}{dy^2} - \lambda^2Y = 0 \tag{4.7}$$

and the partial differential equation has been reduced to two ordinary differential equations. Note that the designation of λ^2 as a positive constant was not arbitrary. If a negative value were selected or a value of $\lambda^2 = 0$ was chosen, it is readily shown (Problem 4.1) that it would be impossible to obtain a solution that satisfies the prescribed boundary conditions.

The general solutions to Equations 4.6 and 4.7 are, respectively,

$$X = C_1 \cos \lambda x + C_2 \sin \lambda x$$
$$Y = C_3 e^{-\lambda y} + C_4 e^{+\lambda y}$$

in which case the general form of the two-dimensional solution is

$$\theta = (C_1 \cos \lambda x + C_2 \sin \lambda x)(C_3 e^{-\lambda y} + C_4 e^{\lambda y}) \tag{4.8}$$

Applying the condition that $\theta(0, y) = 0$, it is evident that $C_1 = 0$. In addition from the requirement that $\theta(x, 0) = 0$, we obtain

$$C_2 \sin \lambda x (C_3 + C_4) = 0$$

which may only be satisfied if $C_3 = -C_4$. Although the requirement could also be satisfied by having $C_2 = 0$, this would result in $\theta(x, y) = 0$, which does not satisfy the boundary condition $\theta(x, W) = 1$. If we now invoke the requirement that $\theta(L, y) = 0$, we obtain

$$C_2 C_4 \sin \lambda L (e^{\lambda y} - e^{-\lambda y}) = 0$$

The only way in which this condition may be satisfied (and still have a nonzero solution) is by requiring that λ assume discrete values for which $\sin \lambda L = 0$. These values must then be of the form

$$\lambda = \frac{n\pi}{L} \quad n = 1, 2, 3, \ldots \tag{4.9}$$

where the integer $n = 0$ is precluded, since it implies $\theta(x, y) = 0$. The desired solution may now be expressed as

$$\theta = C_2 C_4 \sin \frac{n\pi x}{L} (e^{n\pi y/L} - e^{-n\pi y/L}) \tag{4.10}$$

Combining constants and acknowledging that the new constant may depend on n, we obtain

$$\theta(x, y) = C_n \sin \frac{n\pi x}{L} \sinh \frac{n\pi y}{L}$$

where we have also used the fact that $(e^{n\pi y/L} - e^{-n\pi y/L}) = 2 \sinh (n\pi y/L)$. In this form we have really obtained an infinite number of solutions that satisfy the differential equation and boundary conditions. However, since the problem is linear, a more general solution may be obtained from a superposition of the form

$$\theta(x, y) = \sum_{n=1}^{\infty} C_n \sin \frac{n\pi x}{L} \sinh \frac{n\pi y}{L} \tag{4.11}$$

To determine C_n we now apply the remaining boundary condition, which is of the form

$$\theta(x, W) = 1 = \sum_{n=1}^{\infty} C_n \sin \frac{n\pi x}{L} \sinh \frac{n\pi W}{L} \tag{4.12}$$

Although Equation 4.12 would seem to be an extremely complicated relation for evaluating C_n, a standard method is available. It involves writing an infinite series expansion in terms of *orthogonal functions.* An infinite set of functions $g_1(x), g_2(x), \ldots, g_n(x), \ldots$ is said to be orthogonal in the domain $a \leq x \leq b$ if

$$\int_a^b g_m(x) g_n(x) dx = 0 \quad m \neq n \tag{4.13}$$

Many functions exhibit orthogonality, including the trigonometric functions sin $(n\pi x/L)$ and $\cos(n\pi x/L)$ for $0 \leq x \leq L$. Their utility in the present problem rests with the fact that any function $f(x)$ may be expressed in terms of an infinite series of orthogonal functions

$$f(x) = \sum_{n=1}^{\infty} A_n g_n(x) \tag{4.14}$$

The form of the coefficients A_n in this series may be determined by multiplying each side of the equation by $g_m(x)$ and integrating between the limits a and b.

$$\int_a^b f(x)g_m(x)\,dx = \int_a^b g_m(x) \sum_{n=1}^{\infty} A_n g_n(x)\,dx \tag{4.15}$$

However, from Equation 4.13 it is evident that all but one of the terms on the right-hand side of Equation 4.15 must be zero, leaving us with

$$\int_a^b f(x)g_m(x)\,dx = A_m \int_a^b g_m^2(x)\,dx$$

Hence, solving for A_m, and recognizing that this holds for any A_n by switching m to n:

$$A_n = \frac{\int_a^b f(x)g_n(x)\,dx}{\int_a^b g_n^2(x)\,dx} \tag{4.16}$$

The properties of orthogonal functions may be used to solve Equation 4.12 for C_n by formulating an infinite series for the appropriate form of $f(x)$. From Equation 4.14 it is evident that we should choose $f(x) = 1$ and the orthogonal function $g_n(x) = \sin(n\pi x/L)$. Substituting into Equation 4.16 we obtain

$$A_n = \frac{\int_0^L \sin\dfrac{n\pi x}{L}\,dx}{\int_0^L \sin^2\dfrac{n\pi x}{L}\,dx} = \frac{2}{\pi}\frac{(-1)^{n+1}+1}{n}$$

Hence from Equation 4.14, we have

$$1 = \sum_{n=1}^{\infty} \frac{2}{\pi}\frac{(-1)^{n+1}+1}{n}\sin\frac{n\pi x}{L} \tag{4.17}$$

which is simply the expansion of unity in a Fourier series. Comparing Equations 4.12 and 4.17 we obtain

$$C_n = \frac{2[(-1)^{n+1}+1]}{n\pi \sinh(n\pi W/L)} \qquad n = 1, 2, 3, \ldots \tag{4.18}$$

Substituting Equation 4.18 into Equation 4.11, we then obtain for the final solution

$$\theta(x, y) = \frac{2}{\pi}\sum_{n=1}^{\infty}\frac{(-1)^{n+1}+1}{n}\sin\frac{n\pi x}{L}\frac{\sinh(n\pi y/L)}{\sinh(n\pi W/L)} \tag{4.19}$$

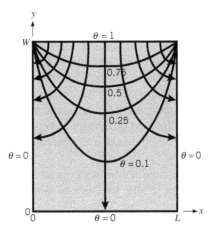

FIGURE 4.3 Isotherms and heat flow lines for two-dimensional conduction in a rectangular plate.

Equation 4.19 is a convergent series, from which the value of θ may be computed for any x and y. Representative results are shown in the form of isotherms for a schematic of the rectangular plate (Figure 4.3). The temperature T corresponding to a value of θ may be obtained from Equation 4.2, and components of the heat flux may be determined by using Equation 4.19 with Equation 2.6. The heat flux components determine the heat flow lines, which are shown in the figure. We note that the temperature distribution is symmetric about $x = L/2$, with $\partial T/\partial x = 0$ at that location. Hence, from Equation 2.6, we know the symmetry plane at $x = L/2$ is adiabatic and therefore is a heat flow line. However, note that the discontinuities prescribed at the upper corners of the plate are physically untenable. In reality, large temperature gradients could be maintained in proximity to the corners, but discontinuities could not exist.

Exact solutions have been obtained for a variety of other geometries and boundary conditions, including cylindrical and spherical systems. Such solutions are presented in specialized books on conduction heat transfer [1–5].

4.3 *The Conduction Shape Factor and the Dimensionless Conduction Heat Rate*

In general, finding analytical solutions to the two- or three-dimensional heat equation is time-consuming and, in many cases, not possible. Therefore, a different approach is often taken. For example, in many instances, two- or three-dimensional conduction problems may be rapidly solved by utilizing *existing* solutions to the heat diffusion equation. These solutions are reported in terms of a *shape factor S* or a steady-state *dimensionless conduction heat rate*, q_{ss}^*. The shape factor is defined such that

$$q = Sk\Delta T_{1-2} \tag{4.20}$$

where ΔT_{1-2} is the temperature difference between boundaries, as shown in, for example, Figure 4.2. It also follows that a two-dimensional conduction resistance may be expressed as

$$R_{t,\text{cond(2D)}} = \frac{1}{Sk} \tag{4.21}$$

Shape factors have been obtained analytically for numerous two- and three-dimensional systems, and results are summarized in Table 4.1 for some common configurations. Results are also available for other configurations [6–9]. In cases 1 through 8 and case 11, two-dimensional conduction is presumed to occur between the boundaries that are maintained at uniform temperatures, with $\Delta T_{1-2} = T_1 - T_2$. In case 9, three-dimensional conduction exists in the corner region, while in case 10 conduction occurs between an isothermal disk (T_1) and a semi-infinite medium of uniform temperature (T_2) at locations well removed from the disk. Shape factors may also be defined for one-dimensional geometries, and from the results of Table 3.3, it follows that for plane, cylindrical, and spherical walls, respectively, the shape factors are A/L, $2\pi L/\ln(r_2/r_1)$, and $4\pi r_1 r_2/(r_2 - r_1)$.

Cases 12 through 15 are associated with conduction from objects held at an isothermal temperature (T_1) that are embedded within an infinite medium of uniform temperature (T_2)

Shape factors for two-dimensional geometries may also be estimated with the graphical method that is described in Section 4S.1.

TABLE 4.1   Conduction shape factors and dimensionless conduction heat rates for selected systems.

(a) **Shape factors** $[q = Sk(T_1 - T_2)]$

System	Schematic	Restrictions	Shape Factor
Case 1 Isothermal sphere buried in a semi-infinite medium		$z > D/2$	$\dfrac{2\pi D}{1 - D/4z}$
Case 2 Horizontal isothermal cylinder of length L buried in a semi-infinite medium		$L \gg D$	$\dfrac{2\pi L}{\cosh^{-1}(2z/D)}$
		$L \gg D$ $z > 3D/2$	$\dfrac{2\pi L}{\ln(4z/D)}$
Case 3 Vertical cylinder in a semi-infinite medium		$L \gg D$	$\dfrac{2\pi L}{\ln(4L/D)}$
Case 4 Conduction between two cylinders of length L in infinite medium		$L \gg D_1, D_2$ $L \gg w$	$\dfrac{2\pi L}{\cosh^{-1}\left(\dfrac{4w^2 - D_1^2 - D_2^2}{2D_1 D_2}\right)}$

TABLE 4.1 *Continued*

System	Schematic	Restrictions	Shape Factor
Case 5 Horizontal circular cylinder of length L midway between parallel planes of equal length and infinite width		$z \gg D/2$ $L \gg z$	$\dfrac{2\pi L}{\ln(8z/\pi D)}$
Case 6 Circular cylinder of length L centered in a square solid of equal length		$w > D$ $L \gg w$	$\dfrac{2\pi L}{\ln(1.08\, w/D)}$
Case 7 Eccentric circular cylinder of length L in a cylinder of equal length		$D > d$ $L \gg D$	$\dfrac{2\pi L}{\cosh^{-1}\left(\dfrac{D^2 + d^2 - 4z^2}{2Dd}\right)}$
Case 8 Conduction through the edge of adjoining walls		$D > 5L$	$0.54D$
Case 9 Conduction through corner of three walls with a temperature difference ΔT_{1-2} across the walls		$L \ll$ length and width of wall	$0.15L$
Case 10 Disk of diameter D and temperature T_1 on a semi-infinite medium of thermal conductivity k and temperature T_2		None	$2D$
Case 11 Square channel of length L		$\dfrac{W}{w} < 1.4$ $\dfrac{W}{w} > 1.4$ $L \gg W$	$\dfrac{2\pi L}{0.785 \ln(W/w)}$ $\dfrac{2\pi L}{0.930 \ln(W/w) - 0.050}$

TABLE 4.1   *Continued*

(*b*) **Dimensionless conduction heat rates** $[q = q_{ss}^* kA_s(T_1 - T_2)/L_c; \; L_c \equiv (A_s/4\pi)^{1/2}]$

System	Schematic	Active Area, A_s	q_{ss}^*
Case 12 Isothermal sphere of diameter D and temperature T_1 in an infinite medium of temperature T_2		πD^2	1
Case 13 Infinitely thin, isothermal disk of diameter D and temperature T_1 in an infinite medium of temperature T_2		$\dfrac{\pi D^2}{2}$	$\dfrac{2\sqrt{2}}{\pi} = 0.900$
Case 14 Infinitely thin rectangle of length L, width w, and temperature T_1 in an infinite medium of temperature T_2		$2wL$	0.932
Case 15 Cuboid shape of height d with a square footprint of width D and temperature T_1 in an infinite medium of temperature T_2		$2D^2 + 4Dd$	$\begin{array}{cc} d/D & q_{ss}^* \\ \hline 0.1 & 0.943 \\ 1.0 & 0.956 \\ 2.0 & 0.961 \\ 10 & 1.111 \end{array}$

at locations removed from the object. For these infinite medium cases, useful results may be obtained by defining a *characteristic length*

$$L_c \equiv (A_s/4\pi)^{1/2} \tag{4.22}$$

where A_s is the surface area of the object. Conduction heat transfer rates from the object to the infinite medium may then be reported in terms of a *dimensionless conduction heat rate* [10]

$$q_{ss}^* \equiv qL_c/kA_s(T_1 - T_2) \tag{4.23}$$

From Table 4.1, it is evident that the values of q_{ss}^*, which have been obtained analytically and numerically, are similar for a wide range of geometrical configurations. As a consequence of this similarity, values of q_{ss}^* may be *estimated* for configurations that are *similar* to those for which q_{ss}^* is known. For example, dimensionless conduction heat rates from cuboid shapes (case 15) over the range $0.1 \leq d/D \leq 10$ may be closely approximated by interpolating the values of q_{ss}^* reported in Table 4.1. Additional procedures that may be exploited to estimate values of q_{ss}^* for other geometries are explained in [10]. Note that results for q_{ss}^* in Table 4.1*b* may be converted to expressions for S listed in Table 4.1*a*. For example, the shape factor of case 10 may be derived from the dimensionless conduction heat rate of case 13 (recognizing that the infinite medium can be viewed as two adjacent semi-infinite media).

The shape factors and dimensionless conduction heat rates reported in Table 4.1 are associated with objects that are held at uniform temperatures. For uniform heat flux conditions, the object's temperature is no longer uniform but varies spatially with the coolest temperatures located near the periphery of the hot object. Hence, the temperature difference that is used to define S or q_{ss}^* is replaced by a temperature difference involving the *spatially averaged* surface temperature of the object $(\overline{T}_1 - T_2)$ or by the difference between the *maximum* surface temperature of the hot object and the far field temperature of the surrounding medium, $(T_{1,max} - T_2)$. For the *uniformly heated* geometry of case 10 (a disk of diameter D in contact with a semi-infinite medium of thermal conductivity k and temperature T_2), the values of S are $3\pi^2 D/16$ and $\pi D/2$ for temperature differences based on the average and maximum disk temperatures, respectively.

EXAMPLE 4.1

A metallic electrical wire of diameter $d = 5$ mm is to be coated with insulation of thermal conductivity $k = 0.35$ W/m · K. It is expected that, for the typical installation, the coated wire will be exposed to conditions for which the total coefficient associated with convection and radiation is $h = 15$ W/m² · K. To minimize the temperature rise of the wire due to ohmic heating, the insulation thickness is specified so that the *critical insulation radius* is achieved (see Example 3.6). During the wire coating process, however, the insulation thickness sometimes varies around the periphery of the wire, resulting in eccentricity of the wire relative to the coating. Determine the change in the thermal resistance of the insulation due to an eccentricity that is 50% of the critical insulation thickness.

SOLUTION

Known: Wire diameter, convective conditions, and insulation thermal conductivity.

Find: Thermal resistance of the wire coating associated with peripheral variations in the coating thickness.

Schematic:

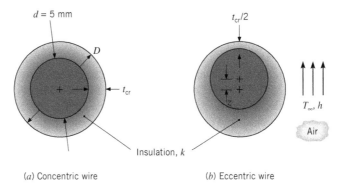

(a) Concentric wire   (b) Eccentric wire

Assumptions:
1. Steady-state conditions.
2. Two-dimensional conduction.

3. Negligible radiation exchange with surroundings.

4. Constant properties.

5. Both the exterior and interior surfaces of the coating are at uniform temperatures.

Analysis: From Example 3.6, the critical insulation radius is

$$r_{cr} = \frac{k}{h} = \frac{0.35 \text{ W/m} \cdot \text{K}}{15 \text{ W/m}^2 \cdot \text{K}} = 0.023 \text{ m} = 23 \text{ mm}$$

Therefore, the critical insulation thickness is

$$t_{cr} = r_{cr} - d/2 = 0.023 \text{ m} - \frac{0.005 \text{ m}}{2} = 0.021 \text{ m} = 21 \text{ mm}$$

The thermal resistance of the coating associated with the concentric wire may be evaluated using Equation 3.33 and is

$$R'_{t,cond} = \frac{\ln[r_{cr}/(d/2)]}{2\pi k} = \frac{\ln[0.023 \text{ m}/(0.005 \text{ m}/2)]}{2\pi(0.35 \text{ W/m} \cdot \text{K})} = 1.0 \text{ m} \cdot \text{K/W}$$

For the eccentric wire, the thermal resistance of the insulation may be evaluated using case 7 of Table 4.1, where the eccentricity is $z = 0.5 \times t_{cr} = 0.5 \times 0.021 \text{ m} = 0.010 \text{ m}$

$$R'_{t,cond(2D)} = \frac{1}{Sk} = \frac{\cosh^{-1}\left(\dfrac{D^2 + d^2 - 4z^2}{2Dd}\right)}{2\pi k}$$

$$= \frac{\cosh^{-1}\left(\dfrac{(2 \times 0.023 \text{ m})^2 + (0.005 \text{ m})^2 - 4(0.010 \text{ m})^2}{2 \times (2 \times 0.023 \text{ m}) \times 0.005 \text{ m}}\right)}{2\pi \times 0.35 \text{ W/m} \cdot \text{K}}$$

$$= 0.91 \text{ m} \cdot \text{K/W}$$

Therefore, the reduction in the thermal resistance of the insulation is $0.10 \text{ m} \cdot \text{K/W}$, or 10%.   ◁

Comments:

1. Reduction in the local insulation thickness leads to a smaller local thermal resistance of the insulation. Conversely, locations associated with thicker coatings have increased local thermal resistances. These effects offset each other, but not exactly; the maximum resistance is associated with the concentric wire case. For this application, eccentricity of the wire relative to the coating provides *enhanced* thermal performance relative to the concentric wire case.

2. The interior surface of the coating will be at nearly uniform temperature if the thermal conductivity of the wire is high relative to that of the insulation. Such is the case for metallic wire. However, the exterior surface temperature of the coating will not be perfectly uniform due to the variation in the local insulation thickness.

4.4 *Finite-Difference Equations*

As discussed in Sections 4.1 and 4.2, analytical methods may be used, in certain cases, to effect exact mathematical solutions to steady, two-dimensional conduction problems. These solutions have been generated for an assortment of simple geometries and boundary conditions and are well documented in the literature [1–5]. However, more often than not, two-dimensional problems involve geometries and/or boundary conditions that preclude such solutions. In these cases, the best alternative is often one that uses a *numerical* technique such as the *finite-difference*, *finite-element*, or *boundary-element* method. Another strength of numerical methods is that they can be readily extended to three-dimensional problems. Because of its ease of application, the finite-difference method is well suited for an introductory treatment of numerical techniques.

4.4.1 The Nodal Network

In contrast to an analytical solution, which allows for temperature determination at *any* point of interest in a medium, a numerical solution enables determination of the temperature at only *discrete* points. The first step in any numerical analysis must therefore be to select these points. Referring to Figure 4.4, this may be done by subdividing the medium of interest into a number of small regions and assigning to each a reference point that is at its center.

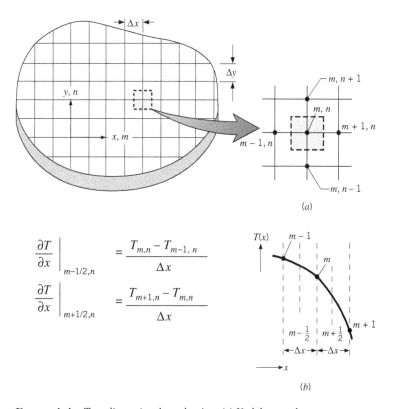

$$\left.\frac{\partial T}{\partial x}\right|_{m-1/2,n} = \frac{T_{m,n} - T_{m-1,n}}{\Delta x}$$

$$\left.\frac{\partial T}{\partial x}\right|_{m+1/2,n} = \frac{T_{m+1,n} - T_{m,n}}{\Delta x}$$

FIGURE 4.4 Two-dimensional conduction. (*a*) Nodal network. (*b*) Finite-difference approximation.

The reference point is frequently termed a *nodal point* (or simply a *node*), and the aggregate of points is termed a *nodal network*, *grid*, or *mesh*. The nodal points are designated by a numbering scheme that, for a two-dimensional system, may take the form shown in Figure 4.4*a*. The *x* and *y* locations are designated by the *m* and *n* indices, respectively.

Each node represents a certain region, and its temperature is a measure of the *average* temperature of the region. For example, the temperature of the node (*m*, *n*) of Figure 4.4*a* may be viewed as the average temperature of the surrounding shaded area. The selection of nodal points is rarely arbitrary, depending often on matters such as geometric convenience and the desired accuracy. The numerical accuracy of the calculations depends strongly on the number of designated nodal points. If this number is large (a *fine mesh*), accurate solutions can be obtained.

4.4.2 Finite-Difference Form of the Heat Equation: No Generation and Constant Properties

Determination of the temperature distribution numerically dictates that an appropriate conservation equation be written for *each* of the nodal points of unknown temperature. The resulting set of equations may then be solved simultaneously for the temperature at each node. For *any interior* node of a two-dimensional system with no generation and uniform thermal conductivity, the *exact* form of the energy conservation requirement is given by the heat equation, Equation 4.1. However, if the system is characterized in terms of a nodal network, it is necessary to work with an *approximate*, or *finite-difference*, form of this equation.

A finite-difference equation that is suitable for the interior nodes of a two-dimensional system may be inferred directly from Equation 4.1. Consider the second derivative, $\partial^2 T/\partial x^2$. From Figure 4.4*b*, the value of this derivative at the (*m*, *n*) nodal point may be approximated as

$$\left.\frac{\partial^2 T}{\partial x^2}\right|_{m,n} \approx \frac{\left.\partial T/\partial x\right|_{m+1/2,n} - \left.\partial T/\partial x\right|_{m-1/2,n}}{\Delta x} \tag{4.24}$$

The temperature gradients may in turn be expressed as a function of the nodal temperatures. That is,

$$\left.\frac{\partial T}{\partial x}\right|_{m+1/2,n} \approx \frac{T_{m+1,n} - T_{m,n}}{\Delta x} \tag{4.25}$$

$$\left.\frac{\partial T}{\partial x}\right|_{m-1/2,n} \approx \frac{T_{m,n} - T_{m-1,n}}{\Delta x} \tag{4.26}$$

Substituting Equations 4.25 and 4.26 into 4.24, we obtain

$$\left.\frac{\partial^2 T}{\partial x^2}\right|_{m,n} \approx \frac{T_{m+1,n} + T_{m-1,n} - 2T_{m,n}}{(\Delta x)^2} \tag{4.27}$$

Proceeding in a similar fashion, it is readily shown that

$$\left.\frac{\partial^2 T}{\partial y^2}\right|_{m,n} \approx \frac{T_{m,n+1} + T_{m,n-1} - 2T_{m,n}}{(\Delta y)^2} \tag{4.28}$$

Using a network for which $\Delta x = \Delta y$ and substituting Equations 4.27 and 4.28 into Equation 4.1, we obtain

$$T_{m,n+1} + T_{m,n-1} + T_{m+1,n} + T_{m-1,n} - 4T_{m,n} = 0 \qquad (4.29)$$

Hence for the (m, n) node, the heat equation, which is an *exact differential equation*, is reduced to an *approximate algebraic equation*. This approximate, *finite-difference form of the heat equation* may be applied to any interior node that is equidistant from its four neighboring nodes. It requires simply that the temperature of an interior node be equal to the average of the temperatures of the four neighboring nodes.

4.4.3 Finite-Difference Form of the Heat Equation: The Energy Balance Method

In many cases, it is desirable to develop the finite-difference equations by an alternative method called the *energy balance method*. As will become evident, this approach enables one to analyze many different phenomena such as problems involving multiple materials, embedded heat sources, or exposed surfaces that do not align with an axis of the coordinate system. In the energy balance method, the finite-difference equation for a node is obtained by applying conservation of energy to a control volume about the nodal region. Since the actual direction of heat flow (into or out of the node) is often unknown, it is convenient to formulate the energy balance by *assuming* that *all* the heat flow is *into the control volume* represented by (m, n). Such a condition is, of course, impossible, but if the rate equations are expressed in a manner consistent with this assumption, the correct form of the finite-difference equation is obtained. For steady-state conditions with generation, the appropriate form of Equation 1.12c is then

$$\dot{E}_{\text{in}} + \dot{E}_g = 0 \qquad (4.30)$$

Consider applying Equation 4.30 to a control volume about the interior node (m, n) of Figure 4.5. For two-dimensional conditions, energy exchange is influenced by conduction between (m, n) and its four adjoining nodes, as well as by generation. Hence Equation 4.30 reduces to

$$\sum_{i=1}^{4} q_{(i)\rightarrow(m,n)} + \dot{q}(\Delta x \cdot \Delta y \cdot 1) = 0$$

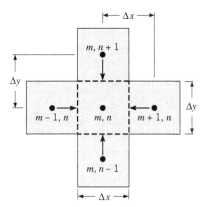

FIGURE 4.5 Conduction to an interior node from its adjoining nodes.

where i refers to the neighboring nodes, $q_{(i) \rightarrow (m,n)}$ is the conduction rate between nodes, and unit depth is assumed. To evaluate the conduction rate terms, we *assume* that conduction transfer occurs exclusively through *lanes* that are oriented in either the x- or y-direction. Simplified forms of Fourier's law may therefore be used. For example, the rate at which energy is transferred by conduction from node $(m - 1, n)$ to (m, n) may be expressed as

$$q_{(m-1,n) \rightarrow (m,n)} = k(\Delta y \cdot 1)\frac{T_{m-1,n} - T_{m,n}}{\Delta x} \tag{4.31}$$

The quantity $(\Delta y \cdot 1)$ is the heat transfer area, and the term $(T_{m-1,n} - T_{m,n})/\Delta x$ is the finite-difference approximation to the temperature gradient at the boundary between the two nodes. The remaining conduction rates may be expressed as

$$q_{(m+1,n) \rightarrow (m,n)} = k(\Delta y \cdot 1)\frac{T_{m+1,n} - T_{m,n}}{\Delta x} \tag{4.32}$$

$$q_{(m,n+1) \rightarrow (m,n)} = k(\Delta x \cdot 1)\frac{T_{m,n+1} - T_{m,n}}{\Delta y} \tag{4.33}$$

$$q_{(m,n-1) \rightarrow (m,n)} = k(\Delta x \cdot 1)\frac{T_{m,n-1} - T_{m,n}}{\Delta y} \tag{4.34}$$

Note that, in evaluating each conduction rate, we have subtracted the temperature of the (m, n) node from the temperature of its adjoining node. This convention is necessitated by the assumption of heat flow toward (m, n) and is consistent with the direction of the arrows shown in Figure 4.5. Substituting Equations 4.31 through 4.34 into the energy balance and remembering that $\Delta x = \Delta y$, it follows that the finite-difference equation for an interior node with generation is

$$T_{m,n+1} + T_{m,n-1} + T_{m+1,n} + T_{m-1,n} + \frac{\dot{q}(\Delta x)^2}{k} - 4T_{m,n} = 0 \tag{4.35}$$

If there is no internally distributed source of energy ($\dot{q} = 0$), this expression reduces to Equation 4.29.

It is important to note that a finite-difference equation is needed for each nodal point at which the temperature is unknown. However, it is not always possible to classify all such points as interior and hence to use Equation 4.29 or 4.35. For example, the temperature may be unknown at an insulated surface or at a surface that is exposed to convective conditions. For points on such surfaces, the finite-difference equation must be obtained by applying the energy balance method.

To further illustrate this method, consider the node corresponding to the internal corner of Figure 4.6. This node represents the three-quarter shaded section and exchanges energy by convection with an adjoining fluid at T_∞. Conduction to the nodal region (m, n) occurs along four different lanes from neighboring nodes in the solid. The conduction heat rates q_{cond} may be expressed as

$$q_{(m-1,n) \rightarrow (m,n)} = k(\Delta y \cdot 1)\frac{T_{m-1,n} - T_{m,n}}{\Delta x} \tag{4.36}$$

FIGURE 4.6 Formulation of the finite-difference equation for an internal corner of a solid with surface convection.

$$q_{(m,n+1) \to (m,n)} = k(\Delta x \cdot 1)\frac{T_{m,n+1} - T_{m,n}}{\Delta y} \qquad (4.37)$$

$$q_{(m+1,n) \to (m,n)} = k\left(\frac{\Delta y}{2} \cdot 1\right)\frac{T_{m+1,n} - T_{m,n}}{\Delta x} \qquad (4.38)$$

$$q_{(m,n-1) \to (m,n)} = k\left(\frac{\Delta x}{2} \cdot 1\right)\frac{T_{m,n-1} - T_{m,n}}{\Delta y} \qquad (4.39)$$

Note that the areas for conduction from nodal regions $(m-1, n)$ and $(m, n+1)$ are proportional to Δy and Δx, respectively, whereas conduction from $(m+1, n)$ and $(m, n-1)$ occurs along lanes of width $\Delta y/2$ and $\Delta x/2$, respectively.

Conditions in the nodal region (m, n) are also influenced by convective exchange with the fluid, and this exchange may be viewed as occurring along half-lanes in the x- and y-directions. The total convection rate q_{conv} may be expressed as

$$q_{(\infty) \to (m,n)} = h\left(\frac{\Delta x}{2} \cdot 1\right)(T_\infty - T_{m,n}) + h\left(\frac{\Delta y}{2} \cdot 1\right)(T_\infty - T_{m,n}) \qquad (4.40)$$

Implicit in this expression is the assumption that the exposed surfaces of the corner are at a uniform temperature corresponding to the nodal temperature $T_{m,n}$. This assumption is consistent with the concept that the entire nodal region is characterized by a single temperature, which represents an average of the actual temperature distribution in the region. In the absence of transient, three-dimensional, and generation effects, conservation of energy, Equation 4.30, requires that the sum of Equations 4.36 through 4.40 be zero. Summing these equations and rearranging, we therefore obtain

$$T_{m-1,n} + T_{m,n+1} + \frac{1}{2}(T_{m+1,n} + T_{m,n-1}) + \frac{h\Delta x}{k}T_\infty - \left(3 + \frac{h\Delta x}{k}\right)T_{m,n} = 0 \qquad (4.41)$$

where again the mesh is such that $\Delta x = \Delta y$.

Nodal energy balance equations pertinent to several common configurations for situations where there is no internal energy generation are presented in Table 4.2.

TABLE 4.2 Summary of nodal finite-difference equations

Configuration	Finite-Difference Equation for $\Delta x = \Delta y$

$$T_{m,n+1} + T_{m,n-1} + T_{m+1,n} + T_{m-1,n} - 4T_{m,n} = 0 \qquad (4.29)$$

Case 1. Interior node

$$2(T_{m-1,n} + T_{m,n+1}) + (T_{m+1,n} + T_{m,n-1})$$
$$+ 2\frac{h\,\Delta x}{k}T_\infty - 2\left(3 + \frac{h\,\Delta x}{k}\right)T_{m,n} = 0 \qquad (4.41)$$

Case 2. Node at an internal corner with convection

$$(2T_{m-1,n} + T_{m,n+1} + T_{m,n-1}) + \frac{2h\,\Delta x}{k}T_\infty - 2\left(\frac{h\,\Delta x}{k} + 2\right)T_{m,n} = 0 \qquad (4.42)^a$$

Case 3. Node at a plane surface with convection

$$(T_{m,n-1} + T_{m-1,n}) + 2\frac{h\,\Delta x}{k}T_\infty - 2\left(\frac{h\,\Delta x}{k} + 1\right)T_{m,n} = 0 \qquad (4.43)$$

Case 4. Node at an external corner with convection

$$(2T_{m-1,n} + T_{m,n+1} + T_{m,n-1}) + \frac{2q''\,\Delta x}{k} - 4T_{m,n} = 0 \qquad (4.44)^b$$

Case 5. Node at a plane surface with uniform heat flux

^{a,b}To obtain the finite-difference equation for an adiabatic surface (or surface of symmetry), simply set h or q'' equal to zero.

EXAMPLE 4.2

Using the energy balance method, derive the finite-difference equation for the (m, n) nodal point located on a plane, insulated surface of a medium with uniform heat generation.

SOLUTION

Known: Network of nodal points adjoining an insulated surface.

Find: Finite-difference equation for the surface nodal point.

Schematic:

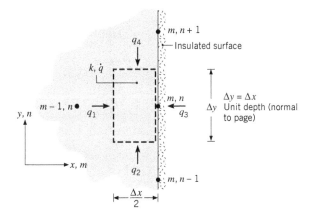

Assumptions:

1. Steady-state conditions.
2. Two-dimensional conduction.
3. Constant properties.
4. Uniform internal heat generation.

Analysis: Applying the energy conservation requirement, Equation 4.30, to the control surface about the region $(\Delta x/2 \cdot \Delta y \cdot 1)$ associated with the (m, n) node, it follows that, with volumetric heat generation at a rate $\dot{q}$,

$$q_1 + q_2 + q_3 + q_4 + \dot{q}\left(\frac{\Delta x}{2} \cdot \Delta y \cdot 1\right) = 0$$

where

$$q_1 = k(\Delta y \cdot 1)\frac{T_{m-1,n} - T_{m,n}}{\Delta x}$$

$$q_2 = k\left(\frac{\Delta x}{2} \cdot 1\right)\frac{T_{m,n-1} - T_{m,n}}{\Delta y}$$

$$q_3 = 0$$

$$q_4 = k\left(\frac{\Delta x}{2} \cdot 1\right)\frac{T_{m,n+1} - T_{m,n}}{\Delta y}$$

Substituting into the energy balance and dividing by $k/2$, it follows that

$$2T_{m-1,n} + T_{m,n-1} + T_{m,n+1} - 4T_{m,n} + \frac{\dot{q}(\Delta x \cdot \Delta y)}{k} = 0 \qquad \triangleleft$$

Comments:

1. The same result could be obtained by using the symmetry condition, $T_{m+1,n} = T_{m-1,n}$, with the finite-difference equation (Equation 4.35) for an interior nodal point. If $\dot{q} = 0$, the desired result could also be obtained by setting $h = 0$ in Equation 4.42 (Table 4.2).

2. As an application of the foregoing finite-difference equation, consider the following two-dimensional system within which thermal energy is uniformly generated at an unknown rate $\dot{q}$. The thermal conductivity of the solid is known, as are convection conditions at one of the surfaces. In addition, temperatures have been measured at locations corresponding to the nodal points of a finite-difference mesh.

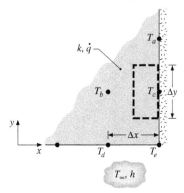

$$T_a = 235.9°C \qquad T_b = 227.6°C$$
$$T_c = 230.9°C \qquad T_d = 220.1°C$$
$$T_e = 222.4°C \qquad T_\infty = 200.0°C$$
$$h = 50 \text{ W/m}^2 \cdot \text{K} \quad k = 1 \text{ W/m} \cdot \text{K}$$
$$\Delta x = 10 \text{ mm} \qquad \Delta y = 10 \text{ mm}$$

The generation rate can be determined by applying the finite-difference equation to node *c*.

$$2T_b + T_e + T_a - 4T_c + \frac{\dot{q}(\Delta x \cdot \Delta y)}{k} = 0$$

$$(2 \times 227.6 + 222.4 + 235.9 - 4 \times 230.9)°C + \frac{\dot{q}(0.01 \text{ m})^2}{1 \text{ W/m} \cdot \text{K}} = 0$$

$$\dot{q} = 1.01 \times 10^5 \text{ W/m}^3$$

From the prescribed thermal conditions and knowledge of $\dot{q}$, we can also determine whether the conservation of energy requirement is satisfied for node *e*. Applying an energy balance to a control volume about this node, it follows that

$$q_1 + q_2 + q_3 + q_4 + \dot{q}(\Delta x/2 \cdot \Delta y/2 \cdot 1) = 0$$

$$k(\Delta x/2 \cdot 1)\frac{T_c - T_e}{\Delta y} + 0 + h(\Delta x/2 \cdot 1)(T_\infty - T_e) + k(\Delta y/2 \cdot 1)\frac{T_d - T_e}{\Delta x}$$

$$+ \dot{q}(\Delta x/2 \cdot \Delta y/2 \cdot 1) = 0$$

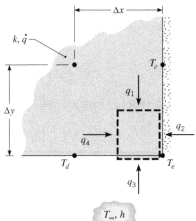

If the energy balance is satisfied, the left-hand side of this equation will be identically equal to zero. Substituting values, we obtain

$$1 \text{ W/m} \cdot \text{K}(0.005 \text{ m}^2)\frac{(230.9 - 222.4)°\text{C}}{0.010 \text{ m}}$$

$$+ 0 + 50 \text{ W/m}^2 \cdot \text{K}(0.005 \text{ m}^2)(200 - 222.4)°\text{C}$$

$$+ 1 \text{ W/m} \cdot \text{K}(0.005 \text{ m}^2)\frac{(220.1 - 222.4)°\text{C}}{0.010 \text{ m}} + 1.01 \times 10^5 \text{ W/m}^3(0.005)^2 \text{ m}^3 = 0(?)$$

$$4.250 \text{ W} + 0 - 5.600 \text{ W} - 1.150 \text{ W} + 2.525 \text{ W} = 0(?)$$

$$0.025 \text{ W} \approx 0$$

The inability to precisely satisfy the energy balance is attributable to temperature measurement errors, the approximations employed in developing the finite-difference equations, and the use of a relatively coarse mesh.

The Energy Balance Method with Thermal Resistances It is useful to note that heat rates between adjoining nodes may also be formulated in terms of the corresponding thermal resistances. Referring, for example, to Figure 4.6, the rate of heat transfer by conduction from node $(m - 1, n)$ to (m, n) may be expressed as

$$q_{(m-1,n) \to (m,n)} = \frac{T_{m-1,n} - T_{m,n}}{R_{t,\text{cond}}} = \frac{T_{m-1,n} - T_{m,n}}{\Delta x/k(\Delta y \cdot 1)}$$

yielding a result that is equivalent to that of Equation 4.36. Similarly, the rate of heat transfer by convection to (m, n) may be expressed as

$$q_{(\infty) \to (m,n)} = \frac{T_\infty - T_{m,n}}{R_{t,\text{conv}}} = \frac{T_\infty - T_{m,n}}{\{h[(\Delta x/2) \cdot 1 + (\Delta y/2) \cdot 1]\}^{-1}}$$

which is equivalent to Equation 4.40.

As an example of the utility of resistance concepts, consider an interface that separates two dissimilar materials and is characterized by a thermal contact resistance $R''_{t,c}$ (Figure 4.7). The rate of heat transfer from node (m, n) to $(m, n - 1)$ may be expressed as

$$q_{(m,n) \to (m,n-1)} = \frac{T_{m,n} - T_{m,n-1}}{R_{\text{tot}}} \tag{4.45}$$

where, for a unit depth,

$$R_{\text{tot}} = \frac{\Delta y/2}{k_A(\Delta x \cdot 1)} + \frac{R''_{t,c}}{\Delta x \cdot 1} + \frac{\Delta y/2}{k_B(\Delta x \cdot 1)} \tag{4.46}$$

FIGURE 4.7 Conduction between adjoining, dissimilar materials with an interface contact resistance.

4.5 *Solving the Finite-Difference Equations*

Once the nodal network has been established and an appropriate finite-difference equation has been written for each node, the temperature distribution may be determined. The problem reduces to one of solving a system of linear, algebraic equations. In this section, we formulate the system of linear, algebraic equations as a matrix equation and briefly discuss its solution by the matrix inversion method. We also present some considerations for verifying the accuracy of the solution.

4.5.1 Formulation as a Matrix Equation

Consider a system of N finite-difference equations corresponding to N unknown temperatures. Identifying the nodes by a single integer subscript, rather than by the double subscript (m, n), the procedure for performing a matrix inversion begins by expressing the equations as

$$
\begin{aligned}
a_{11}T_1 + a_{12}T_2 + a_{13}T_3 + \cdots + a_{1N}T_N &= C_1 \\
a_{21}T_1 + a_{22}T_2 + a_{23}T_3 + \cdots + a_{2N}T_N &= C_2 \\
\vdots \qquad \vdots \qquad \vdots \qquad \vdots \qquad \vdots \qquad & \quad \vdots \\
a_{N1}T_1 + a_{N2}T_2 + a_{N3}T_3 + \cdots + a_{NN}T_N &= C_N
\end{aligned}
\tag{4.47}
$$

where the quantities $a_{11}, a_{12}, \ldots, C_1, \ldots$ are known coefficients and constants involving quantities such as Δx, k, h, and T_∞. Using matrix notation, these equations may be expressed as

$$
[A][T] = [C]
\tag{4.48}
$$

where

$$
A \equiv
\begin{bmatrix}
a_{11} & a_{12} & \cdots & a_{1N} \\
a_{21} & a_{22} & \cdots & a_{2N} \\
\vdots & \vdots & & \vdots \\
a_{N1} & a_{N2} & \cdots & a_{NN}
\end{bmatrix},
\quad
T \equiv
\begin{bmatrix}
T_1 \\
T_2 \\
\vdots \\
T_N
\end{bmatrix},
\quad
C \equiv
\begin{bmatrix}
C_1 \\
C_2 \\
\vdots \\
C_N
\end{bmatrix}
$$

The *coefficient matrix* $[A]$ is square $(N \times N)$, and its *elements* are designated by a double subscript notation, for which the first and second subscripts refer to rows and columns, respectively. The matrices $[T]$ and $[C]$ have a single column and are known as *column vectors*. Typically, they are termed the *solution* and *right-hand side vectors*, respectively. If the matrix multiplication implied by the left-hand side of Equation 4.48 is performed, Equations 4.47 are obtained.

Numerous mathematical methods are available for solving systems of linear, algebraic equations [11, 12], and many computational software programs have the built-in capability to solve Equation 4.48 for the solution vector $[T]$. For small matrices, the solution can be found using a programmable calculator or by hand. One method suitable for hand or computer calculation is the Gauss–Seidel method, which is presented in Appendix D.

4.5.2 Verifying the Accuracy of the Solution

It is good practice to verify that a numerical solution has been correctly formulated by performing an energy balance on a control surface surrounding all nodal regions whose temperatures have been evaluated. The temperatures should be substituted into the energy balance equation, and if the balance is not satisfied to a high degree of precision, the finite-difference equations should be checked for errors.

Even when the finite-difference equations have been properly formulated and solved, the results may still represent a coarse approximation to the actual temperature field. This behavior is a consequence of the finite spacings (Δx, Δy) between nodes and of finite-difference approximations, such as $k(\Delta y \cdot 1)(T_{m-1,n} - T_{m,n})/\Delta x$, to Fourier's law of conduction, $-k(dy \cdot 1)\,\partial T/\partial x$. The finite-difference approximations become more accurate as the nodal network is refined (Δx and Δy are reduced). Hence, if accurate results are desired, grid studies should be performed, whereby results obtained for a fine grid are compared with those obtained for a coarse grid. One could, for example, reduce Δx and Δy by a factor of 2, thereby increasing the number of nodes and finite-difference equations by a factor of 4. If the agreement is unsatisfactory, further grid refinements could be made until the computed temperatures no longer depend significantly on the choice of Δx and Δy. Such *grid-independent* results would provide an accurate solution to the physical problem.

Another option for validating a numerical solution involves comparing results with those obtained from an exact solution. For example, a finite-difference solution of the physical problem described in Figure 4.2 could be compared with the exact solution given by Equation 4.19. Although we seldom seek numerical solutions to problems for which there exist exact solutions, it is often useful to test our finite-difference procedures by applying them to a simpler version of the problem that does have an exact solution.

EXAMPLE 4.3

As discussed in Section 1.6, a major objective in advancing gas turbine technologies is to increase the temperature limit associated with operation of the turbine blades. For example, it is common to use internal cooling by including flow channels within the blades and routing air through the channels. We wish to assess the effect of such a scheme by approximating the blade as a rectangular solid with rectangular channels. The blade, which has a thermal conductivity of $k = 25$ W/m · K, is 6 mm thick, and each channel has a 2 mm × 6 mm rectangular cross section, with a 4-mm spacing between adjoining channels.

Under operating conditions for which $h_o = 1000 \ \text{W/m}^2 \cdot \text{K}$, $T_{\infty,o} = 1700 \ \text{K}$, $h_i = 200 \ \text{W/m}^2 \cdot \text{K}$, and $T_{\infty,i} = 400 \ \text{K}$, determine the temperature field in the turbine blade and the rate of heat transfer per unit length to the channel. At what location is the temperature a maximum?

SOLUTION

Known: Dimensions and operating conditions for a gas turbine blade with embedded channels.

Find: Temperature field in the blade, including a location of maximum temperature. Rate of heat transfer per unit length to the channel.

Schematic:

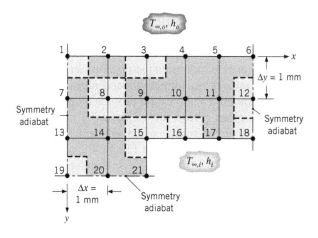

Assumptions:

1. Steady-state, two-dimensional conduction.
2. Constant properties.

Analysis: Adopting a grid spacing of $\Delta x = \Delta y = 1 \ \text{mm}$ and identifying the three lines of symmetry, the foregoing nodal network is constructed. The corresponding finite-difference equations may be obtained by applying the energy balance method to nodes 1, 6, 18, 19, and 21 and by using the results of Table 4.2 for the remaining nodes.

Heat transfer to node 1 occurs by conduction from nodes 2 and 7, as well as by convection from the outer fluid. Since there is no heat transfer from the region beyond the symmetry adiabat, application of an energy balance to the one-quarter section associated with node 1 yields a finite-difference equation of the form

$$\text{Node 1:} \quad T_2 + T_7 - \left(2 + \frac{h_o \Delta x}{k}\right) T_1 = -\frac{h_o \Delta x}{k} T_{\infty,o}$$

A similar result may be obtained for nodal region 6, which is characterized by equivalent surface conditions (2 conduction, 1 convection, 1 adiabatic). Nodes 2 to 5 correspond to case 3 of Table 4.2, and choosing node 3 as an example, it follows that

$$\text{Node 3:} \quad T_2 + T_4 + 2T_9 - 2\left(\frac{h_o \Delta x}{k} + 2\right) T_3 = -\frac{2h_o \Delta x}{k} T_{\infty,o}$$

Nodes 7, 12, 13, and 20 correspond to case 5 of Table 4.2, with $q'' = 0$, and choosing node 12 as an example, it follows that

Node 12: $T_6 + 2T_{11} + T_{18} - 4T_{12} = 0$

Nodes 8 to 11 and 14 are interior nodes (case 1), in which case the finite-difference equation for node 8 is

Node 8: $T_2 + T_7 + T_9 + T_{14} - 4T_8 = 0$

Node 15 is an internal corner (case 2) for which

Node 15: $2T_9 + 2T_{14} + T_{16} + T_{21} - 2\left(3 + \dfrac{h_i \Delta x}{k}\right)T_{15} = -2\dfrac{h_i \Delta x}{k}T_{\infty,i}$

while nodes 16 and 17 are situated on a plane surface with convection (case 3):

Node 16: $2T_{10} + T_{15} + T_{17} - 2\left(\dfrac{h_i \Delta x}{k} + 2\right)T_{16} = -\dfrac{2h_i \Delta x}{k}T_{\infty,i}$

In each case, heat transfer to nodal regions 18 and 21 is characterized by conduction from two adjoining nodes and convection from the internal flow, with no heat transfer occurring from an adjoining adiabat. Performing an energy balance for nodal region 18, it follows that

Node 18: $T_{12} + T_{17} - \left(2 + \dfrac{h_i \Delta x}{k}\right)T_{18} = -\dfrac{h_i \Delta x}{k}T_{\infty,i}$

The last special case corresponds to nodal region 19, which has two adiabatic surfaces and experiences heat transfer by conduction across the other two surfaces.

Node 19: $T_{13} + T_{20} - 2T_{19} = 0$

The equations for nodes 1 through 21 may be solved simultaneously using *IHT*, another commercial code, or a handheld calculator. The following results are obtained:

T_1	T_2	T_3	T_4	T_5	T_6
1526.0 K	1525.3 K	1523.6 K	1521.9 K	1520.8 K	1520.5 K

T_7	T_8	T_9	T_{10}	T_{11}	T_{12}
1519.7 K	1518.8 K	1516.5 K	1514.5 K	1513.3 K	1512.9 K

T_{13}	T_{14}	T_{15}	T_{16}	T_{17}	T_{18}
1515.1 K	1513.7 K	1509.2 K	1506.4 K	1505.0 K	1504.5 K

T_{19}	T_{20}	T_{21}
1513.4 K	1511.7 K	1506.0 K

The temperature field may also be represented in the form of isotherms, and four such lines of constant temperature are shown schematically. Also shown are heat flux lines that have been carefully drawn so that they are everywhere perpendicular to the isotherms and coincident with the symmetry adiabat. The surfaces that are exposed to the combustion gases and air are not isothermal, and therefore the heat flow lines are not perpendicular to these boundaries.

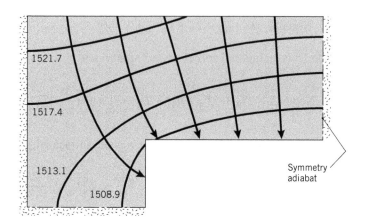

As expected, the maximum temperature exists at the location farthest removed from the coolant, which corresponds to node 1.

The rate of heat transfer per unit length of channel may be calculated in two ways. Based on heat transfer from the blade to the air, it is

$$q' = 4h_i[(\Delta y/2)(T_{21} - T_{\infty,i}) + (\Delta y/2 + \Delta x/2)(T_{15} - T_{\infty,i})$$
$$+ (\Delta x)(T_{16} - T_{\infty,i}) + \Delta x(T_{17} - T_{\infty,i}) + (\Delta x/2)(T_{18} - T_{\infty,i})]$$

Alternatively, based on heat transfer from the combustion gases to the blade, it is

$$q' = 4h_o[(\Delta x/2)(T_{\infty,o} - T_1) + (\Delta x)(T_{\infty,o} - T_2) + (\Delta x)(T_{\infty,o} - T_3)$$
$$+ (\Delta x)(T_{\infty,o} - T_4) + (\Delta x)(T_{\infty,o} - T_5) + (\Delta x/2)(T_{\infty,o} - T_6)]$$

where the factor of 4 originates from the symmetry conditions. In both cases, we obtain

$$q' = 3540.6 \text{ W/m} \qquad \lhd$$

Comments:

1. In matrix notation, following Equation 4.48, the equations for nodes 1 through 21 are of the form $[A][T] = [C]$, where

$$[A] = \begin{bmatrix}
-a & 1 & 0 & 0 & 0 & 0 & 1 & 0 & 0 & 0 & 0 & 0 & 0 & 0 & 0 & 0 & 0 & 0 & 0 & 0 & 0 \\
1 & -b & 1 & 0 & 0 & 0 & 0 & 2 & 0 & 0 & 0 & 0 & 0 & 0 & 0 & 0 & 0 & 0 & 0 & 0 & 0 \\
0 & 1 & -b & 1 & 0 & 0 & 0 & 0 & 2 & 0 & 0 & 0 & 0 & 0 & 0 & 0 & 0 & 0 & 0 & 0 & 0 \\
0 & 0 & 1 & -b & 1 & 0 & 0 & 0 & 0 & 2 & 0 & 0 & 0 & 0 & 0 & 0 & 0 & 0 & 0 & 0 & 0 \\
0 & 0 & 0 & 1 & -b & 1 & 0 & 0 & 0 & 0 & 2 & 0 & 0 & 0 & 0 & 0 & 0 & 0 & 0 & 0 & 0 \\
0 & 0 & 0 & 0 & 1 & -a & 0 & 0 & 0 & 0 & 0 & 1 & 0 & 0 & 0 & 0 & 0 & 0 & 0 & 0 & 0 \\
1 & 0 & 0 & 0 & 0 & 0 & -4 & 2 & 0 & 0 & 0 & 0 & 1 & 0 & 0 & 0 & 0 & 0 & 0 & 0 & 0 \\
0 & 1 & 0 & 0 & 0 & 0 & 1 & -4 & 1 & 0 & 0 & 0 & 0 & 1 & 0 & 0 & 0 & 0 & 0 & 0 & 0 \\
0 & 0 & 1 & 0 & 0 & 0 & 0 & 1 & -4 & 1 & 0 & 0 & 0 & 0 & 1 & 0 & 0 & 0 & 0 & 0 & 0 \\
0 & 0 & 0 & 1 & 0 & 0 & 0 & 0 & 1 & -4 & 1 & 0 & 0 & 0 & 0 & 1 & 0 & 0 & 0 & 0 & 0 \\
0 & 0 & 0 & 0 & 1 & 0 & 0 & 0 & 0 & 1 & -4 & 1 & 0 & 0 & 0 & 0 & 1 & 0 & 0 & 0 & 0 \\
0 & 0 & 0 & 0 & 0 & 1 & 0 & 0 & 0 & 0 & 2 & -4 & 0 & 0 & 0 & 0 & 0 & 1 & 0 & 0 & 0 \\
0 & 0 & 0 & 0 & 0 & 0 & 1 & 0 & 0 & 0 & 0 & 0 & -4 & 2 & 0 & 0 & 0 & 0 & 1 & 0 & 0 \\
0 & 0 & 0 & 0 & 0 & 0 & 0 & 1 & 0 & 0 & 0 & 0 & 1 & -4 & 1 & 0 & 0 & 0 & 0 & 1 & 0 \\
0 & 0 & 0 & 0 & 0 & 0 & 0 & 0 & 2 & 0 & 0 & 0 & 0 & 2 & -c & 1 & 0 & 0 & 0 & 0 & 1 \\
0 & 0 & 0 & 0 & 0 & 0 & 0 & 0 & 0 & 2 & 0 & 0 & 0 & 0 & 1 & -d & 1 & 0 & 0 & 0 & 0 \\
0 & 0 & 0 & 0 & 0 & 0 & 0 & 0 & 0 & 0 & 2 & 0 & 0 & 0 & 0 & 1 & -d & 1 & 0 & 0 & 0 \\
0 & 0 & 0 & 0 & 0 & 0 & 0 & 0 & 0 & 0 & 0 & 1 & 0 & 0 & 0 & 0 & 1 & -e & 0 & 0 & 0 \\
0 & 0 & 0 & 0 & 0 & 0 & 0 & 0 & 0 & 0 & 0 & 0 & 1 & 0 & 0 & 0 & 0 & 0 & -2 & 1 & 0 \\
0 & 0 & 0 & 0 & 0 & 0 & 0 & 0 & 0 & 0 & 0 & 0 & 0 & 2 & 0 & 0 & 0 & 0 & 1 & -4 & 1 \\
0 & 0 & 0 & 0 & 0 & 0 & 0 & 0 & 0 & 0 & 0 & 0 & 0 & 0 & 1 & 0 & 0 & 0 & 0 & 1 & -e
\end{bmatrix} \quad [C] = \begin{bmatrix} -f \\ -2f \\ -2f \\ -2f \\ -2f \\ -f \\ 0 \\ 0 \\ 0 \\ 0 \\ 0 \\ 0 \\ 0 \\ 0 \\ -2g \\ -2g \\ -2g \\ -g \\ 0 \\ 0 \\ -g \end{bmatrix}$$

With $h_o \Delta x/k = 0.04$ and $h_i \Delta x/k = 0.008$, the following coefficients in the equations can be calculated: $a = 2.04$, $b = 4.08$, $c = 6.016$, $d = 4.016$, $e = 2.008$, $f = 68$, and $g = 3.2$. By framing the equations as a matrix equation, standard tools for solving matrix equations may be used.

2. To ensure that no errors have been made in formulating and solving the finite-difference equations, the calculated temperatures should be used to verify that conservation of energy is satisfied for a control surface surrounding all nodal regions. This check has already been performed, since it was shown that the heat transfer rate from the combustion gases to the blade is equal to that from the blade to the air.

3. The accuracy of the finite-difference solution may be improved by refining the grid. If, for example, we halve the grid spacing ($\Delta x = \Delta y = 0.5$ mm), thereby increasing the number of unknown nodal temperatures to 65, we obtain the following results for selected temperatures and the heat rate:

$$T_1 = 1525.9 \text{ K}, \qquad T_6 = 1520.5 \text{ K}, \qquad T_{15} = 1509.2 \text{ K},$$

$$T_{18} = 1504.5 \text{ K}, \qquad T_{19} = 1513.5 \text{ K}, \qquad T_{21} = 1505.7 \text{ K},$$

$$q' = 3539.9 \text{ W/m}$$

Agreement between the two sets of results is excellent. Of course, use of the finer mesh increases setup and computation time, and in many cases the results obtained from a coarse grid are satisfactory. Selection of the appropriate grid is a judgment that the engineer must make.

4. In the gas turbine industry, there is great interest in adopting measures that reduce blade temperatures. Such measures could include use of a different alloy of larger thermal conductivity and/or increasing coolant flow through the channel, thereby

increasing h_i. Using the finite-difference solution with $\Delta x = \Delta y = 1$ mm, the following results are obtained for parametric variations of k and h_i:

k (W/m · K)	h_i (W/m² · K)	T_1 (K)	q' (W/m)
25	200	1526.0	3540.6
50	200	1523.4	3563.3
25	1000	1154.5	11,095.5
50	1000	1138.9	11,320.7

Why do increases in k and h_i reduce temperature in the blade?

5. Note that, because the exterior surface of the blade is at an extremely high temperature, radiation losses to its surroundings may be significant. In the finite-difference analysis, such effects could be considered by linearizing the radiation rate equation (see Equations 1.8 and 1.9) and treating radiation in the same manner as convection. However, because the radiation coefficient h_r depends on the surface temperature, an iterative finite-difference solution would be necessary to ensure that the resulting surface temperatures correspond to the temperatures at which h_r is evaluated at each nodal point.

6. See Example 4.3 in *IHT*. This problem can also be solved using *Tools, Finite-Difference Equations* in the *Advanced* section of *IHT*.

4.6 Summary

The primary objective of this chapter was to develop an appreciation for the nature of a two-dimensional conduction problem and the methods that are available for its solution. When confronted with a two-dimensional problem, one should first determine whether an exact solution is known. This may be done by examining some of the excellent references in which exact solutions to the heat equation are obtained [1–5]. One may also want to determine whether the shape factor or dimensionless conduction heat rate is known for the system of interest [6–10]. However, often, conditions are such that the use of a shape factor, dimensionless conduction heat rate, or an exact solution is not possible, and it is necessary to use a finite-difference or finite-element solution. You should therefore appreciate the inherent nature of the *discretization process* and know how to formulate and solve the finite-difference equations for the discrete points of a nodal network. You may test your understanding of related concepts by addressing the following questions.

- What is an *isotherm*? What is a *heat flow line*? How are the two lines related geometrically?
- What is an *adiabat*? How is it related to a line of symmetry? How is it intersected by an isotherm?
- What parameters characterize the effect of geometry on the relationship between the heat rate and the overall temperature difference for steady conduction in a two-dimensional system? How are these parameters related to the conduction resistance?
- What is represented by the temperature of a *nodal point*, and how does the accuracy of a nodal temperature depend on prescription of the *nodal network*?

References

1. Schneider, P. J., *Conduction Heat Transfer*, Addison-Wesley, Reading, MA, 1955.

2. Carslaw, H. S., and J. C. Jaeger, *Conduction of Heat in Solids*, Oxford University Press, London, 1959.

3. Özisik, M. N., *Heat Conduction*, Wiley Interscience, New York, 1980.

4. Kakac, S., and Y. Yener, *Heat Conduction*, Hemisphere Publishing, New York, 1985.

5. Poulikakos, D., *Conduction Heat Transfer*, Prentice-Hall, Englewood Cliffs, NJ, 1994.

6. Sunderland, J. E., and K. R. Johnson, *Trans. ASHRAE*, **10**, 237–241, 1964.

7. Kutateladze, S. S., *Fundamentals of Heat Transfer*, Academic Press, New York, 1963.

8. General Electric Co. (Corporate Research and Development), *Heat Transfer Data Book*, Section 502, General Electric Company, Schenectady, NY, 1973.

9. Hahne, E., and U. Grigull, *Int. J. Heat Mass Transfer*, **18**, 751–767, 1975.

10. Yovanovich, M. M., in W. M. Rohsenow, J. P. Hartnett, and Y. I. Cho, Eds., *Handbook of Heat Transfer*, McGraw-Hill, New York, 1998, pp. 3.1–3.73.

11. Gerald, C. F., and P. O. Wheatley, *Applied Numerical Analysis*, Pearson Education, Upper Saddle River, NJ, 1998.

12. Hoffman, J. D., *Numerical Methods for Engineers and Scientists*, McGraw-Hill, New York, 1992.

Problems

Exact Solutions

4.1 In the method of separation of variables (Section 4.2) for two-dimensional, steady-state conduction, the separation constant λ^2 in Equations 4.6 and 4.7 must be a positive constant. Show that a negative or zero value of λ^2 will result in solutions that cannot satisfy the prescribed boundary conditions.

4.2 A two-dimensional rectangular plate is subjected to prescribed boundary conditions. Using the results of the exact solution for the heat equation presented in Section 4.2, calculate the temperature at the midpoint $(1, 0.5)$ by considering the first five nonzero terms of the infinite series that must be evaluated. Assess the error resulting from using only the first three terms of the infinite series. Plot the temperature distributions $T(x, 0.5)$ and $T(1.0, y)$.

4.3 Consider the two-dimensional rectangular plate of Problem 4.2 having a thermal conductivity of 50 W/m · K. Beginning with the exact solution for the temperature distribution, derive an expression for the heat transfer rate per unit thickness from the plate along the lower surface $(0 \leq x \leq 2, y = 0)$. Evaluate the heat rate considering the first five nonzero terms of the infinite series.

4.4 A two-dimensional rectangular plate is subjected to the boundary conditions shown. Derive expressions for the steady-state temperature distribution $T(x, y)$ and the steady-state heat flux distribution $q''(x, y = b)$ for $0 \leq x \leq a$.

4.5 A two-dimensional rectangular plate is subjected to prescribed temperature boundary conditions on three sides and a uniform heat flux *into* the plate at the top surface. Using the general approach of Section 4.2, derive an expression for the temperature distribution in the plate.

Shape Factors and Dimensionless Conduction Heat Rates

4.6 Free convection heat transfer is sometimes quantified by writing Equation 4.20 as $q_{conv} = Sk_{eff} \Delta T_{1-2}$, where k_{eff} is an *effective* thermal conductivity. The ratio k_{eff}/k is greater than unity because of fluid motion driven by buoyancy forces, as represented by the dashed streamlines.

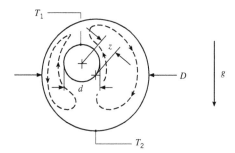

An experiment for the configuration shown yields a heat transfer rate per unit length of $q'_{conv} = 89$ W/m for surface temperatures of $T_1 = 50°C$ and $T_2 = 20°C$, respectively. For inner and outer cylinders of diameters $d = 20$ mm and $D = 60$ mm, and an eccentricity factor of $z = 10$ mm, determine the value of k_{eff}. The actual thermal conductivity of the fluid is $k = 0.27$ W/m · K.

4.7 Consider Problem 4.5 for the case where the plate is of square cross section, $W = L$.

(a) Derive an expression for the shape factor, S_{max}, associated with the *maximum* top surface temperature, such that $q = S_{max} k (T_{2,max} - T_1)$ where $T_{2,max}$ is the maximum temperature along $y = W$.

(b) Derive an expression for the shape factor, S_{avg}, associated with the *average* top surface temperature, $q = S_{avg}k(\bar{T}_2 - T_1)$ where $\bar{T}_2$ is the average temperature along $y = W$.

(c) Evaluate the shape factors that can be used to determine the maximum and average temperatures along $y = W$. Evaluate the maximum and average temperatures for $T_1 = 0°C$, $L = W = 10$ mm, $k = 20$ W/m · K, and $q''_s = 1000$ W/m^2.

4.8 Radioactive wastes are temporarily stored in a spherical container, the center of which is buried a distance of 10 m below the earth's surface. The outside diameter of the container is 2 m, and 500 W of heat are released as a result of radioactive decay. If the soil surface temperature is 20°C, what is the outside surface temperature of the container under steady-state conditions? On a sketch of the soil–container system drawn to scale, show representative isotherms and heat flow lines in the soil.

4.9 Based on the dimensionless conduction heat rates for cases 12–15 in Table 4.1*b*, find shape factors for the following objects having temperature T_1, located at the surface of a semi-infinite medium having temperature T_2. The surface of the semi-infinite medium is adiabatic.

(a) A buried hemisphere, flush with the surface.

(b) A disk on the surface. Compare your result to Table 4.1*a*, case 10.

(c) A square on the surface.

(d) A buried cube, flush with the surface.

(a) (b) and (c) (d)

4.10 Determine the heat transfer rate between two particles of diameter $D = 80$ μm and temperatures $T_1 = 300.1$ K and $T_2 = 299.9$ K, respectively. The particles are in contact and are surrounded by air.

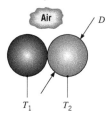

4.11 A two-dimensional object is subjected to isothermal conditions at its left and right surfaces, as shown in the schematic. Both diagonal surfaces are adiabatic and the depth of the object is $L = 100$ mm.

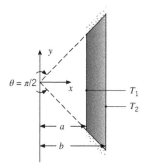

(a) Determine the two-dimensional shape factor for the object for $a = 10$ mm, $b = 12$ mm.

(b) Determine the two-dimensional shape factor for the object for $a = 10$ mm, $b = 15$ mm.

(c) Use the alternative conduction analysis of Section 3.2 to estimate the shape factor for parts (a) and (b). Compare the values of the approximate shape factors of the alternative conduction analysis to the two-dimensional shape factors of parts (a) and (b).

(d) For $T_1 = 100°C$ and $T_2 = 60°C$, determine the heat transfer rate per unit depth for $k = 15$ W/m · K for parts (a) and (b).

4.12 An electrical heater 80 mm long and 4 mm in diameter is inserted into a hole drilled normal to the surface of a large block of material having a thermal conductivity of 15 W/m·K. Estimate the temperature reached by the heater when dissipating 200 W with the surface of the block at a temperature of 25°C.

4.13 Two parallel pipelines spaced 0.5 m apart are buried in soil having a thermal conductivity of 0.5 W/m·K. The pipes have outer diameters of 100 and 75 mm with surface temperatures of 175°C and 5°C, respectively. Estimate the heat transfer rate per unit length between the two pipelines.

4.14 A tube of diameter 50 mm having a surface temperature of 85°C is embedded in the center plane of a concrete slab 0.1 m thick with upper and lower surfaces at 20°C. Using the appropriate tabulated relation for this configuration, find the shape factor. Determine the heat transfer rate per unit length of the tube.

4.15 Pressurized steam at 450 K flows through a long, thin-walled pipe of 0.5-m diameter. The pipe is enclosed in a concrete casing that is of square cross section and 1.5 m on a side. The axis of the pipe is centered in the casing, and the outer surfaces of the casing are maintained at 300 K. What is the heat loss per unit length of pipe?

4.16 The temperature distribution in laser-irradiated materials is determined by the power, size, and shape of the laser beam, along with the properties of the material being irradiated. The beam shape is typically Gaussian, and the local beam irradiation flux (often referred to as the laser *fluence*) is

$$q''(x, y) = q''(x = y = 0)\exp(-x/r_b)^2\exp(-y/r_b)^2$$

The x- and y-coordinates determine the location of interest on the surface of the irradiated material. Consider the case where the center of the beam is located at $x = y = r = 0$. The beam is characterized by a radius r_b, defined as the radial location where the local fluence is $q''(r_b) = q''(r = 0)/e \approx 0.368q''(r = 0)$.

A shape factor for Gaussian heating is $S = 2\pi^{1/2}r_b$, where S is defined in terms of $T_{1,\max} - T_2$ [Nissin, Y. I., A. Lietoila, R. G. Gold, and J. F. Gibbons, *J. Appl. Phys.*, **51**, 274, 1980]. Calculate the maximum steady-state surface temperature associated with irradiation of a material of thermal conductivity $k = 27$ W/m · K

and absorptivity $\alpha = 0.45$ by a Gaussian beam with $r_b = 0.1$ mm and power $P = 1$ W. Compare your result with the maximum temperature that would occur if the irradiation was from a circular beam of the same diameter and power, but characterized by a uniform fluence (a *flat* beam). Also calculate the average temperature of the irradiated surface for the uniform fluence case. The temperature far from the irradiated spot is $T_2 = 25°C$.

4.17 Hot water at 85°C flows through a thin-walled copper tube of 30-mm diameter. The tube is enclosed by an eccentric cylindrical shell that is maintained at 35°C and has a diameter of 120 mm. The eccentricity, defined as the separation between the centers of the tube and shell, is 20 mm. The space between the tube and shell is filled with an insulating material having a thermal conductivity of 0.05 W/m · K. Calculate the heat loss per unit length of the tube, and compare the result with the heat loss for a concentric arrangement.

4.18 A furnace of cubical shape, with external dimensions of 0.70 m, is constructed from a refractory brick (fireclay). If the wall thickness is 100 mm, the inner surface temperature is 620°C, and the outer surface temperature is 40°C, calculate the heat loss from the furnace.

Shape Factors with Thermal Circuits

4.19 A double-glazed window consists of two sheets of glass separated by an $L = 0.2$-mm-thick gap. The gap is evacuated, eliminating conduction and convection across the gap. Small cylindrical pillars, each $L = 0.2$ mm long and $D = 0.15$ mm in diameter, are inserted between the glass sheets to ensure that the glass does not break due to stresses imposed by the pressure difference across each glass sheet. A contact resistance of $R''_{t,c} = 1.5 \times 10^{-6}$ m² · K/W exists between the pillar and the sheet. For nominal glass temperatures of $T_1 = 20°C$ and $T_2 = -10°C$, determine the conduction heat transfer through an individual AISI 302 stainless steel pillar.

4.20 A pipeline, used for the transport of crude oil, is buried in the earth such that its centerline is a distance of 3 m below the surface. The pipe has an outer diameter of 1 m and is insulated with a layer of cellular glass 200 mm thick. What is the heat loss per unit length of pipe when heated oil at 110°C flows through the pipe and the surface of the earth is at a temperature of −10°C?

4.21 A long power transmission cable is buried at a depth (ground-to-cable-centerline distance) of 2 m. The cable is encased in a thin-walled pipe of 0.1-m diameter, and, to render the cable *superconducting* (with essentially zero power dissipation), the space between the cable and pipe is filled with liquid nitrogen at 77 K. If the pipe is covered with a superinsulator ($k_i = 0.005$ W/m · K) of 0.05-m thickness and the surface of the earth ($k_g = 1.2$ W/m · K) is at 300 K, what is the cooling load (W/m) that must be maintained by a cryogenic refrigerator per unit pipe length?

4.22 A small device is used to measure the surface temperature of an object. A thermocouple bead of diameter $D = 120$ μm is positioned a distance $z = 100$ μm from the surface of interest. The two thermocouple wires, each of diameter $d = 25$ μm and length $L = 300$ μm, are held by a large manipulator that is at a temperature of $T_m = 23$°C.

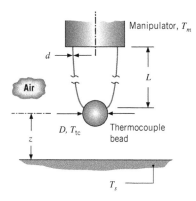

If the thermocouple registers a temperature of $T_{tc} = 29$°C, what is the surface temperature? The thermal conductivities of the chromel and alumel thermocouple wires are $k_{Ch} = 19$ W/m · K and $k_{Al} = 29$ W/m · K, respectively. You may neglect radiation and convection effects.

4.23 Two thick walls are separated by a vacuum gap of thickness L. A cylinder of diameter D runs between the walls. All surfaces are highly polished (their emissivity is small). The walls are at temperatures $T_{b,1}$ and $T_{b,2}$ at locations far from the cylinder.

(a) Draw the thermal resistance network.

(b) Derive an expression for the shape factor, S, associated with conduction between $T_{b,1}$ and $T_{b,2}$.

(c) Determine the value of the shape factor for $D = 0.01$ m, $L = 0.5$ m, and $k = 23$ W/m · K.

4.24 A hot fluid passes through circular channels of a cast iron platen (A) of thickness $L_A = 30$ mm which is in poor contact with the cover plates (B) of thickness $L_B = 7.5$ mm. The channels are of diameter $D = 15$ mm with a centerline spacing of $L_o = 60$ mm. The thermal conductivities of the materials are $k_A = 20$ W/m · K and $k_B = 75$ W/m · K, while the contact resistance between

the two materials is $R''_{t,c} = 2.0 \times 10^{-4}$ m² · K/W. The hot fluid is at $T_i = 150$°C, and the convection coefficient is 1000 W/m² · K. The cover plate is exposed to ambient air at $T_\infty = 25$°C with a convection coefficient of 200 W/m² · K. The shape factor between one channel and the platen top and bottom surfaces is 4.25.

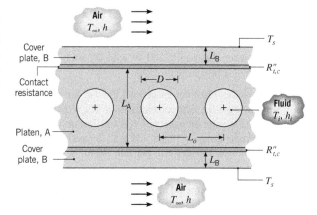

(a) Determine the heat rate from a single channel per unit length of the platen normal to the page, q'_i.

(b) Determine the outer surface temperature of the cover plate, T_s.

(c) Comment on the effects that changing the centerline spacing will have on q'_i and T_s. How would insulating the lower surface affect q'_i and T_s?

4.25 An aluminum heat sink ($k = 240$ W/m·K), used to cool an array of electronic chips, consists of a square channel of inner width $w = 25$ mm, through which liquid flow may be assumed to maintain a uniform surface temperature of $T_1 = 20$°C. The outer width and length of the channel are $W = 40$ mm and $L = 160$ mm, respectively.

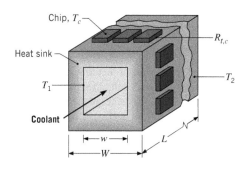

If $N = 120$ chips attached to the outer surfaces of the heat sink maintain an approximately uniform surface temperature of $T_2 = 50$°C and all of the heat dissipated by the chips is assumed to be transferred to the coolant, what is the heat dissipation per chip? If the contact resistance between each chip and the heat sink is $R_{t,c} = 0.2$ K/W, what is the chip temperature?

4.26 Hot water is transported from a cogeneration power station to commercial and industrial users through steel pipes of diameter $D = 150$ mm, with each pipe centered in concrete ($k = 1.4$ W/m · K) of square cross section ($w = 300$ mm). The outer surfaces of the concrete are exposed to ambient air for which $T_\infty = 0°C$ and $h = 25$ W/m² · K.

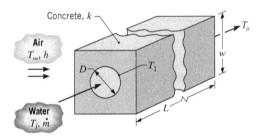

(a) If the inlet temperature of water flowing through the pipe is $T_i = 90°C$, what is the rate of heat loss per unit length of pipe in proximity to the inlet? The temperature of the pipe T_1 may be assumed to be that of the inlet water.

(b) If the difference between the inlet and outlet temperatures of water flowing through a 100-m-long pipe is not to exceed 5°C, estimate the minimum allowable flow rate $\dot{m}$. A value of $c = 4207$ J/kg · K may be used for the specific heat of the water.

4.27 A long constantan wire of 1-mm diameter is butt welded to the surface of a large copper block, forming a thermocouple junction. The wire behaves as a fin, permitting heat to flow from the surface, thereby depressing the sensing junction temperature T_j below that of the block T_o.

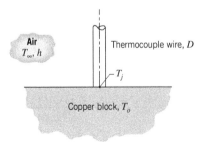

(a) If the wire is in air at 25°C with a convection coefficient of 25 W/m² · K, estimate the measurement error $(T_j - T_o)$ for the thermocouple when the block is at 125°C.

(b) For convection coefficients of 5, 10, and 25 W/m² · K, plot the measurement error as a function of the

thermal conductivity of the block material over the range 15 to 400 W/m · K. Under what circumstances is it advantageous to use smaller diameter wire?

4.28 A hole of diameter $D = 0.375$ m is drilled through the center of a solid block of square cross section with $w = 1.5$ m on a side. The hole is drilled along the length, $l = 4$ m, of the block, which has a thermal conductivity of $k = 27$ W/m·K. The four outer surfaces are exposed to ambient air, with $T_{\infty,2} = 25°C$ and $h_2 = 4$ W/m²·K, while hot oil flowing through the hole is characterized by $T_{\infty,1} = 300°C$ and $h_1 = 50$ W/m²·K. Determine the corresponding heat rate and surface temperatures.

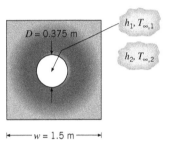

4.29 An igloo is built in the shape of a hemisphere, with an inner radius of 1.8 m and walls of compacted snow that are 0.5 m thick. On the inside of the igloo, the surface heat transfer coefficient is 6 W/m²·K; on the outside, under normal wind conditions, it is 15 W/m²·K. The thermal conductivity of compacted snow is 0.15 W/m·K. The temperature of the ice cap on which the igloo sits is −20°C and has the same thermal conductivity as the compacted snow.

(a) Assuming that the occupants' body heat provides a continuous source of 320 W within the igloo, calculate the inside air temperature when the outside air temperature is $T_\infty = -40°C$. Be sure to consider heat losses through the floor of the igloo.

(b) Using the thermal circuit of part (a), perform a parameter sensitivity analysis to determine which variables have a significant effect on the inside air temperature. For instance, for very high wind conditions, the outside convection coefficient could double or even triple. Does it make sense to construct the igloo with walls half or twice as thick?

4.30 In Chapter 3 we assumed that, whenever fins are attached to a base material, the base temperature is unchanged. What in fact happens is that, if the temperature of the base material exceeds the fluid temperature, attachment of a fin depresses the junction temperature T_j below the original temperature of the base, and heat flow from the base material to the fin is two-dimensional.

Consider conditions for which a long aluminum pin fin of diameter $D = 5$ mm is attached to a base material whose temperature far from the junction is maintained at $T_b = 100°C$. Fin convection conditions correspond to $h = 50$ W/m$^2 \cdot$ K and $T_\infty = 25°C$.

(a) What are the fin heat rate and junction temperature when the base material is (i) aluminum ($k = 240$ W/m $\cdot$ K) and (ii) stainless steel ($k = 15$ W/m $\cdot$ K)?

(b) Repeat the foregoing calculations if a thermal contact resistance of $R''_{t,j} = 3 \times 10^{-5}$ m$^2 \cdot$ K/W is associated with the method of joining the pin fin to the base material.

(c) Considering the thermal contact resistance, plot the heat rate as a function of the convection coefficient over the range $10 \le h \le 100$ W/m$^2 \cdot$ K for each of the two materials.

4.31 A cylinder of diameter $D = 10$ mm and temperature $T_1 = 100°C$ is centrally located in an extruded square coating of bakelite of dimension $w = 11$ mm. The ambient temperature is $T_\infty = 25°C$ and a convection coefficient of value $h = 100$ W/m$^2 \cdot$ K exists at the outer surfaces of the bakelite. Determine the heat transfer rate per unit length of the cylinder. Plot the heat transfer rate per length over the range 10 mm $\le w \le 100$ mm. Explain the dependence of the heat transfer rate on the bakelite thickness that is evident in your plot. Also derive the critical value of w that maximizes heat transfer from the cylinder.

4.32 The elemental unit of an air heater consists of a long circular rod of diameter D, which is encapsulated by a finned sleeve and in which thermal energy is generated by ohmic heating. The N fins of thickness t and length L are integrally fabricated with the square sleeve of width w. Under steady-state operating conditions, the rate of thermal energy generation corresponds to the rate of heat transfer to airflow over the sleeve.

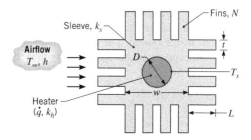

(a) Under conditions for which a uniform surface temperature T_s is maintained around the circumference of the heater and the temperature T_∞ and convection coefficient h of the airflow are known, obtain an expression for the rate of heat transfer per unit length to the air. Evaluate the heat rate for $T_s = 300°C$, $D = 20$ mm, an aluminum sleeve ($k_s = 240$ W/m $\cdot$ K), $w = 40$ mm, $N = 16$, $t = 4$ mm, $L = 20$ mm, $T_\infty = 50°C$, and $h = 500$ W/m$^2 \cdot$ K.

(b) For the foregoing heat rate and a copper heater of thermal conductivity $k_h = 400$ W/m $\cdot$ K, what is the required volumetric heat generation within the heater and its corresponding centerline temperature?

(c) With all other quantities unchanged, explore the effect of variations in the fin parameters (N, L, t) on the heat rate, subject to the constraint that the fin thickness and the spacing between fins cannot be less than 2 mm.

4.33 For a small heat source attached to a large substrate, the *spreading* resistance associated with multidimensional conduction in the substrate may be approximated by the expression [Yovanovich, M. M., and V. W. Antonetti, in *Adv. Thermal Modeling Elec. Comp. and Systems*, Vol. 1, A. Bar-Cohen and A. D. Kraus, Eds., Hemisphere, NY, 79–128, 1988]

$$R_{t(sp)} = \frac{1 - 1.410\, A_r + 0.344\, A_r^3 + 0.043\, A_r^5 + 0.034\, A_r^7}{4 k_{sub} A_{s,h}^{1/2}}$$

where $A_r = A_{s,h}/A_{s,sub}$ is the ratio of the heat source area to the substrate area. Consider application of the expression to an in-line array of square chips of width $L_h = 5$ mm on a side and pitch $S_h = 10$ mm. The interface between the chips and a large substrate of thermal conductivity $k_{sub} = 80$ W/m $\cdot$ K is characterized by a thermal contact resistance of $R''_{t,c} = 0.5 \times 10^{-4}$ m$^2 \cdot$ K/W.

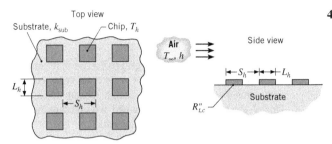

If a convection heat transfer coefficient of $h = 100$ W/m² · K is associated with airflow ($T_\infty = 15°$C) over the chips and substrate, what is the maximum allowable chip power dissipation if the chip temperature is not to exceed $T_h = 85°$C?

Finite-Difference Equations: Derivations

4.34 Consider nodal configuration 2 of Table 4.2. Derive the finite-difference equations under steady-state conditions for the following situations.

    (a) The horizontal boundary of the internal corner is perfectly insulated and the vertical boundary is subjected to the convection process (T_∞, h).

    (b) Both boundaries of the internal corner are perfectly insulated. How does this result compare with Equation 4.41?

4.35 Consider nodal configuration 4 of Table 4.2. Derive the finite-difference equations under steady-state conditions for the following situations.

    (a) The upper boundary of the external corner is perfectly insulated and the side boundary is subjected to the convection process (T_∞, h).

    (b) Both boundaries of the external corner are perfectly insulated. How does this result compare with Equation 4.43?

4.36 One of the strengths of numerical methods is their ability to handle complex boundary conditions. In the sketch, the boundary condition changes from specified heat flux q_s'' (into the domain) to convection, at the location of the node (m, n). Write the steady-state, two-dimensional finite difference equation at this node.

4.37 Determine expressions for $q_{(m-1,n) \to (m,n)}$, $q_{(m+1,n) \to (m,n)}$, $q_{(m,n+1) \to (m,n)}$ and $q_{(m,n-1) \to (m,n)}$ for conduction associated with a control volume that spans two different materials. There is no contact resistance at the interface between the materials. The control volumes are L units long into the page. Write the finite difference equation under steady-state conditions for node (m, n).

4.38 Consider heat transfer in a one-dimensional (radial) cylindrical coordinate system under steady-state conditions with volumetric heat generation.

    (a) Derive the finite-difference equation for any interior node m.

    (b) Derive the finite-difference equation for the node n located at the external boundary subjected to the convection process (T_∞, h).

4.39 In a two-dimensional cylindrical configuration, the radial (Δr) and angular ($\Delta\phi$) spacings of the nodes are uniform. The boundary at $r = r_i$ is of uniform temperature T_i. The boundaries in the radial direction are adiabatic (insulated) and exposed to surface convection (T_∞, h), as illustrated. Derive the finite-difference equations for (i) node 2, (ii) node 3, and (iii) node 1.

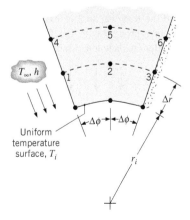

4.40 Upper and lower surfaces of a bus bar are convectively cooled by air at T_∞, with $h_u \neq h_i$. The sides are cooled by maintaining contact with heat sinks at T_o, through a thermal contact resistance of $R''_{t,c}$. The bar is of thermal conductivity k, and its width is twice its thickness L.

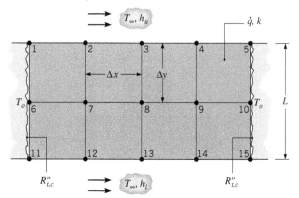

Consider steady-state conditions for which heat is uniformly generated at a volumetric rate $\dot{q}$ due to passage of an electric current. Using the energy balance method, derive finite-difference equations for nodes 1 and 13.

4.41 Derive the nodal finite-difference equations for the following configurations.

(a) Node (m, n) on a diagonal boundary subjected to convection with a fluid at T_∞ and a heat transfer coefficient h. Assume $\Delta x = \Delta y$.

(b) Node (m, n) at the tip of a cutting tool with the upper surface exposed to a constant heat flux q''_o, and the diagonal surface exposed to a convection cooling process with the fluid at T_∞ and a heat transfer coefficient h. Assume $\Delta x = \Delta y$.

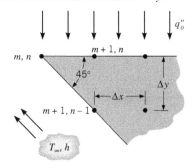

4.42 Consider the cutting tool of Problem 4.41 but with a tip angle of 60°. Derive the nodal finite difference equation for node (m, n).

4.43 Consider the two-dimensional grid ($\Delta x = \Delta y$) representing steady-state conditions with no internal volumetric generation for a system with thermal conductivity k. One of the boundaries is maintained at a constant temperature T_s while the others are adiabatic.

Derive an expression for the heat rate per unit length normal to the page crossing the isothermal boundary (T_s).

4.44 Consider a one-dimensional fin of uniform cross-sectional area, insulated at its tip, $x = L$. (See Table 3.4, case B). The temperature at the base of the fin T_b and of the adjoining fluid T_∞, as well as the heat transfer coefficient h and the thermal conductivity k, are known.

(a) Derive the finite-difference equation for any interior node m.

(b) Derive the finite-difference equation for a node n located at the insulated tip.

Finite-Difference Equations: Analysis

4.45 Consider the network for a two-dimensional system without internal volumetric generation having nodal temperatures shown below. If the grid spacing is 125 mm and the thermal conductivity of the material is 50 W/m·K, calculate the heat rate per unit length normal to the page from the isothermal surface (T_s).

Node	T_i (°C)
1	120.55
2	120.64
3	121.29
4	123.89
5	134.57
6	150.49
7	147.14

4.46 An ancient myth describes how a wooden ship was destroyed by soldiers who reflected sunlight from their polished bronze shields onto its hull, setting the ship ablaze. To test the validity of the myth, a group of college students are given mirrors and they reflect sunlight onto a 100 mm × 100 mm area of a $t = $ 10-mm-thick plywood mockup characterized by $k = 0.8$ W/m·K. The bottom of the mockup is in water at $T_w = 20°C$, while the air temperature is $T_\infty = 25°C$. The surroundings are at $T_{sur} = 23°C$. The wood has an emissivity of $\varepsilon = 0.90$; both the front and back surfaces of the plywood are characterized by $h = 5$ W/m²·K. The absorbed irradiation from the N students' mirrors is $G_{S,N} = 70,000$ W/m² on the front surface of the mockup.

(a) A debate ensues concerning where the beam should be focused, location A or location B. Using a finite difference method with $\Delta x = \Delta y = 100$ mm and treating the wood as a two-dimensional extended surface (Figure 3.17a), enlighten the students as to whether location A or location B will be more effective in igniting the wood by determining the maximum local steady-state temperature.

(b) Some students wonder whether the same technique can be used to melt a stainless steel hull. Repeat part (a) considering a stainless steel mockup of the same dimensions with $k = 15$ W/m ·K and $\varepsilon = 0.2$. The value of the absorbed irradiation is the same as in part (a).

4.47 Square channels of dimension $L_c = \sqrt{2} \cdot 10$ mm $= 14.14$ mm are evenly spaced at $S = 25$ mm along the centerline of a plate of thickness $L_p = 40$ mm. Hot and cold fluids flow through the channels in an alternating pattern as shown. Determine the maximum and minimum temperatures within the solid plate and the heat transfer rate per unit plate length. There are $N = 50$ total channels with $T_{\infty,h} = 120°C$, $T_{\infty,c} = 20°C$, $h = 40$ W/m² · K, and $k = 14$ W/m · K. Use $\Delta x = \Delta y = 5$ mm and the computational domain identified in the sketch.

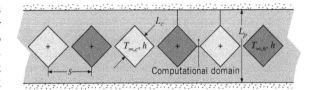

4.48 A long conducting rod of rectangular cross section (20 mm × 30 mm) and thermal conductivity $k = 20$ W/m · K experiences uniform heat generation at a rate $\dot{q} = 5 \times 10^7$ W/m³, while its surfaces are maintained at 300 K.

(a) Using a finite-difference method with a grid spacing of 5 mm, determine the temperature distribution in the rod.

(b) With the boundary conditions unchanged, what heat generation rate will cause the midpoint temperature to reach 600 K?

4.49 A flue passing hot exhaust gases has a square cross section, 300 mm to a side. The walls are constructed of refractory brick 150 mm thick with a thermal conductivity of 0.85 W/m·K. Calculate the heat loss from the flue per unit length when the interior and exterior surfaces are maintained at 350 and 25°C, respectively. Use a grid spacing of 75 mm.

4.50 Steady-state temperatures (K) at three nodal points of a long rectangular rod are as shown. The rod experiences a uniform volumetric generation rate of 5×10^7 W/m³ and has a thermal conductivity of 20 W/m · K. Two of its sides are maintained at a constant temperature of 300 K, while the others are insulated.

(a) Determine the temperatures at nodes 1, 2, and 3.

(b) Calculate the heat transfer rate per unit length (W/m) from the rod using the nodal temperatures. Compare this result with the heat rate calculated from knowledge of the volumetric generation rate and the rod dimensions.

4.51 Functionally graded materials are intentionally fabricated to establish a spatial distribution of properties in the final product. Consider an $L \times L$ two-dimensional object with $L = 20$ mm. The thermal conductivity distribution of the functionally graded material is $k(x) = 20$ W/m·K $+ (7070$ W/m$^{5/2}$·K$)x^{3/2}$. Two sets of boundary conditions, denoted as cases 1 and 2, are applied.

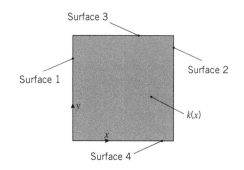

Surface 3

Surface 1

Surface 2

$k(x)$

Surface 4

Case	Surface	Boundary Condition
1	1	$T = 100°C$
—	2	$T = 50°C$
—	3	Adiabatic
—	4	Adiabatic
2	1	Adiabatic
—	2	Adiabatic
—	3	$T = 50°C$
—	4	$T = 100°C$

(a) Determine the spatially averaged value of the thermal conductivity $\bar{k}$. Use this value to estimate the heat rate per unit depth for cases 1 and 2.

(b) Using a grid spacing of 2 mm, determine the heat rate per unit depth for case 1. Compare your result to the estimated value calculated in part (a).

(c) Using a grid spacing of 2 mm, determine the heat rate per unit depth for case 2. Compare your result to the estimated value calculated in part (a).

4.52 Steady-state temperatures at selected nodal points of the symmetrical section of a flow channel are known to be $T_2 = 95.47°C$, $T_3 = 117.3°C$, $T_5 = 79.79°C$, $T_6 = 77.29°C$, $T_8 = 87.28°C$, and $T_{10} = 77.65°C$. The wall experiences uniform volumetric heat generation of $\dot{q} = 10^6$ W/m^3 and has a thermal conductivity of $k = 10$ W/m·K. The inner and outer surfaces of the channel experience convection with fluid temperatures of

$T_{\infty,i} = 50°C$ and $T_{\infty,o} = 25°C$ and convection coefficients of $h_i = 500$ W/m^2·K and $h_o = 250$ W/m^2·K.

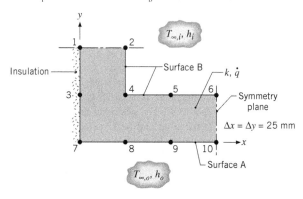

(a) Determine the temperatures at nodes 1, 4, 7, and 9.

(b) Calculate the heat rate per unit length (W/m) from the outer surface A to the adjacent fluid.

(c) Calculate the heat rate per unit length from the inner fluid to surface B.

(d) Verify that your results are consistent with an overall energy balance on the channel section.

4.53 Consider an aluminum heat sink ($k = 240$ W/m·K), such as that shown schematically in Problem 4.25. The inner and outer widths of the square channel are $w = 20$ mm and $W = 40$ mm, respectively, and an outer surface temperature of $T_s = 50°C$ is maintained by the array of electronic chips. In this case, it is not the inner surface temperature that is known, but conditions (T_∞, h) associated with coolant flow through the channel, and we wish to determine the rate of heat transfer to the coolant per unit length of channel. For this purpose, consider a symmetrical section of the channel and a two-dimensional grid with $\Delta x = \Delta y = 5$ mm.

(a) For $T_\infty = 20°C$ and $h = 5000$ W/m^2·K, determine the unknown temperatures, $T_1, \ldots, T_7$, and the rate of heat transfer per unit length of channel, q'.

(b) Assess the effect of variations in h on the unknown temperatures and the heat rate.

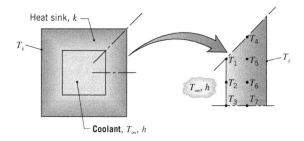

Heat sink, k

T_s

Coolant, T_∞, h

T_{∞}, h

4.54 Conduction within relatively complex geometries can sometimes be evaluated using the finite-difference methods of this text that are applied to *subdomains* and *patched* together. Consider the two-dimensional domain formed by rectangular and cylindrical subdomains patched at the common, dashed control surface. Note that, along the dashed control surface, temperatures in the two subdomains are identical and local conduction heat fluxes to the cylindrical subdomain are identical to local conduction heat fluxes from the rectangular subdomain.

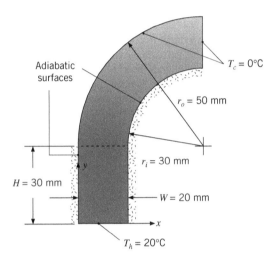

Calculate the heat transfer rate per unit depth into the page, q', using $\Delta x = \Delta y = \Delta r = 10$ mm and $\Delta \phi = \pi/8$. The base of the rectangular subdomain is held at $T_h = 20°C$, while the vertical surface of the cylindrical subdomain and the surface at outer radius r_o are at $T_c = 0°C$. The remaining surfaces are adiabatic, and the thermal conductivity is $k = 10$ W/m·K.

4.55 Consider the two-dimensional tube of a noncircular cross section formed by rectangular and semicylindrical subdomains patched at the common dashed control surfaces in a manner similar to that described in Problem 4.54. Note that, along the dashed control surfaces, temperatures in the two subdomains are identical and local conduction heat fluxes to the semicylindrical subdomain are identical to local conduction heat fluxes from the rectangular subdomain. The bottom of the domain is held at $T_s = 100°C$ by condensing steam, while the flowing fluid is characterized by the temperature and convection coefficient shown in the sketch. The remaining surfaces are insulated, and the thermal conductivity is $k = 15$ W/m·K.

Find the heat transfer rate per unit length of tube, q', using $\Delta x = \Delta y = \Delta r = 10$ mm and $\Delta \phi = \pi/8$. *Hint*: Take advantage of the symmetry of the problem by considering only half of the entire domain.

4.56 The steady-state temperatures (°C) associated with selected nodal points of a two-dimensional system having a thermal conductivity of 1.5 W/m·K are shown on the accompanying grid.

(a) Determine the temperatures at nodes 1, 2, and 3.

(b) Calculate the heat transfer rate per unit thickness normal to the page from the system to the fluid.

4.57 Consider two-dimensional, steady-state conduction in a square cross section with prescribed surface temperatures.

(a) Determine the temperatures at nodes 1, 2, 3, and 4. Estimate the midpoint temperature.

(b) Reducing the mesh size by a factor of 2, determine the corresponding nodal temperatures. Compare your results with those from the coarser grid.

(c) From the results for the finer grid, plot the 75, 150, and 250°C isotherms.

4.58 Consider a long bar of square cross section (0.8 m to the side) and of thermal conductivity 2 W/m · K. Three of these sides are maintained at a uniform temperature of 300°C. The fourth side is exposed to a fluid at 100°C for which the convection heat transfer coefficient is 10 W/m² · K.

(a) Using an appropriate numerical technique with a grid spacing of 0.2 m, determine the midpoint temperature and heat transfer rate between the bar and the fluid per unit length of the bar.

(b) Reducing the grid spacing by a factor of 2, determine the midpoint temperature and heat transfer rate. Plot the corresponding temperature distribution across the surface exposed to the fluid. Also, plot the 200 and 250°C isotherms.

4.59 Consider a two-dimensional, straight triangular fin of length $L = 50$ mm and base thickness $t = 20$ mm. The thermal conductivity of the fin is $k = 25$ W/m · K. The base temperature is $T_b = 50$°C, and the fin is exposed to convection conditions characterized by $h = 50$ W/m² · K, $T_\infty = 20$°C. Using a finite difference mesh with $\Delta x = 10$ mm and $\Delta y = 2$ mm, and taking advantage of symmetry, determine the fin efficiency, η_f. Compare your value of the fin efficiency with that reported in Figure 3.19.

4.60 A common arrangement for heating a large surface area is to move warm air through rectangular ducts below the surface. The ducts are square and located midway between the top and bottom surfaces that are exposed to room air and insulated, respectively.

For the condition when the floor and duct temperatures are 30 and 80°C, respectively, and the thermal conductivity of concrete is 1.4 W/m · K, calculate the heat rate from each duct, per unit length of duct. Use a grid spacing with $\Delta x = 2 \Delta y$, where $\Delta y = 0.125L$ and $L = 150$ mm.

4.61 A long, solid cylinder of diameter $D = 25$ mm is formed of an insulating core that is covered with a *very thin* ($t = 50$ μm), highly polished metal sheathing of thermal conductivity $k = 25$ W/m · K. Electric current flows through the stainless steel from one end of the cylinder to the other, inducing uniform volumetric heating within the sheathing of $\dot{q} = 5 \times 10^6$ W/m³. As will become evident in Chapter 6, values of the convection coefficient between the surface and air for this situation are spatially nonuniform, and for the airstream conditions of the experiment, the convection heat transfer coefficient varies with the angle θ as $h(\theta) = 26 + 0.637\theta - 8.92\theta^2$ for $0 \leq \theta < \pi/2$ and $h(\theta) = 5$ for $\pi/2 \leq \theta \leq \pi$.

(a) Neglecting conduction in the θ-direction within the stainless steel, plot the temperature distribution $T(\theta)$ for $0 \leq \theta \leq \pi$ for $T_\infty = 25$°C.

(b) Accounting for θ-direction conduction in the stainless steel, determine temperatures in the stainless steel at increments of $\Delta\theta = \pi/20$ for $0 \leq \theta \leq \pi$. Compare the temperature distribution with that of part (a).

Hint: The temperature distribution is symmetrical about the horizontal centerline of the cylinder.

4.62 A bar of thermal conductivity $k = 140$ W/m · K is of a trapezoidal cross section as shown in the schematic. The left and right faces are at temperatures $T_h = 100$°C and $T_c = 0$°C, respectively. Determine the heat transfer rate per unit bar length using a finite difference approach with $\Delta x = \Delta y = 10$ mm. Compare the heat rate to that of a bar of a 20 mm × 30 mm rectangular cross section where the height of the domain is 20 mm.

4.63 A thin metallic foil of thickness 0.25 mm with a pattern of extremely small holes serves as an acceleration grid to control the electrical potential of an ion beam. Such a grid is used in a chemical vapor deposition (CVD) process for the fabrication of semiconductors. The top surface of the grid is exposed to a uniform heat flux caused by absorption of the ion beam, $q_s'' = 600$ W/m^2. The edges of the foil are thermally coupled to water-cooled sinks maintained at 300 K. The upper and lower surfaces of the foil experience radiation exchange with the vacuum enclosure walls maintained at 300 K. The effective thermal conductivity of the foil material is 40 W/m·K, and its emissivity is 0.45.

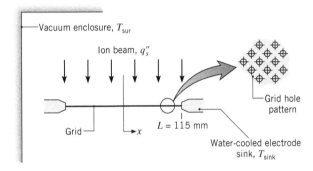

Assuming one-dimensional conduction and using a finite-difference method representing the grid by 10 nodes in the *x*-direction, estimate the temperature distribution for the grid. *Hint:* For each node requiring an energy balance, use the linearized form of the radiation rate equation, Equation 1.8, with the radiation coefficient h_r, Equation 1.9, evaluated for each node.

4.64 A long bar of rectangular cross section, 0.4 m × 0.6 m on a side and having a thermal conductivity of 1.5 W/m·K, is subjected to the boundary conditions shown.

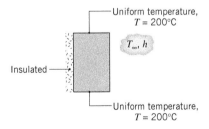

Two of the sides are maintained at a uniform temperature of 200°C. One of the sides is adiabatic, and the remaining side is subjected to a convection process with $T_\infty = 30$°C and $h = 50$ W/m^2·K. Using an appropriate numerical technique with a grid spacing of 0.1 m, determine the temperature distribution in the bar and the heat transfer rate between the bar and the fluid per unit length of the bar.

4.65 The top surface of a plate, including its grooves, is maintained at a uniform temperature of $T_1 = 200$°C. The lower surface is at $T_2 = 20$°C, the thermal conductivity is 15 W/m·K, and the groove spacing is 0.16 m.

(a) Using a finite-difference method with a mesh size of $\Delta x = \Delta y = 40$ mm, calculate the unknown nodal temperatures and the heat transfer rate per width of groove spacing (*w*) and per unit length normal to the page.

(b) With a mesh size of $\Delta x = \Delta y = 10$ mm, repeat the foregoing calculations, determining the temperature field and the heat rate. Also, consider conditions for which the bottom surface is not at a uniform temperature T_2 but is exposed to a fluid at $T_\infty = 20$°C. With $\Delta x = \Delta y = 10$ mm, determine the temperature field and heat rate for values of $h = 5$, 200, and 1000 W/m^2·K, as well as for $h \to \infty$.

4.66 Refer to the two-dimensional rectangular plate of Problem 4.2. Using an appropriate numerical method with $\Delta x = \Delta y = 0.25$ m, determine the temperature at the midpoint (1, 0.5).

4.67 The shape factor for conduction through the edge of adjoining walls for which $D > L/5$, where D and L are the wall depth and thickness, respectively, is shown in Table 4.1. The two-dimensional symmetrical element of the edge, which is represented by inset (*a*), is bounded by the diagonal symmetry adiabat and a section of the wall thickness over which the temperature distribution is assumed to be linear between T_1 and T_2.

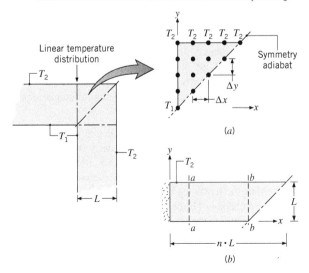

(a) Using the nodal network of inset (*a*) with $L = 40$ mm, determine the temperature distribution in the element for $T_1 = 100°C$ and $T_2 = 0°C$. Evaluate the heat rate per unit depth ($D = 1$ m) if $k = 1$ W/m · K. Determine the corresponding shape factor for the edge, and compare your result with that from Table 4.1.

(b) Choosing a value of $n = 1$ or $n = 1.5$, establish a nodal network for the trapezoid of inset (*b*) and determine the corresponding temperature field. Assess the validity of assuming linear temperature distributions across sections *a–a* and *b–b*.

4.68 The diagonal of a long triangular bar is well insulated, while sides of equivalent length are maintained at uniform temperatures T_a and T_b.

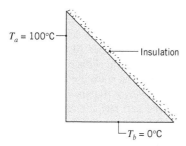

(a) Establish a nodal network consisting of five nodes along each of the sides. For one of the nodes on the diagonal surface, define a suitable control volume and derive the corresponding finite-difference equation. Using this form for the diagonal nodes and appropriate equations for the interior nodes, find the temperature distribution for the bar. On a scale drawing of the shape, show the 25, 50, and 75°C isotherms.

(b) An alternate and simpler procedure to obtain the finite-difference equations for the diagonal nodes follows from recognizing that the insulated diagonal surface is a symmetry plane. Consider a square 5 × 5 nodal network, and represent its diagonal as a symmetry line. Recognize which nodes on either side of the diagonal have identical temperatures. Show that you can treat the diagonal nodes as "interior" nodes and write the finite-difference equations by inspection.

4.69 A simplified representation for cooling in very large-scale integration (VLSI) of microelectronics is shown in the sketch. A silicon chip is mounted in a dielectric substrate, and one surface of the system is convectively cooled, while the remaining surfaces are well insulated from the surroundings. The problem is rendered two-dimensional by assuming the system to be very long in the direction perpendicular to the paper. Under steady-state operation,

electric power dissipation in the chip provides for uniform volumetric heating at a rate of $\dot{q}$. However, the heating rate is limited by restrictions on the maximum temperature that the chip is allowed to achieve.

For the conditions shown on the sketch, will the maximum temperature in the chip exceed 85°C, the maximum allowable operating temperature set by industry standards? A grid spacing of 3 mm is suggested.

4.70 A plate ($k = 10$ W/m · K) is stiffened by a series of longitudinal ribs having a rectangular cross section with length $L = 8$ mm and width $w = 4$ mm. The base of the plate is maintained at a uniform temperature $T_b = 45°C$, while the rib surfaces are exposed to air at a temperature of $T_\infty = 25°C$ and a convection coefficient of $h = 600$ W/m² · K.

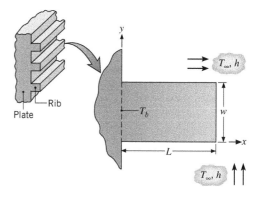

(a) Using a finite-difference method with $\Delta x = \Delta y = 2$ mm and a total of 5 × 3 nodal points and regions, estimate the temperature distribution and the heat rate from the base. Compare these results with those obtained by assuming that heat transfer in the rib is one-dimensional, thereby approximating the behavior of a fin.

(b) The grid spacing used in the foregoing finite-difference solution is coarse, resulting in poor precision for estimates of temperatures and the heat rate. Investigate the effect of grid refinement by reducing the nodal spacing to $\Delta x = \Delta y = 1$ mm (a 9×3 grid) considering symmetry of the center line.

(c) Investigate the nature of two-dimensional conduction in the rib and determine a criterion for which the one-dimensional approximation is reasonable. Do so by extending your finite-difference analysis to determine the heat rate from the base as a function of the length of the rib for the range $1.5 \leq L/w \leq 10$, keeping the length L constant. Compare your results with those determined by approximating the rib as a fin.

4.71 The bottom half of an I-beam providing support for a furnace roof extends into the heating zone. The web is well insulated, while the flange surfaces experience convection with hot gases at $T_\infty = 400°C$ and a convection coefficient of $h = 150$ W/m$^2 \cdot$K. Consider the symmetrical element of the flange region (inset a), assuming that the temperature distribution across the web is uniform at $T_w = 100°C$. The beam thermal conductivity is 10 W/m·K, and its dimensions are $w_f = 80$ mm, $w_w = 30$ mm, and $L = 30$ mm.

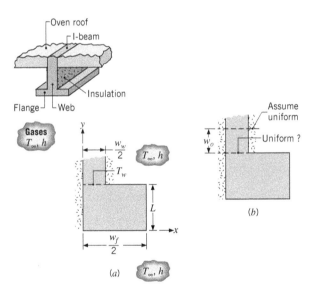

(a) Calculate the heat transfer rate per unit length to the beam using a 5×4 nodal network.

(b) Is it reasonable to assume that the temperature distribution across the web–flange interface is uniform? Consider the L-shaped domain of inset (b) and use a fine grid to obtain the temperature distribution across the web–flange interface. Make the distance $w_o \geq w_w/2$.

4.72 A long trapezoidal bar is subjected to uniform temperatures on two surfaces, while the remaining surfaces are well insulated. If the thermal conductivity of the material is 20 W/m·K, estimate the heat transfer rate per unit length of the bar using a finite-difference method. Use the Gauss–Seidel method of solution with a space increment of 10 mm.

4.73 Small-diameter electrical heating elements dissipating 50 W/m (length normal to the sketch) are used to heat a ceramic plate of thermal conductivity 2 W/m·K. The upper surface of the plate is exposed to ambient air at 30°C with a convection coefficient of 100 W/m$^2 \cdot$K, while the lower surface is well insulated.

(a) Using the Gauss–Seidel method with a grid spacing of $\Delta x = 6$ mm and $\Delta y = 2$ mm, obtain the temperature distribution within the plate.

(b) Using the calculated nodal temperatures, sketch four isotherms to illustrate the temperature distribution in the plate.

(c) Calculate the heat loss by convection from the plate to the fluid. Compare this value with the element dissipation rate.

(d) What advantage, if any, is there in not making $\Delta x = \Delta y$ for this situation?

(e) With $\Delta x = \Delta y = 2$ mm, calculate the temperature field within the plate and the rate of heat transfer from the plate. Under no circumstances may the temperature at any location in the plate exceed 400°C. Would this limit be exceeded if the airflow were terminated and heat transfer to the air were by natural convection with $h = 10$ W/m$^2 \cdot$K?

Transient Conduction

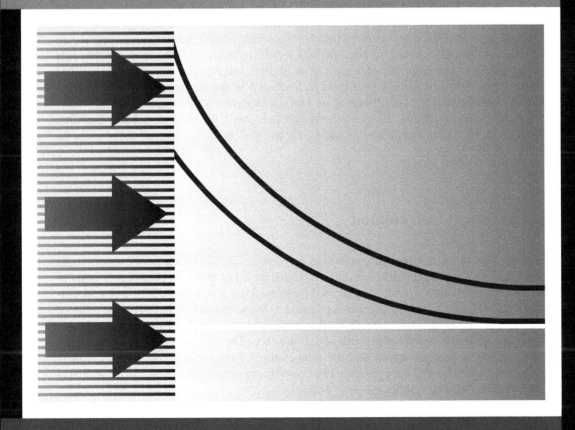

In our treatment of conduction we have gradually considered more complicated conditions. We began with the simple case of one-dimensional, steady-state conduction with no internal generation, and we subsequently considered more realistic situations involving multidimensional and generation effects. However, we have not yet considered situations for which conditions change with time.

We now recognize that many heat transfer problems are time dependent. Such *unsteady,* or *transient,* problems typically arise when the boundary conditions of a system are changed. For example, if the surface temperature of a system is altered, the temperature at each point in the system will also begin to change. The changes will continue to occur until, as is often the case, a *steady-state* temperature distribution is ultimately reached. Consider a hot metal billet that is removed from a furnace and exposed to a cool airstream. Energy is transferred by convection and radiation from its surface to the surroundings. Energy transfer by conduction also occurs from the interior of the metal to the surface, and the temperature at each point in the billet decreases until a steady-state condition is reached.

Our objective in this chapter is to develop procedures for determining the time dependence of the temperature distribution within a solid during a transient process, as well as for determining heat transfer between the solid and its surroundings. The nature of the procedure depends on assumptions that may be made for the process. If, for example, temperature gradients within the solid may be neglected, a comparatively simple approach, termed the *lumped capacitance method*, may be used to determine the variation of temperature with time. The method is developed in Sections 5.1 through 5.3.

Under conditions for which temperature gradients are not negligible, but heat transfer within the solid is one-dimensional, exact solutions to the heat equation may be used to compute the dependence of temperature on both location and time. Such solutions are considered for *finite solids* (plane walls, long cylinders and spheres) in Sections 5.4 through 5.6 and for *semi-infinite solids* in Section 5.7. Section 5.8 presents the transient thermal response of a variety of objects subject to a step change in either surface temperature or surface heat flux. In Section 5.9, the response of a semi-infinite solid to periodic heating conditions at its surface is explored. For more complex conditions, finite-difference or finite-element methods must be used to predict the time dependence of temperatures within the solid, as well as heat rates at its boundaries (Section 5.10).

5.1 *The Lumped Capacitance Method*

A simple, yet common, transient conduction problem is one for which a solid experiences a sudden change in its thermal environment. Consider a hot metal forging that is initially at a uniform temperature T_i and is quenched by immersing it in a liquid of lower temperature $T_\infty < T_i$ (Figure 5.1). If the quenching is said to begin at time $t = 0$, the temperature of the solid will decrease for time $t > 0$, until it eventually reaches T_∞. This reduction is due to convection heat transfer at the solid–liquid interface. The essence of the lumped capacitance method is the assumption that the temperature of the solid is *spatially uniform* at any instant during the transient process. This assumption implies that temperature gradients within the solid are negligible.

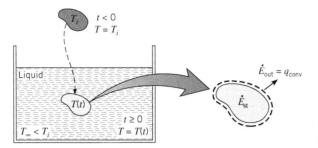

FIGURE 5.1 Cooling of a hot metal forging.

From Fourier's law, heat conduction in the absence of a temperature gradient implies the existence of infinite thermal conductivity. Such a condition is clearly impossible. However, the condition is closely approximated if the resistance to conduction within the solid is small compared with the resistance to heat transfer between the solid and its surroundings. For now we assume that this is, in fact, the case.

In neglecting temperature gradients within the solid, we can no longer consider the problem from within the framework of the heat equation, since the heat equation is a differential equation governing the spatial temperature distribution within the solid. Instead, the transient temperature response is determined by formulating an overall energy balance on the entire solid. This balance must relate the rate of heat loss at the surface to the rate of change of the internal energy. Applying Equation 1.12c to the control volume of Figure 5.1, this requirement takes the form

$$-\dot{E}_{\text{out}} = \dot{E}_{\text{st}} \tag{5.1}$$

or

$$-hA_s(T - T_\infty) = \rho Vc \frac{dT}{dt} \tag{5.2}$$

Introducing the temperature difference

$$\theta \equiv T - T_\infty \tag{5.3}$$

and recognizing that $(d\theta/dt) = (dT/dt)$ if T_∞ is constant, it follows that

$$\frac{\rho Vc}{hA_s} \frac{d\theta}{dt} = -\theta$$

Separating variables and integrating from the initial condition, for which $t = 0$ and $T(0) = T_i$, we then obtain

$$\frac{\rho Vc}{hA_s} \int_{\theta_i}^{\theta} \frac{d\theta}{\theta} = -\int_0^t dt$$

where

$$\theta_i \equiv T_i - T_\infty \tag{5.4}$$

Evaluating the integrals, it follows that

$$\frac{\rho Vc}{hA_s} \ln \frac{\theta_i}{\theta} = t \tag{5.5}$$

or

$$\frac{\theta}{\theta_i} = \frac{T - T_\infty}{T_i - T_\infty} = \exp\left[-\left(\frac{hA_s}{\rho Vc}\right)t\right] \tag{5.6}$$

Equation 5.5 may be used to determine the time required for the solid to reach some temperature T, or, conversely, Equation 5.6 may be used to compute the temperature reached by the solid at some time t.

The foregoing results indicate that the difference between the solid and fluid temperatures decays exponentially to zero as t approaches infinity. This behavior is shown in Figure 5.2. From Equation 5.6 it is also evident that the quantity $(\rho Vc/hA_s)$ may be interpreted as a *thermal time constant* expressed as

$$\tau_t = \left(\frac{1}{hA_s}\right)(\rho Vc) = R_t C_t \tag{5.7}$$

where, from Equation 3.9, R_t is the resistance to convection heat transfer and C_t is the *lumped thermal capacitance* of the solid. Any increase in R_t or C_t will cause the solid to respond more slowly to changes in its thermal environment. This behavior is analogous to the voltage decay that occurs when a capacitor is discharged through a resistor in an electrical *RC* circuit.

To determine the total energy transfer Q occurring up to some time t, we simply write

$$Q = \int_0^t q \, dt = hA_s \int_0^t \theta \, dt$$

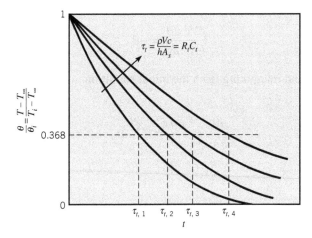

FIGURE 5.2 Transient temperature response of lumped capacitance solids for different thermal time constants τ_t.

Substituting for θ from Equation 5.6 and integrating, we obtain

$$Q = (\rho V c)\theta_i \left[1 - \exp\left(-\frac{t}{\tau_t}\right)\right] \qquad (5.8a)$$

The quantity Q is, of course, related to the change in the internal energy of the solid, and from Equation 1.12b

$$-Q = \Delta E_{st} \qquad (5.8b)$$

For quenching, Q is positive and the solid experiences a decrease in energy. Equations 5.5, 5.6, and 5.8a also apply to situations where the solid is heated ($\theta < 0$), in which case Q is negative and the internal energy of the solid increases.

5.2 *Validity of the Lumped Capacitance Method*

The foregoing results demonstrate that the lumped capacitance method is a simple and convenient method for solving transient heating and cooling problems. In this section, we determine under what conditions it may be used with reasonable accuracy.

To develop a suitable criterion consider steady-state conduction through the plane wall of area A (Figure 5.3). Although we are assuming steady-state conditions, the following criterion is readily extended to transient processes. One surface is maintained at a temperature $T_{s,1}$ and the other surface is exposed to a fluid of temperature $T_\infty < T_{s,1}$. The temperature of this surface will be some intermediate value $T_{s,2}$, for which $T_\infty < T_{s,2} < T_{s,1}$. Hence under steady-state conditions the surface energy balance, Equation 1.13, reduces to

$$\frac{kA}{L}(T_{s,1} - T_{s,2}) = hA(T_{s,2} - T_\infty)$$

where k is the thermal conductivity of the solid. Rearranging, we then obtain

$$\frac{T_{s,1} - T_{s,2}}{T_{s,2} - T_\infty} = \frac{(L/kA)}{(1/hA)} = \frac{R_{t,\text{cond}}}{R_{t,\text{conv}}} = \frac{hL}{k} \equiv Bi \qquad (5.9)$$

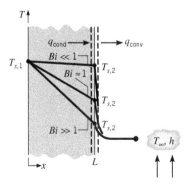

FIGURE 5.3 Effect of Biot number on steady-state temperature distribution in a plane wall with surface convection.

The quantity (hL/k) appearing in Equation 5.9 is a *dimensionless parameter.* It is termed the *Biot number,* and it plays a fundamental role in conduction problems that involve surface convection effects. According to Equation 5.9 and as illustrated in Figure 5.3, the Biot number provides a measure of the temperature drop in the solid relative to the temperature difference between the solid's surface and the fluid. From Equation 5.9, it is also evident that the Biot number may be interpreted as a ratio of thermal resistances. *In particular, if Bi ≪ 1, the resistance to conduction within the solid is much less than the resistance to convection across the fluid boundary layer. Hence, the assumption of a uniform temperature distribution within the solid is reasonable if the Biot number is small.*

Although we have discussed the Biot number in the context of steady-state conditions, we are reconsidering this parameter because of its significance to transient conduction problems. Consider the plane wall of Figure 5.4, which is initially at a uniform temperature T_i and experiences convection cooling when it is immersed in a fluid of $T_\infty < T_i$. The problem may be treated as one-dimensional in x, and we are interested in the temperature variation with position and time, $T(x, t)$. This variation is a strong function of the Biot number, and three conditions are shown in Figure 5.4. Again, for $Bi \ll 1$ the temperature gradients in the solid are small and the assumption of a uniform temperature distribution, $T(x, t) \approx T(t)$ is reasonable. Virtually all the temperature difference is between the solid and the fluid, and the solid temperature remains nearly uniform as it decreases to T_∞. For moderate to large values of the Biot number, however, the temperature gradients within the solid are significant. Hence $T = T(x, t)$. Note that for $Bi \gg 1$, the temperature difference across the solid is much larger than that between the surface and the fluid.

We conclude this section by emphasizing the importance of the lumped capacitance method. Its inherent simplicity renders it the preferred method for solving transient heating and cooling problems. Hence, when confronted with such a problem, *the very first thing that one should do is calculate the Biot number.* If the following condition is satisfied

$$Bi = \frac{hL_c}{k} < 0.1 \qquad (5.10)$$

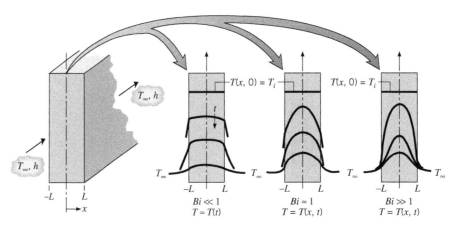

FIGURE 5.4   Transient temperature distributions for different Biot numbers in a plane wall symmetrically cooled by convection.

the error associated with using the lumped capacitance method is small. For convenience, it is customary to define the *characteristic length* of Equation 5.10 as the ratio of the solid's volume to surface area $L_c \equiv V/A_s$. Such a definition facilitates calculation of L_c for solids of complicated shape and reduces to the half-thickness L for a plane wall of thickness $2L$ (Figure 5.4), to $r_o/2$ for a long cylinder, and to $r_o/3$ for a sphere. However, if one wishes to implement the criterion in a conservative fashion, L_c should be associated with the length scale corresponding to the maximum spatial temperature difference. Accordingly, for a symmetrically heated (or cooled) plane wall of thickness $2L$, L_c would remain equal to the half-thickness L. However, for a long cylinder or sphere, L_c would equal the actual radius r_o, rather than $r_o/2$ or $r_o/3$.

Finally, we note that, with $L_c \equiv V/A_s$, the exponent of Equation 5.6 may be expressed as

$$\frac{hA_st}{\rho Vc} = \frac{ht}{\rho cL_c} = \frac{hL_c}{k}\frac{k}{\rho c}\frac{t}{L_c^2} = \frac{hL_c}{k}\frac{\alpha t}{L_c^2}$$

or

$$\frac{hA_st}{\rho Vc} = Bi \cdot Fo \tag{5.11}$$

where

$$Fo \equiv \frac{\alpha t}{L_c^2} \tag{5.12}$$

is termed the Fourier number. It is a *dimensionless time,* which, with the Biot number, characterizes transient conduction problems. Substituting Equation 5.11 into 5.6, we obtain

$$\frac{\theta}{\theta_i} = \frac{T - T_\infty}{T_i - T_\infty} = \exp(-Bi \cdot Fo) \tag{5.13}$$

EXAMPLE 5.1

A thermocouple junction, which may be approximated as a sphere, is to be used for temperature measurement in a gas stream. The convection coefficient between the junction surface and the gas is $h = 400$ W/m$^2 \cdot$ K, and the junction thermophysical properties are $k = 20$ W/m $\cdot$ K, $c = 400$ J/kg $\cdot$ K, and $\rho = 8500$ kg/m^3. Determine the junction diameter needed for the thermocouple to have a time constant of 1 s. If the junction is at 25°C and is placed in a gas stream that is at 200°C, how long will it take for the junction to reach 199°C?

SOLUTION

Known: Thermophysical properties of thermocouple junction used to measure temperature of a gas stream.

Find:

1. Junction diameter needed for a time constant of 1 s.
2. Time required to reach 199°C in gas stream at 200°C.

Schematic:

Assumptions:

1. Temperature of junction is uniform at any instant.
2. Radiation exchange with the surroundings is negligible.
3. Losses by conduction through the leads are negligible.
4. Constant properties.

Analysis:

1. Because the junction diameter is unknown, it is not possible to begin the solution by determining whether the criterion for using the lumped capacitance method, Equation 5.10, is satisfied. However, a reasonable approach is to use the method to find the diameter and to then determine whether the criterion is satisfied. From Equation 5.7 and the fact that $A_s = \pi D^2$ and $V = \pi D^3/6$ for a sphere, it follows that

$$\tau_t = \frac{1}{h\pi D^2} \times \frac{\rho \pi D^3}{6} c$$

Rearranging and substituting numerical values,

$$D = \frac{6h\tau_t}{\rho c} = \frac{6 \times 400 \text{ W/m}^2 \cdot \text{K} \times 1 \text{ s}}{8500 \text{ kg/m}^3 \times 400 \text{ J/kg} \cdot \text{K}} = 7.06 \times 10^{-4} \text{ m} \qquad \triangleleft$$

With $L_c = r_o/3$ it then follows from Equation 5.10 that

$$Bi = \frac{h(r_o/3)}{k} = \frac{400 \text{ W/m}^2 \cdot \text{K} \times 3.53 \times 10^{-4} \text{ m}}{3 \times 20 \text{ W/m} \cdot \text{K}} = 2.35 \times 10^{-3}$$

Accordingly, Equation 5.10 is satisfied (for $L_c = r_o$, as well as for $L_c = r_o/3$) and the lumped capacitance method may be used to an excellent approximation.

2. From Equations 5.5 and 5.6 the time required for the junction to reach a temperature T can be written as

$$t = \frac{\rho V c}{h A_s} \ln\left(\frac{T_i - T_\infty}{T - T_\infty}\right) = \tau_t \ln\left(\frac{T_i - T_\infty}{T - T_\infty}\right)$$

Thus, the time required to reach $T = 199°C$ is

$$t = \tau_t \ln\left(\frac{25 - 200}{199 - 200}\right) = 5.2\tau_t = 5.2 \times 1 \text{ s} = 5.2 \text{ s} \qquad \triangleleft$$

Comments: Heat transfer due to radiation exchange between the junction and the surroundings and conduction through the leads would affect the time response of the junction and would, in fact, yield an equilibrium temperature that differs from T_∞.

5.3 *General Lumped Capacitance Analysis*

Although transient conduction in a solid is commonly initiated by convection heat transfer to or from an adjoining fluid, other processes may induce transient thermal conditions within the solid. For example, a solid may be separated from large surroundings by a gas or vacuum. If the temperatures of the solid and surroundings differ, radiation exchange could cause the internal thermal energy, and hence the temperature, of the solid to change. Temperature changes could also be induced by applying a heat flux at a portion, or all, of the surface or by initiating thermal energy generation within the solid. Surface heating could, for example, be applied by attaching a film or sheet electrical heater to the surface, while thermal energy could be generated by passing an electrical current through the solid.

Figure 5.5 depicts the general situation for which thermal conditions within a solid may be influenced simultaneously by convection, radiation, an applied surface heat flux, and internal energy generation. It is presumed that, initially ($t = 0$), the temperature of the solid T_i differs from that of the fluid T_∞ and the surroundings T_{sur}, and that both surface and volumetric heating (q''_s and $\dot{q}$) are initiated. The imposed heat flux q''_s and the convection–radiation heat transfer occur at mutually exclusive portions of the surface, $A_{s(h)}$ and $A_{s(c,r)}$, respectively, and convection–radiation transfer is presumed to be *from* the surface. Moreover, although convection and radiation have been prescribed for the same surface, the surfaces may, in fact, differ ($A_{s,c} \neq A_{s,r}$). Applying conservation of energy at any instant t, it follows from Equation 1.12c that

$$q''_s A_{s,h} + \dot{E}_g - (q''_{conv} + q''_{rad})A_{s(c,r)} = \rho V c \frac{dT}{dt} \tag{5.14}$$

or, from Equations 1.3a and 1.7,

$$q''_s A_{s,h} + \dot{E}_g - [h(T - T_\infty) + \varepsilon\sigma(T^4 - T_{sur}^4)]A_{s(c,r)} = \rho V c \frac{dT}{dt} \tag{5.15}$$

Equation 5.15 is a nonlinear, first-order, nonhomogeneous, ordinary differential equation that cannot be integrated to obtain an exact solution.[1] However, exact solutions may be obtained for simplified versions of the equation.

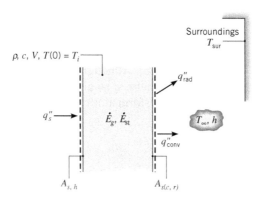

Figure 5.5 Control surface for general lumped capacitance analysis.

---

[1]An approximate, finite-difference solution may be obtained by *discretizing* the time derivative (Section 5.10) and *marching* the solution out in time.

5.3.1   Radiation Only

If there is no imposed heat flux or generation and convection is either nonexistent (a vacuum) or negligible relative to radiation, Equation 5.15 reduces to

$$\rho V c \frac{dT}{dt} = -\varepsilon A_{s,r} \sigma (T^4 - T_{sur}^4) \tag{5.16}$$

Separating variables and integrating from the initial condition to any time t, it follows that

$$\frac{\varepsilon A_{s,r} \sigma}{\rho V c} \int_0^t dt = \int_{T_i}^T \frac{dT}{T_{sur}^4 - T^4} \tag{5.17}$$

Evaluating both integrals and rearranging, the time required to reach the temperature T becomes

$$t = \frac{\rho V c}{4\varepsilon A_{s,r} \sigma T_{sur}^3} \left\{ \ln \left| \frac{T_{sur} + T}{T_{sur} - T} \right| - \ln \left| \frac{T_{sur} + T_i}{T_{sur} - T_i} \right| \right.$$
$$\left. + 2 \left[\tan^{-1} \left(\frac{T}{T_{sur}} \right) - \tan^{-1} \left(\frac{T_i}{T_{sur}} \right) \right] \right\} \tag{5.18}$$

This expression can be used to determine the time required for the solid to reach some temperature, T. Equation 5.17 may also be integrated for the limiting case of $T_{sur} = 0$ (radiation to deep space) yielding

$$t = \frac{\rho V c}{3\varepsilon A_{s,r} \sigma} \left(\frac{1}{T^3} - \frac{1}{T_i^3} \right) \tag{5.19}$$

5.3.2   Negligible Radiation

An exact solution to Equation 5.15 may also be obtained if radiation may be neglected and all quantities (other than T, of course) are independent of time. Introducing a temperature difference $\theta \equiv T - T_\infty$, where $d\theta/dt = dT/dt$, Equation 5.15 reduces to a linear, first-order, nonhomogeneous differential equation of the form

$$\frac{d\theta}{dt} + a\theta - b = 0 \tag{5.20}$$

where $a \equiv (hA_{s,c}/\rho V c)$ and $b \equiv [(q_s'' A_{s,h} + \dot{E}_g)/\rho V c]$. Although Equation 5.20 may be solved by summing its homogeneous and particular solutions, an alternative approach is to eliminate the nonhomogeneity by introducing the transformation

$$\theta' \equiv \theta - \frac{b}{a} \tag{5.21}$$

Recognizing that $d\theta'/dt = d\theta/dt$, Equation 5.21 may be substituted into (5.20) to yield

$$\frac{d\theta'}{dt} + a\theta' = 0 \tag{5.22}$$

Separating variables and integrating from 0 to t (θ_i' to θ'), it follows that

$$\frac{\theta'}{\theta_i'} = \exp(-at) \tag{5.23}$$

or substituting for θ' and θ,

$$\frac{T - T_\infty - (b/a)}{T_i - T_\infty - (b/a)} = \exp(-at) \tag{5.24}$$

Hence

$$\frac{T - T_\infty}{T_i - T_\infty} = \exp(-at) + \frac{b/a}{T_i - T_\infty}[1 - \exp(-at)] \tag{5.25}$$

As it must, Equation 5.25 reduces to Equation 5.6 when $b = 0$ and yields $T = T_i$ at $t = 0$. As $t \rightarrow \infty$, Equation 5.25 reduces to $(T - T_\infty) = (b/a)$, which could also be obtained by performing an energy balance on the control surface of Figure 5.5 for steady-state conditions.

5.3.3 Convection Only with Variable Convection Coefficient

In some cases, such as those involving free convection or boiling, the convection coefficient h varies with the temperature difference between the object and the fluid. In these situations, the convection coefficient can often be approximated with an expression of the form

$$h = C(T - T_\infty)^n \tag{5.26}$$

where n is a constant and the parameter C has units of $\text{W/m}^2 \cdot \text{K}^{(1+n)}$. If radiation, surface heating, and volumetric generation are negligible, Equation 5.15 may be written as

$$-C(T - T_\infty)^n A_{s,c}(T - T_\infty) = -CA_{s,c}(T - T_\infty)^{1+n} = \rho Vc \frac{dT}{dt} \tag{5.27}$$

Substituting θ and $d\theta/dt = dT/dt$ into the preceding expression, separating variables and integrating yields

$$\frac{\theta}{\theta_i} = \left[\frac{nCA_{s,c}\theta_i^n}{\rho Vc}t + 1 \right]^{-1/n} \tag{5.28}$$

It can be shown that Equation 5.28 reduces to Equation 5.6 if the heat transfer coefficient is independent of temperature, $n = 0$.

5.3.4 Additional Considerations

In some cases the ambient or surroundings temperature may vary with time. For example, if the container of Figure 5.1 is insulated and of finite volume, the liquid temperature will increase as the metal forging is cooled. An analytical solution for the time-varying solid (and liquid) temperature is presented in Example 11.8. As evident in Examples 5.2 through 5.4, the lumped capacitance differential equation can be solved numerically for a wide variety of situations involving variable properties or time-varying boundary conditions, internal energy generation rates, or surface heating or cooling.

EXAMPLE 5.2   IHT

Consider the thermocouple and convection conditions of Example 5.1, but now allow for radiation exchange with the walls of a duct that encloses the gas stream. If the duct walls are at 400°C and the emissivity of the thermocouple bead is 0.9, calculate the steady-state temperature of the junction. Also, determine the time for the junction temperature to increase from an initial condition of 25°C to a temperature that is within 1°C of its steady-state value.

SOLUTION

Known: Thermophysical properties and diameter of the thermocouple junction used to measure temperature of a gas stream passing through a duct with hot walls.

Find:

1. Steady-state temperature of the junction.

2. Time required for the thermocouple to reach a temperature that is within 1°C of its steady-state value.

Schematic:

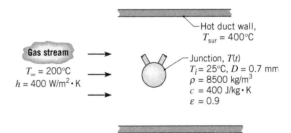

Assumptions: Same as Example 5.1, but radiation transfer is no longer treated as negligible and is approximated as exchange between a small surface and large surroundings.

Analysis:

1. For steady-state conditions, the energy balance on the thermocouple junction has the form

$$\dot{E}_{\text{in}} - \dot{E}_{\text{out}} = 0$$

Recognizing that net radiation to the junction must be balanced by convection from the junction to the gas, the energy balance may be expressed as

$$[\varepsilon\sigma(T_{\text{sur}}^{4} - T^{4}) - h(T - T_{\infty})]A_{s} = 0$$

Substituting numerical values, we obtain

$$T = 218.7°C$$ ◁

2. The temperature–time history, $T(t)$, for the junction, initially at $T(0) = T_i = 25°C$, follows from the energy balance for transient conditions,

$$\dot{E}_{in} - \dot{E}_{out} = \dot{E}_{st}$$

From Equation 5.15, the energy balance may be expressed as

$$-[h(T - T_\infty) + \varepsilon\sigma(T^4 - T_{sur}^4)]A_s = \rho V c \frac{dT}{dt}$$

The solution to this first-order differential equation can be obtained by numerical integration, giving the result, $T(4.9 \text{ s}) = 217.7°C$. Hence, the time required to reach a temperature that is within 1°C of the steady-state value is

$$t = 4.9 \text{ s.}$$ ◁

Comments:

1. The effect of radiation exchange with the hot duct walls is to increase the junction temperature, such that the thermocouple indicates an erroneous gas stream temperature that exceeds the actual temperature by 18.7°C. The time required to reach a temperature that is within 1°C of the steady-state value is slightly less than the result of Example 5.1, which only considers convection heat transfer. Why is this so?

2. The response of the thermocouple and the indicated gas stream temperature depend on the velocity of the gas stream, which in turn affects the magnitude of the convection coefficient. Temperature–time histories for the thermocouple junction are shown in the following graph for values of $h = 200$, 400, and 800 $W/m^2 \cdot K$.

The effect of increasing the convection coefficient is to cause the junction to indicate a temperature closer to that of the gas stream. Further, the effect is to reduce the time required for the junction to reach the near-steady-state condition. What physical explanation can you give for these results?

3. For problems involving convection and radiation heating or cooling, Equation 1.9 may be used to define an *effective* heat transfer coefficient $h_{eff} \equiv h + h_r$ for testing the validity of

the lumped capacitance method. As a first approximation, we may employ the lumped capacitance method if $Bi_{max} = h_{eff,max}L_c/k < 0.1$. For this problem, $T_{max} = 218.7°C = 491.7$ K and $h_{eff,max} = h + \varepsilon\sigma(T_{max} + T_{sur})(T_{max}^2 + T_{sur}^2) = 400$ W/m$^2 \cdot$ K $+ 0.9 \times 5.67 \times 10^{-8}$ W/m$^2 \cdot$ K^4 $(491.7 + 673)$K$(491.7^2 + 673^2)$K$^2 = 400$ W/m$^2 \cdot$ K $+ 41$ W/m$^2 \cdot$ K $= 441$ W/m$^2 \cdot$ K. Thus, $Bi_{max} = h_{eff,max}(r_o/3)/k = 2.6 \times 10^{-3} < 0.1$. This inequality *suggests* that the lumped capacitance approximation remains valid when radiation is included in the analysis.

4. The *IHT* software includes an integral function, Der(T, t), that can be used to represent the temperature–time derivative and to integrate first-order differential equations.

EXAMPLE 5.3

A 3-mm-thick panel of aluminum alloy ($k = 177$ W/m $\cdot$ K, $c = 875$ J/kg $\cdot$ K, and $\rho = 2770$ kg/m^3) is finished on both sides with an epoxy coating that must be cured at or above $T_c = 150°C$ for at least 5 min. The production line for the curing operation involves two steps: (1) heating in a large oven with air at $T_{\infty,o} = 175°C$ and a convection coefficient of $h_o = 40$ W/m$^2 \cdot$ K, and (2) cooling in a large chamber with air at $T_{\infty,c} = 25°C$ and a convection coefficient of $h_c = 10$ W/m$^2 \cdot$ K. The heating portion of the process is conducted over a time interval t_e, which exceeds the time t_c required to reach 150°C by 5 min ($t_e = t_c + 300$ s). The coating has an emissivity of $\varepsilon = 0.8$, and the temperatures of the oven and chamber walls are 175 and 25°C, respectively. If the panel is placed in the oven at an initial temperature of 25°C and removed from the chamber at a *safe-to-touch* temperature of 37°C, what is the total elapsed time for the two-step curing operation?

SOLUTION

Known: Operating conditions for a two-step heating/cooling process in which a coated aluminum panel is maintained at or above a temperature of 150°C for at least 5 min.

Find: Total time t_t required for the two-step process.

Schematic:

Assumptions:

1. Panel temperature is uniform at any instant.
2. Thermal resistance of epoxy is negligible.
3. Constant properties.

Analysis: To assess the validity of the lumped capacitance approximation, we begin by calculating Biot numbers for the heating and cooling processes, both of which are affected by convection and radiation. Following Comment 3 of Example 5.2, a representative value of the radiation heat transfer coefficient may be determined from Equation 1.9 using the maximum possible plate temperature, $T_{\text{max},o} = 175°C = 448$ K, and assuming $T_{\text{sur},o} = T_{\infty,o}$ in which case

$$h_{\text{eff,max},o} = h_o + h_{r,\text{max},o} = h_o + \varepsilon\sigma(T_{\text{max},o} + T_{\text{sur},o})(T_{\text{max},o}^2 + T_{\text{sur},o}^2)$$

$$= 40 \text{ W/m}^2 \cdot \text{K} + 0.8 \times 5.67 \times 10^{-8} \text{ W/m}^2 \cdot \text{K}^4 (448 + 448)\text{K}(448^2 + 448^2)\text{K}^2$$

$$= 40 \text{ W/m}^2 \cdot \text{K} + 16 \text{ W/m}^2 \cdot \text{K} = 56 \text{ W/m}^2 \cdot \text{K}$$

Again following Comment 3 of Example 5.2, the Biot number for heating is

$$Bi_o = \frac{h_{\text{eff,max},o}L_c}{k} = \frac{(56 \text{ W/m}^2 \cdot \text{K})(0.0015 \text{ m})}{177 \text{ W/m} \cdot \text{K}} = 4.8 \times 10^{-4}$$

where L_c is the panel half-thickness. Using $T_{\text{max},c} = 175°C$ and $T_{\text{sur},c} = 25°C$ for the cooling step, we also obtain $h_{\text{eff,max},c} = 49.8 \text{ W/m}^2 \cdot \text{K}$ and $Bi_c = 4.2 \times 10^{-4}$. Since both Bi_o and Bi_c are less than 0.1, the lumped capacitance approximation is valid for the entire heating/cooling process.

With $V = 2LA_s$ and $A_{s,c} = A_{s,r} = 2A_s$, Equation 5.15 may be expressed as

$$-[h(T - T_{\infty}) + \varepsilon\sigma(T^4 - T_{\text{sur}}^4)] = \rho c L \frac{dT}{dt}$$

where h, T_{∞}, and T_{sur} take on different values during the two process steps.

Selecting a suitable time increment, Δt, the equation may be integrated numerically to obtain the panel temperature at $t = \Delta t$, $2\Delta t$, $3\Delta t$, and so on. Selecting $\Delta t = 10$ s, calculations for the heating process are extended to $t_e = t_c + 300$ s, which is 5 min beyond the time required for the panel to reach $T_c = 150°C$. At t_e the cooling process is initiated and continued until the panel temperature reaches $37°C$ at $t = t_t$. The integration was performed using *IHT*, and results of the calculations are plotted as follows:

The total time for the two-step process is

$$t_t = 989 \text{ s} \qquad \lhd$$

with intermediate times of $t_c = 124$ s and $t_e = 424$ s.

Comments:

1. The duration of the two-step process may be reduced by increasing the convection coefficients and/or by reducing the period of extended heating. The second option is viable because during the initial portion of the cooling period, the panel temperature exceeds 150°C. Hence, to satisfy the cure requirement, it is not necessary to extend heating for 5 min beyond $t = t_c$. If the convection coefficients are increased to $h_o = h_c = 100 \text{ W/m}^2 \cdot \text{K}$ and an extended heating period of 300 s is maintained, the numerical integration yields $t_c = 58$ s and $t_t = 445$ s. The corresponding time interval over which the panel temperature exceeds 150°C is $\Delta t_{(T>150°C)} = 306$ s (58 s $\leq t \leq$ 364 s). If the extended heating period is reduced to 294 s, the numerical integration yields $t_c = 58$ s, $t_t = 439$ s, and $\Delta t_{(T>150°C)} = 300$ s. Hence the total process time is reduced, while the curing requirement is still satisfied.

2. Generally, the accuracy of a numerical integration improves with decreasing Δt, but at the expense of increased computation time. In this case, however, results obtained for $\Delta t = 1$ s are virtually identical to those obtained for $\Delta t = 10$ s, indicating that the larger time interval is sufficient to accurately depict the temperature history.

3. The complete solution for this example is provided as a ready-to-solve model in the *Advanced* section of *IHT*, using *Models, Lumped Capacitance*. The model can be used to check the results of Comment 1 or to independently explore modifications of the cure process.

4. If the Biot numbers were not small, it would be inappropriate to apply the lumped capacitance method. For moderate or large Biot numbers, temperatures near the solid's centerline would continue to increase for some time after the conclusion of heating, as thermal energy near the solid's surface propagates inward. The temperatures near the centerline would subsequently reach a maximum and would then decrease to the steady-state value. Correlations for the maximum temperature experienced at the panel's centerline, along with the time at which these maximum temperatures are reached, have been developed for a broad range of Bi_h and Bi_c values [1].

EXAMPLE 5.4

Air to be supplied to a hospital operating room is first purified by forcing it through a single-stage compressor. As it travels through the compressor, the air temperature initially increases due to compression, then decreases as it is returned to atmospheric pressure. Harmful pathogen particles in the air will also be heated and subsequently cooled, and they will be destroyed if their maximum temperature exceeds a *lethal* temperature T_d. Consider spherical pathogen particles ($D = 10$ μm, $\rho = 900 \text{ kg/m}^3$, $c = 1100 \text{ J/kg} \cdot \text{K}$, and $k = 0.2 \text{ W/m} \cdot \text{K}$) that are dispersed in unpurified air. During the process, the air temperature may be described by an expression of the form $T_\infty(t) = 125°C - 100°C \cdot \cos(2\pi t/t_p)$, where t_p is the process time associated with flow through the compressor. If $t_p = 0.004$ s, and the initial and lethal pathogen temperatures are $T_i = 25°C$ and $T_d = 220°C$, respectively, will the pathogens be destroyed? The value of the convection heat transfer coefficient associated with the pathogen particles is $h = 4600 \text{ W/m}^2 \cdot \text{K}$.

SOLUTION

Known: Air temperature versus time, convection heat transfer coefficient, pathogen geometry, size, and properties.

Find: Whether the pathogens are destroyed for $t_p = 0.004$ s.

Schematic:

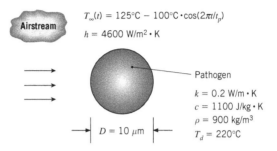

$T_\infty(t) = 125°C - 100°C \cdot \cos(2\pi t/t_p)$

$h = 4600$ W/m² · K

Airstream

Pathogen

$k = 0.2$ W/m · K
$c = 1100$ J/kg · K
$\rho = 900$ kg/m³
$T_d = 220°C$

$D = 10 \ \mu m$

Assumptions:

1. Constant properties.
2. Negligible radiation.

Analysis: The Biot number associated with a spherical pathogen particle is

$$Bi = \frac{h(D/6)}{k} = \frac{4600 \text{ W/m}^2 \cdot \text{K} \times (10 \times 10^{-6} \text{ m/6})}{0.2 \text{ W/m} \cdot \text{K}} = 0.038$$

Hence, the lumped capacitance approximation is valid and we may apply Equation 5.2.

$$\frac{dT}{dt} = -\frac{hA_s}{\rho Vc}[T - T_\infty(t)] = -\frac{6h}{\rho cD}[T - 125°C + 100°C \cdot \cos(2\pi t/t_p)] \tag{1}$$

The solution to this first-order differential equation may be obtained analytically, or by numerical integration.

Numerical Integration A numerical solution of Equation 1 may be obtained by specifying the initial particle temperature, T_i, and using *IHT* or an equivalent numerical solver to integrate the equation. The plot of the numerical solution follows.

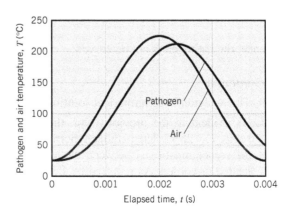

Inspection of the predicted pathogen temperatures yields

$$T_{max} = 212°C < 220°C$$

Hence, the pathogen is not destroyed.   ◁

Analytical Solution   Equation 1 is a linear nonhomogeneous differential equation, therefore the solution can be found as the sum of a homogeneous and a particular solution, $T = T_h + T_p$. The homogeneous part, T_h, corresponds to the homogeneous differential equation, $dT_h/dt = -(6h/\rho cD)T_h$, which has the familiar solution, $T_h = C_0 \exp(-6ht/\rho cD)$. The particular solution, T_p, can then be found using the method of undetermined coefficients; for a nonhomogeneous term that includes a cosine function and a constant term, the particular solution is assumed to be of the form $T_p = C_1 \cos(2\pi t/t_p) + C_2 \sin(2\pi t/t_p) + C_3$. Substituting this expression into Equation 1 yields values for the coefficients, resulting in

$$T_p = 125°C - 100°C \times A\left[\cos\left(\frac{2\pi t}{t_p}\right) + \frac{2\pi\rho cD}{6ht_p}\sin\left(\frac{2\pi t}{t_p}\right)\right] \qquad (2)$$

where

$$A = \frac{(6h/\rho cD)^2}{(6h/\rho cD)^2 + (2\pi/t_p)^2}$$

The initial condition, $T(0) = T_i$, is then applied to the complete solution, $T = T_h + T_p$, to yield $C_0 = 100°C(A - 1)$. Thus, the particle temperature is

$$T(t) = 125°C + 100°C\left\{(A - 1)\exp\left(-\frac{6ht}{\rho cD}\right) - A\left[\cos\left(\frac{2\pi t}{t_p}\right) + \frac{2\pi\rho cD}{6ht_p}\sin\left(\frac{2\pi t}{t_p}\right)\right]\right\} \qquad (3)$$

To find the maximum pathogen temperature, we could differentiate Equation 3 and set the result equal to zero. This yields a lengthy, implicit equation for the critical time t_{crit} at which the maximum temperature is reached. The maximum temperature may then be found by substituting $t = t_{crit}$ into Equation 3. Alternatively, Equation 3 can be plotted or $T(t)$ may be tabulated to find

$$T_{max} = 212°C < 220°C$$

Hence, the pathogen is not destroyed.   ◁

Comments:

1. The analytical and numerical solutions agree, as they must.

2. As evident in the previous plot, the air and pathogen temperatures are initially the same, $T_i = 25°C$. The pathogen thermal response lags that of the air since a temperature difference must exist between the air and the particle in order for the pathogen to be heated or cooled. As required by Equation 1 and as evident in the plot, the maximum particle temperature is reached when there is no temperature difference between the air and the pathogen.

3. The maximum pathogen temperature may be increased by extending the duration of the process. For a process time of $t_p = 0.008$ s, the air and pathogen particle temperatures are as follows.

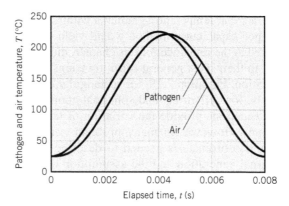

The maximum particle temperature is now $T_{max} = 221°C > T_d = 220°C$, and the pathogen would be killed. However, because the duration of the cycle is twice as long as originally specified, approximately half of the air could be supplied to the operating room compared to the $t_p = 0.004$ s case. A trade-off exists between the amount of air that can be delivered to the operating room and its purity.

4. The maximum possible radiation heat transfer coefficient may be calculated based on the extreme temperatures of the problem and by assuming a particle emissivity of unity. Hence,

$$h_{r,max} = \sigma(T_{max} + T_{min})(T_{max}^2 + T_{min}^2)$$

$$= 5.67 \times 10^{-8} \text{ W/m}^2 \cdot \text{K}^4 \times (498 + 298)\text{K} \times (498^2 + 298^2)\text{K}^2 = 15.2 \text{ W/m}^2 \cdot \text{K}$$

Since $h_{r,max} \ll h$, radiation heat transfer is negligible.

5. The Der(T, t) function of the *IHT* software was used to generate the numerical solution for this problem. See Comment 4 of Example 5.2. If one is familiar with a numerical solver such as *IHT*, it is often much faster to obtain a numerical solution than an analytical solution, as is the case in this example. Moreover, if one seeks maximum or minimum values of a dependent variable, such as the pathogen temperature in this example, it is often faster to determine the maxima or minima by inspection, rather than with an analytical solution. However, analytical solutions often explicitly show parameter dependencies and can provide insights that numerical solutions might obscure.

6. A time increment of $\Delta t = 10$ μs was used to generate the numerical solutions. Generally, the accuracy of a numerical integration improves with decreasing Δt, but at the expense of increased computation time. For this example, results for $\Delta t = 5$ μs are virtually identical to those obtained for the larger time increment, indicating that either increment is sufficient to accurately depict the temperature history and to determine the maximum particle temperature.

7. Assumption of instantaneous pathogen death at the lethal temperature is an approximation. Pathogen destruction also depends on the duration of exposure to the high temperatures [2].

5.4   *Spatial Effects*

Situations frequently arise for which the Biot number is not small, and we must cope with the fact that temperature gradients within the medium are no longer negligible. Use of the lumped capacitance method would yield incorrect results, so alternative approaches, presented in the remainder of this chapter, must be utilized.

In their most general form, transient conduction problems are described by the heat equation, Equation 2.19, for rectangular coordinates or Equations 2.26 and 2.29, respectively, for cylindrical and spherical coordinates. The solutions to these partial differential equations provide the variation of temperature with both time and the spatial coordinates. However, in many problems, such as the plane wall of Figure 5.4, only one spatial coordinate is needed to describe the internal temperature distribution. With no internal generation and the assumption of constant thermal conductivity, Equation 2.19 then reduces to

$$\frac{\partial^2 T}{\partial x^2} = \frac{1}{\alpha}\frac{\partial T}{\partial t} \tag{5.29}$$

To solve Equation 5.29 for the temperature distribution $T(x, t)$, it is necessary to specify an *initial* condition and two *boundary conditions*. For the typical transient conduction problem of Figure 5.4, the initial condition is

$$T(x, 0) = T_i \tag{5.30}$$

and the boundary conditions are

$$\left.\frac{\partial T}{\partial x}\right|_{x=0} = 0 \tag{5.31}$$

and

$$-k\left.\frac{\partial T}{\partial x}\right|_{x=L} = h[T(L, t) - T_\infty] \tag{5.32}$$

Equation 5.30 presumes a uniform temperature distribution at time $t = 0$; Equation 5.31 reflects the *symmetry requirement* for the midplane of the wall; and Equation 5.32 describes the surface condition experienced for time $t > 0$. From Equations 5.29 through 5.32, it is evident that, in addition to depending on x and t, temperatures in the wall also depend on a number of physical parameters. In particular

$$T = T(x, t, T_i, T_\infty, L, k, \alpha, h) \tag{5.33}$$

The foregoing problem may be solved analytically or numerically. These methods will be considered in subsequent sections, but first it is important to note the advantages of *nondimensionalizing* the governing equations. This may be done by arranging the relevant variables into suitable *groups*. Consider the dependent variable T. If the temperature difference $\theta \equiv T - T_\infty$ is divided by the *maximum possible temperature difference* $\theta_i \equiv T_i - T_\infty$, a dimensionless form of the dependent variable may be defined as

$$\theta^* \equiv \frac{\theta}{\theta_i} = \frac{T - T_\infty}{T_i - T_\infty} \tag{5.34}$$

Accordingly, θ^* must lie in the range $0 \le \theta^* \le 1$. A dimensionless spatial coordinate may be defined as

$$x^* \equiv \frac{x}{L} \tag{5.35}$$

where L is the half-thickness of the plane wall, and a dimensionless time may be defined as

$$t^* \equiv \frac{\alpha t}{L^2} \equiv Fo \tag{5.36}$$

where t^* is equivalent to the dimensionless *Fourier number,* Equation 5.12.

Substituting the definitions of Equations 5.34 through 5.36 into Equations 5.29 through 5.32, the heat equation becomes

$$\frac{\partial^2 \theta^*}{\partial x^{*2}} = \frac{\partial \theta^*}{\partial Fo} \tag{5.37}$$

and the initial and boundary conditions become

$$\theta^*(x^*, 0) = 1 \tag{5.38}$$

$$\left.\frac{\partial \theta^*}{\partial x^*}\right|_{x^*=0} = 0 \tag{5.39}$$

and

$$\left.\frac{\partial \theta^*}{\partial x^*}\right|_{x^*=1} = -Bi\ \theta^*(1, t^*) \tag{5.40}$$

where the *Biot number* is $Bi \equiv hL/k$. In dimensionless form the functional dependence of Equation 5.33 may now be expressed as

$$\theta^* = f(x^*, Fo, Bi) \tag{5.41}$$

Recall that a similar functional dependence, without the x^* variation, was obtained for the lumped capacitance method, as shown in Equation 5.13.

Comparing Equations 5.33 and 5.41, the considerable advantage associated with casting the problem in dimensionless form becomes apparent. Equation 5.41 implies that *for a prescribed geometry, the transient temperature distribution is a universal function of x^*, Fo, and Bi.* That is, the *dimensionless solution* has a prescribed form that does not depend on the particular value of T_i, T_∞, L, k, α, or h. Since this generalization greatly simplifies the presentation and utilization of transient solutions, the dimensionless variables are used extensively in subsequent sections.

5.5 *The Plane Wall with Convection*

Exact, analytical solutions to transient conduction problems have been obtained for many simplified geometries and boundary conditions and are well documented [3–6]. Several mathematical techniques, including the method of separation of variables (Section 4.2), may be used for this purpose, and typically the solution for the dimensionless temperature distribution, Equation 5.41, is in the form of an infinite series. However, except for very small values of the Fourier number, this series may be approximated by a single term, considerably simplifying its evaluation.

5.5.1 Exact Solution

Consider the *plane wall* of thickness $2L$ (Figure 5.6a). If the thickness is small relative to the width and height of the wall, it is reasonable to assume that conduction occurs exclusively in the x-direction. If the wall is initially at a uniform temperature, $T(x, 0) = T_i$, and is suddenly immersed in a fluid of $T_\infty \neq T_i$, the resulting temperatures may be obtained by solving Equation 5.37 subject to the conditions of Equations 5.38 through 5.40. Since the convection conditions for the surfaces at $x^* = \pm 1$ are the same, the temperature distribution at any instant must be symmetrical about the midplane ($x^* = 0$). An exact solution to this problem is of the form [4]

$$\theta^* = \sum_{n=1}^{\infty} C_n \exp(-\zeta_n^2 Fo)\cos(\zeta_n x^*) \tag{5.42a}$$

where $Fo = \alpha t/L^2$, the coefficient C_n is

$$C_n = \frac{4\sin\zeta_n}{2\zeta_n + \sin(2\zeta_n)} \tag{5.42b}$$

and the discrete values of ζ_n (*eigenvalues*) are positive roots of the transcendental equation

$$\zeta_n \tan\zeta_n = Bi \tag{5.42c}$$

The first four roots of this equation are given in Appendix B.3. The exact solution given by Equation 5.42a is valid for any time, $0 \leq Fo \leq \infty$.

5.5.2 Approximate Solution

It can be shown (Problem 5.34) that for values of $Fo > 0.2$, the infinite series solution, Equation 5.42a, can be approximated by the first term of the series, $n = 1$. Invoking this approximation, the dimensionless form of the temperature distribution becomes

$$\theta^* = C_1 \exp(-\zeta_1^2 Fo)\cos(\zeta_1 x^*) \tag{5.43a}$$

or

$$\theta^* = \theta_o^* \cos(\zeta_1 x^*) \tag{5.43b}$$

where $\theta_o^* \equiv (T_o - T_\infty)/(T_i - T_\infty)$ represents the midplane ($x^* = 0$) temperature

$$\theta_o^* = C_1 \exp(-\zeta_1^2 Fo) \tag{5.44}$$

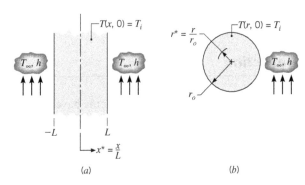

FIGURE 5.6 One-dimensional systems with an initial uniform temperature subjected to sudden convection conditions: (a) Plane wall. (b) Infinite cylinder or sphere.

Graphical representations of the one-term approximations are presented in Section 5S.1.

An important implication of Equation 5.43b is that *the time dependence of the temperature at any location within the wall is the same as that of the midplane temperature.* The coefficients C_1 and ζ_1 are evaluated from Equations 5.42b and 5.42c, respectively, and are given in Table 5.1 for a range of Biot numbers.

TABLE 5.1 Coefficients used in the one-term approximation to the series solutions for transient one-dimensional conduction

Bi^a	Plane Wall		Infinite Cylinder		Sphere	
	ζ_1 (rad)	C_1	ζ_1 (rad)	C_1	ζ_1 (rad)	C_1
0.01	0.0998	1.0017	0.1412	1.0025	0.1730	1.0030
0.02	0.1410	1.0033	0.1995	1.0050	0.2445	1.0060
0.03	0.1723	1.0049	0.2440	1.0075	0.2991	1.0090
0.04	0.1987	1.0066	0.2814	1.0099	0.3450	1.0120
0.05	0.2218	1.0082	0.3143	1.0124	0.3854	1.0149
0.06	0.2425	1.0098	0.3438	1.0148	0.4217	1.0179
0.07	0.2615	1.0114	0.3709	1.0173	0.4551	1.0209
0.08	0.2791	1.0130	0.3960	1.0197	0.4860	1.0239
0.09	0.2956	1.0145	0.4195	1.0222	0.5150	1.0268
0.10	0.3111	1.0161	0.4417	1.0246	0.5423	1.0298
0.15	0.3779	1.0237	0.5376	1.0365	0.6609	1.0445
0.20	0.4328	1.0311	0.6170	1.0483	0.7593	1.0592
0.25	0.4801	1.0382	0.6856	1.0598	0.8447	1.0737
0.30	0.5218	1.0450	0.7465	1.0712	0.9208	1.0880
0.4	0.5932	1.0580	0.8516	1.0932	1.0528	1.1164
0.5	0.6533	1.0701	0.9408	1.1143	1.1656	1.1441
0.6	0.7051	1.0814	1.0184	1.1345	1.2644	1.1713
0.7	0.7506	1.0919	1.0873	1.1539	1.3525	1.1978
0.8	0.7910	1.1016	1.1490	1.1724	1.4320	1.2236
0.9	0.8274	1.1107	1.2048	1.1902	1.5044	1.2488
1.0	0.8603	1.1191	1.2558	1.2071	1.5708	1.2732
2.0	1.0769	1.1785	1.5994	1.3384	2.0288	1.4793
3.0	1.1925	1.2102	1.7887	1.4191	2.2889	1.6227
4.0	1.2646	1.2287	1.9081	1.4698	2.4556	1.7202
5.0	1.3138	1.2402	1.9898	1.5029	2.5704	1.7870
6.0	1.3496	1.2479	2.0490	1.5253	2.6537	1.8338
7.0	1.3766	1.2532	2.0937	1.5411	2.7165	1.8673
8.0	1.3978	1.2570	2.1286	1.5526	1.7654	1.8920
9.0	1.4149	1.2598	2.1566	1.5611	2.8044	1.9106
10.0	1.4289	1.2620	2.1795	1.5677	2.8363	1.9249
20.0	1.4961	1.2699	2.2881	1.5919	2.9857	1.9781
30.0	1.5202	1.2717	2.3261	1.5973	3.0372	1.9898
40.0	1.5325	1.2723	2.3455	1.5993	3.0632	1.9942
50.0	1.5400	1.2727	2.3572	1.6002	3.0788	1.9962
100.0	1.5552	1.2731	2.3809	1.6015	3.1102	1.9990
∞	1.5708	1.2733	2.4050	1.6018	3.1415	2.0000

$^a Bi = hL/k$ for the plane wall and hr_o/k for the infinite cylinder and sphere. See Figure 5.6.

5.5.3   Total Energy Transfer: Approximate Solution

In many situations it is useful to know the total energy that has left (or entered) the wall up to any time t in the transient process. The conservation of energy requirement, Equation 1.12b, may be applied for the time interval bounded by the initial condition ($t = 0$) and any time $t > 0$

$$E_{in} - E_{out} = \Delta E_{st} \tag{5.45}$$

Equating the energy transferred from the wall Q to E_{out} and setting $E_{in} = 0$ and $\Delta E_{st} = E(t) - E(0)$, it follows that

$$Q = -[E(t) - E(0)] \tag{5.46a}$$

or

$$Q = -\int \rho c[T(x, t) - T_i]\, dV \tag{5.46b}$$

where the integration is performed over the volume of the wall. It is convenient to nondimensionalize this result by introducing the quantity

$$Q_o = \rho c V(T_i - T_\infty) \tag{5.47}$$

which may be interpreted as the initial internal energy of the wall relative to the fluid temperature. It is also the *maximum* amount of energy transfer that could occur if the process were continued to time $t = \infty$. Hence, assuming constant properties, the ratio of the total energy transferred from the wall over the time interval t to the maximum possible transfer is

$$\frac{Q}{Q_o} = \int \frac{-[T(x, t) - T_i]}{T_i - T_\infty} \frac{dV}{V} = \frac{1}{V} \int (1 - \theta^*)\, dV \tag{5.48}$$

Employing the approximate form of the temperature distribution for the plane wall, Equation 5.43b, the integration prescribed by Equation 5.48 can be performed to obtain

$$\frac{Q}{Q_o} = 1 - \frac{\sin \zeta_1}{\zeta_1} \theta_o^* \tag{5.49}$$

where θ_o^* can be determined from Equation 5.44, using Table 5.1 for values of the coefficients C_1 and ζ_1.

5.5.4   Additional Considerations

Because the mathematical problem is precisely the same, the foregoing results may also be applied to a plane wall of thickness L that is insulated on one side ($x^* = 0$) and experiences convective transport on the other side ($x^* = +1$). This equivalence is a consequence of the fact that, regardless of whether a symmetrical or an adiabatic requirement is prescribed at $x^* = 0$, the boundary condition is of the form $\partial \theta^*/\partial x^* = 0$.

Also note that the foregoing results may be used to determine the transient response of a plane wall to a sudden change in *surface* temperature. The process is equivalent to having an infinite convection coefficient, in which case the Biot number is infinite ($Bi = \infty$) and the fluid temperature T_∞ is replaced by the prescribed surface temperature T_s.

5.6 *Radial Systems with Convection*

For an infinite cylinder or sphere of radius r_o (Figure 5.6*b*), which is at an initial uniform temperature and experiences a change in convective conditions, results similar to those of Section 5.5 may be developed. That is, an exact series solution may be obtained for the time dependence of the radial temperature distribution, and a one-term approximation may be used for some conditions. The infinite cylinder is an idealization that permits the assumption of one-dimensional conduction in the radial direction. It is a reasonable approximation for cylinders having $L/r_o \gtrsim 10$.

5.6.1 Exact Solutions

For a uniform initial temperature and convective boundary conditions, the exact solutions [4], applicable at any time ($Fo > 0$), are as follows.

Infinite Cylinder In dimensionless form, the temperature is

$$\theta^* = \sum_{n=1}^{\infty} C_n \exp(-\zeta_n^2 Fo) J_0(\zeta_n r^*) \tag{5.50a}$$

where $Fo = \alpha t/r_o^2$ and $r^* = r/r_o$,

$$C_n = \frac{2}{\zeta_n} \frac{J_1(\zeta_n)}{J_0^2(\zeta_n) + J_1^2(\zeta_n)} \tag{5.50b}$$

and the discrete values of ζ_n are positive roots of the transcendental equation

$$\zeta_n \frac{J_1(\zeta_n)}{J_0(\zeta_n)} = Bi \tag{5.50c}$$

where $Bi = hr_o/k$. The quantities J_1 and J_0 are Bessel functions of the first kind, and their values are tabulated in Appendix B.4. Roots of the transcendental equation (5.50c) are tabulated by Schneider [4].

Sphere Similarly, for the sphere

$$\theta^* = \sum_{n=1}^{\infty} C_n \exp(-\zeta_n^2 Fo) \frac{1}{\zeta_n r^*} \sin(\zeta_n r^*) \tag{5.51a}$$

where $Fo = \alpha t/r_o^2$ and $r^* = r/r_o$,

$$C_n = \frac{4[\sin(\zeta_n) - \zeta_n \cos(\zeta_n)]}{2\zeta_n - \sin(2\zeta_n)} \tag{5.51b}$$

and the discrete values of ζ_n are positive roots of the transcendental equation

$$1 - \zeta_n \cot \zeta_n = Bi \qquad (5.51c)$$

where $Bi = hr_o/k$. Roots of the transcendental equation are tabulated by Schneider [4].

5.6.2   Approximate Solutions

For the infinite cylinder and sphere, the foregoing series solutions can again be approximated by a single term, $n = 1$, for $Fo > 0.2$. Hence, as for the case of the plane wall, the time dependence of the temperature at any location within the radial system is the same as that of the centerline or centerpoint.

Infinite Cylinder   The one-term approximation to Equation 5.50a is

$$\theta^* = C_1 \exp(-\zeta_1^2 Fo) J_0(\zeta_1 r^*) \qquad (5.52a)$$

or

$$\theta^* = \theta_o^* J_0(\zeta_1 r^*) \qquad (5.52b)$$

where θ_o^* represents the centerline temperature and is of the form

$$\theta_o^* = C_1 \exp(-\zeta_1^2 Fo) \qquad (5.52c)$$

Values of the coefficients C_1 and ζ_1 have been determined and are listed in Table 5.1 for a range of Biot numbers.

Sphere   From Equation 5.51a, the one-term approximation is

$$\theta^* = C_1 \exp(-\zeta_1^2 Fo) \frac{1}{\zeta_1 r^*} \sin(\zeta_1 r^*) \qquad (5.53a)$$

or

$$\theta^* = \theta_o^* \frac{1}{\zeta_1 r^*} \sin(\zeta_1 r^*) \qquad (5.53b)$$

where θ_o^* represents the center temperature and is of the form

$$\theta_o^* = C_1 \exp(-\zeta_1^2 Fo) \qquad (5.53c)$$

Values of the coefficients C_1 and ζ_1 have been determined and are listed in Table 5.1 for a range of Biot numbers.

5.6.3   Total Energy Transfer: Approximate Solutions

As in Section 5.5.3, an energy balance may be performed to determine the total energy transfer from the infinite cylinder or sphere over the time interval $\Delta t = t$. Substituting from

---

 Graphical representations of the one-term approximations are presented in Section 5S.1.

the approximate solutions, Equations 5.52b and 5.53b, and introducing Q_o from Equation 5.47, the results are as follows.

Infinite Cylinder

$$\frac{Q}{Q_o} = 1 - \frac{2\theta_o^*}{\zeta_1} J_1(\zeta_1) \tag{5.54}$$

Sphere

$$\frac{Q}{Q_o} = 1 - \frac{3\theta_o^*}{\zeta_1^3} [\sin(\zeta_1) - \zeta_1 \cos(\zeta_1)] \tag{5.55}$$

Values of the center temperature θ_o^* are determined from Equation 5.52c or 5.53c, using the coefficients of Table 5.1 for the appropriate system.

5.6.4 Additional Considerations

As for the plane wall, the foregoing results may be used to predict the transient response of long cylinders and spheres subjected to a sudden change in *surface* temperature. Namely, an infinite Biot number would be prescribed, and the fluid temperature T_∞ would be replaced by the constant surface temperature T_s.

EXAMPLE 5.5

Consider a steel pipeline (AISI 1010) that is 1 m in diameter and has a wall thickness of 40 mm. The pipe is heavily insulated on the outside, and, before the initiation of flow, the walls of the pipe are at a uniform temperature of −20°C. With the initiation of flow, hot oil at 60°C is pumped through the pipe, creating a convective condition corresponding to $h = 500 \text{ W/m}^2 \cdot \text{K}$ at the inner surface of the pipe.

1. What are the values of the Biot number and the Fourier number 8 min after the initiation of flow?

2. At $t = 8$ min, what is the temperature of the exterior pipe surface covered by the insulation?

3. What is the heat flux $q''(\text{W/m}^2)$ to the pipe from the oil at $t = 8$ min?

4. How much energy per meter of pipe length has been transferred from the oil to the pipe at $t = 8$ min?

SOLUTION

Known: Wall subjected to sudden change in convective surface condition.

Find:

1. Biot number and Fourier number after 8 min.

2. Temperature of exterior pipe surface after 8 min.

3. Heat flux to the wall at 8 min.

4. Energy transferred to pipe per unit length after 8 min.

Schematic:

Assumptions:

1. Pipe wall can be approximated as plane wall, since thickness is much less than diameter.
2. Constant properties.
3. Outer surface of pipe is adiabatic.

Properties: Table A.1, steel type AISI 1010 [$T = (-20 + 60)°C/2 \approx 300$ K]: $\rho = 7832$ kg/m^3, $c = 434$ J/kg · K, $k = 63.9$ W/m · K, $\alpha = 18.8 \times 10^{-6}$ m^2/s.

Analysis:

1. At $t = 8$ min, the Biot and Fourier numbers are computed from Equations 5.10 and 5.12, respectively, with $L_c = L$. Hence

$$Bi = \frac{hL}{k} = \frac{500 \text{ W/m}^2 \cdot \text{K} \times 0.04 \text{ m}}{63.9 \text{ W/m} \cdot \text{K}} = 0.313 \qquad \lhd$$

$$Fo = \frac{\alpha t}{L^2} = \frac{18.8 \times 10^{-6} \text{ m}^2/\text{s} \times 8 \text{ min} \times 60 \text{ s/min}}{(0.04 \text{ m})^2} = 5.64 \qquad \lhd$$

2. With $Bi = 0.313$, use of the lumped capacitance method is inappropriate. However, since $Fo > 0.2$ and transient conditions in the insulated pipe wall of thickness L correspond to those in a plane wall of thickness $2L$ experiencing the same surface condition, the desired results may be obtained from the one-term approximation for a plane wall. The midplane temperature can be determined from Equation 5.44

$$\theta_o^* = \frac{T_o - T_\infty}{T_i - T_\infty} = C_1 \exp(-\zeta_1^2 Fo)$$

where, with $Bi = 0.313$, $C_1 = 1.047$ and $\zeta_1 = 0.531$ rad from Table 5.1. With $Fo = 5.64$,

$$\theta_o^* = 1.047 \exp [-(0.531 \text{ rad})^2 \times 5.64] = 0.214$$

Hence after 8 min, the temperature of the exterior pipe surface, which corresponds to the midplane temperature of the plane wall solution, is

$$T(0, 8 \text{ min}) = T_\infty + \theta_o^*(T_i - T_\infty) = 60°C + 0.214(-20 - 60)°C = 42.9°C \qquad \lhd$$

3. Heat transfer to the inner surface at $x = L$ is by convection, and at any time t the heat flux may be obtained from Newton's law of cooling. Hence at $t = 480$ s,

$$q_x''(L, \ 480 \text{ s}) \equiv q_L'' = h[T(L, \ 480 \text{ s}) - T_\infty]$$

Using the one-term approximation for the surface temperature, Equation 5.43b with $x^* = 1$ has the form

$$\theta^* = \theta_o^* \cos(\zeta_1)$$

$$T(L, \ t) = T_\infty + (T_i - T_\infty)\theta_o^* \cos(\zeta_1)$$

$$T(L, \ 8 \text{ min}) = 60°\text{C} + (-20 - 60)°\text{C} \times 0.214 \times \cos(0.531 \text{ rad})$$

$$T(L, \ 8 \text{ min}) = 45.2°\text{C}$$

The heat flux at $t = 8$ min is then

$$q_L'' = 500 \text{ W/m}^2 \cdot \text{K} \, (45.2 - 60)°\text{C} = -7400 \text{ W/m}^2 \qquad \lhd$$

4. The energy transfer to the pipe wall over the 8-min interval may be obtained from Equations 5.47 and 5.49. With

$$\frac{Q}{Q_o} = 1 - \frac{\sin(\zeta_1)}{\zeta_1}\theta_o^*$$

$$\frac{Q}{Q_o} = 1 - \frac{\sin(0.531 \text{ rad})}{0.531 \text{ rad}} \times 0.214 = 0.80$$

it follows that

$$Q = 0.80 \, \rho c V(T_i - T_\infty)$$

or with a volume per unit pipe length of $V' = \pi DL$,

$$Q' = 0.80 \, \rho c \pi DL(T_i - T_\infty)$$

$$Q' = 0.80 \times 7832 \text{ kg/m}^3 \times 434 \text{ J/kg} \cdot \text{K}$$

$$\times \pi \times 1 \text{ m} \times 0.04 \text{ m} \, (-20 - 60)°\text{C}$$

$$Q' = -2.73 \times 10^7 \text{ J/m} \qquad \lhd$$

Comments:

1. The minus sign associated with q'' and Q' simply implies that the direction of heat transfer is from the oil to the pipe (into the pipe wall).

2. The solution for this example is provided as a ready-to-solve model in the *Advanced* section of *IHT*, which uses the *Models, Transient Conduction, Plane Wall* option. Since the *IHT* model uses a multiple-term approximation to the series solution, the results are more accurate than those obtained from the foregoing one-term approximation. *IHT Models* for *Transient Conduction* are also provided for the radial systems treated in Section 5.6.

EXAMPLE 5.6

A new process for treatment of a special material is to be evaluated. The material, a sphere of radius $r_o = 5$ mm, is initially in equilibrium at 400°C in a furnace. It is suddenly removed from the furnace and subjected to a two-step cooling process.

Step 1   Cooling in air at 20°C for a period of time t_a until the center temperature reaches a critical value, $T_a(0, t_a) = 335$°C. For this situation, the convection heat transfer coefficient is $h_a = 10$ W/m² · K.

After the sphere has reached this critical temperature, the second step is initiated.

Step 2   Cooling in a well-stirred water bath at 20°C, with a convection heat transfer coefficient of $h_w = 6000$ W/m² · K.

The thermophysical properties of the material are $\rho = 3000$ kg/m³, $k = 20$ W/m · K, $c = 1000$ J/kg · K, and $\alpha = 6.66 \times 10^{-6}$ m²/s.

1. Calculate the time t_a required for step 1 of the cooling process to be completed.
2. Calculate the time t_w required during step 2 of the process for the center of the sphere to cool from 335°C (the condition at the completion of step 1) to 50°C.

SOLUTION

Known: Temperature requirements for cooling a sphere.

Find:

1. Time t_a required to accomplish desired cooling in air.
2. Time t_w required to complete cooling in water bath.

Schematic:

Assumptions:

1. One-dimensional conduction in r.
2. Constant properties.

Analysis:

1. To determine whether the lumped capacitance method can be used, the Biot number is calculated. From Equation 5.10, with $L_c = r_o/3$,

$$Bi = \frac{h_a r_o}{3k} = \frac{10 \text{ W/m}^2 \cdot \text{K} \times 0.005 \text{ m}}{3 \times 20 \text{ W/m} \cdot \text{K}} = 8.33 \times 10^{-4}$$

Accordingly, the lumped capacitance method may be used, and the temperature is nearly uniform throughout the sphere. From Equation 5.5 it follows that

$$t_a = \frac{\rho V c}{h_a A_s} \ln \frac{\theta_i}{\theta_a} = \frac{\rho r_o c}{3 h_a} \ln \frac{T_i - T_\infty}{T_a - T_\infty}$$

where $V = (4/3)\pi r_o^3$ and $A_s = 4\pi r_o^2$. Hence

$$t_a = \frac{3000 \text{ kg/m}^3 \times 0.005 \text{ m} \times 1000 \text{ J/Kg} \cdot \text{K}}{3 \times 10 \text{ W/m}^2 \cdot \text{K}} \ln \frac{400 - 20}{335 - 20} = 94 \text{ s} \qquad \triangleleft$$

2. To determine whether the lumped capacitance method may also be used for the second step of the cooling process, the Biot number is again calculated. In this case

$$Bi = \frac{h_w r_o}{3k} = \frac{6000 \text{ W/m}^2 \cdot \text{K} \times 0.005 \text{ m}}{3 \times 20 \text{ W/m} \cdot \text{K}} = 0.50$$

and the lumped capacitance method is not appropriate. However, to an excellent approximation, the temperature of the sphere is uniform at $t = t_a$, and, as will become evident, the one-term approximation may be used for the calculations. The time t_w at which the center temperature reaches 50°C, that is, $T(0, t_w) = 50°C$, can be obtained by rearranging Equation 5.53c

$$Fo = -\frac{1}{\zeta_1^2} \ln \left[\frac{\theta_o^*}{C_1} \right] = -\frac{1}{\zeta_1^2} \ln \left[\frac{1}{C_1} \times \frac{T(0, t_w) - T_\infty}{T_i - T_\infty} \right]$$

where $t_w = Fo \, r_o^2/\alpha$. With the Biot number now defined as

$$Bi = \frac{h_w r_o}{k} = \frac{6000 \text{ W/m}^2 \cdot \text{K} \times 0.005 \text{ m}}{20 \text{ W/m} \cdot \text{K}} = 1.50$$

Table 5.1 yields $C_1 = 1.376$ and $\zeta_1 = 1.800$ rad. It follows that

$$Fo = -\frac{1}{(1.800 \text{ rad})^2} \ln \left[\frac{1}{1.376} \times \frac{(50 - 20)°C}{(335 - 20)°C} \right] = 0.82$$

and

$$t_w = Fo \frac{r_o^2}{\alpha} = 0.82 \frac{(0.005 \text{ m})^2}{6.66 \times 10^{-6} \text{ m}^2/\text{s}} = 3.1 \text{ s} \qquad \triangleleft$$

Note that, with $Fo = 0.82$, use of the one-term approximation is justified.

Comments:

1. If the temperature distribution in the sphere at the conclusion of step 1 were not uniform, the calculations of step 2 could not be performed using either the one-term approximation or the exact series solution, since both assume a uniform initial temperature.

2. The surface temperature of the sphere at the conclusion of step 2 may be obtained from Equation 5.53b. With $\theta_o^* = 0.095$ and $r^* = 1$,

$$\theta^*(r_o) = \frac{T(r_o) - T_\infty}{T_i - T_\infty} = \frac{0.095}{1.800 \text{ rad}} \sin(1.800 \text{ rad}) = 0.0514$$

and

$$T(r_o) = 20°C + 0.0514(335 - 20)°C = 36°C$$

The infinite series, Equation 5.51a, and its one-term approximation, Equation 5.53b, may be used to compute the temperature at any location in the sphere and at any time $t > t_a$. For $(t - t_a) < 0.2(0.005 \text{ m})^2/6.66 \times 10^{-6} \text{ m}^2/\text{s} = 0.75$ s, a sufficient number of terms must be retained to ensure convergence of the series. For $(t - t_a) > 0.75$ s, satisfactory convergence is provided by the one-term approximation. Computing and plotting the temperature histories for $r = 0$ and $r = r_o$, we obtain the following results for $0 \leq (t - t_a) \leq 5$ s:

3. The *IHT Models, Transient Conduction, Sphere* option could be used to analyze the cooling processes experienced by the sphere in air and water, steps 1 and 2. The *IHT Models, Lumped Capacitance* option may only be used to analyze the air-cooling process, step 1.

5.7   *The Semi-Infinite Solid*

An important simple geometry for which analytical solutions may be obtained is the *semi-infinite solid*. Since, in principle, such a solid extends to infinity in all but one direction, it is characterized by a single identifiable surface (Figure 5.7). If a sudden change of conditions is imposed at this surface, transient, one-dimensional conduction will occur within the

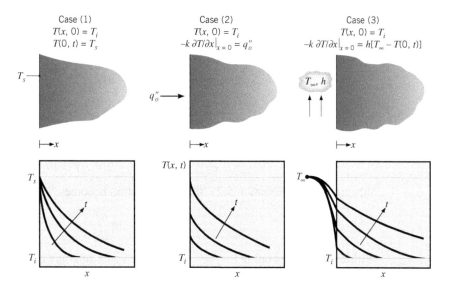

FIGURE 5.7 Transient temperature distributions in a semi-infinite solid for three surface conditions: constant surface temperature, constant surface heat flux, and surface convection.

solid. The semi-infinite solid provides a *useful idealization* for many practical problems. It may be used to determine the transient response of a finite solid, such as a thick slab. For this situation the semi-infinite solid approximation would be reasonable for the early portion of the transient, during which temperatures in the slab interior (well removed from the surface) are essentially uninfluenced by the change in surface conditions. These early portions of the transient might correspond to very small Fourier numbers, and the approximate solutions of Sections 5.5 and 5.6 would not be valid. Although the exact solutions of the preceding sections could be used to determine the temperature distributions, many terms might be required to evaluate the infinite series expressions. The following semi-infinite solid solutions often eliminate the need to evaluate the cumbersome infinite series exact solutions at small *Fo*. It will be shown that a plane wall of thickness 2*L* can be accurately approximated as a semi-infinite solid for $Fo = \alpha t/L^2 \lesssim 0.2$.

The heat equation for transient conduction in a semi-infinite solid is given by Equation 5.29. The initial condition is prescribed by Equation 5.30, and the interior boundary condition is of the form

$$T(x \rightarrow \infty, t) = T_i \tag{5.56}$$

Closed-form solutions have been obtained for three important surface conditions, instantaneously applied at $t = 0$ [3, 4]. These conditions are shown in Figure 5.7. They include application of a constant surface temperature $T_s \neq T_i$, application of a constant surface heat flux q_o'', and exposure of the surface to a fluid characterized by $T_\infty \neq T_i$ and the convection coefficient *h*.

The solution for case 1 may be obtained by recognizing the existence of a *similarity variable* η, through which the heat equation may be transformed from a partial differential equation, involving two independent variables (*x* and *t*), to an ordinary differential equation expressed in terms of the single similarity variable. To confirm that such a requirement is satisfied by $\eta \equiv x/(4\alpha t)^{1/2}$, we first transform the pertinent differential operators, such that

$$\frac{\partial T}{\partial x} = \frac{dT}{d\eta}\frac{\partial \eta}{\partial x} = \frac{1}{(4\alpha t)^{1/2}}\frac{dT}{d\eta}$$

$$\frac{\partial^2 T}{\partial x^2} = \frac{d}{d\eta}\left[\frac{\partial T}{\partial x}\right]\frac{\partial \eta}{\partial x} = \frac{1}{4\alpha t}\frac{d^2 T}{d\eta^2}$$

$$\frac{\partial T}{\partial t} = \frac{\partial T}{\partial \eta}\frac{\partial \eta}{\partial t} = -\frac{x}{2t(4\alpha t)^{1/2}}\frac{dT}{d\eta}$$

Substituting into Equation 5.29, the heat equation becomes

$$\frac{d^2 T}{d\eta^2} = -2\eta\frac{dT}{d\eta} \tag{5.57}$$

With $x = 0$ corresponding to $\eta = 0$, the surface condition may be expressed as

$$T(\eta = 0) = T_s \tag{5.58}$$

and with $x \rightarrow \infty$, as well as $t = 0$, corresponding to $\eta \rightarrow \infty$, both the initial condition and the interior boundary condition correspond to the single requirement that

$$T(\eta \rightarrow \infty) = T_i \tag{5.59}$$

Since the transformed heat equation and the initial/boundary conditions are independent of x and t, $\eta \equiv x/(4\alpha t)^{1/2}$ is, indeed, a similarity variable. Its existence implies that, irrespective of the values of x and t, the temperature may be represented as a unique function of η.

The specific form of the temperature dependence, $T(\eta)$, may be obtained by separating variables in Equation 5.57, such that

$$\frac{d(dT/d\eta)}{(dT/d\eta)} = -2\eta\, d\eta$$

Integrating, it follows that

$$\ln(dT/d\eta) = -\eta^2 + C_1'$$

or

$$\frac{dT}{d\eta} = C_1 \exp(-\eta^2)$$

Integrating a second time, we obtain

$$T = C_1 \int_0^\eta \exp(-u^2)\,du + C_2$$

where u is a dummy variable. Applying the boundary condition at $\eta = 0$, Equation 5.58, it follows that $C_2 = T_s$ and

$$T = C_1 \int_0^\eta \exp(-u^2)\,du + T_s$$

From the second boundary condition, Equation 5.59, we obtain

$$T_i = C_1 \int_0^\infty \exp(-u^2)\, du + T_s$$

or, evaluating the definite integral,

$$C_1 = \frac{2(T_i - T_s)}{\pi^{1/2}}$$

Hence the temperature distribution may be expressed as

$$\frac{T - T_s}{T_i - T_s} = (2/\pi^{1/2}) \int_0^\eta \exp(-u^2)\, du \equiv \operatorname{erf} \eta \tag{5.60}$$

where the *Gaussian error function,* erf η, is a standard mathematical function that is tabulated in Appendix B. Note that erf η asymptotically approaches unity as η becomes infinite. Thus, at any nonzero time, temperatures everywhere are predicted to have changed from T_i (become closer to T_s).[2] The surface heat flux may be obtained by applying Fourier's law at $x = 0$, in which case

$$q_s'' = -k\frac{\partial T}{\partial x}\bigg|_{x=0} = -k(T_i - T_s)\frac{d(\operatorname{erf} \eta)}{d\eta}\frac{\partial \eta}{\partial x}\bigg|_{\eta=0}$$

$$q_s'' = k(T_s - T_i)(2/\pi^{1/2})\exp(-\eta^2)(4\alpha t)^{-1/2}\big|_{\eta=0}$$

$$q_s'' = \frac{k(T_s - T_i)}{(\pi\alpha t)^{1/2}} = \frac{(k\rho c)^{1/2}}{(\pi t)^{1/2}}(T_s - T_i) \tag{5.61}$$

The quantity $(k\rho c)^{1/2}$ appears in a variety of transient heat conduction problems and is sometimes referred to as the *thermal effusivity* of a material. Analytical solutions may also be obtained for the case 2 and case 3 surface conditions, and results for all three cases are summarized as follows.

Case 1   **Constant Surface Temperature:** $T(0, t) = T_s$

$$\frac{T(x, t) - T_s}{T_i - T_s} = \operatorname{erf}\left(\frac{x}{2\sqrt{\alpha t}}\right) \tag{5.60}$$

$$q_s''(t) = \frac{k(T_s - T_i)}{\sqrt{\pi\alpha t}} \tag{5.61}$$

Case 2   **Constant Surface Heat Flux:** $q_s'' = q_o''$

$$T(x, t) - T_i = \frac{2q_o''(\alpha t/\pi)^{1/2}}{k}\exp\left(\frac{-x^2}{4\alpha t}\right) - \frac{q_o''x}{k}\operatorname{erfc}\left(\frac{x}{2\sqrt{\alpha t}}\right) \tag{5.62}$$

---

[2]The infinite speed at which boundary-condition information propagates into the semi-infinite solid is physically unrealistic, but this limitation of Fourier's law is not important except at extremely small time scales, as discussed in Section 2.3.

Case 3   Surface Convection: $-k \left. \dfrac{\partial T}{\partial x} \right|_{x=0} = h[T_\infty - T(0, t)]$

$$\frac{T(x, t) - T_i}{T_\infty - T_i} = \mathrm{erfc}\left(\frac{x}{2\sqrt{\alpha t}} \right)$$

$$-\left[\exp\left(\frac{hx}{k} + \frac{h^2 \alpha t}{k^2} \right) \right] \left[\mathrm{erfc}\left(\frac{x}{2\sqrt{\alpha t}} + \frac{h\sqrt{\alpha t}}{k} \right) \right] \qquad (5.63)$$

The *complementary error function,* erfc w, is defined as erfc $w \equiv 1 - \mathrm{erf}\ w$.

Temperature histories for the three cases are shown in Figure 5.7, and distinguishing features should be noted. With a step change in the surface temperature, case 1, temperatures within the medium monotonically approach T_s with increasing t, while the magnitude of the surface temperature gradient, and hence the surface heat flux, decreases as $t^{-1/2}$. A *thermal penetration depth* δ_p can be defined as the depth to which significant temperature effects propagate within a medium. For example, defining δ_p as the x-location at which $(T - T_s)/(T_i - T_s) = 0.90$, Equation 5.60 results in $\delta_p = 2.3\sqrt{\alpha t}$.[3] Hence, the penetration depth increases as $t^{1/2}$ and is larger for materials with higher thermal diffusivity. For a fixed surface heat flux (case 2), Equation 5.62 reveals that $T(0, t) = T_s(t)$ increases monotonically as $t^{1/2}$. For surface convection (case 3), the surface temperature and temperatures within the medium approach the fluid temperature T_∞ with increasing time. As T_s approaches T_∞, there is, of course, a reduction in the surface heat flux, $q_s''(t) = h[T_\infty - T_s(t)]$.

Specific temperature histories computed from Equation 5.63 are plotted in Figure 5.8. The result corresponding to $h = \infty$ is equivalent to that associated with a sudden change in surface temperature, case 1. That is, for $h = \infty$, the surface instantaneously achieves the imposed fluid temperature ($T_s = T_\infty$), and with the second term on the right-hand side of Equation 5.63 reducing to zero, the result is equivalent to Equation 5.60.

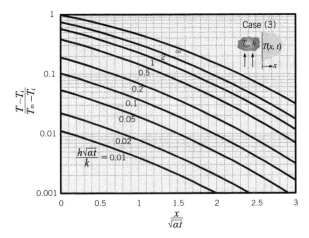

FIGURE 5.8 Temperature histories in a semi-infinite solid with surface convection.

---

[3]To apply the semi-infinite approximation to a plane wall of thickness $2L$, it is necessary that $\delta_p < L$. Substituting $\delta_p = L$ into the expression for the thermal penetration depth yields $Fo = 0.19 \approx 0.2$. Hence, a plane wall of thickness $2L$ can be accurately approximated as a semi-infinite solid for $Fo = \alpha t/L^2 \lesssim 0.2$. This restriction will also be demonstrated in Section 5.8.

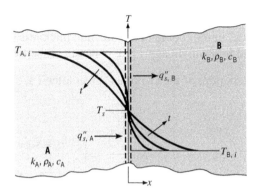

FIGURE 5.9 Interfacial contact between two semi-infinite solids at different initial temperatures.

An interesting permutation of case 1 occurs when two semi-infinite solids, initially at uniform temperatures $T_{A,i}$ and $T_{B,i}$, are placed in contact at their surfaces (Figure 5.9). If the contact resistance is negligible, the requirement of thermal equilibrium dictates that, at the instant of contact ($t = 0$), both surfaces must assume the same temperature T_s, for which $T_{B,i} < T_s < T_{A,i}$. Since T_s does not change with time, it follows that the transient thermal response and the surface heat flux of each of the solids are determined by Equations 5.60 and 5.61, respectively.

The equilibrium surface temperature of Figure 5.9 may be determined from a surface energy balance, which requires that

$$q''_{s,A} = q''_{s,B} \tag{5.64}$$

Substituting from Equation 5.61 for $q''_{s,A}$ and $q''_{s,B}$ and recognizing that the x-coordinate of Figure 5.9 requires a sign change for $q''_{s,A}$, it follows that

$$\frac{-k_A(T_s - T_{A,i})}{(\pi \alpha_A t)^{1/2}} = \frac{k_B(T_s - T_{B,i})}{(\pi \alpha_B t)^{1/2}} \tag{5.65}$$

or, solving for T_s,

$$T_s = \frac{(k\rho c)_A^{1/2} T_{A,i} + (k\rho c)_B^{1/2} T_{B,i}}{(k\rho c)_A^{1/2} + (k\rho c)_B^{1/2}} \tag{5.66}$$

Hence the thermal effusivity, $m \equiv (k\rho c)^{1/2}$, plays the role of a weighting factor that determines whether T_s will more closely approach $T_{A,i} (m_A > m_B)$ or $T_{B,i} (m_B > m_A)$.

EXAMPLE 5.7

On a hot and sunny day, the concrete deck surrounding a swimming pool is at a temperature of $T_d = 55°C$. A swimmer walks across the dry deck to the pool. The soles of the swimmer's dry feet are characterized by an $L_{sf} = 3$-mm-thick skin/fat layer of thermal conductivity $k_{sf} = 0.3$ W/m · K. Consider two types of concrete decking; (i) a dense stone mix and (ii) a lightweight aggregate characterized by density, specific heat, and thermal conductivity of $\rho_{lw} = 1495$ kg/m³, $c_{lw} = 880$ J/kg · K, and $k_{lw} = 0.28$ W/m · K, respectively. The density and specific heat of the skin/fat layer may be approximated to be those of liquid water, and the skin/fat layer is at an initial temperature of $T_{sf,i} = 37°C$. What is the temperature of the bottom of the swimmer's feet after an elapsed time of $t = 1$ s?

SOLUTION

Known: Concrete temperature, initial foot temperature, and thickness of skin/fat layer on the sole of the foot. Skin/fat and lightweight aggregate concrete properties.

Find: The temperature of the bottom of the swimmer's feet after 1 s.

Schematic:

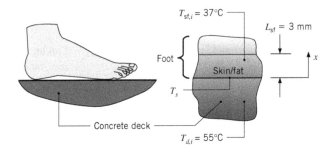

Assumptions:

1. One-dimensional conduction in the x-direction.
2. Constant and uniform properties.
3. Negligible contact resistance.

Properties: Table A.3 stone mix concrete ($T = 300$ K): $\rho_{sm} = 2300$ kg/m^3, $k_{sm} = 1.4$ W/m · K, $c_{sm} = 880$ J/kg · K. Table A.6 water ($T = 310$ K): $\rho_{sf} = 993$ kg/m^3, $c_{sf} = 4178$ J/kg · K.

Analysis: If the skin/fat layer and the deck are both semi-infinite media, from Equation 5.66 the surface temperature T_s is constant when the swimmer's foot is in contact with the deck. For the lightweight aggregate concrete decking, the thermal penetration depth at $t = 1$ s is

$$\delta_{p,\text{lw}} = 2.3\sqrt{\alpha_{\text{lw}}t} = 2.3\sqrt{\frac{k_{\text{lw}}t}{\rho_{\text{lw}}c_{\text{lw}}}} = 2.3\sqrt{\frac{0.28 \text{ W/m} \cdot \text{K} \times 1 \text{ s}}{1495 \text{ kg/m}^3 \times 880 \text{ J/kg} \cdot \text{K}}}$$

$$= 1.06 \times 10^{-3} \text{ m} = 1.06 \text{ mm}$$

Since the thermal penetration depth is relatively small, it is reasonable to assume that the lightweight aggregate deck behaves as a semi-infinite medium. Similarly, the thermal penetration depth in the stone mix concrete is $\delta_{p,\text{sm}} = 1.91$ mm, and the thermal penetration depth associated with the skin/fat layer of the foot is $\delta_{p,\text{sf}} = 0.62$ mm. Hence, it is reasonable to assume that the stone mix concrete deck responds as a semi-infinite medium, and, since $\delta_{p,\text{sf}} < L_{\text{sf}}$, it is also correct to assume that the skin/fat layer behaves as a semi-infinite medium. Therefore, Equation 5.66 may be used to determine the surface temperature of the swimmer's foot for exposure to the two types of concrete decking. For the lightweight aggregate,

$$T_{s,\text{lw}} = \frac{(k\rho c)_{\text{lw}}^{1/2} T_{d,i} + (k\rho c)_{\text{sf}}^{1/2} T_{\text{sf},i}}{(k\rho c)_{\text{lw}}^{1/2} + (k\rho c)_{\text{sf}}^{1/2}}$$

$$= \frac{\begin{bmatrix} (0.28 \text{ W/m} \cdot \text{K} \times 1495 \text{ kg/m}^3 \times 880 \text{ J/kg} \cdot \text{K})^{1/2} \times 55°\text{C} \\ + (0.3 \text{ W/m} \cdot \text{K} \times 993 \text{ kg/m}^3 \times 4178 \text{ J/kg} \cdot \text{K})^{1/2} \times 37°\text{C} \end{bmatrix}}{\begin{bmatrix} (0.28 \text{ W/m} \cdot \text{K} \times 1495 \text{ kg/m}^3 \times 880 \text{ J/kg} \cdot \text{K})^{1/2} \\ + (0.3 \text{ W/m} \cdot \text{K} \times 993 \text{ kg/m}^3 \times 4178 \text{ J/kg} \cdot \text{K})^{1/2} \end{bmatrix}} = 43.3°\text{C} \quad \triangleleft$$

Repeating the calculation for the stone mix concrete gives $T_{s,sm} = 47.8°C$. ◁

Comments:

1. The lightweight aggregate concrete feels cooler to the swimmer, relative to the stone mix concrete. Specifically, the temperature rise from the initial skin/fat temperature that is associated with the stone mix concrete is $\Delta T_{sm} = T_{sm} - T_{sf,i} = 47.8°C - 37°C = 10.8°C$, whereas the temperature rise associated with the lightweight aggregate is $\Delta T_{lw} = T_{lw} - T_{sf,i} = 43.3°C - 37°C = 6.3°C$.

2. The thermal penetration depths associated with an exposure time of $t = 1$ s are small. Stones and air pockets within the concrete may be of the same size as the thermal penetration depth, making the uniform property assumption somewhat questionable. The predicted foot temperatures should be viewed as representative values.

5.8 *Objects with Constant Surface Temperatures or Surface Heat Fluxes*

In Sections 5.5 and 5.6, the transient thermal response of plane walls, cylinders, and spheres to an applied convection boundary condition was considered in detail. It was pointed out that the solutions in those sections may be used for cases involving a step change in surface temperature by allowing the Biot number to be infinite. In Section 5.7, the response of a semi-infinite solid to a step change in surface temperature, or to an applied constant heat flux, was determined. This section will conclude our discussion of transient heat transfer in one-dimensional objects experiencing constant surface temperature or constant surface heat flux boundary conditions. A variety of approximate solutions are presented.

5.8.1 Constant Temperature Boundary Conditions

In the following discussion, the transient thermal response of objects to a step change in surface temperature is considered.

Semi-Infinite Solid   Insight into the thermal response of objects to an applied constant temperature boundary condition may be obtained by casting the heat flux in Equation 5.61 into the nondimensional form

$$q^* \equiv \frac{q_s'' L_c}{k(T_s - T_i)} \tag{5.67}$$

where L_c is a *characteristic length* and q^* is the *dimensionless conduction heat rate* that was introduced in Section 4.3. Substituting Equation 5.67 into Equation 5.61 yields

$$q^* = \frac{1}{\sqrt{\pi Fo}} \tag{5.68}$$

where the Fourier number is defined as $Fo \equiv \alpha t / L_c^2$. When Equations 5.67 and 5.68 are combined, we note that the value of q_s'' is independent of the choice of the characteristic length, as it must be for a semi-infinite solid. Equation 5.68 is plotted in Figure 5.10*a*, and since $q^* \propto Fo^{-1/2}$, the slope of the line is $-1/2$ on the log-log plot.

Interior Heat Transfer: Plane Wall, Cylinder, and Sphere   Results for heat transfer to the *interior* of a plane wall, cylinder, and sphere are also shown in Figure 5.10*a*. These results are generated by using Fourier's law in conjunction with Equations 5.42, 5.50, and 5.51 for $Bi \rightarrow \infty$. As in Sections 5.5 and 5.6, the characteristic length is $L_c = L$ or r_o for a plane wall of thickness $2L$ or a cylinder (or sphere) of radius r_o, respectively. For each geometry, q^* initially follows the semi-infinite solid solution but at some point decreases rapidly as the objects approach their equilibrium temperature and $q_s''(t \rightarrow \infty) \rightarrow 0$. The value of q^* is expected to decrease more rapidly for geometries that possess large surface area to volume ratios, and this trend is evident in Figure 5.10*a*.

Exterior Heat Transfer: Various Geometries   Additional results are shown in Figure 5.10*a* for objects that are embedded in an exterior (surrounding) medium of infinite extent. The infinite medium is initially at temperature T_i, and the surface temperature of the object

(*a*)

(*b*)

FIGURE 5.10   Transient dimensionless conduction heat rates for a variety of geometries. (*a*) Constant surface temperature. Results for the geometries of Table 4.1 lie within the shaded region and are from Yovanovich [7]. (*b*) Constant surface heat flux.

is suddenly changed to T_s. For the exterior cases, L_c is the characteristic length used in Section 4.3, namely $L_c = (A_s/4\pi)^{1/2}$. For the sphere in a surrounding infinite medium, the exact solution for $q^*(Fo)$ is [7]

$$q^* = \frac{1}{\sqrt{\pi Fo}} + 1 \tag{5.69}$$

As seen in the figure, for all of the exterior cases q^* closely mimics that of the sphere when the appropriate length scale is used in its definition, regardless of the object's shape. Moreover, in a manner consistent with the interior cases, q^* initially follows the semi-infinite solid solution. In contrast to the interior cases, q^* eventually reaches the nonzero, steady-state value of q_{ss}^* that is listed in Table 4.1. Note that q_s'' in Equation 5.67 is the *average* surface heat flux for geometries that have nonuniform surface heat flux.

As seen in Figure 5.10*a*, *all* of the thermal responses collapse to that of the semi-infinite solid for early times, that is, for Fo less than approximately 10^{-3}. This remarkable consistency reflects the fact that temperature variations are confined to thin layers adjacent to the surface of any object at early times, regardless of whether internal or external heat transfer is of interest. At early times, therefore, Equations 5.60 and 5.61 may be used to predict the temperatures and heat transfer rates within the thin regions adjacent to the boundaries of any object. For example, predicted local heat fluxes and local dimensionless temperatures using the semi-infinite solid solutions are within approximately 5% of the predictions obtained from the exact solutions for the interior and exterior heat transfer cases involving spheres when $Fo \le 10^{-3}$.

5.8.2 Constant Heat Flux Boundary Conditions

When a constant surface heat flux is applied to an object, the resulting surface temperature history is often of interest. In this situation, the heat flux in the numerator of Equation 5.67 is now constant, and the temperature difference in the denominator, $T_s - T_i$, increases with time.

Semi-Infinite Solid   In the case of a semi-infinite solid, the surface temperature history can be found by evaluating Equation 5.62 at $x = 0$, which may be rearranged and combined with Equation 5.67 to yield

$$q^* = \frac{1}{2}\sqrt{\frac{\pi}{Fo}} \tag{5.70}$$

As for the constant temperature case, $q^* \propto Fo^{-1/2}$, but with a different coefficient. Equation 5.70 is presented in Figure 5.10*b*.

Interior Heat Transfer: Plane Wall, Cylinder, and Sphere   A second set of results is shown in Figure 5.10*b* for the *interior* cases of the plane wall, cylinder, and sphere. As for the constant surface temperature results of Figure 5.10*a*, q^* initially follows the semi-infinite solid solution and subsequently decreases more rapidly, with the decrease occurring first for the sphere, then the cylinder, and finally the plane wall. Compared to the constant surface temperature case, the rate at which q^* decreases is not as dramatic, since steady-state conditions are never reached; the surface temperature must increase with time as more heat is added to these objects. At *late* times (large Fo), the surface temperature increases linearly with time, yielding $q^* \propto Fo^{-1}$, with a slope of -1 on the log-log plot.

Exterior Heat Transfer: Various Geometries Results for heat transfer between a sphere and an exterior infinite medium are also presented in Figure 5.10*b*. The exact solution for the embedded sphere is

$$q^* = [1 - \exp(Fo)\,\mathrm{erfc}(Fo^{1/2})]^{-1} \tag{5.71}$$

As in the constant surface temperature case of Figure 5.10*a*, this solution approaches steady-state conditions, with $q''_{ss} = 1$. For objects of other shapes that are embedded within an infinite medium, q^* would follow the semi-infinite solid solution at small *Fo*. At larger *Fo*, q^* must asymptotically approach the value of q''_{ss} given in Table 4.1 where T_s in Equation 5.67 is the *average* surface temperature for geometries that have nonuniform surface temperatures.

5.8.3 Approximate Solutions

Simple expressions have been developed for $q^*(Fo)$ [8]. These expressions may be used to approximate all the results included in Figure 5.10 over the *entire range of Fo*. These expressions are listed in Table 5.2, along with the corresponding exact solutions. Table 5.2*a* is for the constant surface temperature case corresponding to Figure 5.10*a*, while Table 5.2*b* is for the constant surface heat flux situation of Figure 5.10*b*. For each of the geometries listed in the left-hand column, the tables provide the length scale to be used in the definition of both *Fo* and q^*, the exact solution for $q^*(Fo)$, the approximation solutions for early times ($Fo < 0.2$) and late times ($Fo \geq 0.2$), and the maximum percentage error associated with use of the approximations (which occurs at $Fo \approx 0.2$ for all results except the external sphere with constant heat flux).

EXAMPLE 5.8

Derive an expression for the ratio of the total energy transferred from the isothermal surfaces of a plane wall to the interior of the plane wall, Q/Q_o, that is valid for $Fo < 0.2$. Express your results in terms of the Fourier number *Fo*.

SOLUTION
Known: Plane wall with constant surface temperatures.

Find: Expression for Q/Q_o as a function of $Fo = \alpha t/L^2$.

Schematic:

TABLE 5.2a Summary of transient heat transfer results for constant surface temperature cases[a] [8]

Geometry	Length Scale, L_c	Exact Solutions	$q^*(Fo)$ Approximate Solutions		Maximum Error (%)
			$Fo < 0.2$	$Fo \geq 0.2$	
Semi-infinite	L (arbitrary)	$\dfrac{1}{\sqrt{\pi Fo}}$	Use exact solution.		None
Interior Cases					
Plane wall of thickness 2L	L	$2\sum\limits_{n=1}^{\infty}\exp(-\zeta_n^2 Fo)\quad \zeta_n = (n-\tfrac{1}{2})\pi$	$\dfrac{1}{\sqrt{\pi Fo}}$	$2\exp(-\zeta_1^2 Fo)\quad \zeta_1 = \pi/2$	1.7
Infinite cylinder	r_o	$2\sum\limits_{n=1}^{\infty}\exp(-\zeta_n^2 Fo)\quad J_0(\zeta_n)=0$	$\dfrac{1}{\sqrt{\pi Fo}} - 0.50 - 0.65\,Fo$	$2\exp(-\zeta_1^2 Fo)\quad \zeta_1 = 2.4050$	0.8
Sphere	r_o	$2\sum\limits_{n=1}^{\infty}\exp(-\zeta_n^2 Fo)\quad \zeta_n = n\pi$	$\dfrac{1}{\sqrt{\pi Fo}} - 1$	$2\exp(-\zeta_1^2 Fo)\quad \zeta_1 = \pi$	6.3
Exterior Cases					
Sphere	r_o	$\dfrac{1}{\sqrt{\pi Fo}} + 1$	Use exact solution.		None
Various shapes (Table 4.1, cases 12–15)	$(A_s/4u)^{1/2}$	None	$\dfrac{1}{\sqrt{\pi Fo}} + q_{ss}^*,\ q_{ss}^*$ from Table 4.1		7.1

[a]$q^* \equiv q_s'' L_c / k(T_s - T_i)$ and $Fo \equiv \alpha t/l_c^2$, where L_c is the length scale given in the table, T_s is the object surface temperature, and T_i is (a) the initial object temperature for the interior cases and (b) the temperature of the infinite medium for the exterior cases.

TABLE 5.2b **Summary of transient heat transfer results for constant surface heat flux cases[a] [8]**

Geometry	Length Scale, L_c	$q*(Fo)$			Maximum Error (%)
		Exact Solutions	Approximate Solutions		
			$Fo < 0.2$	$Fo \geq 0.2$	
Semi-infinite	L (arbitrary)	$\dfrac{1}{2}\sqrt{\dfrac{\pi}{Fo}}$	Use exact solution.		None
Interior Cases					
Plane wall of thickness $2L$	L	$\left[Fo + \dfrac{1}{3} - 2\displaystyle\sum_{n=1}^{\infty}\dfrac{\exp(-\zeta_n^2 Fo)}{\zeta_n^2}\right]^{-1}\ \zeta_n = n\pi$	$\dfrac{1}{2}\sqrt{\dfrac{\pi}{Fo}}$	$\left[Fo + \dfrac{1}{3}\right]^{-1}$	5.3
Infinite cylinder	r_o	$\left[2Fo + \dfrac{1}{4} - 2\displaystyle\sum_{n=1}^{\infty}\dfrac{\exp(-\zeta_n^2 Fo)}{\zeta_n^2}\right]^{-1}\ J_1(\zeta_n) = 0$	$\dfrac{1}{2}\sqrt{\dfrac{\pi}{Fo}} - \dfrac{\pi}{8}$	$\left[2Fo + \dfrac{1}{4}\right]^{-1}$	2.1
Sphere	r_o	$\left[3Fo + \dfrac{1}{5} - 2\displaystyle\sum_{n=1}^{\infty}\dfrac{\exp(-\zeta_n^2 Fo)}{\zeta_n^2}\right]^{-1}\ \tan(\zeta_n) = \zeta_n$	$\dfrac{1}{2}\sqrt{\dfrac{\pi}{Fo}} - \dfrac{\pi}{4}$	$\left[3Fo + \dfrac{1}{5}\right]^{-1}$	4.5
Exterior Cases					
Sphere	r_o	$[1 - \exp(Fo)\operatorname{erfc}(Fo^{1/2})]^{-1}$	$\dfrac{1}{2}\sqrt{\dfrac{\pi}{Fo}} + \dfrac{\pi}{4}$	$\dfrac{0.77}{\sqrt{Fo}} + 1$	3.2
Various shapes (Table 4.1, cases 12–15)	$(A_s/4\pi)^{1/2}$	None	$\dfrac{1}{2}\sqrt{\dfrac{\pi}{Fo}} + \dfrac{\pi}{4}$	$\dfrac{0.77}{\sqrt{Fo}} + q_{ss}^*$	Unknown

[a] $q* \equiv q_s'' L_c /k(T_s - T_i)$ and $Fo \equiv \alpha t/L_c^2$, where L_c is the length scale given in the table, T_s is the object surface temperature, and T_i is (a) the initial object temperature for the interior cases and (b) the temperature of the infinite medium for the exterior cases.

Assumptions:

1. One-dimensional conduction.
2. Constant properties.
3. Validity of the approximate solution of Table 5.2*a*.

Analysis: From Table 5.2*a* for a plane wall of thickness 2*L* and $Fo < 0.2$,

$$q^* = \frac{q_s'' L}{k(T_s - T_i)} = \frac{1}{\sqrt{\pi Fo}} \quad \text{where} \quad Fo = \frac{\alpha t}{L^2}$$

Combining the preceding equations yields

$$q_s'' = \frac{k(T_s - T_i)}{\sqrt{\pi \alpha t}}$$

Recognizing that Q is the accumulated heat that has entered the wall up to time t, we can write

$$\frac{Q}{Q_o} = \frac{\int_{t=0}^{t} q_s'' dt}{L\rho c(T_s - T_i)} = \frac{\alpha}{L\sqrt{\pi \alpha}} \int_{t=0}^{t} t^{-1/2}\, dt = \frac{2}{\sqrt{\pi}}\sqrt{Fo} \qquad \triangleleft$$

Comments:

1. The exact solution for Q/Q_o at small Fourier number involves many terms that would need to be evaluated in the infinite series expression. Use of the approximate solution simplifies the evaluation of Q/Q_o considerably.

2. At $Fo = 0.2$, $Q/Q_o \approx 0.5$. Half of the total possible change in thermal energy of the plane wall occurs during $Fo \leq 0.2$. The solution to many problems would preclude use of the one-term approximation to the exact solution, Equation 5.49, which is valid only for $Fo > 0.2$.

EXAMPLE 5.9

A proposed cancer treatment utilizes small, composite *nanoshells* whose size and composition are carefully specified so that the particles efficiently absorb laser irradiation at particular wavelengths [9]. Prior to treatment, antibodies are attached to the nanoscale particles. The doped particles are then injected into the patient's bloodstream and are distributed throughout the body. The antibodies are attracted to malignant sites, and therefore carry and adhere the nanoshells only to cancerous tissue. After the particles have come to rest within the tumor, a laser beam penetrates through the tissue between the skin and the cancer, is absorbed by the nanoshells, and, in turn, heats and destroys the cancerous tissues.

Consider an approximately spherical tumor of diameter $D_t = 3$ mm that is uniformly infiltrated with nanoshells that are highly absorptive of incident radiation from a laser located outside the patient's body.

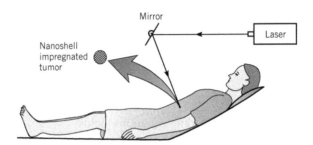

1. Estimate the heat transfer rate from the tumor to the surrounding healthy tissue for a steady-state treatment temperature of $T_{t,ss} = 55°C$ at the surface of the tumor. The thermal conductivity of healthy tissue is approximately $k = 0.5$ W/m · K, and the body temperature is $T_b = 37°C$.

2. Find the laser power necessary to sustain the tumor surface temperature at $T_{t,ss} = 55°C$ if the tumor is located $d = 20$ mm beneath the surface of the skin, and the laser heat flux decays exponentially, $q_l''(x) = q_{l,o}''(1 - \rho)e^{-\kappa x}$, between the surface of the body and the tumor. In the preceding expression, $q_{l,o}''$ is the laser heat flux outside the body, $\rho = 0.05$ is the reflectivity of the skin surface, and $\kappa = 0.02$ mm^{-1} is the *extinction coefficient* of the tissue between the tumor and the surface of the skin. The laser beam has a diameter of $D_l = 5$ mm.

3. Neglecting heat transfer to the surrounding tissue, estimate the time at which the tumor temperature is within 3°C of $T_{t,ss} = 55°C$ for the laser power found in part 2. Assume the tissue's density and specific heat are that of water.

4. Neglecting the thermal mass of the tumor but accounting for heat transfer to the surrounding tissue, estimate the time needed for the surface temperature of the tumor to reach $T_t = 52°C$.

SOLUTION

Known: Size of a spherical tumor; thermal conductivity, reflectivity, and extinction coefficient of tissue; depth of sphere below the surface of the skin.

Find:

1. Heat transfer rate from the tumor to maintain its surface temperature at $T_{t,ss} = 55°C$.

2. Laser power needed to sustain the tumor surface temperature at $T_{t,ss} = 55°C$.

3. Time for the tumor to reach $T_t = 52°C$ when heat transfer to the surrounding tissue is neglected.

4. Time for the tumor to reach $T_t = 52°C$ when heat transfer to the surrounding tissue is considered and the thermal mass of the tumor is neglected.

Schematic:

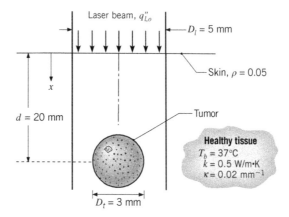

Assumptions:

1. One-dimensional conduction in the radial direction.
2. Constant properties.
3. Healthy tissue can be treated as an infinite medium.
4. The treated tumor absorbs all irradiation incident from the laser.
5. Lumped capacitance behavior for the tumor.
6. Neglect potential nanoscale heat transfer effects.
7. Neglect the effect of blood perfusion.

Properties: Table A.6, water (320 K, assumed): $\rho = v_f^{-1} = 989.1 \text{ kg/m}^3$, $c_p = 4180 \text{ J/kg} \cdot \text{K}$.

Analysis:

1. The steady-state rate of heat loss from the spherical tumor may be determined by evaluating the dimensionless heat rate from case 12 of Table 4.1:

$$q = 2\pi k D_t (T_{t,\text{ss}} - T_b) = 2 \times \pi \times 0.5 \text{ W/m} \cdot \text{K} \times 3 \times 10^{-3} \text{ m} \times (55 - 37)°\text{C}$$
$$= 0.170 \text{ W} \qquad \triangleleft$$

2. The laser irradiation will be absorbed over the projected area of the tumor, $\pi D_t^2/4$. To determine the laser power corresponding to $q = 0.170$ W, we first write an energy balance for the sphere. For a control surface about the sphere, the energy absorbed from the laser irradiation is offset by heat conduction to the healthy tissue, $q = 0.170 \text{ W} \approx q_l''(x = d)\pi D_t^2/4$, where, $q_l''(x = d) = q_{l,o}''(1 - \rho)e^{-\kappa d}$ and the laser power is $P_l = q_{l,o}'' \pi D_l^2/4$. Hence,

$$P_l = q D_l^2 e^{\kappa d}/[(1 - \rho)D_t^2]$$
$$= 0.170 \text{ W} \times (5 \times 10^{-3} \text{ m})^2 \times e^{(0.02 \text{ mm}^{-1} \times 20 \text{ mm})}/[(1 - 0.05) \times (3 \times 10^{-3} \text{ m})^2]$$
$$= 0.74 \text{ W} \qquad \triangleleft$$

3. The general lumped capacitance energy balance, Equation 5.14, may be written

$$q_l''(x = d)\pi D_t^2/4 = q = \rho V c_p \frac{dT}{dt}$$

Separating variables and integrating between appropriate limits,

$$\frac{q}{\rho V c_p} \int_{t=0}^{t} dt = \int_{T_b}^{T_t} dT$$

yields

$$t = \frac{\rho V c_p}{q}(T_t - T_b) = \frac{989.1 \text{ kg/m}^3 \times (\pi/6) \times (3 \times 10^{-3}\text{m})^3 \times 4180 \text{ J/kg} \cdot \text{K}}{0.170 \text{ W}}$$

$$\times (52°\text{C} - 37°\text{C})$$

or

$$t = 5.16 \text{ s} \qquad \triangleleft$$

4. Using Equation 5.71,

$$q/2\pi k D_t(T_t - T_b) = q^* = [1 - \exp(Fo)\text{erfc}(Fo^{1/2})]^{-1}$$

which may be solved by trial-and-error to yield $Fo = 10.3 = 4\alpha t/D_t^2$. Then, with $\alpha = k/\rho c_p = 0.50 \text{ W/m} \cdot \text{K}/(989.1 \text{ kg/m}^3 \times 4180 \text{ J/kg} \cdot \text{K}) = 1.21 \times 10^{-7} \text{ m}^2/\text{s}$, we find

$$t = FoD_t^2/4\alpha = 10.3 \times (3 \times 10^{-3} \text{ m})^2/(4 \times 1.21 \times 10^{-7} \text{ m}^2/\text{s}) = 192 \text{ s} \qquad \triangleleft$$

Comments:

1. The analysis does not account for blood perfusion. The flow of blood would lead to advection of warmed fluid away from the tumor (and relatively cool blood to the vicinity of the tumor), increasing the power needed to reach the desired treatment temperature.

2. The laser power needed to treat various-sized tumors, calculated as in parts 1 and 2 of the problem solution, is shown below. Note that as the tumor becomes smaller, a higher-powered laser is needed, which may seem counterintuitive. The power required to heat

the tumor, which is the same as the rate of heat loss calculated in part 1, increases in direct proportion to the diameter, as might be expected. However, since the laser power flux remains constant, a smaller tumor cannot absorb as much energy (the energy absorbed has a D_t^2 dependence). A smaller fraction of the laser power is utilized to heat the tumor, and the required laser power increases for smaller tumors.

3. To determine the actual time needed for the tumor temperature to approach steady-state conditions, a numerical solution of the heat diffusion equation applied to the surrounding tissue, *coupled* with a solution for the temperature history within the tumor, would be required. However, we see that significantly more time is needed for the surrounding tissue to reach steady-state conditions than to increase the temperature of the isolated spherical tumor (5.16 s). Hence, the actual time to heat *both* the tumor and the surrounding tissue will be slightly greater than 192 s.

5.9 Periodic Heating

In the preceding discussion of transient heat transfer, we have considered objects that experience constant surface temperature or constant surface heat flux boundary conditions. In many practical applications the boundary conditions are not constant, and analytical solutions have been obtained for situations where the conditions vary with time. One situation involving nonconstant boundary conditions is periodic heating, which describes various applications, such as thermal processing of materials using pulsed lasers, and occurs naturally in situations such as those involving the collection of solar energy.

Consider, for example, the semi-infinite solid of Figure 5.11a. For a surface temperature history described by $T(0, t) = T_i + \Delta T \sin \omega t$, the solution of Equation 5.29 subject to the interior boundary condition $T(x \rightarrow \infty, t) = T_i$ is

$$\frac{T(x, t) - T_i}{\Delta T} = \exp[-x\sqrt{\omega/2\alpha}]\sin[\omega t - x\sqrt{\omega/2\alpha}] \tag{5.72}$$

This solution applies after sufficient time has passed to yield a *quasi*-steady state for which all temperatures fluctuate periodically about a time-invariant mean value. At locations in the solid, the fluctuations have a time lag relative to the surface temperature. In addition, the amplitude of the fluctuations within the material decays exponentially with distance from the surface. Consistent with the earlier definition of the thermal penetration depth, δ_p can be defined as the x-location at which the amplitude of the temperature fluctuation is reduced by approximately 90% relative to that of the surface. This results

(a)

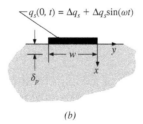

(b)

FIGURE 5.11 Schematic of (a) a periodically heated, one-dimensional semi-infinite solid and (b) a periodically heated strip attached to a semi-infinite solid.

in $\delta_p = 4\sqrt{\alpha/\omega}$. The heat flux at the surface may be determined by applying Fourier's law at $x = 0$, yielding

$$q_s''(t) = k\Delta T\sqrt{\omega/\alpha}\,\sin(\omega t + \pi/4) \tag{5.73}$$

Equation 5.73 reveals that the surface heat flux is periodic, with a time-averaged value of zero.

Periodic heating can also occur in two- or three-dimensional arrangements, as shown in Figure 5.11*b*. Recall that for this geometry, a steady state can be attained with constant heating of the strip placed upon a semi-infinite solid (Table 4.1, case 13). In a similar manner, a quasi-steady state may be achieved when sinusoidal heating ($q_s = \Delta q_s + \Delta q_s \sin \omega t$) is applied to the strip. Again, all temperatures fluctuate about a time-invariant mean value in the quasi-steady state.

The solution of the two-dimensional, transient heat diffusion equation for the two-dimensional configuration shown in Figure 5.11*b* has been obtained, and the relationship between the amplitude of the applied sinusoidal heating and the amplitude of the temperature response of the heated strip can be approximated as [10]

$$\Delta T \approx \frac{\Delta q_s}{L\pi k}\left[-\frac{1}{2}\ln(\omega/2) - \ln(w^2/4\alpha) + C_1\right] = \frac{\Delta q_s}{L\pi k}\left[-\frac{1}{2}\ln(\omega/2) + C_2\right] \tag{5.74}$$

where the constant C_1 depends on the thermal contact resistance at the interface between the heated strip and the underlying material. Note that the amplitude of the temperature fluctuation, ΔT, corresponds to the spatially averaged temperature of the rectangular strip of length L and width w. The heat flux from the strip to the semi-infinite medium is assumed to be spatially uniform. The approximation is valid for $L \gg w$. For the system of Figure 5.11*b*, the thermal penetration depth is smaller than that of Figure 5.11*a* because of the lateral spreading of thermal energy and is $\delta_p \approx \sqrt{\alpha/\omega}$.

EXAMPLE 5.10

A nanostructured dielectric material has been fabricated, and the following method is used to measure its thermal conductivity. A long metal strip 3000 angstroms thick, $w = 100\ \mu$m wide, and $L = 3.5$ mm long is deposited by a photolithography technique on the top surface of a $d = 300$-μm-thick sample of the new material. The strip is heated periodically by an electric current supplied through two connector pads. The heating rate is $q_s(t) = \Delta q_s + \Delta q_s \times \sin(\omega t)$, where Δq_s is 3.5 mW. The instantaneous, spatially averaged temperature of the metal strip is found experimentally by measuring the time variation of its electrical resistance, $R(t) = E(t)/I(t)$, and by knowing how the electrical resistance of the metal varies with temperature. The measured temperature of the metal strip is periodic; it has an amplitude of $\Delta T = 1.37$ K at a relatively low heating frequency of $\omega = 2\pi$ rad/s and 0.71 K at a frequency of 200π rad/s. Determine the thermal conductivity of the nanostructured dielectric material and the thermal penetration depths at the two frequencies. The density and specific heats of the conventional version of the material are 3100 kg/m^3 and 820 J/kg · K, respectively.

SOLUTION

Known: Dimensions of a thin metal strip, frequency and amplitude of the electric power dissipated in the strip, amplitude of the induced oscillating strip temperature, and thickness of the underlying sample.

Find: The thermal conductivity of the nanostructured material.

Schematic:

Assumptions:

1. Two-dimensional transient conduction in the x- and y-directions.
2. Density and specific heat are constant and unaffected by nanostructuring.
3. Negligible radiation and convection losses from the metal strip and top surface of the sample.
4. The nanostructured material sample is a semi-infinite solid.
5. Uniform heat flux at the interface between the heated strip and the sample.

Analysis: Substitution of $\Delta T = 1.37$ K at $\omega = 2\pi$ rad/s and $\Delta T = 0.71$ K at $\omega = 200\pi$ rad/s into Equation 5.74 results in two equations that may be solved simultaneously to yield

$$C_2 = 5.35 \qquad k = 1.11 \text{ W/m} \cdot \text{K} \hspace{2cm} \lhd$$

The thermal diffusivity is $\alpha = 4.37 \times 10^{-7}$ m²/s, while the thermal penetration depths are estimated by $\delta_p \approx \sqrt{\alpha/\omega}$, resulting in

$$\delta_p = 260 \ \mu\text{m}; \delta_p = 26 \ \mu\text{m} \hspace{2cm} \lhd$$

at $\omega = 2\pi$ rad/s and $\omega = 200\pi$ rad/s, respectively.

Comments:

1. The foregoing experimental technique is widely used, and is referred to as the 3 ω *method*. The technique is insensitive to a thermal contact or thermal boundary resistance that may exist at the interface between the sensing strip and the underlying sample, since these effects cancel when measurements are made at different excitation frequencies [10].
2. The thermal penetration depths are less than the sample thickness. Therefore, treating the sample as a semi-infinite solid is valid. Thinner samples could be used if higher heating frequencies were employed.

5.10 *Finite-Difference Methods*

Analytical solutions to transient problems are restricted to simple geometries and boundary conditions, such as the one-dimensional cases considered in the preceding sections. For some simple two- and three-dimensional geometries, analytical solutions are still possible. However, in many cases the geometry and/or boundary conditions preclude the use of analytical techniques, and recourse must be made to *finite-difference* (or *finite-element*) methods. Such methods, introduced in Section 4.4 for steady-state conditions, are readily extended to transient problems. In this section we consider *explicit* and *implicit* forms of finite-difference solutions to transient conduction problems.

5.10.1 Discretization of the Heat Equation: The Explicit Method

Once again consider the two-dimensional system of Figure 4.4. Under transient conditions with constant properties and no internal generation, the appropriate form of the heat equation, Equation 2.21, is

$$\frac{1}{\alpha}\frac{\partial T}{\partial t} = \frac{\partial^2 T}{\partial x^2} + \frac{\partial^2 T}{\partial y^2} \tag{5.75}$$

To obtain the finite-difference form of this equation, we may use the *central-difference* approximations to the spatial derivatives prescribed by Equations 4.27 and 4.28. Once again the *m* and *n* subscripts may be used to designate the *x*- and *y*-locations of *discrete nodal points*. However, in addition to being discretized in space, the problem must be discretized in time. The integer *p* is introduced for this purpose, where

$$t = p\Delta t \tag{5.76}$$

and the finite-difference approximation to the time derivative in Equation 5.75 is expressed as

$$\left.\frac{\partial T}{\partial t}\right|_{m,n} \approx \frac{T_{m,n}^{p+1} - T_{m,n}^{p}}{\Delta t} \tag{5.77}$$

The superscript *p* is used to denote the time dependence of *T*, and the time derivative is expressed in terms of the difference in temperatures associated with the *new* (*p* + 1) and *previous* (*p*) times. Hence calculations must be performed at successive times separated by the interval Δ*t*, and just as a finite-difference solution restricts temperature determination to discrete points in space, it also restricts it to discrete points in time.

If Equation 5.77 is substituted into Equation 5.75, the nature of the finite-difference solution will depend on the specific time at which temperatures are evaluated in the finite-difference approximations to the spatial derivatives. In the *explicit method* of solution, these temperatures are evaluated at the *previous* (*p*) time. Hence Equation 5.77 is considered to be a *forward-difference* approximation to the time derivative. Evaluating terms on

---

Analytical solutions for some simple two- and three-dimensional geometries are found in Section 5S.2.

the right-hand side of Equations 4.27 and 4.28 at p and substituting into Equation 5.75, the explicit form of the finite-difference equation for the interior node (m, n) is

$$\frac{1}{\alpha} \frac{T_{m,n}^{p+1} - T_{m,n}^{p}}{\Delta t} = \frac{T_{m+1,n}^{p} + T_{m-1,n}^{p} - 2T_{m,n}^{p}}{(\Delta x)^2} + \frac{T_{m,n+1}^{p} + T_{m,n-1}^{p} - 2T_{m,n}^{p}}{(\Delta y)^2} \qquad (5.78)$$

Solving for the nodal temperature at the new $(p + 1)$ time and assuming that $\Delta x = \Delta y$, it follows that

$$T_{m,n}^{p+1} = Fo(T_{m+1,n}^{p} + T_{m-1,n}^{p} + T_{m,n+1}^{p} + T_{m,n-1}^{p}) + (1 - 4Fo)T_{m,n}^{p} \qquad (5.79)$$

where Fo is a finite-difference form of the Fourier number

$$Fo = \frac{\alpha \Delta t}{(\Delta x)^2} \qquad (5.80)$$

This approach can easily be extended to one- or three-dimensional systems. If the system is one-dimensional in x, the explicit form of the finite-difference equation for an interior node m reduces to

$$T_m^{p+1} = Fo(T_{m+1}^{p} + T_{m-1}^{p}) + (1 - 2Fo)T_m^{p} \qquad (5.81)$$

Equations 5.79 and 5.81 are *explicit* because *unknown* nodal temperatures for the new time are determined exclusively by *known* nodal temperatures at the previous time. Hence calculation of the unknown temperatures is straightforward. Since the temperature of each interior node is known at $t = 0$ $(p = 0)$ from prescribed initial conditions, the calculations begin at $t = \Delta t$ $(p = 1)$, where Equation 5.79 or 5.81 is applied to each interior node to determine its temperature. With temperatures known for $t = \Delta t$, the appropriate finite-difference equation is then applied at each node to determine its temperature at $t = 2 \Delta t$ $(p = 2)$. In this way, the transient temperature distribution is obtained by *marching out in time,* using intervals of Δt.

The accuracy of the finite-difference solution may be improved by decreasing the values of Δx and Δt. Of course, the number of interior nodal points that must be considered increases with decreasing Δx, and the number of time intervals required to carry the solution to a prescribed final time increases with decreasing Δt. Hence the computation time increases with decreasing Δx and Δt. The choice of Δx is typically based on a compromise between accuracy and computational requirements. Once this selection has been made, however, the value of Δt may not be chosen independently. It is, instead, determined by *stability* requirements.

An undesirable feature of the explicit method is that it is not *unconditionally stable*. In a transient problem, the solution for the nodal temperatures should continuously approach final (steady-state) values with increasing time. However, with the explicit method, the solution may be characterized by numerically induced oscillations, which are physically impossible. The oscillations may become unstable, causing the numerical predictions to diverge from the actual solution. To prevent divergence, the prescribed value of Δt must be maintained below a certain limit, which depends on Δx and other parameters of the system. This dependence is termed a *stability criterion,* which may be obtained mathematically or demonstrated from a thermodynamic argument (see Problem 5.86). For the problems of interest in this text, *the criterion is determined by requiring that the* *coefficient associated with the combined terms involving $T_{m,n}^p$ is greater than or equal to zero.* With Equations 5.79 and 5.81 already expressed in the desired form, it follows that the stability criterion for a one-dimensional interior node is $(1 - 2Fo) \geq 0$, or

$$Fo \leq \frac{1}{2} \tag{5.82}$$

and for a two-dimensional node, it is $(1 - 4Fo) \geq 0$, or

$$Fo \leq \frac{1}{4} \tag{5.83}$$

For prescribed values of Δx and α, these criteria may be used to determine upper limits to the value of Δt.

Equations 5.79 and 5.81 may also be derived by applying the energy balance method of Section 4.4.3 to a control volume about the interior node. Accounting for changes in thermal energy storage, a general form of the energy balance equation may be expressed as

$$\dot{E}_{\text{in}} + \dot{E}_g = \dot{E}_{\text{st}} \tag{5.84}$$

As in Section 4.4.3, we will assume that all heat flow is *into* the node in the following derivations of the finite difference equations.

To illustrate application of Equation 5.84, consider the surface node of the one-dimensional system shown in Figure 5.12. With equal spacing between nodes, the thickness associated with the surface node is one-half that of the interior nodes. Assuming convection transfer from an adjoining fluid and no generation, it follows from Equation 5.84 that

$$hA(T_\infty - T_0^p) + \frac{kA}{\Delta x}(T_1^p - T_0^p) = \rho cA \frac{\Delta x}{2} \frac{T_0^{p+1} - T_0^p}{\Delta t}$$

or, solving for the surface temperature at $t + \Delta t$,

$$T_0^{p+1} = \frac{2h\,\Delta t}{\rho c\,\Delta x}(T_\infty - T_0^p) + \frac{2\alpha\,\Delta t}{\Delta x^2}(T_1^p - T_0^p) + T_0^p$$

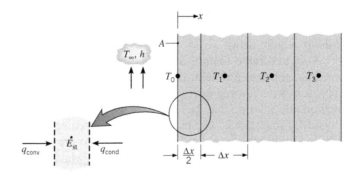

FIGURE 5.12 Surface node with convection and one-dimensional transient conduction.

Recognizing that $(2h\Delta t/\rho c\Delta x) = 2(h\Delta x/k)(\alpha\Delta t/\Delta x^2) = 2\,Bi\,Fo$ and grouping terms involving T_0^p, it follows that

$$T_0^{p+1} = 2Fo(T_1^p + Bi\,T_\infty) + (1 - 2Fo - 2Bi\,Fo)T_0^p \tag{5.85}$$

The finite-difference form of the Biot number is

$$Bi = \frac{h\,\Delta x}{k} \tag{5.86}$$

Recalling the procedure for determining the stability criterion, we require that the coefficient for T_0^p be greater than or equal to zero. Hence

$$1 - 2Fo - 2Bi\,Fo \geq 0$$

or

$$Fo(1 + Bi) \leq \frac{1}{2} \tag{5.87}$$

Since the complete finite-difference solution requires the use of Equation 5.81 for the interior nodes, as well as Equation 5.85 for the surface node, Equation 5.87 must be contrasted with Equation 5.82 to determine which requirement is more stringent. Since $Bi \geq 0$, it is apparent that the limiting value of Fo for Equation 5.87 is less than that for Equation 5.82. To ensure stability for all nodes, Equation 5.87 should therefore be used to select the maximum allowable value of Fo, and hence Δt, to be used in the calculations.

Forms of the explicit finite-difference equation for several common geometries are presented in Table 5.3*a*. Each equation may be derived by applying the energy balance method to a control volume about the corresponding node. To develop confidence in your ability to apply this method, you should attempt to verify at least one of these equations.

EXAMPLE 5.11

A fuel element of a nuclear reactor is in the shape of a plane wall of thickness $2L = 20$ mm and is convectively cooled at both surfaces, with $h = 1100$ W/m² · K and $T_\infty = 250°C$. At normal operating power, heat is generated uniformly within the element at a volumetric rate of $\dot{q}_1 = 10^7$ W/m³. A departure from the steady-state conditions associated with normal operation will occur if there is a change in the generation rate. Consider a sudden change to $\dot{q}_2 = 2 \times 10^7$ W/m³, and use the explicit finite-difference method to determine the fuel element temperature distribution after 1.5 s. The fuel element thermal properties are $k = 30$ W/m · K and $\alpha = 5 \times 10^{-6}$ m²/s.

SOLUTION

Known: Conditions associated with heat generation in a rectangular fuel element with surface cooling.

Find: Temperature distribution 1.5 s after a change in operating power.

TABLE 5.3 Transient, two-dimensional finite-difference equations ($\Delta x = \Delta y$)

Configuration	(a) Explicit Method — Finite-Difference Equation	(a) Explicit Method — Stability Criterion	(b) Implicit Method
1. Interior node	$T_{m,n}^{p+1} = Fo(T_{m+1,n}^p + T_{m-1,n}^p + T_{m,n+1}^p + T_{m,n-1}^p) + (1 - 4Fo)T_{m,n}^p$ (5.79)	$Fo \leq \frac{1}{4}$ (5.83)	$(1 + 4Fo)T_{m,n}^{p+1} - Fo(T_{m+1,n}^{p+1} + T_{m-1,n}^{p+1} + T_{m,n+1}^{p+1} + T_{m,n-1}^{p+1}) = T_{m,n}^p$ (5.95)
2. Node at interior corner with convection	$T_{m,n}^{p+1} = \frac{2}{3}Fo(T_{m+1,n}^p + 2T_{m-1,n}^p + 2T_{m,n+1}^p + T_{m,n-1}^p + 2Bi\,T_\infty^p) + (1 - 4Fo - \frac{4}{3}Bi\,Fo)T_{m,n}^p$ (5.88)	$Fo(3 + Bi) \leq \frac{3}{4}$ (5.89)	$(1 + 4Fo(1 + \frac{1}{3}Bi))T_{m,n}^{p+1} - \frac{2}{3}Fo \cdot (T_{m+1,n}^{p+1} + 2T_{m-1,n}^{p+1} + 2T_{m,n+1}^{p+1} + T_{m,n-1}^{p+1}) = T_{m,n}^p + \frac{4}{3}Bi\,Fo\,T_\infty$ (5.98)
3. Node at plane surface with convection[a]	$T_{m,n}^{p+1} = Fo(2T_{m-1,n}^p + T_{m,n+1}^p + T_{m,n-1}^p + 2Bi\,T_\infty^p) + (1 - 4Fo - 2Bi\,Fo)T_{m,n}^p$ (5.90)	$Fo(2 + Bi) \leq \frac{1}{2}$ (5.91)	$(1 + 2Fo(2 + Bi))T_{m,n}^{p+1} - Fo(2T_{m-1,n}^{p+1} + T_{m,n+1}^{p+1} + T_{m,n-1}^{p+1}) = T_{m,n}^p + 2Bi\,Fo\,T_\infty$ (5.99)
4. Node at exterior corner with convection	$T_{m,n}^{p+1} = 2Fo(T_{m-1,n}^p + T_{m,n-1}^p + 2Bi\,T_\infty^p) + (1 - 4Fo - 4Bi\,Fo)T_{m,n}^p$ (5.92)	$Fo(1 + Bi) \leq \frac{1}{4}$ (5.93)	$(1 + 4Fo(1 + Bi))T_{m,n}^{p+1} - 2Fo(T_{m-1,n}^{p+1} + T_{m,n-1}^{p+1}) = T_{m,n}^p + 4Bi\,Fo\,T_\infty$ (5.100)

[a]To obtain the finite-difference equation and/or stability criterion for an adiabatic surface (or surface of symmetry), simply set Bi equal to zero.

Schematic:

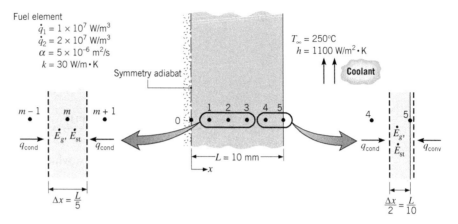

Assumptions:

1. One-dimensional conduction in x.
2. Uniform generation.
3. Constant properties.

Analysis: A numerical solution will be obtained using a space increment of $\Delta x = 2$ mm. Since there is symmetry about the midplane, the nodal network yields six unknown nodal temperatures. Using the energy balance method, Equation 5.84, an explicit finite-difference equation may be derived for any interior node m.

$$kA\frac{T_{m-1}^{p} - T_{m}^{p}}{\Delta x} + kA\frac{T_{m+1}^{p} - T_{m}^{p}}{\Delta x} + \dot{q}A\,\Delta x = \rho A\,\Delta x c\frac{T_{m}^{p+1} - T_{m}^{p}}{\Delta t}$$

Solving for T_{m}^{p+1} and rearranging,

$$T_{m}^{p+1} = Fo\left[T_{m-1}^{p} + T_{m+1}^{p} + \frac{\dot{q}(\Delta x)^{2}}{k}\right] + (1 - 2Fo)T_{m}^{p} \qquad (1)$$

Accounting for the symmetry about $x = 0$, this equation may also be used for the $m = 0$ node, with $T_{-1}^{p} = T_{1}^{p}$. Applying energy conservation to a control volume about node 5,

$$hA(T_{\infty} - T_{5}^{p}) + kA\frac{T_{4}^{p} - T_{5}^{p}}{\Delta x} + \dot{q}A\frac{\Delta x}{2} = \rho A\frac{\Delta x}{2}c\frac{T_{5}^{p+1} - T_{5}^{p}}{\Delta t}$$

or

$$T_{5}^{p+1} = 2Fo\left[T_{4}^{p} + Bi\,T_{\infty} + \frac{\dot{q}(\Delta x)^{2}}{2k}\right] + (1 - 2Fo - 2Bi\,Fo)T_{5}^{p} \qquad (2)$$

Since the most restrictive stability criterion is associated with Equation 2, we select Fo from the requirement that

$$Fo(1 + Bi) \leq \frac{1}{2}$$

Hence, with

$$Bi = \frac{h\,\Delta x}{k} = \frac{1100 \text{ W/m}^2 \cdot \text{K}(0.002 \text{ m})}{30 \text{ W/m} \cdot \text{K}} = 0.0733$$

it follows that

$$Fo \leq 0.466$$

or

$$\Delta t = \frac{Fo(\Delta x^2)}{\alpha} \leq \frac{0.466(2 \times 10^{-3} \text{ m})^2}{5 \times 10^{-6} \text{ m}^2/\text{s}} = 0.373 \text{ s}$$

To be well within the stability limit, we select $\Delta t = 0.3$ s, which corresponds to

$$Fo = \frac{5 \times 10^{-6} \text{ m}^2/\text{s}(0.3 \text{ s})}{(2 \times 10^{-3} \text{ m})^2} = 0.375$$

Substituting numerical values, including $\dot{q} = \dot{q}_2 = 2 \times 10^7$ W/m^3, the finite-difference equations become

$$T_0^{p+1} = 0.375(2T_1^p + 2.67) + 0.250T_0^p$$

$$T_m^{p+1} = 0.375(T_{m-1}^p + T_{m+1}^p + 2.67) + 0.250T_m^p \quad m = 1,2,3,4$$

$$T_5^{p+1} = 0.750(T_4^p + 19.67) + 0.195T_5^p$$

The initial temperature distribution must be known. From Equation 3.51, with $\dot{q} = \dot{q}_1$,

$$T_5 = T_s = T_\infty + \frac{\dot{q}_1 L}{h} = 250°\text{C} + \frac{10^7 \text{ W/m}^3 \times 0.01 \text{ m}}{1100 \text{ W/m}^2 \cdot \text{K}} = 340.91°\text{C}$$

and from Equation 3.47

$$T(x) = 16.67\left(1 - \frac{x^2}{L^2}\right) + 340.91°\text{C}$$

Initial temperatures for the nodal points are shown in the first row of the accompanying table.

Using the finite-difference equations, the nodal temperatures may be sequentially calculated with a time increment of 0.3 s until the desired final time is reached. The results are illustrated in rows 2 through 6 of the table and may be contrasted with the new steady-state condition (row 7), which was obtained by using Equations 3.47 and 3.51 with $\dot{q} = \dot{q}_2$.

Tabulated Nodal Temperatures (°C)

p	$t(s)$	T_0	T_1	T_2	T_3	T_4	T_5
0	0	357.58	356.91	354.91	351.58	346.91	340.91
1	0.3	358.08	357.41	355.41	352.08	347.41	341.41
2	0.6	358.58	357.91	355.91	352.58	347.91	341.88
3	0.9	359.08	358.41	356.41	353.08	348.41	342.35
4	1.2	359.58	358.91	356.91	353.58	348.89	342.82
5	1.5	360.08	359.41	357.41	354.07	349.37	343.27
∞	∞	465.15	463.82	459.82	453.15	443.82	431.82

Comments:

1. It is evident that, at 1.5 s, the wall is in the early stages of the transient process and that many additional calculations would have to be made to reach steady-state conditions with the finite-difference solution. The computation time could be reduced slightly by using the maximum allowable time increment ($\Delta t = 0.373$ s), but with some loss of accuracy. In the interest of maximizing accuracy, the time interval should be reduced until the computed results become independent of further reductions in Δt.

   Extending the finite-difference solutions to $t = 400$ s, the temperature histories computed for the midplane (0) and surface (5) nodes are shown in the figure below.

   It is evident that the new equilibrium condition is reached within approximately 250 s of the step change in operating power.

2. This problem can be solved using *Tools, Finite-Difference Equations, One-Dimensional, Transient* in the *Advanced* section of *IHT*.

5.10.2 Discretization of the Heat Equation: The Implicit Method

In the *explicit* finite-difference scheme, the temperature of any node at $t + \Delta t$ may be calculated from knowledge of temperatures at the same and neighboring nodes for the *preceding time t*. Hence determination of a nodal temperature at some time is *independent* of temperatures at other nodes for the *same time*. Although the method offers computational convenience, it suffers from limitations on the selection of Δt. For a given space increment, the time interval must be compatible with stability requirements. Frequently, this dictates the use of extremely small values of Δt, and a very large number of time intervals may be necessary to obtain a solution.

A reduction in computation time may often be realized by employing an *implicit*, rather than explicit, finite-difference scheme. The implicit form of a finite-difference equation may be derived by again using Equation 5.77 to approximate the time derivative, but evaluating all other temperatures at the *new* ($p + 1$) time, instead of the previous (p) time. Equation 5.77 is then considered to provide a *backward-difference* approximation to the time derivative. In contrast to Equation 5.78, the implicit form of the finite-difference equation for the interior node of a two-dimensional system is then

$$\frac{1}{\alpha} \frac{T_{m,n}^{p+1} - T_{m,n}^{p}}{\Delta t} = \frac{T_{m+1,n}^{p+1} + T_{m-1,n}^{p+1} - 2T_{m,n}^{p+1}}{(\Delta x)^2}$$

$$+ \frac{T_{m,n+1}^{p+1} + T_{m,n-1}^{p+1} - 2T_{m,n}^{p+1}}{(\Delta y)^2} \tag{5.94}$$

Rearranging and assuming $\Delta x = \Delta y$, it follows that

$$(1 + 4Fo)T_{m,n}^{p+1} - Fo(T_{m+1,n}^{p+1} + T_{m-1,n}^{p+1} + T_{m,n+1}^{p+1} + T_{m,n-1}^{p+1}) = T_{m,n}^{p} \qquad (5.95)$$

From Equation 5.95 it is evident that the *new* temperature of the (m, n) node depends on the *new* temperatures of its adjoining nodes, which are, in general, unknown. Hence, to determine the unknown nodal temperatures at time $p + 1$, all nodal equations must be *solved simultaneously*. Such a solution may be effected by using matrix inversion or Gauss–Seidel iteration as discussed in Section 4.5 and Appendix D. The *marching solution* would then involve simultaneously solving the nodal equations at each time $t = \Delta t, 2\Delta t, \ldots$, until the desired final time was reached.

Relative to the explicit method, the implicit formulation has the important advantage of being *unconditionally stable*. That is, all of the coefficients of $T_{m,n}^{p+1}$ are positive, and the solution remains stable for all space and time intervals, in which case there are no restrictions on Δx and Δt. Since larger values of Δt may therefore be used with an implicit method, computation times may often be reduced, with little loss of accuracy. Nevertheless, to maximize accuracy, Δt should be sufficiently small to ensure that the results are independent of further reductions in its value.

The implicit form of a finite-difference equation may also be derived from the energy balance method. For the surface node of Figure 5.12, it is readily shown that

$$(1 + 2Fo + 2Fo\,Bi)T_0^{p+1} - 2FoT_1^{p+1} = 2Fo\,Bi\,T_\infty + T_0^{p} \qquad (5.96)$$

For any interior node of Figure 5.12, it may also be shown that

$$(1 + 2Fo)T_m^{p+1} - Fo(T_{m-1}^{p+1} + T_{m+1}^{p+1}) = T_m^{p} \qquad (5.97)$$

Forms of the implicit finite-difference equation for other common geometries are presented in Table 5.3*b*. Each equation may be derived by applying the energy balance method.

EXAMPLE 5.12

A thick slab of copper initially at a uniform temperature of 20°C is suddenly exposed to radiation at one surface such that the net heat flux is maintained at a constant value of 3×10^5 W/m². Using the explicit and implicit finite-difference techniques with a space increment of $\Delta x = 75$ mm, determine the temperature at the irradiated surface and at an interior point that is 150 mm from the surface after 2 min have elapsed. Compare the results with those obtained from an appropriate analytical solution.

SOLUTION

Known: Thick slab of copper, initially at a uniform temperature, is subjected to a constant net heat flux at one surface.

Find:

1. Using the explicit finite-difference method, determine temperatures at the surface and 150 mm from the surface after an elapsed time of 2 min.

2. Repeat the calculations using the implicit finite-difference method.

3. Determine the same temperatures analytically.

Schematic:

Assumptions:

1. One-dimensional conduction in x.

2. Constant properties.

Properties: Table A.1, copper (300 K): $k = 401$ W/m · K, $\alpha = 117 \times 10^{-6}$ m^2/s.

Analysis:

1. An explicit form of the finite-difference equation for the surface node may be obtained by applying an energy balance to a control volume about the node.

$$q_o'' A + kA \frac{T_1^p - T_0^p}{\Delta x} = \rho A \frac{\Delta x}{2} c \frac{T_0^{p+1} - T_0^p}{\Delta t}$$

or

$$T_0^{p+1} = 2Fo\left(\frac{q_0'' \Delta x}{k} + T_1^p\right) + (1 - 2Fo)T_0^p$$

The finite-difference equation for any interior node is given by Equation 5.81. Both the surface and interior nodes are governed by the stability criterion

$$Fo \le \frac{1}{2}$$

Noting that the finite-difference equations are simplified by choosing the maximum allowable value of Fo, we select $Fo = \frac{1}{2}$. Hence

$$\Delta t = Fo\frac{(\Delta x)^2}{\alpha} = \frac{1}{2}\frac{(0.075 \text{ m})^2}{117 \times 10^{-6} \text{ m}^2/\text{s}} = 24 \text{ s}$$

With

$$\frac{q_o'' \Delta x}{k} = \frac{3 \times 10^5 \text{ W/m}^2 \ (0.075 \text{ m})}{401 \text{ W/m} \cdot \text{K}} = 56.1°\text{C}$$

the finite-difference equations become

$$T_0^{p+1} = 56.1°\text{C} + T_1^p \qquad \text{and} \qquad T_m^{p+1} = \frac{T_{m+1}^p + T_{m-1}^p}{2}$$

for the surface and interior nodes, respectively. Performing the calculations, the results are tabulated as follows:

Explicit Finite-Difference Solution for $Fo = \frac{1}{2}$

p	t(s)	T_0	T_1	T_2	T_3	T_4
0	0	20	20	20	20	20
1	24	76.1	20	20	20	20
2	48	76.1	48.1	20	20	20
3	72	104.2	48.1	34.0	20	20
4	96	104.2	69.1	34.0	27.0	20
5	120	125.2	69.1	48.1	27.0	23.5

After 2 min, the surface temperature and the desired interior temperature are $T_0 = 125.2°C$ and $T_2 = 48.1°C$.

It can be seen from the table that, with each successive time step, an additional nodal temperature changes from its initial value. For this reason, it is not necessary to formally implement the second boundary condition $T(x \rightarrow \infty) = T_i$. Also note that calculation of identical temperatures at successive times for the same node is an idiosyncrasy corresponding to use of the maximum allowable value of Fo with the explicit technique. The actual response is, of course, one in which the temperature changes continuously with time.

To determine the extent to which the accuracy may be improved by reducing Fo, let us redo the calculations for $Fo = \frac{1}{4}$ ($\Delta t = 12$ s). The finite-difference equations are then of the form

$$T_0^{p+1} = \frac{1}{2}(56.1°C + T_1^p) + \frac{1}{2}T_0^p \quad \text{and} \quad T_m^{p+1} = \frac{1}{4}(T_{m+1}^p + T_{m-1}^p) + \frac{1}{2}T_m^p$$

and the results of the calculations are tabulated as follows:

Explicit Finite-Difference Solution for $Fo = \frac{1}{4}$

p	t(s)	T_0	T_1	T_2	T_3	T_4	T_5	T_6	T_7	T_8
0	0	20	20	20	20	20	20	20	20	20
1	12	48.1	20	20	20	20	20	20	20	20
2	24	62.1	27.0	20	20	20	20	20	20	20
3	36	72.6	34.0	21.8	20	20	20	20	20	20
4	48	81.4	40.6	24.4	20.4	20	20	20	20	20
5	60	89.0	46.7	27.5	21.3	20.1	20	20	20	20
6	72	95.9	52.5	30.7	22.5	20.4	20.0	20	20	20
7	84	102.3	57.9	34.1	24.1	20.8	20.1	20.0	20	20
8	96	108.1	63.1	37.6	25.8	21.5	20.3	20.0	20.0	20
9	108	113.6	67.9	41.0	27.6	22.2	20.5	20.1	20.0	20.0
10	120	118.8	72.6	44.4	29.6	23.2	20.8	20.2	20.0	20.0

After 2 min, the temperatures of interest are $T_0 = 118.8°C$ and $T_2 = 44.4°C$ for $Fo = \frac{1}{4}$. It is clear that by reducing Fo we have eliminated the recurring temperature idiosyncracy. An assessment of the improvement in accuracy will be given later, by comparison with an exact solution. In any case, the value of Fo should be successively reduced until the results become essentially independent of Fo.

2. Performing an energy balance on a control volume about the surface node, the implicit form of the finite-difference equation is

$$q_o'' + k\frac{T_1^{p+1} - T_0^{p+1}}{\Delta x} = \rho\frac{\Delta x}{2}c\frac{T_0^{p+1} - T_0^p}{\Delta t}$$

or

$$(1 + 2Fo)T_0^{p+1} - 2FoT_1^{p+1} = \frac{2\alpha q_o'' \Delta t}{k\,\Delta x} + T_0^p$$

Arbitrarily choosing $Fo = \frac{1}{2}$ ($\Delta t = 24$ s), it follows that

$$2T_0^{p+1} - T_1^{p+1} = 56.1 + T_0^p$$

From Equation 5.97, the finite-difference equation for any interior node is then of the form

$$-T_{m-1}^{p+1} + 4T_m^{p+1} - T_{m+1}^{p+1} = 2T_m^p$$

In contrast to the explicit method, the implicit method requires the simultaneous solution of the finite-difference equations for all nodes at time $p + 1$. Hence, the number of nodes under consideration must be specified, and a boundary condition must be applied at the last node. From the results of the explicit method, it is probable that we are safe in choosing nine nodes corresponding to $T_0, T_1, \ldots, T_8$. We are thereby assuming that, at $t = 120$ s, there has been no change in T_9, and the boundary condition is implemented numerically as $T_9 = 20°C$.

We now have a set of nine equations that must be solved simultaneously for each time increment. We can express the equations in the form $[A][T] = [C]$, where

$$[A] = \begin{bmatrix} 2 & -1 & 0 & 0 & 0 & 0 & 0 & 0 & 0 \\ -1 & 4 & -1 & 0 & 0 & 0 & 0 & 0 & 0 \\ 0 & -1 & 4 & -1 & 0 & 0 & 0 & 0 & 0 \\ 0 & 0 & -1 & 4 & -1 & 0 & 0 & 0 & 0 \\ 0 & 0 & 0 & -1 & 4 & -1 & 0 & 0 & 0 \\ 0 & 0 & 0 & 0 & -1 & 4 & -1 & 0 & 0 \\ 0 & 0 & 0 & 0 & 0 & -1 & 4 & -1 & 0 \\ 0 & 0 & 0 & 0 & 0 & 0 & -1 & 4 & -1 \\ 0 & 0 & 0 & 0 & 0 & 0 & 0 & -1 & 4 \end{bmatrix}$$

$$[C] = \begin{bmatrix} 56.1 + T_0^p \\ 2T_1^p \\ 2T_2^p \\ 2T_3^p \\ 2T_4^p \\ 2T_5^p \\ 2T_6^p \\ 2T_7^p \\ 2T_8^p + T_9^{p+1} \end{bmatrix}$$

Note that numerical values for the components of [C] are determined from previous values of the nodal temperatures. Note also how the finite-difference equation for node 8 appears in matrices [A] and [C], with $T_9^{p+1} = 20°C$, as indicated previously.

A table of nodal temperatures may be compiled, beginning with the first row ($p = 0$) corresponding to the prescribed initial condition. To obtain nodal temperatures for subsequent times, the matrix equation must be solved. At each time step $p + 1$, [C] is updated using the previous time step (p) values. The process is carried out five times to determine the nodal temperatures at 120 s. The desired temperatures are $T_0 = 114.7°C$ and $T_2 = 44.2°C$.

Implicit Finite-Difference Solution for $Fo = \frac{1}{2}$

p	$t(s)$	T_0	T_1	T_2	T_3	T_4	T_5	T_6	T_7	T_8
0	0	20.0	20.0	20.0	20.0	20.0	20.0	20.0	20.0	20.0
1	24	52.4	28.7	22.3	20.6	20.2	20.0	20.0	20.0	20.0
2	48	74.0	39.5	26.6	22.1	20.7	20.2	20.1	20.0	20.0
3	72	90.2	50.3	32.0	24.4	21.6	20.6	20.2	20.1	20.0
4	96	103.4	60.5	38.0	27.4	22.9	21.1	20.4	20.2	20.1
5	120	114.7	70.0	44.2	30.9	24.7	21.9	20.8	20.3	20.1

3. Approximating the slab as a semi-infinite medium, the appropriate analytical expression is given by Equation 5.62, which may be applied to any point in the slab.

$$T(x, t) - T_i = \frac{2q_o''(\alpha t/\pi)^{1/2}}{k}\exp\left(-\frac{x^2}{4\alpha t}\right) - \frac{q_o''x}{k}\text{erfc}\left(\frac{x}{2\sqrt{\alpha t}}\right)$$

At the surface, this expression yields

$$T(0, 120\text{ s}) - 20°C = \frac{2 \times 3 \times 10^5 \text{ W/m}^2}{401 \text{ W/m} \cdot \text{K}}(117 \times 10^{-6} \text{ m}^2/\text{s} \times 120 \text{ s}/\pi)^{1/2}$$

or

$$T(0, 120\text{ s}) = 120.0°C \qquad \triangleleft$$

At the interior point ($x = 0.15$ m)

$$T(0.15\text{ m}, 120\text{ s}) - 20°C = \frac{2 \times 3 \times 10^5 \text{ W/m}^2}{401 \text{ W/m} \cdot \text{K}}$$

$$\times (117 \times 10^{-6} \text{ m}^2/\text{s} \times 120 \text{ s}/\pi)^{1/2}$$

$$\times \exp\left[-\frac{(0.15\text{ m})^2}{4 \times 117 \times 10^{-6} \text{ m}^2/\text{s} \times 120 \text{ s}}\right]$$

$$-\frac{3 \times 10^5 \text{ W/m}^2 \times 0.15 \text{ m}}{401 \text{ W/m} \cdot \text{K}}$$

$$\times \left[1 - \text{erf}\left(\frac{0.15 \text{ m}}{2\sqrt{117 \times 10^{-6} \text{ m}^2/\text{s} \times 120 \text{ s}}}\right)\right]$$

$$T(0.15\text{ m}, 120\text{ s}) = 45.4°C \qquad \triangleleft$$

Comments:

1. The accuracy of the foregoing calculations is adversely affected by the coarse grid ($\Delta x = 75$ mm) and large time steps ($\Delta t = 24$ s, 12 s). Results of the implicit method with $\Delta x = 18.75$ mm and $\Delta t = 6$ s ($Fo = 2.0$) are compared to the earlier results in the table below and are seen to be in good agreement with the exact solution.

Method	$T_0 = T(0, 120 \text{ s})$	$T_2 = T(0.15 \text{ m}, 120 \text{ s})$
Explicit ($Fo = \frac{1}{2}$)	125.2	48.1
Explicit ($Fo = \frac{1}{4}$)	118.8	44.4
Implicit ($Fo = \frac{1}{2}$)	114.7	44.2
Implicit ($Fo = 2$)	119.2	45.3
Exact	120.0	45.4

2. Temperature distributions within the copper slab at $t = 60$ and 120 s are plotted below.

Note that, at $t = 120$ s, the assumption of a semi-infinite medium would remain valid if the thickness of the slab exceeded approximately 500 mm.

3. A more general radiative heating condition would be one in which the surface is suddenly exposed to large surroundings at an elevated temperature T_{sur} (Problem 5.103). The net rate at which radiation is transferred to the surface may then be calculated from Equation 1.7. Allowing for convection heat transfer to the surface, application of conservation of energy to the surface node yields an explicit finite-difference equation of the form

$$\varepsilon\sigma[T_{\text{sur}}^4 - (T_0^p)^4] + h(T_\infty - T_0^p) + k\frac{T_1^p - T_0^p}{\Delta x} = \rho\frac{\Delta x}{2}c\frac{T_0^{p+1} - T_0^p}{\Delta t}$$

Use of this finite-difference equation in a numerical solution is complicated by the fact that it is *nonlinear*. However, the equation may be *linearized* by introducing the radiation heat transfer coefficient h_r defined by Equation 1.9, and the finite-difference equation is

$$h_r^p(T_{\text{sur}} - T_0^p) + h(T_\infty - T_0^p) + k\frac{T_1^p - T_0^p}{\Delta x} = \rho\frac{\Delta x}{2}c\frac{T_0^{p+1} - T_0^p}{\Delta t}$$

The solution may proceed in the usual manner, although the effect of a radiative Biot number ($Bi_r \equiv h_r\,\Delta x/k$) must be included in the stability criterion, and the value of h_r must be updated at each step in the calculations. If the implicit method is used, h_r is calculated at $p + 1$, in which case an iterative calculation must be made at each time step.

4. This problem can be solved using *Tools, Finite-Difference Equations, One-Dimensional, Transient* in the *Advanced* section of *IHT*. The Der(T,t) function of *IHT* can be used to represent the temperature-time derivative of Equation 5.75. See Problem 5.93.

5.11 *Summary*

Transient conduction occurs in numerous engineering applications and may be treated using different methods. There is certainly much to be said for simplicity, in which case, when confronted with a transient problem, the first thing you should do is calculate the Biot number. If this number is much less than unity, you may use the lumped capacitance method to obtain accurate results with minimal computational requirements. However, if the Biot number is not much less than unity, spatial effects must be considered, and an alternative method must be used. Analytical results and approximate solutions are available in convenient form for the plane wall, the infinite cylinder, the sphere, and the semi-infinite solid. You should know when and how to use these results. If geometrical complexities and/or the form of the boundary conditions preclude their use, recourse must be made to a numerical technique, such as the finite-difference method.

You may test your understanding of key concepts by addressing the following questions:

- Under what conditions may the *lumped capacitance method* be used to predict the transient response of a solid to a change in its thermal environment?
- What is the physical interpretation of the *Biot number*?
- Is the lumped capacitance method of analysis likely to be more applicable for a hot solid being cooled by forced convection in air or in water? By forced convection in air or natural convection in air?
- Is the lumped capacitance method of analysis likely to be more applicable for cooling of a hot solid made of copper or aluminum? For silicon nitride or glass?
- What parameters determine the *time constant* associated with the transient thermal response of a lumped capacitance solid? Is this response accelerated or decelerated by an increase in the convection coefficient? By an increase in the density or specific heat of the solid?
- For one-dimensional, transient conduction in a plane wall, a long cylinder, or a sphere with surface convection, what dimensionless parameters may be used to simplify the representation of thermal conditions? How are these parameters defined?
- Why is the semi-infinite solution applicable to any geometry at early times?
- What is the physical interpretation of the *Fourier number*?
- What requirement must be satisfied for use of a *one-term approximation* to determine the transient thermal response of a plane wall, a long cylinder, or a sphere experiencing one-dimensional conduction due to a change in surface conditions? At what stage of a transient process is the requirement not satisfied?
- What does transient heating or cooling of a plane wall with equivalent convection conditions at opposite surfaces have in common with a plane wall heated or cooled by convection at one surface and well insulated at the other surface?
- How may a one-term approximation be used to determine the transient thermal response of a plane wall, long cylinder, or sphere subjected to a sudden change in surface temperature?
- For one-dimensional, transient conduction, what is implied by the idealization of a *semi-infinite* solid? Under what conditions may the idealization be applied to a plane wall? A long cylinder? A sphere? An object of arbitrary geometry?
- What differentiates an *explicit*, finite-difference solution to a transient conduction problem from an *implicit* solution?
- What is meant by characterization of the implicit finite-difference method as *unconditionally stable*? What constraint is placed on the explicit method to ensure a stable solution?

References

1. Bergman, T. L., *J. Heat Transfer,* **130**, 094503, 2008.

2. Peleg, M., *Food Res. Int.,* **33**, 531–538, 2000.

3. Carslaw, H. S., and J. C. Jaeger, *Conduction of Heat in Solids,* 2nd ed., Oxford University Press, London, 1986.

4. Schneider, P. J., *Conduction Heat Transfer,* Addison-Wesley, Reading, MA, 1957.

5. Kakac, S., and Y. Yener, *Heat Conduction,* Taylor & Francis, Washington, DC, 1993.

6. Poulikakos, D., *Conduction Heat Transfer,* Prentice-Hall, Englewood Cliffs, NJ, 1994.

7. Yovanovich, M. M., "Conduction and Thermal Contact Resistances (Conductances)," in W. M. Rohsenow, J. P. Hartnett, and Y. I. Cho, Eds., *Handbook of Heat Transfer,* McGraw-Hill, New York, 1998, pp. 3.1–3.73.

8. Lavine, A. S., and T. L. Bergman, *J. Heat Transfer,* **130**, 101302, 2008.

9. Hirsch, L. R., R. J. Stafford, J. A. Bankson, S. R. Sershen, B. Rivera, R. E. Price, J. D. Hazle, N. J. Halas, and J. L. West, *Proc. Nat. Acad. Sciences of the U.S.,* **100**, 13549–13554, 2003.

10. Cahill, D. G., *Rev. Sci. Instrum.,* **61**, 802–808, 1990.

Problems

Qualitative Considerations

5.1 Consider a thin electrical heater attached to a plate and backed by insulation. Initially, the heater and plate are at the temperature of the ambient air, T_∞. Suddenly, the power to the heater is activated, yielding a constant heat flux $q_o''(\text{W/m}^2)$ at the inner surface of the plate.

(a) Sketch and label, on $T - x$ coordinates, the temperature distributions: initial, steady-state, and at two intermediate times.

(b) Sketch the heat flux at the outer surface $q_x''(L, t)$ as a function of time.

5.2 The inner surface of a plane wall is insulated while the outer surface is exposed to an airstream at T_∞. The wall is at a uniform temperature corresponding to that of the airstream. Suddenly, a radiation heat source is switched on, applying a uniform flux q_o'' to the outer surface.

(a) Sketch and label, on $T - x$ coordinates, the temperature distributions: initial, steady-state, and at two intermediate times.

(b) Sketch the heat flux at the outer surface $q_x''(L, t)$ as a function of time.

5.3 A microwave oven operates on the principle that application of a high-frequency field causes electrically polarized molecules in food to oscillate. The net effect is a nearly *uniform generation* of thermal energy within the food. Consider the process of cooking a slab of beef of thickness $2L$ in a microwave oven and compare it with cooking in a conventional oven, where *each side* of the slab is *heated by radiation*. In each case the meat is to be heated from 0°C to a *minimum* temperature of 90°C. Base your comparison on a sketch of the temperature distribution at selected times for each of the cooking processes. In particular, consider the time t_0 at which heating is initiated, a time t_1 during the heating process, the time t_2 corresponding to the conclusion of heating, and a time t_3 well into the subsequent cooling process.

5.4 A thin stainless steel disk of thickness b and outer radius r_o has been heat treated to a high, uniform initial temperature of T_i. The disk is then placed upon a small stand and allowed to cool. The ambient air and surroundings are at T_∞ and T_{sur}, respectively. The convection coefficients on the top and bottom disk surfaces, h_t and h_b, are known, as is the plate emissivity, ε. Derive a differential equation that could be solved to determine the transient thermal response of the disk, $T(r, t)$. List the appropriate initial and boundary conditions.

Top view

Side view

Lumped Capacitance Method

5.5 For each of the following cases, determine an appropriate characteristic length L_c and the corresponding Biot number Bi that is associated with the transient thermal response of the solid object. State whether the lumped capacitance approximation is valid. If temperature information is not provided, evaluate properties at $T = 300$ K.

(a) A toroidal shape of diameter $D = 65$ mm and cross-sectional area $A_c = 7$ mm^2 is of thermal conductivity $k = 2.3$ W/m · K. The surface of the torus is exposed to a coolant corresponding to a convection coefficient of $h = 50$ W/m^2 · K.

(b) A long, hot AISI 302 stainless steel bar of rectangular cross section has dimensions $w = 5$ mm, $W = 7$ mm, and $L = 150$ mm. The bar is subjected to a coolant that provides a heat transfer coefficient of $h = 10$ W/m^2 · K at all exposed surfaces.

(c) A long extruded aluminum (Alloy 2024) tube of inner and outer dimensions $w = 25$ mm and $W = 30$ mm, respectively, is suddenly submerged in water, resulting in a convection coefficient of $h = 40$ W/m^2 · K at the four exterior tube surfaces. The tube is plugged at both ends, trapping stagnant air inside the tube.

(d) An $L = 300$-mm-long solid stainless steel rod of diameter $D = 13$ mm and mass $M = 0.328$ kg is exposed to a convection coefficient of $h = 30$ W/m^2 · K.

(e) A solid sphere of diameter $D = 12$ mm and thermal conductivity $k = 120$ W/m · K is suspended in a large vacuum oven with internal wall temperatures of $T_{sur} = 20°$C. The initial sphere temperature is $T_i = 100°$C, and its emissivity is $\varepsilon = 0.73$.

(f) A long cylindrical rod of diameter $D = 20$ mm, density $\rho = 2300$ kg/m^3, specific heat $c_p = 1750$ J/kg · K, and thermal conductivity $k = 16$ W/m · K is suddenly exposed to convective conditions with $T_\infty = 20°$C. The rod is initially at a uniform temperature of $T_i = 200°$C and reaches a spatially averaged temperature of $T = 100°$C at $t = 225$ s.

(g) Repeat part (f) but now consider a rod diameter of $D = 200$ mm.

Case (a)   Case (b)

Case (c)   Case (d)

Case (e)   Cases (f,g)

5.6 Steel balls 12 mm in diameter are annealed by heating to 1150 K and then slowly cooling to 400 K in an air environment for which $T_\infty = 325$ K and $h = 20$ W/m^2 · K. Assuming the properties of the steel to be $k = 40$ W/m · K, $\rho = 7800$ kg/m^3, and $c = 600$ J/kg · K, estimate the time required for the cooling process.

5.7 Consider the steel balls of Problem 5.6, except now the air temperature increases with time as $T_\infty(t) = 325$ K $+ at$, where $a = 0.1875$ K/s.

(a) Sketch the ball temperature versus time for $0 \le t \le 1$ h. Also show the ambient temperature, T_∞, in your graph. Explain special features of the ball temperature behavior.

(b) Find an expression for the ball temperature as a function of time $T(t)$, and plot the ball temperature for $0 \le t \le 1$ h. Was your sketch correct?

5.8 The heat transfer coefficient for air flowing over a sphere is to be determined by observing the temperature–time history of a sphere fabricated from pure copper. The sphere, which is 12.7 mm in diameter, is at 66°C before it is inserted into an airstream having a temperature of 27°C. A thermocouple on the outer surface of the sphere indicates 55°C 69 s after the sphere is inserted into the airstream. Assume and then justify that the sphere behaves as a spacewise isothermal object and calculate the heat transfer coefficient.

5.9 A solid steel sphere (AISI 1010), 300 mm in diameter, is coated with a dielectric material layer of thickness 2 mm and thermal conductivity 0.04 W/m · K. The coated sphere is initially at a uniform temperature of 500°C and is suddenly quenched in a large oil bath for which $T_\infty = 100°C$ and $h = 3300$ W/m² · K. Estimate the time required for the coated sphere temperature to reach 140°C. *Hint:* Neglect the effect of energy storage in the dielectric material, since its thermal capacitance (ρcV) is small compared to that of the steel sphere.

5.10 A flaked cereal is of thickness $2L = 1.4$ mm. The density, specific heat, and thermal conductivity of the flake are $\rho = 700$ kg/m³, $c_p = 2400$ J/kg · K, and $k = 0.34$ W/m · K, respectively. The product is to be baked by increasing its temperature from $T_i = 20°C$ to $T_f = 220°C$ in a convection oven, through which the product is carried on a conveyor. If the oven is $L_o = 3$ m long and the convection heat transfer coefficient at the product surface and oven air temperature are $h = 55$ W/m² · K and $T_\infty = 300°C$, respectively, determine the required conveyor velocity, V. An engineer suggests that if the flake thickness is reduced to $2L = 1.0$ mm the conveyor velocity can be increased, resulting in higher productivity. Determine the required conveyor velocity for the thinner flake.

5.11 The base plate of an iron has a thickness of $L = 7$ mm and is made from an aluminum alloy ($\rho = 2800$ kg/m³, $c = 900$ J/kg · K, $k = 180$ W/m · K, $\varepsilon = 0.80$). An electric resistance heater is attached to the inner surface of the plate, while the outer surface is exposed to ambient air and large surroundings at $T_\infty = T_{sur} = 25°C$. The areas of both the inner and outer surfaces are $A_s = 0.040$ m².

If an approximately uniform heat flux of $q''_h = 1.5 \times 10^4$ W/m² is applied to the inner surface of the base plate and the convection coefficient at the outer surface is $h = 10$ W/m² · K, estimate the time required for the plate to reach a temperature of 135°C. *Hint:* Numerical integration is suggested in order to solve the problem.

5.12 A tool used for fabricating semiconductor devices consists of a chuck (thick metallic, cylindrical disk) onto which a very thin silicon wafer ($\rho = 2700$ kg/m³, $c = 875$ J/kg · K, $k = 177$ W/m · K) is placed by a robotic arm. Once in position, an electric field in the chuck is energized, creating an electrostatic force that holds the wafer firmly to the chuck. To ensure a reproducible thermal contact resistance between the chuck and the wafer from cycle to cycle, pressurized helium gas is introduced at the center of the chuck and flows (very slowly) radially outward between the asperities of the interface region.

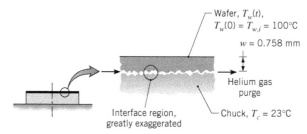

An experiment has been performed under conditions for which the wafer, initially at a uniform temperature $T_{w,i} = 100°C$, is suddenly placed on the chuck, which is at a uniform and constant temperature $T_c = 23°C$. With the wafer in place, the electrostatic force and the helium gas flow are applied. After 15 s, the temperature of the wafer is determined to be 33°C. What is the thermal contact resistance $R''_{t,c}$ (m² · K/W) between the wafer and chuck? Will the value of $R''_{t,c}$ increase, decrease, or remain the same if air, instead of helium, is used as the purge gas?

5.13 A copper sheet of thickness $2L = 2$ mm has an initial temperature of $T_i = 118°C$. It is suddenly quenched in liquid water, resulting in boiling at its two surfaces. For boiling, Newton's law of cooling is expressed as $q'' = h(T_s - T_{sat})$, where T_s is the solid surface temperature and T_{sat} is the saturation temperature of the fluid (in this case $T_{sat} = 100°C$). The convection heat transfer coefficient may be expressed as $h = 1010 \text{ W/m}^2 \cdot \text{K}^3 (T_s - T_{sat})^2$. Determine the time needed for the sheet to reach a temperature of $T = 102°C$. Plot the copper temperature versus time for $0 \le t \le 0.5$ s. On the same graph, plot the copper temperature history assuming the heat transfer coefficient is constant, evaluated at the average copper temperature $\bar{T} = 100°C$. Assume lumped capacitance behavior.

5.14 Carbon steel (AISI 1010) shafts of 80-mm diameter are heat treated in a gas-fired furnace whose gases are at 1200 K and provide a convection coefficient of $125 \text{ W/m}^2 \cdot \text{K}$. If the shafts enter the furnace at 300 K, how long must they remain in the furnace to achieve a centerline temperature of 800 K?

5.15 Small spherical particles of diameter $D = 50 \text{ } \mu m$ contain a fluorescent material that, when irradiated with white light, emits at a wavelength corresponding to the material's temperature. Hence the color of the particle varies with its temperature. Because the small particles are *neutrally buoyant* in liquid water, a researcher wishes to use them to measure instantaneous local water temperatures in a turbulent flow by observing their emitted color. If the particles are characterized by a density, specific heat, and thermal conductivity of $\rho = 999 \text{ kg/m}^3$, $k = 1.2 \text{ W/m} \cdot \text{K}$, and $c_p = 1200 \text{ J/kg} \cdot \text{K}$, respectively, determine the time constant of the particles. *Hint*: Since the particles travel with the flow, heat transfer between the particle and the fluid occurs by conduction. Assume lumped capacitance behavior.

5.16 An electronic device, such as a power transistor mounted on a finned heat sink, can be modeled as a spatially isothermal object with internal heat generation and an external convection resistance.

(a) Consider such a system of mass M, specific heat c, and surface area A_s, which is initially in equilibrium with the environment at T_∞. Suddenly, the electronic device is energized such that a constant heat generation $\dot{E}_g(\text{W})$ occurs. Show that the temperature response of the device is

$$\frac{\theta}{\theta_i} = \exp\left(-\frac{t}{RC}\right)$$

where $\theta \equiv T - T(\infty)$ and $T(\infty)$ is the steady-state temperature corresponding to $t \to \infty$; $\theta_i = T_i - T(\infty)$; T_i = initial temperature of device; R = thermal resistance $1/\bar{h}A_s$; and C = thermal capacitance Mc.

(b) An electronic device, which generates 60 W of heat, is mounted on an aluminum heat sink weighing 0.31 kg and reaches a temperature of 100°C in ambient air at 20°C under steady-state conditions. If the device is initially at 20°C, what temperature will it reach 5 min after the power is switched on?

5.17 A thermal energy storage unit consists of a large rectangular channel, which is well insulated on its outer surface and encloses alternating layers of the storage material and the flow passage.

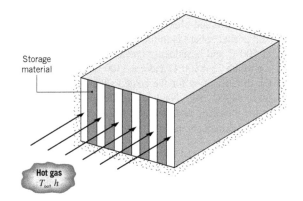

Each layer of the storage material is an aluminum slab of width $W = 0.05$ m, which is at an initial temperature of 25°C. Consider conditions for which the storage unit is charged by passing a hot gas through the passages, with the gas temperature and the convection coefficient assumed to have constant values of $T_\infty = 600°C$ and $h = 100 \text{ W/m}^2 \cdot \text{K}$ throughout the channel. How long will it take to achieve 75% of the maximum possible energy storage? What is the temperature of the aluminum at this time?

5.18 A spherical vessel used as a reactor for producing pharmaceuticals has a 5-mm-thick stainless steel wall ($k = 17 \text{ W/m} \cdot \text{K}$) and an inner diameter of $D_i = 1.0$ m. During production, the vessel is filled with reactants for which $\rho = 1100 \text{ kg/m}^3$ and $c = 2400 \text{ J/kg} \cdot \text{K}$, while exothermic reactions release energy at a volumetric rate of $\dot{q} = 10^4 \text{ W/m}^3$. As first approximations, the reactants may be assumed to be well stirred and the thermal capacitance of the vessel may be neglected.

(a) The exterior surface of the vessel is exposed to ambient air ($T_\infty = 25°C$) for which a convection coefficient of $h = 6 \text{ W/m}^2 \cdot \text{K}$ may be assumed. If the initial temperature of the reactants is 25°C, what is the temperature of the reactants after 5 h of process time? What is the corresponding temperature at the outer surface of the vessel?

(b) Explore the effect of varying the convection coefficient on transient thermal conditions within the reactor.

5.19 *Molecular electronics* is an emerging field associated with computing and data storage utilizing energy transfer at the molecular scale. At this scale, thermal energy is associated exclusively with the vibration of molecular chains. The primary resistance to energy transfer in these proposed devices is the contact resistance at metal-molecule interfaces. To measure the contact resistance, individual molecules are *self-assembled* in a regular pattern onto a very thin gold substrate. The substrate is suddenly heated by a short pulse of laser irradiation, simultaneously transferring thermal energy to the molecules. The molecules vibrate rapidly in their "hot" state, and their vibrational intensity can be measured by detecting the randomness of the electric field produced by the molecule tips, as indicated by the dashed, circular lines in the schematic.

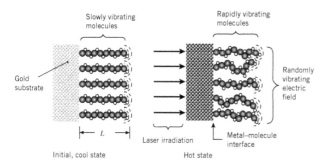

Molecules that are of density $\rho = 180$ kg/m^3 and specific heat $c_p = 3000$ J/kg·K have an initial, relaxed length of $L = 2$ nm. The intensity of the molecular vibration increases exponentially from an initial value of I_i to a steady-state value of $I_{ss} > I_i$ with the time constant associated with the exponential response being $\tau_I = 5$ ps. Assuming the intensity of the molecular vibration represents temperature on the molecular scale and that each molecule can be viewed as a cylinder of initial length L and cross-sectional area A_c, determine the thermal contact resistance, $R''_{t,c}$, at the metal-molecule interface.

5.20 A plane wall of a furnace is fabricated from plain carbon steel ($k = 60$ W/m·K, $\rho = 7850$ kg/m^3, $c = 430$ J/kg·K) and is of thickness $L = 10$ mm. To protect it from the corrosive effects of the furnace combustion gases, one surface of the wall is coated with a thin ceramic film that, for a unit surface area, has a thermal resistance of $R''_{t,f} = 0.01$ m^2·K/W. The opposite surface is well insulated from the surroundings.

At furnace start-up the wall is at an initial temperature of $T_i = 300$ K, and combustion gases at $T_\infty = 1300$ K enter the furnace, providing a convection coefficient of $h = 25$ W/m^2·K at the ceramic film. Assuming the film to have negligible thermal capacitance, how long will it take for the inner surface of the steel to achieve a temperature of $T_{s,i} = 1200$ K? What is the temperature $T_{s,o}$ of the exposed surface of the ceramic film at this time?

5.21 Plasma spray-coating processes are often used to provide surface protection for materials exposed to hostile environments, which induce degradation through factors such as wear, corrosion, or outright thermal failure. *Ceramic* coatings are commonly used for this purpose. By injecting ceramic powder through the nozzle (anode) of a plasma torch, the particles are entrained by the plasma jet, within which they are then accelerated and heated.

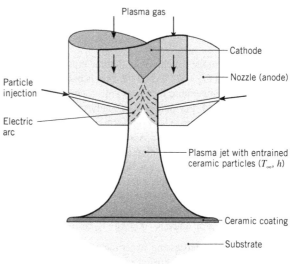

During their *time-in-flight*, the ceramic particles must be heated to their melting point and experience complete conversion to the liquid state. The coating is formed as the molten droplets impinge (*splat*) on the substrate material and experience rapid solidification. Consider conditions for which spherical alumina (Al_2O_3) particles of diameter $D_p = 50\ \mu m$, density $\rho_p = 3970\ kg/m^3$, thermal conductivity $k_p = 10.5\ W/m \cdot K$, and specific heat $c_p = 1560\ J/kg \cdot K$ are injected into an arc plasma, which is at $T_\infty = 10{,}000\ K$ and provides a coefficient of $h = 30{,}000\ W/m^2 \cdot K$ for convective heating of the particles. The melting point and latent heat of fusion of alumina are $T_{mp} = 2318\ K$ and $h_{sf} = 3577\ kJ/kg$, respectively.

(a) Neglecting radiation, obtain an expression for the time-in-flight, t_{i-f}, required to heat a particle from its initial temperature T_i to its melting point T_{mp}, and, once at the melting point, for the particle to experience complete melting. Evaluate t_{i-f} for $T_i = 300\ K$ and the prescribed heating conditions.

(b) Assuming alumina to have an emissivity of $\varepsilon_p = 0.4$ and the particles to exchange radiation with large surroundings at $T_{sur} = 300\ K$, assess the validity of neglecting radiation.

5.22 A chip that is of length $L = 5\ mm$ on a side and thickness $t = 1\ mm$ is encased in a ceramic substrate, and its exposed surface is convectively cooled by a dielectric liquid for which $h = 150\ W/m^2 \cdot K$ and $T_\infty = 20°C$.

In the off-mode the chip is in thermal equilibrium with the coolant ($T_i = T_\infty$). When the chip is energized, however, its temperature increases until a new steady state is established. For purposes of analysis, the energized chip is characterized by uniform volumetric heating with $\dot{q} = 9 \times 10^6\ W/m^3$. Assuming an infinite contact resistance between the chip and substrate and negligible conduction resistance within the chip, determine the steady-state chip temperature T_f. Following activation of the chip, how long does it take to come within 1°C of this temperature? The chip density and specific heat are $\rho = 2000\ kg/m^3$ and $c = 700\ J/kg \cdot K$, respectively.

5.23 In a material processing experiment conducted aboard the space shuttle, a coated niobium sphere of 10-mm diameter is removed from a furnace at 900°C and cooled to a temperature of 300°C. Although properties of the niobium vary over this temperature range, constant values may be assumed to a reasonable approximation, with $\rho = 8600\ kg/m^3$, $c = 290\ J/kg \cdot K$, and $k = 63\ W/m \cdot K$.

(a) If cooling is implemented in a large evacuated chamber whose walls are at 25°C, determine the time required to reach the final temperature if the coating is polished and has an emissivity of $\varepsilon = 0.1$. How long would it take if the coating is oxidized and $\varepsilon = 0.6$?

(b) To reduce the time required for cooling, consideration is given to immersion of the sphere in an inert gas stream for which $T_\infty = 25°C$ and $h = 200\ W/m^2 \cdot K$. Neglecting radiation, what is the time required for cooling?

(c) Considering the effect of both radiation and convection, what is the time required for cooling if $h = 200\ W/m^2 \cdot K$ and $\varepsilon = 0.6$? Explore the effect on the cooling time of independently varying h and ε.

5.24 A long wire of diameter $D = 2\ mm$ is submerged in an oil bath of temperature $T_\infty = 25°C$. The wire has an electrical resistance per unit length of $R'_e = 0.01\ \Omega/m$. If a current of $I = 100\ A$ flows through the wire and the convection coefficient is $h = 400\ W/m^2 \cdot K$, what is the steady-state temperature of the wire? From the time the current is applied, how long does it take for the wire to reach a temperature that is within 1°C of the steady-state value? The properties of the wire are $\rho = 8000\ kg/m^3$, $c = 500\ J/kg \cdot K$, and $k = 20\ W/m \cdot K$.

5.25 Consider the system of Problem 5.1 where the temperature of the plate is spacewise isothermal during the transient process.

(a) Obtain an expression for the temperature of the plate as a function of time $T(t)$ in terms of q''_o, T_∞, h, L, and the plate properties ρ and c.

(b) Determine the thermal time constant and the steady-state temperature for a 12-mm-thick plate of pure copper when $T_\infty = 27°C$, $h = 50\ W/m^2 \cdot K$, and $q''_o = 5000\ W/m^2$. Estimate the time required to reach steady-state conditions.

(c) For the conditions of part (b), as well as for $h = 100$ and $200\ W/m^2 \cdot K$, compute and plot the corresponding temperature histories of the plate for $0 \le t \le 2500\ s$.

5.26 As noted in Problem 5.3, microwave ovens operate by rapidly aligning and reversing water molecules within the food, resulting in volumetric energy generation and, in turn, cooking of the food. When the food is initially frozen, however, the water molecules do not readily oscillate in response to the microwaves, and the volumetric generation rates are between one and two orders of magnitude lower than if the water were in liquid form. (Microwave power that is not absorbed in the food is reflected back to the microwave generator, where it must be dissipated in the form of heat to prevent damage to the generator.)

(a) Consider a frozen, 1-kg spherical piece of ground beef at an initial temperature of $T_i = -20°C$ placed in a microwave oven with $T_\infty = 30°C$ and $h = 15 \text{ W/m}^2 \cdot \text{K}$. Determine how long it will take the beef to reach a uniform temperature of $T = 0°C$, with all the water in the form of ice. Assume the properties of the beef are the same as ice, and assume 3% of the oven power ($P = 1 \text{ kW}$ total) is absorbed in the food.

(b) After all the ice is converted to liquid, determine how long it will take to heat the beef to $T_f = 80°C$ if 95% of the oven power is absorbed in the food. Assume the properties of the beef are the same as liquid water.

(c) When thawing food in microwave ovens, one may observe that some of the food may still be frozen while other parts of the food are overcooked. Explain why this occurs. Explain why most microwave ovens have thaw cycles that are associated with very low oven powers.

5.27 A metal sphere of diameter D, which is at a uniform temperature T_i, is suddenly removed from a furnace and suspended from a fine wire in a large room with air at a uniform temperature T_∞ and the surrounding walls at a temperature T_{sur}.

(a) Neglecting heat transfer by radiation, obtain an expression for the time required to cool the sphere to some temperature T.

(b) Neglecting heat transfer by convection, obtain an expression for the time required to cool the sphere to the temperature T.

(c) How would you go about determining the time required for the sphere to cool to the temperature T if both convection and radiation are of the same order of magnitude?

(d) Consider an anodized aluminum sphere ($\varepsilon = 0.75$) 50 mm in diameter, which is at an initial temperature of $T_i = 800 \text{ K}$. Both the air and surroundings are at 300 K, and the convection coefficient is $10 \text{ W/m}^2 \cdot \text{K}$. For the conditions of parts (a), (b), and (c), determine the time required for the sphere to cool to 400 K. Plot the corresponding temperature histories. Repeat the calculations for a polished aluminum sphere ($\varepsilon = 0.1$).

5.28 As permanent space stations increase in size, there is an attendant increase in the amount of electrical power they dissipate. To keep station compartment temperatures from exceeding prescribed limits, it is necessary to transfer the dissipated heat to space. A novel heat rejection scheme that has been proposed for this purpose is termed a Liquid Droplet Radiator (LDR). The heat is first transferred to a high vacuum oil, which is then injected into outer space as a stream of small droplets. The stream is allowed to traverse a distance L, over which it cools by radiating energy to outer space at absolute zero temperature. The droplets are then collected and routed back to the space station.

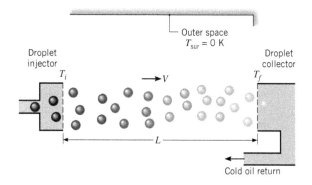

Consider conditions for which droplets of emissivity $\varepsilon = 0.95$ and diameter $D = 0.5 \text{ mm}$ are injected at a temperature of $T_i = 450 \text{ K}$ and a velocity of $V = 0.1 \text{ m/s}$. Properties of the oil are $\rho = 885 \text{ kg/m}^3$, $c = 1900 \text{ J/kg} \cdot \text{K}$, and $k = 0.145 \text{ W/m} \cdot \text{K}$. Assuming each drop to radiate to deep space at $T_{sur} = 4 \text{ K}$, determine the distance L required for the droplets to impact the collector at a final temperature of $T_f = 300 \text{ K}$. What is the amount of thermal energy rejected by each droplet?

5.29 Thin film coatings characterized by high resistance to abrasion and fracture may be formed by using microscale composite particles in a plasma spraying process. A spherical particle typically consists of a *ceramic core*, such as tungsten carbide (WC), and a *metallic shell*, such as cobalt (Co). The ceramic provides the thin film coating with its desired hardness at elevated temperatures, while the metal serves to coalesce the particles on the coated surface and to inhibit crack formation. In the plasma spraying process, the particles are injected into a plasma gas jet that heats them to a temperature above the melting point of the metallic casing and melts the casing before the particles impact the surface.

Consider spherical particles comprised of a WC core of diameter $D_i = 16 \text{ } \mu\text{m}$, which is encased in a Co shell of outer diameter $D_o = 20 \text{ } \mu\text{m}$. If the particles flow in a plasma gas at $T_\infty = 10,000 \text{ K}$ and the coefficient associated with convection from the gas to the particles is $h = 20,000 \text{ W/m}^2 \cdot \text{K}$, how long does it take to heat the particles from an initial temperature of $T_i = 300$ K to the melting point of cobalt, $T_{mp} = 1770$ K? The density and specific heat of WC (the core of the particle) are $\rho_c = 16{,}000$ kg/m^3 and $c_c = 300$ J/kg · K, while the corresponding values for Co (the outer shell) are $\rho_s = 8900$ kg/m^3 and $c_s = 750$ J/kg · K. Once having reached the melting point, how much additional time is required to completely melt the cobalt if its latent heat of fusion is $h_{sf} = 2.59 \times 10^5$ J/kg? You may use the lumped capacitance method of analysis and neglect radiation exchange between the particle and its surroundings.

5.30 In a manufacturing process, a spherical ceramic particle of diameter $D = 300$ μm and initial temperature $T_i = 1100$ K is injected into a chamber containing a hot gas stream at $T_\infty = 1200$ K, $h = 500$ W/m^2 · K. The particulate material has a density of $\rho = 2500$ kg/m^3, specific heat $c = 750$ J/kg·K, thermal conductivity $k = 1.4$ W/m · K, and emissivity $\varepsilon = 0.94$. The surroundings temperature is controlled by externally heating or cooling the chamber walls. Determine the particle temperature at $t = 0.1$ s for $T_{sur} = 750$ K, 850 K, and 950 K. Plot the particle temperatures at times $0 \leq t \leq 0.5$ s for the three surroundings temperatures.

5.31 A long, highly polished aluminum rod of diameter $D = 35$ mm is hung horizontally in a large room. The initial rod temperature is $T_i = 90°$C, and the room air is $T_\infty = 20°$C. At time $t_1 = 1250$ s, the rod temperature is $T_1 = 65°$C, and, at time $t_2 = 6700$ s, the rod temperature is $T_2 = 30°$C. Determine the values of the constants C and n that appear in Equation 5.26. Plot the rod temperature versus time for $0 \leq t \leq 10{,}000$ s. On the same graph, plot the rod temperature versus time for a constant value of the convection heat transfer coefficient, evaluated at a rod temperature of $\bar{T} = (T_i + T_\infty)/2$. For all cases, evaluate properties at $\bar{T} = (T_i + T_\infty)/2$.

5.32 The objective of this problem is to develop thermal models for estimating the steady-state temperature and the transient temperature history of the electrical transformer shown.

The external transformer geometry is approximately cubical, with a length of 32 mm to a side. The combined mass of the iron and copper in the transformer is 0.28 kg, and its weighted-average specific heat is 400 J/kg · K. The transformer dissipates 4.0 W and is operating in ambient air at $T_\infty = 20°$C, with a convection coefficient of 10 W/m^2 · K. List and justify the assumptions made in your analysis, and discuss limitations of the models.

(a) Beginning with a properly defined control volume, develop a model for estimating the steady-state temperature of the transformer, $T(\infty)$. Evaluate $T(\infty)$ for the prescribed operating conditions.

(b) Develop a model for estimating the thermal response (temperature history) of the transformer if it is initially at a temperature of $T_i = T_\infty$ and power is suddenly applied. Determine the time required for the transformer to come within 5°C of its steady-state operating temperature.

5.33 The melting of water initially at the fusion temperature, $T_f = 0°$C, was considered in Example 1.5. Freezing of water often occurs at 0°C. However, pure liquids that undergo a cooling process can remain in a *supercooled* liquid state well below their equilibrium freezing temperature, T_f, particularly when the liquid is not in contact with any solid material. Droplets of liquid water in the atmosphere have a supercooled freezing temperature, $T_{f,sc}$, that can be well correlated to the droplet diameter by the expression $T_{f,sc} = -28 + 0.87 \ln(D_p)$ in the diameter range $10^{-7} < D_p < 10^{-2}$ m, where $T_{f,sc}$ has units of degrees Celsius and D_p is expressed in units of meters. For a droplet of diameter $D = 50$ μm and initial temperature $T_i = 10°$C subject to ambient conditions of $T_\infty = -40°$C and $h = 900$ W/m^2 · K, compare the time needed to completely solidify the droplet for case A, when the droplet solidifies at $T_f = 0°$C, and case B, when the droplet starts to freeze at $T_{f,sc}$. Sketch the temperature histories from the initial time to the time when the droplets are completely solid. *Hint:* When the droplet reaches $T_{f,sc}$ in case B, rapid solidification occurs during which the latent energy released by the freezing water is absorbed by the remaining liquid in the drop. As soon as any ice is formed within the droplet, the remaining liquid is in contact with a solid (the ice) and the freezing temperature immediately shifts from $T_{f,sc}$ to $T_f = 0°$C.

One-Dimensional Conduction: The Plane Wall

5.34 Consider the series solution, Equation 5.42, for the plane wall with convection. Calculate midplane ($x^* = 0$) and surface ($x^* = 1$) temperatures θ^* for $Fo = 0.1$ and 1, using $Bi = 0.1$, 1, and 10. Consider only the first four eigenvalues. Based on these results, discuss the validity of the approximate solutions, Equations 5.43 and 5.44.

5.35 Copper-coated, epoxy-filled fiberglass circuit boards are treated by heating a stack of them under high pressure, as shown in the sketch. The purpose of the pressing–heating operation is to cure the epoxy that bonds the fiberglass sheets, imparting stiffness to the boards. The stack, referred to as a *book,* is comprised of 10 *boards* and 11 *pressing plates,* which prevent epoxy from flowing between the boards and impart a smooth finish to the cured boards. In order to perform simplified thermal analyses, it is reasonable to approximate the book as having an effective thermal conductivity (k) and an effective thermal capacitance (ρc_p). Calculate the effective properties if each of the boards and plates has a thickness of 2.36 mm and the following thermophysical properties: board (b) $\rho_b = 1000$ kg/m³, $c_{p,b} = 1500$ J/kg · K, $k_b = 0.30$ W/m · K; plate (p) $\rho_p = 8000$ kg/m³, $c_{p,p} = 480$ J/kg · K, $k_p = 12$ W/m · K.

5.36 Circuit boards are treated by heating a stack of them under high pressure, as illustrated in Problem 5.35. The platens at the top and bottom of the stack are maintained at a uniform temperature by a circulating fluid. The purpose of the pressing–heating operation is to cure the epoxy, which bonds the fiberglass sheets, and impart stiffness to the boards. The cure condition is achieved when the epoxy has been maintained at or above 170°C for at least 5 min. The effective thermophysical properties of the stack or *book* (boards and metal pressing plates) are $k = 0.613$ W/m · K and $\rho c_p = 2.73 \times 10^6$ J/m³ · K.

(a) If the book is initially at 15°C and, following application of pressure, the platens are suddenly brought to a uniform temperature of 190°C, calculate the elapsed time t_e required for the midplane of the book to reach the cure temperature of 170°C.

(b) If, at this instant of time, $t = t_e$, the platen temperature were reduced suddenly to 15°C, how much energy would have to be removed from the book by the coolant circulating in the platen, in order to return the stack to its initial uniform temperature?

5.37 Annealing is a process by which steel is reheated and then cooled to make it less brittle. Consider the reheat stage for a 125-mm-thick steel plate ($\rho = 7830$ kg/m³, $c = 550$ J/kg · K, $k = 48$ W/m · K), which is initially at a uniform temperature of $T_i = 150$°C and is to be heated to a minimum temperature of 500°C. Heating is effected in a gas-fired furnace, where products of combustion at $T_\infty = 850$°C maintain a convection coefficient of $h = 200$ W/m² · K on both surfaces of the plate. How long should the plate be left in the furnace?

5.38 Consider an acrylic sheet of thickness $L = 4$ mm that is used to coat a hot, isothermal metal substrate at $T_h = 300$°C. The properties of the acrylic are $\rho = 1990$ kg/m³, $c = 1470$ J/kg · K, and $k = 0.21$ W/m · K. Neglecting the thermal contact resistance between the acrylic and the metal substrate, determine how long it will take for the insulated back side of the acrylic to reach its softening temperature, $T_{soft} = 90$°C. The initial acrylic temperature is $T_i = 20$°C.

5.39 Steel is sequentially heated and cooled (*annealed*) to relieve stresses and to make it less brittle. Consider a 125-mm-thick plate ($k = 45$ W/m · K, $\rho = 7800$ kg/m³, $c_p = 500$ J/kg · K) that is initially at a uniform temperature of 300°C and is heated (on both sides) in a gas-fired furnace for which $T_\infty = 750$°C and $h = 400$ W/m² · K. How long will it take for a minimum temperature of 550°C to be reached in the plate?

5.40 Stone mix concrete slabs are used to absorb thermal energy from flowing air that is carried from a large concentrating solar collector. The slabs are heated during the day and release their heat to cooler air at night. If the daytime airflow is characterized by a temperature and convection heat transfer coefficient of $T_\infty = 200$°C and $h = 35$ W/m² · K, respectively, determine the slab thickness $2L$ required to transfer a total amount of energy such that $Q/Q_o = 0.90$ over a $t = 8$-h period. The initial concrete temperature is $T_i = 40$°C.

5.41 During transient operation, the steel nozzle of a rocket engine must not exceed a maximum allowable operating temperature of 1500 K when exposed to combustion gases characterized by a temperature of 2300 K and a convection coefficient of 5000 W/$m^2 \cdot K$. To extend the duration of engine operation, it is proposed that a ceramic *thermal barrier coating* ($k = 10$ W/m $\cdot$ K, $\alpha = 6 \times 10^{-6}$ m²/s) be applied to the interior surface of the nozzle.

(a) If the ceramic coating is 10 mm thick and at an initial temperature of 300 K, obtain a conservative estimate of the maximum allowable duration of engine operation. The nozzle radius is much larger than the combined wall and coating thickness.

(b) Compute and plot the inner and outer surface temperatures of the coating as a function of time for $0 \leq t \leq 150$ s. Repeat the calculations for a coating thickness of 40 mm.

5.42 A plate of thickness $2L = 25$ mm at a temperature of 600°C is removed from a hot pressing operation and must be cooled rapidly to achieve the required physical properties. The process engineer plans to use air jets to control the rate of cooling, but she is uncertain whether it is necessary to cool both sides (case 1) or only one side (case 2) of the plate. The concern is not just for the *time-to-cool*, but also for the maximum temperature difference within the plate. If this temperature difference is too large, the plate can experience significant warping.

Case 1: cooling, both sides   Case 2: cooling, one side only

The air supply is at 25°C, and the convection coefficient on the surface is 400 W/$m^2 \cdot K$. The thermophysical properties of the plate are $\rho = 3000$ kg/m³, $c = 750$ J/kg $\cdot$ K, and $k = 15$ W/m $\cdot$ K.

(a) Using the *IHT* software, calculate and plot on one graph the temperature histories for cases 1 and 2 for a 500-s cooling period. Compare the times required for the maximum temperature in the plate to reach 100°C. Assume no heat loss from the unexposed surface of case 2.

(b) For both cases, calculate and plot on one graph the variation with time of the maximum temperature difference in the plate. Comment on the relative magnitudes of the temperature gradients within the plate as a function of time.

5.43 Two plates of the same material and thickness L are at different initial temperatures $T_{i,1}$ and $T_{i,2}$, where $T_{i,2} > T_{i,1}$. Their faces are suddenly brought into contact. The external surfaces of the two plates are insulated.

(a) Let a dimensionless temperature be defined as $T^*(Fo) \equiv (T - T_{i,1})/(T_{i,2} - T_{i,1})$. Neglecting the thermal contact resistance at the interface between the plates, what are the steady-state dimensionless temperatures of each of the two plates, $T_{ss,1}^*$ and $T_{ss,2}^*$? What is the dimensionless interface temperature T_{int}^* at any time?

(b) An effective overall heat transfer coefficient between the two plates can be defined based on the instantaneous, spatially averaged dimensionless plate temperatures, $U_{eff}^* \equiv q^*/(\overline{T}_2^* - \overline{T}_1^*)$. Noting that a dimensionless heat transfer rate to or from either of the two plates may be expressed as $q^* = d(Q/Q_o)/d\,Fo$, determine an expression for U_{eff}^* for $Fo > 0.2$.

5.44 In a tempering process, glass plate, which is initially at a uniform temperature T_i, is cooled by suddenly reducing the temperature of both surfaces to T_s. The plate is 20 mm thick, and the glass has a thermal diffusivity of 6×10^{-7} m²/s.

(a) How long will it take for the midplane temperature to achieve 75% of its maximum possible temperature reduction?

(b) If $(T_i - T_s) = 300$°C, what is the maximum temperature gradient in the glass at the time calculated in part (a)?

5.45 The strength and stability of tires may be enhanced by heating both sides of the rubber ($k = 0.14$ W/m $\cdot$ K, $\alpha = 6.35 \times 10^{-8}$ m²/s) in a steam chamber for which $T_\infty = 200$°C. In the heating process, a 20-mm-thick rubber wall (assumed to be untreaded) is taken from an initial temperature of 25°C to a midplane temperature of 150°C.

(a) If steam flow over the tire surfaces maintains a convection coefficient of $h = 200$ W/$m^2 \cdot K$, how long will it take to achieve the desired midplane temperature?

(b) To accelerate the heating process, it is recommended that the steam flow be made sufficiently vigorous to maintain the tire surfaces at 200°C throughout the process. Compute and plot the midplane and surface temperatures for this case, as well as for the conditions of part (a).

5.46 A plastic coating is applied to wood panels by first depositing molten polymer on a panel and then cooling the surface of the polymer by subjecting it to airflow at 25°C. As first approximations, the heat of reaction associated with solidification of the polymer may be neglected and the polymer/wood interface may be assumed to be adiabatic.

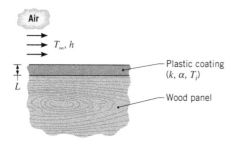

Air
T_∞, h
Plastic coating (k, α, T_i)
L
Wood panel

If the thickness of the coating is $L = 2$ mm and it has an initial uniform temperature of $T_i = 200°C$, how long will it take for the surface to achieve a *safe-to-touch* temperature of 42°C if the convection coefficient is $h = 200$ W/m² · K? What is the corresponding value of the interface temperature? The thermal conductivity and diffusivity of the plastic are $k = 0.25$ W/m · K and $\alpha = 1.20 \times 10^{-7}$ m²/s, respectively.

One-Dimensional Conduction: The Long Cylinder

5.47 A long rod of 60-mm diameter and thermophysical properties $\rho = 8000$ kg/m³, $c = 500$ J/kg · K, and $k = 50$ W/m · K is initially at a uniform temperature and is heated in a forced convection furnace maintained at 750 K. The convection coefficient is estimated to be 1000 W/m² · K.

(a) What is the centerline temperature of the rod when the surface temperature is 550 K?

(b) In a heat-treating process, the centerline temperature of the rod must be increased from $T_i = 300$ K to $T = 500$ K. Compute and plot the centerline temperature histories for $h = 100$, 500, and 1000 W/m² · K. In each case the calculation may be terminated when $T = 500$ K.

5.48 A long cylinder of 30-mm diameter, initially at a uniform temperature of 1000 K, is suddenly quenched in a large, constant-temperature oil bath at 350 K. The cylinder properties are $k = 1.7$ W/m · K, $c = 1600$ J/kg · K, and $\rho = 400$ kg/m³, while the convection coefficient is 50 W/m² · K.

(a) Calculate the time required for the surface of the cylinder to reach 500 K.

(b) Compute and plot the surface temperature history for $0 \leq t \leq 300$ s. If the oil were agitated, providing a convection coefficient of 250 W/m² · K, how would the temperature history change?

5.49 A cylindrical stone mix concrete beam of diameter $D = 0.5$ m initially at $T_i = 20°C$ is exposed to hot gases at $T_\infty = 500°C$. The convection coefficient is $h = 10$ W/m² · K.

(a) Determine the centerline temperature of the beam after an exposure time of $t = 6$ h.

(b) Determine the centerline temperature of a second beam that is of the same size and exposed to the same conditions as in part (a) but fabricated of lightweight aggregate concrete with density $\rho = 1495$ kg/m³, thermal conductivity $k = 0.789$ W/m · K, and specific heat $c_p = 880$ J/kg · K.

5.50 A long rod 35 mm in diameter, fabricated from sapphire (aluminum oxide) and initially at a uniform temperature of 800 K, is suddenly cooled by a fluid at 300 K having a heat transfer coefficient of 1500 W/m² · K. After 30 s, the rod is wrapped in insulation and experiences no heat losses. What will be the temperature of the rod after a long period of time?

5.51 A long plastic rod of 30-mm diameter ($k = 0.3$ W/m · K and $\rho c_p = 1040$ kJ/m³ · K) is uniformly heated in an oven as preparation for a pressing operation. For best results, the temperature in the rod should not be less than 200°C. To what uniform temperature should the rod be heated in the oven if, for the worst case, the rod sits on a conveyor for 3 min while exposed to convection cooling with ambient air at 25°C and with a convection coefficient of 8 W/m² · K? A further condition for good results is a maximum–minimum temperature difference of less than 10°C. Is this condition satisfied? If not, what could you do to satisfy it?

5.52 As part of a heat treatment process, cylindrical, 304 stainless steel rods of 125-mm diameter are cooled from an initial temperature of 500°C by suspending them in an oil bath at 30°C. If a convection coefficient of 400 W/m² · K is maintained by circulation of the oil, how long does it take for the centerline of a rod to reach a temperature of 50°C, at which point it is withdrawn from the bath? If 10 rods of length $L = 1$ m are processed per hour, what is the nominal rate at which energy must be extracted from the bath (the cooling load)?

5.53 In a manufacturing process, long rods of different diameters are at a uniform temperature of 400°C in a curing oven, from which they are removed and cooled by forced convection in air at 25°C. One of the line operators has observed that it takes 280 s for a 40-mm-diameter rod to cool to a *safe-to-handle* temperature of 60°C. For an equivalent convection coefficient, how long will it take for an 80-mm-diameter rod to cool to the same temperature? The thermophysical properties of the rod are $\rho = 2500$ kg/m³, $c = 900$ J/kg · K, and $k = 15$ W/m · K. Comment on your result. Did you anticipate this outcome?

5.54 The density and specific heat of a particular material are known ($\rho = 1200$ kg/m³, $c_p = 1250$ J/kg · K), but its thermal conductivity is unknown. To determine the thermal conductivity, a long cylindrical specimen of diameter $D = 40$ mm is machined, and a thermocouple is inserted through a small hole drilled along the centerline.

The thermal conductivity is determined by performing an experiment in which the specimen is heated to a uniform temperature of $T_i = 100°C$ and then cooled by passing air at $T_\infty = 25°C$ in cross flow over the cylinder. For the prescribed air velocity, the convection coefficient is $h = 55$ W/m² · K.

(a) If a centerline temperature of $T(0, t) = 40°C$ is recorded after $t = 1136$ s of cooling, verify that the material has a thermal conductivity of $k = 0.30$ W/m · K.

(b) For air in cross flow over the cylinder, the prescribed value of $h = 55$ W/m² · K corresponds to a velocity of $V = 6.8$ m/s. If $h = CV^{0.618}$, where the constant C has units of W · s$^{0.618}$/m$^{2.618}$ · K, how does the centerline temperature at $t = 1136$ s vary with velocity for $3 \le V \le 20$ m/s? Determine the centerline temperature histories for $0 \le t \le 1500$ s and velocities of 3, 10, and 20 m/s.

One-Dimensional Conduction: The Sphere

5.55 In heat treating to harden steel ball bearings ($c = 500$ J/kg · K, $\rho = 7800$ kg/m³, $k = 50$ W/m · K), it is desirable to increase the surface temperature for a short time without significantly warming the interior of the ball. This type of heating can be accomplished by sudden immersion of the ball in a molten salt bath with $T_\infty = 1300$ K and $h = 5000$ W/m² · K. Assume that any location within the ball whose temperature exceeds 1000 K will be hardened. Estimate the time required to harden the outer millimeter of a ball of diameter 20 mm, if its initial temperature is 300 K.

5.56 Stainless steel (AISI 304) ball bearings, which have uniformly been heated to 850°C, are hardened by quenching them in an oil bath that is maintained at 40°C. The ball diameter is 25 mm, and the convection coefficient associated with the oil bath is 800 W/m² · K.

(a) If quenching is to occur until the surface temperature of the balls reaches 100°C, how long must the balls be kept in the oil? What is the center temperature at the conclusion of the cooling period?

(b) If 10,000 balls are to be quenched per hour, what is the rate at which energy must be removed by the oil bath cooling system in order to maintain its temperature at 40°C?

5.57 A $D = 100$ mm diameter solid sphere, initially at a uniform temperature of $T_i = 50°C$, is placed in a flowing fluid at $T_\infty = 20°C$. The sphere properties are $k = 2$ W/m · K, $\rho = 2500$ kg/m³, and $c_p = 750$ J/kg · K. The convection heat transfer coefficient is $h = 30$ W/m² · K. At what time does the sphere's surface temperature reach 26°C? How much energy has been transferred from the sphere at this time?

5.58 A cold air chamber is proposed for quenching steel ball bearings of diameter $D = 0.2$ m and initial temperature $T_i = 400°C$. Air in the chamber is maintained at −15°C by a refrigeration system, and the steel balls pass through the chamber on a conveyor belt. Optimum bearing production requires that 70% of the initial thermal energy content of the ball above −15°C be removed. Radiation effects may be neglected, and the convection heat transfer coefficient within the chamber is 1000 W/m² · K. Estimate the residence time of the balls within the chamber, and recommend a drive velocity of the conveyor. The following properties may be used for the steel: $k = 50$ W/m · K, $\alpha = 2 \times 10^{-5}$ m²/s, and $c = 450$ J/kg · K.

5.59 A sphere 30 mm in diameter initially at 800 K is quenched in a large bath having a constant temperature of 320 K with a convection heat transfer coefficient of 75 W/m² · K. The thermophysical properties of the sphere material are: $\rho = 400$ kg/m³, $c = 1600$ J/kg · K, and $k = 1.7$ W/m · K.

(a) Show, in a qualitative manner on $T - t$ coordinates, the temperatures at the center and at the surface of the sphere as a function of time.

(b) Calculate the time required for the surface of the sphere to reach 415 K.

(c) Determine the heat flux (W/m²) at the outer surface of the sphere at the time determined in part (b).

(d) Determine the energy (J) that has been lost by the sphere during the process of cooling to the surface temperature of 415 K.

(e) At the time determined by part (b), the sphere is quickly removed from the bath and covered with perfect insulation, such that there is no heat loss from the surface of the sphere. What will be the temperature of the sphere after a long period of time has elapsed?

(f) Compute and plot the center and surface temperature histories over the period $0 \leq t \leq 150$ s. What effect does an increase in the convection coefficient to $h = 200$ W/m² · K have on the foregoing temperature histories? For $h = 75$ and 200 W/m² · K, compute and plot the surface heat flux as a function of time for $0 \leq t \leq 150$ s.

5.60 Spheres of 40-mm diameter heated to a uniform temperature of 400°C are suddenly removed from the oven and placed in a forced-air bath operating at 25°C with a convection coefficient of 300 W/m² · K on the sphere surfaces. The thermophysical properties of the sphere material are $\rho = 3000$ kg/m³, $c = 850$ J/kg · K, and $k = 15$ W/m · K.

(a) How long must the spheres remain in the air bath for 80% of the thermal energy to be removed?

(b) The spheres are then placed in a packing carton that prevents further heat transfer to the environment. What uniform temperature will the spheres eventually reach?

5.61 To determine which parts of a spider's brain are triggered into neural activity in response to various optical stimuli, researchers at the University of Massachusetts Amherst desire to examine the brain as it is shown images that might evoke emotions such as fear or hunger. Consider a spider at $T_i = 20$°C that is shown a frightful scene and is then immediately immersed in liquid nitrogen at $T_\infty = 77$ K. The brain is subsequently dissected in its frozen state and analyzed to determine which parts of the brain reacted to the stimulus. Using your knowledge of heat transfer, determine how much time elapses before the spider's brain begins to freeze. Assume the brain is a sphere of diameter $D_b = 1$ mm, centrally located in the spider's cephalothorax, which may be approximated as a spherical shell of diameter $D_c = 3$ mm. The brain and cephalothorax properties correspond to those of liquid water. Neglect the effects of the latent heat of fusion and assume the heat transfer coefficient is $h = 100$ W/m² · K.

5.62 The convection coefficient for flow over a solid sphere may be determined by submerging the sphere, which is initially at 25°C, into the flow, which is at 75°C, and measuring its surface temperature at some time during the transient heating process.

(a) If the sphere has a diameter of 0.1 m, a thermal conductivity of 15 W/m · K, and a thermal diffusivity of 10^{-5} m²/s, at what time will a surface temperature of 60°C be recorded if the convection coefficient is 300 W/m² · K?

(b) Assess the effect of thermal diffusivity on the thermal response of the material by computing center and surface temperature histories for $\alpha = 10^{-6}$, 10^{-5}, and 10^{-4} m²/s. Plot your results for the period $0 \leq t \leq 300$ s. In a similar manner, assess the effect of thermal conductivity by considering values of $k = 1.5$, 15, and 150 W/m · K.

5.63 Consider the sphere of Example 5.6, which is initially at a uniform temperature when it is suddenly removed from the furnace and subjected to a two-step cooling process. Use the *Transient Conduction, Sphere* model of *IHT* to obtain the following solutions.

(a) For step 1, calculate the time required for the center temperature to reach $T(0, t) = 335$°C, while cooling in air at 20°C with a convection coefficient of 10 W/m² · K. What is the Biot number for this cooling process? Do you expect radial temperature gradients to be appreciable? Compare your results to those of the example.

(b) For step 2, calculate the time required for the center temperature to reach $T(0, t) = 50$°C, while cooling in a water bath at 20°C with a convection coefficient of 6000 W/m² · K.

(c) For the step 2 cooling process, calculate and plot the temperature histories, $T(r, t)$, for the center and surface of the sphere. Identify and explain key features of the histories. When do you expect the temperature gradients in the sphere to be the largest?

Semi-Infinite Media

5.64 Two large blocks of different materials, such as copper and concrete, have been sitting in a room (23°C) for a very long time. Which of the two blocks, if either, will feel colder to the touch? Assume the blocks to be semi-infinite solids and your hand to be at a temperature of 37°C.

5.65 Asphalt pavement may achieve temperatures as high as 50°C on a hot summer day. Assume that such a temperature exists throughout the pavement, when suddenly a rainstorm reduces the surface temperature to 20°C. Calculate the total amount of energy (J/m²) that will be transferred from the asphalt over a 30-min period in which the surface is maintained at 20°C.

5.66 A thick steel slab ($\rho = 7800 \text{ kg/m}^3$, $c = 480 \text{ J/kg} \cdot \text{K}$, $k = 50 \text{ W/m} \cdot \text{K}$) is initially at 300°C and is cooled by water jets impinging on one of its surfaces. The temperature of the water is 30°C, and the jets maintain an extremely large, approximately uniform convection coefficient at the surface. Assuming that the surface is maintained at the temperature of the water throughout the cooling, how long will it take for the temperature to reach 50°C at a distance of 20 mm from the surface?

5.67 A tile-iron consists of a massive plate maintained at 150°C by an embedded electrical heater. The iron is placed in contact with a tile to soften the adhesive, allowing the tile to be easily lifted from the subflooring. The adhesive will soften sufficiently if heated above 50°C for at least 2 min, but its temperature should not exceed 120°C to avoid deterioration of the adhesive. Assume the tile and subfloor to have an initial temperature of 25°C and to have equivalent thermophysical properties of $k = 0.15 \text{ W/m} \cdot \text{K}$ and $\rho c_p = 1.5 \times 10^6 \text{ J/m}^3 \cdot \text{K}$.

Tile, 4–mm thickness

Subflooring

(a) How long will it take a worker using the tile-iron to lift a tile? Will the adhesive temperature exceed 120°C?

(b) If the tile-iron has a square surface area 254 mm to the side, how much energy has been removed from it during the time it has taken to lift the tile?

5.68 An insurance company has hired you as a consultant to improve their understanding of burn injuries. They are especially interested in injuries induced when a portion of a worker's body comes into contact with machinery that is at elevated temperatures in the range of 50 to 100°C. Their medical consultant informs them that irreversible thermal injury (cell death) will occur in any living tissue that is maintained at $T \geq 48°C$ for a duration $\Delta t \geq 10$ s. They want information concerning the extent of irreversible tissue damage (as measured by distance from the skin surface) as a function of the machinery temperature and the time during which contact is made between the skin and the machinery. Assume that living tissue has a normal temperature of 37°C, is isotropic, and has constant properties equivalent to those of liquid water.

(a) To assess the seriousness of the problem, compute locations in the tissue at which the temperature will reach 48°C after 10 s of exposure to machinery at 50°C and 100°C.

(b) For a machinery temperature of 100°C and $0 \leq t \leq 30$ s, compute and plot temperature histories at tissue locations of 0.5, 1, and 2 mm from the skin.

5.69 The manufacturer of a heat flux gage like the one shown below claims the time constant for a 63.2% response to be $\tau = (4d^2 \rho c_p)/\pi^2 k$, where ρ, c_p, and k are the thermophysical properties of the gage material and d is its thickness. Not knowing the origin of this relation, your task is to model the gage considering the two extreme cases illustrated below. In both cases, the gage, initially at a uniform temperature T_i, is exposed to a sudden change in surface temperature, $T(0, t) = T_s$. For case (a) the backside of the gage is insulated, and for case (b) the gage is imbedded in a semi-infinite solid having the same thermophysical properties as those of the gage.

Develop relationships for predicting the time constant of the gage for the two cases and compare them to the manufacturer's relation. What conclusion can you draw from this analysis regarding the transient response of gages for different applications?

5.70 A very thick slab with thermal diffusivity $5.6 \times 10^{-6} \, m^2/s$ and thermal conductivity $20 \, W/m \cdot K$ is initially at a uniform temperature of $325°C$. Suddenly, the surface is exposed to a coolant at $15°C$ for which the convection heat transfer coefficient is $100 \, W/m^2 \cdot K$.

(a) Determine temperatures at the surface and at a depth of 45 mm after 3 min have elapsed.

(b) Compute and plot temperature histories ($0 \leq t \leq 300 \, s$) at $x = 0$ and $x = 45 \, mm$ for the following parametric variations: (i) $\alpha = 5.6 \times 10^{-7}$, 5.6×10^{-6}, and $5.6 \times 10^{-5} \, m^2/s$; and (ii) $k = 2, 20$, and $200 \, W/m \cdot K$.

5.71 A generous amount of gel is applied to the surface of the window of an ultrasound probe prior to its use. The window is made of a polycarbonate material of density $\rho = 1200 \, kg/m^3$, thermal conductivity $k = 0.2 \, W/m \cdot K$, and specific heat $c_p = 1200 \, J/kg \cdot K$. The probe is initially at a temperature of $T_i = 20°C$. To ensure that the gel remains adhered to the probe until the probe is brought into contact with the patient's skin, its temperature-dependent viscosity must not fall below a critical value. This corresponds to a requirement that the maximum gel temperature remain below $12°C$ prior to making contact with the patient. Determine the required initial gel temperature. The thermal conductivity, density, and specific heat of the gel may be assumed to be those of liquid water.

5.72 A thick oak wall, initially at $25°C$, is suddenly exposed to combustion products for which $T_\infty = 800°C$ and $h = 22 \, W/m^2 \cdot K$.

(a) Determine the time of exposure required for the surface to reach the ignition temperature of $400°C$.

(b) Plot the temperature distribution $T(x)$ in the medium at $t = 325 \, s$. The distribution should extend to a location for which $T \approx 25°C$.

5.73 Standards for firewalls may be based on their thermal response to a prescribed radiant heat flux. Consider a 0.25-m-thick concrete wall ($\rho = 2300 \, kg/m^3$, $c = 880$ $J/kg \cdot K$, $k = 1.4 \, W/m \cdot K$), which is at an initial temperature of $T_i = 25°C$ and irradiated at one surface by lamps that provide a uniform heat flux of $q_s'' = 10^4 \, W/m^2$. The absorptivity of the surface to the irradiation is $\alpha_s = 1.0$. If building code requirements dictate that the temperatures of the irradiated and back surfaces must not exceed $325°C$ and $25°C$, respectively, after 30 min of heating, will the requirements be met?

5.74 It is well known that, although two materials are at the same temperature, one may feel cooler to the touch than the other. Consider thick plates of copper and glass, each at an initial temperature of 300 K. Assuming your finger to be at an initial temperature of 310 K and to have thermophysical properties of $\rho = 1000 \, kg/m^3$, $c = 4180 \, J/kg \cdot K$, and $k = 0.625 \, W/m \cdot K$, determine whether the copper or the glass will feel cooler to the touch.

5.75 Two stainless steel plates ($\rho = 8000 \, kg/m^3$, $c = 500 \, J/kg \cdot K$, $k = 15 \, W/m \cdot K$), each 25 mm thick and insulated on one surface, are initially at 400 and 300 K when they are pressed together at their uninsulated surfaces. What is the temperature of the insulated surface of the hot plate after 1 min has elapsed?

5.76 Special coatings are often formed by depositing thin layers of a molten material on a solid substrate. Solidification begins at the substrate surface and proceeds until the thickness S of the solid layer becomes equal to the thickness δ of the deposit.

(a) Consider conditions for which molten material at its fusion temperature T_f is deposited on a *large* substrate that is at an initial uniform temperature T_i. With $S = 0$ at $t = 0$, develop an expression for estimating the time t_d required to completely solidify the deposit if it remains at T_f throughout the solidification process. Express your result in terms of the substrate thermal conductivity and thermal diffusivity (k_s, α_s), the density and latent heat of fusion of the deposit (ρ, h_{sf}), the deposit thickness δ, and the relevant temperatures (T_f, T_i).

(b) The plasma spray deposition process of Problem 5.21 is used to apply a thin ($\delta = 2 \, mm$) alumina coating on a thick tungsten substrate. The substrate has a uniform initial temperature of $T_i = 300 \, K$, and its thermal conductivity and thermal diffusivity may be approximated as $k_s = 120 \, W/m \cdot K$ and $\alpha_s = 4.0 \times 10^{-5} \, m^2/s$, respectively. The density and latent heat of fusion of the alumina are $\rho = 3970 \, kg/m^3$ and $h_{sf} = 3577 \, kJ/kg$, respectively, and the alumina solidifies at its fusion temperature ($T_f = 2318 \, K$). Assuming that the molten layer is instantaneously deposited on the substrate, estimate the time required for the deposit to solidify.

5.77 When a molten metal is cast in a mold that is a poor conductor, the dominant resistance to heat flow is within the mold wall. Consider conditions for which a liquid metal is solidifying in a thick-walled mold of thermal conductivity k_w and thermal diffusivity α_w. The density and latent heat of fusion of the metal are designated as ρ and h_{sf}, respectively, and in both its molten and solid states, the thermal conductivity of the metal is very much larger than that of the mold.

Just before the start of solidification ($S = 0$), the mold wall is everywhere at an initial uniform temperature T_i and the molten metal is everywhere at its fusion (melting point) temperature of T_f. Following the start of solidification, there is conduction heat transfer into the mold wall and the thickness of the solidified metal S increases with time t.

(a) Sketch the one-dimensional temperature distribution, $T(x)$, in the mold wall and the metal at $t = 0$ and at two subsequent times during the solidification. Clearly indicate any underlying assumptions.

(b) Obtain a relation for the variation of the solid layer thickness S with time t, expressing your result in terms of appropriate parameters of the system.

Objects with Constant Surface Temperatures or Surface Heat Fluxes and Periodic Heating

5.78 A rewritable optical disc (DVD) is formed by sandwiching a 15-nm-thick binary compound storage material between two 1-mm-thick polycarbonate sheets. Data are *written* to the opaque storage medium by irradiating it from below with a relatively high-powered laser beam of diameter 0.4 μm and power 1 mW, resulting in rapid heating of the compound material (the polycarbonate is transparent to the laser irradiation). If the temperature of the storage medium exceeds 900 K, a noncrystalline, amorphous material forms at the heated spot when the laser irradiation is curtailed and the spot is allowed to cool rapidly. The resulting spots of amorphous material have a different reflectivity from the surrounding crystalline material, so they can subsequently be *read* by irradiating them with a sec-

ond, low-power laser and detecting the changes in laser radiation transmitted through the entire DVD thickness. Determine the irradiation (write) time needed to raise the storage medium temperature from an initial value of 300 K to 1000 K. The absorptivity of the storage medium is 0.8. The polycarbonate properties are $\rho = 1200$ kg/m^3, $k = 0.21$ W/m $\cdot$ K, and $c_p = 1260$ J/kg $\cdot$ K.

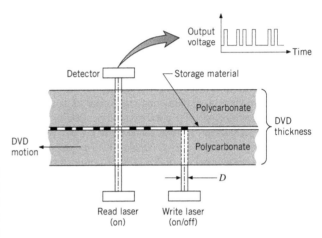

5.79 Ground source heat pumps operate by using the soil, rather than ambient air, as the heat source (or sink) for heating (or cooling) a building. A liquid transfers energy from (to) the soil by way of buried plastic tubing. The tubing is at a depth for which annual variations in the temperature of the soil are much less than those of the ambient air. For example, at a location such as South Bend, Indiana, deep-ground temperatures may remain at approximately 11°C, while annual excursions in the ambient air temperature may range from –25°C to +37°C. Consider the tubing to be laid out in a *closely spaced* serpentine arrangement.

To what depth should the tubing be buried so that the soil can be viewed as an infinite medium at constant temperature over a 12-month period? Account for the periodic cooling (heating) of the soil due to both annual changes in ambient conditions and variations in heat pump operation from the winter heating to the summer cooling mode.

5.80 To enable cooking a wider range of foods in microwave ovens, thin, metallic packaging materials have been developed that will readily absorb microwave energy. As the packaging material is heated by the microwaves, conduction simultaneously occurs from the hot packaging material to the cold food. Consider the spherical piece of frozen ground beef of Problem 5.26 that is now wrapped in the thin microwave-absorbing packaging material. Determine the time needed for the beef that is immediately adjacent to the packaging material to reach $T = 0°C$ if 50% of the oven power ($P = 1$ kW total) is absorbed in the packaging material.

5.81 (a) Derive an expression for the ratio of the total energy transferred from the isothermal surface of an infinite cylinder to the interior of the cylinder, Q/Q_o, that is valid for $Fo < 0.2$. Express your results in terms of the Fourier number Fo.

    (b) Derive an expression for the ratio of the total energy transferred from the isothermal surface of a sphere to the interior of the sphere Q/Q_o that is valid for $Fo < 0.2$. Express your result in terms of the Fourier number, Fo.

5.82 Consider the plane wall of thickness $2L$, the infinite cylinder of radius r_o, and the sphere of radius r_o. Each configuration is subjected to a constant surface heat flux q_s''. Using the approximate solutions of Table 5.2b for $Fo \geq 0.2$, derive expressions for each of the three geometries for the quantity $(T_{s,\text{act}} - T_i)/(T_{s,\text{lc}} - T_i)$. In this expression, $T_{s,\text{act}}$ is the actual surface temperature as determined by the relations of Table 5.2b, and $T_{s,\text{lc}}$ is the temperature associated with lumped capacitance behavior. Determine criteria associated with $(T_{s,\text{act}} - T_i)/(T_{s,\text{lc}} - T_i) \leq 1.1$, that is, determine when the lumped capacitance approximation is accurate to within 10%.

5.83 Consider the experimental measurement of Example 5.10. It is desired to measure the thermal conductivity of an extremely thin sample of the same nanostructured material having the same length and width. To minimize experimental uncertainty, the experimenter wishes to keep the amplitude of the temperature response, ΔT, above a value of 0.1°C. What is the minimum sample thickness that can be measured? Assume the properties of the thin sample and the magnitude of the applied heating rate are the same as those measured and used in Example 5.10.

5.84 Consider a packed bed of spheres, each of diameter $D = 2$ mm, at an initial temperature of $T_i = 90°C$.

Cold water at $T_w = 5°C$ is poured onto the bed at $t = 0$, producing a large convection coefficient about each sphere. Determine the rate (in °C/s) at which an individual sphere near the top surface of the bed is cooled at $t = 0.1, 0.25,$ and 1 s. The spheres have thermophysical properties similar to Pyrex.

5.85 During the cold winter months, an office building is partially heated by extracting thermal energy stored in soil beneath the building using a heat pump. During the summer, the building is partially cooled by sending thermal energy to the same soil with an air conditioner. The heat pump–air conditioning system creates a periodic soil surface temperature variation $T(0,t)$ characterized by an amplitude of $\Delta T = 30°C$. Determine the thermal penetration depth within the soil and the maximum soil surface heat flux.

Finite-Difference Equations: Derivations

5.86 The stability criterion for the explicit method requires that the coefficient of the T_m^p term of the one-dimensional, finite-difference equation be zero or positive. Consider the situation for which the temperatures at the two neighboring nodes (T_{m-1}^p, T_{m+1}^p) are 100°C while the center node (T_m^p) is at 50°C. Show that for values of $Fo > \frac{1}{2}$ the finite-difference equation will predict a value of T_m^{p+1} that violates the second law of thermodynamics.

5.87 A thin rod of diameter D is initially in equilibrium with its surroundings, a large vacuum enclosure at temperature T_{sur}. Suddenly an electrical current I (A) is passed through the rod having an electrical resistivity ρ_e and emissivity ε. Other pertinent thermophysical properties are identified in the sketch. Derive the transient, finite-difference equation for node m.

5.88 A plane wall ($\rho = 4000$ kg/m³, $c_p = 500$ J/kg · K, $k = 10$ W/m · K) of thickness $L = 20$ mm initially has a linear, steady-state temperature distribution with boundaries maintained at $T_1 = 0°C$ and $T_2 = 100°C$. Suddenly, an electric current is passed through the wall, causing uniform energy generation at a rate $\dot{q} = 2 \times 10^7$ W/m³. The boundary conditions T_1 and T_2 remain fixed.

$L = 20$ mm

(a) On $T - x$ coordinates, sketch temperature distributions for the following cases: (i) initial condition ($t \leq 0$); (ii) steady-state conditions ($t \to \infty$), assuming that the maximum temperature in the wall exceeds T_2; and (iii) for two intermediate times. Label all important features of the distributions.

(b) For the system of three nodal points shown schematically (1, m, 2), define an appropriate control volume for node m and, identifying all relevant processes, derive the corresponding finite-difference equation using either the *explicit* or *implicit* method.

(c) With a time increment of $\Delta t = 5$ s, use the finite-difference method to obtain values of T_m for the first 45 s of elapsed time. Determine the corresponding heat fluxes at the boundaries, that is, q''_x (0, 45 s) and q''_x (20 mm, 45 s).

(d) To determine the effect of mesh size, repeat your analysis using grids of 5 and 11 nodal points ($\Delta x = 5.0$ and 2.0 mm, respectively).

5.89 A round solid cylinder made of a plastic material ($\alpha = 6 \times 10^{-7}$ m²/s) is initially at a uniform temperature of 20°C and is well insulated along its lateral surface and at one end. At time $t = 0$, heat is applied to the left boundary causing T_0 to increase linearly with time at a rate of 1°C/s.

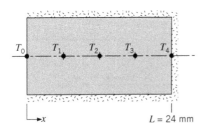

$L = 24$ mm

(a) Using the explicit method with $Fo = \frac{1}{2}$, derive the finite-difference equations for nodes 1, 2, 3, and 4.

(b) Format a table with headings of p, t(s), and the nodal temperatures T_0 to T_4. Determine the surface temperature T_0 when $T_4 = 35$°C.

5.90 Derive the explicit finite-difference equation for an interior node for three-dimensional transient conduction.

Also determine the stability criterion. Assume constant properties and equal grid spacing in all three directions.

5.91 Derive the transient, two-dimensional finite-difference equation for the temperature at nodal point 0 located on the boundary between two different materials.

Finite-Difference Solutions: One-Dimensional Systems

5.92 A wall 0.12 m thick having a thermal diffusivity of 1.5×10^{-6} m²/s is initially at a uniform temperature of 85°C. Suddenly one face is lowered to a temperature of 20°C, while the other face is perfectly insulated.

(a) Using the explicit finite-difference technique with space and time increments of 30 mm and 300 s, respectively, determine the temperature distribution at $t = 45$ min.

(b) With $\Delta x = 30$ mm and $\Delta t = 300$ s, compute $T(x, t)$ for $0 \leq t \leq t_{ss}$, where t_{ss} is the time required for the temperature at each nodal point to reach a value that is within 1°C of the steady-state temperature. Repeat the foregoing calculations for $\Delta t = 75$ s. For each value of Δt, plot temperature histories for each face and the midplane.

5.93 The Der(T,t) function of *IHT* can be used to represent the temperature-time derivative of Equation 5.75. In this problem we will use the Der function to solve Example 5.12.

(a) Show, using the energy balance method, that the implicit form of the finite-difference equations for nodes $1 \leq m \leq 8$ of Example 5.12 can be written as

$$\rho c \Delta x \, \text{Der}(T_m^{p+1}, t) = \frac{k}{\Delta x}(T_{m-1}^{p+1} - T_m^{p+1})$$
$$+ \frac{k}{\Delta x}(T_{m+1}^{p+1} - T_m^{p+1})$$

Note: *IHT* makes use of the implicit finite-difference scheme.

(b) Derive the implicit form of the finite-difference equation for node $m = 0$ of Example 5.12.

(c) For $T_9 = 20$°C and using a time step of $\Delta t = 24$ s, determine the temperatures at nodes 0 through 8 at $t = 120$ s.

5.94 Consider the solar collector of Problem 3.92. After reaching a steady-state operating condition with $q''_{rad} = 800 \text{ W/m}^2$, the net radiation flux to the surface is suddenly reduced to $q''_{rad} = 0$ due to the passage of a cloud.

(a) With $q''_{rad} = 800 \text{ W/m}^2$, determine the steady-state temperature distribution within the absorber plate and the rate of thermal energy delivered to a single tube per unit length, q', using a finite difference model with $\Delta x = 10$ mm. Take advantage of the symmetry about $x = 0$ (located directly above a representative tube) and $L/2$ to minimize the number of finite-difference equations. What is the maximum steady-state absorber plate temperature? Evaluate the plate properties at $T = 325$ K.

(b) Using the temperature distribution of part (a) as the initial condition, determine the temperature distribution within the absorber plate $T(x,t)$ when $q''_{rad} = 0$ over the time period $0 \le t \le 500$ s. Use a grid spacing of $\Delta x = 10$ mm and a time step of $\Delta t = 1$ s. Plot the maximum absorber plate temperature over the time period $0 \le t \le 500$ s. What is the maximum absorber plate temperature at $t = 0$, 10, and 100 s? *Hint:* The solution will be expedited by using the Der(T,t) function of *IHT*. See Problem 5.93.

(c) Plot the heat transfer rate to the tube per unit length over the time period $0 \le t \le 500$ s. How long can the cloud cover last before the rate of heat transfer to the tube from the absorber plate is reduced by 50 percent?

5.95 A molded plastic product ($\rho = 1200 \text{ kg/m}^3$, $c = 1500$ J/kg $\cdot$ K, $k = 0.30$ W/m $\cdot$ K) is cooled by exposing one surface to an array of air jets, while the opposite surface is well insulated. The product may be approximated as a slab of thickness $L = 60$ mm, which is initially at a uniform temperature of $T_i = 80°C$. The air jets are at a temperature of $T_\infty = 20°C$ and provide a uniform convection coefficient of $h = 100 \text{ W/m}^2 \cdot$ K at the cooled surface.

Using a finite-difference solution with a space increment of $\Delta x = 6$ mm, determine temperatures at the cooled and insulated surfaces after 1 h of exposure to the gas jets.

5.96 Consider a one-dimensional plane wall at a uniform initial temperature T_i. The wall is 10 mm thick, and has a thermal

diffusivity of $\alpha = 6 \times 10^{-7} \text{ m}^2/\text{s}$. The left face is insulated, and suddenly the right face is lowered to a temperature $T_{s,r}$.

(a) Using the implicit finite-difference technique with $\Delta x = 2$ mm and $\Delta t = 2$ s, determine how long it will take for the temperature at the left face $T_{s,l}$ to achieve 50% of its maximum possible temperature reduction.

(b) At the time determined in part (a), the right face is suddenly returned to the initial temperature. Determine how long it will take for the temperature at the left face to recover to a 20% temperature reduction, that is, $T_i - T_{s,l} = 0.2(T_i - T_{s,r})$.

5.97 The plane wall of Problem 2.46 ($k = 50$ W/m $\cdot$ K, $\alpha = 1.5 \times 10^{-6} \text{ m}^2/\text{s}$) has a thickness of $L = 40$ mm and an initial uniform temperature of $T_o = 25°C$. Suddenly, the boundary at $x = L$ experiences heating by a fluid for which $T_\infty = 50°C$ and $h = 1000 \text{ W/m}^2 \cdot$ K, while heat is uniformly generated within the wall at $\dot{q} = 1 \times 10^7 \text{ W/m}^3$. The boundary at $x = 0$ remains at T_o.

(a) With $\Delta x = 4$ mm and $\Delta t = 1$ s, plot temperature distributions in the wall for (i) the initial condition, (ii) the steady-state condition, and (iii) two intermediate times.

(b) On $q''_x - t$ coordinates, plot the heat flux at $x = 0$ and $x = L$. At what elapsed time is there zero heat flux at $x = L$?

5.98 Consider two plates, A and B, that are each initially isothermal and each of thickness $L = 5$ mm. The faces of the plates are suddenly brought into contact in a joining process. Material A is acrylic, initially at $T_{i,A} = 20°C$ with $\rho_A = 1990 \text{ kg/m}^3$, $c_A = 1470$ J/kg $\cdot$ K, and $k_A = 0.21$ W/m $\cdot$ K. Material B is steel initially at $T_{i,B} = 300°C$ with $\rho_B = 7800 \text{ kg/m}^3$, $c_B = 500$ J/kg $\cdot$ K, and $k_B = 45$ W/m $\cdot$ K. The external (back) surfaces of the acrylic and steel are insulated. Neglecting the thermal contact resistance between the plates, determine how long it will take for the external surface of the acrylic to reach its softening temperature, $T_{soft} = 90°C$. Plot the acrylic's external surface temperature as well as the average temperatures of both materials over the time span $0 \le t \le 300$ s. Use 20 equally spaced nodal points.

5.99 Consider the fuel element of Example 5.11, which operates at a uniform volumetric generation rate of $\dot{q} = 10^7 \text{ W/m}^3$, until the generation rate suddenly changes to $\dot{q} = 2 \times 10^7 \text{ W/m}^3$. Use the *Finite-Difference Equations, One-Dimensional, Transient* conduction model builder of *IHT* to obtain the implicit form of the finite-difference equations for the 6 nodes, with $\Delta x = 2$ mm, as shown in the example.

(a) Calculate the temperature distribution 1.5 s after the change in operating power, and compare your results with those tabulated in the example.

(b) Use the *Explore* and *Graph* options of *IHT* to calculate and plot temperature histories at the midplane (00) and surface (05) nodes for $0 \le t \le 400$ s. What are the steady-state temperatures, and approximately how long does it take to reach the new equilibrium condition after the step change in operating power?

5.100 In a thin-slab, continuous casting process, molten steel leaves a mold with a thin solid shell, and the molten material solidifies as the slab is quenched by water jets en route to a section of rollers. Once fully solidified, the slab continues to cool as it is brought to an acceptable handling temperature. It is this portion of the process that is of interest.

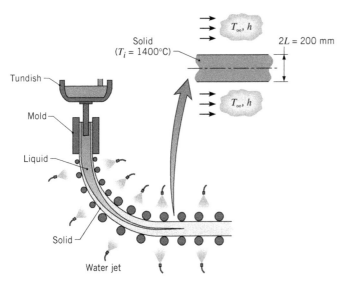

Consider a 200-mm-thick solid slab of steel ($\rho = 7800$ kg/m³, $c = 700$ J/kg · K, $k = 30$ W/m · K), initially at a uniform temperature of $T_i = 1400$°C. The slab is cooled at its top and bottom surfaces by water jets ($T_\infty = 50$°C), which maintain an approximately uniform convection coefficient of $h = 5000$ W/m² · K at both surfaces. Using a finite-difference solution with a space increment of $\Delta x = 1$ mm, determine the time required to cool the surface of the slab to 200°C. What is the corresponding temperature at the midplane of the slab? If the slab moves at a speed of $V = 15$ mm/s, what is the required length of the cooling section?

5.101 A very thick plate with thermal diffusivity 5.6×10^{-6} m²/s and thermal conductivity 20 W/m · K is initially at a uniform temperature of 325°C. Suddenly, the surface is exposed to a coolant at 15°C for which the convection heat transfer coefficient is 100 W/m² · K. Using the finite-difference method with a space increment of $\Delta x = 15$ mm and a time increment of 18 s, determine temperatures at the surface and at a depth of 45 mm after 3 min have elapsed.

5.102 Consider the nacelle of the wind turbine in Example 1.3. The nacelle is formed of a thermoplastic composite material of thickness $L = 20$ mm. The thermal conductivity, density, and specific heat of the nacelle material are $k = 1.5$ W/m · K, $\rho = 1250$ kg/m³, and $c = 1500$ J/kg · K. In reality, the heat transfer coefficient at the exterior nacelle surface is not constant, but varies as the blades rotate about the nacelle leading to a sinusoidal variation in the heat transfer coefficient, $h(t) = \bar{h} + \Delta h \sin(\omega t)$. The values of $\bar{h}$ and Δh are 35 W/m² · K and 15 W/m² · K, respectively. The frequency ω may be determined from the number of blades (3) and the rotational speed of the blades (17 rpm).

At time $t = 0$, the temperature distribution through the wall is given by $T(x, t = 0) = Ax + B$, where $x = 0$ corresponds to the interior nacelle surface, $A = -3460$°C/m, and $B = 212$°C. Using this initial condition, determine the temperature distribution within the wall $T(x,t)$ over the time period $40 \le t \le 50$ s. Assume that the inner surface of the nacelle is maintained at its initial value of 212°C. Use a grid spacing of $\Delta x = 2$ mm and a time step of $\Delta t = 0.05$ s. Plot the exterior nacelle surface temperature, the exterior surface convection heat flux, and the exterior surface radiation heat flux over the time period $40 \le t \le 50$ s. Explain the physical basis for the phenomena you observe from your solution. *Hint:* The solution will be expedited by using the Der(T,t) function of *IHT*. See Problem 5.93.

5.103 Referring to Example 5.12, Comment 3, consider a sudden exposure of the surface to large surroundings at an elevated temperature (T_{sur}) and to convection (T_∞, h).

(a) Derive the explicit, finite-difference equation for the surface node in terms of *Fo*, *Bi*, and Bi_r.

(b) Obtain the stability criterion for the surface node. Does this criterion change with time? Is the criterion more restrictive than that for an interior node?

(c) A thick slab of material ($k = 1.5$ W/m · K, $\alpha = 7 \times 10^{-7}$ m²/s, $\varepsilon = 0.9$), initially at a uniform temperature of 27°C, is suddenly exposed to large surroundings at 1000 K. Neglecting convection and using a space increment of 10 mm, determine temperatures at the surface and 30 mm from the surface after an elapsed time of 1 min.

5.104 Consider the thick slab of copper in Example 5.12, which is initially at a uniform temperature of 20°C and is suddenly exposed to a net radiant flux of 3×10^5 W/m^2. Use the *Finite-Difference Equations/One-Dimensional/Transient* conduction model builder of *IHT* to obtain the implicit form of the finite-difference equations for the interior nodes. In your analysis, use a space increment of $\Delta x = 37.5$ mm with a total of 17 nodes (00–16), and a time increment of $\Delta t = 1.2$ s. For the surface node 00, use the finite-difference equation derived in Section 2 of the Example.

(a) Calculate the 00 and 04 nodal temperatures at $t = 120$ s, that is, $T(0, 120\,\text{s})$ and $T(0.15\,\text{m}, 120\,\text{s})$, and compare the results with those given in Comment 1 for the exact solution. Will a time increment of 0.12 s provide more accurate results?

(b) Plot temperature histories for $x = 0$, 150, and 600 mm, and explain key features of your results.

5.105 A constant-property, one-dimensional plane wall of width $2L$, at an initial uniform temperature T_i, is heated convectively (both surfaces) with an ambient fluid at $T_\infty = T_{\infty,1}$, $h = h_1$. At a later instant in time, $t = t_1$, heating is curtailed, and convective cooling is initiated. Cooling conditions are characterized by $T_\infty = T_{\infty,2} = T_i$, $h = h_2$.

(a) Write the heat equation as well as the initial and boundary conditions in their dimensionless form for the heating phase (Phase 1). Express the equations in terms of the dimensionless quantities θ^*, x^*, Bi_1, and Fo, where Bi_1 is expressed in terms of h_1.

(b) Write the heat equation as well as the initial and boundary conditions in their dimensionless form for the cooling phase (Phase 2). Express the equations in terms of the dimensionless quantities θ^*, x^*, Bi_2, Fo_1, and Fo where Fo_1 is the dimensionless time associated with t_1, and Bi_2 is expressed in terms of h_2. To be consistent with part (a), express the dimensionless temperature in terms of $T_\infty = T_{\infty,1}$.

(c) Consider a case for which $Bi_1 = 10$, $Bi_2 = 1$, and $Fo_1 = 0.1$. Using a finite-difference method with $\Delta x^* = 0.1$ and $\Delta Fo = 0.001$, determine the transient thermal response of the surface ($x^* = 1$), midplane ($x^* = 0$), and quarter-plane ($x^* = 0.5$) of the slab. Plot these three dimensionless temperatures as a function of dimensionless time over the range $0 \leq Fo \leq 0.5$.

(d) Determine the minimum dimensionless temperature at the midplane of the wall, and the dimensionless time at which this minimum temperature is achieved.

5.106 Steel-reinforced concrete pillars are used in the construction of large buildings. Structural failure can occur at high temperatures due to a fire because of softening of the metal core. Consider a 200-mm-thick composite pillar consisting of a central steel core (50 mm thick) sandwiched between two 75-mm-thick concrete walls. The pillar is at a uniform initial temperature of $T_i = 27$°C and is suddenly exposed to combustion products at $T_\infty = 900$°C, $h = 40$ W/m$^2 \cdot$ K on both exposed surfaces. The surroundings temperature is also 900°C.

(a) Using an implicit finite difference method with $\Delta x = 10$ mm and $\Delta t = 100$ s, determine the temperature of the exposed concrete surface and the center of the steel plate at $t = 10,000$ s. Steel properties are: $k_s = 55$ W/m $\cdot$ K, $\rho_s = 7850$ kg/m^3, and $c_s = 450$ J/kg $\cdot$ K. Concrete properties are: $k_c = 1.4$ W/m $\cdot$ K, $\rho_c = 2300$ kg/m^3, $c_c = 880$ J/kg $\cdot$ K, and $\varepsilon = 0.90$. Plot the maximum and minimum concrete temperatures along with the maximum and minimum steel temperatures over the duration $0 \leq t \leq 10,000$ s.

(b) Repeat part (a) but account for a thermal contact resistance of $R_{t,c}'' = 0.20$ m$^2 \cdot$ K/W at the concrete-steel interface.

(c) At $t = 10,000$ s, the fire is extinguished, and the surroundings and ambient temperatures return to $T_\infty = T_{\text{sur}} = 27$°C. Using the same convection heat transfer coefficient and emissivity as in parts (a) and (b), determine the maximum steel temperature and the *critical time* at which the maximum steel temperature occurs for cases with and without the contact resistance. Plot the concrete surface temperature, the concrete temperature adjacent to the steel, and the steel temperatures over the duration $10,000 \leq t \leq 20,000$ s.

5.107 One end of a stainless steel (AISI 316) rod of diameter 10 mm and length 0.16 m is inserted into a fixture maintained at 200°C. The rod, covered with an insulating sleeve, reaches a uniform temperature throughout its length. When the sleeve is removed, the rod is subjected to ambient air at 25°C such that the convection heat transfer coefficient is 30 W/m$^2 \cdot$ K.

(a) Using the explicit finite-difference technique with a space increment of $\Delta x = 0.016$ m, estimate the time required for the midlength of the rod to reach 100°C.

(b) With $\Delta x = 0.016$ m and $\Delta t = 10$ s, compute $T(x, t)$ for $0 \leq t \leq t_1$, where t_1 is the time required for the midlength of the rod to reach 50°C. Plot the temperature distribution for $t = 0$, 200 s, 400 s, and t_1.

5.108 Consider the acceleration-grid foil ($k = 40$ W/m $\cdot$ K, $\alpha = 3 \times 10^{-5}$ m^2/s, $\varepsilon = 0.45$) of Problem 4.63. Develop an implicit, finite-difference model of the foil, which can be used for the following purposes.

(a) Assuming the foil to be at a uniform temperature of 300 K when the ion beam source is activated, obtain a plot of the midspan temperature–time history. At what elapsed time does this point on the foil reach a temperature within 1 K of the steady-state value?

(b) The foil is operating under steady-state conditions when, suddenly, the ion beam is deactivated. Obtain a plot of the subsequent midspan temperature–time history. How long does it take for the hottest point on the foil to cool to 315 K, a *safe-to-touch* condition?

5.109 A tantalum rod of diameter 3 mm and length 120 mm is supported by two electrodes within a large vacuum enclosure. Initially the rod is in equilibrium with the electrodes and its surroundings, which are maintained at 300 K. Suddenly, an electrical current, $I = 80$ A, is passed through the rod. Assume the emissivity of the rod is 0.1 and the electrical resistivity is 95×10^{-8} $\Omega \cdot$ m. Use Table A.1 to obtain the other thermophysical properties required in your solution. Use a finite-difference method with a space increment of 10 mm.

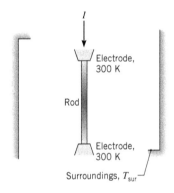

(a) Estimate the time required for the midlength of the rod to reach 1000 K.

(b) Determine the steady-state temperature distribution and estimate approximately how long it will take to reach this condition.

Finite-Difference Equations: Cylindrical Coordinates

5.110 A thin circular disk is subjected to induction heating from a coil, the effect of which is to provide a uniform heat generation within a ring section as shown. Convection occurs at the upper surface, while the lower surface is well insulated.

(a) Derive the transient, finite-difference equation for node m, which is within the region subjected to induction heating.

(b) On $T - r$ coordinates sketch, in qualitative manner, the steady-state temperature distribution, identifying important features.

5.111 An electrical cable, experiencing uniform volumetric generation $\dot{q}$, is half buried in an insulating material while the upper surface is exposed to a convection process (T_∞, h).

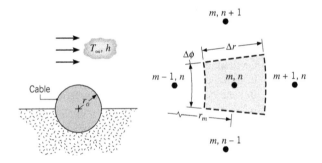

(a) Derive the explicit, finite-difference equations for an interior node (m, n), the center node $(m = 0)$, and the outer surface nodes (M, n) for the convection and insulated boundaries.

(b) Obtain the stability criterion for each of the finite-difference equations. Identify the most restrictive criterion.

Finite-Difference Solutions: Two-Dimensional Systems

5.112 Two very long (in the direction normal to the page) bars having the prescribed initial temperature distributions are to be soldered together. At time $t = 0$, the

$m = 3$ face of the copper (pure) bar contacts the $m = 4$ face of the steel (AISI 1010) bar. The solder and flux act as an interfacial layer of negligible thickness and effective contact resistance $R''_{t,c} = 2 \times 10^{-5} \ m^2 \cdot K/W$.

Initial Temperatures (K)

n/m	1	2	3	4	5	6
1	700	700	700	1000	900	800
2	700	700	700	1000	900	800
3	700	700	700	1000	900	800

(a) Derive the explicit, finite-difference equation in terms of Fo and $Bi_c = \Delta x/kR''_{t,c}$ for $T_{4,2}$ and determine the corresponding stability criterion.

(b) Using $Fo = 0.01$, determine $T_{4,2}$ one time step after contact is made. What is Δt? Is the stability criterion satisfied?

5.113 A flue passing hot exhaust gases has a square cross section, 300 mm to a side. The walls are constructed of refractory brick 150 mm thick with a thermal conductivity of 0.85 W/m · K and thermal diffusivity of $\alpha = 5.5 \times 10^{-7} \ m^2/s$. Initially with no flue gases flowing, the walls are at a uniform temperature of 25°C. The interior surface is suddenly exposed to hot gases at 350°C with a convection coefficient of 100 W/m² · K, while the exterior surface experiences convection with air at 25°C and a convection coefficient of 5 W/m² · K. Using the implicit, finite-difference method with a grid spacing of 50 mm and a time increment of 1 h, find the temperature distribution in the wall and the rate of heat loss by convection from the exterior of the flue at 5, 10, 50, and 100 h after introduction of the flue gases.

5.114 Small-diameter electrical heating elements dissipating 50 W/m (length normal to the sketch) are used to heat a ceramic plate of thermal conductivity 2 W/m · K. The upper surface of the plate is exposed to ambient air at 30°C with a convection coefficient of 100 W/m² · K, while the lower surface is well insulated.

Initially, the ceramic plate ($\alpha = 1.5 \times 10^{-6} \ m^2/s$) is at a uniform temperature of 30°C, and suddenly the electrical heating elements are energized. Using the implicit finite-difference method, estimate the time required for the difference between the surface immediately above a heating element and initial temperatures to reach 95% of the difference for steady-state conditions. Use a grid spacing of $\Delta x = 6$ mm and $\Delta y = 2$ mm and a time increment of 1 s.

Introduction to Convection

6

Thus far we have focused on heat transfer by conduction and have considered convection only to the extent that it provides a possible boundary condition for conduction problems. In Section 1.2.2 we used the term *convection* to describe energy transfer between a surface and a fluid moving over the surface. Convection includes energy transfer by both the bulk fluid motion (advection) and the random motion of fluid molecules (conduction or diffusion).

In our treatment of convection, we have two major objectives. In addition to obtaining an understanding of the physical mechanisms that underlie convection transfer, we wish to develop the means to perform convection transfer calculations. This chapter and the material of Appendix E are devoted primarily to achieving the former objective. Physical origins are discussed, and relevant dimensionless parameters, as well as important analogies, are developed.

A unique feature of this chapter is the manner in which convection mass transfer effects are introduced by analogy to those of convection heat transfer. In mass transfer by convection, gross fluid motion combines with diffusion to promote the transfer of a species for which there exists a concentration gradient. In this text, we focus on convection mass transfer that occurs at the surface of a volatile solid or liquid due to motion of a gas over the surface.

With conceptual foundations established, subsequent chapters are used to develop useful tools for quantifying convection effects. Chapters 7 and 8 present methods for computing the coefficients associated with forced convection in external and internal flow configurations, respectively. Chapter 9 describes methods for determining these coefficients in free convection, and Chapter 10 considers the problem of convection with phase change (boiling and condensation). Chapter 11 develops methods for designing and evaluating the performance of heat exchangers, devices that are widely used in engineering practice to effect heat transfer between fluids.

Accordingly, we begin by developing our understanding of the nature of convection.

6.1 *The Convection Boundary Layers*

The concept of boundary layers is central to the understanding of convection heat and mass transfer between a surface and a fluid flowing past it. In this section, velocity, thermal, and concentration boundary layers are described, and their relationships to the friction coefficient, convection heat transfer coefficient, and convection mass transfer coefficient are introduced.

6.1.1 **The Velocity Boundary Layer**

To introduce the concept of a boundary layer, consider flow over the flat plate of Figure 6.1. When fluid particles make contact with the surface, their velocity is reduced significantly relative to the fluid velocity upstream of the plate, and for most situations it is valid to assume that the particle velocity is zero at the wall.[1] These particles then act to retard the motion of particles in the adjoining fluid layer, which act to retard the motion of particles in the next layer, and so on until, at a distance $y = \delta$ from the surface, the effect becomes negligible.

---

[1]This is an approximation of the situation discussed in Section 3.7.3, wherein fluid molecules or particles continually collide with and are reflected from the surface. The momentum of an individual fluid particle will change in response to its collision with the surface. This effect may be described by *momentum accommodation coefficients*, as will be discussed in Section 8.8. In this chapter, we assume that nano- and microscale effects are not important, in which case the assumption of zero fluid velocity at the wall is valid.

FIGURE 6.1 Velocity boundary layer development on a flat plate.

This retardation of fluid motion is associated with *shear stresses* τ acting in planes that are parallel to the fluid velocity (Figure 6.1). With increasing distance y from the surface, the x velocity component of the fluid, u, must then increase until it approaches the free stream value u_∞. The subscript ∞ is used to designate conditions in the *free stream* outside the boundary layer.

The quantity δ is termed the *boundary layer thickness*, and it is typically defined as the value of y for which $u = 0.99u_\infty$. The *boundary layer velocity profile* refers to the manner in which u varies with y through the boundary layer. Accordingly, the fluid flow is characterized by two distinct regions, a thin fluid layer (the boundary layer) in which velocity gradients and shear stresses are large and a region outside the boundary layer in which velocity gradients and shear stresses are negligible. With increasing distance from the leading edge, the effects of viscosity penetrate farther into the free stream and the boundary layer grows (δ increases with x).

Because it pertains to the fluid velocity, the foregoing boundary layer may be referred to more specifically as the *velocity boundary layer*. It develops whenever there is fluid flow over a surface, and it is of fundamental importance to problems involving convection transport. In fluid mechanics its significance to the engineer stems from its relation to the surface shear stress τ_s, and hence to surface frictional effects. For external flows it provides the basis for determining the local *friction coefficient*

$$C_f \equiv \frac{\tau_s}{\rho u_\infty^2 / 2} \tag{6.1}$$

a key dimensionless parameter from which the surface frictional drag may be determined. Assuming a *Newtonian fluid*, the surface shear stress may be evaluated from knowledge of the velocity gradient at the surface

$$\tau_s = \mu \frac{\partial u}{\partial y}\bigg|_{y=0} \tag{6.2}$$

where μ is a fluid property known as the *dynamic viscosity*. In a velocity boundary layer, the velocity gradient at the surface depends on the distance x from the leading edge of the plate. Therefore, the surface shear stress and friction coefficient also depend on x.

6.1.2 The Thermal Boundary Layer

Just as a velocity boundary layer develops when there is fluid flow over a surface, a *thermal boundary layer* must develop if the fluid free stream and surface temperatures differ. Consider flow over an isothermal flat plate (Figure 6.2). At the leading edge the *temperature*

FIGURE 6.2   Thermal boundary layer development on an isothermal flat plate.

profile is uniform, with $T(y) = T_\infty$. However, fluid particles that come into contact with the plate achieve thermal equilibrium at the plate's surface temperature.[2] In turn, these particles exchange energy with those in the adjoining fluid layer, and temperature gradients develop in the fluid. The region of the fluid in which these temperature gradients exist is the thermal boundary layer, and its thickness δ_t is typically defined as the value of y for which the ratio $[(T_s - T)/(T_s - T_\infty)] = 0.99$. With increasing distance from the leading edge, the effects of heat transfer penetrate farther into the free stream and the thermal boundary layer grows.

The relation between conditions in this boundary layer and the convection heat transfer coefficient may readily be demonstrated. At any distance x from the leading edge, the *local* surface heat flux may be obtained by applying Fourier's law to the *fluid* at $y = 0$. That is,

$$q_s'' = -k_f \frac{\partial T}{\partial y}\bigg|_{y=0} \tag{6.3}$$

The subscript s has been used to emphasize that this is the surface heat flux, but it will be dropped in later sections. This expression is appropriate because, *at the surface, there is no fluid motion and energy transfer occurs only by conduction.* Recalling Newton's law of cooling, we see that

$$q_s'' = h(T_s - T_\infty) \tag{6.4}$$

and combining this with Equation 6.3, we obtain

$$h = \frac{-k_f \partial T/\partial y\big|_{y=0}}{T_s - T_\infty} \tag{6.5}$$

Hence, conditions in the thermal boundary layer, which strongly influence the wall temperature gradient $\partial T/\partial y\big|_{y=0}$, determine the rate of heat transfer from the surface. Since $(T_s - T_\infty)$ is a constant, independent of x, while δ_t increases with increasing x, temperature gradients in the boundary layer must decrease with increasing x. Accordingly, the magnitude of $\partial T/\partial y\big|_{y=0}$ decreases with increasing x, and it follows that q_s'' and h decrease with increasing x.

---

[2]Micro- and nanoscale effects are assumed to be negligible in this chapter. Hence, the thermal accommodation coefficient of Section 3.7.3 attains a value of unity, in which case the fluid particles achieve thermal equilibrium with the surface of the plate. Micro- and nanoscale effects will be discussed in Section 8.8.

FIGURE 6.3 Species concentration boundary layer development on a flat plate.

6.1.3 The Concentration Boundary Layer

When air moves past the surface of a pool of water, the liquid water will evaporate, and water vapor will be transferred into the airstream. This is an example of convection mass transfer. More generally, consider a *binary mixture* of chemical species A and B that flows over a surface (Figure 6.3). The molar concentration (kmol/m^3) of species A at the surface is $C_{A,s}$, and in the free stream it is $C_{A,\infty}$. If $C_{A,s}$ differs from $C_{A,\infty}$, transfer of species A by convection will occur. For example, species A could be a vapor that is transferred into a gas stream (species B) due to *evaporation* at a liquid surface (as in the water example) or due to *sublimation* at a solid surface. In this situation, a *concentration boundary layer* will develop that is similar to the velocity and thermal boundary layers. The concentration boundary layer is the region of the fluid in which concentration gradients exist, and its thickness δ_c is typically defined as the value of y for which $[(C_{A,s} - C_A)/(C_{A,s} - C_{A,\infty})] = 0.99$. With increasing distance from the leading edge, the effects of species transfer penetrate farther into the free stream and the concentration boundary layer grows.

Species transfer by convection between the surface and the free stream fluid is determined by conditions in the boundary layer, and we are interested in determining the rate at which this transfer occurs. In particular, we are interested in the molar flux of species A, N_A'' (kmol/s · m^2). It is helpful to recognize that the molar flux associated with species transfer *by diffusion* is determined by an expression that is analogous to Fourier's law. For the conditions of interest in this chapter, the expression, which is termed *Fick's law*, has the form

$$N_A'' = -D_{AB}\frac{\partial C_A}{\partial y} \qquad (6.6)[3]$$

where D_{AB} is a property of the binary mixture known as the *binary diffusion coefficient*. At any point corresponding to $y > 0$ in the concentration boundary layer of Figure 6.3, species transfer is due to both bulk fluid motion (*advection*) and diffusion. However, absent nano- or microscale effects and the influence of species diffusion on the velocity normal to the surface, fluid motion at the surface can be neglected.[4] Accordingly, species transfer at the surface is only by diffusion, and applying Fick's law at $y = 0$, the molar flux is

$$N_{A,s}'' = -D_{AB}\frac{\partial C_A}{\partial y}\bigg|_{y=0} \qquad (6.7)$$

---

[3]This expression is an approximation of a more general form of Fick's law of diffusion (Section 14.1.3) when the total molar concentration of the mixture, $C = C_A + C_B$, is a constant.

[4]The basis for neglecting the effects of diffusion on bulk fluid motion is considered in Sections 14.2 and 14.3.

The subscript s has been used to emphasize that this is the molar flux at the surface, but it will be dropped in later sections. Analogous to Newton's law of cooling, an equation can be written that relates the molar flux to the concentration difference across the boundary layer, as

$$N''_{A,s} = h_m(C_{A,s} - C_{A,\infty}) \tag{6.8}$$

where h_m (m/s) is the *convection mass transfer coefficient*, analogous to the convection heat transfer coefficient. Combining Equations 6.7 and 6.8, it follows that

$$h_m = \frac{-D_{AB} \partial C_A / \partial y \big|_{y=0}}{C_{A,s} - C_{A,\infty}} \tag{6.9}$$

Therefore, conditions in the concentration boundary layer, which strongly influence the surface concentration gradient $\partial C_A / \partial y \big|_{y=0}$, also influence the convection mass transfer coefficient and hence the rate of species transfer from the surface.

6.1.4   Significance of the Boundary Layers

For flow over any surface, there will always exist a velocity boundary layer and hence surface friction. Likewise, a thermal boundary layer, and hence convection heat transfer, will always exist if the surface and free stream temperatures differ. Similarly, a concentration boundary layer and convection mass transfer will exist if the fluid's species concentration at the surface differs from its species concentration in the free stream. The velocity boundary layer is of extent $\delta(x)$ and is characterized by the presence of velocity gradients and shear stresses. The thermal boundary layer is of extent $\delta_t(x)$ and is characterized by temperature gradients and heat transfer. Finally, the concentration boundary layer is of extent $\delta_c(x)$ and is characterized by concentration gradients and species transfer. Situations can arise in which all three boundary layers are present. In such cases, the boundary layers rarely grow at the same rate, and the values of δ, δ_t, and δ_c at a given location are not the same.

For the engineer, the principal manifestations of the three boundary layers are, respectively, *surface friction*, *convection heat transfer*, and *convection mass transfer*. The key boundary layer parameters are then the *friction coefficient* C_f and the *heat* and *mass transfer convection coefficients* h and h_m, respectively. We now turn our attention to examining these key parameters, which are central to the analysis of convection heat and mass transfer problems.

6.2   *Local and Average Convection Coefficients*

6.2.1   Heat Transfer

Consider the conditions of Figure 6.4*a*. A fluid of velocity V and temperature T_∞ flows over a surface of arbitrary shape and of area A_s. The surface is presumed to be at a uniform temperature, T_s, and if $T_s \neq T_\infty$, we know that convection heat transfer will occur. From Section 6.1.2, we also know that the surface heat flux and convection heat transfer coefficient

FIGURE 6.4 Local and total convection heat transfer. (*a*) Surface of arbitrary shape. (*b*) Flat plate.

both vary along the surface. The *total heat transfer rate q* may be obtained by integrating the local flux over the entire surface. That is,

$$q = \int_{A_s} q'' dA_s \tag{6.10}$$

or, from Equation 6.4,

$$q = (T_s - T_\infty) \int_{A_s} h dA_s \tag{6.11}$$

Defining an *average convection coefficient* $\bar{h}$ for the entire surface, the total heat transfer rate may also be expressed as

$$q = \bar{h} A_s (T_s - T_\infty) \tag{6.12}$$

Equating Equations 6.11 and 6.12, it follows that the average and local convection coefficients are related by an expression of the form

$$\bar{h} = \frac{1}{A_s} \int_{A_s} h dA_s \tag{6.13}$$

Note that for the special case of flow over a flat plate (Figure 6.4*b*), h varies only with the distance x from the leading edge and Equation 6.13 reduces to

$$\bar{h} = \frac{1}{L} \int_0^L h dx \tag{6.14}$$

The Problem of Convection Heat Transfer The local flux and/or the total transfer rate are of paramount importance in any convection problem. These quantities may be determined from the rate equations, Equations 6.4 and 6.12, which depend on knowledge of the local (h) and average ($\bar{h}$) convection coefficients. It is for this reason that determination of these coefficients is viewed as *the problem of convection*. However, the problem is not a simple one, for in addition to depending on numerous *fluid properties* such as density, viscosity, thermal conductivity, and specific heat, the coefficients depend on the *surface geometry* and the *flow conditions*. This multiplicity of independent variables is attributable to the dependence of convection transfer on the boundary layers that develop on the surface.

6.2.2 Mass Transfer

Similar results may be obtained for convection mass transfer. If a fluid of species molar concentration $C_{A,\infty}$ flows over a surface at which the species concentration is maintained at some uniform value $C_{A,s} \neq C_{A,\infty}$ (Figure 6.5*a*), transfer of the species by convection will occur. From Section 6.1.3, we know that the surface molar flux and convection mass

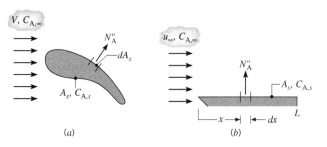

FIGURE 6.5 Local and total convection species transfer. (*a*) Surface of arbitrary shape. (*b*) Flat plate.

transfer coefficient both vary along the surface. The total molar transfer rate for an entire surface, N_A (kmol/s), may then be expressed as

$$N_A = \bar{h}_m A_s (C_{A,s} - C_{A,\infty}) \tag{6.15}$$

where the average and local mass transfer convection coefficients are related by an equation of the form

$$\bar{h}_m = \frac{1}{A_s} \int_{A_s} h_m \, dA_s \tag{6.16}$$

For the flat plate of Figure 6.5*b*, it follows that

$$\bar{h}_m = \frac{1}{L} \int_0^L h_m \, dx \tag{6.17}$$

Species transfer may also be expressed as a mass flux, n_A'' (kg/s · m^2), or as a mass transfer rate, n_A (kg/s), by multiplying both sides of Equations 6.8 and 6.15, respectively, by the molecular weight $\mathcal{M}_A$ (kg/kmol) of species A. Accordingly,

$$n_A'' = h_m (\rho_{A,s} - \rho_{A,\infty}) \tag{6.18}$$

and

$$n_A = \bar{h}_m A_s (\rho_{A,s} - \rho_{A,\infty}) \tag{6.19}$$

where ρ_A (kg/m^3) is the mass density of species A.[5] We can also write Fick's law on a mass basis by multiplying Equation 6.7 by $\mathcal{M}_A$ to yield

$$n_{A,s}'' = -D_{AB} \frac{\partial \rho_A}{\partial y} \bigg|_{y=0} \tag{6.20}$$

---

[5]Although the foregoing nomenclature is well suited for characterizing mass transfer processes of interest in this text, there is by no means a standard nomenclature, and it is often difficult to reconcile the results from different publications. A review of the different ways in which driving potentials, fluxes, and convection coefficients may be formulated is provided by Webb [1].

Furthermore, multiplying the numerator and denominator of Equation 6.9 by $\mathcal{M}_A$ yields an alternative expression for h_m:

$$h_m = \frac{-D_{AB}\,\partial\rho_A/\partial y\big|_{y=0}}{\rho_{A,s} - \rho_{A,\infty}} \qquad (6.21)$$

To perform a convection mass transfer calculation, it is necessary to determine the value of $C_{A,s}$ or $\rho_{A,s}$. Such a determination may be made by assuming thermodynamic equilibrium at the interface between the gas and the liquid or solid phase. One implication of equilibrium is that the temperature of the vapor at the interface is equal to the surface temperature T_s. A second implication is that the vapor is in a *saturated state*, in which case thermodynamic tables, such as Table A.6 for water, may be used to obtain its density from knowledge of T_s. To a good approximation, the molar concentration of the vapor at the surface may also be determined from the vapor pressure through application of the equation of state for an ideal gas. That is,

$$C_{A,s} = \frac{p_{\text{sat}}(T_s)}{\mathcal{R}T_s} \qquad (6.22)$$

where $\mathcal{R}$ is the universal gas constant and $p_{\text{sat}}(T_s)$ is the vapor pressure corresponding to saturated conditions at T_s. Note that the vapor mass density and molar concentration are related by $\rho_A = \mathcal{M}_A C_A$.

The Problem of Convection Mass Transfer The local mass flux and/or the total mass transfer rate can be determined from Equations 6.8 and 6.15, which depend on knowledge of the local (h_m) and average ($\bar{h}_m$) mass transfer convection coefficients. The *problem of convection mass transfer* is then to determine these coefficients, which depend on *fluid properties* such as density, viscosity, and diffusion coefficient, the *surface geometry*, and the *flow conditions*.

EXAMPLE 6.1

Experimental results for the local heat transfer coefficient h_x for flow over a flat plate with an extremely rough surface were found to fit the relation

$$h_x(x) = ax^{-0.1}$$

where a is a coefficient (W/m$^{1.9}$ · K) and x (m) is the distance from the leading edge of the plate.

1. Develop an expression for the ratio of the average heat transfer coefficient $\bar{h}_x$ over the region from 0 to x to the local heat transfer coefficient h_x at x.
2. Plot the variation of h_x and $\bar{h}_x$ as a function of x.

SOLUTION

Known: Variation of the local heat transfer coefficient, $h_x(x)$.

Find:

1. The ratio of the average heat transfer coefficient $\bar{h}(x)$ to the local value $h_x(x)$.
2. Plot of the variation of h_x and $\bar{h}_x$ with x.

Schematic:

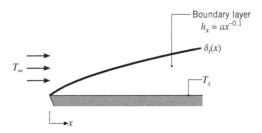

Analysis:

1. From Equation 6.14 the average value of the convection heat transfer coefficient, $h_x(x) = ax^{-0.1}$, from 0 to x is

$$\bar{h}_x = \frac{1}{x}\int_0^x ax^{-0.1}\,dx = \frac{a}{x}\int_0^x x^{-0.1}\,dx = \frac{a}{x}\left(\frac{x^{+0.9}}{0.9}\right) = 1.11ax^{-0.1}$$

or

$$\bar{h}_x = 1.11h_x \qquad \triangleleft$$

2. The variation of h_x and $\bar{h}_x$ with x is as follows:

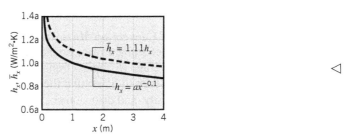

$\triangleleft$

Comments: Boundary layer development causes both the local and average coefficients to decrease with increasing distance from the leading edge. The average coefficient up to x must therefore exceed the local value at x.

EXAMPLE 6.2

A long circular cylinder 20 mm in diameter is fabricated from solid naphthalene, a common moth repellant, and is exposed to an airstream that provides for an average convection mass transfer coefficient of $\bar{h}_m = 0.05$ m/s. The molar concentration of naphthalene vapor at the cylinder surface is 5×10^{-6} kmol/m³, and its molecular weight is 128.16 kg/kmol. What is the mass sublimation rate per unit length of cylinder?

SOLUTION

Known: Saturated vapor concentration of naphthalene.

Find: Sublimation rate per unit length, n'_A (kg/s · m).

Schematic:

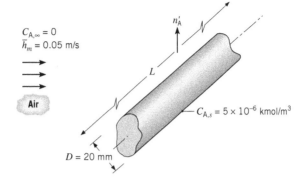

$C_{A,\infty} = 0$
$\bar{h}_m = 0.05$ m/s

Air

n'_A

L

$C_{A,s} = 5 \times 10^{-6}$ kmol/m³

$D = 20$ mm

Assumptions:

1. Steady-state conditions.

2. Negligible concentration of naphthalene in free stream of air.

Analysis: Naphthalene is transported to the air by convection, and from Equation 6.15, the molar transfer rate for the cylinder is

$$N_A = \bar{h}_m\, \pi D L (C_{A,s} - C_{A,\infty})$$

With $C_{A,\infty} = 0$ and $N'_A = N_A/L$, it follows that

$$N'_A = (\pi D)\bar{h}_m C_{A,s} = \pi \times 0.02 \text{ m} \times 0.05 \text{ m/s} \times 5 \times 10^{-6} \text{ kmol/m}^3$$

$$N'_A = 1.57 \times 10^{-8} \text{ kmol/s} \cdot \text{m}$$

The mass sublimation rate is then

$$n'_A = \mathcal{M}_A N'_A = 128.16 \text{ kg/kmol} \times 1.57 \times 10^{-8} \text{ kmol/s} \cdot \text{m}$$

$$n'_A = 2.01 \times 10^{-6} \text{ kg/s} \cdot \text{m} \qquad \triangleleft$$

EXAMPLE 6.3

At some location on the surface of a pan of water in air at one atmosphere, measurements of the partial pressure of water vapor p_A (atm) are made as a function of the distance y from the surface, and the results are as follows:

Determine the convection mass transfer coefficient $h_{m,x}$ at this location.

SOLUTION

Known: Partial pressure p_A of water vapor as a function of distance y at a particular location on the surface of a water layer.

Find: Convection mass transfer coefficient at the prescribed location.

Schematic:

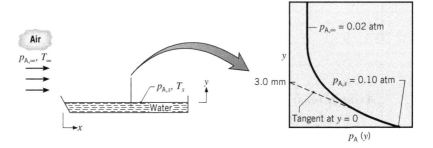

Assumptions:
1. Water vapor may be approximated as an ideal gas.
2. Conditions are isothermal.

Properties: Table A.6, saturated vapor (0.1 atm = 0.101 bar): $T_s = 319$ K. Table A.8, water vapor–air (319 K): D_{AB} (319 K) = D_{AB} (298 K) × (319 K/298 K)$^{3/2}$ = 0.288 × 10^{-4} m^2/s.

Analysis: From Equation 6.21 the local convection mass transfer coefficient is

$$h_{m,x} = \frac{-D_{AB}\, \partial \rho_A / \partial y \big|_{y=0}}{\rho_{A,s} - \rho_{A,\infty}}$$

or, approximating the vapor as an ideal gas

$$p_A = \rho_A R T$$

with constant T (isothermal conditions),

$$h_{m,x} = \frac{-D_{AB}\, \partial p_A / \partial y \big|_{y=0}}{p_{A,s} - p_{A,\infty}}$$

From the measured vapor pressure distribution

$$\frac{\partial p_A}{\partial y}\bigg|_{y=0} = \frac{(0 - 0.1)\ \text{atm}}{(0.003 - 0)\ \text{m}} = -33.3\ \text{atm/m}$$

Hence

$$h_{m,x} = \frac{-0.288 \times 10^{-4}\ \text{m}^2/\text{s}\ (-33.3\ \text{atm/m})}{(0.1 - 0.02)\ \text{atm}} = 0.0120\ \text{m/s} \qquad \triangleleft$$

Comments: From thermodynamic equilibrium at the liquid–vapor interface, the interfacial temperature, $T_s = 319$ K, was determined from Table A.6.

6.3 *Laminar and Turbulent Flow*

In the discussion of convection so far, we have not addressed the significance of the *flow conditions*. An essential step in the treatment of any convection problem is to determine whether the boundary layer is *laminar* or *turbulent*. Surface friction and the convection transfer rates depend strongly on which of these conditions exists.

6.3.1 Laminar and Turbulent Velocity Boundary Layers

Boundary layer *development* on a flat plate is illustrated in Figure 6.6. In many cases, laminar and turbulent flow conditions both occur, with the laminar section preceding the turbulent section. For either condition, the fluid motion is characterized by velocity components in the *x*- and *y*-directions. Fluid motion away from the surface is necessitated by the slowing of the fluid near the wall as the boundary layer grows in the *x*-direction. Figure 6.6 shows that there are sharp differences between laminar and turbulent flow conditions, as described in the following paragraphs.

In the laminar boundary layer, the fluid flow is highly ordered and it is possible to identify streamlines along which fluid particles move. From Section 6.1.1 we know that the boundary layer thickness grows and that velocity gradients at $y = 0$ decrease in the streamwise (increasing *x*) direction. From Equation 6.2, we see that the local surface shear stress τ_s also decreases with increasing *x*. The highly ordered behavior continues until a *transition* zone is reached, across which a conversion from laminar to turbulent conditions occurs. Conditions within the transition zone change with time, with the flow sometimes exhibiting laminar behavior and sometimes exhibiting the characteristics of turbulent flow.

Flow in the fully turbulent boundary layer is, in general, highly irregular and is characterized by random, three-dimensional motion. Mixing within the boundary layer carries high-speed fluid toward the solid surface and transfers slower-moving fluid farther into the free stream. Much of the mixing is promoted by streamwise vortices called *streaks* that are generated intermittently near the flat plate, where they rapidly grow and decay. Recent analytical and experimental studies have suggested that these and other *coherent structures* within the turbulent flow can travel in *waves* at velocities that can exceed u_∞, interact nonlinearly, and spawn the chaotic conditions that characterize turbulent flow [2].

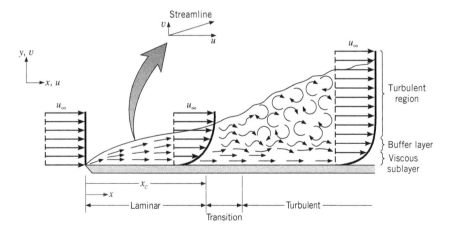

FIGURE 6.6 Velocity boundary layer development on a flat plate.

As a result of the interactions that lead to chaotic flow conditions, velocity and pressure fluctuations occur at any point within the turbulent boundary layer. Three different regions may be delineated within the turbulent boundary layer as a function of distance from the surface. We may speak of a *viscous sublayer* in which transport is dominated by diffusion and the velocity profile is nearly linear. There is an adjoining *buffer layer* in which diffusion and turbulent mixing are comparable, and there is a *turbulent zone* in which transport is dominated by turbulent mixing. A comparison of the laminar and turbulent boundary layer profiles for the *x*-component of the velocity, provided in Figure 6.7, shows that the turbulent velocity profile is relatively flat due to the mixing that occurs within the buffer layer and turbulent region, giving rise to large velocity gradients within the viscous sublayer. Hence, τ_s is generally larger in the turbulent portion of the boundary layer of Figure 6.6 than in the laminar portion.

The transition from laminar to turbulent flow is ultimately due to *triggering mechanisms*, such as the interaction of unsteady flow structures that develop naturally within the fluid or small disturbances that exist within many typical boundary layers. These disturbances may originate from fluctuations in the free stream, or they may be induced by surface roughness or minute surface vibrations. The onset of turbulence depends on whether the triggering mechanisms are amplified or attenuated in the direction of fluid flow, which in turn depends on a dimensionless grouping of parameters called the *Reynolds number*,

$$Re_x = \frac{\rho u_\infty x}{\mu} \tag{6.23}$$

where, for a flat plate, the characteristic length is *x*, the distance from the leading edge. It will be shown later that the Reynolds number represents the ratio of the inertia to viscous forces. If the Reynolds number is small, inertia forces are insignificant relative to viscous forces. The disturbances are then dissipated, and the flow remains laminar. For a large Reynolds number, however, the inertia forces can be sufficient to amplify the triggering mechanisms, and a transition to turbulence occurs.

In determining whether the boundary layer is laminar or turbulent, it is frequently reasonable to assume that transition begins at some location x_c, as shown in Figure 6.6. This location is determined by the *critical* Reynolds number, $Re_{x,c}$. For flow over a flat plate,

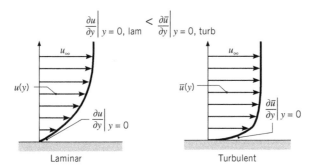

FIGURE 6.7 Comparison of laminar and turbulent velocity boundary layer profiles for the same free stream velocity.[6]

---

[6]Since velocity fluctuates with time in turbulent flow, the time-averaged velocity, $\bar{u}$, is plotted in Figure 6.7.

$Re_{x,c}$ is known to vary from approximately 10^5 to 3×10^6, depending on surface roughness and the turbulence level of the free stream. A representative value of

$$Re_{x,c} \equiv \frac{\rho u_\infty x_c}{\mu} = 5 \times 10^5 \qquad (6.24)$$

is often assumed for boundary layer calculations and, unless otherwise noted, is used for the calculations of this text that involve a flat plate.

6.3.2 Laminar and Turbulent Thermal and Species Concentration Boundary Layers

Since the velocity distribution determines the advective component of thermal energy or chemical species transfer within the boundary layer, the nature of the flow also has a profound effect on convective heat and mass transfer rates. For laminar conditions, we have seen that the thermal and species boundary layers grow in the streamwise (increasing x) direction. Hence, temperature and species concentration gradients in the fluid at $y = 0$ decrease in the streamwise direction, and, from Equations 6.5 and 6.9, the heat and mass transfer coefficients also decrease with increasing x.

Just as it induces large velocity gradients at $y = 0$, as shown in Figure 6.7, turbulent mixing promotes large temperature and species concentration gradients adjacent to the solid surface as well as a corresponding increase in the heat and mass transfer coefficients across the transition region.

When laminar and turbulent conditions both occur, the velocity boundary layer thickness δ and the local convection heat transfer coefficient h vary as illustrated in Figure 6.8. Because turbulence induces mixing, which in turn reduces the importance of conduction and diffusion in determining the thermal and species boundary layer thicknesses, *differences* in the thicknesses of the velocity, thermal, and species boundary layers tend to be much smaller in turbulent flow than in laminar flow. As is evident in Equation 6.24, the presence of heat and/or mass transfer can affect the location of the transition from laminar to turbulent flow x_c since the density and dynamic viscosity of the fluid can depend on the temperature or species concentration.

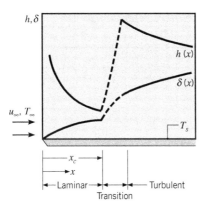

FIGURE 6.8 Variation of velocity boundary layer thickness δ and the local heat transfer coefficient h for flow over an isothermal flat plate.

EXAMPLE 6.4

Water flows at a velocity $u_\infty = 1$ m/s over a flat plate of length $L = 0.6$ m. Consider two cases, one for which the water temperature is approximately 300 K and the other for an approximate water temperature of 350 K. In the laminar and turbulent regions, experimental measurements show that the local convection coefficients are well described by

$$h_{\text{lam}}(x) = C_{\text{lam}}x^{-0.5} \quad h_{\text{turb}}(x) = C_{\text{turb}}x^{-0.2}$$

where x has units of m. At 300 K,

$$C_{\text{lam},300} = 395 \text{ W/m}^{1.5}\cdot\text{K} \quad C_{\text{turb},300} = 2330 \text{ W/m}^{1.8}\cdot\text{K}$$

while at 350 K,

$$C_{\text{lam},350} = 477 \text{ W/m}^{1.5}\cdot\text{K} \quad C_{\text{turb},350} = 3600 \text{ W/m}^{1.8}\cdot\text{K}$$

As is evident, the constant C depends on the nature of the flow as well as the water temperature because of the thermal dependence of various properties of the fluid.

Determine the average convection coefficient, $\overline{h}$, over the entire plate for the two water temperatures.

SOLUTION

Known: Water flow over a flat plate, expressions for the dependence of the local convection coefficient with distance from the plate's leading edge x, and approximate temperature of the water.

Find: Average convection coefficient, $\overline{h}$.

Schematic:

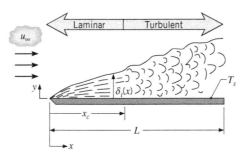

Assumptions:

1. Steady-state conditions.
2. Transition occurs at a critical Reynolds number of $Re_{x,c} = 5 \times 10^5$.

Properties: Table A.6, water ($\overline{T} \approx 300$ K): $\rho = v_f^{-1} = 997$ kg/m^3, $\mu = 855 \times 10^{-6}$ N · s/m^2. Table A.6 ($\overline{T} \approx 350$ K): $\rho = v_f^{-1} = 974$ kg/m^3, $\mu = 365 \times 10^{-6}$ N · s/m^2.

Analysis: The local convection coefficient is highly dependent on whether laminar or turbulent conditions exist. Therefore, we first determine the extent to which these conditions exist by finding the location where transition occurs, x_c. From Equation 6.24, we know that at 300 K,

$$x_c = \frac{Re_{x,c}\mu}{\rho u_\infty} = \frac{5 \times 10^5 \times 855 \times 10^{-6} \text{ N} \cdot \text{s/m}^2}{997 \text{ kg/m}^3 \times 1 \text{ m/s}} = 0.43 \text{ m}$$

while at 350 K,

$$x_c = \frac{Re_{x,c}\mu}{\rho u_\infty} = \frac{5 \times 10^5 \times 365 \times 10^{-6} \text{ N} \cdot \text{s/m}^2}{974 \text{ kg/m}^3 \times 1 \text{ m/s}} = 0.19 \text{ m}$$

From Equation 6.14 we know that

$$\overline{h} = \frac{1}{L} \int_0^L h\,dx = \frac{1}{L}\left[\int_0^{x_c} h_{\text{lam}}\,dx + \int_{x_c}^L h_{\text{turb}}\,dx \right]$$

or

$$\overline{h} = \frac{1}{L}\left[\frac{C_{\text{lam}}}{0.5} x^{0.5} \Big|_0^{x_c} + \frac{C_{\text{turb}}}{0.8} x^{0.8} \Big|_{x_c}^L \right]$$

At 300 K,

$$\overline{h} = \frac{1}{0.6 \text{ m}}\left[\frac{395 \text{ W/m}^{1.5} \cdot \text{K}}{0.5} \times (0.43^{0.5}) \text{ m}^{0.5} + \frac{2330 \text{ W/m}^{1.8} \cdot \text{K}}{0.8} \right.$$

$$\left. \times (0.6^{0.8} - 0.43^{0.8}) \text{ m}^{0.8} \right] = 1620 \text{ W/m}^2 \cdot \text{K} \qquad \triangleleft$$

while at 350 K,

$$\overline{h} = \frac{1}{0.6 \text{ m}}\left[\frac{477 \text{ W/m}^{1.5} \cdot \text{K}}{0.5} \times (0.19^{0.5}) \text{ m}^{0.5} + \frac{3600 \text{ W/m}^{1.8} \cdot \text{K}}{0.8} \right.$$

$$\left. \times (0.6^{0.8} - 0.19^{0.8}) \text{ m}^{0.8} \right] = 3710 \text{ W/m}^2 \cdot \text{K} \qquad \triangleleft$$

The local and average convection coefficient distributions for the plate are shown in the following figure.

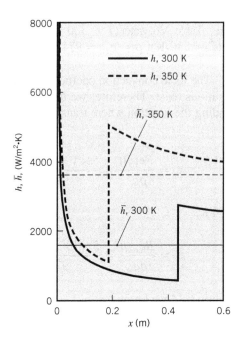

Comments:

1. The average convection coefficient at $T \approx 350$ K is over twice as large as the value at $T \approx 300$ K. This strong temperature dependence is due primarily to the shift of x_c that is associated with the smaller viscosity of the water at the higher temperature. Careful consideration of the temperature dependence of fluid properties is crucial when performing a convection heat transfer analysis.

2. Spatial variations in the local convection coefficient are significant. The largest local convection coefficients occur at the leading edge of the flat plate, where the laminar thermal boundary layer is extremely thin, and just downstream of x_c, where the turbulent boundary layer is thinnest.

6.4 *The Boundary Layer Equations*

We can improve our understanding of boundary layer behavior and further illustrate its relevance to convection transport by considering the equations that govern boundary layer conditions, such as those illustrated in Figure 6.9.

As discussed in Section 6.1, the velocity boundary layer results from the difference between the free stream velocity and the zero velocity at the wall, while the thermal boundary layer results from a difference between the free stream and surface temperatures. The fluid is considered to be a binary mixture of species A and B, and the concentration boundary layer originates from a difference between the free stream and surface concentrations ($C_{A,\infty} \neq C_{A,s}$). Illustration of the relative thicknesses ($\delta_t > \delta_c > \delta$) in Figure 6.9 is arbitrary, for the moment, and the factors that influence relative boundary layer development are discussed later in this chapter.

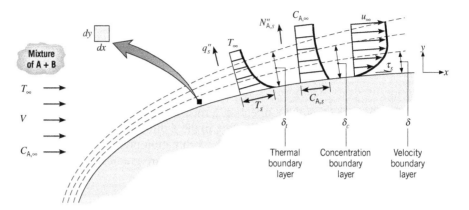

FIGURE 6.9 Development of the velocity, thermal, and concentration boundary layers for an arbitrary surface.

Our objective in this section is to examine the differential equations that govern the velocity, temperature, and species concentration fields that are applicable to boundary layer flow with heat and species transfer. Section 6.4.1 presents the laminar boundary layer equations, and Appendix F gives the corresponding equations for turbulent conditions. In Section 6.5, these equations will be used to determine important dimensionless parameters associated with convection that will be used extensively in subsequent chapters.

6.4.1 Boundary Layer Equations for Laminar Flow

 Motion of a fluid in which there are coexisting velocity, temperature, and concentration gradients must comply with several *fundamental laws of nature*. In particular, at each point in the fluid, *conservation of mass*, *energy*, and *chemical species*, as well as *Newton's second law of motion*, must be satisfied. Equations representing these requirements are derived by applying the laws to a stationary differential control volume situated in the flow. The resulting equations, in Cartesian coordinates, for the *steady, two-dimensional flow* of an *incompressible fluid* with *constant properties* are given in Appendix E. These equations serve as starting points for our analysis of laminar boundary layers. Note that turbulent flows are inherently unsteady, and the equations governing them are presented in Appendix F.

We begin by restricting attention to applications for which *body forces are negligible* ($X = Y = 0$ in Equations E.2 and E.3), there is *no thermal energy generation in the fluid* ($\dot{q} = 0$ in Equation E.4), and the *flow is nonreacting* ($\dot{N}_A = 0$ in Equation E.6). Additional simplifications may be made by invoking approximations pertinent to conditions in the velocity, thermal, and concentration boundary layers. Specifically, the boundary layer thicknesses are typically very small relative to the size of the object upon which they form, and the x-direction velocity, temperature, and concentration must change from their surface to their free stream values over these very small distances. Therefore, gradients normal to the object's surface are much larger than those along the surface. As a result, we can

---

 These equations are derived in Section 6S.1.

neglect terms that represent x-direction diffusion of momentum, thermal energy, and chemical species, relative to their y-direction counterparts. That is [3, 4]:

$$\frac{\partial^2 u}{\partial x^2} \ll \frac{\partial^2 u}{\partial y^2} \qquad \frac{\partial^2 T}{\partial x^2} \ll \frac{\partial^2 T}{\partial y^2} \qquad \frac{\partial^2 C_A}{\partial x^2} \ll \frac{\partial^2 C_A}{\partial y^2} \tag{6.25}$$

By neglecting the x-direction terms, we are assuming the net shear stress, conduction heat flux, and species diffusion flux in the x-direction to be negligible. Furthermore, because the boundary layers are so thin, the x-direction pressure gradient within a boundary layer can be approximated as the free stream pressure gradient:

$$\frac{\partial p}{\partial x} \approx \frac{dp_\infty}{dx} \tag{6.26}$$

The form of $p_\infty(x)$ depends on the surface geometry and may be obtained from a separate consideration of flow conditions in the free stream where shear stresses are negligible [5]. Hence, the pressure gradient may be treated as a known quantity.

With the foregoing simplifications and approximations, the overall continuity equation is unchanged from Equation E.1:

$$\frac{\partial u}{\partial x} + \frac{\partial v}{\partial y} = 0 \tag{6.27}$$

This equation is an outgrowth of applying conservation of mass to the differential, $dx \cdot dy \cdot 1$ control volume shown in Figure 6.9. The two terms represent the *net* outflow (outflow minus inflow) of mass in the x- and y-directions, the sum of which must be zero for steady flow.

The x-momentum equation (Equation E.2) reduces to:

$$u\frac{\partial u}{\partial x} + v\frac{\partial u}{\partial y} = -\frac{1}{\rho}\frac{dp_\infty}{dx} + v\frac{\partial^2 u}{\partial y^2} \tag{6.28}$$

This equation results from application of Newton's second law of motion in the x-direction to the $dx \cdot dy \cdot 1$ differential control volume in the fluid. The left-hand side represents the net rate at which x-momentum leaves the control volume due to fluid motion across its boundaries. The first term on the right-hand side represents the net pressure force, and the second term represents the net force due to viscous shear stresses.

The energy equation (Equation E.4) reduces to:

$$u\frac{\partial T}{\partial x} + v\frac{\partial T}{\partial y} = \alpha\frac{\partial^2 T}{\partial y^2} + \frac{v}{c_p}\left(\frac{\partial u}{\partial y}\right)^2 \tag{6.29}$$

This equation results from application of conservation of energy to the $dx \cdot dy \cdot 1$ differential control volume in the flowing fluid. Terms on the left-hand side account for the net rate at which thermal energy leaves the control volume due to bulk fluid motion (advection). The first term on the right-hand side accounts for the net inflow of thermal energy due to y-direction conduction. The last term on the right-hand side is what remains of the viscous dissipation, Equation E.5, when it is acknowledged that, in a boundary layer, the velocity component in the direction along the surface, u, is much larger than that normal to the surface, v, and gradients normal to the surface are much larger than those along the surface. In many situations this term may be neglected relative to those that account for advection and conduction. However, aerodynamic heating that accompanies high-speed (especially supersonic) flight is a noteworthy situation in which this term is important.

The species conservation equation (Equation E.6) reduces to:

$$u\frac{\partial C_A}{\partial x} + v\frac{\partial C_A}{\partial y} = D_{AB}\frac{\partial^2 C_A}{\partial y^2} \tag{6.30}$$

This equation is obtained by applying conservation of chemical species to the $dx \cdot dy \cdot 1$ differential control volume in the flow. Terms on the left-hand side account for net transport of species A due to bulk fluid motion (advection), while the right-hand side represents the net inflow due to y-direction diffusion.

After specifying appropriate boundary conditions, Equations 6.27 through 6.30 may be solved to determine the spatial variations of u, v, T, and C_A in the different laminar boundary layers. For incompressible, constant property flow, Equations 6.27 and 6.28 are *uncoupled* from Equations 6.29 and 6.30. That is, Equations 6.27 and 6.28 may be solved for the *velocity field*, $u(x, y)$ and $v(x, y)$, without consideration of Equations 6.29 and 6.30. From knowledge of $u(x, y)$, the velocity gradient $(\partial u/\partial y)|_{y=0}$ could then be evaluated, and the wall shear stress could be obtained from Equation 6.2. In contrast, through the appearance of u and v in Equations 6.29 and 6.30, the temperature and species concentration are *coupled* to the velocity field. Hence $u(x, y)$ and $v(x, y)$ must be known before Equations 6.29 and 6.30 may be solved for $T(x, y)$ and $C_A(x, y)$. Once $T(x, y)$ and $C_A(x, y)$ have been obtained from such solutions, the convection heat and mass transfer coefficients may be determined from Equations 6.5 and 6.9, respectively. It follows that these coefficients depend strongly on the velocity field.[7]

Because boundary layer solutions generally involve mathematics beyond the scope of this text, our treatment of such solutions will be restricted to the analysis of laminar parallel flow over a flat plate (Section 7.2 and Appendix G). However, other analytical solutions are discussed in advanced texts on convection [7–9], and detailed boundary layer solutions may be obtained by using numerical (finite-difference or finite-element) techniques [10]. It is also essential to recognize that a wide array of situations of engineering relevance involve turbulent convective heat transfer, which is both mathematically and physically more complex than laminar convection. The boundary layer equations for turbulent flow are included in Appendix F.

---

[7]Special attention must be given to the effect of species transfer on the velocity boundary layer. Recall that velocity boundary layer development is generally characterized by the existence of zero fluid velocity *at the surface*. This condition pertains to the velocity component v normal to the surface, as well as to the velocity component u along the surface. However, if there is simultaneous mass transfer to or from the surface, it is evident that v can no longer be zero at the surface. Nevertheless, for the mass transfer problems of interest in this text, it is reasonable to assume that $v = 0$ at the surface, which is equivalent to assuming that mass transfer has a negligible effect on the velocity boundary layer. The assumption is appropriate for many problems involving evaporation or sublimation from gas–liquid or gas–solid interfaces, respectively. It is not appropriate, however, for *mass transfer cooling* problems that involve large surface mass transfer rates [6]. In addition, we note that, with mass transfer, the boundary layer fluid is a binary mixture of species A and B, and its properties should be those of the mixture. However, in all problems in this text, $C_A \ll C_B$ and it is reasonable to assume that the boundary layer properties (such as k, μ, c_p, etc.) are those of species B.

It is important to stress that we have not developed the laminar boundary layer equations primarily for the purpose of obtaining solutions. Rather, we have been motivated mainly by two other considerations. One motivation has been to obtain an appreciation for the physical processes that occur in boundary layers. These processes affect wall friction as well as energy and species transfer in the boundary layers. A second important motivation arises from the fact that the equations may be used to identify key *boundary layer similarity parameters*, as well as important *analogies* between *momentum, heat*, and *mass* transfer that have numerous practical applications. The laminar boundary layer equations will be used for this purpose in Sections 6.5 through 6.7, but the same key parameters and analogies hold true for turbulent conditions as well.

6.4.2 Compressible Flow

The equations of the foregoing section and Appendix E are restricted to incompressible flows, that is, for cases where the fluid density can be treated as constant.[8] Flows in which the fluids experience significant density changes as a result of *pressure variations* associated with the fluid motion are deemed to be compressible. The treatment of convection heat transfer associated with *compressible flow* is beyond the scope of this text. Although liquids may nearly always be treated as incompressible, density variations in flowing gases should be considered when the velocity approaches or exceeds the speed of sound. Specifically, a gradual transition from incompressible to compressible flow in gases occurs at a critical Mach number of $Ma_c \approx 0.3$, where $Ma \equiv V/a$ and V and a are the gas velocity and speed of sound, respectively [11, 12]. For an ideal gas, $a = \sqrt{\gamma R T}$ where γ is the ratio of specific heats, $\gamma \equiv c_p/c_v$, R is the gas constant, and the temperature is expressed in kelvins. As an example, for air at $T = 300$ K and $p = 1$ atm, we may assume ideal gas behavior. The gas constant is $R \equiv \mathcal{R}/\mathcal{M} = 8315$ J/kmol · K/28.7 kg/kmol = 287 J/kg · K and $c_v \equiv c_p - R = 1007$ J/kg · K − 287 J/kg · K = 720 J/kg · K. The ratio of specific heats is therefore $\gamma = c_p/c_v = 1007$ J/kg · K/720 J/kg · K = 1.4, and the speed of sound is $a = \sqrt{1.4 \times 287 \text{ J/kg} \cdot \text{K} \times 300 \text{ K}} = 347$ m/s. Hence air flowing at 300 K must be treated as being compressible if $V > 0.3 \times 347$ m/s $\cong 100$ m/s.

Since the material in Chapters 6 through 9 is restricted to incompressible or *low-speed* flow, it is important to confirm that compressibility effects are not important when utilizing the material to solve a convection heat transfer problem.[9]

6.5 *Boundary Layer Similarity: The Normalized Boundary Layer Equations*

If we examine Equations 6.28, 6.29, and 6.30 we note a strong similarity. In fact, if the pressure gradient appearing in Equation 6.28 and the viscous dissipation term of Equation 6.29 are negligible, the three equations are of the same form. *Each equation is characterized by advection terms on the left-hand side and a diffusion term on the right-hand side.* This

---

[8]Chapter 9 addresses flows that arise due to the variation of density with temperature. These *free convection* flows can nearly always be treated as if the fluid is incompressible but with an extra term in the momentum equation to account for buoyancy forces.

[9]Turbulence and compressibility often coincide, since large velocities can lead to large Reynolds and Mach numbers. It can be shown (Problem 6.25) that, for sufficiently small geometries, any flow that is turbulent is also compressible.

situation describes *low-speed, forced convection flows,* which are found in many engineering applications. Implications of this similarity may be developed in a rational manner by first *nondimensionalizing* the governing equations.

6.5.1 Boundary Layer Similarity Parameters

The boundary layer equations are *normalized* by first defining dimensionless independent variables of the forms

$$x^* \equiv \frac{x}{L} \quad \text{and} \quad y^* \equiv \frac{y}{L} \tag{6.31}$$

where L is a *characteristic length* for the surface of interest (e.g., the length of a flat plate). Moreover, dependent dimensionless variables may also be defined as

$$u^* \equiv \frac{u}{V} \quad \text{and} \quad v^* \equiv \frac{v}{V} \tag{6.32}$$

where V is the velocity upstream of the surface (Figure 6.9), and as

$$T^* \equiv \frac{T - T_s}{T_\infty - T_s} \tag{6.33}$$

$$C_A^* \equiv \frac{C_A - C_{A,s}}{C_{A,\infty} - C_{A,s}} \tag{6.34}$$

The dimensional variables may then be written in terms of the new dimensionless variables (for example, from Equation 6.31 $x \equiv x^*L$ and $y \equiv y^*L$) and the resulting expressions for x, y, u, v, T, and C_A may be substituted into Equations 6.28, 6.29, and 6.30 to obtain the dimensionless forms of the conservation equations shown in Table 6.1. Note that viscous dissipation has been neglected and that $p^* \equiv (p_\infty/\rho V^2)$ is a dimensionless pressure. The y-direction boundary conditions required to solve the equations are also shown in the table.

By normalizing the boundary layer equations, three very important dimensionless *similarity parameters* evolve and are introduced in Table 6.1. They are the Reynolds number, Re_L; Prandtl number, Pr; and Schmidt number, Sc. Such similarity parameters are important because they allow us to apply results obtained for a surface experiencing one set of convective conditions to *geometrically similar* surfaces experiencing entirely different conditions. These conditions may vary, for example, with the fluid, the fluid velocity as described by the free stream value V, and/or the size of the surface as described by the characteristic length L. As long as the similarity parameters *and* dimensionless boundary conditions are the same for two sets of conditions, the solutions of the differential equations of Table 6.1 for the nondimensional velocity, temperature, and species concentration will be *identical*. This concept will be amplified in the remainder of this section.

6.5.2 Dependent Dimensionless Parameters

Equations 6.35 through 6.43 in Table 6.1 are extremely useful from the standpoint of suggesting how important boundary layer results may be simplified and generalized. The momentum equation (6.35) suggests that, although conditions in the velocity boundary layer depend on the fluid properties ρ and μ, the velocity V, and the length scale L, this

TABLE 6.1 The boundary layer equations and their *y*-direction boundary conditions in nondimensional form

Boundary Layer	Conservation Equation	Boundary Conditions		Similarity Parameter(s)
		Wall	Free Stream	
Velocity	$u^* \dfrac{\partial u^*}{\partial x^*} + v^* \dfrac{\partial u^*}{\partial y^*} = -\dfrac{dp^*}{dx^*} + \dfrac{1}{Re_L}\dfrac{\partial^2 u^*}{\partial y^{*2}}$ (6.35)	$u^*(x^*,0) = 0$	$u^*(x^*,\infty) = \dfrac{u_\infty(x^*)}{V}$ (6.38)	$Re_L = \dfrac{VL}{\nu}$ (6.41)
Thermal	$u^* \dfrac{\partial T^*}{\partial x^*} + v^* \dfrac{\partial T^*}{\partial y^*} = \dfrac{1}{Re_L\, Pr}\dfrac{\partial^2 T^*}{\partial y^{*2}}$ (6.36)	$T^*(x^*,0) = 0$	$T^*(x^*,\infty) = 1$ (6.39)	$Re_L,\, Pr = \dfrac{\nu}{\alpha}$ (6.42)
Concentration	$u^* \dfrac{\partial C_A^*}{\partial x^*} + v^* \dfrac{\partial C_A^*}{\partial y^*} = \dfrac{1}{Re_L\, Sc}\dfrac{\partial^2 C_A^*}{\partial y^{*2}}$ (6.37)	$C_A^*(x^*,0) = 0$	$C_A^*(x^*,\infty) = 1$ (6.40)	$Re_L,\, Sc = \dfrac{\nu}{D_{AB}}$ (6.43)

Convection Similarity Parameters Re_L, Pr, and Sc

dependence may be simplified by grouping these variables in the form of the Reynolds number. We therefore anticipate that the solution to Equation 6.35 will be of the functional form

$$u^* = f\left(x^*, y^*, Re_L, \frac{dp^*}{dx^*}\right) \tag{6.44}$$

Since the pressure distribution $p^*(x^*)$ depends on the surface geometry and may be obtained independently by considering flow conditions in the free stream, the appearance of dp^*/dx^* in Equation 6.44 represents the influence of geometry on the velocity distribution.

From Equations 6.2, 6.31, and 6.32, the shear stress at the surface, $y^* = 0$, may be expressed as

$$\tau_s = \mu \frac{\partial u}{\partial y}\bigg|_{y=0} = \left(\frac{\mu V}{L}\right) \frac{\partial u^*}{\partial y^*}\bigg|_{y^*=0}$$

and from Equations 6.1 and 6.41, it follows that the friction coefficient is

Friction Coefficient:

$$C_f = \frac{\tau_s}{\rho V^2/2} = \frac{2}{Re_L} \frac{\partial u^*}{\partial y^*}\bigg|_{y^*=0} \tag{6.45}$$

From Equation 6.44 we also know that

$$\frac{\partial u^*}{\partial y^*}\bigg|_{y^*=0} = f\left(x^*, Re_L, \frac{dp^*}{dx^*}\right)$$

Hence, *for a prescribed geometry* Equation 6.45 may be expressed as

$$C_f = \frac{2}{Re_L} f(x^*, Re_L) \tag{6.46}$$

The significance of this result should not be overlooked. Equation 6.46 states that the friction coefficient, a dimensionless parameter of considerable importance to the engineer, may be expressed exclusively in terms of a dimensionless space coordinate and the Reynolds number. Hence, for a prescribed geometry we expect the function that relates C_f to x^* and Re_L to be *universally* applicable. That is, we expect it to apply to different fluids and over a wide range of values for V and L.

Similar results may be obtained for the convection coefficients of heat and mass transfer. Intuitively, we might anticipate that h depends on the fluid properties (k, c_p, μ, and ρ), the fluid velocity V, the length scale L, and the surface geometry. However, Equation 6.36 suggests the manner in which this dependence may be simplified. In particular, the solution to this equation may be expressed in the form

$$T^* = f\left(x^*, y^*, Re_L, Pr, \frac{dp^*}{dx^*}\right) \tag{6.47}$$

where the dependence on dp^*/dx^* originates from the influence of the geometry on the fluid motion (u^* and v^*), which, in turn, affects the thermal conditions. Once again the term dp^*/dx^*

represents the effect of surface geometry. From the definition of the convection coefficient, Equation 6.5, and the dimensionless variables, Equations 6.31 and 6.33, we also obtain

$$
h = -\frac{k_f}{L} \frac{(T_\infty - T_s)}{(T_s - T_\infty)} \frac{\partial T^*}{\partial y^*}\bigg|_{y^*=0} = +\frac{k_f}{L} \frac{\partial T^*}{\partial y^*}\bigg|_{y^*=0}
$$

This expression suggests defining an important dependent dimensionless parameter termed the Nusselt number.

Nusselt Number:

$$
Nu \equiv \frac{hL}{k_f} = +\frac{\partial T^*}{\partial y^*}\bigg|_{y^*=0} \tag{6.48}
$$

This parameter is equal to the dimensionless temperature gradient at the surface, and it provides a measure of the convection heat transfer occurring at the surface. From Equation 6.47 it follows that, *for a prescribed geometry,*

$$
Nu = f(x^*, Re_L, Pr) \tag{6.49}
$$

The Nusselt number is to the thermal boundary layer what the friction coefficient is to the velocity boundary layer. Equation 6.49 implies that for a given geometry, the Nusselt number must be some *universal function* of x^*, Re_L, and Pr. If this function were known, it could be used to compute the value of Nu for different fluids and for different values of V and L. From knowledge of Nu, the local convection coefficient h may be found and the *local* heat flux may then be computed from Equation 6.4. Moreover, since the *average* heat transfer coefficient is obtained by integrating over the surface of the body, it must be independent of the spatial variable x^*. Hence the functional dependence of the *average* Nusselt number is

$$
\overline{Nu} = \frac{\overline{h}L}{k_f} = f(Re_L, Pr) \tag{6.50}
$$

Similarly, it may be argued that, for mass transfer in a gas flow over an evaporating liquid or a sublimating solid, the convection mass transfer coefficient h_m depends on the properties D_{AB}, ρ, and μ, the velocity V, and the characteristic length L. However, Equation 6.37 suggests that this dependence may be simplified. The solution to this equation must be of the form

$$
C_A^* = f\left(x^*, y^*, Re_L, Sc, \frac{dp^*}{dx^*}\right) \tag{6.51}
$$

where the dependence on dp^*/dx^* again originates from the influence of the surface geometry. From the definition of the convection coefficient, Equation 6.9, and the dimensionless variables, Equations 6.31 and 6.34, we know that

$$
h_m = -\frac{D_{AB}}{L} \frac{(C_{A,\infty} - C_{A,s})}{(C_{A,s} - C_{A,\infty})} \frac{\partial C_A^*}{\partial y^*}\bigg|_{y^*=0} = +\frac{D_{AB}}{L} \frac{\partial C_A^*}{\partial y^*}\bigg|_{y^*=0}
$$

Hence we may define a dependent dimensionless parameter termed the Sherwood number (*Sh*).

Sherwood Number:

$$Sh \equiv \frac{h_m L}{D_{AB}} = + \frac{\partial C_A^*}{\partial y^*}\bigg|_{y^*=0} \tag{6.52}$$

This parameter is equal to the dimensionless concentration gradient at the surface, and it provides a measure of the convection mass transfer occurring at the surface. From Equation 6.51 it follows that, *for a prescribed geometry,*

$$Sh = f(x^*, Re_L, Sc) \tag{6.53}$$

The Sherwood number is to the concentration boundary layer what the Nusselt number is to the thermal boundary layer, and Equation 6.53 implies that it must be a universal function of x^*, Re_L, and Sc. As for the Nusselt number, it is also possible to work with an *average* Sherwood number that depends on only Re_L and Sc.

$$\overline{Sh} = \frac{\overline{h}_m L}{D_{AB}} = f(Re_L, Sc) \tag{6.54}$$

From the foregoing development we have obtained the relevant dimensionless parameters for low-speed, forced-convection boundary layers. We have done so by nondimensionalizing the differential equations that describe the physical processes occurring within the boundary layers. An alternative approach could have involved the use of dimensional analysis in the form of the Buckingham pi theorem [5]. However, the success of that method depends on one's ability to select, largely from intuition, the various parameters that influence a problem. For example, knowing beforehand that $\overline{h} = f(k, c_p, \rho, \mu, V, L)$, one could use the Buckingham pi theorem to obtain Equation 6.50. However, having begun with the differential form of the conservation equations, we have eliminated the guesswork and have established the similarity parameters in a rigorous fashion.

The importance of an expression such as Equation 6.50 should be fully appreciated. It states that values of the average heat transfer coefficient $\overline{h}$, whether obtained theoretically, experimentally, or numerically, can be completely represented in terms of only three dimensionless groups, instead of the original seven dimensional parameters. The convenience and power afforded by such simplification will become evident in Chapters 7 through 10. Moreover, once the form of the functional dependence of Equation 6.50 has been obtained for a particular surface geometry, let us say from laboratory measurements, it is known to be *universally* applicable. By this we mean that it may be applied for different fluids, velocities, and length scales, as long as the assumptions implicit in the originating boundary layer equations remain valid (e.g., negligible viscous dissipation and body forces).

EXAMPLE 6.5

Experimental tests using air as the working fluid are conducted on a portion of the turbine blade shown in the sketch. The heat flux to the blade at a particular point (x^*) on the surface is measured to be $q'' = 95{,}000 \text{ W/m}^2$. To maintain a steady-state surface temperature of 800°C, heat transferred to the blade is removed by circulating a coolant inside the blade.

Original
conditions

1. Determine the heat flux to the blade at x^* if its temperature is reduced to $T_{s,1} = 700°C$ by increasing the coolant flow.

2. Determine the heat flux at the same dimensionless location x^* for a similar turbine blade having a chord length of $L = 80$ mm, when the blade operates in an airflow at $T_\infty = 1150°C$ and $V = 80$ m/s, with $T_s = 800°C$.

SOLUTION

Known: Operating conditions of an internally cooled turbine blade.

Find:

1. Heat flux to the blade at a point x^* when the surface temperature is reduced.

2. Heat flux at the same dimensionless location to a larger turbine blade of the same shape with reduced air velocity.

Schematic:

Case 1   Case 2

Assumptions:

1. Steady-state, incompressible flow.

2. Constant air properties.

Analysis:

1. When the surface temperature is 800°C, the local convection heat transfer coefficient between the surface and the air at x^* can be obtained from Newton's law of cooling:

$$q'' = h(T_\infty - T_s)$$

Thus,

$$h = \frac{q''}{(T_\infty - T_s)}$$

We proceed without calculating the value for now. From Equation 6.49, it follows that, for the prescribed geometry,

$$Nu = \frac{hL}{k} = f(x^*, Re_L, Pr)$$

Hence, since constant air properties are assumed, x^*, Re_L, and Pr are equal for the two cases, and the local Nusselt number is unchanged. Moreover, since L and k are unchanged, the local convection coefficient remains the same. Thus, when the surface temperature is reduced to 700°C, the heat flux may be obtained from Newton's law of cooling, using the same local convection coefficient:

$$q_1'' = h(T_\infty - T_{s,1}) = \frac{q''}{(T_\infty - T_s)}(T_\infty - T_{s,1}) = \frac{95{,}000 \text{ W/m}^2}{(1150 - 800)°\text{C}}(1150 - 700)°\text{C}$$

$$= 122{,}000 \text{ W/m}^2 \qquad \triangleleft$$

2. To determine the heat flux at x^* associated with the larger blade and the reduced airflow (case 2), we first note that, although L has increased by a factor of 2, the velocity has decreased by the same factor and the Reynolds number has not changed. That is,

$$Re_{L,2} = \frac{V_2 L_2}{v} = \frac{VL}{v} = Re_L$$

Accordingly, since x^* and Pr are also unchanged, the local Nusselt number remains the same.

$$Nu_2 = Nu$$

Because the characteristic length is different, however, the local convection coefficient changes, where

$$\frac{h_2 L_2}{k} = \frac{hL}{k} \qquad \text{or} \qquad h_2 = h\frac{L}{L_2} = \frac{q''}{(T_\infty - T_s)}\frac{L}{L_2}$$

The heat flux at x^* is then

$$q_2'' = h_2(T_\infty - T_s) = q''\frac{(T_\infty - T_s)}{(T_\infty - T_s)}\frac{L}{L_2}$$

$$q_2'' = 95{,}000 \text{ W/m}^2 \times \frac{0.04 \text{ m}}{0.08 \text{ m}} = 47{,}500 \text{ W/m}^2 \qquad \triangleleft$$

Comments:

1. If the Reynolds numbers for the two situations of part 2 differed, that is, $Re_{L,2} \neq Re_L$, the local heat flux q_2'' could be obtained only if the particular functional dependence of Equation 6.49 were known. Such forms are provided for many different shapes in subsequent chapters.

2. Air temperatures in the boundary layer range from the blade surface temperature T_s to the ambient value T_∞. Hence, as will become evident in Section 7.1 representative air properties could be evaluated at arithmetic mean or *film* temperatures of $T_{f,1} = (T_{s,1} + T_\infty)/2 = (700°\text{C} + 1150°\text{C})/2 = 925°\text{C}$ and $T_{f,2} = (800°\text{C} + 1150°\text{C})/2 = 975°\text{C}$, respectively. Based

on properties corresponding to these film temperatures, the Reynolds (and Nusselt) numbers for the two cases would be slightly different. However, the difference would not be large enough to significantly change the calculated value of the local heat flux for case 2.

3. At $T = 1150°C = 1423$ K, $c_v \equiv c_p - R = 1167$ J/kg·K $- 287$ J/kg·K $= 880$ J/kg·K, and the specific heat ratio is $\gamma = c_p/c_v = 1167$ J/kg·K/800 J/kg·K $= 1.33$. Assuming the air behaves as an ideal gas, the speed of sound in the air is $a = \sqrt{\gamma RT} = \sqrt{1.33 \times 287 \text{J/kg·K} \times 1423 \text{ K}} = 736$ m/s. Therefore $Ma = V/a = 0.22$ and 0.11 for cases 1 and 2, respectively. Hence the flow is incompressible in both cases. If the flow were to be compressible for either case, the Nusselt number would also depend on the Mach number, and the two thermal boundary layers would not be similar.

EXAMPLE 6.6

Consider convective cooling of a two-dimensional streamlined strut of characteristic length $L_{H_2} = 40$ mm. The strut is exposed to hydrogen flowing at $p_{H_2} = 2$ atm, $V_{H_2} = 8.1$ m/s, and $T_{\infty,H_2} = -30°C$. Of interest is the value of the average heat transfer coefficient $\bar{h}_{H_2}$, when the surface temperature is $T_{s,H_2} = -15°C$. Rather than conducting expensive experiments involving pressurized hydrogen, an engineer proposes to take advantage of similarity by performing wind tunnel experiments using air at atmospheric pressure with $T_{\infty,\text{Air}} = 23°C$. A geometrically similar strut of characteristic length $L_{\text{Air}} = 60$ mm and perimeter $P = 150$ mm is placed in the wind tunnel. Measurements reveal a surface temperature of $T_{s,\text{Air}} = 30°C$ when the rate of heat loss per unit object length (into the page) is $q'_{\text{Air}} = 50$ W/m. Determine the required air velocity in the wind tunnel experiment V_{Air} and the average convective heat transfer coefficient in the hydrogen $\bar{h}_{H_2}$.

SOLUTION

Known: Flow across a strut. Hydrogen pressure, velocity, and temperature. Air temperature and pressure, as well as rate of heat loss per unit length. Surface temperatures of the strut in hydrogen and in air.

Find: Air velocity and average convective heat transfer coefficient for the strut that is exposed to hydrogen.

Schematic:

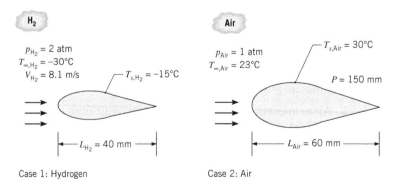

Case 1: Hydrogen

Case 2: Air

Assumptions:

1. Steady-state, incompressible boundary layer behavior.
2. Ideal gas behavior.
3. Constant properties.
4. Negligible viscous dissipation.

Properties: Table A.4, air ($p = 1$ atm, $T_f = (23°C + 30°C)/2 = 26.5°C \approx 300$ K): $Pr = 0.707$, $\nu = 15.89 \times 10^{-6}$ m²/s, $k = 26.3 \times 10^{-3}$ W/m · K. Table A.4 hydrogen ($p = 1$ atm, $T_f = -22.5°C \approx 250$ K): $Pr = 0.707$, $\nu = 81.4 \times 10^{-6}$ m²/s, $k = 157 \times 10^{-3}$ W/m · K.

The properties k, Pr, c_p, and μ may be assumed to be independent of pressure to an excellent approximation. However, for a gas, the kinematic viscosity $\nu = \mu/\rho$ will vary with pressure through its dependence on density. From the ideal gas law, $\rho = p/RT$, it follows that the ratio of kinematic viscosities for a gas at the same temperature but at different pressures, p_1 and p_2, is $(\nu_1/\nu_2) = (p_2/p_1)$. Hence, the kinematic viscosity of hydrogen at 250 K and 2 atm is $\nu_{H_2} = 81.4 \times 10^{-6}$ m²/s $\times$ 1 atm/2 atm $= 40.7 \times 10^{-6}$ m²/s. Since Pr is independent of pressure, $Pr_{H_2}(p = 2$ atm, $T_f = -22.5°C) = Pr_{Air}(p = 1$ atm, $T_f = 26.5°C) = 0.707$.

Analysis: From Equation 6.50, we know that the average Nusselt numbers are related to the Reynolds and Prandtl numbers by the functional dependence

$$\overline{Nu}_{H_2} = \frac{\overline{h}_{H_2} L_{H_2}}{k_{H_2}} = f(Re_{L,H_2}, Pr_{H_2}) \quad \text{and} \quad \overline{Nu}_{Air} = \frac{\overline{h}_{Air} L_{Air}}{k_{Air}} = f(Re_{L,Air}, Pr_{Air})$$

Since $Pr_{H_2} = Pr_{Air}$, similarity exists if $Re_{L,Air} = Re_{L,H_2}$, in which case the average Nusselt numbers for the air and hydrogen will be identical. Equating the Reynolds numbers for the hydrogen and air yields an expression for the required air velocity

$$V_{Air} = \frac{Re_{L,Air} \nu_{Air}}{L_{Air}} = \frac{Re_{L,H_2} \nu_{Air}}{L_{Air}} = \frac{V_{H_2} L_{H_2} \nu_{Air}}{\nu_{H_2} L_{Air}}$$

$$= \frac{8.1 \text{ m/s} \times 0.04 \text{ m} \times 15.89 \times 10^{-6} \text{ m}^2/\text{s}}{40.7 \times 10^{-6} \text{ m}^2/\text{s} \times 0.06 \text{ m}} = 2.10 \text{ m/s} \qquad \triangleleft$$

With $Re_{L,Air} = Re_{L,H_2}$ and $Pr_{Air} = Pr_{H_2}$, we may equate the Nusselt numbers for the hydrogen and air, and incorporate Newton's law of cooling. Doing so gives

$$\overline{h}_{H_2} = \overline{h}_{Air} \frac{L_{Air} k_{H_2}}{L_{H_2} k_{Air}} = \frac{q'_{Air}}{P(T_{s,Air} - T_{\infty,Air})} \times \frac{L_{Air} k_{H_2}}{L_{H_2} k_{Air}}$$

$$= \frac{50 \text{ W/m}}{150 \times 10^{-3} \text{ m} \times (30 - 23)°C} \times \frac{0.06 \text{ m} \times 0.157 \text{ W/m} \cdot \text{K}}{0.04 \text{ m} \times 0.0263 \text{ W/m} \cdot \text{K}} = 426 \text{ W/m}^2 \cdot \text{K} \qquad \triangleleft$$

Comments:

1. The fluid properties are evaluated at the arithmetic mean of the free stream and surface temperatures. As will become evident in Section 7.1, the temperature dependence of fluid properties is often accounted for by evaluating properties at the film temperature, $T_f = (T_s + T_\infty)/2$.

2. Experiments involving pressurized hydrogen can be relatively expensive because care must be taken to prevent leakage of this small-molecule, flammable gas.

6.6 *Physical Interpretation of the Dimensionless Parameters*

All of the foregoing dimensionless parameters have physical interpretations that relate to conditions in the flow, not only for boundary layers but also for other flow types, such as the internal flows we will see in Chapter 8. Consider the *Reynolds number* Re_L (Equation 6.41), which may be interpreted as the *ratio of inertia to viscous forces* in a region of characteristic dimension L. Inertia forces are associated with an increase in the momentum of a moving fluid. From Equation 6.28, it is evident that these forces (per unit mass) are of the form $u\partial u/\partial x$, in which case an order-of-magnitude approximation gives $F_I \approx V^2/L$. Similarly, the net shear force (per unit mass) is found on the right-hand side of Equation 6.28 as $\nu(\partial^2 u/\partial y^2)$ and may be approximated as $F_s \approx \nu V/L^2$. Therefore, the ratio of forces is

$$\frac{F_I}{F_s} \approx \frac{\rho V^2/L}{\mu V/L^2} = \frac{\rho V L}{\mu} = Re_L$$

We therefore expect inertia forces to dominate for large values of Re_L and viscous forces to dominate for small values of Re_L.

There are several important implications of the foregoing expectation. Recall from Section 6.3.1 that the Reynolds number determines the existence of laminar or turbulent flow. We should also expect the magnitude of the Reynolds number to influence the velocity boundary layer thickness δ. With increasing Re_L at a fixed location on a surface, we expect viscous forces to become less influential relative to inertia forces. Hence the effects of viscosity do not penetrate as far into the free stream, and the value of δ diminishes.

The *Prandtl number* is defined as the ratio of the kinematic viscosity, also referred to as the momentum diffusivity, ν, to the thermal diffusivity α. It is therefore a fluid property. The Prandtl number provides a *measure of the relative effectiveness of momentum and energy transport by diffusion in the velocity and thermal boundary layers*, respectively. From this interpretation it follows that the value of Pr strongly influences the relative growth of the velocity and thermal boundary layers. In fact for laminar boundary layers (in which transport by diffusion is *not* overshadowed by turbulent mixing), it is reasonable to expect that

$$\frac{\delta}{\delta_t} \approx Pr^n \tag{6.55}$$

where n is a positive exponent. From Table A.4 we see that the Prandtl number of gases is near unity, in which case $\delta_t \approx \delta$. In a liquid metal (Table A.7), $Pr \ll 1$ and $\delta_t \gg \delta$. The opposite is true for oils (Table A.5), for which $Pr \gg 1$ and $\delta_t \ll \delta$.

Similarly, the *Schmidt number,* which is defined in Equation 6.43, is a fluid property and provides *a measure of the relative effectiveness of momentum and mass transport by diffusion in the velocity and concentration boundary layers*, respectively. For convection mass transfer in laminar flows, it therefore determines the relative velocity and concentration boundary layer thicknesses, where

$$\frac{\delta}{\delta_c} \approx Sc^n \tag{6.56}$$

Another dimensionless fluid property, which is related to Pr and Sc, is the *Lewis number* (Le). It is defined as

$$Le = \frac{\alpha}{D_{AB}} = \frac{Sc}{Pr} \tag{6.57}$$

and is relevant to any situation involving simultaneous heat and mass transfer by convection. From Equations 6.55 through 6.57, it then follows that

$$\frac{\delta_t}{\delta_c} \approx Le^n \tag{6.58}$$

The Lewis number is therefore a measure of the relative thermal and concentration boundary layer thicknesses. For most applications it is reasonable to assume a value of $n = 1/3$ in Equations 6.55, 6.56, and 6.58.

Table 6.2 lists the dimensionless groups that appear frequently in heat and mass transfer. The list includes groups already considered, as well as those yet to be introduced for special conditions. As a new group is confronted, its definition and interpretation should be committed to memory. Note that the *Grashof number* provides a measure of the ratio of buoyancy forces to viscous forces in the velocity boundary layer. Its role in free convection (Chapter 9) is much the same as that of the Reynolds number in forced convection. The *Eckert number* provides a measure of the kinetic energy of the flow relative to the enthalpy difference across the thermal boundary layer. It plays an important role in high-speed flows for which viscous dissipation is significant. Note also that, although similar in form, the Nusselt and Biot numbers differ in both definition and interpretation. Whereas the Nusselt number is

TABLE 6.2 Selected dimensionless groups of heat and mass transfer

Group	Definition	Interpretation
Biot number (Bi)	$\dfrac{hL}{k_s}$	Ratio of the internal thermal resistance of a solid to the boundary layer thermal resistance
Mass transfer Biot number (Bi_m)	$\dfrac{h_m L}{D_{AB}}$	Ratio of the internal species transfer resistance to the boundary layer species transfer resistance
Bond number (Bo)	$\dfrac{g(\rho_l - \rho_v)L^2}{\sigma}$	Ratio of gravitational and surface tension forces
Coefficient of friction (C_f)	$\dfrac{\tau_s}{\rho V^2/2}$	Dimensionless surface shear stress
Eckert number (Ec)	$\dfrac{V^2}{c_p(T_s - T_\infty)}$	Kinetic energy of the flow relative to the boundary layer enthalpy difference
Fourier number (Fo)	$\dfrac{\alpha t}{L^2}$	Ratio of the heat conduction rate to the rate of thermal energy storage in a solid. Dimensionless time
Mass transfer Fourier number (Fo_m)	$\dfrac{D_{AB}t}{L^2}$	Ratio of the species diffusion rate to the rate of species storage. Dimensionless time
Friction factor (f)	$\dfrac{\Delta p}{(L/D)(\rho u_m^2/2)}$	Dimensionless pressure drop for internal flow

(*continued*)

TABLE 6.2 *Continued*

Group	Definition	Interpretation
Grashof number (Gr_L)	$\dfrac{g\beta(T_s - T_\infty)L^3}{v^2}$	Measure of the ratio of buoyancy forces to viscous forces
Colburn j factor (j_H)	$St\,Pr^{2/3}$	Dimensionless heat transfer coefficient
Colburn j factor (j_m)	$St_m\,Sc^{2/3}$	Dimensionless mass transfer coefficient
Jakob number (Ja)	$\dfrac{c_p(T_s - T_{\text{sat}})}{h_{fg}}$	Ratio of sensible to latent energy absorbed during liquid–vapor phase change
Lewis number (Le)	$\dfrac{\alpha}{D_{AB}}$	Ratio of the thermal and mass diffusivities
Mach number (Ma)	$\dfrac{V}{a}$	Ratio of velocity to speed of sound
Nusselt number (Nu_L)	$\dfrac{hL}{k_f}$	Ratio of convection to pure conduction heat transfer
Peclet number (Pe_L)	$\dfrac{VL}{\alpha} = Re_L\,Pr$	Ratio of advection to conduction heat transfer rates
Prandtl number (Pr)	$\dfrac{c_p\mu}{k} = \dfrac{v}{\alpha}$	Ratio of the momentum and thermal diffusivities
Reynolds number (Re_L)	$\dfrac{VL}{v}$	Ratio of the inertia and viscous forces
Schmidt number (Sc)	$\dfrac{v}{D_{AB}}$	Ratio of the momentum and mass diffusivities
Sherwood number (Sh_L)	$\dfrac{h_m L}{D_{AB}}$	Ratio of convection to pure diffusion mass transfer
Stanton number (St)	$\dfrac{h}{\rho V c_p} = \dfrac{Nu_L}{Re_L\,Pr}$	Modified Nusselt number
Mass transfer Stanton number (St_m)	$\dfrac{h_m}{V} = \dfrac{Sh_L}{Re_L\,Sc}$	Modified Sherwood number
Weber number (We)	$\dfrac{\rho V^2 L}{\sigma}$	Ratio of inertia to surface tension forces

defined in terms of the thermal conductivity of the fluid, the Biot number, Equation 5.9, is based on the solid thermal conductivity and is generally not used in convection analysis.

6.7 Boundary Layer Analogies

As engineers, our interest in boundary layer behavior is directed principally toward the dimensionless parameters C_f, Nu, and Sh. From knowledge of these parameters, we may compute the wall shear stress and the convection heat and mass transfer rates. It is therefore understandable that expressions that relate C_f, Nu, and Sh to each other are useful in convection analysis. Such expressions are available in the form of *boundary layer analogies*.

6.7.1 The Heat and Mass Transfer Analogy

If two or more processes are governed by dimensionless equations of the same form, the processes are said to be *analogous*. Clearly, then, from Equations 6.36 and 6.37 and the boundary conditions, Equations 6.39 and 6.40, of Table 6.1, convection heat and mass transfer are analogous. Moreover, as shown in Equations 6.36 and 6.37, each equation is related to the velocity field through Re_L, and the parameters Pr and Sc assume analogous roles. One implication of this analogy is that dimensionless relations that govern thermal boundary layer behavior must be of the same form as those that govern the concentration boundary layer. Hence the boundary layer temperature and concentration profiles must also be of the same functional form if the applied boundary conditions are analogous.

Recalling the discussion of Section 6.5.2, the features of which are summarized in Table 6.3, an important result of the heat and mass transfer analogy may be obtained. From the foregoing paragraph, it follows that Equation 6.47 must be of the same functional form as Equation 6.51. From Equations 6.48 and 6.52, it then follows that the dimensionless temperature and concentration gradients evaluated at the surface, and therefore the values of Nu and Sh, are analogous. Similarly, expressions for the average Nusselt and Sherwood numbers, Equations 6.50 and 6.54, respectively, are also of the same form. *Accordingly, heat and mass transfer relations for a particular geometry are interchangeable.* If, for example, one has performed a set of heat transfer experiments to determine the functional form of Equation 6.49 for a particular surface geometry, the results may be used for convection mass transfer involving the same geometry, simply by replacing Nu with Sh and Pr with Sc.

The analogy may also be used to directly relate the two convection coefficients. In subsequent chapters we will find that Nu and Sh are generally proportional to Pr^n and Sc^n, respectively, where n is a positive exponent less than 1. Anticipating this dependence, we use Equations 6.49 and 6.53 to obtain

$$Nu = f(x^*, Re_L)Pr^n \qquad \text{and} \qquad Sh = f(x^*, Re_L)Sc^n$$

in which case, with equivalent functions, $f(x^*, Re_L)$,

$$\frac{Nu}{Pr^n} = \frac{Sh}{Sc^n} \tag{6.59}$$

TABLE 6.3 Functional relations pertinent to the boundary layer analogies

Fluid Flow		Heat Transfer		Mass Transfer	
$u^* = f\left(x^*, y^*, Re_L, \dfrac{dp^*}{dx^*}\right)$	(6.44)	$T^* = f\left(x^*, y^*, Re_L, Pr, \dfrac{dp^*}{dx^*}\right)$	(6.47)	$C_A^* = f\left(x^*, y^*, Re_L, Sc, \dfrac{dp^*}{dx^*}\right)$	(6.51)
$C_f = \dfrac{2}{Re_L}\left.\dfrac{\partial u^*}{\partial y^*}\right|_{y^*=0}$	(6.45)	$Nu = \dfrac{hL}{k} = +\left.\dfrac{\partial T^*}{\partial y^*}\right|_{y^*=0}$	(6.48)	$Sh = \dfrac{h_m L}{D_{AB}} = +\left.\dfrac{\partial C_A^*}{\partial y^*}\right|_{y^*=0}$	(6.52)
$C_f = \dfrac{2}{Re_L} f(x^*, Re_L)$	(6.46)	$Nu = f(x^*, Re_L, Pr)$	(6.49)	$Sh = f(x^*, Re_L, Sc)$	(6.53)
		$\overline{Nu} = f(Re_L, Pr)$	(6.50)	$\overline{Sh} = f(Re_L, Sc)$	(6.54)

Substituting from Equations 6.48 and 6.52 we then obtain

$$\frac{hL/k}{Pr^n} = \frac{h_m L/D_{AB}}{Sc^n}$$

or, from Equation 6.57,

$$\frac{h}{h_m} = \frac{k}{D_{AB} Le^n} = \rho c_p Le^{1-n} \tag{6.60}$$

This result may often be used to determine one convection coefficient, for example, h_m, from knowledge of the other coefficient. The same relation may be applied to the average coefficients $\bar{h}$ and $\bar{h}_m$, and it may be used in turbulent, as well as laminar, flow. For most applications it is reasonable to assume a value of $n = \frac{1}{3}$.

EXAMPLE 6.7

A solid of arbitrary shape is suspended in atmospheric air having a free stream temperature and velocity of 20°C and 100 m/s, respectively. The solid has a characteristic length of 1 m, and its surface is maintained at 80°C. Under these conditions measurements of the convection heat flux at a particular point (x^*) on the surface and of the temperature in the boundary layer above this point (x^*, y^*) reveal values of 10^4 W/m^2 and 60°C, respectively. A mass transfer operation is to be effected for a second solid having the same shape but a characteristic length of 2 m. In particular, a thin film of water on the solid is to be evaporated in dry atmospheric air having a free stream velocity of 50 m/s, with the air and the solid both at a temperature of 50°C. What are the molar concentration and the species molar flux of the water vapor at the same nondimensional locations where the temperature and heat flux measurements were made in the first case?

SOLUTION

Known: A boundary layer temperature and heat flux at a location on a solid in an airstream of prescribed temperature and velocity.

Find: Water vapor concentration and flux associated with the same location on a larger surface of the same shape.

Schematic:

Case 1: heat transfer   Case 2: mass transfer

Assumptions:

1. Steady-state, two-dimensional, incompressible boundary layer behavior; constant properties.
2. Boundary layer approximations are valid.
3. Negligible viscous dissipation.
4. Mole fraction of water vapor in concentration boundary layer is much less than unity.

Properties: Table A.4, air (50°C): $\nu = 18.2 \times 10^{-6}$ m^2/s, $k = 28 \times 10^{-3}$ W/m · K, $Pr = 0.70$. Table A.6, saturated water vapor (50°C): $\rho_{A,sat} = \nu_g^{-1} = 0.082$ kg/m^3. Table A.8, water vapor–air (50°C): $D_{AB} \approx 0.26 \times 10^{-4}$ m^2/s.

Analysis: The desired molar concentration and flux may be determined by invoking the analogy between heat and mass transfer. From Equations 6.47 and 6.51, we know that

$$T^* \equiv \frac{T - T_s}{T_\infty - T_s} = f\left(x^*, y^*, Re_L, Pr, \frac{dp^*}{dx^*}\right)$$

and

$$C_A^* \equiv \frac{C_A - C_{A,s}}{C_{A,\infty} - C_{A,s}} = f\left(x^*, y^*, Re_L, Sc, \frac{dp^*}{dx^*}\right)$$

For case 1

$$Re_{L,1} = \frac{V_1 L_1}{\nu} = \frac{100 \text{ m/s} \times 1 \text{ m}}{18.2 \times 10^{-6} \text{ m}^2/\text{s}} = 5.5 \times 10^6, \qquad Pr = 0.70$$

while for case 2

$$Re_{L,2} = \frac{V_2 L_2}{\nu} = \frac{50 \text{ m/s} \times 2 \text{ m}}{18.2 \times 10^{-6} \text{ m}^2/\text{s}} = 5.5 \times 10^6$$

$$Sc = \frac{\nu}{D_{AB}} = \frac{18.2 \times 10^{-6} \text{ m}^2/\text{s}}{26 \times 10^{-6} \text{ m}^2/\text{s}} = 0.70$$

Since $Re_{L,1} = Re_{L,2}$, $Pr = Sc$, $x_1^* = x_2^*$, $y_1^* = y_2^*$, and the geometries are the same, it follows that the temperature and concentration distributions have the same functional form. Hence

$$\frac{C_A(x^*, y^*) - C_{A,s}}{C_{A,\infty} - C_{A,s}} = \frac{T(x^*, y^*) - T_s}{T_\infty - T_s} = \frac{60 - 80}{20 - 80} = 0.33$$

or, with $C_{A,\infty} = 0$,

$$C_A(x^*, y^*) = C_{A,s}(1 - 0.33) = 0.67 C_{A,s}$$

With

$$C_{A,s} = C_{A,sat}(50°C) = \frac{\rho_{A,sat}}{\mathcal{M}_A} = \frac{0.082 \text{ kg/m}^3}{18 \text{ kg/kmol}} = 0.0046 \text{ kmol/m}^3$$

it follows that

$$C_A(x^*, y^*) = 0.67 \,(0.0046 \text{ kmol/m}^3) = 0.0031 \text{ kmol/m}^3 \qquad \triangleleft$$

The molar flux may be obtained from Equation 6.8

$$N_A''(x^*) = h_m(C_{A,s} - C_{A,\infty})$$

with h_m evaluated from the analogy. From Equations 6.49 and 6.53 we know that, since $x_1^* = x_2^*$, $Re_{L,1} = Re_{L,2}$, and $Pr = Sc$, the corresponding functional forms are equivalent. Hence

$$Sh = \frac{h_m L_2}{D_{AB}} = Nu = \frac{hL_1}{k}$$

With $h = q''/(T_s - T_\infty)$ from Newton's law of cooling,

$$h_m = \frac{L_1}{L_2} \times \frac{D_{AB}}{k} \times \frac{q''}{(T_s - T_\infty)} = \frac{1}{2} \times \frac{0.26 \times 10^{-4} \text{ m}^2/\text{s}}{0.028 \text{ W/m} \cdot \text{K}} \times \frac{10^4 \text{ W/m}^2}{(80-20)^\circ\text{C}}$$

$$h_m = 0.077 \text{ m/s}$$

Hence

$$N_A''(x^*) = 0.077 \text{ m/s} \,(0.0046 - 0.0) \text{ kmol/m}^3$$

or

$$N_A''(x^*) = 3.54 \times 10^{-4} \text{ kmol/s} \cdot \text{m}^2 \qquad \triangleleft$$

Comments:

1. Recognize that, since the mole fraction of water vapor in the concentration boundary layer is small, the kinematic viscosity of air (ν_B) may be used to evaluate $Re_{L,2}$.

2. Air properties for case 1 are evaluated at the film temperature, $T_f = (T_s + T_\infty)/2 = (80^\circ\text{C} + 20^\circ\text{C})/2 = 50^\circ\text{C}$.

6.7.2 Evaporative Cooling

An important application of the heat and mass transfer analogy is to the process of *evaporative cooling*, which occurs when a gas flows over a liquid (Figure 6.10). Evaporation occurs from the liquid surface, and the energy associated with the phase change is the latent heat of vaporization of the liquid. Evaporation occurs when liquid molecules near the surface experience collisions that increase their energy above that needed to overcome the surface binding energy. The energy required to sustain the evaporation must come from the internal energy of the liquid, which would then experience a reduction in temperature (the cooling effect). However, if steady-state conditions are to be maintained, the latent energy lost by the liquid because of evaporation must be replenished by energy transfer to the liquid from its surroundings. Neglecting radiation effects, this transfer may be due to the convection of sensible energy from the gas or to heat addition by other means, as, for example, by an electrical

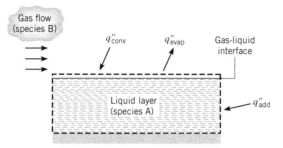

FIGURE 6.10 Latent and sensible heat exchange at a gas–liquid interface.

heater submerged in the liquid. Applying conservation of energy to a control surface about the liquid (Equation 1.12c), it follows that, for a unit surface area,

$$q''_{conv} + q''_{add} = q''_{evap} \tag{6.61}$$

where q''_{evap} may be approximated as the product of the evaporative mass flux and the latent heat of vaporization

$$q''_{evap} = n''_A h_{fg} \tag{6.62}$$

If there is no heat addition by other means, Equation 6.61 reduces to a balance between convection heat transfer from the gas and the evaporative heat loss from the liquid. Substituting from Equations 6.4, 6.18, and 6.62, Equation 6.61 may then be expressed as

$$h(T_\infty - T_s) = h_{fg} h_m [\rho_{A,sat}(T_s) - \rho_{A,\infty}] \tag{6.63}$$

where the vapor density at the surface is that associated with saturated conditions at T_s. Hence the magnitude of the cooling effect may be expressed as

$$(T_\infty - T_s) = h_{fg} \left(\frac{h_m}{h} \right) [\rho_{A,sat}(T_s) - \rho_{A,\infty}] \tag{6.64}$$

Substituting for (h_m/h) from Equation 6.60 and for the vapor densities from the ideal gas law, the cooling effect may also be expressed as

$$(T_\infty - T_s) = \frac{\mathcal{M}_A h_{fg}}{\mathcal{R} \rho c_p Le^{2/3}} \left[\frac{p_{A,sat}(T_s)}{T_s} - \frac{p_{A,\infty}}{T_\infty} \right] \tag{6.65}$$

In the interest of accuracy, the gas (species B) properties ρ, c_p, and Le should be evaluated at the arithmetic mean, or film, temperature of the thermal boundary layer, $T_{am} = T_f = (T_s + T_\infty)/2$, while the latent heat of vaporization of species A, h_{fg}, should be evaluated at the surface temperature T_s. A representative value of $n = \frac{1}{3}$ has been assumed for the Pr and Sc exponent of Equation 6.60.

Numerous environmental and industrial applications of the foregoing results arise for situations in which the gas is *air* and the liquid is *water*.

EXAMPLE 6.8

A container, which is wrapped in a fabric that is continually moistened with a highly volatile liquid, may be used to keep beverages cool in hot, arid regions. Suppose that the container is placed in dry ambient air at 40°C, with heat and mass transfer between the wetting
agent and the air occurring by forced convection. The wetting agent is known to have a
molecular weight of 200 kg/kmol and a latent heat of vaporization of 100 kJ/kg. Its saturated
vapor pressure for the prescribed conditions is approximately 5000 N/m², and the diffusion
coefficient of the vapor in air is 0.2×10^{-4} m²/s. What is the steady-state temperature of the
beverage?

SOLUTION

Known: Properties of wetting agent used to evaporatively cool a beverage container.

Find: Steady-state temperature of beverage.

Schematic:

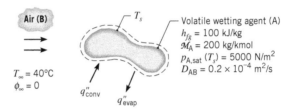

Assumptions:

1. Heat and mass transfer analogy is applicable.
2. Vapor displays ideal gas behavior.
3. Radiation effects are negligible.
4. Air properties may be evaluated at a mean boundary layer temperature assumed to be
   $T_f = 300$ K.

Properties: Table A.4, air (300 K): $\rho = 1.16$ kg/m³, $c_p = 1.007$ kJ/kg · K, $\alpha = 22.5 \times 10^{-6}$ m²/s.

Analysis: Subject to the foregoing assumptions, the evaporative cooling effect is given
by Equation 6.65.

$$(T_\infty - T_s) = \frac{\mathcal{M}_A h_{fg}}{\mathcal{R} \rho c_p Le^{2/3}} \left[\frac{p_{A,\text{sat}}(T_s)}{T_s} - \frac{p_{A,\infty}}{T_\infty} \right]$$

Setting $p_{A,\infty} = 0$ and rearranging, it follows that

$$T_s^2 - T_\infty T_s + B = 0$$

where the coefficient B is

$$B = \frac{\mathcal{M}_A h_{fg} p_{A,\text{sat}}}{\mathcal{R} \rho c_p Le^{2/3}}$$

or

$$B = [200 \text{ kg/kmol} \times 100 \text{ kJ/kg} \times 5000 \text{ N/m}^2 \times 10^{-3} \text{ kJ/N} \cdot \text{m}]$$

$$\div \left[8.314 \text{ kJ/kmol} \cdot \text{K} \times 1.16 \text{ kg/m}^3 \times 1.007 \text{ kJ/kg} \cdot \text{K} \right.$$

$$\left. \times \left(\frac{22.5 \times 10^{-6} \text{ m}^2/\text{s}}{20 \times 10^{-6} \text{ m}^2/\text{s}} \right)^{2/3} \right] = 9520 \text{ K}^2$$

Hence

$$T_s = \frac{T_\infty \pm \sqrt{T_\infty^2 - 4B}}{2} = \frac{313 \text{ K} \pm \sqrt{(313)^2 - 4(9520)} \text{ K}}{2}$$

Rejecting the minus sign on physical grounds (T_s must equal T_∞ if there is no evaporation, in which case $p_{A,\text{sat}} = 0$ and $B = 0$), it follows that

$$T_s = 278.9 \text{ K} = 5.9°\text{C} \qquad \triangleleft$$

Comments: The result is independent of the shape of the container as long as the heat and mass transfer analogy may be used.

6.7.3 The Reynolds Analogy

A second boundary layer analogy may be obtained by noting from Table 6.1 that, for $dp^*/dx^* = 0$ and $Pr = Sc = 1$, the boundary layer equations, Equations 6.35 through 6.37, are of precisely the same form. For a flat plate parallel to the incoming flow, we have $dp^*/dx^* = 0$ and there is no variation in the free stream velocity outside the boundary layer. With $u_\infty = V$, Equations 6.38 through 6.40 also have the same form. Hence the functional forms of the solutions for u^*, T^*, and C_A^*, Equations 6.44, 6.47, and 6.51, must be equivalent. From Equations 6.45, 6.48, and 6.52, it follows that

$$C_f \frac{Re_L}{2} = Nu = Sh \qquad (6.66)$$

Replacing Nu and Sh by the *Stanton number* (St) and the *mass transfer Stanton number* (St_m), respectively,

$$St \equiv \frac{h}{\rho V c_p} = \frac{Nu}{Re \, Pr} \qquad (6.67)$$

$$St_m \equiv \frac{h_m}{V} = \frac{Sh}{Re \, Sc} \qquad (6.68)$$

Equation 6.66 may also be expressed in the form

$$\frac{C_f}{2} = St = St_m \qquad (6.69)$$

Equation 6.69 is known as the *Reynolds analogy,* and it relates the key engineering parameters of the velocity, thermal, and concentration boundary layers. If the velocity parameter is known, the analogy may be used to obtain the other parameters, and vice versa. However, there are restrictions associated with using this result. In addition to relying on the validity of the boundary layer approximations, the accuracy of Equation 6.69 depends on having Pr and $Sc \approx 1$ and $dp^*/dx^* \approx 0$. However, it has been shown that the analogy may be applied over a wide range of Pr and Sc, if certain corrections are added. In particular the *modified Reynolds,* or *Chilton–Colburn, analogies* [13, 14], have the form

$$\frac{C_f}{2} = St\, Pr^{2/3} \equiv j_H \qquad 0.6 < Pr < 60 \qquad (6.70)$$

$$\frac{C_f}{2} = St_m\, Sc^{2/3} \equiv j_m \qquad 0.6 < Sc < 3000 \qquad (6.71)$$

where j_H and j_m are the *Colburn j factors* for heat and mass transfer, respectively. For laminar flow Equations 6.70 and 6.71 are only appropriate when $dp^*/dx^* < 0$, but in turbulent flow, conditions are less sensitive to the effect of pressure gradients and these equations remain approximately valid. If the analogy is applicable at every point on a surface, it may be applied to the surface average coefficients.

6.8 *Summary*

In this chapter we have considered several fundamental issues related to convection transport phenomena. In the process, however, you should not lose sight of what remains *the problem of convection.* Our primary objective is still one of developing means to determine the convection coefficients h and h_m. Although these coefficients may be obtained by solving the boundary layer equations, it is only for simple flow situations that such solutions are readily effected. The more practical approach frequently involves calculating h and h_m from empirical relations of the form given by Equations 6.49 and 6.53. The particular form of these equations is obtained by *correlating* measured convection heat and mass transfer results in terms of appropriate dimensionless groups. It is this approach that is emphasized in the chapters that follow.

To test your comprehension of the material, you should challenge yourself with appropriate questions.

- What is the difference between a *local* convection heat transfer coefficient and an *average* coefficient? What are their units? What is the difference between local and average convection coefficients for species transfer? What are their units?
- What are the forms of *Newton's law of cooling* for a *heat flux* and a *heat rate*? What are the analogous forms for convection mass transfer, expressed in molar and mass units?

- Provide some examples for which species transfer by convection is pertinent.
- What is *Fick's law*?
- What are the *velocity, thermal,* and *concentration boundary layers*? Under what conditions do they develop?
- What quantities change with location in a *velocity boundary layer*? A *thermal boundary layer*? A *concentration boundary layer*?
- Recognizing that convection heat (mass) transfer is strongly influenced by conditions associated with fluid flow over a surface, how is it that we may determine the convection heat (species) flux by applying Fourier's (Fick's) law to the fluid at the surface?
- Do we expect heat and mass transfer to change with transition from a laminar to a turbulent boundary layer? If so, how?
- What laws of nature are embodied by the *convection transfer equations*?
- What physical processes are represented by the terms of the x-momentum equation (6.28)? By the energy equation (6.29)? By the species conservation equation (6.30)?
- What special approximations may be made for conditions within *thin* velocity, thermal, and concentration boundary layers?
- What is the *film* temperature and how is it used?
- How is the *Reynolds number* defined? What is its physical interpretation? What role is played by the *critical Reynolds number*?
- What is the definition of the *Prandtl number*? How does its value affect relative growth of the velocity and thermal boundary layers for laminar flow over a surface? What are representative room-temperature values of the Prandtl number for a liquid metal, a gas, water, and an oil?
- What is the definition of the *Schmidt number*? The *Lewis number*? What are their physical interpretations, and how do they influence relative velocity, thermal, and concentration boundary layer development for laminar flow over a surface?
- What is the *friction coefficient*? The *Nusselt number*? The *Sherwood number*? For flow over a prescribed geometry, what are the independent parameters that determine local and average values of these quantities?
- Under what conditions may velocity, thermal, and concentration boundary layers be termed *analogous*? What is the physical basis of analogous behavior?
- What important boundary layer parameters are linked by the *heat and mass transfer analogy*?
- What is the physical basis of the *evaporative cooling effect*? Have you ever experienced the effect?
- What important boundary layer parameters are linked by the *Reynolds analogy*?
- What physical features distinguish a turbulent flow from a laminar flow?

References

1. Webb, R. L., *Int. Comm. Heat Mass Trans.,* **17,** 529, 1990.

2. Hof, B., C. W. H. van Doorne, J. Westerweel, F. T. M. Nieuwstadt, H. Faisst, B. Eckhardt, H. Wedin, R. R. Kerswell, and F. Waleffe, *Science,* **305,** 1594, 2004.

3. Schlichting, H., and K. Gersten, *Boundary Layer Theory,* 8th ed., Springer-Verlag, New York, 1999.

4. Bird, R. B., W. E. Stewart, and E. N. Lightfoot, *Transport Phenomena,* 2nd ed., Wiley, New York, 2002.

5. Fox, R. W., A. T. McDonald, and P. J. Pritchard, *Introduction to Fluid Mechanics,* 6th ed., Wiley, Hoboken, NJ, 2003.

6. Hartnett, J. P., "Mass Transfer Cooling," in W. M. Rohsenow and J. P. Hartnett, Eds., *Handbook of Heat Transfer,* McGraw-Hill, New York, 1973.

7. Kays, W. M., M. E. Crawford, and B. Weigand, *Convective Heat and Mass Transfer,* 4th ed., McGraw-Hill Higher Education, Boston, 2005.

8. Burmeister, L. C., *Convective Heat Transfer,* 2nd ed., Wiley, New York, 1993.

9. Kaviany, M., *Principles of Convective Heat Transfer,* Springer-Verlag, New York, 1994.

10. Patankar, S. V., *Numerical Heat Transfer and Fluid Flow,* Hemisphere Publishing, New York, 1980.

11. Oosthuizen, P. H., and W. E. Carscallen, *Compressible Fluid Flow*, McGraw-Hill, New York, 1997.

12. John, J. E. A., and T. G. Keith, *Gas Dynamics*, 3rd ed., Pearson Prentice Hall, Upper Saddle River, NJ, 2006.

13. Colburn, A. P., *Trans. Am. Inst. Chem. Eng., **29,*** 174, 1933.

14. Chilton, T. H., and A. P. Colburn, *Ind. Eng. Chem.,* **26,** 1183, 1934.

Problems

Boundary Layer Profiles

6.1 The temperature distribution within a laminar thermal boundary layer associated with flow over an isothermal flat plate is shown in the sketch. The temperature distribution shown is located at $x = x_2$.

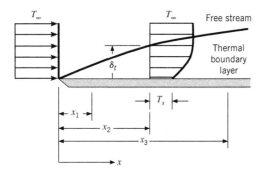

(a) Is the plate being heated or cooled by the fluid?

(b) Carefully sketch the temperature distributions at $x = x_1$ and $x = x_3$. Based on your sketch, at which of the three x-locations is the local heat flux largest? At which location is the local heat flux smallest?

(c) As the free stream velocity increases, the velocity and thermal boundary layers both become thinner. Carefully sketch the temperature distributions at $x = x_2$ for (i) a low free stream velocity and (ii) a high free stream velocity. Based on your sketch, which velocity condition will induce the larger local convective heat flux?

6.2 In flow over a surface, velocity and temperature profiles are of the forms

$$u(y) = Ay + By^2 - Cy^3 \quad \text{and}$$
$$T(y) = D + Ey + Fy^2 - Gy^3$$

where the coefficients A through G are constants. Obtain expressions for the friction coefficient C_f and the convection coefficient h in terms of u_∞, T_∞, and appropriate profile coefficients and fluid properties.

6.3 In a particular application involving airflow over a surface, the boundary layer temperature distribution may be approximated as

$$\frac{T - T_s}{T_\infty - T_s} = 1 - \exp\left(-Pr \frac{u_\infty y}{\nu}\right)$$

where y is the distance normal to the surface and the Prandtl number, $Pr = c_p \mu / k = 0.7$, is a dimensionless fluid property. If $T_\infty = 400$ K, $T_s = 300$ K, and $u_\infty/\nu = 5000$ m^{-1}, what is the surface heat flux?

6.4 Water at a temperature of $T_\infty = 25°C$ flows over one of the surfaces of a steel wall (AISI 1010) whose temperature is $T_{s,1} = 40°C$. The wall is 0.35 m thick, and its other surface temperature is $T_{s,2} = 100°C$. For steady-state conditions what is the convection coefficient associated with the water flow? What is the temperature gradient in the wall and in the water that is in contact with the wall? Sketch the temperature distribution in the wall and in the adjoining water.

Heat Transfer Coefficients

6.5 For laminar free convection from a heated vertical surface, the local convection coefficient may be expressed as $h_x = Cx^{-1/4}$, where h_x is the coefficient at a distance x from the leading edge of the surface and the quantity C, which depends on the fluid properties, is independent of x. Obtain an expression for the ratio $\bar{h}_x/h_x$, where $\bar{h}_x$ is the average coefficient between the leading edge ($x = 0$) and the x-location. Sketch the variation of h_x and $\bar{h}_x$ with x.

6.6 Parallel flow of atmospheric air over a flat plate of length $L = 3$ m is disrupted by an array of stationary rods placed in the flow path over the plate.

Laboratory measurements of the local convection coefficient at the surface of the plate are made for a prescribed value of V and $T_s > T_\infty$. The results are correlated by an expression of the form $h_x = 0.7 + 13.6x - 3.4x^2$, where h_x has units of $W/m^2 \cdot K$ and x is in meters. Evaluate the average convection coefficient $\bar{h}_L$ for the entire plate and the ratio $\bar{h}_L/h_L$ at the trailing edge.

6.7 A circular, hot gas jet at T_∞ is directed normal to a circular plate that has radius r_o and is maintained at a uniform temperature T_s. Gas flow over the plate is axisymmetric, causing the local convection coefficient to have a radial dependence of the form $h(r) = a + br^n$, where a, b, and n are constants. Determine the rate of heat transfer to the plate, expressing your result in terms of T_∞, T_s, r_o, a, b, and n.

6.8 Laminar flow normally persists on a smooth flat plate until a critical Reynolds number value is reached. However, the flow can be *tripped* to a turbulent state by adding roughness to the leading edge of the plate. For a particular situation, experimental results show that the local heat transfer coefficients for laminar and turbulent conditions are

$$h_{\text{lam}}(x) = 1.74 \text{ W/m}^{1.5} \cdot \text{K } x^{-0.5}$$
$$h_{\text{turb}}(x) = 3.98 \text{ W/m}^{1.8} \cdot \text{K } x^{-0.2}$$

Calculate the average heat transfer coefficients for laminar and turbulent conditions for plates of length $L = 0.1$ m and 1 m.

6.9 Experiments have been conducted to determine local heat transfer coefficients for flow perpendicular to a long, isothermal bar of rectangular cross section. The bar is of width c parallel to the flow, and height d normal to the flow. For Reynolds numbers in the range $10^4 \le Re_d \le 5 \times 10^4$, the *face-averaged* Nusselt numbers are well correlated by an expression of the form

$$\overline{Nu}_d = \bar{h}d/k = C Re_d^m Pr^{1/3}$$

The values of C and m for the front face, side faces, and back face of the rectangular rod are found to be the following:

Face	c/d	C	m
Front	$0.33 \le c/d \le 1.33$	0.674	1/2
Side	0.33	0.153	2/3
Side	1.33	0.107	2/3
Back	0.33	0.174	2/3
Back	1.33	0.153	2/3

Determine the value of the average heat transfer coefficient for the entire exposed surface (that is, averaged over all four faces) of a $c = 40$-mm-wide,

$d = 30$-mm-tall rectangular rod. The rod is exposed to air in cross flow at $V = 10$ m/s, $T_\infty = 300$ K. Provide a plausible explanation of the relative values of the face-averaged heat transfer coefficients on the front, side, and back faces.

6.10 A concentrating solar collector consists of a parabolic reflector and a collector tube of diameter D, through which flows a working fluid that is heated with concentrated solar irradiation. Throughout the day, the reflector is slowly repositioned to track the sun. For wind conditions characterized by a steady, horizontal flow normal to the tube axis, the local heat transfer coefficient on the tube surface varies, as shown in the schematic for various reflector positions.

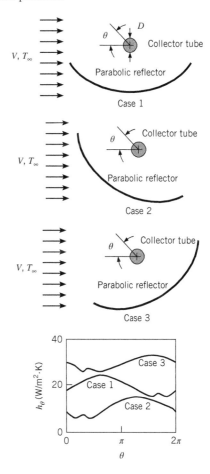

(a) Estimate the value of the average heat transfer coefficient over the entire collector tube surface for each of the three cases.

(b) Assuming the tube receives the same amount of solar irradiation in each case, which case would have the highest collector efficiency?

6.11 Helium at a free stream temperature of $T_\infty = 25°C$ is in parallel flow over a flat plate of length $L = 3$ m and temperature $T_s = 85°C$. However, obstacles placed in the flow intensify mixing with increasing distance x from the leading edge, and the spatial variation of temperatures measured in the boundary layer is correlated by an expression of the form $T(°C) = 25 + 60 \exp(-600xy)$, where x and y are in meters. Determine and plot the manner in which the local convection coefficient h varies with x. Evaluate the average convection coefficient $\bar{h}$ for the plate.

6.12 Experiments to determine the local convection heat transfer coefficient for uniform flow normal to a heated circular disk have yielded a radial Nusselt number distribution of the form

$$Nu_D = \frac{h(r)D}{k} = Nu_o\left[1 + a\left(\frac{r}{r_o}\right)^n\right]$$

where both n and a are positive. The Nusselt number at the stagnation point is correlated in terms of the Reynolds ($Re_D = VD/\nu$) and Prandtl numbers

$$Nu_o = \frac{h(r=0)D}{k} = 0.814\,Re_D^{1/2}Pr^{0.36}$$

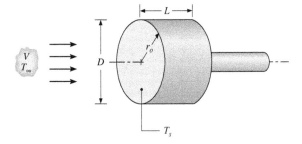

Obtain an expression for the average Nusselt number, $\overline{Nu}_D = \bar{h}D/k$, corresponding to heat transfer from an isothermal disk. Typically, boundary layer development from a stagnation point yields a decaying convection coefficient with increasing distance from the stagnation point. Provide a plausible explanation for why the opposite trend is observed for the disk.

6.13 An experimental procedure for validating results of Problem 6.12 involves preheating a copper disk to an initial elevated temperature T_i and recording its temperature history $T(t)$ as it is subsequently cooled by the impinging flow to a final temperature T_f. The measured temperature decay may then be compared with predictions based on the correlation for $\overline{Nu}_D$. Assume that values of $a = 0.30$ and $n = 2$ are associated with the correlation.

Consider experimental conditions for which a disk of diameter $D = 50$ mm and length $L = 25$ mm is preheated to $T_i = 1000$ K and cooled to $T_f = 400$ K by an impinging airflow at $T_\infty = 300$ K. The cooled surface of the disk has an emissivity of $\varepsilon = 0.8$ and is exposed to large, isothermal surroundings for which $T_{sur} = T_\infty$. The remaining surfaces of the disk are well insulated, and heat transfer through the supporting rod may be neglected. Using results from Problem 6.12, compute and plot temperature histories corresponding to air velocities of $V = 4, 20$, and 50 m/s. Constant properties may be assumed for the copper ($\rho = 8933$ kg/m³, $c_p = 425$ J/kg · K, $k = 386$ W/m · K) and air ($\nu = 38.8 \times 10^{-6}$ m²/s, $k = 0.0407$ W/m · K, $Pr = 0.684$).

6.14 If laminar flow is induced at the surface of a disk due to rotation about its axis, the local convection coefficient is known to be a constant, $h = C$, independent of radius. Consider conditions for which a disk of radius $r_o = 100$ mm is rotating in stagnant air at $T_\infty = 20°C$ and a value of $C = 20$ W/m² · K is maintained.

If an embedded electric heater maintains a surface temperature of $T_s = 50°C$, what is the local heat flux at the top surface of the disk? What is the total electric power requirement? What can you say about the nature of boundary layer development on the disk?

6.15 Consider the rotating disk of Problem 6.14. A disk-shaped, stationary plate is placed a short distance away from the rotating disk, forming a gap of width g. The stationary plate and ambient air are at $T_\infty = 20°C$. If the flow is laminar and the gap-to-radius ratio, $G = g/r_o$, is small, the local radial Nusselt number distribution is of the form

$$Nu_r = \frac{h(r)r}{k} = 70(1 + e^{-140G})\,Re_{r_o}^{-0.456}Re_r^{0.478}$$

where $Re_r = \Omega r^2/\nu$ [Pelle J., and S. Harmand, *Exp. Thermal Fluid Science*, **31**, 165, 2007]. Determine the value of the average Nusselt number, $\overline{Nu}_D = \bar{h}D/k$ where $D = 2r_o$.

If the rotating disk temperature is $T_s = 50°C$, what is the total heat flux from the disk's top surface for $g = 1$ mm, $\Omega = 150$ rad/s? What is the total electric power requirement? What can you say about the nature of the flow between the disks?

Boundary Layer Transition

6.16 Consider airflow over a flat plate of length $L = 1$ m under conditions for which transition occurs at $x_c = 0.5$ m based on the critical Reynolds number, $Re_{x,c} = 5 \times 10^5$.

(a) Evaluating the thermophysical properties of air at 350 K, determine the air velocity.

(b) In the laminar and turbulent regions, the local convection coefficients are, respectively,

$$h_{\text{lam}}(x) = C_{\text{lam}}x^{-0.5} \quad \text{and} \quad h_{\text{turb}} = C_{\text{turb}}x^{-0.2}$$

where, at $T = 350$ K, $C_{\text{lam}} = 8.845$ W/m$^{3/2} \cdot$ K, $C_{\text{turb}} = 49.75$ W/m$^{1.8} \cdot$ K, and x has units of m. Develop an expression for the average convection coefficient, $\bar{h}_{\text{lam}}(x)$, as a function of distance from the leading edge, x, for the laminar region, $0 \leq x \leq x_c$.

(c) Develop an expression for the average convection coefficient, $\bar{h}_{\text{turb}}(x)$, as a function of distance from the leading edge, x, for the turbulent region, $x_c \leq x \leq L$.

(d) On the same coordinates, plot the local and average convection coefficients, h_x and $\bar{h}_x$, respectively, as a function of x for $0 \leq x \leq L$.

6.17 A fan that can provide air speeds up to 50 m/s is to be used in a low-speed wind tunnel with atmospheric air at 25°C. If one wishes to use the wind tunnel to study flat-plate boundary layer behavior up to Reynolds numbers of $Re_x = 10^8$, what is the minimum plate length that should be used? At what distance from the leading edge would transition occur if the critical Reynolds number were $Re_{x,c} = 5 \times 10^5$?

6.18 Consider the flow conditions of Example 6.4 for two situations, one in which the flow is completely laminar, and the second for flow that is tripped to turbulence at the leading edge of the plate. Determine whether there is a plate length L for which the average convection coefficient for laminar flow is the same as the average convection coefficient for turbulent flow. Assume a water temperature of 300 K.

6.19 For the situation described in Example 6.4, the boundary layer can be *tripped* into a turbulent state by applying roughness to the surface of the flat plate at a particular x-location. Hence the location where transition occurs, x_c, can be moved upstream relative to the transition location associated with the smooth plate of

the example. Calculate and plot the average convection coefficient over the entire plate $\bar{h}$ for roughness applied over the range $0 \leq x_r \leq L$. What values of x_r provide the minimum and maximum values of $\bar{h}$? Assume the water temperature is 300 K.

6.20 To a good approximation, the dynamic viscosity μ, the thermal conductivity k, and the specific heat c_p are independent of pressure. In what manner do the kinematic viscosity ν and thermal diffusivity α vary with pressure for an incompressible liquid and an ideal gas? Determine α of air at 350 K for pressures of 1, 5, and 10 atm. Assuming a transition Reynolds number of 5×10^5, determine the distance from the leading edge of a flat plate at which transition will occur for air at 350 K at pressures of 1, 5, and 10 atm with $u_\infty = 2$ m/s.

Similarity and Dimensionless Parameters

6.21 Consider a laminar boundary layer developing over a flat plate. The flow is incompressible.

(a) Substitute Equations 6.31 and 6.32 into Equation 6.38 to determine the boundary conditions in dimensional form associated with flow over a flat plate of length L.

(b) Substitute Equations 6.31, 6.32, as well as the definition of Re_L into Equation 6.35, and compare the resulting expression with Equation 6.28. Note that for a flat plate, $dp/dx = 0$ and $u_\infty = V$.

6.22 Consider laminar boundary layer flow over a flat plate. For identical flow conditions, determine the entries not shown in the table below, using the fact that $\delta/\delta_t = Pr^n$. The pressure is $p = 1$ atm for each case. *Hint*: See Appendices A.4 through A.6.

Fluid	Temperature (K)	δ/δ_t
Air	300	0.89
Water (liquid)	350	
Engine oil		15.4
		45.3

6.23 Experiments have shown that, for airflow at $T_\infty = 35°C$ and $V_1 = 100$ m/s, the rate of heat transfer from a turbine blade of characteristic length $L_1 = 0.15$ m and surface temperature $T_{s,1} = 300°C$ is $q_1 = 1500$ W. What would be the heat transfer rate from a second turbine blade of characteristic length $L_2 = 0.3$ m operating at $T_{s,2} = 400°C$ in airflow of $T_\infty = 35°C$ and $V_2 = 50$ m/s? The surface area of the blade may be assumed to be directly proportional to its characteristic length.

6.24 An object of irregular shape has a characteristic length of $L = 1$ m and is maintained at a uniform surface temperature of $T_s = 400$ K. When placed in atmospheric air at a temperature of $T_\infty = 300$ K and moving with a velocity of $V = 100$ m/s, the average heat flux from the surface to the air is 20,000 W/m². If a second object of the same shape, but with a characteristic length of $L = 5$ m, is maintained at a surface temperature of $T_s = 400$ K and is placed in atmospheric air at $T_\infty = 300$ K, what will the value of the average convection coefficient be if the air velocity is $V = 20$ m/s?

6.25 Experiments have shown that the transition from laminar to turbulent conditions for flow normal to the axis of a long cylinder occurs at a critical Reynolds number of $Re_{D,c} \approx 2 \times 10^5$, where D is the cylinder diameter. Moreover, the transition from incompressible to compressible flow occurs at a critical Mach number of $Ma_c \approx 0.3$. For air at a pressure of $p = 1$ atm and temperature $T = 27°C$, determine the critical cylinder diameter D_c below which, if the flow is turbulent, compressibility effects are likely to be important.

6.26 Experimental measurements of the convection heat transfer coefficient for a square bar in cross flow yielded the following values:

$$\bar{h}_1 = 50 \text{ W/m}^2 \cdot \text{K} \quad \text{when} \quad V_1 = 20 \text{ m/s}$$
$$\bar{h}_2 = 40 \text{ W/m}^2 \cdot \text{K} \quad \text{when} \quad V_2 = 15 \text{ m/s}$$

Assume that the functional form of the Nusselt number is $\overline{Nu} = C\, Re^m Pr^n$, where C, m, and n are constants.

(a) What will be the convection heat transfer coefficient for a similar bar with $L = 1$ m when $V = 15$ m/s?

(b) What will be the convection heat transfer coefficient for a similar bar with $L = 1$ m when $V = 30$ m/s?

(c) Would your results be the same if the side of the bar, rather than its diagonal, were used as the characteristic length?

6.27 To assess the efficacy of different liquids for cooling an object of given size and shape by forced convection, it is convenient to introduce a *figure of merit*, F_F, which combines the influence of all pertinent fluid properties on the convection coefficient. If the Nusselt number is governed by an expression of the form, $\overline{Nu}_L \sim Re_L^m Pr^n$, obtain the corresponding relationship between F_F and the fluid properties. For representative values of

$m = 0.80$ and $n = 0.33$, calculate values of F_F for air ($k = 0.026$ W/m · K, $v = 1.6 \times 10^{-5}$ m²/s, $Pr = 0.71$), water ($k = 0.600$ W/m · K, $v = 10^{-6}$ m²/s, $Pr = 5.0$), and a dielectric liquid ($k = 0.064$ W/m · K, $v = 10^{-6}$ m²/s, $Pr = 25$). Which fluid is the most effective cooling agent?

6.28 Experimental results for heat transfer over a flat plate with an extremely rough surface were found to be correlated by an expression of the form

$$Nu_x = 0.04\, Re_x^{0.9}\, Pr^{1/3}$$

where Nu_x is the local value of the Nusselt number at a position x measured from the leading edge of the plate. Obtain an expression for the ratio of the average heat transfer coefficient $\bar{h}_x$ to the local coefficient h_x.

6.29 Consider conditions for which a fluid with a free stream velocity of $V = 1$ m/s flows over a surface with a characteristic length of $L = 1$ m, providing an average convection heat transfer coefficient of $\bar{h} = 100$ Wm² · K. Calculate the dimensionless parameters $\overline{Nu}_L$, Re_L, Pr, and $\bar{j}_H$ for the following fluids: air, engine oil, mercury, and water. Assume the fluids to be at 300 K.

6.30 For flow over a flat plate of length L, the local heat transfer coefficient h_x is known to vary as $x^{-1/2}$, where x is the distance from the leading edge of the plate. What is the ratio of the average Nusselt number for the entire plate ($\overline{Nu}_L$) to the local Nusselt number at $x = L$ (Nu_L)?

6.31 For laminar boundary layer flow over a flat plate with air at 20°C and 1 atm, the thermal boundary layer thickness δ_t is approximately 13% larger than the velocity boundary layer thickness δ. Determine the ratio δ/δ_t if the fluid is ethylene glycol under the same flow conditions.

6.32 Consider parallel flow over a flat plate for air at 300 K and engine oil at 380 K. The free stream velocity is $u_\infty = 2$ m/s. The temperature difference between the surface and the free stream is the same in both cases, with $T_s > T_\infty$.

(a) Determine the location where transition to turbulence occurs, x_c, for both fluids.

(b) For laminar flow over a flat plate, the velocity boundary layer thickness is given by

$$\frac{\delta}{x} = \frac{5}{\sqrt{Re_x}}$$

Calculate and plot the velocity boundary layer thickness δ over the range $0 \le x \le x_c$ for each fluid.

(c) Calculate and plot the thermal boundary layer thickness δ_t for the two fluids over the same range of x used in part (b). At an x-location where both fluids experience laminar flow conditions, explain which fluid has the largest temperature gradient at the plate surface, $-\partial T/\partial y|_{y=0}$. Which fluid is associated with the largest local Nusselt number Nu? Which fluid is associated with the largest local heat transfer coefficient h?

6.33 Forced air at $T_\infty = 25°C$ and $V = 10$ m/s is used to cool electronic elements on a circuit board. One such element is a chip, 4 mm × 4 mm, located 120 mm from the leading edge of the board. Experiments have revealed that flow over the board is disturbed by the elements and that convection heat transfer is correlated by an expression of the form

$$Nu_x = 0.04\, Re_x^{0.85} Pr^{1/3}$$

Estimate the surface temperature of the chip if it is dissipating 30 mW.

6.34 Consider the electronic elements that are cooled by forced convection in Problem 6.33. The cooling system is designed and tested at sea level ($p \approx 1$ atm), but the circuit board is sold to a customer in Mexico City, with an elevation of 2250 m and atmospheric pressure of 76.5 kPa.

(a) Estimate the surface temperature of the chip located 120 mm from the leading edge of the board when the board is operated in Mexico City. The dependence of various thermophysical properties on pressure is noted in Problem 6.20.

(b) It is desirable for the chip operating temperature to be independent of the location of the customer. What air velocity is required for operation in Mexico City if the chip temperature is to be the same as at sea level?

6.35 Consider the chip on the circuit board of Problem 6.33. To ensure reliable operation over extended periods, the chip temperature should not exceed 85°C. Assuming the availability of forced air at $T_\infty = 25°C$ and applicability of the prescribed heat transfer correlation, compute and plot the maximum allowable chip power dissipation P_c as a function of air velocity for $1 \leq V \leq 25$ m/s. If the chip surface has an emissivity of 0.80 and the board is mounted in a large enclosure whose walls are at 25°C, what is the effect of radiation on the $P_c - V$ plot?

6.36 A major contributor to product defects in electronic modules relates to stresses induced during thermal cycling (intermittent heating and cooling). For example, in circuit cards having active and passive components

with materials of different thermal expansion coefficients, thermal stresses are the principal source of failure in component joints, such as soldered and wired connections. Although concern is generally for fatigue failure resulting from numerous excursions during the life of a product, it is possible to identify defective joints by performing accelerated thermal stress tests before the product is released to the customer. In such cases, it is important to achieve rapid thermal cycling to minimize disruptions to production schedules.

A manufacturer of circuit cards wishes to develop an apparatus for imposing rapid thermal transients on the cards by subjecting them to forced convection characterized by a relation of the form $\overline{Nu}_L = C\, Re_L^m Pr^n$, where $m = 0.8$ and $n = 0.33$. However, he does not know whether to use air ($k = 0.026$ W/m·K, $\nu = 1.6 \times 10^{-5}$ m²/s, $Pr = 0.71$) or a dielectric liquid ($k = 0.064$ W/m·K, $\nu = 10^{-6}$ m²/s, $Pr = 25$) as the working fluid. Assuming equivalent air and liquid velocities and validity of the lumped capacitance model for the components, obtain a quantitative estimate of the ratio of the thermal time constants for the two fluids. What fluid provides the faster thermal response?

6.37 A microscale detector monitors a steady flow ($T_\infty = 27°C$, $V = 10$ m/s) of air for the possible presence of small, hazardous particulate matter that may be suspended in the room. The sensor is heated to a slightly higher temperature to induce a chemical reaction associated with certain substances of interest that might impinge on the sensor's active surface. The active surface produces an electric current if such surface reactions occur; the electric current is then sent to an alarm. To maximize the sensor head's surface area and, in turn, the probability of capturing and detecting a particle, the sensor head is designed with a very complex shape. The value of the average heat transfer coefficient associated with the heated sensor must be known so that the required electrical power to the sensor can be determined.

Consider a sensor with a characteristic dimension of $L_s = 80$ μm. A scale model of the sensor is placed in a recirculating (closed) wind tunnel using hydrogen as the working fluid. If the wind tunnel operates at a hydrogen absolute pressure of 0.5 atm and velocity of $V = 0.5$ m/s, find the required hydrogen temperature and characteristic dimension of the scale model, L_m.

Reynolds Analogy

6.38 A thin, flat plate that is 0.2 m × 0.2 m on a side is oriented parallel to an atmospheric airstream having a velocity of 40 m/s. The air is at a temperature of $T_\infty = 20°C$, while the plate is maintained at $T_s = 120°C$. The air flows over the top and bottom surfaces of the plate, and measurement of the drag force reveals a value of 0.075 N. What is the rate of heat transfer from both sides of the plate to the air?

6.39 Atmospheric air is in parallel flow ($u_\infty = 15$ m/s, $T_\infty = 15°C$) over a flat heater surface that is to be maintained at a temperature of 140°C. The heater surface area is 0.25 m², and the airflow is known to induce a drag force of 0.25 N on the heater. What is the electrical power needed to maintain the prescribed surface temperature?

6.40 For flow over a flat plate with an extremely rough surface, convection heat transfer effects are known to be correlated by the expression of Problem 6.28. For airflow at 50 m/s, what is the surface shear stress at $x = 1$ m from the leading edge of the plate? Assume the air to be at a temperature of 300 K.

6.41 A thin, flat plate that is 0.2 m × 0.2 m on a side with rough top and bottom surfaces is placed in a wind tunnel so that its surfaces are parallel to an atmospheric airstream having a velocity of 30 m/s. The air is at a temperature of $T_\infty = 20°C$ while the plate is maintained at $T_s = 80°C$. The plate is rotated 45° about its center point, as shown in the schematic. Air flows over the top and bottom surfaces of the plate, and measurement of the heat transfer rate is 2000 W. What is the drag force on the plate?

Top view of thin, flat plate

6.42 As a means of preventing ice formation on the wings of a small, private aircraft, it is proposed that electric resistance heating elements be installed within the wings. To determine representative power requirements, consider nominal flight conditions for which the plane moves at 100 m/s in air that is at a temperature of −23°C. If the characteristic length of the airfoil is $L = 2$ m and wind tunnel measurements indicate an average friction coefficient of $\bar{C}_f = 0.0025$ for the nominal conditions, what is the average heat flux needed to maintain a surface temperature of $T_s = 5°C$?

6.43 A circuit board with a dense distribution of integrated circuits (ICs) and dimensions of 120 mm × 120 mm on a side is cooled by the parallel flow of atmospheric air with a velocity of 2 m/s.

From wind tunnel tests under the same flow conditions, the average frictional shear stress on the upper surface is determined to be 0.0625 N/m². What is the allowable power dissipation from the upper surface of the board if the average surface temperature of the ICs must not exceed the ambient air temperature by more than 25°C? Evaluate the thermophysical properties of air at 300 K.

Mass Transfer Coefficients

6.44 On a summer day the air temperature is 27°C and the relative humidity is 30%. Water evaporates from the surface of a lake at a rate of 0.10 kg/h per square meter of water surface area. The temperature of the water is also 27°C. Determine the value of the convection mass transfer coefficient.

6.45 It is observed that a 230-mm-diameter pan of water at 23°C has a mass loss rate of 1.5×10^{-5} kg/s when the ambient air is dry and at 23°C.

  (a) Determine the convection mass transfer coefficient for this situation.

  (b) Estimate the evaporation mass loss rate when the ambient air has a relative humidity of 50%.

  (c) Estimate the evaporation mass loss rate when the water and ambient air temperatures are 47°C, assuming that the convection mass transfer coefficient remains unchanged and the ambient air is dry.

6.46 The dryer section of a paper mill consists of 30 hot, 1.5-m-diameter cylindrical rollers of length 3 m. A 225° arc of each roller is in contact with the moist paper sheet, while the remaining roller surface is covered with a water film that is directly exposed to high-velocity air at a temperature of $T_\infty = 35°C$ and relative humidity of 95%. Determine the area-averaged mass transfer coefficient associated with the exposed roller surfaces if the water temperature is 90°C and the total evaporation rate from the exposed surfaces is 80 kg/s.

6.47 Photosynthesis, as it occurs in the leaves of a green plant, involves the transport of carbon dioxide (CO_2) from the atmosphere to the chloroplasts of the leaves. The rate of photosynthesis may be quantified in terms of the rate of CO_2 assimilation by the chloroplasts. This assimilation is strongly influenced by CO_2 transfer through the boundary layer that develops on the leaf surface. Under conditions for which the density of CO_2 is 6×10^{-4} kg/m³ in the air and 5×10^{-4} kg/m³ at the leaf surface and the convection mass transfer coefficient is 10^{-2} m/s, what is the rate of photosynthesis in terms of kilograms of CO_2 assimilated per unit time and area of leaf surface?

6.48 Species A is evaporating from a flat surface into species B. Assume that the concentration profile for species A in the concentration boundary layer is of the form $C_A(y) = Dy^2 + Ey + F$, where D, E, and F are constants at any x-location and y is measured along a normal from the surface. Develop an expression for the mass transfer convection coefficient h_m in terms of these constants, the concentration of A in the free stream $C_{A,\infty}$ and the mass diffusivity D_{AB}. Write an expression for the molar flux of mass transfer by convection for species A.

Similarity and Heat–Mass Transfer Analogy

6.49 Consider conditions for which a fluid with a free stream velocity of $V = 1$ m/s flows over an evaporating or subliming surface with a characteristic length of $L = 1$ m, providing an average mass transfer convection coefficient of $\overline{h}_m = 10^{-2}$ m/s. Calculate the dimensionless parameters $\overline{Sh}_L$, Re_L, Sc, and $\overline{j}_m$ for the following combinations: airflow over water, airflow over naphthalene, and warm glycerol over ice. Assume a fluid temperature of 300 K and a pressure of 1 atm.

6.50 An object of irregular shape 1 m long maintained at a constant temperature of 100°C is suspended in an airstream having a free stream temperature of 0°C, a pressure of 1 atm, and a velocity of 120 m/s. The air temperature measured at a point near the object in the airstream is 80°C. A second object having the same shape is 2 m long and is suspended in an airstream in the same manner. The air free stream velocity is 60 m/s. Both the air and the object are at 50°C, and the total pressure is 1 atm. A plastic coating on the surface of the object is being dried by this process. The molecular weight of the vapor is 82, and the saturation pressure at 50°C for the plastic material is 0.0323 atm. The mass diffusivity for the vapor in air at 50°C is 2.60×10^{-5} m²/s.

(a) For the second object, at a location corresponding to the point of measurement on the first object, determine the vapor concentration and partial pressure.

(b) If the average heat flux q'' is 2000 W/m² for the first object, determine the average mass flux n''_A (kg/s · m²) for the second object.

6.51 On a cool day in April a scantily clothed runner is known to lose heat at a rate of 500 W because of convection to the surrounding air at $T_\infty = 10$°C. The runner's skin remains dry and at a temperature of $T_s = 30$°C. Three months later, the runner is moving at the same speed, but the day is warm and humid with a temperature of $T_\infty = 30$°C and a relative humidity of $\phi_\infty = 60\%$. The runner is now drenched in sweat and has a uniform surface temperature of 35°C. Under both conditions constant air properties may be assumed with $\nu = 1.6 \times 10^{-5}$ m²/s, $k = 0.026$ W/m · K, $Pr = 0.70$, and D_{AB} (water vapor–air) $= 2.3 \times 10^{-5}$ m²/s.

(a) What is the rate of water loss due to evaporation on the summer day?

(b) What is the total convective heat loss on the summer day?

6.52 An experiment involves the flow of air at atmospheric pressure and 300 K over an object of characteristic length $L = 1$ m that is coated with species A. The measured average mass transfer coefficients are shown in the table. Determine the Schmidt, Reynolds, and Sherwood numbers for each case. If $\overline{Sh}$ is related to Re and Sc by an expression of the form $\overline{Sh} = C Re_L^m Sc^n$, determine the values of C, m, and n by analyzing the experimental results. Based on the values you have determined for the expression, calculate the area-averaged mass transfer coefficient for evaporation of benzene in air flowing at a velocity of 3.0 m/s over the object.

Species A	V (m/s)	$\overline{h}_m \times 10^3$ (m/s)
Water	2	12
Naphthalene	1.5	3.6
Acetone	2.5	8.0

6.53 An industrial process involves the evaporation of water from a liquid film that forms on a contoured surface. Dry air is passed over the surface, and from laboratory measurements the convection heat transfer correlation is of the form

$$\overline{Nu}_L = 0.43\, Re_L^{0.58}\, Pr^{0.4}$$

(a) For an air temperature and velocity of 27°C and 10 m/s, respectively, what is the rate of evaporation from a surface of 1-m² area and characteristic length $L = 1$ m? Approximate the density of saturated vapor as $\rho_{A,sat} = 0.0077$ kg/m³.

(b) What is the steady-state temperature of the liquid film?

6.54 The *naphthalene sublimation technique* involves the use of a mass transfer experiment coupled with an analysis based on the heat and mass transfer analogy to obtain local or average convection heat transfer coefficients for complex surface geometries. A coating of naphthalene, which is a volatile solid at room temperature, is applied to the surface and is then subjected to airflow in a wind tunnel. Alternatively, solid objects may be cast from liquid naphthalene. Over a designated time interval, Δt, there is a discernible loss of naphthalene due to sublimation, and by measuring the surface recession at locations of interest or the mass loss of the sample, local or average mass transfer coefficients may be determined.

Consider a rectangular rod of naphthalene exposed to air in cross flow at $V = 10$ m/s, $T_\infty = 300$ K, as in Problem 6.9, except now $c = 10$ mm and $d = 30$ mm. Determine the change in mass of the $L = 500$-mm-long rod over a time period of $\Delta t = 30$ min. Naphthalene's solid–vapor saturation pressure at 27°C and 1 atm is $p_{A,sat} = 1.33 \times 10^{-4}$ bar.

6.55 Consider application of the naphthalene sublimation technique (Problem 6.54) to a gas turbine blade that is coated with naphthalene and has a surface area of $A_s = 0.05$ m^2.

To determine the average convection heat transfer coefficient for a representative operating condition, an experiment is performed in which the coated blade is exposed for 30 min to atmospheric air at the desired velocity and a temperature of $T_\infty = 27$°C. During the experiment the surface temperature is $T_s = 27$°C, and at its conclusion the mass of the blade is reduced by $\Delta m = 8$ g. What is the average convection heat transfer coefficient associated with the operating condition?

6.56 A streamlined strut supporting a bearing housing is exposed to a hot airflow from an engine exhaust. It is necessary to run experiments to determine the average convection heat transfer coefficient $\overline{h}$ from the air to the strut in order to be able to cool the strut to the desired surface temperature T_s. It is decided to run mass transfer experiments on an object of the same shape and to obtain the desired heat transfer results by using the heat and mass transfer analogy.

The mass transfer experiments were conducted using a half-size model strut constructed from naphthalene exposed to an airstream at 27°C. Mass transfer measurements yielded these results:

Re_L	$\overline{Sh}_L$
60,000	282
120,000	491
144,000	568
288,000	989

(a) Using the mass transfer experimental results, determine the coefficients C and m for a correlation of the form $\overline{Sh}_L = C\,Re_L^m\,Sc^{1/3}$.

(b) Determine the average convection heat transfer coefficient $\overline{h}$ for the full-sized strut, $L_H = 60$ mm, when exposed to a free stream airflow with $V = 60$ m/s, $T_\infty = 184$°C, and $p_\infty = 1$ atm when $T_s = 70$°C.

(c) The surface area of the strut can be expressed as $A_s = 2.2L_H \cdot l$, where l is the length normal to the page. For the conditions of part (b), what is the change in the rate of heat transfer to the strut if the characteristic length L_H is doubled?

6.57 Consider the conditions of Problem 6.6, for which a heat transfer experiment yielded the prescribed distribution of the local convection coefficient, $h_x(x)$. The experiment was performed for surface and free stream temperatures of 310 and 290 K, respectively. Now consider repeating the experiment under conditions for which the surface is coated with a thin layer of naphthalene and both the surface and air are at 300 K. What is the corresponding value of the average convection mass transfer coefficient, $\overline{h}_{m,L}$?

6.58 To reduce the threat of predators, the sand grouse, a bird of Kenya, will lay its eggs in locations well removed from sources of groundwater. To bring water to its chicks, the grouse will then fly to the nearest source and, by submerging the lower part of its body, will entrain water within its plumage. The grouse will then return to its nest, and the chicks will imbibe water from the plumage. Of course, if the time of flight is too long, evaporative losses could cause a significant reduction in the water content of the plumage, and the chicks could succumb to dehydration.

To gain a better understanding of convective transfer during flight, wind tunnel studies were performed using molded models of the grouse. By heating the portion of the model that corresponds to the water-encapsulating plumage, an average convection heat transfer coefficient was determined. Results for different air speeds and model sizes were then used to develop an empirical correlation of the form

$$\overline{Nu}_L = 0.034\, Re_L^{4/5}\, Pr^{1/3}$$

The effective surface area of the water-encapsulating portion of the plumage is designated as A_s, and the characteristic length is defined as $L = (A_s)^{1/2}$.

Consider conditions for which a grouse has entrained 0.05 kg of water within plumage of $A_s = 0.04\ m^2$ and is returning to its nest at a constant speed of $V = 30$ m/s. The ambient air is stagnant and at a temperature and relative humidity of $T_\infty = 37°C$ and $\phi_\infty = 25\%$, respectively. If, throughout the flight, the surface A_s is covered with a liquid water film at $T_s = 32°C$, what is the maximum allowable distance of the nest from the water source, if the bird must return with at least 50% of its initial water supply?

6.59 In the spring, concrete surfaces such as sidewalks and driveways are sometimes very wet in the morning, even when it has not rained during the night. Typical nighttime conditions are shown in the sketch.

(a) Determine the heat fluxes associated with convection, q''_{conv}, evaporation, q''_{evap}, and radiation exchange with the sky, q''_{rad}.

(b) Do your calculations suggest why the concrete is wet instead of dry? Explain briefly.

(c) Is heat flowing from the liquid layer to the concrete? Or from the concrete to the liquid layer? Determine the heat flux by conduction into or out of the concrete.

6.60 Dry air at 32°C flows over a wetted (water) plate of 0.2 m² area. The average convection coefficient is $\overline{h} = 20\ W/m^2\cdot K$, and the heater power required to maintain the plate at a temperature of 27°C is 432 W. Estimate the power required to maintain the wetted plate at a temperature of 37°C in dry air at 32°C if the convection coefficients remain unchanged.

6.61 Dry air at 32°C flows over a wetted plate of length 200 mm and width 1 m (case A). An embedded electrical heater supplies 432 W and the surface temperature is 27°C.

(a) What is the evaporation rate of water from the plate (kg/h)?

(b) After a long period of operation, all the water is evaporated from the plate and its surface is dry (case B). For the same free stream conditions and the same heater power as case A, estimate the temperature of the plate, T_s. Assume $\overline{Nu}_L = C\, Re_L^{1/2}\, Pr^{1/3}$.

Evaporative Cooling

6.62 A successful California engineer has installed a circular hot tub in his backyard and finds that, for the typical operating conditions shown in the sketch, water must be added at a rate of 0.001 kg/s to maintain a fixed water level in the tub.

If the tub is well insulated on its sides and bottom and if the temperature of the makeup water is equal to that of the tub water, at what rate must electrical heaters supply energy to maintain the tub water at 310 K?

6.63 It is known that on clear nights the air temperature need not drop below 0°C before a thin layer of water on the ground will freeze. Consider such a layer of water on a clear night for which the effective sky temperature is −30°C and the convection heat transfer coefficient due to wind motion is $h = 25$ W/m²·K. The water may be assumed to have an emissivity of 1.0 and to be insulated from the ground as far as conduction is concerned.

(a) Neglecting evaporation, determine the lowest temperature the air can have without the water freezing.

(b) For the conditions given, estimate the mass transfer coefficient for water evaporation h_m (m/s).

(c) Accounting now for the effect of evaporation, what is the lowest temperature the air can have without the water freezing? Assume the air to be dry.

6.64 A mist cooler is used to provide relief for a fatigued athlete. Water at $T_i = 10$°C is injected as a mist into a fan airstream with ambient temperature of $T_\infty = 32$°C. The droplet diameters are $D = 100$ μm. For small droplets the average Nusselt number is correlated by an expression of the form

$$\overline{Nu}_D = \bar{h}D/k = 2$$

(a) At the initial time, calculate the rate of convection heat transfer to the droplet, the rate of evaporative heat loss, and the rate of change of temperature of the droplet for two values of the relative humidity of the fan airstream, $\phi_\infty = 0.20$ and 0.95. Explain what is happening to the droplet in each case.

(b) Calculate the steady-state droplet temperature for each of the two relative humidity values in part (a).

6.65 An expression for the actual water vapor partial pressure in terms of wet-bulb and dry-bulb temperatures, referred to as the Carrier equation, is given as

$$p_v = p_{gw} - \frac{(p - p_{gw})(T_{db} - T_{wb})}{1810 - T_{wb}}$$

where p_v, p_{gw}, and p are the actual partial pressure, the saturation pressure at the wet-bulb temperature, and the total pressure (all in bars), while T_{db} and T_{wb} are the dry- and wet-bulb temperatures in kelvins. Consider air at 1 atm and 37.8°C flowing over a wet-bulb thermometer that indicates 21.1°C.

(a) Using Carrier's equation, calculate the partial pressure of the water vapor in the free stream. What is the relative humidity?

(b) Refer to a psychrometric chart and obtain the relative humidity directly for the conditions indicated. Compare the result with part (a).

(c) Use Equation 6.65 to determine the relative humidity. Compare the result to parts (a) and (b).

6.66 A 2-mm-thick layer of water on an electrically heated plate is maintained at a temperature of $T_w = 340$ K, as dry air at $T_\infty = 300$ K flows over the surface of the water (case A). The arrangement is in large surroundings that are also at 300 K.

(a) If the evaporative flux from the surface of the water to the air is $n_A'' = 0.030$ kg/s·m², what is the corresponding value of the convection mass transfer coefficient? How long will it take for the water to completely evaporate?

(b) What is the corresponding value of the convection heat transfer coefficient and the rate at which electrical power must be supplied per unit area of the plate to maintain the prescribed temperature of the water? The emissivity of water is $\varepsilon_w = 0.95$.

(c) If the electrical power determined in part (b) is maintained after complete evaporation of the water (case B), what is the resulting temperature of the plate, whose emissivity is $\varepsilon_p = 0.60$?

6.67 An experiment is conducted to determine the average mass transfer convection coefficient of a small droplet using a heater controlled to operate at a constant temperature. The power history required to completely evaporate the droplet at a temperature of 37°C is shown in the sketch. It was observed that, as the droplet dried, its wetted diameter on the heater surface remained nearly constant at a value of 4 mm.

(a) Calculate the average mass transfer convection coefficient based on the wetted area during the evaporation process when the droplet, heater, and the *dry* ambient air are at 37°C.

(b) How much energy will be required to evaporate the droplet if the *dry* ambient air temperature is 27°C, while the droplet–heater temperature remains at 37°C?

6.68 It is desired to develop a simple model for predicting the temperature–time history of a plate during the drying cycle in a dishwasher. Following the wash cycle the plate is at $T_p(t) = T_p(0) = 65°C$ and the air in the dishwasher is completely saturated ($\phi_\infty = 1.0$) at $T_\infty = 55°C$. The values of the plate surface area A_s, mass m, and specific heat c are such that $mc/A_s = 1600 \text{ J/m}^2 \cdot \text{K}$.

(a) Assuming the plate is completely covered by a thin film of water and neglecting the thermal resistances of the film and plate, derive a differential equation for predicting the plate temperature as a function of time.

(b) For the initial conditions ($t = 0$) estimate the change in plate temperature with time, dT/dt (°C/s), assuming that the average heat transfer coefficient on the plate is 3.5 W/m^2 · K.

External Flow

7

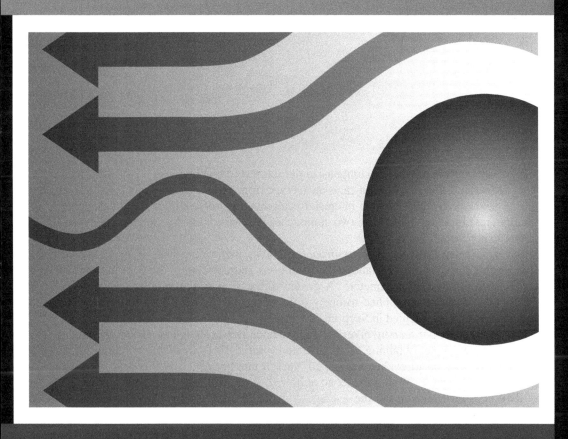

In this chapter we focus on the problem of computing heat and mass transfer rates to or from a surface in *external flow*. In such a flow boundary layers develop freely, without constraints imposed by adjacent surfaces. Accordingly, there will always exist a region of the flow outside the boundary layer in which velocity, temperature, and/or concentration gradients are negligible. Examples include fluid motion over a flat plate (inclined or parallel to the free stream velocity) and flow over curved surfaces such as a sphere, cylinder, airfoil, or turbine blade.

For the moment we confine our attention to problems of *low-speed, forced convection* with *no phase change* occurring within the fluid. In addition, we will not consider potential micro- or nanoscale effects within the fluid, as described in Section 2.2, in this chapter. In *forced* convection, the relative motion between the fluid and the surface is maintained by external means, such as a fan or a pump, and not by buoyancy forces due to temperature gradients in the fluid (*natural* convection). *Internal flows, natural convection,* and *convection with phase change* are treated in Chapters 8, 9, and 10, respectively.

Our primary objective is to determine convection coefficients for different flow geometries. In particular, we wish to obtain specific forms of the functions that represent these coefficients. By nondimensionalizing the boundary layer equations in Chapter 6, we found that the local and average convection coefficients may be correlated by equations of the form

Heat Transfer:

$$Nu_x = f(x^*, Re_x, Pr) \tag{6.49}$$

$$\overline{Nu}_x = f(Re_x, Pr) \tag{6.50}$$

Mass Transfer:

$$Sh_x = f(x^*, Re_x, Sc) \tag{6.53}$$

$$\overline{Sh}_x = f(Re_x, Sc) \tag{6.54}$$

The subscript x has been added to emphasize our interest in conditions at a particular location on the surface. The overbar indicates an average from $x^* = 0$, where the boundary layer begins to develop, to the location of interest. Recall that *the problem of convection* is one of obtaining these functions. There are two approaches that we could take, one theoretical and the other experimental.

The *experimental* or *empirical approach* involves performing heat or mass transfer measurements under controlled laboratory conditions and correlating the data in terms of appropriate dimensionless parameters. A general discussion of the approach is provided in Section 7.1. It has been applied to many different geometries and flow conditions, and important results are presented in Sections 7.2 through 7.8.

The *theoretical approach* involves solving the boundary layer equations for a particular geometry. For example, obtaining the temperature profile T^* from such a solution, Equation 6.48 may be used to evaluate the local Nusselt number Nu_x, and therefore the local convection coefficient h_x. With knowledge of how h_x varies over the surface, Equation 6.13 may then be used to determine the average convection coefficient $\overline{h}_x$, and therefore the Nusselt number $\overline{Nu}_x$. In Section 7.2.1 this approach is illustrated by using the *similarity* *method* to obtain an *exact solution* of the boundary layer equations for a flat plate in parallel, laminar flow [1–3]. An *approximate solution* to the same problem is obtained in Appendix G by using the *integral method* [4].

7.1 The Empirical Method

The manner in which a convection heat transfer correlation may be obtained experimentally is illustrated in Figure 7.1. If a prescribed geometry, such as the flat plate in parallel flow, is heated electrically to maintain $T_s > T_\infty$, convection heat transfer occurs from the surface to the fluid. It would be a simple matter to measure T_s and T_∞, as well as the electrical power, $E \cdot I$, which is equal to the total heat transfer rate q. The convection coefficient $\bar{h}_L$, which is an average associated with the entire plate, could then be computed from Newton's law of cooling, Equation 6.12. Moreover, from knowledge of the characteristic length L and the fluid properties, the Nusselt, Reynolds, and Prandtl numbers could be computed from their definitions, Equations 6.50, 6.41, and 6.42, respectively.

The foregoing procedure could be repeated for a variety of test conditions. We could vary the velocity u_∞ and the plate length L, as well as the nature of the fluid, using, for example, air, water, and engine oil, which have substantially different Prandtl numbers. We would then be left with many different values of the Nusselt number corresponding to a wide range of Reynolds and Prandtl numbers, and the results could be plotted on a *log–log* scale, as shown in Figure 7.2a. Each symbol represents a unique set of test conditions. As is often the case, the results associated with a given fluid, and hence a fixed Prandtl number, fall close to a straight line, indicating a power law dependence of the Nusselt number on the Reynolds number. Considering all the fluids, the data may then be represented by an algebraic expression of the form

$$\overline{Nu}_L = C\,Re_L^m\,Pr^n \tag{7.1}$$

Since the values of C, m, and n are often independent of the nature of the fluid, the family of straight lines corresponding to different Prandtl numbers can be collapsed to a single line by plotting the results in terms of the ratio, $\overline{Nu}_L/Pr^n$, as shown in Figure 7.2b.

Because Equation 7.1 is inferred from experimental measurements, it is termed an *empirical correlation.* The specific values of the coefficient C and the exponents m and n vary with the nature of the surface geometry and the type of flow.

We will use expressions of the form given by Equation 7.1 for many special cases, and it is important to note that the assumption of *constant fluid properties* is often implicit in the results. However, we know that the fluid properties vary with temperature across the boundary layer and that this variation can certainly influence the heat transfer rate. This

FIGURE 7.1 Experiment for measuring the average convection heat transfer coefficient $\bar{h}_L$.

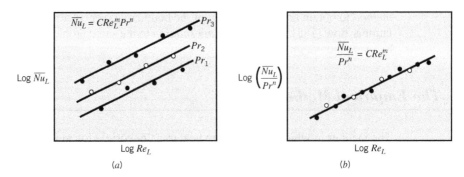

FIGURE 7.2   Dimensionless representation of convection heat transfer measurements.

influence may be handled in one of two ways. In one method, Equation 7.1 is used with all properties evaluated at a mean boundary layer temperature T_f, termed the *film temperature.*

$$T_f \equiv \frac{T_s + T_\infty}{2} \tag{7.2}$$

The alternate method is to evaluate all properties at T_∞ and to multiply the right-hand side of Equation 7.1 by an additional parameter to account for the property variations. The parameter is commonly of the form $(Pr_\infty/Pr_s)^r$ or $(\mu_\infty/\mu_s)^r$, where the subscripts ∞ and s designate evaluation of the properties at the free stream and surface temperatures, respectively. Both methods are used in the results that follow.

Finally, we note that experiments may also be performed to obtain convection mass transfer correlations. However, under conditions for which the heat and mass transfer analogy (Section 6.7.1) may be applied, the mass transfer correlation assumes the same form as the corresponding heat transfer correlation. Accordingly, we anticipate correlations of the form

$$\overline{Sh}_L = C\,Re_L^m\,Sc^n \tag{7.3}$$

where, for a given geometry and flow condition, the values of C, m, and n are the same as those appearing in Equation 7.1.

7.2 *The Flat Plate in Parallel Flow*

Despite its simplicity, parallel flow over a flat plate (Figure 7.3) occurs in numerous engineering applications. As discussed in Section 6.3, laminar boundary layer development begins at the leading edge ($x = 0$) and transition to turbulence may occur at a downstream location (x_c) for which a critical Reynolds number $Re_{x,c}$ is achieved. We begin by analytically determining the velocity, temperature, and concentration distributions in the laminar boundary layers that are shown qualitatively in Figures 6.1, 6.2, and 6.3, respectively. From knowledge of these distributions, we will determine expressions for the local and average friction coefficients, Nusselt numbers, and Sherwood numbers. Subsequently, we will report experimentally determined correlations for the friction coefficient, Nusselt numbers, and Sherwood numbers for turbulent boundary layers.

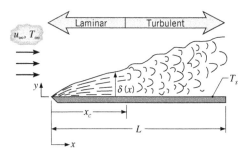

FIGURE 7.3 The flat plate in parallel flow.

7.2.1 Laminar Flow over an Isothermal Plate: A Similarity Solution

The major convection parameters may be obtained by solving the appropriate form of the boundary layer equations. Assuming *steady, incompressible, laminar* flow with *constant fluid properties* and *negligible viscous dissipation* and recognizing that $dp/dx = 0$, the boundary layer equations (6.27, 6.28, 6.29, and 6.30) reduce to

Continuity:

$$\frac{\partial u}{\partial x} + \frac{\partial v}{\partial y} = 0 \tag{7.4}$$

Momentum:

$$u\frac{\partial u}{\partial x} + v\frac{\partial u}{\partial y} = \nu\frac{\partial^2 u}{\partial y^2} \tag{7.5}$$

Energy:

$$u\frac{\partial T}{\partial x} + v\frac{\partial T}{\partial y} = \alpha\frac{\partial^2 T}{\partial y^2} \tag{7.6}$$

Species:

$$u\frac{\partial \rho_A}{\partial x} + v\frac{\partial \rho_A}{\partial y} = D_{AB}\frac{\partial^2 \rho_A}{\partial y^2} \tag{7.7}$$

Solution of these equations is simplified by the fact that for constant properties, conditions in the velocity (hydrodynamic) boundary layer are independent of temperature and species concentration. Hence we may begin by solving the hydrodynamic problem, Equations 7.4 and 7.5, to the exclusion of Equations 7.6 and 7.7. Once the hydrodynamic problem has been solved, solutions to Equations 7.6 and 7.7, which depend on u and v, may be obtained.

Hydrodynamic Solution The hydrodynamic solution follows the method of Blasius [1, 2]. The first step is to define a stream function $\psi(x, y)$, such that

$$u \equiv \frac{\partial \psi}{\partial y} \quad \text{and} \quad v \equiv -\frac{\partial \psi}{\partial x} \tag{7.8}$$

Equation 7.4 is then automatically satisfied and hence is no longer needed. New dependent and independent variables, f and η, respectively, are then defined such that

$$f(\eta) \equiv \frac{\psi}{u_\infty \sqrt{\nu x/u_\infty}} \tag{7.9}$$

$$\eta \equiv y\sqrt{u_\infty/\nu x} \tag{7.10}$$

As we will find, use of these variables simplifies matters by reducing the partial differential equation, Equation 7.5, to an ordinary differential equation.

The Blasius solution is termed a *similarity solution,* and η is a *similarity variable.* This terminology is used because, despite growth of the boundary layer with distance x from the leading edge, the velocity profile u/u_∞ remains *geometrically similar.* This similarity is of the functional form

$$\frac{u}{u_\infty} = \phi\left(\frac{y}{\delta}\right)$$

where δ is the boundary layer thickness. We will find from the Blasius solution that δ varies as $(\nu x/u_\infty)^{1/2}$; thus, it follows that

$$\frac{u}{u_\infty} = \phi(\eta) \tag{7.11}$$

Hence the velocity profile is uniquely determined by the similarity variable η, which depends on both x and y.

From Equations 7.8 through 7.10 we obtain

$$u = \frac{\partial \psi}{\partial y} = \frac{\partial \psi}{\partial \eta}\frac{\partial \eta}{\partial y} = u_\infty \sqrt{\frac{\nu x}{u_\infty}}\frac{df}{d\eta}\sqrt{\frac{u_\infty}{\nu x}} = u_\infty \frac{df}{d\eta} \tag{7.12}$$

and

$$v = -\frac{\partial \psi}{\partial x} = -\left(u_\infty\sqrt{\frac{\nu x}{u_\infty}}\frac{\partial f}{\partial x} + \frac{u_\infty}{2}\sqrt{\frac{\nu}{u_\infty x}}f\right)$$

$$v = \frac{1}{2}\sqrt{\frac{\nu u_\infty}{x}}\left(\eta\frac{df}{d\eta} - f\right) \tag{7.13}$$

By differentiating the velocity components, it may also be shown that

$$\frac{\partial u}{\partial x} = -\frac{u_\infty}{2x}\eta\frac{d^2 f}{d\eta^2} \tag{7.14}$$

$$\frac{\partial u}{\partial y} = u_\infty\sqrt{\frac{u_\infty}{\nu x}}\frac{d^2 f}{d\eta^2} \tag{7.15}$$

$$\frac{\partial^2 u}{\partial y^2} = \frac{u_\infty^2}{\nu x}\frac{d^3 f}{d\eta^3} \tag{7.16}$$

Substituting these expressions into Equation 7.5, we then obtain

$$2\frac{d^3f}{d\eta^3} + f\frac{d^2f}{d\eta^2} = 0 \tag{7.17}$$

Hence the hydrodynamic boundary layer problem is reduced to one of solving a nonlinear, third-order ordinary differential equation. The appropriate boundary conditions are

$$u(x, 0) = v(x, 0) = 0 \quad \text{and} \quad u(x, \infty) = u_\infty$$

or, in terms of the similarity variables,

$$\left.\frac{df}{d\eta}\right|_{\eta=0} = f(0) = 0 \quad \text{and} \quad \left.\frac{df}{d\eta}\right|_{\eta\to\infty} = 1 \tag{7.18}$$

The solution to Equation 7.17, subject to the conditions of Equations 7.18, may be obtained by a series expansion [2] or by numerical integration [3]. Selected results are presented in Table 7.1, from which useful information may be extracted. The x-component velocity distribution from the third column of the table is plotted in Figure 7.4a. We also note that, to a good approximation, $(u/u_\infty) = 0.99$ for $\eta = 5.0$. Defining the boundary layer thickness δ as that value of y for which $(u/u_\infty) = 0.99$, it follows from Equation 7.10 that

$$\delta = \frac{5.0}{\sqrt{u_\infty/\nu x}} = \frac{5x}{\sqrt{Re_x}} \tag{7.19}$$

TABLE 7.1 Flat plate laminar boundary layer functions [3]

$\eta = y\sqrt{\dfrac{u_\infty}{\nu x}}$	f	$\dfrac{df}{d\eta} = \dfrac{u}{u_\infty}$	$\dfrac{d^2f}{d\eta^2}$
0	0	0	0.332
0.4	0.027	0.133	0.331
0.8	0.106	0.265	0.327
1.2	0.238	0.394	0.317
1.6	0.420	0.517	0.297
2.0	0.650	0.630	0.267
2.4	0.922	0.729	0.228
2.8	1.231	0.812	0.184
3.2	1.569	0.876	0.139
3.6	1.930	0.923	0.098
4.0	2.306	0.956	0.064
4.4	2.692	0.976	0.039
4.8	3.085	0.988	0.022
5.2	3.482	0.994	0.011
5.6	3.880	0.997	0.005
6.0	4.280	0.999	0.002
6.4	4.679	1.000	0.001
6.8	5.079	1.000	0.000

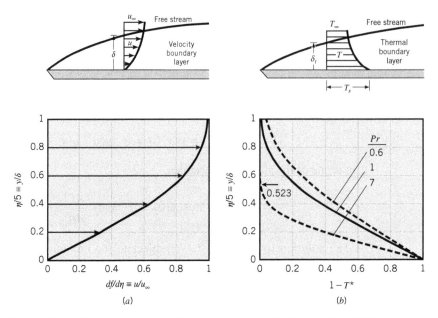

FIGURE 7.4 Similarity solution for laminar flow over an isothermal plate. (*a*) The *x*-component of the velocity. (*b*) Temperature distributions for $Pr = 0.6$, 1, and 7.

From Equation 7.19 it is clear that δ increases with increasing x and v but decreases with increasing u_∞ (the larger the free stream velocity, the *thinner* the boundary layer). In addition, from Equation 7.15 the wall shear stress may be expressed as

$$\tau_s = \mu \frac{\partial u}{\partial y}\bigg|_{y=0} = \mu u_\infty \sqrt{u_\infty/vx}\; \frac{d^2 f}{d\eta^2}\bigg|_{\eta=0}$$

Hence from Table 7.1

$$\tau_s = 0.332 u_\infty \sqrt{\rho \mu u_\infty/x}$$

From Equation 6.45, the *local* friction coefficient is then

$$C_{f,x} \equiv \frac{\tau_{s,x}}{\rho u_\infty^2 /2} = 0.664\, Re_x^{-1/2} \tag{7.20}$$

Heat Transfer Solution   From knowledge of conditions in the velocity boundary layer, the energy equation may now be solved. We begin by introducing the dimensionless temperature $T^* \equiv [(T - T_s)/(T_\infty - T_s)]$ and assume a similarity solution of the form $T^* = T^*(\eta)$. Making the necessary substitutions, Equation 7.6 reduces to

$$\frac{d^2 T^*}{d\eta^2} + \frac{Pr}{2}\, f\, \frac{dT^*}{d\eta} = 0 \tag{7.21}$$

Note the dependence of the thermal solution on hydrodynamic conditions through appearance of the variable f in Equation 7.21. The appropriate boundary conditions are

$$T^*(0) = 0 \quad \text{and} \quad T^*(\infty) = 1 \tag{7.22}$$

Subject to the conditions of Equation 7.22, Equation 7.21 may be solved by numerical integration for different values of the Prandtl number; representative temperature distributions for $Pr = 0.6$, 1, and 7 are shown in Figure 7.4b. The temperature distribution is identical to the velocity distribution for $Pr = 1$. Thermal effects penetrate farther into the velocity boundary layer with decreasing Prandtl number and transcend the velocity boundary layer for $Pr < 1$. A practical consequence of this solution is that, for $Pr \gtrsim 0.6$, results for the surface temperature gradient $dT^*/d\eta|_{\eta=0}$ may be correlated by the following relation:

$$\left.\frac{dT^*}{d\eta}\right|_{\eta=0} = 0.332 \, Pr^{1/3}$$

Expressing the local convection coefficient as

$$h_x = \frac{q_s''}{T_s - T_\infty} = -\frac{T_\infty - T_s}{T_s - T_\infty} k \left.\frac{\partial T^*}{\partial y}\right|_{y=0}$$

$$h_x = k \left(\frac{u_\infty}{\nu x}\right)^{1/2} \left.\frac{dT^*}{d\eta}\right|_{\eta=0}$$

it follows from Equation 6.48 that the *local* Nusselt number is

$$Nu_x \equiv \frac{h_x x}{k} = 0.332 \, Re_x^{1/2} Pr^{1/3} \qquad Pr \gtrsim 0.6 \tag{7.23}$$

From the solution to Equation 7.21, it also follows that, for $Pr \gtrsim 0.6$, the ratio of the velocity to thermal boundary layer thickness is

$$\frac{\delta}{\delta_t} \approx Pr^{1/3} \tag{7.24}$$

where δ is given by Equation 7.19. For example, for $Pr = 7$, $\delta/\delta_t = 1.91$ ($\delta_t/\delta = 0.523$), as shown in Figure 7.4b.

Mass Transfer Solution   The species boundary layer equation, Equation 7.7, is of the same form as the energy boundary layer equation, Equation 7.6, with D_{AB} replacing α. Introducing a normalized species density $\rho_A^* = [(\rho_A - \rho_{A,s})/(\rho_{A,\infty} - \rho_{A,s})]$ and noting that, for a fixed surface species concentration

$$\rho_A^*(0) = 0 \quad \text{and} \quad \rho_A^*(\infty) = 1 \tag{7.25}$$

we also see that the species boundary conditions are of the same form as the temperature boundary conditions given in Equation 7.22. Therefore, as discussed in Section 6.7.1, the heat and mass transfer analogy may be applied since the differential equation and boundary conditions for the species concentration are of the same form as for temperature. Hence, with reference to Equation 7.23, application of the analogy yields

$$Sh_x \equiv \frac{h_{m,x}x}{D_{AB}} = 0.332\,Re_x^{1/2}\,Sc^{1/3} \qquad Sc \gtrsim 0.6 \qquad (7.26)$$

By analogy to Equation 7.24, it also follows that the ratio of boundary layer thicknesses is

$$\frac{\delta}{\delta_c} \approx Sc^{1/3} \qquad (7.27)$$

The foregoing results may be used to compute important *laminar* boundary layer parameters for $0 < x < x_c$, where x_c is the distance from the leading edge at which transition begins. Equations 7.20, 7.23, and 7.26 imply that $\tau_{s,x}$, h_x, and $h_{m,x}$ are, in principle, infinite at the leading edge and decrease as $x^{-1/2}$ in the flow direction. Equations 7.24 and 7.27 also imply that, for values of Pr and Sc close to unity, which is the case for most gases, the three boundary layers experience nearly identical growth.

Average Boundary Layer Parameters for Laminar Conditions   From the foregoing local results, average boundary layer parameters may be determined. With the average friction coefficient defined as

$$\bar{C}_{f,x} \equiv \frac{\bar{\tau}_{s,x}}{\rho u_\infty^2/2} \qquad (7.28)$$

where

$$\bar{\tau}_{s,x} \equiv \frac{1}{x}\int_0^x \tau_{s,x}\,dx$$

the form of $\tau_{s,x}$ may be substituted from Equation 7.20 and the integration performed to obtain

$$\bar{C}_{f,x} = 1.328\,Re_x^{-1/2} \qquad (7.29)$$

Moreover, from Equations 6.14 and 7.23, the *average* heat transfer coefficient for laminar flow is

$$\bar{h}_x = \frac{1}{x}\int_0^x h_x\,dx = 0.332\left(\frac{k}{x}\right)Pr^{1/3}\left(\frac{u_\infty}{\nu}\right)^{1/2}\int_0^x \frac{dx}{x^{1/2}}$$

Integrating and substituting from Equation 7.23, it follows that $\bar{h}_x = 2h_x$. Hence

$$\overline{Nu}_x \equiv \frac{\bar{h}_x x}{k} = 0.664\,Re_x^{1/2}\,Pr^{1/3} \qquad Pr \gtrsim 0.6 \qquad (7.30)$$

Employing the heat and mass transfer analogy, it follows that

$$\overline{Sh}_x \equiv \frac{\bar{h}_{m,x}x}{D_{AB}} = 0.664\,Re_x^{1/2}\,Sc^{1/3} \qquad Sc \gtrsim 0.6 \qquad (7.31)$$

If the flow is laminar over the entire surface, the subscript x may be replaced by L, and Equations 7.29 through 7.31 may be used to predict average conditions for the entire surface.

From the foregoing expressions we see that, for laminar flow over a flat plate, the *average* friction and convection coefficients from the leading edge to a point x on the surface are *twice* the *local* coefficients at that point. We also note that, in using these expressions, the effect of variable properties can be treated by evaluating all properties at the *film temperature*, Equation 7.2.

Liquid Metals For fluids of small Prandtl number, namely, *liquid metals*, Equation 7.23 does not apply. However, for this case the thermal boundary layer development is much more rapid than that of the velocity boundary layer ($\delta_t \gg \delta$), and it is reasonable to assume uniform velocity ($u = u_\infty$) throughout the thermal boundary layer. From a solution to the thermal boundary layer equation based on this assumption [5], it may then be shown that

$$Nu_x = 0.564 \, Pe_x^{1/2} \quad Pr \lesssim 0.05, \quad Pe_x \gtrsim 100 \tag{7.32}$$

where $Pe_x \equiv Re_x Pr$ is the *Peclet number* (Table 6.2). Despite the corrosive and reactive nature of liquid metals, their unique properties (low melting point and vapor pressure, as well as high thermal capacity and conductivity) render them attractive as coolants in applications requiring high heat transfer rates.

A single correlating equation, which applies for all Prandtl numbers, has been recommended by Churchill and Ozoe [6]. For laminar flow over an isothermal plate, the local convection coefficient may be obtained from

$$Nu_x = \frac{0.3387 \, Re_x^{1/2} \, Pr^{1/3}}{[1 + (0.0468/Pr)^{2/3}]^{1/4}} \quad Pe_x \gtrsim 100 \tag{7.33}$$

with $\overline{Nu_x} = 2Nu_x$.

7.2.2 Turbulent Flow over an Isothermal Plate

It is not possible to obtain exact analytical solutions for turbulent boundary layers, which are inherently unsteady. From experiment [2] it is known that, for turbulent flows with Reynolds numbers up to approximately 10^8, the *local* friction coefficient is correlated to within 15% accuracy by an expression of the form

$$C_{f,x} = 0.0592 \, Re_x^{-1/5} \quad Re_{x,c} \lesssim Re_x \lesssim 10^8 \tag{7.34}$$

Moreover, it is known that, to a reasonable approximation, the velocity boundary layer thickness may be expressed as

$$\delta = 0.37x \, Re_x^{-1/5} \tag{7.35}$$

Comparing these results with those for the laminar boundary layer, Equations 7.19 and 7.20, we see that turbulent boundary layer growth is much more rapid (δ varies as $x^{4/5}$ in contrast to $x^{1/2}$ for laminar flow) and that the decay in the friction coefficient is more gradual ($x^{-1/5}$ versus $x^{-1/2}$). For turbulent flow, boundary layer development is influenced strongly by random fluctuations in the fluid and not by molecular diffusion. Hence relative boundary layer growth does not depend on the value of Pr or Sc, and Equation 7.35 may be used to obtain the thermal and concentration, as well as the velocity, boundary layer thicknesses. That is, for turbulent flow, $\delta \approx \delta_t \approx \delta_c$.

Using Equation 7.34 with the modified Reynolds, or Chilton–Colburn, analogy, Equations 6.70 and 6.71, the *local* Nusselt number for turbulent flow is

$$Nu_x = St\,Re_x\,Pr = 0.0296\,Re_x^{4/5}\,Pr^{1/3} \quad 0.6 \lesssim Pr \lesssim 60 \qquad (7.36)$$

and the *local* Sherwood number is

$$Sh_x = St_m\,Re_x\,Sc = 0.0296\,Re_x^{4/5}\,Sc^{1/3} \quad 0.6 \lesssim Sc \lesssim 3000 \qquad (7.37)$$

Enhanced mixing causes the turbulent boundary layer to grow more rapidly than the laminar boundary layer and to have larger friction and convection coefficients.

Expressions for the average coefficients may now be determined. However, since the turbulent boundary layer is generally preceded by a laminar boundary layer, we first consider *mixed* boundary layer conditions.

7.2.3 Mixed Boundary Layer Conditions

For laminar flow over the entire plate, Equations 7.29 through 7.31 may be used to compute the average coefficients. Moreover, if transition occurs toward the rear of the plate, for example, in the range $0.95 \lesssim (x_c/L) \lesssim 1$, these equations may be used to compute the average coefficients to a reasonable approximation. However, when transition occurs sufficiently upstream of the trailing edge, $(x_c/L) \lesssim 0.95$, the surface average coefficients will be influenced by conditions in both the laminar and turbulent boundary layers.

In the mixed boundary layer situation (Figure 7.3), Equation 6.14 may be used to obtain the average convection heat transfer coefficient for the entire plate. Integrating over the laminar region ($0 \le x \le x_c$) and then over the turbulent region ($x_c < x \le L$), this equation may be expressed as

$$\overline{h}_L = \frac{1}{L}\left(\int_0^{x_c} h_{\text{lam}}\,dx + \int_{x_c}^{L} h_{\text{turb}}\,dx \right)$$

where it is assumed that transition occurs abruptly at $x = x_c$. Substituting from Equations 7.23 and 7.36 for h_{lam} and h_{turb}, respectively, we obtain[1]

$$\overline{h}_L = \left(\frac{k}{L}\right)\left[0.332\left(\frac{u_\infty}{\nu}\right)^{1/2}\int_0^{x_c}\frac{dx}{x^{1/2}} + 0.0296\left(\frac{u_\infty}{\nu}\right)^{4/5}\int_{x_c}^{L}\frac{dx}{x^{1/5}} \right]Pr^{1/3}$$

---

[1]In addition to assuming an abrupt transition at $x = x_c$, it is assumed that the virtual origin of the turbulent boundary layer is at $x = 0$. Relaxation of these assumptions leads to slightly modified variations of Equation 7.38 [7].

Integrating, we then obtain

$$\overline{Nu}_L = (0.037 \, Re_L^{4/5} - A) \, Pr^{1/3} \tag{7.38}$$
$$\begin{bmatrix} 0.6 \lesssim Pr \lesssim 60 \\ Re_{x,c} \lesssim Re_L \lesssim 10^8 \end{bmatrix}$$

where the bracketed relations indicate the range of applicability and the constant A is determined by the value of the critical Reynolds number, $Re_{x,c}$. That is,

$$A = 0.037 \, Re_{x,c}^{4/5} - 0.664 \, Re_{x,c}^{1/2} \tag{7.39}$$

Similarly, the average friction coefficient may be found using the expression

$$\overline{C}_{f,L} = \frac{1}{L} \left(\int_0^{x_c} C_{f,x,\text{lam}} \, dx + \int_{x_c}^L C_{f,x,\text{turb}} \, dx \right)$$

Substituting expressions for $C_{f,x,\text{lam}}$ and $C_{f,x,\text{turb}}$ from Equations 7.20 and 7.34, respectively, and carrying out the integration provides an expression of the form

$$\overline{C}_{f,L} = 0.074 \, Re_L^{-1/5} - \frac{2A}{Re_L} \tag{7.40}$$
$$[Re_{x,c} \lesssim Re_L \lesssim 10^8]$$

Applying the heat and mass transfer analogy to Equation 7.38 yields

$$\overline{Sh}_L = (0.037 \, Re_L^{4/5} - A) \, Sc^{1/3} \tag{7.41}$$
$$\begin{bmatrix} 0.6 \lesssim Sc \lesssim 60 \\ Re_{x,c} \lesssim Re_L \lesssim 10^8 \end{bmatrix}$$

For a completely turbulent boundary layer ($Re_{x,c} = 0$), $A = 0$. Such a condition may be realized by *tripping* the boundary layer at the leading edge, using a fine wire or some other turbulence promoter. For a transition Reynolds number of $Re_{x,c} = 5 \times 10^5$, $A = 871$.

All of the foregoing correlations require evaluation of the fluid properties at the film temperature, Equation 7.2.

7.2.4 Unheated Starting Length

All the foregoing Nusselt number expressions are restricted to situations for which the surface temperature T_s is uniform. A common exception involves existence of an *unheated starting length* ($T_s = T_\infty$) upstream of a heated section ($T_s \neq T_\infty$). As shown in Figure 7.5, velocity boundary layer growth begins at $x = 0$, while thermal boundary layer development begins at $x = \xi$. Hence there is no heat transfer for $0 \leq x \leq \xi$. Through use of an integral boundary layer solution [5], it is known that, for laminar flow,

$$Nu_x = \frac{Nu_x|_{\xi=0}}{[1 - (\xi/x)^{3/4}]^{1/3}} \tag{7.42}$$

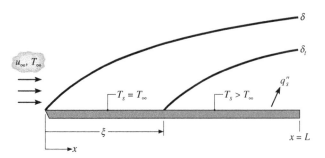

FIGURE 7.5   Flat plate in parallel flow with unheated starting length.

where $Nu_x|_{\xi=0}$, is given by Equation 7.23. In both Nu_x and $Nu_x|_{\xi=0}$, the characteristic length x is measured from the leading edge of the unheated starting length. It has also been found that, for turbulent flow,

$$Nu_x = \frac{Nu_x|_{\xi=0}}{[1-(\xi/x)^{9/10}]^{1/9}} \qquad (7.43)$$

where $Nu_x|_{\xi=0}$ is given by Equation 7.36. Analogous mass transfer results are obtained by replacing (Nu_x, Pr) with (Sh_x, Sc).

By using Equation 6.14 with local convection coefficients given by the foregoing relations, expressions may be obtained for the *average Nusselt number* of an isothermal plate with an unheated starting length [8]. For a plate of total length L, with laminar *or* turbulent flow over the entire surface, the expressions are of the form

$$\overline{Nu}_L = \overline{Nu}_L|_{\xi=0} \frac{L}{L-\xi}[1-(\xi/L)^{(p+1)/(p+2)}]^{p/(p+1)} \qquad (7.44)$$

where $p = 2$ for laminar flow and $p = 8$ for turbulent flow. The quantity $\overline{Nu}_L|_{\xi=0}$ is the average Nusselt number for a plate of length L when heating starts at the leading edge of the plate. For laminar flow, it can be obtained from Equation 7.30 (with x replaced by L); for turbulent flow it is given by Equation 7.38 with $A = 0$ (assuming turbulent flow over the entire surface). Note that $\overline{Nu}_L$ is equal to $\overline{h}L/k$, where $\overline{h}$ is averaged over the heated portion of the plate only, which is of length $(L - \xi)$. The corresponding value of $\overline{h}_L$ must therefore be multiplied by the area of the heated section to determine the total heat rate from the plate.

7.2.5   Flat Plates with Constant Heat Flux Conditions

It is also possible to have a uniform surface heat flux, rather than a uniform temperature, imposed at the plate. For laminar flow, it may be shown that [5]

$$Nu_x = 0.453\, Re_x^{1/2}\, Pr^{1/3} \qquad Pr \gtrsim 0.6 \qquad (7.45)$$

while for turbulent flow

$$Nu_x = 0.0308\, Re_x^{4/5}\, Pr^{1/3} \qquad 0.6 \lesssim Pr \lesssim 60 \qquad (7.46)$$

Hence the Nusselt number is 36% and 4% larger than the constant surface temperature result for laminar and turbulent flow, respectively. Correction for the effect of an unheated starting length may be made by using Equations 7.45 and 7.46 with Equations 7.42 and 7.43, respectively. If the heat flux is known, the convection coefficient may be used to determine the local surface temperature

$$T_s(x) = T_\infty + \frac{q_s''}{h_x} \tag{7.47}$$

Since the total heat rate is readily determined from the product of the uniform flux and the surface area, $q = q_s'' A_s$, it is not necessary to introduce an average convection coefficient for the purpose of determining q. However, one may still wish to determine an *average surface temperature* from an expression of the form

$$(\overline{T_s - T_\infty}) = \frac{1}{L} \int_0^L (T_s - T_\infty) dx = \frac{q_s''}{L} \int_0^L \frac{x}{k\, Nu_x} dx$$

where Nu_x is obtained from the appropriate convection correlation. Substituting from Equation 7.45, it follows that

$$(\overline{T_s - T_\infty}) = \frac{q_s'' L}{k\, \overline{Nu}_L} \tag{7.48}$$

where

$$\overline{Nu}_L = 0.680\, Re_L^{1/2}\, Pr^{1/3} \tag{7.49}$$

This result is only 2% larger than that obtained by evaluating Equation 7.30 at $x = L$. Differences are even smaller for turbulent flow, suggesting that any of the $\overline{Nu}_L$ results obtained for a uniform surface temperature may be used with Equation 7.48 to evaluate $(\overline{T_s - T_\infty})$. Expressions for the average temperature of a plate that is subjected to a uniform heat flux downstream of an unheated starting section have been obtained by Ameel [8].

7.2.6 Limitations on Use of Convection Coefficients

Although the equations of this section are suitable for most engineering calculations, in practice they rarely provide exact values for the convection coefficients. Conditions vary according to free stream turbulence and surface roughness, and errors as large as 25% may be incurred by using the expressions. A detailed description of free stream turbulence effects is provided by Blair [9].

7.3 *Methodology for a Convection Calculation*

Although we have only discussed correlations for parallel flow over a flat plate, selection and application of a convection correlation for *any flow situation* are facilitated by following a few simple rules.

1. *Become immediately cognizant of the flow geometry.* For example, does the problem involve flow over a flat plate, a sphere, or a cylinder? The specific form of the convection correlation depends, of course, on the geometry.

2. *Specify the appropriate reference temperature and evaluate the pertinent fluid properties at that temperature.* For moderate boundary layer temperature differences, the film temperature, Equation 7.2, may be used for this purpose. However, we will also consider correlations that require property evaluation at the free stream temperature and include a property ratio to account for the nonconstant property effect.

3. *In mass transfer problems the pertinent fluid properties are those of species B.* In our treatment of convection mass transfer, we are only concerned with *dilute, binary mixtures.* That is, the problems involve transport of some species A for which $x_A \ll 1$. To a good approximation the properties of the mixture may then be assumed to be the properties of species B. The Schmidt number, for example, would be $Sc = \nu_B/D_{AB}$ and the Reynolds number would be $Re_L = (VL/\nu_B)$.

4. *Calculate the Reynolds number.* Boundary layer conditions are strongly influenced by this parameter. If the geometry is a flat plate in parallel flow, determine whether the flow is laminar or turbulent.

5. *Decide whether a local or surface average coefficient is required.* Recall that for constant surface temperature or vapor density, the local coefficient is used to determine the flux at a particular point on the surface, whereas the average coefficient determines the transfer rate for the entire surface.

6. *Select the appropriate correlation.*

EXAMPLE 7.1   IHT

Air at a pressure of 6 kN/m^2 and a temperature of 300°C flows with a velocity of 10 m/s over a flat plate 0.5 m long. Estimate the cooling rate per unit width of the plate needed to maintain it at a surface temperature of 27°C.

SOLUTION

Known: Airflow over an isothermal flat plate.

Find: Cooling rate per unit width of the plate, q' (W/m).

Schematic:

Assumptions:

1. Steady-state, incompressible flow conditions.

2. Negligible radiation effects.

Properties: Table A.4, air ($T_f = 437$ K, $p = 1$ atm): $\nu = 30.84 \times 10^{-6}$ m²/s, $k = 36.4 \times 10^{-3}$ W/m·K, $Pr = 0.687$. As noted in Example 6.6, the properties k, Pr, c_p, and μ may be assumed to be independent of pressure. However, for an ideal gas, the kinematic viscosity is inversely proportional to pressure. Hence the kinematic viscosity of air at 437 K and $p_\infty = 6 \times 10^3$ N/m² is

$$\nu = 30.84 \times 10^{-6} \text{ m}^2/\text{s} \times \frac{1.0133 \times 10^5 \text{ N/m}^2}{6 \times 10^3 \text{ N/m}^2} = 5.21 \times 10^{-4} \text{ m}^2/\text{s}$$

Analysis: For a plate of unit width, it follows from Newton's law of cooling that the rate of convection heat transfer *to* the plate is

$$q' = \bar{h}L(T_\infty - T_s)$$

To determine the appropriate convection correlation for computing $\bar{h}$, the Reynolds number must first be determined

$$Re_L = \frac{u_\infty L}{\nu} = \frac{10 \text{ m/s} \times 0.5 \text{ m}}{5.21 \times 10^{-4} \text{ m}^2/\text{s}} = 9597$$

Hence the flow is laminar over the entire plate, and the appropriate correlation is given by Equation 7.30.

$$\overline{Nu}_L = 0.664 \, Re_L^{1/2} \, Pr^{1/3} = 0.664(9597)^{1/2}(0.687)^{1/3} = 57.4$$

The average convection coefficient is then

$$\bar{h} = \frac{\overline{Nu}_L k}{L} = \frac{57.4 \times 0.0364 \text{ W/m·K}}{0.5 \text{ m}} = 4.18 \text{ W/m}^2 \cdot \text{K}$$

and the required cooling rate per unit width of plate is

$$q' = 4.18 \text{ W/m}^2 \cdot \text{K} \times 0.5 \text{ m} (300 - 27)°\text{C} = 570 \text{ W/m} \qquad \lhd$$

Comments:

1. The results of Table A.4 apply to gases at atmospheric pressure.

2. Example 7.1 in *IHT* demonstrates how to use the *Correlations* and *Properties* tools, which can facilitate performing convection calculations.

EXAMPLE 7.2

A flat plate of width $w = 1$ m is maintained at a uniform surface temperature of $T_s = 230°$C by using independently controlled heated segments, each of which is $L_h = 50$ mm long. If atmospheric air at 25°C flows over the plate at a velocity of 60 m/s, which segment requires the largest heater power, and what is the value of this power?

SOLUTION

Known: Airflow over a flat plate with segmented heaters.

Find: Maximum heater power requirement.

Schematic:

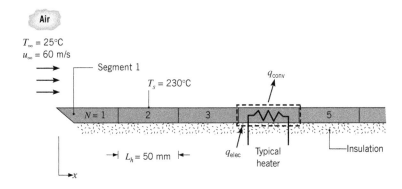

Assumptions:

1. Steady-state, incompressible flow conditions.
2. Negligible radiation effects.
3. Bottom surfaces of heated segments are adiabatic.

Properties: Table A.4, air ($T_f = 400$ K, $p = 1$ atm): $\nu = 26.41 \times 10^{-6}$ m^2/s, $k = 0.0338$ W/m · K, $Pr = 0.690$.

Analysis: The Reynolds number based upon the length, L_h, of the first heated segment is

$$Re_1 = \frac{u_\infty L_h}{\nu} = \frac{60 \text{ m/s} \times 0.05 \text{ m}}{26.41 \times 10^{-6} \text{ m}^2/\text{s}} = 1.14 \times 10^5$$

If the transition Reynolds number is assumed to be $Re_{x,c} = 5 \times 10^5$, it follows that transition will occur at

$$x_c = \frac{\nu}{u_\infty} Re_{x,c} = \frac{26.41 \times 10^{-6} \text{ m}^2/\text{s}}{60 \text{ m/s}} 5 \times 10^5 = 0.22 \text{ m}$$

which is within the fifth heated segment. Knowing how the local convection coefficient varies with distance from the leading edge of the plate, there are three possibilities regarding which segment will have the maximum power requirement:

1. Segment 1, since it corresponds to the largest local, laminar convection coefficient.
2. Segment 5, since it corresponds to the largest local, turbulent convection coefficient.
3. Segment 6, since turbulent conditions exist over the entire segment.

In general, for heater segment N, the power requirement is

$$q_{\text{elec},N} = q_{\text{conv},N} = \bar{h}_N L_h w (T_s - T_\infty) \tag{1}$$

Applying the conservation of energy principle, the power requirement for segment N may be determined by subtracting the rate of heat loss associated with the first $N - 1$ segments from the rate of heat loss associated with all N segments. With $\overline{h}_{1-N}$ defined as the average convection heat transfer coefficient over segments 1 through N, the power requirement for segment N is equal to the rate of convection heat transfer from the segment, which may be expressed as

$$q_{\text{conv},N} = \overline{h}_{1-N}(NL_h)w(T_s - T_\infty) - \overline{h}_{1-(N-1)}[(N - 1)L_h]w(T_s - T_\infty)$$

$$= [N\overline{h}_{1-N} - (N - 1)\overline{h}_{1-(N-1)}]\, L_h w(T_s - T_\infty) \tag{2}$$

Combining Equations 1 and 2 yields

$$\overline{h}_N = N\overline{h}_{1-N} - (N - 1)h_{1-(N-1)} \tag{3}$$

Segment 1: The flow is laminar, and the average convection coefficient, $\overline{h}_1$, may be determined from Equation 7.30,

$$\overline{Nu}_1 = 0.664\, Re_1^{1/2}\, Pr^{1/3} = 0.664(1.14 \times 10^5)^{1/2}(0.69)^{1/3} = 198$$

yielding

$$\overline{h}_1 = \frac{\overline{Nu}_1 k}{L_h} = \frac{198 \times 0.0338\ \text{W/m} \cdot \text{K}}{0.05\ \text{m}} = 134\ \text{W/m}^2 \cdot \text{K}$$

Segment 5: Mixed conditions exist. The average Nusselt number for segments 1 through 5 may be obtained from Equation 7.38 with $A = 871$ and $Re_5 = 5Re_1 = 5.68 \times 10^5$:

$$\overline{Nu}_5 = (0.037Re_5^{4/5} - 871)Pr^{1/3} = [0.037(5.68 \times 10^5)^{4/5} - 871](0.69)^{1/3} = 542$$

Therefore

$$\overline{h}_{1-5} = \frac{\overline{Nu}_5 k}{5L_h} = \frac{542 \times 0.0338\ \text{W/m} \cdot \text{K}}{0.25\ \text{m}} = 73.3\ \text{W/m}^2 \cdot \text{K}$$

The value of $\overline{Nu}_4$ may be obtained from Equation 7.30. With $Re_4 = 4Re_1 = 4.54 \times 10^5$,

$$\overline{Nu}_4 = 0.664(4.54 \times 10^5)^{1/2}(0.69)^{1/3} = 396$$

Therefore

$$\overline{h}_{1-4} = \frac{\overline{Nu}_4 k}{4L_h} = \frac{396 \times 0.0338\ \text{W/m} \cdot \text{K}}{0.2\ \text{m}} = 66.8\ \text{W/m}^2 \cdot \text{K}$$

Evaluation of Equation 3 yields

$$\overline{h}_5 = 5\overline{h}_{1-5} - 4\overline{h}_{1-4} = 5 \times 73.3\ \text{W/m}^2 \cdot \text{K} - 4 \times 66.8\ \text{W/m}^2 \cdot \text{K} = 99.3\ \text{W/m}^2 \cdot \text{K}$$

Segment 6: The value of $\overline{Nu}_6$ may be obtained from Equation 7.38. With $Re_6 = 6Re_1 = 6.84 \times 10^5$,

$$\overline{Nu}_6 = [0.037(6.82 \times 10^5)^{4/5} - 871](0.69)^{1/3} = 748$$

Therefore

$$\overline{h}_{1-6} = \frac{\overline{Nu}_6 k}{6L_h} = \frac{748 \times 0.0338 \text{ W/m} \cdot \text{K}}{0.3 \text{ m}} = 84.3 \text{ W/m}^2 \cdot \text{K}$$

Equation 3 yields

$$\overline{h}_6 = 6\overline{h}_{1-6} - 5\overline{h}_{1-5} = 6 \times 84.3 \text{ W/m}^2 \cdot \text{K} - 5 \times 73.3 \text{ W/m}^2 \cdot \text{K} = 139 \text{ W/m}^2 \cdot \text{K}$$

Since $\overline{h}_6 > \overline{h}_1 > \overline{h}_5$, the maximum power requirement is associated with segment 6 and is

$$q_{conv,6} = \overline{h}_6 L_h w(T_s - T_\infty) = 139 \text{ W/m}^2 \cdot \text{K} \times 0.05 \text{ m} \times 1 \text{ m} \times (230°\text{C} - 25°\text{C}) = 1430 \text{ W} \quad \triangleleft$$

Comments:

1. A less accurate, alternative approach is to assume that the average heat transfer coefficient for a particular segment N is well represented by the value of the local heat transfer coefficient at the middle of the segment, $x_{mid,N}$. The following results were obtained using this approach.

Segment	$x_{mid,N}$ (m)	Flow	Correlation	$h_{x,mid}$ (W/m^2 · K)
1	0.025	Laminar	Equation 7.23	$95 \neq \overline{h}_1$
5	0.225	Turbulent	Equation 7.36	$145 \neq \overline{h}_5$
6	0.275	Turbulent	Equation 7.36	$139 \approx \overline{h}_6$

With this approach, we not only predict incorrect values of the average heat transfer coefficient for each segment, but we also incorrectly identify segment 5 as having the largest power requirement. This procedure yields reasonable results only when spatial variation of the convection coefficient is gradual, such as in regions of turbulent flow that are not in the vicinity of the flow transition.

2. This example is solved in the *Advanced* section of *IHT*.

EXAMPLE 7.3

Drought conditions in the southwestern United States have prompted officials to question whether the operation of residential swimming pools should be permitted. As the chief engineer of a city that has a large number of pools, you have been asked to estimate the daily water loss due to pool evaporation. For representative conditions, you may assume water and ambient air temperatures of 25°C and an ambient relative humidity of 50%. Typical pool surface dimensions are 6 m × 12 m. There is a 1.5-m-wide deck around the pool, which is raised relative to the surrounding ground. Wind blows in the direction of the long side of the pool, with velocity 2 m/s. You may assume the free stream turbulence of the air to be negligible, the surface of the water to be smooth and level with the pool deck, and the deck to be dry. What is the water loss for the pool in kilograms per day?

SOLUTION

Known: Ambient air conditions above a swimming pool, pool and deck dimensions.

Find: Daily evaporative water loss.

Schematic:

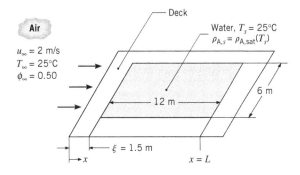

Assumptions:

1. Steady-state, incompressible flow conditions.
2. Smooth water surface and negligible free stream turbulence.
3. Deck is dry.
4. Heat and mass transfer analogy applicable.
5. Flow is tripped to turbulence by the leading edge of the deck.
6. Ideal gas behavior for water vapor in free stream.

Properties: Table A.4, air (25°C): $\nu = 15.7 \times 10^{-6}$ m²/s. Table A.8, water vapor–air (25°C): $D_{AB} = 0.26 \times 10^{-4}$ m²/s, $Sc = \nu/D_{AB} = 0.60$. Table A.6, saturated water vapor (25°C): $\rho_{A,sat} = v_g^{-1} = 0.0226$ kg/m³.

Analysis: The leading edge of the velocity boundary layer is at the edge of the deck; therefore, the trailing edge of the pool is at a distance of $L = 13.5$ m from the leading edge. The Reynolds number at that point is

$$Re_L = \frac{u_\infty L}{\nu} = \frac{2 \text{ m/s} \times 13.5 \text{ m}}{15.7 \times 10^{-6} \text{ m}^2/\text{s}} = 1.72 \times 10^6$$

Applying the heat and mass transfer analogy to Equation 7.44 yields

$$\overline{Sh}_L = \overline{Sh}_L\big|_{\xi=0} \frac{L}{L - \xi}[1 - (\xi/L)^{(p+1)/(p+2)}]^{p/(p+1)} \tag{1}$$

The average Sherwood number, $\overline{Sh}_L\big|_{\xi=0}$, is evaluated from Equation 7.41 with $A = 0$ because the boundary layer is tripped to turbulent conditions by the leading edge of the deck

$$\overline{Sh}_L\big|_{\xi=0} = 0.037 \, Re_L^{4/5} Sc^{1/3}$$

$$\overline{Sh}_L\big|_{\xi=0} = 0.037(1.72 \times 10^6)^{4/5} \times (0.60)^{1/3} = 3040$$

With $p = 8$ for turbulent flow, Equation 1 may be evaluated as

$$\overline{Sh}_L = 3040 \frac{13.5 \text{ m}}{(13.5 \text{ m} - 1.5 \text{ m})}[1 - (1.5 \text{ m}/13.5 \text{ m})^{(8+1)/(8+2)}]^{8/(8+1)} = 2990$$

It follows that

$$\bar{h}_{m,L} = \overline{Sh}_L \left(\frac{D_{AB}}{L} \right) = 2990 \frac{0.26 \times 10^{-4}\,\text{m}^2/\text{s}}{13.5\,\text{m}} = 5.77 \times 10^{-3}\,\text{m/s}$$

The evaporation rate for the pool is then

$$n_A = \bar{h}_m A (\rho_{A,s} - \rho_{A,\infty})$$

where A is the pool area (not including the deck). With the free stream vapor assumed to be an ideal gas,

$$\phi_\infty = \frac{\rho_{A,\infty}}{\rho_{A,\text{sat}}(T_\infty)}$$

and with $\rho_{A,s} = \rho_{A,\text{sat}}(T_s)$,

$$n_A = \bar{h}_m A [\rho_{A,\text{sat}}(T_s) - \phi_\infty \rho_{A,\text{sat}}(T_\infty)]$$

Since $T_s = T_\infty = 25°C$, it follows that

$$n_A = \bar{h}_m A \rho_{A,\text{sat}}(25°C)[1 - \phi_\infty]$$

Hence

$$n_A = 5.77 \times 10^{-3}\,\text{m/s} \times 72\,\text{m}^2 \times 0.0226\,\text{kg/m}^3 \times 0.5 \times 86{,}400\,\text{s/day}$$

$$n_A = 405\,\text{kg/day} \qquad \qquad \triangleleft$$

Comments:

1. The water surface temperature is likely to be slightly less than the air temperature because of the evaporative cooling effect.

2. The volume loss, with a water density of 996 kg/m³, is $n_A/\rho = 0.4$ m³/day. This would mean a drop in the pool level of 6 mm per day. Of course the loss would be greater in summer when the air temperature is higher.

3. The influence of the deck length on the daily evaporation rate is shown in the figure. As the length of the deck is increased, the overall evaporation rate is reduced due to the extension of the leading edge of the velocity boundary layer farther away from the pool.

7.4 *The Cylinder in Cross Flow*

7.4.1 Flow Considerations

Another common external flow involves fluid motion normal to the axis of a circular cylinder. As shown in Figure 7.6, the free stream fluid is brought to rest at the *forward stagnation point,* with an accompanying rise in pressure. From this point, the pressure decreases with increasing x, the streamline coordinate, and the boundary layer develops under the influence of a *favorable pressure gradient* ($dp/dx < 0$). However, the pressure must eventually reach a minimum, and toward the rear of the cylinder further boundary layer development occurs in the presence of an *adverse pressure gradient* ($dp/dx > 0$).

In Figure 7.6, the distinction between the upstream velocity V and the free stream velocity u_∞ should be noted. Unlike conditions for the flat plate in parallel flow, these velocities differ, with u_∞ now depending on the distance x from the stagnation point. From Euler's equation for an inviscid flow [10], $u_\infty(x)$ must exhibit behavior opposite to that of $p(x)$. That is, from $u_\infty = 0$ at the stagnation point, the fluid accelerates because of the favorable pressure gradient ($du_\infty/dx > 0$ when $dp/dx < 0$), reaches a maximum velocity when $dp/dx = 0$, and decelerates because of the adverse pressure gradient ($du_\infty/dx < 0$ when $dp/dx > 0$). As the fluid decelerates, the velocity gradient at the surface, $\partial u/\partial y|_{y=0}$, eventually becomes zero (Figure 7.7). At this location, termed the *separation point,* fluid near the surface lacks sufficient momentum to overcome the pressure gradient, and continued downstream movement is impossible. Since the oncoming fluid also precludes flow back upstream, *boundary layer separation* must occur. This is a condition for which the boundary layer detaches from the surface, and a *wake* is formed in the downstream region. Flow in this region is characterized by vortex formation and is highly irregular. An excellent review of flow conditions in the wake of a circular cylinder is provided by Coutanceau and Defaye [11].

The occurrence of *boundary layer transition,* which depends on the Reynolds number, strongly influences the position of the separation point. For the circular cylinder the characteristic length is the diameter, and the Reynolds number is defined as

$$Re_D \equiv \frac{\rho VD}{\mu} = \frac{VD}{\nu}$$

Since the momentum of fluid in a turbulent boundary layer is larger than in the laminar boundary layer, it is reasonable to expect transition to delay the occurrence of separation.

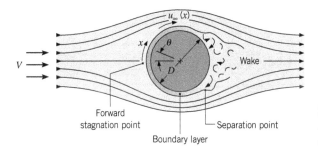

Forward
stagnation point

Boundary layer

Separation point

Wake

$u_\infty(x)$

FIGURE 7.6 Boundary layer formation and separation on a circular cylinder in cross flow.

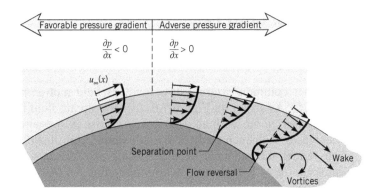

FIGURE 7.7   Velocity profile associated with separation on a circular
cylinder in cross flow.

If $Re_D \lesssim 2 \times 10^5$, the boundary layer is laminar, and separation occurs at $\theta \approx 80°$ (Figure 7.8). However, if $Re_D \gtrsim 2 \times 10^5$, boundary layer transition occurs, and separation is delayed to $\theta \approx 140°$.

The foregoing processes strongly influence the drag force, F_D, acting on the cylinder. This force has two components, one of which is due to the boundary layer surface shear stress (*friction drag*). The other component is due to a pressure differential in the flow direction resulting from formation of the wake (*form,* or *pressure, drag*). A dimensionless *drag coefficient C_D* may be defined as

$$C_D \equiv \frac{F_D}{A_f(\rho V^2/2)} \tag{7.50}$$

where A_f is the cylinder frontal area (the area projected perpendicular to the free stream velocity). The drag coefficient is a function of the Reynolds number and results are presented in Figure 7.9. For $Re_D \lesssim 2$ separation effects are negligible, and conditions are dominated by friction drag. However, with increasing Reynolds number, the effect of separation, and therefore form drag, becomes more important. The large reduction in C_D that occurs for $Re_D \gtrsim 2 \times 10^5$ is due to boundary layer transition, which delays separation, thereby reducing the extent of the wake region and the magnitude of the form drag.

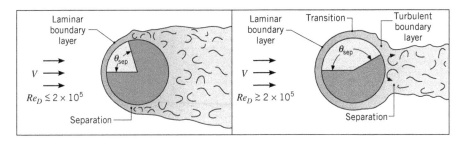

FIGURE 7.8   The effect of turbulence on separation.

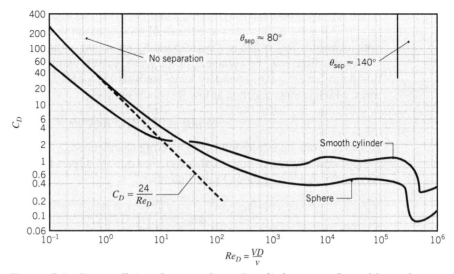

FIGURE 7.9 Drag coefficients for a smooth circular cylinder in cross flow and for a sphere. Boundary layer separation angles are for a cylinder. Based on Schlichting, H., and K. Gersten, *Boundary Layer Theory*, Springer, New York, 2000.

7.4.2 Convection Heat and Mass Transfer

Experimental results for the variation of the local Nusselt number with θ are shown in Figure 7.10 for the cylinder in a cross flow of air. Not unexpectedly, the results are strongly influenced by the nature of boundary layer development on the surface. Consider conditions for $Re_D \lesssim 10^5$. Starting at the stagnation point, Nu_θ decreases with increasing θ as a result of laminar boundary layer development. However, a minimum is reached at $\theta \approx 80°$, where separation occurs and Nu_θ increases with θ due to mixing associated with vortex formation in the wake. In contrast, for $Re_D \gtrsim 10^5$ the variation of Nu_θ with θ is characterized by two minima. The decline in Nu_θ from the value at the stagnation point is again due to laminar boundary layer development, but the sharp increase that occurs between 80° and 100° is now due to boundary layer transition to turbulence. With further development

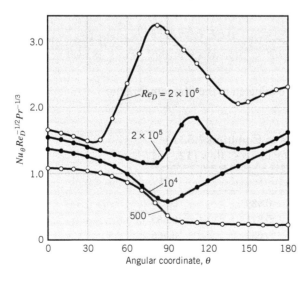

FIGURE 7.10 Local Nusselt number for airflow normal to a circular cylinder. (Adapted with permission from Zukauskas, A., "Convective Heat Transfer in Cross Flow," in S. Kakac, R. K. Shah, and W. Aung, Eds., *Handbook of Single-Phase Convective Heat Transfer*, Wiley, New York, 1987.)

of the turbulent boundary layer, Nu_θ again begins to decline. Eventually separation occurs ($\theta \approx 140°$), and Nu_θ increases as a result of mixing in the wake region. The increase in Nu_θ with increasing Re_D is due to a corresponding reduction in the boundary layer thickness.

Correlations may be obtained for the local Nusselt number, and at the forward stagnation point for $Pr \gtrsim 0.6$, boundary layer analysis [5] yields an expression of the following form, which is most accurate at low Reynolds number:

$$Nu_D(\theta = 0) = 1.15\, Re_D^{1/2}\, Pr^{1/3} \tag{7.51}$$

However, from the standpoint of engineering calculations, we are more interested in overall average conditions. An empirical correlation due to Hilpert [12] that has been modified to account for fluids of various Prandtl numbers,

$$\overline{Nu}_D \equiv \frac{\bar{h}D}{k} = C\,Re_D^m\, Pr^{1/3} \tag{7.52}$$

is widely used for $Pr \gtrsim 0.7$, where the constants C and m are listed in Table 7.2. Equation 7.52 may also be used for flow over cylinders of noncircular cross section, with the characteristic length D and the constants obtained from Table 7.3. In working with Equations 7.51 and 7.52 all properties are evaluated at the film temperature.

Other correlations have been suggested for the circular cylinder in cross flow [16, 17, 18]. The correlation due to Zukauskas [17] is of the form

$$\overline{Nu}_D = C\,Re_D^m\, Pr^n \left(\frac{Pr}{Pr_s}\right)^{1/4} \tag{7.53}$$

$$\begin{bmatrix} 0.7 \lesssim Pr \lesssim 500 \\ 1 \lesssim Re_D \lesssim 10^6 \end{bmatrix}$$

where all properties are evaluated at T_∞, except Pr_s, which is evaluated at T_s. Values of C and m are listed in Table 7.4. If $Pr \lesssim 10$, $n = 0.37$; if $Pr \gtrsim 10$, $n = 0.36$. Churchill and Bernstein [18] have proposed a single comprehensive equation that covers the entire range of Re_D for which data are available, as well as a wide range of Pr. The equation is recommended for all $Re_D\, Pr \gtrsim 0.2$ and has the form

$$\overline{Nu}_D = 0.3 + \frac{0.62\,Re_D^{1/2}\,Pr^{1/3}}{[1 + (0.4/Pr)^{2/3}]^{1/4}} \left[1 + \left(\frac{Re_D}{282,000}\right)^{5/8}\right]^{4/5} \tag{7.54}$$

where all properties are evaluated at the film temperature.

TABLE 7.2 Constants of Equation 7.52 for the circular cylinder in cross flow [12, 13]

Re_D	C	m
0.4–4	0.989	0.330
4–40	0.911	0.385
40–4000	0.683	0.466
4000–40,000	0.193	0.618
40,000–400,000	0.027	0.805

TABLE 7.3   Constants of Equation 7.52 for noncircular cylinders in cross flow of a gas [14, 15][a]

Geometry			Re_D	C	m
Square					
$V \rightarrow$ ◇	D		6000–60,000	0.304	0.59
$V \rightarrow$ ▢	D		5000–60,000	0.158	0.66
Hexagon					
$V \rightarrow$ ⬡	D		5200–20,400	0.164	0.638
			20,400–105,000	0.039	0.78
$V \rightarrow$ ⬡	D		4500–90,700	0.150	0.638
Thin plate perpendicular to flow					
$V \rightarrow$ ▯	D	Front	10,000–50,000	0.667	0.500
		Back	7000–80,000	0.191	0.667

[a]These tabular values are based on the recommendations of Sparrow et al. [15] for air, with extension to other fluids through the $Pr^{1/3}$ dependence of Equation 7.52. A Prandtl number of $Pr = 0.7$ was assumed for the experimental results for air that are described in [15].

Again we caution the reader not to view any of the foregoing correlations as sacrosanct. Each correlation is reasonable over a certain range of conditions, but for most engineering calculations one should not expect accuracy to much better than 20%. Because they are based on more recent results encompassing a wide range of conditions, Equations 7.53 and 7.54 are generally used for the calculations of this text. Detailed reviews of the many correlations that have been developed for the circular cylinder have been provided by Sparrow et al. [15] as well as Morgan [19]. Finally, we note that by invoking the heat and mass transfer analogy, Equations 7.51 through 7.54 may be applied to problems involving convection mass transfer from a cylinder in cross flow. It is simply a matter of replacing $\overline{Nu}_D$ by $\overline{Sh}_D$ and Pr by Sc. In mass transfer problems, boundary layer property variations are typically small. Hence, when using the mass transfer analog of Equation 7.53, the property ratio, which accounts for nonconstant property effects, may be neglected.

TABLE 7.4   Constants of Equation 7.53 for the circular cylinder in cross flow [18]

Re_D	C	m
1–40	0.75	0.4
40–1000	0.51	0.5
10^3–2×10^5	0.26	0.6
2×10^5–10^6	0.076	0.7

EXAMPLE 7.4

Experiments have been conducted using a metallic cylinder 12.7 mm in diameter and 94 mm long. The cylinder is heated internally by an electrical heater and is subjected to a cross flow of air in a low-speed wind tunnel. Under a specific set of operating conditions for which the upstream air velocity and temperature were maintained at $V = 10$ m/s and 26.2°C, respectively, the heater power dissipation was measured to be $P = 46$ W, while the average cylinder surface temperature was determined to be $T_s = 128.4$°C.

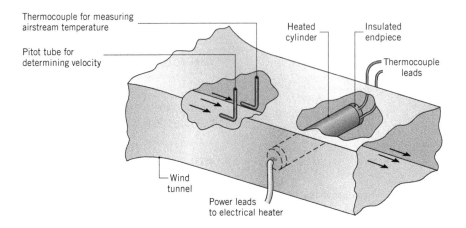

It is estimated that 15% of the power dissipation is lost through the cumulative effects of surface radiation and conduction through the endpieces. The cumulative uncertainty associated with (*i*) the air velocity and temperature measurements, (*ii*) estimating the heat losses by radiation and from the cylinder ends, and (*iii*) averaging the cylinder surface temperature, which varies axially and circumferentially, renders the experimental value of the convection coefficient accurate to no better than 20%.

1. Determine the convection heat transfer coefficient from the experimental observations.

2. Compare the experimental result with the convection coefficient computed from an appropriate correlation.

SOLUTION
Known: Operating conditions for a heated cylinder.

Find:

1. Convection coefficient associated with the operating conditions.

2. Convection coefficient from an appropriate correlation.

Schematic:

Assumptions:

1. Steady-state, incompressible flow conditions.
2. Uniform cylinder surface temperature.

Properties: Table A.4, air ($T_\infty = 26.2°C \approx 300$ K): $\nu = 15.89 \times 10^{-6}$ m²/s, $k = 26.3 \times 10^{-3}$ W/m·K, $Pr = 0.707$. Table A.4, air ($T_f \approx 350$ K): $\nu = 20.92 \times 10^{-6}$ m²/s, $k = 30 \times 10^{-3}$ W/m·K, $Pr = 0.700$. Table A.4, air ($T_s = 128.4°C = 401$ K): $Pr = 0.690$.

Analysis:

1. The convection heat transfer coefficient may be determined from the data by using Newton's law of cooling. That is,

$$\bar{h} = \frac{q}{A(T_s - T_\infty)}$$

With $q = 0.85P$ and $A = \pi DL$, it follows that

$$\bar{h} = \frac{0.85 \times 46 \text{ W}}{\pi \times 0.0127 \text{ m} \times 0.094 \text{ m} (128.4 - 26.2)°C} = 102 \text{ W/m}^2 \cdot \text{K} \qquad \triangleleft$$

2. Working with the Zukauskas relation, Equation 7.53,

$$\overline{Nu}_D = C\, Re_D^m\, Pr^n \left(\frac{Pr}{Pr_s}\right)^{1/4}$$

all properties, except Pr_s, are evaluated at T_∞. Accordingly,

$$Re_D = \frac{VD}{\nu} = \frac{10 \text{ m/s} \times 0.0127 \text{ m}}{15.89 \times 10^{-6} \text{ m}^2/\text{s}} = 7992$$

Hence, from Table 7.4, $C = 0.26$ and $m = 0.6$. Also, since $Pr < 10$, $n = 0.37$. It follows that

$$\overline{Nu}_D = 0.26(7992)^{0.6}(0.707)^{0.37}\left(\frac{0.707}{0.690}\right)^{0.25} = 50.5$$

$$\bar{h} = \overline{Nu}_D \frac{k}{D} = 50.5\frac{0.0263 \text{ W/m}\cdot\text{K}}{0.0127 \text{ m}} = 105 \text{ W/m}^2 \cdot \text{K} \qquad \triangleleft$$

Using the Churchill relation, Equation 7.54,

$$\overline{Nu}_D = 0.3 + \frac{0.62\, Re_D^{1/2}\, Pr^{1/3}}{[1 + (0.4/Pr)^{2/3}]^{1/4}}\left[1 + \left(\frac{Re_D}{282,000}\right)^{5/8}\right]^{4/5}$$

With all properties evaluated at T_f, $Pr = 0.70$ and

$$Re_D = \frac{VD}{\nu} = \frac{10 \text{ m/s} \times 0.0127 \text{ m}}{20.92 \times 10^{-6} \text{ m}^2/\text{s}} = 6071$$

Hence the Nusselt number and the convection coefficient are

$$\overline{Nu}_D = 0.3 + \frac{0.62(6071)^{1/2}(0.70)^{1/3}}{[1 + (0.4/0.70)^{2/3}]^{1/4}}\left[1 + \left(\frac{6071}{282{,}000}\right)^{5/8}\right]^{4/5} = 40.6$$

$$\overline{h} = \overline{Nu}_D\frac{k}{D} = 40.6\frac{0.030\ \text{W/m}\cdot\text{K}}{0.0127\ \text{m}} = 96.0\ \text{W/m}^2\cdot\text{K} \qquad \triangleleft$$

Alternatively, from the Hilpert correlation, Equation 7.52,

$$\overline{Nu}_D = C\,Re_D^m\,Pr^{1/3}$$

With all properties evaluated at the film temperature, $Re_D = 6071$ and $Pr = 0.70$. Hence, from Table 7.2, $C = 0.193$ and $m = 0.618$. The Nusselt number and the convection coefficient are then

$$\overline{Nu}_D = 0.193(6071)^{0.618}(0.700)^{1/3} = 37.3$$

$$\overline{h} = \overline{Nu}_D\frac{k}{D} = 37.3\frac{0.030\ \text{W/m}\cdot\text{K}}{0.0127\ \text{m}} = 88.2\ \text{W/m}^2\cdot\text{K} \qquad \triangleleft$$

Comments:

1. Calculations based on the three correlations are within the range of the measured value of the convection heat transfer coefficient, $\overline{h} = 102 \pm 20\ \text{W/m}^2\cdot\text{K}$.

2. Recognize the importance of using the proper temperature when evaluating fluid properties.

EXAMPLE 7.5

Hydrogen is often stored by *adsorbing* it into a metal hydride powder. The hydrogen can be desorbed as needed, by heating the metal hydride throughout its volume. Consider a hydrogen fuel cell–powered automobile cruising at a speed of $V = 25$ m/s. The car consumes $\dot{m}_{H_2} = 1.35 \times 10^{-4}$ kg/s of hydrogen, which is supplied from a cylindrical, stainless steel canister loaded with metal hydride powder. The canister is of inside diameter $D_i = 0.1$ m, length $L = 0.8$ m, and wall thickness $t = 5$ mm, and is subject to air in cross flow at $V = 25$ m/s, $T_\infty = 23°C$.

    In order for desorption to occur, the metal hydride must be maintained at an operating temperature of at least 275 K. The desorption process is an endothermic reaction corresponding to a thermal generation rate expressed as

$$\dot{E}_g = -\dot{m}_{H_2} \times (29.5 \times 10^3\ \text{kJ/kg})$$

where $\dot{m}_{H_2}$ is the hydrogen desorption rate (kg/s). Determine how much additional heating, beyond that due to convection from the air, should be supplied to the canister to maintain the required operating temperature.

SOLUTION

Known: Size and shape of a hydrogen storage canister, hydrogen desorption rate, required hydrogen operating pressure, velocity and temperature of air in cross flow.

Find: The rate of convective heat transfer to the canister and the additional heating needed to sustain $p_{H_2} > p_{fc}$.

Schematic:

Assumptions:

1. Steady-state, incompressible flow conditions.
2. Uniform cylinder surface temperature.
3. Negligible heat gain through the ends of the cylinder.
4. Uniform metal hydride temperature.
5. Negligible contact resistance between the canister wall and the metal hydride.

Properties: Table A.4, air ($T_f \approx 285$ K): $\nu = 14.56 \times 10^{-6}$ m²/s, $k = 25.2 \times 10^{-3}$ W/m · K, $Pr = 0.712$. Table A.1, AISI 316 stainless steel ($T_{ss} \approx 300$ K): $k_{ss} = 13.4$ W/m · K.

Analysis: The thermal energy generation rate associated with the desorption of hydrogen from the metal hydride at the required flow rate is

$$\dot{E}_g = -(1.35 \times 10^{-4} \text{ kg/s}) \times (29.5 \times 10^6 \text{ J/kg}) = -3982 \text{ W}$$

To determine the convective heat transfer rate, we begin by calculating the Reynolds number:

$$Re_D = \frac{V(D_i + 2t)}{\nu} = \frac{23 \text{ m/s} \times (0.1 \text{ m} + 2 \times 0.005 \text{ m})}{14.56 \times 10^{-6} \text{ m}^2/\text{s}} = 173,760$$

Use of Equation 7.54

$$\overline{Nu}_D = 0.3 + \frac{0.62 \, Re_D^{1/2} \, Pr^{1/3}}{[1 + (0.4/Pr)^{2/3}]^{1/4}} \left[1 + \left(\frac{Re_D}{282,000}\right)^{5/8}\right]^{4/5}$$

yields

$$\overline{Nu}_D = 0.3 + \frac{0.62(173,760)^{1/2}(0.712)^{1/3}}{[1 + (0.4/0.712)^{2/3}]^{1/4}} \left[1 + \left(\frac{173,760}{282,000}\right)^{5/8}\right]^{4/5} = 315.8$$

Therefore, the average convection heat transfer coefficient is

$$\overline{h} = \overline{Nu}_D \frac{k}{(D_i + 2t)} = 315.8 \times \frac{25.3 \times 10^{-3} \text{ W/m} \cdot \text{K}}{(0.1 \text{ m} + 2 \times 0.005 \text{ m})} = 72.6 \text{ W/m}^2 \cdot \text{K}$$

Simplifying Equation 3.34, we find

$$q_{\text{conv}} = \frac{T_\infty - T_i}{\dfrac{1}{\pi L(D_i + 2t)\overline{h}} + \dfrac{\ln[(D_i + 2t)/D_i]}{2\pi k_{\text{ss}}L}}$$

or, substituting values,

$$q_{\text{conv}} = \frac{296 \text{ K} - 275.2 \text{ K}}{\dfrac{1}{\pi(0.8 \text{ m})(0.1 \text{ m} + 2 \times 0.005 \text{ m})(72.6 \text{ W/m}^2 \cdot \text{K})} + \dfrac{\ln[(0.1 \text{ m} + 2 \times 0.005 \text{ m})/0.1 \text{ m}]}{2\pi(13.4 \text{ W/m} \cdot \text{K})(0.8 \text{ m})}}$$

$$= 406 \text{ W}$$

The additional thermal energy, q_{add}, that must be supplied to the canister to maintain the steady-state operating temperature may be found from an energy balance, $q_{\text{add}} + q_{\text{conv}} + \dot{E}_g = 0$. Therefore,

$$q_{\text{add}} = -q_{\text{conv}} - \dot{E}_g = -406 \text{ W} + 3982 \text{ W} = 3576 \text{ W} \qquad \triangleleft$$

Comments:

1. Additional heating will occur due to radiation, conduction from the canister mounting hardware and fuel lines, and possibly condensation of water vapor on the cool canister.

2. The thermal resistances associated with conduction in the canister wall and convection are 0.0014 K/W and 0.053 K/W, respectively. The convection resistance dominates and can be reduced by adding fins to the exterior of the canister.

3. The amount of additional heating that is required will increase if the automobile is operated at a higher speed, since the hydrogen consumption scales as V^3, while the convective heat transfer coefficient increases as $V^{0.7}$ to $V^{0.8}$. Additional heating is also needed when the automobile is operated in a cooler climate.

7.5 *The Sphere*

Boundary layer effects associated with flow over a sphere are much like those for the circular cylinder, with transition and separation playing prominent roles. Results for the drag coefficient, which is defined by Equation 7.50, are presented in Figure 7.9. In the limit of very small Reynolds numbers (*creeping flow*), the coefficient is inversely proportional to the Reynolds number and the specific relation is termed *Stokes' law*

$$C_D = \frac{24}{Re_D} \qquad Re_D \lesssim 0.5 \qquad (7.55)$$

Numerous heat transfer correlations have been proposed, and Whitaker [16] recommends an expression of the form

$$\overline{Nu}_D = 2 + (0.4\,Re_D^{1/2} + 0.06\,Re_D^{2/3})Pr^{0.4}\left(\frac{\mu}{\mu_s}\right)^{1/4} \qquad (7.56)$$

$$\begin{bmatrix} 0.71 \lesssim Pr \lesssim 380 \\ 3.5 \lesssim Re_D \lesssim 7.6 \times 10^4 \\ 1.0 \lesssim (\mu/\mu_s) \lesssim 3.2 \end{bmatrix}$$

All properties except μ_s are evaluated at T_∞, and the result may be applied to mass transfer problems simply by replacing $\overline{Nu}_D$ and Pr with $\overline{Sh}_D$ and Sc, respectively. A special case of convection heat and mass transfer from spheres relates to transport from freely falling liquid drops, and the correlation of Ranz and Marshall [20] is often used

$$\overline{Nu}_D = 2 + 0.6\,Re_D^{1/2}\,Pr^{1/3} \qquad (7.57)$$

In the limit $Re_D \rightarrow 0$, Equations 7.56 and 7.57 reduce to $\overline{Nu}_D = 2$, which corresponds to heat transfer by conduction from a spherical surface to a stationary, infinite medium around the surface, as may be derived from Case 1 of Table 4.1.

EXAMPLE 7.6

Electrical circuitry is written onto a photovoltaic panel by depositing a stream of small ($D = 55$ μm) droplets of electrically conducting ink from a thermal inkjet printer. The drops are at an initial temperature of $T_i = 200°C$, and it is desirable for them to strike the panel at a temperature of $T_{final} = 50°C$. The quiescent air and surroundings are at $T_\infty = T_{sur} = 25°C$, and the drops are ejected from the print head at their terminal velocity. Determine the required standoff distance L between the printer and the photovoltaic panel. The properties of the electrically conducting ink drop are $\rho_d = 2400$ kg/m^3, $c_d = 800$ J/kg · K, and $k_d = 5.0$ W/m · K.

SOLUTION

Known: Droplet size and properties, initial and desired final droplet temperature. Droplet injected at its terminal velocity.

Find: Required standoff distance between the printer and the photovoltaic panel.

Schematic:

Assumptions:

1. Constant air properties evaluated at 25°C.
2. Negligible radiation effects.
3. Negligible temperature variation within the droplets (lumped capacitance approximation).

Properties: Table A.4, air ($T_f = 75°C$): $\rho = 1.002$ kg/m^3, $\nu = 20.72 \times 10^{-6}$ m^2/s. Table A.4, air ($T_\infty = 25°C$): $\nu = 15.71 \times 10^{-6}$ m^2/s, $k = 0.0261$ W/m · K, $Pr = 0.708$.

Analysis: Since the droplets travel at their terminal velocities, the net force on each drop must be zero. Hence the weight of the drop is offset by the buoyancy force associated with the displaced air and the drag force:

$$\rho_d g\left(\pi \frac{D^3}{6} \right) = \rho g\left(\pi \frac{D^3}{6} \right) + C_D \left(\frac{\pi D^2}{4} \right)\left(\rho \frac{V^2}{2} \right) \quad (1)$$

where Equation 7.50 has been used to express the drag force F_D. Since the droplets are small, we anticipate that the Reynolds number will also be small. If this is the case, Stokes' law, Equation 7.55, may be used to express the drag coefficient as

$$C_D = \frac{24}{Re_D} = \frac{24v}{VD} \tag{2}$$

Substituting Equation (2) into Equation (1) and solving for the velocity,

$$V = \frac{gD^2}{18v\rho}(\rho_d - \rho) = \frac{9.8 \text{ m/s}^2 \times (55 \times 10^{-6} \text{ m})^2}{18 \times 20.72 \times 10^{-6} \text{ m}^2/\text{s} \times 1.002 \text{ kg/m}^3} \times (2400 - 1.002) \text{kg/m}^3$$

$$= 0.190 \text{ m/s} = 190 \text{ mm/s}$$

Therefore, the Reynolds number is $Re_D = VD/v = 0.190 \text{ m/s} \times 55 \times 10^{-6} \text{ m}/20.72 \times 10^{-6}$ $\text{m}^2/\text{s} = 0.506$, and use of Stokes' law is appropriate. The Nusselt number and convection coefficient can be calculated from the Ranz and Marshall correlation, Equation 7.57, using properties evaluated at the free stream temperature (see Table 7.7):

$$\overline{Nu}_D = 2 + 0.6 \, Re_D^{1/2} \, Pr^{1/3} = 2 + 0.6 \times \left(\frac{0.190 \text{ m/s} \times 55 \times 10^{-6} \text{ m}}{15.71 \times 10^{-6} \text{ m}^2/\text{s}} \right)^{1/2} \times 0.708^{1/3} = 2.44$$

$$\overline{h} = \frac{\overline{Nu}_D k}{D} = \frac{2.44 \times 0.0261 \text{ W/m} \cdot \text{K}}{55 \times 10^{-6} \text{ m}} = 1160 \text{ W/m}^2 \cdot \text{K}$$

Applying the lumped capacitance method, Equation 5.5, the required *time-of-flight* is then

$$t = \frac{\rho_d V c_d}{\overline{h} A_s} \ln\left(\frac{\theta_i}{\theta_{\text{final}}} \right) = \frac{\rho_d c_d D}{6\overline{h}} \ln\left(\frac{T_i - T_\infty}{T_{\text{final}} - T_\infty} \right)$$

$$= \frac{2400 \text{ kg/m}^3 \times 800 \text{ J/kg} \cdot \text{K} \times 55 \times 10^{-6} \text{ m}}{6 \times 1160 \text{ W/m}^2 \cdot \text{K}} \ln\left(\frac{(200 - 25)°\text{C}}{(50 - 25)°\text{C}} \right)$$

$$= 0.030 \text{ s}$$

and the standoff distance is

$$L = Vt = 0.190 \text{ m/s} \times 0.030 \text{ s} = 0.0056 \text{ m} = 5.6 \text{ mm} \qquad \triangleleft$$

Comments:

1. The validity of the lumped capacitance method may be determined by calculating the Biot number. Applying Equation 5.10 in the conservative fashion with $L_c = D/2$,

$$Bi = \frac{\overline{h}(D/2)}{k_p} = \left(\frac{1160 \text{ W/m}^2 \cdot \text{K} \times 55 \times 10^{-6} \text{ m}}{2} \right) / 5.0 \text{ W/m} \cdot \text{K} = 0.006 < 0.1$$

and the criterion is satisfied.

2. Use of Equation 7.55, Stokes' law, to describe the Reynolds number dependence of the drag coefficient is valid since $Re_D \lesssim 0.5$. For larger particles, Figure 7.9 would need to be consulted to determine the relationship between C_D and Re_D.

3. If the particles were not injected at their terminal velocity, they would either accelerate or decelerate during flight, complicating the analysis.

4. Assuming blackbody behavior and using the maximum (initial) temperature of the particle, $T_s = 473$ K, the maximum radiation heat transfer coefficient is $h_r = \sigma(T_s + T_{sur})$ $(T_s^2 + T_{sur}^2) = 5.67 \times 10^{-8}$ W/m$^2 \cdot$ K^4 × (473 K + 298 K) × [(473 K)^2 + (298 K)^2] = 13.7 W/m$^2 \cdot$ K. Since $h_r \ll \bar{h}$, radiation heat transfer is negligible.

7.6 *Flow Across Banks of Tubes*

Heat transfer to or from a bank (or bundle) of tubes in cross flow is relevant to numerous industrial applications, such as steam generation in a boiler or air cooling in the coil of an air conditioner. The geometric arrangement is shown schematically in Figure 7.11. Typically, one fluid moves over the tubes, while a second fluid at a different temperature passes through the tubes. In this section we are specifically interested in the convection heat transfer associated with cross flow over the tubes.

The tube rows of a bank can be either *aligned* or *staggered* in the direction of the fluid velocity V (Figure 7.12). The configuration is characterized by the tube diameter D and by the *transverse pitch* S_T and *longitudinal pitch* S_L measured between tube centers. Flow

FIGURE 7.11 Schematic of a tube bank in cross flow.

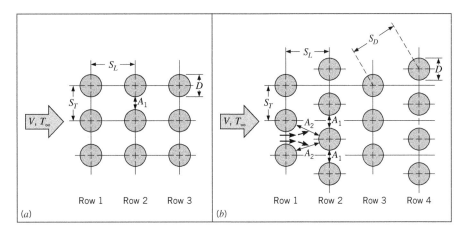

FIGURE 7.12 Tube arrangements in a bank. (*a*) Aligned. (*b*) Staggered.

conditions within the bank are dominated by boundary layer separation effects and by wake interactions, which in turn influence convection heat transfer.

Flow around the tubes in the first row of a tube bank is similar to that for a single (isolated) cylinder in cross flow. Correspondingly, the heat transfer coefficient for a tube in the first row is approximately equal to that for a single tube in cross flow. For downstream rows, flow conditions depend strongly on the tube bank arrangement (Figure 7.13). Aligned tubes beyond the first row reside in the wakes of upstream tubes, and for moderate values of S_L convection coefficients associated with downstream rows are enhanced by mixing, or turbulation, of the flow. Typically, the convection coefficient of a row increases with increasing row number until approximately the fifth row, after which there is little change in flow conditions and hence in the convection coefficient. For large S_L, the influence of upstream rows decreases, and heat transfer in the downstream rows is not enhanced. For this reason, operation of aligned tube banks with $S_T/S_L < 0.7$ is undesirable. For the staggered tube array, the path of the main flow is more tortuous, and mixing of the cross-flowing fluid is increased relative to the aligned tube arrangement. In general, heat transfer enhancement is favored by the more tortuous flow of a staggered arrangement, particularly for small Reynolds numbers ($Re_D \lesssim 100$).

Typically, we wish to know the *average* heat transfer coefficient for the *entire* tube bank. Zukauskas [17] has proposed a correlation of the form

$$\overline{Nu}_D = C_1 \, Re_{D,\text{max}}^m \, Pr^{0.36} \left(\frac{Pr}{Pr_s} \right)^{1/4} \qquad (7.58)$$

$$\begin{bmatrix} N_L \geq 20 \\ 0.7 \lesssim Pr \lesssim 500 \\ 10 \lesssim Re_{D,\text{max}} \lesssim 2 \times 10^6 \end{bmatrix}$$

where N_L is the number of tube rows, all properties except Pr_s are evaluated at the arithmetic mean of the fluid inlet ($T_i = T_\infty$) and outlet (T_o) temperatures, and the constants C_1 and m are listed in Table 7.5. The need to evaluate fluid properties at the arithmetic mean of the inlet and outlet temperatures is dictated by the fact that the fluid temperature will decrease or increase, respectively, due to heat transfer to or from the tubes. If the change of the mean

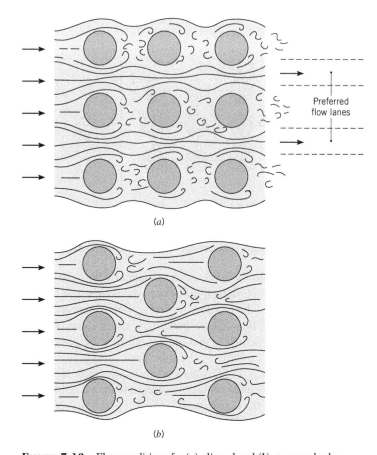

FIGURE 7.13   Flow conditions for (*a*) aligned and (*b*) staggered tubes.

TABLE 7.5   Constants of Equation 7.58 for the tube bank
in cross flow [17]

Configuration	$Re_{D,\max}$	C_1	m
Aligned	10–10^2	0.80	0.40
Staggered	10–10^2	0.90	0.40
Aligned	10^2–10^3	Approximate as a single	
Staggered	10^2–10^3	(isolated) cylinder	
Aligned $(S_T/S_L > 0.7)^a$	10^3–2×10^5	0.27	0.63
Staggered $(S_T/S_L < 2)$	10^3–2×10^5	$0.35(S_T/S_L)^{1/5}$	0.60
Staggered $(S_T/S_L > 2)$	10^3–2×10^5	0.40	0.60
Aligned	2×10^5–2×10^6	0.021	0.84
Staggered	2×10^5–2×10^6	0.022	0.84

^{a}For $S_T/S_L < 0.7$, heat transfer is inefficient and aligned tubes should not be used.

fluid temperature, $|T_i - T_o|$, is large, significant error could result from the evaluation of the properties at the inlet temperature.

If there are 20 or fewer rows of tubes, $N_L \lesssim 20$, the average heat transfer coefficient is typically reduced relative to banks with more tube rows, and a correction factor may be applied such that

$$\overline{Nu}_D\big|_{(N_L<20)} = C_2\overline{Nu}_D\big|_{(N_L\geq20)} \tag{7.59}$$

where C_2 is given in Table 7.6.

The Reynolds number $Re_{D,max}$ for the foregoing correlation is based on the *maximum fluid velocity* occurring within the tube bank, $Re_{D,max} \equiv \rho V_{max}D/\mu$. For the aligned arrangement, V_{max} occurs at the transverse plane A_1 of Figure 7.12a, and from the mass conservation requirement for a constant density fluid

$$V_{max} = \frac{S_T}{S_T - D}V \tag{7.60}$$

For the staggered configuration, the maximum velocity may occur at either the transverse plane A_1 or the diagonal plane A_2 of Figure 7.12b. It will occur at A_2 if the rows are spaced such that

$$2(S_D - D) < (S_T - D)$$

The factor of 2 results from the bifurcation experienced by the fluid moving from the A_1 to the A_2 planes. Hence V_{max} occurs at A_2 if

$$S_D = \left[S_L^2 + \left(\frac{S_T}{2}\right)^2\right]^{1/2} < \frac{S_T + D}{2}$$

in which case it is given by

$$V_{max} = \frac{S_T}{2(S_D - D)}V \tag{7.61}$$

If V_{max} occurs at A_1 for the staggered configuration, it may again be computed from Equation 7.60.

TABLE 7.6   **Correction factor C_2 of Equation 7.59 for $N_L < 20$ $(Re_{D,max} \gtrsim 10^3)$ [17]**

N_L	1	2	3	4	5	7	10	13	16
Aligned	0.70	0.80	0.86	0.90	0.92	0.95	0.97	0.98	0.99
Staggered	0.64	0.76	0.84	0.89	0.92	0.95	0.97	0.98	0.99

Since the fluid may experience a large change in temperature as it moves through the tube bank, the heat transfer rate could be significantly overpredicted by using $\Delta T = T_s - T_\infty$ as the temperature difference in Newton's law of cooling. As the fluid moves through the bank, its temperature approaches T_s and $|\Delta T|$ decreases. In Chapter 8, the appropriate form of ΔT is shown to be a *log-mean temperature difference*,

$$\Delta T_{lm} = \frac{(T_s - T_i) - (T_s - T_o)}{\ln\left(\dfrac{T_s - T_i}{T_s - T_o}\right)} \tag{7.62}$$

where T_i and T_o are temperatures of the fluid as it enters and leaves the bank, respectively. The outlet temperature, which is needed to determine ΔT_{lm}, may be estimated from

$$\frac{T_s - T_o}{T_s - T_i} = \exp\left(-\frac{\pi D N \bar{h}}{\rho V N_T S_T c_p}\right) \tag{7.63}$$

where N is the total number of tubes in the bank and N_T is the number of tubes in each row. Once ΔT_{lm} is known, the heat transfer rate per unit length of the tubes may be computed from

$$q' = N(\bar{h}\pi D \Delta T_{lm}) \tag{7.64}$$

Additional results, obtained for specific values of S_T/D and S_L/D are reported by Zukauskas [17] and Grimison [21]. The results of Grimison are restricted to air as the cross-flowing fluid, and predicted values of the average Nusselt numbers generated by the correlations of the two references agree to within approximately 15% over a broad range of $Re_{D,max}$. The foregoing results may be used to determine mass transfer rates associated with evaporation or sublimation from the surfaces of a bank of cylinders in cross flow. Once again it is only necessary to replace $\overline{Nu}_D$ and Pr by $\overline{Sh}_D$ and Sc, respectively.

We close by recognizing that there is generally as much interest in the pressure drop associated with flow across a tube bank as in the overall heat transfer rate. The power required to move the fluid across the bank is often a major operating expense and is directly proportional to the pressure drop, which may be expressed as [17]

$$\Delta p = N_L \chi \left(\frac{\rho V_{max}^2}{2}\right) f \tag{7.65}$$

The friction factor f and the correction factor χ are plotted in Figures 7.14 and 7.15. Figure 7.14 pertains to a square, in-line tube arrangement for which the dimensionless longitudinal and transverse pitches, $P_L \equiv S_L/D$ and $P_T \equiv S_T/D$, respectively, are equal. The correction factor χ, plotted in the inset, is used to apply the results to other in-line arrangements. Similarly, Figure 7.15 applies to a staggered arrangement of tubes in the form of an equilateral triangle ($S_T = S_D$), and the correction factor enables extension of the results to other staggered arrangements. Note that the Reynolds number appearing in Figures 7.14 and 7.15 is based on the maximum fluid velocity V_{max}.

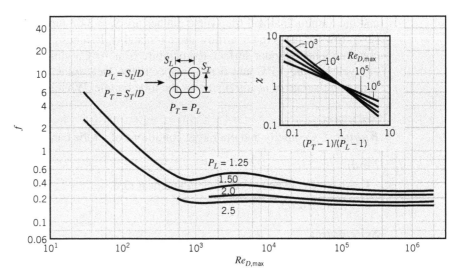

FIGURE 7.14 Friction factor f and correction factor χ for Equation 7.65. In-line tube bundle arrangement [17]. (Used with permission.)

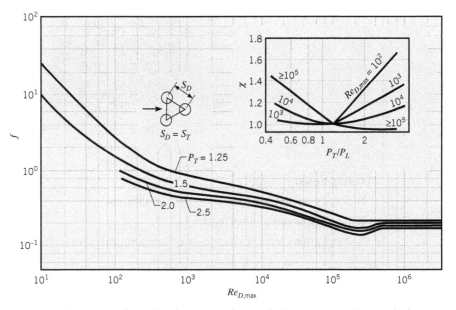

FIGURE 7.15 Friction factor f and correction factor χ for Equation 7.65. Staggered tube bundle arrangement [17]. (Used with permission.)

EXAMPLE 7.7

Pressurized water is often available at elevated temperatures and may be used for space heating or industrial process applications. In such cases it is customary to use a tube bundle in which the water is passed through the tubes, while air is passed in cross flow over the tubes. Consider a staggered arrangement for which the tube outside diameter is 16.4 mm and the longitudinal and transverse pitches are $S_L = 34.3$ mm and $S_T = 31.3$ mm. There are seven rows of tubes in the airflow direction and eight tubes per row. Under typical operating conditions the cylinder surface temperature is at 70°C, while the air upstream temperature and velocity are 15°C and 6 m/s, respectively. Determine the air-side convection coefficient and the rate of heat transfer for the tube bundle. What is the air-side pressure drop?

SOLUTION

Known: Geometry and operating conditions of a tube bank.

Find:

1. Air-side convection coefficient and heat rate.
2. Pressure drop.

Schematic:

Assumptions:

1. Steady-state, incompressible flow conditions.
2. Negligible radiation effects.
3. Negligible effect of change in air temperature across tube bank on air properties.

Properties: Table A.4, air $((T_i + T_o)/2 \approx 15°C)$: $\rho = 1.217$ kg/m^3, $c_p = 1007$ J/kg · K, $\nu = 14.82 \times 10^{-6}$ m^2/s, $k = 0.0253$ W/m · K, $Pr = 0.710$. Table A.4, air $(T_s = 70°C)$: $Pr = 0.701$.

Analysis:

1. From Equations 7.58 and 7.59, the air-side Nusselt number is

$$\overline{Nu}_D = C_2 C_1 Re_{D,\max}^m \, Pr^{0.36} \left(\frac{Pr}{Pr_s} \right)^{1/4}$$

Since $S_D = [S_L^2 + (S_T/2)^2]^{1/2} = 37.7$ mm is greater than $(S_T + D)/2$, the maximum velocity occurs on the transverse plane, A_1, of Figure 7.12. Hence from Equation 7.60

$$V_{\max} = \frac{S_T}{S_T - D} V = \frac{31.3 \text{ mm}}{(31.3 - 16.4) \text{ mm}} 6 \text{ m/s} = 12.6 \text{ m/s}$$

With

$$Re_{D,\max} = \frac{V_{\max}D}{\nu} = \frac{12.6 \text{ m/s} \times 0.0164 \text{ m}}{14.82 \times 10^{-6} \text{ m}^2/\text{s}} = 13,943$$

and

$$\frac{S_T}{S_L} = \frac{31.3 \text{ mm}}{34.3 \text{ mm}} = 0.91 < 2$$

it follows from Tables 7.5 and 7.6 that

$$C_1 = 0.35\left(\frac{S_T}{S_L}\right)^{1/5} = 0.34, \quad m = 0.60, \quad \text{and} \quad C_2 = 0.95$$

Hence

$$\overline{Nu}_D = 0.95 \times 0.34(13,943)^{0.60}(0.71)^{0.36}\left(\frac{0.710}{0.701}\right)^{0.25} = 87.9$$

and

$$\bar{h} = \overline{Nu}_D \frac{k}{D} = 87.9 \times \frac{0.0253 \text{ W/m} \cdot \text{K}}{0.0164 \text{ m}} = 135.6 \text{ W/m}^2 \cdot \text{K} \qquad \triangleleft$$

From Equation 7.63

$$T_s - T_o = (T_s - T_i) \exp\left(-\frac{\pi D N \bar{h}}{\rho V N_T S_T c_p}\right)$$

$$T_s - T_o = (55°\text{C}) \exp\left(-\frac{\pi(0.0164 \text{ m}) \, 56 \, (135.6 \text{ W/m}^2 \cdot \text{K})}{1.217 \text{ kg/m}^3 (6 \text{ m/s}) \, 8 \, (0.0313 \text{ m}) \, 1007 \text{ J/kg} \cdot \text{K}}\right)$$

$$T_s - T_o = 44.5°\text{C}$$

Hence from Equations 7.62 and 7.64

$$\Delta T_{\text{lm}} = \frac{(T_s - T_i) - (T_s - T_o)}{\ln\left(\dfrac{T_s - T_i}{T_s - T_o}\right)} = \frac{(55 - 44.5)°\text{C}}{\ln\left(\dfrac{55}{44.5}\right)} = 49.6°\text{C}$$

and

$$q' = N(\bar{h}\pi D \Delta T_{\text{lm}}) = 56\pi \times 135.6 \text{ W/m}^2 \cdot \text{K} \times 0.0164 \text{ m} \times 49.6°\text{C}$$

$$q' = 19.4 \text{ kW/m} \qquad \triangleleft$$

2. The pressure drop may be obtained from Equation 7.65.

$$\Delta p = N_L \chi \left(\frac{\rho V_{\max}^2}{2}\right) f$$

With $Re_{D,\max} = 13,943$, $P_T = (S_T/D) = 1.91$, $P_L = (S_L/D) = 2.09$, and $(P_T/P_L) = 0.91$, it follows from Figure 7.15 that $\chi \approx 1.04$ and $f \approx 0.35$. Hence with $N_L = 7$

$$\Delta p = 7 \times 1.04 \left[\frac{1.217 \text{ kg/m}^3 (12.6 \text{ m/s})^2}{2}\right] 0.35$$

$$\Delta p = 246 \text{ N/m}^2 = 2.46 \times 10^{-3} \text{ bars} \qquad \triangleleft$$

Comments:

1. Had $\Delta T_i \equiv T_s - T_i$ been used in lieu of ΔT_{lm} in Equation 7.64, the heat rate would have been overpredicted by 11%.

2. Since the air temperature is predicted to increase by only 10.5°C, evaluation of the air properties at $T_i = 15$°C is a reasonable approximation. However, if improved accuracy is desired, the calculations could be repeated with the properties reevaluated at $(T_i + T_o)/2 = 20.25$°C. An exception pertains to the density ρ in the exponential term of Equation 7.63. As it appears in the denominator of this term, ρ is matched with the inlet velocity to provide a product (ρV) that is linked to the mass flow rate of air entering the tube bank. Hence, in this term, ρ should be evaluated at T_i.

3. The air outlet temperature and heat rate may be increased by increasing the number of tube rows, and for a fixed number of rows, they may be varied by adjusting the air velocity. For $5 \leq N_L \leq 25$ and $V = 6$ m/s, parametric calculations based on Equations 7.58, 7.59, and 7.62 through 7.64 yield the following results:

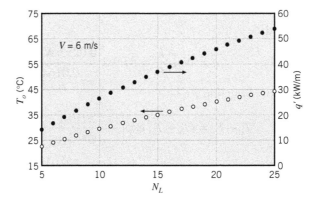

The air outlet temperature would asymptotically approach the surface temperature with increasing N_L, at which point the heat rate approaches a constant value and there is no advantage to adding more tube rows. Note that Δp increases linearly with increasing N_L. For $N_L = 25$ and $1 \leq V \leq 20$ m/s, we obtain

Although the heat rate increases with increasing V, the air outlet temperature decreases, approaching T_i as $V \to \infty$.

7.7 *Impinging Jets*

A single gas jet or an array of such jets, impinging normally on a surface, may be used to achieve enhanced coefficients for convective heating, cooling, or drying. Applications include tempering of glass plate, annealing of metal sheets, drying of textile and paper products, cooling of heated components in gas turbine engines, and deicing of aircraft systems.

7.7.1 Hydrodynamic and Geometric Considerations

As shown in Figure 7.16, gas jets are typically discharged into a quiescent ambient from a round nozzle of diameter D or a slot (rectangular) nozzle of width W. Typically, the jet is turbulent and, at the nozzle exit, is characterized by a uniform velocity profile. However, with increasing distance from the exit, momentum exchange between the jet and the ambient causes the free boundary of the jet to broaden and the *potential core,* within which the uniform exit velocity is retained, to contract. Downstream of the potential core the velocity profile is nonuniform over the entire jet cross section and the maximum (center) velocity decreases with increasing distance from the nozzle exit. The region of the flow over which conditions are unaffected by the impingement (target) surface is termed the *free jet.*

Within the *stagnation* or *impingement zone,* flow is influenced by the target surface and is decelerated and accelerated in the normal (z) and transverse $(r$ or $x)$ directions, respectively. However, since the flow continues to entrain zero momentum fluid from the ambient, transverse acceleration cannot continue indefinitely and accelerating flow in the stagnation zone is transformed to a decelerating *wall jet.* Hence, with increasing r or x, velocity components parallel to the surface increase from a value of zero to some maximum and

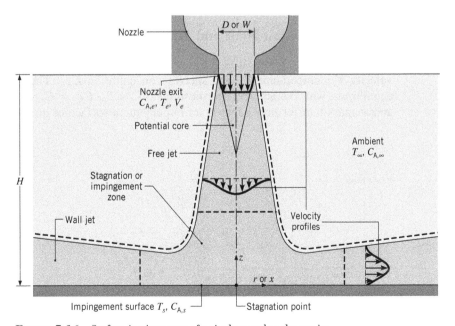

FIGURE 7.16 Surface impingement of a single round or slot gas jet.

subsequently decay to zero. Velocity profiles within the wall jet are characterized by zero velocity at both the impingement and free surfaces. If $T_s \neq T_e$ and/or $C_{A,s} \neq C_{A,e}$, convection heat and/or mass transfer occurs in both the stagnation and wall jet regions.

Many impingement heat (mass) transfer schemes involve an array of jets, as, for example, the array of slot jets shown in Figure 7.17. In addition to flow from each nozzle exhibiting free jet, stagnation, and wall jet regions, secondary stagnation zones result from the interaction of adjoining wall jets. In many such schemes the jets are discharged into a restricted volume bounded by the target surface and the nozzle plate from which the jets originate. The overall rate of heat (mass) transfer depends strongly on the manner in which *spent gas,* whose temperature (species concentration) is between values associated with the nozzle exit and the impingement surface, is vented from the system. For the configuration of Figure 7.17, spent gas cannot flow upward between the nozzles but must instead flow symmetrically in the $\pm y$-directions. As the temperature (surface cooling) or species concentration (surface evaporation) of the gas increases with increasing $|y|$, the local surface-to-gas temperature or concentration difference decreases, causing a reduction in local convection fluxes. A preferable situation is one for which the space between adjoining nozzles is open to the ambient, thereby permitting continuous upflow and direct discharge of the spent gas.

Plan (top) views of single round and slot nozzles, as well as regular arrays of round and slot nozzles, are shown in Figure 7.18. For the isolated nozzles (Figures 7.18*a, d*), local and average convection coefficients are associated with any $r > 0$ and $x > 0$. For the arrays, with discharge of the spent gas in the vertical (z) direction, symmetry dictates equivalent local and average values for each of the unit cells delineated by dashed lines. For a large number of square-in-line (Figure 7.18*b*) or equilaterally staggered (Figure 7.18*c*) round jets, the unit cells correspond to a square or hexagon, respectively. A pertinent geometric parameter is the relative nozzle area, which is defined as the ratio of the nozzle exit cross-sectional area to the surface area of the cell ($A_r \equiv A_{c,e}/A_{cell}$). In each case, S represents the pitch of the array.

7.7.2 Convection Heat and Mass Transfer

In the results that follow, it is presumed that the gas jet exits its nozzle with a uniform velocity V_e, temperature T_e, and species concentration $C_{A,e}$. Thermal and compositional equilibrium with the ambient are presumed ($T_e = T_\infty$, $C_{A,e} = C_{A,\infty}$), while convection heat and/or mass transfer may occur at an impingement surface of uniform temperature and/or

FIGURE 7.17 Surface impingement of an array of slot jets.

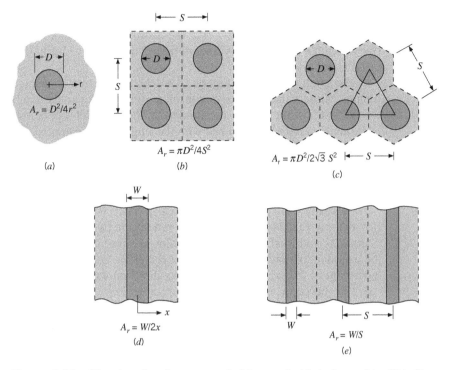

FIGURE 7.18 Plan view of pertinent geometrical features for (*a*) single round jet, (*b*) in-line array of round jets, (*c*) staggered array of round jets, (*d*) single slot jet, and (*e*) array of slot jets.

species composition ($T_s \neq T_e$, $C_{A,s} \neq C_{A,e}$). Newton's law of cooling and its mass transfer analog are then

$$q'' = h(T_s - T_e) \tag{7.66}$$

$$N_A'' = h_m(C_{A,s} - C_{A,e}) \tag{7.67}$$

Conditions are presumed to be uninfluenced by the level of turbulence at the nozzle exit, and the surface is presumed to be stationary. However, this requirement may be relaxed for surface velocities which are much less than the jet velocity.

An extensive review of available convection coefficient data for impinging gas jets has been performed by Martin [22], and for a single round or slot nozzle, distributions of the *local* Nusselt number have the characteristic forms shown in Figure 7.19. The characteristic length is the *hydraulic diameter* of the nozzle, which is defined as four times its cross-sectional area divided by its wetted perimeter ($D_h \equiv 4A_{c,e}/P$). Hence the characteristic length is the diameter of a round nozzle, and assuming $L \gg W$, it is twice the width of a slot nozzle. It follows

FIGURE 7.19 Distribution of local Nusselt number associated with a single round or slot nozzle for (*a*) large and (*b*) small relative nozzle-to-plate spacings.

that $Nu = hD/k$ for a round nozzle and $Nu = h(2W/k)$ for a slot nozzle. For large nozzle-to-plate separations, Figure 7.19*a*, the distribution is characterized by a bell-shaped curve for which Nu monotonically decays from a maximum value at the *stagnation point, r/D or x/2W* equal to zero.

For small separations, Figure 7.19*b*, the distribution is characterized by a second maximum, whose value increases with increasing jet Reynolds number and may exceed that of the first maximum. The threshold separation of $H/D \approx 5$, below which there is a second maximum, is loosely associated with the length of the potential core (Figure 7.16). Appearance of the second maximum is attributed to a sharp rise in the turbulence level which accompanies the transition from an accelerating stagnation region flow to a decelerating wall jet [22]. Additional maxima have been observed and attributed to the formation of vortices in the stagnation zone, as well as transition to a turbulent wall jet [23].

Secondary maxima in Nu are also associated with the interaction of adjoining wall jets for an array [22, 24]. However, distributions are two-dimensional, exhibiting, for example, variations with both x and y for the slot jet array of Figure 7.17. Variations with x could be expected to yield maxima at the jet centerline and halfway between adjoining jets, while constraint of the exhaust flow to the $\pm y$-direction would induce acceleration with increasing $|y|$ and hence a monotonically increasing Nu with $|y|$. However, variations with y decrease with increasing cross-sectional area of the outflow and may be neglected if $S \times H \gtrsim W \times L$ [22].

Average Nusselt numbers may be obtained by integrating local results over the appropriate surface area. The resulting correlations are reported in the form

$$\overline{Nu} = f(Re, Pr, A_r, H/D_h) \tag{7.68}$$

where

$$\overline{Nu} \equiv \frac{\bar{h}D_h}{k} \tag{7.69}$$

$$Re = \frac{V_e D_h}{\nu} \tag{7.70}$$

and $D_h = D$ (round nozzle) or $D_h = 2W$ (slot nozzle).

Round Nozzles   Having assessed data from several sources, Martin [22] recommends the following correlation for a *single round nozzle* $(A_r = D^2/4r^2)$

$$\frac{\overline{Nu}}{Pr^{0.42}} = G\left(A_r, \frac{H}{D}\right)[2\, Re^{1/2}(1 + 0.005\, Re^{0.55})^{1/2}] \tag{7.71}$$

where

$$G = 2A_r^{1/2}\frac{1 - 2.2A_r^{1/2}}{1 + 0.2(H/D - 6)A_r^{1/2}} \tag{7.72}$$

The ranges of validity are

$$\begin{bmatrix} 2000 \lesssim Re \lesssim 400{,}000 \\ 2 \lesssim H/D \lesssim 12 \\ 0.004 \lesssim A_r \lesssim 0.04 \end{bmatrix}$$

For $A_r \gtrsim 0.04$, results for $\overline{Nu}$ are available in graphical form [22].

For an *array of round nozzles* ($A_r = \pi D^2/4S^2$ or $\pi D^2/2\sqrt{3}S^2$ for in-line and staggered arrays, respectively),

$$\frac{\overline{Nu}}{Pr^{0.42}} = 0.5\, K\left(A_r, \frac{H}{D}\right) G\left(A_r, \frac{H}{D}\right) Re^{2/3} \qquad (7.73)$$

where

$$K = \left[1 + \left(\frac{H/D}{0.6/A_r^{1/2}}\right)^6\right]^{-0.05} \qquad (7.74)$$

and G is the single nozzle function given by Equation 7.72. The function K accounts for the fact that, for $H/D \gtrsim 0.6/A_r^{1/2}$, the average Nusselt number for the array decays more rapidly with increasing H/D than that for the single nozzle. The correlation is valid over the ranges

$$\left[\begin{array}{l} 2000 \lesssim Re \lesssim 100{,}000 \\ 2 \lesssim H/D \lesssim 12 \\ 0.004 \lesssim A_r \lesssim 0.04 \end{array}\right]$$

Slot Nozzles For a *single slot nozzle* ($A_r = W/2x$), the recommended correlation is

$$\frac{\overline{Nu}}{Pr^{0.42}} = \frac{3.06}{0.5/A_r + H/W + 2.78} Re^m \qquad (7.75)$$

where

$$m = 0.695 - \left[\left(\frac{1}{4A_r}\right) + \left(\frac{H}{2W}\right)^{1.33} + 3.06\right]^{-1} \qquad (7.76)$$

and the ranges of validity are

$$\left[\begin{array}{l} 3000 \lesssim Re \lesssim 90{,}000 \\ 2 \lesssim H/W \lesssim 10 \\ 0.025 \lesssim A_r \lesssim 0.125 \end{array}\right]$$

As a *first approximation*, Equation 7.75 may be used for $A_r \gtrsim 0.125$, yielding predictions for the stagnation point ($x = 0$, $A_r \rightarrow \infty$) that are within 40% of measured values.

For an *array of slot nozzles* ($A_r = W/S$), the recommended correlation is

$$\frac{\overline{Nu}}{Pr^{0.42}} = \frac{2}{3} A_{r,o}^{3/4}\left(\frac{2\,Re}{A_r/A_{r,o} + A_{r,o}/A_r}\right)^{2/3} \qquad (7.77)$$

where

$$A_{r,o} = \left[60 + 4\left(\frac{H}{2W} - 2\right)^2\right]^{-1/2} \qquad (7.78)$$

The correlation pertains to conditions for which the outflow of spent gas is restricted to the $\pm y$-directions of Figure 7.17 and the outflow area is large enough to satisfy the requirement $(S \times H)/(W \times L) \gtrsim 1$. Additional restrictions are

$$
\begin{bmatrix}
1500 \lesssim Re \lesssim 40{,}000 \\
2 \lesssim H/W \lesssim 80 \\
0.008 \lesssim A_r \lesssim 2.5 A_{r,o}
\end{bmatrix}
$$

An *optimal* arrangement of nozzles would be one for which the values of H, S, and D_h yielded the largest value of $\overline{Nu}$ for a prescribed total gas flow rate per unit surface area of the target. For fixed H and for arrays of both round and slot nozzles, optimal values of D_h and S have been found to be [22]

$$D_{h,\text{op}} \approx 0.2H \tag{7.79}$$

$$S_{\text{op}} \approx 1.4H \tag{7.80}$$

The optimum value of $(D_h/H)^{-1} \approx 5$ coincides approximately with the length of the potential core. Beyond the potential core, the midline jet velocity decays, causing an attendant reduction in convection coefficients.

Invoking the heat and mass transfer analogy by substituting $\overline{Sh}/Sc^{0.42}$ for $\overline{Nu}/Pr^{0.42}$, the foregoing correlations may also be applied to convection mass transfer. However, for both heat and mass transfer, application of the equations should be restricted to conditions for which they were developed. For example, in their present form, the correlations may not be used if the jets emanate from sharp-edged orifices instead of bell-shaped nozzles. The orifice jet is strongly affected by a flow contraction phenomenon that alters convection heat or mass transfer [22, 23]. In the case of convection heat transfer, conditions are also influenced by differences between the jet exit and ambient temperatures ($T_e \neq T_\infty$). The exit temperature is then an inappropriate temperature in Newton's law of cooling, Equation 7.66, and should be replaced by what is commonly termed the recovery, or adiabatic wall, temperature [25, 26]. Finally, care should be taken in situations involving small nozzle diameters or narrow slot widths. For these situations, high jet velocities are necessary for Reynolds numbers to be within the range of application of Equations 7.71, 7.73, 7.75, or 7.77. When the Mach number based on the jet velocity exceeds 0.3 ($V_e/a \gtrsim 0.3$), compressibility effects may be significant [27], invalidating use of the correlations of this section. Additional correlations for liquid jet impingement, flame jet impingement, and impingement on nonflat surfaces are available [28–30].

7.8 *Packed Beds*

Gas flow through a *packed bed* of solid particles (Figure 7.20) is relevant to many industrial processes, which include the transfer and storage of thermal energy, heterogeneous catalytic reactions, and drying. The term *packed bed* refers to a condition for which the position of the particles is *fixed*. In contrast, a *fluidized bed* is one for which the particles are in motion due to advection with the fluid.

For a packed bed a large amount of heat or mass transfer surface area can be obtained in a small volume, and the irregular flow that exists in the voids of the bed enhances transport

FIGURE 7.20   Gas flow through a packed bed of solid particles.

through mixing. Many correlations that have been developed for different particle shapes, sizes, and packing densities are described in the literature [31–34]. One such correlation, which has been recommended for gas flow in a bed of spheres, is of the form

$$\varepsilon \bar{j}_H = \varepsilon \bar{j}_m = 2.06 \, Re_D^{-0.575} \begin{bmatrix} Pr(\text{or } Sc) \approx 0.7 \\ 90 \lesssim Re_D \lesssim 4000 \end{bmatrix} \tag{7.81}$$

where $\bar{j}_H$ and $\bar{j}_m$ are the Colburn j factors defined by Equations 6.70 and 6.71. The Reynolds number $Re_D = VD/\nu$ is defined in terms of the sphere diameter and the upstream velocity V that would exist in the empty channel without the packing. The quantity ε is the *porosity*, or *void fraction*, of the bed (volume of void space per unit volume of bed), and its value typically ranges from 0.30 to 0.50. The correlation may be applied to packing materials other than spheres by multiplying the right-hand side by an appropriate correction factor. For a bed of uniformly sized cylinders, with length-to-diameter ratio of 1, the factor is 0.79; for a bed of cubes it is 0.71.

In using Equation 7.81, properties should be evaluated at the arithmetic mean of the fluid temperatures entering and leaving the bed. If the particles are at a uniform temperature T_s, the heat transfer rate for the bed may be computed from

$$q = \bar{h} A_{p,t} \Delta T_{\text{lm}} \tag{7.82}$$

where $A_{p,t}$ is the total surface area of the particles and ΔT_{lm} is the log-mean temperature difference defined by Equation 7.62. The outlet temperature, which is needed to compute ΔT_{lm}, may be estimated from

$$\frac{T_s - T_o}{T_s - T_i} = \exp\left(-\frac{\bar{h} A_{p,t}}{\rho V A_{c,b} c_p}\right) \tag{7.83}$$

where ρ and V are the inlet density and velocity, respectively, and $A_{c,b}$ is the bed (channel) cross-sectional area.

7.9   *Summary*

In this chapter we have considered *forced convection* heat and mass transfer for an important class of problems involving *external flow* at low-to-moderate speeds. Consideration was given to several common geometries, for which convection coefficients depend on the nature of boundary layer development. You should test your understanding of related concepts by addressing the following questions.

- What is an *external flow*?
- What is an empirical heat or mass transfer coefficient?
- What are the inherent dimensionless parameters for forced convection?
- How does the velocity boundary layer thickness vary with distance from the leading edge for laminar flow over a flat plate? For turbulent flow? What determines the relative velocity, thermal, and concentration boundary layer thicknesses for laminar flow? For turbulent flow?
- How does the *local* convection heat or mass transfer coefficient vary with distance from the leading edge for *laminar flow* over a flat plate? For *turbulent flow*? For flow in which *transition* to turbulence occurs on the plate?
- How is local heat transfer from the surface of a flat plate affected by the existence of an *unheated starting length*?
- What are the manifestations of *boundary layer separation* from the surface of a circular cylinder in cross flow? How is separation influenced by whether the upstream flow is laminar or turbulent?
- How is variation of the local convection coefficient on the surface of a circular cylinder in cross flow affected by boundary layer separation? By boundary layer transition? Where do local maxima and minima in the convection coefficient occur on the surface?
- How does the average convection coefficient of a tube vary with its location in a tube bank?
- For jet impingement on a surface, what are distinguishing features of the *free jet*? The *potential core*? The *impingement zone*? The *wall jet*?
- At what location on the surface of an impinging jet will a maximum in the convection coefficient always exist? Under what conditions will there be a secondary maximum?
- For an *array* of impinging jets, how are flow and heat transfer affected by the manner in which *spent fluid* is discharged from the system?
- What is the difference between a *packed bed* and a *fluidized bed* of solid particles?
- What is the *film temperature*?
- What temperature difference must be used when computing the total rate of heat transfer from a bank of tubes or a packed bed?

In this chapter we have also compiled convection correlations that may be used to estimate convection transfer rates for a variety of external flow conditions. For simple surface geometries these results may be derived from a boundary layer analysis, but in most cases they are obtained from generalizations based on experiment. You should know when and how to use the various expressions, and you should be familiar with the general methodology of a convection calculation. To facilitate their use, the correlations are summarized in Table 7.7.

TABLE 7.7   Summary of convection heat transfer correlations for external flow[a,b]

Correlation		Geometry	Conditions[c]
$\delta = 5x\,Re_x^{-1/2}$	(7.19)	Flat plate	Laminar, T_f
$C_{f,x} = 0.664\,Re_x^{-1/2}$	(7.20)	Flat plate	Laminar, local, T_f
$Nu_x = 0.332\,Re_x^{1/2}\,Pr^{1/3}$	(7.23)	Flat plate	Laminar, local, T_f, $Pr \gtrsim 0.6$
$\delta_t = \delta\,Pr^{-1/3}$	(7.24)	Flat plate	Laminar, T_f
$\overline{C}_{f,x} = 1.328\,Re_x^{-1/2}$	(7.29)	Flat plate	Laminar, average, T_f

TABLE 7.7 *(Continued)*

Correlation		Geometry	Conditions[c]
$\overline{Nu}_x = 0.664\,Re_x^{1/2}\,Pr^{1/3}$	(7.30)	Flat plate	Laminar, average, T_f, $Pr \gtrsim 0.6$
$Nu_x = 0.564\,Pe_x^{1/2}$	(7.32)	Flat plate	Laminar, local, T_f, $Pr \lesssim 0.05$, $Pe_x \gtrsim 100$
$C_{f,x} = 0.0592\,Re_x^{-1/5}$	(7.34)	Flat plate	Turbulent, local, T_f, $Re_x \lesssim 10^8$
$\delta = 0.37x\,Re_x^{-1/5}$	(7.35)	Flat plate	Turbulent, T_f, $Re_x \lesssim 10^8$
$Nu_x = 0.0296\,Re_x^{4/5}\,Pr^{1/3}$	(7.36)	Flat plate	Turbulent, local, T_f, $Re_x \lesssim 10^8$, $0.6 \lesssim Pr \lesssim 60$
$\overline{C}_{f,L} = 0.074\,Re_L^{-1/5} - 1742\,Re_L^{-1}$	(7.40)	Flat plate	Mixed, average, T_f, $Re_{x,c} = 5 \times 10^5$, $Re_L \lesssim 10^8$
$\overline{Nu}_L = (0.037\,Re_L^{4/5} - 871)\,Pr^{1/3}$	(7.38)	Flat plate	Mixed, average, T_f, $Re_{x,c} = 5 \times 10^5$, $Re_L \lesssim 10^8$, $0.6 \lesssim Pr \lesssim 60$
$\overline{Nu}_D = C\,Re_D^m\,Pr^{1/3}$ (Table 7.2)	(7.52)	Cylinder	Average, T_f, $0.4 \lesssim Re_D \lesssim 4 \times 10^5$, $Pr \gtrsim 0.7$
$\overline{Nu}_D = C\,Re_D^m\,Pr^n(Pr/Pr_s)^{1/4}$ (Table 7.4)	(7.53)	Cylinder	Average, T_∞, $1 \lesssim Re_D \lesssim 10^6$, $0.7 \lesssim Pr \lesssim 500$
$\overline{Nu}_D = 0.3 + [0.62\,Re_D^{1/2}\,Pr^{1/3}$ $\times[1 + (0.4/Pr)^{2/3}]^{-1/4}]$ $\times[1 + (Re_D/282{,}000)^{5/8}]^{4/5}$	(7.54)	Cylinder	Average, T_f, $Re_D\,Pr \gtrsim 0.2$
$\overline{Nu}_D = 2 + (0.4\,Re_D^{1/2}$ $+ 0.06\,Re_D^{2/3})Pr^{0.4}$ $\times(\mu/\mu_s)^{1/4}$	(7.56)	Sphere	Average, T_∞, $3.5 \lesssim Re_D \lesssim 7.6 \times 10^4$, $0.71 \lesssim Pr \lesssim 380$, $1.0 \lesssim (\mu/\mu_s) \lesssim 3.2$
$\overline{Nu}_D = 2 + 0.6\,Re_D^{1/2}\,Pr^{1/3}$	(7.57)	Falling drop	Average, T_∞
$\overline{Nu}_D = C_1 C_2\,Re_{D,\text{max}}^m\,Pr^{0.36}(Pr/Pr_s)^{1/4}$ (Tables 7.5, 7.6)	(7.58), (7.59)	Tube bank[d]	Average, $\overline{T}$, $10 \lesssim Re_D \lesssim 2 \times 10^6$, $0.7 \lesssim Pr \lesssim 500$
Single round nozzle	(7.71)	Impinging jet	Average, T_f, $2000 \lesssim Re \lesssim 4 \times 10^5$, $2 \lesssim (H/D) \lesssim 12$, $2.5 \lesssim (r/D) \lesssim 7.5$
Single slot nozzle	(7.75)	Impinging jet	Average, T_f, $3000 \lesssim Re \lesssim 9 \times 10^4$, $2 \lesssim (H/W) \lesssim 10$, $4 \lesssim (x/W) \lesssim 20$
Array of round nozzles	(7.73)	Impinging jet	Average, T_f, $2000 \lesssim Re \lesssim 10^5$, $2 \lesssim (H/D) \lesssim 12$, $0.004 \lesssim A_r \lesssim 0.04$
Array of slot nozzles	(7.77)	Impinging jet	Average, T_f, $1500 \lesssim Re \lesssim 4 \times 10^4$, $2 \lesssim (H/W) \lesssim 80$, $0.008 \lesssim A_r \lesssim 2.5A_{r,o}$
$\varepsilon\overline{j}_H = \varepsilon\overline{j}_m = 2.06\,Re_D^{-0.575}$	(7.81)	Packed bed of spheres[d]	Average, $\overline{T}$, $90 \lesssim Re_D \lesssim 4000$, Pr (or Sc) ≈ 0.7

[a]Correlations in this table pertain to isothermal surfaces; for special cases involving an unheated starting length or a uniform surface heat flux, see Section 7.2.4 or 7.2.5.

[b]When the heat and mass transfer analogy is applicable, the corresponding mass transfer correlations may be obtained by replacing Nu and Pr by Sh and Sc, respectively.

[c]The temperature listed under "Conditions" is the temperature at which properties should be evaluated.

[d]For tube banks and packed beds, properties are evaluated at the average fluid temperature, $\overline{T} = (T_i + T_o)/2$.

References

1. Blasius, H., *Z. Math. Phys.*, **56**, 1, 1908. English translation in National Advisory Committee for Aeronautics Technical Memo No. 1256.

2. Schlichting, H., and K. Gersten, *Boundary Layer Theory*, Springer, New York, 2000.

3. Howarth, L., *Proc. R. Soc. Lond., Ser. A*, **164**, 547, 1938.

4. Pohlhausen, E., *Z. Angew. Math. Mech.*, **1**, 115, 1921.

5. Kays, W. M., M. E. Crawford, and B. Weigand, *Convective Heat and Mass Transfer*, 4th ed. McGraw-Hill Higher Education, Boston, 2005.

6. Churchill, S. W., and H. Ozoe, *J. Heat Transfer*, **95**, 78, 1973.

7. Brewster, M. Q., *J. Heat Transfer*, **136**, 114501–1, 2014.

8. Ameel, T. A., *Int. Comm. Heat Mass Transfer*, **24**, 1113, 1997.

9. Blair, M. F., *J. Heat Transfer*, **105**, 33 and 41, 1983.

10. Fox, R. W., A. T. McDonald, and P. J. Pritchard, *Introduction to Fluid Mechanics,* 6th ed., Wiley, New York, 2003.

11. Coutanceau, M., and J.-R. Defaye, *Appl. Mech. Rev.*, **44**, 255, 1991.

12. Hilpert, R., *Forsch. Geb. Ingenieurwes.*, **4**, 215, 1933.

13. Knudsen, J. D., and D. L. Katz, *Fluid Dynamics and Heat Transfer*, McGraw-Hill, New York, 1958.

14. Jakob, M., *Heat Transfer*, Vol. 1, Wiley, New York, 1949.

15. Sparrow, E. M., J. P. Abraham, and J. C. K. Tong, *Int. J. Heat Mass Transfer*, **47**, 5285, 2004.

16. Whitaker, S., *AIChE J.*, **18**, 361, 1972.

17. Zukauskas, A., "Heat Transfer from Tubes in Cross Flow," in J. P. Hartnett and T. F. Irvine, Jr., Eds., *Advances in Heat Transfer*, Vol. 8, Academic Press, New York, 1972.

18. Churchill, S. W., and M. Bernstein, *J. Heat Transfer*, **99**, 300, 1977.

19. Morgan, V. T., "The Overall Convective Heat Transfer from Smooth Circular Cylinders," in T. F. Irvine, Jr. and J. P. Hartnett, Eds., *Advances in Heat Transfer*, Vol. 11, Academic Press, New York, 1975.

20. Ranz, W., and W. Marshall, *Chem. Eng. Prog.*, **48**, 141, 1952.

21. Grimison, E. D., *Trans. ASME*, **59**, 583, 1937.

22. Martin, H., "Heat and Mass Transfer between Impinging Gas Jets and Solid Surfaces," in J. P. Hartnett and T. F. Irvine, Jr., Eds., *Advances in Heat Transfer*, Vol. 13, Academic Press, New York, 1977.

23. Popiel, Cz. O., and L. Bogusiawski, "Mass or Heat Transfer in Impinging Single Round Jets Emitted by a Bell-Shaped Nozzle and Sharp-Ended Orifice," in C. L. Tien, V. P. Carey, and J. K. Ferrell, Eds., *Heat Transfer 1986*, Vol. 3, Hemisphere Publishing, New York, 1986.

24. Goldstein, R. J., and J. F. Timmers, *Int. J. Heat Mass Transfer*, **25**, 1857, 1982.

25. Hollworth, B. R., and L. R. Gero, *J. Heat Transfer*, **107**, 910, 1985.

26. Goldstein, R. J., A. I. Behbahani, and K. K. Heppelman, *Int. J. Heat Mass Transfer*, **29**, 1227, 1986.

27. Pence, D. V., P. A. Boeschoten, and J. A. Liburdy, *J. Heat Transfer*, **125**, 447, 2003.

28. Webb, B. W., and C.-F. Ma, in J. P. Hartnett, T. F. Irvine, Y. I. Cho, and G. A. Greene, Eds., *Advances in Heat Transfer*, Vol. 26, Academic Press, New York, 1995.

29. Baukal, C. E., and B. Gebhart, *Int. J. Heat and Fluid Flow*, **4**, 386, 1996.

30. Chander, S., and A. Ray, *Energy Conversion and Management*, **46**, 2803, 2005.

31. Bird, R. B., W. E. Stewart, and E. N. Lightfoot, *Transport Phenomena*, 2nd ed., Wiley, New York, 2002.

32. Jakob, M., *Heat Transfer*, Vol. 2, Wiley, New York, 1957.

33. Geankopplis, C. J., *Mass Transport Phenomena*, Holt, Rinehart & Winston, New York, 1972.

34. Sherwood, T. K., R. L. Pigford, and C. R. Wilkie, *Mass Transfer*, McGraw-Hill, New York, 1975.

Problems

Flat Plate in Parallel Flow

7.1 Consider the following fluids at a film temperature of 300 K in parallel flow over a flat plate with velocity of 1 m/s: atmospheric air, water, engine oil, and mercury.

(a) For each fluid, determine the velocity and thermal boundary layer thicknesses at a distance of 40 mm from the leading edge.

(b) For each of the prescribed fluids and on the same coordinates, plot the boundary layer thicknesses as a function of distance from the leading edge to a plate length of 40 mm.

7.2 Consider steady, parallel flow of atmospheric air over a flat plate. The air has a temperature and free stream velocity of 300 K and 15 m/s.

   (a) Evaluate the boundary layer thickness at distances of $x = 1$, 10, and 100 mm from the leading edge. If a second plate were installed parallel to and at a distance of 4 mm from the first plate, what is the distance from the leading edge at which boundary layer merger would occur?

   (b) Evaluate the surface shear stress and the y-velocity component at the outer edge of the boundary layer for the single plate at $x = 1$, 10, and 100 mm.

   (c) Comment on the validity of the boundary layer approximations.

7.3 Consider a liquid metal ($Pr \ll 1$), with free stream conditions u_∞ and T_∞, in parallel flow over an isothermal flat plate at T_s. Assuming that $u = u_\infty$ throughout the thermal boundary layer, write the corresponding form of the boundary layer energy equation. Applying appropriate initial ($x = 0$) and boundary conditions, solve this equation for the boundary layer temperature field, $T(x, y)$. Use the result to obtain an expression for the local Nusselt number Nu_x. *Hint:* This problem is analogous to one-dimensional heat transfer in a semi-infinite medium with a sudden change in surface temperature.

7.4 Consider the velocity boundary layer profile for flow over a flat plate to be of the form $u = C_1 + C_2 y$. Applying appropriate boundary conditions, obtain an expression for the velocity profile in terms of the boundary layer thickness δ and the free stream velocity u_∞. Using the integral form of the boundary layer momentum equation (Appendix G), obtain expressions for the boundary layer thickness and the local friction coefficient, expressing your result in terms of the local Reynolds number. Compare your results with those obtained from the exact solution (Section 7.2.1) and the integral solution with a cubic profile (Appendix G).

7.5 Consider a steady, turbulent boundary layer on an isothermal flat plate of temperature T_s. The boundary layer is "tripped" at the leading edge $x = 0$ by a fine wire. Assume constant physical properties and velocity and temperature profiles of the form

$$\frac{u}{u_\infty} = \left(\frac{y}{\delta}\right)^{1/7} \quad \text{and} \quad \frac{T - T_\infty}{T_s - T_\infty} = 1 - \left(\frac{y}{\delta_t}\right)^{1/7}$$

   (a) From experiment it is known that the surface shear stress is related to the boundary layer thickness by an expression of the form

$$\tau_s = 0.0228 \rho u_\infty^2 \left(\frac{u_\infty \delta}{\nu}\right)^{-1/4}$$

Beginning with the momentum integral equation (Appendix G), show that

$$\delta/x = 0.376 \, Re_x^{-1/5}.$$

Determine the average friction coefficient $\overline{C}_{f,x}$.

   (b) Beginning with the energy integral equation, obtain an expression for the local Nusselt number Nu_x and use this result to evaluate the average Nusselt number $\overline{Nu}_x$.

7.6 Consider flow over a flat plate for which it is desired to determine the average heat transfer coefficient over the short span x_1 to x_2, $\overline{h}_{1-2}$, where $(x_2 - x_1) \ll L$.

Provide three different expressions that can be used to evaluate $\overline{h}_{1-2}$ in terms of (a) the local coefficient at $x = (x_1 + x_2)/2$, (b) the local coefficients at x_1 and x_2, and (c) the average coefficients at x_1 and x_2. Indicate which of the expressions is approximate. Considering whether the flow is laminar, turbulent, or mixed, indicate when it is appropriate or inappropriate to use each of the equations.

7.7 Consider atmospheric air at 25°C and a velocity of 25 m/s flowing over both surfaces of a 1-m-long flat plate that is maintained at 125°C. Determine the rate of heat transfer per unit width from the plate for values of the critical Reynolds number corresponding to 10^5, 5×10^5, and 10^6.

7.8 A flat plate of width 1 m is maintained at a uniform surface temperature of $T_s = 150°C$ by using independently controlled, heat-generating rectangular modules of thickness $a = 10$ mm and length $b = 50$ mm. Each module is insulated from its neighbors, as well as on its back side. Atmospheric air at 25°C flows over the plate at a velocity of 30 m/s. The thermophysical properties of the module are $k = 5.2$ W/m · K, $c_p = 320$ J/kg · K, and $\rho = 2300$ kg/m³.

(a) Find the required power generation, $\dot{q}$ (W/m^3), in a module positioned at a distance 700 mm from the leading edge.

(b) Find the maximum temperature T_{max} in the heat-generating module.

7.9 An electric air heater consists of a horizontal array of thin metal strips that are each 10 mm long in the direction of an airstream that is in parallel flow over the top of the strips. Each strip is 0.2 m wide, and 25 strips are arranged side by side, forming a continuous and smooth surface over which the air flows at 2 m/s. During operation, each strip is maintained at 500°C and the air is at 25°C.

(a) What is the rate of convection heat transfer from the first strip? The fifth strip? The tenth strip? All the strips?

(b) For air velocities of 2, 5, and 10 m/s, determine the convection heat rates for all the locations of part (a). Represent your results in tabular or bar graph form.

(c) Repeat part (b), but under conditions for which the flow is fully turbulent over the entire array of strips.

7.10 Consider a flat plate subject to parallel flow (top and bottom) characterized by $u_\infty = 5$ m/s, $T_\infty = 20$°C.

(a) Determine the average convection heat transfer coefficient, convective heat transfer rate, and drag force associated with an $L = 2$-m-long, $w = 2$-m-wide flat plate for airflow and surface temperatures of $T_s = 50$°C and 80°C.

(b) Determine the average convection heat transfer coefficient, convective heat transfer rate, and drag force associated with an $L = 0.1$-m-long, $w = 0.1$-m-wide flat plate for water flow and surface temperatures of $T_s = 50$°C and 80°C.

7.11 Explain under what conditions the total rate of heat transfer from an isothermal flat plate of dimensions $L \times 2L$ would be the same, independent of whether parallel flow over the plate is directed along the side of length L or $2L$. With a critical Reynolds number of 5×10^5, for what values of Re_L would the total heat transfer be independent of orientation?

7.12 In Example 7.2, it was determined that the sixth segment of the flat plate has the maximum power requirement. Which segment has the minimum power requirement if conditions are identical except the air temperature is raised to $T_\infty = 190$°C? What is this power requirement?

7.13 Consider water at 27°C in parallel flow over an isothermal, 1-m-long flat plate with a velocity of 2 m/s.

(a) Plot the variation of the local heat transfer coefficient, $h_x(x)$, with distance along the plate for three flow conditions corresponding to transition Reynolds numbers of (i) 5×10^5, (ii) 3×10^5, and (iii) 0 (the flow is fully turbulent).

(b) Plot the variation of the average heat transfer coefficient $\bar{h}_x(x)$ with distance for the three flow conditions of part (a).

(c) What are the average heat transfer coefficients for the entire plate $\bar{h}_L$ for the three flow conditions of part (a)?

7.14 Air at a pressure of 1 atm and a temperature of 50°C is in parallel flow over the top surface of a flat plate that is heated to a uniform temperature of 100°C. The plate has a length of 0.20 m (in the flow direction) and a width of 0.10 m. The Reynolds number based on the plate length is 40,000. What is the rate of heat transfer from the plate to the air? If the free stream velocity of the air is doubled and the pressure is increased to 10 atm, what is the rate of heat transfer?

7.15 Concentration of sunlight onto photovoltaic cells is desired since the concentrating mirrors and lenses are less expensive than the photovoltaic material. Consider the solar photovoltaic cell of Example 3.3. A 100 mm × 100 mm photovoltaic cell is irradiated with concentrated solar energy. Since the concentrating lens is glass, it absorbs 10% of the irradiation instead of the top surface of the solar cell, as in Example 3.3. The remaining irradiation is reflected from the system (7%) or is absorbed in the silicon semiconductor material of the photovoltaic cell (83%). The photovoltaic cell is cooled by air directed parallel to its top and bottom surfaces. The air temperature and velocity are 25°C and 5 m/s, respectively, and the bottom surface is coated with a high-emissivity paint, $\varepsilon_b = 0.95$.

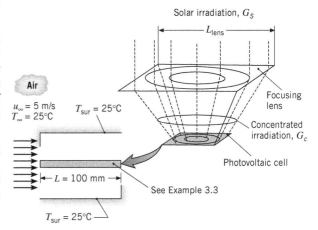

(a) Determine the electric power produced by the photovoltaic cell and the silicon temperature for a square concentrating lens with $L_{lens} = 400$ mm, which focuses the irradiation falling on the lens to the smaller area of the photovoltaic cell. Assume the concentrating lens temperature is 25°C and does not interfere with boundary layer development over the photovoltaic cell's top surface. The top and bottom boundary layers are both tripped to turbulent conditions at the leading edge of the photovoltaic material.

(b) Determine the electric power output of the photovoltaic cell and the silicon temperature over the range 100 mm ≤ L_{lens} ≤ 600 mm.

7.16 The roof of a refrigerated truck compartment is of composite construction, consisting of a layer of foamed urethane insulation (t_2 = 50 mm, k_i = 0.026 W/m · K) sandwiched between aluminum alloy panels (t_1 = 5 mm, k_p = 180 W/m · K). The length and width of the roof are L = 10 m and W = 3.5 m, respectively, and the temperature of the inner surface is $T_{s,i}$ = −10°C. Consider conditions for which the truck is moving at a speed of V = 105 km/h, the air temperature is $T_∞$ = 32°C, and the solar irradiation is G_S = 750 W/m². Turbulent flow may be assumed over the entire length of the roof.

(a) For equivalent values of the solar absorptivity and the emissivity of the outer surface ($α_S$ = $ε$ = 0.5), estimate the average temperature $T_{s,o}$ of the outer surface. What is the corresponding heat load imposed on the refrigeration system?

(b) A special finish ($α_S$ = 0.15, $ε$ = 0.8) may be applied to the outer surface. What effect would such an application have on the surface temperature and the heat load?

(c) If, with $α_S$ = $ε$ = 0.5, the roof is not insulated (t_2 = 0), what are the corresponding values of the surface temperature and the heat load?

7.17 The top surface of a compartment consists of very smooth (A) and highly roughened (B) portions, and the surface is placed in an atmospheric airstream. In the interest of minimizing total convection heat transfer to

the surface, which orientation, (1) or (2), is preferred? If T_s = 110°C, $T_∞$ = 30°C, and $u_∞$ = 20 m/s, what is the rate of convection heat transfer to the entire surface for this orientation?

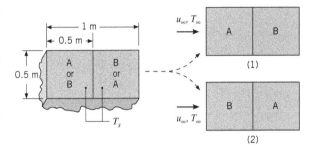

7.18 Calculate the value of the average heat transfer coefficient for the plate of Problem 7.17 when the entire plate is rotated 90° so that half of the leading edge consists of a very smooth portion (A) and the other half consists of a highly roughened portion (B).

7.19 Steel (AISI 1010) plates of thickness $δ$ = 6 mm and length L = 1 m on a side are conveyed from a heat treatment process and are concurrently cooled by atmospheric air of velocity $u_∞$ = 10 m/s and $T_∞$ = 30°C in parallel flow over the plates.

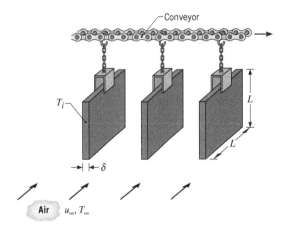

For an initial plate temperature of T_i = 310°C, what is the rate of heat transfer from the plate? What is the corresponding rate of change of the plate temperature? The velocity of the air is much larger than that of the plate.

7.20 Consider weather conditions for which the prevailing wind blows past the penthouse tower on a tall building. The tower length in the wind direction is 10 m and there are 10 window panels.

(a) Calculate the average convection coefficient for the first, third, and tenth window panels when the wind speed is 5 m/s. Use a film temperature of 300 K to evaluate the thermophysical properties required of the correlation. Would this be a suitable value of the film temperature for ambient air temperatures in the range $-15 \leq T_\infty \leq 38°C$?

(b) For the first, third, and tenth windows, on one graph, plot the variation of the average convection coefficient with wind speed for the range $5 \leq u_\infty \leq 100$ km/h. Explain the major features of each curve and their relative magnitudes.

7.21 The Weather Channel reports that it is a hot, muggy day with an air temperature of 90°F, a 10 mph breeze out of the southwest, and bright sunshine with a solar insolation of 400 W/m². Consider the wall of a metal building over which the prevailing wind blows. The length of the wall in the wind direction is 10 m, and the emissivity is 0.93. Assume that all the solar irradiation is absorbed, that irradiation from the sky is negligible, and that flow is fully turbulent over the wall. Estimate the average wall temperature.

7.22 In the production of sheet metals or plastics, it is customary to cool the material before it leaves the production process for storage or shipment to the customer. Typically, the process is continuous, with a sheet of thickness δ and width W cooled as it transits the distance L between two rollers at a velocity V. In this problem, we consider cooling of an aluminum alloy (2024-T6) by an airstream moving at a velocity u_∞ in counter flow over the top surface of the sheet. A turbulence promoter is used to provide turbulent boundary layer development over the entire surface.

(a) By applying conservation of energy to a differential control surface of length dx, which either moves with the sheet *or* is stationary and through which the sheet passes, derive a differential equation that governs the temperature distribution along the sheet. Because of the low emissivity of the aluminum, radiation effects may be neglected. Express

your result in terms of the velocity, thickness, and properties of the sheet (V, δ, ρ, c_p), the local convection coefficient h_x associated with the counter flow, and the air temperature. For a known temperature of the sheet (T_i) at the onset of cooling and a negligible effect of the sheet velocity on boundary layer development, solve the equation to obtain an expression for the outlet temperature T_o.

(b) For $\delta = 1$ mm, $V = 0.15$ m/s, $L = 5$ m, $W = 0.5$ m, $u_\infty = 20$ m/s, $T_\infty = 30°C$, and $T_i = 290°C$, what is the outlet temperature T_o?

7.23 An array of electronic chips is mounted within a sealed rectangular enclosure, and cooling is implemented by attaching an aluminum heat sink ($k = 180$ W/m · K). The base of the heat sink has dimensions of $w_1 = w_2 = 100$ mm, while the 6 fins are of thickness $t = 10$ mm and pitch $S = 18$ mm. The fin length is $L_f = 50$ mm, and the base of the heat sink has a thickness of $L_b = 10$ mm.

If cooling is implemented by water flow through the heat sink, with $u_\infty = 3$ m/s and $T_\infty = 17°C$, what is the base temperature T_b of the heat sink when power dissipation by the chips is $P_{elec} = 1800$ W? The average convection coefficient for surfaces of the fins and the exposed base may be estimated by assuming parallel flow over a flat plate. Properties of the water may be approximated as $k = 0.62$ W/m · K, $\rho = 995$ kg/m³, $c_p = 4178$ J/kg · K, $\nu = 7.73 \times 10^{-7}$ m²/s, and $Pr = 5.2$.

7.24 Consider the concentrating photovoltaic apparatus of Problem 7.15. The apparatus is to be installed in a desert environment, so the space between the concentrating lens and top of the photovoltaic cell is enclosed to protect the cell from sand abrasion in windy conditions. Since convection cooling from the top of the cell is reduced by the enclosure, an engineer proposes to cool the photovoltaic cell by attaching an aluminum heat sink to its bottom surface. The heat sink dimensions and material are the same as those of Problem 7.23. A contact resistance of 0.5×10^{-4} m² · K/W exists at the photovoltaic cell/heat sink interface and a dielectric liquid ($k = 0.064$ W/m · K, $\rho = 1400$ kg/m³, $c_p = 1300$ J/kg · K, $\nu = 10^{-6}$ m²/s, $Pr = 25$) flows between the heat sink fins at $u_\infty = 3$ m/s, $T_\infty = 25$°C.

(a) Determine the electric power produced by the photovoltaic cell and the silicon temperature for a square concentrating lens with $L_{lens} = 400$ mm.

(b) Compare the electric power produced by the photovoltaic cell with the heat sink in place and with the bottom surface cooled directly by the dielectric fluid (i.e., no heat sink) for $L_{lens} = 1.5$ m.

(c) Determine the electric power output and the silicon temperature over the range 100 mm ≤ L_{lens} < 3000 mm with the aluminum heat sink in place.

7.25 In the production of sheet metals or plastics, it is customary to cool the material before it leaves the production process for storage or shipment to the customer. Typically, the process is continuous, with a sheet of thickness δ and width W cooled as it transits the distance L between two rollers at a velocity V. In this problem, we consider cooling of plain carbon steel by an airstream moving at a velocity u_∞ in cross flow over the top and bottom surfaces of the sheet. A turbulence promoter is used to provide turbulent boundary layer development over the entire surface.

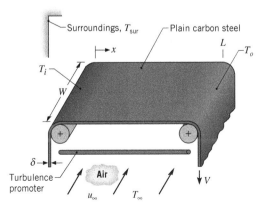

(a) By applying conservation of energy to a differential control surface of length dx, which either moves with the sheet *or* is stationary and through which the sheet passes, and assuming a uniform sheet temperature in the direction of airflow, derive a differential equation that governs the temperature distribution, $T(x)$, along the sheet. Consider the effects of radiation, as well as convection, and express your result in terms of the velocity, thickness, and properties of the sheet ($V, \delta, \rho, c_p, \varepsilon$), the average convection coefficient $\bar{h}_w$ associated with the cross flow, and the environmental temperatures (T_∞, T_{sur}).

(b) Neglecting radiation, obtain a closed form solution to the foregoing equation. For $\delta = 3$ mm, $V = 0.10$ m/s, $L = 10$ m, $W = 1$ m, $u_\infty = 20$ m/s, $T_\infty = 20$°C, and a sheet temperature of $T_i = 500$°C at the onset of cooling, what is the outlet temperature T_o? Assume a negligible effect of the sheet velocity on boundary layer development in the direction of airflow. The density and specific heat of the steel are $\rho = 7850$ kg/m³ and $c_p = 620$ J/kg · K, while properties of the air may be taken to be $k = 0.044$ W/m · K, $\nu = 4.5 \times 10^{-5}$ m²/s, $Pr = 0.68$.

(c) Accounting for the effects of radiation, with $\varepsilon = 0.70$ and $T_{sur} = 20$°C, numerically integrate the differential equation derived in part (a) to determine the temperature of the sheet at $L = 10$ m. Explore the effect of V on the temperature distribution along the sheet.

7.26 A steel strip emerges from the hot roll section of a steel mill at a speed of 20 m/s and a temperature of 1200 K. Its length and thickness are $L = 75$ m and $\delta = 0.002$ m, respectively, and its density and specific heat are 7900 kg/m³ and 640 J/kg · K, respectively.

Accounting for heat transfer from the top and bottom surfaces and neglecting radiation and strip conduction effects, determine the time rate of change of the strip temperature at a distance of 0.1 m from the leading edge and at the trailing edge. Determine the distance from the leading edge at which the minimum cooling rate is achieved.

7.27 A flat plate of width 1 m and length 0.2 m is maintained at a temperature of 32°C. Ambient fluid at 22°C flows across the top of the plate in parallel flow. Determine the average heat transfer coefficient, the convection heat transfer rate from the top of the plate, and the drag force on the plate for the following:

(a) The fluid is water flowing at a velocity of 0.5 m/s.

(b) The nanofluid of Example 2.2 is flowing at a velocity of 0.5 m/s.

(c) Water is flowing at a velocity of 2.5 m/s.

(d) The nanofluid of Example 2.2 is flowing at a velocity of 2.5 m/s.

7.28 One hundred electrical components, each dissipating 40 W, are attached to one surface of a square (0.25 m × 0.25 m) copper plate, and all the dissipated energy is transferred to water in parallel flow over the opposite surface. A protuberance at the leading edge of the plate acts to *trip* the boundary layer, and the plate itself may be assumed to be isothermal. The water velocity and temperature are $u_\infty = 2$ m/s and $T_\infty = 17°C$, and the water's thermophysical properties may be approximated as $\nu = 0.96 \times 10^{-6}$ m²/s, $k = 0.620$ W/m · K, and $Pr = 5.2$.

(a) What is the temperature of the copper plate?

(b) If each component has a plate contact surface area of 1 cm² and the corresponding contact resistance is 2×10^{-4} m² · K/W, what is the component temperature? Neglect the temperature variation across the thickness of the copper plate.

7.29 The boundary layer associated with parallel flow over an isothermal plate may be tripped at any *x*-location by using a fine wire that is stretched across the width of the plate. Determine the value of the critical Reynolds number $Re_{x,c,op}$ that is associated with the optimal location of the trip wire from the leading edge that will result in maximum heat transfer from the warm plate to the cool fluid.

7.30 Air at 27°C with a free stream velocity of 10 m/s is used to cool electronic devices mounted on a printed circuit board. Each device, 4 mm × 4 mm, dissipates 40 mW, which is removed from the top surface. A turbulator is located at the leading edge of the board, causing the boundary layer to be turbulent.

(a) Estimate the surface temperature of the fourth device located 15 mm from the leading edge of the board.

(b) Generate a plot of the surface temperature of the first four devices as a function of the free stream velocity for $5 \leq u_\infty \leq 15$ m/s.

(c) What is the minimum free stream velocity if the surface temperature of the hottest device is not to exceed 80°C?

7.31 Forced air at 25°C and 10 m/s is used to cool electronic elements mounted on a circuit board. Consider a chip of length 4 mm and width 4 mm located 80 mm from the leading edge. Because the board surface is irregular, the flow is disturbed and the appropriate convection correlation is of the form $Nu_x = 0.04 \, Re_x^{0.85} \, Pr^{0.33}$.

Estimate the surface temperature of the chip, T_s, if its heat dissipation rate is 40 mW.

7.32 Consider a thin, 50 mm × 50 mm fuel cell similar to that of Example 1.5, with air in parallel flow over its surfaces. Very small-diameter wires are stretched across both sides of the fuel cell at a distance $x = x_c$ from the leading edge in order to trip the flow into turbulent conditions. Using an appropriate correlation from Chapter 7, determine the minimum velocity needed to sustain the fuel cell at $T_c = 77°C$, and the associated location of the wire. The air and large surroundings are at $T_\infty = T_{sur} = 27°C$ and the fuel cell dissipates $\dot{E}_g = 11$ W. The fuel cell emissivity is $\varepsilon = 0.85$.

7.33 The cover plate of a flat-plate solar collector is at 15°C, while ambient air at 10°C is in parallel flow over the plate, with $u_\infty = 2$ m/s.

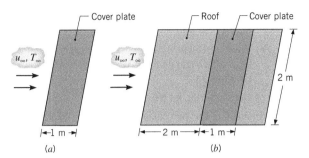

(a) (b)

(a) What is the rate of convective heat loss from the plate?

(b) If the plate is installed 2 m from the leading edge of a roof and flush with the roof surface, what is the rate of convective heat loss?

7.34 An array of 10 silicon chips, each of length $L = 10$ mm on a side, is insulated on one surface and cooled on the opposite surface by atmospheric air in parallel flow with $T_\infty = 20°C$ and $u_\infty = 30$ m/s. When in use, the same electrical power is dissipated in each chip, maintaining a uniform heat flux over the entire cooled surface.

If the temperature of each chip may not exceed 80°C, what is the maximum allowable power per chip? What is the maximum allowable power if a turbulence promoter is used to trip the boundary layer at the leading edge? Would it be preferable to orient the array normal, instead of parallel, to the airflow?

7.35 A square (10 mm × 10 mm) silicon chip is insulated on one side and cooled on the opposite side by atmospheric air in parallel flow at $u_\infty = 20$ m/s and $T_\infty = 30°C$. When in use, electrical power dissipation within the chip maintains a uniform heat flux at the cooled surface. If the chip temperature may not exceed 80°C at any point on its surface, what is the maximum allowable power? What is the maximum allowable power if the chip is flush mounted in a substrate that provides for an unheated starting length of 20 mm?

Cylinder in Cross Flow

7.36 Consider the following fluids, each with a velocity of $V = 5$ m/s and a temperature of $T_\infty = 20°C$, in cross flow over a 10-mm-diameter cylinder maintained at 50°C: atmospheric air, saturated water, and engine oil.

(a) Calculate the rate of heat transfer per unit length, q', using the Churchill–Bernstein correlation.

(b) Generate a plot of q' as a function of fluid velocity for $0.5 \leq V \leq 10$ m/s.

7.37 A circular pipe of 25-mm outside diameter is placed in an airstream at 25°C and 1-atm pressure. The air moves in cross flow over the pipe at 15 m/s, while the outer surface of the pipe is maintained at 100°C. What is the drag force exerted on the pipe per unit length? What is the rate of heat transfer from the pipe per unit length?

7.38 A long, cylindrical, electrical heating element of diameter $D = 10$ mm, thermal conductivity $k = 240$ W/m · K, density $\rho = 2700$ kg/m^3, and specific heat $c_p = 900$ J/kg · K is installed in a duct for which air moves in cross flow over the heater at a temperature and velocity of 27°C and 10 m/s, respectively.

(a) Neglecting radiation, estimate the steady-state surface temperature when, per unit length of the heater, electrical energy is being dissipated at a rate of 1000 W/m.

(b) If the heater is activated from an initial temperature of 27°C, estimate the time required for the surface temperature to come within 10°C of its steady-state value.

7.39 Consider the conditions of Problem 7.38, but now allow for radiation exchange between the surface of the heating element ($\varepsilon = 0.8$) and the walls of the duct, which form a large enclosure at 27°C.

(a) Evaluate the steady-state surface temperature.

(b) If the heater is activated from an initial temperature of 27°C, estimate the time required for the surface temperature to come within 10°C of the steady-state value.

(c) To guard against overheating due to unanticipated excursions in the blower output, the heater controller is designed to maintain a fixed surface temperature of 275°C. Determine the power dissipation required to maintain this temperature for air velocities in the range $5 \leq V \leq 10$ m/s.

7.40 Determine the convection heat transfer coefficient, thermal resistance for convection, and the convection heat transfer rate that are associated with air at atmospheric pressure in cross flow over a cylinder of diameter $D = 100$ mm and length $L = 2$ m. The cylinder temperature is $T_s = 70°C$ while the air velocity and temperature are $V = 3$ m/s and $T_\infty = 20°C$, respectively. Plot the convection heat transfer coefficient and the heat transfer rate from the cylinder over the range 0.05 m ≤ D ≤ 0.5 m.

7.41 A pin fin of 8-mm diameter dissipates 50 W by forced convection to air in cross flow with a Reynolds number of 3000. If the diameter of the fin is doubled and all other conditions remain the same, estimate the fin heat rate. Assume the pin to be infinitely long.

7.42 A long, thin metal plate is hung vertically after a heat treating process. The plate, of width $W = 0.2$ m, vertical dimension of 3 m, and an initial temperature of $T_s = 320°C$, is cooled by atmospheric air of velocity $V = 4$ m/s and $T_\infty = 30°C$. Determine the initial rate of heat loss from the plate if the air is in parallel flow over the plate. Account for heat loss from both surfaces of the thin plate. Determine the initial rate of heat loss if the plate is turned so that its width is perpendicular to the air flow. Account for heat loss from both the front surface and the back surface of the plate. Which orientation will maximize the cooling rate?

7.43 Pin fins are to be specified for use in an industrial cooling application. The fins will be subjected to a gas in cross flow at $V = 10$ m/s. The cylindrical fin has a diameter of $D = 15$ mm, and the cross-sectional area is the same for each configuration shown in the sketch.

Configuration A   Configuration B   Configuration C

Cross sections of cylindrical and square fins in cross flow

For fins of equal length and therefore equal mass, which fin has the largest heat transfer rate? Assume the gas properties are those of air at $T = 350$ K. *Hint:* Assume the fins can be treated as infinitely long and apply the Hilpert correlation to the fin of circular cross section.

7.44 Air at 24°C and a velocity of 5 m/s passes over the small region A_s (20 mm × 20 mm) on a large surface, which is maintained at $T_s = 130°C$. For these conditions, 0.7 W is removed from the surface A_s. To increase the heat removal rate, a stainless steel (AISI 304) pin fin of diameter 5 mm is affixed to A_s, which is assumed to remain at $T_s = 130°C$.

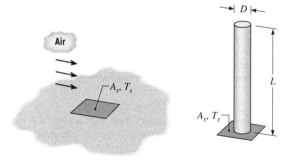

(a) Determine the maximum possible heat removal rate through the fin.

(b) What fin length would provide a close approximation to the heat rate found in part (a)? *Hint:* Refer to Example 3.9.

(c) Determine the fin effectiveness, ε_f.

(d) What is the percentage increase in the heat rate from A_s due to installation of the fin?

7.45 Hot water at 50°C is routed from one building in which it is generated to an adjoining building in which it is used for space heating. Transfer between the buildings occurs in a steel pipe ($k = 60$ W/m · K) of 100-mm outside diameter and 8-mm wall thickness. During the winter, representative environmental conditions involve air at $T_\infty = -5°C$ and $V = 3$ m/s in cross flow over the pipe.

(a) If the cost of producing the hot water is $0.10 per kW · h, what is the representative daily cost of heat loss from an uninsulated pipe to the air per meter of pipe length? The convection resistance associated with water flow in the pipe may be neglected.

(b) Determine the savings associated with application of a 10-mm-thick coating of urethane insulation ($k = 0.026$ W/m · K) to the outer surface of the pipe.

7.46 To enhance heat transfer from a silicon chip of width $W = 4$ mm on a side, a copper pin fin is brazed to the surface of the chip. The pin length and diameter are $L = 12$ mm and $D = 2$ mm, respectively, and atmospheric air at $V = 10$ m/s and $T_\infty = 300$ K is in cross flow over the pin. The surface of the chip, and hence the base of the pin, are maintained at a temperature of $T_b = 350$ K.

(a) Assuming the chip to have a negligible effect on flow over the pin, what is the average convection coefficient for the surface of the pin?

(b) Neglecting radiation and assuming the convection coefficient at the pin tip to equal that calculated in part (a), determine the pin heat transfer rate.

(c) Neglecting radiation and assuming the convection coefficient at the exposed chip surface to equal that calculated in part (a), determine the total rate of heat transfer from the chip.

(d) Independently determine and plot the effect of increasing velocity ($10 \leq V \leq 40$ m/s) and pin diameter ($2 \leq D \leq 4$ mm) on the total rate of heat transfer from the chip. What is the heat rate for $V = 40$ m/s and $D = 4$ mm?

7.47 A fine wire of diameter D is positioned across a passage to determine flow velocity from heat transfer characteristics. Current is passed through the wire to heat it, and the heat is dissipated to the flowing fluid by convection. The resistance of the wire is determined from electrical measurements, and the temperature is known from the resistance.

(a) For a fluid of arbitrary Prandtl number, develop an expression for its velocity in terms of the difference between the temperature of the wire and the free stream temperature of the fluid.

(b) What is the velocity of an airstream at 1 atm and 20°C, if a wire of 1-mm diameter achieves a temperature of 45°C while dissipating 75 W/m?

7.48 To determine air velocity changes, it is proposed to measure the electric current required to maintain a platinum wire of 0.5-mm diameter at a constant temperature of 77°C in a stream of air at 27°C.

(a) Assuming Reynolds numbers in the range $40 < Re_D < 1000$, develop a relationship between the wire current and the velocity of the air that is in cross flow over the wire. Use this result to establish a relation between fractional changes in the current, $\Delta I/I$, and the air velocity, $\Delta V/V$.

(b) Calculate the current required when the air velocity is 10 m/s and the electrical resistivity of the platinum wire is 17.1×10^{-5} $\Omega \cdot$ m.

7.49 A computer code is being developed to analyze a temperature sensor of 12.5-mm diameter experiencing cross flow of water with a free stream temperature of 80°C and variable velocity. Derive an expression for the convection heat transfer coefficient as a function of the sensor surface temperature T_s for the range $20 < T_s < 80$°C and for velocities V in the range $0.005 < V < 0.20$ m/s. Use the Zukauskas correlation for the range $40 < Re_D < 1000$ and assume that the Prandtl number of water has a linear temperature dependence.

7.50 A 25-mm-diameter, high-tension line has an electrical resistance of 10^{-4} Ω/m and is transmitting a current of 1000 A.

(a) If ambient air at 10°C and 5 m/s is in cross flow over the line, what is its surface temperature?

(b) If the line may be approximated as a solid copper rod, what is its centerline temperature?

(c) Generate a plot that depicts the variation of the surface temperature with air velocity for $1 \leq V \leq 10$ m/s.

7.51 A temperature sensor of 10.5-mm diameter experiences cross flow of water with a free stream temperature of 80°C and variable velocity. Derive an expression for the convection heat transfer coefficient as a function of the sensor surface temperature T_s for the range $20 < T_s < 80$°C and for velocities V in the range $0.005 < V < 0.20$ m/s. Use the Zukauskas correlation for the range $40 < Re_D < 1000$ and assume that the Prandtl number of water has a linear temperature dependence.

7.52 An aluminum transmission line with a diameter of 20 mm has an electrical resistance of $R'_{elec} = 2.636 \times 10^{-4}$ Ω/m and carries a current of 700 A. The line is subjected to frequent and severe cross winds, increasing the probability of contact between adjacent lines, thereby causing sparks and creating a potential fire hazard for nearby vegetation. The remedy is to insulate the line, but with the adverse effect of increasing the conductor operating temperature.

(a) Calculate the conductor temperature when the air temperature is 20°C and the line is subjected to cross flow with a velocity of 10 m/s.

(b) Calculate the conductor temperature for the same conditions, but with a 2-mm-thick insulation having a thermal conductivity of 0.15 W/m · K.

(c) Calculate and plot the temperatures of the bare and insulated conductors for wind velocities in the range from 2 to 20 m/s. Comment on features of the curves and the effect of the wind velocity on the conductor temperatures.

7.53 Fluid velocities can be measured using hot-film sensors, and a common design is one for which the sensing element forms a thin film about the circumference of a quartz rod. The film is typically comprised of a thin (~100 nm) layer of platinum, whose electrical resistance is proportional to its temperature. Hence, when submerged in a fluid stream, an electric current may be passed through the film to maintain its temperature above that of the fluid. The temperature of the film is controlled by monitoring its electric resistance, and with concurrent measurement of the electric current, the power dissipated in the film may be determined.

Fluid
$V, T_\infty = 20$°C

Quartz rod, $D = 1.5$ mm
$k = 1.4$ W/m·K

Power leads, thick films, no power dissipation

Hot-film sensor, $T_s = 50$°C

$w = 0.3$ mm

Proper operation is assured only if the heat generated in the film is transferred to the fluid, rather than conducted from the film into the quartz rod. Thermally, the film should therefore be strongly coupled to the fluid and weakly coupled to the quartz rod. This condition is satisfied if the Biot number is very large, $Bi = \bar{h}D/2k \gg 1$, where $\bar{h}$ is the convection coefficient between the fluid and the film and k is the thermal conductivity of the rod.

(a) For the following fluids and velocities, calculate and plot the convection coefficient as a function of velocity: (i) water, $0.5 \leq V \leq 5$ m/s; (ii) air, $1 \leq V \leq 20$ m/s.

(b) Comment on the suitability of using this hot-film sensor for the foregoing conditions.

7.54 An uninsulated steam pipe is used to transport high-temperature steam from one building to another. The pipe is of 0.5-m diameter, has a surface temperature of 150°C, and is exposed to ambient air at −10°C. The air moves in cross flow over the pipe with a velocity of 5 m/s.

(a) What is the rate of heat loss per unit length of pipe?

(b) Consider the effect of insulating the pipe with a rigid urethane foam ($k = 0.026$ W/m · K). Evaluate and plot the rate of heat loss as a function of the thickness δ of the insulation layer for $0 \leq \delta \leq 50$ mm.

7.55 A thermocouple is inserted into a hot air duct to measure the air temperature. The thermocouple (T_1) is soldered to the tip of a steel *thermocouple well* of length $L = 0.1$ m and inner and outer diameters of $D_i = 5$ mm and $D_o = 8$ mm. A second thermocouple (T_2) is used to measure the duct wall temperature.

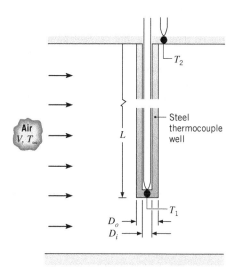

Consider conditions for which the air velocity in the duct is $V = 5$ m/s and the two thermocouples register temperatures of $T_1 = 450$ K and $T_2 = 375$ K. Neglecting radiation, determine the air temperature T_∞. Assume that, for steel, $k = 35$ W/m · K, and, for air, $\rho = 0.774$ kg/m³, $\mu = 251 \times 10^{-7}$ N · s/m², $k = 0.0373$ W/m · K, and $Pr = 0.686$.

7.56 Consider conditions for which a mercury-in-glass thermometer of 4-mm diameter is inserted to a length L through the wall of a duct in which air at 77°C is flowing. If the stem of the thermometer at the duct wall is at the wall temperature $T_w = 15$°C, conduction heat transfer through the glass causes the bulb temperature to be lower than that of the airstream.

(a) Develop a relationship for the *immersion* error, $\Delta T_i = T(L) − T_\infty$, as a function of air velocity, thermometer diameter, and insertion length L.

(b) To what length L must the thermometer be inserted if the immersion error is not to exceed 0.25°C when the air velocity is 10 m/s?

(c) Using the insertion length determined in part (b), calculate and plot the immersion error as a function of air velocity for the range 2 to 20 m/s.

(d) For a given insertion length, will the immersion error increase or decrease if the diameter of the thermometer is increased? Is the immersion error more sensitive to the diameter or air velocity?

7.57 The objective of an experiment performed by our students is to determine the effect of pin fins on the thermal resistance between a flat plate and an airstream. A 25.9-mm-square polished aluminum plate is subjected to an airstream in parallel flow at $T_\infty = 20$°C and $u_\infty = 6$ m/s. An electrical heating patch is attached to the backside of the plate and dissipates 15.5 W under all conditions. Pin fins of diameter $D = 4.8$ mm and length $L = 25.4$ mm are fabricated from brass and can be firmly attached to the plate at various locations over its surface. Thermocouples are attached to the plate surface and the tip of one of the fins.

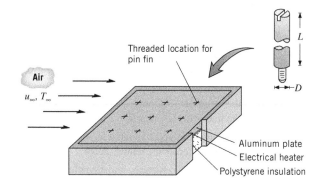

Measured temperatures for five pin-fin configurations are tabulated.

Number of Pin Fins	Temperature (°C)	
	Fin Tip	Plate Base
0	—	70.2
1	40.6	67.4
2	39.5	64.7
5	36.4	57.4
8	34.2	52.1

(a) Using the experimental observations and neglecting the effect of flow interactions between pins, determine the thermal resistance between the plate and the airstream for the five configurations.

(b) Develop a model of the plate–pin fin system and using appropriate convection correlations, predict the thermal resistances for the five configurations. Compare your predictions with the observations and explain any differences.

(c) Use your model to predict the thermal resistances when the airstream velocity is doubled.

Spheres

7.58 Air at 25°C flows over a 10-mm-diameter sphere with a velocity of 25 m/s, while the surface of the sphere is maintained at 75°C.

(a) What is the drag force on the sphere?

(b) What is the rate of heat transfer from the sphere?

(c) Generate a plot of the heat transfer from the sphere as a function of the air velocity for the range 1 to 25 m/s.

7.59 Consider a sphere with a diameter of 25 mm and a surface temperature of 60°C that is immersed in a fluid at a temperature of 30°C and a velocity of 2 m/s. Calculate the drag force and the heat rate when the fluid is (a) water and (b) air at atmospheric pressure. Explain why the results for the two fluids are so different.

7.60 A spherical, underwater instrument pod used to make soundings and to measure conditions in the water has a diameter of 85 mm and dissipates 300 W.

(a) Estimate the surface temperature of the pod when suspended in a bay where the current is 1 m/s and the water temperature is 15°C.

(b) Inadvertently, the pod is hauled out of the water and suspended in ambient air without deactivating the power. Estimate the surface temperature of the pod if the air temperature is 15°C and the wind speed is 3 m/s.

7.61 Oil is cooled by spraying a mist of the hot liquid through cool air at $T_\infty = 27°C$. Two droplets, each of diameter $D = 100$ μm and temperature 250°C, suddenly collide and coalesce into a single droplet. Determine the heat transfer rate from the oil to the air before and after coalescence of the droplets. The velocity of the droplets is $V = 5$ m/s.

7.62 Worldwide, over a billion solder balls must be manufactured daily for assembling electronics packages. The *uniform droplet spray* method uses a piezoelectric device to vibrate a shaft in a pot of molten solder that, in turn, ejects small droplets of solder through a precision-machined nozzle. As they traverse a collection chamber, the droplets cool and solidify. The collection chamber is flooded with an inert gas such as nitrogen to prevent oxidation of the solder ball surfaces.

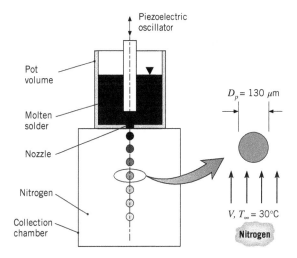

(a) Molten solder droplets of diameter 130 μm are ejected at a velocity of 2 m/s at an initial temperature of 225°C into gaseous nitrogen that is at 30°C and slightly above atmospheric pressure. Determine the terminal velocity of the particles and the distance the particles have traveled when they become completely solidified. The solder properties are $\rho = 8230$ kg/m^3, $c = 240$ J/kg · K, $k = 38$ W/m · K, $h_{sf} = 42$ kJ/kg. The solder's melting temperature is 183°C.

(b) The piezoelectric device oscillates at 1.8 kHz, producing 1800 particles per second. Determine the separation distance between the particles as they traverse the nitrogen gas and the pot volume needed in order to produce the solder balls continuously for one week.

7.63 Consider the material processing experiment of Problem 5.23, with atmospheric nitrogen used to implement cooling by convection. However, instead of using a prescribed value of the convection coefficient, compute the coefficient from an appropriate correlation.

(a) Neglecting radiation, determine the time required to cool the sphere from 900°C to 300°C if the velocity and temperature of the nitrogen are $V = 5$ m/s and $T_\infty = 25$°C.

(b) Accounting for the effects of both convection and radiation, with $\varepsilon = 0.6$ and $T_{sur} = 25$°C, determine the time required to cool the sphere. Explore the effects of the flow velocity on your result.

7.64 Copper spheres of 20-mm diameter are quenched by being dropped into a tank of water that is maintained at 280 K. The spheres may be assumed to reach the terminal velocity on impact and to drop freely through the water. Estimate the terminal velocity by equating the drag and gravitational forces acting on the sphere. What is the approximate height of the water tank needed to cool the spheres from an initial temperature of 360 K to a center temperature of 320 K?

7.65 For the conditions of Problem 7.64, what are the terminal velocity and the tank height if engine oil at 300 K, rather than water, is used as the coolant?

7.66 Consider the plasma spray coating process of Problem 5.21. In addition to the prescribed conditions, the argon plasma jet is known to have a mean velocity of $V = 400$ m/s, while the *initial* velocity of the injected alumina particles may be approximated as zero. The nozzle exit and the substrate are separated by a distance of $L = 100$ mm, and pertinent properties of the argon plasma may be approximated as $k = 0.671$ W/m·K, $c_p = 1480$ J/kg·K, $\mu = 2.70 \times 10^{-4}$ kg/s·m, and $\nu = 5.6 \times 10^{-3}$ m²/s.

(a) Assuming the motion of particles entrained by the plasma jet to be governed by Stokes' law, derive expressions for the particle velocity, $V_p(t)$, and its distance of travel from the nozzle exit, $x_p(t)$, as a function of time, t, where $t = 0$ corresponds to particle injection. Evaluate the time-in-flight required for a particle to traverse the separation distance, $x_p = L$, and the velocity V_p at this time.

(b) Assuming an average relative velocity of $(\overline{V - V_p}) = 315$ m/s during the time-of-flight, estimate the convection coefficient associated with heat transfer from the plasma to the particle. Using this coefficient and assuming an initial particle temperature of $T_i = 300$ K, estimate the time-in-flight required to heat a particle to its melting point, T_{mp}, and, once at T_{mp}, for the particle to experience complete melting. Is the prescribed value of L sufficient to ensure complete particle melting before surface impact?

7.67 Highly reflective aluminum coatings may be formed on the surface of a substrate by impacting the surface with molten drops of aluminum. The droplets are discharged from an injector, proceed through an inert gas (helium), and must still be in a molten state at the time of impact.

Consider conditions for which droplets with a diameter, velocity, and initial temperature of $D = 100$ μm, $V = 15$ m/s, and $T_i = 1100$ K, respectively, traverse a stagnant layer of atmospheric helium that is at a temperature of $T_\infty = 300$ K. What is the maximum allowable thickness of the helium layer needed to ensure that the temperature of droplets impacting the substrate is greater than or equal to the melting point of aluminum ($T_f \geq T_{mp} = 933$ K)? Properties of the molten aluminum may be approximated as $\rho = 2500$ kg/m³, $c = 1200$ J/kg·K, and $k = 200$ W/m·K.

7.68 Tissue engineering involves the development of biological substitutes that restore or improve tissue function. Once manufactured, engineered organs can be implanted and grow within the patient, obviating chronic shortages of natural organs that arise when traditional organ transplant procedures are used. Artificial organ manufacture involves two major steps. First, a porous *scaffold* is fabricated with a specific pore size and pore distribution, as well as overall shape and size. Second, the top surface of the scaffold is seeded with human cells that grow into the pores of the scaffold. The scaffold material is biodegradable and is eventually replaced with healthy tissue. The artificial organ is then ready to be implanted in the patient.

   The complex pore shapes, small pore sizes, and unusual organ shapes preclude use of traditional manufacturing methods to fabricate the scaffolds. A method that has been used with success is a *solid freeform fabrication* technique whereby small spherical drops are directed to a substrate. The drops are initially molten and solidify when they impact the room-temperature substrate. By controlling the location of the droplet deposition, complex scaffolds can be built up, one drop at a time. A device similar to that of Problem 7.62 is used to generate uniform, 75-μm-diameter drops at an initial temperature of $T_i = 150$°C. The particles are sent through quiescent air at $T_\infty = 25$°C. The droplet properties are $\rho = 2200$ kg/m³, $c = 700$ J/kg·K.

Droplet generator

Exit nozzle

$D = 75 \ \mu m$
$T_i = 150°C$

Air $T_\infty = 25°C, V$

L

Tissue scaffold

(a) It is desirable for the droplets to exit the nozzle at their terminal velocity. Determine the terminal velocity of the drops.

(b) It is desirable for the droplets to impact the structure at a temperature of $T_2 = 120°C$. What is the required distance between the exit nozzle and the structure, L?

7.69 A silicon chip ($k = 150 \ \text{W/m} \cdot \text{K}$, $\rho = 2300 \ \text{kg/m}^3$, $c_p = 700 \ \text{J/kg} \cdot \text{K}$), 10 mm on a side and 1 mm thick, is connected to a substrate by solder balls ($k = 40 \ \text{W/m} \cdot \text{K}$, $\rho = 10,000 \ \text{kg/m}^3$, $c_p = 150 \ \text{J/kg} \cdot \text{K}$) of 1-mm diameter, and during an accelerated thermal stress test, the system is exposed to the flow of a dielectric liquid ($k = 0.064 \ \text{W/m} \cdot \text{K}$, $\nu = 10^{-6} \ \text{m}^2/\text{s}$, $Pr = 25$). As first approximations, treat the top and bottom surfaces of the chip as flat plates in turbulent, parallel flow and assume the substrate and lower chip surfaces to have a negligible effect on flow over the solder balls. Also assume point contact between the chip and the solder, thereby neglecting heat transfer by conduction between the components.

$t = 1 \text{ mm}$

Silicon chip, T_i

$L = 10$ mm

Solder ball
$D = 1$ mm

V
T_∞

Dielectric fluid

Substrate

(a) The stress test begins with the components at ambient temperature ($T_i = 20°C$) and proceeds with heating by the fluid at $T_\infty = 80°C$. If the fluid velocity is $V = 0.2$ m/s, estimate the ratio of the time constant of the chip to that of a solder ball. Which component responds more rapidly to the heating process?

(b) The thermal stress acting on the solder joint is proportional to the chip-to-solder temperature difference. What is this temperature difference 0.25 s after the start of heating?

7.70 A thermocouple junction is inserted in a large duct to measure the temperature of hot gases flowing through the duct.

Duct T_s

Hot gases

T_g

Thermocouple junction

(a) If the duct surface temperature T_s is less than the gas temperature T_g, will the thermocouple sense a temperature that is less than, equal to, or greater than T_g? Justify your answer on the basis of a simple analysis.

(b) A thermocouple junction in the shape of a 2-mm-diameter sphere with a surface emissivity of 0.60 is placed in a gas stream moving at 3 m/s. If the thermocouple senses a temperature of 320°C when the duct surface temperature is 175°C, what is the actual gas temperature? The gas may be assumed to have the properties of air at atmospheric pressure.

(c) How would changes in velocity and emissivity affect the temperature measurement error? Determine the measurement error for velocities in the range $1 \le V \le 25$ m/s ($\varepsilon = 0.6$) and for emissivities in the range $0.1 \le \varepsilon \le 1.0$ ($V = 3$ m/s).

7.71 Consider temperature measurement in a gas stream using the thermocouple junction described in Problem 7.70 ($D = 2$ mm, $\varepsilon = 0.60$). If the gas velocity and temperature are 3 m/s and 500°C, respectively, what temperature will be indicated by the thermocouple if the duct surface temperature is 200°C? The gas may be assumed to have the properties of atmospheric air. What temperature will be indicated by the thermocouple if the gas pressure is doubled and all other conditions remain the same?

Tube Banks

7.72 Repeat Example 7.7 for a more compact tube bank in which the longitudinal and transverse pitches are $S_L = S_T = 20.5$ mm. All other conditions remain the same.

7.73 A preheater involves the use of condensing steam at 100°C on the inside of a bank of tubes to heat air that enters at 1 atm and 25°C. The air moves at 5 m/s in cross flow over the tubes. Each tube is 1 m long and has an outside diameter of 10 mm. The bank consists of 196 tubes in a square, aligned array for which $S_T = S_L = 15$ mm. What is the total rate of heat transfer to the air? What is the pressure drop associated with the airflow?

7.74 Consider the in-line tube bank of Problem 7.73 ($D = 10$ mm, $L = 1$ m, and $S_T = S_L = 15$ mm), with condensing steam used to heat atmospheric air entering the tube bank at $T_i = 25°C$ and $V = 5$ m/s. In this case, however, the desired outlet temperature, not the number of tube rows, is known. What is the minimum value of N_L needed to achieve an outlet temperature of $T_o ≥ 75°C$? What is the corresponding pressure drop across the tube bank?

7.75 A tube bank uses an aligned arrangement of 10-mm-diameter tubes with $S_T = S_L = 20$ mm. There are 10 rows of tubes with 50 tubes in each row. Consider an application for which cold water flows through the tubes, maintaining the outer surface temperature at 27°C, while flue gases at 427°C and a velocity of 5 m/s are in cross flow over the tubes. The properties of the flue gas may be approximated as those of atmospheric air at 427°C. What is the total rate of heat transfer per unit length of the tubes in the bank?

7.76 An air duct heater consists of an aligned array of electrical heating elements in which the longitudinal and transverse pitches are $S_L = S_T = 24$ mm. There are 3 rows of elements in the flow direction ($N_L = 3$) and 4 elements per row ($N_T = 4$). Atmospheric air with an upstream velocity of 12 m/s and a temperature of 25°C moves in cross flow over the elements, which have a diameter of 12 mm, a length of 250 mm, and are maintained at a surface temperature of 350°C.

(a) Determine the total rate of heat transfer to the air and the temperature of the air leaving the duct heater.

(b) Determine the pressure drop across the element bank and the fan power requirement.

(c) Compare the average convection coefficient obtained in your analysis with the value for an isolated (single) element. Explain the difference between the results.

(d) What effect would increasing the longitudinal and transverse pitches to 30 mm have on the exit temperature of the air, the total heat rate, and the pressure drop?

7.77 A tube bank uses an aligned arrangement of 30-mm-diameter tubes with $S_T = S_L = 60$ mm and a tube length of 1 m. There are 10 tube rows in the flow direction ($N_L = 7$) and 7 tubes per row ($N_T = 10$). Air with upstream conditions of $T_∞ = 27°C$ and $V = 10.5$ m/s is in cross flow over the tubes, while a tube wall temperature of 100°C is maintained by steam condensation inside the tubes. Determine the temperature of air leaving the tube bank, the pressure drop across the bank, and the fan power requirement.

7.78 Electrical components mounted to each of two isothermal plates are cooled by passing atmospheric air between the plates, and an in-line array of aluminum pin fins is used to enhance heat transfer to the air.

The pins are of diameter $D = 2$ mm, length $L = 100$ mm, and thermal conductivity $k = 240$ W/m · K. The longitudinal and transverse pitches are $S_L = S_T = 4$ mm, with a square array of 625 pins ($N_T = N_L = 25$) mounted to square plates that are each of width $W = 100$ mm on a side. Air enters the pin array with a velocity of 10 m/s and a temperature of 300 K.

(a) Evaluating air properties at 300 K, estimate the average convection coefficient for the array of pin fins.

(b) Assuming a uniform convection coefficient over all heat transfer surfaces (plates and pins), use the result of part (a) to determine the air outlet temperature and total heat rate when the plates are maintained at 350 K. *Hint:* The air outlet temperature is governed by an exponential relation of the form $[(T_s − T_o)/(T_s − T_i)] = \exp[−(\bar{h}A_t η_o)/\dot{m}c_p]$, where $\dot{m} = ρVLN_TS_T$ is the mass flow rate of air passing through the array, A_t is the total heat transfer surface area (plates and fins), and $η_o$ is the overall surface efficiency defined by Equation 3.107.

7.79 An air-cooled steam condenser is operated with air in cross flow over a square, in-line array of 400 tubes ($N_L = N_T = 20$), with an outside tube diameter of 20 mm and longitudinal and transverse pitches of $S_L = 60$ mm and $S_T = 30$ mm, respectively. Saturated steam at a pressure of 2.455 bars enters the tubes, and a uniform tube outer surface temperature of $T_s = 390$ K may be assumed to be maintained as condensation occurs within the tubes.

(a) If the temperature and velocity of the air upstream of the array are $T_i = 300$ K and $V = 4$ m/s, what is the temperature T_o of the air that leaves the array? As a first approximation, evaluate the properties of air at 300 K.

(b) If the tubes are 2 m long, what is the total heat transfer rate for the array? What is the rate at which steam is condensed in kg/s?

(c) Assess the effect of increasing N_L by a factor of 2, while reducing S_L to 30 mm. For this configuration, explore the effect of changes in the air velocity.

Impinging Jets

7.80 Heating and cooling with *miniature* impinging jets has been proposed for numerous applications. For a single round jet, determine the minimum jet diameter for which Equation 7.71 may be applied for air at atmospheric pressure (a) at $T_e = 0°C$ and (b) at $T_e = 500°C$.

7.81 A circular transistor of 10-mm diameter is cooled by impingement of an air jet exiting a 2-mm-diameter round nozzle with a velocity of 20 m/s and a temperature of 15°C. The jet exit and the exposed surface of the transistor are separated by a distance of 10 mm.

If the transistor is well insulated at all but its exposed surface and the surface temperature is not to exceed 85°C, what is the transistor's maximum allowable operating power?

7.82 A long rectangular plate of AISI 302 stainless steel is initially at 1000 K and is cooled by an array of slot jets (see Figure 7.17). The nozzle width and pitch are $W = 10$ mm and $S = 100$ mm, respectively, and the nozzle-to-plate separation is $H = 200$ mm. The plate thickness and width are $t = 5$ mm and $L = 1$ m, respectively. If air exits the nozzles at a temperature of 600 K and a velocity of 30 m/s, what is the initial cooling rate of the plate?

7.83 A cryogenic probe is used to treat cancerous skin tissue. The probe consists of a single round jet of diameter $D_e = 2$ mm that issues from a nozzle concentrically situated within a larger, enclosed cylindrical tube of outer diameter $D_o = 15$ mm. The wall thickness of the AISI 302 stainless steel probe is $t = 2$ mm, and the separation distance between the nozzle and the inner surface of the probe is $H = 5$ mm.

Assuming the cancerous skin tissue to be a semi-infinite medium with $k_c = 0.20$ W/m · K and $T_c = 37°C$ far from the probe location, determine the surface temperature T_s. Neglect the contact resistance between the probe and the tissue. Cold nitrogen exits the jet at $T_e = 100$ K, $V_e = 20$ m/s. *Hint:* Due to the probe walls, the jet is confined and behaves as if it were one in an array such as in Figure 7.18c.

7.84 A 25-mm-diameter hot surface at $T_s = 85°C$ is cooled by an air jet exiting a 5-mm-diameter round nozzle with a velocity of 35 m/s and temperature of 25°C. The nozzle exit is 25 mm from the hot surface. Determine the percentage change in the average heat transfer coefficient at the hot surface if the air is replaced with carbon dioxide or helium.

7.85 Air at 10 m/s and 15°C is used to cool a square hot molded plastic plate 0.5 m to a side having a surface temperature of 140°C. To increase the throughput of the production process, it is proposed to cool the plate using an array of slotted nozzles with width and pitch of 4 mm and 56 mm, respectively, and a nozzle-to-plate separation of 40 mm. The air exits the nozzle at a temperature of 15°C and a velocity of 10 m/s.

(a) Determine the improvement in cooling rate that can be achieved using the slotted nozzle arrangement in lieu of turbulated air at 10 m/s and 15°C in parallel flow over the plate.

(b) Would the heat rates for both arrangements change significantly if the air velocities were increased by a factor of 2?

(c) What is the air mass rate requirement for the slotted nozzle arrangement?

7.86 You have been asked to determine the feasibility of using an impinging jet in a soldering operation for electronic assemblies. The schematic illustrates the use of a single, round nozzle to direct high-velocity, hot air to a location where a *surface mount* joint is to be formed.

For your study, consider a round nozzle with a diameter of 1 mm located a distance of 2 mm from the region of the surface mount, which has a diameter of 2.5 mm.

(a) For an air jet velocity of 70 m/s and a temperature of 500°C, estimate the average convection coefficient over the area of the surface mount.

(b) Assume that the surface mount region on the printed circuit board (PCB) can be modeled as a semi-infinite medium, which is initially at a uniform temperature of 25°C and suddenly experiences convective heating by the jet. Estimate the time required for the surface to reach 183°C. The thermophysical properties of a typical solder are $\rho = 8333 \text{ kg/m}^3$, $c_p = 188 \text{ J/kg} \cdot \text{K}$, and $k = 51 \text{ W/m} \cdot \text{K}$.

(c) For each of three air jet temperatures of 500, 600, and 700°C, calculate and plot the surface temperature as a function of time for $0 \le t \le 150$ s. On this plot, identify important temperature limits for the soldering process: the lower limit corresponding to the solder's eutectic temperature, $T_{\text{sol}} = 183°\text{C}$, and the upper limit corresponding to the glass transition temperature, $T_{\text{gl}} = 250°\text{C}$, at which the PCB becomes plastic. Comment on the outcome of your study, the appropriateness of the assumptions, and the feasibility of using the jet for a soldering application.

Packed Beds

7.87 Thermal energy storage systems commonly involve a packed bed of solid spheres, through which a hot gas flows if the system is being charged, or a cold gas if it is being discharged. In a charging process, heat transfer from the hot gas increases thermal energy stored within the colder spheres; during discharge, the stored energy decreases as heat is transferred from the warmer spheres to the cooler gas.

Consider the packed bed of aluminum spheres described above under conditions for which the bed is charged by hot air with an inlet velocity of $V = 1$ m/s and temperature of $T_{g,i} = 300°\text{C}$, but for which the convection coefficient is not prescribed. If the porosity of the bed is $\varepsilon = 0.40$ and the initial temperature of the spheres is $T_i = 25°\text{C}$, how long does it take a sphere near the inlet of the bed to accumulate 90% of its maximum possible energy?

7.88 The use of rock pile thermal energy storage systems has been considered for solar energy and industrial process heat applications. A particular system involves a cylindrical container, 2 m long by 1 m in diameter, in which nearly spherical rocks of 0.03-m diameter are packed. The bed has a void space of 0.42, and the density and specific heat of the rock are $\rho = 2300 \text{ kg/m}^3$ and $c_p = 879 \text{ J/kg} \cdot \text{K}$, respectively. Consider conditions for which atmospheric air is supplied to the rock pile at a steady flow rate of 1 kg/s and a temperature of 90°C. The air flows in the axial direction through the container. If the rock is at a temperature of 25°C, what is the total rate of heat transfer from the air to the rock pile?

7.89 The cylindrical chamber of a *pebble bed nuclear reactor* is of length $L = 10$ m, and diameter $D = 3$ m. The chamber is filled with spherical uranium oxide pellets of core diameter $D_p = 50$ mm. Each pellet generates thermal energy in its core at a rate of $\dot{E}_g$ and is coated with a layer of non-heat-generating graphite, which is of uniform thickness $\delta = 5$ mm, to form a pebble. The uranium oxide and graphite each have a thermal conductivity of 2 W/m · K. The packed bed has a porosity of $\varepsilon = 0.4$. Pressurized helium at 40 bars is used to absorb the thermal energy from the pebbles. The helium enters the packed bed at $T_i = 450°\text{C}$ with a velocity of 3.2 m/s. The properties of the helium may be assumed to be

$c_p = 5193$ J/kg · K, $k = 0.3355$ W/m · K, $\rho = 2.1676$ kg/m³, $\mu = 4.214 \times 10^{-5}$ kg/s · m, $Pr = 0.654$.

(a) For a desired overall thermal energy transfer rate of $q = 125$ MW, determine the mean outlet temperature of the helium leaving the bed, T_o, and the amount of thermal energy generated by each pellet, $\dot{E}_g$.

(b) The amount of energy generated by the fuel decreases if a maximum operating temperature of approximately 2100°C is exceeded. Determine the maximum internal temperature of the hottest pellet in the packed bed. For Reynolds numbers in the range $4000 \leq Re_D \leq 10,000$, Equation 7.81 may be replaced by $\varepsilon \bar{j}_H = 2.876 \, Re_D^{-1} + 0.3023 \, Re_D^{-0.35}$.

7.90 *Latent heat capsules* consist of a thin-walled spherical shell within which a solid-liquid, phase-change material (PCM) of melting point T_{mp} and latent heat of fusion h_{sf} is enclosed. As shown schematically, the capsules may be packed in a cylindrical vessel through which there is fluid flow. If the PCM is in its solid state and $T_{mp} < T_i$, heat is transferred from the fluid to the capsules and latent energy is stored in the PCM as it melts. Conversely, if the PCM is a liquid and $T_{mp} > T_i$, energy is released from the PCM as it freezes and heat is transferred to the fluid. In either situation, all of the capsules within the packed bed would remain at T_{mp} through much of the phase change process, in which case the fluid outlet temperature would remain at a fixed value T_o.

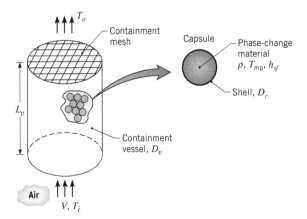

Consider an application for which air at atmospheric pressure is chilled by passing it through a packed bed ($\varepsilon = 0.5$) of capsules ($D_c = 50$ mm) containing an organic compound with a melting point of $T_{mp} = 4°C$. The air enters a cylindrical vessel ($L_v = D_v = 0.40$ m) at $T_i = 25°C$ and $V = 1.0$ m/s.

(a) If the PCM in each capsule is in the solid state at T_{mp} as melting occurs within the capsule, what is the outlet temperature of the air? If the density and latent heat of fusion of the PCM are $\rho = 1200$ kg/m³ and $h_{sf} = 165$ kJ/kg, what is the mass rate (kg/s) at which the PCM is converted from solid to liquid in the vessel?

(b) Explore the effect of the inlet air velocity and capsule diameter on the outlet temperature.

(c) At what location in the vessel will complete melting of the PCM in a capsule first occur? Once complete melting begins to occur, how will the outlet temperature vary with time and what is its asymptotic value?

7.91 The porosity of a packed bed can be decreased by vibrating the containment vessel as the vessel is filled with the particles. The vibration promotes particle settling.

(a) Consider the air chilling process of Problem 7.90a. Determine the outlet air temperature T_o and mass rate at which the PCM is melted for $\varepsilon = 0.30$. Assume the total mass of PCM and the mass flow rate of air are unchanged. The length of the containment vessel L_v is decreased to compensate for the reduced porosity.

(b) Determine T_o and the PCM melting rate for the case where the diameter of the containment vessel D_v is decreased to compensate for the reduced porosity. Which containment vessel configuration is preferred?

7.92 Consider the packed bed ($\varepsilon = 0.5$) of latent heat capsules ($D_c = 50$ mm) described in Problem 7.90, but now for an application in which ambient air is to be heated by passing it through the bed. In this case the capsules contain an organic compound with a melting point of $T_{mp} = 50°C$, and the air enters the vessel ($L_v = D_v = 0.40$ m) at $T_i = 20°C$ and $V = 1.0$ m/s.

(a) If the PCM in each capsule is in the liquid state at T_{mp} as solidification occurs within the capsule, what is the outlet temperature of the air? If the density and latent heat of fusion of the PCM are $\rho = 900$ kg/m³ and $h_{sf} = 200$ kJ/kg, what is the mass rate (kg/s) at which the PCM is converted from liquid to solid in the vessel?

(b) Explore the effect of the inlet air velocity and capsule diameter on the outlet temperature.

(c) At what location in the vessel will complete freezing of the PCM in a capsule first occur? Once complete freezing begins to occur, how will the outlet temperature vary with time and what is its asymptotic value?

7.93 Packed beds of spherical particles can be *sintered* at high temperature to form permeable, rigid foams. A foam sheet of thickness $t = 10$ mm is comprised of sintered bronze spheres, each of diameter $D = 0.6$ mm. The metal foam has a porosity of $\varepsilon = 0.25$, and the foam sheet fills the cross section of an $L = 40$ mm $\times W = 40$ mm wind tunnel. The upper and lower surfaces of the foam are at temperatures $T_s = 80°C$, and the two other foam edges (the front edge shown in the schematic and the corresponding back edge) are insulated. Air flows in the wind tunnel at an upstream temperature and velocity of $T_i = 20°C$ and $V = 10$ m/s, respectively.

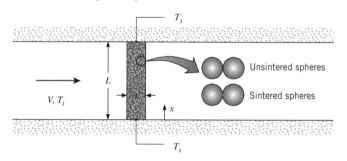

(a) Assuming the foam is at a uniform temperature T_s, *estimate* the convection heat transfer rate to the air. Do you expect the actual heat transfer rate to be equal to, less than, or greater than your estimated value?

(b) Assuming one-dimensional conduction in the *x*-direction, use an extended surface analysis to *estimate* the heat transfer rate to the air. To do so, show that the *effective* perimeter associated with Equation 3.70 is $P_{eff} = A_{p,t}/L$. Determine the effective thermal conductivity of the foam k_{eff} by using Equation 3.25. Do you expect the actual heat transfer rate to be equal to, less than, or greater than your estimated value?

Heat and Mass Transfer

7.94 Consider mass loss from a smooth wet flat plate due to forced convection at atmospheric pressure. The plate is 0.5 m long and 3 m wide. Dry air at 300 K and a free stream velocity of 35 m/s flows over the surface, which

is also at a temperature of 300 K. Estimate the average mass transfer coefficient $\bar{h}_m$ and determine the water vapor mass loss rate (kg/s) from the plate.

7.95 Consider dry, atmospheric air in parallel flow over a 1-m-long plate whose surface is wetted. The air velocity is 35 m/s, and the air and water are each at a temperature of 300 K. Estimate the heat loss and evaporation rates per unit width of the plate, q' and n'_A, respectively.

7.96 A flat plate coated with a volatile substance (species A) is exposed to dry, atmospheric air in parallel flow with $T_\infty = 20°C$ and $u_\infty = 8$ m/s. The plate is maintained at a constant temperature of 134°C by an electrical heating element, and the substance evaporates from the surface. The plate has a width of 0.25 m (normal to the plane of the sketch) and is well insulated on the bottom.

The molecular weight and the latent heat of vaporization of species A are $\mathcal{M}_A = 150$ kg/kmol and $h_{fg} = 5.44 \times 10^6$ J/kg, respectively, and the mass diffusivity is $D_{AB} = 7.75 \times 10^{-7}$ m²/s. If the saturated vapor pressure of the substance is 0.12 atm at 134°C, what is the electrical power required to maintain steady-state conditions?

7.97 Dry air at atmospheric pressure and 350 K, with a free stream velocity of 25 m/s, flows over a smooth, porous plate 1 m long.

(a) Assuming the plate to be saturated with liquid water at 350 K, estimate the mass rate of evaporation per unit width of the plate, n'_A (kg/s · m).

(b) For air and liquid water temperatures of 300, 325, and 350 K, generate plots of n'_A as a function of velocity for the range from 1 to 25 m/s.

7.98 A scheme for dissipating heat from an array of $N = 100$ integrated circuits involves joining the circuits to the bottom of a plate and exposing the top of the plate to a water bath. The water container is of length $L = 100$ mm on a side and is exposed to airflow at its top surface. The flow is turbulated by the protruding lip of the side wall.

Air

$u_\infty = 10$ m/s
$T_\infty = 300$ K
$\phi_\infty = 0.5$

Water bath, T_b — Plate

Integrated circuits

|← L →|

If the sides and bottom of the container are well insulated from the surroundings and heat is uniformly dissipated in each circuit, at what rate may heat be dissipated from each circuit when the water temperature is maintained at $T_b = 350$ K?

7.99 Dry air at 35°C and a velocity of 20 m/s flows over a wetted plate of length 500 mm and width 150 mm. An embedded electrical heater supplies power to maintain the plate surface temperature at 20°C.

(a) What is the evaporation rate (kg/h) of water from the plate? What electrical power is required to maintain steady-state conditions?

(b) After a long period of operation, all the water is evaporated from the plate and its surface is dry. For the same free stream conditions and heater power of part (a), estimate the temperature of the plate.

7.100 A minivan traveling 90 km/h has just passed through a thunderstorm that left a film of water 0.1 mm thick on the top of the van. The top of the van can be assumed to be a flat plate 6 m long. Assume isothermal conditions at 27°C, an ambient air relative humidity of 80%, and turbulent flow over the entire surface. What location on the van top will be the last to dry? What is the water evaporation rate per unit area (kg/s · m²) at the trailing edge of the van top?

7.101 Atmospheric air of 40% relative humidity and temperature $T_\infty = 300$ K is in parallel flow over a series of water-filled trays, with $u_\infty = 12$ m/s.

Air

u_∞
T_∞
ϕ_∞

Tray 1   Tray 2   Tray 3

x   0.5 m   1.0 m   1.5 m

What is the rate at which energy must be supplied to each of the first three trays to maintain the water at 300 K?

7.102 An electric power plant generates 500 MW of electric power and operates at a thermal efficiency of 38 percent. The waste heat from the power plant is transferred to the environment from a cooling pond that is 2000 m wide and 2000 m long. Assuming conditions are steady, an ambient temperature of $T_\infty = 20$°C, and a relative humidity of 50 percent, determine the pond water temperature if a breeze of 3 m/s flows parallel to a side of the pond. What is the heat loss rate from the pond by convection? What is the heat loss rate due to evaporation? Conditions are turbulent everywhere. Assume radiation heat transfer is negligible.

7.103 A 20-mm-diameter, 1-m-long, thin-walled copper tube contains a material that generates thermal energy at a volumetric rate of $\dot{q} = 0.2$ MW/m³. The rod is cooled by air at atmospheric pressure in cross flow at a velocity of 2 m/s. It is proposed to decrease the temperature of the material by spraying a thin, porous coating of thickness $t = 2$ mm onto the exterior of the copper tube. The coating is then saturated with liquid water, leading to evaporative cooling at the exposed surface of the coating. The thermal conductivity of the water-saturated coating is $k_c = 3.5$ W/m · K. Determine the temperature of the copper tube with and without the coating. The ambient conditions are characterized by a temperature of 25°C and a relative humidity of 45 percent.

7.104 Condenser cooling water for a power plant is stored in a cooling pond that is 1000 m long by 500 m wide. However, because of evaporative losses, it is necessary to periodically add "makeup" water to the pond in order to maintain a suitable water level. Assuming isothermal conditions at 27°C for the water and the air, that the free stream air is dry and moving at a velocity of 2 m/s in the direction of the 1000-m pond length, and that the boundary layer on the water surface is everywhere turbulent, determine the amount of makeup water that should be added to the pond daily.

7.105 In a paper-drying process, the paper moves on a conveyor belt at 0.2 m/s, while dry air from an in-line array of round jets (Figure 7.18b) impinges normal to its surface. The nozzle diameter and pitch are $D = 20$ mm and $S = 100$ mm, respectively, and the nozzle-to-paper separation is $H = 200$ mm. Air exits the nozzle at a velocity and temperature of 20 m/s and 300 K, while the wet paper is maintained at 300 K. In kg/s · m², what is the average drying rate of the paper?

7.106 In a paper mill drying process, a sheet of paper slurry (water–fiber mixture) has a linear velocity of 5 m/s as it is rolled. Radiant heaters maintain a sheet temperature of $T_s = 330$ K, as evaporation occurs to dry, ambient air at 300 K above and below the sheet.

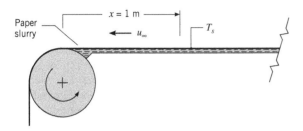

(a) What is the evaporative flux at a distance of $x = 1$ m from the leading edge of the roll? What is the corresponding value of the radiant flux (irradiation, G) that must be supplied to the sheet to maintain its temperature at 330 K? The sheet has an absorptivity of $\alpha = 1$.

(b) To accelerate the drying and paper production processes, the velocity and temperature of the strip are increased to 10 m/s and 340 K, respectively. To maintain a uniform strip temperature, the irradiation G must be varied with x along the strip. For $0 \le x \le 1$ m, compute and plot the variations $h_{m,x}(x)$, $N_A''(x)$, and $G(x)$.

7.107 A channel of triangular cross section, which is 25 m long and 1 m deep, is used for the storage of water.

The water and the surrounding air are each at a temperature of 25°C, and the relative humidity of the air is 50%.

(a) If the air moves at a velocity of 5 m/s along the length of the channel, what is the rate of water loss due to evaporation from the surface?

(b) Obtain an expression for the rate at which the water depth would decrease with time due to evaporation. For the above conditions, how long would it take for all the water to evaporate?

7.108 Dry air at 20°C and a velocity of 15 m/s flows over a 20-mm-diameter rod covered with a thin porous coating that is saturated with water. The rod ($k = 175$ W/m · K) is 250 mm long and its ends are attached to heat sinks maintained at 35°C.

Perform a steady-state, finite-difference analysis of the rod–porous coating system, considering conduction in the rod as well as energy transfer from the surface by convection heat and mass transfer. Use the analysis to estimate the temperature at the midspan of the rod and the evaporation rate from the surface. (*Suggestions:* Use 10 nodes to represent the half-length of the system. Estimate the overall average convection heat transfer coefficient based on an average film temperature for the system, and use the heat–mass transfer analogy to determine the average convection mass transfer coefficient. Validate your code by using it to predict a temperature distribution that agrees with the analytical solution for a fin without evaporation.)

7.109 Dry air at 35°C and a velocity of 15 m/s flows over a long cylinder of 20-mm diameter. The cylinder is covered with a thin porous coating saturated with water, and an embedded electrical heater supplies power to maintain the coating surface temperature at 20°C.

(a) What is the evaporation rate of water from the cylinder per unit length (kg/h · m)? What electrical power per unit length of the cylinder (W/m) is required to maintain steady-state conditions?

(b) After a long period of operation, all the water is evaporated from the coating and its surface is dry. For the same free stream conditions and heater power of part (a), estimate the temperature of the surface.

7.110 Approximate the human form as an unclothed vertical cylinder of 0.3-m diameter and 1.75-m length with a surface temperature of 30°C.

(a) Calculate the rate of heat loss in a 10-m/s wind at 20°C.

(b) What is the rate of heat loss if the skin is covered with a thin layer of water at 30°C and the relative humidity of the air is 60%?

7.111 In the first stage of a paper-drying process, a cylinder of diameter 0.15 m is covered by moisture-soaked paper. The temperature of the paper is maintained at 70°C by embedded electrical heaters. Dry air at a velocity of 10 m/s and temperature of 20°C flows over the cylinder.

(a) Calculate the required electrical power and the evaporation rate per unit length of the cylinder, q' and n'_A, respectively.

(b) Generate plots of q' and n'_A as a function of the dry air velocity for $5 \leq V \leq 20$ m/s and for paper temperatures of 65, 70, and 75°C.

7.112 Cylindrical dry-bulb and wet-bulb thermometers are installed in a large-diameter duct to obtain the temperature T_∞ and the relative humidity ϕ_∞ of moist air flowing through the duct at a velocity V. The dry-bulb thermometer has a bare glass surface of diameter D_{db} and emissivity ε_g. The wet-bulb thermometer is covered with a thin wick that is saturated with water flowing continuously by capillary action from a bottom reservoir. Its diameter and emissivity are designated as D_{wb} and ε_w. The duct inside surface is at a known temperature T_s, which is less than T_∞. Develop expressions that may be used to obtain T_∞ and ϕ_∞ from knowledge of the dry-bulb and wet-bulb temperatures T_{db} and T_{wb} and the foregoing parameters. Determine T_∞ and ϕ_∞ when $T_{db} = 45°C$, $T_{wb} = 25°C$, $T_s = 35°C$, $p = 1$ atm, $V = 5$ m/s, $D_{db} = 3$ mm, $D_{wb} = 4$ mm, and $\varepsilon_g = \varepsilon_w = 0.95$. As a first approximation, evaluate the dry- and wet-bulb air properties at 45 and 25°C, respectively.

7.113 The thermal pollution problem is associated with discharging warm water from an electrical power plant or from an industrial source to a natural body of water. Methods for alleviating this problem involve cooling the warm water before allowing the discharge to occur. Two such methods, involving wet cooling towers or spray ponds, rely on heat transfer from the warm water in droplet form to the surrounding atmosphere. To develop an understanding of the mechanisms that contribute to this cooling, consider a spherical droplet of diameter D and temperature T, which is moving at a velocity V relative to air at a temperature T_∞ and relative humidity ϕ_∞. The surroundings are characterized by the temperature T_{sur}. Develop expressions for the droplet evaporation and cooling rates. Calculate the evaporation rate (kg/s) and cooling rate (K/s) when $D = 3$ mm, $V = 7$ m/s, $T = 40°C$, $T_\infty = 25°C$, $T_{sur} = 15°C$, and $\phi_\infty = 0.60$. The emissivity of water is $\varepsilon_w = 0.96$.

7.114 Cranberries are harvested by flooding the bogs in which they are grown and raking them into troughs for transport. At the processing plant, the surface moisture on the berries is removed as they roll over a fine screen through which warm air is blown. The berries have an average diameter of 15 mm, and the thickness of the water layer is 0.2 mm.

If the velocity and temperature of the heated air are 2 m/s and 30°C, respectively, estimate the time required to dry the berries. Assume that the water film on the berries is also at 30°C.

7.115 A spherical drop of water, 0.5 mm in diameter, is falling at a velocity of 2.15 m/s through dry, still air at 1-atm pressure. Estimate the instantaneous rate of evaporation from the drop if the drop surface is at 60°C and the air is at 100°C.

7.116 In a home furnace humidification system, water droplets of diameter D are discharged in a direction opposing the motion of warm air emerging from the heater. The air is humidified by evaporation from the droplets, and the excess water is collected on a *splash plate,* from which it is routed to a drain.

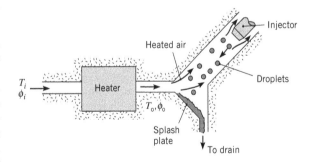

Consider conditions for which air enters the heater at a temperature and relative humidity of 17°C and 70%, respectively, and leaves the heater at a temperature of 47°C. The droplet diameter is 1 mm, and the relative velocity between the droplets and the heated air is 15 m/s. During the time-of-flight, the change in droplet diameter may be neglected and the droplet temperature may be assumed to remain at 47°C. What is the rate of evaporation from a single droplet?

7.117 Motile bacteria are equipped with flagella that are rotated by tiny, biological electrochemical engines which, in turn, propel the bacteria through a host liquid. Consider a nominally spherical *Escherichia coli* bacterium that is of diameter $D = 2\ \mu$m. The bacterium is in a water-based solution at 37°C containing a nutrient which is characterized by a binary diffusion coefficient of $D_{AB} = 0.7 \times 10^{-9}$ m²/s and a food energy value of $\mathcal{N} = 16{,}000$ kJ/kg. There is a nutrient density difference between the fluid and the shell of the bacterium of $\Delta\rho_A = 860 \times 10^{-12}$ kg/m³. Assuming a propulsion efficiency of $\eta = 0.5$, determine the maximum speed of the *E. coli*. Report your answer in body diameters per second.

7.118 Evaporation of liquid fuel droplets is often studied in the laboratory by using a porous sphere technique in which the fuel is supplied at a rate just sufficient to maintain a completely wetted surface on the sphere.

Consider the use of kerosene at 300 K with a porous sphere of 1-mm diameter. At this temperature the kerosene has a saturated vapor density of 0.015 kg/m³ and a latent heat of vaporization of 300 kJ/kg. The mass diffusivity for the vapor–air mixture is 10^{-5} m²/s. If dry, atmospheric air at $V = 15$ m/s and $T_\infty = 300$ K flows over the sphere, what is the minimum mass rate at which kerosene must be supplied to maintain a wetted surface? For this condition, by how much must T_∞ actually exceed T_s to maintain the wetted surface at 300 K?

7.119 Consider an air-conditioning system composed of a bank of tubes arranged normal to air flowing in a duct at a mass rate of $\dot{m}_a$ (kg/s). A coolant flowing through the tubes is able to maintain the surface temperature of the tubes at a constant value of $T_s < T_{a,i}$, where $T_{a,i}$ is the inlet air temperature (upstream of the tube bank). It has been suggested that air cooling may be enhanced if a thin, uniform film of water is maintained on the outer surface of each of the tubes.

(a) Assuming the water film to be at the temperature T_s, develop an expression for the ratio of the amount of cooling that occurs with the water film to the amount of cooling that occurs without the film. The amount of cooling may be defined as $T_{a,i} - T_{a,o}$, where $T_{a,o}$ is the outlet air temperature (downstream of the tube bank). The upstream air may be assumed to be dry, and the driving potentials for convection heat and mass transfer may be approximated as $(T_{a,i} - T_s)$ and $\rho_{A,sat}(T_s)$, respectively. *Note:* The total rate of heat loss from the air may be expressed as $q = \dot{m}_a c_{p,a}(T_{a,i} - T_{a,o})$. Estimate the value of this ratio under conditions for which $T_{a,i} = 35$°C and $T_s = 10$°C.

(b) Consider a tube bank that is 5 rows deep, with 12 tubes in a row. Each tube is 0.5 m long, with an outside diameter of 8 mm, and a staggered arrangement is used for which $S_T = S_L = 24$ mm. Under conditions for which $\dot{m}_a = 0.5$ kg/s, $V = 3$ m/s, $T_{a,i} = 35$°C, and $T_s = 10$°C, what is the value of $T_{a,o}$ if the tubes are wetted? What is the specific humidity of the air leaving the tube bank?

CHAPTER

Internal Flow

Having acquired the means to compute convection transfer rates for external flow, we now consider the convection transfer problem for *internal flow*. Recall that an external flow is one for which boundary layer development on a surface is allowed to continue without external constraints, as for the flat plate of Figure 6.6. In contrast, an internal flow, such as flow in a pipe, is one for which the fluid is *confined* by a surface. Hence the boundary layer is unable to develop without eventually being constrained. The internal flow configuration represents a convenient geometry for heating and cooling fluids used in chemical processing, environmental control, and energy conversion technologies.

Our objectives are to develop an appreciation for the physical phenomena associated with internal flow and to obtain convection coefficients for flow conditions of practical importance. As in Chapter 7, we will restrict attention to problems of low-speed, forced convection with no phase change occurring in the fluid. We begin by considering velocity (hydrodynamic) effects pertinent to internal flows, focusing on certain unique features of boundary layer development. Thermal boundary layer effects are considered next, and an overall energy balance is applied to determine fluid temperature variations in the flow direction. Finally, correlations for estimating the convection heat transfer coefficient are presented for a variety of internal flow conditions.

8.1 *Hydrodynamic Considerations*

When considering external flow, it is necessary to ask only whether the flow is laminar or turbulent. However, for an internal flow we must also be concerned with the existence of *entrance* and *fully developed* regions.

8.1.1 Flow Conditions

Consider laminar flow in a circular tube of radius r_o (Figure 8.1), where fluid enters the tube with a uniform velocity. We know that when the fluid makes contact with the surface, viscous effects become important, and a boundary layer develops with increasing x. This development occurs at the expense of a shrinking inviscid flow region and concludes with boundary layer merger at the centerline. Following this merger, viscous effects extend over

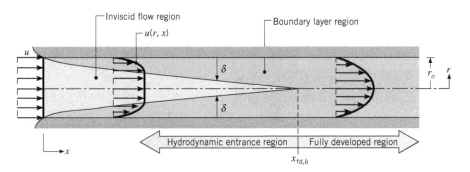

FIGURE 8.1 Laminar, hydrodynamic boundary layer development in a circular tube.

the entire cross section and the velocity profile no longer changes with increasing x. The flow is then said to be *fully developed,* and the distance from the entrance at which this condition is achieved is termed the *hydrodynamic entry length,* $x_{\text{fd},h}$. As shown in Figure 8.1, the *fully developed velocity profile* is parabolic for laminar flow in a circular tube. For turbulent flow, the profile is *flatter* due to turbulent mixing in the radial direction.

When dealing with internal flows, it is important to be cognizant of the extent of the entry region, which depends on whether the flow is laminar or turbulent. The Reynolds number for flow in a circular tube is defined as

$$Re_D \equiv \frac{\rho u_m D}{\mu} = \frac{u_m D}{\nu} \tag{8.1}$$

where u_m is the mean fluid velocity over the tube cross section and D is the tube diameter. In a fully developed flow, the critical Reynolds number corresponding to the *onset* of turbulence is

$$Re_{D,c} \approx 2300 \tag{8.2}$$

although much larger Reynolds numbers ($Re_D \approx 10{,}000$) are needed to achieve fully turbulent conditions. The transition to turbulence is likely to begin in the developing boundary layer of the entrance region [1].

For laminar flow ($Re_D \lesssim 2300$), the hydrodynamic entry length may be obtained from an expression of the form [2]

$$\left(\frac{x_{\text{fd},h}}{D}\right)_{\text{lam}} \approx 0.05\, Re_D \tag{8.3}$$

This expression is based on the presumption that fluid enters the tube from a rounded converging nozzle and is hence characterized by a nearly uniform velocity profile at the entrance (Figure 8.1). Although there is no satisfactory general expression for the entry length in turbulent flow, we know that it is approximately independent of Reynolds number and that, as a first approximation [3],

$$10 \lesssim \left(\frac{x_{\text{fd},h}}{D}\right)_{\text{turb}} \lesssim 60 \tag{8.4}$$

For the purposes of this text, we shall assume fully developed turbulent flow for $(x/D) > 10$.

8.1.2 The Mean Velocity

Because the velocity varies over the cross section and there is no well-defined free stream, it is necessary to work with a mean velocity u_m when dealing with internal flows. This velocity is defined such that, when multiplied by the fluid density ρ and the cross-sectional area of the tube A_c, it provides the rate of mass flow through the tube. Hence

$$\dot{m} = \rho u_m A_c \tag{8.5}$$

For steady, incompressible flow in a tube of uniform cross-sectional area, $\dot{m}$ and u_m are constants independent of x. From Equations 8.1 and 8.5 it is evident that, for flow in a *circular tube* ($A_c = \pi D^2/4$), the Reynolds number reduces to

$$Re_D = \frac{4\dot{m}}{\pi D \mu} \tag{8.6}$$

Since the mass flow rate may also be expressed as the integral of the mass flux (ρu) over the cross section

$$\dot{m} = \int_{A_c} \rho u(r, x) \, dA_c \tag{8.7}$$

it follows that, for *incompressible* flow in a *circular* tube,

$$u_m = \frac{\int_{A_c} \rho u(r, x) \, dA_c}{\rho A_c} = \frac{2\pi\rho}{\rho\pi r_o^2} \int_0^{r_o} u(r, x) r \, dr = \frac{2}{r_o^2} \int_0^{r_o} u(r, x) r \, dr \tag{8.8}$$

The foregoing expression may be used to determine u_m at any axial location x from knowledge of the velocity profile $u(r)$ at that location.

8.1.3   Velocity Profile in the Fully Developed Region

The form of the velocity profile may readily be determined for the *laminar flow* of an *incompressible, constant property fluid* in the *fully developed region* of a *circular tube*. An important feature of hydrodynamic conditions in the fully developed region is that both the radial velocity component v and the gradient of the axial velocity component ($\partial u/\partial x$) are everywhere zero.

$$v = 0 \quad \text{and} \quad \left(\frac{\partial u}{\partial x}\right) = 0 \tag{8.9}$$

Hence the axial velocity component depends only on r, $u(x, r) = u(r)$.

The radial dependence of the axial velocity may be obtained by solving the appropriate form of the x-momentum equation. This form is determined by first recognizing that, for the conditions of Equation 8.9, the net momentum flux is everywhere zero in the fully developed region. Hence the momentum conservation requirement reduces to a simple balance between shear and pressure forces in the flow. For the annular differential element of Figure 8.2, this force balance may be expressed as

$$\tau_r(2\pi r \, dx) - \left\{ \tau_r(2\pi r \, dx) + \frac{d}{dr}[\tau_r(2\pi r \, dx)] dr \right\}$$

$$+ \, p(2\pi r \, dr) - \left\{ p(2\pi r \, dr) + \frac{d}{dx}[p(2\pi r \, dr)] dx \right\} = 0$$

which reduces to

$$-\frac{d}{dr}(r\tau_r) = r\frac{dp}{dx} \tag{8.10}$$

FIGURE 8.2 Force balance on a differential element for laminar, fully developed flow in a circular tube.

With $y = r_o - r$, Newton's law of viscosity, Equation 6S.10, assumes the form

$$\tau_r = -\mu \frac{du}{dr} \qquad (8.11)$$

and Equation 8.10 becomes

$$\frac{\mu}{r} \frac{d}{dr}\left(r \frac{du}{dr}\right) = \frac{dp}{dx} \qquad (8.12)$$

Since the axial pressure gradient is independent of r, Equation 8.12 may be solved by integrating twice to obtain

$$r \frac{du}{dr} = \frac{1}{\mu}\left(\frac{dp}{dx}\right)\frac{r^2}{2} + C_1$$

and

$$u(r) = \frac{1}{\mu}\left(\frac{dp}{dx}\right)\frac{r^2}{4} + C_1 \ln r + C_2$$

The integration constants may be determined by invoking the boundary conditions

$$u(r_o) = 0 \qquad \text{and} \qquad \left.\frac{\partial u}{\partial r}\right|_{r=0} = 0$$

which, respectively, impose the requirements of zero slip at the tube surface and radial symmetry about the centerline. It is a simple matter to evaluate the constants, and it follows that

$$u(r) = -\frac{1}{4\mu}\left(\frac{dp}{dx}\right)r_o^2\left[1 - \left(\frac{r}{r_o}\right)^2\right] \qquad (8.13)$$

Hence the fully developed velocity profile is *parabolic*, as illustrated in Figure 8.2. Note that the pressure gradient must always be negative.

The foregoing result may be used to determine the mean velocity of the flow. Substituting Equation 8.13 into Equation 8.8 and integrating, we obtain

$$u_m = -\frac{r_o^2}{8\mu}\frac{dp}{dx} \qquad (8.14)$$

Substituting this result into Equation 8.13, the velocity profile is then

$$\frac{u(r)}{u_m} = 2\left[1 - \left(\frac{r}{r_o}\right)^2\right] \tag{8.15}$$

Since u_m can be computed from knowledge of the mass flow rate, Equation 8.14 can be used to determine the pressure gradient.

8.1.4 Pressure Gradient and Friction Factor in Fully Developed Flow

The engineer is frequently interested in the pressure drop needed to sustain an internal flow because this parameter determines pump or fan power requirements. To determine the pressure drop, it is convenient to work with the *Moody* (or Darcy) *friction factor*, which is a dimensionless parameter defined as

$$f \equiv \frac{-(dp/dx)D}{\rho u_m^2/2} \tag{8.16}$$

This quantity is not to be confused with the *friction coefficient*, sometimes called the Fanning friction factor, which is defined as

$$C_f \equiv \frac{\tau_s}{\rho u_m^2/2} \tag{8.17}$$

Since $\tau_s = -\mu(du/dr)_{r=r_o}$, it follows from Equation 8.13 that

$$C_f = \frac{f}{4} \tag{8.18}$$

Substituting Equations 8.1 and 8.14 into 8.16, it follows that, for fully developed laminar flow,

$$f = \frac{64}{Re_D} \tag{8.19}$$

For fully developed turbulent flow, the analysis is much more complicated, and we must ultimately rely on experimental results. In addition to depending on the Reynolds number, the friction factor is a function of the tube surface condition and increases with surface roughness e. Measured friction factors covering a wide range of conditions have been correlated by Colebrook [4, 5] and are described by the transcendental expression

$$\frac{1}{\sqrt{f}} = -2.0 \log\left[\frac{e/D}{3.7} + \frac{2.51}{Re_D\sqrt{f}}\right] \tag{8.20}$$

A correlation for the smooth surface condition that encompasses a large Reynolds number range has been developed by Petukhov [6] and is of the form

$$f = (0.790 \ln Re_D - 1.64)^{-2} \qquad 3000 \lesssim Re_D \lesssim 5 \times 10^6 \tag{8.21}$$

Equations 8.19 and 8.20 are plotted in the *Moody diagram* [7] of Figure 8.3.

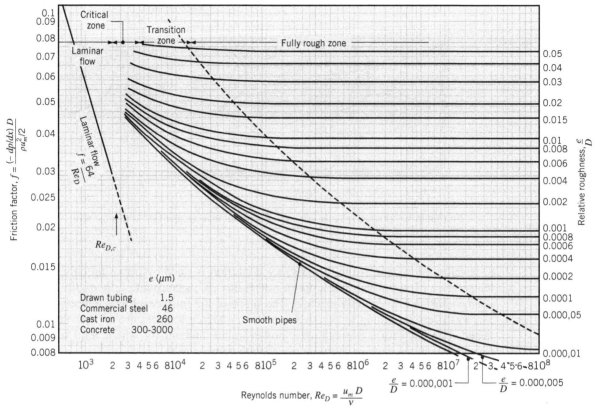

FIGURE 8.3 Friction factor for fully developed flow in a circular tube. Used with permission from Moody, L. F., *Trans. ASME*, **66**, 671, 1944.

Note that f, hence dp/dx, is a constant in the fully developed region. From Equation 8.16 the pressure drop $\Delta p = p_1 - p_2$ associated with fully developed flow from the axial position x_1 to x_2 may then be expressed as

$$\Delta p = -\int_{p_1}^{p_2} dp = f \frac{\rho u_m^2}{2D} \int_{x_1}^{x_2} dx = f \frac{\rho u_m^2}{2D}(x_2 - x_1) \tag{8.22a}$$

where f is obtained from Figure 8.3 or from Equation 8.19 for laminar flow and from Equation 8.20 or 8.21 for turbulent flow. The pump or fan power required to overcome the resistance to flow associated with this pressure drop may be expressed as

$$P = (\Delta p)\dot{\forall} \tag{8.22b}$$

where the volumetric flow rate $\dot{\forall}$ may, in turn, be expressed as $\dot{\forall} = \dot{m}/\rho$ for an incompressible fluid.

8.2 Thermal Considerations

Having reviewed the fluid mechanics of internal flow, we now consider thermal effects. If fluid enters the tube of Figure 8.4 at a uniform temperature $T(r, 0)$ that is less than the surface temperature, convection heat transfer occurs and a *thermal boundary layer* begins

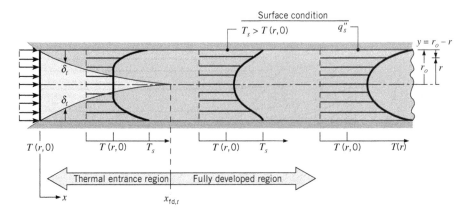

FIGURE 8.4 Thermal boundary layer development in a circular tube with a hot wall.

to develop. Moreover, if the tube *surface* condition is fixed by imposing either a uniform temperature (T_s is constant) or a uniform heat flux (q_s'' is constant), a *thermally fully developed condition* is eventually reached. The shape of the fully developed temperature profile $T(r, x)$ differs according to whether a uniform surface temperature or heat flux is maintained. For both surface conditions, however, the amount by which fluid temperatures exceed the entrance temperature increases with increasing x.

For laminar flow the *thermal entry length* may be expressed as [3]

$$\left(\frac{x_{\text{fd},t}}{D}\right)_{\text{lam}} \approx 0.05\, Re_D\, Pr \tag{8.23}$$

Comparing Equations 8.3 and 8.23, it is evident that, if $Pr > 1$, the hydrodynamic boundary layer develops more rapidly than the thermal boundary layer ($x_{\text{fd},h} < x_{\text{fd},t}$), while the inverse is true for $Pr < 1$. For large Prandtl number fluids such as oils, $x_{\text{fd},h}$ is much smaller than $x_{\text{fd},t}$ and it is reasonable to assume a fully developed velocity profile throughout the thermal entry region. In contrast, for turbulent flow, conditions are nearly independent of Prandtl number, and to a first approximation, we shall assume ($x_{\text{fd},t}/D$) = 10.

Thermal conditions in the fully developed region are characterized by several interesting and useful features. Before we can consider these features (Section 8.2.3), however, it is necessary to introduce the concept of a mean temperature and the appropriate form of Newton's law of cooling.

8.2.1 The Mean Temperature

Just as the absence of a free stream velocity requires use of a mean velocity to describe an internal flow, the absence of a fixed free stream temperature necessitates using a *mean (or bulk) temperature*. To provide a definition of the mean temperature, we begin by returning to Equation 1.12e:

$$q = \dot{m}c_p(T_{\text{out}} - T_{\text{in}}) \tag{1.12e}$$

Recall that the terms on the right-hand side represent the thermal energy for an incompressible liquid or the enthalpy (thermal energy plus flow work) for an ideal gas, which is carried by the fluid. In developing this equation, it was implicitly assumed that the temperature was uniform across the inlet and outlet cross-sectional areas. In reality, this is not true if convection heat transfer occurs, and we *define* the mean temperature so that the term $\dot{m}c_pT_m$ is equal to the true rate of thermal energy (or enthalpy) advection integrated over the cross section. This true advection rate may be obtained by integrating the product of mass flux (ρu) and the thermal energy (or enthalpy) per unit mass, c_pT, over the cross section. Therefore, we define T_m from

$$\dot{m}c_pT_m = \int_{A_c} \rho u c_p T\, dA_c \tag{8.24}$$

or

$$T_m = \frac{\int_{A_c} \rho u c_p T\, dA_c}{\dot{m}c_p} \tag{8.25}$$

For flow in a circular tube with constant ρ and c_p, it follows from Equations 8.5 and 8.25 that

$$T_m = \frac{2}{u_m r_o^2} \int_0^{r_o} u T r\, dr \tag{8.26}$$

It is important to note that, when multiplied by the mass flow rate and the specific heat, T_m provides the rate at which thermal energy (or enthalpy) is advected with the fluid as it moves along the tube.

8.2.2 Newton's Law of Cooling

The mean temperature T_m is a convenient reference temperature for internal flows, playing much the same role as the free stream temperature T_∞ for external flows. Accordingly, Newton's law of cooling may be expressed as

$$q_s'' = h(T_s - T_m) \tag{8.27}$$

where h is the *local* convection heat transfer coefficient. However, there is an essential difference between T_m and T_∞. Whereas T_∞ is constant in the flow direction, T_m must vary in this direction. That is, dT_m/dx is never zero if heat transfer is occurring. The value of T_m increases with x if heat transfer is from the surface to the fluid ($T_s > T_m$); it decreases with x if the opposite is true ($T_s < T_m$).

8.2.3 Fully Developed Conditions

Since the existence of convection heat transfer between the surface and the fluid dictates that the fluid temperature must continue to change with x, one might legitimately question whether fully developed thermal conditions can ever be reached. The situation is certainly different from the hydrodynamic case, for which $(\partial u/\partial x) = 0$ in the fully developed region. In contrast, if there is heat transfer, (dT_m/dx), as well as $(\partial T/\partial x)$ at any radius r, is not zero.

Accordingly, the temperature profile $T(r)$ is continuously changing with x, and it would seem that a fully developed condition could never be reached. This apparent contradiction may be reconciled by working with a dimensionless form of the temperature, as was done for transient conduction (Chapter 5) and the energy conservation equation (Chapter 6).

Introducing a dimensionless temperature difference of the form $(T_s - T)/(T_s - T_m)$, conditions for which this ratio becomes independent of x are known to exist [3]. That is, although the temperature profile $T(r)$ continues to change with x, the *relative* shape of the profile no longer changes and the flow is said to be *thermally fully developed*. The requirement for such a condition is formally stated as

$$\frac{\partial}{\partial x}\left[\frac{T_s(x) - T(r, x)}{T_s(x) - T_m(x)}\right]_{\text{fd},t} = 0 \tag{8.28}$$

where T_s is the tube surface temperature, T is the local fluid temperature, and T_m is the mean temperature of the fluid over the cross section of the tube.

The condition given by Equation 8.28 is eventually reached in a tube for which there is either a *uniform surface heat flux* (q_s'' is constant) *or a uniform surface temperature* (T_s is constant). These surface conditions arise in many engineering applications. For example, a constant surface heat flux would exist if the tube wall were heated electrically or if the outer surface were uniformly irradiated. In contrast, a constant surface temperature would exist if a phase change (due to boiling or condensation) were occurring at the outer surface. Note that it is impossible to *simultaneously* impose the conditions of constant surface heat flux and constant surface temperature. If q_s'' is constant, T_s must vary with x; conversely, if T_s is constant, q_s'' must vary with x.

Several important features of thermally developed flow may be inferred from Equation 8.28. Since the temperature ratio is independent of x, the derivative of this ratio with respect to r must also be independent of x. Evaluating this derivative at the tube surface (note that T_s and T_m are constants insofar as differentiation with respect to r is concerned), we then obtain

$$\frac{\partial}{\partial r}\left(\frac{T_s - T}{T_s - T_m}\right)\bigg|_{r=r_o} = \frac{-\partial T/\partial r|_{r=r_o}}{T_s - T_m} \neq f(x)$$

Substituting for $\partial T/\partial r$ from Fourier's law, which, from Figure 8.4, is of the form

$$q_s'' = -k\frac{\partial T}{\partial y}\bigg|_{y=0} = k\frac{\partial T}{\partial r}\bigg|_{r=r_o}$$

and for q_s'' from Newton's law of cooling, Equation 8.27, we obtain

$$\frac{h}{k} \neq f(x) \tag{8.29}$$

Hence *in the thermally fully developed flow* of a fluid *with constant properties,* the *local convection coefficient is a constant, independent of x.*

Equation 8.28 is not satisfied in the entrance region, where h varies with x, as shown in Figure 8.5. Because the thermal boundary layer thickness is zero at the tube entrance, the convection coefficient is extremely large at $x = 0$. However, h decays rapidly as the thermal boundary layer develops, until the constant value associated with fully developed conditions is reached.

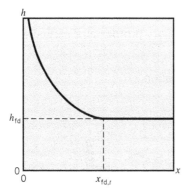

FIGURE 8.5   Axial variation of the convection heat transfer coefficient for flow in a tube.

Additional simplifications are associated with the special case of *uniform surface heat flux*. Since both h and q''_s are constant in the fully developed region, it follows from Equation 8.27 that

$$\left.\frac{dT_s}{dx}\right|_{\text{fd},t} = \left.\frac{dT_m}{dx}\right|_{\text{fd},t} \qquad q''_s = \text{constant} \tag{8.30}$$

If we expand Equation 8.28 and solve for $\partial T/\partial x$, it also follows that

$$\left.\frac{\partial T}{\partial x}\right|_{\text{fd},t} = \left.\frac{dT_s}{dx}\right|_{\text{fd},t} - \frac{(T_s - T)}{(T_s - T_m)}\left.\frac{dT_s}{dx}\right|_{\text{fd},t} + \frac{(T_s - T)}{(T_s - T_m)}\left.\frac{dT_m}{dx}\right|_{\text{fd},t} \tag{8.31}$$

Substituting from Equation 8.30, we then obtain

$$\left.\frac{\partial T}{\partial x}\right|_{\text{fd},t} = \left.\frac{dT_m}{dx}\right|_{\text{fd},t} \qquad q''_s = \text{constant} \tag{8.32}$$

Hence the axial temperature gradient is independent of the radial location.

For the case of *constant surface temperature* ($dT_s/dx = 0$), it also follows from Equation 8.31 that

$$\left.\frac{\partial T}{\partial x}\right|_{\text{fd},t} = \frac{(T_s - T)}{(T_s - T_m)}\left.\frac{dT_m}{dx}\right|_{\text{fd},t} \qquad T_s = \text{constant} \tag{8.33}$$

in which case the value of $\partial T/\partial x$ depends on the radial coordinate.

From the foregoing results, it is evident that the mean temperature is a very important variable for internal flows. To describe such flows, its variation with x must be known. This variation may be obtained by applying an *overall energy balance* to the flow, as will be shown in the next section.

EXAMPLE 8.1

For flow of a liquid metal through a circular tube, the velocity and temperature profiles at a particular axial location may be approximated as being uniform and parabolic, respectively. That is, $u(r) = C_1$ and $T(r) - T_s = C_2[1 - (r/r_o)^2]$, where C_1 and C_2 are constants. What is the value of the Nusselt number Nu_D at this location?

SOLUTION

Known: Form of the velocity and temperature profiles at a particular axial location for flow in a circular tube.

Find: Nusselt number at the prescribed location.

Schematic:

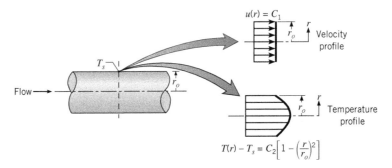

Assumptions: Incompressible, constant property flow.

Analysis: The Nusselt number may be obtained by first determining the convection coefficient, which, from Equation 8.27, is given as

$$h = \frac{q_s''}{T_s - T_m}$$

From Equation 8.26, the mean temperature is

$$T_m = \frac{2}{u_m r_o^2} \int_0^{r_o} u T r\, dr = \frac{2C_1}{u_m r_o^2} \int_0^{r_o} \left\{ T_s + C_2 \left[1 - \left(\frac{r}{r_o} \right)^2 \right] \right\} r\, dr$$

or, since $u_m = C_1$ from Equation 8.8,

$$T_m = \frac{2}{r_o^2} \int_0^{r_o} \left\{ T_s + C_2 \left[1 - \left(\frac{r}{r_o} \right)^2 \right] \right\} r\, dr$$

$$T_m = \frac{2}{r_o^2} \left[T_s \frac{r^2}{2} + C_2 \frac{r^2}{2} - \frac{C_2}{4} \frac{r^4}{r_o^2} \right]\Bigg|_0^{r_o}$$

$$T_m = \frac{2}{r_o^2} \left(T_s \frac{r_o^2}{2} + \frac{C_2}{2} r_o^2 - \frac{C_2}{4} r_o^2 \right) = T_s + \frac{C_2}{2}$$

The heat flux may be obtained from Fourier's law, in which case

$$q_s'' = k \frac{\partial T}{\partial r}\Bigg|_{r=r_o} = -k C_2 2 \frac{r}{r_o^2}\Bigg|_{r=r_o} = -2C_2 \frac{k}{r_o}$$

Hence

$$h = \frac{q_s''}{T_s - T_m} = \frac{-2C_2(k/r_o)}{-C_2/2} = \frac{4k}{r_o}$$

and

$$Nu_D = \frac{hD}{k} = \frac{(4k/r_o) \times 2r_o}{k} = 8 \qquad \triangleleft$$

8.3 *The Energy Balance*

8.3.1 General Considerations

Because the flow in a tube is completely enclosed, an energy balance may be applied to determine how the mean temperature $T_m(x)$ varies with position along the tube and how the total convection heat transfer rate q_{conv} is related to the difference in temperatures at the tube inlet and outlet. Consider the tube flow of Figure 8.6. Fluid moves at a constant flow rate $\dot{m}$, and convection heat transfer occurs at the inner surface. Typically, it will be reasonable to make one of the four assumptions in Section 1.3 that leads to the simplified steady-flow thermal energy equation, Equation 1.12e. For example, it is often the case that viscous dissipation is negligible (see Problem 8.9) and that the fluid can be modeled as either an incompressible liquid or an ideal gas with negligible pressure variation. In addition, it is usually reasonable to neglect net heat transfer by conduction in the axial direction, so the heat transfer term in Equation 1.12e includes only q_{conv}. Therefore, Equation 1.12e may be written in the form

$$q_{conv} = \dot{m}c_p(T_{m,o} - T_{m,i}) \qquad (8.34)$$

for a tube of finite length. This simple overall energy balance relates three important thermal variables (q_{conv}, $T_{m,o}$, $T_{m,i}$). *It is a general expression that applies irrespective of the nature of the surface thermal or tube flow conditions.*

Applying Equation 1.12e to the differential control volume of Figure 8.6 and recalling that the mean temperature is defined such that $\dot{m}c_pT_m$ represents the true rate of thermal energy (or enthalpy) advection integrated over the cross section, we obtain

$$dq_{conv} = \dot{m}c_p[(T_m + dT_m) - T_m] \qquad (8.35)$$

or

$$dq_{conv} = \dot{m}c_p dT_m \qquad (8.36)$$

FIGURE 8.6 Control volume for internal flow in a tube.

Equation 8.36 may be cast in a convenient form by expressing the rate of convection heat transfer to the differential element as $dq_{conv} = q_s'' P\,dx$, where P is the surface perimeter ($P = \pi D$ for a circular tube). Substituting from Equation 8.27, it follows that

$$\frac{dT_m}{dx} = \frac{q_s'' P}{\dot{m}c_p} = \frac{P}{\dot{m}c_p}h(T_s - T_m) \tag{8.37}$$

This expression is an extremely useful result, from which the axial variation of T_m may be determined. If $T_s > T_m$, heat is transferred to the fluid and T_m increases with x; if $T_s < T_m$, the opposite is true.

The manner in which quantities on the right-hand side of Equation 8.37 vary with x should be noted. Although P may vary with x, most commonly it is a constant (a tube of constant cross-sectional area). Hence the quantity $(P/\dot{m}c_p)$ is a constant. In the fully developed region, the convection coefficient h is also constant, although it decreases with x in the entrance region (Figure 8.5). Finally, although T_s may be constant, T_m must always vary with x (except for the trivial case of no heat transfer, $T_s = T_m$).

The solution to Equation 8.37 for $T_m(x)$ depends on the surface thermal condition. Recall that the two special cases of interest are *constant surface heat flux* and *constant surface temperature*. It is common to find one of these conditions existing to a reasonable approximation.

8.3.2 Constant Surface Heat Flux

For constant surface heat flux we first note that it is a simple matter to determine the total heat transfer rate q_{conv}. Since q_s'' is independent of x, it follows that

$$q_{conv} = q_s''(P \cdot L) \tag{8.38}$$

This expression could be used with Equation 8.34 to determine the fluid temperature change, $T_{m,o} - T_{m,i}$.

For constant q_s'' it also follows that the middle expression in Equation 8.37 is a constant independent of x. Hence

$$\frac{dT_m}{dx} = \frac{q_s'' P}{\dot{m}c_p} \neq f(x) \tag{8.39}$$

Integrating from $x = 0$, it follows that

$$T_m(x) = T_{m,i} + \frac{q_s'' P}{\dot{m}c_p}x \qquad q_s'' = \text{constant} \tag{8.40}$$

Accordingly, the mean temperature varies *linearly* with x along the tube (Figure 8.7a). Moreover, from Equation 8.27 and Figure 8.5 we also expect the temperature difference $(T_s - T_m)$ to vary with x, as shown in Figure 8.7a. This difference is initially small (due to the large value of h near the entrance) but increases with increasing x due to the decrease in h that occurs as the boundary layer develops. However, in the fully developed region we

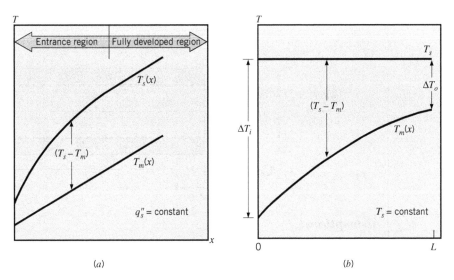

FIGURE 8.7 Axial temperature variations for heat transfer in a tube. (*a*) Constant surface heat flux. (*b*) Constant surface temperature.

know that h is independent of x. Hence from Equation 8.27 it follows that $(T_s - T_m)$ must also be independent of x in this region.

It should be noted that, if the heat flux is not constant but is, instead, a known function of x, Equation 8.37 may still be integrated to obtain the variation of the mean temperature with x. Similarly, the total heat rate may be obtained from the requirement that $q_{\text{conv}} = \int_0^L q_s''(x) P \, dx$.

EXAMPLE 8.2

A system for heating water from an inlet temperature of $T_{m,i} = 20°C$ to an outlet temperature of $T_{m,o} = 60°C$ involves passing the water through a thick-walled tube having inner and outer diameters of 20 and 40 mm. The outer surface of the tube is well insulated, and electrical heating within the wall provides for a uniform generation rate of $\dot{q} = 10^6 \, \text{W/m}^3$.

1. For a water mass flow rate of $\dot{m} = 0.1 \, \text{kg/s}$, how long must the tube be to achieve the desired outlet temperature?

2. If the inner surface temperature of the tube is $T_s = 70°C$ at the outlet, what is the local convection heat transfer coefficient at the outlet?

SOLUTION

Known: Internal flow through thick-walled tube having uniform heat generation.

Find:

1. Length of tube needed to achieve the desired outlet temperature.

2. Local convection coefficient at the outlet.

Schematic:

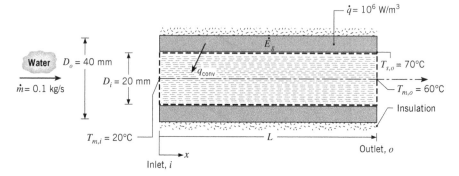

Assumptions:

1. Steady-state conditions.
2. Uniform heat flux.
3. Incompressible liquid and negligible viscous dissipation.
4. Constant properties.
5. Adiabatic outer tube surface.

Properties: Table A.6, water ($\overline{T}_m$ = 313 K): c_p = 4179 J/kg · K.

Analysis:

1. Since the outer surface of the tube is adiabatic, the rate at which energy is generated within the tube wall must equal the rate at which it is convected to the water.

$$\dot{E}_g = q_{\text{conv}}$$

With

$$\dot{E}_g = \dot{q}\frac{\pi}{4}(D_o^2 - D_i^2)L$$

it follows from Equation 8.34 that

$$\dot{q}\frac{\pi}{4}(D_o^2 - D_i^2)L = \dot{m}c_p(T_{m,o} - T_{m,i})$$

or

$$L = \frac{4\dot{m}c_p}{\pi(D_o^2 - D_i^2)\dot{q}}(T_{m,o} - T_{m,i})$$

$$L = \frac{4 \times 0.1 \text{ kg/s} \times 4179 \text{ J/kg} \cdot \text{K}}{\pi(0.04^2 - 0.02^2)\text{m}^2 \times 10^6 \text{ W/m}^3}(60 - 20)°\text{C} = 17.7 \text{ m} \qquad \triangleleft$$

2. From Newton's law of cooling, Equation 8.27, the local convection coefficient at the tube exit is

$$h_o = \frac{q_s''}{T_{s,o} - T_{m,o}}$$

Assuming that uniform heat generation in the wall provides a constant surface heat flux, with

$$q''_s = \frac{\dot{E}_g}{\pi D_i L} = \frac{\dot{q}}{4} \frac{D_o^2 - D_i^2}{D_i}$$

$$q''_s = \frac{10^6 \ \text{W/m}^3}{4} \frac{(0.04^2 - 0.02^2)\,\text{m}^2}{0.02 \ \text{m}} = 1.5 \times 10^4 \ \text{W/m}^2$$

it follows that

$$h_o = \frac{1.5 \times 10^4 \ \text{W/m}^2}{(70 - 60)°\text{C}} = 1500 \ \text{W/m}^2 \cdot \text{K} \qquad \triangleleft$$

Comments:

1. If conditions are fully developed over the entire tube, the local convection coefficient and the temperature difference $(T_s - T_m)$ are independent of x. Hence $h = 1500 \ \text{W/m}^2 \cdot \text{K}$ and $(T_s - T_m) = 10°\text{C}$ over the entire tube. The inner surface temperature at the tube inlet is then $T_{s,i} = 30°\text{C}$.

2. The required tube length L could have been computed by applying the expression for $T_m(x)$, Equation 8.40, at $x = L$.

8.3.3 Constant Surface Temperature

Results for the total heat transfer rate and the axial distribution of the mean temperature are entirely different for the *constant surface temperature* condition. Defining ΔT as $T_s - T_m$, Equation 8.37 may be expressed as

$$\frac{dT_m}{dx} = -\frac{d(\Delta T)}{dx} = \frac{P}{\dot{m}c_p} h \Delta T$$

Separating variables and integrating from the tube inlet to the outlet,

$$\int_{\Delta T_i}^{\Delta T_o} \frac{d(\Delta T)}{\Delta T} = -\frac{P}{\dot{m}c_p} \int_0^L h \, dx$$

or

$$\ln \frac{\Delta T_o}{\Delta T_i} = -\frac{PL}{\dot{m}c_p} \left(\frac{1}{L} \int_0^L h \, dx \right)$$

From the definition of the average convection heat transfer coefficient, Equation 6.13, it follows that

$$\ln \frac{\Delta T_o}{\Delta T_i} = -\frac{PL}{\dot{m}c_p} \bar{h}_L \qquad T_s = \text{constant} \qquad (8.41a)$$

where $\bar{h}_L$, or simply $\bar{h}$, is the average value of h for the entire tube. Rearranging,

$$\frac{\Delta T_o}{\Delta T_i} = \frac{T_s - T_{m,o}}{T_s - T_{m,i}} = \exp\left(-\frac{PL}{\dot{m}c_p}\bar{h}\right) \qquad T_s = \text{constant} \qquad (8.41b)$$

Had we integrated from the tube inlet to some axial position x within the tube, we would have obtained the similar, but more general, result that

$$\frac{T_s - T_m(x)}{T_s - T_{m,i}} = \exp\left(-\frac{Px}{\dot{m}c_p}\bar{h}(x)\right) \qquad T_s = \text{constant} \qquad (8.42)$$

where $\bar{h}(x)$ is the average value of h from the tube inlet to x. This result tells us that the temperature difference $(T_s - T_m)$ *decays in an exponential fashion* with distance along the tube. The axial surface and mean temperature distributions are therefore as shown in Figure 8.7*b*.

Determination of an expression for the total heat transfer rate q_{conv} is complicated by the nature of the temperature decay. Expressing Equation 8.34 in the form

$$q_{\text{conv}} = \dot{m}c_p[(T_s - T_{m,i}) - (T_s - T_{m,o})] = \dot{m}c_p(\Delta T_i - \Delta T_o)$$

and substituting for $\dot{m}c_p$ from Equation 8.41a, we obtain

$$q_{\text{conv}} = \bar{h}A_s\Delta T_{\text{lm}} \qquad T_s = \text{constant} \qquad (8.43)$$

where A_s is the tube surface area ($A_s = P \cdot L$) and ΔT_{lm} is the *log mean temperature difference*,

$$\Delta T_{\text{lm}} \equiv \frac{\Delta T_o - \Delta T_i}{\ln(\Delta T_o/\Delta T_i)} \qquad (8.44)$$

Equation 8.43 is a form of Newton's law of cooling for the entire tube, and ΔT_{lm} is the appropriate *average* of the temperature difference over the tube length. The logarithmic nature of this average temperature difference [in contrast, e.g., to an *arithmetic mean temperature* difference of the form $\Delta T_{\text{am}} = (\Delta T_i + \Delta T_o)/2$] is due to the exponential nature of the temperature decay.

Before concluding this section, it is important to note that, in many applications, it is the temperature of an *external* fluid, rather than the tube surface temperature, that is fixed (Figure 8.8). In such cases, it is readily shown that the results of this section may still be used if T_s is replaced by T_∞ (the free stream temperature of the external fluid) and $\bar{h}$ is replaced by $\bar{U}$ (the average overall heat transfer coefficient). For such cases, it follows that

$$\frac{\Delta T_o}{\Delta T_i} = \frac{T_\infty - T_{m,o}}{T_\infty - T_{m,i}} = \exp\left(-\frac{\bar{U}A_s}{\dot{m}c_p}\right) \qquad (8.45a)$$

and

$$q = \bar{U}A_s\,\Delta T_{\text{lm}} \qquad (8.46a)$$

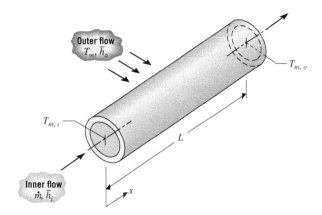

FIGURE 8.8　Heat transfer between fluid flowing over a tube and fluid passing through the tube.

The overall heat transfer coefficient is defined in Section 3.3.1, and for this application it would include contributions due to convection at the tube inner and outer surfaces. For a thick-walled tube of small thermal conductivity, it would also include the effect of conduction across the tube wall. Note that the product $\bar{U}A_s$ yields the same result, irrespective of whether it is defined in terms of the inner ($\bar{U}_i A_{s,i}$) or outer ($\bar{U}_o A_{s,o}$) surface areas of the tube (see Equation 3.37). Also note that ($\bar{U}A_s$)$^{-1}$ is equivalent to the total thermal resistance between the two fluids, in which case Equations 8.45a and 8.46a may be expressed as

$$\frac{\Delta T_o}{\Delta T_i} = \frac{T_\infty - T_{m,o}}{T_\infty - T_{m,i}} = \exp\left(-\frac{1}{\dot{m}c_p R_{\text{tot}}}\right) \tag{8.45b}$$

and

$$q = \frac{\Delta T_{\text{lm}}}{R_{\text{tot}}} \tag{8.46b}$$

A common variation of the foregoing conditions is one for which the uniform temperature of an *outer* surface, $T_{s,o}$, rather than the free stream temperature of an external fluid, T_∞, is known. In the foregoing equations, T_∞ is then replaced by $T_{s,o}$, and the total resistance embodies the convection resistance associated with the internal flow, as well as the resistance due to conduction between the inner surface of the tube and the surface corresponding to $T_{s,o}$.

EXAMPLE 8.3

Steam condensing on the outer surface of a thin-walled circular tube of diameter $D = 50$ mm and length $L = 6$ m maintains a uniform outer surface temperature of 100°C. Water flows through the tube at a rate of $\dot{m} = 0.25$ kg/s, and its inlet and outlet temperatures are $T_{m,i} = 15$°C and $T_{m,o} = 57$°C. What is the average convection coefficient associated with the water flow?

SOLUTION

Known: Flow rate and inlet and outlet temperatures of water flowing through a tube of prescribed dimensions and surface temperature.

Find: Average convection heat transfer coefficient.

Schematic:

Assumptions:

1. Negligible tube wall conduction resistance.
2. Incompressible liquid and negligible viscous dissipation.
3. Constant properties.

Properties: Table A.6, water ($\overline{T}_m = 36°C$): $c_p = 4178$ J/kg · K.

Analysis: Combining the energy balance, Equation 8.34, with the rate equation, Equation 8.43, the average convection coefficient is given by

$$\overline{h} = \frac{\dot{m}c_p}{\pi DL} \frac{(T_{m,o} - T_{m,i})}{\Delta T_{\mathrm{lm}}}$$

From Equation 8.44

$$\Delta T_{\mathrm{lm}} = \frac{(T_s - T_{m,o}) - (T_s - T_{m,i})}{\ln[(T_s - T_{m,o})/(T_s - T_{m,i})]}$$

$$\Delta T_{\mathrm{lm}} = \frac{(100 - 57) - (100 - 15)}{\ln[(100 - 57)/(100 - 15)]} = 61.6°C$$

Hence

$$\overline{h} = \frac{0.25 \text{ kg/s} \times 4178 \text{ J/kg} \cdot \text{K}}{\pi \times 0.05 \text{ m} \times 6 \text{ m}} \frac{(57 - 15)°C}{61.6°C}$$

or

$$\overline{h} = 755 \text{ W/m}^2 \cdot \text{K} \qquad \triangleleft$$

Comments: If conditions were fully developed over the entire tube, the local convection coefficient would be everywhere equal to 755 W/m² · K.

8.4 *Laminar Flow in Circular Tubes: Thermal Analysis and Convection Correlations*

To use many of the foregoing results, the convection coefficients must be known. In this section we outline the manner in which such coefficients may be obtained theoretically for laminar flow in a circular tube. In subsequent sections we consider empirical correlations pertinent to turbulent flow in a circular tube, as well as to flows in tubes of noncircular cross section.

8.4.1 The Fully Developed Region

Here, the problem of heat transfer in *laminar flow* of an *incompressible, constant property fluid* in the *fully developed region* of a *circular tube* is treated theoretically. The resulting temperature distribution is used to determine the convection coefficient.

A differential equation governing the temperature distribution is determined by applying the simplified, steady-flow, thermal energy equation, Equation 1.12e [$q = \dot{m}c_p(T_{\text{out}} - T_{\text{in}})$], to the annular differential element of Figure 8.9. If we neglect the effects of net axial conduction, the heat input, q, is due only to conduction through the radial surfaces. Since the radial velocity is zero in the fully developed region, there is no advection of thermal energy through the radial control surfaces, and the only advection is in the axial direction. Thus, Equation 1.12e leads to Equation 8.47, which expresses a balance between radial conduction and axial advection:

$$q_r - q_{r+dr} = (d\dot{m})c_p\left[\left(T + \frac{\partial T}{\partial x}dx\right) - T\right] \tag{8.47a}$$

or

$$(d\dot{m})c_p\frac{\partial T}{\partial x}dx = q_r - \left(q_r + \frac{\partial q_r}{\partial r}dr\right) = -\frac{\partial q_r}{\partial r}dr \tag{8.47b}$$

The differential mass flow rate in the axial direction is $d\dot{m} = \rho u 2\pi r dr$, and the radial heat transfer rate is $q_r = -k(\partial T/\partial r)2\pi r dx$. If we assume constant properties, Equation 8.47b becomes

$$u\frac{\partial T}{\partial x} = \frac{\alpha}{r}\frac{\partial}{\partial r}\left(r\frac{\partial T}{\partial r}\right) \tag{8.48}$$

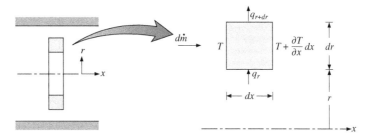

FIGURE 8.9 Thermal energy balance on a differential element for laminar, fully developed flow in a circular tube.

We will now proceed to solve for the temperature distribution for the case of *constant surface heat flux*. In this case, the assumption of negligible net axial conduction is exactly satisfied, that is, $(\partial^2 T/\partial x^2) = 0$. Substituting for the axial temperature gradient from Equation 8.32 and for the axial velocity component, u, from Equation 8.15, the energy equation, Equation 8.48, reduces to

$$\frac{1}{r}\frac{\partial}{\partial r}\left(r\frac{\partial T}{\partial r}\right) = \frac{2u_m}{\alpha}\left(\frac{dT_m}{dx}\right)\left[1 - \left(\frac{r}{r_o}\right)^2\right] \qquad q_s'' = \text{constant} \qquad (8.49)$$

where $T_m(x)$ varies linearly with x and $(2u_m/\alpha)(dT_m/dx)$ is a constant. Separating variables and integrating twice, we obtain an expression for the radial temperature distribution:

$$T(r, x) = \frac{2u_m}{\alpha}\left(\frac{dT_m}{dx}\right)\left[\frac{r^2}{4} - \frac{r^4}{16r_o^2}\right] + C_1 \ln r + C_2$$

The constants of integration may be evaluated by applying appropriate boundary conditions. From the requirement that the temperature remain finite at $r = 0$, it follows that $C_1 = 0$. From the requirement that $T(r_o) = T_s$, where T_s varies with x, it also follows that

$$C_2 = T_s(x) - \frac{2u_m}{\alpha}\left(\frac{dT_m}{dx}\right)\left(\frac{3r_o^2}{16}\right)$$

Accordingly, for the fully developed region with constant surface heat flux, the temperature profile is of the form

$$T(r, x) = T_s(x) - \frac{2u_m r_o^2}{\alpha}\left(\frac{dT_m}{dx}\right)\left[\frac{3}{16} + \frac{1}{16}\left(\frac{r}{r_o}\right)^4 - \frac{1}{4}\left(\frac{r}{r_o}\right)^2\right] \qquad (8.50)$$

From knowledge of the temperature profile, all other thermal parameters may be determined. For example, if the velocity and temperature profiles, Equations 8.15 and 8.50, respectively, are substituted into Equation 8.26 and the integration over r is performed, the mean temperature is found to be

$$T_m(x) = T_s(x) - \frac{11}{48}\left(\frac{u_m r_o^2}{\alpha}\right)\left(\frac{dT_m}{dx}\right) \qquad (8.51)$$

From Equation 8.39, where $P = \pi D$ and $\dot{m} = \rho u_m(\pi D^2/4)$, we then obtain

$$T_m(x) - T_s(x) = -\frac{11}{48}\frac{q_s''D}{k} \qquad (8.52)$$

Combining Newton's law of cooling, Equation 8.27, and Equation 8.52, it follows that

$$h = \frac{48}{11}\left(\frac{k}{D}\right)$$

or

$$Nu_D \equiv \frac{hD}{k} = 4.36 \qquad q_s'' = \text{constant} \qquad (8.53)$$

Hence in a *circular tube* characterized by *uniform surface heat flux* and *laminar, fully developed conditions*, the *Nusselt number is a constant*, independent of Re_D, Pr, and axial location.

For *laminar, fully developed conditions* with a *constant surface temperature,* the assumption of negligible net axial conduction is often reasonable. Substituting for the velocity profile from Equation 8.15 and for the axial temperature gradient from Equation 8.33, the energy equation becomes

$$\frac{1}{r}\frac{\partial}{\partial r}\left(r\frac{\partial T}{\partial r}\right) = \frac{2u_m}{\alpha}\left(\frac{dT_m}{dx}\right)\left[1 - \left(\frac{r}{r_o}\right)^2\right]\frac{T_s - T}{T_s - T_m} \qquad T_s = \text{constant} \qquad (8.54)$$

A solution to this equation may be obtained by an iterative procedure, which involves making successive approximations to the temperature profile. The resulting profile is not described by a simple algebraic expression, but the resulting Nusselt number may be shown to be [3]

$$Nu_D = 3.66 \qquad T_s = \text{constant} \qquad (8.55)$$

Note that in using Equation 8.53 or 8.55 to determine h, the thermal conductivity should be evaluated at T_m.[1]

EXAMPLE 8.4

In the human body, blood flows from the heart into a series of branching blood vessels having successively smaller diameters. Typical diameters, lengths, and average blood velocities for three different types of vessels are given in the table. The perfusion term in the bioheat equation (Eq. 3.112), is based on the assumption that blood enters the vessel at the arterial temperature, T_a, and exits at the temperature of the surrounding tissue, T (referred to here as T_t). For each of the three types of vessels, assume an entrance temperature $T_{m,i} = T_a$, and determine the *equilibration length L* required for the mean blood temperature to approach the tissue temperature, specifically, to satisfy the criterion $(T_t - T_{m,o})/(T_t - T_{m,i}) = 0.05$ [8, 9]. Heat transfer between the vessel wall and surrounding tissue can be described by an effective heat transfer coefficient, $h_t = k_t/D$, where $k_t = 0.5$ W/m · K.

Vessel	Vessel Diameter, D (mm)	Vessel Length, L_v (mm)	Blood Velocity, u_m (mm/s)
Large artery	3	200	130
Arteriole	0.02	2	3
Capillary	0.008	1	0.7

SOLUTION

Known: Blood vessel diameter and average blood velocity. Tissue thermal conductivity and effective heat transfer coefficient.

Find: Equilibration length, L, required for the mean blood temperature to approach the tissue temperature.

---

[1]If heat transfer occurs in a liquid that is characterized by a highly temperature-dependent viscosity, as is the case for many oils, experimental results have shown that the Nusselt numbers of Equations 8.53 or 8.55 may be corrected to account for this property variation as described in Section 8.4.3.

Schematic:

Assumptions:

1. Steady-state conditions.
2. Constant properties.
3. Negligible blood vessel wall thermal resistance.
4. Thermal properties of blood can be approximated by those of water.
5. Blood is an incompressible liquid with negligible viscous dissipation.
6. Tissue temperature is fixed.
7. Effects of pulsation of flow are negligible.

Properties: Table A.6, water ($\bar{T}_m$ = 310 K) $\rho = v_f^{-1}$ = 993 kg/m³, c_p = 4178 J/kg · K, $\mu = 695 \times 10^{-6}$ N · s/m², $k = 0.628$ W/m · K, $Pr = 4.62$.

Analysis: Since the tissue temperature T_t is fixed and heat transfer between the blood vessel wall and the tissue can be represented by an effective heat transfer coefficient, Equation 8.45a is applicable, with the free stream temperature T_∞ replaced by T_t. This equation can be used to find the length L that satisfies the criterion. However, we must first find $\bar{U}$, which requires knowledge of the heat transfer coefficient for the blood flow, h_b.

For the large artery, the Reynolds number is

$$Re_D = \frac{\rho u_m D}{\mu} = \frac{993 \text{ kg/m}^3 \times 130 \times 10^{-3} \text{ m/s} \times 3 \times 10^{-3} \text{ m}}{695 \times 10^{-6} \text{ N} \cdot \text{s/m}^2} = 557$$

so the flow is laminar. Since the other vessels have smaller diameters and velocities, their flows will also be laminar. We begin by assuming fully developed conditions. Moreover, because the situation is neither one of constant surface temperature nor constant surface heat flux, we will approximate the Nusselt number as $Nu_D \approx 4$, in which case $h_b = 4k_b/D$. Neglecting the thermal resistance of the vessel wall, for the large artery

$$\frac{1}{\bar{U}} = \frac{1}{h_b} + \frac{1}{h_t} = \frac{D}{4k_b} + \frac{D}{k_t} = \frac{3 \times 10^{-3} \text{ m}}{4 \times 0.628 \text{ W/m} \cdot \text{K}} + \frac{3 \times 10^{-3} \text{ m}}{0.5 \text{ W/m} \cdot \text{K}}$$

$$= 7.2 \times 10^{-3} \text{ m}^2 \cdot \text{K/W}$$

or

$$\bar{U} = 140 \text{ W/m}^2 \cdot \text{K}$$

The large artery length needed to satisfy the criterion can be found by solving Equation 8.45a, with $\dot{m} = \rho u_m \pi D^2/4$:

$$L = -\frac{\rho u_m D c_p}{4\bar{U}} \ln\left(\frac{T_t - T_{m,o}}{T_t - T_{m,i}}\right)$$

$$= -\frac{993 \text{ kg/m}^3 \times 130 \times 10^{-3} \text{ m/s} \times 3 \times 10^{-3} \text{ m} \times 4178 \text{ J/kg} \cdot \text{K}}{4 \times 140 \text{ W/m}^2 \cdot \text{K}} \ln(0.05)$$

$$= 8.7 \text{ m}$$

Using Equations 8.3 and 8.23:

$$x_{\text{fd},h} = 0.05 \, Re_D \, D = 0.05 \times 557 \times 3 \times 10^{-3} \text{ m} = 0.08 \text{ m}$$

$$x_{\text{fd},t} = x_{\text{fd},h} \, Pr = 0.08 \text{ m} \times 4.62 = 0.4 \text{ m}$$

Therefore, the flow would become fully developed well within the length of 8.7 m. The calculations can be repeated for the other two cases, and the results are tabulated below. ◁

Vessel	Re_D	$\bar{U}$ (W/m$^2 \cdot$ K)	L (m)	$x_{\text{fd},h}$ (m)	$x_{\text{fd},t}$ (m)
Large artery	557	140	8.7	8×10^{-2}	0.4
Arteriole	8.6×10^{-2}	2.1×10^4	8.9×10^{-6}	9×10^{-8}	4×10^{-7}
Capillary	8.0×10^{-3}	5.2×10^4	3.3×10^{-7}	3×10^{-9}	1×10^{-8}

The large value of L for the large artery suggests that the temperature remains close to the inlet arterial blood temperature. This is due to its relatively large diameter, which leads to a small overall heat transfer coefficient. In the intermediate arterioles, the blood temperature approaches the tissue temperature within a length on the order of 10 μm. Since arterioles are on the order of millimeters in length, the blood temperature exiting them and entering the capillaries would be approximately equal to the tissue temperature. There could then be no further temperature drop in the capillaries. Thus, it is in the arterioles and slightly larger vessels in which the blood temperature equilibrates to the tissue temperature, not in the small capillaries as Pennes assumed. ◁

Comments:

1. Since the length of an artery ($L_v \approx 200$ mm) is small compared to the calculated equilibration length L, the blood temperature remains close to the inlet arterial blood temperature, and the assumption inherent in the perfusion term of Eq. 3.112 is not satisfied. The blood that exits the large arteries enters the arterioles at a temperature close to T_a and the length of an arteriole ($L_v \approx 2$ mm) is much greater than the equilibration length. Therefore, the blood temperature exits the arterioles close to the tissue temperature, satisfying the assumption of the perfusion term. Since blood travels from the arterioles to the capillaries, the blood temperature is already at the tissue temperature as it enters the capillaries. Hence, the assumption of Eq. 3.112 is not satisfied in the capillaries.

2. The properties of blood are moderately close to those of water. The property that differs most is the viscosity, as blood is more viscous than water. However, this discrepancy would have no effect on the foregoing results and conclusions. Since the Reynolds number would be even smaller for the higher-viscosity blood, the flow would still be laminar and the heat transfer would be unaffected.

3. Blood cells have dimensions on the order of the capillary diameter. Thus, for the capillaries, an accurate model of blood flow would account for the individual cells surrounded by plasma.

8.4.2   The Entry Region

The results of the preceding section are valid only when *both* the velocity and temperature profiles are fully developed, as determined by the entry length expressions of Equations 8.3 and 8.23. If either or both profiles are not fully developed, the flow is said to be in the *entry region*. The energy equation for the entry region is more complicated than Equation 8.48 because there would be a radial advection term (since $v \neq 0$ in the entry region). In addition, both velocity and temperature now depend on x, as well as r, and the axial temperature gradient $\partial T/\partial x$ may no longer be simplified through Equation 8.32 or 8.33. However, two different entry length solutions have been obtained. The simplest solution is for the *thermal entry length problem,* and it is based on assuming that thermal conditions develop in the presence of a *fully developed velocity profile.* Such a situation would exist if the location at which heat transfer begins were preceded by an *unheated starting length.* It could also be assumed to a reasonable approximation for large Prandtl number fluids, such as oils. Even in the absence of an unheated starting length, velocity boundary layer development would occur far more rapidly than thermal boundary layer development for large Prandtl number fluids, and a thermal entry length approximation could be made. In contrast, the *combined* (thermal and velocity) *entry length problem* corresponds to the case for which the temperature and velocity profiles develop simultaneously. It would never be the case that thermal conditions are fully developed and hydrodynamic conditions are developing. Since the temperature distribution depends on the velocity distribution, as long as the velocity is still changing, thermal conditions cannot be fully developed.

Solutions have been obtained for both the thermal and combined entry length conditions [3], and selected results are shown in Figure 8.10. As evident in Figure 8.10*a*, local

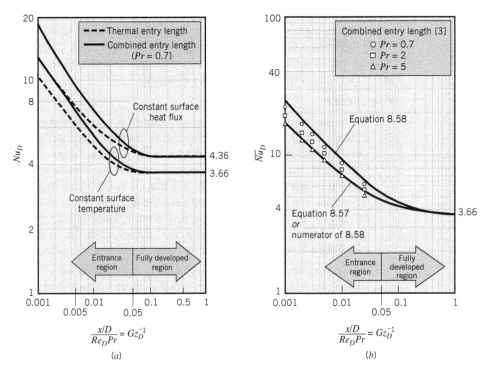

FIGURE 8.10   Results obtained from entry length solutions for laminar flow in a circular tube with constant surface temperature: (*a*) Local Nusselt numbers. (*b*) Average Nusselt numbers.

Nusselt numbers Nu_D are, in principle, infinite at $x = 0$ and decay to their asymptotic (fully developed) values with increasing x. These results are plotted against the dimensionless parameter $x\alpha/(u_m D^2) = x/(D\,Re_D\,Pr)$, which is the reciprocal of the *Graetz* number,

$$Gz_D \equiv (D/x)\,Re_D\,Pr \tag{8.56}$$

The manner in which Nu_D varies with Gz_D^{-1} is independent of Pr for the thermal entry problem, since the fully developed velocity profile, given by Equation 8.13, is independent of the fluid viscosity. In contrast, for the combined entry length problem, results depend on the manner in which the velocity distribution develops, which is sensitive to the fluid viscosity. Hence, heat transfer results depend on the Prandtl number for the combined entry length case and are presented in Figure 8.10*a* for $Pr = 0.7$, which is representative of most gases. At any location within the entry region, Nu_D decreases with increasing Pr and approaches the thermal entry length condition as $Pr \to \infty$. Note that fully developed conditions are reached for $[(x/D)/Re_D\,Pr] \approx 0.05$.

For the *constant surface temperature* condition, it is desirable to know the *average* convection coefficient for use with Equation 8.42 or 8.43. Selection of the appropriate correlation depends on whether a thermal or combined entry length exists.

For the *thermal entry length problem*, Kays [10] presents a correlation attributed to Hausen [11], which is of the form

$$\overline{Nu}_D = 3.66 + \frac{0.0668\,Gz_D}{1 + 0.04\,Gz_D^{2/3}} \tag{8.57}$$

$$\left[\begin{array}{l} T_s = \text{constant} \\ \text{thermal entry length} \\ or \\ \text{combined entry length with } Pr \gtrsim 5 \end{array}\right]$$

where $\overline{Nu}_D \equiv \bar{h}D/k$, and $\bar{h}$ is the heat transfer coefficient averaged from the tube inlet to x. Equation 8.57 is applicable to all situations where the velocity profile is fully developed. However, from Figure 8.10*b*, it is apparent that, for $Pr \gtrsim 5$, the thermal entry length approximation is reasonable since it agrees well with the combined entry length solution [3].

For the *combined entry problem*, the average Nusselt number depends on the Prandtl and Graetz numbers. Baehr and Stephan [12] recommend a correlation of the form

$$\overline{Nu}_D = \frac{\dfrac{3.66}{\tanh[2.264\,Gz_D^{-1/3} + 1.7\,Gz_D^{-2/3}]} + 0.0499\,Gz_D\,\tanh(Gz_D^{-1})}{\tanh(2.432\,Pr^{1/6}\,Gz_D^{-1/6})} \tag{8.58}$$

$$\left[\begin{array}{l} T_s = \text{constant} \\ \text{combined entry length} \\ Pr \gtrsim 0.1 \end{array}\right]$$

Equation 8.58, evaluated for $Pr = 0.7$, is shown in Figure 8.10*b* and agrees well with the data points obtained by solving the governing equations for the combined entry problem [3]. As $Pr \to \infty$, the denominator of Equation 8.58 approaches unity. Therefore, the numerator of

Equation 8.58 corresponds to the $Pr \to \infty$, thermal entry length problem and yields values of $\overline{Nu}_D$ that are within 3% of the Hausen correlation for $0.006 \leq Gz_D^{-1} \leq 1$, also shown in Figure 8.10*b*. All properties appearing in Equations 8.57 and 8.58 should be evaluated at the average mean temperature, $\overline{T}_m = (T_{m,i} + T_{m,o})/2$.

The subject of laminar flow in ducts has been studied extensively, and numerous results are available for a variety of duct cross sections and surface conditions. Representative results have been compiled in a monograph by Shah and London [13] and in a review by Shah and Bhatti [14]. Correlations for the combined entry region for non-circular ducts have been developed by Muzychka and Yovanovich [15].

8.4.3 Temperature-Dependent Properties

When differences between the surface temperature T_s and the mean temperature T_m correspond to large fluid property variations, the Nusselt number calculated from Equations 8.53, 8.55, 8.57, or 8.58 can be affected. For gases, this effect is usually small. For liquids, however, the viscosity variation may be particularly important. This is especially true for oils. Viscosity variation changes the radial velocity distribution, which affects the radial temperature distribution and ultimately alters the Nusselt number. Kays et al. [3] recommend applying the following correction factor to the Nusselt number for liquids:

$$\frac{Nu_{D,c}}{Nu_D} = \frac{\overline{Nu}_{D,c}}{\overline{Nu}_D} = \left(\frac{\mu}{\mu_s}\right)^{0.14} \tag{8.59}$$

In this expression, $Nu_{D,c}$ and $\overline{Nu}_{D,c}$ are the corrected Nusselt numbers, while Nu_D and $\overline{Nu}_D$ are Nusselt numbers found from Equations 8.53, 8.55, 8.57, or 8.58. All properties in Equation 8.59 are evaluated at $\overline{T}_m$ except for μ_s, which is evaluated at the surface temperature T_s. This correction factor can be applied to laminar flow of a liquid in a circular tube, regardless of whether the flow is fully developed or in the entry length region. The correction factor may also be applied to tubes of noncircular cross section in the absence of other alternatives [16].

8.5 *Convection Correlations: Turbulent Flow in Circular Tubes*

Since the analysis of turbulent flow conditions is more involved than for laminar conditions, empirical correlations are emphasized here. For *fully developed (hydrodynamically and thermally) turbulent flow* in a *smooth circular tube*, the *local* Nusselt number may be obtained from the *Dittus–Boelter equation*[2] [17]:

$$Nu_D = 0.023 \, Re_D^{4/5} Pr^n \tag{8.60}$$

[2]Although it has become common practice to refer to Equation 8.60 as the *Dittus–Boelter equation*, the original Dittus–Boelter equations are actually of the form
$$Nu_D = 0.0243 \, Re_D^{4/5} Pr^{0.4} \quad \text{(Heating)}$$
$$Nu_D = 0.0265 \, Re_D^{4/5} Pr^{0.3} \quad \text{(Cooling)}$$
The historical origins of Equation 8.60 are discussed by Winterton [17].

where $n = 0.4$ for heating $(T_s > T_m)$ and 0.3 for cooling $(T_s < T_m)$. These equations have been confirmed experimentally for the range of conditions

$$\left[\begin{array}{l} 0.6 \lesssim Pr \lesssim 160 \\ Re_D \gtrsim 10{,}000 \\ \dfrac{L}{D} \gtrsim 10 \end{array}\right]$$

The equations may be used for small to moderate temperature differences, $T_s - T_m$, with all properties evaluated at T_m. For flows characterized by large property variations, the following equation, due to Sieder and Tate [18], is recommended:

$$Nu_D = 0.027\, Re_D^{4/5}\, Pr^{1/3} \left(\frac{\mu}{\mu_s}\right)^{0.14} \tag{8.61}$$

$$\left[\begin{array}{l} 0.7 \lesssim Pr \lesssim 16{,}700 \\ Re_D \gtrsim 10{,}000 \\ \dfrac{L}{D} \gtrsim 10 \end{array}\right]$$

where all properties except μ_s are evaluated at T_m. *To a good approximation, the foregoing correlations may be applied for both the uniform surface temperature and heat flux conditions.*

Although Equations 8.60 and 8.61 are easily applied and are certainly satisfactory for the purposes of this text, errors as large as 25% may result from their use. Such errors may be reduced to less than 10% through the use of more recent, but generally more complex, correlations [6, 19]. One correlation, valid for *smooth tubes* over a large Reynolds number range including the transition region, is provided by Gnielinski [20]:

$$Nu_D = \frac{(f/8)(Re_D - 1000)\, Pr}{1 + 12.7\,(f/8)^{1/2}\,(Pr^{2/3} - 1)} \tag{8.62}$$

where the friction factor may be obtained from the Moody diagram or from Equation 8.21. The correlation is valid for $0.5 \lesssim Pr \lesssim 2000$ and $3000 \lesssim Re_D \lesssim 5 \times 10^6$. In using Equation 8.62, which applies for both uniform surface heat flux and temperature, properties should be evaluated at T_m. If temperature differences are large, additional consideration must be given to variable-property effects and available options are reviewed by Kakac [21].

We note that, unless specifically developed for the transition region $(2300 < Re_D < 10^4)$, caution should be exercised when applying a turbulent flow correlation for $Re_D < 10^4$. If the correlation was developed for fully turbulent conditions $(Re_D > 10^4)$, it may be used as a first approximation at smaller Reynolds numbers, with the understanding that the convection coefficient will be overpredicted. If a higher level of accuracy is desired, the Gnielinski correlation, Equation 8.62, may be used. A comprehensive discussion of heat transfer in the transition region is provided by Ghajar and Tam [22].

We also note that Equations 8.60 through 8.62 pertain to smooth tubes. For turbulent flow in *rough tubes*, the heat transfer coefficient increases with wall roughness, and, as a first approximation, it may be computed by using Equation 8.62 with friction factors obtained from Equation 8.20 or the Moody diagram, Figure 8.3. However, although the general trend is one of increasing h with increasing f, the increase in f is proportionately larger, and when f is approximately four times larger than the corresponding value for a smooth surface, h no longer changes with additional increases in f [23]. Procedures for estimating the effect of wall roughness on convection heat transfer in fully developed turbulent flow are discussed by Bhatti and Shah [19].

Since entry lengths for turbulent flow are typically short, $10 \lesssim (x_{fd}/D) \lesssim 60$, it is often reasonable to assume that the average Nusselt number for the entire tube is equal to the value associated with the fully developed region, $\overline{Nu}_D \approx Nu_{D,fd}$. However, for short tubes $\overline{Nu}_D$ will exceed $Nu_{D,fd}$ and may be calculated from an expression of the form

$$\frac{\overline{Nu}_D}{Nu_{D,fd}} = 1 + \frac{C}{(x/D)^m} \tag{8.63}$$

where C and m depend on the nature of the inlet (e.g., sharp-edged or nozzle) and entry region (thermal or combined), as well as on the Prandtl and Reynolds numbers [3, 19, 24]. Typically, errors of less than 15% are associated with assuming $\overline{Nu}_D = Nu_{D,fd}$ for $(L/D) > 60$. When determining $\overline{Nu}_D$, all fluid properties should be evaluated at the arithmetic average of the mean temperature, $\bar{T}_m \equiv (T_{m,i} + T_{m,o})/2$.

Finally, we note that the foregoing correlations do not apply to *liquid metals*. For fully developed turbulent flow in smooth circular tubes with constant surface heat flux, Skupinski et al. [25] recommend a correlation of the form

$$Nu_D = 4.82 + 0.0185\, Pe_D^{0.827} \quad q_s'' = \text{constant} \tag{8.64}$$

$$\left[\begin{array}{l} 3 \times 10^{-3} \lesssim Pr \lesssim 5 \times 10^{-2} \\ 3.6 \times 10^3 \lesssim Re_D \lesssim 9.05 \times 10^5 \\ 10^2 \lesssim Pe_D \lesssim 10^4 \end{array} \right]$$

Similarly, for constant surface temperature Seban and Shimazaki [26] recommend the following correlation for $Pe_D \gtrsim 100$:

$$Nu_D = 5.0 + 0.025\, Pe_D^{0.8} \quad T_s = \text{constant} \tag{8.65}$$

Extensive data and additional correlations are available in the literature [27].

EXAMPLE 8.5

A method to generate electric power from solar irradiation involves concentrating sunlight onto absorber tubes that are placed at the focal points of parabolic reflectors. The absorber tubes carry a liquid *concentrator fluid* that is heated as it flows through the tubes. After it leaves the concentrating field, the fluid enters a heat exchanger, where it transfers thermal energy to the *working fluid* of a Rankine cycle. The cooled concentrator fluid is returned to the concentrator field after it exits the heat exchanger. A power plant consists of many concentrators.

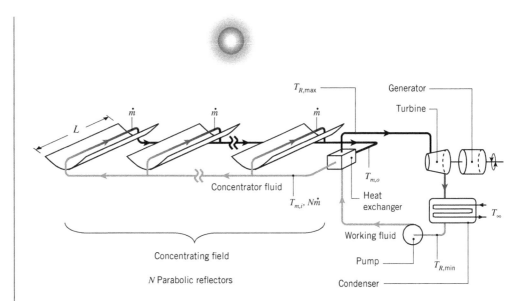

Consider conditions for which a concentrated heat flux of $q_s'' = 20{,}000 \text{ W/m}^2$, assumed to be uniform over the tube surface, heats a concentrator fluid of density, thermal conductivity, specific heat, and viscosity of $\rho = 700 \text{ kg/m}^3$, $k = 0.078 \text{ W/m} \cdot \text{K}$, $c_p = 2590 \text{ J/kg} \cdot \text{K}$, and $\mu = 0.15 \times 10^{-3} \text{ N} \cdot \text{s/m}^2$, respectively. The tube diameter is $D = 70$ mm, and the mass flow rate of the fluid in a single concentrator tube is $\dot{m} = 2.5$ kg/s.

1. If the concentrator fluid enters each tube at $T_{m,i} = 400°C$ and exits at $T_{m,o} = 450°C$, what is the required concentrator length, L? At what rate, q, is heat transferred to the concentrator fluid in a single concentrator-tube arrangement?

2. What is the surface temperature of the tube at the exit of a concentrator, $T_s(L)$?

3. The maximum and minimum temperatures of the entire power plant are the exit temperature of the concentrator fluid $T_{m,o}$ and the ambient temperature T_∞, respectively. If a temperature difference of $\Delta T = T_{m,o} - T_{R,\text{max}} = 20°C$ occurs across the heat exchanger and a second temperature difference of $\Delta T = T_{R,\text{min}} - T_\infty = 20°C$ exists across the condenser, where $T_\infty = 20°C$, determine the minimum number of concentrators N, each of length L, needed to generate $P = 20$ MW of electric power.

SOLUTION

Known: Tube diameter, surface heat flux, fluid mass flow rate and properties. Inlet and outlet mean temperatures of fluid in concentrator. Temperature differences across the heat exchanger and condenser.

Find:

1. Length of concentrator to achieve required temperature increase and the corresponding heat transfer rate q.

2. Tube surface temperature at the end of the concentrator.

3. Minimum number of concentrators needed to generate $P = 20$ MW of electric power.

Schematic:

Assumptions:

1. Steady-state conditions.
2. Incompressible liquid with negligible viscous dissipation.
3. Constant properties.
4. Thin tube wall.

Analysis:

1. For constant heat flux conditions, Equation 8.38 may be used with the appropriate energy balance, Equation 8.34, to obtain

$$A_s = \pi DL = \frac{\dot{m}c_p \left(T_{m,o} - T_{m,i}\right)}{q_s''}$$

or

$$L = \frac{\dot{m}c_p}{\pi D q_s''}\left(T_{m,o} - T_{m,i}\right)$$

where L is the concentrator tube length. Hence

$$L = \frac{2.5 \text{ kg/s} \times 2590 \text{ J/kg} \cdot \text{K}}{\pi \times 0.070 \text{ m} \times 20{,}000 \text{ W/m}^2}(450°\text{C} - 400°\text{C}) = 73.6 \text{ m} \qquad \lhd$$

The heat transfer rate is

$$q = q_s''A = q_s''\pi DL = 20{,}000 \text{ W/m}^2 \times \pi \times 0.070 \text{ m} \times 73.6 \text{ m}$$

$$= 0.324 \times 10^6 \text{ W} = 0.324 \text{ MW} \qquad \lhd$$

2. The tube surface temperature at the end of the concentrator may be obtained from Newton's law of cooling, Equation 8.27, where

$$T_s(L) = \frac{q_s''}{h} + T_{m,o}$$

To find the local convection coefficient at the tube outlet, the nature of the flow condition must first be established. From Equation 8.6,

$$Re_D = \frac{4\dot{m}}{\pi D \mu} = \frac{4 \times 2.5 \text{ kg/s}}{\pi \times 0.070 \text{ m} \times 0.15 \times 10^{-3} \text{ N} \cdot \text{s/m}^2} = 3.03 \times 10^5$$

Hence the flow is turbulent. The Prandtl number of the concentrator fluid may be determined from

$$Pr = \frac{\nu}{\alpha} = \frac{\mu c_p}{k} = \frac{0.15 \times 10^{-3} \text{ N} \cdot \text{s/m}^2 \times 2590 \text{ J/kg} \cdot \text{K}}{0.078 \text{ W/m} \cdot \text{K}} = 4.98$$

Since $L/D = 73.6 \text{ m}/0.070 \text{ m} = 1050$, we conclude from Equation 8.4 that conditions are fully developed within the tube at the end of the concentrator. The local Nusselt number at $x = L$ is obtained from Equation 8.60

$$Nu_D = 0.023 \, Re_D^{4/5} \, Pr^{0.4} = 0.023 \times (3.03 \times 10^5)^{4/5} \times 4.98^{0.4} = 1113$$

from which the local convection heat transfer coefficient is

$$h = \frac{k}{D} Nu_D = \frac{0.078 \text{ W/m} \cdot \text{K}}{0.070 \text{ m}} \times 1113 = 1240 \text{ W/m}^2 \cdot \text{K}$$

The tube surface temperature at the end of the concentrator is

$$T_s(L) = \frac{20,000 \text{ W/m}^2}{1240 \text{ W/m}^2 \cdot \text{K}} + 450°\text{C} = 466°\text{C} \qquad \triangleleft$$

3. The *minimum* number of concentrators may be determined by first calculating the corresponding minimum amount of thermal energy required to generate $P = 20$ MW of electricity. The maximum possible (Carnot) efficiency is $\eta_C = 1 - T_{R,\min}/T_{R,\max} = 1 - (T_\infty + \Delta T)/(T_{m,o} - \Delta T) = 1 - (293 \text{ K} + 20 \text{ K})/(723 \text{ K} - 20 \text{ K}) = 0.555$. Hence the minimum thermal energy required is

$$q_{\min} = \frac{P}{\eta_C} = \frac{20 \text{ MW}}{0.555} = 36.1 \text{ MW}$$

Correspondingly, the minimum number of concentrators required is

$$N = \frac{q_{\min}}{q} = \frac{36.1 \text{ MW}}{0.324 \text{ MW}} = 111 \qquad \triangleleft$$

Comments:

1. If temperature differences within the heat exchanger and the condenser could be eliminated ($\Delta T = 0°\text{C}$), the Carnot efficiency would be $\eta_C = 1 - T_{R,\min}/T_{R,\max} = 1 - T_\infty/T_{m,o} = 1 - (293 \text{ K})/(723 \text{ K}) = 0.595$. This yields $q_{\min} = 33.6$ MW and $N = 104$. Minimizing thermal resistances in the heat exchanger and condenser reduces the number of concentrators required and can reduce the capital cost of the plant.

2. Actual thermal efficiencies are less than the Carnot efficiency, and a nominal value of 38% is associated with parabolic trough *Solar Electric Generating Stations* (SEGS)

operating in Southern California since the mid-1980s. However, the *overall efficiency* of power plants using concentrating solar collectors is typically defined as a ratio of the rate of power generation to the rate at which solar energy is intercepted by the collectors. With a nominal efficiency of 40% for conversion of solar energy to thermal energy, the overall efficiency of the SEGS systems is approximately 15%.

3. A contemporary research challenge is to develop concentrator fluids that do not boil during periods of high solar irradiation and that resist freezing at night. Moreover, development of inexpensive and safe liquids capable of withstanding even higher temperatures will lead to higher maximum Rankine cycle temperatures, $T_{R,\max}$, and, in turn, increased plant efficiency.

EXAMPLE 8.6

Hot air flows with a mass rate of $\dot{m} = 0.050$ kg/s through an uninsulated sheet metal duct of diameter $D = 0.15$ m, which is in the crawlspace of a house. The hot air enters at 103°C and, after a distance of $L = 5$ m, cools to 85°C. The heat transfer coefficient between the duct outer surface and the ambient air at $T_\infty = 0°C$ is known to be $h_o = 6$ W/m² · K.

1. Calculate the rate of heat loss (W) from the duct over the length L.

2. Determine the heat flux and the duct surface temperature at $x = L$.

SOLUTION

Known: Hot air flowing in a duct.

Find:

1. Rate of heat loss from the duct over the length L.

2. Heat flux and surface temperature at $x = L$.

Schematic:

Assumptions:

1. Steady-state conditions.

2. Constant properties.

3. Ideal gas behavior.

4. Negligible viscous dissipation and negligible pressure variations.

5. Negligible duct wall thermal resistance.

6. Uniform convection coefficient at outer surface of duct.

7. Negligible radiation.

Properties: Table A.4, air ($\overline{T}_m = 367$ K): $c_p = 1011$ J/kg · K. Table A.4, air ($T_{m,L} = 358$ K): $k = 0.0306$ W/m · K, $\mu = 211.7 \times 10^{-7}$ N · s/m^2, $Pr = 0.698$.

Analysis:

1. From the energy balance for the entire tube, Equation 8.34,

$$q = \dot{m}c_p(T_{m,L} - T_{m,0})$$

$$q = 0.05 \text{ kg/s} \times 1011 \text{ J/kg} \cdot \text{K} (85 - 103)°\text{C} = -910 \text{ W} \qquad \lhd$$

2. An expression for the heat flux at $x = L$ may be inferred from the resistance network

where $h_x(L)$ is the inside convection heat transfer coefficient at $x = L$. Hence

$$q_s''(L) = \frac{T_{m,L} - T_\infty}{1/h_x(L) + 1/h_o}$$

The inside convection coefficient may be obtained from knowledge of the Reynolds number. From Equation 8.6

$$Re_D = \frac{4\dot{m}}{\pi D \mu} = \frac{4 \times 0.05 \text{ kg/s}}{\pi \times 0.15 \text{ m} \times 211.7 \times 10^{-7} \text{ N} \cdot \text{s/m}^2} = 20{,}050$$

Hence the flow is turbulent. Moreover, with $(L/D) = (5/0.15) = 33.3$, it is reasonable to assume fully developed conditions at $x = L$. From Equation 8.60, with $n = 0.3$,

$$Nu_D = \frac{h_x(L)D}{k} = 0.023 \, Re_D^{4/5} \, Pr^{0.3} = 0.023(20{,}050)^{4/5}(0.698)^{0.3} = 56.4$$

$$h_x(L) = Nu_D \frac{k}{D} = 56.4 \frac{0.0306 \text{ W/m} \cdot \text{K}}{0.15 \text{ m}} = 11.5 \text{ W/m}^2 \cdot \text{K}$$

Therefore

$$q_s''(L) = \frac{(85 - 0)°\text{C}}{(1/11.5 + 1/6.0)\text{m}^2 \cdot \text{K/W}} = 335 \text{ W/m}^2$$

Referring back to the network, it also follows that

$$q_s''(L) = \frac{T_{m,L} - T_{s,L}}{1/h_x(L)}$$

in which case

$$T_{s,L} = T_{m,L} - \frac{q_s''(L)}{h_x(L)} = 85°\text{C} - \frac{335 \text{ W/m}^2}{11.5 \text{ W/m}^2 \cdot \text{K}} = 55.9°\text{C} \qquad \lhd$$

Comments:

1. In using the energy balance of part 1 for the entire tube, properties (in this case, only c_p) are evaluated at $\overline{T}_m = (T_{m,0} + T_{m,L})/2$. However, in using the correlation for a local heat transfer coefficient, Equation 8.60, properties are evaluated at the local mean temperature, $T_{m,L} = 85°C$.

2. The overall average heat transfer coefficient $\overline{U}$ may be determined from Equation 8.45a, which can be rearranged to yield

$$\overline{U} = -\frac{\dot{m}c_p}{\pi DL}\ln\left[\frac{T_\infty - T_{m,o}}{T_\infty - T_{m,i}}\right] = -\frac{0.05 \text{ kg/s} \times 1011 \text{ J/kg} \cdot \text{K}}{\pi \times 0.15 \text{ m} \times 5 \text{ m}}\ln\left[\frac{-85°C}{-103°C}\right] = 4.12 \text{ W/m}^2 \cdot \text{K}$$

It follows from Assumption 6 that $\overline{h}_o = h_o$ and that $\overline{h}_i = 1/(1/\overline{U} - 1/\overline{h}_o) = 13.2 \text{ W/m}^2 \cdot \text{K}$. The average inside convection heat transfer coefficient is larger than $h_x(L)$, as expected from Equation 8.63.

3. This problem is characterized neither by constant surface temperature nor by constant surface heat flux. It would therefore be erroneous to presume that the total rate of heat loss from the tube is given by $q_s''(L)\pi DL = 790$ W. This result is substantially less than the actual heat loss rate of 910 W because $q_s''(x)$ decreases with increasing x. This decrease in $q_s''(x)$ is due to reductions in both $h_x(x)$ and $[T_m(x) - T_\infty]$ with increasing x.

8.6 *Convection Correlations: Noncircular Tubes and the Concentric Tube Annulus*

Although we have thus far restricted our consideration to internal flows of circular cross section, many engineering applications involve convection transport in *noncircular tubes*. At least to a first approximation, however, many of the circular tube results may be applied by using *an effective diameter* as the characteristic length. It is termed the *hydraulic diameter* and is defined as

$$D_h \equiv \frac{4A_c}{P} \tag{8.66}$$

where A_c and P are the *flow* cross-sectional area and the *wetted perimeter,* respectively. It is this diameter that should be used in calculating parameters such as Re_D and Nu_D.

For turbulent flow, which still occurs if $Re_D \gtrsim 2300$, it is reasonable to use the correlations of Section 8.5 for $Pr \gtrsim 0.7$. However, in a noncircular tube the convection coefficients vary around the periphery, approaching zero in the corners. Hence in using a circular tube correlation, the coefficient is presumed to be an average over the perimeter.

For laminar flow, the use of circular tube correlations is less accurate, particularly with cross sections characterized by sharp corners. For such cases the Nusselt number corresponding to fully developed conditions may be obtained from Table 8.1, which is based on solutions of the differential momentum and energy equations for flow through the different duct cross sections. As for the circular tube, results differ according to the surface thermal

TABLE 8.1 Nusselt numbers and friction factors for fully developed laminar flow in tubes of differing cross section

Cross Section	$\dfrac{b}{a}$	$Nu_D \equiv \dfrac{hD_h}{k}$ (Uniform q_s'')	$Nu_D \equiv \dfrac{hD_h}{k}$ (Uniform T_s)	fRe_{D_h}
◯	—	4.36	3.66	64
□	1.0	3.61	2.98	57
▭	1.43	3.73	3.08	59
▭	2.0	4.12	3.39	62
▭	3.0	4.79	3.96	69
▭	4.0	5.33	4.44	73
▭	8.0	6.49	5.60	82
▭	∞	8.23	7.54	96
Heated / Insulated	∞	5.39	4.86	96
△	—	3.11	2.49	53

Adapted from W. M. Kays and M. E. Crawford, *Convection Heat and Mass Transfer*, 3rd ed. McGraw-Hill, New York, 1993.

condition. Nusselt numbers tabulated for a uniform surface heat flux presume a constant flux in the axial (flow) direction, but a constant temperature around the perimeter at any cross section. This condition is typical of highly conductive tube wall materials. Results tabulated for a uniform surface temperature apply when the temperature is constant in both the axial and peripheral directions. Care should be taken when comparing the values of the Nusselt numbers associated with different cross-sectional shapes. Specifically, a cross section that is characterized by a larger Nusselt number does not necessarily imply more effective convection heat transfer, since both the hydraulic diameter and the wetted perimeter are cross section–dependent.

Although the foregoing procedures are generally satisfactory, exceptions do exist. Detailed discussions of heat transfer in noncircular tubes are provided in several sources [13, 14, 28].

Many internal flow problems involve heat transfer in a *concentric tube annulus* (Figure 8.11). Fluid passes through the space (annulus) formed by the concentric tubes, and convection heat transfer may occur to or from both the inner and outer tube surfaces. It is possible to independently specify the heat flux or temperature, that is, the thermal

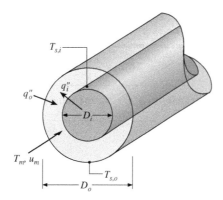

$T_{s,i}$

q_o'' q_i''

D_i

T_m, u_m

$T_{s,o}$

D_o

FIGURE 8.11   The concentric tube annulus.

condition, at each of these surfaces. In any case the heat flux from each surface may be computed with expressions of the form

$$q_i'' = h_i(T_{s,i} - T_m) \tag{8.67}$$

$$q_o'' = h_o(T_{s,o} - T_m) \tag{8.68}$$

Note that separate convection coefficients are associated with the inner and outer surfaces. The corresponding Nusselt numbers are of the form

$$Nu_i \equiv \frac{h_i D_h}{k} \tag{8.69}$$

$$Nu_o \equiv \frac{h_o D_h}{k} \tag{8.70}$$

where, from Equation 8.66, the hydraulic diameter D_h is

$$D_h = \frac{4(\pi/4)(D_o^2 - D_i^2)}{\pi D_o + \pi D_i} = D_o - D_i \tag{8.71}$$

For the case of fully developed laminar flow with one surface insulated and the other surface at a constant temperature, Nu_i or Nu_o may be obtained from Table 8.2. Note that in

TABLE 8.2   Nusselt number for fully developed laminar flow in a circular tube annulus with one surface insulated and the other at constant temperature

D_i/D_o	Nu_i	Nu_o	*Comments*
0	—	3.66	See Equation 8.55
0.05	17.46	4.06	
0.10	11.56	4.11	
0.25	7.37	4.23	
0.50	5.74	4.43	
≈1.00	4.86	4.86	See Table 8.1, $b/a \to \infty$

Used with permission from W. M. Kays and H. C. Perkins, in W. M. Rohsenow and J. P. Hartnett, Eds., *Handbook of Heat Transfer,* Chap. 7, McGraw-Hill, New York, 1973.

TABLE 8.3 Influence coefficients for fully developed
laminar flow in a circular tube annulus with
uniform heat flux maintained at both surfaces

D_i/D_o	Nu_{ii}	Nu_{oo}	θ_i^*	θ_o^*
0	—	4.364^a	∞	0
0.05	17.81	4.792	2.18	0.0294
0.10	11.91	4.834	1.383	0.0562
0.20	8.499	4.883	0.905	0.1041
0.40	6.583	4.979	0.603	0.1823
0.60	5.912	5.099	0.473	0.2455
0.80	5.58	5.24	0.401	0.299
1.00	5.385	5.385^b	0.346	0.346

Used with permission from W. M. Kays and H. C. Perkins, in W. M. Rohsenow and
J. P. Hartnett, Eds., *Handbook of Heat Transfer,* Chap. 7, McGraw-Hill, New York, 1973.

^{a}See Equation 8.53.
^{b}See Table 8.1 for $b/a \rightarrow \infty$ with one surface insulated.

such cases we would be interested only in the convection coefficient associated with the isothermal (nonadiabatic) surface.

If uniform heat flux conditions exist at both surfaces, the Nusselt numbers may be computed from expressions of the form

$$Nu_i = \frac{Nu_{ii}}{1 - (q_o''/q_i'')\theta_i^*} \tag{8.72}$$

$$Nu_o = \frac{Nu_{oo}}{1 - (q_i''/q_o'')\theta_o^*} \tag{8.73}$$

The influence coefficients (Nu_{ii}, Nu_{oo}, θ_i^*, and θ_o^*) appearing in these equations may be obtained from Table 8.3. Note that q_i'' and q_o'' may be positive or negative, depending on whether heat transfer is to or from the fluid, respectively. Moreover, situations may arise for which the values of h_i and h_o are negative. Such results, when used with the sign convention implicit in Equations 8.67 and 8.68, reveal the relative magnitudes of T_s and T_m.

For fully developed turbulent flow, the influence coefficients are a function of the Reynolds and Prandtl numbers [28]. However, to a first approximation the inner and outer convection coefficients may be assumed to be equal, and they may be evaluated by using the hydraulic diameter, Equation 8.71, with the Dittus–Boelter equation, Equation 8.60.

8.7 *Heat Transfer Enhancement*

Several options are available for enhancing heat transfer associated with internal flows. Enhancement may be achieved by increasing the convection coefficient and/or by increasing the convection surface area. For example, h may be increased by introducing surface roughness to enhance turbulence, as, for example, through machining or insertion of a coil-spring wire. The wire insert (Figure 8.12a) provides a helical roughness element

FIGURE 8.12 Internal flow heat transfer enhancement schemes: (*a*) longitudinal section and end view of coil-spring wire insert, (*b*) longitudinal section and cross-sectional view of twisted tape insert, (*c*) cut-away section and end view of longitudinal fins, and (*d*) longitudinal section and end view of helical ribs.

in contact with the tube inner surface. Alternatively, the convection coefficient may be increased by inducing swirl through insertion of a twisted tape (Figure 8.12*b*). The insert consists of a thin strip that is periodically twisted through 360°. Introduction of a tangential velocity component increases the speed of the flow, particularly near the tube wall. The heat transfer area may be increased by manufacturing a tube with a grooved inner surface (Figure 8.12*c*), while both the convection coefficient and area may be increased by using spiral fins or ribs (Figure 8.12*d*). In evaluating any heat transfer enhancement scheme, attention must also be given to the attendant increase in pressure drop and hence fan or pump power requirements. Comprehensive assessments of enhancement options have been published [29–32], and the *Journal of Enhanced Heat Transfer* provides access to recent developments in the field.

By coiling a tube (Figure 8.13), heat transfer may be enhanced without turbulence or additional heat transfer surface area. In this case, centrifugal forces within the fluid induce

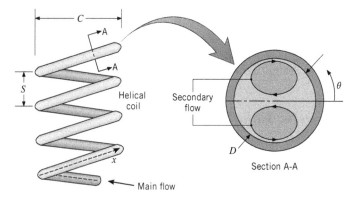

FIGURE 8.13 Schematic of helically coiled tube and secondary flow in enlarged cross-sectional view.

a *secondary flow* consisting of a pair of longitudinal vortices that, in contrast to conditions in a straight tube, can result in highly nonuniform local heat transfer coefficients around the periphery of the tube. Hence, local heat transfer coefficients vary with θ as well as x. If constant heat flux conditions are applied, the mean fluid temperature, $T_m(x)$, may be calculated using the conservation of energy principle, Equation 8.40. For situations where the fluid is heated, maximum fluid temperatures occur at the tube wall, but calculation of the maximum local temperature is not straightforward because of the θ-dependence of the heat transfer coefficient. Therefore, correlations for the *peripherally averaged* Nusselt number are of little use if constant heat flux conditions are applied. In contrast, correlations for the peripherally averaged Nusselt number for constant wall temperature boundary conditions are useful, and the relationships recommended by Shah and Joshi [33] are provided in the next paragraphs.

The secondary flow increases friction losses and heat transfer rates. In addition, the secondary flow decreases entrance lengths and reduces the difference between laminar and turbulent heat transfer rates, relative to the straight tube cases considered previously in this chapter. Pressure drops and heat transfer rates exhibit little dependence on the coil pitch, S. The critical Reynolds number corresponding to the onset of turbulence for the helical tube, $Re_{D,c,h}$, is

$$Re_{D,c,h} = Re_{D,c}[1 + 12(D/C)^{0.5}] \tag{8.74}$$

where $Re_{D,c}$ is given in Equation 8.2 and C is defined in Figure 8.13. Strong secondary flow associated with tightly wound helically coiled tubes delays the transition to turbulence.

For fully developed laminar flow with $C/D \gtrsim 3$, the friction factor is

$$f = \frac{64}{Re_D} \qquad Re_D(D/C)^{1/2} \lesssim 30 \tag{8.19}$$

$$f = \frac{27}{Re_D^{0.725}}(D/C)^{0.1375} \qquad 30 \lesssim Re_D(D/C)^{1/2} \lesssim 300 \tag{8.75a}$$

$$f = \frac{7.2}{Re_D^{0.5}}(D/C)^{0.25} \qquad 300 \lesssim Re_D(D/C)^{1/2} \tag{8.75b}$$

For cases where $C/D \lesssim 3$, recommendations provided in Shah and Joshi [33] should be followed. The heat transfer coefficient for use in Equation 8.27 may be evaluated from a correlation of the form

$$Nu_D = \left[\left(3.66 + \frac{4.343}{a} \right)^3 + 1.158 \left(\frac{Re_D(D/C)^{1/2}}{b} \right)^{3/2} \right]^{1/3} \left(\frac{\mu}{\mu_s} \right)^{0.14} \tag{8.76}$$

where

$$a = \left(1 + \frac{927(C/D)}{Re_D^2\, Pr} \right) \qquad \text{and} \qquad b = 1 + \frac{0.477}{Pr} \tag{8.77a,b}$$

$$\begin{bmatrix} 0.005 \lesssim Pr \lesssim 1600 \\ 1 \lesssim Re_D(D/C)^{1/2} \lesssim 1000 \end{bmatrix}$$

Friction factor correlations for turbulent flow in coiled tubes are based on limited data. Furthermore, heat transfer augmentation due to the secondary flow is minor when the flow is turbulent and is less than 10% for $C/D \gtrsim 20$. As such, helically coiled tubes are typically employed to enhance heat transfer only for laminar flow conditions. In laminar flow, the entrance length is 20% to 50% shorter than that of a straight tube, while the flow becomes fully developed within the first half-turn of the helically coiled tube under turbulent conditions. Therefore, the entrance region may be neglected in most engineering calculations. A compilation of additional correlations is available [34].

When a gas or liquid is heated in a straight tube, a fluid parcel that enters near the centerline of the tube will exit the tube faster, and will be cooler, than a fluid parcel that enters near the tube wall. Hence, the *time-temperature histories* of individual fluid parcels, processed in the same heating tube, can be dramatically different. In addition to augmenting heat transfer, the secondary flow associated with the helically coiled tube serves to *mix* the fluid relative to laminar flow in a straight tube, resulting in similar time-temperature histories for all the fluid parcels. It is for this reason that coiled tubes are routinely used to process and manufacture highly viscous, high value-added fluids, such as pharmaceuticals, cosmetics, and personal care products [34].

8.8 *Forced Convection in Small Channels*

The tubes and channels considered thus far have been of conventional size. However, many technologies involve forced convection in channels of relatively small dimension. A motivation for using *microfluidic devices* is evident from inspection of Equations 8.53 and 8.55, as well as Tables 8.1 through 8.3. That is, convection coefficients are inversely proportional to the hydraulic diameter, and when small channels are used, can attain values that far exceed the typical values listed in Table 1.1 [35].

8.8.1 Microscale Convection in Gases ($0.1 \ \mu m \lesssim D_h \lesssim 100 \ \mu m$)

In most situations, unrealistically high gas velocities are required to achieve turbulent flow in channels or tubes characterized by $D_h \lesssim 100 \ \mu$m. Therefore, turbulent microscale convection involving gases is rarely encountered. However, standard laminar convection correlations may not apply because the interaction of gas molecules with the tube or channel wall may affect the convection process. As discussed in Section 3.7.3, conduction through a gas layer can be altered when the characteristic length of the gas volume is of the same magnitude as the mean free path of the gas, λ_{mfp}. Convection processes can be similarly affected in channels of small hydraulic diameter. More specifically, the internal flow convection correlations presented in the preceding sections are not expected to apply for gases when $D_h/\lambda_{\text{mfp}} \lesssim 100$. For air at atmospheric temperature and pressure, this limit corresponds to $D_h \lesssim 10 \ \mu$m.

If the situation shown in Figure 3.25 were to involve a bulk flow in the vertical direction, the manner in which individual gas molecules scatter from the two solid walls would also affect the transfer of momentum throughout the gas and, in turn, the velocity distribution within the gas. Since the gas temperature distribution depends on the gas velocity, a *momentum accommodation coefficient* α_p will influence convection heat transfer rates, as will the thermal accommodation coefficient α_t of Equation 3.130 [36]. Values of the momentum accommodation coefficient are in the range $0 \leq \alpha_p \leq 1$. Specifically, specular reflection (where the speed of the molecule is unchanged, and the angle of reflection from the surface is equal to the angle of incidence on the surface) corresponds to $\alpha_p = 0$. On the other hand, diffuse reflection (with no preferred angle of reflection) corresponds to $\alpha_p = 1$.

Values of α_p for air interacting with most engineering surfaces range from 0.87 to unity, while for nitrogen, argon, or CO_2 in silicon channels $0.75 \lesssim \alpha_p \lesssim 0.85$ [36].

Convection heat transfer in microscale internal gas flow has been analyzed, accounting for both thermal and momentum interactions between the gas molecules and the solid walls. For laminar, fully developed flow in a circular tube of diameter D with a uniform surface heat flux, the Nusselt number may be expressed as [37]

$$Nu_D = \frac{hD}{k} = \frac{48}{11 - 6\zeta + \zeta^2 + 48\Gamma_t} \tag{8.78a}$$

where

$$\zeta = 8\Gamma_p/(1 + 8\Gamma_p) \tag{8.78b}$$

$$\Gamma_p = \frac{2 - \alpha_p}{\alpha_p}\left[\frac{\lambda_{\text{mfp}}}{D}\right] \tag{8.78c}$$

$$\Gamma_t = \frac{2 - \alpha_t}{\alpha_t}\frac{2\gamma}{\gamma + 1}\left[\frac{\lambda_{\text{mfp}}}{Pr\,D}\right] \tag{8.78d}$$

The term $\gamma \equiv c_p/c_v$ is the ratio of specific heats of the gas. For large tube diameters ($\lambda_{\text{mfp}}/D \to 0$), $Nu_D \to 48/11 = 4.36$, in agreement with Equation 8.53.

Similarly, for laminar, fully developed flow in a channel formed by large plates separated by a spacing a, the Nusselt number for uniform and equal plate heat fluxes is [38]

$$Nu_D = \frac{hD_h}{k} = \frac{140}{17 - 6\zeta + (2/3)\zeta^2 + 70\Gamma_t} \tag{8.79a}$$

where

$$\zeta = 6\Gamma_p/(1 + 6\Gamma_p) \tag{8.79b}$$

The parameters Γ_p and Γ_t are defined as in Equations 8.78c,d with $D = D_h$. For infinitely large plates the hydraulic diameter is $D_h = 2a$ and for large plate spacing, $\lambda_{\text{mfp}}/D_h \to 0$, $Nu_D \to 140/17 = 8.23$, in agreement with Table 8.1.

The preceding relations may be applied only when the gas flow can be treated as incompressible, that is, when the Mach number is small ($Ma \lesssim 0.3$).

8.8.2   Microscale Convection in Liquids

Experiments have shown that Equations 8.19 and 8.22a may be applied to laminar liquid flows in tubes with diameters as small as 17 μm [39, 40].[3] These equations are expected to be valid for most liquids for hydraulic diameters as small as 1 μm [39, 41]. Convection heat transfer in microscale internal flows involving liquids is the subject of ongoing research. The analytical results of Chapters 6 and 8 should be used with caution for liquids when $D_h \lesssim 1$ μm.

---

[3]From the discussion of Section 6.3.1, one might anticipate that, since turbulence is characterized by the motion of relatively large parcels of fluid in devices of conventional size, Equations 8.2, 8.20, and 8.21 would not be applicable for flow in microfluidic devices because the volume of the fluid parcel is restricted by the hydraulic diameter of the channel. Nonetheless, careful measurements using various liquids have shown that Equation 8.2 does indeed hold for liquid flow in tubes with diameters at least as small as 17 μm [40].

8.8.3   Nanoscale Convection ($D_h \lesssim 100$ nm)

As the hydraulic diameter approaches $D_h \approx 0.1$ μm = 100 nm, molecular interactions must, in general, be accounted for both in the fluid and in the solid wall. Nanoscale convection, including liquid convection within hollow carbon nanotubes, is an area of current research [42].

EXAMPLE 8.7

Combinatorial chemistry is used in the pharmaceutical industry to produce large populations, or *libraries*, of compounds that are subsequently screened to identify candidates of therapeutic value. A *microreactor chip* subjects two reactants, A and B, to N unique time-temperature histories, resulting in N new compounds for testing. The 20 mm × 20 mm chip is fabricated by first coating a 1-mm-thick glass microscope slide with a photoresist material of thickness $a = 40$ μm. Channels of width $b = 160$ μm and length $L = 20$ mm are subsequently etched into the photoresist, and a second microscope slide is adhered to the top of the structure. The spacing between channels is $s = 40$ μm, so that $N = L/(b + s) = 100$ parallel channels are formed. Flow is induced through each channel by applying a pressure difference of $\Delta p = 500$ kPa from the entrance to the exit of the microreactor. The reactants enter each channel at $T_{m,i} = 5$°C, and the N products of reaction are cooled back to 5°C upon exiting the chip. The walls of each of the N channels are maintained at unique temperatures by holding the edges of the chip at temperatures $T_1 = 125$°C and $T_2 = 25$°C. For the warmest (125°C) and coolest (25°C) microchannels, estimate the time during which the liquid is within 1 degree Celsius of the wall temperature. Assume the liquid has thermophysical properties similar to ethylene glycol.

SOLUTION

Known: Dimensions and operating conditions for reactants flowing in a microreactor.

Find: Time-at-temperature associated with the warmest and coolest channels.

Schematic:

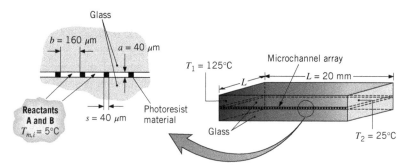

Assumptions:

1. Linear temperature distribution across the width of the microreactor.
2. Steady-state conditions.
3. Incompressible liquid with constant properties.
4. Negligible viscous dissipation.
5. Uniform wall temperatures for each channel.

Properties: Table A.5, ethylene glycol ($\overline{T}_m = 288$ K): $\rho = 1120.2$ kg/m^3, $c_p = 2359$ J/kg · K, $\mu = 2.82 \times 10^{-2}$ N · s/m^2, $k = 247 \times 10^{-3}$ W/m · K, $Pr = 269$. ($\overline{T}_m = 338$ K): $\rho = 1085$ kg/m^3, $c_p = 2583$ J/kg · K, $\mu = 0.427 \times 10^{-2}$ N · s/m^2, $k = 261 \times 10^{-3}$ W/m · K, $Pr = 45.2$.

Analysis: The hydraulic diameter of each microchannel is found from Equation 8.66 and is

$$D_h = \frac{4A_c}{P} = \frac{4ab}{(2a + 2b)} = \frac{4 \times 40 \times 10^{-6} \text{ m} \times 160 \times 10^{-6} \text{ m}}{(80 \times 10^{-6} \text{ m} \times 320 \times 10^{-6} \text{ m})} = 64 \times 10^{-6} \text{ m}$$

We begin by assuming laminar flow and a short entrance length, to be verified later, so the flow rate may be estimated by using the friction factor for fully developed conditions. From Table 8.1 for $b/a = 4$, $f = 73/Re_{D_h}$. Substituting this expression into Equation 8.22a, rearranging terms, and using properties at $T_m = 338$ K for the 125°C microchannel (in this equation and those following) results in

$$u_m = \frac{2}{73} \frac{D_h^2 \Delta p}{\mu L} = \frac{2}{73} \times \frac{(64 \times 10^{-6} \text{ m})^2 \times 500 \times 10^3 \text{ N/m}^2}{0.427 \times 10^{-2} \text{ N} \cdot \text{s/m}^2 \times 20 \times 10^{-3} \text{ m}} = 0.657 \text{ m/s}$$

Hence, the Reynolds number is

$$Re_D = \frac{u_m D_h \rho}{\mu} = \frac{0.657 \text{ m/s} \times 64 \times 10^{-6} \text{ m} \times 1085 \text{ kg/m}^3}{0.427 \times 10^{-2} \text{ N} \cdot \text{s/m}^2} = 10.7$$

and the flow is deep in the laminar regime. Equation 8.3 may be used to determine the hydrodynamic entrance length, which is

$$x_{\text{fd},h} \approx 0.05 \, D_h \, Re_D = 0.05 \times 64 \times 10^{-6} \text{ m} \times 10.7 = 34.2 \times 10^{-6} \text{ m}$$

and the thermal entrance length may be obtained from Equation 8.23, yielding

$$x_{\text{fd},t} \approx x_{\text{fd},h} Pr = 34.2 \times 10^{-6} \text{ m} \times 45.2 = 1.55 \times 10^{-3} \text{ m}$$

Both entrance lengths occupy less than 10% of the total microchannel length, $L = 20$ mm. Therefore, use of fully developed values of f are justified, and the mass flow rate for the $T = 125$°C microchannel is

$$\dot{m} = \rho A_c u_m = \rho a b u_m = 1085 \text{ kg/m}^3 \times 40 \times 10^{-6} \text{ m} \times 160 \times 10^{-6} \text{ m} \times 0.657 \text{ m/s}$$
$$= 4.56 \times 10^{-6} \text{ kg/s}$$

Equation 8.42 may now be used to determine the distance from the entrance of the microchannel to the location, x_c, where $T_{m,c} = 124$°C, that is, within 1°C of the surface temperature. The average heat transfer coefficient, $\overline{h}$, is replaced by the fully developed value of the heat transfer coefficient, h, because of the relatively short entrance length. From Table 8.1, $Nu_D = hD_h/k = 4.44$ for $b/a = 4$. Therefore,

$$\overline{h} \approx h = Nu_D \frac{k}{D_h} = 4.44 \times \frac{0.261 \text{ W/m} \cdot \text{K}}{64 \times 10^{-6} \text{ m}} = 1.81 \times 10^4 \text{ W/m}^2 \cdot \text{K}$$

As expected, the convection coefficient is large.

Rearranging Equation 8.42 yields

$$x_c = \frac{\dot{m}c_p}{Ph} \ln\left[\frac{T_s - T_{m,i}}{T_s - T_{m,c}}\right] = \frac{4.56 \times 10^{-6} \text{ kg/s} \times 2583 \text{ J/kg} \cdot \text{K}}{0.4 \times 10^{-3} \text{ m} \times 1.81 \times 10^4 \text{ W/m}^2 \cdot \text{K}} \ln\left[\frac{(125-5)°\text{C}}{(125-124)°\text{C}}\right]$$

$$= 7.79 \times 10^{-3} \text{ m}$$

Therefore, the time-at-temperature is

$$t_T = (L - x_c)/u_m = (20 \times 10^{-3} - 7.79 \times 10^{-3}) \text{ m}/0.657 \text{ m/s} = 0.019 \text{ s} \qquad \lhd$$

Repeating the calculations for the microchannel associated with the smallest processing temperature of 25°C yields $u_m = 0.0995$ m/s, $Re_D = 0.253$, $x_{\text{fd},h} = 8.09 \times 10^{-7}$ m, $x_{\text{fd},t} = 0.218 \times 10^{-3}$ m, $h = 1.71 \times 10^4$ W/m$^2 \cdot$ K, $x_c = 0.73 \times 10^{-3}$ m, and $t_T = 0.19$ s. $\qquad \lhd$

Comments:

1. The total thickness of the glass (2 mm) is 50 times greater than the depth of each microchannel, while the thermal conductivity of the glass, $k_{\text{glass}} \approx 1.4$ W/m · K (Table A.3), is 5 times greater than that of the fluid. The presence of such a small amount of fluid is expected to have a negligible effect on the linear temperature distribution that is established across the chip. The temperature difference across the bottom or top surface of each channel is approximately $\Delta T = (T_1 - T_2)b/L = (125 - 25)°\text{C} \times (160 \times 10^{-6} \text{ m})/(20 \times 10^{-3} \text{ m}) = 0.8°\text{C}$.

2. As discussed in Section 8.7, the non-uniform velocity and temperature distributions within a heated straight channel lead to shorter residence times and cooler temperatures for fluid parcels near the centerline of the channel, relative to residence times and temperatures associated with parcels near the channel walls. Achieving uniform time-at-temperature histories in fluids flowing in microreactors is made difficult by the laminar flow conditions. However, inducing swirl within the liquids by using curved microchannels, or by fabricating helical rib patterns onto the interior microchannel walls, can promote mixing and more uniform time-temperature histories for the flowing fluids [43].

3. The mean temperature distributions of the hottest and coldest channels are shown below.

8.9 *Convection Mass Transfer*

Mass transfer by convection may also occur for internal flows. For example, a gas may flow through a tube whose surface has been wetted or is sublimable. Evaporation or sublimation will then occur, and a concentration boundary layer will develop. Just as the mean temperature is the appropriate reference temperature for heat transfer, the mean species concentration $\rho_{A,m}$ plays an analogous role for mass transfer. Defining the mass flow rate of species A in a duct of arbitrary cross section A_c as $\dot{m}_A = \rho_{A,m} u_m A_c = \int_{A_c} (\rho_A u) dA_c$, the mean species density is, accordingly,

$$\rho_{A,m} = \frac{\int_{A_c} (\rho_A u) dA_c}{u_m A_c} \tag{8.80a}$$

or for a circular tube,

$$\rho_{A,m} = \frac{2}{u_m r_o^2} \int_0^{r_o} (\rho_A u r) \, dr \tag{8.80b}$$

The concentration boundary layer development is characterized by entrance and fully developed regions, and Equation 8.23 may be used (with Pr replaced by Sc) to determine the *concentration entry length* $x_{fd,c}$ for laminar flow. Equation 8.4 may again be used as a first approximation for turbulent flow. Moreover, by analogy to Equation 8.28, for both laminar and turbulent flows, fully developed conditions exist when

$$\frac{\partial}{\partial x} \left[\frac{\rho_{A,s} - \rho_A(r, x)}{\rho_{A,s} - \rho_{A,m}(x)} \right]_{fd,c} = 0 \tag{8.81}$$

where a uniform species concentration $\rho_{A,s}$ is presumed to exist at the surface.

The local mass flux of species A from the surface may be computed from an expression of the form

$$n_A'' = \bar{h}_m (\rho_{A,s} - \rho_{A,m}) \tag{8.82}$$

while the total rate of species transfer for a duct of surface area A_s may be expressed as

$$n_A = \bar{h}_m A_s \Delta \rho_{A,lm} \tag{8.83}$$

where the *log mean concentration difference*

$$\Delta \rho_{A,lm} = \frac{\Delta \rho_{A,o} - \Delta \rho_{A,i}}{\ln(\Delta \rho_{A,o}/\Delta \rho_{A,i})} \tag{8.84}$$

is analogous to the log mean temperature difference of Equation 8.44 and the concentration difference is defined as $\Delta \rho_A = \rho_{A,s} - \rho_{A,m}$. From application of conservation of species A to a control volume about the duct, the total rate of species transfer may also be expressed as

$$n_A = \frac{\dot{m}}{\rho} (\rho_{A,o} - \rho_{A,i}) \tag{8.85}$$

where ρ and $\dot{m}$ are the total mass density and flow rate, respectively, and $\dot{m}/\rho = u_m A_c$. Equations 8.83 and 8.85 are the mass transfer equivalents of Equations 8.43 and 8.34, respectively, for heat transfer. In addition, the analog to Equation 8.42 that characterizes the variation of the mean vapor density with distance x from the duct entrance may be expressed as

$$\frac{\rho_{A,s} - \rho_{A,m}(x)}{\rho_{A,s} - \rho_{A,m,i}} = \exp\left(-\frac{\bar{h}_m(x)\rho P}{\dot{m}}x\right) \tag{8.86}$$

where P is the duct perimeter.

The convection mass transfer coefficients, h_m and $\bar{h}_m$, may be obtained from appropriate correlations for the corresponding Sherwood numbers, which are defined as $Sh_D = h_m D/D_{AB}$ and $\overline{Sh}_D = \bar{h}_m D/D_{AB}$. The specific form of a correlation may be inferred from the foregoing heat transfer results by invoking the heat and mass transfer analogy, with Sh_D and Sc substituted for Nu_D and Pr, respectively. For example, with a uniform vapor density at the surface of a circular duct and fully developed laminar flow through the duct,

$$Sh_D = 3.66 \tag{8.87}$$

For fully developed turbulent flow, the mass transfer analog to the Dittus–Boelter equation is

$$Sh_D = 0.023\, Re_D^{4/5}\, Sc^{0.4} \tag{8.88}$$

Microscale conditions for mass transfer are similar to those discussed for heat transfer in Section 8.8.

EXAMPLE 8.8

A thin liquid film of ammonia (NH_3), which has formed on the inner surface of a tube of diameter $D = 10$ mm and length $L = 1$ m, is removed by passing dry air through the tube at a flow rate of 3×10^{-4} kg/s. The tube and the air are at 25°C. What is the average mass transfer convection coefficient?

SOLUTION

Known: Liquid ammonia on the inner surface of a tube is removed by evaporation into an airstream.

Find: Average mass transfer convection coefficient for the tube.

Schematic:

Assumptions:

1. Thin ammonia film with smooth surface.

2. Heat and mass transfer analogy is applicable.

Properties: Table A.4, air (25°C): $\nu = 15.7 \times 10^{-6}$ m²/s, $\mu = 183.6 \times 10^{-7}$ N · s/m². Table A.8, ammonia–air (25°C): $D_{AB} = 0.28 \times 10^{-4}$ m²/s, $Sc = (\nu/D_{AB}) = 0.56$.

Analysis: From Equation 8.6

$$Re_D = \frac{4 \times 3 \times 10^{-4} \text{ kg/s}}{\pi \times 0.01 \text{ m} \times 183.6 \times 10^{-7} \text{ N} \cdot \text{s/m}^2} = 2080$$

in which case the flow is laminar. From Equation 8.3, the hydrodynamic entrance length is

$$x_{\text{fd},h} \approx 0.05 \, D \, Re_D = 0.05 \times 0.01 \text{ m} \times 2080 = 1.04 \text{ m}$$

The concentration entrance length may be obtained from the mass transfer analog of Equation 8.23, yielding

$$x_{\text{fd},c} \approx x_{\text{fd},h} \, Sc = 1.04 \text{ m} \times 0.56 = 0.58 \text{ m}$$

Since $x_{\text{fd},h} > L$, the combined entry length condition holds even though $x_{\text{fd},c} < L$. A constant ammonia vapor concentration is maintained at the surface of the film, which is analogous to a constant surface temperature. The mass transfer analog of Equation 8.58 is appropriate, with $Gz_D = (D/L) \, Re_D \, Sc = (0.01 \text{ m}/1 \text{ m}) \times 2080 \times 0.56 = 11.6$. Thus

$$
\overline{Sh}_D = \frac{\dfrac{3.66}{\tanh[2.264 \, Gz_D^{-1/3} + 1.7 \, Gz_D^{-2/3}]} + 0.0499 \, Gz_D \, \tanh(Gz_D^{-1})}{\tanh(2.432 \, Sc^{1/6} \, Gz_D^{-1/6})}
$$

$$
= \frac{\dfrac{3.66}{\tanh[2.264 \times 11.6^{-1/3} + 1.7 \times 11.6^{-2/3}]} + 0.0499 \times 11.6 \times \tanh(11.6^{-1})}{\tanh(2.432 \times 0.56^{1/6} \times 11.6^{-1/6})}
$$

$$= 4.74$$

Finally,

$$\overline{h}_m = \overline{Sh}_D \frac{D_{AB}}{D} = \frac{4.74 \times 0.28 \times 10^{-4} \text{ m}^2/\text{s}}{0.01 \text{ m}} = 0.013 \text{ m/s} \qquad \triangleleft$$

Comments: An assumption of fully developed conditions over the entire tube would provide a value of $\overline{Sh}_D = 3.66$, which is 23% less than the preceding result.

8.10 *Summary*

In this chapter we have considered forced convection heat and mass transfer for an important class of problems involving *internal flow*. Such flows are encountered in numerous applications, and you should be able to perform engineering calculations that involve an energy balance and appropriate convection correlations. The methodology involves determining whether the flow is laminar or turbulent and establishing the length of the entry region. After deciding whether you are interested in local conditions (at a particular axial location) or in average conditions (for the entire tube), the convection correlation may be selected and used with the appropriate form of the energy balance to solve the problem. A summary of the correlations is provided in Table 8.4.

You should test your understanding of related concepts by addressing the following questions.

- What are the salient features of a *hydrodynamic entry region*? A *thermal entry region*? Are hydrodynamic and thermal entry lengths equivalent? If not, on what do the relative lengths depend?
- What are the salient *hydrodynamic* features of *fully developed flow*? How is the friction factor for fully developed flow affected by wall roughness?
- To what important characteristic of an internal flow is the *mean* or *bulk temperature* linked?
- What are the salient *thermal* features of *fully developed flow*?
- If fluid enters a tube at a uniform temperature and there is heat transfer to or from the surface of the tube, how does the convection coefficient vary with distance along the tube?
- For fluid flow through a tube with a uniform surface heat flux, how does the mean temperature of the fluid vary with distance from the tube entrance in (a) the entrance region and (b) the fully developed region? How does the surface temperature vary with distance in the entrance and fully developed regions?
- For heat transfer to or from a fluid flowing through a tube with a uniform surface temperature, how does the mean temperature of the fluid vary with distance from the entrance? How does the surface heat flux vary with distance from the entrance?
- Why is a *log mean temperature difference*, rather than an arithmetic mean temperature difference, used to calculate the total rate of heat transfer to or from a fluid flowing through a tube with a constant surface temperature?
- What two equations may be used to calculate the total heat rate to a fluid flowing through a tube with a uniform surface heat flux? What two equations may be used to calculate the total heat rate to or from a fluid flowing through a tube with a uniform surface temperature?
- Under what conditions is the Nusselt number associated with internal flow equal to a constant value, independent of Reynolds number and Prandtl number?
- Is the average Nusselt number associated with flow through a tube larger than, equal to, or less than the Nusselt number for fully developed conditions? Why?
- How is the characteristic length defined for a noncircular tube?
- What are the salient features of a *concentration entry region*?
- What are the salient features of fully developed flow for mass transfer?
- How may convection mass transfer correlations be inferred?

Several features that complicate internal flows have not been considered in this chapter. For example, a situation may exist for which there is a prescribed axial variation in T_s or q_s'', rather than uniform surface conditions. Among other things, such a variation would preclude the existence of a fully developed region. There may also exist surface roughness effects, circumferential heat flux or temperature variations, widely varying fluid properties, or transition flow conditions. For a complete discussion of these effects, the literature should be consulted [13, 14, 19, 21, 28].

TABLE 8.4   Summary of convection correlations for flow in a circular tube[a,b,e]

Correlation		Conditions
$f = 64/Re_D$	(8.19)	Laminar, fully developed
$Nu_D = 4.36$	(8.53)	Laminar, fully developed, uniform q_s''
$Nu_D = 3.66$	(8.55)	Laminar, fully developed, uniform T_s
$\overline{Nu}_D = 3.66 + \dfrac{0.0668\,Gz_D}{1 + 0.04\,Gz_D^{2/3}}$	(8.57)	Laminar, thermal entry (or combined entry with $Pr \gtrsim 5$), uniform T_s, $Gz_D = (D/x)\,Re_D\,Pr$
$\overline{Nu}_D = \dfrac{\dfrac{3.66}{\tanh[2.264\,Gz_D^{-1/3} + 1.7\,Gz_D^{-2/3}]} + 0.0499\,Gz_D\tanh(Gz_D^{-1})}{\tanh(2.432\,Pr^{1/6}Gz_D^{-1/6})}$	(8.58)	Laminar, combined entry, $Pr \gtrsim 0.1$, uniform T_s, $Gz_D = (D/x)\,Re_D\,Pr$
$\dfrac{1}{\sqrt{f}} = -2.0\log\left[\dfrac{e/D}{3.7} + \dfrac{2.51}{Re_D\sqrt{f}}\right]$	(8.20)[c]	Turbulent, fully developed
$f = (0.790\ln Re_D - 1.64)^{-2}$	(8.21)[c]	Turbulent, fully developed, smooth walls, $3000 \lesssim Re_D \lesssim 5\times10^6$
$Nu_D = 0.023\,Re_D^{4/5}\,Pr^n$	(8.60)[d]	Turbulent, fully developed, $0.6 \lesssim Pr \lesssim 160$, $Re_D \gtrsim 10{,}000$, $(L/D) \gtrsim 10$, $n = 0.4$ for $T_s > T_m$ and $n = 0.3$ for $T_s < T_m$
$Nu_D = 0.027\,Re_D^{4/5}\,Pr^{1/3}\left(\dfrac{\mu}{\mu_s}\right)^{0.14}$	(8.61)[d]	Turbulent, fully developed, $0.7 \lesssim Pr \lesssim 16{,}700$, $Re_D \gtrsim 10{,}000$, $L/D \gtrsim 10$
$Nu_D = \dfrac{(f/8)(Re_D - 1000)\,Pr}{1 + 12.7(f/8)^{1/2}(Pr^{2/3} - 1)}$	(8.62)[d]	Turbulent, fully developed, $0.5 \lesssim Pr \lesssim 2000$, $3000 \lesssim Re_D \lesssim 5\times10^6$, $(L/D) \gtrsim 10$
$Nu_D = 4.82 + 0.0185(Re_D\,Pr)^{0.827}$	(8.64)	Liquid metals, turbulent, fully developed, uniform q_s'', $3.6\times10^3 \lesssim Re_D \lesssim 9.05\times10^5$, $3\times10^{-3} \lesssim Pr \lesssim 5\times10^{-2}$, $10^2 \lesssim Re_D\,Pr \lesssim 10^4$
$Nu_D = 5.0 + 0.025(Re_D\,Pr)^{0.8}$	(8.65)	Liquid metals, turbulent, fully developed, uniform T_s, $Re_D\,Pr \gtrsim 100$

[a]The mass transfer correlations may be obtained by replacing Nu_D and Pr by Sh_D and Sc, respectively.

[b]Properties in Equations 8.53, 8.55, 8.60, 8.61, 8.62, 8.64, and 8.65 are based on T_m; properties in Equations 8.19, 8.20, and 8.21 are based on $T_f = (T_s + T_m)/2$; properties in Equations 8.57 and 8.58 are based on $\overline{T}_m = (T_{m,i} + T_{m,o})/2$.

[c]Equation 8.20 pertains to smooth or rough tubes. Equation 8.21 pertains to smooth tubes.

[d]As a first approximation, Equations 8.60, 8.61, or 8.62 may be used to evaluate the average Nusselt number $\overline{Nu}_D$ over the entire tube length, if $(L/D) \gtrsim 10$. The properties should then be evaluated at the average of the mean temperature, $\overline{T}_m = (T_{m,i} + T_{m,o})/2$.

[e]For tubes of noncircular cross section, $Re_D \equiv D_h u_m/\nu$, $D_h \equiv 4A_c/P$, and $u_m = \dot{m}/\rho A_c$. Results for fully developed laminar flow are provided in Table 8.1. For turbulent flow, Equation 8.60 may be used as a first approximation.

References

1. Abraham, J. P., E. M. Sparrow, and J. C. K. Tong, *Int. J. Heat Mass Transfer*, **52**, 557, 2009.

2. Langhaar, H. L., *J. Appl. Mech.*, **64**, A-55, 1942.

3. Kays, W. M., M. E. Crawford, and B. Weigand, *Convective Heat and Mass Transfer*, 4th ed., McGraw-Hill Higher Education, Boston, 2005.

4. Munson, B. R., D. F. Young, T. H. Okiishi, and W. W. Huebsch, *Fundamentals of Fluid Mechanics*, 6th ed. Wiley, Hoboken, NJ, 2009.

5. Fox, R. W., P. J. Pritchard, and A. T. McDonald, *Introduction to Fluid Mechanics*, 7th ed., Wiley, Hoboken, NJ, 2009.

6. Petukhov, B. S., in T. F. Irvine and J. P. Hartnett, Eds., *Advances in Heat Transfer*, Vol. 6, Academic Press, New York, 1970.

7. Moody, L. F., *Trans. ASME*, **66**, 671, 1944.

8. Chen, M. M., and K. R. Holmes, *Ann. N. Y. Acad. Sci.*, **335**, 137, 1980.

9. Chato, J. C., *J. Biomech. Eng.*, **102**, 110, 1980.

10. Kays, W. M., *Trans. ASME*, **77**, 1265, 1955.

11. Hausen, H., *Z. VDI Beih. Verfahrenstech.*, **4**, 91, 1943.

12. Baehr, H. D., and K. Stephan, *Heat Transfer*, 2nd ed., Springer, Berlin, 2006.

13. Shah, R. K., and A. L. London, *Laminar Flow Forced Convection in Ducts*, Academic Press, New York, 1978.

14. Shah, R. K., and M. S. Bhatti, in S. Kakac, R. K. Shah, and W. Aung, Eds., *Handbook of Single-Phase Convective Heat Transfer*, Chap. 3, Wiley-Interscience, Hoboken, NJ, 1987.

15. Muzychka, Y. S., and M. M. Yovanovich, *J. Heat Transfer*, **126**, 54, 2004.

16. Burmeister, L. C., *Convective Heat Transfer*, 2nd ed., Wiley, Hoboken, NJ, 1993.

17. Winterton, R. H. S., *Int. J. Heat Mass Transfer*, **41**, 809, 1998.

18. Sieder, E. N., and G. E. Tate, *Ind. Eng. Chem.*, **28**, 1429, 1936.

19. Bhatti, M. S., and R. K. Shah, in S. Kakac, R. K. Shah, and W. Aung, Eds., *Handbook of Single-Phase Convective Heat Transfer*, Chap. 4, Wiley-Interscience, Hoboken, NJ, 1987.

20. Gnielinski, V., *Int. Chem. Eng.*, **16**, 359, 1976.

21. Kakac, S., in S. Kakac, R. K. Shah, and W. Aung, Eds., *Handbook of Single-Phase Convective Heat Transfer*, Chap. 18, Wiley-Interscience, Hoboken, NJ, 1987.

22. Ghajar, A. J., and L.-M. Tam, *Exp. Thermal and Fluid Science*, **8**, 79, 1994.

23. Norris, R. H., in A. E. Bergles and R. L. Webb, Eds., *Augmentation of Convective Heat and Mass Transfer*, ASME, New York, 1970.

24. Molki, M., and E. M. Sparrow, *J. Heat Transfer*, **108**, 482, 1986.

25. Skupinski, E. S., J. Tortel, and L. Vautrey, *Int. J. Heat Mass Transfer*, **8**, 937, 1965.

26. Seban, R. A., and T. T. Shimazaki, *Trans. ASME*, **73**, 803, 1951.

27. Reed, C. B., in S. Kakac, R. K. Shah, and W. Aung, Eds., *Handbook of Single-Phase Convective Heat Transfer*, Chap. 8, Wiley-Interscience, Hoboken, NJ, 1987.

28. Kays, W. M., and H. C. Perkins, in W. M. Rohsenow, J. P. Hartnett, and E. N. Ganic, Eds., *Handbook of Heat Transfer, Fundamentals*, Chap. 7, McGraw-Hill, New York, 1985.

29. Bergles, A. E., "Principles of Heat Transfer Augmentation," *Heat Exchangers, Thermal-Hydraulic Fundamentals and Design*, Hemisphere Publishing, New York, 1981, pp. 819–842.

30. Webb, R. L., in S. Kakac, R. K. Shah, and W. Aung, Eds., *Handbook of Single-Phase Convective Heat Transfer*, Chap. 17, Wiley-Interscience, Hoboken, NJ, 1987.

31. Webb, R. L., *Principles of Enhanced Heat Transfer*, Wiley, Hoboken, NJ, 1993.

32. Manglik, R. M., and A. E. Bergles, in J. P. Hartnett, T. F. Irvine, Y. I. Cho, and R. E. Greene, Eds., *Advances in Heat Transfer*, Vol. 36, Academic Press, New York, 2002.

33. Shah, R. K., and S. D. Joshi, in *Handbook of Single-Phase Convective Heat Transfer*, Chap. 5, Wiley-Interscience, Hoboken, NJ, 1987.

34. Vashisth, S., V. Kumar, and K. D. P. Nigam, *Ind. Eng. Chem. Res.*, **47**, 3291, 2008.

35. Jensen, K. F., *Chem. Eng. Sci.*, **56**, 293, 2001.

36. Zhang, Z. M., *Nano/Microscale Heat Transfer*, McGraw-Hill, New York, 2007.

37. Sparrow, E. M., and S. H. Lin, *J. Heat Transfer*, **84**, 363, 1962.

38. Inman, R., *Laminar Slip Flow Heat Transfer in a Parallel Plate Channel or a Round Tube with Uniform Wall Heating*, NASA TN D-2393, 1964.

39. Sharp, K. V., and R. J. Adrian, *Exp. Fluids*, **36**, 741, 2004.

40. Rands, C., B. W. Webb, and D. Maynes, *Int. J. Heat Mass Transfer*, **49**, 2924, 2006.

41. Travis, K. P., B. D. Todd, and D. J. Evans, *Phys. Rev. E*, **55**, 4288, 1997.

42. Whitby, M., and N. Quirke, in K. D. Sattler, Ed., *Handbook of Nanophysics*, Chap. 11, CRC Press, Boca Raton, FL, 2011.

43. Stroock, A. D., S. K. W. Deteringer, A. Ajdar, I. Mezić, H. A. Stone, and G. M. Whitesides, *Science*, **295**, 647, 2002.

Problems

Hydrodynamic Considerations

8.1 Fully developed conditions are known to exist for water flowing through a 25-mm-diameter tube at 0.01 kg/s and 27°C. What is the maximum velocity of the water in the tube? What is the pressure gradient associated with the flow?

8.2 Water at 35°C is pumped through a horizontal, 200-m-long, 30-mm-diameter tube at 0.25 kg/s. Over time, a 2-mm-thick layer of scale of surface roughness $e = 200$ μm is deposited on the interior tube wall. Determine the pressure drop from the entrance to the exit of the tube and the required pump power for the clean and fouled conditions.

8.3 What is the pressure drop associated with water at 27°C flowing with a mean velocity of 0.2 m/s through a 600-m-long cast iron pipe of 0.15-m inside diameter?

8.4 Water at 27°C flows with a mean velocity of 1 m/s through a 1-km-long pipe of 0.25-m inside diameter.

(a) Determine the pressure drop over the pipe length and the corresponding pump power requirement, if the pipe surface is smooth.

(b) If the pipe is made of cast iron and its surface is clean, determine the pressure drop and pump power requirement.

(c) For the smooth pipe condition, generate a plot of pressure drop and pump power requirement for mean velocities in the range from 0.05 to 1.5 m/s.

8.5 For fully developed laminar flow through a parallel-plate channel, the x-momentum equation has the form

$$\mu\left(\frac{d^2u}{dy^2}\right) = \frac{dp}{dx} = \text{constant}$$

The purpose of this problem is to develop expressions for the velocity distribution and pressure gradient analogous to those for the circular tube in Section 8.1.

(a) Show that the velocity profile, $u(y)$, is parabolic and of the form

$$u(y) = \frac{3}{2}u_m\left[1 - \frac{y^2}{(a/2)^2}\right]$$

where u_m is the mean velocity

$$u_m = -\frac{a^2}{12\mu}\left(\frac{dp}{dx}\right)$$

and $-dp/dx = \Delta p/L$, where Δp is the pressure drop across the channel of length L.

(b) Write an expression defining the friction factor, f, using the hydraulic diameter D_h as the characteristic length. What is the hydraulic diameter for the parallel-plate channel?

(c) The friction factor is estimated from the expression $f = C/Re_{D_h}$, where C depends upon the flow cross section, as shown in Table 8.1. What is the coefficient C for the parallel-plate channel?

(d) Airflow in a parallel-plate channel with a separation of 5 mm and a length of 200 mm experiences a pressure drop of $\Delta p = 3.75$ N/m². Calculate the mean velocity and the Reynolds number for air at atmospheric pressure and 300 K. Is the assumption of fully developed flow reasonable for this application? If not, what is the effect on the estimate for u_m?

Thermal Entry Length and Energy Balance Considerations

8.6 Consider pressurized water, engine oil (unused), and NaK (22%/78%) flowing in a 20-mm-diameter tube.

(a) Determine the mean velocity, the hydrodynamic entry length, and the thermal entry length for each of the fluids when the fluid temperature is 366 K and the flow rate is 0.01 kg/s.

(b) Determine the mass flow rate, the hydrodynamic entry length, and the thermal entry length for water and engine oil at 300 and 400 K and a mean velocity of 0.02 m/s.

8.7 At a particular axial station, velocity and temperature profiles for laminar flow in a parallel plate channel have the form

$$u(y) = 0.75[1 - (y/y_o)^2]$$

$$T(y) = 5.0 + 95.66(y/y_o)^2 - 47.83(y/y_o)^4$$

with units of m/s and °C, respectively.

Determine corresponding values of the mean velocity, u_m, and mean (or bulk) temperature, T_m. Plot the velocity and temperature distributions. Do your values of u_m and T_m appear reasonable?

8.8 In Chapter 1, it was stated that for incompressible liquids, flow work could usually be neglected in the steady-flow energy equation (Equation 1.12d). In the trans-Alaska pipeline, the high viscosity of the oil and long distances cause significant pressure drops, and it is reasonable to question whether flow work would be significant. Consider an $L = 100$ km length of pipe of diameter $D = 1.2$ m, with oil flow rate $\dot{m} = 500$ kg/s. The oil properties are $\rho = 900$ kg/m^3, $c_p = 2000$ J/kg · K, $\mu = 0.765$ N · s/m^2. Calculate the pressure drop, the flow work, and the temperature rise caused by the flow work.

8.9 When viscous dissipation is included, Equation 8.48 (multiplied by ρc_p) becomes

$$\rho c_p u \frac{\partial T}{\partial x} = \frac{k}{r} \frac{\partial}{\partial r}\left(r \frac{\partial T}{\partial r}\right) + \mu \left(\frac{du}{dr}\right)^2$$

This problem explores the importance of viscous dissipation. The conditions under consideration are laminar,

fully developed flow in a circular pipe, with u given by Equation 8.15.

(a) By integrating the left-hand side over a section of a pipe of length L and radius r_o, show that this term yields the right-hand side of Equation 8.34.

(b) Integrate the viscous dissipation term over the same volume.

(c) Find the temperature rise caused by viscous dissipation by equating the two terms calculated above. Use the same conditions as in Problem 8.8.

8.10 Consider a circular tube of diameter D and length L, with a mass flow rate of $\dot{m}$.

(a) For constant heat flux conditions, derive an expression for the ratio of the temperature difference between the tube wall at the tube exit and the inlet temperature, $T_s(x = L) - T_{m,i}$, to the total heat transfer rate to the fluid q. Express your result in terms of $\dot{m}$, L, the local Nusselt number at the tube exit $Nu_D(x = L)$, and relevant fluid properties.

(b) Repeat part (a) for constant surface temperature conditions. Express your result in terms of $\dot{m}$, L, the average Nusselt number from the tube inlet to the tube exit $\overline{Nu}_D$, and relevant fluid properties.

8.11 Dry, compressed air at $T_{m,i} = 75$°C, $p = 10$ atm, with a mass flow rate of $\dot{m} = 0.001$ kg/s, enters a 30-mm-diameter, 5-m-long tube whose surface is at $T_s = 25$°C.

(a) Determine the thermal entry length, the mean temperature of the air at the tube outlet, the rate of heat transfer from the air to the tube wall, and the power required to flow the air through the tube. For these conditions the fully developed heat transfer coefficient is $h = 3.58$ W/m^2 · K.

(b) In an effort to reduce the capital cost of the installation it is proposed to use a smaller, 28-mm-diameter tube. Determine the thermal entry length, the mean temperature of the air at the tube outlet, the heat transfer rate, and the required power for the smaller tube. For laminar flow conditions it is known that the value of the fully developed heat transfer coefficient is inversely proportional to the tube diameter.

8.12 Atmospheric air enters the heated section of a circular tube at a flow rate of 0.005 kg/s and a temperature of 20°C. The tube is of diameter $D = 50$ mm, and fully developed conditions with $h = 25$ W/m^2 · K exist over the entire length of $L = 3$ m.

(a) For the case of uniform surface heat flux at $q_s'' = 1000$ W/m^2, determine the total heat transfer rate q and the mean temperature of the air leaving the tube $T_{m,o}$. What is the value of the surface temperature at the tube inlet $T_{s,i}$ and outlet $T_{s,o}$? Sketch the axial variation of T_s and T_m. On the same figure, also sketch (qualitatively) the axial variation of T_s and T_m for the more realistic case in which the local convection coefficient varies with x.

(b) If the surface heat flux varies linearly with x, such that q_s'' (W/m^2) = 500x (m), what are the values of q, $T_{m,o}$, $T_{s,i}$, and $T_{s,o}$? Sketch the axial variation of T_s and T_m. On the same figure, also sketch (qualitatively) the axial variation of T_s and T_m for the more realistic case in which the local convection coefficient varies with x.

(c) For the two heating conditions of parts (a) and (b), plot the mean fluid and surface temperatures, $T_m(x)$ and $T_s(x)$, respectively, as functions of distance along the tube. What effect will a fourfold increase in the convection coefficient have on the temperature distributions?

(d) For each type of heating process, what heat fluxes are required to achieve an air outlet temperature of 125°C? Plot the temperature distributions.

8.13 Consider flow in a circular tube. Within the test section length (between 1 and 2) a constant heat flux q_s'' is maintained.

(a) For the following two cases, sketch the surface temperature $T_s(x)$ and the fluid mean temperature $T_m(x)$ as a function of distance along the test section x. In case A, flow is hydrodynamically and thermally fully developed. In case B, flow is not developed.

(b) Assuming that the surface flux q_s'' and the inlet mean temperature $T_{m,1}$ are identical for both cases, will the exit mean temperature $T_{m,2}$ for case A be greater than, equal to, or less than $T_{m,2}$ for case B? Briefly explain why.

8.14 A flat-plate solar collector is used to heat atmospheric air flowing through a rectangular channel. The bottom surface of the channel is well insulated, while the top surface is subjected to a uniform heat flux q_o'', which is due to the net effect of solar radiation absorption and heat exchange between the absorber and cover plates.

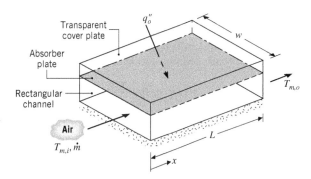

(a) Beginning with an appropriate differential control volume, obtain an equation that could be used to determine the mean air temperature $T_m(x)$ as a function of distance along the channel. Solve this equation to obtain an expression for the mean temperature of the air leaving the collector.

(b) With air inlet conditions of $\dot{m}$ = 0.1 kg/s and $T_{m,i}$ = 40°C, what is the air outlet temperature if L = 3 m, w = 1 m, and q_o'' = 700 W/m^2? The specific heat of air is c_p = 1008 J/kg · K.

8.15 Water enters a tube at 27°C with a flow rate of 450 kg/h. The rate of heat transfer from the tube wall to the fluid is given as q_s'(W/m) = ax, where the coefficient a is 20 W/m^2 and x (m) is the axial distance from the tube entrance.

(a) Beginning with a properly defined differential control volume in the tube, derive an expression for the temperature distribution $T_m(x)$ of the water.

(b) What is the outlet temperature of the water for a heated section 30 m long?

(c) Sketch the mean fluid temperature, $T_m(x)$, and the tube wall temperature, $T_s(x)$, as a function of distance along the tube for fully developed *and* developing flow conditions.

(d) What value of a uniform wall heat flux, q_s'' (instead of q_s' = ax), would provide the same fluid outlet temperature as that determined in part (b)? For this type of heating, sketch the temperature distributions requested in part (c).

8.16 Slug flow is an idealized tube flow condition for which the velocity is assumed to be uniform over the entire tube cross section. For the case of laminar slug flow with a uniform surface heat flux, determine the form of the fully developed temperature distribution $T(r)$ and the Nusselt number Nu_D.

8.17 Superimposing a control volume that is differential in x on the tube flow conditions of Figure 8.8, derive Equation 8.45a.

8.18 An experimental nuclear core simulation apparatus consists of a long thin-walled metallic tube of diameter D and length L, which is electrically heated to produce the sinusoidal heat flux distribution

$$q_s''(x) = q_o'' \sin\left(\frac{\pi x}{L}\right)$$

where x is the distance measured from the tube inlet. Fluid at an inlet temperature $T_{m,i}$ flows through the tube at a rate of $\dot{m}$. Assuming the flow is turbulent and fully developed over the entire length of the tube, develop expressions for:

(a) the total rate of heat transfer, q, from the tube to the fluid;

(b) the fluid outlet temperature, $T_{m,o}$;

(c) the axial distribution of the wall temperature, $T_s(x)$; and

(d) the magnitude and position of the highest wall temperature.

(e) Consider a 40-mm-diameter tube of 4-m length with a sinusoidal heat flux distribution for which $q_o'' = 10{,}000 \text{ W/m}^2$. Fluid passing through the tube has a flow rate of 0.025 kg/s, a specific heat of 4180 J/kg · K, an entrance temperature of 25°C, and a convection coefficient of 1000 W/m² · K. Plot the mean fluid and surface temperatures as a function of distance along the tube. Identify important features of the distributions. Explore the effect of ±25% changes in the convection coefficient and the heat flux on the distributions.

Heat Transfer Correlations: Circular Tubes

8.19 Engine oil is heated by flowing through a circular tube of diameter $D = 50$ mm and length $L = 25$ m and whose surface is maintained at 150°C.

(a) If the flow rate and inlet temperature of the oil are 0.5 kg/s and 20°C, what is the outlet temperature $T_{m,o}$? What is the total heat transfer rate q for the tube?

(b) For flow rates in the range $0.5 \leq \dot{m} \leq 2.0$ kg/s, compute and plot the variations of $T_{m,o}$ and q with $\dot{m}$. For what flow rate(s) are q and $T_{m,o}$ maximized? Explain your results.

8.20 In the final stages of production, a pharmaceutical is sterilized by heating it from 25 to 75°C as it moves at 0.2 m/s through a straight thin-walled stainless steel tube of 12.7-mm diameter. A uniform heat flux is maintained by an electric resistance heater wrapped around the outer surface of the tube. If the tube is 10 m long, what is the required heat flux? If fluid enters the tube with a fully developed velocity profile and a uniform temperature profile, what is the surface temperature at the tube exit and at a distance of 0.5 m from the entrance? Fluid properties may be approximated as $\rho = 1000 \text{ kg/m}^3$, $c_p = 4000 \text{ J/kg} \cdot \text{K}$, $\mu = 2 \times 10^{-3} \text{ kg/s} \cdot \text{m}$, $k = 0.8 \text{ W/m} \cdot \text{K}$, and $Pr = 10$.

8.21 Consider the laminar thermal boundary layer development near the entrance of the tube shown in Figure 8.4. When the hydrodynamic boundary layer is thin relative to the tube diameter, the inviscid flow region has a uniform velocity that is approximately equal to the mean velocity u_m. Hence the boundary layer development is similar to what would occur for a flat plate.

(a) Beginning with Equation 7.23, derive an expression for the local Nusselt number Nu_D, as a function of the Prandtl number Pr and the inverse Graetz number Gz_D^{-1}. Plot the expression using the coordinates shown in Figure 8.10a for $Pr = 0.7$.

(b) Beginning with Equation 7.30, derive an expression for the average Nusselt number $\overline{Nu}_D$, as a function of the Prandtl number Pr and the inverse Graetz number Gz_D^{-1}. Compare your results with the Nusselt number for the combined entrance length in the limit of small x.

8.22 Fluid enters a tube with a flow rate of 0.015 kg/s and an inlet temperature of 20°C. The tube, which has a length of 6 m and diameter of 15 mm, has a surface temperature of 30°C.

(a) Determine the heat transfer rate to the fluid if it is water.

(b) Determine the heat transfer rate for the nanofluid of Example 2.2.

8.23 An oil preheater consists of a single tube of 10-mm diameter and 5-m length, with its surface maintained at 175°C by swirling combustion gases. The engine oil (new) enters at 75°C. What flow rate must be supplied to maintain an oil outlet temperature of 100°C? What is the corresponding heat transfer rate?

8.24 Engine oil flows at a rate of 1 kg/s through a 5-mm-diameter straight tube. The oil has an inlet temperature of 45°C and it is desired to heat the oil to a mean temperature of 80°C at the exit of the tube. The surface of the tube is maintained at 150°C. Determine the required length of the tube. *Hint:* Calculate the Reynolds numbers at the entrance and exit of the tube before proceeding with your analysis.

8.25 Air at $p = 1$ atm enters a thin-walled ($D = 5$-mm diameter) long tube ($L = 2$ m) at an inlet temperature of $T_{m,i} = 100°C$. A constant heat flux is applied to the air from the tube surface. The air mass flow rate is $\dot{m} = 135 \times 10^{-6}$ kg/s.

(a) If the tube surface temperature at the exit is $T_{s,o} = 160°C$, determine the heat rate entering the tube. Evaluate properties at $T = 400$ K.

(b) If the tube length of part (a) were reduced to $L = 0.2$ m, how would flow conditions at the tube exit be affected? Would the value of the heat transfer coefficient at the tube exit be greater than, equal to, or smaller than the heat transfer coefficient for part (a)?

(c) If the flow rate of part (a) were increased by a factor of 10, would there be a difference in flow conditions at the tube exit? Would the value of the heat transfer coefficient at the tube exit be greater than, equal to, or smaller than the heat transfer coefficient for part (a)?

8.26 To cool a summer home without using a vapor-compression refrigeration cycle, air is routed through a plastic pipe ($k = 0.15$ W/m·K, $D_i = 0.15$ m, $D_o = 0.17$ m) that is submerged in an adjoining body of water. The water temperature is nominally at $T_\infty = 17°C$, and a convection coefficient of $h_o \approx 1500$ W/m²·K is maintained at the outer surface of the pipe.

If air from the home enters the pipe at a temperature of $T_{m,i} = 29°C$ and a volumetric flow rate of $\dot{V}_i = 0.025$ m³/s, what pipe length L is needed to provide a discharge temperature of $T_{m,o} = 21°C$? What is the fan power required to move the air through this length of pipe if its inner surface is smooth?

8.27 *Batch processes* are often used in chemical and pharmaceutical operations to achieve a desired chemical composition for the final product. Related heat transfer processes are typically transient, involving a liquid of fixed volume that may be heated from room temperature to a desired process temperature, or cooled from the process temperature to room temperature. Consider a batch process for which a pharmaceutical (the cold fluid, c) is poured into an insulated, highly agitated vessel (a *stirred reactor*) and heated by passing a hot fluid (h) through a submerged heat exchanger coil of thin-walled tubing and surface area A_s. The flow rate, $\dot{m}_h$, mean inlet temperature, $T_{h,i}$, and specific heat, $c_{p,h}$, of the hot fluid are known, as are the initial temperature, $T_{c,i} < T_{h,i}$, the volume, V_c, mass density, ρ_c, and specific heat, $c_{v,c}$, of the pharmaceutical. Heat transfer from the hot fluid to the pharmaceutical is governed by an overall heat transfer coefficient U.

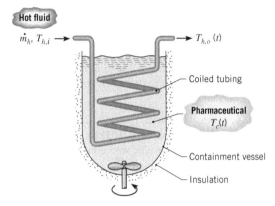

(a) Starting from basic principles, derive expressions that can be used to determine the variation of T_c and $T_{h,o}$ with time during the heating process. *Hint:* Two equations may be written for the rate of heat transfer, $q(t)$, to the pharmaceutical, one based on the log-mean temperature difference and the other on an energy balance for flow of the hot fluid through the tube. Equate these expressions to determine $T_{h,o}(t)$ as a function of $T_c(t)$ and prescribed parameters. Use the expression for $T_{h,o}(t)$ and the energy balance for flow through the tube with an energy balance for a control volume containing the pharmaceutical to obtain an expression for $T_c(t)$.

(b) Consider a pharmaceutical of volume $V_c = 1$ m³, density $\rho_c = 1100$ kg/m³, specific heat $c_{v,c} = 2000$ J/kg·K, and an initial temperature of $T_{c,i} = 25°C$. A coiled tube of length $L = 40$ m, diameter $D = 50$ mm, and coil diameter $C = 500$ mm is submerged in the vessel, and hot fluid enters the tubing at $T_{h,i} = 200°C$ and $\dot{m}_h = 2.4$ kg/s. The convection coefficient at the outer surface of the tubing may be approximated as $h_o = 1000$ W/m²·K, and the fluid properties are $c_{p,h} = 2500$ J/kg·K, $\mu_h = 0.002$ N·s/m², $k_h = 0.260$ W/m·K, and $Pr_h = 20$. For the foregoing conditions, compute and plot the pharmaceutical temperature T_c and the outlet temperature $T_{h,o}$ as a function of time over the range $0 \leq t \leq 3600$ s. How long does it take to reach a batch temperature of $T_c = 160°C$? The process operator may control the heating time by varying $\dot{m}_h$. For $1 \le \dot{m}_h \le 5$ kg/s, explore the effect of the flow rate on the time t_c required to reach a value of $T_c = 160°C$.

8.28 Water flowing at 2 kg/s through a 40-mm-diameter tube is to be heated from 25 to 75°C by maintaining the tube surface temperature at 100°C.

(a) What is the required tube length for these conditions?

(b) To design a water heating system, we wish to consider using tube diameters in the range from 30 to 50 mm. What are the required tube lengths for water flow rates of 1, 2, and 3 kg/s? Represent this design information graphically.

(c) Plot the pressure gradient as a function of tube diameter for the three flow rates. Assume the tube wall is smooth.

8.29 Dry, compressed air at $T_{m,i} = 55°C$, $p = 15$ atm, with a mass flow rate of $\dot{m} = 0.05$ kg/s, enters a 50-mm-diameter, 2.5-m-long tube whose surface temperature is $T_s = 25°C$.

(a) Determine the mean temperature of the air at the tube outlet, the rate of heat transfer from the air to the tube wall, and the power required to flow the air through the tube.

(b) In an effort to reduce the capital cost of the installation it is proposed to use a smaller, 40-mm-diameter tube. Determine the required tube length and the required power for the smaller tube in order to achieve the same heat transfer rate as for the larger tube.

8.30 Consider the conditions associated with the hot water pipe of Problem 7.44, but now account for the convection resistance associated with water flow at a mean velocity of $u_m = 0.5$ m/s in the pipe. What is the corresponding daily cost of heat loss per meter of the uninsulated pipe?

8.31 Water flows through a thick-walled tube with an inner diameter of 12 mm and a length of 8 m. The tube is immersed in a well-stirred, hot reaction tank maintained at 85°C, and the conduction resistance of the tube wall (based on the inner surface area) is $R''_{cd} = 0.002$ m² · K/W. The inlet temperature of the process fluid is $T_{m,i} = 20°C$, and the flow rate is 33 kg/h.

(a) Estimate the outlet temperature of the process fluid, $T_{m,o}$. Assume, and then justify, fully developed flow and thermal conditions within the tube.

(b) Do you expect $T_{m,o}$ to increase or decrease if combined thermal and hydrodynamic entry conditions exist within the tube? Estimate the outlet temperature of the water for this condition.

8.32 Consider fully developed conditions in a circular tube with constant surface temperature $T_s < T_m$. Determine whether a small- or large-diameter tube is more effective in minimizing heat loss from the flowing fluid characterized by a mass flow rate of $\dot{m}$. Consider both laminar and turbulent conditions.

8.33 Atmospheric air enters a 10-m-long, 150-mm-diameter uninsulated heating duct at 60°C and 0.04 kg/s. The duct surface temperature is approximately constant at $T_s = 15°C$.

(a) What are the outlet air temperature, the heat rate q, and pressure drop Δp for these conditions?

(b) To illustrate the tradeoff between heat transfer rate and pressure drop considerations, calculate q and Δp for diameters in the range from 0.1 to 0.2 m. In your analysis, maintain the total surface area, $A_s = \pi DL$, at the value computed for part (a). Plot q, Δp, and L as a function of the duct diameter.

8.34 NaK (45%/55%), which is an alloy of sodium and potassium, is used to cool fast neutron nuclear reactors. The NaK flows at a rate of $\dot{m} = 1$ kg/s through a $D = 50$-mm-diameter tube that has a surface temperature of $T_s = 450$ K. The NaK enters the tube at $T_{m,i} = 332$ K and exits at an outlet temperature of $T_{m,o} = 400$ K. Determine the tube length L and the local convective heat flux at the tube exit.

8.35 Consider the encased pipe of Problem 4.26, but now allow for the difference between the mean temperature of the fluid, which changes along the pipe length, and that of the pipe.

(a) For the prescribed values of k, D, w, h, and T_∞ and a pipe of length $L = 100$ m, what is the outlet temperature $T_{m,o}$ of water that enters the pipe at a temperature of $T_{m,i} = 90°C$ and a flow rate of $\dot{m} = 2$ kg/s?

(b) What is the pressure drop of the water and the corresponding pump power requirement?

(c) Subject to the constraint that the width of the duct is fixed at $w = 0.30$ m, explore the effects of the flow rate and the pipe diameter on the outlet temperature.

8.36 Liquid mercury at 0.5 kg/s is to be heated from 300 to 400 K by passing it through a 50-mm-diameter tube whose surface is maintained at 450 K. Calculate the required tube length by using an appropriate liquid metal convection heat transfer correlation. Compare your result with that which would have been obtained by using a correlation appropriate for $Pr \gtrsim 0.7$.

8.37 The surface of a 50-mm-diameter, thin-walled tube is maintained at 100°C. In one case air is in cross flow over the tube with a temperature of 25°C and a velocity of 30 m/s. In another case air is in fully developed flow through the tube with a temperature of 25°C and a mean velocity of 30 m/s. Compare the heat flux from the tube to the air for the two cases.

8.38 Consider a horizontal, thin-walled circular tube of diameter $D = 0.025$ m submerged in a container of *n*-octadecane (paraffin), which is used to store thermal energy. As hot water flows through the tube, heat is transferred to the paraffin, converting it from the solid to liquid state at the phase change temperature of $T_\infty = 27.4°C$. The latent heat of fusion and density of paraffin are $h_{sf} = 244$ kJ/kg and $\rho = 770$ kg/m^3, respectively, and thermophysical properties of the water may be taken as $c_p = 4.185$ kJ/kg · K, $k = 0.653$ W/m · K, $\mu = 467 \times 10^{-6}$ kg/s · m, and $Pr = 2.99$.

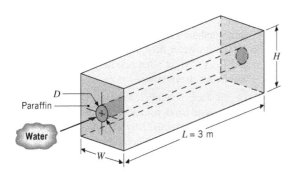

(a) Assuming the tube surface to have a uniform temperature corresponding to that of the phase change, determine the water outlet temperature and total heat transfer rate for a water flow rate of 0.1 kg/s and an inlet temperature of 60°C. If $H = W = 0.25$ m, how long would it take to completely liquefy the paraffin, from an initial state for which all the paraffin is solid and at 27.4°C?

(b) The liquefaction process can be accelerated by increasing the flow rate of the water. Compute and plot the heat rate and outlet temperature as a function of flow rate for $0.1 \leq \dot{m} \leq 0.5$ kg/s. How long would it take to melt the paraffin for $\dot{m} = 0.5$ kg/s?

8.39 Compressed air at $p = 20$ atm enters a 20-mm-diameter tube at $T_{m,i} = 20°C$ and a mass flow rate of $\dot{m} = 6 \times 10^{-4}$ kg/s. The air is warmed by a constant surface heat flux so that its exit mean temperature is $T_{m,o} = 50°C$. Determine the required surface heat flux and the temperature of the tube wall at the tube exit for $L = 0.15, 1.5,$ and 15 m.

8.40 Cooling water flows through the 25.4-mm-diameter thin-walled tubes of a steam condenser at 1 m/s, and a surface temperature of 350 K is maintained by the

condensing steam. The water inlet temperature is 290 K, and the tubes are 5 m long.

(a) What is the water outlet temperature? Evaluate water properties at an assumed average mean temperature, $\bar{T}_m = 300$ K.

(b) Was the assumed value for $\bar{T}_m$ reasonable? If not, repeat the calculation using properties evaluated at a more appropriate temperature.

(c) A range of tube lengths from 4 to 7 m is available to the engineer designing this condenser. Generate a plot to show what coolant mean velocities are possible if the water outlet temperature is to remain at the value found for part (b). All other conditions remain the same.

8.41 Air at 200 kPa enters a 2-m-long, thin-walled tube of 25-mm diameter at 150°C and 6 m/s. Steam at 20 bars condenses on the outer surface.

(a) Determine the outlet temperature and pressure drop of the air, as well as the rate of heat transfer to the air.

(b) Calculate the parameters of part (a) if the pressure of the air is doubled.

8.42 Heated air required for a food-drying process is generated by passing ambient air at 20°C through long, circular tubes ($D = 50$ mm, $L = 5$ m) housed in a steam condenser. Saturated steam at atmospheric pressure condenses on the outer surface of the tubes, maintaining a uniform surface temperature of 100°C.

(a) If an airflow rate of 0.01 kg/s is maintained in each tube, determine the air outlet temperature $T_{m,o}$ and the total heat rate q for the tube.

(b) The air outlet temperature may be controlled by adjusting the tube mass flow rate. Compute and plot $T_{m,o}$ as a function of $\dot{m}$ for $0.005 \leq \dot{m} \leq 0.050$ kg/s. If a particular drying process requires approximately 1 kg/s of air at 75°C, what design and operating conditions should be prescribed for the air heater, subject to the constraint that the tube diameter and length be fixed at 50 mm and 5 m, respectively?

8.43 A common procedure for cooling a high-performance computer chip involves joining the chip to a heat sink within which circular microchannels are machined. During operation, the chip produces a uniform heat flux q_c'' at its interface with the heat sink, while a liquid coolant (water) is routed through the channels. Consider a square chip and heat sink, each L on a side, with microchannels of diameter D and pitch $S = C_1 D$, where the constant C_1 is greater than unity. Water is supplied at an inlet temperature $T_{m,i}$ and a total mass flow rate $\dot{m}$ (for the entire heat sink).

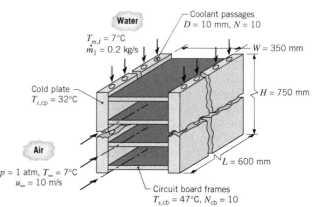

(a) Assuming that q_c'' is dispersed in the heat sink such that a uniform heat flux q_s'' is maintained at the surface of each channel, obtain expressions for the longitudinal distributions of the mean fluid, $T_m(x)$, and surface, $T_s(x)$, temperatures in each channel. Assume laminar, fully developed flow throughout each channel, and express your results in terms of $\dot{m}$, q_c'', C_1, D, and/or L, as well as appropriate thermophysical properties.

(b) For $L = 12$ mm, $D = 1$ mm, $C_1 = 2$, $q_c'' = 20$ W/cm^2, $\dot{m} = 0.010$ kg/s, and $T_{m,i} = 290$ K, compute and plot the temperature distributions $T_m(x)$ and $T_s(x)$.

(c) A common objective in designing such heat sinks is to maximize q_c'' while maintaining the heat sink at an acceptable temperature. Subject to prescribed values of $L = 12$ mm and $T_{m,i} = 290$ K and the constraint that $T_{s,\max} \leq 50°C$, explore the effect on q_c'' of variations in heat sink design and operating conditions.

8.44 One way to cool chips mounted on the circuit boards of a computer is to encapsulate the boards in metal frames that provide efficient pathways for conduction to supporting *cold plates*. Heat generated by the chips is then dissipated by transfer to water flowing through passages drilled in the plates. Because the plates are made from a metal of large thermal conductivity (typically aluminium or copper), they may be assumed to be at a temperature, $T_{s,cp}$.

(a) Consider circuit boards attached to cold plates of height $H = 750$ mm and width $L = 600$ mm, each with $N = 10$ holes of diameter $D = 10$ mm. If operating conditions maintain plate temperatures of $T_{s,cp} = 32°C$ with water flow at $\dot{m}_1 = 0.2$ kg/s per passage and $T_{m,i} = 7°C$, at what rate may heat be generated within the circuit boards?

(b) To enhance cooling, thereby allowing increased power generation without an attendant increase in system temperatures, a hybrid cooling scheme may be used. The scheme involves forced airflow over the encapsulated circuit boards, as well as water flow through the cold plates. Consider conditions for which $N_{cb} = 10$ circuit boards of width $W = 350$ mm are attached to the cold plates and their average surface temperature is $T_{s,cb} = 47°C$ when $T_{s,cp} = 32°C$. If air is in parallel flow over the plates with $u_\infty = 10$ m/s and $T_\infty = 7°C$, how much of the heat generated by the circuit boards is transferred to the air?

8.45 Refrigerant-134a is being transported at 0.1 kg/s through a Teflon tube of inside diameter $D_i = 25$ mm and outside diameter $D_o = 28$ mm, while atmospheric air at $V = 25$ m/s and 300 K is in cross flow over the tube. What is the heat transfer per unit length of tube to Refrigerant-134a at 240 K?

8.46 Oil at 150°C flows *slowly* through a long, thin-walled pipe of 30-mm inner diameter. The pipe is suspended in a room for which the air temperature is 20°C and the convection coefficient at the outer tube surface is 11 W/m^2 · K. Estimate the heat loss per unit length of tube.

8.47 Exhaust gases from a wire processing oven are discharged into a tall stack, and the gas and stack surface temperatures at the outlet of the stack must be estimated.

Knowledge of the outlet gas temperature $T_{m,o}$ is useful for predicting the dispersion of effluents in the thermal plume, while knowledge of the outlet stack surface temperature $T_{s,o}$ indicates whether condensation of the gas products will occur. The thin-walled, cylindrical stack is 0.5 m in diameter and 6.0 m high. The exhaust gas flow rate is 0.5 kg/s, and the inlet temperature is 600°C.

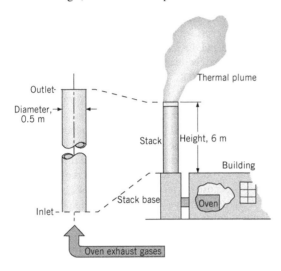

(a) Consider conditions for which the ambient air temperature and wind velocity are 4°C and 5 m/s, respectively. Approximating the thermophysical properties of the gas as those of atmospheric air, estimate the outlet gas and stack surface temperatures for the given conditions.

(b) The gas outlet temperature is sensitive to variations in the ambient air temperature and wind velocity. For $T_\infty = -25°C$, 5°C, and 35°C, compute and plot the gas outlet temperature as a function of wind velocity for $2 \leq V \leq 10$ m/s.

8.48 The problem of heat losses from a fluid moving through a buried pipeline has received considerable attention. Practical applications include the trans-Alaska pipeline, as well as power plant steam and water distribution lines. Consider a steel pipe of diameter D that is used to transport oil flowing at a rate $\dot{m}_o$ through a cold region. The pipe is covered with a layer of insulation of thickness t and thermal conductivity k_i and is buried in soil to a depth z (distance from the soil surface to the pipe centerline). Each section of pipe is of length L and extends between pumping stations in which the oil is heated to ensure low viscosity and hence low pump power requirements. The temperature of the oil entering the pipe from a pumping station and the temperature of the ground above the pipe are designated as $T_{m,i}$ and T_s, respectively, and are known.

Consider conditions for which the oil (o) properties may be approximated as $\rho_o = 900$ kg/m³, $c_{p,o} = 2000$ J/kg · K, $\nu_o = 8.5 \times 10^{-4}$ m²/s, $k_o = 0.140$ W/m · K, $Pr_o = 10^4$; the oil flow rate is $\dot{m}_o = 500$ kg/s; and the pipe diameter is 1.2 m.

(a) Expressing your results in terms of D, L, z, t, $\dot{m}_o$, $T_{m,i}$, and T_s, as well as the appropriate oil (o), insulation (i), and soil (s) properties, obtain all the expressions needed to estimate the temperature $T_{m,o}$ of the oil leaving the pipe.

(b) If $T_s = -40°C$, $T_{m,i} = 120°C$, $t = 0.15$ m, $k_i = 0.05$ W/m · K, $k_s = 0.5$ W/m · K, $z = 3$ m, and $L = 100$ km, what is the value of $T_{m,o}$? What is the total rate of heat transfer q from a section of the pipeline?

(c) The operations manager wants to know the trade-off between the burial depth of the pipe and insulation thickness on the heat loss from the pipe. Develop a graphical representation of this design information.

8.49 Consider a thin-walled tube of 10-mm diameter and 2-m length. Water enters the tube from a large reservoir at $\dot{m} = 0.2$ kg/s and $T_{m,i} = 47°C$.

(a) If the tube surface is maintained at a uniform temperature of 27°C, what is the outlet temperature of the water, $T_{m,o}$? To obtain the properties of water, assume an average mean temperature of $\overline{T}_m = 300$ K.

(b) What is the exit temperature of the water if it is heated by passing air at $T_\infty = 100°C$ and $V = 10$ m/s in cross flow over the tube? The properties of air may be evaluated at an assumed film temperature of $T_f = 350$ K.

(c) In the foregoing calculations, were the assumed values of $\overline{T}_m$ and T_f appropriate? If not, use properly evaluated properties and recompute $T_{m,o}$ for the conditions of part (b).

8.50 Water at a flow rate of $\dot{m} = 0.215$ kg/s is cooled from 70°C to 30°C by passing it through a thin-walled tube of diameter $D = 50$ mm and maintaining a coolant at $T_\infty = 15°C$ in cross flow over the tube.

(a) What is the required tube length if the coolant is air and its velocity is $V = 20$ m/s?

(b) What is the tube length if the coolant is water and $V = 2$ m/s?

8.51 A circular tube of diameter $D = 2$ mm and length $L = 100$ mm imposes a constant heat flux of $q'' = 200,000$ W/m² on a fluid with a mass flow rate of $\dot{m} = 10$ g/s. For an inlet temperature of $T_{m,i} = 29°C$, determine the tube wall temperature at $x = L$ for pure water. Evaluate fluid properties at $\bar{T} = 300\,\text{K}$. For the same conditions, determine the tube wall temperature at $x = L$ for the nanofluid of Example 2.2.

8.52 Heat is to be removed from a reaction vessel operating at 75°C by supplying water at 27°C and 0.12 kg/s through a thin-walled tube of 15-mm diameter. The convection coefficient between the tube outer surface and the fluid in the vessel is 3000 W/m² · K.

(a) If the outlet water temperature cannot exceed 47°C, what is the maximum rate of heat transfer from the vessel?

(b) What tube length is required to accomplish the heat transfer rate of part (a)?

8.53 A heating contractor must heat 0.2 kg/s of water from 15°C to 35°C using hot gases in cross flow over a thin-walled tube.

Your assignment is to develop a series of design graphs that can be used to demonstrate acceptable combinations of tube dimensions (D and L) and of hot gas conditions (T_∞ and V) that satisfy this requirement. In your analysis, consider the following parameter ranges: $D = 20$, 30, or 40 mm; $L = 3$, 4, or 6 m; $T_\infty = 250$, 375, or 500°C; and $20 \le V \le 40$ m/s.

8.54 A thin-walled tube with a diameter of 6 mm and length of 20 m is used to carry exhaust gas from a smoke stack to the laboratory in a nearby building for analysis. The gas enters the tube at 200°C and with a mass flow rate of 0.003 kg/s. Autumn winds at a temperature of 15°C blow directly across the tube at a velocity of 5 m/s. Assume the thermophysical properties of the exhaust gas are those of air.

(a) Estimate the average heat transfer coefficient for the exhaust gas flowing inside the tube.

(b) Estimate the heat transfer coefficient for the air flowing across the outside of the tube.

(c) Estimate the overall heat transfer coefficient U and the temperature of the exhaust gas when it reaches the laboratory.

8.55 A thin-walled, uninsulated 0.3-m-diameter duct is used to route chilled air at 0.05 kg/s through the attic of a large commercial building. The attic air is at 37°C, and natural circulation provides a convection coefficient of 2 W/m² · K at the outer surface of the duct. If chilled air enters a 15-m-long duct at 7°C, what is its exit temperature and the rate of heat gain? Properties of the chilled air may be evaluated at an assumed average temperature of 300 K.

8.56 Pressurized water at $T_{m,i} = 200°C$ is pumped at $\dot{m} = 2$ kg/s from a power plant to a nearby industrial user through a thin-walled, round pipe of inside diameter $D = 1$ m. The pipe is covered with a layer of insulation of thickness $t = 0.15$ m and thermal conductivity $k = 0.05$ W/m · K. The pipe, which is of length $L = 500$ m, is exposed to a cross flow of air at $T_\infty = -10°C$ and $V = 4$ m/s. Obtain a differential equation that could be used to solve for the variation of the mixed mean temperature of the water $T_m(x)$ with the axial coordinate. As a first approximation, the internal flow may be assumed to be fully developed throughout the pipe. Express your results in terms of $\dot{m}$, V, T_∞, D, t, k, and appropriate water (w) and air (a) properties. Evaluate the rate of heat loss per unit length of the pipe at the inlet. What is the mean temperature of the water at the outlet?

8.57 Water at 290 K and 0.2 kg/s flows through a Teflon tube ($k = 0.35$ W/m · K) of inner and outer radii equal to 10 and 13 mm, respectively. A thin electrical heating tape wrapped around the outer surface of the tube delivers a uniform surface heat flux of 2000 W/m², while a convection coefficient of 25 W/m² · K is maintained on the outer surface of the tape by ambient air at 300 K. What is the fraction of the power dissipated by the tape, which is transferred to the water? What is the outer surface temperature of the Teflon tube?

8.58 The temperature of flue gases flowing through the large stack of a boiler is measured by means of a thermocouple enclosed within a cylindrical tube as shown. The tube axis is oriented normal to the gas flow, and the thermocouple senses a temperature T_t corresponding to that of the tube surface. The gas flow rate and temperature are designated as $\dot{m}_g$ and T_g, respectively, and the gas flow may be assumed to be fully developed. The stack is fabricated from sheet metal that is at a uniform temperature T_s and is exposed to ambient air at T_∞ and large surroundings at T_{sur}. The convection coefficient associated with the outer surface of the duct is designated as h_o, while those associated with the inner surface of the duct and the tube surface are designated as h_i and h_t, respectively. The tube and duct surface emissivities are designated as ε_t and ε_s, respectively.

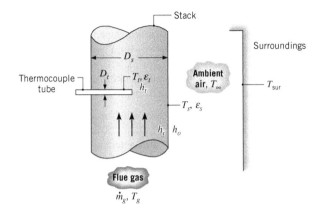

(a) Neglecting conduction losses along the thermocouple tube, develop an analysis that could be used to predict the error $(T_g - T_t)$ in the temperature measurement.

(b) Assuming the flue gas to have the properties of atmospheric air, evaluate the error for $T_t = 300°C$, $D_s = 0.6$ m, $D_t = 10$ mm, $\dot{m}_g = 1$ kg/s, $T_\infty = T_{sur} = 27°C$, $\varepsilon_t = \varepsilon_s = 0.8$, and $h_o = 25$ W/m$^2 \cdot$ K.

8.59 A 50-mm-diameter, thin-walled metal pipe covered with a 25-mm-thick layer of insulation (0.085 W/m · K) and carrying superheated steam at atmospheric pressure is suspended from the ceiling of a large room. The steam temperature entering the pipe is 120°C, and the air temperature is 20°C. The convection heat transfer coefficient on the outer surface of the covered pipe is 10 W/m$^2 \cdot$ K. If the velocity of the steam is 10 m/s, at what point along the pipe will the steam begin condensing?

8.60 In a biomedical supplies manufacturing process, a requirement exists for a large platen that is to be maintained at 45 ± 0.25°C. The proposed design features the attachment of heating tubes to the platen at a relative spacing S. The thick-walled, copper tubes have an inner diameter of $D_i = 8$ mm and are attached to the platen with a high thermal conductivity solder, which provides a contact width of $2D_i$. The heating fluid (ethylene glycol) flows through each tube at a fixed rate of $\dot{m} = 0.06$ kg/s. The platen has a thickness of $w = 25$ mm and is fabricated from a stainless steel with a thermal conductivity of 15 W/m · K.

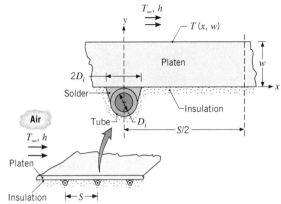

Considering the two-dimensional cross section of the platen shown in the inset, perform an analysis to determine the heating fluid temperature T_m and the tube spacing S required to maintain the surface temperature of the platen, $T(x, w)$, at 45 ± 0.25°C, when the ambient temperature is 25°C and the convection coefficient is 100 W/m$^2 \cdot$ K.

Noncircular Ducts

8.61 Air at 3×10^{-4} kg/s and 27°C enters a rectangular duct that is 1 m long and 4 mm × 16 mm on a side. A uniform heat flux of 600 W/m^2 is imposed on the duct surface. What is the temperature of the air and of the duct surface at the outlet?

8.62 Air at 25°C flows at 30×10^{-6} kg/s within 100-mm-long channels used to cool a high thermal conductivity metal mold. Assume the flow is hydrodynamically and thermally fully developed.

(a) Determine the rate of heat transferred to the air for a circular channel ($D = 10$ mm) when the mold temperature is 50°C (case A).

(b) Using new manufacturing methods (see Problem 8.80), channels of complex cross section can be readily fabricated within metal objects, such as molds. Consider air flowing under the same conditions as in case A, except now the channel is segmented into six smaller triangular sections. The flow area of case A is equal to the total flow area of case B. Determine the rate of heat transferred to the air for the segmented channel.

(c) Compare the pressure drops for cases A and B.

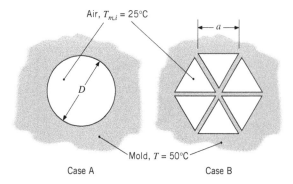

Case A   Case B

8.63 A *cold plate* is an active cooling device that is attached to a heat-generating system in order to dissipate the heat while maintaining the system at an acceptable temperature. It is typically fabricated from a material of high thermal conductivity, k_{cp}, within which channels are machined and a coolant is passed. Consider a copper cold plate of height H and width W on a side, within which water passes through square channels of width $w = h$. The transverse spacing between channels δ is twice the spacing between the sidewall of an outer channel and the sidewall of the cold plate.

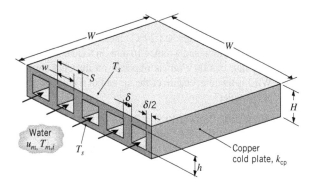

Consider conditions for which *equivalent* heat-generating systems are attached to the top and bottom of the cold plate, maintaining the corresponding surfaces at the same temperature T_s. The mean velocity and inlet temperature of the coolant are u_m and $T_{m,i}$, respectively.

(a) Assuming fully developed turbulent flow throughout each channel, obtain a system of equations that may be used to evaluate the total rate of heat transfer to the cold plate, q, and the outlet temperature of the water, $T_{m,o}$, in terms of the specified parameters.

(b) Consider a cold plate of width $W = 100$ mm and height $H = 10$ mm, with 10 square channels of width $w = 6$ mm and a spacing of $\delta = 4$ mm between channels. Water enters the channels at

a temperature of $T_{m,i} = 300$ K and a velocity of $u_m = 2$ m/s. If the top and bottom cold plate surfaces are at $T_s = 360$ K, what is the outlet water temperature and the total rate of heat transfer to the cold plate? The thermal conductivity of the copper is 400 W/m · K, while average properties of the water may be taken to be $\rho = 984$ kg/m³, $c_p = 4184$ J/kg · K, $\mu = 489 \times 10^{-6}$ N · s/m², $k = 0.65$ W/m · K, and $Pr = 3.15$. Is this a good cold plate design? How could its performance be improved?

8.64 Air at 1 atm and 285 K enters a 2-m-long rectangular duct with cross section 75 mm × 150 mm. The duct is maintained at a constant surface temperature of 400 K, and the air mass flow rate is 0.10 kg/s. Determine the heat transfer rate from the duct to the air and the air outlet temperature.

8.65 The cold plate design of Problem 8.63 has not been optimized with respect to selection of the channel width, and we wish to explore conditions for which the rate of heat transfer may be enhanced. Assume that the width and height of the copper cold plate are fixed at $W = 100$ mm and $H = 10$ mm, while the channel height and spacing between channels are fixed at $h = 6$ mm and $\delta = 4$ mm. The mean velocity and inlet temperature of the water are maintained at $u_m = 2$ m/s and $T_{m,i} = 300$ K, while equivalent heat-generating systems attached to the top and bottom of the cold plate maintain the corresponding surfaces at 360 K. Evaluate the effect of changing the channel width, and hence the number of channels, on the rate of heat transfer to the cold plate. Include consideration of the limiting case for which $w = 96$ mm (one channel).

8.66 A device that recovers heat from high-temperature combustion products involves passing the combustion gas between parallel plates, each of which is maintained at 350 K by water flow on the opposite surface. The plate separation is 40 mm, and the gas flow is fully developed. The gas may be assumed to have the properties of atmospheric air, and its mean temperature and velocity are 1000 K and 60 m/s, respectively.

(a) What is the heat flux at the plate surface?

(b) If a third plate, 20 mm thick, is suspended midway between the original plates, what is the surface heat flux for the original plates? Assume the temperature and *flow rate* of the gas to be unchanged and radiation effects to be negligible.

8.67 You have been asked to perform a feasibility study on the design of a blood warmer to be used during the transfusion of blood to a patient. This exchanger is to heat blood taken from the bank at 10°C to 37°C at a flow rate of 200 ml/min. The blood passes through a rectangular cross-section tube, 6.4 mm × 1.6 mm, which is sandwiched between two plates held at a constant temperature of 40°C.

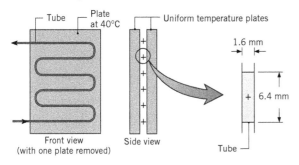

Front view
(with one plate removed)   Side view   Tube

(a) Compute the length of the tubing required to achieve the desired outlet conditions at the specified flow rate. Assume the flow is fully developed and the blood has the same properties as water.

(b) Assess your assumptions and indicate whether your analysis over- or underestimates the necessary length.

8.68 An electronic circuit board dissipating 50 W is sandwiched between two ducted, forced-air-cooled heat sinks. The sinks are 150 mm in length and have 20 rectangular passages 6 mm × 25 mm. Atmospheric air at a volumetric flow rate of 0.060 m³/s and 27°C is drawn through the sinks by a blower. Estimate the operating temperature of the board and the pressure drop across the sinks.

8.69 Rectangular flow channels are formed by milling long rectangular slots into flat metal plates, and brazing the plates together as shown below. Either large channels (Configuration A) or twice as many small channels (Configuration B) can be formed by shifting the

plates relative to one another prior to brazing. Consider isothermal plates at $T_s = 67°C$ with ethylene glycol at $T_m = 27°C$ flowing at a rate of 0.1 kg/s in each of the large channels. The flow rate in each small channel is 0.05 kg/s. For dimensions $a = 10$ mm and $b = 30$ mm, determine the heat transfer rate per unit channel length for one large channel and for a pair of small channels. Also determine the pressure gradient in the fluid associated with Configurations A and B.

8.70 To slow down large prime movers like locomotives, a process termed dynamic electric braking is used to switch the traction motor to a generator mode in which mechanical power from the drive wheels is absorbed and used to generate electrical current. As shown in the schematic, the electric power is passed through a resistor grid (a), which consists of an array of metallic blades electrically connected in series (b). The blade material is a high-temperature, high electrical resistivity alloy, and the electrical power is dissipated as heat by internal volumetric generation. To cool the blades, a motor-fan moves high-velocity air through the grid.

(a) Treating the space between the blades as a rectangular channel of 220-mm × 4-mm cross section and 70-mm length, estimate the heat removal rate per blade if the airstream has an inlet temperature and velocity of 25°C and 50 m/s, respectively, while the blade has an operating temperature of 600°C.

(b) On a locomotive pulling a 10-car train, there may be 2000 of these blades. Based on your result from part (a), how long will it take to slow a train whose total mass is 10^6 kg from a speed of 120 km/h to 50 km/h using dynamic electric braking?

8.71 Water at $\dot{m} = 0.02$ kg/s and $T_{m,i} = 20°C$ enters an annular region formed by an inner tube of diameter $D_i = 25$ mm and an outer tube of diameter $D_o = 100$ mm. Saturated steam flows through the inner tube, maintaining its surface at a uniform temperature of $T_{s,i} = 100°C$, while the outer surface of the outer tube is well insulated.

(a) If fully developed conditions may be assumed throughout the annulus, how long must the system be to provide an outlet water temperature of 75°C?

(b) What is the heat flux from the inner tube at the outlet?

(c) How long must the annulus be if the water flow rate is 0.30 kg/s instead of 0.02 kg/s?

8.72 Referring to Figure 8.11, consider conditions in an annulus having an outer surface that is insulated ($q''_o = 0$) and a uniform heat flux q''_i at the inner surface. Fully developed, laminar flow may be assumed to exist.

(a) Determine the velocity profile $u(r)$ in the annular region.

(b) Determine the temperature profile $T(r)$ and obtain an expression for the Nusselt number Nu_i associated with the inner surface.

8.73 Consider a concentric tube annulus for which the inner and outer diameters are 25 and 50 mm. Water enters the annular region at 0.04 kg/s and 25°C. If the inner tube wall is heated electrically at a rate (per unit length) of q' = 4000 W/m, while the outer tube wall is insulated, how long must the tubes be for the water to achieve an outlet temperature of 85°C? What is the inner tube surface temperature at the outlet, where fully developed conditions may be assumed?

8.74 It is common practice to recover waste heat from an oil- or gas-fired furnace by using the exhaust gases to preheat the combustion air. A device commonly used for this purpose consists of a concentric pipe arrangement for which the exhaust gases are passed through the inner pipe, while the cooler combustion air flows through an annular passage around the pipe.

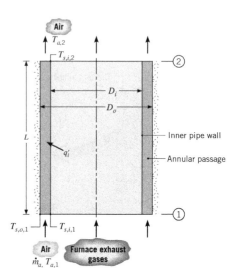

Consider conditions for which there is a uniform heat transfer rate per unit length, $q'_i = 1.25 \times 10^5$ W/m, from the exhaust gases to the pipe inner surface, while air flows through the annular passage at a rate of $\dot{m}_a = 2.1$ kg/s. The thin-walled inner pipe is of diameter $D_i = 2$ m, while the outer pipe, which is well insulated from the surroundings, is of diameter $D_o = 2.05$ m. The air properties may be taken to be $c_p = 1030$ J/kg · K, $\mu = 270 \times 10^{-7}$ N · s/m^2, $k = 0.041$ W/m · K, and $Pr = 0.68$.

(a) If air enters at $T_{a,1} = 300$ K and $L = 7$ m, what is the air outlet temperature $T_{a,2}$?

(b) If the airflow is fully developed throughout the annular region, what is the temperature of the inner pipe at the inlet ($T_{s,i,1}$) and outlet ($T_{s,i,2}$) sections of the device? What is the outer surface temperature $T_{s,o,1}$ at the inlet?

Heat Transfer Enhancement

8.75 Consider the air cooling system and conditions of Problem 8.26, but with a prescribed pipe length of $L = 15$ m.

(a) What is the air outlet temperature, $T_{m,o}$? What is the fan power requirement?

(b) The convection coefficient associated with airflow in the pipe may be increased twofold by inserting a coiled spring along the length of the pipe to disrupt flow conditions near the inner surface. If such a heat transfer enhancement scheme is adopted, what is the attendant value of $T_{m,o}$? Use of the insert would not come without a corresponding increase in the fan power requirement. What is the power requirement if the friction factor is increased by 50%?

(c) After extended exposure to the water, a thin coating of organic matter forms on the outer surface of the pipe, and its thermal resistance (for a unit area of the outer surface) is $R''_{t,o} = 0.050 \text{ m}^2 \cdot \text{K/W}$. What is the corresponding value of $T_{m,o}$ without the insert of part (b)?

8.76 Consider sterilization of the pharmaceutical product of Problem 8.20. To avoid any possibility of heating the product to an unacceptably high temperature, atmospheric steam is condensed on the exterior of the tube instead of using the resistance heater, providing a uniform surface temperature, $T_s = 100°C$.

(a) For the conditions of Problem 8.20, determine the required length of straight tube, L_s, that would be needed to increase the mean temperature of the pharmaceutical product from 25°C to 75°C.

(b) Consider replacing the straight tube with a coiled tube characterized by a coil diameter $C = 100 \text{ mm}$ and a coil pitch $S = 25 \text{ mm}$. Determine the overall length of the coiled tube, L_{cl} (i.e., the product of the tube pitch and the number of coils), necessary to increase the mean temperature of the pharmaceutical to the desired value.

(c) Calculate the pressure drop through the straight tube and through the coiled tube.

(d) Calculate the steam condensation rate.

8.77 An engineer proposes to insert a solid rod of diameter D_i into a circular tube of diameter D_o to enhance heat transfer from the flowing fluid of temperature T_m to the outer tube wall of temperature $T_{s,o}$. Assuming laminar flow, calculate the ratio of the heat flux from the fluid to the outer tube wall with the rod to the heat flux without the rod, $q''_o/q''_{o,\text{wo}}$, for $D_i/D_o = 0$, 0.10, 0.25 and 0.50. The rod is placed concentrically within the tube.

8.78 An electrical power transformer of diameter 230 mm and height 500 mm dissipates 1000 W. It is desired to maintain its surface temperature at 47°C by supplying ethylene glycol at 24°C through thin-walled tubing of 20-mm diameter welded to the lateral surface of the transformer. All the heat dissipated by the transformer is assumed to be transferred to the ethylene glycol.

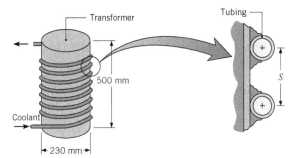

Assuming the maximum allowable temperature rise of the coolant to be 6°C, determine the required coolant flow rate, the total length of tubing, and the coil pitch S between turns of the tubing.

8.79 A *bayonet cooler* is used to reduce the temperature of a pharmaceutical fluid. The pharmaceutical fluid flows through the cooler, which is fabricated of 10-mm-diameter, thin-walled tubing with two 250-mm-long straight sections and a coil with six and a half turns and a coil diameter of 75 mm. A coolant flows outside the cooler, with a convection coefficient at the outside surface of $h_o = 500 \text{ W/m}^2 \cdot \text{K}$ and a coolant temperature of 20°C. Consider the situation where the pharmaceutical fluid enters at 90°C with a mass flow rate of 0.005 kg/s. The pharmaceutical has the following properties: $\rho = 1200 \text{ kg/m}^3$, $\mu = 4 \times 10^{-3} \text{ N} \cdot \text{s/m}^2$, $c_p = 2000 \text{ J/kg} \cdot \text{K}$, and $k = 0.5 \text{ W/m} \cdot \text{K}$.

(a) Determine the outlet temperature of the pharmaceutical fluid.

(b) It is desired to further reduce the outlet temperature of the pharmaceutical. However, because the cooling process is just one part of an intricate processing operation, flow rates cannot be changed. A young engineer suggests that the outlet temperature might be reduced by inserting stainless steel coiled springs into the straight sections of the cooler with the notion that the springs will disturb the flow adjacent to the inner tube wall and, in turn, increase the heat transfer coefficient at the inner tube wall. A senior engineer asserts that insertion of the springs should double the heat transfer coefficient at the straight inner tube walls. Determine the outlet temperature of the pharmaceutical fluid with the springs inserted into the tubes, assuming the senior engineer is correct in his assertion.

(c) Would you expect the outlet temperature of the pharmaceutical to depend on whether the springs have a left-hand or right-hand spiral? Why?

8.80 The mold used in an injection molding process consists of a top half and a bottom half. Each half is 60 mm × 60 mm × 20 mm and is constructed of metal ($\rho = 7800$ kg/m^3, $c = 450$ J/kg · K). The cold mold (100°C) is to be heated to 200°C with pressurized water (available at 275°C and a total flow rate of 0.02 kg/s) prior to injecting the thermoplastic material. The injection takes only a fraction of a second, and the hot mold (200°C) is subsequently cooled with cold water (available at 25°C and a total flow rate of 0.02 kg/s) prior to ejecting the molded part. After part ejection, which also takes a fraction of a second, the process is repeated.

(a) In conventional mold design, straight cooling (heating) passages are bored through the mold in a location where the passages will not interfere with the molded part. Determine the initial heating rate and the initial cooling rate of the mold when five 5-mm-diameter, 60-mm-long passages are bored in each half of the mold (10 passages total). The velocity distribution of the water is fully developed at the entrance of each passage in the hot (or cold) mold.

(b) New additive manufacturing processes, known as *selective freeform fabrication*, or *SFF*, are used to construct molds that are configured with *conformal cooling passages*. Consider the same mold as before, but now a 5-mm-diameter, coiled, conformal cooling passage is designed within each half of the SFF-manufactured mold. Each of the two coiled passages has $N = 2$ turns. The coiled passage does not interfere with the molded part. The conformal channels have a coil diameter $C = 50$ mm. The total water flow remains the same as in part (a) (0.01 kg/s per coil). Determine the initial heating rate and the initial cooling rate of the mold.

(c) Compare the surface areas of the conventional and conformal cooling passages. Compare the rate at which the mold temperature changes for molds configured with the conventional and conformal heating and cooling passages. Which cooling passage, conventional or conformal, will enable production of more parts per day? Neglect the presence of the thermoplastic material.

8.81 Consider the pharmaceutical product of Problem 8.20. Prior to finalizing the manufacturing process, test trials are performed to experimentally determine the dependence of the shelf life of the drug as a function of the sterilization temperature. Hence, the sterilization temperature must be carefully controlled in the trials. To promote good mixing of the pharmaceutical and, in turn, relatively uniform outlet temperatures across the exit tube area, experiments are performed using a device that is constructed of two interwoven coiled tubes, each of 10-mm

diameter. The thin-walled tubing is welded to a solid high thermal conductivity rod of diameter $D_r = 40$ mm. One tube carries the pharmaceutical product at a mean velocity of $u_p = 0.1$ m/s and inlet temperature of 25°C, while the second tube carries pressurized liquid water at $u_w = 0.12$ m/s with an inlet temperature of 127°C. The tubes do not contact each other but are each welded to the solid metal rod, with each tube making 20 turns around the rod. The exterior of the apparatus is well insulated.

Heating fluid in

Pharmaceutical in

|← 40 mm →|

(a) Determine the outlet temperature of the pharmaceutical product. Evaluate the liquid water properties at 380 K.

(b) Investigate the sensitivity of the pharmaceutical's outlet temperature to the velocity of the pressurized water over the range $0.10 < u_w < 0.25$ m/s.

Flow in Small Channels

8.82 An extremely effective method of cooling high-power-density silicon chips involves etching microchannels in the back (noncircuit) surface of the chip. The channels are covered with a silicon cap, and cooling is maintained by passing water through the channels.

H $\quad$ W

Circuits
Chip, T_s
Microchannels
Cap, T_s

Consider a chip that is 10 mm × 10 mm on a side and in which fifty 10-mm-long rectangular microchannels, each of width $W = 50$ μm and height $H = 200$ μm, have been etched. Consider operating conditions for which water enters each microchannel at a temperature of 290 K and a flow rate of 10^{-4} kg/s, while the chip and cap are at a uniform temperature of 350 K. Assuming fully developed flow in the channel and that all the heat dissipated by the circuits is transferred to the water, determine the water outlet temperature and the chip power dissipation. Water properties may be evaluated at 300 K.

8.83 Due to its comparatively large thermal conductivity, water is a preferred fluid for convection cooling. However, in applications involving electronic devices, water must not come into contact with the devices, and dielectric fluids such as air are commonly used in lieu of water.

Consider the microchannel chip cooling application of Problem 8.82, but with air as the coolant and a flow rate of $\dot{m}_1 = 10^{-6}$ kg/s per channel. Determine the air outlet temperature and the chip power dissipation. Evaluate air properties at $T = 300$ K, $p = 1$ atm. Use thermal and momentum accommodation coefficients of $\alpha_t = 0.8$ and $\alpha_p = 0.9$, respectively. Assume that $Nu_D = hD_h/k$ for the uniform wall temperature condition is increased due to microscale effects in the same proportion as for flow in a circular tube with a uniform heat flux condition.

8.84 An ideal gas flows within a small diameter tube. Derive an expression for the transition density of the gas ρ_c below which microscale effects must be accounted for. Express your result in terms of the gas molecule diameter, universal gas constant, Boltzmann's constant, and the tube diameter. Evaluate the transition density for a $D = 10$-μm-diameter tube for hydrogen, air, and carbon dioxide. Compare the calculated transition densities with the gas density at atmospheric pressure and $T = 23°C$.

8.85 Consider the microchannel cooling arrangement of Problem 8.82. However, instead of assuming the entire chip and cap to be at a uniform temperature, adopt a more conservative (and realistic) approach that prescribes a temperature of $T_s = 350$ K at the base of the channels ($x = 0$) and allows for a decrease in temperature with increasing x along the side walls of each channel.

(a) For the operating conditions prescribed in Problem 8.82 and a chip thermal conductivity of $k_{ch} = 140$ W/m · K, determine the water outlet temperature and the chip power dissipation. Heat transfer from the sides of the chip to the surroundings and from the side walls of a channel to the cap may be neglected. Note that the spacing between channels, $\delta = S - W$, is twice the spacing between the side wall of an outer channel and the outer surface of the chip. The channel pitch is $S = L/N$, where $L = 10$ mm is the chip width and $N = 50$ is the number of channels.

(b) The channel geometry prescribed in Problem 8.82 and considered in part (a) is not optimized, and larger heat rates may be dissipated by adjusting related dimensions. Consider the effect of reducing the pitch to a value of $S = 100$ μm, while retaining a width of $W = 50$ μm and a flow rate per channel of $\dot{m}_1 = 10^{-4}$ kg/s.

8.86 The onset of turbulence in a gas flowing within a circular tube occurs at $Re_{D,c} \approx 2300$, while a transition from incompressible to compressible flow occurs at a critical Mach number of $Ma_c \approx 0.3$. Determine the critical tube diameter D_c, below which incompressible turbulent flow and heat transfer cannot exist for (i) air, (ii) CO_2, (iii) He. Evaluate properties at atmospheric pressure and a temperature of $T = 300$ K.

8.87 Due to its comparatively large thermal conductivity, water is a preferred fluid for convection cooling. However, in applications involving electronic devices, water must not come into contact with the devices, which would therefore have to be hermetically sealed. To circumvent related design and operational complexities and to ensure that the devices are not rendered inoperable by contact with the coolant, a dielectric fluid is commonly used in lieu of water. Many gases have excellent dielectric characteristics, and despite its poor heat transfer properties, air is the common choice for electronic cooling. However, there is an alternative, which involves a class of *perfluorinated liquids* that are excellent dielectrics and have heat transfer properties superior to those of gases.

Consider the microchannel chip cooling application of Problem 8.85 but now for a perfluorinated liquid with properties of $c_p = 1050$ J/kg · K, $k = 0.065$ W/m · K, $\mu = 0.0012$ N · s/m^2, and $Pr = 15$.

(a) For channel dimensions of $H = 200$ μm, $W = 50$ μm, and $S = 20$ μm, a chip thermal conductivity of $k_{ch} = 140$ W/m · K and width $L = 10$ mm, a channel base temperature ($x = 0$) of $T_s = 350$ K, a channel inlet temperature of $T_{m,i} = 290$ K, and a flow rate of $\dot{m}_1 = 10^{-4}$ kg/s per channel, determine the outlet temperature and the chip power dissipation for the dielectric liquid.

(b) Consider the foregoing conditions, but with air at a flow rate of $\dot{m}_1 = 10^{-6}$ kg/s used as the coolant. Using properties of $c_p = 1007$ J/kg · K, $k = 0.0263$ W/m · K, and $\mu = 185 \times 10^{-7}$ N · s/m^2, determine the air outlet temperature and the chip power dissipation.

8.88 A novel scheme for dissipating heat from the chips of a multichip array involves machining coolant channels in the ceramic substrate to which the chips are attached. The square chips ($L_c = 5$ mm) are aligned above each of the channels, with longitudinal and transverse pitches of $S_L = S_T = 20$ mm. Water flows through the square cross section ($W = 5$ mm) of each channel with a mean velocity of $u_m = 1$ m/s, and its properties may be approximated as $\rho = 1000$ kg/m³, $c_p = 4180$ J/kg · K, $\mu = 855 \times 10^{-6}$ kg/s · m, $k = 0.610$ W/m · K, and $Pr = 5.8$. Symmetry in the transverse direction dictates the existence of equivalent conditions for each substrate section of length L_s and width S_T.

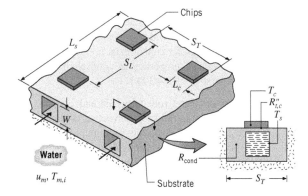

(a) Consider a substrate whose length in the flow direction is $L_s = 200$ mm, thereby providing a total of $N_L = 10$ chips attached in-line above each flow channel. To a good approximation, all the heat dissipated by the chips above a channel may be assumed to be transferred to the water flowing through the channel. If each chip dissipates 5 W, what is the temperature rise of the water passing through the channel?

(b) The chip-substrate contact resistance is $R''_{t,c} = 0.5 \times 10^{-4}$ m² · K/W, and the three-dimensional conduction resistance for the $L_s \times S_T$ substrate section is $R_{cond} = 0.120$ K/W. If water enters the substrate at 25°C and is in fully developed flow, estimate the temperature T_c of the chips and the temperature T_s of the substrate channel surface.

8.89 An experiment is designed to study microscale forced convection. Water at $T_{m,i} = 300$ K is to be heated in a straight, circular glass tube with a 50-μm inner diameter and a wall thickness of 1 mm. Warm water at $T_\infty = 350$ K, $V = 2$ m/s is in cross flow over the exterior tube surface. The experiment is to be designed to cover the operating range $1 \leq Re_D \leq 2000$, where Re_D is the Reynolds number associated with the internal flow.

(a) Determine the tube length L that meets a design requirement that the tube be twice as long as the thermal entrance length associated with the highest Reynolds number of interest. Evaluate water properties at 305 K.

(b) Determine the water outlet temperature, $T_{m,o}$, that is expected to be associated with $Re_D = 2000$. Evaluate the heating water (water in cross flow over the tube) properties at 330 K.

(c) Calculate the pressure drop from the entrance to the exit of the tube for $Re_D = 2000$.

(d) Based on the calculated flow rate and pressure drop in the tube, estimate the height of a column of water (at 300 K) needed to supply the necessary pressure at the tube entrance and the time needed to collect 0.1 liter of water. Discuss how the outlet temperature of the water flowing from the tube, $T_{m,o}$, might be measured.

8.90 Determine the tube diameter that corresponds to a 10% reduction in the convection heat transfer coefficient for thermal and momentum accommodation coefficients of $\alpha_t = 0.92$ and $\alpha_p = 0.89$, respectively. Determine the channel spacing, a, that is associated with a 10% reduction in h using the same accommodation coefficients. The gas is air at $T = 350$ K and atmospheric pressure for both the tube and the parallel plate configurations. The flow is laminar and fully developed with constant surface heat flux.

8.91 An experiment is devised to measure liquid flow and convective heat transfer rates in microscale channels. The mass flow rate through a channel is determined by measuring the amount of liquid that has flowed through the channel and dividing by the duration of the experiment. The mean temperature of the outlet fluid is also measured. To minimize the time needed to perform the experiment (that is, to collect a significant amount of liquid so that its mass and temperature can be accurately measured), *arrays* of microchannels are typically used. Consider an array of microchannels of circular cross section, each with a nominal diameter of 50 μm, fabricated into a copper block. The channels are 20 mm long, and the block is held at 310 K. Water at an inlet temperature of 300 K is forced into the channels from a pressurized plenum, so that a pressure difference of 2.5×10^6 Pa exists from the entrance to the exit of each channel.

In many microscale systems, the characteristic dimensions are similar to the tolerances that can be controlled during the manufacture of the experimental apparatus. Hence, careful consideration of the effect of machining tolerances must be made when interpreting the experimental results.

(a) Consider the case in which three microchannels are machined in the copper block. The channel diameters exhibit some deviation due to manufacturing constraints and are of actual diameter 45 μm, 50 μm, and 55 μm, respectively. Calculate the mass flow rate through each of the three channels, along with the mean outlet temperature of each channel.

(b) If the water exiting each of the three channels is collected and mixed in a single container, calculate the average flow rate through each of the three channels and the average mixed temperature of the water that is collected from all three channels.

(c) The enthusiastic experimentalist uses the average flow rate and the average mixed outlet temperature to analyze the performance of the average (50 μm) diameter channel and concludes that flow rates and heat transfer coefficients are increased and decreased, respectively, by about 5% when forced convection occurs in microchannels. Comment on the validity of the experimentalist's conclusion.

Mass Transfer

8.92 Air at 300 K and a flow rate of 3 kg/h passes upward through a 30-mm tube, as shown in the sketch. A thin film of water, also at 300 K, slowly falls downward on the inner surface of the tube. Determine the convection mass transfer coefficient for this situation.

8.93 In the processing of very long plastic tubes of 2-mm inside diameter, air flows inside the tubing with a Reynolds number of 1000. The interior layer of the plastic material evaporates into the air under fully developed conditions. Both plastic and air are at 400 K, and the Schmidt number for the mixture of the plastic vapor and air is 2.0. Determine the convection mass transfer coefficient.

8.94 What is the convection mass transfer coefficient associated with fully developed atmospheric airflow at 27°C and 0.04 kg/s through a 50-mm-diameter tube whose surface has been coated with a thin layer of naphthalene? Determine the velocity and concentration entry lengths.

8.95 Air is forced through holes that are bored through a 0.5-m-thick block of solid (of molecular weight 95 kg/kmol) that sublimates into the flowing air. Consider air at 320 K and a pressure of 2 atm flowing through a 10-mm-diameter hole at a flow rate of 3×10^{-4} kg/s. The vapor pressure of the sublimating material at the solid-vapor interface is 10 mm Hg. The diffusion coefficient of the vapor in air at 1 atm and 320 K is 2.5×10^{-5} m²/s. Assuming there is only a trace concentration of sublimated vapor in the air at any axial location, determine the local mass fluxes from the solid at $x = 0.1$, 0.25, and 0.5 m. Based on the calculated fluxes, estimate the hole diameters at the preceding x-locations after 30 minutes of operation. The density of the solid material is 1500 kg/m³.

8.96 Dry air at 35°C and a velocity of 10 m/s flows over a thin-walled tube of 20-mm diameter and 200-mm length, having a fibrous coating that is water-saturated.

To maintain an approximately uniform surface temperature of 27°C, water at a prescribed flow rate and temperature passes through the tube.

(a) Considering the heat and mass transfer processes on the external surface of the tube, determine the heat rate from the tube.

(b) For a flow rate of 0.025 kg/s, determine the inlet temperature, $T_{m,i}$, at which water must be supplied to the tube.

8.97 Consider gas flow of mass density ρ and rate $\dot{m}$ through a tube whose inner surface is coated with a liquid or a sublimable solid of uniform vapor density $\rho_{A,s}$. Derive Equation 8.86 for variation of the mean vapor density, $\rho_{A,m}$, with distance x from the tube entrance and Equation 8.83 for the total rate of vapor transfer from a tube of length L.

8.98 Atmospheric air at 25°C and 3×10^{-4} kg/s flows through a 10-mm-diameter, 1-m-long circular tube whose inner surface is wetted with a water film also maintained at 25°C. Determine the water vapor density at the tube outlet, assuming the inlet air to be dry. What is the rate at which vapor is added to the air?

8.99 A humidifier consists of a bundle of vertical tubes, each of 20-mm diameter, through which dry atmospheric air is in fully developed flow at 10^{-3} kg/s and 298 K. The inner tube surface is wetted with a water film. Determine the tube length required for the water vapor to reach 99% of saturation. What is the rate at which energy must be supplied to each tube to maintain its temperature at 298 K?

8.100 The final step of a manufacturing process in which a protective coating is applied to the inner surface of a circular tube involves passage of dry, atmospheric air through the tube to remove a residual liquid associated with the process. Consider a coated 5-m-long tube with an inner diameter of 50 mm. The tube is maintained at a temperature of 300 K, and the residual liquid exists as a thin film whose corresponding vapor pressure is 15 mm Hg. The molecular weight and diffusion coefficient of the vapor are $\mathcal{M}_A = 70$ kg/kmol and $D_{AB} = 10^{-5}$ m²/s, respectively. Air enters the tube at a mean velocity of 0.5 m/s and a temperature of 300 K.

(a) Estimate the partial pressure and mass density of vapor in the air exiting the tube.

(b) What is the rate of liquid removal from the tube in kg/s?

8.101 Dry air is inhaled at a rate of 10 liter/min through a trachea with a diameter of 20 mm and a length of 125 mm. Both the air and the inner surface of the trachea are at a normal body temperature of 37°C and the trachea may be assumed to be saturated with water.

Trachea

(a) Assuming steady, fully developed flow in the trachea, estimate the mass transfer convection coefficient.

(b) Estimate the daily water loss (liter/day) associated with evaporation in the trachea.

8.102 A mass transfer operation is preceded by laminar flow of a gaseous species B through a circular tube that is sufficiently long to achieve a fully developed velocity profile. Once the fully developed condition is reached, the gas enters a section of the tube that is wetted with a liquid film (A). The film maintains a uniform vapor density $\rho_{A,s}$ along the tube surface.

(a) Write the differential equation and boundary conditions that govern the species A mass density distribution, $\rho_A(x, r)$, for $x > 0$.

(b) What is the heat transfer analog to this problem? From this analog, write an expression for the average Sherwood number associated with mass exchange over the region $0 \leq x \leq L$.

(c) Beginning with application of conservation of species to a differential control volume of extent $\pi r_o^2 dx$, derive an expression (Equation 8.86) that may be used to determine the mean vapor density $\rho_{A,m,o}$ at $x = L$.

(d) Consider conditions for which species B is air at 25°C and 1 atm and the liquid film consists of water, also at 25°C. The flow rate is $\dot{m} = 2.5 \times 10^{-4}$ kg/s, and the tube diameter is $D = 10$ mm. What is the mean vapor density at the tube outlet if $L = 1$ m?

CHAPTER

Free Convection

In preceding chapters we considered convection transfer in fluid flows that originate from an *external forcing* condition. For example, fluid motion may be induced by a fan or a pump, or it may result from propulsion of a solid through the fluid. In the presence of a temperature gradient, *forced convection* heat transfer will occur.

Now we consider situations for which there is no *forced* velocity, yet convection currents exist within the fluid. Such situations are referred to as *free* or *natural convection,* and they originate when a *body force* acts on a fluid in which there are *density gradients.* The net effect is a *buoyancy force,* which induces free convection currents. In the most common case, the density gradient is due to a temperature gradient, and the body force is due to the gravitational field.

Since free convection flow velocities are generally much smaller than those associated with forced convection, the corresponding convection transfer rates are also smaller. It is perhaps tempting to therefore attach less significance to free convection processes. This temptation should be resisted. In many systems involving multimode heat transfer effects, free convection provides the largest resistance to heat transfer and therefore plays an important role in the design or performance of the system. Moreover, when it is desirable to minimize heat transfer rates or to minimize operating cost, free convection is often preferred to forced convection.

There are, of course, many applications. Free convection strongly influences the operating temperatures of power generating and electronic devices. It plays a major role in a vast array of thermal manufacturing applications. Free convection is important in establishing temperature distributions within buildings and in determining heat losses or heat loads for heating, ventilating, and air conditioning systems. Free convection distributes the poisonous products of combustion during fires and is relevant to the environmental sciences, where it drives oceanic and atmospheric motions, as well as the related heat transfer and mass transfer processes.

In this chapter our objectives are to obtain an appreciation for the physical origins and nature of buoyancy-driven flows and to acquire tools for performing related heat transfer calculations.

9.1 *Physical Considerations*

In free convection fluid motion is due to buoyancy forces within the fluid, while in forced convection it is externally imposed. *Buoyancy is due to the combined presence of a fluid density gradient* and *a body force that is proportional to density.* In practice, the body force is usually *gravitational,* although it may be a centrifugal force in rotating fluid machinery or a Coriolis force in atmospheric and oceanic rotational motions. There are also several ways in which a mass density gradient may arise in a fluid, but for the most common situation it is due to the presence of a temperature gradient. We know that the density of gases and liquids depends on temperature, generally decreasing (due to fluid expansion) with increasing temperature ($\partial \rho / \partial T < 0$).

In this text we focus on free convection problems in which the density gradient is due to a temperature gradient and the body force is gravitational. However, the presence of a fluid density gradient in a gravitational field does not ensure the existence of free convection currents. Consider the conditions of Figure 9.1. A fluid is enclosed by two large, horizontal plates of different temperature ($T_1 \neq T_2$). In case *a* the temperature of the lower

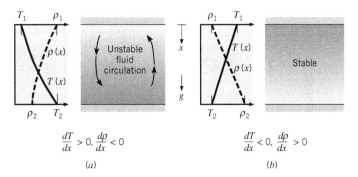

FIGURE 9.1 Conditions in a fluid between large horizontal plates at different temperatures: (*a*) Unstable temperature gradient. (*b*) Stable temperature gradient.

plate exceeds that of the upper plate, and the density decreases in the direction of the gravitational force. If the temperature difference exceeds a critical value, conditions are *unstable* and buoyancy forces are able to overcome the retarding influence of viscous forces. The gravitational force on the denser fluid in the upper layers exceeds that acting on the lighter fluid in the lower layers, and the designated circulation pattern will exist. The heavier fluid will descend, being warmed in the process, while the lighter fluid will rise, cooling as it moves. However, this condition does not characterize case *b*, for which $T_1 > T_2$ and the density no longer decreases in the direction of the gravitational force. Conditions are now *stable,* and there is no bulk fluid motion. In case *a* heat transfer occurs from the bottom to the top surface by free convection; for case *b* heat transfer (from top to bottom) occurs by conduction.

Free convection flows may be classified according to whether the flow is bounded by a surface. In the absence of an adjoining surface, *free boundary flows* may occur in the form of a *plume* or a *buoyant jet* (Figure 9.2). A plume is associated with fluid rising from a submerged hot object. Consider the hot wire of Figure 9.2*a*, which is immersed in an *extensive, quiescent*

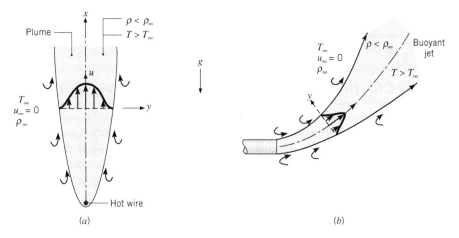

FIGURE 9.2 Buoyancy-driven free boundary layer flows in an extensive, quiescent medium. (*a*) Plume formation above a hot wire. (*b*) Buoyant jet associated with a hot discharge.

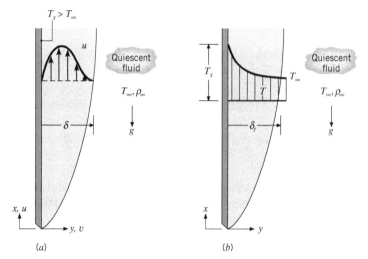

FIGURE 9.3   Boundary layer development on a hot vertical plate: (*a*) Velocity boundary layer. (*b*) Thermal boundary layer.

fluid.[1] Fluid that is heated by the wire rises due to buoyancy forces, entraining fluid from the quiescent region. Although the width of the plume increases with distance from the wire, the plume itself will eventually dissipate as a result of viscous effects and a reduction in the buoyancy force caused by cooling of the fluid in the plume. The distinction between a plume and a buoyant jet is generally made on the basis of the *initial* fluid velocity. This velocity is zero for the plume, but non-zero for the buoyant jet. Figure 9.2*b* shows a hot fluid being discharged as a horizontal jet into a quiescent medium of lower temperature. The vertical motion that the jet begins to assume is due to the buoyancy force. Such a condition occurs when warm water from the condenser of a central power station is discharged into a reservoir of cooler water. Free boundary flows are discussed in considerable detail by Jaluria [1] and Gebhart et al. [2].

In this text we focus on free convection flows bounded by a surface, and a classic example relates to boundary layer development on a hot vertical plate (Figure 9.3). The plate is immersed in an extensive, quiescent fluid, and with $T_s > T_\infty$ the fluid close to the plate is less dense than fluid that is further removed. Buoyancy forces therefore induce a free convection boundary layer in which the fluid that is heated by the plate rises vertically, entraining fluid from the quiescent region. The resulting velocity distribution is unlike that associated with forced convection boundary layers. In particular, the velocity is zero as $y \rightarrow \infty$, as well as at $y = 0$. A free convection boundary layer also develops if $T_s < T_\infty$. In this case, however, fluid motion is downward.

9.2   *The Governing Equations for Laminar Boundary Layers*

As for forced convection, the equations that describe momentum and energy transfer in free convection originate from the related conservation principles. Moreover, the specific processes are much like those that dominate in forced convection. Inertia and viscous forces remain important, as does energy transfer by advection and diffusion. The difference between the two flows is that, in free convection, a major role is played by buoyancy forces. Such forces, in fact, drive the flow.

---

[1]An extensive medium is, in principle, an infinite medium. Since a quiescent fluid is one that is otherwise at rest, the velocity of fluid far from the hot wire is zero.

Consider a laminar boundary layer flow (Figure 9.3) that is driven by buoyancy forces. Assume steady, two-dimensional, constant property conditions in which the gravity force acts in the negative x-direction. Also, with one exception, assume the fluid to be incompressible. The exception involves accounting for the effect of variable density only in the buoyancy force, since it is this variation that induces fluid motion. Finally, assume that the boundary layer approximations of Section 6.4.1 are valid.

With the foregoing simplifications the x-momentum equation (Equation E.2) reduces to the boundary layer equation (Equation 6.28), except that the body force term X is retained. If the only contribution to this force is made by gravity, the body force per unit volume is $X = -\rho g$, where g is the local acceleration due to gravity. The appropriate form of the x-momentum equation is then

$$u\frac{\partial u}{\partial x} + v\frac{\partial u}{\partial y} = -\frac{1}{\rho}\frac{dp_\infty}{dx} - g + v\frac{\partial^2 u}{\partial y^2} \tag{9.1}$$

where dp_∞/dx is the free stream pressure gradient in the quiescent region *outside* the boundary layer. In this region, $u = 0$ and Equation 9.1 reduces to

$$\frac{dp_\infty}{dx} = -\rho_\infty g$$

Substituting this equation into 9.1, we obtain the following expression:

$$u\frac{\partial u}{\partial x} + v\frac{\partial u}{\partial y} = g(\Delta\rho/\rho) + v\frac{\partial^2 u}{\partial y^2} \tag{9.2}$$

where $\Delta\rho = \rho_\infty - \rho$. This expression must apply at every point in the free convection boundary layer.

The first term on the right-hand side of Equation 9.2 is the buoyancy force per unit mass, and flow originates because the density ρ is a variable. If density variations are due only to temperature variations, the term may be related to a fluid property known as the *volumetric thermal expansion coefficient*

$$\beta = -\frac{1}{\rho}\left(\frac{\partial\rho}{\partial T}\right)_p \tag{9.3}$$

This *thermodynamic* property of the fluid provides a measure of the amount by which the density changes in response to a change in temperature at constant pressure. If it is expressed in the following approximate form,

$$\beta \approx -\frac{1}{\rho}\frac{\Delta\rho}{\Delta T} = -\frac{1}{\rho}\frac{\rho_\infty - \rho}{T_\infty - T}$$

it follows that

$$(\rho_\infty - \rho) \approx \rho\beta(T - T_\infty)$$

This simplification is known as the *Boussinesq approximation*, and substituting into Equation 9.2, the x-momentum equation becomes

$$u\frac{\partial u}{\partial x} + v\frac{\partial u}{\partial y} = g\beta(T - T_\infty) + v\frac{\partial^2 u}{\partial y^2} \tag{9.4}$$

where it is now apparent how the buoyancy force, which drives the flow, is related to the temperature difference.

Since buoyancy effects are confined to the momentum equation (9.4), the mass and energy conservation equations are unchanged from forced convection. Equations 6.27 and 6.29 may then be used to complete the problem formulation. The set of governing equations is then

$$\frac{\partial u}{\partial x} + \frac{\partial v}{\partial y} = 0 \tag{9.5}$$

$$u\frac{\partial T}{\partial x} + v\frac{\partial T}{\partial y} = \alpha \frac{\partial^2 T}{\partial y^2} \tag{9.6}$$

and Equation 9.4. Note that viscous dissipation has been neglected in the energy equation, (9.6), an assumption that is certainly reasonable for the small velocities associated with free convection. In the mathematical sense the appearance of the buoyancy term in Equation 9.4 complicates matters. No longer may the hydrodynamic problem, given by Equations 9.4 and 9.5, be uncoupled from and solved to the exclusion of the thermal problem, given by Equation 9.6. The solution to the momentum equation depends on knowledge of T, and hence on the solution to the energy equation. Equations 9.4 through 9.6 are therefore strongly coupled and must be solved simultaneously.

As evident in Equation 9.4, the fluid's volumetric thermal expansion coefficient, β, plays a direct role in driving fluid motion in free convection. For an ideal gas, $\rho = p/RT$, and from Equation 9.3,

$$\beta = -\frac{1}{\rho}\left(\frac{\partial \rho}{\partial T}\right)_p = \frac{1}{\rho}\frac{p}{RT^2} = \frac{1}{T} \tag{9.7}$$

where T is the *absolute* temperature. For liquids and nonideal gases, β must be obtained from appropriate property tables (Appendix A).

9.3 *Similarity Considerations*

Let us now consider the dimensionless parameters that govern free convective flow and heat transfer for the vertical plate. As for forced convection (Chapter 6), the parameters may be obtained by nondimensionalizing the governing equations. Introducing

$$x^* \equiv \frac{x}{L} \qquad y^* \equiv \frac{y}{L}$$

$$u^* \equiv \frac{u}{u_0} \qquad v^* \equiv \frac{v}{u_0} \qquad T^* \equiv \frac{T - T_\infty}{T_s - T_\infty}$$

where L is a characteristic length and u_0 is a reference velocity,[2] the x-momentum and energy equations (9.4 and 9.6) reduce to

$$u^*\frac{\partial u^*}{\partial x^*} + v^*\frac{\partial u^*}{\partial y^*} = \frac{g\beta(T_s - T_\infty)L}{u_0^2}T^* + \frac{1}{Re_L}\frac{\partial^2 u^*}{\partial y^{*2}} \tag{9.8}$$

$$u^*\frac{\partial T^*}{\partial x^*} + v^*\frac{\partial T^*}{\partial y^*} = \frac{1}{Re_L\,Pr}\frac{\partial^2 T^*}{\partial y^{*2}} \tag{9.9}$$

---

[2]Since free stream conditions are quiescent in free convection, there is no logical external reference velocity (V or u_∞), as in forced convection.

The dimensionless parameter in the first term on the right-hand side of Equation 9.8 is a direct consequence of the buoyancy force. The reference velocity u_0 can be specified to simplify the form of the equation. It is convenient to choose $u_0^2 = g\beta(T_s - T_\infty)L$, so that the term multiplying T^* becomes unity. Then, Re_L becomes $[g\beta(T_s - T_\infty)L^3/v^2]^{1/2}$. It is customary to define the *Grashof number* Gr_L as the square of this Reynolds number:

$$Gr_L \equiv \frac{g\beta(T_s - T_\infty)L^3}{v^2} \tag{9.10}$$

As a result, Re_L in Equations 9.8 and 9.9 is replaced by $Gr_L^{1/2}$, and we see that the Grashof number (or more precisely, $Gr_L^{1/2}$) plays a role in free convection that is similar to the role played by the Reynolds number in forced convection. That is, while the Reynolds number is a measure of *inertial* forces relative to viscous forces, the Grashof number is a measure of *buoyancy* forces relative to viscous forces. We therefore expect heat transfer correlations for free convection to be of the form

$$\overline{Nu}_L = f(Gr_L, Pr) \tag{9.11}$$

With the exception of zero gravity conditions, inertial *and* buoyancy forces exist in every forced convection heat transfer scenario. Similarly, buoyancy *and* inertial forces will be present in the boundary layer of Figure 9.3 if the free stream velocity, u_∞, is non-zero. In the case of Figure 9.3 with $u_\infty \neq 0$, it is convenient to choose u_∞ as the characteristic velocity, resulting in $u^*(y^* \to \infty) \to 1$. The T^* term in Equation 9.8 will then be multiplied by Gr_L/Re_L^2, and the resulting Nusselt number expressions will be of the form

$$\overline{Nu}_L = f(Re_L, Gr_L, Pr) \tag{9.12}$$

In general, fluid motion may be driven primarily by inertial forces, primarily by buoyancy forces, or by a combination of inertial and buoyancy forces. The dominant effect is determined by the value of Gr_L/Re_L^2, which is a measure of buoyancy forces relative to inertial forces. The various convection processes are summarized in Table 9.1. Note that, in Chapters 6 through 8, we implicitly assumed that the inequality $Gr_L/Re_L^2 \ll 1$ was satisfied. We will assume $Gr_L/Re_L^2 \gg 1$ in Sections 9.1 through 9.8. Discussion of combined (mixed) free and forced convection is provided in Section 9.9.

9.4 *Laminar Free Convection on a Vertical Surface*

Numerous solutions to the laminar free convection boundary layer equations have been obtained, and a special case that has received much attention involves free convection from

TABLE 9.1 Free, forced, and mixed convection processes, and the corresponding correlation forms

Process	Measure of buoyancy relative to inertial forces	Form of correlation	
Forced convection	$Gr_L/Re_L^2 \ll 1$	$\overline{Nu}_L = f(Re_L, Pr)$	(6.50)
Free convection	$Gr_L/Re_L^2 \gg 1$	$\overline{Nu}_L = f(Gr_L, Pr)$	(9.11)
Mixed convection	$Gr_L/Re_L^2 \approx 1$	$\overline{Nu}_L = f(Re_L, Gr_L, Pr)$	(9.12)

an isothermal vertical surface in an extensive quiescent medium (Figure 9.3). For this geometry Equations 9.4 through 9.6 must be solved subject to boundary conditions of the form[3]

$$y = 0: \qquad u = v = 0 \qquad T = T_s$$
$$y \to \infty: \qquad u \to 0 \qquad T \to T_\infty$$

A similarity solution to the foregoing problem has been obtained by Ostrach [3]. The solution involves transforming variables by introducing a similarity variable of the form

$$\eta \equiv \frac{y}{x}\left(\frac{Gr_x}{4}\right)^{1/4} \tag{9.13}$$

and representing the velocity components in terms of a stream function defined as

$$\psi(x,y) \equiv f(\eta)\left[4v\left(\frac{Gr_x}{4}\right)^{1/4}\right] \tag{9.14}$$

With the foregoing definition of the stream function, the x-velocity component may be expressed as

$$u = \frac{\partial \psi}{\partial y} = \frac{\partial \psi}{\partial \eta}\frac{\partial \eta}{\partial y} = 4v\left(\frac{Gr_x}{4}\right)^{1/4}f'(\eta)\frac{1}{x}\left(\frac{Gr_x}{4}\right)^{1/4}$$

$$= \frac{2v}{x}Gr_x^{1/2}f'(\eta) \tag{9.15}$$

where primed quantities indicate differentiation with respect to η. Hence $f'(\eta) \equiv df/d\eta$. Evaluating the y-velocity component $v = -\partial\psi/\partial x$ in a similar fashion and introducing the dimensionless temperature

$$T^* \equiv \frac{T - T_\infty}{T_s - T_\infty} \tag{9.16}$$

the three original partial differential equations (9.4 through 9.6) may then be reduced to two ordinary differential equations of the form

$$f''' + 3ff'' - 2(f')^2 + T^* = 0 \tag{9.17}$$

$$T^{*''} + 3Prf\,T^{*'} = 0 \tag{9.18}$$

where f and T^* are functions only of η and the double and triple primes, respectively, refer to second and third derivatives with respect to η. Note that f is the key dependent variable for the velocity boundary layer and that the continuity equation (9.5) is automatically satisfied through introduction of the stream function.

The transformed boundary conditions required to solve the momentum and energy equations (9.17 and 9.18) are of the form

$$\eta = 0: \qquad f = f' = 0 \qquad T^* = 1$$
$$\eta \to \infty: \qquad f' \to 0 \qquad T^* \to 0$$

---

[3]The boundary layer approximations are assumed in using Equations 9.4 through 9.6. However, the approximations are only valid for $(Gr_x Pr) \gtrsim 10^4$. Below this value (close to the leading edge), the boundary layer thickness is too large relative to the characteristic length x to ensure the validity of the approximations.

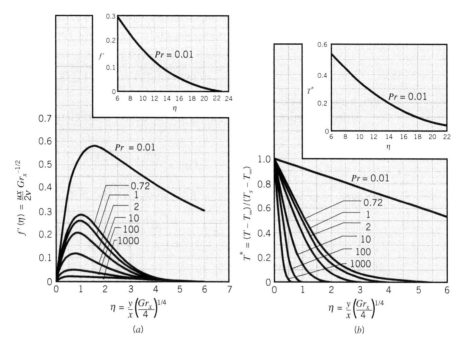

FIGURE 9.4 Laminar, free convection boundary layer conditions on an isothermal, vertical surface. (*a*) Velocity profiles. (*b*) Temperature profiles. Adapted from Ostrach, S., NACA Report 1111, 1953.

A numerical solution has been obtained by Ostrach [3], and selected results are shown in Figure 9.4. Values of the x-velocity component u and the temperature T at any value of x and y may be obtained from Figure 9.4a and Figure 9.4b, respectively.

Figure 9.4b may also be used to infer the appropriate form of the heat transfer correlation. Using Newton's law of cooling for the local convection coefficient h, the local Nusselt number may be expressed as

$$Nu_x = \frac{hx}{k} = \frac{[q_s''/(T_s - T_\infty)]x}{k}$$

Using Fourier's law to obtain q_s'' and expressing the surface temperature gradient in terms of η, Equation 9.13, and T^*, Equation 9.16, it follows that

$$q_s'' = -k\frac{\partial T}{\partial y}\bigg|_{y=0} = -\frac{k}{x}(T_s - T_\infty)\left(\frac{Gr_x}{4}\right)^{1/4}\frac{dT^*}{d\eta}\bigg|_{\eta=0}$$

Hence

$$Nu_x = \frac{hx}{k} = -\left(\frac{Gr_x}{4}\right)^{1/4}\frac{dT^*}{d\eta}\bigg|_{\eta=0} = \left(\frac{Gr_x}{4}\right)^{1/4}g(Pr) \qquad (9.19)$$

which acknowledges that the dimensionless temperature gradient at the surface is a function of the Prandtl number $g(Pr)$. This dependence is evident from Figure 9.4b and has been determined numerically for selected values of Pr [3]. The results have been correlated to within 0.5% by an interpolation formula of the form [4]

$$g(Pr) = \frac{0.75\, Pr^{1/2}}{(0.609 + 1.221\, Pr^{1/2} + 1.238\, Pr)^{1/4}} \tag{9.20}$$

which applies for $0 \le Pr \le \infty$.

Using Equation 9.19 for the local convection coefficient and substituting for the local Grashof number,

$$Gr_x = \frac{g\beta(T_s - T_\infty)x^3}{v^2}$$

the average convection coefficient for a surface of length L is then

$$\bar{h} = \frac{1}{L}\int_0^L h\, dx = \frac{k}{L}\left[\frac{g\beta(T_s - T_\infty)}{4v^2}\right]^{1/4} g(Pr) \int_0^L \frac{dx}{x^{1/4}}$$

Integrating, it follows that

$$\overline{Nu_L} = \frac{\bar{h}L}{k} = \frac{4}{3}\left(\frac{Gr_L}{4}\right)^{1/4} g(Pr) \tag{9.21}$$

or substituting from Equation 9.19, with $x = L$,

$$\overline{Nu_L} = \tfrac{4}{3} Nu_L \tag{9.22}$$

The foregoing results apply irrespective of whether $T_s > T_\infty$ or $T_s < T_\infty$. If $T_s < T_\infty$, conditions are inverted from those of Figure 9.3. The leading edge is at the top of the plate, and positive x is defined in the direction of the gravity force.

9.5 The Effects of Turbulence

It is important to note that free convection boundary layers are not restricted to laminar flow. As with forced convection, *hydrodynamic instabilities* may arise. That is, disturbances in the flow may be amplified, leading to transition from laminar to turbulent flow. This process is shown schematically in Figure 9.5 for a hot vertical plate.

Transition in a free convection boundary layer depends on the relative magnitude of the buoyancy and viscous forces in the fluid. It is customary to correlate its occurrence in

FIGURE 9.5 Free convection boundary layer transition on a vertical plate.

terms of the *Rayleigh number,* which is simply the product of the Grashof and Prandtl numbers. For vertical plates the critical Rayleigh number is

$$Ra_{x,c} = Gr_{x,c} \, Pr = \frac{g\beta(T_s - T_\infty)x^3}{\nu\alpha} \approx 10^9 \tag{9.23}$$

An extensive discussion of stability and transition effects is provided by Gebhart et al. [2].

As in forced convection, transition to turbulence has a strong effect on heat transfer. Hence the results of the foregoing section apply only if $Ra_L \lesssim 10^9$. To obtain appropriate correlations for turbulent flow, emphasis is placed on experimental results.

EXAMPLE 9.1

Consider a 0.25-m-long vertical plate that is at 70°C. The plate is suspended in quiescent air that is at 25°C. Estimate the velocity boundary layer thickness and maximum upward velocity at the trailing edge of the plate. How does the boundary layer thickness compare with the thickness that would exist if the air were flowing over the plate at a free stream velocity of 5 m/s?

SOLUTION

Known: Vertical plate is in quiescent air at a lower temperature.

Find: Velocity boundary layer thickness and maximum upward velocity at trailing edge. Compare boundary layer thickness with value corresponding to an air speed of 5 m/s.

Schematic:

Assumptions:

1. Ideal gas.
2. Constant properties.

Properties: Table A.4, air ($T_f = 320.5$ K): $\nu = 17.95 \times 10^{-6}$ m²/s, $Pr = 0.7$, $\beta = T_f^{-1} = 3.12 \times 10^{-3}$ K⁻¹.

Analysis: Equation 9.10 gives

$$Gr_L = \frac{g\beta(T_s - T_\infty)L^3}{\nu^2}$$

$$= \frac{9.8 \text{ m/s}^2 \times (3.12 \times 10^{-3} \text{ K}^{-1})(70 - 25)°\text{C}(0.25 \text{ m})^3}{(17.95 \times 10^{-6} \text{ m}^2\text{/s})^2} = 6.69 \times 10^7$$

Hence $Ra_L = Gr_L\, Pr = 4.68 \times 10^7$ and, from Equation 9.23, the free convection boundary layer is laminar. The analysis of Section 9.4 is therefore applicable. From the results of Figure 9.4a, it follows that, for $Pr = 0.7$, $\eta \approx 6.0$ at the edge of the boundary layer, that is, at $y \approx \delta$. Hence

$$\delta_L \approx \frac{6L}{(Gr_L/4)^{1/4}} = \frac{6(0.25\ \text{m})}{(1.67 \times 10^7)^{1/4}} = 0.024\ \text{m} \qquad \triangleleft$$

From Figure 9.4a, it can be seen that the maximum velocity corresponds to $f'(\eta) \approx 0.28$ and the velocity is

$$u = \frac{2vf'(\eta)Gr_L^{1/2}}{L} \approx \frac{2 \times 17.95 \times 10^{-6}\ \text{m}^2/\text{s} \times 0.28 \times (6.69 \times 10^7)^{1/2}}{0.25\ \text{m}} = 0.33\ \text{m/s} \qquad \triangleleft$$

For airflow at $u_\infty = 5$ m/s

$$Re_L = \frac{u_\infty L}{v} = \frac{(5\ \text{m/s}) \times 0.25\ \text{m}}{17.95 \times 10^{-6}\ \text{m}^2/\text{s}} = 6.97 \times 10^4$$

and $Gr_L/Re_L^2 = 6.69 \times 10^7/(6.97 \times 10^4)^2 = 0.014 \ll 1$ so an analysis using the results of Chapter 7 is applicable. Since $Re_L < 5 \times 10^5$, the boundary layer is laminar, and from Equation 7.19,

$$\delta_L \approx \frac{5L}{Re_L^{1/2}} = \frac{5(0.25\ \text{m})}{(6.97 \times 10^4)^{1/2}} = 0.0047\ \text{m} \qquad \triangleleft$$

Comments: Free convection boundary layers typically have smaller velocities than in forced convection, which leads to thicker boundary layers. In turn, free convection boundary layers typically pose a larger resistance to heat transfer than forced convection boundary layers.

9.6 *Empirical Correlations: External Free Convection Flows*

In the preceding sections, we considered free convection associated with laminar boundary layer development adjacent to a hot vertical plate and transition of the laminar flow to a turbulent state. In doing so, we introduced two dimensionless parameters, the Grashof number Gr and the Rayleigh number Ra, which also appear in empirical correlations for free convection involving both laminar and turbulent flow conditions and in geometries other than a flat plate.

In this section we summarize empirical correlations that have been developed for common *immersed* (external flow) geometries. The correlations are suitable for many engineering calculations and are often of the form

$$\overline{Nu}_L = \frac{\bar{h}L}{k} = C\, Ra_L^n \qquad (9.24)$$

where the Rayleigh number,

$$Ra_L = Gr_L Pr = \frac{g\beta(T_s - T_\infty)L^3}{\nu\alpha} \tag{9.25}$$

is based on the characteristic length L of the geometry. Typically, $n = \frac{1}{4}$ and $\frac{1}{3}$ for laminar and turbulent flows, respectively. For turbulent flow it then follows that $\bar{h}_L$ is independent of L. Note that all properties are evaluated at the film temperature, $T_f \equiv (T_s + T_\infty)/2$.

9.6.1 The Vertical Plate

Expressions of the form given by Equation 9.24 have been developed for the vertical plate [5–7]. For laminar flow ($10^4 \lesssim Ra_L \lesssim 10^9$), $C = 0.59$ and $n = 1/4$, and for combined laminar and turbulent flow ($10^9 \lesssim Ra_L \lesssim 10^{13}$), $C = 0.10$ and $n = 1/3$. A correlation that may be applied over the *entire* range of Ra_L has been recommended by Churchill and Chu [8] and is of the form

$$\overline{Nu}_L = \left\{ 0.825 + \frac{0.387\, Ra_L^{1/6}}{[1 + (0.492/Pr)^{9/16}]^{8/27}} \right\}^2 \tag{9.26}$$

where L is the plate length. Although Equation 9.26 is suitable for most engineering calculations, slightly better accuracy may be obtained for laminar flow by using [8]

$$\overline{Nu}_L = 0.68 + \frac{0.670\, Ra_L^{1/4}}{[1 + (0.492/Pr)^{9/16}]^{4/9}} \qquad Ra_L \lesssim 10^9 \tag{9.27}$$

When the Rayleigh number is moderately large, the second term on the right-hand side of Equations 9.26 and 9.27 dominates, and the correlations are the same form as Equation 9.24, except that the constant, C, is replaced by a function of Pr. Equation 9.27 is then in excellent quantitative agreement with the analytical solution given by Equations 9.21 and 9.20. In contrast, when the Rayleigh number is small, the first term on the right-hand side of Equations 9.26 and 9.27 dominates, and the equations yield the same behavior since $0.825^2 \approx 0.68$. The presence of leading constants in Equations 9.26 and 9.27 accounts for the fact that, for small Rayleigh number, the boundary layer assumptions become invalid and conduction parallel to the plate is important.

It is important to recognize that the foregoing results have been obtained for an isothermal plate (constant T_s). If the surface condition is, instead, one of uniform heat flux (constant q_s''), the temperature difference ($T_s - T_\infty$) will vary with x, increasing from the leading edge. An approximate procedure for determining this variation may be based on results [8, 9] showing that $\overline{Nu}_L$ correlations obtained for the isothermal plate may still be used to an excellent approximation, if $\overline{Nu}_L$ and Ra_L are defined in terms of the temperature difference at the midpoint of the plate, $\Delta T_{L/2} = T_s(L/2) - T_\infty$. Hence, with $\bar{h} \equiv q_s''/\Delta T_{L/2}$, a correlation such as Equation 9.27 could be used to determine $\Delta T_{L/2}$ (for example, using a trial-and-error technique), and hence the midpoint surface temperature $T_s(L/2)$. If it is assumed that $Nu_x \propto Ra_x^{1/4}$ over the entire plate, it follows that

$$\frac{q_s'' x}{k\Delta T} \propto \Delta T^{1/4} x^{3/4}$$

or

$$\Delta T \propto x^{1/5}$$

Hence the temperature difference at any x is

$$\Delta T_x \approx \frac{x^{1/5}}{(L/2)^{1/5}} \Delta T_{L/2} = 1.15 \left(\frac{x}{L} \right)^{1/5} \Delta T_{L/2} \tag{9.28}$$

A more detailed discussion of constant heat flux results is provided by Churchill [10].

The foregoing results may also be applied to *vertical* cylinders of height L, if the boundary layer thickness δ is much less than the cylinder diameter D. This condition is known to be satisfied [11] when

$$\frac{D}{L} \gtrsim \frac{35}{Gr_L^{1/4}}$$

Cebeci [12] and Minkowycz and Sparrow [13] present results for slender, vertical cylinders not meeting this condition, where transverse curvature influences boundary layer development and enhances the rate of heat transfer.

EXAMPLE 9.2

A glass-door firescreen, used to reduce exfiltration of room air through a chimney, has a height of 0.71 m and a width of 1.02 m and reaches a temperature of 232°C. If the room temperature is 23°C, estimate the convection heat rate from the fireplace to the room.

SOLUTION

Known: Glass screen situated in fireplace opening.

Find: Rate of heat transfer by convection between screen and room air.

Schematic:

Assumptions:
1. Screen is at a uniform temperature T_s.
2. Room air is quiescent.
3. Ideal gas.
4. Constant properties.

Properties: Table A.4, air ($T_f = 400$ K): $k = 33.8 \times 10^{-3}$ W/m · K, $\nu = 26.4 \times 10^{-6}$ m²/s, $\alpha = 38.3 \times 10^{-6}$ m²/s, $Pr = 0.690$, $\beta = (1/T_f) = 0.0025$ K^{-1}.

Analysis: The rate of heat transfer by free convection from the panel to the room is given by Newton's law of cooling

$$q = \bar{h}A_s(T_s - T_\infty)$$

where $\bar{h}$ may be obtained from knowledge of the Rayleigh number. Using Equation 9.25,

$$Ra_L = \frac{g\beta(T_s - T_\infty)L^3}{\alpha v}$$

$$= \frac{9.8 \text{ m/s}^2 \times 0.0025 \text{ K}^{-1} \times (232 - 23)°C \times (0.71 \text{ m})^3}{38.3 \times 10^{-6} \text{ m}^2/\text{s} \times 26.4 \times 10^{-6} \text{ m}^2/\text{s}} = 1.813 \times 10^9$$

and from Equation 9.23 it follows that transition to turbulence occurs on the panel. The appropriate correlation is then given by Equation 9.26

$$\overline{Nu}_L = \left\{ 0.825 + \frac{0.387 \, Ra_L^{1/6}}{[1 + (0.492/Pr)^{9/16}]^{8/27}} \right\}^2$$

$$\overline{Nu}_L = \left\{ 0.825 + \frac{0.387(1.813 \times 10^9)^{1/6}}{[1 + (0.492/0.690)^{9/16}]^{8/27}} \right\}^2 = 147$$

Hence

$$\bar{h} = \frac{\overline{Nu}_L \cdot k}{L} = \frac{147 \times 33.8 \times 10^{-3} \text{ W/m} \cdot \text{K}}{0.71 \text{ m}} = 7.0 \text{ W/m}^2 \cdot \text{K}$$

and

$$q = 7.0 \text{ W/m}^2 \cdot \text{K}(1.02 \times 0.71) \text{ m}^2(232 - 23)°C = 1060 \text{ W} \qquad \triangleleft$$

Comments:

1. Radiation heat transfer effects are often significant relative to free convection. Using Equation 1.7 and assuming $\varepsilon = 1.0$ for the glass surface and $T_{sur} = 23°C$, the net rate of radiation heat transfer between the glass and the surroundings is

$$q_{rad} = \varepsilon A_s \sigma(T_s^4 - T_{sur}^4)$$

$$q_{rad} = 1(1.02 \times 0.71)\text{m}^2 \times 5.67 \times 10^{-8} \text{ W/m}^2 \cdot \text{K}^4(505^4 - 296^4)\text{K}^4$$

$$q_{rad} = 2355 \text{ W}$$

Hence in this case radiation heat transfer exceeds free convection heat transfer by more than a factor of 2.

2. The effects of radiation and free convection on heat transfer from the glass depend strongly on its temperature. With $q \propto T_s^4$ for radiation and $q \propto T_s^n$ for free convection, where $1.25 < n < 1.33$, we expect the relative influence of radiation to increase with increasing temperature. This behavior is revealed by computing and plotting the heat rates as a function of temperature for $50 \leq T_s \leq 250°C$.

For each value of T_s used to generate the foregoing free convection results, air properties were determined at the corresponding value of T_f.

9.6.2   Inclined and Horizontal Plates

For a vertical plate that is hot (or cold) relative to an ambient fluid, the plate is aligned with the gravitational vector, and the buoyancy force acts exclusively to induce fluid motion in the upward (or downward) direction. However, if the plate is inclined with respect to gravity, the buoyancy force has a component normal, as well as parallel, to the plate surface. With a reduction in the buoyancy force parallel to the surface, there is a reduction in fluid velocities along the plate, and one might expect there to be an attendant reduction in convection heat transfer. Whether, in fact, there is such a reduction depends on whether one is interested in heat transfer from the top or bottom surface of the plate.

As shown in Figure 9.6a, if the plate is cold, the y-component of the buoyancy force, which is normal to the plate, acts to maintain the descending boundary layer flow in contact with the top surface of the plate. Since the x-component of the gravitational acceleration is reduced to $g \cos \theta$, fluid velocities along the plate are reduced and there is an attendant reduction in convection heat transfer to the top surface. However, at the bottom surface, the y-component of the buoyancy force acts to move fluid from the surface, and boundary layer development is interrupted by the discharge of parcels of cool fluid from the surface (Figure 9.6a). The resulting flow is three-dimensional, and, as shown by the spanwise (z-direction) variations of Figure 9.6b, the cool fluid discharged from the bottom surface is continuously replaced by the warmer ambient fluid. The displacement of cool boundary layer fluid by the warmer ambient and the attendant reduction in the thermal boundary layer thickness act to increase convection heat transfer to the bottom surface. In fact, heat transfer enhancement due to the three-dimensional flow typically exceeds the reduction associated with the reduced x-component of g, and the combined effect is to increase heat transfer to the bottom surface. Similar trends characterize a hot plate (Figure 9.6c,d), and the three-dimensional flow is now associated with the upper surface, from which parcels of warm fluid are discharged. Such flows have been observed by several investigators [14–16].

In an early study of heat transfer from inclined plates, Rich [17] suggested that convection coefficients could be determined from vertical plate correlations, if g is replaced by $g \cos \theta$ in computing the plate Rayleigh number. Since then, however, it has been determined that this approach is only satisfactory for the top and bottom surfaces of cold and hot plates, respectively. It is not appropriate for the top and bottom surfaces of hot and cold plates,

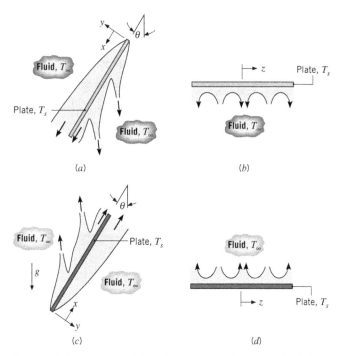

FIGURE 9.6 Buoyancy-driven flows on an inclined plate: (*a*) Side view of flows at top and bottom surfaces of a cold plate ($T_s < T_\infty$). (*b*) End view of flow at bottom surface of cold plate. (*c*) Side view of flows at top and bottom surfaces of a hot plate ($T_s > T_\infty$). (*d*) End view of flow at top surface of hot plate.

respectively, where the three-dimensionality of the flow has limited the ability to develop generalized correlations. At the top and bottom surfaces of cold and hot inclined plates, respectively, it is therefore recommended that, for $0 \leq \theta \lesssim 60°$, g be replaced by $g \cos \theta$ and that Equation 9.26 or 9.27 be used to compute the average Nusselt number. For the opposite surfaces, no recommendations are made, and the literature should be consulted [14–16].

If the plate is horizontal, the buoyancy force is exclusively normal to the surface. As for the inclined plate, flow patterns and heat transfer depend strongly on whether the surface is cold or hot and on whether it is facing upward or downward. For a cold surface facing upward (Figure 9.7*a*) and a hot surface facing downward (Figure 9.7*d*), the tendency of the fluid to descend and ascend, respectively, is impeded by the plate. The flow must move horizontally before it can descend or ascend from the edges of the plate, and convection heat transfer is somewhat ineffective. In contrast, for a cold surface facing downward (Figure 9.7*b*) and a hot surface facing upward (Figure 9.7*c*), flow is driven by descending and ascending parcels of fluid, respectively. Conservation of mass dictates that cold (warm) fluid descending (ascending) from a surface be replaced by ascending (descending) warmer (cooler) fluid from the ambient, and heat transfer is much more effective.

For horizontal plates of various shapes (for example, squares, rectangles, or circles), there is a need to define the characteristic length for use in the Nusselt and Rayleigh numbers. Experiments have shown [18, 19] that a single set of correlations can be used for a variety of different plate shapes when the characteristic length is defined as

$$L \equiv \frac{A_s}{P} \tag{9.29}$$

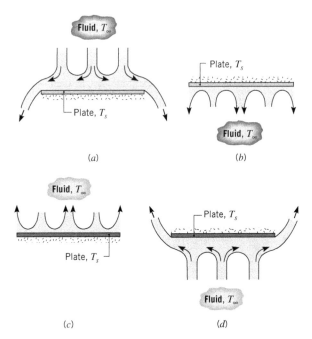

FIGURE 9.7 Buoyancy-driven flows on horizontal cold ($T_s < T_\infty$) and hot ($T_s > T_\infty$) plates: (*a*) Top surface of cold plate. (*b*) Bottom surface of cold plate. (*c*) Top surface of hot plate. (*d*) Bottom surface of hot plate.

where A_s and P are the plate surface area (one side) and perimeter, respectively. Using this characteristic length, the recommended correlations for the average Nusselt number are

Upper Surface of Hot Plate or Lower Surface of Cold Plate [19]:

$$\overline{Nu}_L = 0.54\, Ra_L^{1/4} \qquad (10^4 \lesssim Ra_L \lesssim 10^7,\ Pr \gtrsim 0.7) \tag{9.30}$$

$$\overline{Nu}_L = 0.15\, Ra_L^{1/3} \qquad (10^7 \lesssim Ra_L \lesssim 10^{11},\ \text{all } Pr) \tag{9.31}$$

Lower Surface of Hot Plate or Upper Surface of Cold Plate [20]:

$$\overline{Nu}_L = 0.52\, Ra_L^{1/5} \qquad (10^4 \lesssim Ra_L \lesssim 10^9,\ Pr \gtrsim 0.7) \tag{9.32}$$

Additional correlations can be found in [21].

EXAMPLE 9.3

Airflow through a long rectangular heating duct that is 0.75 m wide and 0.3 m high maintains the outer duct surface at 45°C. If the duct is uninsulated and exposed to air at 15°C in the crawlspace beneath a home, what is the rate of heat loss from the duct per meter of length?

SOLUTION

Known: Surface temperature of a long rectangular duct.

Find: Rate of heat loss from duct per meter of length.

Schematic:

Assumptions:

1. Ambient air is quiescent.
2. Surface radiation effects are negligible.
3. Ideal gas.
4. Constant properties.

Properties: Table A.4, air ($T_f = 303$ K): $\nu = 16.2 \times 10^{-6}$ m²/s, $\alpha = 22.9 \times 10^{-6}$ m²/s, $k = 0.0265$ W/m · K, $\beta = 0.0033$ K⁻¹, $Pr = 0.71$.

Analysis: Surface heat loss is by free convection from the vertical sides and the horizontal top and bottom. From Equation 9.25

$$Ra_L = \frac{g\beta(T_s - T_\infty)L^3}{\nu\alpha} = \frac{(9.8 \text{ m/s}^2)(0.0033 \text{ K}^{-1})(30 \text{ K}) L^3 \text{ (m}^3)}{(16.2 \times 10^{-6} \text{ m}^2/\text{s})(22.9 \times 10^{-6} \text{ m}^2/\text{s})}$$

$$Ra_L = 2.62 \times 10^9 \, L^3$$

For the two sides, $L = H = 0.3$ m. Hence $Ra_L = 7.07 \times 10^7$. The free convection boundary layer is therefore laminar, and from Equation 9.27

$$\overline{Nu}_L = 0.68 + \frac{0.670 \, Ra_L^{1/4}}{[1 + (0.492/Pr)^{9/16}]^{4/9}}$$

The convection coefficient associated with the sides is then

$$\overline{h}_s = \frac{k}{H} \overline{Nu}_L$$

$$\overline{h}_s = \frac{0.0265 \text{ W/m} \cdot \text{K}}{0.3 \text{ m}} \left\{ 0.68 + \frac{0.670(7.07 \times 10^7)^{1/4}}{[1 + (0.492/0.71)^{9/16}]^{4/9}} \right\} = 4.23 \text{ W/m}^2 \cdot \text{K}$$

For the top and bottom, $L = (A_s/P) \approx (w/2) = 0.375$ m. Hence $Ra_L = 1.38 \times 10^8$, and from Equations 9.31 and 9.32, respectively,

$$\overline{h}_t = [k/(w/2)] \times 0.15 \, Ra_L^{1/3} = \frac{0.0265 \text{ W/m} \cdot \text{K}}{0.375 \text{ m}} \times 0.15(1.38 \times 10^8)^{1/3}$$

$$= 5.47 \text{ W/m}^2 \cdot \text{K}$$

$$\overline{h}_b = [k/(w/2)] \times 0.52 \, Ra_L^{1/5} = \frac{0.0265 \text{ W/m} \cdot \text{K}}{0.375 \text{ m}} \times 0.52(1.38 \times 10^8)^{1/5}$$

$$= 1.56 \text{ W/m}^2 \cdot \text{K}$$

The rate of heat loss per unit length of duct is then

$$q' = 2q'_s + q'_t + q'_b = (2\overline{h}_s \cdot H + \overline{h}_t \cdot w + \overline{h}_b \cdot w)(T_s - T_\infty)$$

$$q' = (2 \times 4.23 \times 0.3 + 5.47 \times 0.75 + 1.56 \times 0.75)(45 - 15) \text{ W/m}$$

$$q' = 234 \text{ W/m} \qquad \triangleleft$$

Comments:

1. The heat loss may be reduced by insulating the duct. We consider this option for a 25-mm-thick layer of blanket insulation ($k = 0.035$ W/m · K) that is wrapped around the duct.

The rate of heat loss at each surface may be expressed as

$$q' = \frac{T_{s,1} - T_\infty}{R'_{\text{cond}} + R'_{\text{conv}}}$$

where R'_{conv} is associated with free convection from the outer surface and hence depends on the unknown temperature $T_{s,2}$. This temperature may be determined by applying an energy balance to the outer surface, from which it follows that

$$q''_{\text{cond}} = q''_{\text{conv}}$$

or

$$\frac{(T_{s,1} - T_{s,2})}{(t/k)} = \frac{(T_{s,2} - T_\infty)}{(1/\overline{h})}$$

Since different convection coefficients are associated with the sides, top, and bottom ($\overline{h}_s$, $\overline{h}_t$, and $\overline{h}_b$), a separate solution to this equation must be obtained for each of the three surfaces. The solutions are iterative, since the properties of air and the convection coefficients depend on T_s. Performing the calculations, we obtain

Sides $T_{s,2} = 24°\text{C}$, $\overline{h}_s = 3.18$ W/m² · K

Top $T_{s,2} = 23°\text{C}$, $\overline{h}_t = 3.66$ W/m² · K

Bottom $T_{s,2} = 30°\text{C}$, $\overline{h}_b = 1.36$ W/m² · K

Neglecting heat loss through the corners of the insulation, the total heat rate per unit length of duct is then

$$q' = 2q'_s + q'_t + q'_b$$

$$q' = \frac{2H(T_{s,1} - T_\infty)}{(t/k) + (1/\overline{h}_s)} + \frac{w(T_{s,1} - T_\infty)}{(t/k) + (1/\overline{h}_t)} + \frac{w(T_{s,1} - T_\infty)}{(t/k) + (1/\overline{h}_b)}$$

which yields

$$q' = (17.5 + 22.8 + 15.5) \text{ W/m} = 55.8 \text{ W/m}$$

The insulation therefore provides a 76% reduction in heat loss to the ambient air by natural convection.

2. Although they have been neglected, radiation losses may still be significant. From Equation 1.7 with ε assumed to be unity and $T_{\text{sur}} = 288$ K, $q'_{\text{rad}} = 398$ W/m for the uninsulated duct. Inclusion of radiation effects in the energy balance for the insulated duct would reduce the outer surface temperatures, thereby reducing the convection heat rates. With radiation, however, the total heat rate ($q'_{\text{conv}} + q'_{\text{rad}}$) would increase.

9.6.3 The Long Horizontal Cylinder

This important geometry has been studied extensively, and many existing correlations have been reviewed by Morgan [22]. For an isothermal cylinder, Morgan suggests an expression of the form

$$\overline{Nu}_D = \frac{\overline{h}D}{k} = C \, Ra_D^n \tag{9.33}$$

where C and n are given in Table 9.2 and Ra_D and $\overline{Nu}_D$ are based on the cylinder diameter. In contrast, Churchill and Chu [23] have recommended a single correlation for a wide Rayleigh number range:

$$\overline{Nu}_D = \left\{ 0.60 + \frac{0.387 \, Ra_D^{1/6}}{[1 + (0.559/Pr)^{9/16}]^{8/27}} \right\}^2 \qquad Ra_D \lesssim 10^{12} \tag{9.34}$$

The foregoing correlations provide the average Nusselt number over the entire circumference of an isothermal cylinder. As shown in Figure 9.8 for a hot cylinder, local Nusselt numbers are influenced by boundary layer development, which begins at $\theta = 0$ and concludes at $\theta < \pi$ with formation of a plume ascending from the cylinder. If the flow remains laminar over the entire surface, the distribution of the local Nusselt number with θ is characterized by a maximum at $\theta = 0$ and a monotonic decay with increasing θ. This decay would be disrupted at Rayleigh numbers sufficiently large ($Ra_D \gtrsim 10^9$) to permit transition to turbulence within the boundary layer. If the cylinder is cold relative to the ambient fluid, boundary layer development begins at $\theta = \pi$, the local Nusselt number is a maximum at this location, and the plume descends from the cylinder.

TABLE 9.2   Constants of Equation 9.33 for free convection
on a horizontal circular cylinder [22]

Ra_D	C	n
10^{-10}–10^{-2}	0.675	0.058
10^{-2}–10^2	1.02	0.148
10^2–10^4	0.850	0.188
10^4–10^7	0.480	0.250
10^7–10^{12}	0.125	0.333

FIGURE 9.8   Boundary layer development and Nusselt number distribution on a hot horizontal cylinder.

EXAMPLE 9.4

The fluid of Example 2.2 is characterized by a thermal conductivity, density, specific heat, and dynamic viscosity of 0.705 W/m · K, 1146 kg/m³, 3587 J/kg · K, and 962 × 10⁻⁶ N · s/m², respectively. An experiment is conducted in which a long aluminum rod of diameter $D = 20$ mm and initial temperature $T_i = 32°C$ is suddenly immersed horizontally into a large bath of the fluid at a temperature of $T_\infty = 22°C$. At $t = 65$ s, the measured temperature of the rod is $T_f = 23°C$. Determine the thermal expansion coefficient of the fluid β.

SOLUTION

Known: Initial and final temperatures of aluminum rod of known diameter. Temperature and properties of the fluid.

Find: Thermal expansion coefficient of the fluid.

Schematic:

Assumptions:

1. Constant properties.
2. Spatially uniform rod temperature (applicability of the lumped capacitance approximation).

Properties: Table A.1, aluminum ($\overline{T} = 300$ K): $\rho_s = 2702$ kg/m³, $c_{p,s} = 903$ J/kg · K, $k_s = 237$ W/m · K.

Analysis: Because the temperature difference between the rod and the fluid decreases with time, we expect the convection heat transfer coefficient to decrease as cooling proceeds. Since $\bar{h}$ depends on buoyancy forces established by temperature differences, the analysis of Section 5.3.3 may be applied. From Equations 5.28 and 5.26

$$\frac{\theta}{\theta_i} = \left[\frac{nC_1 A_{s,c}\theta_i^n}{\rho_s V c_{p,s}} t + 1 \right]^{-1/n} \tag{1}$$

where $\bar{h} = C_1(T_s - T_\infty)^n$ and $\theta = T_s - T_\infty$. From Equation 9.33, $\overline{Nu}_D = C\,Ra_D^n$. Substituting definitions of the Nusselt and Rayleigh numbers into Equation 9.33 yields

$$\bar{h} = C\frac{k_l}{D}\left[\frac{g\beta D^3}{\nu_l \alpha_l} \right]^n (T_s - T_\infty)^n \tag{2}$$

From a comparison of Equation 2 with the expression $\bar{h} = C_1(T_s - T_\infty)^n$, it is evident that

$$C_1 = C\frac{k_l}{D}\left[\frac{g\beta D^3}{\nu_l \alpha_l} \right]^n \tag{3}$$

Defining a final excess temperature as $\theta_f = T_{s,f} - T_\infty$ at $t = t_f$ and noting that $\nu_l = \mu_l/\rho_l$, $\alpha_l = k_l/\rho_l c_{p,l}$, and $A_{s,c}/V = 4/D$, Equation 3 may be substituted into Equation 1, yielding

$$\beta = \frac{k_l \mu_l}{c_{p,l}\rho_l^2 g D^3}\left\{ \frac{\rho_s c_{p,s} D^2}{4 k_l C n t_f \theta_i^n}\left[\left(\frac{\theta_f}{\theta_i}\right)^{-n} - 1 \right] \right\}^{1/n} \tag{4}$$

For now we will assume the Rayleigh number falls in the range $10^4 \leq Ra_D \leq 10^7$, for which $C = 0.480$ and $n = 0.250$ from Table 9.2. Hence the thermal expansion coefficient is

$$\beta = \frac{0.705\ \text{W/m} \cdot \text{K} \times 962 \times 10^{-6}\ \text{N} \cdot \text{s/m}^2}{3587\ \text{J/kg} \cdot \text{K} \times (1146\ \text{kg/m}^3)^2 \times 9.8\ \text{m/s}^2 \times (20 \times 10^{-3}\ \text{m})^3}$$

$$\times \left\{ \frac{2702\ \text{kg/m}^3 \times 903\ \text{J/kg} \cdot \text{K} \times (20 \times 10^{-3}\ \text{m})^2}{4 \times 0.705\ \text{W/m} \cdot \text{K} \times 0.480 \times 0.25 \times 65\ \text{s} \times (10\ \text{K})^{0.25}}\left[\left(\frac{1.0}{10}\right)^{-0.25} - 1 \right] \right\}^{1/0.25}$$

$$= 261 \times 10^{-6}\ \text{K}^{-1} \qquad\qquad\qquad \lhd$$

Using this value of the thermal expansion coefficient, the Rayleigh number based on the initial temperature difference is

$$Ra_{D,\text{max}} = \frac{g\beta\theta_i D^3}{\nu_l \alpha_l} = \frac{g\beta\rho^2 c_{p,l}\theta_i D^3}{\mu_l k_l}$$

$$= \frac{9.8\ \text{m/s}^2 \times 261 \times 10^{-6}\ \text{K}^{-1} \times (1146\ \text{kg/m}^3)^2 \times 3587\ \text{J/kg} \cdot \text{K} \times 10\ \text{K} \times (20 \times 10^{-3}\ \text{m})^3}{962 \times 10^{-6}\ \text{N} \cdot \text{s/m}^2 \times 0.705\ \text{W/m} \cdot \text{K}}$$

$$= 1.42 \times 10^6$$

Since $\theta_f = \theta_i/10$, the minimum value of the Rayleigh number during the cooling process is $Ra_{D,\text{min}} = Ra_{D,\text{max}}/10 = 1.42 \times 10^5$. Therefore, $10^4 < Ra_{D,\text{min}} < Ra_{D,\text{max}} < 10^7$, and the values of C and n selected from Table 9.2 are appropriate. Hence the foregoing value of the thermal expansion coefficient is correct.

Comments:

1. The manner in which the rod temperature decreases during the cooling process may be determined from Equation 1 with C_1 obtained from Equation 3. The rod temperature history is shown in the figure.

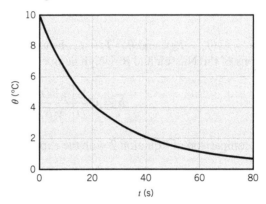

2. Since the temperature difference between the rod and the fluid decreases with time, the Rayleigh number also decreases as cooling proceeds. This leads to a gradual reduction in the convection heat transfer coefficient during cooling, as can be determined by solving Equation 2 once $\theta(t)$ is known.

3. The maximum value of the convection heat transfer coefficient is $\bar{h}_{max} = 584$ W/m$^2 \cdot$K. This corresponds to a maximum Biot number of $Bi_{max} = \bar{h}_{max}(D/2)/k_s = 584$ W/m$^2 \cdot$K $\times$ $(20 \times 10^{-3}$ m/2)/237 W/m$\cdot$K $= 0.025$ when the criterion of Equation 5.10 is applied in a conservative fashion. Since $Bi_{max} < 0.1$, we conclude that the lumped capacitance approximation is valid.

4. Because the rod temperature continually decreases, the buoyancy forces within the fluid decrease with time. Hence fluid velocities continually evolve as the temperature difference between the rod and the fluid slowly decays. Equation 9.33 is strictly applicable only for steady-state conditions. In applying the correlation here, we have implicitly assumed that the instantaneous heat transfer rate from the rod is the same as the steady-state heat transfer rate *if* the same temperature difference exists between the rod and the fluid. This assumption often yields predictions of acceptable accuracy and is referred to as the *quasi-steady approximation*.

9.6.4 Spheres

The following correlation due to Churchill [10] is recommended for spheres in fluids of $Pr \gtrsim 0.7$ and for $Ra_D \lesssim 10^{11}$.

$$\overline{Nu}_D = 2 + \frac{0.589 \, Ra_D^{1/4}}{[1 + (0.469/Pr)^{9/16}]^{4/9}} \tag{9.35}$$

In the limit as $Ra_D \to 0$, Equation 9.35 reduces to $\overline{Nu}_D = 2$, which corresponds to heat transfer by conduction between a spherical surface and a stationary infinite medium, in a manner consistent with Equations 7.56 and 7.57.

Recommended correlations from this section are summarized in Table 9.3. Results for other immersed geometries and special conditions are presented in comprehensive reviews by Churchill [10] and Raithby and Hollands [21].

TABLE 9.3 Summary of free convection empirical correlations for immersed geometries

Geometry	Recommended Correlation	Restrictions
1. Vertical plates[a]	Equation 9.26	None
2. Inclined plates Cold surface up or hot surface down	Equation 9.26 $g \to g \cos \theta$	$0 \le \theta \lesssim 60°$
3. Horizontal plates (a) Hot surface up or cold surface down	Equation 9.30 Equation 9.31	$10^4 \lesssim Ra_L \lesssim 10^7, Pr \gtrsim 0.7$ $10^7 \lesssim Ra_L \lesssim 10^{11}$
(b) Cold surface up or hot surface down	Equation 9.32	$10^4 \lesssim Ra_L \lesssim 10^9, Pr \gtrsim 0.7$

TABLE 9.3   *Continued*

Geometry	Recommended Correlation	Restrictions
4. Horizontal cylinder	Equation 9.34	$Ra_D \lesssim 10^{12}$
5. Sphere	Equation 9.35	$Ra_D \lesssim 10^{11}$ $Pr \gtrsim 0.7$

[a] The correlation may be applied to a vertical cylinder if $(D/L) \gtrsim (35/Gr_L^{1/4})$.

9.7   *Free Convection Within Parallel Plate Channels*

A common free convection geometry involves vertical (or inclined) parallel plate channels that are open to the ambient at opposite ends (Figure 9.9). The plates could constitute a fin array used to enhance free convection heat transfer from a base surface to which the fins are attached, or they could constitute an array of circuit boards with heat-dissipating electronic components. Surface thermal conditions may be idealized as being isothermal or isoflux and symmetrical ($T_{s,1} = T_{s,2}$; $q''_{s,1} = q''_{s,2}$) or asymmetrical ($T_{s,1} \neq T_{s,2}$; $q''_{s,1} \neq q''_{s,2}$).

For vertical channels ($\theta = 0$) buoyancy acts exclusively to induce motion in the streamwise (x) direction and, beginning at $x = 0$, boundary layers develop on each surface. For short channels and/or large spacings (small L/S), independent boundary layer development occurs at each surface and conditions correspond to those for an *isolated plate* in an infinite, quiescent medium. For large L/S, however, boundary layers developing on opposing surfaces eventually merge to yield a fully developed condition. If the channel is inclined, there is a component of the buoyancy force normal, as well as parallel, to the streamwise direction, and conditions may be strongly influenced by development of a three-dimensional, secondary flow.

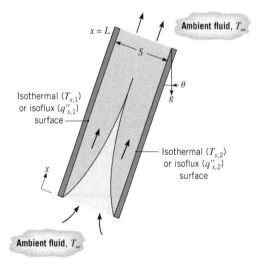

FIGURE 9.9   Free convection flow between hot parallel plates with opposite ends exposed to a quiescent fluid.

9.7.1 Vertical Channels

Beginning with the benchmark paper by Elenbaas [24], the vertical orientation has been studied extensively for plates with symmetric and asymmetric isothermal or isoflux surface conditions. For *isothermal plates of the same temperature,* Elenbaas obtained the following semiempirical correlation:

$$\overline{Nu}_S = \frac{1}{24} Ra_S \left(\frac{S}{L}\right) \left\{ 1 - \exp\left[-\frac{35}{Ra_S(S/L)}\right] \right\}^{3/4} \tag{9.36}$$

where the average Nusselt and Rayleigh numbers are defined as

$$\overline{Nu}_S = \left(\frac{q/A}{T_s - T_\infty}\right) \frac{S}{k} \tag{9.37}$$

and

$$Ra_S = \frac{g\beta(T_s - T_\infty)S^3}{\nu\alpha} \tag{9.38}$$

Equation 9.36 was developed for air as the working fluid, and its range of applicability is

$$\left[10^{-1} \lesssim \frac{S}{L} Ra_S \lesssim 10^5\right]$$

Knowledge of the average Nusselt number for a plate therefore permits determination of the total heat rate for the plate. In the fully developed limit ($S/L \to 0$), Equation 9.36 reduces to

$$\overline{Nu}_{S(\text{fd})} = \frac{Ra_S(S/L)}{24} \tag{9.39}$$

Retention of the L dependence results from defining $\overline{Nu}_S$ in terms of the fixed inlet (ambient) temperature and not in terms of the fluid mixed-mean temperature, which is not explicitly known. For the common condition corresponding to adjoining isothermal ($T_{s,1}$) and insulated ($q''_{s,2} = 0$) plates, the fully developed limit yields the following expression for the isothermal surface [25]:

$$\overline{Nu}_{S(\text{fd})} = \frac{Ra_S(S/L)}{12} \tag{9.40}$$

For isoflux surfaces, it is more convenient to define a local Nusselt number as

$$Nu_{S,L} = \left(\frac{q''_s}{T_{s,L} - T_\infty}\right) \frac{S}{k} \tag{9.41}$$

and to correlate results in terms of a modified Rayleigh number defined as

$$Ra_S^* = \frac{g\beta q''_s S^4}{k\nu\alpha} \tag{9.42}$$

The subscript L refers to conditions at $x = L$, where the plate temperature is a maximum. For symmetric, isoflux plates the fully developed limit corresponds to [25]

$$Nu_{S,L(\text{fd})} = 0.144 \, [Ra_S^*(S/L)]^{1/2} \tag{9.43}$$

and for asymmetric isoflux conditions with one surface insulated ($q_{s,2}'' = 0$) the limit is

$$Nu_{S,L(\text{fd})} = 0.204 \, [Ra_S^*(S/L)]^{1/2} \tag{9.44}$$

Combining the foregoing relations for the fully developed limit with available results for the isolated plate limit, Bar-Cohen and Rohsenow [25] obtained Nusselt number correlations applicable to the complete range of S/L. For isothermal and isoflux conditions, respectively, the correlations are of the form

$$\overline{Nu}_S = \left[\frac{C_1}{(Ra_S \, S/L)^2} + \frac{C_2}{(Ra_S \, S/L)^{1/2}} \right]^{-1/2} \tag{9.45}$$

$$Nu_{S,L} = \left[\frac{C_1}{Ra_S^* \, S/L} + \frac{C_2}{(Ra_S^* \, S/L)^{2/5}} \right]^{-1/2} \tag{9.46}$$

where the constants C_1 and C_2 are given in Table 9.4 for the different surface thermal conditions. In each case the fully developed and isolated plate limits correspond to Ra_S (or $Ra_S^*)S/L \lesssim 10$ and Ra_S (or $Ra_S^*)S/L \gtrsim 100$, respectively.

Bar-Cohen and Rohsenow [25] used the foregoing correlations to infer the optimum plate spacing S_{opt} for maximizing heat transfer from an array of isothermal plates, as well as the spacing S_{max} needed to maximize heat transfer from each plate in the array. Existence of an optimum for the array results from the fact that, although heat transfer from each plate decreases with decreasing S, the number of plates that may be placed in a prescribed volume increases. Hence S_{opt} maximizes heat transfer from the array by yielding a maximum for the product of $\overline{h}$ and the total plate surface area. In contrast, to maximize heat transfer from each plate, S_{max} must be large enough to preclude overlap of adjoining boundary layers, such that the isolated plate limit remains valid over the entire plate.

Consideration of the optimum plate spacing is particularly important for vertical parallel plates used as fins to enhance heat transfer by natural convection from a base surface of fixed width W. With the temperature of the fins exceeding that of the ambient fluid, flow between the fins is induced by buoyancy forces. However, resistance to the flow is associated with viscous forces imposed by the surface of the fins, and the rate of mass flow between adjoining fins is governed by a balance between buoyancy and viscous forces. Since viscous forces increase with decreasing S, there is an accompanying reduction in the

TABLE 9.4 Heat transfer parameters for free convection between vertical parallel plates

Surface Condition	C_1	C_2	S_{opt}	$S_{\text{max}}/S_{\text{opt}}$
Symmetric isothermal plates ($T_{s,1} = T_{s,2}$)	576	2.87	$2.71(Ra_S/S^3L)^{-1/4}$	1.71
Symmetric isoflux plates ($q_{s,1}'' = q_{s,2}''$)	48	2.51	$2.12(Ra_S^*/S^4L)^{-1/5}$	4.77
Isothermal/adiabatic plates ($T_{s,1}, q_{s,2}'' = 0$)	144	2.87	$2.15(Ra_S/S^3L)^{-1/4}$	1.71
Isoflux/adiabatic plates ($q_{s,1}'', q_{s,2}'' = 0$)	24	2.51	$1.69(Ra_S^*/S^4L)^{-1/5}$	4.77

flow rate, and hence $\bar{h}$. However, for fixed W, the attendant increase in the number of fins increases the total surface area A_s and yields a maximum in $\bar{h}A_s$ for $S = S_{opt}$. For $S < S_{opt}$, the amount by which $\bar{h}$ is diminished by viscous effects exceeds the increase in A_s; for $S > S_{opt}$, the amount by which A_s is diminished exceeds the increase in $\bar{h}$.

For isoflux plates, the total volumetric heat rate simply increases with decreasing S. However, the need to maintain T_s below prescribed limits precludes reducing S to extremely small values. Hence S_{opt} may be defined as that value of S which yields the maximum volumetric heat dissipation per unit temperature difference, $T_s(L) - T_\infty$. The spacing S_{max} that yields the lowest possible surface temperature for a prescribed heat flux, without regard to volumetric considerations, is again the value of S that precludes boundary layer merger. Values of S_{opt} and S_{max}/S_{opt} are presented in Table 9.4 for plates of negligible thickness.

In using the foregoing correlations, fluid properties are evaluated at average temperatures of $\bar{T} = (T_s + T_\infty)/2$ for isothermal surfaces and $\bar{T} = (T_{s,L} + T_\infty)/2$ for isoflux surfaces.

9.7.2　Inclined Channels

Experiments have been performed by Azevedo and Sparrow [16] for inclined channels in water. Symmetric isothermal plates and isothermal-insulated plates were considered for $0 \leq \theta \leq 45°$ and conditions within the isolated plate limit, $Ra_S(S/L) > 200$. Although three-dimensional secondary flows were observed at the lower plate, when it was hot, data for all experimental conditions were correlated to within ±10% by

$$\overline{Nu}_S = 0.645[Ra_S(S/L)]^{1/4} \tag{9.47}$$

Departures of the data from the correlation were most pronounced at large tilt angles with bottom surface heating and were attributed to heat transfer enhancement by the three-dimensional secondary flow. Fluid properties are evaluated at $\bar{T} = (T_s + T_\infty)/2$.

9.8　*Empirical Correlations: Enclosures*

The foregoing results pertain to free convection between a surface and an extensive fluid medium. However, engineering applications frequently involve heat transfer between surfaces that are at different temperatures and are separated by an *enclosed* fluid. In this section we present correlations that are pertinent to several common geometries.

9.8.1　Rectangular Cavities

The rectangular cavity (Figure 9.10) has been studied extensively, and comprehensive reviews of both experimental and theoretical results are available [26, 27]. Two of the opposing walls are maintained at different temperatures ($T_1 > T_2$), while the remaining walls are insulated from the surroundings. The tilt angle τ between the hot and cold surfaces

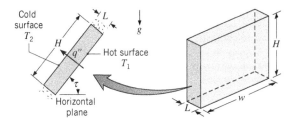

FIGURE 9.10　Free convection in a rectangular cavity.

and the horizontal can vary from $0°$ (*horizontal cavity* with bottom surface hot) to $90°$ (*vertical cavity* with hot and cold sidewalls) to $180°$ (*horizontal cavity* with hot top surface). The heat flux across the cavity, which is expressed as

$$q'' = h(T_1 - T_2) \tag{9.48}$$

can depend strongly on the aspect ratio H/L, as well as the value of τ. For large values of the aspect ratio w/L, its dependence on w/L is small and may be neglected for the purposes of this text.

The horizontal cavity heated from below ($\tau = 0$) has been considered by many investigators. For H/L, $w/L \gg 1$, and Rayleigh numbers less than a critical value of $Ra_{L,c} = 1708$, buoyancy forces cannot overcome the resistance imposed by viscous forces and there is no advection within the cavity. Hence heat transfer from the bottom to the top surface occurs by conduction or, for a gas, by conduction and radiation. Since conditions correspond to one-dimensional conduction through a plane fluid layer, the convection coefficient is $h = k/L$ and $Nu_L = 1$. However, for

$$Ra_L \equiv \frac{g\beta(T_1 - T_2)L^3}{\nu\alpha} > 1708$$

conditions are thermally unstable and there is advection within the cavity. For Rayleigh numbers in the range $1708 < Ra_L \lesssim 5 \times 10^4$, fluid motion consists of regularly spaced roll cells (Figure 9.11), while for larger Rayleigh numbers, the cells break down and the fluid motion evolves through many different patterns before becoming turbulent.

As a first approximation, convection coefficients for the horizontal cavity heated from below may be obtained from the following correlation proposed by Globe and Dropkin [28]:

$$\overline{Nu_L} = \frac{\overline{h}L}{k} = 0.069\, Ra_L^{1/3}\, Pr^{0.074} \qquad 3 \times 10^5 \lesssim Ra_L \lesssim 7 \times 10^9 \tag{9.49}$$

where all properties are evaluated at the average temperature, $\overline{T} \equiv (T_1 + T_2)/2$. The correlation applies for values of L/H sufficiently small to ensure a negligible effect of the sidewalls. More detailed correlations, which apply over a wider range of Ra_L, have been proposed [29, 30]. In concluding the discussion of horizontal cavities, it is noted that in the absence of radiation, for heating from above ($\tau = 180°$), heat transfer from the top to the bottom surface is exclusively by conduction ($Nu_L = 1$), irrespective of the value of Ra_L.

In the vertical rectangular cavity ($\tau = 90°$), the vertical surfaces are hot and cold, while the horizontal surfaces are adiabatic. As shown in Figure 9.12, fluid motion is characterized by a recirculating or cellular flow for which fluid ascends along the hot wall and descends along the cold wall. For small Rayleigh numbers, $Ra_L \lesssim 10^3$, the buoyancy-driven flow is weak and, in the absence of radiation, heat transfer is primarily by conduction across the fluid. Hence, from Fourier's law, the Nusselt number is again $Nu_L = 1$. With increasing Rayleigh number,

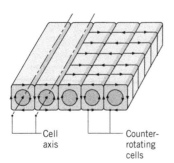

Cell axis   Counter-rotating cells

FIGURE 9.11   Longitudinal roll cells characteristic of advection in a horizontal fluid layer heated from below ($1708 < Ra_L \lesssim 5 \times 10^4$).

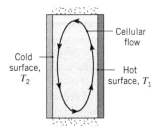

FIGURE 9.12 Cellular flow in a vertical cavity with different sidewall temperatures.

the cellular flow intensifies and becomes concentrated in thin boundary layers adjoining the sidewalls. The core becomes nearly stagnant, although additional cells can develop in the corners and the sidewall boundary layers eventually undergo transition to turbulence. For aspect ratios in the range $1 \lesssim (H/L) \lesssim 10$, the following correlations have been suggested [27]:

$$\overline{Nu}_L = 0.22 \left(\frac{Pr}{0.2 + Pr} Ra_L \right)^{0.28} \left(\frac{H}{L} \right)^{-1/4} \qquad (9.50)$$

$$\left[\begin{array}{l} 2 \lesssim \dfrac{H}{L} \lesssim 10 \\[2mm] Pr \lesssim 10^5 \\[2mm] 10^3 \lesssim Ra_L \lesssim 10^{10} \end{array} \right]$$

$$\overline{Nu}_L = 0.18 \left(\frac{Pr}{0.2 + Pr} Ra_L \right)^{0.29} \qquad (9.51)$$

$$\left[\begin{array}{l} 1 \lesssim \dfrac{H}{L} \lesssim 2 \\[2mm] 10^{-3} \lesssim Pr \lesssim 10^5 \\[2mm] 10^3 \lesssim \dfrac{Ra_L \, Pr}{0.2 + Pr} \end{array} \right]$$

while for larger aspect ratios, the following correlations have been proposed [31]:

$$\overline{Nu}_L = 0.42 \, Ra_L^{1/4} \, Pr^{0.012} \left(\frac{H}{L} \right)^{-0.3} \qquad \left[\begin{array}{l} 10 \lesssim \dfrac{H}{L} \lesssim 40 \\[2mm] 1 \lesssim Pr \lesssim 2 \times 10^4 \\[2mm] 10^4 \lesssim Ra_L \lesssim 10^7 \end{array} \right] \qquad (9.52)$$

$$\overline{Nu}_L = 0.046 \, Ra_L^{1/3} \qquad \left[\begin{array}{l} 1 \lesssim \dfrac{H}{L} \lesssim 40 \\[2mm] 1 \lesssim Pr \lesssim 20 \\[2mm] 10^6 \lesssim Ra_L \lesssim 10^9 \end{array} \right] \qquad (9.53)$$

TABLE 9.5 Critical angle for inclined rectangular cavities

(H/L)	1	3	6	12	>12
τ^*	25°	53°	60°	67°	70°

Convection coefficients computed from the foregoing expressions are to be used with Equation 9.48. Again, all properties are evaluated at the mean temperature, $(T_1 + T_2)/2$.

Studies of free convection in tilted cavities are often stimulated by applications involving flat-plate solar collectors [32–37]. For such cavities, the fluid motion consists of a combination of the roll structure of Figure 9.11 and the cellular structure of Figure 9.12. Typically, transition between the two types of fluid motion occurs at a critical tilt angle, τ^*, with a corresponding change in the value of $\overline{Nu}_L$. For large aspect ratios, $(H/L) \gtrsim 12$, and tilt angles less than the critical value τ^* given in Table 9.5, the following correlation due to Hollands et al. [37] is in excellent agreement with available data:

$$\overline{Nu}_L = 1 + 1.44 \left[1 - \frac{1708}{Ra_L \cos \tau} \right]^{\cdot} \left[1 - \frac{1708 (\sin 1.8\tau)^{1.6}}{Ra_L \cos \tau} \right]$$

$$+ \left[\left(\frac{Ra_L \cos \tau}{5830} \right)^{1/3} - 1 \right]^{\cdot} \quad \begin{bmatrix} \dfrac{H}{L} \gtrsim 12 \\ 0 < \tau \lesssim \tau^* \end{bmatrix} \tag{9.54}$$

The notation [ ]˙ implies that, if the quantity in brackets is negative, it must be set equal to zero. The implication is that, if the Rayleigh number is less than a critical value $Ra_{L,c} = 1708/\cos \tau$, there is no flow within the cavity. For small aspect ratios Catton [27] suggests that reasonable results may be obtained from a correlation of the form

$$\overline{Nu}_L = \overline{Nu}_L(\tau = 0) \left[\frac{\overline{Nu}_L(\tau = 90°)}{\overline{Nu}_L(\tau = 0)} \right]^{\tau/\tau^*} (\sin \tau^*)^{(\tau/4\tau^*)} \quad \begin{bmatrix} \dfrac{H}{L} \lesssim 12 \\ 0 < \tau \lesssim \tau^* \end{bmatrix} \tag{9.55}$$

Beyond the critical tilt angle, the following correlations due to Ayyaswamy and Catton [32] and Arnold et al. [35], respectively, have been recommended [27] for all aspect ratios (H/L):

$$\overline{Nu}_L = \overline{Nu}_L(\tau = 90°)(\sin \tau)^{1/4} \qquad \tau^* \lesssim \tau < 90° \tag{9.56}$$

$$\overline{Nu}_L = 1 + [\overline{Nu}_L(\tau = 90°) - 1] \sin \tau \qquad 90° < \tau < 180° \tag{9.57}$$

9.8.2 Concentric Cylinders

Free convection heat transfer in the annular space between *long,* horizontal concentric cylinders (Figure 9.13) has been considered by Raithby and Hollands [38]. Flow in the annular region is characterized by two cells that are symmetric about the vertical midplane. If the inner cylinder is hot and the outer cylinder is cold ($T_i > T_o$), fluid ascends and descends along the inner and outer cylinders, respectively. If $T_i < T_o$, the cellular flows are

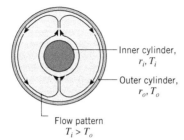

FIGURE 9.13   Free convection flow in the annular space between long, horizontal, concentric cylinders or concentric spheres of inner radius r_i and outer radius r_o.

reversed. The convection heat transfer rate (W) between the two cylinders, each of length L, is expressed by Equation 3.32 (with an *effective thermal conductivity, k_{eff}*, replacing the molecular thermal conductivity, k) as

$$q = \frac{2\pi L k_{eff}(T_i - T_o)}{\ln(r_o/r_i)} \tag{9.58}$$

We see that the effective conductivity of a fictitious *stationary* fluid will transfer the same amount of heat as the actual *moving* fluid. The suggested correlation for k_{eff} is

$$\frac{k_{eff}}{k} = 0.386\left(\frac{Pr}{0.861 + Pr}\right)^{1/4} Ra_c^{1/4} \tag{9.59}$$

where the length scale in Ra_c is given by

$$L_c = \frac{2[\ln(r_o/r_i)]^{4/3}}{(r_i^{-3/5} + r_o^{-3/5})^{5/3}} \tag{9.60}$$

Equation 9.59 may be used for the range $0.7 \lesssim Pr \lesssim 6000$ and $Ra_c \lesssim 10^7$. Properties are evaluated at the mean temperature, $T_m = (T_i + T_o)/2$. Of course, the minimum heat transfer rate between the cylinders cannot fall below the conduction limit; therefore, $k_{eff} = k$ if the value of k_{eff}/k predicted by Equation 9.59 is less than unity. A more detailed correlation, which accounts for cylinder eccentricity effects, has been developed by Kuehn and Goldstein [39].

9.8.3   Concentric Spheres

Raithby and Hollands [38] have also considered free convection heat transfer between concentric spheres (Figure 9.13) and express the convection heat transfer rate by Equation 3.40 (with an effective thermal conductivity, k_{eff}, replacing the molecular thermal conductivity, k) as

$$q = \frac{4\pi k_{eff}(T_i - T_o)}{(1/r_i) - (1/r_o)} \tag{9.61}$$

The effective thermal conductivity is

$$\frac{k_{eff}}{k} = 0.74\left(\frac{Pr}{0.861 + Pr}\right)^{1/4} Ra_s^{1/4} \tag{9.62}$$

where the length scale in Ra_s is given by

$$L_s = \frac{\left(\dfrac{1}{r_i} - \dfrac{1}{r_o}\right)^{4/3}}{2^{1/3}(r_i^{-7/5} + r_o^{-7/5})^{5/3}} \qquad (9.63)$$

The result may be used to a reasonable approximation for $0.7 \lesssim Pr \lesssim 4000$ and $Ra_s \lesssim 10^4$. Properties are evaluated at $T_m = (T_i + T_o)/2$, and $k_{\text{eff}} = k$ if the value of k_{eff}/k predicted by Equation 9.62 is less than unity.

EXAMPLE 9.5

A long tube of 0.1-m diameter is maintained at 120°C by passing steam through its interior. A radiation shield is installed concentric to the tube with an air gap of 10 mm. If the shield is at 35°C, estimate the rate of heat transfer by free convection from the tube per unit length. What is the rate of heat loss if the space between the tube and the shield is filled with glass-fiber blanket insulation?

SOLUTION

Known: Temperatures and diameters of a steam tube and a concentric radiation shield.

Find:

1. Rate of heat loss per unit length of tube.

2. Rate of heat loss if air space is filled with glass-fiber blanket insulation.

Schematic:

Assumptions:

1. Radiation heat transfer may be neglected.

2. Contact resistance with insulation is negligible.

3. Ideal gas.

4. Constant properties.

Properties: Table A.4, air [$T = (T_i + T_o)/2 = 350$ K]: $k = 0.030$ W/m · K, $\nu = 20.92 \times 10^{-6}$ m²/s, $\alpha = 29.9 \times 10^{-6}$ m²/s, $Pr = 0.70$, $\beta = 0.00285$ K^{-1}. Table A.3, insulation, glass-fiber ($T \approx 300$ K): $k = 0.038$ W/m · K.

Analysis:

1. From Equation 9.58, the rate of heat loss per unit length by free convection is

$$q' = \frac{2\pi k_{\text{eff}}(T_i - T_o)}{\ln(r_o/r_i)}$$

where k_{eff} may be obtained from Equations 9.59 and 9.60. With

$$L_c = \frac{2[\ln(r_o/r_i)]^{4/3}}{(r_i^{-3/5} + r_o^{-3/5})^{5/3}} = \frac{2[\ln(0.06 \text{ m}/0.05 \text{ m})]^{4/3}}{(0.05^{-3/5} + 0.06^{-3/5})^{5/3} \text{ m}^{-1}} = 0.00117 \text{ m}$$

we find

$$Ra_c = \frac{g\beta(T_i - T_o)L_c^3}{\nu\alpha} = \frac{9.8 \text{ m/s}^2 \times 0.00285 \text{ K}^{-1} \times (120 - 35)°\text{C} \times (0.00117 \text{ m})^3}{20.92 \times 10^{-6} \text{ m}^2/\text{s} \times 29.9 \times 10^{-6} \text{ m}^2/\text{s}}$$

$$= 171$$

The effective thermal conductivity is then

$$k_{\text{eff}} = 0.386k\left(\frac{Pr}{0.861 + Pr}\right)^{1/4} Ra_c^{1/4}$$

$$= 0.386 \times 0.030 \text{ W/m} \cdot \text{K}\left(\frac{0.70}{0.861 + 0.70}\right)^{1/4}(171)^{1/4} = 0.0343 \text{ W/m} \cdot \text{K}$$

and the rate of heat loss is

$$q' = \frac{2\pi k_{\text{eff}}(T_i - T_o)}{\ln(r_o/r_i)} = \frac{2\pi(0.0343 \text{ W/m} \cdot \text{K})}{\ln(0.06 \text{ m}/0.05 \text{ m})}(120 - 35)°\text{C} = 100 \text{ W/m} \qquad \triangleleft$$

2. With insulation in the space between the tube and the shield, heat loss is by conduction; comparing Equation 3.32 and Equation 9.58,

$$q'_{\text{ins}} = q'\frac{k_{\text{ins}}}{k_{\text{eff}}} = 100 \text{ W/m}\frac{0.038 \text{ W/m} \cdot \text{K}}{0.0343 \text{ W/m} \cdot \text{K}} = 111 \text{ W/m} \qquad \triangleleft$$

Comments: Although there is slightly more heat loss by conduction through the insulation than by free convection across the air space, the total heat loss across the air space may exceed that through the insulation because of the effects of radiation. The heat loss due to radiation may be minimized by using a radiation shield of low emissivity, and the means for calculating the loss will be developed in Chapter 13.

9.9 *Combined Free and Forced Convection*

In dealing with forced convection (Chapters 6 through 8), we ignored the effects of free convection. This was, of course, an assumption; for, as we now know, buoyancy forces develop in fluids as a result of temperature gradients, and unstable gradients might induce free convection. Similarly, in the preceding sections of this chapter, we assumed that forced convection was negligible. It is now time to acknowledge that situations may arise for which free and forced convection effects are comparable, in which case it is inappropriate to neglect either process. In Section 9.3 we indicated that free convection is negligible if $(Gr_L/Re_L^2) \ll 1$ and that forced convection is negligible if $(Gr_L/Re_L^2) \gg 1$. Hence the *combined free* and *forced* (or *mixed*) *convection* regime is generally one for which $(Gr_L/Re_L^2) \approx 1$.

The effect of buoyancy on heat transfer in a forced flow is strongly influenced by the direction of the buoyancy force relative to that of the flow. Three special cases that have been studied extensively correspond to buoyancy-induced and forced motions having the same direction (*assisting* flow), opposite directions (*opposing* flow), and perpendicular directions (*transverse* flow). Upward and downward forced motions over a hot vertical plate are examples of assisting and opposing flows, respectively. Examples of transverse flow include horizontal motion over a hot cylinder, sphere, or horizontal plate. In assisting and transverse flows, buoyancy acts to enhance the rate of heat transfer associated with pure forced convection; in opposing flows, it acts to decrease this rate.

It has become common practice to correlate mixed convection heat transfer results for external and internal flows by an expression of the form

$$Nu^n = Nu_F^n \pm Nu_N^n \tag{9.64}$$

For the specific geometry of interest, the Nusselt numbers Nu_F and Nu_N are determined from correlations for pure forced and natural (free) convection, respectively. The plus sign on the right-hand side of Equation 9.64 applies for assisting and transverse flows, while the minus sign applies for opposing flow. The best correlation of data is often obtained for $n = 3$, although values of 7/2 and 4 may be better suited for transverse flows involving horizontal plates and cylinders (or spheres), respectively.

Equation 9.64 should be viewed as a first approximation, and any serious treatment of a mixed convection problem should be accompanied by an examination of the open literature. Mixed convection flows received considerable attention in the late 1970s to the 1990s, and comprehensive literature reviews are available [40–43]. The flows are endowed with a variety of rich and unusual features that can complicate heat transfer predictions. For example, in a horizontal, parallel-plate channel, three-dimensional flows in the form of longitudinal vortices are induced by bottom heating, and the longitudinal variation of the Nusselt number is characterized by a decaying oscillation [44, 45]. Moreover, in channel flows, significant asymmetries may be associated with convection heat transfer at top and bottom surfaces [46]. Finally, we note that, although buoyancy effects can significantly enhance heat transfer for laminar forced convection flows, enhancement is typically negligible if the forced flow is turbulent [47].

9.10 *Convection Mass Transfer*

The buoyancy term on the right-hand side of Equation 9.2 is due to density variations in the fluid, which may arise from species concentration gradients, as well as temperature gradients. Hence, a more general form of the Grashof number, Equation 9.10, is

$$Gr_L = \frac{g(\Delta\rho/\rho)L^3}{\nu^2} = \frac{g(\rho_s - \rho_\infty)L^3}{\rho\nu^2} \tag{9.65}$$

which may be applied to natural convection flows driven by concentration gradients and/or temperature gradients. As shown in Section 9.2, if density variations are due only to temperature gradients, $(\Delta\rho/\rho) = -\beta\Delta T$. However, if there are no temperature gradients, motion may still be induced by spatial variations in the species composition, and similarity considerations lead to the conclusion that $Sh_L = f(Gr_L, Sc)$. Moreover, correlations for convection mass transfer may be inferred from those for heat transfer by invoking the heat and mass transfer analogy. For example, if species A is evaporating or sublimating from a vertical surface into a quiescent ambient fluid B, the convection mass transfer coefficient may be obtained from the analogous form of Equation 9.24. That is,

$$\overline{Sh}_L = \frac{\bar{h}_m L}{D_{AB}} = C(Gr_L Sc)^n \tag{9.66}$$

where Gr_L is given by Equation 9.65. If the molecular weight of species A is less than that of species B, $\rho_s < \rho_\infty$ and the buoyancy-induced flow is upward along the surface. If the opposite is true, $\rho_s > \rho_\infty$ and the flow is descending.

The analogy may only be applied in the foregoing manner for isothermal conditions. If there are gradients in *both* temperature and species concentration, heat and mass transfer will occur concurrently by natural convection. Similarity considerations then yield $\overline{Nu}_L = f(Gr_L, Pr, Sc)$ and $\overline{Sh}_L = f(Gr_L, Sc, Pr)$, where the density difference $\Delta\rho$ is due to both temperature and concentration variations. As a first approximation, existing correlations of the form $\overline{Nu}_L = f(Gr_L, Pr)$ and $\overline{Sh}_L = f(Gr_L, Sc)$ may be used to determine the convection transfer coefficients, so long as the value of $\Delta\rho = \rho_s - \rho_\infty$ is calculated by including the effects of both temperature and concentration variations on ρ_s and ρ_∞ and $Le = Pr/Sc \approx 1$. In a binary mixture of species A and B, the surface and free stream densities are defined as $\rho_s = \rho_{s,A} + \rho_{s,B}$ and $\rho_\infty = \rho_{\infty,A} + \rho_{\infty,B}$, respectively, where the species densities depend on the surface and free stream temperatures. The average density across the boundary layer(s) is $\rho = (\rho_s + \rho_\infty)/2$. For cases involving fluids for which Le is much less or much greater than unity, the literature should be consulted.

9.11 *Summary*

We have considered convective flows that originate in part or exclusively from buoyancy forces, and we have introduced the dimensionless parameters needed to characterize such flows. You should be able to discern when free convection effects are important and to quantify the associated heat transfer rates. An assortment of empirical correlations has been provided for this purpose.

To test your understanding of related concepts, consider the following questions.

• What is an *extensive, quiescent* fluid?
• What conditions are required for a buoyancy-driven flow?
• How does the velocity profile in the free convection boundary layer on a hot vertical plate differ from the velocity profile in the boundary layer associated with forced, parallel flow over a flat plate?
• What is the general form of the buoyancy term in the x-momentum equation for a free convection boundary layer? How may it be approximated if the flow is due to temperature variations? What is the name of the approximation?

- What is the physical interpretation of the *Grashof number*? What is the *Rayleigh number*? How does each parameter depend on the characteristic length?
- For a hot horizontal plate in quiescent air, do you expect heat transfer to be larger for the top or bottom surface? Why? For a cold horizontal plate in quiescent air, do you expect heat transfer to be larger for the top or bottom surface? Why?
- For free convection within a vertical parallel plate channel, what kind of force balance governs the flow rate in the channel?
- For a vertical channel with isothermal plates, what is the physical basis for existence of an optimum spacing?
- What is the nature of flow in a cavity whose vertical surfaces are hot and cold? What is the nature of flow in an annular space between concentric cylindrical surfaces that are hot and cold?
- What is meant by the term *mixed convection*? How can one determine if mixed convection effects should be considered in a heat transfer analysis? Under what conditions is heat transfer enhanced by mixed convection? Under what conditions is it reduced?
- Consider transport of species A from a horizontal surface facing upward in a quiescent fluid B. If $T_s = T_\infty$ and the molecular weight of A is less than that of B, what is the analogous heat transfer problem? What is the analogous heat transfer problem if the molecular weight of A exceeds that of B?

References

1. Jaluria, Y., *Natural Convection Heat and Mass Transfer,* Pergamon Press, New York, 1980.

2. Gebhart, B., Y. Jaluria, R. L. Mahajan, and B. Sammakia, *Buoyancy-Induced Flows and Transport,* Hemisphere Publishing, Washington, DC, 1988.

3. Ostrach, S., "An Analysis of Laminar Free Convection Flow and Heat Transfer About a Flat Plate Parallel to the Direction of the Generating Body Force," National Advisory Committee for Aeronautics, Report 1111, 1953.

4. LeFevre, E. J., "Laminar Free Convection from a Vertical Plane Surface," *Proc. Ninth Int. Congr. Appl. Mech.,* Brussels, Vol. 4, 168, 1956.

5. McAdams, W. H., *Heat Transmission,* 3rd ed., McGraw-Hill, New York, 1954, Chap. 7.

6. Warner, C. Y., and V. S. Arpaci, *Int. J. Heat Mass Transfer,* **11,** 397, 1968.

7. Bayley, F. J., *Proc. Inst. Mech. Eng.,* **169,** 361, 1955.

8. Churchill, S. W., and H. H. S. Chu, *Int. J. Heat Mass Transfer,* **18,** 1323, 1975.

9. Sparrow, E. M., and J. L. Gregg, *Trans. ASME,* **78,** 435, 1956.

10. Churchill, S. W., "Free Convection Around Immersed Bodies," in G. F. Hewitt, Exec. Ed., *Heat Exchanger Design Handbook,* Section 2.5.7, Begell House, New York, 2002.

11. Sparrow, E. M., and J. L. Gregg, *Trans. ASME,* **78,** 1823, 1956.

12. Cebeci, T., "Laminar-Free-Convective Heat Transfer from the Outer Surface of a Vertical Slender Circular Cylinder," *Proc. Fifth Int. Heat Transfer Conf.,* Paper NC1.4, pp. 15–19, 1974.

13. Minkowycz, W. J., and E. M. Sparrow, *J. Heat Transfer,* **96,** 178, 1974.

14. Vliet, G. C., *Trans. ASME,* **91C,** 511, 1969.

15. Fujii, T., and H. Imura, *Int. J. Heat Mass Transfer,* **15,** 755, 1972.

16. Azevedo, L. F. A., and E. M. Sparrow, *J. Heat Transfer,* **107,** 893, 1985.

17. Rich, B. R., *Trans. ASME,* **75,** 489, 1953.

18. Goldstein, R. J., E. M. Sparrow, and D. C. Jones, *Int. J. Heat Mass Transfer,* **16,** 1025, 1973.

19. Lloyd, J. R., and W. R. Moran, *J. Heat Transfer,* **96,** 443, 1974.

20. Radziemska, E., and W. M. Lewandowski, *Applied Energy*, **68,** 347, 2001.

21. Raithby, G. D., and K. G. T. Hollands, in W. M. Rohsenow, J. P. Hartnett, and Y. I. Cho, Eds., *Handbook of Heat Transfer Fundamentals,* Chap. 4, McGraw-Hill, New York, 1998.

22. Morgan, V. T., "The Overall Convective Heat Transfer from Smooth Circular Cylinders," in T. F. Irvine and J. P. Hartnett, Eds., *Advances in Heat Transfer,* Vol. 11, Academic Press, New York, 1975, pp. 199–264.

23. Churchill, S. W., and H. H. S. Chu, *Int. J. Heat Mass Transfer,* **18,** 1049, 1975.

24. Elenbaas, W., *Physica,* **9,** 1, 1942.

25. Bar-Cohen, A., and W. M. Rohsenow, *J. Heat Transfer,* **106,** 116, 1984.

26. Ostrach, S., "Natural Convection in Enclosures," in J. P. Hartnett and T. F. Irvine, Eds., *Advances in Heat Transfer,* Vol. 8, Academic Press, New York, 1972, pp. 161–227.

27. Catton, I., "Natural Convection in Enclosures," *Proc. 6th Int. Heat Transfer Conf.,* Toronto, Canada, 1978, Vol. 6, pp. 13–31.

28. Globe, S., and D. Dropkin, *J. Heat Transfer,* **81C,** 24, 1959.

29. Hollands, K. G. T., G. D. Raithby, and L. Konicek, *Int. J. Heat Mass Transfer,* **18,** 879, 1975.

30. Churchill, S. W., "Free Convection in Layers and Enclosures," in G. F. Hewitt, Exec. Ed., *Heat Exchanger Design Handbook,* Section 2.5.8, Begell House, New York, 2002.

31. MacGregor, R. K., and A. P. Emery, *J. Heat Transfer,* **91,** 391, 1969.

32. Ayyaswamy, P. S., and I. Catton, *J. Heat Transfer,* **95,** 543, 1973.

33. Catton, I., P. S. Ayyaswamy, and R. M. Clever, *Int. J. Heat Mass Transfer,* **17,** 173, 1974.

34. Clever, R. M., *J. Heat Transfer,* **95,** 407, 1973.

35. Arnold, J. N., I. Catton, and D. K. Edwards, "Experimental Investigation of Natural Convection in Inclined Rectangular Regions of Differing Aspect Ratios," ASME Paper 75-HT-62, 1975.

36. Buchberg, H., I. Catton, and D. K. Edwards, *J. Heat Transfer,* **98,** 182, 1976.

37. Hollands, K. G. T., S. E. Unny, G. D. Raithby, and L. Konicek, *J. Heat Transfer,* **98,** 189, 1976.

38. Raithby, G. D., and K. G. T. Hollands, "A General Method of Obtaining Approximate Solutions to Laminar and Turbulent Free Convection Problems," in T. F. Irvine and J. P. Hartnett, Eds., *Advances in Heat Transfer,* Vol. 11, Academic Press, New York, 1975, pp. 265–315.

39. Kuehn, T. H., and R. J. Goldstein, *Int. J. Heat Mass Transfer,* **19,** 1127, 1976.

40. Churchill, S. W., "Combined Free and Forced Convection around Immersed Bodies," in G. F. Hewitt, Exec. Ed., *Heat Exchanger Design Handbook*, Section 2.5.9, Begell House, New York, 2002.

41. Churchill, S. W., "Combined Free and Forced Convection in Channels," in G. F. Hewitt, Exec. Ed., *Heat Exchanger Design Handbook*, Section 2.5.10, Begell House, New York, 2002.

42. Chen, T. S., and B. F. Armaly, in S. Kakac, R. K. Shah, and W. Aung, Eds., *Handbook of Single-Phase Convective Heat Transfer,* Chap. 14, Wiley-Interscience, New York, 1987.

43. Aung, W., in S. Kakac, R. K. Shah, and W. Aung, Eds., *Handbook of Single-Phase Convective Heat Transfer,* Chap. 15, Wiley-Interscience, New York, 1987.

44. Incropera, F. P., A. J. Knox, and J. R. Maughan, *J. Heat Transfer,* **109,** 434, 1987.

45. Maughan, J. R., and F. P. Incropera, *Int. J. Heat Mass Transfer,* **30,** 1307, 1987.

46. Osborne, D. G., and F. P. Incropera, *Int. J. Heat Mass Transfer,* **28,** 207, 1985.

47. Osborne, D. G., and F. P. Incropera, *Int. J. Heat Mass Transfer,* **28,** 1337, 1985.

Problems

Properties and General Considerations

9.1 The one-dimensional plane wall of Figure 3.1 is of thickness $L = 75$ mm and thermal conductivity $k = 5$ W/m · K. The fluid temperatures are $T_{\infty,1} = 200°C$ and $T_{\infty,2} = 100°C$, respectively. Using the minimum and maximum typical values of the convection heat transfer coefficients listed in Table 1.1, determine the minimum and maximum steady-state heat fluxes through the wall for (i) free convection in gases, (ii) free convection in liquids, (iii) forced convection in gases, (iv) forced convection in liquids, and (v) convection with phase change.

9.2 Using the values of density for water in Table A.6, calculate the volumetric thermal expansion coefficient at 300 K from its definition, Equation 9.3, and compare your result with the tabulated value.

9.3 An object of characteristic length $L = 0.5$ m is 10°C warmer than the surrounding fluid, which is flowing at a velocity of 0.5 m/s. Graph the Reynolds number, the Grashof number, and the ratio Gr_L/Re_L^2 over the range $10 \leq T_f \leq 90°C$, where T_f is the film temperature. Consider the fluid to be either air at atmospheric pressure or liquid water. Use of *IHT* is recommended.

9.4 Consider an object of characteristic length 0.01 m and a situation for which the temperature difference is 30°C. Evaluating thermophysical properties at the prescribed conditions, determine the Rayleigh number for the following fluids: air (1 atm, 400 K), helium (1 atm, 400 K), glycerin (285 K), and water (310 K).

9.5 To assess the efficacy of different liquids for cooling by natural convection, it is convenient to introduce a *figure of merit*, F_N, which combines the influence of all pertinent fluid properties on the convection coefficient. If the Nusselt number is governed by an expression of the form, $Nu_L \sim Ra^n$, obtain the corresponding relationship between F_N and the fluid properties. For a representative value of $n = 0.33$, calculate values of F_N for air ($k = 0.026$ W/m · K, $\beta = 0.0035$ K^{-1}, $\nu = 1.5 \times 10^{-5}$ m^2/s, $Pr = 0.70$), water ($k = 0.600$ W/m · K, $\beta = 2.7 \times 10^{-4}$ K^{-1}, $\nu = 10^{-6}$ m^2/s, $Pr = 5.0$), and a dielectric liquid ($k = 0.064$ W/m · K, $\beta = 0.0014$ K^{-1}, $\nu = 10^{-6}$ m^2/s, $Pr = 25$). What fluid is the most effective cooling agent?

9.6 In many cases, we are concerned with free convection involving gases that are contained within sealed enclosures. Consider air at 27°C and pressures of 1, 10, and 100 bars. Determine the *figure of merit* described in Problem 9.5 for each of these three pressures. Which air pressure will provide the most effective cooling? *Hint:* See Problem 6.20.

Vertical Plates

9.7 Consider a large vertical plate with a uniform surface temperature of 130°C suspended in quiescent air at 25°C and atmospheric pressure.

(a) Estimate the boundary layer thickness at a location 0.25 m measured from the lower edge.

(b) What is the maximum velocity in the boundary layer at this location and at what position in the boundary layer does the maximum occur?

(c) Using the similarity solution result, Equation 9.19, determine the heat transfer coefficient 0.25 m from the lower edge.

(d) At what location on the plate measured from the lower edge will the boundary layer become turbulent?

9.8 The heat transfer rate due to free convection from a vertical surface, 1 m high and 0.6 m wide, to quiescent air that is 20 K colder than the surface is known. What is the ratio of the heat transfer rate for that situation to the rate corresponding to a vertical surface, 0.6 m high and 1 m wide, when the quiescent air is 20 K warmer than the surface? Neglect heat transfer by radiation and any influence of temperature on the relevant thermophysical properties of air.

9.9 Consider an array of vertical rectangular fins, which is to be used to cool an electronic device mounted in quiescent, atmospheric air at $T_\infty = 27°C$. Each fin has $L = 20$ mm and $H = 150$ mm and operates at an approximately uniform temperature of $T_s = 77°C$.

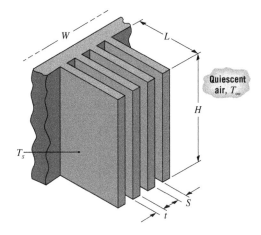

(a) Viewing each fin surface as a vertical plate in an infinite, quiescent medium, briefly describe why there exists an optimum fin spacing S. Using Figure 9.4, estimate the optimum value of S for the prescribed conditions.

(b) For the optimum value of S and a fin thickness of $t = 1.5$ mm, estimate the rate of heat transfer from the fins for an array of width $W = 355$ mm.

9.10 A number of thin plates are to be cooled by vertically suspending them in a water bath at a temperature of 20°C. If the plates are initially at 54°C and are 0.15 m long, what minimum spacing would prevent interference between their free convection boundary layers?

9.11 Beginning with the free convection correlation of the form given by Equation 9.24, show that for air at atmospheric pressure and a film temperature of 400 K, the average heat transfer coefficient for a vertical plate can be expressed as

$$\bar{h}_L = 1.40 \left(\frac{\Delta T}{L} \right)^{1/4} \qquad 10^4 < Ra_L < 10^9$$

$$\bar{h}_L = 0.98 \Delta T^{1/3} \qquad 10^9 < Ra_L < 10^{13}$$

9.12 A square aluminum plate 5 mm thick and 200 mm on a side is heated while vertically suspended in quiescent air at 40°C. Determine the average heat transfer coefficient for the plate when its temperature is 15°C by two methods: using results from the similarity solution to the boundary layer equations, and using results from an empirical correlation.

9.13 An aluminum alloy (2024) plate, heated to a uniform temperature of 227°C, is allowed to cool while vertically suspended in a room where the ambient air and surroundings are at 27°C. The plate is 0.3 m square with a thickness of 15 mm and an emissivity of 0.25.

The plate has been used in an experiment to determine the free convection heat transfer coefficient. At an instant of time when the plate temperature was 127°C, the time rate of change of this temperature was observed to be −0.0465 K/s. What is the corresponding free convection heat transfer coefficient? Compare this result with an estimate based on a standard empirical correlation.

9.14 Determine the average convection heat transfer coefficient for the 2.5-m-high vertical walls of a home having respective interior air and wall surface temperatures of (a) 20 and 10°C and (b) 27 and 37°C.

9.15 Show that the laminar, free convection correlations for a vertical, isothermal flat plate of height L can be applied to a vertical, isothermal cylinder of diameter D and height L if the boundary layer thickness at the top of the cylinder is approximately one-fourth the cylinder diameter. Determine the critical cylinder diameter, D_{crit}, below which the flat plate correlations cannot be applied when the surface is at $T_s = 35$°C, the surrounding, quiescent air is at $T_\infty = 25$°C, and $L = 0.75$ m. Calculate the rate of free convection heat transfer from the vertical cylinder wall corresponding to the critical cylinder diameter.

9.16 Consider a vertical plate of dimension 0.25 m × 0.50 m that is at $T_s = 100$°C in a quiescent environment at $T_\infty = 20$°C. In the interest of minimizing heat transfer from the plate, which orientation, (A) or (B), is preferred? Estimate the rate of heat transfer by free convection from the front surface of the plate when it is in the preferred orientation.

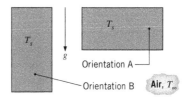

9.17 During a winter day, the window of a patio door with a height of 1.8 m and width of 1.0 m shows a frost line near its base. The room wall and air temperatures are 15°C.

(a) Explain why the window would show a frost layer at the base rather than at the top.

(b) Estimate the rate of heat loss through the window due to free convection and radiation. Assume the window has a uniform temperature of 0°C and the emissivity of the glass surface is 0.94. If the room has electric baseboard heating, estimate the corresponding daily cost of the window heat loss for a utility rate of 0.21 \$/kW · h.

9.18 A vertical, thin pane of window glass that is 1 m on a side separates quiescent room air at $T_{\infty,i} = 20$°C from quiescent ambient air at $T_{\infty,o} = -20$°C. The walls of the room and the external surroundings (landscape, buildings, etc.) are also at $T_{sur,i} = 20$°C and $T_{sur,o} = -20$°C, respectively.

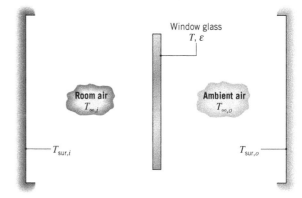

If the glass has an emissivity of $\varepsilon = 1$, what is its temperature T? What is the rate of heat loss through the glass?

9.19 Consider the conditions of Problem 9.18, but now allow for a difference between the inner and outer surface temperatures, $T_{s,i}$ and $T_{s,o}$, of the window. For a glass thickness and thermal conductivity of $t_g = 12$ mm and $k_g = 1.4$ W/m · K, respectively, evaluate $T_{s,i}$ and $T_{s,o}$. What is the rate of heat loss through the window?

9.20 Consider laminar flow about a vertical isothermal plate of length L, providing an average heat transfer coefficient of $\bar{h}_L$. If the plate is divided into N smaller plates, each of length $L_N = L/N$, determine an expression for the ratio of the heat transfer coefficient averaged over the N plates to the heat transfer coefficient averaged over the single plate, $\bar{h}_{L,N}/\bar{h}_{L,1}$.

9.21 A household oven door of 0.5-m height and 0.7-m width reaches an average surface temperature of 32°C during operation. Estimate the heat loss to the room with ambient air at 22°C. If the door has an emissivity of 1.0 and the surroundings are also at 22°C, comment on the heat loss by free convection relative to that by radiation.

9.22 Consider a vertical, single-pane window of equivalent width and height ($W = L = 1$ m). The interior surface is exposed to the air and walls of a room, which are each at 18°C.

(a) Under cold ambient conditions for which a thin layer of frost has formed on the inner surface, what is the heat loss through the window?

(b) How would your analysis be affected by a frost layer whose thickness is not negligible?

(c) During incipience of frost formation, where would you expect the frost to begin to develop on the window? The frost may be assumed to have an emissivity of $\varepsilon = 0.90$.

9.23 A thin-walled container with a hot process fluid at 50°C is placed in a quiescent, cold water bath at 10°C. Heat transfer at the inner and outer surfaces of the container may be approximated by free convection from a vertical plate.

9.24 Consider an experiment to investigate the transition to turbulent flow in a free convection boundary layer that develops along a vertical plate suspended in a large room. The plate is constructed of a thin heater that is sandwiched between two aluminum plates and may be assumed to be isothermal. The heated plate is 1 m high and 2 m wide. The quiescent air and the surroundings are both at 25°C.

(a) The exposed surfaces of the aluminum plate are painted with a very thin coating of high emissivity ($\varepsilon = 0.95$) paint. Determine the electrical power that must be supplied to the heater to sustain the plate at a temperature of $T_s = 35$°C. How much of the plate is exposed to turbulent conditions in the free convection boundary layer?

(b) The experimentalist speculates that the roughness of the paint is affecting the transition to turbulence in the boundary layer and decides to remove the paint and polish the aluminum surface ($\varepsilon = 0.05$). If the same power is supplied to the plate as in part (a), what is the steady-state plate temperature? How much of the plate is exposed to turbulent conditions in the free convection boundary layer?

9.25 The vertical rear window of an automobile is of thickness $L = 8$ mm and height $H = 0.5$ m and contains fine-meshed heating wires that can induce nearly uniform volumetric heating, $\dot{q}$(W/m^3).

(a) Consider steady-state conditions for which the interior surface of the window is exposed to quiescent air at 10°C, while the exterior surface is exposed to air at −10°C moving in parallel flow over the surface with a velocity of 20 m/s. Determine the volumetric heating rate needed to maintain the interior window surface at $T_{s,i} = 15$°C.

(b) The interior and exterior window temperatures, $T_{s,i}$ and $T_{s,o}$, depend on the compartment and ambient temperatures, $T_{\infty,i}$ and $T_{\infty,o}$, as well as on the velocity u_∞ of air flowing over the exterior surface and the volumetric heating rate $\dot{q}$. Subject to the constraint that $T_{s,i}$ is to be maintained at 15°C, we wish to develop guidelines for varying the heating rate in response to changes in $T_{\infty,i}$, $T_{\infty,o}$, and/or u_∞. If $T_{\infty,i}$ is maintained at 10°C, how will $\dot{q}$ and $T_{s,o}$ vary with $T_{\infty,o}$ for $-25 \leq T_{\infty,o} \leq 5$°C and $u_\infty = 10, 20,$ and 30 m/s?

(a) Determine the overall heat transfer coefficient between the hot process fluid and the cold water bath. Assume the properties of the hot process fluid are those of water.

(b) Generate a plot of the overall heat transfer coefficient as a function of the hot process fluid temperature $T_{\infty,h}$ for the range 20 to 60°C, with all other conditions remaining the same.

If a constant vehicle speed is maintained, such that $u_\infty = 30$ m/s, how will $\dot{q}$ and $T_{s,o}$ vary with $T_{\infty,i}$ for $5 \leq T_{\infty,i} \leq 20°C$ and $T_{\infty,o} = -25, -10$, and $5°C$?

9.26 Circuit boards are mounted to interior vertical surfaces of a rectangular duct of height $H = 400$ mm and length $L = 800$ mm. Although the boards are cooled by forced convection heat transfer to air flowing through the duct, not all of the heat dissipated by the electronic components is transferred to the flow. Some of the heat is instead transferred by conduction to the vertical walls of the duct and then by natural convection and radiation to the ambient (atmospheric) air and surroundings, which are at equivalent temperatures of $T_\infty = T_{sur} = 20°C$. The walls are metallic and, to a first approximation, may be assumed to be isothermal at a temperature T_s.

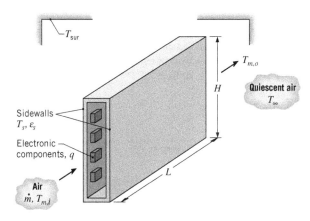

(a) Consider conditions for which the electronic components dissipate 200 W and air enters the duct at a flow rate of $\dot{m} = 0.015$ kg/s and a temperature of $T_{m,i} = 20°C$. If the emissivity of the side walls is $\varepsilon_s = 0.15$ and the outlet temperature of the air is $T_{m,o} = 30°C$, what is the surface temperature T_s?

(b) To reduce the temperature of the electronic components, it is desirable to enhance heat transfer from the side walls. Assuming no change in the airflow conditions, what is the effect on T_s of applying a high emissivity coating ($\varepsilon_s = 0.90$) to the side walls?

(c) If there is a loss of airflow while power continues to be dissipated, what are the resulting values of T_s for $\varepsilon_s = 0.15$ and $\varepsilon_s = 0.90$?

9.27 A refrigerator door has a height and width of $H = 1$ m and $W = 0.65$ m, respectively, and is situated in a large room for which the air and walls are at $T_\infty = T_{sur} = 25°C$. The door consists of a layer of polystyrene insulation ($k = 0.03$ W/m · K) sandwiched between thin sheets of steel ($\varepsilon = 0.6$) and polypropylene. Under normal operating conditions, the inner surface of the door is maintained at a fixed temperature of $T_{s,i} = 5°C$.

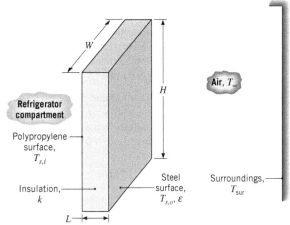

(a) Estimate the rate of heat gain through the door for the worst case condition corresponding to no insulation ($L = 0$).

(b) Compute and plot the rate of heat gain and the outer surface temperature $T_{s,o}$ as a function of insulation thickness for $0 \leq L \leq 25$ mm.

9.28 Air at 3 atm and 100°C is discharged from a compressor into a vertical receiver of 2.5-m height and 0.75-m diameter. Assume that the receiver wall has negligible thermal resistance, is at a uniform temperature, and that heat transfer at its inner and outer surfaces is by free convection from a vertical plate. Neglect radiation exchange and any losses from the top.

(a) Estimate the receiver wall temperature and the heat transfer to the ambient air at 25°C. To facilitate use of the free convection correlations with appropriate film temperatures, assume that the receiver wall temperature is 60°C.

(b) Were the assumed film temperatures of part (a) reasonable? If not, use an iteration procedure to find consistent values.

(c) Now consider two features of the receiver neglected in the previous analysis: (i) radiation exchange from the exterior surface of emissivity 0.85 to large surroundings, also at 25°C; and (ii) the thermal resistance of a 20-mm-thick wall with a thermal conductivity of 0.25 W/m · K. Represent the system by a thermal circuit and estimate the wall temperatures and the heat transfer rate.

9.29 Determine the maximum allowable uniform heat flux that may be imposed at a wall heating panel 1 m high if the maximum temperature is not to exceed 37°C when the ambient air temperature is 25°C.

Horizontal and Inclined Plates

9.30 Airflow through a long, 0.2-m-square air conditioning duct maintains the outer duct surface temperature at 10°C. If the horizontal duct is uninsulated and exposed to air at 35°C in the crawlspace beneath a home, what is the heat gain per unit length of the duct?

9.31 Consider the transformer of Problem 8.78, whose lateral surface is being maintained at 47°C by a forced convection coolant line removing 1000 W. It is desired to explore cooling of the transformer by free convection and radiation, assuming the surface to have an emissivity of 0.80.

(a) Determine how much power could be removed by free convection and radiation from the lateral *and* the upper horizontal surfaces when the ambient temperature and the surroundings are at 27°C.

(b) Vertical fins, 5 mm thick, 75 mm wide, and 500 mm long, can easily be welded to the lateral surface. What is the heat removal rate by free convection if 30 such fins are attached?

9.32 An electrical heater in the form of a horizontal disk of 400-mm diameter is used to heat the bottom of a tank filled with engine oil at a temperature of 5°C. Calculate the power required to maintain the heater surface temperature at 70°C.

9.33 Consider a horizontal 5-mm-thick, 100-mm-long straight fin fabricated from plain carbon steel ($k = 57$ W/m · K, $\varepsilon = 0.5$). The base of the fin is maintained at 100°C, while the quiescent ambient air and the surroundings are at 25°C. Assume the fin tip is adiabatic, and estimate the fin heat rate per unit width, q_f'. Use an average fin surface temperature of 80°C to estimate the free convection coefficient and the linearized radiation coefficient. How sensitive is your estimate to the choice of the average fin surface temperature?

9.34 A hot, 0.5-m-diameter, 35-mm-thick aluminum alloy disk is quenched from an initial temperature of $T_i = 400°C$ in a large oil bath of temperature $T_\infty = 35°C$. As it is quenched, the disk rests horizontally on the bottom of the well-insulated bath container. Use the results of Section 5.3.3 to determine the time needed to reduce the disk temperature to 100°C. The metal properties are $\rho = 1000$ kg/m³, $k = 185$ W/m · K, and $c = 775$ J/kg · K.

9.35 The thermal conductivity and surface emissivity of a material may be determined by heating its bottom surface and exposing its top surface to quiescent air and large surroundings of equivalent temperatures, $T_\infty = T_{sur} = 25°C$. The remaining surfaces of the sample/heater are well insulated.

Consider a sample of thickness $L = 25$ mm and a square planform of width $W = 250$ mm. In an experiment performed under steady-state conditions, temperature measurements made at the lower and upper surface of the sample yield values of $T_1 = 150°C$ and $T_2 = 100°C$, respectively, for a power input of $P_{elec} = 70$ W. What are the thermal conductivity and emissivity of the sample?

9.36 Many laptop computers are equipped with thermal management systems that involve liquid cooling of the central processing unit (CPU), transfer of the heated liquid to the back of the laptop screen assembly, and dissipation of heat from the back of the screen assembly by way of a flat, isothermal heat spreader. The cooled liquid is recirculated to the CPU and the process continues. Consider an aluminum heat spreader that is of width $w = 275$ mm and height $L = 175$ mm. The screen assembly is oriented at an angle $\theta = 30°$ from the vertical direction, and the heat spreader is attached to the $t = 3$-mm-thick plastic housing with a thermally conducting adhesive. The plastic housing has a thermal conductivity of $k = 0.21$ W/m·K and emissivity of $\varepsilon = 0.85$. The contact resistance associated with the heat spreader-housing interface is $R''_{t,c} = 2.0 \times 10^{-4}$ m²·K/W. If the CPU generates, on average, 15 W of thermal energy, what is the temperature of the heat spreader when $T_\infty = T_{sur} = 23°C$? Which thermal resistance (contact, conduction, radiation, or free convection) is the largest?

9.37 A circular grill of diameter 0.25 m and emissivity 0.9 is maintained at a constant surface temperature of 130°C. What electrical power is required when the room air and surroundings are at 24°C?

9.38 Consider the roof of the refrigerated truck compartment described in Problem 7.16, but under conditions for which the truck is parked ($V = 0$). All other conditions remain unchanged. For $\alpha_S = \varepsilon = 0.5$, determine the outer surface temperature, $T_{s,o}$, and the heat load imposed on the refrigeration system. *Hint*: Assume $T_{s,o} > T_\infty$ and $Ra_L > 10^7$.

9.39 The 4 m × 4 m horizontal roof of an uninsulated aluminum melting furnace is comprised of a 0.08-m-thick fireclay brick refractory covered by a 5-mm-thick steel (AISI 1010) plate. The refractory surface exposed to the furnace gases is maintained at 1700 K during operation, while the outer surface of the steel is exposed to the air and walls of a large room at 25°C. The emissivity of the steel is $\varepsilon = 0.3$.

(a) What is the rate of heat loss from the roof?

(b) If a 20-mm-thick layer of alumina-silica insulation (64 kg/m³) is placed between the refractory and the steel, what is the new rate of heat loss from the roof? What is the temperature at the inner surface of the insulation?

(c) One of the process engineers claims that the temperature at the inner surface of the insulation found in part (b) is too high for safe, long-term operation. What thickness of fireclay brick would reduce this temperature to 1350 K?

9.40 A 200-mm-square, 10-mm-thick tile has the thermophysical properties of Pyrex ($\varepsilon = 0.80$) and emerges from a curing process at an initial temperature of $T_i = 140°C$. The backside of the tile is insulated while the upper surface is exposed to ambient air and surroundings at 25°C.

(a) Estimate the time required for the tile to cool to a final, safe-to-touch temperature of $T_f = 40°C$. Use an average tile surface temperature of $\bar{T} = (T_i + T_f)/2$ to estimate the average free convection coefficient and the linearized radiation coefficient. How sensitive is your estimate to the assumed value for $\bar{T}$?

(b) Estimate the required cooling time if ambient air is blown in parallel flow over the tile with a velocity of 10 m/s.

9.41 Integrated circuit (IC) boards are stacked within a duct and dissipate a total of 500 W. The duct has a square cross section with $w = H = 150$ mm and a length of 0.5 m. Air flows into the duct at 25°C and 1.2 m³/min, and the convection coefficient between the air and the inner surfaces of the duct is $\bar{h}_i = 50$ W/m²·K. The entire outer surface of the duct, which is anodized with an emissivity of 0.5, is exposed to ambient air and large surroundings at 25°C.

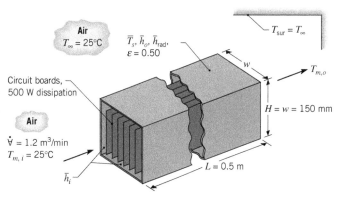

Your assignment is to develop a model to estimate the outlet temperature of the air, $T_{m,o}$, and the average surface temperature of the duct, $\bar{T}_s$.

(a) Assuming a surface temperature of 37°C, estimate the average free convection coefficient, $\bar{h}_o$, for the outer surface of the duct.

(b) Assuming a surface temperature of 37°C, estimate the average linearized radiation coefficient, $\bar{h}_{rad}$, for the outer surface of the duct.

(c) Perform an energy balance on the duct by considering the dissipation of electrical power in the ICs, the rate of change in the energy of air flowing through the duct, and the rate of heat transfer from the air in the duct to the surroundings. Express the last process in terms of thermal resistances between the mean temperature, $\bar{T}_m$, of the air in the duct and the temperature of the ambient air and the surroundings.

(d) Substitute numerical values into the expression of part (c) and calculate the air outlet temperature, $T_{m,o}$. Estimate the corresponding value of $\bar{T}_s$. Comment on your results and the assumptions inherent in your model.

9.42 Certain wood stove designs rely exclusively on heat transfer by radiation and natural convection to the surroundings. Consider a stove that forms a cubical enclosure, $L_s = 1$ m on a side, in a large room. The exterior walls of the stove have an emissivity of $\varepsilon = 0.8$ and are at an operating temperature of $T_{s,s} = 500$ K.

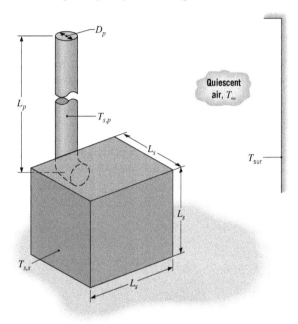

The stove pipe, which may be assumed to be isothermal at an operating temperature of $T_{s,p} = 400$ K, has a diameter of $D_p = 0.25$ m and a height of $L_p = 2$ m, extending from stove to ceiling. The stove is in a large room whose air and walls are at $T_\infty = T_{sur} = 300$ K. Neglecting heat transfer from the small horizontal section of the pipe and radiation exchange between the pipe and stove, estimate the rate at which heat is transferred from the stove and pipe to the surroundings.

9.43 A plate 1 m × 1 m, inclined at an angle of 45°, is exposed to a net radiation heat flux of 350 W/m² at its bottom surface. If the top surface of the plate is well insulated, estimate the temperature the plate reaches when the ambient air is quiescent and at a temperature of 0°C.

Horizontal Cylinders and Spheres

9.44 A horizontal uninsulated steam pipe passes through a large room whose walls and ambient air are at 300 K. The pipe of 150-mm diameter has an emissivity of 0.85 and an outer surface temperature of 400 K. Calculate the heat loss per unit length from the pipe.

9.45 Beverage in cans 150 mm long and 60 mm in diameter is initially at 27°C and is to be cooled by placement in a refrigerator compartment at 4°C. In the interest of maximizing the cooling rate, should the cans be laid horizontally or vertically in the compartment? As a first approximation, neglect heat transfer from the ends.

9.46 A long, uninsulated steam line with a diameter of 100 mm and a surface emissivity of 0.8 transports steam at 150°C and is exposed to atmospheric air and large surroundings at an equivalent temperature of 20°C.

(a) Calculate the rate of heat loss per unit length for a calm day.

(b) Calculate the rate of heat loss on a breezy day when the wind speed is 8 m/s.

(c) For the conditions of part (a), calculate the rate of heat loss with a 20-mm-thick layer of insulation ($k = 0.08$ W/m · K). Would the heat loss change significantly with an appreciable wind speed?

9.47 A horizontal tube of 12.5-mm diameter with an outer surface temperature of 240°C is located in a room with an air temperature of 20°C. Estimate the heat transfer rate per unit length of the tube due to free convection.

9.48 Saturated steam at 4 bars absolute pressure with a mean velocity of 3 m/s flows through a horizontal pipe whose inner and outer diameters are 55 and 65 mm, respectively. The heat transfer coefficient for the steam flow is known to be 11,000 W/m² · K.

(a) If the pipe is covered with a 25-mm-thick layer of 85% magnesia insulation and is exposed to atmospheric air at 25°C, determine the rate of heat transfer by free convection to the room per unit length of the pipe. If the steam is saturated at the inlet of the pipe, estimate its quality at the outlet of a pipe 30 m long.

(b) Net radiation to the surroundings also contributes to heat loss from the pipe. If the insulation has a surface emissivity of $\varepsilon = 0.8$ and the surroundings are at $T_{sur} = T_\infty = 25°C$, what is the rate of heat transfer to the room per unit length of pipe? What is the quality of the outlet flow?

(c) The heat loss may be reduced by increasing the insulation thickness and/or reducing its emissivity. What is the effect of increasing the insulation thickness to 50 mm if $\varepsilon = 0.8$? Of decreasing the emissivity to 0.2 if the insulation thickness is 25 mm? Of reducing the emissivity to 0.2 and increasing the insulation thickness to 50 mm?

9.49 An electric immersion heater, 10 mm in diameter and 300 mm long, is rated at 550 W. If the heater is horizontally positioned in a large tank of water at 20°C, estimate its surface temperature. Estimate the surface temperature if the heater is accidentally operated in air at 20°C.

9.50 The maximum surface temperature of the 20-mm-diameter shaft of a motor operating in ambient air at 27°C should not exceed 87°C. Because of power dissipation within the motor housing, it is desirable to reject as much heat as possible through the shaft to the ambient air. In this problem, we will investigate several methods for heat removal.

(a) For rotating cylinders, a suitable correlation for estimating the convection coefficient is of the form

$$\overline{Nu}_D = 0.133\, Re_D^{2/3}\, Pr^{1/3}$$
$$(Re_D < 4.3 \times 10^5, \quad 0.7 < Pr < 670)$$

where $Re_D \equiv \Omega D^2/\nu$ and Ω is the rotational velocity (rad/s). Determine the convection coefficient and the maximum heat rate per unit length as a function of rotational speed in the range from 5000 to 15,000 rpm.

(b) Estimate the free convection coefficient and the maximum heat rate per unit length for the stationary shaft. Mixed free and forced convection effects may become significant for $Re_D < 4.7(Gr_D^3/Pr)^{0.137}$. Are free convection effects important for the range of rotational speeds designated in part (a)?

(c) Assuming the emissivity of the shaft is 0.8 and the surroundings are at the ambient air temperature, is radiation exchange important?

(d) If ambient air is in cross flow over the shaft, what air velocities are required to remove the heat rates determined in part (a)?

9.51 Consider a horizontal pin fin of 6-mm diameter and 60-mm length fabricated from plain carbon steel ($k = 57$ W/m · K, $\varepsilon = 0.5$). The base of the fin is maintained at 150°C, while the quiescent ambient air and the surroundings are at 25°C. Assume the fin tip is adiabatic.

(a) Estimate the fin heat rate, q_f. Use an average fin surface temperature of 125°C in estimating the free convection coefficient and the linearized radiation coefficient. How sensitive is this estimate to your choice of the average fin surface temperature?

(b) Use the finite-difference method of solution to obtain q_f when the convection and radiation coefficients are based on local, rather than average, temperatures for the fin. How does your result compare with the analytical solution of part (a)?

9.52 The concept of a critical insulation radius was introduced in Example 3.6. Consider the thin-walled copper tube and insulation of the example. Now, the tube temperature is −10°C and it is suspended horizontally in quiescent air at 25°C. Neglecting radiation, determine the heat transfer rate per unit tube length for an insulation thickness of 10 mm. Graph the conduction, convection, and total thermal resistances for a unit length for insulation thickness $r - r_i$ over the range $0 \le r - r_i \le 50$ mm, and compare your results to that of the example. Repeat the calculation and the graphing exercise for a tube diameter of 1 mm. For each case, does a critical insulation radius exist? Why or why not? Evaluate air properties at 285 K.

9.53 A horizontal electrical cable of 25-mm diameter has a heat dissipation rate of 30 W/m. If the ambient air temperature is 27°C, estimate the surface temperature of the cable.

9.54 Hot air flows from a furnace through a 0.15-m-diameter, thin-walled steel duct with a velocity of 3 m/s. The duct passes through the crawlspace of a house, and its uninsulated exterior surface is exposed to quiescent air and surroundings at 0°C.

(a) At a location in the duct for which the mean air temperature is 70°C, determine the heat loss per unit duct length and the duct wall temperature. The duct outer surface has an emissivity of 0.5.

(b) If the duct is wrapped with a 25-mm-thick layer of 85% magnesia insulation ($k = 0.050$ W/m · K) having a surface emissivity of $\varepsilon = 0.60$, what are the duct wall temperature, the outer surface temperature, and the heat loss per unit length?

9.55 A computer code is being developed to analyze a 12.5-mm-diameter, cylindrical sensor used to determine ambient air temperature. The sensor experiences free convection while positioned horizontally in quiescent air at $T_\infty = 27$°C. For the temperature range from 30 to 80°C, derive an expression for the convection coefficient as a function of only $\Delta T = T_s - T_\infty$, where T_s is the sensor temperature. Evaluate properties at an appropriate film temperature and show what effect this approximation has on the convection coefficient estimate.

9.56 A thin-walled tube of 20-mm diameter passes hot fluid at a mean temperature of 45°C in an experimental flow loop. The tube is mounted horizontally in quiescent air at a temperature of 15°C. To satisfy the stringent temperature control requirements of the experiment, it was decided to wind thin electrical heating tape on the outer surface of the tube to prevent heat loss from the hot fluid to the ambient air.

(a) Neglecting radiation heat loss, calculate the heat flux q_e'' that must be supplied by the electrical tape to ensure a uniform fluid temperature.

(b) Assuming the emissivity of the tape is 0.95 and the surroundings are also at 15°C, calculate the required heat flux.

(c) The heat loss may be reduced by wrapping the heating tape in a layer of insulation. For 85% magnesia insulation ($k = 0.050$ W/m · K) having a surface emissivity of $\varepsilon = 0.60$, compute and plot the required heat flux q_e'' as a function of insulation thickness in the range from 0 to 20 mm. For this range, compute and plot the convection and radiation heat rates per unit tube length as a function of insulation thickness.

9.57 Long stainless steel rods of 50-mm diameter are preheated to a uniform temperature of 1000 K before being suspended from an overhead conveyor for transport to a hot forming operation. The conveyor is in a large room whose walls and air are at 300 K.

(a) Assuming the linear motion of the rod to have a negligible effect on convection heat transfer from its surface, determine the average convection coefficient at the start of the transport process.

(b) If the surface emissivity of the rod is $\varepsilon = 0.40$, what is the effective radiation heat transfer coefficient at the start of the transport process?

(c) Assuming a constant cumulative (radiation plus convection) heat transfer coefficient corresponding to the results of parts (a) and (b), what is the maximum allowable conveyor transit time, if the centerline temperature of the rod must exceed 900 K for the forming operation? Properties of the steel are $k = 25$ W/m · K and $\alpha = 5.2 \times 10^{-6}$ m²/s.

(d) Heat transfer by convection and radiation are actually decreasing during the transfer operation. Accounting for this reduction, reconsider the conditions of part (c) and obtain a more accurate estimate of the maximum allowable conveyor transit time.

9.58 Consider a batch process in which 200 L of a pharmaceutical are heated from 25°C to 70°C by saturated steam condensing at 2.455 bars as it flows through a coiled tube of 15-mm diameter and 15-m length. At any time during the process, the liquid may be approximated as an infinite, quiescent medium of uniform temperature and may be assumed to have constant properties of $\rho = 1100$ kg/m³, $c = 2000$ J/kg · K, $k = 0.25$ W/m · K, $v = 4.0 \times 10^{-6}$ m²/s, $Pr = 10$, and $\beta = 0.002$ K⁻¹. The thermal resistances of the condensing steam and tube wall may be neglected.

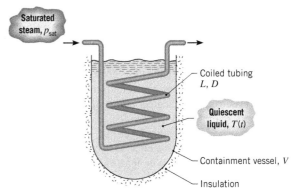

(a) What is the initial rate of heat transfer to the pharmaceutical?

(b) Neglecting heat transfer between the tank and its surroundings, how long does it take to heat the pharmaceutical to 70°C? Plot the corresponding variation with time of the fluid temperature and the convection coefficient at the outer surface of the tube. How much steam is condensed during the heating process?

9.59 A hot fluid at 35°C is to be transported through a tube horizontally positioned in quiescent air at 25°C. Which of the tube shapes, each of equal cross-sectional area, would you use in order to minimize heat losses to the ambient air by free convection?

(1)   (2)   (3)   (4)

Use the following correlation of Lienhard [*Int. J. Heat Mass Transfer,* **16**, 2121, 1973] to approximate the laminar convection coefficient for an immersed body on which the boundary layer does not separate from the surface,

$$\overline{Nu}_l = 0.52\, Ra_l^{1/4}$$

The characteristic length l is the length of travel of the fluid in the boundary layer across the shape surface.

9.60 Consider a 2-mm-diameter sphere immersed in a fluid at 300 K and 1 atm.

(a) If the fluid around the sphere is quiescent and extensive, show that the conduction limit of heat transfer from the sphere can be expressed as $Nu_{D,\text{cond}} = 2$. *Hint:* Begin with the expression for the thermal resistance of a hollow sphere, Equation 3.41, letting $r_2 \rightarrow \infty$, and then expressing the result in terms of the Nusselt number.

(b) Considering free convection, at what surface temperature will the Nusselt number be twice that for the conduction limit? Consider air and water as the fluids.

(c) Considering forced convection, at what velocity will the Nusselt number be twice that for the conduction limit? Consider air and water as the fluids.

9.61 A sphere of 25-mm diameter contains an embedded electrical heater. Calculate the power required to maintain the surface temperature at 94°C when the sphere is exposed to a quiescent medium at 20°C for: (a) air at atmospheric pressure, (b) water, and (c) ethylene glycol.

Parallel Plate Channels

9.62 Consider two long vertical plates maintained at uniform temperatures $T_{s,1} > T_{s,2}$. The plates are open at their ends and are separated by the distance $2L$.

(a) Sketch the velocity distribution in the space between the plates.

(b) Write appropriate forms of the continuity, momentum, and energy equations for laminar flow between the plates.

(c) Evaluate the temperature distribution, and express your result in terms of the mean temperature, $T_m = (T_{s,1} + T_{s,2})/2$.

(d) Estimate the vertical pressure gradient by assuming the density to be a constant ρ_m corresponding to T_m. Substituting from the Boussinesq approximation, obtain the resulting form of the momentum equation.

(e) Determine the velocity distribution.

9.63 Consider the conditions of Problem 9.9, but now view the problem as one involving free convection between vertical, parallel plate channels. What is the optimum fin spacing S? For this spacing and the prescribed values of t and W, what is the rate of heat transfer from the fins?

9.64 A vertical array of circuit boards is immersed in quiescent ambient air at $T_\infty = 20$°C. Although the components protrude from their substrates, it is reasonable, as a first approximation, to assume *flat* plates with uniform surface heat flux q_s''. Consider boards of length and width $L = W = 0.4$ m and spacing $S = 25$ mm. If the maximum allowable board temperature is 80°C, what is the maximum allowable power dissipation per board?

9.65 Determined to reduce the $7 per week cost associated with heat loss through their patio window by convection and radiation, the tenants of Problem 9.17 cover the inside of the window with a 50-mm-thick sheet of extruded insulation. Because they are not very handy around the house, the insulation is installed poorly, resulting in an $S = 5$-mm gap between the extruded insulation and the window pane, allowing the room air to infiltrate into the space between the pane and the insulation.

(a) Determine the rate of heat loss through the window and associated weekly cost with the ill-fitting insulation in place. The insulation will significantly reduce the radiation losses through the window. Losses will be due almost entirely to convection.

(b) Plot the rate of heat loss through the patio window as a function of the gap spacing for 1 mm ≤ S ≤ 20 mm.

9.66 The front door of a dishwasher of width 580 mm has a vertical air vent that is 500 mm in height with a 20-mm spacing between the inner tub operating at 55°C and an outer plate that is thermally insulated.

(a) Determine the rate of heat loss from the tub surface when the ambient air is 30°C.

(b) A change in the design of the door provides the opportunity to increase or decrease the 20-mm spacing by 10 mm. What recommendations would you offer with regard to how the change in spacing will alter heat losses?

9.67 A natural convection air heater consists of an array of parallel, equally spaced vertical plates, which may be maintained at a fixed temperature T_s by embedded electrical heaters. The plates are of length and width $L = W = 300$ mm and are in quiescent, atmospheric air at $T_\infty = 25$°C. The total width of the array cannot exceed a value of $W_{ar} = 150$ mm.

For $T_s = 80$°C, what is the plate spacing S that maximizes heat transfer from the array? For this spacing, how many plates comprise the array and what is the corresponding rate of heat transfer from the array?

9.68 A bank of drying ovens is mounted on a rack in a room with an ambient air temperature of 27°C. The cubical ovens are 500 mm to a side, and the spacing between the ovens is 15 mm.

(a) Estimate the rate of heat loss from a facing side of an oven when its surface temperature is 47°C.

(b) Explore the effect of the spacing on the rate of heat loss. At what spacing is the rate of heat loss a maximum? Describe the boundary layer behavior for this condition. Can this condition be analyzed by treating the side of an oven as an isolated vertical plate?

9.69 A solar collector consists of a parallel plate channel that is connected to a water storage plenum at the bottom and to a heat sink at the top. The channel is inclined $\theta = 30°$ from the vertical and has a transparent cover plate. Solar radiation transmitted through the cover plate and the water maintains the isothermal absorber plate at a temperature $T_s = 67$°C, while water returned to the reservoir from the heat sink is at $T_\infty = 27$°C. The system operates as a *thermosyphon,* for which water flow is driven exclusively by buoyancy forces. The plate spacing and length are $S = 15$ mm and $L = 1.5$ m.

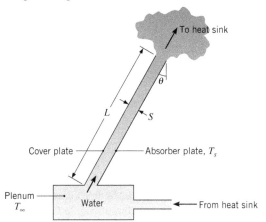

Assuming the cover plate to be adiabatic with respect to convection heat transfer to or from the water, estimate the rate of heat transfer per unit width normal to the flow direction (W/m) from the absorber plate to the water.

Rectangular Cavities

9.70 As is evident from the property data of Tables A.3 and A.4, the thermal conductivity of glass at room temperature is more than 50 times larger than that of air. It is therefore desirable to use windows of double-pane construction, for which the two panes of glass enclose an air space. If heat transfer across the air space is by conduction, the corresponding thermal resistance may be increased by increasing the thickness L of the space. However, there are limits to the efficacy of such a measure, since convection currents are induced if L exceeds a critical value, beyond which the thermal resistance decreases.

Consider atmospheric air enclosed by vertical panes at temperatures of $T_1 = 22°C$ and $T_2 = -20°C$. If the critical Rayleigh number for the onset of convection is $Ra_L \approx 2000$, what is the maximum allowable spacing for conduction across the air? How is this spacing affected by the temperatures of the panes? How is it affected by the pressure of the air, as, for example, by partial evacuation of the space?

9.71 A building window pane that is 1.2 m high and 0.9 m wide is separated from the ambient air by a storm window of the same height and width. The air space between the two windows is 0.06 m thick. If the building and storm windows are at 20 and $-10°C$, respectively, what is the rate of heat loss by free convection across the air space?

9.72 To reduce heat losses, a horizontal rectangular duct that is $W = 0.80$ m wide and $H = 0.3$ m high is encased in a metal radiation shield. The duct wall and shield are separated by an air gap of thickness $t = 0.06$ m. For a duct wall temperature of $T_d = 45°C$ and a shield temperature of $T_{sh} = 25°C$, determine the rate of convection heat loss per unit length from the duct.

9.73 A rectangular cavity consists of two parallel, 0.5-m-square plates separated by a distance of 50 mm, with the lateral boundaries insulated. The heated plate is maintained at 325 K and the cooled plate at 275 K. Estimate the heat flux between the surfaces for three orientations of the cavity using the notation of Figure 9.6: vertical with $\tau = 90°$, horizontal with $\tau = 0°$, and horizontal with $\tau = 180°$.

9.74 Consider a thermal storage system in which the phase change material (paraffin) is housed in a large container whose bottom, horizontal surface is maintained at $T_s = 50°C$ by warm water delivered from a solar collector.

(a) Neglecting the change in sensible energy of the liquid phase, estimate the amount of paraffin that is melted over a five-hour period beginning with an initial liquid layer at the bottom of the container of thickness $s_i = 10$ mm. The paraffin of Problem 8.38 is used as the phase change material and is initially at the phase change temperature, $T_{mp} = 27.4°C$. The bottom surface area of the container is $A = 2.5$ m^2.

(b) Compare the amount of energy needed to melt the paraffin to the amount of energy required to increase the temperature of the same amount of liquid from the phase change temperature to the average liquid temperature, $(T_s + T_{mp})/2$.

(c) Neglecting the change in sensible energy of the liquid phase, estimate the amount of paraffin that would melt over a five-hour time period if the hot plate is placed at the top of the container and $s_i = 10$ mm.

9.75 Consider a horizontal flat roof section having the same dimensions as a vertical wall section. For both sections, the surfaces exposed to the air gap are at 18°C (inside) and $-10°C$ (outside).

(a) Estimate the ratio of the convection heat rate for the horizontal section to that of the vertical section.

(b) What effect will inserting a baffle at the mid-height of the vertical section have on the convection heat rate for that section?

9.76 The space between the panes of a double-glazed window can be filled with either air or carbon dioxide at atmospheric pressure. The window is 1.5 m high and the spacing between the panes can be varied. Develop an analysis to predict the convection heat transfer rate across the window as a function of pane spacing and determine, under otherwise identical conditions, whether air or carbon dioxide will yield the smaller rate. Illustrate the results of your analysis for two surface-temperature conditions: winter (−10°C, 20°C) and summer (35°C, 25°C).

9.77 The top surface (0.5 m × 0.5 m) of an oven is 60°C for a particular operating condition when the room air is 23°C. To reduce heat loss from the oven and to minimize burn hazard, it is proposed to create a 50-mm air space by adding a cover plate.

(a) Assuming the same oven surface temperature T_s for both situations, estimate the reduction in the convection heat loss resulting from installation of the cover plate. What is the temperature of the cover plate?

(b) Explore the effect of the cover plate spacing on the convection heat loss and the cover plate temperature for spacings in the range $5 \leq L \leq 50$ mm. Is there an optimum spacing?

9.78 A solar water heater consists of a flat-plate collector that is coupled to a storage tank. The collector consists of a transparent cover plate and an absorber plate that are separated by an air gap.

Although much of the solar energy collected by the absorber plate is transferred to a working fluid passing through a coiled tube brazed to the back of the absorber, some of the energy is lost by free convection and net radiation transfer across the air gap. In Chapter 13, we will evaluate the contribution of radiation exchange to this loss. For now, we restrict our attention to the free convection effect.

(a) Consider a collector that is inclined at an angle of $\tau = 60°$ and has dimensions of $H = w = 2$ m on a side, with an air gap of $L = 30$ mm. If the absorber and cover plates are at $T_1 = 70°C$ and $T_2 = 30°C$, respectively, what is the rate of heat transfer by free convection from the absorber plate?

(b) The heat loss by free convection depends on the spacing between the plates. Compute and plot the rate of heat loss as a function of spacing for $5 \leq L \leq 50$ mm. Is there an optimum spacing?

Concentric Cylinders and Spheres

9.79 Consider the cylindrical, 0.12-m-diameter radiation shield of Example 9.5 that is installed concentric with a 0.10-m-diameter tube carrying steam. The spacing provides an air gap of $L = 10$ mm.

(a) Calculate the rate of heat loss per unit length of the tube by convection when a second shield of 0.14-m diameter is installed, with the second shield maintained at 35°C. Compare the result to that for the single shield of the example.

(b) In the two-shield configuration of part (a), the air gaps formed by the annular concentric tubes are $L = 10$ mm. Calculate the rate of heat loss per unit length if the gap dimension is $L = 15$ mm. Do you expect the heat loss to increase or decrease?

9.80 The effective thermal conductivity k_{eff} for concentric cylinders and concentric spheres is provided in Equations 9.59 and 9.62, respectively. Derive expressions for the critical Rayleigh numbers associated with the cylindrical and spherical geometries, $Ra_{c,\text{crit}}$ and $Ra_{s,\text{crit}}$, respectively, below which k_{eff} is minimized. Evaluate $Ra_{c,\text{crit}}$ and $Ra_{s,\text{crit}}$ for air, water, and glycerin at a mean temperature of 300 K. For specified inner and outer surface temperatures and inner cylinder or sphere radii, comment on the heat transfer rate for outer cylinder or sphere radii corresponding to $Ra_{c,\text{crit}}$ and $Ra_{s,\text{crit}}$, respectively.

9.81 A solar collector design consists of an inner tube enclosed concentrically in an outer tube that is transparent to solar radiation. The tubes are thin walled with inner and outer diameters of 0.10 and 0.15 m, respectively. The annular space between the tubes is completely enclosed and filled with air at atmospheric pressure. Under operating conditions for which the inner and outer tube surface temperatures are 70 and 30°C, respectively, what is the convective heat loss per meter of tube length across the air space?

9.82 Graph the heat loss per unit length from the solar collector of Problem 9.81 over the range $0.1 \leq D_o \leq 0.25$ m assuming (i) conduction heat transfer across the annular space and (ii) convection heat transfer across the annular space. Determine the outer diameter that yields the minimum heat loss across the air space per unit collector length, and the value of this loss.

9.83 A proposed method to reduce heat losses from a horizontal, isothermal cylinder placed within a large room is to encase it within a larger cylinder, as shown in the schematic, with all surfaces painted with a low emissivity coating.

Air at atmospheric pressure exists within the annular region. For a concentrically located inner cylinder of surface temperature T_i = 70°C and radius r_i = 20 mm, and for an ambient temperature of

T_∞ = 30°C, determine the optimal radius of the outer cylinder r_o that will minimize the convection heat loss. Compare the convection heat loss from the inner cylinder with the optimally sized outer cylinder in place to the heat loss without the outer cylinder. Is the approach effective?

9.84 The surfaces of two long, horizontal, concentric thin-walled tubes having radii of 100 and 125 mm are maintained at 300 and 400 K, respectively. If the annular space is pressurized with nitrogen at 5 atm, estimate the convection heat transfer rate per unit length of the tubes.

9.85 Liquid nitrogen is stored in a thin-walled spherical vessel of diameter D_i = 1 m. The vessel is positioned concentrically within a larger, thin-walled spherical container of diameter D_o = 1.10 m, and the intervening cavity is filled with atmospheric helium.

Under normal operating conditions, the inner and outer surface temperatures are T_i = 77 K and T_o = 283 K. If the latent heat of vaporization of nitrogen is 2×10^5 J/kg, what is the mass rate $\dot{m}$ (kg/s) at which gaseous nitrogen is vented from the system?

9.86 The human eye contains aqueous humor, which separates the external cornea and the internal iris–lens structure. It is hypothesized that, in some individuals, small flakes of pigment are intermittently liberated from the iris and migrate to, and subsequently damage, the cornea. Approximating the geometry of the enclosure formed by the cornea and iris–lens structure as a pair of concentric hemispheres of outer radius r_o = 10 mm and inner radius r_i = 7 mm, respectively, investigate whether free convection can occur in the aqueous humor by evaluating the effective thermal conductivity ratio, k_{eff}/k. If free convection can occur, it is possible that the damaging particles are advected from the iris to the cornea. The iris–lens structure is at the core temperature, T_i = 37°C, while the cornea temperature is measured to be T_o = 34°C. The properties of the aqueous humor are ρ = 990 kg/m³, k = 0.58 W/m · K, c_p = 4.2×10^3 J/kg · K, μ = 7.1×10^{-4} N · s/m², and β = 3.2×10^{-4} K⁻¹.

Mixed Convection

9.87 A horizontal, 25-mm diameter cylinder is maintained at a uniform surface temperature of 35°C. A fluid with a velocity of 0.05 m/s and temperature of 20°C is in cross flow over the cylinder. Determine whether heat transfer by free convection will be significant for (i) air, (ii) water, (iii) engine oil, and (iv) mercury.

9.88 According to experimental results for parallel air-flow over a uniform temperature, heated vertical plate, the effect of free convection on the heat transfer convection coefficient will be 5% when $Gr_L/Re_L^2 = 0.08$. Consider a heated vertical plate 0.3 m long, maintained at a surface temperature of 60°C in atmospheric air at 25°C. What is the minimum vertical velocity required of the airflow such that free convection effects will be less than 5% of the heat transfer rate?

9.89 A vertical array of circuit boards of 150-mm height is to be air cooled such that the board temperature does not exceed 60°C when the ambient temperature is 25°C.

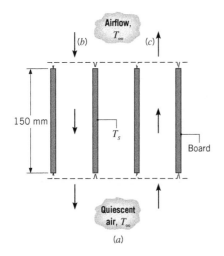

Assuming isothermal surface conditions, determine the allowable electrical power dissipation per board for the cooling arrangements:

(a) Free convection only (no forced airflow).

(b) Airflow with a downward velocity of 0.6 m/s.

(c) Airflow with an upward velocity of 0.3 m/s.

(d) Airflow with a velocity (upward or downward) of 5 m/s.

9.90 A horizontal, 25-mm-diameter cylinder is maintained at a uniform surface temperature of 35°C. Air at atmospheric pressure with a velocity of 0.10 m/s and temperature 20°C is in cross flow over the cylinder. Calculate the convection heat rate per unit cylinder length. How might the cylinder be maintained at the warm temperature if its surface is adiabatic?

9.91 A probe, used to measure the velocity of air in a low-speed wind tunnel, is fabricated of an $L = 100$ mm long, $D = 8$-mm outside diameter horizontal aluminum tube. Power resistors are inserted into the stationary tube and dissipate $P = 1.5$ W. The surface temperature of the tube is determined experimentally by measuring the emitted radiation from the exterior of the tube. To maximize surface emission, the exterior of the tube is painted with flat black paint having an emissivity of $\varepsilon = 0.95$.

(a) For air at a temperature and cross flow velocity of $T_\infty = 25°C$, $V = 0.1$ m/s, respectively, determine the surface temperature of the tube. The surroundings temperature is $T_{sur} = 25°C$.

(b) For the conditions of part (a), plot the tube surface temperature versus the cross flow velocity over the range 0.05 m/s $\leq V \leq 1$ m/s.

9.92 Square panels (250 mm × 250 mm) with a decorative, highly reflective plastic finish are cured in an oven at 125°C and cooled in quiescent air at 29°C. Quality considerations dictate that the panels remain horizontal and that the cooling rate be controlled. To increase productivity in the plant, it is proposed to replace the batch cooling method with a conveyor system having a velocity of 0.5 m/s.

Compare the initial (immediately after leaving the oven) convection heat transfer rates for the two methods.

9.93 An experiment involves heating a very small sphere that is suspended by a fine string in air with a laser beam in order to induce the highest sphere temperature possible. After inspecting Equation 9.64, a research assistant suggests inducing a uniform downward airflow to *exactly* offset free convection from the sphere, thereby minimizing heat losses and maximizing the steady-state sphere temperature. In the limiting case of a *very small* sphere, what is the minimum value of the convection heat transfer coefficient expressed in terms of the sphere diameter and thermal conductivity of the air?

Mass Transfer

9.94 A garment soaked with water is hung up to dry in a warm room at atmospheric pressure. The still air is dry and at a temperature of 40°C. The garment may be assumed to have a temperature of 25°C and a characteristic length of 1m in the vertical direction. Estimate the drying rate per unit width of the garment.

9.95 A 205 mm × 245 mm sheet of wetted fabric is hung up to dry on a warm, sunny day. The still air is at a temperature of 30°C and relative humidity of 40%. To maximize the drying rate, should the sheet be hung with its long dimension or short dimension in the vertical direction? Determine the maximum drying rate when the sheet temperature is 26°C.

9.96 A water bath is used to maintain canisters containing experimental biological reactions at a uniform temperature of 37°C. The top of the bath has a width and length of 0.25 m and 0.50 m, respectively, and is uncovered to allow easy access for removal or insertion of the canisters. The bath is located in a draft-free laboratory with air at atmospheric pressure, a temperature of 20°C, and a relative humidity of 60%. The walls of the laboratory are at a uniform temperature of 25°C.

(a) Estimate the rate of heat loss from the surface of the bath by radiation exchange with the surroundings.

(b) Calculate the Grashof number using Equation 9.65, which can be applied to natural convection flows driven by temperature and concentration gradients. Use a characteristic length L that is appropriate for the exposed surface of the water bath.

(c) Estimate the free convection heat transfer coefficient using the result for Gr_L obtained in part (b).

(d) Invoke the heat and mass transfer analogy and use an appropriate correlation to estimate the mass transfer coefficient using Gr_L. Calculate the water evaporation rate on a daily basis and the rate of heat loss by evaporation.

(e) Calculate the total rate of heat loss from the surface, and compare the relative contributions of the sensible, latent, and radiative effects. Review the assumptions made in your analysis, especially those relating to the heat and mass transfer analogy.

9.97 Fuel cells similar to the PEM cell of Example 1.5 operate with a mixture of liquid water and methanol instead of hydrogen; the anode is placed in direct contact with the liquid fuel. Oxygen (species A) is delivered to the exposed cathode by free convection. Hence, no fans or pumps are needed to operate the device. The power output of *passive, direct methanol fuel cells* (DMFCs) can become *mass transfer limited*, since the electric current produced by the DMFC is related to the rate at which oxygen is consumed at the cathode by the expression $I = 4n_A \mathcal{F} \mathcal{M}_A$, where $\mathcal{F}$ is Faraday's constant, $\mathcal{F} = 96{,}489$ coulombs/mol. Consider a passive DMFC with a 120 mm × 120 mm membrane. Determine the maximum possible electric current produced by the DMFC when the oxygen mass fraction at the cathode is $m_{A,s} = 0.10$ for cases where the cathode is facing up or is vertical. As a first approximation and to illustrate the sensitivity of the device to its orientation relative to the vertical direction, assume buoyancy forces are dominated by the difference in density associated with the change in the oxygen mass fraction between the cathode surface and the quiescent environment, which is atmospheric air at $T_\infty = 25$°C. Assume the quiescent air is composed of nitrogen and oxygen, with an oxygen mass fraction $m_{A,\infty} = 0.233$.

Quiescent air
$T_\infty = 20$°C
$\phi_\infty = 0.60$

Room surroundings
$T_{sur} = 25$°C

Water, 37°C

Thermostated water bath

Experiment canister

Electrical heating elements

Boiling and Condensation

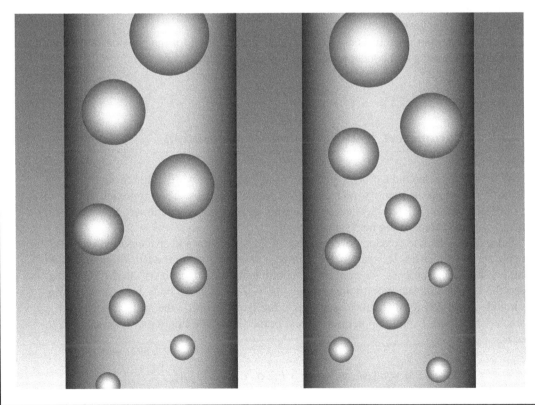

In this chapter we focus on convection processes associated with the change in phase of a fluid. In particular, we consider processes that can occur at a solid–liquid or solid–vapor interface, namely, *boiling* and *condensation*. For these cases *latent* heat effects associated with the phase change are significant. The change from the liquid to the vapor state due to boiling is sustained by heat transfer from the solid surface; conversely, condensation of a vapor to the liquid state results in heat transfer to the solid surface.

Since they involve fluid motion, boiling and condensation are classified as forms of the convection mode of heat transfer. However, they are characterized by unique features. Because there is a phase change, heat transfer to or from the fluid can occur without influencing the fluid temperature. In fact, through boiling or condensation, large heat transfer rates may be achieved with small temperature differences. In addition to the *latent heat* h_{fg}, two other parameters are important in characterizing the processes, namely, the *surface tension* σ at the liquid–vapor interface and the *density difference* between the two phases. This difference induces a *buoyancy force*, which is proportional to $g(\rho_l - \rho_v)$. Because of combined latent heat and buoyancy-driven flow effects, boiling and condensation heat transfer coefficients and rates are generally much larger than those characteristic of convection heat transfer without phase change.

Many engineering applications that are characterized by high heat fluxes involve boiling and condensation. In a closed-loop power cycle, pressurized liquid is converted to vapor in a *boiler*. After expansion in a turbine, the vapor is restored to its liquid state in a *condenser*, whereupon it is pumped to the boiler to repeat the cycle. *Evaporators*, in which the boiling process occurs, and condensers are also essential components in vapor-compression refrigeration cycles. The high heat transfer coefficients associated with boiling make it attractive to consider for purposes of managing the thermal performance of advanced electronics equipment. The rational design of such components dictates that the associated phase change processes be well understood.

In this chapter our objectives are to develop an appreciation for the physical conditions associated with boiling and condensation and to provide a basis for performing related heat transfer calculations.

10.1 *Dimensionless Parameters in Boiling and Condensation*

In our treatment of boundary layer phenomena (Chapter 6), we nondimensionalized the governing equations to identify relevant dimensionless groups. This approach enhanced our understanding of related physical mechanisms and suggested simplified procedures for generalizing and representing heat transfer results.

Since it is difficult to develop governing equations for boiling and condensation processes, the appropriate dimensionless parameters can be obtained by using the Buckingham pi theorem [1]. For either process, the convection coefficient could depend on the difference between the surface and saturation temperatures, $\Delta T = |T_s - T_{\text{sat}}|$, the body force arising from the liquid–vapor density difference, $g(\rho_l - \rho_v)$, the latent heat h_{fg}, the surface tension σ, a characteristic length L, and the thermophysical properties of the liquid or vapor: ρ, c_p, k, μ. That is,

$$h = h\,[\Delta T, g(\rho_l - \rho_v), h_{fg}, \sigma, L, \rho, c_p, k, \mu] \tag{10.1}$$

Since there are 10 variables in 5 dimensions (m, kg, s, J, K), there are $(10 - 5) = 5$ pi-groups, which can be expressed in the following forms:

$$\frac{hL}{k} = f\left[\frac{\rho g(\rho_l - \rho_v)L^3}{\mu^2}, \frac{c_p \Delta T}{h_{fg}}, \frac{\mu c_p}{k}, \frac{g(\rho_l - \rho_v)L^2}{\sigma}\right] \qquad (10.2a)$$

or, defining the dimensionless groups,

$$Nu_L = f\left[\frac{\rho g(\rho_l - \rho_v)L^3}{\mu^2}, Ja, Pr, Bo\right] \qquad (10.2b)$$

The Nusselt and Prandtl numbers are familiar from our earlier single-phase convection analyses. The new dimensionless parameters are the Jakob number Ja, the Bond number Bo, and a nameless parameter that bears a strong resemblance to the Grashof number (see Equation 9.10 and recall that $\beta\Delta T \approx \Delta\rho/\rho$). This unnamed parameter represents the effect of buoyancy-induced fluid motion on heat transfer. The Jakob number is the ratio of the maximum sensible energy absorbed by the liquid (vapor) to the latent energy absorbed by the liquid (vapor) during condensation (boiling). In many applications, the sensible energy is much less than the latent energy and Ja has a small numerical value. The Bond number is the ratio of the buoyancy force to the surface tension force. In subsequent sections, we will delineate the role of these parameters in boiling and condensation.

10.2 *Boiling Modes*

When evaporation occurs at a solid–liquid interface, it is termed *boiling*. The process occurs when the temperature of the surface T_s exceeds the saturation temperature T_{sat} corresponding to the liquid pressure. Heat is transferred from the solid surface to the liquid, and the appropriate form of Newton's law of cooling is

$$q_s'' = h(T_s - T_{sat}) = h\,\Delta T_e \qquad (10.3)$$

where $\Delta T_e \equiv T_s - T_{sat}$ is termed the *excess temperature*. The process is characterized by the formation of vapor bubbles, which grow and subsequently detach from the surface. Vapor bubble growth and dynamics depend, in a complicated manner, on the excess temperature, the nature of the surface, and thermophysical properties of the fluid, such as its surface tension. In turn, the dynamics of vapor bubble formation affect liquid motion near the surface and therefore strongly influence the heat transfer coefficient.

Boiling may occur under various conditions. For example, in *pool boiling* the liquid is quiescent and its motion near the surface is due to free convection and to mixing induced by bubble growth and detachment. In contrast, for *forced convection boiling,* fluid motion is induced by external means, as well as by free convection and bubble-induced mixing. Boiling may also be classified according to whether it is *subcooled* or *saturated.* In subcooled boiling, the temperature of most of the liquid is below the saturation temperature and bubbles formed at the surface may condense in the liquid. In contrast, the temperature of the liquid slightly exceeds the saturation temperature in *saturated boiling*. Bubbles formed at the surface are then propelled through the liquid by buoyancy forces, eventually escaping from a free surface.

10.3 *Pool Boiling*

Saturated pool boiling, as shown in Figure 10.1, has been studied extensively. Although there is a sharp increase in the liquid temperature close to the solid surface, the temperature through most of the liquid remains slightly above saturation. Bubbles generated at the liquid–solid interface rise to the liquid–vapor interface, where the vapor is ultimately transported across the interface. An appreciation for the underlying physical mechanisms may be obtained by examining the *boiling curve.*

10.3.1 The Boiling Curve

Nukiyama [2] was the first to identify different regimes of pool boiling using the apparatus of Figure 10.2. The heat flux from a horizontal *nichrome* wire to saturated water was determined by measuring the current flow I and potential drop E. The wire temperature was determined from knowledge of the manner in which its electrical resistance varied with temperature. This arrangement is termed *power-controlled* heating, wherein the wire temperature T_s (hence the excess temperature ΔT_e) is the dependent variable and the power setting (hence the heat flux q_s'') is the independent variable. Following the arrows of the *heating curve* of Figure 10.3, it is evident that as power is applied, the heat flux increases, at first slowly and then very rapidly, with excess temperature.

Nukiyama observed that boiling, as evidenced by the presence of bubbles, did not begin until $\Delta T_e \approx 5°C$. With further increase in power, the heat flux increased to very high levels until *suddenly,* for a value slightly larger than q_{max}'', the wire temperature jumped to the melting point and burnout occurred. However, repeating the experiment with a *platinum* wire having a higher melting point (2045 K vs. 1500 K), Nukiyama was able to maintain heat fluxes above q_{max}'' without burnout. When he subsequently reduced the power, the variation of ΔT_e with q_s'' followed the *cooling curve* of Figure 10.3. When the heat flux reached the minimum point q_{min}'', a further decrease in power caused the excess temperature to drop abruptly, and the process followed the original heating curve back to the saturation point.

Nukiyama believed that the hysteresis effect of Figure 10.3 was a consequence of the power-controlled method of heating, where ΔT_e is a dependent variable. He also believed that by using a heating process permitting the independent control of ΔT_e, the missing (dashed) portion of the curve could be obtained. His conjecture was subsequently confirmed by Drew and Mueller [3]. By condensing steam inside a tube at

Figure 10.1 Temperature distribution in saturated pool boiling with a liquid–vapor interface.

FIGURE 10.2   Nukiyama's power-controlled heating apparatus for demonstrating the boiling curve.

different pressures, they were able to control the value of ΔT_e for boiling of a low boiling point organic fluid at the tube outer surface and thereby obtain the missing portion of the boiling curve.

10.3.2   Modes of Pool Boiling

An appreciation for the underlying physical mechanisms may be obtained by examining the different modes, or regimes, of pool boiling. These regimes are identified in the boiling curve of Figure 10.4. The specific curve pertains to water at 1 atm, although similar trends characterize the behavior of other fluids. From Equation 10.3 we note that q''_s depends on the convection coefficient h, as well as on the excess temperature ΔT_e. Different boiling regimes may be delineated according to the value of ΔT_e.

Free Convection Boiling   Free convection boiling is said to exist if $\Delta T_e \leq \Delta T_{e,A}$, where $\Delta T_{e,A} \approx 5°C$. The surface temperature must be somewhat above the saturation temperature in order to sustain bubble formation. As the excess temperature is increased, bubble inception will eventually occur, but below point A (referred to as the *onset of nucleate boiling*, ONB), fluid motion is determined principally by free convection effects. According to

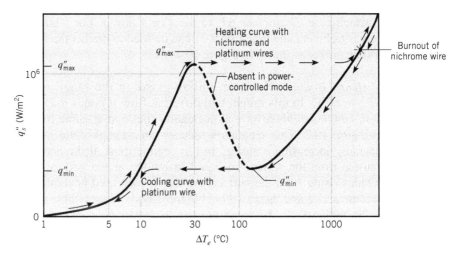

FIGURE 10.3   Nukiyama's boiling curve for saturated water at atmospheric pressure.

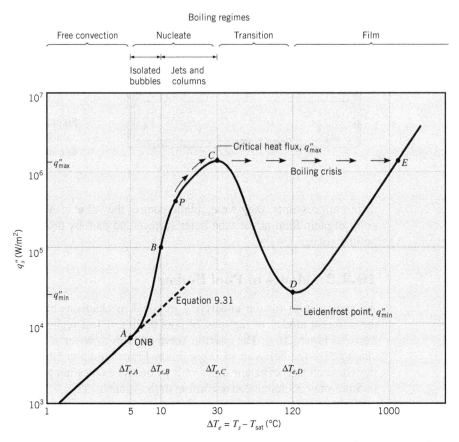

FIGURE 10.4   Typical boiling curve for water at 1 atm: surface heat flux q_s'' as a function of excess temperature, $\Delta T_e \equiv T_s - T_{sat}$.

whether the flow is laminar or turbulent, h varies as ΔT_e to the $\frac{1}{4}$ or $\frac{1}{3}$ power, respectively, in which case q_s'' varies as ΔT_e to the $\frac{5}{4}$ or $\frac{4}{3}$ power. For a large horizontal plate, the fluid flow is turbulent and Equation 9.31 can be used to predict the free convection portion of the boiling curve, as shown in Figure 10.4.

Nucleate Boiling   Nucleate boiling exists in the range $\Delta T_{e,A} \leq \Delta T_e \leq \Delta T_{e,C}$, where $\Delta T_{e,C} \approx 30°C$. In this range, two different flow regimes may be distinguished. In region *A–B, isolated bubbles* form at nucleation sites and separate from the surface, as illustrated in Figure 10.2. This separation induces considerable fluid mixing near the surface, substantially increasing h and q_s''. In this regime most of the heat exchange is through direct transfer from the surface to liquid in motion at the surface, and not through the vapor bubbles rising from the surface. As ΔT_e is increased beyond $\Delta T_{e,B}$, more nucleation sites become active and increased bubble formation causes bubble interference and coalescence. In the region *B–C,* the vapor escapes as *jets* or *columns,* which subsequently merge into slugs of the vapor. This condition is illustrated in Figure 10.5*a.* Interference between the densely populated bubbles inhibits the motion of liquid near the surface. Point *P* of Figure 10.4 represents a change in the behavior of the boiling curve. Before point *P,* the

(a)

(b)

FIGURE 10.5   Boiling of methanol on a horizontal tube. (a) Nucleate boiling in the jets and columns regime. (b) Film boiling. (Photographs courtesy of Professor J. W. Westwater, University of Illinois at Champaign-Urbana.)

boiling curve can be approximated as a straight line on a log–log plot, meaning that $q_s'' \propto \Delta T_e^n$. Beyond this point, the heat flux increases more slowly as ΔT_e is increased. At some point between P and C, the decaying increase of the heat flux leads to a reduction of the heat transfer coefficient $h = q_s''/\Delta T_e$. The maximum heat flux, $q_{s,C}'' = q_{max}''$, is usually termed the *critical heat flux,* and in water at atmospheric pressure it exceeds 1 MW/m². At the point of this maximum, considerable vapor is being formed, making it difficult for liquid to continuously wet the surface.

Because high heat transfer rates and convection coefficients are associated with small values of the excess temperature, it is desirable to operate many engineering devices in the nucleate boiling regime. The approximate magnitude of the convection coefficient may be inferred by using Equation 10.3 with the boiling curve of Figure 10.4. Dividing q_s'' by ΔT_e, it is evident that convection coefficients in excess of 10^4 W/m² · K are characteristic of this regime. These values are considerably larger than those normally corresponding to convection with no phase change.

Transition Boiling   The region corresponding to $\Delta T_{e,C} \leq \Delta T_e \leq \Delta T_{e,D}$, where $\Delta T_{e,D} \approx$ 120°C, is termed *transition boiling, unstable film boiling,* or *partial film boiling.* Bubble formation is now so rapid that a vapor film or blanket begins to form on the surface. At any point on the surface, conditions may oscillate between film and nucleate boiling, but the fraction of the total surface covered by the film increases with increasing ΔT_e. Because the thermal conductivity of the vapor is much less than that of the liquid, h (and q_s'') must decrease with increasing ΔT_e.

Film Boiling   Film boiling exists for $\Delta T_e \geq \Delta T_{e,D}$. At point D of the boiling curve, referred to as the *Leidenfrost point*,[1] the heat flux is a minimum, $q_{s,D}'' = q_{\min}''$, and the surface is completely covered by a *vapor blanket*. Heat transfer from the surface to the liquid occurs by conduction and radiation through the vapor. As the surface temperature is increased, radiation through the vapor film becomes more significant and the heat flux increases with increasing ΔT_e. Figure 10.5*b* illustrates the nature of the vapor formation and bubble dynamics associated with film boiling. The photographs of Figure 10.5 were obtained for the boiling of methanol on a horizontal tube.

Although the foregoing discussion of the boiling curve assumes that control may be maintained over T_s, it is important to remember the Nukiyama experiment and be mindful of the many applications that involve controlling q_s'' (e.g., in a nuclear reactor or in an electric resistance heater) rather than ΔT_e. Consider starting at point P in Figure 10.4 and gradually increasing q_s''. The value of ΔT_e, and hence the value of T_s, will also increase, following the boiling curve to point C. However, any increase in q_s'' beyond point C will induce a sharp increase from $\Delta T_{e,C} \approx 30°C$ to $\Delta T_{e,E} \equiv T_{s,E} - T_{\text{sat}} \approx 1100°C$. Because $T_{s,E}$ may exceed the melting point of the solid, system failure may occur. For this reason point C is often termed the *burnout point* or the *boiling crisis,* and accurate knowledge of the critical heat flux (CHF), $q_{s,C}'' \equiv q_{\max}''$, is important. Although we may want to operate a heat transfer surface close to the CHF, we would rarely want to exceed it.

10.4 *Pool Boiling Correlations*

From the shape of the boiling curve and the fact that various physical mechanisms characterize the different regimes, it is no surprise that a multiplicity of heat transfer correlations exist for the boiling process. For the region below $\Delta T_{e,A}$ of the boiling curve (Figure 10.4), appropriate free convection correlations from Chapter 9 can be used to estimate heat transfer coefficients and heat rates. In this section we review some of the more widely used correlations for nucleate and film boiling.

10.4.1 Nucleate Pool Boiling

The analysis of nucleate boiling requires prediction of the number of surface nucleation sites and the rate at which bubbles originate from each site. While mechanisms associated with this boiling regime have been studied extensively, complete and reliable mathematical models have yet to be developed. Yamagata et al. [4] were the first to show the influence of nucleation sites on the heat rate and to demonstrate that q_s'' is approximately proportional to ΔT_e^3. It is desirable to develop correlations that reflect this relationship between the surface heat flux and the excess temperature.

---

[1]It was Leidenfrost who in 1756 observed that water droplets supported by the vapor film slowly boil away as they move about a hot surface.

In Section 10.3.2 we noted that within region *A-B* of Figure 10.4, most of the heat exchange is due to direct transfer from the hot surface to the liquid. Hence, the boiling phenomena in this region may be thought of as a type of liquid phase forced convection in which the fluid motion is induced by the rising bubbles. We have seen that forced convection correlations are generally of the form

$$\overline{Nu}_L = C_{fc}\, Re_L^{m_{fc}}\, Pr^{n_{fc}} \tag{7.1}$$

and Equation 7.1 may provide insight into how pool boiling data can be correlated, provided that a length scale and a characteristic velocity can be identified for inclusion in the Nusselt and Reynolds numbers. The subscript *fc* is added to the constants that appear in Equation 7.1 to remind us that they apply to this *forced convection* expression. As we saw in Chapter 7, these constants are determined experimentally for complicated flows. Because it is postulated that the rising bubbles mix the liquid, an appropriate length scale for relatively large heater surfaces is the bubble diameter, D_b. The diameter of the bubble upon its departure from the hot surface may be determined from a force balance in which the buoyancy force (which promotes bubble departure and is proportional to D_b^3) is equal to the surface tension force (which adheres the bubble to the surface and is proportional to D_b), resulting in the expression

$$D_b \propto \sqrt{\frac{\sigma}{g(\rho_l - \rho_v)}} \tag{10.4a}$$

The constant of proportionality depends on the angle of contact between the liquid, its vapor, and the solid surface; the contact angle depends on the particular liquid and solid surface that is considered. The subscripts *l* and *v* denote the saturated liquid and vapor states, respectively, and σ (N/m) is the surface tension.

A characteristic velocity for the agitation of the liquid may be found by dividing the distance the liquid travels to fill in behind a departing bubble (proportional to D_b) by the time between bubble departures, t_b. The time t_b is equal to the energy it takes to form a vapor bubble (proportional to D_b^3), divided by the rate at which heat is added over the solid–vapor contact area (proportional to D_b^2). Thus,

$$V \propto \frac{D_b}{t_b} \propto \frac{D_b}{\left(\dfrac{\rho_l h_{fg} D_b^3}{q_s'' D_b^2}\right)} \propto \frac{q_s''}{\rho_l h_{fg}} \tag{10.4b}$$

Substituting Equations 10.4a and 10.4b into Equation 7.1, absorbing the proportionalities into the constant C_{fc}, and substituting the resulting expression for *h* into Equation 10.3 provides the following expression, where the constants $C_{s,f}$ and *n* are newly introduced and the exponent m_{fc} in Equation 7.1 has an experimentally determined value of 2/3:

$$q_s'' = \mu_l h_{fg} \left[\frac{g(\rho_l - \rho_v)}{\sigma}\right]^{1/2} \left(\frac{c_{p,l}\, \Delta T_e}{C_{s,f}\, h_{fg}\, Pr_l^n}\right)^3 \tag{10.5}$$

Equation 10.5 was developed by Rohsenow [5] and is the first and most widely used correlation for nucleate boiling. All properties are for the liquid, except for ρ_v, and all should be evaluated at T_{sat}. The coefficient $C_{s,f}$ and the exponent *n* depend on the solid–fluid combination, and representative experimentally determined values are presented in Table 10.1. Values for other surface–fluid combinations may be obtained from the literature [6–8].

TABLE 10.1   Values of $C_{s,f}$ for various surface–fluid combinations [5–7]

Surface–Fluid Combination	$C_{s,f}$	n
Water–copper		
Scored	0.0068	1.0
Polished	0.0128	1.0
Water–stainless steel		
Chemically etched	0.0133	1.0
Mechanically polished	0.0132	1.0
Ground and polished	0.0080	1.0
Water–brass	0.0060	1.0
Water–nickel	0.006	1.0
Water–platinum	0.0130	1.0
n-Pentane–copper		
Polished	0.0154	1.7
Lapped	0.0049	1.7
Benzene–chromium	0.0101	1.7
Ethyl alcohol–chromium	0.0027	1.7

Values of the surface tension and the latent heat of vaporization of water are presented in Table A.6 and for selected fluids in Table A.5. Values for additional fluids may be obtained from any recent edition of the *Handbook of Chemistry and Physics*. If Equation 10.5 is rewritten in terms of a Nusselt number based on an arbitrary length scale L, it will be in the form $Nu_L \propto Ja^2 \, Pr^{1-3n} \, Bo^{1/2}$. Comparing with Equation 10.2b, we see that only the first dimensionless parameter does not appear. If the Nusselt number is based on the characteristic bubble diameter given in Equation 10.4a, the expression reduces to the simpler form $Nu_{D_b} \propto Ja^2 \, Pr^{1-3n}$.

The Rohsenow correlation applies only for clean surfaces. When it is used to estimate the heat flux, errors can amount to $\pm 100\%$. However, since $\Delta T_e \propto (q_s'')^{1/3}$, this error is reduced by a factor of 3 when the expression is used to estimate ΔT_e from knowledge of q_s''. Also, since $q_s'' \propto h_{fg}^{-2}$ and h_{fg} decreases with increasing saturation pressure (temperature), the nucleate boiling heat flux will increase as the liquid is pressurized.

10.4.2   Critical Heat Flux for Nucleate Pool Boiling

We recognize that the critical heat flux, $q_{s,C}'' = q_{max}''$, represents an important point on the boiling curve. We may wish to operate a boiling process close to this point, but we appreciate the danger of dissipating heat in excess of this amount. Kutateladze [9], through dimensional analysis, and Zuber [10], through a hydrodynamic stability analysis, obtained an expression which can be approximated as

$$q_{max}'' = C h_{fg} \rho_v \left[\frac{\sigma g (\rho_l - \rho_v)}{\rho_v^2} \right]^{1/4} \tag{10.6}$$

which is independent of surface material and is weakly dependent upon the hot surface geometry through the leading constant, C. For large horizontal cylinders, for spheres, and for many large finite surfaces, use of a leading constant with the value $C = \pi/24 \approx 0.131$ (the Zuber constant) agrees with experimental data to within 16% [11]. For large horizontal plates, a value of $C = 0.149$ gives better agreement with experimental data. The properties in Equation 10.6 are evaluated at the saturation temperature. Equation 10.6 applies when the characteristic length of the hot surface, L, is large relative to the bubble diameter, D_b. However, when the hot surface is small, such that the Confinement number, $Co = \sqrt{\sigma/(g[\rho_l - \rho_v])}/L = Bo^{-1/2}$ [12], is greater than approximately 0.2, a correction factor must be applied to account for the small size of the surface. Lienhard [11] reports correction factors for various geometries, including horizontal plates, cylinders, spheres, and vertically and horizontally oriented ribbons.

It is important to note that the critical heat flux depends strongly on pressure, mainly through the pressure dependence of surface tension and the heat of vaporization. Cichelli and Bonilla [13] have experimentally demonstrated that the peak flux increases with pressure up to one-third of the critical pressure, after which it falls to zero at the critical pressure.

10.4.3 Minimum Heat Flux

The transition boiling regime is of little practical interest, as it may be obtained only by controlling the surface temperature. While no adequate theory has been developed for this regime, conditions can be characterized by periodic, *unstable* contact between the liquid and the hot surface. However, the upper limit of this regime is of interest because it corresponds to formation of a *stable* vapor blanket or film and to a minimum heat flux condition. If the heat flux drops below this minimum, the film will collapse, causing the surface to cool and nucleate boiling to be reestablished.

Zuber [10] used stability theory to derive the following expression for the minimum heat flux, $q''_{s,D} = q''_{\min}$, from a large horizontal plate.

$$q''_{\min} = C\rho_v h_{fg}\left[\frac{g\sigma(\rho_l - \rho_v)}{(\rho_l + \rho_v)^2}\right]^{1/4} \tag{10.7}$$

where the properties are evaluated at the saturation temperature. The constant, $C = 0.09$, has been experimentally determined by Berenson [14]. This result is accurate to approximately 50% for most fluids at moderate pressures but provides poorer estimates at higher pressures [15]. A similar result has been obtained for horizontal cylinders [16].

10.4.4 Film Pool Boiling

At excess temperatures beyond the Leidenfrost point, a continuous vapor film blankets the surface and there is no contact between the liquid phase and the surface. Because conditions in the stable vapor film bear a strong resemblance to those of laminar film condensation (Section 10.7), it is customary to base film boiling correlations on results obtained from condensation theory. One such result, which applies to film boiling on a cylinder or sphere of diameter D, is of the form

$$\overline{Nu}_D = \frac{\overline{h}_{\mathrm{conv}}D}{k_v} = C\left[\frac{g(\rho_l - \rho_v)h'_{fg}D^3}{\nu_v k_v(T_s - T_{\mathrm{sat}})}\right]^{1/4} \tag{10.8}$$

The correlation constant C is 0.62 for horizontal cylinders [17] and 0.67 for spheres [11]. The corrected latent heat h'_{fg} accounts for the sensible energy required to maintain temperatures within the vapor blanket above the saturation temperature. Although it may be approximated as $h'_{fg} = h_{fg} + 0.80c_{p,v}(T_s - T_{sat})$, it is known to depend weakly on the Prandtl number of the vapor [18]. Vapor properties are evaluated at the system pressure and the film temperature, $T_f = (T_s + T_{sat})/2$, whereas ρ_l and h_{fg} are evaluated at the saturation temperature.

At elevated surface temperatures ($T_s \gtrsim 300°C$), radiation heat transfer across the vapor film can become significant. Since radiation acts to increase the film thickness, it is not reasonable to assume that the radiative and convective processes are simply additive. Bromley [17] investigated film boiling from the outer surface of horizontal tubes and suggested calculating the total heat transfer coefficient from a transcendental equation of the form

$$\bar{h}^{4/3} = \bar{h}_{conv}^{4/3} + \bar{h}_{rad}\bar{h}^{1/3} \tag{10.9}$$

If $\bar{h}_{rad} < \bar{h}_{conv}$, a simpler form may be used:

$$\bar{h} = \bar{h}_{conv} + \tfrac{3}{4}\bar{h}_{rad} \tag{10.10}$$

The effective radiation coefficient $\bar{h}_{rad}$ is expressed as

$$\bar{h}_{rad} = \frac{\varepsilon\sigma(T_s^4 - T_{sat}^4)}{T_s - T_{sat}} \tag{10.11}$$

where ε is the emissivity of the solid (Table A.11) and σ is the Stefan–Boltzmann constant.

Note that the analogy between film boiling and film condensation does not hold for small surfaces with high curvature because of the large disparity between vapor and liquid film thicknesses for the two processes. The analogy is also questionable for a vertical surface, although satisfactory predictions have been obtained for limited conditions.

10.4.5 Parametric Effects on Pool Boiling

In this section we briefly consider other parameters that can affect pool boiling, confining our attention to the gravitational field, liquid subcooling, and solid surface conditions.

The influence of the *gravitational field* on boiling must be considered in applications involving space travel and rotating machinery. This influence is evident from appearance of the gravitational acceleration g in the foregoing expressions. Siegel [19], in his review of low gravity effects, confirms that the $g^{1/4}$ dependence in Equations 10.6, 10.7, and 10.8 (for the maximum and minimum heat fluxes and for film boiling) is correct for values of g as low as 0.10 m/s². For nucleate boiling, however, evidence indicates that the heat flux is nearly independent of gravity, which is in contrast to the $g^{1/2}$ dependence of Equation 10.5. Above-normal gravitational forces show similar effects, although near the ONB, gravity can influence bubble-induced convection.

If liquid in a pool boiling system is maintained at a temperature that is less than the saturation temperature, the liquid is said to be *subcooled*, where $\Delta T_{sub} \equiv T_{sat} - T_l$. In the natural convection regime, the heat flux increases typically as $(T_s - T_l)^n$ or $(\Delta T_e + \Delta T_{sub})^n$, where $5/4 \leq n \leq 4/3$ depending on the geometry of the hot surface. In contrast, for nucleate boiling, the influence of subcooling is considered to be negligible, although the maximum

FIGURE 10.6 Formation of nucleation sites. (*a*) Wetted cavity with no trapped vapor. (*b*) Reentrant cavity with trapped vapor. (*c*) Enlarged profile of a roughened surface.

and minimum heat fluxes, q''_{max} and q''_{min}, are known to increase linearly with ΔT_{sub}. For film boiling, the heat flux increases strongly with increasing ΔT_{sub}.

The influence of *surface roughness* (by machining, grooving, scoring, or sandblasting) on the maximum and minimum heat fluxes and on film boiling is negligible. However, as demonstrated by Berenson [20], increased surface roughness can cause a large increase in heat flux for the nucleate boiling regime. As Figure 10.6 illustrates, a roughened surface has numerous cavities that serve to trap vapor, providing more and larger sites for bubble growth. It follows that the nucleation site density for a rough surface can be substantially larger than that for a smooth surface. However, under prolonged boiling, the effects of surface roughness generally diminish, indicating that the new, large sites created by roughening are not stable sources of vapor entrapment.

Special surface arrangements that provide stable *augmentation* (*enhancement*) of nucleate boiling are available commercially and have been reviewed by Webb [21]. *Enhancement surfaces* are of two types: (1) coatings of very porous material formed by sintering, brazing, flame spraying, electrolytic deposition, or foaming, and (2) mechanically machined or formed double-reentrant cavities to ensure continuous vapor trapping (see Figure 10.7). Such surfaces provide for continuous renewal of vapor at the nucleation sites and heat transfer augmentation by more than an order of magnitude. Active augmentation techniques, such as surface wiping–rotation, surface vibration, fluid vibration, and electrostatic fields, have also been reviewed by Bergles [22, 23]. However, because such techniques complicate the boiling system and, in many instances, impair reliability, they have found little practical application.

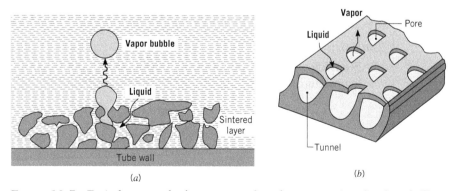

FIGURE 10.7 Typical structured enhancement surfaces for augmentation of nucleate boiling. (*a*) Sintered metallic coating. (*b*) Mechanically formed double-reentrant cavity.

EXAMPLE 10.1

The bottom of a copper pan, 0.3 m in diameter, is maintained at 118°C by an electric heater. Estimate the power required to boil water in this pan. What is the evaporation rate? Estimate the critical heat flux.

SOLUTION

Known: Water boiling in a copper pan of prescribed surface temperature.

Find:

1. Power required by electric heater to cause boiling.
2. Rate of water evaporation due to boiling.
3. Critical heat flux corresponding to the burnout point.

Schematic:

Assumptions:

1. Steady-state conditions.
2. Water exposed to standard atmospheric pressure, 1.01 bar.
3. Water at uniform temperature $T_{\text{sat}} = 100°C$.
4. Large pan bottom surface of polished copper.
5. Negligible losses from heater and pan to surroundings.

Properties: Table A.6, saturated water, liquid (100°C): $\rho_l = 1/v_f = 957.9$ kg/m³, $c_{p,l} = c_{p,f} = 4.217$ kJ/kg · K, $\mu_l = \mu_f = 279 \times 10^{-6}$ N · s/m², $Pr_l = Pr_f = 1.76$, $h_{fg} = 2257$ kJ/kg, $\sigma = 58.9 \times 10^{-3}$ N/m. Table A.6, saturated water, vapor (100°C): $\rho_v = 1/v_g = 0.5956$ kg/m³.

Analysis:

1. From knowledge of the saturation temperature T_{sat} of water boiling at 1 atm and the temperature of the copper surface T_s, the excess temperature ΔT_e is

$$\Delta T_e \equiv T_s - T_{\text{sat}} = 118°C - 100°C = 18°C$$

According to the boiling curve of Figure 10.4, nucleate pool boiling will occur and the recommended correlation for estimating the surface heat flux is given by Equation 10.5.

$$q_s'' = \mu_l h_{fg} \left[\frac{g(\rho_l - \rho_v)}{\sigma} \right]^{1/2} \left(\frac{c_{p,l}\,\Delta T_e}{C_{s,f}\, h_{fg}\, Pr_l^n} \right)^3$$

The values of $C_{s,f}$ and n corresponding to the polished copper surface–water combination are determined from the experimental results of Table 10.1, where $C_{s,f} = 0.0128$ and $n = 1.0$. Substituting numerical values, the boiling heat flux is

$$q_s'' = 279 \times 10^{-6} \text{ N} \cdot \text{s/m}^2 \times 2257 \times 10^3 \text{ J/kg}$$

$$\times \left[\frac{9.8 \text{ m/s}^2 \, (957.9 - 0.5956) \text{ kg/m}^3}{58.9 \times 10^{-3} \text{ N/m}} \right]^{1/2}$$

$$\times \left(\frac{4.217 \times 10^3 \text{ J/kg} \cdot \text{K} \times 18°\text{C}}{0.0128 \times 2257 \times 10^3 \text{ J/kg} \times 1.76} \right)^3 = 836 \text{ kW/m}^2$$

Hence the boiling heat transfer rate is

$$q_s = q_s'' \times A = q_s'' \times \frac{\pi D^2}{4}$$

$$q_s = 8.36 \times 10^5 \text{ W/m}^2 \times \frac{\pi (0.30 \text{ m})^2}{4} = 59.1 \text{ kW} \qquad \triangleleft$$

2. Neglecting losses to the surroundings, all heat addition to the pan will result in water evaporation from the pan. Hence

$$q_s = \dot{m}_b h_{fg}$$

where $\dot{m}_b$ is the rate at which water evaporates from the free surface to the room. It follows that

$$\dot{m}_b = \frac{q_s}{h_{fg}} = \frac{5.91 \times 10^4 \text{ W}}{2257 \times 10^3 \text{ J/kg}} = 0.0262 \text{ kg/s} = 94 \text{ kg/h} \qquad \triangleleft$$

3. The critical heat flux for nucleate pool boiling can be estimated from Equation 10.6:

$$q_{max}'' = 0.149 h_{fg} \rho_v \left[\frac{\sigma g (\rho_l - \rho_v)}{\rho_v^2} \right]^{1/4}$$

Substituting the appropriate numerical values,

$$q_{max}'' = 0.149 \times 2257 \times 10^3 \text{ J/kg} \times 0.5956 \text{ kg/m}^3$$

$$\times \left[\frac{58.9 \times 10^{-3} \text{ N/m} \times 9.8 \text{ m/s}^2 \, (957.9 - 0.5956) \text{ kg/m}^3}{(0.5956 \text{ kg/m}^3)^2} \right]^{1/4}$$

$$q_{max}'' = 1.26 \text{ MW/m}^2 \qquad \triangleleft$$

Comments:

1. Note that the critical heat flux $q_{max}'' = 1.26 \text{ MW/m}^2$ represents the maximum heat flux for nucleate boiling of water at normal atmospheric pressure. Operation of the heater at $q_s'' = 0.836 \text{ MW/m}^2$ is therefore below the critical condition.

2. Using Equation 10.7, the minimum heat flux at the Leidenfrost point is $q_{min}'' = 18.9 \text{ kW/m}^2$. Note from Figure 10.4 that, for this condition, $\Delta T_e \approx 120°\text{C}$.

EXAMPLE 10.2

A metal-clad heating element of 6-mm diameter and emissivity $\varepsilon = 1$ is horizontally immersed in a water bath. The surface temperature of the metal is 255°C under steady-state boiling conditions. Estimate the power dissipation per unit length of the heater.

SOLUTION

Known: Boiling from outer surface of horizontal cylinder in water.

Find: Power dissipation per unit length for the cylinder, q_s'.

Schematic:

Ambient air
$p = 1$ atm

Water
$T_{sat} = 100°C$ Electrical heater
$D = 6$ mm

$T_s = 255°C$

Assumptions:

1. Steady-state conditions.
2. Water exposed to standard atmospheric pressure and at uniform temperature T_{sat}.

Properties: Table A.6, saturated water, liquid (100°C): $\rho_l = 1/v_f = 957.9$ kg/m³, $h_{fg} = 2257$ kJ/kg. Table A.4, water vapor at atmospheric pressure ($T_f \approx 450$ K): $\rho_v = 0.4902$ kg/m³, $c_{p,v} = 1.980$ kJ/kg · K, $k_v = 0.0299$ W/m · K, $\mu_v = 15.25 \times 10^{-6}$ N · s/m².

Analysis: The excess temperature is

$$\Delta T_e = T_s - T_{sat} = 255°C - 100°C = 155°C$$

According to the boiling curve of Figure 10.4, film pool boiling conditions are achieved, in which case heat transfer is due to both convection and radiation. The heat transfer rate follows from Equation 10.3, written on a per unit length basis for a cylindrical surface of diameter D:

$$q_s' = q_s'' \pi D = \bar{h} \pi D \, \Delta T_e$$

The heat transfer coefficient $\bar{h}$ is calculated from Equation 10.9,

$$\bar{h}^{4/3} = \bar{h}_{conv}^{4/3} + \bar{h}_{rad} \bar{h}^{1/3}$$

where the convection and radiation heat transfer coefficients follow from Equations 10.8 and 10.11, respectively. For the convection coefficient:

$$\bar{h}_{\text{conv}} = 0.62 \left[\frac{k_v^3 \rho_v (\rho_l - \rho_v) g (h_{fg} + 0.8 c_{p,v} \Delta T_e)}{\mu_v D \, \Delta T_e} \right]^{1/4}$$

$$\bar{h}_{\text{conv}} = 0.62$$

$$\times \left[\frac{(0.0299 \text{ W/m} \cdot \text{K})^3 \times 0.4902 \text{ kg/m}^3 \, (957.9 - 0.4902) \text{ kg/m}^3 \times 9.8 \text{ m/s}^2}{1} \right.$$

$$\left. \times \frac{(2257 \times 10^3 \text{ J/kg} + 0.8 \times 1.98 \times 10^3 \text{ J/kg} \cdot \text{K} \times 155°\text{C})}{15.25 \times 10^{-6} \text{ N} \cdot \text{s/m}^2 \times 6 \times 10^{-3} \text{ m} \times 155°\text{C}} \right]^{1/4}$$

$$\bar{h}_{\text{conv}} = 238 \text{ W/m}^2 \cdot \text{K}$$

For the radiation heat transfer coefficient:

$$\bar{h}_{\text{rad}} = \frac{\varepsilon \sigma (T_s^4 - T_{\text{sat}}^4)}{T_s - T_{\text{sat}}}$$

$$\bar{h}_{\text{rad}} = \frac{5.67 \times 10^{-8} \text{ W/m}^2 \cdot \text{K}^4 (528^4 - 373^4) \text{K}^4}{(528 - 373) \text{ K}} = 21.3 \text{ W/m}^2 \cdot \text{K}$$

Solving Equation 10.9 by trial and error,

$$\bar{h}^{4/3} = 238^{4/3} + 21.3 \bar{h}^{1/3}$$

it follows that

$$\bar{h} = 254.1 \text{ W/m}^2 \cdot \text{K}$$

Hence the heat transfer rate per unit length of heater element is

$$q_s' = 254.1 \text{ W/m}^2 \cdot \text{K} \times \pi \times 6 \times 10^{-3} \text{ m} \times 155°\text{C} = 742 \text{ W/m} \qquad \triangleleft$$

Comments: Equation 10.10 is also appropriate for estimating $\bar{h}$; it provides a value of $254.0 \text{ W/m}^2 \cdot \text{K}$.

10.5 *Forced Convection Boiling*

In *pool boiling*, fluid flow is due primarily to the buoyancy-driven motion of bubbles originating from the hot surface. In contrast, for *forced convection boiling,* flow is due to a directed (bulk) motion of the fluid, as well as to buoyancy effects. Conditions depend strongly on geometry, which may involve *external* flow over hot plates and cylinders or *internal* (duct) flow. Internal, forced convection boiling is commonly referred to as *two-phase flow* and is characterized by rapid changes from liquid to vapor in the flow direction.

10.5.1   External Forced Convection Boiling

For external flow over a hot plate, the heat flux can be estimated by standard forced convection correlations up to the inception of boiling. As the temperature of the plate is increased, nucleate boiling will occur, causing the heat flux to increase. If vapor generation is not extensive and the liquid is subcooled, Bergles and Rohsenow [24] suggest a method for estimating the total heat flux in terms of components associated with pure forced convection and pool boiling.

Both forced convection and subcooling are known to increase the critical heat flux q''_{max} for nucleate boiling. Experimental values as high as 35 MW/m^2 (compared with 1.3 MW/m^2 for pool boiling of water at 1 atm) have been reported [25]. For a liquid of velocity V moving in cross flow over a cylinder of diameter D, Lienhard and Eichhorn [26] have developed the following expressions for low- and high-velocity flows, where properties are evaluated at the saturation temperature.

Low Velocity:

$$\frac{q''_{max}}{\rho_v h_{fg} V} = \frac{1}{\pi}\left[1 + \left(\frac{4}{We_D}\right)^{1/3}\right] \tag{10.12}$$

High Velocity:

$$\frac{q''_{max}}{\rho_v h_{fg} V} = \frac{(\rho_l/\rho_v)^{3/4}}{169\pi} + \frac{(\rho_l/\rho_v)^{1/2}}{19.2\pi\, We_D^{1/3}} \tag{10.13}$$

The Weber number We_D is the ratio of inertia to surface tension forces and has the form

$$We_D \equiv \frac{\rho_v V^2 D}{\sigma} \tag{10.14}$$

The high- and low-velocity regions, respectively, are determined by whether the heat flux parameter $q''_{max}/\rho_v h_{fg} V$ is less than or greater than $[(0.275/\pi)\,(\rho_l/\rho_v)^{1/2} + 1]$. In most cases, Equations 10.12 and 10.13 correlate q''_{max} data within 20%.

10.5.2   Two-Phase Flow

Internal forced convection boiling is associated with bubble formation at the inner surface of a hot tube through which a liquid is flowing. Bubble growth and separation are strongly influenced by the flow velocity, and hydrodynamic effects differ significantly from those corresponding to pool boiling. The process is accompanied by the existence of a variety of two-phase flow patterns.

Consider flow development in a vertical tube that is subjected to a constant surface heat flux, through which fluid is moving in the upward direction, as shown in Figure 10.8. Heat transfer to the subcooled liquid that enters the tube is initially by *single-phase forced convection* and may be predicted using the correlations of Chapter 8. Farther down the tube, the wall temperature exceeds the saturation temperature of the liquid, and vaporization is initiated in the *subcooled flow boiling region*. This region is characterized by large radial temperature gradients, with bubbles forming adjacent to the hot wall and subcooled liquid flowing near the center of the tube. The thickness of the bubble region increases farther

FIGURE 10.8   Flow regimes for forced convection boiling in a tube.

downstream, and eventually, the core of the liquid reaches the saturation temperature of the fluid. Bubbles can then exist at any radial location, and the time-averaged mass fraction of vapor in the fluid,[2] X, exceeds zero at any radial location. This marks the beginning of the *saturated flow boiling region*. Within the saturated flow boiling region, the mean vapor mass fraction defined as

$$\overline{X} \equiv \frac{\int_{A_c} \rho u(r,x) X dA_c}{\dot{m}}$$

increases and, due to the large density difference between the vapor and liquid phases, the mean velocity of the fluid, u_m, increases substantially.

  The first stage of the saturated flow boiling region corresponds to the *bubbly flow regime*. As $\overline{X}$ increases further, individual bubbles coalesce to form slugs of vapor. This *slug-flow regime* is followed by an *annular-flow regime* in which the liquid forms a film on the tube wall. This film moves along the inner surface of the tube, while vapor moves at a larger velocity through the core of the tube. Dry spots eventually appear on the inner surface

---

[2]This term is often referred to as the *quality* of a two-phase fluid.

of the tube and grow in size within a *transition regime*. Eventually, the entire tube surface is completely dry, and all remaining liquid is in the form of droplets that travel at high velocity within the core of the tube in the *mist regime*. After the droplets are completely vaporized, the fluid consists of superheated vapor in a *second* single-phase forced convection region. The increase in the vapor fraction along the tube length, along with the large difference in the densities of the liquid and vapor phases, increases the mean velocity of the fluid by several orders of magnitude between the first and the second single-phase forced convection regions.

The local heat transfer coefficient varies significantly as $\bar{X}$ and u_m decrease and increase, respectively, along the length of the tube, x. In general, the heat transfer coefficient can increase by approximately an order of magnitude through the subcooled flow boiling region. Heat transfer coefficients are further increased in the early stages of the saturated flow boiling region. Conditions become more complex deeper in the saturated flow boiling region since the convection coefficient, defined in Equation 10.3, *either* increases or decreases with increasing $\bar{X}$, depending on the fluid and tube wall material. Typically, the smallest convection coefficients exist in the second (vapor) forced convection region owing to the low thermal conductivity of the vapor relative to that of the liquid.

The following correlation has been developed for the saturated flow boiling region in smooth circular tubes [27, 28]:

$$\frac{h}{h_{sp}} = 0.6683\left(\frac{\rho_l}{\rho_v}\right)^{0.1} \bar{X}^{0.16}(1-\bar{X})^{0.64} f(Fr) + 1058\left(\frac{q_s''}{\dot{m}''h_{fg}}\right)^{0.7}(1-\bar{X})^{0.8}G_{s,f} \quad (10.15a)$$

or

$$\frac{h}{h_{sp}} = 1.136\left(\frac{\rho_l}{\rho_v}\right)^{0.45} \bar{X}^{0.72}(1-\bar{X})^{0.08} f(Fr) + 667.2\left(\frac{q_s''}{\dot{m}''h_{fg}}\right)^{0.7}(1-\bar{X})^{0.8}G_{s,f} \quad (10.15b)$$

$$0 < \bar{X} \lesssim 0.8$$

where $\dot{m}'' = \dot{m}/A_c$ is the mass flow rate per unit cross-sectional area. In utilizing Equation 10.15, the larger value of the heat transfer coefficient, h, should be used. In this expression, the liquid phase *Froude number* is $Fr = (\dot{m}''/\rho_l)^2/gD$ and the coefficient $G_{s,f}$ depends on the surface–fluid combination, with representative values given in Table 10.2. Equation 10.15 applies for horizontal as well as vertical tubes, where the *stratification parameter*, $f(Fr)$, accounts for stratification of the liquid and vapor phases that may occur for horizontal tubes. Its value is unity for vertical tubes and for horizontal tubes with $Fr \gtrsim 0.04$. For horizontal tubes with $Fr \lesssim 0.04$, $f(Fr) = 2.63\,Fr^{0.3}$. All properties are evaluated at the saturation temperature, T_{sat}. The single-phase convection coefficient, h_{sp}, is associated with the liquid forced convection region of Figure 10.8 and is obtained from Equation 8.62 with properties evaluated at T_{sat}. Because Equation 8.62 is for turbulent flow, it is recommended that Equation 10.15 not be applied to situations where the liquid single-phase convection is laminar. Equation 10.15 is applicable when the channel dimension is large relative to the bubble diameter, that is, for Confinement numbers, $Co = \sqrt{\sigma/(g[\rho_l - \rho_v])}/D_h \lesssim 1/2$ [3].

In order to use Equation 10.15, the mean vapor mass fraction, $\bar{X}$, must be known. For negligible changes in the fluid's kinetic and potential energy as well as negligible work, Equation 1.12d may be rearranged to yield

$$\bar{X}(x) = \frac{q_s''\pi Dx}{\dot{m}h_{fg}} \quad (10.16)$$

TABLE 10.2   Values of $G_{s,f}$ for various
surface–fluid combinations [27, 28]

Fluid in Commercial Copper Tubing	$G_{s,f}$
Kerosene	0.488
Refrigerant R-134a	1.63
Refrigerant R-152a	1.10
Water	1.00
For stainless steel tubing, use $G_{s,f} = 1$.	

where the origin of the x-coordinate, $x = 0$, corresponds to the axial location where $\overline{X}$ begins to exceed zero, and the change in enthalpy, $u_t + pv$, is equal to the change in $\overline{X}$ multiplied by the enthalpy of vaporization, h_{fg}.

Correlations for the subcooled flow boiling region and annular as well as mist regimes are available in the literature [28]. For constant heat flux conditions, critical heat fluxes may occur in the subcooled flow boiling region, in the saturated flow boiling region where $\overline{X}$ is large, or in the vapor forced convection region. Critical heat flux conditions may lead to melting of the tube material in extreme conditions [29]. Additional discussions of flow boiling are available in the literature [7, 30–33]. Extensive databases consisting of thousands of experimentally measured values of the critical heat flux for wide ranges of operating conditions are also available [34, 35].

10.5.3 Two-Phase Flow in Microchannels

Two-phase microchannels feature forced convection boiling of a liquid through circular or non-circular tubes having hydraulic diameters ranging from 10 to 1000 μm, resulting in extremely high heat transfer rates [36, 37]. In these situations, the characteristic bubble size can occupy a significant fraction of the tube diameter and the Confinement number can become very large ($Co \gtrsim 1/2$). Hence, different types of flow regimes exist, including regimes where the bubbles occupy nearly the full diameter of the tube [38]. This can lead to a dramatic increase in the convection coefficient, h, corresponding to the peak in Figure 10.8. Thereafter, h decreases with increasing x as it does in Figure 10.8. Equation 10.15 cannot be used to predict correct values of the heat transfer coefficient and does not even predict correct trends for microchannel flow boiling cases. Recourse must be made to more sophisticated modeling [36, 39].

10.6   *Condensation: Physical Mechanisms*

Condensation occurs when the temperature of a vapor is reduced below its saturation temperature. In industrial equipment, the process commonly results from contact between the vapor and a cool *surface* (Figures 10.9a, b). The latent energy of the vapor is released, heat is transferred to the surface, and the condensate is formed. Other common modes are *homogeneous* condensation (Figure 10.9c), where vapor condenses out as droplets suspended in a gas phase to form a fog, and *direct contact* condensation (Figure 10.9d), which occurs when vapor is brought into contact with a cold liquid. In this chapter we will consider only surface condensation.

As shown in Figures 10.9a, b, condensation may occur in one of two ways, depending on the condition of the surface. The dominant form of condensation is one in which a liquid film

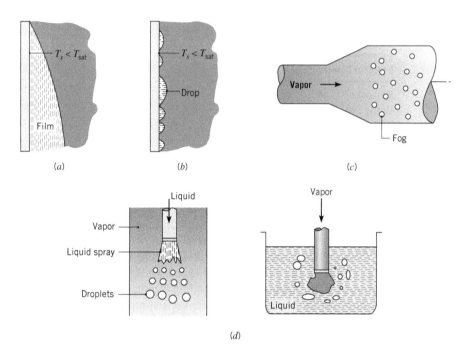

FIGURE 10.9   Modes of condensation. (*a*) Film. (*b*) Dropwise condensation on a surface. (*c*) Homogeneous condensation or fog formation resulting from increased pressure due to expansion. (*d*) Direct contact condensation.

covers the entire condensing surface, and under the action of gravity the film flows continuously from the surface. *Film condensation* is generally characteristic of clean, uncontaminated surfaces. However, if the surface is coated with a substance that inhibits wetting, it is possible to maintain *dropwise condensation*. The drops form in cracks, pits, and cavities on the surface and may grow and coalesce through continued condensation. Typically, more than 90% of the surface is covered by drops, ranging from a few micrometers in diameter to agglomerations visible to the naked eye. The droplets flow from the surface due to the action of gravity. Film and dropwise condensation of steam on a vertical copper surface are shown in Figure 10.10. A thin coating of cupric oleate was applied to the left-hand portion of the surface to promote the dropwise condensation. A thermocouple probe of 1-mm diameter extends across the photograph.

Regardless of whether it is in the form of a film or droplets, the condensate provides a resistance to heat transfer between the vapor and the surface. Because this resistance increases with condensate thickness, which increases in the flow direction, it is desirable to use short vertical surfaces or horizontal cylinders in situations involving film condensation. Most condensers therefore consist of horizontal tube bundles through which a liquid coolant flows and around which the vapor to be condensed is circulated. In terms of maintaining high condensation and heat transfer rates, droplet formation is superior to film formation. In dropwise condensation most of the heat transfer is through drops of less than 100-μm diameter, and transfer rates that are more than an order of magnitude larger than those associated with film condensation may be achieved. It is therefore common practice to use surface coatings that inhibit wetting, and hence stimulate dropwise condensation. Silicones, Teflon, and an assortment of waxes and fatty acids are often used for this purpose. However, such coatings gradually lose their effectiveness due to oxidation, fouling, or outright removal, and film condensation eventually occurs.

(a)   (b)

FIGURE 10.10   Condensation on a vertical surface. (a) Dropwise. (b) Film. (Photograph courtesy of Professor J. W. Westwater, University of Illinois at Champaign-Urbana.)

Although it is desirable to achieve dropwise condensation in industrial applications, it is often difficult to maintain this condition. For this reason and because the convection coefficients for film condensation are smaller than those for the dropwise case, condenser design calculations are often based on the assumption of film condensation. In the remaining sections of this chapter, we focus on film condensation and mention only briefly results available for dropwise condensation.

10.7 *Laminar Film Condensation on a Vertical Plate*

As shown in Figure 10.11*a*, there may be several complicating features associated with film condensation. The film originates at the top of the plate and flows downward under the influence of gravity. The thickness δ and the condensate mass flow rate $\dot{m}$ increase with increasing x because of continuous condensation at the liquid–vapor interface, which is at T_{sat}. There is then heat transfer from this interface through the film to the surface, which is maintained at $T_s < T_{\text{sat}}$. In the most general case the vapor may be superheated ($T_{v,\infty} > T_{\text{sat}}$) and may be part of a mixture containing one or more noncondensable gases. Moreover, there exists a finite shear stress at the liquid–vapor interface, contributing to a velocity gradient in the vapor, as well as in the film [40, 41].

Despite the complexities associated with film condensation, useful results may be obtained by making assumptions that originated with an analysis by Nusselt [42].

1. Laminar flow and constant properties are assumed for the liquid film.

2. The gas is assumed to be a pure vapor and at a uniform temperature equal to T_{sat}. With no temperature gradient in the vapor, heat transfer to the liquid–vapor interface can occur only by condensation at the interface and not by conduction from the vapor.

3. The shear stress at the liquid–vapor interface is assumed to be negligible, in which case $\partial u/\partial y|_{y=\delta} = 0$. With this assumption and the foregoing assumption of a uniform vapor temperature, there is no need to consider the vapor velocity or thermal boundary layers shown in Figure 10.11*a*.

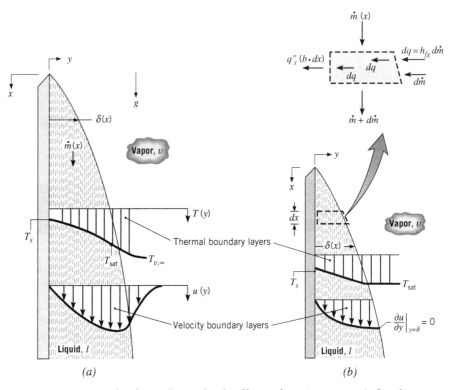

FIGURE 10.11   Boundary layer effects related to film condensation on a vertical surface. (*a*) Without approximation. (*b*) With assumptions associated with Nusselt's analysis, for a vertical plate of width *b*.

4. Momentum and energy transfer by advection in the condensate film are assumed to be negligible. This assumption is reasonable by virtue of the low velocities associated with the film. It follows that heat transfer across the film occurs only by conduction, in which case the liquid temperature distribution is linear.

Film conditions resulting from the assumptions are shown in Figure 10.11*b*.

The *x*-momentum equation for the film can be found from Equation 9.1, with $\rho = \rho_l$ and $\nu = \nu_l$ for the liquid, and with the sign of the gravity term changed since *x* is now in the direction of gravity. The pressure gradient is obtained under free stream conditions and is $dp_\infty/dx = +\rho_v g$, since the free stream density is the vapor density. From the fourth approximation, momentum advection terms may be neglected, and the *x*-momentum equation may be expressed as

$$\frac{\partial^2 u}{\partial y^2} = -\frac{g}{\mu_l}(\rho_l - \rho_v) \tag{10.17}$$

Integrating twice and applying boundary conditions of the form $u(0) = 0$ and $\partial u/\partial y\big|_{y=\delta} = 0$, the velocity profile in the film becomes

$$u(y) = \frac{g(\rho_l - \rho_v)\delta^2}{\mu_l}\left[\frac{y}{\delta} - \frac{1}{2}\left(\frac{y}{\delta}\right)^2\right] \tag{10.18}$$

From this result the condensate mass flow rate per unit width $\Gamma(x)$ may be obtained in terms of an integral involving the velocity profile:

$$\frac{\dot{m}(x)}{b} = \int_0^{\delta(x)} \rho_l u(y)\,dy \equiv \Gamma(x) \tag{10.19}$$

Substituting from Equation 10.18, it follows that

$$\Gamma(x) = \frac{g\rho_l(\rho_l - \rho_v)\delta^3}{3\mu_l} \tag{10.20}$$

The specific variation with x of δ, and hence of Γ, may be obtained by first applying the conservation of energy requirement to the differential element shown in Figure 10.11b. At a portion of the liquid–vapor interface of unit width and length dx, the rate of heat transfer into the film, dq, must equal the rate of energy release due to condensation at the interface. Hence

$$dq = h_{fg}\,d\dot{m} \tag{10.21}$$

Since advection is neglected, it also follows that the rate of heat transfer across the interface must equal the rate of heat transfer to the surface. Hence

$$dq = q_s''(b \cdot dx) \tag{10.22}$$

Since the liquid temperature distribution is linear, Fourier's law may be used to express the surface heat flux as

$$q_s'' = \frac{k_l(T_{\mathrm{sat}} - T_s)}{\delta} \tag{10.23}$$

Combining Equations 10.19 and 10.21 through 10.23, we then obtain

$$\frac{d\Gamma}{dx} = \frac{k_l(T_{\mathrm{sat}} - T_s)}{\delta h_{fg}} \tag{10.24}$$

Differentiating Equation 10.20, we also obtain

$$\frac{d\Gamma}{dx} = \frac{g\rho_l(\rho_l - \rho_v)\delta^2}{\mu_l}\frac{d\delta}{dx} \tag{10.25}$$

Combining Equations 10.24 and 10.25, it follows that

$$\delta^3 d\delta = \frac{k_l\mu_l(T_{\mathrm{sat}} - T_s)}{g\rho_l(\rho_l - \rho_v)h_{fg}}\,dx$$

Integrating from $x = 0$, where $\delta = 0$, to any x-location of interest on the surface,

$$\delta(x) = \left[\frac{4k_l\mu_l(T_{\mathrm{sat}} - T_s)x}{g\rho_l(\rho_l - \rho_v)h_{fg}}\right]^{1/4} \tag{10.26}$$

This result may then be substituted into Equation 10.20 to obtain $\Gamma(x)$.

An improvement to the foregoing result for $\delta(x)$ was made by Nusselt [42] and Rohsenow [43], who showed that, with the inclusion of thermal advection effects, a term is added to the latent heat of vaporization. In lieu of h_{fg}, Rohsenow recommended using a modified latent heat of the form $h'_{fg} = h_{fg} + 0.68 c_{p,l}(T_{\text{sat}} - T_s)$, or in terms of the Jakob number,

$$h'_{fg} = h_{fg}(1 + 0.68\,Ja) \tag{10.27}$$

More recently, Sadasivan and Lienhard [18] have shown that the modified latent heat depends weakly on the Prandtl number of the liquid.

The surface heat flux may be expressed as

$$q''_s = h_x(T_{\text{sat}} - T_s) \tag{10.28}$$

Substituting from Equation 10.23, the local convection coefficient is then

$$h_x = \frac{k_l}{\delta} \tag{10.29}$$

or, from Equation 10.26, with h_{fg} replaced by h'_{fg},

$$h_x = \left[\frac{g\rho_l(\rho_l - \rho_v)k_l^3 h'_{fg}}{4\mu_l(T_{\text{sat}} - T_s)x} \right]^{1/4} \tag{10.30}$$

Since h_x depends on $x^{-1/4}$, it follows that the average convection coefficient for the entire plate is

$$\overline{h}_L = \frac{1}{L}\int_0^L h_x\,dx = \frac{4}{3}h_L$$

or

$$\overline{h}_L = 0.943\left[\frac{g\rho_l(\rho_l - \rho_v)k_l^3 h'_{fg}}{\mu_l(T_{\text{sat}} - T_s)L} \right]^{1/4} \tag{10.31}$$

The average Nusselt number then has the form

$$\overline{Nu}_L = \frac{\overline{h}_L L}{k_l} = 0.943\left[\frac{\rho_l g(\rho_l - \rho_v)h'_{fg}L^3}{\mu_l k_l(T_{\text{sat}} - T_s)} \right]^{1/4} \tag{10.32}$$

In using this equation in conjunction with Equation 10.27, all liquid properties should be evaluated at the film temperature $T_f = (T_{\text{sat}} + T_s)/2$. The vapor density ρ_v and latent heat of vaporization h_{fg} should be evaluated at T_{sat}.

A more detailed boundary layer analysis of film condensation on a vertical plate has been performed by Sparrow and Gregg [40]. Their results, confirmed by Chen [44], indicate that errors associated with using Equation 10.32 are less than 3% for $Ja \leq 0.1$ and $1 \leq Pr \leq 100$. Dhir and Lienhard [45] have also shown that Equation 10.32 may be used for inclined plates, if g is replaced by $g \cdot \cos\theta$, where θ is the angle between the vertical and the surface. However, it must be used with caution for large values of θ and does not apply if $\theta = \pi/2$. The expression may be used for condensation on the inner or outer surface of a vertical tube of radius R, if $R \gg \delta$.

The total rate of heat transfer to the surface may be obtained by using Equation 10.31 with the following form of Newton's law of cooling:

$$q = \bar{h}_L A(T_{sat} - T_s) \tag{10.33}$$

The total condensation rate may then be determined from the relation

$$\dot{m} = \frac{q}{h'_{fg}} = \frac{\bar{h}_L A(T_{sat} - T_s)}{h'_{fg}} \tag{10.34}$$

Equations 10.33 and 10.34 are generally applicable to any surface geometry, although the form of $\bar{h}_L$ will vary according to geometry and flow conditions.

10.8 *Turbulent Film Condensation*

As for all previously discussed convection phenomena, turbulent flow conditions may exist in film condensation. Consider the vertical surface of Figure 10.12*a*. The transition criterion may be expressed in terms of a Reynolds number defined as

$$Re_\delta \equiv \frac{4\Gamma}{\mu_l} \tag{10.35}$$

With the condensate mass flow rate given by $\dot{m} = \rho_l u_m b \delta$, the Reynolds number may be expressed as

$$Re_\delta = \frac{4\dot{m}}{\mu_l b} = \frac{4\rho_l u_m \delta}{\mu_l} \tag{10.36}$$

where u_m is the average velocity in the film and δ, the film thickness, is the characteristic length. As in the case of single-phase boundary layers, the Reynolds number is an indicator of flow conditions. As shown in Figure 10.12*b*, for $Re_\delta \lesssim 30$, the film is laminar and wave free. For increased Re_δ, ripples or waves form on the condensate film, and at $Re_\delta \approx 1800$ the transition from laminar to turbulent flow is complete.

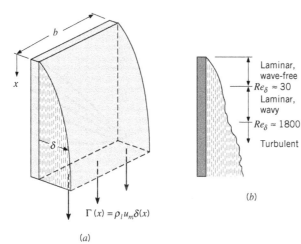

(a)

(b)

FIGURE 10.12 Film condensation on a vertical plate. (*a*) Condensate rate for plate of width *b*. (*b*) Flow regimes.

For the wave-free laminar regime ($Re_\delta \lesssim 30$), Equations 10.35 and 10.20 may be combined to yield

$$Re_\delta = \frac{4g\rho_l(\rho_l - \rho_v)\delta^3}{3\mu_l^2} \tag{10.37}$$

Assuming $\rho_l \gg \rho_v$, Equations 10.26, 10.31, and 10.37 may be combined to provide an expression for an average modified Nusselt number associated with condensation in the wave-free laminar regime:

$$\overline{Nu}_L = \frac{\overline{h}_L(v_l^2/g)^{1/3}}{k_l} = 1.47 Re_\delta^{-1/3} \qquad Re_\delta \lesssim 30 \tag{10.38}$$

where the average heat transfer coefficient $\overline{h}_L$ is associated with condensation over the entire plate. When the flow at the bottom of the plate is in the laminar, wavy regime, Kutateladze [46] recommends a correlation of the form

$$\overline{Nu}_L = \frac{\overline{h}_L(v_l^2/g)^{1/3}}{k_l} = \frac{Re_\delta}{1.08 Re_\delta^{1.22} - 5.2} \qquad 30 \lesssim Re_\delta \lesssim 1800 \tag{10.39}$$

and when the flow at the bottom of the plate is in the turbulent regime, Labuntsov [47] recommends

$$\overline{Nu}_L = \frac{\overline{h}_L(v_l^2/g)^{1/3}}{k_l} = \frac{Re_\delta}{8750 + 58 Pr_l^{-0.5}(Re_\delta^{0.75} - 253)} \qquad Re_\delta \gtrsim 1800, Pr_l \geq 1 \tag{10.40}$$

Graphical representation of the foregoing correlations is provided in Figure 10.13, and the trends have been verified experimentally by Gregorig et al. [48] for water over the range $1 < Re_\delta < 7200$. All properties are evaluated as for laminar film condensation, as explained beneath Equation 10.32.

The Reynolds number in Equations 10.38 through 10.40 is associated with the film thickness δ that exists at the bottom of the condensing surface, $x = L$. If δ is

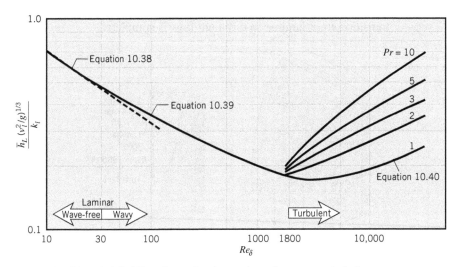

FIGURE 10.13 Modified Nusselt number for condensation on a vertical plate.

unknown, it is preferable to rewrite these equations in a form that eliminates Re_δ. To do so, Equations 10.34 and 10.36 may be combined with the definition of the average Nusselt number to provide

$$Re_\delta = 4P \frac{\overline{h}_L (v_l^2/g)^{1/3}}{k_l} = 4P\,\overline{Nu}_L \tag{10.41}$$

where the dimensionless parameter P is

$$P = \frac{k_l L (T_{\text{sat}} - T_s)}{\mu_l h'_{fg} (v_l^2/g)^{1/3}} \tag{10.42}$$

Substituting Equation 10.41 into Equations 10.38, 10.39, and 10.40, we can solve for the average Nusselt numbers in terms of P to yield

$$\overline{Nu}_L = \frac{\overline{h}_L (v_l^2/g)^{1/3}}{k_l} = 0.943\,P^{-1/4} \qquad\qquad P \lesssim 15.8 \tag{10.43}$$

$$\overline{Nu}_L = \frac{\overline{h}_L (v_l^2/g)^{1/3}}{k_l} = \frac{1}{P}(0.68P + 0.89)^{0.82} \qquad 15.8 \lesssim P \lesssim 2530 \tag{10.44}$$

$$\overline{Nu}_L = \frac{\overline{h}_L (v_l^2/g)^{1/3}}{k_l} = \frac{1}{P}[(0.024P - 53)Pr_l^{1/2} + 89]^{4/3} \qquad P \gtrsim 2530,\ Pr_l \geq 1 \tag{10.45}$$

Equation 10.43 is identical to Equation 10.32 with $\rho_l \gg \rho_v$.

For a particular problem, P may be determined from Equation 10.42, after which the average Nusselt number or average heat transfer coefficient may be found from Equation 10.43, 10.44, or 10.45.

EXAMPLE 10.3

The outer surface of a vertical tube, which is 1 m long and has an outer diameter of 80 mm, is exposed to saturated steam at atmospheric pressure and is maintained at 50°C by the flow of cool water through the tube. What is the rate of heat transfer to the coolant, and what is the rate at which steam is condensed at the surface?

SOLUTION

Known: Dimensions and temperature of a vertical tube experiencing condensation of steam at its outer surface.

Find: Heat transfer and condensation rates.

Schematic:

$D = 0.08$ m

Condensate

$L = 1$ m

Saturated steam
$p = 1$ atm

$T_s = 50°C$

Assumptions:

1. The condensate film thickness is small relative to the cylinder diameter.
2. Negligible concentration of noncondensable gases in the steam.

Properties: Table A.6, saturated water, vapor ($p = 1.0133$ bars): $T_{sat} = 100°C$, $\rho_v = 1/v_g = 0.596$ kg/m^3, $h_{fg} = 2257$ kJ/kg. Table A.6, saturated water, liquid ($T_f = 75°C$): $\rho_l = 1/v_f = 975$ kg/m^3, $\mu_l = 375 \times 10^{-6}$ N · s/m^2, $k_l = 0.668$ W/m · K, $c_{p,l} = 4193$ J/kg · K, $v_l = \mu_l/\rho_l = 385 \times 10^{-9}$ m^2/s.

Analysis: Since we assume the film thickness is small relative to the cylinder diameter, we may use the correlations of Sections 10.7 and 10.8. With

$$Ja = \frac{c_{p,l}(T_{sat} - T_s)}{h_{fg}} = \frac{4193 \text{ J/kg} \cdot \text{K}(100 - 50) \text{ K}}{2257 \times 10^3 \text{ J/kg}} = 0.0929$$

it follows that

$$h'_{fg} = h_{fg}(1 + 0.68\, Ja) = 2257 \text{ kJ/kg} \,(1.0632) = 2400 \text{ kJ/kg}$$

From Equation 10.42,

$$P = \frac{k_l L(T_{sat} - T_s)}{\mu_l h'_{fg}(v_l^2/g)^{1/3}}$$

$$= \frac{0.668 \text{ W/m} \cdot \text{K} \times 1 \text{ m} \times (100 - 50) \text{ K}}{375 \times 10^{-6} \text{ N} \cdot \text{s/m}^2 \times 2.4 \times 10^6 \text{ J/kg}\left[\dfrac{(385 \times 10^{-9} \text{ m}^2/\text{s})^2}{9.8 \text{ m/s}^2}\right]^{1/3}} = 1501$$

Therefore, Equation 10.44 applies:

$$\overline{Nu}_L = \frac{1}{P}(0.68P + 0.89)^{0.82} = \frac{1}{1501}(0.68 \times 1501 + 0.89)^{0.82} = 0.20$$

Then

$$\overline{h}_L = \frac{\overline{Nu}_L k_l}{(v_l^2/g)^{1/3}} = \frac{0.20 \times 0.668 \text{ W/m} \cdot \text{K}}{\left[\dfrac{(385 \times 10^{-9} \text{ m}^2/\text{s})^2}{9.8 \text{ m/s}^2}\right]^{1/3}} = 5300 \text{ W/m}^2 \cdot \text{K}$$

and from Equations 10.33 and 10.34

$$q = \bar{h}_L (\pi DL)(T_{sat} - T_s) = 5300 \text{ W/m}^2 \cdot \text{K} \times \pi \times 0.08 \text{ m} \times 1 \text{ m} \, (100 - 50) \text{ K} = 66.6 \text{ kW} \quad \lhd$$

$$\dot{m} = \frac{q}{h'_{fg}} = \frac{66.6 \times 10^3 \text{ W}}{2.4 \times 10^6 \text{ J/kg}} = 0.0276 \text{ kg/s} \quad \lhd$$

Note that using Equation 10.26, with the corrected latent heat, the film thickness at the bottom of the tube $\delta(L)$ for the wave-free laminar assumption is

$$\delta(L) = \left[\frac{4 k_l \mu_l (T_{sat} - T_s) L}{g \rho_l (\rho_l - \rho_v) h'_{fg}} \right]^{1/4}$$

$$\delta(L) = \left[\frac{4 \times 0.668 \text{ W/m} \cdot \text{K} \times 375 \times 10^{-6} \text{ kg/s} \cdot \text{m} \, (100 - 50) \text{ K} \times 1 \text{ m}}{9.8 \text{ m/s}^2 \times 975 \text{ kg/m}^3 \, (975 - 0.596) \text{ kg/m}^3 \times 2.4 \times 10^6 \text{ J/kg}} \right]^{1/4}$$

$$\delta(L) = 2.18 \times 10^{-4} \text{ m} = 0.218 \text{ mm}$$

Hence $\delta(L) \ll (D/2)$, and use of the vertical plate correlation for a vertical cylinder is justified.

Comments:

1. The condensation heat and mass rates may be increased by increasing the length of the tube. For $1 \leq L \leq 2$ m, the calculations yield the variations shown below, for which $1000 \leq Re_\delta \leq 2330$ or $1500 \leq P \leq 3010$. The foregoing calculations were performed by using the wavy-laminar correlation, Equation 10.44, for $P \leq 2530$ ($L \leq 1.68$ m) and Equation 10.45, for $P > 2530$ ($L > 1.68$ m). Note, however, that the correlations do not provide equivalent results at $P = 2530$. In particular, Equation 10.45 is a function of Pr, whereas Equation 10.44 is not.

2. If a noncondensable gas such as air is mixed with the steam, heat transfer and condensation rates can be reduced significantly. This is due to multiple effects [36]. For example, q and $\dot{m}$ can drop by 65% if the steam contains only 1% air by weight. Steam condensers that operate at subatmospheric pressure, such as those utilized in Rankine cycles, must be meticulously designed to prevent infiltration of air.

10.9 *Film Condensation on Radial Systems*

The Nusselt analysis of Section 10.7 may be extended to laminar film condensation on the outer surface of a sphere or a horizontal tube (Figures 10.14*a,b*) and the average Nusselt number has the form

$$\overline{Nu}_D = \frac{\overline{h}_D D}{k_l} = C \left[\frac{\rho_l g (\rho_l - \rho_v) h'_{fg} D^3}{\mu_l k_l (T_{\text{sat}} - T_s)} \right]^{1/4} \tag{10.46}$$

where $C = 0.826$ for the sphere [49] and 0.729 for the tube [45]. The properties in this equation and in Equations 10.48 and 10.49 below are evaluated as explained beneath Equation 10.32.

When a liquid–vapor interface is curved, such as those of Figure 10.14, pressure differences are established across the interface by the effects of surface tension. This pressure difference is described by the *Young–Laplace* equation, which for a two-dimensional system may be written

$$\Delta p = p_v - p_l = \frac{\sigma}{r_c} \tag{10.47}$$

where r_c is the local radius of curvature of the liquid–vapor interface. If r_c varies along the interface (and the vapor pressure p_v is constant), the pressure on the liquid side of the interface is nonuniform, influencing the velocity distribution within the liquid and the heat transfer rate. For the unfinned tube of Figure 10.14*b*, the interface curvature is relatively large, $r_c \approx D/2$, except where the liquid sheet departs from the bottom of the tube. Hence $p_l \approx p_v$ along nearly the entire liquid–vapor interface, and the surface tension does not influence the condensation rate.

FIGURE 10.14 Film condensation on (*a*) a sphere, (*b*) a single horizontal tube, (*c*) a vertical tier of horizontal tubes with a continuous condensate sheet, and (*d*) with dripping condensate.

FIGURE 10.15   Condensation on a horizontal finned tube.

Condensation on a tube with annular fins is shown in Figure 10.15. In this case, the sharp corners of the finned tube lead to large variations in the liquid–vapor interface curvature, and surface tension effects can be important. For the finned tube, surface tension forces tend to increase heat transfer rates near the fin tips by reducing the film thickness and decrease heat transfer rates in the inter-fin region by retaining condensate. Just as the liquid layer is thicker on the bottom of a sphere (Figure 10.14*a*) or unfinned horizontal tube (Figure 10.14*b*), there is more retained condensate on the underside of the horizontal finned tube.

Heat transfer rates for the finned tube q_{ft} may be related to those for a corresponding unfinned tube q_{uft} by an *enhancement ratio*, $\varepsilon_{ft} = q_{ft}/q_{uft}$. The degree of enhancement depends primarily on the fluid, the ambient pressure, and the fin geometry, and is weakly dependent on the difference between the tube and ambient temperatures [50]. Small fins, relative to those commonly used for single-phase convection, promote a highly curved liquid surface and, in turn, can enhance heat transfer significantly. The small fins can be fabricated by, for example, removing material from a tube of radius r_2 as shown in Figure 10.15, thereby eliminating contact resistances at the tube–fin interface. Moreover, when manufactured from a metal of high thermal conductivity such as copper, it is often reasonable to assume the tube and small fins have the same, uniform temperature.

Heat transfer correlations for finned tubes tend to be cumbersome and have restricted ranges of application [51]. For design purposes, however, correlations derived by Rose [50] may be used to estimate the *minimum* enhancement associated with the use of a finned tube. This minimum enhancement occurs when condensate is retained in the *entire* inter-fin region, and is

$$\varepsilon_{ft,min} = \frac{q_{ft,min}}{q_{uft}} = \frac{tr_2}{Sr_1}\left[\frac{r_1}{r_2} + 1.02\frac{\sigma r_1}{(\rho_l - \rho_v)gt^3}\right]^{1/4} \tag{10.48}$$

where ρ_l and ρ_v are evaluated as described below Equation 10.32 and the surface tension σ is evaluated at T_{sat}. Actual enhancements exceed $\varepsilon_{ft,min}$ and have been reported to be in the range $2 \leq \varepsilon_{ft} \leq 4$ for water [50]. Procedures for estimating the heat transfer rates associated with nonisothermal fins are provided by Briggs and Rose [52].

For vertically aligned tubes with a continuous condensate sheet, as shown in Figure 10.14*c*, the heat transfer rate associated with the lower tubes is less than that of the top tube because the films on the lower tubes are thicker than on the top tube. For a vertical tier of N horizontal *unfinned* tubes the average coefficient (over all N tubes) may be expressed as

$$\bar{h}_{D,N} = \bar{h}_D N^n \qquad (10.49)$$

where $\bar{h}_D$ is the heat transfer coefficient for the top tube given by Equation 10.46. The Nusselt analysis may be extended to account for the increasing tube-to-tube film thickness, yielding $n = -1/4$. However, an empirical value of $n = -1/6$ is often found to be more appropriate [53].

The discrepancy between the analytical and empirical values of n may be attributed to several effects. The analysis is based on the assumption of a continuous adiabatic sheet of condensate spanning the tubes, as illustrated in Figure 10.14c. However, heat transfer to the liquid sheet and its increase in momentum as it falls freely under gravity also increase overall heat transfer rates. Chen [54] accounted for these influences and reported their effects in terms of the Jakob number and the number of tubes in the tier, N. For $Ja < 0.1$, heat transfer was predicted to be enhanced by less than 15%. Larger measured values of $\bar{h}_{D,N}$ might also be attributed to condensate dripping, as illustrated in Figure 10.14d. As individual drops impinge on the lower tube, turbulence and waves propagate throughout the film, enhancing heat transfer. For tubes with annular fins, lateral propagation of condensate is hindered by the fins, directly exposing more of the lower tube surface to vapor and resulting in values of n in the range $-1/6 < n \lesssim 0$ [53].

If the length-to-diameter ratio of an unfinned tube exceeds $1.8 \tan \theta$ [55], Equations 10.46 and 10.49 may be applied to inclined tubes by replacing g with $g \cos \theta$, where the angle θ is measured from the horizontal position. For either finned or unfinned tubes, the presence of noncondensable gases will decrease the convection coefficients relative to values obtained from the foregoing correlations.

EXAMPLE 10.4

The tube bank of a steam condenser consists of a square array of 400 tubes, each of diameter $D = 2r_1 = 6$ mm.

1. If horizontal, unfinned tubes are exposed to saturated steam at a pressure of $p = 0.15$ bar and the tube surface is maintained at $T_s = 25°C$, what is the rate at which steam is condensed per unit length of the tube bank?

2. If annular fins of height $h = r_2 - r_1 = 1$ mm, thickness $t = 1$ mm, and pitch $S = 2$ mm are added, determine the minimum condensation rate per unit length of tubing.

SOLUTION

Known: Configuration and surface temperature of unfinned and finned condenser tubes exposed to saturated steam at 0.15 bar.

Find:

1. Condensation rate per unit length of unfinned tubing.

2. Minimum condensation rate per unit length of finned tubing.

Schematic:

Assumptions:

1. Spatially uniform cylinder and fin temperature.
2. Average heat transfer coefficient varies with tube row with $n = -1/6$ in Equation 10.49.
3. Negligible concentration of noncondensable gases in the steam.

Properties: Table A.6, saturated water, vapor ($p = 0.15$ bar): $T_{sat} = 327$ K = 54°C, $\rho_v = 1/v_g = 0.098$ kg/m^3, $h_{fg} = 2373$ kJ/kg, $\sigma = 0.0671$ N/m. Table A.6, saturated water, liquid ($T_f = 312.5$ K): $\rho_l = 1/v_f = 992$ kg/m^3, $\mu_l = 663 \times 10^{-6}$ N · s/m^2, $k_l = 0.631$ W/m · K, $c_{p,l} = 4178$ J/kg · K.

Analysis:

1. Equation 10.46 may be rearranged to yield an expression for the convection coefficient for the top, unfinned tube which is of the form

$$\bar{h}_D = C \left[\frac{\rho_l g(\rho_l - \rho_v)k_l^3 h'_{fg}}{\mu_l(T_{sat} - T_s)D} \right]^{1/4}$$

where $C = 0.729$ for a tube and

$$
\begin{aligned}
h'_{fg} &= h_{fg}(1 + 0.68\,Ja) = h_{fg} + 0.68c_{p,l}(T_{sat} - T_s) \\
&= 2373 \times 10^3 \text{ J/kg} + 0.68 \times 4178 \text{ J/kg} \cdot \text{K} \times (327 - 298) \text{ K} \\
&= 2455 \text{ kJ/kg}
\end{aligned}
$$

Therefore

$$\bar{h}_D = 0.729 \left[\frac{
\begin{array}{c}
992 \text{ kg/m}^3 \times 9.8 \text{ m/s}^2 \times (992 - 0.098) \text{ kg/m}^3 \\
\times (0.631 \text{ W/m} \cdot \text{K})^3 \times 2455 \times 10^3 \text{ J/kg}
\end{array}
}{663 \times 10^{-6} \text{ kg/s} \cdot \text{m} \times (327 - 298) \text{ K} \times 6 \times 10^{-3} \text{ m}} \right]^{1/4}$$

$$= 10{,}980 \text{ W/m}^2 \cdot \text{K}$$

From Equation 10.49 the array-averaged convection coefficient is

$$\overline{h}_{D,N} = \overline{h}_D N^n = 10{,}980 \text{ W/m}^2 \cdot \text{K} \times 20^{-1/6} = 6667 \text{ W/m}^2 \cdot \text{K}$$

From Equation 10.34 the condensation rate per unit length of tubing is

$$\dot{m}'_{\text{uft}} = N \times N \frac{\overline{h}_{D,N}(\pi D)(T_{\text{sat}} - T_s)}{h'_{fg}}$$

$$= 20 \times 20 \times 6667 \text{ W/m}^2 \cdot \text{K} \times \pi \times 6 \times 10^{-3} \text{ m} \times (327 - 298) \text{ K}/2455 \times 10^3 \text{ J/kg}$$

$$= 0.594 \text{ kg/s} \cdot \text{m} \qquad \qquad \triangleleft$$

2. From Equation 10.48, the minimum enhancement attributable to the annular fins is

$$\varepsilon_{\text{ft,min}} = \frac{q_{\text{ft,min}}}{q_{\text{uft}}} = \frac{\dot{m}'_{\text{ft,min}}}{\dot{m}'_{\text{uft}}} = \frac{tr_2}{Sr_1}\left[\frac{r_1}{r_2} + 1.02\frac{\sigma r_1}{(\rho_l - \rho_v)gt^3}\right]^{1/4}$$

$$= \frac{1 \times 4}{2 \times 3}\left[\frac{3}{4} + 1.02\frac{0.0671 \text{ N/m} \times 3 \times 10^{-3} \text{ m}}{(992 - 0.098) \text{ kg/m}^3 \times 9.8 \text{ m/s}^2 \times (1 \times 10^{-3} \text{ m})^3}\right]^{1/4}$$

$$= 1.44$$

Therefore, the minimum condensation rate for the finned tubes is

$$\dot{m}'_{\text{ft,min}} = \varepsilon_{\text{ft,min}}\dot{m}'_{\text{uft}} = 1.44 \times 0.594 \text{ kg/s} \cdot \text{m} = 0.855 \text{ kg/s} \cdot \text{m} \qquad \triangleleft$$

Comment: A value of $n = -1/6$ was used in Equation 10.49. However, for finned tubes the value of n is expected to be between zero and $-1/6$. For $n = 0$, the condensation rate per unit length of tubing would be

$$\dot{m}'_{\text{ft,min}} = \varepsilon_{\text{ft,min}} \times N \times N \frac{\overline{h}_D(\pi D)(T_{\text{sat}} - T_s)}{h'_{fg}}$$

$$= 1.44 \times 20 \times 20 \times 10{,}980 \text{ W/m}^2 \cdot \text{K} \times \pi \times 6 \times 10^{-3} \text{ m} \times (327 - 298) \text{ K}/2455 \times 10^3 \text{ J/kg}$$

$$= 1.41 \text{ kg/s} \cdot \text{m}$$

The preceding rate is for a *nonoptimized* condition where condensate fills the entire inter-fin region. Actual enhancements of $\varepsilon_{\text{ft,max}} \approx 4$ might be expected [50]. For $\varepsilon_{\text{ft,max}} = 4$ and $n = 0$, the condensation rate would be

$$\dot{m}'_{\text{ft}} = \varepsilon_{\text{ft,max}} \times N \times N \frac{\overline{h}_D(\pi D)(T_{\text{sat}} - T_s)}{h'_{fg}}$$

$$= 4 \times 20 \times 20 \times 10{,}980 \text{ W/m}^2 \cdot \text{K} \times \pi \times 6 \times 10^{-3} \text{ m} \times (327 - 298) \text{ K}/2455 \times 10^3 \text{ J/kg}$$

$$= 3.91 \text{ kg/s} \cdot \text{m}$$

Hence the condensation rate could potentially be increased by $100 \times (3.91 - 0.594) \text{ kg/s} \cdot \text{m}/0.594 \text{ kg/s} \cdot \text{m} = 559\%$ by using finned tubes.

10.10 *Condensation in Horizontal Tubes*

Condensers used for refrigeration and air-conditioning systems generally involve vapor condensation inside horizontal or vertical tubes. Conditions within the tube depend strongly on the velocity of the vapor flowing through the tube, the mass fraction of vapor X, which decreases along the tube as condensation occurs, and the fluid properties. If the vapor velocity is small, condensation occurs in the manner depicted in Figure 10.16a for a horizontal tube. The fluid condenses in the upper regions of the tube wall and flows downward to a larger pool of liquid. In turn, the liquid pool is propelled down the length of the tube by shear forces imparted by the flowing vapor. For low vapor velocities such that

$$Re_{v,i} = \left(\frac{\rho_v u_{m,v} D}{\mu_v}\right)_i < 35{,}000 \qquad (10.50)$$

where i refers to the tube inlet, heat transfer occurs predominantly through the falling condensate film. Dobson and Chato [56] recommend use of Equation 10.46 with $C = 0.555$ and $h'_{fg} = h_{fg} + 0.375 c_{p,l}(T_{\text{sat}} - T_s)$. The value of C is less than that recommended for condensation on the outside of a cylinder ($C = 0.729$) because heat transfer associated with the condensate pool is small. Property evaluation is explained beneath Equation 10.32.

At high vapor velocities the two-phase flow becomes turbulent and annular (Figure 10.16b). The vapor occupies the core of the annulus, which diminishes in diameter as the thickness of the outer condensate layer increases in the flow direction. Dobson and Chato [56] recommend an empirical correlation for a local heat transfer coefficient h of the form

$$Nu_D = \frac{hD}{k_l} = 0.023\, Re_{D,l}^{0.8}\, Pr_l^{0.4}\left[1 + \frac{2.22}{X_{tt}^{0.89}}\right] \qquad (10.51a)$$

where $Re_{D,l} = 4\dot{m}(1-X)/(\pi D \mu_l)$, $X \equiv \dot{m}_v/\dot{m}$ is the mass fraction of vapor in the fluid, and X_{tt} is the *Martinelli parameter* corresponding to the existence of turbulent flow in both the liquid and vapor phases

$$X_{tt} = \left(\frac{1-X}{X}\right)^{0.9}\left(\frac{\rho_v}{\rho_l}\right)^{0.5}\left(\frac{\mu_l}{\mu_v}\right)^{0.1} \qquad (10.51b)$$

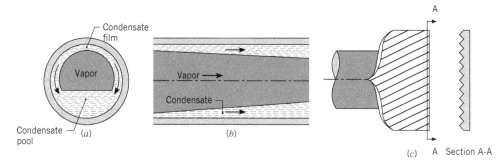

FIGURE 10.16 Film condensation in a horizontal tube. (a) Cross section of condensate flow for low vapor velocities. (b) Longitudinal section of condensate flow for large vapor velocities. (c) Microfins arranged in a helical pattern.

In generating Equations 10.51a,b, Dobson and Chato evaluated all properties at the saturation temperature T_{sat}. The equations are recommended for use when the mass flow rate per unit cross-sectional tube area exceeds 500 kg/s · m^2 [56]. For single-phase liquid convection, $X \rightarrow 0$, $X_{tt} \rightarrow \infty$ and $Re_{D,l} \rightarrow Re_D$. In this case Equation 10.51a reduces to the Dittus–Boelter correlation, Equation 8.60, except for the exponent on the Prandtl number.

Condensation inside tubes at intermediate vapor velocities (or at low vapor mass fractions) is characterized by a variety of complex flow regimes. Heat transfer correlations have been developed for the individual regimes, and recommendations for their use are included in Dobson and Chato [56]. Condensation inside smaller tubes is influenced by surface tension effects and other considerations [36].

Condensation rates can be increased by adding small fins to the interior of the tube. Microfin tubes are typically made of copper with triangular or trapezoidal-shaped fins of height 0.1 to 0.25 mm as shown in Figure 10.16c. Heat transfer is increased due to the increase in the copper surface area, but also by turbulence induced by the fin structure and surface tension effects similar to those discussed for Figure 10.15. The fins are typically arranged in a helical or herringbone pattern down the tube length, with heat transfer rates enhanced by 50 to 180% [51].

10.11 *Dropwise Condensation*

Typically, heat transfer coefficients for dropwise condensation are an order of magnitude larger than those for film condensation. In fact, in heat exchanger applications for which dropwise condensation is promoted, other thermal resistances may be significantly larger than that due to condensation and, therefore, reliable correlations for the condensation process are not needed.

Of the many surface–fluid systems studied [57, 58], most of the data are for steam condensation on well-promoted copper surfaces—that is, surfaces for which wetting is inhibited—and are correlated by expressions of the form [59]

$$\bar{h}_{dc} = 51{,}104 + 2044\, T_{sat}(°C) \qquad 22°C \lesssim T_{sat} \lesssim 100°C \qquad (10.52)$$

$$\bar{h}_{dc} = 255{,}510 \qquad\qquad 100°C \lesssim T_{sat} \qquad (10.53)$$

where the heat transfer coefficient has units of (W/m^2 · K). The heat transfer rate and condensation rate can be calculated from Equations 10.33 and 10.34, where h'_{fg} is given by Equation 10.27, and properties are evaluated as explained beneath Equation 10.32. The effect of subcooling, $T_{sat} - T_s$, on $\bar{h}_{dc}$ is small and may be neglected.

The effect of noncondensable vapors in the steam can be very important and has been studied by Shade and Mikic [60]. In addition, if the condensing surface material does not conduct as well as copper or silver, its thermal resistance becomes a factor. Since all the heat is transferred to the drops, which are very small and widely distributed over the surface, heat flow lines within the surface material near the active areas of condensation will *crowd*, inducing a *constriction* resistance. This effect has been studied by Hannemann and Mikic [61].

10.12 *Summary*

This chapter identifies the essential physical features of boiling and condensation processes and presents correlations suitable for approximate engineering calculations. However, a great deal of additional information is available, and much of it has been summarized in several extensive reviews of the subject [7, 15, 25, 30–33, 36, 51, 56, 58–59, 61–67].

You may test your understanding of heat transfer with phase change by addressing the following questions.

- What is *pool boiling*? *Forced convection boiling*? *Subcooled boiling*? *Saturated boiling*?
- How is the *excess temperature* defined?
- Sketch the *boiling curve* and identify key regimes and features. What is the *critical heat flux*? What is the *Leidenfrost point*? How does progression along the boiling curve occur if the surface heat flux is controlled? What is the nature of the hysteresis effect? How does progression along the boiling curve occur if the surface temperature is controlled?
- How does heat flux depend on the excess temperature in the *nucleate boiling* regime?
- What modes of heat transfer are associated with *film boiling*?
- How is the amount of liquid *subcooling* defined?
- To what extent is the boiling heat flux influenced by the magnitude of the gravitational field, liquid subcooling, and surface roughness?
- How do two-phase flow and heat transfer in microchannels differ from two-phase flow and heat transfer in larger tubes?
- How does *dropwise condensation* differ from *film condensation*? Which mode of condensation is characterized by larger heat transfer coefficients?
- For laminar film condensation on a vertical surface, how do the local and average convection coefficients vary with distance from the leading edge?
- How is the Reynolds number defined for film condensation on a vertical surface? What are the corresponding flow regimes?
- How does surface tension affect condensation on or in finned tubes?

References

1. Fox, R. W., A. T. McDonald, and P. J. Pritchard, *Introduction to Fluid Mechanics*, 6th ed. Wiley, Hoboken, NJ, 2003.

2. Nukiyama, S., *J. Japan Soc. Mech. Eng.*, **37**, 367, 1934 (Translation: *Int. J. Heat Mass Transfer*, **9**, 1419, 1966).

3. Drew, T. B., and C. Mueller, *Trans. AIChE*, **33**, 449, 1937.

4. Yamagata, K., F. Kirano, K. Nishiwaka, and H. Matsuoka, *Mem. Fac. Eng. Kyushu*, **15**, 98, 1955.

5. Rohsenow, W. M., *Trans. ASME*, **74**, 969, 1952.

6. Vachon, R. I., G. H. Nix, and G. E. Tanger, *J. Heat Transfer*, **90**, 239, 1968.

7. Collier, J. G., and J. R. Thome, *Convective Boiling and Condensation*, 3rd ed., Oxford University Press, New York, 1996.

8. Pioro I. L., *Int. J. Heat Mass Transfer*, **42**, 2003, 1999.

9. Kutateladze, S. S., *Kotloturbostroenie*, **3**, 10, 1948.

10. Zuber, N., *Trans. ASME*, **80**, 711, 1958.

11. Lienhard, J. H., *A Heat Transfer Textbook*, 2nd ed., Prentice-Hall, Englewood Cliffs, NJ, 1987.

12. Nakayama, W., A. Yabe, P. Kew, K. Cornwell, S. G. Kandlikar, and V. K. Dhir, in S. G. Kandlikar, M. Shoji, and V. K. Dhir, Eds., *Handbook of Phase Change: Boiling*

and Condensation, Chap. 16, Taylor & Francis, New York, 1999.

13. Cichelli, M. T., and C. F. Bonilla, *Trans. AIChE*, **41**, 755, 1945.

14. Berenson, P. J., *J. Heat Transfer*, **83**, 351, 1961.

15. Hahne, E., and U. Grigull, *Heat Transfer in Boiling*, Hemisphere/Academic Press, New York, 1977.

16. Lienhard, J. H., and P. T. Y. Wong, *J. Heat Transfer*, **86**, 220, 1964.

17. Bromley, L. A., *Chem. Eng. Prog.*, **46**, 221, 1950.

18. Sadasivan, P., and J. H. Lienhard, *J. Heat Transfer*, **109**, 545, 1987.

19. Siegel, R., *Adv. Heat Transfer*, **4**, 143, 1967.

20. Berenson, P. J., *Int. J. Heat Mass Transfer*, **5**, 985, 1962.

21. Webb, R. L., *Heat Transfer Eng.*, **2**, 46, 1981, and *Heat Transfer Eng.*, **4**, 71, 1983.

22. Bergles, A. E., "Enhancement of Heat Transfer," *Heat Transfer 1978*, Vol. 6, pp. 89–108, Hemisphere Publishing, New York, 1978.

23. Bergles, A. E., in G. F. Hewitt, Exec. Ed., *Heat Exchanger Design Handbook*, Section 2.7.9, Begell House, New York, 2002.

24. Bergles, A. E., and W. H. Rohsenow, *J. Heat Transfer*, **86**, 365, 1964.

25. van Stralen, S., and R. Cole, *Boiling Phenomena*, McGraw-Hill/Hemisphere, New York, 1979.

26. Lienhard, J. H., and R. Eichhorn, *Int. J. Heat Mass Transfer*, **19**, 1135, 1976.

27. Kandlikar, S. G., *J. Heat Transfer*, **112**, 219, 1990.

28. Kandlikar, S. G., and H. Nariai, in S. G. Kandlikar, M. Shoji, and V. K. Dhir, Eds., *Handbook of Phase Change: Boiling and Condensation*, Chap. 15, Taylor & Francis, New York, 1999.

29. Celata, G. P., and A. Mariani, in S. G. Kandlikar, M. Shoji, and V. K. Dhir, Eds., *Handbook of Phase Change: Boiling and Condensation*, Chap. 17, Taylor & Francis, New York, 1999.

30. Tong, L. S., and Y. S. Tang, *Boiling Heat Transfer and Two Phase Flow*, 2nd ed., Taylor & Francis, New York, 1997.

31. Rohsenow, W. M., in W. M. Rohsenow and J. P. Hartnett, Eds., *Handbook of Heat Transfer*, Chap. 13, McGraw-Hill, New York, 1973.

32. Griffith, P., in W. M. Rohsenow and J. P. Hartnett, Eds., *Handbook of Heat Transfer*, Chap. 14, McGraw-Hill, New York, 1973.

33. Ginoux, J. N., *Two-Phase Flow and Heat Transfer*, McGraw-Hill/Hemisphere, New York, 1978.

34. Hall, D. D., and I. Mudawar, *Int. J. Heat Mass Transfer*, **43**, 2573, 2000.

35. Hall, D. D., and I. Mudawar, *Int. J. Heat Mass Transfer*, **43**, 2605, 2000.

36. Faghri, A., and Y. Zhang, *Transport Phenomena in Multiphase Systems*, Elsevier, Amsterdam, 2006.

37. Qu, W., and I. Mudawar, *Int. J. Heat Mass Transfer*, **46**, 2755, 2003.

38. Ghiaasiaan, S. M., and S. I. Abdel-Khalik, in J. P. Hartnett, T. F. Irvine, Y. I. Cho, and G. A. Greene, Eds., *Advances in Heat Transfer*, Vol. 34, Academic Press, New York, 2001.

39. Qu, W., and I. Mudawar, *Int. J. Heat Mass Transfer*, **46**, 2773, 2003.

40. Sparrow, E. M., and J. L. Gregg, *J. Heat Transfer*, **81**, 13, 1959.

41. Koh, J. C. Y., E. M. Sparrow, and J. P. Hartnett, *Int. J. Heat Mass Transfer*, **2**, 69, 1961.

42. Nusselt, W., *Z. Ver. Deut. Ing.*, **60**, 541, 1916.

43. Rohsenow, W. M., *Trans. ASME*, **78**, 1645, 1956.

44. Chen, M. M., *J. Heat Transfer*, **83**, 48, 1961.

45. Dhir, V. K., and J. H. Lienhard, *J. Heat Transfer*, **93**, 97, 1971.

46. Kutateladze, S. S., *Fundamentals of Heat Transfer*, Academic Press, New York, 1963.

47. Labuntsov, D. A., *Teploenergetika*, **4**, 72, 1957.

48. Gregorig, R., J. Kern, and K. Turek, *Wärme Stoffübertrag.*, **7**, 1, 1974.

49. Popiel, Cz. O., and L. Boguslawski, *Int. J. Heat Mass Transfer*, **18**, 1486, 1975.

50. Rose, J. W., *Int. J. Heat Mass Transfer*, **37**, 865, 1994.

51. Cavallini, A., G. Censi, D. Del Col, L. Doretti, G. A. Longo, L. Rossetto, and C. Zilio, *Int. J. Refrig.* **26**, 373, 2003.

52. Briggs, A., and J. W. Rose, *Int. J. Heat Mass Transfer*, **37**, 457, 1994.

53. Murase, T., H. S. Wang, and J. W. Rose, *Int. J. Heat Mass Transfer*, **49**, 3180, 2006.

54. Chen, M. M., *J. Heat Transfer*, **83**, 55, 1961.

55. Selin, G., "Heat Transfer by Condensing Pure Vapours Outside Inclined Tubes," *International Developments in Heat Transfer*, Part 2, International Heat Transfer Conference, University of Colorado, pp. 278–289, ASME, New York, 1961.

56. Dobson, M. K., and J. C. Chato, *J. Heat Transfer*, **120**, 193, 1998.

57. Tanner, D. W., D. Pope, C. J. Potter, and D. West, *Int. J. Heat Mass Transfer*, **11**, 181, 1968.

58. Rose, J. W., *Proc. Instn. Mech. Engrs. A: Power and Energy*, **216**, 115, 2001.

59. Griffith, P., in G. F. Hewitt, Exec. Ed., *Heat Exchanger Design Handbook*, Section 2.6.5, Hemisphere Publishing, New York, 1990.

60. Shade, R., and B. Mikic, "The Effects of Noncondensable Gases on Heat Transfer During Dropwise Condensation," Paper 67b presented at the 67th Annual Meeting of the American Institute of Chemical Engineers, Washington, DC, 1974.

61. Hannemann, R., and B. Mikic, *Int. J. Heat Mass Transfer*, **19**, 1309, 1976.

62. Marto, P. J., in W. M. Rohsenow, J. P. Hartnett, and Y. I. Cho, Eds., *Handbook of Heat Transfer*, 3rd ed., Chap. 14, McGraw-Hill, New York, 1998.

63. Collier, J. G., and V. Wadekar, in G. F. Hewitt, Exec. Ed., *Heat Exchanger Design Handbook*, Section 2.7.2, Begell House, New York, 2002.

64. Butterworth, D., in D. Butterworth and G. F. Hewitt, Eds., *Two-Phase Flow and Heat Transfer*, Oxford University Press, London, 1977, pp. 426–462.

65. McNaught, J., and D. Butterworth, in G. F. Hewitt, Exec. Ed., *Heat Exchanger Design Handbook*, Section 2.6.2, Begell House, New York, 2002.

66. Rose, J. W., *Int. J. Heat Mass Transfer*, **24**, 191, 1981.

67. Pioro, L. S., and I. L. Pioro, *Industrial Two-Phase Thermosyphons*, Begell House, New York, 1997.

Problems

General Considerations

10.1 Show that, for water at 1-atm pressure with $T_s - T_{sat} = 10°C$, the Jakob number is much less than unity. What is the physical significance of this result? Verify that this conclusion applies to other fluids.

10.2 The surface of a horizontal, 7-mm-diameter cylinder is maintained at an excess temperature of 5°C in saturated water at 1 atm. Estimate the heat flux using an appropriate free convection correlation and compare your result to the boiling curve of Figure 10.4. Repeat the calculation for a horizontal, 7-μm-diameter wire at the same excess temperature. What can you say about the general applicability of Figure 10.4 to all situations involving boiling of water at 1 atm?

10.3 The role of surface tension in bubble formation can be demonstrated by considering a spherical bubble of pure saturated vapor in *mechanical* and *thermal* equilibrium with its superheated liquid.

  (a) Beginning with an appropriate free-body diagram of the bubble, perform a force balance to obtain an expression of the bubble radius,

$$r_b = \frac{2\sigma}{p_{sat} - p_l}$$

where p_{sat} is the pressure of the saturated vapor and p_l is the pressure of the superheated liquid outside the bubble.

  (b) On a p–v diagram, represent the bubble and liquid states. Discuss what changes in these conditions will cause the bubble to grow or collapse.

  (c) Calculate the bubble size under equilibrium conditions for which the vapor is saturated at 101°C and the liquid pressure corresponds to a saturation temperature of 100°C.

10.4 Estimate the heat transfer coefficient, h, associated with Points A, B, C, D, and E in Figure 10.4. Which point is associated with the largest value of h? Which point corresponds to the smallest value of h? Determine the thickness of the vapor blanket at the Leidenfrost point, neglecting radiation heat transfer through the blanket. Assume the solid is a flat surface.

Nucleate Boiling and Critical Heat Flux

10.5 A long, 1-mm-diameter wire passes an electrical current dissipating 3150 W/m and reaches a surface temperature of 126°C when submerged in water at 1 atm. What is the boiling heat transfer coefficient? Estimate the value of the correlation coefficient $C_{s,f}$.

10.6 Estimate the nucleate pool boiling heat transfer coefficient for water boiling at atmospheric pressure on the outer surface of a platinum-plated 10-mm-diameter tube maintained 10°C above the saturation temperature.

10.7 The hysteresis behavior of the boiling curve becomes less pronounced as the difference between the minimum and maximum (critical) heat fluxes becomes small. Calculate the ratio of the maximum to minimum boiling heat fluxes for water, ethanol, refrigerant R-134a, and refrigerant R-22. Which fluid has the greatest potential for the most severe hysteresis effect?

For which fluid is the critical heat flux largest, and what is its value? The boiling occurs at atmospheric pressure on a large, horizontal surface.

10.8 A simple expression to account for the effect of pressure on the nucleate boiling convection coefficient in water (W/m² · K) is

$$h = C\left(\Delta T_e\right)^n \left(\frac{p}{p_a}\right)^{0.4}$$

where p and p_a are the system pressure and standard atmospheric pressure, respectively. For a horizontal plate and the range $15 < q'' < 235 \text{ kW/m}^2$, $C = 5.56$ and $n = 3$. Units of ΔT_e are kelvins. Compare predictions from this expression with the Rohsenow correlation ($C_{s,f} = 0.013$, $n = 1$) for pressures of 2 and 5 bars with $\Delta T_e = 10°C$.

10.9 In Example 10.1 we considered conditions for which vigorous boiling occurs in a pan of water, and we determined the electric power (heat rate) required to maintain a prescribed temperature for the bottom of the pan. However, the electric power is, in fact, the control (independent) variable, from which the temperature of the pan follows.

(a) For nucleate boiling in the copper pan of Example 10.1, compute and plot the temperature of the pan as a function of the heat rate for $1 \leq q \leq 100 \text{ kW}$.

(b) If the water is initially at room temperature, it must, of course, be heated for a period of time before it will boil. Consider conditions shortly after heating is initiated and the water is at 20°C. Estimate the temperature of the pan bottom for a heat rate of 8 kW.

10.10 Water at atmospheric pressure boils on the surface of a large horizontal copper tube. The heat flux is 90% of the critical value. The tube surface is initially scored; however, over time the effects of scoring diminish and the boiling eventually exhibits behavior similar to that associated with a polished surface. Determine the tube surface temperature immediately after installation and after prolonged service.

10.11 The bottom of a copper pan, 180 mm in diameter, is maintained at 115°C by the heating element of an electric range. Estimate the power required to boil the water in this pan. Determine the evaporation rate. What is the ratio of the surface heat flux to the critical heat flux? What pan temperature is required to achieve the critical heat flux?

10.12 A nickel-coated heater element with a thickness of 15 mm and a thermal conductivity of 50 W/m · K is exposed to saturated water at atmospheric pressure. A thermocouple is attached to the back surface, which is well insulated. Measurements at a particular operating condition yield an electrical power dissipation in the heater element of $6.950 \times 10^7 \text{ W/m}^3$ and a temperature of $T_o = 266.4°C$.

(a) From the foregoing data, calculate the surface temperature, T_s, and the heat flux at the exposed surface.

(b) Using the surface heat flux determined in part (a), estimate the surface temperature by applying an appropriate boiling correlation.

10.13 A 0.3-m-diameter, 5-mm-thick horizontal copper disk is heated from below at a rate of 70 kW. The top of the disk is polished, and exposed to a water bath at atmospheric pressure. Determine the maximum disk temperature when nucleate boiling is occurring. In an effort to reduce the temperature of the copper disk, a 1-mm-thick nickel coating is applied to the top of the disk, which in turn reduces the value of $C_{s,f}$. Calculate the maximum temperature of the coated disk. Does the coating reduce the maximum copper temperature? Neglect any contact resistance between the copper substrate and the nickel coating.

10.14 Advances in very large scale integration (VLSI) of electronic devices on a chip are often restricted by the ability to cool the chip. For mainframe computers, an array of several hundred chips, each of area 25 mm², may be mounted on a ceramic substrate. A method of cooling the array is by immersion in a low boiling point fluid such as refrigerant R-134a. At 1 atm and 247 K, properties of the saturated liquid are $\mu = 1.46 \times 10^{-4}$ N · s/m², $c_p = 1551 \text{ J/kg · K}$, and $Pr = 3.2$. Assume values of $C_{s,f} = 0.004$ and $n = 1.7$.

(a) Estimate the power dissipated by a single chip if it is operating at 50% of the critical heat flux. What is the corresponding value of the chip temperature?

(b) Compute and plot the chip temperature as a function of surface heat flux for $0.25 \leq q_s''/q_{max}'' \leq 0.90$.

10.15 Copper tubes 25 mm in diameter and 0.75 m long are used to boil saturated water at 1 atm.

(a) If the tubes are operated at 75% of the critical heat flux, how many tubes are needed to provide a vapor production rate of 750 kg/h? What is the corresponding tube surface temperature?

(b) Compute and plot the tube surface temperature as a function of heat flux for $0.25 \leq q''/q''_{max} < 0.90$. On the same graph, plot the corresponding number of tubes needed to provide the prescribed vapor production rate.

10.16 Estimate the power (W/m²) required to maintain a brass plate at $\Delta T_e = 15°C$ while boiling saturated water at 1 atm. What is the power requirement if the water is pressurized to 10 atm? At what fraction of the critical heat flux is the plate operating?

10.17 A dielectric fluid at atmospheric pressure is heated with a 0.5-mm-diameter, horizontal platinum wire. Determine the temperature of the wire if the heat flux is 50% of the critical value. The properties of the fluid are $c_{p,l} = 1300$ J/kg · K, $h_{fg} = 142$ kJ/kg, $k_l = 0.075$ W/m · K, $v_l = 0.32 \times 10^{-6}$ m²/s, $\rho_l = 1400$ kg/m³, $\rho_v = 7.2$ kg/m³, $\sigma = 12.4 \times 10^{-3}$ N/m, $T_{sat} = 34°C$. Assume the nucleate boiling constants are $C_{s,f} = 0.005$ and $n = 1.7$. For small horizontal cylinders, the critical heat flux is found by multiplying the value associated with large horizontal cylinders by a correction factor F, where $F = 0.89 + 2.27$ $\exp(-3.44\, Co^{-1/2})$. The Confinement number is based on the radius of the cylinder, and the range of applicability for the correction factor is $1.3 \lesssim Co \lesssim 6.7$ [11].

10.18 It has been demonstrated experimentally that the critical heat flux is highly dependent on pressure, primarily through the pressure dependence of the fluid surface tension and latent heat of vaporization. Using Equation 10.6, calculate values of q''_{max} for water on a large horizontal surface as a function of pressure. Demonstrate that the peak critical heat flux occurs at approximately one-third the critical pressure ($p_c = 221$ bars). Since all common fluids have this characteristic, suggest what coordinates should be used to plot critical heat flux–pressure values to obtain a universal curve.

10.19 What is the critical heat flux for boiling water at 1 atm on a large horizontal surface on the surface of the moon, where the gravitational acceleration is one-sixth that of the earth?

10.20 A heater for boiling a saturated liquid consists of two concentric stainless steel tubes packed with dense boron nitride powder. Electrical current is passed through the inner tube, creating uniform volumetric heating $\dot{q}$ (W/m³). The exposed surface of the outer tube is in contact with the liquid and the boiling heat flux is given as

$$q''_s = C\left(T_s - T_{sat}\right)^3$$

It is feared that under high-power operation the stainless steel tubes would severely oxidize if temperatures exceed $T_{ss,x}$ or that the boron nitride would deteriorate if its temperature exceeds $T_{bn,x}$. Presuming that the saturation temperature of the liquid (T_{sat}) and the boiling surface temperature (T_s) are prescribed, derive expressions for the maximum temperatures in the stainless steel (ss) tubes and in the boron nitride (bn). Express your results in terms of geometric parameters (r_1, r_2, r_3, r_4), thermal conductivities (k_{ss}, k_{bn}), and the boiling parameters (C, T_{sat}, T_s).

10.21 A device for performing boiling experiments consists of a copper bar ($k = 400$ W/m · K), which is exposed to a boiling liquid at one end, encapsulates an electrical heater at the other end, and is well insulated from its surroundings at all but the exposed surface. Thermocouples inserted in the bar are used to measure temperatures at distances of $x_1 = 10$ mm and $x_2 = 25$ mm from the surface.

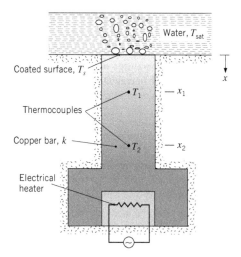

(a) An experiment is performed to determine the boiling characteristics of a special coating applied to the exposed surface. Under steady-state conditions, nucleate boiling is maintained in saturated water at atmospheric pressure, and values of $T_1 = 133.7°C$ and $T_2 = 158.6°C$ are recorded. If $n = 1$, what value of the coefficient $C_{s,f}$ is associated with the Rohsenow correlation?

(b) Assuming applicability of the Rohsenow correlation with the value of $C_{s,f}$ determined from part (a), compute and plot the excess temperature ΔT_e as a function of the boiling heat flux for $10^5 \leq q''_s \leq 10^6$ W/m^2. What are the corresponding values of T_1 and T_2 for $q''_s = 10^6$ W/m^2? If q''_s were increased to 1.5×10^6 W/m^2, could the foregoing results be extrapolated to infer the corresponding values of ΔT_e, T_1, and T_2?

Minimum Heat Flux and Film Boiling

10.22 A small copper sphere, initially at a uniform, elevated temperature $T(0) = T_i$, is suddenly immersed in a large fluid bath maintained at T_{sat}. The initial temperature of the sphere exceeds the Leidenfrost point corresponding to the temperature T_D of Figure 10.4.

(a) Sketch the variation of the average sphere temperature, $\overline{T}(t)$, with time during the quenching process. Indicate on this sketch the temperatures T_i, T_D, and T_{sat}, as well as the regimes of film, transition, and nucleate boiling and the regime of single-phase convection. Identify key features of the temperature history.

(b) At what time(s) in this cooling process do you expect the surface temperature of the sphere to deviate most from its center temperature? Explain your answer.

10.23 Determine the minimum and maximum possible surface temperatures of a polished, horizontal copper tube of diameter $D = 35$ mm, in water at atmospheric pressure, associated with a heat flux of 2×10^4 W/m^2.

10.24 A disk-shaped turbine rotor is heat-treated by quenching in water at $p = 1$ atm. Initially, the rotor is at a uniform temperature of $T_i = 1100°C$ and the water is at its boiling point as the rotor is lowered into the quenching bath by a harness.

(a) Assuming lumped-capacitance behavior and constant properties for the rotor, carefully plot the rotor temperature versus time, pointing out important features of your $T(t)$ curve. The rotor is in Orientation A.

(b) If the rotor is reoriented so that its large surfaces are horizontal (Orientation B), would the rotor temperature decrease more rapidly or less rapidly relative to Orientation A?

10.25 A steel bar, 20 mm in diameter and 200 mm long, with an emissivity of 0.9, is removed from a furnace at 455°C and suddenly submerged horizontally in a water bath under atmospheric pressure. Estimate the initial heat transfer rate from the bar.

10.26 Electrical current passes through a horizontal, 2-mm-diameter conductor of emissivity 0.5 when immersed in water under atmospheric pressure.

(a) Estimate the power dissipation per unit length of the conductor required to maintain the surface temperature at 555°C.

(b) For conductor diameters of 1.5, 2.0, and 2.5 mm, compute and plot the power dissipation per unit length as a function of surface temperature for $250 \leq T_s \leq 650°C$. On a separate figure, plot the percentage contribution of radiation as a function of T_s.

10.27 Consider a horizontal, $D = 1$-mm-diameter platinum wire suspended in saturated water at atmospheric pressure. The wire is heated by an electrical current. Determine the heat flux from the wire at the instant when the surface of the wire reaches its melting point. Determine the corresponding centerline temperature of the wire. Due to oxidation at very high temperature, the wire emissivity is $\varepsilon = 0.80$ when it burns out. The water vapor properties at the film temperature of 1209 K are $\rho_v = 0.189$ kg/m^3, $c_{p,v} = 2404$ J/kg · K, $v_v = 231 \times 10^{-6}$ m^2/s, $k_v = 0.113$ W/m · K.

10.28 A 1-mm-diameter horizontal platinum wire of emissivity $\varepsilon = 0.25$ is operated in saturated water at 1-atm pressure.

(a) What is the surface heat flux if the surface temperature is $T_s = 800$ K?

(b) For emissivities of 0.1, 0.25, and 0.95, generate a log–log plot of the heat flux as a function of surface excess temperature, $\Delta T_e \equiv T_s - T_{sat}$, for $150 \leq \Delta T_e \leq 550$ K. Show the critical heat flux and the Leidenfrost point on your plot. Separately, plot the percentage contribution of radiation to the total heat flux for $150 \leq \Delta T_e \leq 550$ K.

10.29 As strip steel leaves the last set of rollers in a hot rolling mill, it is quenched by planar water jets before being coiled. Due to the large plate temperatures, film boiling is achieved shortly downstream of the jet impingement region.

Consider conditions for which the strip steel beneath the vapor blanket is at a temperature of 907 K and has an emissivity of 0.35. Neglecting the effects of the strip and jet motions and assuming boiling within the film to be approximated by that associated with a large horizontal cylinder of 1-m diameter, estimate the rate of heat transfer per unit surface area from the strip to the wall jet.

10.30 A cylinder of 120-mm diameter at 1000 K is quenched in saturated water at 1 atm. Describe the quenching process and estimate the maximum heat removal rate per unit length during the process.

Forced Convection Boiling

10.31 A tube of 2-mm diameter is used to heat saturated water at 1 atm, which is in cross flow over the tube. Calculate and plot the critical heat flux as a function of water velocity over the range 0 to 2 m/s. On your plot, identify the pool boiling region and the transition region between the low- and high-velocity ranges. *Hint:* Problem 10.17 contains relevant information for pool boiling on small-diameter cylinders.

10.32 Saturated water at 1 atm and velocity 2 m/s flows over a cylindrical heating element of diameter 5 mm. What is the maximum heating rate (W/m) for nucleate boiling?

10.33 A vertical steel tube carries water at a pressure of 10 bars. Saturated liquid water is pumped into the $D = 0.1$-m-diameter tube at its bottom end ($x = 0$) with a mean velocity of $u_m = 0.05$ m/s. The tube is exposed to combusting pulverized coal, providing a uniform heat flux of $q'' = 100{,}000$ W/m^2.

(a) Determine the tube wall temperature and the quality of the flowing water at $x = 15$ m. Assume $G_{s,f} = 1$.

(b) Determine the tube wall temperature at a location beyond $x = 15$ m where single-phase flow of the vapor exists at a mean temperature of T_{sat}.

Assume the vapor at this location is also at a pressure of 10 bars.

(c) Plot the tube wall temperature in the range $-5 \text{ m} \leq x \leq 30 \text{ m}$.

10.34 Consider refrigerant R-134a flowing in a smooth, horizontal, 10-mm-inner-diameter tube of wall thickness 2 mm. The refrigerant is at a saturation temperature of 15°C (for which $\rho_{v,\text{sat}} = 23.75$ kg/m^3) and flows at a rate of 0.01 kg/s. Determine the maximum wall temperature associated with a heat flux of 10^5 W/m^2 at the inner wall at a location 0.4 m downstream from the onset of boiling for tubes fabricated of (a) pure copper and (b) AISI 316 stainless steel.

Film Condensation

10.35 Saturated steam at 0.1 bar condenses with a convection coefficient of 6800 W/m$^2 \cdot$K on the outside of a brass tube having inner and outer diameters of 16.5 and 19 mm, respectively. The convection coefficient for water flowing inside the tube is 5200 W/m$^2 \cdot$K. Estimate the steam condensation rate per unit length of the tube when the mean water temperature is 30°C.

10.36 Consider a container exposed to a saturated vapor, T_{sat}, having a cold bottom surface, $T_s < T_{sat}$, and with insulated sidewalls.

Assuming a linear temperature distribution for the liquid, perform a surface energy balance on the liquid–vapor interface to obtain the following expression for the growth rate of the liquid layer:

$$\delta(t) = \left[\frac{2k_l(T_{\text{sat}} - T_s)}{\rho_l h_{fg}} t \right]^{1/2}$$

Calculate the thickness of the liquid layer formed in 1 h for a 200-mm^2 bottom surface maintained at 80°C and exposed to saturated steam at 1 atm. Compare this result with the condensate formed by a *vertical* plate of the same dimensions for the same period of time.

10.37 Saturated steam at 1.5 bars condenses on the outer surface of a vertical, 100-mm-diameter pipe 1 m long, having a uniform surface temperature of 94°C. Estimate the total condensation rate and the heat transfer rate to the pipe.

10.38 Determine the total condensation rate of water onto the front surface of a vertical plate that is 10 mm high and 1 m in the horizontal direction. The plate is exposed to saturated steam at atmospheric pressure and is maintained at 75°C. Determine the total condensation rate of water onto a 1-m-long horizontal tube that has the same surface area and temperature as that of the vertical plate.

10.39 Consider wave-free laminar condensation on a vertical isothermal plate of length L, providing an average heat transfer coefficient of $\bar{h}_L$. If the plate is divided into N smaller plates, each of length $L_N = L/N$, determine an expression for the ratio of the heat transfer coefficient averaged over the N plates to the heat transfer coefficient averaged over the single plate, $\bar{h}_{L,N}/\bar{h}_{L,1}$.

10.40 A 2 m × 2 m vertical plate is exposed on one side to saturated steam at atmospheric pressure and on the other side to cooling water that maintains a plate temperature of 50°C.

    (a) What is the rate of heat transfer to the coolant? What is the rate at which steam condenses on the plate?

    (b) For plates inclined at an angle θ from the vertical, the average convection coefficient for condensation on the upper surface, $\bar{h}_{L(\text{incl})}$, may be approximated by an expression of the form, $\bar{h}_{L(\text{incl})} \approx (\cos\theta)^{1/4} \cdot \bar{h}_{L(\text{vert})}$, where $\bar{h}_{L(\text{vert})}$ is the average coefficient for the vertical orientation. If the 2 m × 2 m plate is inclined 45° from the normal, what are the rates of heat transfer and condensation?

10.41 A vertical plate 500 mm high and 200 mm wide is to be used to condense saturated steam at 1 atm.

    (a) At what surface temperature must the plate be maintained to achieve a condensation rate of $\dot{m} = 25$ kg/h?

    (b) Compute and plot the surface temperature as a function of condensation rate for $15 \leq \dot{m} \leq 50$ kg/h.

    (c) On the same graph and for the same range of $\dot{m}$, plot the surface temperature as a function of condensation rate if the plate is 200 mm high and 500 mm wide.

10.42 Condensation rates can be restricted by the speed of sound of the vapor phase. Hence, condensation can approach a *sonic limit*. Consider an $S = 2$ mm wide, $L = 200$ mm long vertical slot in a solid metal block. The slot is long into the page, so the fluid flow is two-dimensional. Saturated water vapor at atmospheric pressure condenses on the two large vertical surfaces of the slot, which are maintained at $T_s = 40°C$. As condensation proceeds, liquid water leaves the slot through the bottom opening and is replenished by saturated water vapor flowing upward through the same opening. Estimate the Mach number of the water vapor flowing upward through the slot opening. At what slot width does the Mach number of the water vapor reach unity?

10.43 The condenser of a steam power plant consists of AISI 302 stainless steel tubes ($k_s = 15$ W/m · K), each of outer and inner diameters $D_o = 30$ mm and $D_i = 26$ mm, respectively. Saturated steam at 0.135 bar condenses on the outer surface of a tube, while water at a mean temperature of $T_m = 290$ K is in fully developed flow through the tube.

    (a) For a water flow rate of $\dot{m} = 0.25$ kg/s, what is the outer surface temperature $T_{s,o}$ of the tube and the rates of heat transfer and steam condensation per unit tube length? As a first estimate, you may evaluate the properties of the liquid film at the saturation temperature. If one wishes to increase the transfer rates, what is the limiting factor that should be addressed?

    (b) Explore the effect of the water flow rate on $T_{s,o}$ and the rate of heat transfer per unit length.

10.44 Saturated vapor from a chemical process condenses at a slow rate on the inner surface of a vertical, thin-walled cylindrical container of length L and diameter D. The container wall is maintained at a uniform temperature T_s by flowing cold water across its outer surface.

Chemical process
saturated vapor, T_{sat}

Cold
water

Condensate
film

x

g

T

L

D

Derive an expression for the time, t_f, required to fill the container with condensate, assuming that the condensate film is laminar. Express your result in terms of D, L, $(T_{sat} - T_s)$, g, and appropriate fluid properties.

10.45 The condenser of a steam power plant consists of a square (in-line) array of 900 tubes, each of 25-mm diameter. Consider conditions for which saturated steam at 0.135 bars condenses on the outer surface of each tube, while a tube wall temperature of 23°C is maintained by the flow of cooling water through the tubes. What is the rate of heat transfer to the water per unit length of the tube array? What is the corresponding condensation rate?

10.46 (a) A horizontal tube of 50-mm outer diameter, with a surface temperature of 34°C, is exposed to steam at 0.2 bar. Estimate the condensation rate and heat transfer rate per unit length of the tube.

(b) Now consider that the tube of part (a) is modified by milling sharp-cornered grooves around its periphery, as in Figure 10.15. The 2-mm-deep grooves are each 2 mm wide with a pitch of $S = 4$ mm. Estimate the minimum condensation and heat transfer rates per unit length that would be expected for the modified tube. How much is the performance enhanced relative to the original tube of part (a)?

10.47 A horizontal tube 1 m long with a surface temperature of 70°C is used to condense saturated steam at 1 atm.

(a) What diameter is required to achieve a condensation rate of 125 kg/h?

(b) Plot the condensation rate as a function of surface temperature for $70 \leq T_s \leq 90°C$ and tube diameters of 125, 150, 175 mm.

10.48 A technique for cooling a multichip module involves submerging the module in a saturated fluorocarbon liquid. Vapor generated due to boiling at the module surface is condensed on the outer surface of copper tubing suspended in the vapor space above the liquid. The thin-walled tubing is of diameter $D = 10$ mm and is coiled in a horizontal plane. It is cooled by water that enters at 285 K and leaves at 315 K. All the heat dissipated by the chips within the module is transferred from a 100-mm × 100-mm boiling surface, at which the flux is 10^5 W/m², to the fluorocarbon liquid, which is at $T_{sat} = 57°C$. Liquid properties are $k_l = 0.0537$ W/m · K, $c_{p,l} = 1100$ J/kg · K, $h'_{fg} \approx h_{fg} = 84{,}400$ J/kg, $\rho_l = 1619.2$ kg/m³, $\rho_v = 13.4$ kg/m³, $\sigma = 8.1 \times 0^{-3}$ N/m, $\mu_l = 440 \times 10^{-6}$ kg/m · s, and $Pr_l = 9$.

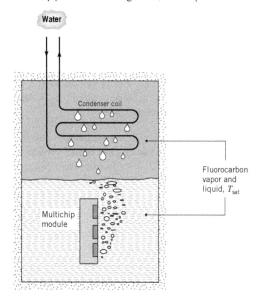

Water

Condenser coil

Fluorocarbon
vapor and
liquid, T_{sat}

Multichip
module

(a) For the prescribed heat dissipation, what is the required condensation rate (kg/s) and water flow rate (kg/s)?

(b) Assuming fully developed flow throughout the tube, determine the tube surface temperature at the coil inlet and outlet.

(c) Assuming a uniform tube surface temperature of $T_s = 53.0°C$, determine the required length of the coil.

10.49 Determine the rate of condensation on a 100-mm-diameter sphere with a surface temperature of 150°C in saturated ethylene glycol vapor at 1 atm. Approximate the liquid properties as those corresponding to saturated conditions at 373 K (Table A.5).

Condensation in Tubes

10.50 The Clean Air Act prohibited the production of chlorofluorocarbons (CFCs) in the United States as of 1996. One widely used CFC, refrigerant R-12, has been replaced by R-134a in many applications because of their similar properties, including a low boiling point at atmospheric pressure, T_{sat} = 243 K and 246.9 K for R-12 and R-134a, respectively. Compare the performance of these two refrigerants under the following conditions. The saturated refrigerant vapor at 310 K is condensed as it flows through a 30-mm-diameter, 0.8-m-long tube whose wall temperature is maintained at 290 K. If vapor enters the tube at a flow rate of 0.010 kg/s, what is the rate of condensation and the flow rate of vapor leaving the tube? The relevant properties of R-12 at T_{sat} = 310 K are ρ_v = 50.1 kg/m³, h_{fg} = 160 kJ/kg, and μ_v = 150 × 10⁻⁷ N · s/m² and those of liquid R-12 at T_f = 300 K are ρ_l = 1306 kg/m³, $c_{p,l}$ = 978 J/kg · K, μ_l = 2.54 × 10⁻⁴ N · s/m², k_l = 0.072 W/m · K. The properties of the saturated R-134a vapor are ρ_v = 46.1 kg/m³, h_{fg} = 166 kJ/kg, and μ_v = 136 × 10⁻⁷ N · s/m².

10.51 Saturated steam at 1.5 bars condenses inside a horizontal, 75-mm-diameter pipe whose surface is maintained at 100°C. Assuming low vapor velocities and film condensation, estimate the heat transfer coefficient and the condensation rate per unit length of the pipe.

10.52 Consider the situation of Problem 10.51 at relatively high vapor velocities, with a fluid mass flow rate of $\dot{m}$ = 2.5 kg/s.

    (a) Determine the heat transfer coefficient and condensation rate per unit length of tube for a mass fraction of vapor of X = 0.2.

    (b) Plot the heat transfer coefficient and the condensation rate for $0.1 \leq X \leq 0.3$.

10.53 Refrigerant R-22 with a mass flow rate of $\dot{m}$ = 8.75 × 10⁻³ kg/s is condensed inside a 7-mm-diameter tube. Annular flow is observed. The saturation temperature of the pressurized refrigerant is T_{sat} = 45°C, and the wall temperature is T_s = 40°C. Vapor properties are ρ_v = 77 kg/m³ and μ_v = 15 × 10⁻⁶ N · s/m².

    (a) Determine the heat transfer coefficient and the heat transfer and condensation rates per unit length at a quality of X = 0.5.

    (b) Plot the condensation rate per unit length over the range $0.2 < X < 0.8$.

Dropwise Condensation

10.54 Consider Problem 10.35. In an effort to increase the condensation rate, an engineer proposes to apply an L = 100-μm-thick Teflon coating to the exterior surface of the brass tube to promote dropwise condensation. Estimate the new condensation convection coefficient and the steam condensation rate per unit length of the tube after application of the coating. Comment on the proposed scheme's effect on the condensation rate (the condensation rate per unit length in Problem 10.35 is approximately 1×10^{-3} kg/s).

10.55 Wetting of some metallic surfaces can be inhibited by means of ion implantation of the surface prior to its use, thereby promoting dropwise condensation. The degree of wetting inhibition and, in turn, the efficacy of the implantation process vary from metal to metal. Consider a vertical metal plate that is exposed to saturated steam at atmospheric pressure. The plate is t = 1 mm thick, and its vertical and horizontal dimensions are L = 250 mm and b = 100 mm, respectively. The temperature of the plate surface that is exposed to the steam is found to be T_s = 90°C when the opposite surface of the metal plate is held at a cold temperature, T_c.

    (a) Determine T_c for 2024-T6 aluminum. Assume the ion-implantation process does not promote dropwise condensation for this metal.

    (b) Determine T_c for AISI 302 stainless steel, assuming the ion-implantation process is effective in promoting dropwise condensation.

Combined Boiling/Condensation

10.56 A passive technique for cooling heat-dissipating integrated circuits involves submerging the ICs in a low boiling point dielectric fluid. Vapor generated in cooling the circuits is condensed on vertical plates suspended in the vapor cavity above the liquid. The temperature of the plates is maintained below the saturation temperature, and during steady-state operation a balance is established between the rate of heat transfer to the condenser plates and the rate of heat dissipation by the ICs.

Consider conditions for which the 25-mm² surface area of each IC is submerged in a fluorocarbon liquid for which $T_{sat} = 50°C$, $\rho_l = 1700$ kg/m³, $c_{p,l} = 1005$ J/kg · K, $\mu_l = 6.80 \times 10^{-4}$ kg/s · m, $k_l = 0.062$ W/m · K, $Pr_l = 11.0$, $\sigma = 0.013$ N/m, $h_{fg} = 1.05 \times 10^5$ J/kg, $C_{s,f} = 0.004$, and $n = 1.7$. If the integrated circuits are operated at a surface temperature of $T_s = 75°C$, what is the rate at which heat is dissipated by each circuit? If the condenser plates are of height $H = 50$ mm and are maintained at a temperature of $T_c = 15°C$ by an internal coolant, how much condenser surface area must be provided to balance the heat generated by 500 integrated circuits?

10.57 A novel scheme for cooling computer chips uses a thermosyphon containing a saturated fluorocarbon. The chip is brazed to the bottom of a cuplike container, within which heat is dissipated by boiling and subsequently transferred to an external coolant (water) via condensation on the inner surface of a thin-walled tube.

The properties of the saturated fluorocarbon are $c_{p,l} = 1100$ J/kg · K, $h_{fg} = 84,400$ J/kg, $\rho_l = 1619.2$ kg/m³, $\rho_v = 13.4$ kg/m³, $\sigma = 8.1 \times 10^{-3}$ N/m, $\mu_l = 440 \times 10^{-6}$ kg/m · s, and $Pr_l = 9.01$. In addition, the nucleate boiling constants are $C_{s,f} = 0.005$ and $n = 1.7$ and $k_l = 0.054$ W/m · K.

(a) If the chip operates under steady-state conditions and its surface heat flux is maintained at 90% of the critical heat flux, what is its temperature T? What is the total power dissipation if the chip width is $L_c = 20$ mm on a side?

(b) If the tube diameter is $D = 30$ mm and its surface is maintained at $T_s = 25°C$ by the water, what tube length L is required to maintain the designated conditions?

10.58 A condenser–boiler section contains a 2-m × 2-m copper plate operating at a uniform temperature of $T_s = 100°C$ and separating saturated steam, which is condensing, from a saturated liquid-X, which experiences nucleate pool boiling. A portion of the boiling curve for liquid-X is shown as follows. Both

saturated steam and saturated liquid-X are supplied to the system, while water condensate and vapor-X are removed by means not shown in the sketch. At a pressure of 1 bar, fluid-X has a saturation temperature and a latent heat of vaporization of $T_{sat} = 80°C$ and $h_{fg} = 700,000$ J/kg, respectively.

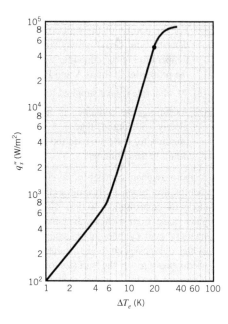

(a) Estimate the rates of evaporation and condensation (kg/s) for the two fluids.

(b) Determine the saturation temperature T_{sat} and pressure p for the steam, assuming that film condensation occurs.

10.59 A thin-walled cylindrical container of diameter D and height L is filled to a height y with a low boiling point liquid (A) at $T_{sat,A}$. The container is located in a large chamber filled with the vapor of a high boiling point fluid (B). Vapor-B condenses into a laminar film on the outer surface of the cylindrical container, extending from the location of the liquid-A free surface. The condensation process sustains nucleate boiling in liquid-A along the container wall according to the relation $q'' = C(T_s - T_{sat})^3$, where C is a known empirical constant.

(a) For the portion of the wall covered with the condensate film, derive an equation for the average temperature of the container wall, T_s. Assume that the properties of fluids A and B are known.

(b) At what rate is heat supplied to liquid-A?

(c) Assuming the container is initially filled completely with liquid, that is, $y = L$, derive an expression for the time required to evaporate all the liquid in the container.

10.60 A thermosyphon consists of a closed container that absorbs heat along its boiling section and rejects heat along its condensation section. Consider a thermosyphon made from a thin-walled mechanically polished stainless steel cylinder of diameter D. Heat supplied to the thermosyphon boils saturated water at atmospheric pressure on the surfaces of the lower boiling section of length L_b and is then rejected by condensing vapor into a thin film, which falls by gravity along the wall of the condensation section of length L_c back into the boiling section. The two sections are separated by an insulated section of length L_i. The top surface of the condensation section may be treated as being insulated. The thermosyphon dimensions are $D = 20$ mm, $L_b = 20$ mm, $L_c = 40$ mm, and $L_i = 40$ mm.

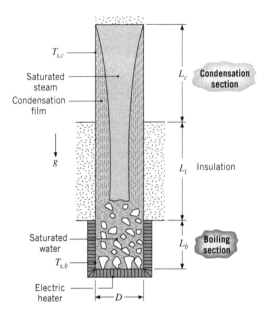

(a) Find the mean surface temperature, $T_{s,b}$, of the boiling surface if the nucleate boiling heat flux is to be maintained at 30% of the critical heat flux.

(b) Find the total condensation flow rate, $\dot{m}$, and the mean surface temperature of the condensation section, $T_{s,c}$.

10.61 It has been proposed that the very hot air trapped inside the attic of a house in the summer may be used as the energy source for a *passive* water heater installed in the attic. Energy costs associated with heating the cool water and air conditioning the house are both reduced. Ten thermosyphons, similar to that of Problem 10.60, are inserted in the bottom of a well-insulated water heater. Each thermosyphon has a condensing section that is $L_c = 50$ mm long, an insulated section that is of length $L_i = 40$ mm, and a boiling section that is $L_b = 30$ mm long. The diameter of each thermosyphon is $D = 20$ mm. The working fluid within the thermosyphons is water at a pressure of $p = 0.047$ bars.

(a) Determine the heating rate delivered by the 10 thermosyphons when boiling occurs at 25% of the CHF. What are the mean temperatures of the boiling and condensing sections?

(b) At night the attic air temperature drops below the temperature of the water. Estimate the rate of heat loss from the hot water tank to the cool attic, assuming losses through the tank insulation are negligible and the stainless steel tube wall thickness of each thermosyphon is very small.

Heat Exchangers

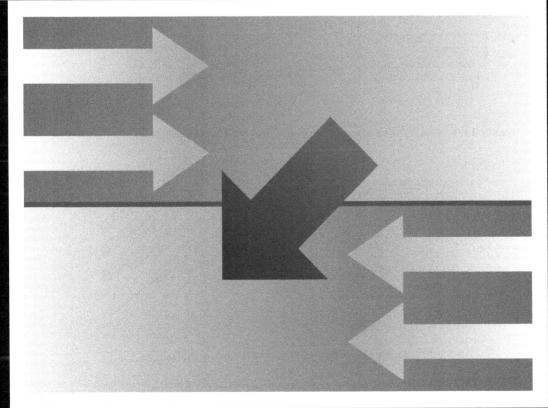

The process of heat exchange between two fluids that are at different temperatures and separated by a solid wall occurs in many engineering applications. The device used to implement this exchange is termed a *heat exchanger*, and specific applications may be found in space heating and air-conditioning, power production, waste heat recovery, and chemical processing.

In this chapter our objectives are to introduce performance parameters for assessing the efficacy of a heat exchanger and to develop methodologies for designing a heat exchanger or for predicting the performance of an existing exchanger operating under prescribed conditions.

11.1 *Heat Exchanger Types*

Heat exchangers are typically classified according to *flow arrangement* and *type of construction*. The simplest heat exchanger is one for which the hot and cold fluids move in the same or opposite directions in a *concentric tube* (or *double-pipe*) construction. In the *parallel-flow* arrangement of Figure 11.1*a*, the hot and cold fluids enter at the same end, flow in the same direction, and leave at the same end. In the *counterflow* arrangement of Figure 11.1*b*, the fluids enter at opposite ends, flow in opposite directions, and leave at opposite ends.

Alternatively, the fluids may move in *cross flow* (perpendicular to each other), as shown by the *finned* and *unfinned* tubular heat exchangers of Figure 11.2. The two configurations

(a)   (b)

FIGURE 11.1 Concentric tube heat exchangers. (*a*) Parallel flow. (*b*) Counterflow.

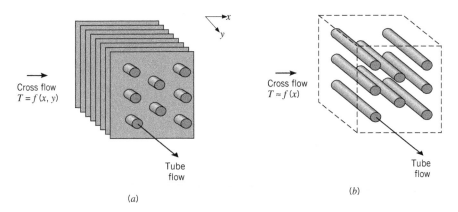

(a)   (b)

FIGURE 11.2 Cross-flow heat exchangers. (*a*) Finned with both fluids unmixed. (*b*) Unfinned with one fluid mixed and the other unmixed.

FIGURE 11.3 Shell-and-tube heat exchanger with one shell pass and one tube pass (cross-counterflow mode of operation).

are typically differentiated by an idealization that treats fluid motion over the tubes as *unmixed* or *mixed*. In Figure 11.2a, the cross-flowing fluid is said to be unmixed because the fins inhibit motion in a direction (y) that is transverse to the main-flow direction (x). In this case the cross-flowing fluid temperature varies with x and y. In contrast, for the unfinned tube bundle of Figure 11.2b, fluid motion, hence mixing, in the transverse direction is possible, and temperature variations are primarily in the main-flow direction. Since the tube flow is unmixed in either heat exchanger, both fluids are unmixed in the finned exchanger, while the cross-flowing fluid is mixed and the tube fluid is unmixed in the unfinned exchanger. The nature of the mixing condition influences heat exchanger performance.

Another common configuration is the *shell-and-tube* heat exchanger [1]. Specific forms differ according to the number of shell-and-tube passes, and the simplest form, which involves single tube and shell *passes*, is shown in Figure 11.3. Baffles are usually installed to increase the convection coefficient of the shell-side fluid by inducing turbulence and a cross-flow velocity component relative to the tubes. In addition, the baffles physically support the tubes, reducing flow-induced tube vibration. Baffled heat exchangers with one shell pass and two tube passes and with two shell passes and four tube passes are shown in Figures 11.4a and 11.4b, respectively.

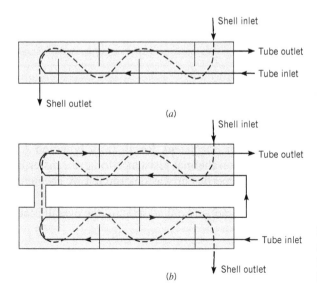

FIGURE 11.4 Shell-and-tube heat exchangers. (a) One shell pass and two tube passes. (b) Two shell passes and four tube passes.

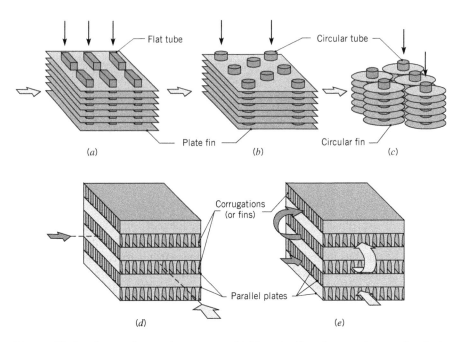

FIGURE 11.5   Compact heat exchanger cores. (*a*) Fin–tube (flat tubes, continuous plate fins).
(*b*) Fin–tube (circular tubes, continuous plate fins). (*c*) Fin–tube (circular tubes, circular fins).
(*d*) Plate–fin (single pass). (*e*) Plate–fin (multipass).

A special and important class of heat exchangers is used to achieve a very large
($\gtrsim 400$ m²/m³ for liquids and $\gtrsim 700$ m²/m³ for gases) heat transfer surface area per unit
volume. Termed *compact heat exchangers*, these devices have dense arrays of finned tubes
or plates and are typically used when at least one of the fluids is a gas, and is hence characterized by a small convection coefficient. The tubes may be *flat* or *circular*, as in Figures
11.5*a* and 11.5*b*, *c*, respectively, and the fins may be *plate* or *circular*, as in Figures 11.5*a*, *b*
and 11.5*c*, respectively. Parallel-plate heat exchangers may be finned or corrugated and
may be used in single-pass (Figure 11.5*d*) or multipass (Figure 11.5*e*) modes of operation.
Flow passages associated with compact heat exchangers are typically small ($D_h \lesssim 5$ mm),
and the flow is usually laminar.

11.2   *The Overall Heat Transfer Coefficient*

An essential, and often the most uncertain, part of any heat exchanger analysis is determination of the *overall heat transfer coefficient*. Recall from Equation 3.19 that this coefficient is defined in terms of the total thermal resistance to heat transfer between two fluids.
In Equations 3.18 and 3.36, the coefficient was determined by accounting for conduction
and convection resistances between fluids separated by composite plane and cylindrical
walls, respectively. For a wall separating two fluid streams, the overall heat transfer coefficient
may be expressed as

$$\frac{1}{UA} = \frac{1}{U_c A_c} = \frac{1}{U_h A_h} = \frac{1}{(hA)_c} + R_w + \frac{1}{(hA)_h} \tag{11.1a}$$

where c and h refer to the cold and hot fluids, respectively. Note that calculation of the UA product does not require designation of the hot or cold side ($U_cA_c = U_hA_h$). However, calculation of an overall coefficient depends on whether it is based on the cold or hot side surface area, since $U_c \neq U_h$ if $A_c \neq A_h$. The conduction resistance R_w is obtained from Equation 3.6 for a plane wall or Equation 3.33 for a cylindrical wall.

It is important to acknowledge, however, that Equation 11.1a applies only to *clean, unfinned* surfaces. During normal heat exchanger operation, surfaces are often subject to fouling by fluid impurities, rust formation, or other reactions between the fluid and the wall material. The subsequent deposition of a film or scale on the surface can greatly increase the resistance to heat transfer between the fluids. This effect can be treated by introducing an additional thermal resistance in Equation 11.1a, termed the *fouling factor, R_f*. Its value depends on the operating temperature, fluid velocity, and length of service of the heat exchanger.

In addition, we know that fins are often added to surfaces exposed to either or both fluids and that, by increasing the surface area, they reduce the overall resistance to heat transfer. Accordingly, with inclusion of surface fouling and fin (extended surface) effects, the overall heat transfer coefficient is modified as follows:

$$\frac{1}{UA} = \frac{1}{(\eta_o h A)_c} + \frac{R''_{f,c}}{(\eta_o A)_c} + R_w + \frac{R''_{f,h}}{(\eta_o A)_h} + \frac{1}{(\eta_o h A)_h} \quad (11.1b)$$

Although representative fouling factors (R''_f) are listed in Table 11.1, the factor is a variable during heat exchanger operation (increasing from zero for a clean surface, as deposits accumulate on the surface). Comprehensive discussions of fouling are provided in References 2 through 4.

The quantity η_o in Equation 11.1b is termed the *overall surface efficiency* or *temperature effectiveness* of a finned surface. It is defined such that, for the hot or cold surface without fouling, the heat transfer rate is

$$q = \eta_o h A (T_b - T_\infty) \quad (11.2)$$

where T_b is the base surface temperature (Figure 3.21) and A is the total (fin plus exposed base) surface area. The quantity was introduced in Section 3.6.5, and the following expression was derived:

$$\eta_o = 1 - \frac{A_f}{A}(1 - \eta_f) \quad (11.3)$$

where A_f is the entire fin surface area and η_f is the efficiency of a single fin. To be consistent with the nomenclature commonly used in heat exchanger analysis, the ratio of fin surface area to the total surface area has been expressed as A_f/A. This representation differs

TABLE 11.1   Representative fouling factors [1]

Fluid	R''_f (m^2 · K/W)
Seawater and treated boiler feedwater (below 50°C)	0.0001
Seawater and treated boiler feedwater (above 50°C)	0.0002
River water (below 50°C)	0.0002–0.001
Fuel oil	0.0009
Refrigerating liquids	0.0002
Steam (nonoil bearing)	0.0001

from that of Section 3.6.5, where the ratio is expressed as NA_f/A_t, with A_f representing the area of a single fin and A_t the total surface area. If a straight or pin fin of length L (Figure 3.17) is used and an adiabatic tip is assumed, Equations 3.81 and 3.91 yield

$$\eta_f = \frac{\tanh(mL)}{mL} \tag{11.4}$$

where $m = (2h/kt)^{1/2}$ and t is the fin thickness. For several common fin shapes, the efficiency may be obtained from Table 3.5.

Note that, as written, Equation 11.2 corresponds to negligible fouling. However, if fouling is significant, the convection coefficient in Equation 11.2 must be replaced by a *partial* overall heat transfer coefficient of the form $U_p = h/(1 + hR_f'')$. In contrast to Equation 11.1b, which provides the overall heat transfer coefficient between the hot and cold fluids, U_p is termed a partial coefficient because it only includes the convection coefficient and fouling factor associated with one fluid and its adjoining surface. Partial coefficients for the hot and cold sides are then $U_{p,h} = h_h/(1 + h_h R_{f,h}'')$ and $U_{p,c} = h_c/(1 + h_c R_{f,c}'')$, respectively. Equation 11.3 may still be used to evaluate η_o for the hot and/or cold side, but U_p must be used in lieu of h to evaluate the corresponding fin efficiency. Moreover, it is readily shown that the second and fourth terms on the right-hand side of Equation 11.1b may be deleted if the convection coefficients in the first and fifth terms are replaced by $U_{p,c}$ and $U_{p,h}$, respectively.

The wall conduction term in Equation 11.1a or 11.1b may often be neglected, since a thin wall of large thermal conductivity is generally used. Also, one of the convection coefficients is often much smaller than the other and hence dominates determination of the overall coefficient. For example, if one of the fluids is a gas and the other is a liquid or a liquid–vapor mixture experiencing boiling or condensation, the gas-side convection coefficient is much smaller. It is in such situations that fins are used to enhance gas-side convection. Representative values of the overall coefficient are summarized in Table 11.2.

For the unfinned, tubular heat exchangers of Figures 11.1 through 11.4, Equation 11.1b reduces to

$$\frac{1}{UA} = \frac{1}{U_i A_i} = \frac{1}{U_o A_o}$$

$$= \frac{1}{h_i A_i} + \frac{R_{f,i}''}{A_i} + \frac{\ln(D_o/D_i)}{2\pi k L} + \frac{R_{f,o}''}{A_o} + \frac{1}{h_o A_o} \tag{11.5}$$

where subscripts i and o refer to inner and outer tube surfaces ($A_i = \pi D_i L$, $A_o = \pi D_o L$), which may be exposed to either the hot or the cold fluid.

TABLE 11.2   Representative values of the overall heat transfer coefficient

Fluid Combination	U (W/m² · K)
Water to water	850–1700
Water to oil	110–350
Steam condenser (water in tubes)	1000–6000
Ammonia condenser (water in tubes)	800–1400
Alcohol condenser (water in tubes)	250–700
Finned-tube heat exchanger (water in tubes, air in cross flow)	25–50

The overall heat transfer coefficient may be determined from knowledge of the hot and cold fluid convection coefficients and fouling factors and from appropriate geometric parameters. For unfinned surfaces, the convection coefficients may be estimated from correlations presented in Chapters 7 and 8. For standard fin configurations, the coefficients may be obtained from results compiled by Kays and London [5].

11.3 *Heat Exchanger Analysis: Use of the Log Mean Temperature Difference*

To design or to predict the performance of a heat exchanger, it is essential to relate the total heat transfer rate to quantities such as the inlet and outlet fluid temperatures, the overall heat transfer coefficient, and the total surface area for heat transfer. Two such relations may readily be obtained by applying overall energy balances to the hot and cold fluids, as shown in Figure 11.6. In particular, if q is the total rate of heat transfer between the hot and cold fluids and there is negligible heat transfer between the exchanger and its surroundings, as well as negligible potential and kinetic energy changes, application of the steady flow energy equation, Equation 1.12d, gives

$$q = \dot{m}_h(i_{h,i} - i_{h,o}) \tag{11.6a}$$

and

$$q = \dot{m}_c(i_{c,o} - i_{c,i}) \tag{11.7a}$$

where i is the fluid enthalpy. The subscripts h and c refer to the hot and cold fluids, whereas the subscripts i and o designate the fluid inlet and outlet conditions. If the fluids are not undergoing a phase change and constant specific heats are assumed, these expressions reduce to

$$q = \dot{m}_h c_{p,h}(T_{h,i} - T_{h,o}) \tag{11.6b}$$

and

$$q = \dot{m}_c c_{p,c}(T_{c,o} - T_{c,i}) \tag{11.7b}$$

where the temperatures appearing in the expressions refer to the *mean* fluid temperatures at the designated locations. Note that Equations 11.6 and 11.7 are independent of the flow arrangement and heat exchanger type.

Another useful expression may be obtained by relating the total heat transfer rate q to the temperature difference ΔT between the hot and cold fluids, where

$$\Delta T \equiv T_h - T_c \tag{11.8}$$

Such an expression would be an extension of Newton's law of cooling, with the overall heat transfer coefficient U used in place of the single convection coefficient h. However,

FIGURE 11.6 Overall energy balances for the hot and cold fluids of a two-fluid heat exchanger.

since ΔT varies with position in the heat exchanger, it is necessary to work with a rate equation of the form

$$q = UA\Delta T_m \qquad (11.9)$$

where ΔT_m is an appropriate *mean* temperature difference. Equation 11.9 may be used with Equations 11.6 and 11.7 to perform a heat exchanger analysis. Before this can be done, however, the specific form of ΔT_m must be established.

11.3.1 The Parallel-Flow Heat Exchanger

The hot and cold mean fluid temperature distributions associated with a parallel-flow heat exchanger are shown in Figure 11.7. The temperature difference ΔT is initially large but decays with increasing x, approaching zero asymptotically. It is important to note that, for such an exchanger, the outlet temperature of the cold fluid never exceeds that of the hot fluid. In Figure 11.7 the subscripts 1 and 2 designate opposite ends of the heat exchanger. This convention is used for all types of heat exchangers considered. For parallel flow, it follows that $T_{h,i} = T_{h,1}$, $T_{h,o} = T_{h,2}$, $T_{c,i} = T_{c,1}$, and $T_{c,o} = T_{c,2}$.

The form of ΔT_m may be determined by applying an energy balance to differential elements in the hot and cold fluids. Each element is of length dx and heat transfer surface area dA, as shown in Figure 11.7. The energy balances and the subsequent analysis are subject to the following assumptions.

1. The heat exchanger is insulated from its surroundings, in which case the only heat exchange is between the hot and cold fluids.
2. Axial conduction along the tubes is negligible.
3. Potential and kinetic energy changes are negligible.
4. The fluid specific heats are constant.
5. The overall heat transfer coefficient is constant.

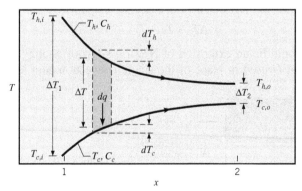

FIGURE 11.7 Temperature distributions for a parallel-flow heat exchanger.

The specific heats may of course change as a result of temperature variations, and the overall heat transfer coefficient may change because of variations in fluid properties and flow conditions. However, in many applications such variations are not significant, and it is reasonable to work with average values of $c_{p,c}$, $c_{p,h}$, and U for the heat exchanger.

Applying an energy balance to each of the differential elements of Figure 11.7, it follows that

$$dq = -\dot{m}_h c_{p,h}\, dT_h \equiv -C_h\, dT_h \tag{11.10}$$

and

$$dq = \dot{m}_c c_{p,c}\, dT_c \equiv C_c\, dT_c \tag{11.11}$$

where C_h and C_c are the hot and cold fluid *heat capacity rates*, respectively. These expressions may be integrated across the heat exchanger to obtain the overall energy balances given by Equations 11.6b and 11.7b. The rate of heat transfer across the surface area dA may also be expressed as

$$dq = U\,\Delta T\, dA \tag{11.12}$$

where $\Delta T = T_h - T_c$ is the *local* temperature difference between the hot and cold fluids.

To determine the integrated form of Equation 11.12, we begin by substituting Equations 11.10 and 11.11 into the differential form of Equation 11.8

$$d(\Delta T) = dT_h - dT_c$$

to obtain

$$d(\Delta T) = -dq\left(\frac{1}{C_h} + \frac{1}{C_c}\right)$$

Substituting for dq from Equation 11.12 and integrating across the heat exchanger, we obtain

$$\int_1^2 \frac{d(\Delta T)}{\Delta T} = -U\left(\frac{1}{C_h} + \frac{1}{C_c}\right)\int_1^2 dA$$

or

$$\ln\left(\frac{\Delta T_2}{\Delta T_1}\right) = -UA\left(\frac{1}{C_h} + \frac{1}{C_c}\right) \tag{11.13}$$

Substituting for C_h and C_c from Equations 11.6b and 11.7b, respectively, it follows that

$$\ln\left(\frac{\Delta T_2}{\Delta T_1}\right) = -UA\left(\frac{T_{h,i} - T_{h,o}}{q} + \frac{T_{c,o} - T_{c,i}}{q}\right)$$

$$= -\frac{UA}{q}[(T_{h,i} - T_{c,i}) - (T_{h,o} - T_{c,o})]$$

Recognizing that, for the parallel-flow heat exchanger of Figure 11.7, $\Delta T_1 = (T_{h,i} - T_{c,i})$ and $\Delta T_2 = (T_{h,o} - T_{c,o})$, we then obtain

$$q = UA\frac{\Delta T_2 - \Delta T_1}{\ln(\Delta T_2/\Delta T_1)}$$

Comparing the above expression with Equation 11.9, we conclude that the appropriate average temperature difference is a *log mean temperature difference,* ΔT_{lm}. Accordingly, we may write

$$q = UA\,\Delta T_{lm} \tag{11.14}$$

where

$$\Delta T_{lm} = \frac{\Delta T_2 - \Delta T_1}{\ln(\Delta T_2/\Delta T_1)} = \frac{\Delta T_1 - \Delta T_2}{\ln(\Delta T_1/\Delta T_2)} \tag{11.15}$$

Remember that, for the *parallel-flow exchanger,*

$$\left[\begin{array}{l} \Delta T_1 \equiv T_{h,1} - T_{c,1} = T_{h,i} - T_{c,i} \\ \Delta T_2 \equiv T_{h,2} - T_{c,2} = T_{h,o} - T_{c,o} \end{array} \right] \tag{11.16}$$

Referring back to Section 8.3.3, it can be seen that there is a strong similarity between the preceding analysis and the analysis of internal tube flow in which heat transfer occurs between the flowing fluid and either a surface at constant temperature or an external fluid at constant temperature. For this reason, internal tube flow is sometimes referred to as a *single stream heat exchanger.* Equations 8.43 and 8.44 or Equations 8.45a and 8.46a are analogous to Equations 11.14 and 11.15.

11.3.2 The Counterflow Heat Exchanger

The hot and cold fluid temperature distributions associated with a counterflow heat exchanger are shown in Figure 11.8. In contrast to the parallel-flow exchanger, this configuration provides for heat transfer between the hotter portions of the two fluids at one end, as well as between the colder portions at the other. For this reason, the change in the temperature difference, $\Delta T = T_h - T_c$, with respect to x is nowhere as large as it is for the inlet region of the parallel-flow exchanger. Note that the outlet temperature of the cold fluid may now exceed the outlet temperature of the hot fluid.

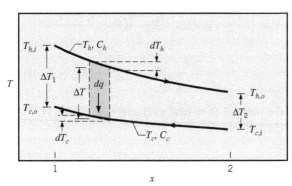

FIGURE 11.8   Temperature distributions for a counterflow heat exchanger.

Equations 11.6b and 11.7b apply to any heat exchanger and hence may be used for the counterflow arrangement. Moreover, from an analysis like that performed in Section 11.3.1, it may be shown that Equations 11.14 and 11.15 also apply. However, for the *counterflow exchanger* the endpoint temperature differences must now be defined as

$$\left[\begin{array}{l} \Delta T_1 \equiv T_{h,1} - T_{c,1} = T_{h,i} - T_{c,o} \\ \Delta T_2 \equiv T_{h,2} - T_{c,2} = T_{h,o} - T_{c,i} \end{array}\right] \tag{11.17}$$

Note that, for the same inlet and outlet temperatures, the log mean temperature difference for counterflow exceeds that for parallel flow, $\Delta T_{lm,CF} > \Delta T_{lm,PF}$. Hence the surface area required to effect a prescribed heat transfer rate q is smaller for the counterflow than for the parallel-flow arrangement, assuming the same value of U. Also note that $T_{c,o}$ can exceed $T_{h,o}$ for counterflow but not for parallel flow.

11.3.3 Special Operating Conditions

It is useful to note certain special conditions under which heat exchangers may be operated. Figure 11.9*a* shows temperature distributions for a heat exchanger in which the hot fluid has a heat capacity rate, $C_h \equiv \dot{m}_h c_{p,h}$, which is much larger than that of the cold fluid, $C_c \equiv \dot{m}_c c_{p,c}$. For this case the temperature of the hot fluid remains approximately constant throughout the heat exchanger, while the temperature of the cold fluid increases. The same condition is achieved if the hot fluid is a condensing vapor. Condensation occurs at constant temperature, and, for all practical purposes, $C_h \rightarrow \infty$. Conversely, in an evaporator or a boiler (Figure 11.9*b*), it is the cold fluid that experiences a change in phase and remains at a nearly uniform temperature ($C_c \rightarrow \infty$). The same effect is achieved without phase change if $C_h \ll C_c$. Note that, with condensation or evaporation, the heat rate is given by Equation 11.6a or 11.7a, respectively. Conditions illustrated in Figure 11.9*a* or 11.9*b* also characterize an internal tube flow (or *single stream heat exchanger*) exchanging heat with a surface at constant temperature or an external fluid at constant temperature.

The third special case (Figure 11.9*c*) involves a counterflow heat exchanger for which the heat capacity rates are equal ($C_h = C_c$). The temperature difference ΔT must then be constant throughout the exchanger, in which case $\Delta T_1 = \Delta T_2 = \Delta T_{lm}$.

Although flow conditions are more complicated in multipass and cross-flow heat exchangers, Equations 11.6, 11.7, 11.14, and 11.15 may still be used if modifications are made to the definition of the log mean temperature difference [6].

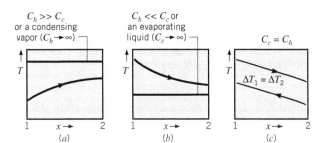

FIGURE 11.9 Special heat exchanger conditions. (*a*) $C_h \gg C_c$ or a condensing vapor. (*b*) An evaporating liquid or $C_h \ll C_c$. (*c*) A counterflow heat exchanger with equivalent fluid heat capacities ($C_h = C_c$).

Methodologies for using the LMTD method for other heat exchanger types are included in Section 11S.1.

> **EXAMPLE 11.1**
>
> A counterflow, concentric tube heat exchanger is used to cool the lubricating oil for a large industrial gas turbine engine. The flow rate of cooling water through the inner tube (D_i = 25 mm) is 0.2 kg/s, while the flow rate of oil through the outer annulus (D_o = 45 mm) is 0.1 kg/s. The oil and water enter at temperatures of 100 and 30°C, respectively. How long must the tube be made if the outlet temperature of the oil is to be 60°C?
>
> **SOLUTION**
>
> **Known:** Fluid flow rates and inlet temperatures for a counterflow, concentric tube heat exchanger of prescribed inner and outer diameter.
>
> **Find:** Tube length to achieve a desired hot fluid outlet temperature.
>
> **Schematic:**
>
>
>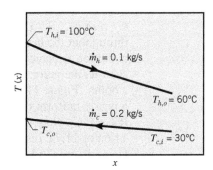
>
> **Assumptions:**
>
> 1. Negligible heat loss to the surroundings.
> 2. Negligible kinetic and potential energy changes.
> 3. Constant properties.
> 4. Negligible tube wall thermal resistance and fouling factors.
> 5. Fully developed conditions for the water and oil (U independent of x).
>
> **Properties:** Table A.5, unused engine oil ($\bar{T}_h$ = 80°C = 353 K): c_p = 2131 J/kg · K, μ = 3.25×10^{-2} N · s/m^2, k = 0.138 W/m · K. Table A.6, water ($\bar{T}_c$ ≈ 35°C): c_p = 4178 J/kg · K, μ = 725×10^{-6} N · s/m^2, k = 0.625 W/m · K, Pr = 4.85.
>
> **Analysis:** The required heat transfer rate may be obtained from the overall energy balance for the hot fluid, Equation 11.6b.
>
> $$q = \dot{m}_h c_{p,h}(T_{h,i} - T_{h,o})$$
>
> $$q = 0.1 \text{ kg/s} \times 2131 \text{ J/kg} \cdot \text{K} \,(100 - 60)°\text{C} = 8524 \text{ W}$$

Applying Equation 11.7b, the water outlet temperature is

$$T_{c,o} = \frac{q}{\dot{m}_c c_{p,c}} + T_{c,i}$$

$$T_{c,o} = \frac{8524 \text{ W}}{0.2 \text{ kg/s} \times 4178 \text{ J/kg} \cdot \text{K}} + 30°\text{C} = 40.2°\text{C}$$

Accordingly, use of $\overline{T}_c = 35°\text{C}$ to evaluate the water properties was a good choice. The required heat exchanger length may now be obtained from Equation 11.14,

$$q = UA \, \Delta T_{\text{lm}}$$

where $A = \pi D_i L$ and from Equations 11.15 and 11.17

$$\Delta T_{\text{lm}} = \frac{(T_{h,i} - T_{c,o}) - (T_{h,o} - T_{c,i})}{\ln\,[(T_{h,i} - T_{c,o})/(T_{h,o} - T_{c,i})]} = \frac{59.8 - 30}{\ln\,(59.8/30)} = 43.2°\text{C}$$

From Equation 11.5 the overall heat transfer coefficient is

$$U = \frac{1}{(1/h_i) + (1/h_o)}$$

For water flow through the tube,

$$Re_D = \frac{4\dot{m}_c}{\pi D_i \mu} = \frac{4 \times 0.2 \text{ kg/s}}{\pi(0.025 \text{ m})725 \times 10^{-6} \text{ N} \cdot \text{s/m}^2} = 14{,}050$$

Accordingly, the flow is turbulent and the convection coefficient may be computed from Equation 8.60

$$Nu_D = 0.023 \, Re_D^{4/5} \, Pr^{0.4}$$

$$Nu_D = 0.023(14{,}050)^{4/5}(4.85)^{0.4} = 90$$

Hence

$$h_i = Nu_D \frac{k}{D_i} = \frac{90 \times 0.625 \text{ W/m} \cdot \text{K}}{0.025 \text{ m}} = 2250 \text{ W/m}^2 \cdot \text{K}$$

For the flow of oil through the annulus, the hydraulic diameter is, from Equation 8.71, $D_h = D_o - D_i = 0.02$ m, and the Reynolds number is

$$Re_D = \frac{\rho u_m D_h}{\mu} = \frac{\rho(D_o - D_i)}{\mu} \times \frac{\dot{m}_h}{\rho\pi(D_o^2 - D_i^2)/4}$$

$$Re_D = \frac{4\dot{m}_h}{\pi(D_o + D_i)\mu} = \frac{4 \times 0.1 \text{ kg/s}}{\pi(0.045 + 0.025) \text{ m} \times 3.25 \times 10^{-2} \text{ kg/s} \cdot \text{m}} = 56.0$$

The annular flow is therefore laminar. Assuming uniform temperature along the inner surface of the annulus and a perfectly insulated outer surface, the convection coefficient at the *inner* surface may be obtained from Nu_i in Table 8.2. With $(D_i/D_o) = 0.56$, linear interpolation provides

$$Nu_i = \frac{h_o D_h}{k} = 5.63$$

and

$$h_o = 5.63 \frac{0.138 \ \text{W/m} \cdot \text{K}}{0.020 \ \text{m}} = 38.8 \ \text{W/m}^2 \cdot \text{K}$$

The overall convection coefficient is then

$$U = \frac{1}{(1/2250 \ \text{W/m}^2 \cdot \text{K}) + (1/38.8 \ \text{W/m}^2 \cdot \text{K})} = 38.1 \ \text{W/m}^2 \cdot \text{K}$$

and from the rate equation it follows that

$$L = \frac{q}{U \pi D_i \, \Delta T_{\text{lm}}} = \frac{8524 \ \text{W}}{38.1 \ \text{W/m}^2 \cdot \text{K} \ \pi (0.025 \ \text{m})(43.2°\text{C})} = 65.9 \ \text{m} \qquad \triangleleft$$

Comments:

1. The hot side convection coefficient controls the rate of heat transfer between the two fluids, and the low value of h_o is responsible for the large value of L. Incorporation of heat transfer enhancement methods, such as described in Section 8.7, could be used to decrease the size of the heat exchanger.

2. Because $h_i \gg h_o$, the tube wall temperature will follow closely that of the coolant water. Accordingly, the assumption of uniform wall temperature, which is inherent in the use of Table 8.2 to obtain h_o, is reasonable.

EXAMPLE 11.2

The counterflow, concentric tube heat exchanger of Example 11.1 is replaced with a compact, plate-type heat exchanger that consists of a stack of thin metal sheets, separated by N gaps of width a. The oil and water flows are subdivided into $N/2$ individual flow streams, with the oil and water moving in opposite directions within alternating gaps. It is desirable for the stack to be of a cubical geometry, with a characteristic exterior dimension L. Determine the exterior dimensions of the heat exchanger as a function of the number of gaps if the flow rates, inlet temperatures, and desired oil outlet temperature are the same as in Example 11.1. Compare the pressure drops of the water and oil streams within the plate-type heat exchanger to the pressure drops of the flow streams in Example 11.1, if 60 gaps are specified.

SOLUTION

Known: Configuration of a plate-type heat exchanger. Fluid flow rates, inlet temperatures, and desired oil outlet temperature.

Find:

1. Exterior dimensions of the heat exchanger.

2. Pressure drops within the plate-type heat exchanger with $N = 60$ gaps, and the concentric tube heat exchanger of Example 11.1.

Schematic:

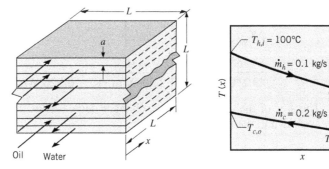

Assumptions:

1. Negligible heat loss to the surroundings.
2. Negligible kinetic and potential energy changes.
3. Constant properties.
4. Negligible plate thermal resistance and fouling factors.
5. Fully developed conditions for the water and oil.
6. Identical gap-to-gap heat transfer coefficients.
7. Heat exchanger exterior dimension is large compared to the gap width.

Properties: See Example 11.1. In addition, Table A.5, unused engine oil ($\overline{T}_h$ = 353 K): ρ = 852.1 kg/m^3. Table A.6, water ($\overline{T}_c$ ≈ 35°C): $\rho = v_f^{-1}$ = 994 kg/m^3.

Analysis:

1. The gap width may be related to the overall dimension of the heat exchanger by the expression $a = L/N$, and the total heat transfer area is $A = L^2 (N - 1)$. Assuming $a \ll L$ and the existence of laminar flow, the Nusselt number for each interior gap is provided in Table 8.1 and is

$$Nu_D = \frac{hD_h}{k} = 7.54$$

From Equation 8.66, the hydraulic diameter is $D_h = 2a$. Combining the preceding expressions yields for the water:

$$h_c = 7.54kN/2L = 7.54 \times 0.625 \text{ W/m} \cdot \text{K} \times N/2L = (2.36 \text{ W/m} \cdot \text{K})N/L$$

Likewise, for the oil:

$$h_h = 7.54kN/2L = 7.54 \times 0.138 \text{ W/m} \cdot \text{K} \times N/2L = (0.520 \text{ W/m} \cdot \text{K})N/L$$

and the overall convection coefficient is

$$U = \frac{1}{1/h_c + 1/h_h}$$

From Example 11.1, the required log mean temperature difference and heat transfer rate are ΔT_{lm} = 43.2°C and q = 8524 W, respectively. From Equation 11.14 it follows that

$$UA = \frac{L^2 (N - 1)}{1/h_c + 1/h_h} = \frac{q}{\Delta T_{lm}}$$

which may be rearranged to yield

$$L = \frac{q}{\Delta T_{\text{lm}}(N-1)}\left[\frac{1}{h_c L} + \frac{1}{h_h L}\right]$$

$$= \frac{8524\ \text{W}}{43.2°\text{C}(N-1)N}\left[\frac{1}{2.36\ \text{W/m} \cdot \text{K}} + \frac{1}{0.520\ \text{W/m} \cdot \text{K}}\right] = \frac{463\ \text{m}}{N(N-1)} \qquad \triangleleft$$

The size of the compact heat exchanger decreases as the number of gaps is increased, as shown in the figure below.

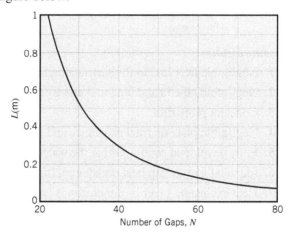

2. For $N = 60$ gaps, the stack dimension is $L = 0.131$ m from the results of part 1, and the gap width is $a = L/N = 0.131\ \text{m}/60 = 0.00218$ m.

The hydraulic diameter is $D_h = 0.00436$ m, and the mean velocity in each water-filled gap is

$$u_m = \frac{\dot{m}}{\rho L^2/2} = \frac{2 \times 0.2\ \text{kg/s}}{994\ \text{kg/m}^3 \times 0.131^2\ \text{m}^2} = 0.0235\ \text{m/s}$$

providing a Reynolds number of

$$Re_D = \frac{\rho u_m D_h}{\mu} = \frac{994\ \text{kg/m}^3 \times 0.0235\ \text{m/s} \times 0.00436\ \text{m}}{725 \times 10^{-6}\ \text{N} \cdot \text{s/m}^2} = 141$$

For the oil-filled gaps

$$u_m = \frac{\dot{m}}{\rho L^2/2} = \frac{2 \times 0.1\ \text{kg/s}}{852.1\ \text{kg/m}^3 \times 0.131^2\ \text{m}^2} = 0.0137\ \text{m/s}$$

yielding a Reynolds number of

$$Re_D = \frac{\rho u_m D_h}{\mu} = \frac{852.1\ \text{kg/m}^3 \times 0.0137\ \text{m/s} \times 0.00436\ \text{m}}{3.25 \times 10^{-2}\ \text{N} \cdot \text{s/m}^2} = 1.57$$

Therefore, the flow is laminar for both fluids, as assumed in part 1. Equations 8.19 and 8.22a may be used to calculate the pressure drop for the water:

$$\Delta p = \frac{64}{Re_D} \cdot \frac{\rho u_m^2}{2D_h} \cdot L = \frac{64}{141} \times \frac{994\ \text{kg/m}^3 \times 0.0235^2\ \text{m}^2/\text{s}^2}{2 \times 0.00436\ \text{m}} = 0.131\ \text{m}$$

$$= 3.76\ \text{N/m}^2 \qquad \triangleleft$$

Similarly, for the oil

$$\Delta p = \frac{64}{Re_D} \cdot \frac{\rho u_m^2}{2D_h} \cdot L = \frac{64}{1.57} \times \frac{852.1 \text{ kg/m}^3 \times 0.0137^2 \text{ m}^2/\text{s}^2}{2 \times 0.00436 \text{ m}} \times 0.131 \text{ m}$$
$$= 98.2 \text{ N/m}^2 \qquad \qquad \triangleleft$$

For Example 11.1, the friction factor associated with the water flow may be calculated using Equation 8.21, and for a smooth surface condition is $f = (0.790 \ln(14,050) - 1.64)^{-2} = 0.0287$. The mean velocity is $u_m = 4\dot{m}/(\rho \pi D_i^2) = 4 \times 0.2 \text{ kg/s}/(994 \text{ kg/m}^3 \times \pi \times 0.025^2 \text{ m}^2) = 0.410 \text{ m/s}$, and the pressure drop is

$$\Delta p = f \cdot \frac{\rho u_m^2}{2D_h} \cdot L = 0.0287 \times \frac{994 \text{ kg/m}^3 \times 0.410^2 \text{ m}^2/\text{s}^2}{2 \times 0.025 \text{ m}} \times 65.9 \text{ m}$$
$$= 6310 \text{ N/m}^2 \qquad \qquad \triangleleft$$

For the oil flowing in the annular region, the mean velocity is $u_m = 4\dot{m}/[\rho \pi (D_o^2 - D_i^2)] = 4 \times 0.1 \text{ kg/s}/[852.1 \text{ kg/m}^3 \times \pi \times (0.045^2 - 0.025^2) \text{m}^2] = 0.107 \text{ m/s}$, and the pressure drop is

$$\Delta p = \frac{64}{Re_D} \cdot \frac{\rho u_m^2}{2D_h} \cdot L = \frac{64}{56} \times \frac{852.1 \text{ kg/m}^3 \times 0.107^2 \text{ m}^2/\text{s}^2}{2 \times 0.020 \text{ m}} \times 65.9 \text{ m}$$
$$= 18,300 \text{ N/m}^2 \qquad \qquad \triangleleft$$

Comments:

1. Increasing the number of gaps increases the *UA* product by simultaneously providing more surface area and increasing the heat transfer coefficients associated with the flow of the fluids through smaller passages.

2. The area-to-volume ratio of the $N = 60$ heat exchanger is $L^2(N-1)/L^3 = (N-1)/L = (60-1)/0.131 \text{ m} = 451 \text{ m}^2/\text{m}^3$.

3. The volume occupied by the concentric tube heat exchanger is $V = \pi D_o^2 L/4 = \pi \times 0.045^2 \text{ m}^2 \times 65.9 \text{ m}/4 = 0.10 \text{ m}^3$, while the volume of the compact plate-type exchanger is $V = L^3 = 0.131^3 \text{ m}^3 = 0.0022 \text{ m}^3$. Use of the plate-type heat exchanger results in a 97.8% reduction in volume relative to the conventional, concentric tube heat exchanger.

4. The pressure drops associated with use of the compact heat exchanger are significantly less than for a conventional concentric tube configuration. Pressure drops are reduced by 99.9% and 99.5% for the water and oil flows, respectively.

5. Fouling of the heat transfer surfaces may result in a decrease in the gap width, as well as an associated reduction in heat transfer rate and increase in pressure drop.

6. Because $h_c > h_h$, the temperatures of the thin metal sheets will follow closely that of the water, and, as in Example 11.1, the assumption of uniform temperature conditions to obtain h_c and h_h is reasonable.

7. One method to fabricate such a heat exchanger is presented in C. F. McDonald, *Appl. Thermal Engin.*, **20**, 471, 2000.

11.4 *Heat Exchanger Analysis: The Effectiveness–NTU Method*

It is a simple matter to use the log mean temperature difference (LMTD) method of heat exchanger analysis when the fluid inlet temperatures are known and the outlet temperatures are specified or readily determined from the energy balance expressions, Equations 11.6b and 11.7b. The value of ΔT_{lm} for the exchanger may then be determined. However, if only the inlet temperatures are known, use of the LMTD method requires a cumbersome iterative procedure. It is therefore preferable to employ an alternative approach termed the *effectiveness*–NTU (or NTU) method.

11.4.1 Definitions

To define the *effectiveness of a heat exchanger*, we must first determine the *maximum possible heat transfer rate*, q_{max}, for the exchanger. This heat transfer rate could, in principle, be achieved in a counterflow heat exchanger (Figure 11.8) of infinite length. In such an exchanger, one of the fluids would experience the maximum possible temperature difference, $T_{h,i} - T_{c,i}$. To illustrate this point, consider a situation for which $C_c < C_h$, in which case, from Equations 11.10 and 11.11, $|dT_c| > |dT_h|$. The cold fluid would then experience the larger temperature change, and since $L \rightarrow \infty$, it would be heated to the inlet temperature of the hot fluid ($T_{c,o} = T_{h,i}$). Accordingly, from Equation 11.7b,

$$C_c < C_h: \qquad q_{max} = C_c(T_{h,i} - T_{c,i})$$

Similarly, if $C_h < C_c$, the hot fluid would experience the larger temperature change and would be cooled to the inlet temperature of the cold fluid ($T_{h,o} = T_{c,i}$). From Equation 11.6b, we then obtain

$$C_h < C_c: \qquad q_{max} = C_h(T_{h,i} - T_{c,i})$$

From the foregoing results we are then prompted to write the general expression

$$q_{max} = C_{min}(T_{h,i} - T_{c,i}) \tag{11.18}$$

where C_{min} is equal to C_c or C_h, whichever is smaller. For prescribed hot and cold fluid inlet temperatures, Equation 11.18 provides the maximum heat transfer rate that could possibly be delivered by an exchanger. A quick mental exercise should convince the reader that the maximum possible heat transfer rate is *not* equal to $C_{max}(T_{h,i} - T_{c,i})$. If the fluid having the larger heat capacity rate were to experience the maximum possible temperature change, conservation of energy in the form $C_c(T_{c,o} - T_{c,i}) = C_h(T_{h,i} - T_{h,o})$ would require that the other fluid experience yet a larger temperature change. For example, if $C_{max} = C_c$ and one argues that it is possible for $T_{c,o}$ to be equal to $T_{h,i}$, it follows that $(T_{h,i} - T_{h,o}) = (C_c/C_h)(T_{h,i} - T_{c,i})$, in which case $(T_{h,i} - T_{h,o}) > (T_{h,i} - T_{c,i})$. Such a condition is clearly impossible.

It is now logical to define the *effectiveness*, ε, as the ratio of the actual heat transfer rate for a heat exchanger to the maximum possible heat transfer rate:

$$\varepsilon \equiv \frac{q}{q_{max}} \tag{11.19}$$

From Equations 11.6b, 11.7b, and 11.18, it follows that

$$\varepsilon = \frac{C_h(T_{h,i} - T_{h,o})}{C_{min}(T_{h,i} - T_{c,i})} \tag{11.20}$$

or

$$\varepsilon = \frac{C_c(T_{c,o} - T_{c,i})}{C_{min}(T_{h,i} - T_{c,i})} \tag{11.21}$$

By definition the effectiveness, which is dimensionless, must be in the range $0 \leq \varepsilon \leq 1$. It is useful because, if ε, $T_{h,i}$, and $T_{c,i}$ are known, the actual heat transfer rate may readily be determined from the expression

$$q = \varepsilon C_{min}(T_{h,i} - T_{c,i}) \tag{11.22}$$

For any heat exchanger it can be shown that [5]

$$\varepsilon = f\left(\text{NTU}, \frac{C_{min}}{C_{max}}\right) \tag{11.23}$$

where C_{min}/C_{max} is equal to C_c/C_h or C_h/C_c, depending on the relative magnitudes of the hot and cold fluid heat capacity rates. The *number of transfer units* (NTU) is a dimensionless parameter that is widely used for heat exchanger analysis and is defined as

$$\text{NTU} \equiv \frac{UA}{C_{min}} \tag{11.24}$$

11.4.2   Effectiveness–NTU Relations

To determine a specific form of the effectiveness–NTU relation, Equation 11.23, consider a *parallel-flow* heat exchanger for which $C_{min} = C_h$. From Equation 11.20 we then obtain

$$\varepsilon = \frac{T_{h,i} - T_{h,o}}{T_{h,i} - T_{c,i}} \tag{11.25}$$

and from Equations 11.6b and 11.7b it follows that

$$\frac{C_{min}}{C_{max}} = \frac{\dot{m}_h c_{p,h}}{\dot{m}_c c_{p,c}} = \frac{T_{c,o} - T_{c,i}}{T_{h,i} - T_{h,o}} \tag{11.26}$$

Now consider Equation 11.13, which may be expressed as

$$\ln\left(\frac{T_{h,o} - T_{c,o}}{T_{h,i} - T_{c,i}}\right) = -\frac{UA}{C_{min}}\left(1 + \frac{C_{min}}{C_{max}}\right)$$

or from Equation 11.24

$$\frac{T_{h,o} - T_{c,o}}{T_{h,i} - T_{c,i}} = \exp\left[-\text{NTU}\left(1 + \frac{C_{min}}{C_{max}}\right)\right] \tag{11.27}$$

Rearranging the left-hand side of this expression as

$$\frac{T_{h,o} - T_{c,o}}{T_{h,i} - T_{c,i}} = \frac{T_{h,o} - T_{h,i} + T_{h,i} - T_{c,o}}{T_{h,i} - T_{c,i}}$$

and substituting for $T_{c,o}$ from Equation 11.26, it follows that

$$\frac{T_{h,o} - T_{c,o}}{T_{h,i} - T_{c,i}} = \frac{(T_{h,o} - T_{h,i}) + (T_{h,i} - T_{c,i}) - (C_{min}/C_{max})(T_{h,i} - T_{h,o})}{T_{h,i} - T_{c,i}}$$

or from Equation 11.25

$$\frac{T_{h,o} - T_{c,o}}{T_{h,i} - T_{c,i}} = -\varepsilon + 1 - \left(\frac{C_{\min}}{C_{\max}}\right)\varepsilon = 1 - \varepsilon\left(1 + \frac{C_{\min}}{C_{\max}}\right)$$

Substituting the above expression into Equation 11.27 and solving for ε, we obtain for the *parallel-flow heat exchanger*

$$\varepsilon = \frac{1 - \exp\{-\text{NTU}[1 + (C_{\min}/C_{\max})]\}}{1 + (C_{\min}/C_{\max})} \tag{11.28a}$$

Since precisely the same result may be obtained for $C_{\min} = C_c$, Equation 11.28a applies for any parallel-flow heat exchanger, irrespective of whether the minimum heat capacity rate is associated with the hot or cold fluid.

Similar expressions have been developed for a variety of heat exchangers [5], and representative results are summarized in Table 11.3, where C_r is the *heat capacity ratio* $C_r \equiv C_{\min}/C_{\max}$. In deriving Equation 11.31a for a shell-and-tube heat exchanger with multiple shell passes, it is assumed that the total NTU is equally distributed between shell passes of the same arrangement, $\text{NTU} = n(\text{NTU})_1$. In order to determine ε, $(\text{NTU})_1$ would first be calculated using the heat transfer area for one shell, ε_1 would then be calculated from Equation 11.30a, and ε would finally be calculated from Equation 11.31a. Note that for $C_r = 0$, as in a boiler, condenser, or single stream heat exchanger, ε is given by Equation 11.35a for

TABLE 11.3   Heat exchanger effectiveness relations [5]

Flow Arrangement	Relation	
Parallel flow	$\varepsilon = \dfrac{1 - \exp[-\text{NTU}(1 + C_r)]}{1 + C_r}$	(11.28a)
Counterflow	$\varepsilon = \dfrac{1 - \exp[-\text{NTU}(1 - C_r)]}{1 - C_r \exp[-\text{NTU}(1 - C_r)]} \quad (C_r < 1)$	
	$\varepsilon = \dfrac{\text{NTU}}{1 + \text{NTU}} \quad (C_r = 1)$	(11.29a)
Shell-and-tube		
One shell pass (2, 4, ... tube passes)	$\varepsilon_1 = 2\left\{1 + C_r + (1 + C_r^2)^{1/2} \times \dfrac{1 + \exp[-(\text{NTU})_1(1 + C_r^2)^{1/2}]}{1 - \exp[-(\text{NTU})_1(1 + C_r^2)^{1/2}]}\right\}^{-1}$	(11.30a)
n shell passes (2n, 4n, ... tube passes)	$\varepsilon = \left[\left(\dfrac{1 - \varepsilon_1 C_r}{1 - \varepsilon_1}\right)^n - 1\right]\left[\left(\dfrac{1 - \varepsilon_1 C_r}{1 - \varepsilon_1}\right)^n - C_r\right]^{-1}$	(11.31a)
Cross-flow (single pass)		
Both fluids unmixed	$\varepsilon = 1 - \exp\left[\left(\dfrac{1}{C_r}\right)(\text{NTU})^{0.22}\{\exp[-C_r(\text{NTU})^{0.78}] - 1\}\right]$	(11.32)
$C_{\max}$ (mixed), $C_{\min}$ (unmixed)	$\varepsilon = \left(\dfrac{1}{C_r}\right)(1 - \exp\{-C_r[1 - \exp(-\text{NTU})]\})$	(11.33a)
$C_{\min}$ (mixed), $C_{\max}$ (unmixed)	$\varepsilon = 1 - \exp(-C_r^{-1}\{1 - \exp[-C_r(\text{NTU})]\})$	(11.34a)
All exchangers ($C_r = 0$)	$\varepsilon = 1 - \exp(-\text{NTU})$	(11.35a)

all flow arrangements. Hence, for this special case, it follows that heat exchanger behavior is independent of flow arrangement. For the cross-flow heat exchanger with both fluids unmixed, Equation 11.32 is exact only for $C_r = 1$. However, it may be used to a good approximation for all $0 < C_r \leq 1$. For $C_r = 0$, Equation 11.35a must be used.

In heat exchanger design calculations (Section 11.5), it is more convenient to work with ε–NTU relations of the form

$$\text{NTU} = f\left(\varepsilon, \frac{C_{min}}{C_{max}}\right)$$

Explicit relations for NTU as a function of ε and C_r are provided in Table 11.4. Note that Equation 11.32 may not be manipulated to yield a direct relationship for NTU as a function of ε and C_r. Note also that to determine the NTU for a shell-and-tube heat exchanger with multiple shell passes, ε would first be calculated for the entire heat exchanger. The variables F and ε_1 would then be calculated using Equations 11.31c and 11.31b, respectively. The parameter E would subsequently be determined from Equation 11.30c and substituted into Equation 11.30b to find $(\text{NTU})_1$. Finally, this result would be multiplied by n to obtain the NTU for the entire exchanger, as indicated in Equation 11.31d.

The foregoing expressions are represented graphically in Figures 11.10 through 11.15. When using Figure 11.13, the abscissa corresponds to the total number of transfer units,

TABLE 11.4 Heat exchanger NTU relations

Flow Arrangement	Relation	
Parallel flow	$\text{NTU} = -\dfrac{\ln[1 - \varepsilon(1 + C_r)]}{1 + C_r}$	(11.28b)
Counterflow	$\text{NTU} = \dfrac{1}{C_r - 1}\ln\left(\dfrac{\varepsilon - 1}{\varepsilon C_r - 1}\right) \qquad (C_r < 1)$	
	$\text{NTU} = \dfrac{\varepsilon}{1 - \varepsilon} \qquad (C_r = 1)$	(11.29b)
Shell-and-tube		
One shell pass (2, 4, . . . tube passes)	$(\text{NTU})_1 = -(1 + C_r^2)^{-1/2} \ln\left(\dfrac{E - 1}{E + 1}\right)$	(11.30b)
	$E = \dfrac{2/\varepsilon_1 - (1 + C_r)}{(1 + C_r^2)^{1/2}}$	(11.30c)
n shell passes (2n, 4n, . . . tube passes)	Use Equations 11.30b and 11.30c with $\varepsilon_1 = \dfrac{F - 1}{F - C_r} \quad F = \left(\dfrac{\varepsilon C_r - 1}{\varepsilon - 1}\right)^{1/n} \quad \text{NTU} = n(\text{NTU})_1$	(11.31b, c, d)
Cross-flow (single pass)		
C_{max} (mixed), C_{min} (unmixed)	$\text{NTU} = -\ln\left[1 + \left(\dfrac{1}{C_r}\right)\ln(1 - \varepsilon C_r)\right]$	(11.33b)
C_{min} (mixed), C_{max} (unmixed)	$\text{NTU} = -\left(\dfrac{1}{C_r}\right)\ln[C_r \ln(1 - \varepsilon) + 1]$	(11.34b)
All exchangers ($C_r = 0$)	$\text{NTU} = -\ln(1 - \varepsilon)$	(11.35b)

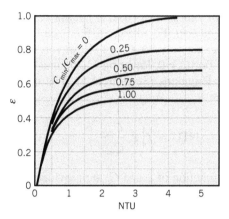

FIGURE 11.10   Effectiveness of a parallel-flow heat exchanger (Equation 11.28).

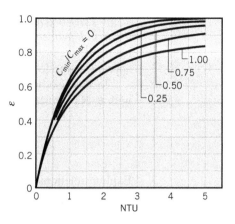

FIGURE 11.11   Effectiveness of a counterflow heat exchanger (Equation 11.29).

FIGURE 11.12   Effectiveness of a shell-and-tube heat exchanger with one shell and any multiple of two tube passes (two, four, etc. tube passes) (Equation 11.30).

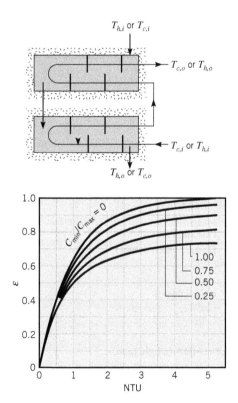

FIGURE 11.13   Effectiveness of a shell-and-tube heat exchanger with two shell passes and any multiple of four tube passes (four, eight, etc. tube passes) (Equation 11.31 with $n = 2$).

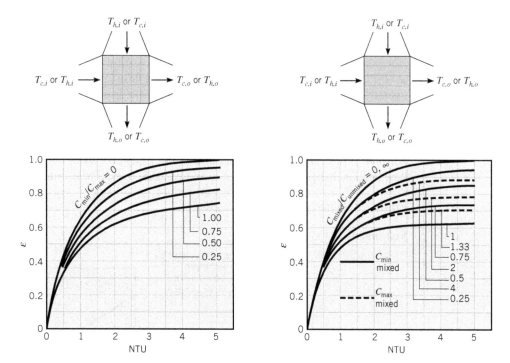

FIGURE 11.14 Effectiveness of a single-pass, cross-flow heat exchanger with both fluids unmixed (Equation 11.32).

FIGURE 11.15 Effectiveness of a single-pass, cross-flow heat exchanger with one fluid mixed and the other unmixed (Equations 11.33, 11.34).

NTU = n(NTU)$_1$. For Figure 11.15 the solid curves correspond to C_{min} mixed and C_{max} unmixed, while the dashed curves correspond to C_{min} unmixed and C_{max} mixed. Note that for $C_r = 0$, all heat exchangers have the same effectiveness, which may be computed from Equation 11.35a. Moreover, if NTU $\lesssim 0.25$, all heat exchangers have approximately the same effectiveness, regardless of the value of C_r, and ε may again be computed from Equation 11.35a. More generally, for $C_r > 0$ and NTU $\gtrsim 0.25$, the counterflow exchanger is the most effective. For any exchanger, maximum and minimum values of the effectiveness are associated with $C_r = 0$ and $C_r = 1$, respectively.

As noted previously, in the context of cross-flow heat exchangers, the terms *mixed* and *unmixed* are idealizations representing limiting cases of actual flow conditions. That is, most flows are neither completely mixed nor unmixed, but exhibit partial degrees of mixing. This issue has been addressed by DiGiovanni and Webb [7], and algebraic expressions have been developed to determine the ε–NTU relationship for arbitrary values of partial mixing.

We also note that both the LMTD and ε–NTU methods approach heat exchanger analysis from a global perspective and provide no information concerning conditions within the exchanger. Although flow and temperature variations within a heat exchanger may be determined using commercial CFD (computational fluid dynamic) computer codes, simpler numerical procedures may be adopted. Such procedures have been applied by Ribando et al. to determine temperature variations in concentric tube and shell-and-tube heat exchangers [8].

EXAMPLE 11.3

Hot exhaust gases, which enter a finned-tube, cross-flow heat exchanger at 300°C and leave at 100°C, are used to heat pressurized water at a flow rate of 1 kg/s from 35 to 125°C. The overall heat transfer coefficient based on the gas-side surface area is $U_h = 100$ W/m² · K. Determine the required gas-side surface area A_h using the NTU method.

SOLUTION

Known: Inlet and outlet temperatures of hot gases and water used in a finned-tube, cross-flow heat exchanger. Water flow rate and gas-side overall heat transfer coefficient.

Find: Required gas-side surface area.

Schematic:

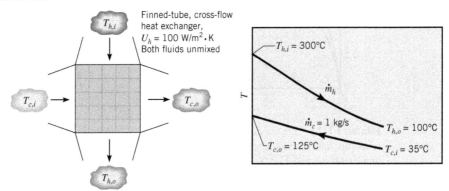

Assumptions:

1. Negligible heat loss to the surroundings and kinetic and potential energy changes.
2. Constant properties.

Properties: Table A.6, water ($\overline{T}_c = 80$°C): $c_{p,c} = 4197$ J/kg · K.

Analysis: The required surface area may be obtained from knowledge of the number of transfer units, which, in turn, may be obtained from knowledge of the ratio of heat capacity rates and the effectiveness. To determine the minimum heat capacity rate, we begin by computing

$$C_c = \dot{m}_c c_{p,c} = 1 \text{ kg/s} \times 4197 \text{ J/kg} \cdot \text{K} = 4197 \text{ W/K}$$

Since $\dot{m}_h$ is not specified, C_h is obtained by combining the overall energy balances, Equations 11.6b and 11.7b:

$$C_h = \dot{m}_h c_{p,h} = C_c \frac{T_{c,o} - T_{c,i}}{T_{h,i} - T_{h,o}} = 4197 \frac{125 - 35}{300 - 100} = 1889 \text{ W/K} = C_{\min}$$

From Equation 11.18

$$q_{\max} = C_{\min}(T_{h,i} - T_{c,i}) = 1889 \text{ W/K} (300 - 35)°\text{C} = 5.00 \times 10^5 \text{ W}$$

From Equation 11.7b the actual heat transfer rate is

$$q = C_c(T_{c,o} - T_{c,i}) = 4197 \text{ W/K } (125 - 35)°C$$

$$q = 3.78 \times 10^5 \text{ W}$$

Hence from Equation 11.19 the effectiveness is

$$\varepsilon = \frac{q}{q_{max}} = \frac{3.78 \times 10^5 \text{ W}}{5.00 \times 10^5 \text{ W}} = 0.755$$

With

$$\frac{C_{min}}{C_{max}} = \frac{1889}{4197} = 0.45$$

it follows from Figure 11.14 that

$$\text{NTU} = \frac{U_h A_h}{C_{min}} \approx 2.0$$

or

$$A_h = \frac{2.0(1889 \text{ W/K})}{100 \text{ W/m}^2 \cdot \text{K}} = 37.8 \text{ m}^2 \qquad \triangleleft$$

Comments:

1. Equation 11.32 may be solved iteratively or by trial and error to yield NTU = 2.0, which is in excellent agreement with the estimate obtained from the charts.

2. With the heat exchanger sized ($A_h = 37.8 \text{ m}^2$) and placed into operation, its actual performance is subject to uncontrolled variations in the exhaust gas inlet temperature ($200 \leq T_{h,i} \leq 400°C$) and to gradual degradation of the heat exchanger surfaces due to fouling (U_h decreasing from 100 to 60 W/m² · K). For a fixed value of $C_{min} = C_h = 1889$ W/K, the reduction in U_h corresponds to a reduction in the NTU (to NTU ≈ 1.20) and hence to a reduction in the heat exchanger effectiveness, which can be computed from Equation 11.32. The effect of the variations on the water outlet temperature has been computed and is plotted as follows:

If the intent is to maintain a fixed water outlet temperature of $T_{c,o} = 125°C$, adjustments in the flow rates, $\dot{m}_c$ and $\dot{m}_h$, could be made to compensate for the variations. The model equations could be used to determine the adjustments and hence as a basis for designing the requisite *controller*.

11.5 *Heat Exchanger Design and Performance Calculations*

Two general types of heat exchanger problems are commonly encountered by the practicing engineer.

In the *heat exchanger design problem*, the fluid inlet temperatures and flow rates, as well as a desired hot or cold fluid outlet temperature, are prescribed. The design problem is then one of specifying a specific heat exchanger type and determining its size—that is, the heat transfer surface area A—required to achieve the desired outlet temperature. The design problem is commonly encountered when a heat exchanger is to be custom-built for a specific application. Alternatively, in a *heat exchanger performance calculation*, an existing heat exchanger is analyzed to determine the heat transfer rate and the fluid outlet temperatures for prescribed flow rates and inlet temperatures. The performance calculation is commonly associated with the use of off-the-shelf heat exchanger types and sizes available from a vendor.

For heat exchanger design problems, the NTU method may be used by first calculating ε and (C_{min}/C_{max}). The appropriate equation (or chart) may then be used to obtain the NTU value, which in turn may be used to determine A. For a performance calculation, the NTU and (C_{min}/C_{max}) values may be computed and ε may then be determined from the appropriate equation (or chart) for a particular exchanger type. Since q_{max} may also be computed from Equation 11.18, it is a simple matter to determine the actual heat transfer rate from the requirement that $q = \varepsilon q_{max}$. Both fluid outlet temperatures may then be determined from Equations 11.6b and 11.7b.

EXAMPLE 11.4

Consider a heat exchanger similar to that of Example 11.3, that is, a finned-tube, cross-flow heat exchanger with a gas-side overall heat transfer coefficient and area of 100 W/m²·K and 40 m², respectively. The water flow rate and inlet temperature remain at 1 kg/s and 35°C. However, a change in operating conditions for the hot gas generator causes the gases to now enter the heat exchanger with a flow rate of 1.5 kg/s and a temperature of 250°C. What is the rate of heat transfer by the exchanger, and what are the gas and water outlet temperatures?

SOLUTION

Known: Hot and cold fluid inlet conditions for a finned-tube, cross-flow heat exchanger of known surface area and overall heat transfer coefficient.

Find: Heat transfer rate and fluid outlet temperatures.

Schematic:

Assumptions:

1. Negligible heat loss to surroundings and kinetic and potential energy changes.
2. Constant properties (unchanged from Example 11.3).

Analysis: The problem may be classified as one requiring a heat exchanger *performance calculation.* The heat capacity rates are

$$C_c = \dot{m}_c c_{p,c} = 1\,\text{kg/s} \times 4197\,\text{J/kg} \cdot \text{K} = 4197\,\text{W/K}$$

$$C_h = \dot{m}_h c_{p,h} = 1.5\,\text{kg/s} \times 1000\,\text{J/kg} \cdot \text{K} = 1500\,\text{W/K} = C_{min}$$

in which case

$$\frac{C_{min}}{C_{max}} = \frac{1500}{4197} = 0.357$$

The number of transfer units is

$$\text{NTU} = \frac{U_h A_h}{C_{min}} = \frac{100\,\text{W/m}^2 \cdot \text{K} \times 40\,\text{m}^2}{1500\,\text{W/K}} = 2.67$$

From Figure 11.14 the heat exchanger effectiveness is then $\varepsilon \approx 0.82$, and from Equation 11.18 the maximum possible heat transfer rate is

$$q_{max} = C_{min}(T_{h,i} - T_{c,i}) = 1500\,\text{W/K}\,(250 - 35)°\text{C} = 3.23 \times 10^5\,\text{W}$$

Accordingly, from the definition of ε, Equation 11.19, the actual heat transfer rate is

$$q = \varepsilon q_{max} = 0.82 \times 3.23 \times 10^5\,\text{W} = 2.65 \times 10^5\,\text{W} \qquad \triangleleft$$

It is now a simple matter to determine the outlet temperatures from the overall energy balances. From Equation 11.6b

$$T_{h,o} = T_{h,i} - \frac{q}{\dot{m}_h c_{p,h}} = 250°\text{C} - \frac{2.65 \times 10^5\,\text{W}}{1500\,\text{W/K}} = 73.3°\text{C} \qquad \triangleleft$$

and from Equation 11.7b

$$T_{c,o} = T_{c,i} + \frac{q}{\dot{m}_c c_{p,c}} = 35°\text{C} + \frac{2.65 \times 10^5 \text{ W}}{4197 \text{ W/K}} = 98.1°\text{C} \qquad \lhd$$

Comments:

1. From Equation 11.32, $\varepsilon = 0.845$, which is in good agreement with the estimate obtained from the charts.

2. The overall heat transfer coefficient has tacitly been assumed to be unaffected by the change in $\dot{m}_h$. In fact, with an approximately 20% reduction in $\dot{m}_h$, there would be a significant, albeit smaller percentage, reduction in U_h.

3. As discussed in the Comment of Example 11.3, flow rate adjustments could be made to maintain a fixed water outlet temperature. If, for example, the outlet temperature must be maintained at $T_{c,o} = 125°\text{C}$, the water flow rate could be reduced to an amount prescribed by Equation 11.7b. That is,

$$\dot{m}_c = \frac{q}{c_{p,c}(T_{c,o} - T_{c,i})} = \frac{2.65 \times 10^5 \text{ W}}{4197 \text{ J/kg} \cdot \text{K} (125 - 35)°\text{C}} = 0.702 \text{ kg/s}$$

The change in flow rate has again been presumed to have a negligible effect on U_h. In this case the assumption is good, since the dominant contribution to U_h is made by the gas-side, and not the water-side, convection coefficient.

EXAMPLE 11.5

The condenser of a large steam power plant is a heat exchanger in which steam is condensed to liquid water. Assume the condenser to be a *shell-and-tube* heat exchanger consisting of a single shell and 30,000 tubes, each executing two passes. The tubes are of thin wall construction with $D = 25$ mm, and steam condenses on their outer surface with an associated convection coefficient of $h_o = 11,000$ W/m$^2 \cdot$ K. The heat transfer rate that must be effected by the exchanger is $q = 2 \times 10^9$ W, and this is accomplished by passing cooling water through the tubes at a rate of 3×10^4 kg/s (the flow rate per tube is therefore 1 kg/s). The water enters at 20°C, while the steam condenses at 50°C. What is the temperature of the cooling water emerging from the condenser? What is the required tube length L per pass?

SOLUTION

Known: Heat exchanger consisting of single shell and 30,000 tubes with two passes each.

Find:

1. Outlet temperature of the cooling water.

2. Tube length per pass to achieve required heat transfer rate.

Schematic:

Assumptions:

1. Negligible heat transfer between exchanger and surroundings and negligible kinetic and potential energy changes.
2. Tube internal flow and thermal conditions fully developed.
3. Negligible thermal resistance of tube material and fouling effects.
4. Constant properties.

Properties: Table A.6, water (assume $\overline{T}_c \approx 27°C = 300$ K): $\rho = 997$ kg/m^3, $c_p = 4179$ J/kg·K, $\mu = 855 \times 10^{-6}$ N·s/m^2, $k = 0.613$ W/m·K, $Pr = 5.83$.

Analysis:

1. The cooling water outlet temperature may be obtained from the overall energy balance, Equation 11.7b. Accordingly,

$$T_{c,o} = T_{c,i} + \frac{q}{\dot{m}_c c_{p,c}} = 20°C + \frac{2 \times 10^9 \text{ W}}{3 \times 10^4 \text{ kg/s} \times 4197 \text{ J/kg·K}}$$

$$T_{c,o} = 36.0°C \qquad \lhd$$

2. The problem may be classified as one requiring a *heat exchanger design calculation*. First, we determine the overall heat transfer coefficient for use in the NTU method.
   From Equation 11.5

$$U = \frac{1}{(1/h_i) + (1/h_o)}$$

where h_i may be estimated from an internal flow correlation. With

$$Re_D = \frac{4\dot{m}}{\pi D \mu} = \frac{4 \times 1 \text{ kg/s}}{\pi(0.025 \text{ m})855 \times 10^{-6} \text{ N·s/m}^2} = 59,567$$

the flow is turbulent and from Equation 8.60

$$Nu_D = 0.023 \, Re_D^{4/5} \, Pr^{0.4} = 0.023(59,567)^{0.8}(5.83)^{0.4} = 308$$

Hence

$$h_i = Nu_D \frac{k}{D} = 308 \frac{0.613 \text{ W/m·K}}{0.025 \text{ m}} = 7543 \text{ W/m}^2 \cdot \text{K}$$

$$U = \frac{1}{[(1/7543) + (1/11,000)] \text{ m}^2 \cdot \text{K/W}} = 4474 \text{ W/m}^2 \cdot \text{K}$$

Using the design calculation methodology, we note that

$$C_h = C_{max} = \infty$$

and

$$C_{min} = \dot{m}_c c_{p,c} = 3 \times 10^4 \text{ kg/s} \times 4179 \text{ J/kg} \cdot \text{K} = 1.25 \times 10^8 \text{ W/K}$$

from which

$$\frac{C_{min}}{C_{max}} = C_r = 0$$

The maximum possible heat transfer rate is

$$q_{max} = C_{min}(T_{h,i} - T_{c,i}) = 1.25 \times 10^8 \text{ W/K} \times (50 - 20) \text{ K} = 3.76 \times 10^9 \text{ W}$$

from which

$$\varepsilon = \frac{q}{q_{max}} = \frac{2 \times 10^9 \text{ W}}{3.76 \times 10^9 \text{ W}} = 0.532$$

From Equation 11.35b or Figure 11.12, we find NTU = 0.759. From Equation 11.24, it follows that the tube length per pass is

$$L = \frac{\text{NTU} \cdot C_{min}}{U(N2\pi D)} = \frac{0.759 \times 1.25 \times 10^8 \text{ W/K}}{4474 \text{ W/m}^2 \cdot \text{K} \, (30{,}000 \times 2 \times \pi \times 0.025 \text{ m})} = 4.51 \text{ m} \qquad \triangleleft$$

Comments:

1. Recognize that L is the tube length per pass, in which case the total length per tube is 9.0 m. The entire length of tubing in the condenser is $N \times L \times 2 = 30{,}000 \times 4.51$ m $\times 2 = 271{,}000$ m or 271 km.

2. Over time, the performance of the heat exchanger would be degraded by fouling on both the inner and outer tube surfaces. A representative maintenance schedule would call for taking the heat exchanger off-line and cleaning the tubes when fouling factors reached values of $R''_{f,i} = R''_{f,o} = 10^{-4}$ m$^2 \cdot$ K/W. To determine the effect of fouling on performance, the ε–NTU method may be used to calculate the total heat rate as a function of the fouling factor, with $R''_{f,o}$ assumed to equal $R''_{f,i}$. The following results are obtained:

To maintain the requirement of $q = 2 \times 10^9$ W with the maximum allowable fouling and the restriction of $\dot{m}_{c,1} = 1$ kg/s, the tube length or the number of tubes would have to be increased. Keeping the length per pass at $L = 4.51$ m, $N = 48,300$ tubes would be needed to transfer 2×10^9 W for $R''_{f,i} = R''_{f,o} = 10^{-4}$ m$^2 \cdot$ K/W. The corresponding increase in the total flow rate to $\dot{m}_c = N\dot{m}_{c,1} = 48,300$ kg/s would have the beneficial effect of reducing the water outlet temperature to $T_{c,o} = 29.9°$C, thereby ameliorating potentially harmful effects associated with discharge into the environment. The additional tube length associated with increasing the number of tubes to $N = 48,300$ is 165 km, which would result in a significant increase in the capital cost of the condenser.

3. The steam plant generates 1250 MW of electricity with a wholesale value of $0.05 per kW $\cdot$ h. If the plant is shut down for 48 hours to clean the condenser tubes, the loss in revenue for the plant's owner is 48 h $\times$ 1250 $\times 10^6$ W $\times$ $0.05/(1 \times 10^3$ W $\cdot$ h) = $3 million.

4. Assuming a smooth surface condition within each tube, the friction factor may be determined from Equation 8.21, $f = (0.790 \ln (59,567) - 1.64)^{-2} = 0.020$. The pressure drop within one tube of length $L = 9$ m may be determined from Equation 8.22a, where $u_m = 4\dot{m}/(\rho \pi D^2) = (4 \times 1 \text{ kg/s})/(997 \text{ kg/m}^3 \times \pi \times 0.025^2 \text{ m}^2) = 2.04$ m/s·

$$\Delta p = f \frac{\rho u_m^2}{2D} L = 0.020 \frac{997 \text{ kg/m}^3 (2.04 \text{ m/s})^2}{2(0.025 \text{ m})} 9.0 \text{ m} = 15,300 \text{ N/m}^2$$

Therefore, the power required to pump the cooling water through the 48,300 tubes may be found by using Equation 8.22b and is

$$p = \frac{\Delta p \dot{m}}{\rho} = \frac{15,300 \text{ N/m}^2 \times 48,300 \text{ kg/s}}{997 \text{ kg/m}^3} = 742,000 \text{ W} = 0.742 \text{ MW}$$

The cooling water pump is driven by an electric motor. If the combined efficiency of the pump and motor is 87%, the annual cost to overcome friction losses in the condenser tubes is 24 h/day $\times$ 365 days/yr $\times$ 0.742 $\times 10^6$ W $\times$ $0.05/1 $\times 10^3$ W $\cdot$ h/0.87 = $374,000.

5. Optimal condenser designs are based on the desired thermal performance and environmental considerations as well as on the capital cost, operating cost, and maintenance cost associated with the device.

EXAMPLE 11.6

A geothermal power plant utilizes pressurized, deep groundwater at $T_G = 147°$C as the heat source for an *organic Rankine cycle*, the operation of which is described further in Comment 2. An evaporator, consisting of a vertically oriented shell-and-tube heat exchanger with one shell pass and one tube pass, transfers heat between the tube side groundwater and the counterflowing shell-side organic fluid of the power cycle. The organic fluid enters the shell side of the evaporator as a subcooled liquid at $T_{c,i} = 27°$C, and exits the evaporator as a saturated vapor of *quality* $X_{R,o} = 1$ and temperature $T_{c,o} = T_{sat} = 122°$C. Within the evaporator, heat transfer occurs between liquid groundwater and the organic fluid in Stage A with $U_A = 900$ W/m$^2 \cdot$ K, and between liquid groundwater and boiling organic fluid in Stage B with $U_B = 1200$ W/m$^2 \cdot$ K. For groundwater and organic fluid flow rates of $\dot{m}_G = 10$ kg/s and $\dot{m}_R = 5.2$ kg/s, respectively, determine the required evaporator heat transfer surface area. The specific heat of the liquid organic fluid of the Rankine cycle is $c_{p,R} = 1300$ J/kg $\cdot$ K, and its latent heat of vaporization is $h_{fg} = 110$ kJ/kg.

SOLUTION

Known: Mass flow rates of groundwater and Rankine cycle organic fluids. Inlet and outlet temperatures and qualities of organic fluid. Inlet temperature of groundwater. Overall heat transfer coefficients for top and bottom stages of evaporator.

Find: Required evaporator heat transfer surface area.

Schematic:

Assumptions:
1. Steady-state conditions.
2. Constant properties.
3. Negligible heat loss to surroundings and kinetic and potential energy changes.

Properties: Table A.6, water (assume $\overline{T} \approx 405$ K): $c_{p,G} = 4267$ J/kg $\cdot$ K.

Analysis: Applying the conservation of energy principle to the organic fluid within the evaporator consisting of Stages A and B yields

$$q = q_A + q_B = \dot{m}_R[c_{p,R}(T_{sat} - T_{c,i}) + h_{fg}]$$

$$= 5.2 \text{ kg/s}[1300 \text{ J/kg} \cdot \text{K } (122 - 27)°\text{C} + 110 \times 10^3 \text{ J/kg}]$$

$$= 642 \times 10^3 \text{ W} + 572 \times 10^3 \text{ W} = 1.214 \times 10^6 \text{ W} = 1.214 \text{ MW}$$

The groundwater temperature exiting the evaporator may be determined from an energy balance on the hot stream

$$T_{h,o} = T_{h,i} - \frac{q}{\dot{m}_G c_{p,G}} = 147°C - \frac{1.214 \times 10^6 \, \text{W}}{10 \, \text{kg/s} \times 4267 \, \text{J/kg} \cdot \text{K}} = 118.5°C$$

The inlet and outlet temperatures for the cold stream are

$$T_{c,i,A} = T_{c,i} = 27°C; \quad T_{c,o,A} = 122°C; \quad T_{c,i,B} = T_{c,o,A} = 122°C; \quad T_{c,o,B} = T_{c,o} = 122°C$$

while for the hot stream

$$T_{h,i,B} = T_{h,i} = 147°C; \; T_{h,o,B} = T_{h,i,B} - \frac{q_B}{\dot{m}_G c_{p,G}} = 147°C - \frac{572 \times 10^3 \, \text{W}}{10 \, \text{kg/s} \times 4267 \, \text{J/kg} \cdot \text{K}} = 133.6°C$$

$$T_{h,i,A} = T_{h,o,B} = 133.6°C; \; T_{h,o,A} = T_{h,o} = 118.5°C$$

The heat capacity rates in the bottom stage (A) of the evaporator are

$$C_h = \dot{m}c_{p,h} = \dot{m}_G c_{p,G} = 10 \, \text{kg/s} \times 4267 \, \text{J/kg} \cdot \text{K} = 42{,}670 \, \text{W/K}$$

$$C_c = \dot{m}c_{p,c} = \dot{m}_R c_{p,R} = 5.2 \, \text{kg/s} \times 1300 \, \text{J/kg} \cdot \text{K} = 6760 \, \text{W/K}$$

$$C_{r,A} = \frac{C_{\text{min},A}}{C_{\text{max},A}} = \frac{6760}{42{,}670} = 0.158$$

Therefore, the effectiveness associated with the bottom stage of the evaporator is

$$\varepsilon_A = \frac{q_A}{C_{\text{min},A}(T_{h,i,A} - T_{c,i,A})} = \frac{642 \times 10^3 \, \text{W}}{6760 \, \text{W/K} \times (133.6 - 27)°C} = 0.891$$

The NTU can be calculated from the relation for a counterflow heat exchanger, Equation 11.29b, as

$$\text{NTU}_A = \frac{1}{C_{r,A} - 1} \ln\left(\frac{\varepsilon_A - 1}{\varepsilon_A C_{r,A} - 1}\right) = \frac{1}{0.158 - 1} \ln\left(\frac{0.891 - 1}{0.891 \times 0.158 - 1}\right) = 2.45$$

Hence the required heat transfer area for Stage A is

$$A_A = \frac{\text{NTU}_A C_{\text{min},A}}{U_A} = \frac{2.45 \times 6760 \, \text{W/K}}{900 \, \text{W/m}^2 \cdot \text{K}} = 18.4 \, \text{m}^2$$

Phase change occurs in the organic fluid in the top (B) stage. Therefore $C_{r,B} = 0$ and $C_{\text{min},B} = 42{,}670 \, \text{W/K}$. The effectiveness for Stage B is

$$\varepsilon_B = \frac{q_B}{C_{\text{min},B}(T_{h,i,B} - T_{c,i,B})} = \frac{572 \times 10^3 \, \text{W}}{42{,}670 \, \text{W/K} \times (147 - 122)°C} = 0.536$$

From Equation 11.35b

$$\text{NTU}_B = -\ln(1 - \varepsilon_B) = -\ln(1 - 0.536) = 0.768$$

and

$$A_B = \frac{\text{NTU}_B C_{\text{min},B}}{U_B} = \frac{0.768 \times 42{,}670 \, \text{W/K}}{1200 \, \text{W/m}^2 \cdot \text{K}} = 27.3 \, \text{m}^2$$

Therefore, the entire heat transfer area is

$$A = A_{\mathrm{A}} + A_{\mathrm{B}} = 18.4 \text{ m}^2 + 27.3 \text{ m}^2 = 45.7 \text{ m}^2 \qquad \triangleleft$$

Comments:

1. Although a baffled, shell-and-tube heat exchanger is used, there is only one tube pass, and it is appropriate to assume counterflow conditions.

2. Thermodynamic cycles can be described in terms of their temperature-entropy, or $T - s$ diagrams [9]. A Rankine cycle with water as the working fluid is shown in the diagram on the left below. Also included is the *vapor dome* of water, under which there exists a two-phase mixture of liquid and vapor. Superheated water vapor exists to the right of the vapor dome. Note that within the turbine the water exists as a two-phase mixture; saturated liquid droplets are mixed with the saturated vapor. The droplets can impinge on the turbine blades, causing failure of the turbine. As such, most water-based Rankine cycles require the addition of an expensive *superheater* to ensure that condensation does not occur within the turbine.

 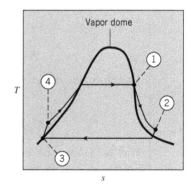

① Evaporator outlet, turbine inlet

② Turbine outlet, condenser inlet

③ Condenser outlet, pump inlet

④ Pump outlet, evaporator inlet

Many organic fluids are characterized by a vapor dome such as that shown in the diagram on the right. Note that, in contrast to using water as the working fluid, condensation cannot occur in the turbine. Hence, a superheater is not needed, making organic Rankine cycles attractive for a broad range of applications such as geothermal energy generation, conversion of waste heat from large turbine or diesel engines to electricity, and concentrating solar power applications as in Example 8.5 where cost reduction is critical [10].

3. The temperatures and mass flow rates in this problem correspond to an electric power generation of 250 kW using pentafluoropropane (R245fa) as the organic working fluid with high and low cycle pressures of $p = 20$ and 1.2 bars, respectively.

11.6 *Additional Considerations*

Because there are many important applications, heat exchanger research and development has had a long history. Such activity is by no means complete, however, as many talented workers continue to seek ways of improving design and performance. In fact, with heightened concern for energy conservation, there has been a steady and substantial increase in activity. A focal point for this work has been *heat transfer enhancement,* which includes the search for special heat exchanger surfaces through which increased heat transfer rates may be achieved. Also, as discussed in Section 11.1 and illustrated in Example 11.2, *compact heat exchangers* are typically used when enhancement is desired and at least one of the fluids is a gas. Many different tubular and plate configurations have been considered, where differences are due primarily to fin design and arrangement.

In addition to application to heat exchanger analysis, the LMTD and NTU methods are powerful tools that may also be applied to similar thermal systems, as illustrated in the following two examples.

EXAMPLE 11.7

A small copper heat sink with dimensions $W_1 = W_2 = 40$ mm, $L_b = 1.0$ mm, $S = 1.6$ mm, $t = 0.8$ mm, and $L_f = 5$ mm has a uniform maximum temperature of $T_h = 50°C$ on its bottom surface. An insulating cap is placed on the top of the heat sink. Water is used as the coolant, entering the heat sink at $T_{m,i} = 30°C$ and $u_m = 1.75$ m/s, providing an average heat transfer coefficient of $\bar{h} = 7590$ W/m$^2 \cdot$ K. Determine the heat transfer rate from the hot surface to the water.

SOLUTION

Known: Dimensions of copper heat sink, maximum heat sink temperature and water inlet temperature, mean velocity, and average heat transfer coefficient.

Find: Heat transfer rate.

Schematic:

A discussion of compact heat exchanger analysis is presented in Section 11S.2.

Assumptions:

1. Steady-state conditions.
2. Adiabatic tips on heat sink fins.
3. Adiabatic heat sink sides, front and back surfaces.
4. Isothermal bottom surface temperature T_h.
5. Constant properties.
6. Negligible axial conduction in the heat sink.

Properties: Table A.1, copper ($T = 300$ K): $k_{Cu} = 401$ W/m·K. Table A.6, water (assume $\bar{T} \approx 310$ K): $\rho = 993$ kg/m³, $c_p = 4178$ J/kg·K.

Analysis: Since the heat sink's bottom surface temperature is spatially uniform, and axial conduction is neglected, the heat sink's thermal behavior corresponds to a single stream heat exchanger as shown in Figure 11.9a. Specifically, the bottom surface temperature does not vary in the x-direction, but the water temperature increases as it flows through the heat sink. Hence, we may use Equation 11.22 to determine the heat transfer rate,

$$q = \varepsilon C_{min}(T_{h,i} - T_{c,i}) \tag{1}$$

where $C_{min} = C_c = \dot{m}_c c_{p,c}$ and $C_r \to 0$. From Section 11.2 and the discussion surrounding Equations 8.45b and 8.46b, we note that the term $1/UA$ used in the definition of NTU corresponds to the overall thermal resistance between the two fluid streams of a heat exchanger. In this example, $UA = 1/R_{tot}$ where R_{tot} is the total thermal resistance between the bottom of the heat sink and the fluid. Therefore, Equation 11.35a may be written as

$$\varepsilon = 1 - \exp(-\text{NTU}) = 1 - \exp\left(-\frac{UA}{C_{min}}\right) = 1 - \exp\left(-\frac{1}{R_{tot}C_{min}}\right) \tag{2}$$

Once C_{min} and R_{tot} are evaluated, the effectiveness can be found from Equation 2, and the heat rate may be determined from Equation 1.

The number of fins is equal to the number of channels and is $N = W_1/S = 40$ mm/1.6 mm = 25. The minimum heat capacity rate is

$$C_{min} = \dot{m}_c c_{p,c} = N \rho u_m L_f (S - t) c_p$$

$$= 25 \times 993 \text{ kg/m}^3 \times 1.75 \text{ m/s} \times 0.005 \text{ m} \times (0.0016 \text{ m} - 0.0008 \text{ m}) \times 4178 \text{ J/kg·K}$$

$$= 726 \text{ W/K}$$

The total thermal resistance is calculated in Comment 4 and is $R_{tot} = 17.8 \times 10^{-3}$ K/W. From Equation 2,

$$\varepsilon = 1 - \exp\left(-\frac{1}{R_{tot}C_{min}}\right) = 1 - \exp\left(-\frac{1}{17.8 \times 10^{-3} \text{ K/W} \times 726 \text{ W/K}}\right) = 0.0745$$

and from Equation 1,

$$q = \varepsilon C_{min}(T_h - T_{c,i}) = 0.0745 \times 726 \text{ W/K} \times (50°\text{C} - 30°\text{C}) = 1080 \text{ W} \qquad \triangleleft$$

Comments:

1. If the water temperature is assumed to be constant as it flows through the heat sink, the heat rate is $q = (T_h - T_{c,i})/R_{tot} = (50°C - 30°C)/17.8 \times 10^{-3}$ K/W = 1120 W. The assumption of constant water temperature leads to an overestimate of the actual heat rate.

2. The outlet temperature of the water is $T_{c,o} = T_{c,i} + q/C_{min} = 30°C + (1080 \text{ W})/(726 \text{ W/K}) = 31.5°C$.

3. From Equations 11.15 and 11.16, $\Delta T_{lm} = [(T_b - T_{c,i}) - (T_b - T_{c,o})]/\ln[(T_b - T_{c,i})/(T_b - T_{c,o})] = [31.5°C - 30°C]/\ln[(50°C - 30°C)/(50°C - 31.5°C)] = 19.2°C$ and $q = \Delta T_{lm}/R_{tot} = 19.2°C/17.8 \times 10^{-3}$ K/W = 1080 W. Therefore, the appropriate mean temperature difference shown in the following thermal circuit, ΔT_m, is the *log mean temperature difference* [11]. As such, this problem could have been solved using an LMTD approach, but an iterative solution would have been required.

4. The total thermal resistance corresponds to the following thermal circuit.

where ΔT_m is the appropriate mean temperature difference between the bottom of the heat sink base and the fluid. The thermal resistance of the base is

$$R_{t,\text{base}} = L_b/(k_{Cu}W_1 W_2) = (0.001 \text{ m})/(401 \text{ W/m} \cdot \text{K} \times 0.040 \text{ m} \times 0.040 \text{ m})$$
$$= 1.56 \times 10^{-3} \text{ K/W}$$

The parallel resistances in the thermal circuit represent the fins and unfinned portion of the base. The combination of these two resistances is the overall thermal resistance of the fin array, as given by Equation 3.108 with Equation 11.3:

$$R_{t,o} = \frac{1}{\eta_o \bar{h} A} = \frac{1}{\bar{h}[A - A_f(1 - \eta_f)]}$$

In this expression, A_f is the surface area of all the fins and $A = A_f + A_b$, where A_b is the area of the unfinned portion of the base. Thus

$$A_f = 2L_f W_2 N = 2 \times 0.005 \text{ m} \times 0.40 \text{ m} \times 25 = 0.01 \text{ m}^2$$

and

$$A = A_f + (W_1 - Nt)W_2 = 0.01 \text{ m}^2 + (0.040 \text{ m} - 25 \times 0.0008 \text{ m}) \times 0.040 \text{ m}$$
$$= 0.0108 \text{ m}^2$$

The quantity η_f is the efficiency of a single fin, given by Equation 11.4. We first calculate

$$mL_f = \sqrt{2\bar{h}/k_{Cu}t} \, L_f$$
$$= \sqrt{2 \times 7590 \text{ W/m}^2 \cdot \text{K}/(401 \text{ W/m} \cdot \text{K} \times 0.0008 \text{ m})} \times 0.005 \text{ m} = 1.09$$

Then

$$\eta_f = \frac{\tanh(mL_f)}{mL_f} = \frac{\tanh(1.09)}{1.09} = 0.732$$

$$R_{t,o} = \frac{1}{\overline{h}[A - A_f(1 - \eta_f)]}$$

$$= \frac{1}{7590 \text{ W/m}^2 \cdot \text{K}[0.0108 \text{ m}^2 - 0.01 \text{ m}^2(1 - 0.732)]} = 0.0162 \text{ K/W}$$

Therefore, the total thermal resistance is

$$R_{\text{tot}} = R_{t,\text{base}} + R_{t,o} = 1.56 \times 10^{-3} \text{ K/W} + 0.0162 \text{ K/W} = 17.8 \times 10^{-3} \text{ K/W}$$

EXAMPLE 11.8

Spherical steel balls of diameter $D = 10$ mm are cooled from an initial temperature of $T_{h,i} = 1000$ K by submersing them in an insulated oil bath initially at $T_{c,i} = 300$ K. The total mass of the balls is $m_h = 200$ kg, while the mass of oil is $m_c = 500$ kg. The convection coefficient associated with the spheres and the oil is $h = 40$ W/m$^2 \cdot$ K, and the steel properties are $k_h = 40$ W/m $\cdot$ K, $\rho_h = 7800$ kg/m^3, and $c_h = 600$ J/kg $\cdot$ K. Determine the steady-state ball and oil temperatures, and the time needed for the balls to reach a temperature of $T_{h,f} = 500$ K.

SOLUTION

Known: Mass, diameter, properties, and initial temperature of steel spheres. Mass and initial temperature of oil bath.

Find: Steady-state ball and oil temperatures, time to cool balls to $T_{h,f} = 500$ K.

Schematic:

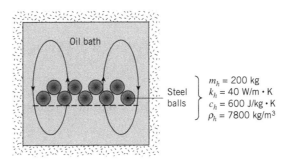

Assumptions:

1. Constant properties.
2. Negligible heat loss from oil bath.

Properties: Table A.5, engine oil (assume $\bar{T} \approx 350$ K): $c_c = 2118$ J/kg · K.

Analysis: We begin by examining whether the lumped capacitance analysis of Chapter 5 may be applied to the balls. With $L_c = r_o/3$, it follows from Equation 5.10 that

$$Bi = \frac{h(r_o/3)}{k_h} = \frac{40 \text{ W/m}^2 \cdot \text{K} \times (0.005 \text{ m/3})}{40 \text{ W/m} \cdot \text{K}} = 0.0017$$

Accordingly, Equation 5.10 is satisfied, and the spheres are nearly isothermal at any instant of time. Treating the steel balls collectively, the sphere and average oil temperatures, T_h and T_c, respectively, may be determined from an energy balance of the form

$$\Delta E_c = -\Delta E_h = \Delta E = C_{t,c}(T_c - T_{c,i}) = C_{t,h}(T_{h,i} - T_h) \tag{1}$$

where $C_{t,c} = m_c c_c = 500$ kg $\times$ 2118 J/kg · K $= 1.06 \times 10^6$ J/K and $C_{t,h} = m_h c_h = 200$ kg $\times$ 600 J/kg · K $= 120 \times 10^3$ J/K are the thermal capacitances of the oil and balls, respectively, as defined in Equation 5.7. The steady-state temperature is achieved when $T_c = T_h = T_{ss}$, and is,

$$T_{ss} = \frac{C_{t,h}T_{h,i} + C_{t,c}T_{c,i}}{C_{t,h} + C_{t,c}} = \frac{120 \times 10^3 \text{ J/K} \times 1000 \text{ K} + 1.06 \times 10^6 \text{ J/K} \times 300 \text{ K}}{120 \times 10^3 \text{ J/K} + 1.06 \times 10^6 \text{ J/K}} = 371 \text{ K} \triangleleft$$

Since the oil temperature increases with time, the lumped capacitance analysis of Chapter 5, which presumes a constant ambient temperature T_∞ is not valid. Rather, heat transfer follows the process described in the schematic below.

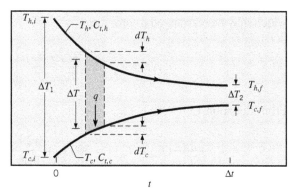

This process is analogous to that of the parallel-flow heat exchanger shown in Figure 11.7, where the total heat transfer area for the N spheres is

$$A = N 4\pi r_o^2 = \frac{m_h}{\rho_h (4/3)\pi r_o^3} 4\pi r_o^2 = \frac{3m_h}{\rho_h r_o} = \frac{3 \times 200 \text{ kg}}{7800 \text{ kg/m}^3 \times 0.005 \text{ m}} = 15.5 \text{ m}^2$$

Applying an energy balance to each of the differential elements shown schematically above yields

$$dE = qdt = -C_{t,h}dT_h \quad \text{and} \quad dE = qdt = C_{t,c}dT_c \qquad (2\text{a, b})$$

where

$$q = hA(T_h - T_c) = UA\Delta T = UA(T_h - T_c) \qquad (3)$$

Substituting expressions for dT_h and dT_c from Equations 2a,b into the expression $d(\Delta T) = dT_h - dT_c$ gives

$$d(\Delta T) = -\frac{qdt}{C_{t,h}} - \frac{qdt}{C_{t,c}} \qquad (4)$$

Combining Equations 3 and 4 yields the relationship

$$d(\Delta T) = -\frac{UA\Delta T}{C_{t,h}}dt - \frac{UA\Delta T}{C_{t,c}}dt$$

Separating variables and integrating,

$$\int_1^2 \frac{d(\Delta T)}{\Delta T} = -UA\left(\frac{1}{C_{t,h}} + \frac{1}{C_{t,c}}\right)\int_{t_1}^{t_2} dt$$

or

$$\ln\left(\frac{\Delta T_2}{\Delta T_1}\right) = -UA\left(\frac{1}{C_{t,h}} + \frac{1}{C_{t,c}}\right)\Delta t \qquad (5)$$

which may be rearranged to provide

$$\Delta t = -\frac{1}{UA}\ln\left[\frac{T_{h,f} - T_{c,f}}{T_{h,i} - T_{c,i}}\right]\Bigg/\left[\frac{1}{C_{t,h}} + \frac{1}{C_{t,c}}\right] \qquad (6)$$

From Equation 1,

$$T_{c,f} = T_{c,i} + \frac{C_{t,h}}{C_{t,c}}(T_{h,i} - T_{h,f}) = 300\text{ K} + \frac{120 \times 10^3\text{ J/K}}{1.06 \times 10^6\text{ J/K}}(1000\text{ K} - 500\text{ K}) = 357\text{ K}$$

Therefore, Equation 6 may be evaluated as

$$\Delta t = -\frac{1}{40\,\text{W/m}^2 \cdot \text{K} \times 15.5\,\text{m}^2}\ln\left[\frac{500\text{ K} - 357\text{ K}}{1000\text{ K} - 300\text{ K}}\right]\Bigg/\left[\frac{1}{120 \times 10^3\text{ K/J}} + \frac{1}{1.06 \times 10^6\text{ K/J}}\right]$$

$$= 278\text{ s} \qquad \triangleleft$$

Comments:

1. The sphere and oil temperature histories are plotted below. Note the asymptotic approach of both temperatures to the steady-state value, $T_{ss} = 371$ K.

2. Equation 5 may be rewritten as

$$\frac{T_{h,f} - T_{c,f}}{T_{h,i} - T_{c,i}} = \exp\left[-UA\left(\frac{1}{C_{t,h}} + \frac{1}{C_{t,c}}\right)\Delta t\right]$$

and for an *infinitely large* oil bath, $C_{t,c} \to \infty$, and $T_{c,f} = T_{c,i} = T_\infty$. Hence,

$$\frac{T_{h,f} - T_\infty}{T_{h,i} - T_\infty} = \exp\left[-\left(\frac{UA}{\rho_h m_h}\right)\Delta t\right]$$

which is equivalent to Equation 5.6 of the lumped capacitance analysis.

3. From Equation 1 we note that $C_{t,c} = \Delta E/(T_{c,f} - T_{c,i})$ and $C_{t,h} = \Delta E/(T_{h,i} - T_{h,f})$. Substituting these two expressions into Equation 6 yields

$$\ln\left(\frac{\Delta T_2}{\Delta T_1}\right) = -UA\Delta t\left(\frac{T_{h,i} - T_{h,f}}{\Delta E} + \frac{T_{c,f} - T_{c,i}}{\Delta E}\right) = -\frac{UA\Delta t}{\Delta E}(\Delta T_1 - \Delta T_2)$$

which can be rearranged to provide an expression involving a *log mean temperature difference* that is of the same form as Equation 11.15

$$\Delta E = UA\Delta t\frac{\Delta T_2 - \Delta T_1}{\ln(\Delta T_2/\Delta T_1)} = UA\Delta t\Delta T_{lm}$$

Applying the LMTD expression to this problem yields

$$\Delta E = 40 \ \mathrm{W/m^2 \cdot K} \times 15.5 \ \mathrm{m^2} \times 278 \ \mathrm{s} \left[\frac{(500 \ \mathrm{K} - 357 \ \mathrm{K}) - (1000 \ \mathrm{K} - 300 \ \mathrm{K})}{\ln\left(\dfrac{500 \ \mathrm{K} - 357 \ \mathrm{K}}{1000 \ \mathrm{K} - 300 \ \mathrm{K}} \right)} \right]$$

$$= 60 \times 10^6 \ \mathrm{J} = 60 \ \mathrm{MJ}$$

which can be verified using Equation 1, with $\Delta E = C_{t,c}(T_{c,f} - T_{c,i}) = C_{t,h}(T_{h,i} - T_{h,f}) = 1.06 \times 10^6 \ \mathrm{J/K} \times (357 \ \mathrm{K} - 300 \ \mathrm{K}) = 120 \times 10^3 \ \mathrm{J/K} \times (1000 \ \mathrm{K} - 500 \ \mathrm{K}) = 60 \ \mathrm{MJ}$.

4. Proceeding in a manner similar to that for the parallel-flow, two-fluid heat exchanger, note that the maximum possible change in thermal energy of the spheres or the oil is

$$\Delta E_{\max} \equiv C_{t,\min}(T_{h,i} - T_{c,i})$$

Likewise, a modified effectiveness and NTU may be defined as

$$\varepsilon^* = \frac{\Delta E}{\Delta E_{\max}} \qquad \text{and} \qquad \mathrm{NTU}^* = \frac{UA\Delta t}{C_{t,\min}}$$

It may be shown that, with $C_{t,r} = C_{t,\min}/C_{t,\max}$,

$$\varepsilon^* = \frac{1 - \exp[-\mathrm{NTU}^*(1 + C_{t,r})]}{1 + C_{t,r}} \qquad \text{and} \qquad \mathrm{NTU}^* = -\frac{\ln[1 - \varepsilon^*(1 + C_{t,r})]}{1 + C_{t,r}}$$

which are of the same form as Equations 11.28a and 11.28b, respectively. For this problem, $C_{t,r} = 120 \times 10^3 \ \mathrm{J/K}/1.06 \times 10^6 \ \mathrm{J/K} = 0.113$ and $\varepsilon^* = \Delta E/\Delta E_{\max} = 60 \times 10^6 \ \mathrm{J/K}/[120 \times 10^3 \ \mathrm{J/K} \times (1000 \ \mathrm{K} - 300 \ \mathrm{K})] = 0.714$. Therefore,

$$\mathrm{NTU}^* = -\frac{\ln[1 - 0.714(1 + 0.113)]}{1 + 0.113} = 1.42$$

and

$$\Delta t = \mathrm{NTU}^* C_{t,\min}/UA = 1.42 \times 120 \times 10^3 \ \mathrm{J/K}/[40 \ \mathrm{W/m^2 \cdot K} \times 15.5 \ \mathrm{m^2}] = 278 \ \mathrm{s}$$

which is in agreement with the problem solution.

5. This problem illustrates the value of recognizing *analogous behavior* that characterizes various thermal systems. In general, the LMTD and ε–NTU heat exchanger analyses can be used to determine the transient thermal responses of two materials between which heat is exchanged, if each material can be characterized by a unique temperature at any time.

11.7 *Summary*

In this chapter we have developed tools that will allow you to perform approximate heat exchanger calculations. More detailed considerations of the subject are available in the literature, including treatment of the uncertainties associated with heat exchanger analysis [3, 4, 7, 12–18].

Although we have restricted attention to heat exchangers involving separation of hot and cold fluids by a stationary wall, there are other important options. For example, *evaporative* heat exchangers enable *direct contact* between a liquid and a gas (there is no separating wall), and because of latent energy effects, large heat transfer rates per unit volume are possible. Also, for gas-to-gas heat exchange, use is often made of *regenerators* in which the same space is alternately occupied by the hot and cold gases. In a fixed regenerator such as a packed bed, the hot and cold gases alternately enter a stationary, porous solid. In a rotary regenerator, the porous solid is a rotating wheel, which alternately exposes its surfaces to the continuously flowing hot and cold gases. Detailed descriptions of such heat exchangers are available in the literature [3, 4, 12, 15, 19–22].

You should test your understanding of fundamental issues by addressing the following questions.

- What are the two possible arrangements for a *concentric tube heat exchanger*? For each arrangement, what restrictions are associated with the fluid outlet temperatures?
- As applied to a *cross-flow heat exchanger*, what is meant by the terms *mixed* and *unmixed*? In what sense are they idealizations of actual conditions?
- Why are baffles used in a *shell-and-tube heat exchanger*?
- What is the principal distinguishing feature of a *compact heat exchanger*?
- What effect does *fouling* have on the overall heat transfer coefficient and hence the performance of a heat exchanger?
- What effect do *finned surfaces* have on the overall heat transfer coefficient and hence the performance of a heat exchanger? When is the use of fins most appropriate?
- When can the overall heat transfer coefficient be expressed as $U = (h_i^{-1} + h_o^{-1})^{-1}$?
- What is the appropriate form of the mean temperature difference for the two fluids of a parallel or counterflow heat exchanger?
- What can be said about the change in temperature of a saturated fluid undergoing evaporation or condensation in a heat exchanger?
- Will the fluid having the minimum or the maximum heat capacity rate experience the largest temperature change in a heat exchanger?
- Why is the maximum possible heat rate for a heat exchanger *not* equal to $C_{max}(T_{h,i} - T_{c,i})$? Can the outlet temperature of the cold fluid ever exceed the inlet temperature of the hot fluid?
- What is the *effectiveness* of a heat exchanger? What is its range of possible values? What is the *number of transfer units*? What is its range of possible values?
- Generally, how does the effectiveness change if the size (surface area) of a heat exchanger is increased? If the overall heat transfer coefficient is increased? If the ratio of heat capacity rates is decreased? As manifested by the number of transfer units, are there limitations to the foregoing trends? What penalty is associated with increasing the size of a heat exchanger? With increasing the overall heat transfer coefficient?

References

1. *Standards of the Tubular Exchange Manufacturers Association*, 6th ed., Tubular Exchanger Manufacturers Association, New York, 1978.

2. Chenoweth, J. M., and M. Impagliazzo, Eds., *Fouling in Heat Exchange Equipment*, American Society of Mechanical Engineers Symposium Volume HTD-17, ASME, New York, 1981.

3. Kakac, S., A. E. Bergles, and F. Mayinger, Eds., *Heat Exchangers*, Hemisphere Publishing, New York, 1981.

4. Kakac, S., R. K. Shah, and A. E. Bergles, Eds., *Low Reynolds Number Flow Heat Exchangers*, Hemisphere Publishing, New York, 1983.

5. Kays, W. M., and A. L. London, *Compact Heat Exchangers*, 3rd ed., McGraw-Hill, New York, 1984.

6. Bowman, R. A., A. C. Mueller, and W. M. Nagle, *Trans. ASME*, **62**, 283, 1940.

7. DiGiovanni, M. A., and R. L. Webb, *Heat Transfer Eng.*, **10**, 61, 1989.

8. Ribando, R. J., G. W. O'Leary, and S. Carlson-Skalak, *Comp. Appl. Eng. Educ.*, **5**, 231, 1997.

9. Moran, M. J., and H. N. Shapiro, *Engineering Thermodynamics*, 6th ed., Wiley, Hoboken, NJ, 2008.

10. Dai, Y., J. Wang, and L. Gao, *Energy Conv. Management*, **50**, 576, 2009.

11. Webb, R. L., *Trans. ASME, J. Heat Transfer*, **129**, 899, 2007.

12. Shah, R. K., C. F. McDonald, and C. P. Howard, Eds., *Compact Heat Exchangers*, American Society of Mechanical Engineers Symposium Volume HTD-10, ASME, New York, 1980.

13. Webb, R. L., in G. F. Hewitt, Exec. Ed., *Heat Exchanger Design Handbook*, Section 3.9, Begell House, New York, 2002.

14. Marner, W. J., A. E. Bergles, and J. M. Chenoweth, *Trans. ASME, J. Heat Transfer*, **105**, 358, 1983.

15. G. F. Hewitt, Exec. Ed., *Heat Exchanger Design Handbook*, Vols. 1–5, Begell House, New York, 2002.

16. Webb, R. L., and N.-H. Kim, *Principles of Enhanced Heat Transfer*, 2nd ed., Taylor & Francis, New York, 2005.

17. Andrews, M. J., and L. S. Fletcher, *ASME/JSME Thermal Eng. Conf.*, **4**, 359, 1995.

18. James, C. A., R. P. Taylor, and B. K. Hodge, *ASME/JSME Thermal Eng. Conf.*, **4**, 337, 1995.

19. Coppage, J. E., and A. L. London, *Trans. ASME*, **75**, 779, 1953.

20. Treybal, R. E., *Mass-Transfer Operations*, 3rd ed., McGraw-Hill, New York, 1980.

21. Sherwood, T. K., R. L. Pigford, and C. R. Wilkie, *Mass Transfer*, McGraw-Hill, New York, 1975.

22. Schmidt, F. W., and A. J. Willmott, *Thermal Energy Storage and Regeneration*, Hemisphere Publishing, New York, 1981.

Problems

Overall Heat Transfer Coefficient

11.1 In a fire-tube boiler, hot products of combustion flowing through an array of thin-walled tubes are used to boil water flowing over the tubes. At the time of installation, the overall heat transfer coefficient was $400 \text{ W/m}^2 \cdot \text{K}$. After 1 year of use, the inner and outer tube surfaces are fouled, with corresponding fouling factors of $R''_{f,i} = 0.0015$ and $R''_{f,o} = 0.0005 \text{ m}^2 \cdot \text{K/W}$, respectively. Should the boiler be scheduled for cleaning of the tube surfaces?

11.2 An industrial boiler consists of tubes inside of which flow hot combustion gases. Water boils on the exterior of the tubes. When installed, the clean boiler has an overall heat transfer coefficient of $300 \text{ W/m}^2 \cdot \text{K}$. Based on experience, it is anticipated that the fouling factors on the inner and outer surfaces will increase linearly with time as $R''_{f,i} = a_i t$ and $R''_{f,o} = a_o t$ where $a_i = 2.5 \times 10^{-11} \text{ m}^2 \cdot \text{K/W} \cdot \text{s}$ and $a_o = 1.0 \times 10^{-11} \text{ m}^2 \cdot \text{K/W} \cdot \text{s}$ for the inner and outer tube surfaces, respectively. If the boiler is to be cleaned when the overall heat transfer coefficient is reduced from its initial value by 25%, how long after installation should the first cleaning be scheduled? The boiler operates continuously between cleanings.

11.3 A type-302 stainless steel tube of inner and outer diameters $D_i = 22 \text{ mm}$ and $D_o = 27 \text{ mm}$, respectively, is used in a cross-flow heat exchanger. The fouling factors,

R''_f, for the inner and outer surfaces are estimated to be 0.0004 and 0.0002 $m^2 \cdot K/W$, respectively.

(a) Determine the overall heat transfer coefficient based on the outside area of the tube, U_o. Compare the thermal resistances due to convection, tube wall conduction, and fouling.

(b) Instead of air flowing over the tube, consider a situation for which the cross-flow fluid is water at 15°C with a velocity of $V_o = 1$ m/s. Determine the overall heat transfer coefficient based on the outside area of the tube, U_o. Compare the thermal resistances due to convection, tube wall conduction, and fouling.

(c) For the water–air conditions of part (a) and mean velocities, $u_{m,i}$, of 0.2, 0.5, and 1.0 m/s, plot the overall heat transfer coefficient as a function of the cross-flow velocity for $5 \le V_o \le 30$ m/s.

(d) For the water–water conditions of part (b) and cross-flow velocities, V_o, of 1, 3, and 8 m/s, plot the overall heat transfer coefficient as a function of the mean velocity for $0.5 \le u_{m,i} \le 2.5$ m/s.

11.4 A steel tube ($k = 50$ W/m $\cdot$ K) of inner and outer diameters $D_i = 20$ mm and $D_o = 26$ mm, respectively, is used to transfer heat from hot gases flowing over the tube ($h_h = 200$ W/m^2 $\cdot$ K) to cold water flowing through the tube ($h_c = 8000$ W/m^2 $\cdot$ K). What is the cold-side overall heat transfer coefficient U_c? To enhance heat transfer, 16 straight fins of rectangular profile are installed longitudinally along the outer surface of the tube. The fins are equally spaced around the circumference of the tube, each having a thickness of 2 mm and a length of 15 mm. What is the corresponding overall heat transfer coefficient U_c?

11.5 A heat recovery device involves transferring energy from the hot flue gases passing through an annular region to pressurized water flowing through the inner

tube of the annulus. The inner tube has inner and outer diameters of 24 and 30 mm and is connected by eight struts to an insulated outer tube of 60-mm diameter. Each strut is 4 mm thick and is integrally fabricated with the inner tube from carbon steel ($k = 50$ W/m $\cdot$ K).

Consider conditions for which water at 300 K flows through the inner tube at 0.161 kg/s while flue gases at 800 K flow through the annulus, maintaining a convection coefficient of 100 W/m^2 $\cdot$ K on both the struts and the outer surface of the inner tube. What is the rate of heat transfer per unit length of tube from gas to the water?

11.6 A novel design for a condenser consists of a tube of thermal conductivity 200 W/m $\cdot$ K with longitudinal fins snugly fitted into a larger tube. Condensing refrigerant at 45°C flows axially through the inner tube, while water at a flow rate of 0.012 kg/s passes through the six channels around the inner tube. The pertinent diameters are $D_1 = 10$ mm, $D_2 = 14$ mm, and $D_3 = 50$ mm, while the fin thickness is $t = 2$ mm. Assume that the convection coefficient associated with the condensing refrigerant is extremely large.

Determine the heat removal rate per unit tube length in a section of the tube for which the water is at 15°C.

11.7 Thin-walled aluminum tubes of diameter $D = 10$ mm are used in the condenser of an air conditioner. Under normal operating conditions, a convection coefficient of $h_i = 5000$ W/m$^2 \cdot$ K is associated with condensation on the inner surface of the tubes, while a coefficient of $h_o = 100$ W/m$^2 \cdot$ K is maintained by airflow over the tubes.

(a) What is the overall heat transfer coefficient if the tubes are unfinned?

(b) What is the overall heat transfer coefficient based on the inner surface, U_i, if aluminum annular fins of thickness $t = 1.5$ mm, outer diameter $D_o = 20$ mm, and pitch $S = 3.5$ mm are added to the outer surface? Base your calculations on a 1-m-long section of tube. Subject to the requirements that $t \geq 1$ mm and $(S - t) \geq 1.5$ mm, explore the effect of variations in t and S on U_i. What combination of t and S would yield the best heat transfer performance?

11.8 A finned-tube, cross-flow heat exchanger is to use the exhaust of a gas turbine to heat pressurized water. Laboratory measurements are performed on a prototype version of the exchanger, which has a surface area of 10 m^2, to determine the overall heat transfer coefficient as a function of operating conditions. Measurements made under particular conditions, for which $\dot{m}_h = 2$ kg/s, $T_{h,i} = 325°C$, $\dot{m}_c = 0.5$ kg/s, and $T_{c,i} = 25°C$, reveal a water outlet temperature of $T_{c,o} = 150°C$. What is the overall heat transfer coefficient of the exchanger?

11.9 Water at a rate of 45,500 kg/h is heated from 80 to 150°C in a heat exchanger having two shell passes and eight tube passes with a total surface area of 930 m^2. Hot exhaust gases having approximately the same thermophysical properties as air enter at 350°C and exit at 175°C. Determine the overall heat transfer coefficient.

11.10 A novel heat exchanger concept consists of a large number of extruded polypropylene sheets ($k = 0.17$ W/m $\cdot$ K), each having a fin-like geometry, that are subsequently stacked and melted together to form the heat exchanger core. Besides being inexpensive, the heat exchanger can be easily recycled at the end of its life. Carbon dioxide at a mean temperature of 10°C and pressure of 2 atm flows in the cool channels at a mean velocity of $u_m = 0.1$ m/s. Air at 30°C and 2 atm flows at 0.2 m/s in the warm channels. Neglecting the thermal contact resistance at the welded interface, determine the product of the overall heat transfer coefficient and heat transfer area, UA, for a heat exchanger core consisting of 200 cool channels and 200 warm channels.

Design and Performance Calculations

11.11 The properties and flow rates for the hot and cold fluids of a heat exchanger are shown in the following table. Which fluid limits the heat transfer rate of the exchanger? Explain your choice.

	Hot fluid	**Cold fluid**
Density, kg/m^3	997	1247
Specific heat, J/kg $\cdot$ K	4179	2564
Thermal conductivity, W/m $\cdot$ K	0.613	0.287
Viscosity, N $\cdot$ s/m^2	8.55×10^{-4}	1.68×10^{-4}
Flow rate, m^3/h	12	18

11.12 A process fluid having a specific heat of 3500 J/kg $\cdot$ K and flowing at 2 kg/s is to be cooled from 80°C to 50°C with chilled water, which is supplied at a temperature of 15°C and a flow rate of 2.5 kg/s. Assuming an overall heat transfer coefficient of 2000 W/m$^2 \cdot$ K, calculate the required heat transfer areas for the following exchanger configurations: (a) parallel flow, (b) counterflow, (c) shell-and-tube, one shell pass and two tube passes, and (d) cross-flow, single pass, both fluids unmixed. Compare the results of your analysis. Your work can be reduced by using *IHT*.

11.13 A shell-and-tube exchanger (two shells, four tube passes) is used to heat 10,000 kg/h of pressurized water from 35 to 120°C with 5000 kg/h pressurized water entering the exchanger at 300°C. If the overall heat transfer coefficient is 1600 W/m$^2 \cdot$ K, determine the required heat exchanger area.

11.14 Consider the heat exchanger of Problem 11.13. After several years of operation, it is observed that the outlet temperature of the cold water reaches only 95°C rather than the desired 120°C for the same flow rates and inlet temperatures of the fluids. Determine the cumulative (inner and outer surface) fouling factor that is the cause of the poorer performance.

11.15 Consider Example 7.7. Determine the heat rate using the ε-NTU method. Assume the air flows in a duct of height $N_T S_T$ where $N_T = 8$ tubes per row and S_T is the transverse tube pitch. The tube length and the duct width are both 1 m.

    The tubes in Example 7.7 were assumed to be isothermal. Now, consider a water mass flow rate of 10 kg/s through the tube bundle. Using the air-side average convection coefficient of the example, determine the overall heat transfer coefficient for the heat exchanger. Evaluate water properties at 70°C. Was the assumption of isothermal tubes valid?

11.16 The hot and cold inlet temperatures to a concentric tube heat exchanger are $T_{h,i} = 200°C$, $T_{c,i} = 100°C$, respectively. The outlet temperatures are $T_{h,o} = 110°C$ and $T_{c,o} = 125°C$. Is the heat exchanger operating in a parallel flow or in a counterflow configuration? What is the heat exchanger effectiveness? What is the NTU? Phase change does not occur in either fluid.

11.17 A counterflow, concentric tube heat exchanger is designed to heat water from 20 to 80°C using hot oil, which is supplied to the annulus at 160°C and discharged at 140°C. The thin-walled inner tube has a diameter of $D_i = 20$ mm, and the overall heat transfer coefficient is 500 W/m² · K. The design condition calls for a total heat transfer rate of 3000 W.

    (a) What is the length of the heat exchanger?

    (b) After 3 years of operation, performance is degraded by fouling on the water side of the exchanger, and the water outlet temperature is only 65°C for the same fluid flow rates and inlet temperatures. What are the corresponding values of the heat transfer rate, the outlet temperature of the oil, the overall heat transfer coefficient, and the water-side fouling factor, $R''_{f,c}$?

11.18 Consider a concentric tube heat exchanger with an area of 50 m² operating under the following conditions:

	Hot fluid	Cold fluid
Heat capacity rate, kW/K	6	3
Inlet temperature, °C	60	30
Outlet temperature, °C	—	54

    (a) Determine the outlet temperature of the hot fluid.

    (b) Is the heat exchanger operating in counterflow or parallel flow, or can't you tell from the available information?

    (c) Calculate the overall heat transfer coefficient.

    (d) Calculate the effectiveness of this exchanger.

    (e) What would be the effectiveness of this exchanger if its length were made very large?

11.19 As part of a senior project, a student was given the assignment to design a heat exchanger that meets the following specifications:

	$\dot{m}$(kg/s)	$T_{m,i}$(°C)	$T_{m,o}$ (°C)
Hot water	28	90	—
Cold water	27	34	60

Like many real-world situations, the customer hasn't revealed, or doesn't know, additional requirements that would allow you to proceed directly to a final configuration. At the outset, it is helpful to make a first-cut design based upon simplifying assumptions, which can be evaluated to determine what additional requirements and trade-offs should be considered by the customer.

    (a) Design a heat exchanger to meet the foregoing specifications. List and explain your assumptions. *Hint:* Begin by finding the required value for *UA* and using representative values of *U* to determine *A*.

    (b) Evaluate your design by identifying what features and configurations could be explored with your customer in order to develop more complete specifications.

11.20 A shell-and-tube heat exchanger must be designed to heat 2.5 kg/s of water from 15 to 85°C. The heating is to be accomplished by passing hot engine oil, which is available at 160°C, through the shell side of the exchanger. The oil is known to provide an average convection coefficient of $h_o = 400$ W/m² · K on the outside of the tubes. Ten tubes pass the water through the shell. Each tube is thin walled, of diameter $D = 25$ mm, and makes eight passes through the shell. If the oil leaves the exchanger at 100°C, what is its flow rate? How long must the tubes be to accomplish the desired heating?

11.21 A concentric tube heat exchanger for cooling lubricating oil is comprised of a thin-walled inner tube of 25-mm diameter carrying water and an outer tube of 45-mm diameter carrying the oil. The exchanger operates in counterflow with an overall heat transfer coefficient of 60 W/m² · K and the tabulated average properties.

Properties	Water	Oil
ρ (kg/m³)	1000	800
c_p (J/kg · K)	4200	1900
v (m²/s)	7×10^{-7}	1×10^{-5}
k (W/m · K)	0.64	0.134
Pr	4.7	140

(a) If the outlet temperature of the oil is 60°C, determine the total heat transfer and the outlet temperature of the water.

(b) Determine the length required for the heat exchanger.

11.22 A counterflow, concentric tube heat exchanger used for engine cooling has been in service for an extended period of time. The heat transfer surface area of the exchanger is 5 m², and the *design value* of the overall convection coefficient is 38 W/m² · K. During a test run, engine oil flowing at 0.1 kg/s is cooled from 110°C to 66°C by water supplied at a temperature of 25°C and a flow rate of 0.2 kg/s. Determine whether fouling has occurred during the service period. If so, calculate the fouling factor, $R_f''\left(\text{m}^2 \cdot \text{K/W}\right)$.

11.23 Hot air for a large-scale drying operation is to be produced by routing the air over a tube bank (unmixed), while products of combustion are routed through the tubes. The surface area of the cross-flow heat exchanger is $A = 25$ m², and for the proposed operating conditions, the manufacturer specifies an overall heat transfer coefficient of $U = 35$ W/m² · K. The air and the combustion gases may each be assumed to have a specific heat of $c_p = 1040$ J/kg · K. Consider conditions for which combustion gases flowing at 1 kg/s enter the heat exchanger at 800 K, while air at 5 kg/s has an inlet temperature of 300 K.

(a) What are the air and gas outlet temperatures?

(b) After extended operation, deposits on the inner tube surfaces are expected to provide a fouling resistance of $R_f'' = 0.004$ m² · K/W. Should operation be suspended in order to clean the tubes?

(c) The heat exchanger performance may be improved by increasing the surface area and/or the overall heat transfer coefficient. Explore the effect of such changes on the air outlet temperature for $500 \leq UA \leq 2500$ W/K.

11.24 A shell-and-tube heat exchanger with one shell pass and two tube passes is used as a *regenerator,* to preheat milk before it is pasteurized. Cold milk enters the regenerator at $T_{c,i} = 5$°C, while hot milk, which has completed the pasteurization process, enters at $T_{h,i} = 70$°C. After leaving the regenerator, the heated milk enters a second heat exchanger, which raises its temperature from $T_{c,o}$ to 70°C.

The density and specific heat of milk are 1030 kg/m³ and 3860 J/kg · K, respectively.

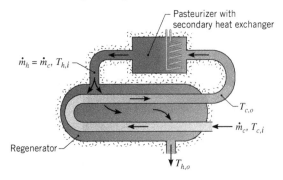

(a) A regenerator is to be used in a pasteurization process for which the flow rate of the milk is $\dot{m}_c = \dot{m}_h = 5$ kg/s. For this flow rate, the manufacturer of the regenerator specifies an overall heat transfer coefficient of 2000 W/m² · K. If the desired effectiveness of the regenerator is 0.5, what is the requisite heat transfer area? What are the corresponding rate of heat recovery and the fluid outlet temperatures?

(b) If the hot fluid in the secondary heat exchanger derives its energy from the combustion of natural gas and the burner has an efficiency of 90%, what would be the annual savings in energy and fuel costs associated with installation of the regenerator? The facility operates continuously throughout the year, and the cost of natural gas is $C_{ng} = \$0.02/\text{MJ}$.

11.25 A twin-tube, counterflow heat exchanger operates with balanced flow rates of 0.003 kg/s for the hot and cold airstreams. The cold stream enters at 280 K and must be heated to 340 K using hot air at 360 K. The average pressure of the airstreams is 1 atm and the maximum allowable pressure drop for the cold air is 10 kPa. The tube walls may be assumed to act as fins, each with an efficiency of 100%.

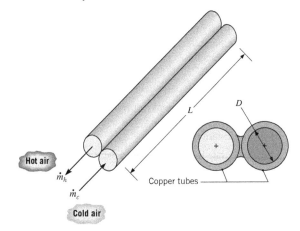

(a) Determine the tube diameter D and length L that satisfy the prescribed heat transfer and pressure drop requirements.

(b) For the diameter D and length L found in part (a), generate plots of the cold stream outlet temperature, the heat transfer rate, and pressure drop as a function of balanced flow rates in the range from 0.002 to 0.004 kg/s. Comment on your results.

11.26 A 5-m-long, twin-tube, counterflow heat exchanger, such as that illustrated in Problem 11.25, is used to heat air for a drying operation. Each tube is made from plain carbon steel ($k = 60$ W/m · K) and has an inner diameter and wall thickness of 50 mm and 4 mm, respectively. The thermal resistance per unit length of the brazed joint connecting the tubes is $R'_t = 0.01$ m · K/W. Consider conditions for which air enters one tube at a pressure of 5 atm, a temperature of 17°C, and flow rate of 0.030 kg/s, while saturated steam at 2.455 bar condenses in the other tube. The convection coefficient for condensation may be approximated as 5000 W/m² · K. What is the air outlet temperature? What is the mass rate at which condensate leaves the system? *Hint:* Account for the effects of circumferential conduction in the tubes by treating them as extended surfaces.

11.27 Hot water for an industrial washing operation is produced by recovering heat from the flue gases of a furnace. A cross-flow heat exchanger is used, with the gases passing over the tubes and the water making a single pass through the tubes. The steel tubes ($k = 60$ W/m · K) have inner and outer diameters of $D_i = 15$ mm and $D_o = 20$ mm, while the staggered tube array has longitudinal and transverse pitches of $S_T = S_L = 40$ mm. The plenum in which the array is installed has a width (corresponding to the tube length) of $W = 2$ m and a height (normal to the tube axis) of $H = 1.2$ m. The number of tubes in the transverse plane is therefore $N_T \approx H/S_T = 30$. The gas properties may be approximated as those of atmospheric air, and the convection coefficient associated with water flow in the tubes may be approximated as 3000 W/m² · K.

(a) If 50 kg/s of water are to be heated from 290 to 350 K by 40 kg/s of flue gases entering the exchanger at 700 K, what is the gas outlet temperature and how many tube rows N_L are required?

(b) The water outlet temperature may be controlled by varying the gas flow rate and/or inlet temperature. For the value of N_L determined in part (a) and the prescribed values of H, W, S_T, $\dot{m}_c$, and $T_{c,i}$, compute and plot $T_{c,o}$ as a function of $\dot{m}_h$ over the range $20 \leq \dot{m}_h \leq 40$ kg/s for values of $T_{h,i} = 500$, 600, and 700 K. Also plot the corresponding variations of $T_{h,o}$. If $T_{h,o}$ must not drop below 400 K to

prevent condensation of corrosive vapors on the heat exchanger surfaces, are there any constraints on $\dot{m}_h$ and $T_{h,i}$?

11.28 The compartment heater of an automobile exchanges heat between warm radiator fluid and cooler outside air. The flow rate of water is large compared to the air, and the effectiveness, ε, of the heater is known to depend on the flow rate of air according to the relation, $\varepsilon \sim \dot{m}_{air}^{-0.2}$.

(a) If the fan is switched to high and $\dot{m}_{air}$ is doubled, determine the percentage increase in the heat added to the car, if fluid inlet temperatures remain the same.

(b) For the low-speed fan condition, the heater warms outdoor air from 0 to 30°C. When the fan is turned to medium, the airflow rate increases 50% and the heat transfer increases 20%. Find the new outlet temperature.

11.29 Consider a *coupled* shell-in-tube heat exchange device consisting of two identical heat exchangers A and B. Air flows on the shell side of heat exchanger A, entering at $T_{h,i,A} = 520$ K and $\dot{m}_{h,A} = 10$ kg/s. Ammonia flows in the shell of heat exchanger B, entering at $T_{c,i,B} = 280$ K, $\dot{m}_{c,B} = 5$ kg/s. The tube-side flow is common to both heat exchangers and consists of water at a flow rate $\dot{m}_{c,A} = \dot{m}_{h,B}$ with two tube passes. The *UA* product increases with water flow rate for heat exchanger A as expressed by the relation $UA_A = a + b\dot{m}_{c,A}$ where $a = 6000$ W/K and $b = 100$ J/kg · K. For heat exchanger B, $UA_B = 1.2UA_A$.

(a) For $\dot{m}_{c,A} = \dot{m}_{h,B} = 1$ kg/s, determine the outlet air and ammonia temperatures, as well as the heat transfer rate.

(b) The plant engineer wishes to fine-tune the heat exchanger performance by installing a variable-speed pump to allow adjustment of the water flow rate. Plot the outlet air and outlet ammonia temperatures versus the water flow rate over the range 0 kg/s $\leq \dot{m}_{c,A} = \dot{m}_{h,B} \leq 2$kg/s.

11.30 A shell-and-tube heat exchanger with one shell and two tube passes is designed to heat liquid water flowing at 3.0 kg/s from 25 to 65°C using hot air. The air enters the shell of the exchanger at 175°C, and when the heat exchanger is new, the air exits at 90°C. The overall heat transfer coefficient is composed of thermal resistances associated with water flow inside the tubes, conduction through the tube walls, and air flow across the tube bank. Its initial value is 200 W/m² · K. Over time, 20% of the tubes develop leaks, and are welded shut during routine maintenance. With the same mass flow rate of water flowing through fewer tubes, would you expect the overall heat transfer coefficient to change significantly? Explain your reasoning. Assuming the overall heat transfer coefficient is unchanged, determine the exit temperature of the water when operating with only 80% of the tubes active.

11.31 For health reasons, public spaces require the continuous exchange of a specified mass of stale indoor air with fresh outdoor air. To conserve energy during the heating season, it is expedient to recover the thermal energy in the exhausted, warm indoor air and transfer it to the incoming, cold fresh air. A *coupled* single-pass, cross-flow heat exchanger with both fluids unmixed is installed in the intake and return ducts of a heating system as shown in the schematic. Water containing an anti-freeze agent is used as the working fluid in the coupled heat exchange device, which is composed of individual heat exchangers A and B. Hence, heat is transferred from the warm stale air to the cold fresh air by way of the pumped water.

Consider a specified air mass flow rate (in each duct) of $\dot{m} = 1.50$ kg/s, an overall heat transfer coefficient–area product of $UA = 2500$ W/K (for each heat exchanger), an outdoor temperature of $T_{c,i,A} = -4°C$ and an indoor temperature of $T_{h,i,B} = 23°C$. Since the warm air has been humidified, excessive heat transfer can result in unwanted condensation in the ductwork. What water flow rate is necessary to maximize heat transfer while ensuring the outlet temperature associated with heat exchanger B does not fall below the dew point temperature, $T_{h,o,B} = T_{dp} = 13°C$? *Hint:* Assume the maximum heat capacity rate is associated with the air.

11.32 Saturated steam at 0.14 bar is condensed in a shell-and-tube heat exchanger with one shell pass and two tube passes consisting of 130 brass tubes, each with a length per pass of 2 m. The tubes have inner and outer diameters of 13.4 and 15.9 mm, respectively. Cooling water enters the tubes at 20°C with a mean velocity of 1.25 m/s. The heat transfer coefficient for condensation on the outer surfaces of the tubes is 13,500 W/m² · K.

(a) Determine the overall heat transfer coefficient, the cooling water outlet temperature, and the steam condensation rate.

(b) With all other conditions remaining the same, but accounting for changes in the overall coefficient, plot the cooling water outlet temperature and the steam condensation rate as a function of the water flow rate for $10 \leq \dot{m}_c \leq 30$ kg/s.

11.33 A feedwater heater that supplies a boiler consists of a shell-and-tube heat exchanger with one shell pass and two tube passes. One hundred thin-walled tubes each have a diameter of 20 mm and a length (per pass) of 2 m. Under normal operating conditions water enters the tubes at 10 kg/s and 290 K and is heated by condensing saturated steam at 1 atm on the outer surface of the tubes. The convection coefficient of the saturated steam is 10,000 W/m² · K.

(a) Determine the water outlet temperature.

(b) With all other conditions remaining the same, but accounting for changes in the overall heat transfer coefficient, plot the water outlet temperature as a function of the water flow rate for $5 \leq \dot{m}_c \leq 20$ kg/s.

(c) On the plot of part (b), generate two additional curves for the water outlet temperature as a function of flow rate for fouling factors of $R''_f = 0.0002$ and 0.0005 m² · K/W.

11.34 Saturated steam at 110°C is condensed in a shell-and-tube heat exchanger (1 shell pass; 2, 4, · · · tube passes) with a *UA* value of 2.5 kW/K. Cooling water enters at 40°C.

(a) Calculate the cooling water flow rate required to maintain a heat rate of 150 kW.

(b) Assuming that *UA* is independent of flow rate, calculate and plot the water flow rate required to provide heat rates over the range from 130 to 160 kW. Comment on the validity of your assumption.

11.35 A shell-and-tube heat exchanger (1 shell pass, 2 tube passes) is to be used to condense 2.73 kg/s of saturated steam at 340 K. Condensation occurs on the outer tube surfaces, and the corresponding convection coefficient is 10,000 W/m² · K. The temperature of the cooling water entering the tubes is 15°C, and the exit temperature is not to exceed 30°C. Thin-walled tubes of 19-mm diameter are specified, and the mean velocity of water flow through the tubes is to be maintained at 0.5 m/s.

(a) What is the minimum number of tubes that should be used, and what is the corresponding tube length per pass?

(b) To reduce the size of the heat exchanger, it is proposed to increase the water-side convection coefficient by inserting a wire mesh in the tubes. If the mesh increases the convection coefficient by a factor of two, what is the required tube length per pass?

11.36 A two-fluid heat exchanger has inlet and outlet temperatures of 65 and 40°C for the hot fluid and 15 and 30°C for the cold fluid. Can you tell whether this exchanger is operating under counterflow or parallel-flow conditions? Determine the effectiveness of the heat exchanger.

11.37 A heat exchanger has inlet and outlet temperatures of 90 and 45°C for the hot fluid and 15 and 42°C for the cold fluid. Can you tell whether this exchanger is operating under counterflow or parallel-flow conditions? Determine the effectiveness of the heat exchanger.

11.38 The human brain is especially sensitive to elevated temperatures. The cool blood in the veins leaving the face and neck and returning to the heart may contribute to thermal regulation of the brain by cooling the arterial blood flowing to the brain. Consider a vein and artery running between the chest and the base of the skull for a distance $L = 250$ mm, with mass flow rates of 3×10^{-3} kg/s in opposite directions in the two vessels. The vessels are of diameter $D = 5$ mm and are separated by a distance $w = 7$ mm. The thermal conductivity of the surrounding tissue is $k_t = 0.5$ W/m · K. If the arterial blood enters at 37°C and the venous blood enters at 27°C, at what temperature will the arterial blood exit? If the arterial blood becomes overheated, and the body responds by halving the blood flow rate, how much hotter can the entering arterial blood be and still maintain its exit temperature below 37°C? *Hint:* If we assume that all the heat leaving the artery enters the vein, then heat transfer between the two vessels can be modeled using a relationship found in Table 4.1. Approximate the blood properties as those of water.

11.39 A plate-fin heat exchanger is used to condense a saturated refrigerant vapor in an air-conditioning system. The vapor has a saturation temperature of 45°C, and a condensation rate of 0.015 kg/s is dictated by system performance requirements. The frontal area of

the condenser is fixed at $A_{fr} = 0.25$ m² by installation requirements, and a value of $h_{fg} = 135$ kJ/kg may be assumed for the refrigerant.

(a) The condenser design is to be based on a nominal air inlet temperature of $T_{c,i} = 30$°C and nominal air inlet velocity of $V = 2$ m/s for which the manufacturer of the heat exchanger core indicates an overall coefficient of $U = 50$ W/m² · K. What is the corresponding value of the heat transfer surface area required to achieve the prescribed condensation rate? What is the air outlet temperature?

(b) From the manufacturer of the heat exchanger core, it is also known that $U \propto V^{0.7}$. During daily operation the air inlet temperature is not controllable and may vary from 27 to 38°C. If the heat exchanger area is fixed by the result of part (a), what is the range of air velocities needed to maintain the prescribed condensation rate? Plot the velocity as a function of the air inlet temperature.

11.40 A shell-and-tube heat exchanger is to heat 10,000 kg/h of water from 16 to 84°C by hot engine oil flowing through the shell. The oil makes a single shell pass, entering at 160°C and leaving at 94°C, with an average heat transfer coefficient of 400 W/m² · K. The water flows through 11 brass tubes of 22.9-mm inside diameter and 25.4-mm outside diameter, with each tube making four passes through the shell.

(a) Assuming fully developed flow for the water, determine the required tube length per pass.

(b) For the tube length found in part (a), plot the effectiveness, fluid outlet temperatures, and water-side convection coefficient as a function of the water flow rate for $5000 \leq \dot{m}_c \leq 15,000$ kg/h, with all other conditions remaining the same.

11.41 In a supercomputer, signal propagation delays are reduced by resorting to high-density circuit arrangements which are cooled by immersing them in a special dielectric liquid. The fluid is pumped in a closed loop through the computer and an adjoining shell-and-tube heat exchanger having one shell and two tube passes.

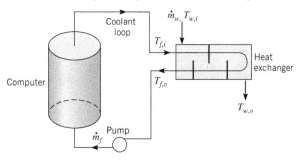

During normal operation, heat generated within the computer is transferred to the dielectric fluid passing through the computer at a flow rate of $\dot{m}_f = 4.81$ kg/s. In turn, the fluid passes through the tubes of the heat exchanger and the heat is transferred to water passing over the tubes. The dielectric fluid may be assumed to have constant properties of $c_p = 1040$ J/kg $\cdot$ K, $\mu = 7.65 \times 10^{-4}$ kg/s $\cdot$ m, $k = 0.058$ W/m $\cdot$ K, and $Pr = 14$. During normal operation, chilled water at a flow rate of $\dot{m}_w = 2.5$ kg/s and an inlet temperature of $T_{w,i} = 5°C$ passes over the tubes. The water has a specific heat of 4200 J/kg $\cdot$ K and provides an average convection coefficient of 10,000 W/m² $\cdot$ K over the outer surface of the tubes.

(a) If the heat exchanger consists of 72 thin-walled tubes, each of 10-mm diameter, and fully developed flow is assumed to exist within the tubes, what is the convection coefficient associated with flow through the tubes?

(b) If the dielectric fluid enters the heat exchanger at $T_{f,i} = 25°C$ and is to leave at $T_{f,o} = 15°C$, what is the required tube length per pass?

(c) For the exchanger with the tube length per pass determined in part (b), plot the outlet temperature of the dielectric fluid as a function of its flow rate for $4 \le \dot{m}_f \le 6$ kg/s. Account for corresponding changes in the overall heat transfer coefficient, but assume all other conditions to remain the same.

(d) The site specialist for the computer facilities is concerned about changes in the performance of the water chiller supplying the cold water ($\dot{m}_w$, $T_{w,i}$) and their effect on the outlet temperature $T_{f,o}$ of the dielectric fluid. With all other conditions remaining the same, determine the effect of a ±10% change in the cold water flow rate on $T_{f,o}$.

(e) Repeat the performance analysis of part (d) to determine the effect of a ±3°C change in the water inlet temperature on $T_{f,o}$, with all other conditions remaining the same.

11.42 A shell-and-tube heat exchanger consists of 135 thin-walled tubes in a double-pass arrangement, each of 12.5-mm diameter with a total surface area of 47.5 m². Water (the tube-side fluid) enters the heat exchanger at 15°C and 6.5 kg/s and is heated by exhaust gas entering at 200°C and 5 kg/s. The gas may be assumed to have the properties of atmospheric air, and the overall heat transfer coefficient is approximately 200 W/m² $\cdot$ K.

(a) What are the gas and water outlet temperatures?

(b) Assuming fully developed flow, what is the tube-side convection coefficient?

(c) With all other conditions remaining the same, plot the effectiveness and fluid outlet temperatures as a function of the water flow rate over the range from 6 to 12 kg/s.

(d) What gas inlet temperature is required for the exchanger to supply 10 kg/s of hot water at an outlet temperature of 42°C, all other conditions remaining the same? What is the effectiveness for this operating condition?

11.43 An ocean thermal energy conversion system is being proposed for electric power generation. Such a system is based on the standard power cycle for which the working fluid is evaporated, passed through a turbine, and subsequently condensed. The system is to be used in very special locations for which the oceanic water temperature near the surface is approximately 300 K, while the temperature at reasonable depths is approximately 280 K. The warmer water is used as a heat source to evaporate the working fluid, while the colder water is used as a heat sink for condensation of the fluid. Consider a power plant that is to generate 2 MW of electricity at an efficiency (electric power output per heat input) of 3%. The evaporator is a heat exchanger consisting of a single shell with many tubes executing two passes. If the working fluid is evaporated at its phase change temperature of 290 K, with ocean water entering at 300 K and leaving at 292 K, what is the heat exchanger area required for the evaporator? What flow rate must be maintained for the water passing through the evaporator? The overall heat transfer coefficient may be approximated as 1200 W/m² $\cdot$ K.

11.44 Using the results shown in Table 11.3, determine an expression for the ratio of the effectiveness of a counterflow heat exchanger to that of a parallel flow heat exchanger. Considering the same heat exchanger used in either counter- or parallel flow mode, what is the limiting value of the effectiveness ratio for extremely small area (low cost) heat exchangers? What is the limiting value of the effectiveness ratio for heat exchangers of infinite area (high cost)? Plot the effectiveness ratio over the range $0 \leq NTU \leq 10$ for $C_r = 0.25$, 0.5, and 0.75. Under what conditions (high or low cost) does the counterflow mode most significantly outperform the parallel flow mode?

11.45 Saturated process steam at 1 atm is condensed in a shell-and-tube heat exchanger (one shell, two tube passes). Cooling water enters the tubes at 15°C with an average velocity of 3.5 m/s. The tubes are thin walled and made of copper with a diameter of 14 mm and length of 0.5 m. The convective heat transfer coefficient for condensation on the outer surface of the tubes is 21,800 W/m² · K.

(a) Find the number of tubes/pass required to condense 2.3 kg/s of steam.

(b) Find the outlet water temperature.

(c) Find the maximum possible condensation rate that could be achieved with this heat exchanger using the same water flow rate and inlet temperature.

(d) Using the heat transfer surface area found in part (a), plot the water outlet temperature and steam condensation rate for water mean velocities in the range from 1 to 5 m/s. Assume that the shell-side convection coefficient remains unchanged.

11.46 The chief engineer at a university that is constructing a large number of new student dormitories decides to install a counterflow concentric tube heat exchanger on each of the dormitory shower drains. The thin-walled copper drains are of diameter $D_i = 50$ mm. Wastewater from the shower enters the heat exchanger at $T_{h,i} = 38°C$ while fresh water enters the dormitory at $T_{c,i} = 10°C$. The wastewater flows down the vertical wall of the drain in a thin, *falling film*, providing $h_h = 10{,}000$ W/m² · K.

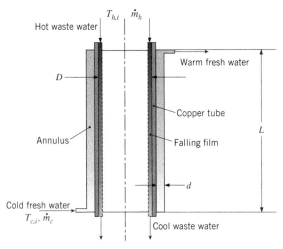

(a) If the annular gap is $d = 10$ mm, the heat exchanger length is $L = 1$ m, and the water flow rate is

$\dot{m} = 10$ kg/min, determine the heat transfer rate and the outlet temperature of the warmed fresh water.

(b) If a helical spring is installed in the annular gap so the fresh water is forced to follow a spiral path from the inlet to the fresh water outlet, resulting in $h_c = 9050$ W/m² · K, determine the heat transfer rate and the outlet temperature of the fresh water.

(c) Based on the result for part (b), calculate the daily savings if 15,000 students each take a 10-minute shower per day and the cost of water heating is $0.07/kW · h.

11.47 A shell-and-tube heat exchanger with one shell pass and 20 tube passes uses hot water on the tube side to heat oil on the shell side. The single copper tube has inner and outer diameters of 20 and 24 mm and a length per pass of 3 m. The water enters at 87°C and 0.2 kg/s and leaves at 27°C. Inlet and outlet temperatures of the oil are 7 and 37°C. What is the average convection coefficient for the tube outer surface?

11.48 The oil in an engine is cooled by air in a cross-flow heat exchanger where both fluids are unmixed. Atmospheric air enters at 30°C and 0.53 kg/s. Oil at 0.026 kg/s enters at 75°C and flows through a tube of 10-mm diameter. Assuming fully developed flow and constant wall heat flux, estimate the oil-side heat transfer coefficient. If the overall convection coefficient is 53 W/m² · K and the total heat transfer area is 1 m², determine the effectiveness. What is the exit temperature of the oil?

11.49 It is proposed that the exhaust gas from a natural gas–powered electric generation plant be used to generate steam in a shell-and-tube heat exchanger with one shell and one tube pass. The steel tubes have a thermal conductivity of 40 W/m · K, an inner diameter of 50 mm, and a wall thickness of 4 mm. The exhaust gas, whose flow rate is 2 kg/s, enters the heat exchanger at 400°C and must leave at 215°C. To limit the pressure drop within the tubes, the tube gas velocity should not exceed 25 m/s. If saturated water at 11.7 bar is supplied to the shell side of the exchanger, determine the required number of tubes and their length. Assume that the properties of the exhaust gas can be approximated as those of atmospheric air and that the water-side thermal resistance is negligible. However, account for fouling on the gas side of the tubes and use a fouling resistance of 0.0015 m² · K/W.

11.50 A recuperator is a heat exchanger that heats the air used in a combustion process by extracting energy from the products of combustion (the flue gas). Consider using a single-pass, cross-flow heat exchanger as a recuperator.

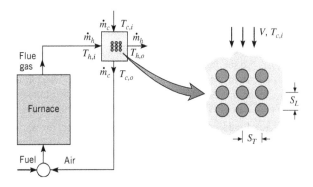

Eighty (80) silicon carbide ceramic tubes ($k = 20$ W/m · K) of inner and outer diameters equal to 55 and 80 mm, respectively, and of length $L = 1.4$ m are arranged as an aligned tube bank of longitudinal and transverse pitches $S_L = 100$ mm and $S_T = 120$ mm, respectively. Cold air is in cross flow over the tube bank with upstream conditions of $V = 1$ m/s and $T_{c,i} = 300$ K, while hot flue gases of inlet temperature $T_{h,i} = 1400$ K pass through the tubes. The tube outer surface is clean, while the inner surface is characterized by a fouling factor of $R_f'' = 2 \times 10^{-4}$ m² · K/W. The air and flue gas flow rates are $\dot{m}_c = 1.0$ kg/s and $\dot{m}_h = 1.05$ kg/s, respectively. As first approximations, (1) evaluate all required air properties at 1 atm and 300 K, (2) assume the flue gas to have the properties of air at 1 atm and 1400 K, and (3) assume the tube wall temperature to be at 800 K for the purpose of treating the effect of variable properties on convection heat transfer.

(a) If there is a 1% fuel savings associated with each 10°C increase in the temperature of the combustion air ($T_{c,o}$) above 300 K, what is the percentage fuel savings for the prescribed conditions?

(b) The performance of the recuperator is strongly influenced by the product of the overall heat transfer coefficient and the total surface area, *UA*. Compute and plot $T_{c,o}$ and the percentage fuel savings as a function of *UA* for $300 \leq UA \leq 600$ W/K. Without changing the flow rates, what measures may be taken to increase *UA*?

11.51 A concentric tube heat exchanger uses water, which is available at 15°C, to cool ethylene glycol from 100 to 60°C. The water and glycol flow rates are each 0.5 kg/s. What are the maximum possible heat transfer rate and effectiveness of the exchanger? Which is preferred, a parallel-flow or counterflow mode of operation?

11.52 Water is used for both fluids (unmixed) flowing through a single-pass, cross-flow heat exchanger. The hot water enters at 80°C and 15,000 kg/h, while the cold water enters at 10°C and 18,000 kg/h. If the effectiveness of the exchanger is 65%, determine the cold water exit temperature.

11.53 A cross-flow heat exchanger consists of a bundle of 32 tubes in a 0.6-m² duct. Hot water at 150°C and a mean velocity of 0.5 m/s enters the tubes having inner and outer diameters of 10.2 and 12.5 mm. Atmospheric air at 10°C enters the exchanger with a volumetric flow rate of 1.0 m³/s. The convection heat transfer coefficient on the tube outer surfaces is 400 W/m² · K. Estimate the fluid outlet temperatures.

11.54 Derive Equation 11.35a. *Hint:* See Section 8.3.3.

11.55 A liquefied natural gas (LNG) regasification facility utilizes a vertical heat exchanger or *vaporizer* that consists of a shell with a single-pass tube bundle used to convert the fuel to its vapor form for subsequent delivery through a land-based pipeline. Pressurized LNG is off-loaded from an oceangoing tanker to the bottom of the vaporizer at $T_{c,i} = -155°C$ and $\dot{m}_{LNG} = 150$ kg/s and flows through the shell. The pressurized LNG has a vaporization temperature of $T_f = -75°C$ and specific heat $c_{p,l} = 4200$ J/kg · K. The specific heat of the vaporized natural gas is $c_{p,v} = 2210$ J/kg · K while the gas has a latent heat of vaporization of $h_{fg} = 575$ kJ/kg. The LNG is heated with seawater flowing through the tubes, also introduced at the bottom of the vaporizer, that is available at $T_{h,i} = 20°C$ with a specific heat of $c_{p,SW} = 3985$ J/kg · K. If the gas is to leave the vaporizer at $T_{c,o} = 8°C$ and the seawater is to exit the device at $T_{h,o} = 10°C$, determine the required vaporizer heat transfer area. *Hint:* Divide the vaporizer into three sections, as shown in the schematic, with $U_A = 150$ W/m² · K, $U_B = 260$ W/m² · K, and $U_C = 40$ W/m² · K.

11.56 Work Problem 11.55 for the situation where the seawater is introduced to the top of the vaporizer, resulting in counterflowing natural gas and seawater.

11.57 A shell-and-tube heat exchanger consisting of one shell pass and two tube passes is used to transfer heat from an ethylene glycol–water solution (shell side) supplied from a rooftop solar collector to pure water (tube side) used for household purposes. The tubes are of inner and outer diameters $D_i = 3.6$ mm and $D_o = 3.8$ mm, respectively. Each of the 100 tubes is 0.8 m long (0.4 m per pass), and the heat transfer coefficient associated with the ethylene glycol–water mixture is $h_o = 11{,}000$ W/m² · K.

(a) For pure copper tubes, calculate the heat transfer rate from the ethylene glycol–water solution ($\dot{m} = 2.5$ kg/s, $T_{h,i} = 80°C$) to the pure water ($\dot{m} = 2.5$ kg/s, $T_{c,i} = 20°C$). Determine the outlet temperatures of both streams of fluid. The density and specific heat of the ethylene glycol–water mixture are 1040 kg/m³ and 3660 J/kg · K, respectively.

(b) It is proposed to replace the copper tube bundle with a bundle composed of high-temperature nylon tubes of the same diameter and tube wall thickness. The nylon is characterized by a thermal conductivity of $k_n = 0.31$ W/m · K. Determine the tube length required to transfer the same amount of energy as in part (a).

11.58 In analyzing thermodynamic cycles involving heat exchangers, it is useful to express the heat rate in terms of an overall thermal resistance R_t and the inlet temperatures of the hot and cold fluids,

$$q = \frac{(T_{h,i} - T_{c,i})}{R_t}$$

The heat transfer rate can also be expressed in terms of the rate equations,

$$q = UA \, \Delta T_{\mathrm{lm}} = \frac{1}{R_{\mathrm{lm}}} \Delta T_{\mathrm{lm}}$$

(a) Derive a relation for R_{lm}/R_t for a *parallel-flow* heat exchanger in terms of a single dimensionless parameter B, which does not involve any fluid temperatures but only U, A, C_h, C_c (or $C_{\min}$, $C_{\max}$).

(b) Calculate and plot R_{lm}/R_t for values of $B = 0.1$, 1.0, and 5.0. What conclusions can be drawn from the plot?

11.59 The power needed to overcome wind and friction drag associated with an automobile traveling at a constant velocity of 25 m/s is 9 kW.

(a) Determine the required heat transfer area of the radiator if the vehicle is equipped with an internal combustion engine operating at an efficiency of 21%. (Assume 79% of the energy generated by the engine is in the form of waste heat removed by the radiator.) The inlet and outlet mean temperatures of the water with respect to the radiator are $T_{m,i} = 400$ K and $T_{m,o} = 330$ K, respectively. Cooling air is available at 3 kg/s and 300 K. The radiator may be analyzed as a cross-flow heat exchanger with both fluids unmixed with an overall heat transfer coefficient of 400 W/m² · K.

(b) Determine the required water mass flow rate and heat transfer area of the radiator if the vehicle is equipped with a fuel cell operating at 50% efficiency. The fuel cell operating temperature is limited to approximately 85°C, so the inlet and outlet mean temperatures of the water with respect to the radiator are $T_{m,i} = 355$ K and $T_{m,o} = 330$ K, respectively. The air inlet temperature is as in part (a). Assume the flow rate of air is proportional to the surface area of the radiator. *Hint:* Iteration is required.

(c) Determine the required heat transfer area of the radiator and the outlet mean temperature of the water for the fuel cell–equipped vehicle if the mass flow rate of the water is the same as in part (a).

11.60 In a Rankine power system, 1.5 kg/s of steam leaves the turbine as saturated vapor at 0.51 bar. The steam is condensed to saturated liquid by passing it over the tubes of a shell-and-tube heat exchanger, while liquid water, having an inlet temperature of $T_{c,i} = 280$ K, is passed through the tubes. The condenser contains 100 thin-walled tubes, each of 10-mm diameter, and the total water flow rate through the tubes is 15 kg/s. The average convection coefficient associated with condensation on the outer surface of the tubes may be approximated as $\bar{h}_o = 5000$ W/m² · K. Appropriate property values for the liquid water are $c_p = 4178$ J/kg · K, $\mu = 700 \times 10^{-6}$ kg/s · m, $k = 0.628$ W/m · K, and $Pr = 4.6$.

(a) What is the water outlet temperature?

(b) What is the required tube length (per tube)?

(c) After extended use, deposits accumulating on the inner and outer tube surfaces provide a cumulative fouling factor of 0.0003 m² · K/W. For the prescribed inlet conditions and the computed tube length, what mass fraction of the vapor is condensed?

(d) For the tube length computed in part (b) and the fouling factor prescribed in part (c), explore the extent to which the water flow rate and inlet temperature may be varied (within physically plausible ranges) to improve the condenser performance. Represent your results graphically, and draw appropriate conclusions.

11.61 Consider a Rankine cycle with saturated steam leaving the boiler at a pressure of 2 MPa and a condenser pressure of 10 kPa.

(a) Calculate the thermal efficiency of the ideal Rankine cycle for these operating conditions.

(b) If the net reversible work for the cycle is 0.5 MW, calculate the required flow rate of cooling water supplied to the condenser at 15°C with an allowable temperature rise of 10°C.

(c) Design a shell-and-tube heat exchanger (one-shell, multiple-tube passes) that will meet the heat rate and temperature conditions required of the condenser. Your design should specify the number of tubes and their diameter and length.

11.62 Consider a concentric tube heat exchanger characterized by a uniform overall heat transfer coefficient and operating under the following conditions:

	$\dot{m}$ (kg/s)	c_p (J/kg · K)	T_i (°C)	T_o (°C)
Cold fluid	0.125	4200	40	95
Hot fluid	0.125	2100	210	—

What is the maximum possible heat transfer rate? What is the heat exchanger effectiveness? Should the heat exchanger be operated in parallel flow or in counterflow? What is the ratio of the required areas for these two flow conditions?

11.63 Consider the influence of a finite sheet thickness in Example 11.2, when there are 40 gaps.

(a) Determine the exterior dimension, L, of the heat exchanger core for a sheet thickness of $t = 0.8$ mm for pure aluminum ($k_{al} = 237$ W/m · K) and polyvinylidene fluoride (PVDF, $k_{pv} = 0.17$ W/m · K) sheets. Neglect the thickness of the top and bottom exterior plates.

(b) Plot the heat exchanger core dimension as a function of the sheet thickness for aluminum and PVDF over the range $0 \le t \le 1$ mm.

11.64 Hot exhaust gases are used in a shell-and-tube exchanger to heat 2.5 kg/s of water from 35 to 85°C. The gases, assumed to have the properties of air, enter at 200°C and leave at 93°C. The overall heat transfer coefficient is 180 W/m² · K. Using the effectiveness–NTU method, calculate the area of the heat exchanger.

11.65 In open heart surgery under hypothermic conditions, the patient's blood is cooled before the surgery and rewarmed afterward. It is proposed that a concentric tube, counterflow heat exchanger of length 0.5 m be used for this purpose, with the thin-walled inner tube having a diameter of 55 mm. The specific heat of the blood is 3500 J/kg · K.

(a) If water at $T_{h,i} = 60$°C and $\dot{m}_h = 0.10$ kg/s is used to heat blood entering the exchanger at $T_{c,i} = 18$°C and $\dot{m}_c = 0.05$ kg/s, what is the temperature of the blood leaving the exchanger? The overall heat transfer coefficient is 500 W/m² · K.

(b) The surgeon may wish to control the heat rate q and the outlet temperature $T_{c,o}$ of the blood by altering the flow rate and/or inlet temperature of the water during the rewarming process. To assist in the development of an appropriate controller for the prescribed values of $\dot{m}_c$ and $T_{c,i}$, compute and plot q and $T_{c,o}$ as a function of $\dot{m}_h$ for $0.05 \le \dot{m}_h \le 0.20$ kg/s and values of $T_{h,i} = 50$, 60, and 70°C. Since the dominant influence on the overall heat transfer coefficient is associated with the blood flow conditions, the value of U may be assumed to remain at 500 W/m² · K. Should certain operating conditions be excluded?

11.66 Ethylene glycol and water, at 60 and 10°C, respectively, enter a shell-and-tube heat exchanger for which the total heat transfer area is 15 m². With ethylene glycol and water flow rates of 2 and 5 kg/s, respectively, the overall heat transfer coefficient is 800 W/m² · K.

(a) Determine the rate of heat transfer and the fluid outlet temperatures.

(b) Assuming all other conditions to remain the same, plot the effectiveness and fluid outlet temperatures as a function of the flow rate of ethylene glycol for $0.5 \le \dot{m}_h \le 5$ kg/s.

11.67 Waste heat from the exhaust gas of an industrial furnace is recovered by mounting a bank of unfinned tubes in the furnace stack. Pressurized water at a flow rate of 0.025 kg/s makes a single pass through *each* of the tubes, while the exhaust gas, which has an upstream velocity of 5.0 m/s, moves in cross flow over the tubes at 2.25 kg/s. The tube bank consists of a square array of 100 thin-walled tubes (10×10), each 25 mm in diameter and 4 m long. The tubes are aligned with a transverse pitch of 50 mm. The inlet temperatures of the water and the exhaust gas are 300 and 800 K, respectively. The water flow is fully developed, and the gas properties may be assumed to be those of atmospheric air.

(a) What is the overall heat transfer coefficient?

(b) What are the fluid outlet temperatures?

(c) Operation of the heat exchanger may vary according to the demand for hot water. For the prescribed heat exchanger design and inlet conditions, compute and plot the rate of heat recovery and the fluid outlet temperatures as a function of water flow rate per tube for $0.02 \leq \dot{m}_{c,1} \leq 0.20$ kg/s.

Additional Considerations

11.68 Derive the expression for the modified effectiveness ε^*, given in Comment 4 of Example 11.8.

11.69 In a manufacturing operation, a 1-mm-thick tubular steel collar is to be shrink-fitted onto a 3-mm-diameter pin of circular cross section. Prior to initiating the shrink-fit operation, the collar is heated to 200°C and the aluminum alloy pin is cooled to −100°C. After these temperatures are achieved, the hot collar is quickly slid over the cold pin. The collar subsequently cools and shrinks as the pin heats and expands, ultimately forming a tight fit between the two. If the tight fit is achieved when the temperature difference between the collar and pin is reduced to 50°C, how much time elapses before formation of the tight fit? Prior to the tight fit, a thermal contact resistance of 5×10^{-4} m² · K/W exists at the rod-collar interface. Plot the steel and aluminum temperatures over the time interval $0 \leq t \leq 2$ s. The aluminum alloy density and specific heat are 2800 kg/m³ and 1000 J/kg · K, respectively, while the corresponding steel properties are 7800 kg/m³ and 510 J/kg · K.

11.70 Work Problem 7.23, taking into account the increase in temperature of the water as it flows through the heat sink. Properties of water are listed in Problem 7.23, along with $\rho = 995$ kg/m³ and $c_p = 4178$ J/kg · K. *Hint:* Assume the water does not escape through the upper surface of the heat sink and that the boundary layers on each fin surface do not merge, allowing evaluation of the heat transfer coefficient using a correlation from Chapter 7. Also, see Problem 11.54.

11.71 The heat sink of Problem 7.23 is considered for an application in which the power dissipation is only 70 W,

and the engineer proposes to use air at $T_\infty = 20$°C for cooling. Taking into account the increase in temperature of the air as it flows through the heat sink, plot the allowable power dissipation and the air exit temperature as a function of the air velocity over the range 1 m/s $\leq u_\infty \leq$ 5 m/s, with the constraint that the base temperature not exceed $T_b = 70$°C. Properties of the air may be approximated as $k = 0.027$ W/m · K, $\nu = 16.4 \times 10^{-6}$ m²/s, $Pr = 0.706$, $\rho = 1.145$ kg/m³, and $c_p = 1007$ J/kg · K. *Hint:* Assume the air does not escape through the upper surface of the heat sink, use a correlation for internal flow, and see Problem 11.54.

11.72 Consider Problem 7.93. Estimate the heat transfer rate to the air, accounting for both the increase in the air temperature as it flows through the foam and the thermal resistance associated with conduction in the foam in the x-direction. Do you expect the actual heat transfer rate to the air to be equal to, less than, or greater than the value you have calculated?

11.73 The metallic foam of Problem 7.93 is brazed to the surface of a silicon chip of width $W = 25$ mm on a side. The foam heat sink is $L = 10$ mm tall. Air at $T_i = 27$°C, $V = 5$ m/s impinges on the foam heat sink while the chip surface is maintained at 70°C. Determine the heat transfer rate from the chip. To calculate a conservative estimate of the heat transfer rate, neglect convection and radiation from the top and sides of the heat sink.

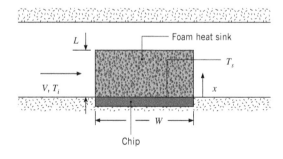

Radiation: Processes and Properties

12

We have come to recognize that heat transfer by conduction and convection requires the presence of a temperature gradient in some form of matter. In contrast, heat transfer by *thermal radiation* requires no matter. It is an extremely important process, and in the physical sense it is perhaps the most interesting of the heat transfer modes. It is relevant to many industrial heating, cooling, and drying processes, as well as to energy conversion methods that involve fossil fuel combustion and solar radiation.

In this chapter our objective is to consider the means by which thermal radiation is generated, the specific nature of the radiation, and the manner in which it interacts with matter. We give particular attention to radiative interactions at a surface and to the properties that must be introduced to describe these interactions. In Chapter 13 we focus on means for computing radiative exchange between two or more surfaces.

12.1 *Fundamental Concepts*

Consider a solid that is initially at a higher temperature T_s than that of its surroundings T_{sur}, but around which there exists a vacuum (Figure 12.1). The presence of the vacuum precludes energy loss from the surface of the solid by conduction or convection. However, our intuition tells us that the solid will cool and eventually achieve thermal equilibrium with its surroundings. This cooling is associated with a reduction in the internal energy stored by the solid and is a direct consequence of the *emission* of thermal radiation from the surface. In turn, the surface will intercept and absorb radiation originating from the surroundings. However, if $T_s > T_{sur}$ the *net* heat transfer rate by radiation $q_{rad,net}$ is *from* the surface, and the surface will cool until T_s reaches T_{sur}.

We associate thermal radiation with the rate at which energy is emitted by matter as a result of its temperature. At this moment thermal radiation is being emitted by all the matter that surrounds you: by the furniture and walls of the room, if you are indoors, or by the ground, buildings, and the atmosphere and sun if you are outdoors. The mechanism of emission is related to energy released as a result of oscillations or transitions of the many electrons that constitute matter. These oscillations are, in turn, sustained by the internal energy, and therefore the temperature, of the matter. Hence we associate the emission of thermal radiation with thermally excited conditions within the matter.

All forms of matter emit radiation. For gases and for semitransparent solids, such as glass and salt crystals at elevated temperatures, emission is a *volumetric phenomenon,* as

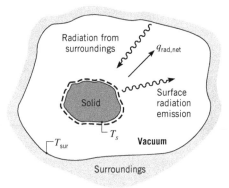

FIGURE 12.1   Radiation cooling of a hot solid.

illustrated in Figure 12.2. That is, radiation emerging from a finite volume of matter is the integrated effect of local emission throughout the volume. However, in this text we concentrate on situations for which radiation can be treated as a *surface phenomenon*. In most solids and liquids, radiation emitted from interior molecules is strongly absorbed by adjoining molecules. Accordingly, radiation that is emitted from a solid or a liquid originates from molecules that are within a distance of approximately 1 μm from the exposed surface. It is for this reason that emission from a solid or a liquid into an adjoining gas or a vacuum can be viewed as a surface phenomenon, except in situations involving nanoscale or microscale devices.

We know that radiation originates due to emission by matter and that its subsequent transport does not require the presence of any matter. But what is the nature of this transport? One theory views radiation as the propagation of a collection of particles termed *photons* or *quanta*. Alternatively, radiation may be viewed as the propagation of *electromagnetic waves*. In any case we wish to attribute to radiation the standard wave properties of frequency ν and wavelength λ. For radiation propagating in a particular medium, the two properties are related by

$$\lambda = \frac{c}{\nu} \tag{12.1}$$

where c is the speed of light in the medium. For propagation in a vacuum, $c_o = 2.998 \times 10^8$ m/s. The unit of wavelength is commonly the micrometer (μm), where 1 μm = 10^{-6} m.

The complete electromagnetic spectrum is delineated in Figure 12.3. The short wavelength gamma rays, X rays, and ultraviolet (UV) radiation are primarily of interest to the high-energy physicist and the nuclear engineer, while the long wavelength microwaves and radio waves ($\lambda > 10^5$ μm) are of concern to the electrical engineer. It is the intermediate portion of the spectrum, which extends from approximately 0.1 to 100 μm and includes a

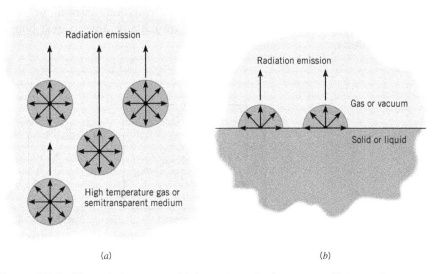

(*a*)   (*b*)

FIGURE 12.2   The emission process. (*a*) As a volumetric phenomenon. (*b*) As a surface phenomenon.

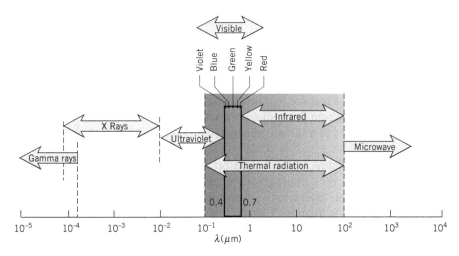

FIGURE 12.3   Spectrum of electromagnetic radiation.

portion of the UV and all of the visible and infrared (IR), that is termed *thermal radiation* because it is both caused by and affects the thermal state or temperature of matter. For this reason, thermal radiation is pertinent to heat transfer.

Thermal radiation emitted by a surface encompasses a range of wavelengths. As shown in Figure 12.4*a*, the magnitude of the radiation varies with wavelength, and the term *spectral* is used to refer to the nature of this dependence. As we will find, both the magnitude of the radiation at any wavelength and the *spectral distribution* vary with the nature and temperature of the emitting surface.

The spectral nature of thermal radiation is one of two features that complicates its description. The second feature relates to its *directionality*. As shown in Figure 12.4*b*, a surface may emit preferentially in certain directions, creating a *directional distribution* of the emitted radiation. To quantify the emission, absorption, reflection, and transmission concepts introduced in Chapter 1, we must be able to treat both spectral and directional effects.

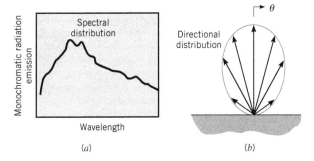

FIGURE 12.4   Radiation emitted by a surface. (*a*) Spectral distribution. (*b*) Directional distribution.

12.2 *Radiation Heat Fluxes*

Various types of heat fluxes are pertinent to the analysis of radiation heat transfer. Table 12.1 lists four distinct radiation fluxes that can be defined at a surface such as the one in Figure 12.2*b*. The *emissive power, E* (W/m^2), is the rate at which radiation is emitted from a surface per unit surface area, over all wavelengths and in all directions. In Chapter 1, this emissive power was related to the behavior of a *blackbody* through the relation $E = \varepsilon \sigma T_s^4$ (Equation 1.5), where ε is a surface property known as the *emissivity*.

Radiation from the surroundings, which may consist of multiple surfaces at various temperatures, is incident upon the surface. The surface might also be irradiated by the sun or by a laser. In any case, we define the *irradiation, G* (W/m^2), as the rate at which radiation is incident upon the surface per unit surface area, over all wavelengths and from all directions. The two remaining heat fluxes of Table 12.1 are readily described once we consider the fate of the irradiation arriving at the surface.

When radiation is incident upon a *semitransparent medium*, portions of the irradiation may be reflected, absorbed, and transmitted, as discussed in Section 1.2.3 and illustrated in Figure 12.5*a*. Transmission refers to radiation passing through the medium, as occurs when a layer of water or a glass plate is irradiated by the sun or artificial lighting. Absorption occurs when radiation interacts with the medium, causing an increase in the internal thermal energy of the medium. Reflection is the process of incident radiation being redirected away from the surface, with no effect on the medium. We define reflectivity ρ as

TABLE 12.1 Radiative fluxes (over all wavelengths and in all directions)

Flux (W/m^2)	Description	Comment
Emissive power, E	Rate at which radiation is emitted from a surface per unit area	$E = \varepsilon \sigma T_s^4$
Irradiation, G	Rate at which radiation is incident upon a surface per unit area	Irradiation can be reflected, absorbed, or transmitted
Radiosity, J	Rate at which radiation leaves a surface per unit area	For an opaque surface $J = E + \rho G$
Net radiative flux, $q''_{\text{rad}} = J - G$	Net rate of radiation leaving a surface per unit area	For an opaque surface $q''_{\text{rad}} = \varepsilon \sigma T_s^4 - \alpha G$

FIGURE 12.5 Radiation at a surface. (*a*) Reflection, absorption, and transmission of irradiation for a semitransparent medium. (*b*) The radiosity for an opaque medium.

the fraction of the irradiation that is reflected, absorptivity α as the fraction of the irradiation that is absorbed, and transmissivity τ as the fraction of the irradiation that is transmitted. Because all of the irradiation must be reflected, absorbed, or transmitted, it follows that

$$\rho + \alpha + \tau = 1 \qquad (12.2)$$

A medium that experiences no transmission ($\tau = 0$) is *opaque,* in which case

$$\rho + \alpha = 1 \qquad (12.3)$$

With this understanding of the partitioning of the irradiation into reflected, absorbed, and transmitted components, two additional and useful radiation fluxes can be defined. The *radiosity, J* (W/m²), of a surface accounts for *all* the radiant energy leaving the surface. For an opaque surface, it includes emission and the reflected portion of the irradiation, as illustrated in Figure 12.5*b*. It is therefore expressed as

$$J = E + G_{\text{ref}} = E + \rho G \qquad (12.4)$$

Radiosity can also be defined at a surface of a semitransparent medium. In that case, the radiosity leaving the top surface of Figure 12.5*a* (not shown) would include radiation transmitted through the medium from below.

Finally, the *net* radiative flux *from* a surface, q''_{rad} (W/m²), is the difference between the outgoing and incoming radiation

$$q''_{\text{rad}} = J - G \qquad (12.5)$$

Combining Equations 12.5, 12.4, 12.3, and 1.4, the net flux for an opaque surface is

$$q''_{\text{rad}} = E + \rho G - G = \varepsilon \sigma T_s^4 - \alpha G \qquad (12.6)$$

A similar expression may be written for a semitransparent surface involving the transmissivity. Because it affects the temperature distribution within the system, the net radiative flux (or net radiation heat transfer rate, $q_{\text{rad}} = q''_{\text{rad}} A$), is an important quantity in heat transfer analysis. As will become evident, the quantities E, G, and J are typically used to determine q''_{rad}, but they are also intrinsically important in applications involving *radiation detection* and *temperature measurement*.

The various fluxes in Table 12.1 may, in general, be quantified only when the directional and spectral nature of the radiation is known. Directional effects are considered by introducing the concept of *radiation intensity* in Section 12.3, while spectral effects are treated by introducing the concept of blackbody radiation in Section 12.4. The emissive power of a real surface will be related to that of the blackbody through the definition of *emissivity* in Section 12.5. The spectral and directional characteristics of the emissivity, absorptivity, reflectivity, and transmissivity of real surfaces are included in Sections 12.5 and 12.6. Sections 12.7 and 12.8 develop the important concept of a diffuse, gray surface, which has the property that $\alpha = \varepsilon$. We have implicitly assumed diffuse, gray surfaces in our treatment of radiation heat transfer so far. Finally, Section 12.9 addresses environmental radiation, or the interaction between solar radiation and radiation emitted by the earth's surface.

12.3   *Radiation Intensity*

Radiation that leaves a surface can propagate in all possible directions (Figure 12.4*b*), and we are often interested in knowing its directional distribution. Also, radiation incident upon a surface may come from different directions, and the manner in which the surface responds to this radiation depends on the direction. Such directional effects can be of primary importance in determining the net radiative heat transfer rate and may be treated by introducing the concept of *radiation intensity*.

12.3.1   Mathematical Definitions

Due to its nature, mathematical treatment of radiation heat transfer involves the extensive use of the spherical coordinate system. From Figure 12.6*a*, we recall that the differential plane angle $d\alpha$ is defined by a region between the rays of a circle and is measured as the ratio of the arc length dl on the circle to the radius r of the circle. Similarly, from Figure 12.6*b*, the differential solid angle $d\omega$ is defined by a region between the rays of a sphere and is measured as the ratio of the area dA_n on the sphere to the sphere's radius squared. Accordingly,

$$d\omega \equiv \frac{dA_n}{r^2} \tag{12.7}$$

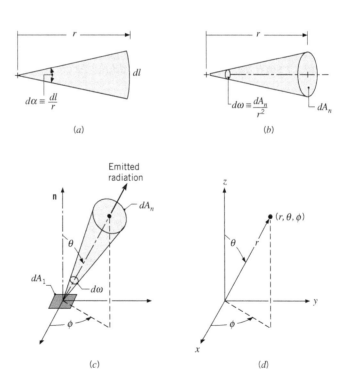

FIGURE 12.6   Mathematical definitions. (*a*) Plane angle. (*b*) Solid angle. (*c*) Emission of radiation from a differential area dA_1 into a solid angle $d\omega$ subtended by dA_n at a point on dA_1. (*d*) The spherical coordinate system.

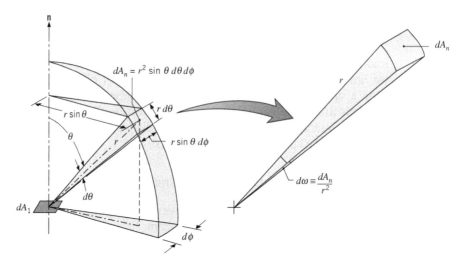

Figure 12.7   The solid angle subtended by dA_n at a point on dA_1 in the spherical coordinate system.

Consider emission in a particular direction from an element of surface area dA_1, as shown in Figure 12.6c. The direction may be specified in terms of the zenith and azimuthal angles, θ and ϕ, respectively, of a spherical coordinate system (Figure 12.6d). The area dA_n, through which the radiation passes, subtends a differential solid angle $d\omega$ when viewed from a point on dA_1. As shown in Figure 12.7, the area dA_n is a rectangle of dimension $r\,d\theta \times r\sin\theta\,d\phi$; thus, $dA_n = r^2\sin\theta\,d\theta\,d\phi$. Accordingly,

$$d\omega = \sin\theta\,d\theta\,d\phi \tag{12.8}$$

When viewed from a point on an *opaque* surface area element dA_1, radiation may be emitted into any direction defined by a hypothetical hemisphere above the surface. The solid angle associated with the entire hemisphere may be obtained by integrating Equation 12.8 over the limits $\phi = 0$ to $\phi = 2\pi$ and $\theta = 0$ to $\theta = \pi/2$. Hence,

$$\int_h d\omega = \int_0^{2\pi}\int_0^{\pi/2}\sin\theta\,d\theta\,d\phi = 2\pi\int_0^{\pi/2}\sin\theta\,d\theta = 2\pi \text{ sr} \tag{12.9}$$

where the symbol h refers to integration over the hemisphere. Note that the unit of the solid angle is the steradian (sr), analogous to radians for plane angles.

12.3.2   Radiation Intensity and Its Relation to Emission

Returning to Figure 12.6c, we now consider the rate at which emission from dA_1 passes through dA_n. This quantity may be expressed in terms of the *spectral intensity* $I_{\lambda,e}$ of the emitted radiation. We formally define $I_{\lambda,e}$ as the *rate at which radiant energy is emitted at the wavelength λ in the (θ, ϕ) direction, per unit area of the emitting surface normal to this direction, per unit solid angle about this direction, and per unit wavelength interval $d\lambda$ about λ*. Note that the area used to define the intensity is the component of dA_1

FIGURE 12.8 The projection of dA_1 normal to the direction of radiation.

perpendicular to the direction of the radiation. From Figure 12.8, we see that this projected area is equal to $dA_1 \cos \theta$. In effect it is how dA_1 would appear to an observer situated on dA_n. The spectral intensity, which has units of $W/m^2 \cdot sr \cdot \mu m$, is then

$$I_{\lambda,e}(\lambda, \theta, \phi) \equiv \frac{dq}{dA_1 \cos \theta \cdot d\omega \cdot d\lambda} \qquad (12.10)$$

where $(dq/d\lambda) \equiv dq_\lambda$ is the rate at which radiation of wavelength λ leaves dA_1 and passes through dA_n. Rearranging Equation 12.10, it follows that

$$dq_\lambda = I_{\lambda,e}(\lambda, \theta, \phi)dA_1 \cos \theta \, d\omega \qquad (12.11)$$

where dq_λ has the units of $W/\mu m$. This important expression allows us to compute the rate at which radiation emitted by a surface propagates into the region of space defined by the solid angle $d\omega$ about the (θ, ϕ) direction. However, to compute this rate, the spectral intensity $I_{\lambda,e}$ of the emitted radiation must be known. The manner in which this quantity may be determined is discussed later, in Sections 12.4 and 12.5. Expressing Equation 12.11 per unit area of the emitting surface and substituting from Equation 12.8, the spectral radiation *flux* associated with dA_1 is

$$dq''_\lambda = I_{\lambda,e}(\lambda, \theta, \phi) \cos \theta \sin \theta \, d\theta \, d\phi \qquad (12.12)$$

If the spectral and directional distributions of $I_{\lambda,e}$ are known, that is, $I_{\lambda,e}(\lambda, \theta, \phi)$ is known, the heat flux associated with emission into any finite solid angle or over any finite wavelength interval may be determined by integrating Equation 12.12. For example, we define the *spectral, hemispherical emissive power* E_λ ($W/m^2 \cdot \mu m$) as the rate at which radiation of wavelength λ is emitted in *all directions* from a surface per unit wavelength interval $d\lambda$ about λ and per unit surface area. Thus, E_λ is the spectral heat flux associated with emission into a hypothetical hemisphere above dA_1, as shown in Figure 12.9, or

$$E_\lambda(\lambda) = q''_\lambda(\lambda) = \int_0^{2\pi} \int_0^{\pi/2} I_{\lambda,e}(\lambda, \theta, \phi) \cos \theta \sin \theta \, d\theta \, d\phi \qquad (12.13)$$

Note that E_λ is a flux based on the *actual* surface area, whereas $I_{\lambda,e}$ is based on the *projected* area. The $\cos \theta$ term appearing in the integrand is a consequence of this difference.

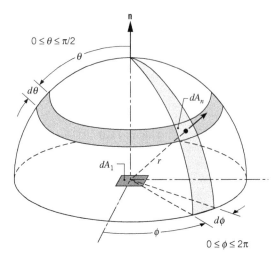

FIGURE 12.9 Emission from a differential element of area dA_1 into a hypothetical hemisphere centered at a point on dA_1.

The *total, hemispherical emissive power, E* (W/m²), is the rate at which radiation is emitted per unit area at all possible wavelengths and in all possible directions. Accordingly,

$$E = \int_0^\infty E_\lambda(\lambda)\, d\lambda \qquad (12.14)$$

or from Equation 12.13

$$E = \int_0^\infty \int_0^{2\pi} \int_0^{\pi/2} I_{\lambda,e}(\lambda, \theta, \phi) \cos\theta \sin\theta\, d\theta\, d\phi\, d\lambda \qquad (12.15)$$

Since the term "emissive power" implies emission in all directions, the adjective "hemispherical" is redundant and is often dropped. One then speaks of the *spectral emissive power E_λ*, or the *total emissive power E* that was first introduced in Equation 1.5 and again in Table 12.1.

Although the directional distribution of surface emission varies according to the nature of the surface, there is a special case that provides a reasonable approximation for many surfaces. We speak of a *diffuse emitter* as a surface for which the intensity of the emitted radiation is independent of direction, in which case $I_{\lambda,e}(\lambda, \theta, \phi) = I_{\lambda,e}(\lambda)$. Removing $I_{\lambda,e}$ from the integrand of Equation 12.13 and performing the integration, it follows that

$$E_\lambda(\lambda) = \pi I_{\lambda,e}(\lambda) \qquad (12.16)$$

Similarly, from Equation 12.15

$$E = \pi I_e \qquad (12.17)$$

where I_e is the *total intensity* of the emitted radiation. Note that the constant appearing in the above expressions is π, not 2π, and has the unit steradians.

EXAMPLE 12.1

A small surface of area $A_1 = 10^{-3}$ m^2 is known to emit diffusely, and from measurements the total intensity associated with emission in the normal direction is $I_n = 7000$ W/m$^2 \cdot$ sr.

Radiation emitted from the surface is intercepted by three other surfaces of area $A_2 = A_3 = A_4 = 10^{-3}$ m^2, which are 0.5 m from A_1 and are oriented as shown. What is the intensity associated with emission in each of the three directions? What are the solid angles subtended by the three surfaces when viewed from A_1? What is the rate at which radiation emitted by A_1 is intercepted by the three surfaces?

SOLUTION

Known: Normal intensity of diffuse emitter of area A_1 and orientation of three surfaces relative to A_1.

Find:

1. Intensity of emission in each of the three directions.
2. Solid angles subtended by the three surfaces.
3. Rate at which radiation is intercepted by the three surfaces.

Schematic:

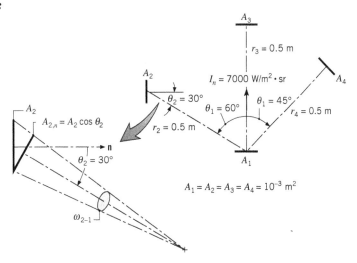

Assumptions:

1. Surface A_1 emits diffusely.
2. A_1, A_2, A_3, and A_4 may be approximated as differential surfaces, $(A_j/r_j^2) \ll 1$.

Analysis:

1. From the definition of a diffuse emitter, we know that the intensity of the emitted radiation is independent of direction. Hence

$$I = 7000 \text{ W/m}^2 \cdot \text{sr} \qquad \triangleleft$$

for each of the three directions.

2. Treating A_2, A_3, and A_4 as differential surface areas, the solid angles may be computed from Equation 12.7

$$d\omega \equiv \frac{dA_n}{r^2}$$

where dA_n is the projection of the surface normal to the direction of the radiation. Since surfaces A_3 and A_4 are normal to the direction of radiation, the solid angles subtended by these surfaces can be directly found from this equation as

$$\omega_{3-1} = \omega_{4-1} = \frac{A_3}{r^2} = \frac{10^{-3} \text{ m}^2}{(0.5 \text{ m})^2} = 4.00 \times 10^{-3} \text{ sr} \qquad \triangleleft$$

Since surface A_2 is not normal to the direction of radiation, we use $dA_{n,2} = dA_2 \cos \theta_2$, where θ_2 is the angle between the surface normal and the direction of the radiation. Thus

$$\omega_{2-1} = \frac{A_2 \cos \theta_2}{r^2} = \frac{10^{-3} \text{ m}^2 \times \cos 30°}{(0.5 \text{ m})^2} = 3.46 \times 10^{-3} \text{ sr} \qquad \triangleleft$$

3. Approximating A_1 as a differential surface, the rate at which radiation is intercepted by each of the three surfaces may be found from Equation 12.11, which, for the total radiation, may be expressed as

$$q_{1-j} = I \times A_1 \cos \theta_1 \times \omega_{j-1}$$

where θ_1 is the angle between the normal to surface 1 and the direction of the radiation. Hence

$$q_{1-2} = 7000 \text{ W/m}^2 \cdot \text{sr} \, (10^{-3} \text{ m}^2 \times \cos 60°) \, 3.46 \times 10^{-3} \text{ sr}$$
$$= 12.1 \times 10^{-3} \text{ W} \qquad \triangleleft$$

$$q_{1-3} = 7000 \text{ W/m}^2 \cdot \text{sr} \, (10^{-3} \text{ m}^2 \times \cos 0°) \, 4.00 \times 10^{-3} \text{ sr}$$
$$= 28.0 \times 10^{-3} \text{ W} \qquad \triangleleft$$

$$q_{1-4} = 7000 \text{ W/m}^2 \cdot \text{sr} \, (10^{-3} \text{ m}^2 \times \cos 45°) \, 4.00 \times 10^{-3} \text{ sr}$$
$$= 19.8 \times 10^{-3} \text{ W} \qquad \triangleleft$$

Comments:

1. Note the different values of θ_1 for the emitting surface and the values of θ_2, θ_3, and θ_4 for the receiving surfaces.

2. If the surfaces were not small relative to the square of the separation distance, the solid angles and radiation heat transfer rates would have to be obtained by integrating Equations 12.8 and 12.11, respectively, over the appropriate surface areas.

3. Any spectral component of the radiation rate can also be obtained using these procedures, if the spectral intensity I_λ is known.

4. Even though the intensity of the emitted radiation is independent of direction, the rate at which radiation is intercepted by the three surfaces differs significantly due to differences in the solid angles and projected areas. For example, consider moving surface A_4 to various θ_1 positions, keeping A_4 normal to the direction of radiation and r_4 constant at 0.5 m as shown in Figure (*a*) below.

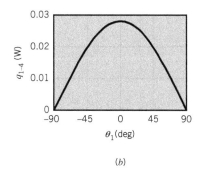

(a)   (b)

Under these conditions $\omega_{4-1} = 4.00 \times 10^{-3}$ sr is constant and

$$q_{1-4} = I \times A_1 \omega_{4-1} \cos \theta_1 = [7000 \text{ W/m}^2 \cdot \text{sr} \times 10^{-3} \text{ m}^2 \times 4.00 \times 10^{-3} \text{ sr}] \times \cos \theta_1$$

The energy that is emitted from A_1 and subsequently intercepted by A_4 is plotted in Figure (*b*) above. Consistent with our intuition, the energy intercepted by surface A_4 is maximum at $\theta_1 = 0°$ ($q_{1-4} = 28.0 \times 10^{-3}$ W) since an observer on A_4 would see the largest projected area of A_1. Also consistent with our intuition, the energy intercepted by A_4 is zero at $\theta_1 = \pm 90°$, even though the intensity of the radiation emitted from A_1 is independent of θ. At $\theta_1 = \pm 90°$, an observer on A_4 would be unable to see A_1 and hence would intercept none of the energy emitted from A_1. Many real surfaces emit radiation in a manner that is approximately diffuse.

12.3.3 Relation to Irradiation

The foregoing concepts may be extended to *incident* radiation (Figure 12.10). Such radiation may originate from emission and reflection occurring at other surfaces and will have spectral and directional distributions determined by the spectral intensity $I_{\lambda,i}(\lambda, \theta, \phi)$. This quantity is defined as the rate at which radiant energy of wavelength λ is incident from the (θ, ϕ) direction, per unit area of the *intercepting surface* normal to this direction, per unit solid angle about this direction, and per unit wavelength interval $d\lambda$ about λ.

The intensity of the incident radiation may be related to the irradiation, which encompasses radiation incident *from all directions*. The *spectral irradiation* $G_\lambda(\text{W/m}^2 \cdot \mu\text{m})$ is defined as the rate at which radiation of wavelength λ is incident on a surface, per unit area of the surface and per unit wavelength interval $d\lambda$ about λ. Accordingly,

$$G_\lambda(\lambda) = \int_0^{2\pi} \int_0^{\pi/2} I_{\lambda,i}(\lambda, \theta, \phi) \cos \theta \sin \theta \, d\theta \, d\phi \tag{12.18}$$

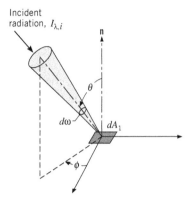

FIGURE 12.10   Directional nature of incident radiation.

where $\sin\theta\,d\theta\,d\phi$ is the unit solid angle. The $\cos\theta$ factor originates because G_λ is a flux based on the actual surface area, whereas $I_{\lambda,i}$ is defined in terms of the projected area. If the *total irradiation G* (W/m^2) represents the rate at which radiation is incident per unit area from all directions and at all wavelengths, it follows that

$$G = \int_0^\infty G_\lambda(\lambda)\,d\lambda \qquad (12.19)$$

or from Equation 12.18

$$G = \int_0^\infty \int_0^{2\pi} \int_0^{\pi/2} I_{\lambda,i}(\lambda,\theta,\phi)\cos\theta\sin\theta\,d\theta\,d\phi\,d\lambda \qquad (12.20)$$

The total irradiation was first introduced in Section 1.2.3 and again in Table 12.1. If the incident radiation is *diffuse*, $I_{\lambda,i}$ is independent of θ and ϕ and it follows that

$$G_\lambda(\lambda) = \pi I_{\lambda,i}(\lambda) \qquad (12.21)$$

and

$$G = \pi I_i \qquad (12.22)$$

EXAMPLE 12.2

The spectral distribution of surface irradiation is as follows:

What is the total irradiation?

SOLUTION

Known: Spectral distribution of surface irradiation.

Find: Total irradiation.

Analysis: The total irradiation may be obtained from Equation 12.19.

$$G = \int_0^\infty G_\lambda \, d\lambda$$

The integral is readily evaluated by breaking it into parts. That is,

$$G = \int_0^{5\,\mu m} G_\lambda \, d\lambda + \int_{5\,\mu m}^{20\,\mu m} G_\lambda \, d\lambda + \int_{20\,\mu m}^{25\,\mu m} G_\lambda \, d\lambda + \int_{25\,\mu m}^\infty G_\lambda \, d\lambda$$

Hence

$$\begin{aligned}
G &= \tfrac{1}{2}(1000 \text{ W/m}^2 \cdot \mu m)(5 - 0) \, \mu m + (1000 \text{ W/m}^2 \cdot \mu m)(20 - 5) \, \mu m \\
&\quad + \tfrac{1}{2}(1000 \text{ W/m}^2 \cdot \mu m)(25 - 20) \, \mu m + 0 \\
&= (2500 + 15{,}000 + 2500) \text{ W/m}^2 \\
G &= 20{,}000 \text{ W/m}^2
\end{aligned}$$

◁

Comments: Generally, radiation sources do not provide such a regular spectral distribution for the irradiation. However, the procedure of computing the total irradiation from knowledge of the spectral distribution remains the same, although evaluation of the integral is likely to involve more detail.

12.3.4 Relation to Radiosity for an Opaque Surface

As discussed in Section 12.2, the radiosity accounts for *all* the radiant energy leaving a surface. Since this radiation includes the *reflected* portion of the irradiation, as well as direct emission (Figure 12.5*b*), the radiosity is generally different from the emissive power. The *spectral radiosity* J_λ (W/m$^2 \cdot \mu m$) represents the rate at which radiation of wavelength λ leaves a unit area of the surface, per unit wavelength interval $d\lambda$ about λ. Since it accounts for radiation leaving in all directions, it is related to the intensity associated with emission and reflection, $I_{\lambda,e+r}(\lambda, \theta, \phi)$, by the expression

$$J_\lambda(\lambda) = \int_0^{2\pi} \int_0^{\pi/2} I_{\lambda,e+r}(\lambda, \theta, \phi) \cos \theta \sin \theta \, d\theta \, d\phi \qquad (12.23)$$

Hence the *total radiosity* J (W/m^2) associated with the entire spectrum is

$$J = \int_0^\infty J_\lambda(\lambda) \, d\lambda \qquad (12.24)$$

or

$$J = \int_0^\infty \int_0^{2\pi} \int_0^{\pi/2} I_{\lambda,e+r}(\lambda,\theta,\phi)\cos\theta\sin\theta\, d\theta\, d\phi\, d\lambda \tag{12.25}$$

This quantity is included in Table 12.1. If the surface is both a *diffuse reflector* and a *diffuse emitter*, $I_{\lambda,e+r}$ is independent of θ and ϕ, and it follows that

$$J_\lambda(\lambda) = \pi I_{\lambda,e+r}(\lambda) \tag{12.26}$$

and

$$J = \pi I_{e+r} \tag{12.27}$$

Again, note that the radiation flux, in this case the radiosity, is based on the actual surface area, while the intensity is based on the projected area.

12.3.5 Relation to the Net Radiative Flux for an Opaque Surface

As expressed in Equation 12.5, the net radiative flux from an opaque surface is equal to the difference between the outgoing radiosity J and the incoming irradiation G. From Equations 12.20 and 12.25, Equation 12.5 may be written in terms of the intensities associated with emission, reflection, and irradiation as

$$
\begin{aligned}
q''_{\text{rad}} &= \int_0^\infty \int_0^{2\pi} \int_0^{\pi/2} I_{\lambda,e+r}(\lambda,\theta,\phi)\cos\theta\sin\theta\, d\theta\, d\phi\, d\lambda \\
&\quad -\int_0^\infty \int_0^{2\pi} \int_0^{\pi/2} I_{\lambda,i}(\lambda,\theta,\phi)\cos\theta\sin\theta\, d\theta\, d\phi\, d\lambda
\end{aligned}
\tag{12.28}
$$

Hence the net radiative heat flux can be determined if the various intensities are known. The formal integration of Equation 12.28 is sometimes carried out in practice but will not be performed here. Rather, as will become evident in Sections 12.4 through 12.7, evaluation of the net radiative flux can be simplified by expressing the various intensities in terms of the intensity associated with a perfectly emitting and absorbing surface, the blackbody, and the emissivity, absorptivity, and reflectivity of the surface.

12.4 *Blackbody Radiation*

To evaluate the emissive power, irradiation, radiosity, or net radiative heat flux of a real opaque surface, we must quantify the spectral intensities used in Equations 12.15, 12.20, 12.25, and 12.28. To do so, it is useful to first introduce the concept of a *blackbody*.

1. *A blackbody absorbs all incident radiation, regardless of wavelength and direction.*
2. *For a prescribed temperature and wavelength, no surface can emit more energy than a blackbody.*
3. *Although the radiation emitted by a blackbody is a function of wavelength and temperature, it is independent of direction. That is, the blackbody is a diffuse emitter.*

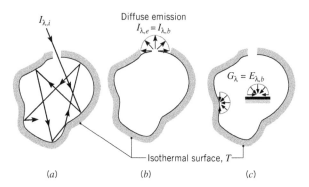

FIGURE 12.11 Characteristics of an isothermal blackbody cavity. (*a*) Complete absorption. (*b*) Diffuse emission from an aperture. (*c*) Diffuse irradiation of interior surfaces.

As the perfect absorber and emitter, the blackbody serves as a *standard* against which the radiative properties of actual surfaces may be compared.

Although closely approximated by some surfaces, it is important to note that no surface has precisely the properties of a blackbody. The closest approximation is achieved by a *cavity* whose inner surface is at a uniform temperature. If radiation enters the cavity through a small aperture (Figure 12.11*a*), it is likely to experience many reflections before reemergence. Since some radiation is absorbed by the inner surface upon each reflection, it is eventually almost entirely absorbed by the cavity, and blackbody behavior is approximated. From thermodynamic principles it may then be argued that radiation leaving the aperture depends only on the surface temperature and corresponds to blackbody emission (Figure 12.11*b*). Since blackbody emission is diffuse, the spectral intensity $I_{\lambda,b}$ of radiation leaving the cavity is independent of direction. Moreover, since the radiation field in the cavity, which is the cumulative effect of emission and reflection from the cavity surface, must be of the same form as the radiation emerging from the aperture, it also follows that a blackbody radiation field exists within the cavity. Accordingly, any small surface in the cavity (Figure 12.11*c*) experiences irradiation for which $G_\lambda = E_{\lambda,b}(\lambda, T)$. This surface is diffusely irradiated, regardless of its orientation. *Blackbody radiation exists within the cavity irrespective of whether the cavity surface is highly reflecting or absorbing.*

12.4.1 The Planck Distribution

The blackbody spectral intensity is well known, having first been determined by Planck [1]. It is

$$I_{\lambda,b}(\lambda, T) = \frac{2hc_o^2}{\lambda^5[\exp(hc_o/\lambda k_B T) - 1]} \tag{12.29}$$

where $h = 6.626 \times 10^{-34}$ J · s and $k_B = 1.381 \times 10^{-23}$ J/K are the universal Planck and Boltzmann constants, respectively, $c_o = 2.998 \times 10^8$ m/s is the speed of light in vacuum, and T is the *absolute* temperature of the blackbody (K). Since the blackbody is a diffuse emitter, it follows from Equation 12.16 that its spectral emissive power is

$$E_{\lambda,b}(\lambda, T) = \pi I_{\lambda,b}(\lambda, T) = \frac{C_1}{\lambda^5[\exp(C_2/\lambda T) - 1]} \tag{12.30}$$

where the first and second radiation constants are $C_1 = 2\pi hc_o^2 = 3.742 \times 10^8$ W · μm^4/m^2 and $C_2 = (hc_o/k_B) = 1.439 \times 10^4$ μm · K.

Equation 12.30, known as the *Planck distribution,* or *Planck's law,* is plotted in Figure 12.12 for selected temperatures. Several important features should be noted.

1. The emitted radiation varies *continuously* with wavelength.[1]

2. At any wavelength the magnitude of the emitted radiation increases with increasing temperature.

3. The spectral region in which the radiation is concentrated depends on temperature, with *comparatively* more radiation appearing at shorter wavelengths as the temperature increases.

4. A significant fraction of the radiation emitted by the sun, which may be approximated as a blackbody at 5800 K, is in the visible region of the spectrum. In contrast, for $T \lesssim 800$ K, emission is predominantly in the infrared region of the spectrum and is not visible to the eye.

12.4.2   Wien's Displacement Law

From Figure 12.12 we see that the blackbody spectral distribution has a maximum and that the corresponding wavelength λ_{max} depends on temperature. The nature of this dependence may be obtained by differentiating Equation 12.30 with respect to λ and setting the result equal to zero. In so doing, we obtain

$$\lambda_{max}T = C_3 \tag{12.31}$$

where the third radiation constant is $C_3 = 2898$ μm · K.

Equation 12.31 is known as *Wien's displacement law,* and the locus of points described by the law is plotted as the dashed line of Figure 12.12. According to this result, the maximum spectral emissive power is displaced to shorter wavelengths with increasing temperature. This emission is in the middle of the visible spectrum ($\lambda \approx 0.50$ μm) for solar radiation, since the sun emits approximately as a blackbody at 5800 K. For a blackbody at 1000 K, peak emission occurs at 2.90 μm, with some of the emitted radiation appearing visible as red light. With increasing temperature, shorter wavelengths become more prominent, until eventually significant emission occurs over the entire visible spectrum. For example, a tungsten filament lamp operating at 2900 K ($\lambda_{max} = 1$ μm) emits white light, although most of the emission remains in the IR region.

12.4.3   The Stefan–Boltzmann Law

Substituting the Planck distribution, Equation 12.30, into Equation 12.14, the total emissive power of a blackbody E_b may be expressed as

$$E_b = \int_0^\infty \frac{C_1}{\lambda^5[\exp(C_2/\lambda T) - 1]} \, d\lambda$$

---

[1]The *continuous* nature of blackbody emission can be determined *only* by considering the *discontinuous* energy states of atomic matter. Planck's derivation of the blackbody intensity distribution is one of the most important discoveries in quantum physics [2].

FIGURE 12.12 Spectral blackbody emissive power.

Performing the integration, it may be shown that

$$E_b = \sigma T^4 \tag{12.32}$$

where the *Stefan–Boltzmann* constant, which depends on C_1 and C_2, has the numerical value

$$\sigma = 5.670 \times 10^{-8} \ \text{W/m}^2 \cdot \text{K}^4$$

This simple, yet important, result is termed the *Stefan–Boltzmann law*. It enables calculation of the amount of radiation emitted in all directions and over all wavelengths simply from knowledge of the temperature of the blackbody. Because this emission is diffuse, it follows from Equation 12.17 that the total intensity associated with blackbody emission is

$$I_b = \frac{E_b}{\pi} \tag{12.33}$$

12.4.4 Band Emission

To account for spectral effects, it is often necessary to know the fraction of the total emission from a blackbody that is in a certain wavelength interval or *band*. For a prescribed

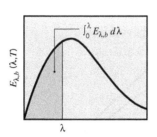

FIGURE 12.13   Radiation emission from a blackbody in the spectral band 0 to λ.

FIGURE 12.14   Fraction of the total blackbody emission in the spectral band from 0 to λ as a function of λT.

temperature and the interval from 0 to λ, this fraction is determined by the ratio of the shaded section to the total area under the curve of Figure 12.13. Hence

$$F_{(0 \to \lambda)} \equiv \frac{\int_0^\lambda E_{\lambda,b}\, d\lambda}{\int_0^\infty E_{\lambda,b}\, d\lambda} = \frac{\int_0^\lambda E_{\lambda,b}\, d\lambda}{\sigma T^4} = \int_0^{\lambda T} \frac{E_{\lambda,b}}{\sigma T^5}\, d(\lambda T) = f(\lambda T) \qquad (12.34)$$

Since the integrand $(E_{\lambda,b}/\sigma T^5)$ is exclusively a function of the wavelength–temperature product λT, the integral of Equation 12.34 may be evaluated to obtain $F_{(0 \to \lambda)}$ as a function of only λT. The results are presented in Table 12.2 and Figure 12.14. They may also be used to obtain the fraction of the radiation between any two wavelengths λ_1 and λ_2, since

$$F_{(\lambda_1 \to \lambda_2)} = \frac{\int_0^{\lambda_2} E_{\lambda,b}\, d\lambda - \int_0^{\lambda_1} E_{\lambda,b}\, d\lambda}{\sigma T^4} = F_{(0 \to \lambda_2)} - F_{(0 \to \lambda_1)} \qquad (12.35)$$

TABLE 12.2   **Blackbody radiation functions**

λT (μm · K)	$F_{(0 \to \lambda)}$	$I_{\lambda,b}(\lambda, T)/\sigma T^5$ (μm · K · sr)$^{-1}$	$\dfrac{I_{\lambda,b}(\lambda, T)}{I_{\lambda,b}(\lambda_{\max}, T)}$
200	0.000000	0.375034×10^{-27}	0.000000
400	0.000000	0.490335×10^{-13}	0.000000
600	0.000000	0.104046×10^{-8}	0.000014
800	0.000016	0.991126×10^{-7}	0.001372
1,000	0.000321	0.118505×10^{-5}	0.016406
1,200	0.002134	0.523927×10^{-5}	0.072534
1,400	0.007790	0.134411×10^{-4}	0.186082
1,600	0.019718	0.249130	0.344904
1,800	0.039341	0.375568	0.519949
2,000	0.066728	0.493432	0.683123
2,200	0.100888	0.589649×10^{-4}	0.816329
2,400	0.140256	0.658866	0.912155
2,600	0.183120	0.701292	0.970891
2,800	0.227897	0.720239	0.997123
2,898	0.250108	0.722318×10^{-4}	1.000000

TABLE 12.2 *Continued*

λT ($\mu m \cdot K$)	$F_{(0 \to \lambda)}$	$I_{\lambda,b}(\lambda, T)/\sigma T^5$ ($\mu m \cdot K \cdot sr)^{-1}$	$\dfrac{I_{\lambda,b}(\lambda, T)}{I_{\lambda,b}(\lambda_{max}, T)}$
3,000	0.273232	0.720254×10^{-4}	0.997143
3,200	0.318102	0.705974	0.977373
3,400	0.361735	0.681544	0.943551
3,600	0.403607	0.650396	0.900429
3,800	0.443382	0.615225×10^{-4}	0.851737
4,000	0.480877	0.578064	0.800291
4,200	0.516014	0.540394	0.748139
4,400	0.548796	0.503253	0.696720
4,600	0.579280	0.467343	0.647004
4,800	0.607559	0.433109	0.599610
5,000	0.633747	0.400813	0.554898
5,200	0.658970	0.370580×10^{-4}	0.513043
5,400	0.680360	0.342445	0.474092
5,600	0.701046	0.316376	0.438002
5,800	0.720158	0.292301	0.404671
6,000	0.737818	0.270121	0.373965
6,200	0.754140	0.249723×10^{-4}	0.345724
6,400	0.769234	0.230985	0.319783
6,600	0.783199	0.213786	0.295973
6,800	0.796129	0.198008	0.274128
7,000	0.808109	0.183534	0.254090
7,200	0.819217	0.170256×10^{-4}	0.235708
7,400	0.829527	0.158073	0.218842
7,600	0.839102	0.146891	0.203360
7,800	0.848005	0.136621	0.189143
8,000	0.856288	0.127185	0.176079
8,500	0.874608	0.106772×10^{-4}	0.147819
9,000	0.890029	0.901463×10^{-5}	0.124801
9,500	0.903085	0.765338	0.105956
10,000	0.914199	0.653279×10^{-5}	0.090442
10,500	0.923710	0.560522	0.077600
11,000	0.931890	0.483321	0.066913
11,500	0.939959	0.418725	0.057970
12,000	0.945098	0.364394×10^{-5}	0.050448
13,000	0.955139	0.279457	0.038689
14,000	0.962898	0.217641	0.030131
15,000	0.968933	0.171866×10^{-5}	0.023794
16,000	0.973814	0.137429	0.019026
18,000	0.980860	0.908240×10^{-6}	0.012574
20,000	0.985602	0.623310	0.008629
25,000	0.992215	0.276474	0.003828
30,000	0.995340	0.140469×10^{-6}	0.001945
40,000	0.997967	0.473891×10^{-7}	0.000656
50,000	0.998953	0.201605	0.000279
75,000	0.999713	0.418597×10^{-8}	0.000058
100,000	0.999905	0.135752	0.000019

Additional blackbody functions are listed in the third and fourth columns of Table 12.2. The third column facilitates calculation of the spectral intensity for a prescribed wavelength and temperature. In lieu of computing this quantity from Equation 12.29, it may be obtained by simply multiplying the tabulated value of $I_{\lambda,b}/\sigma T^5$ by σT^5. The fourth column is used to obtain a quick estimate of the ratio of the spectral intensity at any wavelength to that at λ_{max}.

EXAMPLE 12.3

Determine an expression for the net radiative heat flux at the surface of the small solid object of Figure 12.1 in terms of the surface and surroundings temperatures and the Stefan–Boltzmann constant. The small object is a blackbody.

SOLUTION

Known: Surface temperature of a small blackbody, T_s, and the surroundings temperature, T_{sur}.

Find: Expression for the net radiative flux at the surface of the small object, q''_{rad}.

Assumptions: Small object experiences blackbody irradiation.

Schematic:

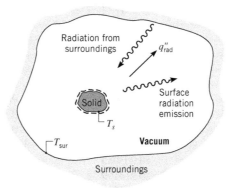

Analysis: Since none of the irradiation is reflected from the small object, Equation 12.28 may be written as

$$q''_{rad} = \int_0^\infty \int_0^{2\pi} \int_0^{\pi/2} I_{\lambda,e}(\lambda, \theta, \phi) \cos\theta \sin\theta \, d\theta \, d\phi \, d\lambda$$
$$- \int_0^\infty \int_0^{2\pi} \int_0^{\pi/2} I_{\lambda,i}(\lambda, \theta, \phi) \cos\theta \sin\theta \, d\theta \, d\phi \, d\lambda \qquad (1)$$

The intensity emitted by the small object corresponds to that of a blackbody. Hence

$$I_{\lambda,e}(\lambda, \theta, \phi) = I_{\lambda,b}(\lambda, T_s) \qquad (2)$$

The intensity corresponding to the irradiation is also black. Therefore

$$I_{\lambda,i}(\lambda, \theta, \phi) = I_{\lambda,b}(\lambda, T_{sur}) \qquad (3)$$

Since the blackbody intensity is diffuse, it is independent of angles θ and ϕ. Therefore, substituting Equations 2 and 3 into Equation 1 yields

$$q''_{rad} = \int_0^{2\pi} \int_0^{\pi/2} \cos\theta \sin\theta \, d\theta \, d\phi \times \int_0^\infty I_{\lambda,b}(\lambda, T_{sur}) \, d\lambda$$
$$- \int_0^{2\pi} \int_0^{\pi/2} \cos\theta \sin\theta \, d\theta \, d\phi \times \int_0^\infty I_{\lambda,b}(\lambda, T_s) \, d\lambda$$
$$= \pi \left[\int_0^\infty I_{\lambda,b}(\lambda, T_{sur}) \, d\lambda - \int_0^\infty I_{\lambda,b}(\lambda, T_s) \, d\lambda \right]$$

Substituting from Equations 12.32 and 12.33 yields

$$q''_{rad} = \sigma(T_s^4 - T_{sur}^4) \qquad \qquad \lhd$$

which is identical to Equation 1.7 with $\varepsilon = 1$.

EXAMPLE 12.4

Consider a large isothermal enclosure that is maintained at a uniform temperature of 2000 K. Calculate the emissive power of the radiation that emerges from a small aperture on the enclosure surface. What is the wavelength λ_1 below which 10% of the emission is concentrated? What is the wavelength λ_2 above which 10% of the emission is concentrated? Determine the maximum spectral emissive power and the wavelength at which this emission occurs. What is the irradiation incident on a small object placed inside the enclosure?

SOLUTION

Known: Large isothermal enclosure at uniform temperature.

Find:

1. Emissive power of a small aperture on the enclosure.
2. Wavelengths below which and above which 10% of the radiation is concentrated.
3. Spectral emissive power and wavelength associated with maximum emission.
4. Irradiation on a small object inside the enclosure.

Schematic:

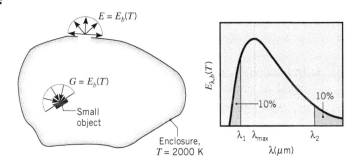

Assumptions: Areas of aperture and object are very small relative to enclosure surface.

Analysis:

1. Emission from the aperture of any isothermal enclosure will have the characteristics of blackbody radiation. Hence, from Equation 12.32,

$$E = E_b(T) = \sigma T^4 = 5.670 \times 10^{-8} \text{ W/m}^2 \cdot \text{K}^4 (2000 \text{ K})^4$$
$$E = 9.07 \times 10^5 \text{ W/m}^2 \qquad \triangleleft$$

2. The wavelength λ_1 corresponds to the upper limit of the spectral band $(0 \rightarrow \lambda_1)$ containing 10% of the emitted radiation. With $F_{(0 \rightarrow \lambda_1)} = 0.10$ it follows from Table 12.2 that $\lambda_1 T = 2195$ μm · K. Hence

$$\lambda_1 = 1.1 \; \mu\text{m} \qquad \triangleleft$$

The wavelength λ_2 corresponds to the lower limit of the spectral band $(\lambda_2 \rightarrow \infty)$ containing 10% of the emitted radiation. With

$$F_{(\lambda_2 \rightarrow \infty)} = 1 - F_{(0 \rightarrow \lambda_2)} = 0.1$$
$$F_{(0 \rightarrow \lambda_2)} = 0.9$$

it follows from Table 12.2 that $\lambda_2 T = 9382$ μm · K. Hence

$$\lambda_2 = 4.69 \; \mu\text{m} \qquad \triangleleft$$

3. From Wien's displacement law, Equation 12.31, $\lambda_{max} T = 2898$ μm · K. Hence

$$\lambda_{max} = 1.45 \; \mu\text{m} \qquad \triangleleft$$

The spectral emissive power associated with this wavelength may be computed from Equation 12.30 or from the third column of Table 12.2. For $\lambda_{max} T = 2898$ μm · K it follows from Table 12.2 that

$$I_{\lambda,b}(1.45 \; \mu\text{m}, \; T) = 0.722 \times 10^{-4} \sigma T^5$$

Hence

$$I_{\lambda,b}(1.45 \; \mu\text{m}, \; 2000 \text{ K}) = 0.722 \times 10^{-4} \; (\mu\text{m} \cdot \text{K} \cdot \text{sr})^{-1} \times 5.67$$
$$\times 10^{-8} \text{ W/m}^2 \cdot \text{K}^4 \; (2000 \text{ K})^5$$

$$I_{\lambda,b}(1.45 \; \mu\text{m}, \; 2000 \text{ K}) = 1.31 \times 10^5 \text{ W/m}^2 \cdot \text{sr} \cdot \mu\text{m}$$

Since the emission is diffuse, it follows from Equation 12.16 that

$$E_{\lambda,b} = \pi I_{\lambda,b} = 4.12 \times 10^5 \text{ W/m}^2 \cdot \mu\text{m} \qquad \triangleleft$$

4. Irradiation of any small object inside the enclosure may be approximated as being equal to emission from a blackbody at the enclosure surface temperature. Hence $G = E_b(T)$, in which case

$$G = 9.07 \times 10^5 \text{ W/m}^2 \qquad \triangleleft$$

EXAMPLE 12.5

A surface emits as a blackbody at 1500 K. What is the rate per unit area (W/m^2) at which it emits radiation over all directions corresponding to $0° \leq \theta \leq 60°$ and over the wavelength interval $2 \ \mu m \leq \lambda \leq 4 \ \mu m$?

SOLUTION

Known: Temperature of a surface that emits as a blackbody.

Find: Rate of emission per unit area over all directions between $\theta = 0°$ and $60°$ and over all wavelengths between $\lambda = 2$ and $4 \ \mu m$.

Schematic:

Blackbody at 1500 K

Assumptions: Surface emits as a blackbody.

Analysis: The desired emission may be inferred from Equation 12.15, with the limits of integration restricted as follows:

$$\Delta E = \int_2^4 \int_0^{2\pi} \int_0^{\pi/3} I_{\lambda,b} \cos \theta \sin \theta \ d\theta \ d\phi \ d\lambda$$

or, since a blackbody emits diffusely,

$$\Delta E = \int_2^4 I_{\lambda,b} \left(\int_0^{2\pi} \int_0^{\pi/3} \cos \theta \sin \theta \ d\theta \ d\phi \right) d\lambda$$

$$\Delta E = \int_2^4 I_{\lambda,b} \left(2\pi \frac{\sin^2\theta}{2} \bigg|_0^{\pi/3} \right) d\lambda = 0.75 \int_2^4 \pi I_{\lambda,b} \ d\lambda$$

Substituting from Equation 12.16 and multiplying and dividing by E_b, this result may be put in a form that allows for use of Table 12.2 in evaluating the spectral integration. In particular,

$$\Delta E = 0.75 E_b \int_2^4 \frac{E_{\lambda,b}}{E_b} d\lambda = 0.75 E_b [F_{(0 \rightarrow 4)} - F_{(0 \rightarrow 2)}]$$

where from Table 12.2

$$\lambda_1 T = 2 \ \mu m \times 1500 \ K = 3000 \ \mu m \cdot K: \qquad F_{(0 \rightarrow 2)} = 0.273$$
$$\lambda_2 T = 4 \ \mu m \times 1500 \ K = 6000 \ \mu m \cdot K: \qquad F_{(0 \rightarrow 4)} = 0.738$$

Hence

$$\Delta E = 0.75(0.738 - 0.273)E_b = 0.75(0.465)E_b$$

From Equation 12.31, it then follows that

$$\Delta E = 0.75(0.465)5.67 \times 10^{-8} \ \text{W/m}^2 \cdot \text{K}^4 (1500 \ \text{K})^4 = 10^5 \ \text{W/m}^2 \qquad \lhd$$

Comments: The total, hemispherical emissive power is reduced by 25% and 53.5% due to the directional and spectral restrictions, respectively.

12.5 *Emission from Real Surfaces*

Having developed the notion of a blackbody to describe *ideal* surface behavior, we may now consider the behavior of *real* surfaces. Recall that the blackbody is an ideal emitter in the sense that no surface can emit more radiation than a blackbody at the same temperature. It is therefore convenient to designate the blackbody as a reference in describing emission from a real surface. A surface radiative property known as the *emissivity*[2] may then be defined as the *ratio* of the radiation emitted by the surface to the radiation emitted by a blackbody at the same temperature.

It is important to acknowledge that the spectral radiation emitted by a real surface differs from the Planck distribution (Figure 12.15*a*). Moreover, the directional distribution (Figure 12.15*b*) may be other than diffuse. Hence the emissivity may assume different values according to whether one is interested in emission at a given wavelength or in a given direction, or in integrated averages over wavelength and direction.

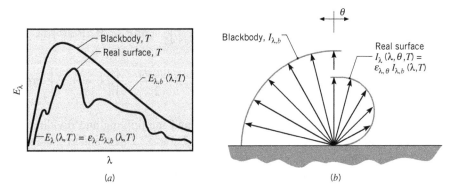

FIGURE 12.15 Comparison of blackbody and real surface emission. (*a*) Spectral distribution. (*b*) Directional distribution.

---

[2]In this text we use the *-ivity*, rather than the *-ance*, ending for material radiative properties (e.g., "emissivity" rather than "emittance"). Although efforts are being made to reserve the *-ivity* ending for optically smooth, uncontaminated surfaces, no such distinction is made in much of the literature, and none is made in this text.

The emissivity that accounts for emission over all wavelengths and in all directions is the *total, hemispherical* emissivity, which is the ratio of the total emissive power of a real surface, $E(T)$, to the total emissive power of a blackbody at the same temperature, $E_b(T)$. That is,

$$\varepsilon(T) \equiv \frac{E(T)}{E_b(T)} \tag{12.36}$$

If the total, hemispherical emissivity of a surface is known, it is a simple matter to express its emissive power in terms of the emissive power of a blackbody by combining Equation 12.36 with Equation 12.32, namely

$$E(T) = \varepsilon(T)E_b(T) = \varepsilon(T)\sigma T^4 \tag{12.37}$$

While Equation 12.37 is simple in form, its simplicity is deceptive in that $\varepsilon(T)$ itself depends on the spectral and directional characteristics of the surface emission. To develop an appropriate understanding of Equation 12.37, we define the *spectral, directional emissivity* $\varepsilon_{\lambda,\theta}(\lambda, \theta, \phi, T)$ of a surface at the temperature T as the ratio of the intensity of the radiation emitted at the wavelength λ and in the direction of θ and ϕ to the intensity of the radiation emitted by a blackbody at the same values of T and λ. Hence

$$\varepsilon_{\lambda,\theta}(\lambda, \theta, \phi, T) \equiv \frac{I_{\lambda,e}(\lambda, \theta, \phi, T)}{I_{\lambda,b}(\lambda, T)} \tag{12.38}$$

Note how the subscripts λ and θ designate interest in a specific wavelength and direction for the emissivity. In contrast, terms appearing within parentheses designate functional dependence on wavelength, direction, and/or temperature. The absence of directional variables in the parentheses of the denominator in Equation 12.38 implies that the intensity is independent of direction, which is, of course, a characteristic of blackbody emission. In like manner a *total, directional emissivity* ε_θ, which represents a spectral average of $\varepsilon_{\lambda,\theta}$, may be defined as

$$\varepsilon_\theta(\theta, \phi, T) \equiv \frac{I_e(\theta, \phi, T)}{I_b(T)} \tag{12.39}$$

For most engineering calculations, it is desirable to work with surface properties that represent directional averages. A *spectral, hemispherical emissivity* is therefore defined as

$$\varepsilon_\lambda(\lambda, T) \equiv \frac{E_\lambda(\lambda, T)}{E_{\lambda,b}(\lambda, T)} \tag{12.40}$$

It may be related to the directional emissivity $\varepsilon_{\lambda,\theta}$ by substituting the expression for the spectral emissive power, Equation 12.13, to obtain

$$\varepsilon_\lambda(\lambda, T) = \frac{\int_0^{2\pi} \int_0^{\pi/2} I_{\lambda,e}(\lambda, \theta, \phi, T) \cos\theta \sin\theta \, d\theta \, d\phi}{\int_0^{2\pi} \int_0^{\pi/2} I_{\lambda,b}(\lambda, T) \cos\theta \sin\theta \, d\theta \, d\phi}$$

In contrast to Equation 12.13, the temperature dependence of emission is now acknowledged. From Equation 12.38 and the fact that $I_{\lambda,b}$ is independent of θ and ϕ, it follows that

$$\varepsilon_\lambda(\lambda, T) = \frac{\int_0^{2\pi} \int_0^{\pi/2} \varepsilon_{\lambda,\theta}(\lambda, \theta, \phi, T) \cos\theta \sin\theta \, d\theta \, d\phi}{\int_0^{2\pi} \int_0^{\pi/2} \cos\theta \sin\theta \, d\theta \, d\phi} \qquad (12.41)$$

Assuming $\varepsilon_{\lambda,\theta}$ to be independent of ϕ, which is a reasonable assumption for most surfaces, and evaluating the denominator, we obtain

$$\varepsilon_\lambda(\lambda, T) = 2\int_0^{\pi/2} \varepsilon_{\lambda,\theta}(\lambda, \theta, T) \cos\theta \sin\theta \, d\theta \qquad (12.42)$$

The *total, hemispherical emissivity,* which represents an average over all possible directions and wavelengths, is defined in Equation 12.36. Substituting Equations 12.14 and 12.40 into Equation 12.36, it follows that

$$\varepsilon(T) = \frac{\int_0^\infty \varepsilon_\lambda(\lambda, T) E_{\lambda,b}(\lambda, T) d\lambda}{E_b(T)} \qquad (12.43)$$

If the emissivities of a surface are known, it is a simple matter to compute its emission characteristics. For example, if $\varepsilon_\lambda(\lambda, T)$ is known, it may be used with Equations 12.30 and 12.40 to compute the spectral emissive power of the surface at any wavelength and temperature,

$$E_\lambda(\lambda, T) = \varepsilon_\lambda(\lambda, T) E_{\lambda,b}(\lambda, T) = \frac{C_1 \varepsilon_\lambda(\lambda, T)}{\lambda^5 [\exp(C_2/\lambda T) - 1]} \qquad (12.44)$$

As noted previously, if $\varepsilon(T)$ is known, it may be used to compute the emissive power of the surface at any temperature, as in Equation 12.37. Measurements have been performed to determine these properties for many different materials and surface coatings.

The directional emissivity of a *diffuse emitter* is a constant, independent of direction. However, although this condition is often a reasonable *approximation,* all surfaces exhibit some departure from diffuse behavior. Representative variations of ε_θ with θ are shown schematically in Figure 12.16 for conducting and nonconducting materials. For conductors

FIGURE 12.16 Representative directional distributions of the total, directional emissivity.

ε_θ is approximately constant over the range $\theta \lesssim 40°$, after which it increases with increasing θ but ultimately decays to zero. In contrast, for nonconductors ε_θ is approximately constant for $\theta \lesssim 70°$, beyond which it decreases sharply with increasing θ. One implication of these variations is that, although there are preferential directions for emission, the hemispherical emissivity ε will not differ markedly from the normal emissivity ε_n, corresponding to $\theta = 0$. In fact the ratio rarely falls outside the range $1.0 \le (\varepsilon/\varepsilon_n) \le 1.3$ for conductors and the range of $0.95 \le (\varepsilon/\varepsilon_n) \le 1.0$ for nonconductors. Hence to a reasonable approximation,

$$\varepsilon \approx \varepsilon_n \tag{12.45}$$

Note that, although the foregoing statements were made for the total emissivity, they also apply to spectral components.

Since the spectral distribution of emission from real surfaces departs from the Planck distribution (Figure 12.15a), we do not expect the value of the spectral emissivity ε_λ to be independent of wavelength. Representative spectral distributions of ε_λ are shown in Figure 12.17. The manner in which ε_λ varies with λ depends on whether the solid is a conductor or nonconductor, as well as on the nature of the surface coating.

Representative values of the total, normal emissivity ε_n are plotted in Figures 12.18 and 12.19 and listed in Table A.11. Several generalizations may be made.

1. The emissivity of metallic surfaces is generally small, achieving values as low as 0.02 for highly polished gold and silver.

2. The presence of oxide layers may significantly increase the emissivity of metallic surfaces. From Figure 12.18, contrast the values of 0.3 and 0.7 for stainless steel at 900 K, depending on whether it is polished or heavily oxidized.

3. The emissivity of nonconductors is comparatively large, generally exceeding 0.6.

4. The emissivity of conductors increases with increasing temperature; however, depending on the specific material, the emissivity of nonconductors may either increase or decrease with increasing temperature. Note that the variations of ε_n with T shown in

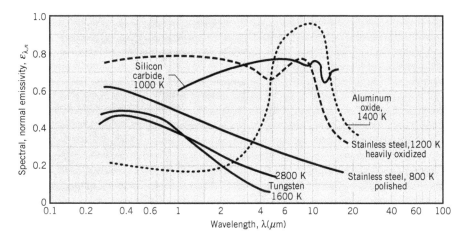

FIGURE 12.17 Spectral dependence of the spectral, normal emissivity $\varepsilon_{\lambda,n}$ of selected materials.

FIGURE 12.18   Temperature dependence of the total, normal emissivity ε_n of selected materials.

Figure 12.18 are consistent with the spectral distributions of $\varepsilon_{\lambda,n}$ shown in Figure 12.17. These trends follow from Equation 12.43. Although the spectral distribution of $\varepsilon_{\lambda,n}$ is approximately independent of temperature, there is proportionately more emission at lower wavelengths with increasing temperature. Hence, if $\varepsilon_{\lambda,n}$ increases with decreasing wavelength for a particular material, ε_n will increase with increasing temperature for that material.

It should be recognized that the emissivity depends strongly on the nature of the surface, which can be influenced by the method of fabrication, thermal cycling, and chemical reaction with its environment. More comprehensive surface emissivity compilations are available in the literature [3–6].

FIGURE 12.19   Representative values of the total, normal emissivity ε_n.

EXAMPLE 12.6

A diffuse surface at 1600 K has the spectral, hemispherical emissivity shown as follows.

Determine the total, hemispherical emissivity and the total emissive power. At what wavelength will the spectral emissive power be a maximum?

SOLUTION

Known: Spectral, hemispherical emissivity of a diffuse surface at 1600 K.

Find:

1. Total, hemispherical emissivity.
2. Total emissive power.
3. Wavelength at which spectral emissive power will be a maximum.

Assumptions: Surface is a diffuse emitter.

Analysis:

1. The total, hemispherical emissivity is given by Equation 12.43, where the integration may be performed in parts as follows:

$$\varepsilon = \frac{\int_0^\infty \varepsilon_\lambda E_{\lambda,b}\, d\lambda}{E_b} = \frac{\varepsilon_1 \int_0^2 E_{\lambda,b}\, d\lambda}{E_b} + \frac{\varepsilon_2 \int_2^5 E_{\lambda,b}\, d\lambda}{E_b}$$

or

$$\varepsilon = \varepsilon_1 F_{(0\to 2\,\mu m)} + \varepsilon_2 [F_{(0\to 5\,\mu m)} - F_{(0\to 2\,\mu m)}]$$

From Table 12.2 we obtain

$$\lambda_1 T = 2\,\mu m \times 1600\,K = 3200\,\mu m \cdot K: \qquad F_{(0\to 2\,\mu m)} = 0.318$$
$$\lambda_2 T = 5\,\mu m \times 1600\,K = 8000\,\mu m \cdot K: \qquad F_{(0\to 5\,\mu m)} = 0.856$$

Hence

$$\varepsilon = 0.4 \times 0.318 + 0.8[0.856 - 0.318] = 0.558 \qquad \triangleleft$$

2. From Equation 12.36 the total emissive power is

$$E = \varepsilon E_b = \varepsilon \sigma T^4$$
$$E = 0.558(5.67 \times 10^{-8}\ \text{W/m}^2 \cdot \text{K}^4)(1600\ \text{K})^4 = 207\ \text{kW/m}^2 \qquad \triangleleft$$

3. If the surface emitted as a blackbody or if its emissivity were a constant, independent of λ, the wavelength corresponding to maximum spectral emission could be obtained from Wien's displacement law. However, because ε_1 varies with λ, it is not immediately obvious where peak emission occurs. From Equation 12.31 we know that

$$\lambda_{\max} = \frac{2898\ \mu\text{m} \cdot \text{K}}{1600\ \text{K}} = 1.81\ \mu\text{m}$$

The spectral emissive power at this wavelength may be obtained by using Equation 12.40 with Table 12.2. That is,

$$E_\lambda(\lambda_{\max},\ T) = \varepsilon_\lambda(\lambda_{\max}) E_{\lambda,b}(\lambda_{\max},\ T)$$

or, since the surface is a diffuse emitter,

$$E_\lambda(\lambda_{\max},\ T) = \pi \varepsilon_\lambda(\lambda_{\max}) I_{\lambda,b}(\lambda_{\max},\ T)$$
$$= \pi \varepsilon_\lambda(\lambda_{\max}) \frac{I_{\lambda,b}(\lambda_{\max},\ T)}{\sigma T^5} \times \sigma T^5$$

$$E_\lambda(1.81\ \mu\text{m},\ 1600\ \text{K}) = \pi \times 0.4 \times 0.722 \times 10^{-4}\ (\mu\text{m} \cdot \text{K} \cdot \text{sr})^{-1} \times 5.67$$

$$\times 10^{-8}\ \text{W/m}^2 \cdot \text{K}^4 \times (1600\ \text{K})^5 = 54\ \text{kW/m}^2 \cdot \mu\text{m}$$

Since $\varepsilon_\lambda = 0.4$ from $\lambda = 0$ to $\lambda = 2\ \mu\text{m}$, the foregoing result provides the maximum spectral emissive power for the region $\lambda < 2\ \mu\text{m}$. However, with the change in ε_λ that occurs at $\lambda = 2\ \mu\text{m}$, the value of E_λ at $\lambda = 2\ \mu\text{m}$ may be larger than that for $\lambda = 1.81\ \mu\text{m}$. To determine whether this is, in fact, the case, we compute

$$E_\lambda(\lambda_1,\ T) = \pi \varepsilon_\lambda(\lambda_1) \frac{I_{\lambda,b}(\lambda_1,\ T)}{\sigma T^5} \times \sigma T^5$$

where, for $\lambda_1 T = 3200\ \mu\text{m} \cdot \text{K}$, $[I_{\lambda,b}(\lambda_1,\ T)/\sigma T^5] = 0.706 \times 10^{-4}\ (\mu\text{m} \cdot \text{K} \cdot \text{sr})^{-1}$. Hence

$$E_\lambda(2\ \mu\text{m},\ 1600\ \text{K}) = \pi \times 0.80 \times 0.706 \times 10^{-4}\ (\mu\text{m} \cdot \text{K} \cdot \text{sr})^{-1} \times 5.67$$
$$\times 10^{-8}\ \text{W/m}^2 \cdot \text{K}^4 (1600\ \text{K})^5$$
$$E_\lambda(2\ \mu\text{m},\ 1600\ \text{K}) = 105.5\ \text{kW/m}^2 \cdot \mu\text{m} > E_\lambda(1.81\ \mu\text{m},\ 1600\ \text{K})$$

and peak emission occurs at

$$\lambda = \lambda_1 = 2\ \mu\text{m} \qquad \triangleleft$$

Comments: For the prescribed spectral distribution of ε_λ, the spectral emissive power will vary with wavelength as shown.

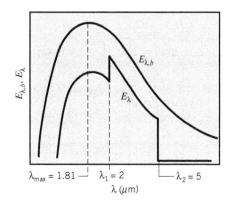

EXAMPLE 12.7

Measurements of the spectral, directional emissivity of a metallic surface at $T = 2000$ K and $\lambda = 1.0$ μm yield a directional distribution that may be approximated as follows:

Determine corresponding values of the spectral, normal emissivity; the spectral, hemispherical emissivity; the spectral intensity of radiation emitted in the normal direction; and the spectral emissive power.

SOLUTION

Known: Directional distribution of $\varepsilon_{\lambda,\theta}$ at $\lambda = 1$ μm for a metallic surface at 2000 K.

Find:

1. Spectral, normal emissivity $\varepsilon_{\lambda,n}$ and spectral, hemispherical emissivity ε_λ.

2. Spectral, normal intensity $I_{\lambda,n}$ and spectral emissive power E_λ.

Analysis:

1. From the measurement of $\varepsilon_{\lambda,\theta}$ at $\lambda = 1$ μm, we see that

$$\varepsilon_{\lambda,n} = \varepsilon_{\lambda,\theta}(1 \ \mu m, \ 0°) = 0.3 \qquad \lhd$$

From Equation 12.42, the spectral, hemispherical emissivity is

$$\varepsilon_\lambda(1 \ \mu m) = 2 \int_0^{\pi/2} \varepsilon_{\lambda,\theta} \cos \theta \sin \theta \ d\theta$$

or

$$\varepsilon_\lambda(1 \ \mu m) = 2 \left[0.3 \int_0^{\pi/3} \cos \theta \ \sin \theta \ d\theta + 0.6 \int_{\pi/3}^{4\pi/9} \cos \theta \ \sin \theta \ d\theta \right]$$

$$\varepsilon_\lambda(1 \ \mu m) = 2 \left[0.3 \frac{\sin^2 \theta}{2} \Big|_0^{\pi/3} + 0.6 \frac{\sin^2 \theta}{2} \Big|_{\pi/3}^{4\pi/9} \right]$$

$$= 2 \left[\frac{0.3}{2}(0.75) + \frac{0.6}{2}(0.97 - 0.75) \right]$$

$$\varepsilon_\lambda(1 \ \mu m) = 0.36 \qquad \qquad \lhd$$

2. From Equation 12.38 the spectral intensity of radiation emitted at $\lambda = 1 \ \mu m$ in the normal direction is

$$I_{\lambda,n}(1 \ \mu m, 0°, 2000 \ K) = \varepsilon_{\lambda,\theta}(1 \ \mu m, \ 0°)I_{\lambda,b}(1 \ \mu m, 2000 \ K)$$

where $\varepsilon_{\lambda,\theta}(1 \ \mu m, 0°) = 0.3$ and $I_{\lambda,b}(1 \ \mu m, 2000 \ K)$ may be obtained from Table 12.2. For $\lambda T = 2000 \ \mu m \cdot K$, $(I_{\lambda,b}/\sigma T^5) = 0.493 \times 10^{-4} \ (\mu m \cdot K \cdot sr)^{-1}$ and

$$I_{\lambda,b} = 0.493 \times 10^{-4} \ (\mu m \cdot K \cdot sr)^{-1} \times 5.67 \times 10^{-8} \ W/m^2 \cdot K^4 \ (2000 \ K)^5$$

$$I_{\lambda,b} = 8.95 \times 10^4 \ W/m^2 \cdot \mu m \cdot sr$$

Hence

$$I_{\lambda,n}(1 \ \mu m, \ 0°, 2000 \ K) = 0.3 \times 8.95 \times 10^4 \ W/m^2 \cdot \mu m \cdot sr$$

$$I_{\lambda,n}(1 \ \mu m, \ 0°, 2000 \ K) = 2.69 \times 10^4 \ W/m^2 \cdot \mu m \cdot sr \qquad \lhd$$

From Equation 12.40 the spectral emissive power for $\lambda = 1 \ \mu m$ and $T = 2000 \ K$ is

$$E_\lambda(1 \ \mu m, 2000 \ K) = \varepsilon_\lambda(1 \ \mu m)E_{\lambda,b}(1 \ \mu m, 2000 \ K)$$

where

$$E_{\lambda,b}(1 \ \mu m, \ 2000 \ K) = \pi I_{\lambda,b}(1 \ \mu m, \ 2000 \ K)$$

$$E_{\lambda,b}(1 \ \mu m, \ 2000 \ K) = \pi \ sr \times 8.95 \times 10^4 \ W/m^2 \cdot \mu m \cdot sr$$

$$= 2.81 \times 10^5 \ W/m^2 \cdot \mu m$$

Hence

$$E_\lambda(1 \ \mu m, 2000 \ K) = 0.36 \times 2.81 \times 10^5 \ W/m^2 \cdot \mu m$$

or

$$E_\lambda(1 \ \mu m, 2000 \ K) = 1.01 \times 10^5 \ W/m^2 \cdot \mu m \qquad \lhd$$

12.6   *Absorption, Reflection, and Transmission by Real Surfaces*

In the preceding section, we learned that emission from a real surface is associated with a surface property termed the emissivity ε. To determine the net radiative heat flux from the surface, it is also necessary to consider properties that determine the absorption, reflection, and transmission of the irradiation. In Section 12.3.3 we defined the *spectral irradiation* $G_\lambda(\text{W/m}^2 \cdot \mu\text{m})$ as the rate at which radiation of wavelength λ is incident on a surface per unit area of the surface and per unit wavelength interval $d\lambda$ about λ. It may be incident from all possible directions, and it may originate from several different sources. The *total irradiation* G (W/m^2) encompasses all spectral contributions and may be evaluated from Equation 12.19.

In the most general situation the irradiation interacts with a *semitransparent medium*, such as a layer of water or a glass plate. As shown in Figure 12.20 for a spectral component of the irradiation, portions of this radiation may be *reflected, absorbed,* and *transmitted.* From a radiation balance on the medium, it follows that

$$G_\lambda = G_{\lambda,\text{ref}} + G_{\lambda,\text{abs}} + G_{\lambda,\text{tr}} \tag{12.46}$$

In general, determination of these components is complex, depending on the upper and lower surface conditions, the wavelength of the radiation, and the composition and thickness of the medium. Moreover, conditions may be strongly influenced by *volumetric* effects occurring within the medium.

In a simpler situation, which pertains to most engineering applications, the medium is *opaque* to the incident radiation. In this case, $G_{\lambda,\text{tr}} = 0$ and the remaining absorption and reflection processes may be treated as *surface phenomena*. That is, they are controlled by processes occurring within a fraction of a micrometer from the irradiated surface. It is therefore appropriate to speak of irradiation being absorbed and reflected *by the surface,* with the relative magnitudes of $G_{\lambda,\text{abs}}$ and $G_{\lambda,\text{ref}}$ depending on λ and the nature of the surface material. There is no net effect of the reflection process on the medium, while absorption has the effect of increasing the internal thermal energy of the medium.

It is interesting to note that surface absorption and reflection are responsible for our perception of *color.* Unless the surface is at a high temperature ($T_s \gtrsim 1000$ K), such that it is *incandescent,* color is in no way due to emission, which is concentrated in the IR region, and is hence imperceptible to the eye. Color is instead due to selective reflection and absorption of the visible portion of the irradiation that is incident from the sun or an artificial source of light. A shirt is "red" because it contains a pigment that preferentially absorbs the blue, green, and yellow components of the incident light. Hence the relative

FIGURE 12.20 Spectral absorption, reflection, and transmission processes associated with a semitransparent medium.

contributions of these components to the reflected light, which is seen, is diminished, and the red component is dominant. Similarly, a leaf is "green" because its cells contain chlorophyll, a pigment that shows strong absorption in the blue and the red and preferential reflection in the green. A surface appears "black" if it absorbs all incident visible radiation, and it is "white" if it reflects this radiation. However, we must be careful how we interpret such *visual* effects. For a prescribed irradiation, the "color" of a surface may not indicate its overall capacity as an absorber or reflector, since much of the irradiation may be in the IR region. A "white" surface such as snow, for example, is highly reflective to visible radiation but strongly absorbs IR radiation, thereby approximating blackbody behavior at long wavelengths.

In Section 12.2, we introduced properties to characterize the absorption, reflection, and transmission processes. In general these properties depend on surface material and finish, surface temperature, and the wavelength and direction of the incident radiation. These properties are considered in the following subsections.

12.6.1 Absorptivity

The absorptivity is a property that determines the fraction of the irradiation absorbed by a surface. Like emissivity, it may be characterized by both a directional and spectral dependence. The *spectral, directional absorptivity, $\alpha_{\lambda,\theta}(\lambda, \theta, \phi)$*, of a surface is defined as the fraction of the spectral intensity incident in the direction of θ and ϕ that is absorbed by the surface. Hence

$$\alpha_{\lambda,\theta}(\lambda, \theta, \phi) \equiv \frac{I_{\lambda,i,\mathrm{abs}}(\lambda, \theta, \phi)}{I_{\lambda,i}(\lambda, \theta, \phi)} \tag{12.47}$$

In this expression, we have neglected any dependence of the absorptivity on the surface temperature. Such a dependence is small for most spectral radiative properties.

It is implicit in the foregoing result that surfaces may exhibit selective absorption with respect to the wavelength and direction of the incident radiation. For most engineering calculations, however, it is desirable to work with surface properties that represent directional averages. We therefore define a *spectral, hemispherical absorptivity $\alpha_\lambda(\lambda)$* as

$$\alpha_\lambda(\lambda) \equiv \frac{G_{\lambda,\mathrm{abs}}(\lambda)}{G_\lambda(\lambda)} \tag{12.48}$$

which, from Equations 12.18 and 12.47, may be expressed as

$$\alpha_\lambda(\lambda) = \frac{\int_0^{2\pi} \int_0^{\pi/2} \alpha_{\lambda,\theta}(\lambda, \theta, \phi) I_{\lambda,i}(\lambda, \theta, \phi) \cos \theta \sin \theta \, d\theta \, d\phi}{\int_0^{2\pi} \int_0^{\pi/2} I_{\lambda,i}(\lambda, \theta, \phi) \cos \theta \sin \theta \, d\theta \, d\phi} \tag{12.49}$$

Hence α_λ depends on the directional distribution of the incident radiation, as well as on the wavelength of the radiation and the nature of the absorbing surface. Note that, if the incident radiation is diffusely distributed and $\alpha_{\lambda,\theta}$ is independent of ϕ, Equation 12.49 reduces to

$$\alpha_\lambda(\lambda) = 2\int_0^{\pi/2}\alpha_{\lambda,\theta}(\lambda,\theta)\cos\theta\sin\theta\,d\theta \tag{12.50}$$

The *total, hemispherical absorptivity*, α, represents an integrated average over both direction and wavelength. It is defined as the fraction of the total irradiation absorbed by a surface

$$\alpha \equiv \frac{G_{\text{abs}}}{G} \tag{12.51}$$

and, from Equations 12.19 and 12.48, it may be expressed as

$$\alpha = \frac{\displaystyle\int_0^\infty \alpha_\lambda(\lambda)G_\lambda(\lambda)\,d\lambda}{\displaystyle\int_0^\infty G_\lambda(\lambda)\,d\lambda} \tag{12.52}$$

Accordingly, α depends on the spectral distribution of the incident radiation, as well as on its directional distribution and the nature of the absorbing surface. Note that, although α is approximately independent of surface temperature, the same may not be said for the total, hemispherical emissivity, ε. From Equation 12.43 it is evident that this property is strongly temperature dependent.

Because α depends on the spectral distribution of the irradiation, its value for a surface exposed to solar radiation may differ appreciably from its value for the same surface exposed to longer wavelength radiation originating from a source of lower temperature. Since the spectral distribution of solar radiation is nearly proportional to that of emission from a blackbody at 5800 K, it follows from Equation 12.52 that the total absorptivity to solar radiation α_S may be approximated as

$$\alpha_S \approx \frac{\displaystyle\int_0^\infty \alpha_\lambda(\lambda)E_{\lambda,b}(\lambda,\,5800\text{ K})\,d\lambda}{\displaystyle\int_0^\infty E_{\lambda,b}(\lambda,\,5800\text{ K})\,d\lambda} \tag{12.53}$$

The integrals appearing in this equation may be evaluated by using the blackbody radiation function $F_{(0\to\lambda)}$ of Table 12.2.

12.6.2 Reflectivity

Surfaces may be idealized as *diffuse* or *specular,* according to the manner in which they reflect radiation (Figure 12.21). Diffuse reflection occurs if, regardless of the direction of the incident radiation, the intensity of the reflected radiation is independent of the reflection angle. In contrast, if all the reflection is in the direction of θ_2, which equals the incident angle θ_1, specular reflection is said to occur. Although no surface is perfectly diffuse or specular, the latter condition is more closely approximated by polished, mirror-like surfaces and the former condition by rough surfaces. The assumption of diffuse reflection is reasonable for most engineering applications.

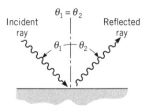

FIGURE 12.21   Diffuse and specular reflection.

The reflectivity is a property that determines the fraction of the incident radiation reflected by a surface. Its specific definition may take several different forms, because the property is inherently *bidirectional* [7] as implied in Figure 12.21. That is, in addition to depending on the direction of the incident radiation, it also depends on the direction of the reflected radiation. We shall avoid this complication by working exclusively with a reflectivity that represents an integrated average over the hemisphere associated with the reflected radiation and therefore provides no information concerning the directional distribution of this radiation. Accordingly, the *spectral, directional reflectivity,* $\rho_{\lambda,\theta}(\lambda, \theta, \phi)$, of a surface is defined as the fraction of the spectral intensity incident in the direction of θ and ϕ, which is reflected by the surface. Hence

$$\rho_{\lambda,\theta}(\lambda, \theta, \phi) \equiv \frac{I_{\lambda,i,\text{ref}}(\lambda, \theta, \phi)}{I_{\lambda,i}(\lambda, \theta, \phi)} \tag{12.54}$$

The *spectral, hemispherical reflectivity* $\rho_\lambda(\lambda)$ is then defined as the fraction of the spectral irradiation that is reflected by the surface. Accordingly,

$$\rho_\lambda(\lambda) \equiv \frac{G_{\lambda,\text{ref}}(\lambda)}{G_\lambda(\lambda)} \tag{12.55}$$

which is equivalent to

$$\rho_\lambda(\lambda) = \frac{\int_0^{2\pi} \int_0^{\pi/2} \rho_{\lambda,\theta}(\lambda, \theta, \phi) I_{\lambda,i}(\lambda, \theta, \phi) \cos\theta \sin\theta \, d\theta \, d\phi}{\int_0^{2\pi} \int_0^{\pi/2} I_{\lambda,i}(\lambda, \theta, \phi) \cos\theta \sin\theta \, d\theta \, d\phi} \tag{12.56}$$

The *total, hemispherical reflectivity* ρ is then defined as

$$\rho \equiv \frac{G_{\text{ref}}}{G} \tag{12.57}$$

in which case

$$\rho = \frac{\int_0^\infty \rho_\lambda(\lambda) G_\lambda(\lambda) \, d\lambda}{\int_0^\infty G_\lambda(\lambda) \, d\lambda} \tag{12.58}$$

12.6.3 Transmissivity

Although advanced treatment of the response of a semitransparent material to incident radiation is a complicated problem [7], reasonable results may often be obtained through the use of hemispherical transmissivities defined as

$$\tau_\lambda = \frac{G_{\lambda,\mathrm{tr}}(\lambda)}{G_\lambda(\lambda)} \qquad (12.59)$$

and

$$\tau = \frac{G_{\mathrm{tr}}}{G} \qquad (12.60)$$

The total transmissivity τ is related to the spectral component τ_λ by

$$\tau = \frac{\int_0^\infty G_{\lambda,\mathrm{tr}}(\lambda)\,d\lambda}{\int_0^\infty G_\lambda(\lambda)\,d\lambda} = \frac{\int_0^\infty \tau_\lambda(\lambda)G_\lambda(\lambda)\,d\lambda}{\int_0^\infty G_\lambda(\lambda)\,d\lambda} \qquad (12.61)$$

12.6.4 Special Considerations

From the radiation balance of Equation 12.46 and the foregoing definitions,

$$\rho_\lambda + \alpha_\lambda + \tau_\lambda = 1 \qquad (12.62)$$

for a *semitransparent* medium. This is analogous to Equation 12.2 but on a spectral basis. Of course, if the medium is *opaque*, there is no transmission, and absorption and reflection are surface processes for which

$$\alpha_\lambda + \rho_\lambda = 1 \qquad (12.63)$$

which is analogous to Equation 12.3. Hence knowledge of one property implies knowledge of the other.

Spectral distributions of the normal reflectivity and absorptivity are plotted in Figure 12.22 for selected *opaque* surfaces. A material such as glass or water, which is semitransparent at short wavelengths, becomes opaque at longer wavelengths. This behavior is shown in Figure 12.23, which presents the spectral transmissivity of several common *semitransparent* materials. Note that the transmissivity of glass is affected by its iron content and that the transmissivity of plastics, such as Tedlar, is greater than that of glass in the IR region. These factors have an important bearing on the selection of cover plate materials for solar collector applications, design and selection of windows for energy conservation, and specification of materials for the optical components of infrared imaging systems. Values for the total transmissivity to solar radiation of common collector cover plate materials are presented in Table A.12, along with surface solar absorptivities and low-temperature emissivities.

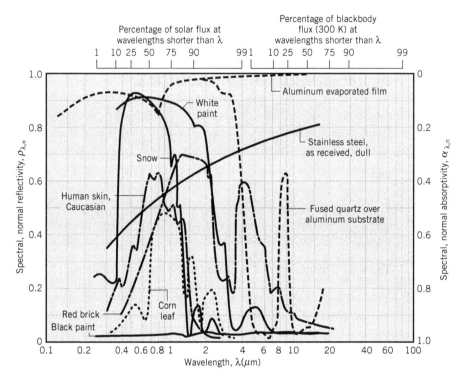

FIGURE 12.22 Spectral dependence of the spectral, normal absorptivity $\alpha_{\lambda,n}$ and reflectivity $\rho_{\lambda,n}$ of selected opaque materials.

FIGURE 12.23 Spectral dependence of the spectral transmissivities τ_λ of selected semitransparent materials.

EXAMPLE 12.8

The spectral, hemispherical absorptivity of an opaque surface and the spectral irradiation at the surface are as shown.

How does the spectral, hemispherical reflectivity vary with wavelength? What is the total, hemispherical absorptivity of the surface? If the surface is initially at 500 K and has a total, hemispherical emissivity of 0.8, how will its temperature change upon exposure to the irradiation?

SOLUTION

Known: Spectral, hemispherical absorptivity and irradiation of a surface. Surface temperature (500 K) and total, hemispherical emissivity (0.8).

Find:

1. Spectral distribution of reflectivity.
2. Total, hemispherical absorptivity.
3. Nature of surface temperature change.

Schematic:

Assumptions:

1. Surface is opaque.
2. Surface convection effects are negligible.
3. Back surface is insulated.

Analysis:

1. From Equation 12.63, $\rho_\lambda = 1 - \alpha_\lambda$. Hence from knowledge of $\alpha_\lambda(\lambda)$, the spectral distribution of ρ_λ is as shown.

2. From Equations 12.51 and 12.52,

$$\alpha = \frac{G_{\text{abs}}}{G} = \frac{\int_0^\infty \alpha_\lambda G_\lambda \, d\lambda}{\int_0^\infty G_\lambda \, d\lambda}$$

or, subdividing the integral into parts,

$$\alpha = \frac{0.2 \int_2^6 G_\lambda \, d\lambda + 500 \int_6^8 \alpha_\lambda \, d\lambda + 1.0 \int_8^{16} G_\lambda \, d\lambda}{\int_2^6 G_\lambda \, d\lambda + \int_6^{12} G_\lambda \, d\lambda + \int_{12}^{16} G_\lambda \, d\lambda}$$

$$\begin{aligned}
\alpha = \Big\{ &0.2(\tfrac{1}{2})500 \text{ W/m}^2 \cdot \mu\text{m}(6-2)\,\mu\text{m} \\
&+ 500 \text{ W/m}^2 \cdot \mu\text{m}[0.2(8-6)\,\mu\text{m} + (1-0.2)(\tfrac{1}{2})(8-6)\,\mu\text{m}] \\
&+ [1 \times 500 \text{ W/m}^2 \cdot \mu\text{m}\,(12-8)\,\mu\text{m} \\
&+ 1(\tfrac{1}{2})500 \text{ W/m}^2 \cdot \mu\text{m}\,(16-12)\,\mu\text{m}]\Big\} \\
&\div [(\tfrac{1}{2})500 \text{ W/m}^2 \cdot \mu\text{m}\,(6-2)\,\mu\text{m} + 500 \text{ W/m}^2 \cdot \mu\text{m}\,(12-6)\,\mu\text{m} \\
&+ (\tfrac{1}{2})500 \text{ W/m}^2 \cdot \mu\text{m}\,(16-12)\,\mu\text{m}]
\end{aligned}$$

Hence

$$\alpha = \frac{G_{\text{abs}}}{G} = \frac{(200 + 600 + 3000) \text{ W/m}^2}{(1000 + 3000 + 1000) \text{ W/m}^2} = \frac{3800 \text{ W/m}^2}{5000 \text{ W/m}^2} = 0.76 \qquad \triangleleft$$

3. Neglecting convection effects, the net heat flux *to* the surface is

$$q''_{\text{net}} = \alpha G - E = \alpha G - \varepsilon \sigma T^4$$

Hence

$$q''_{\text{net}} = 0.76(5000 \text{ W/m}^2) - 0.8 \times 5.67 \times 10^{-8} \text{ W/m}^2 \cdot \text{K}^4 (500 \text{ K})^4$$

$$q''_{\text{net}} = 3800 - 2835 = 965 \text{ W/m}^2$$

Since $q''_{\text{net}} > 0$, the surface temperature will *increase* with time. $\qquad \triangleleft$

EXAMPLE 12.9

The cover glass on a flat-plate solar collector has a low iron content, and its spectral transmissivity may be approximated by the following distribution.

What is the total transmissivity of the cover glass to solar radiation?

SOLUTION

Known: Spectral transmissivity of solar collector cover glass.

Find: Total transmissivity of cover glass to solar radiation.

Assumptions: Spectral distribution of solar irradiation is proportional to that of blackbody emission at 5800 K.

Analysis: From Equation 12.61 the total transmissivity of the cover is

$$\tau = \frac{\int_0^\infty \tau_\lambda G_\lambda \, d\lambda}{\int_0^\infty G_\lambda \, d\lambda}$$

where the irradiation G_λ is due to solar emission. Having assumed that the sun emits as a blackbody at 5800 K, it follows that

$$G_\lambda(\lambda) \propto E_{\lambda,b}(5800 \text{ K})$$

With the proportionality constant canceling from the numerator and denominator of the expression for τ, we obtain

$$\tau = \frac{\int_0^\infty \tau_\lambda E_{\lambda,b}(5800 \text{ K}) \, d\lambda}{\int_0^\infty E_{\lambda,b}(5800 \text{ K}) \, d\lambda}$$

or, for the prescribed spectral distribution of $\tau_\lambda(\lambda)$,

$$\tau = 0.90 \frac{\int_{0.3}^{2.5} E_{\lambda,b}(5800 \text{ K}) \, d\lambda}{E_b(5800 \text{ K})}$$

From Table 12.2

$$\lambda_1 = 0.3 \ \mu m, \ T = 5800 \ \text{K:} \qquad \lambda_1 T = 1740 \ \mu m \cdot \text{K}, \ F_{(0 \to \lambda_1)} = 0.0335$$

$$\lambda_2 = 2.5 \ \mu m, \ T = 5800 \ \text{K:} \qquad \lambda_2 T = 14{,}500 \ \mu m \cdot \text{K}, \ F_{(0 \to \lambda_2)} = 0.9664$$

Hence from Equation 12.35

$$\tau = 0.90[F_{(0 \to \lambda_2)} - F_{(0 \to \lambda_1)}] = 0.90(0.9664 - 0.0335) = 0.84 \qquad \lhd$$

Comments: It is important to recognize that the irradiation at the cover plate is not equal to the emissive power of a blackbody at 5800 K, $G_\lambda \ne E_{\lambda,b}(5800 \ \text{K})$. It is simply assumed to be proportional to this emissive power, in which case it is assumed to have a spectral distribution of the same form. With G_λ appearing in both the numerator and denominator of the expression for τ, it is then possible to replace G_λ by $E_{\lambda,b}$.

12.7 *Kirchhoff's Law*

In the foregoing sections we separately considered real surface properties associated with emission and irradiation. In Sections 12.7 and 12.8 we consider conditions for which the emissivity and absorptivity are equal.

Consider a *large, isothermal enclosure* of surface temperature T_s, within which several small bodies are confined (Figure 12.24). Since these bodies are small relative to the enclosure, they have a negligible influence on the radiation field, which is due to the cumulative effect of emission and reflection by the enclosure surface. Recall that, regardless of its radiative properties, such a surface forms a *blackbody cavity*. Accordingly, regardless of its orientation, the irradiation experienced by any body in the cavity is diffuse and equal to emission from a blackbody at T_s.

$$G = E_b(T_s) \tag{12.64}$$

Under steady-state conditions, *thermal equilibrium* must exist between the bodies and the enclosure. Hence $T_1 = T_2 = \cdots = T_s$, and the net rate of energy transfer to each

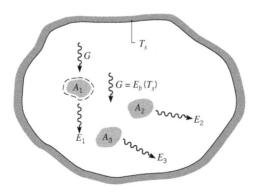

FIGURE 12.24 Radiative exchange in an isothermal enclosure.

surface must be zero. Applying an energy balance to a control surface about body 1, it follows that

$$\alpha_1 G A_1 - E_1(T_s) A_1 = 0$$

or, from Equation 12.64,

$$\frac{E_1(T_s)}{\alpha_1} = E_b(T_s)$$

Since this result must apply to each of the confined bodies, we then obtain

$$\frac{E_1(T_s)}{\alpha_1} = \frac{E_2(T_s)}{\alpha_2} = \cdots = E_b(T_s) \qquad (12.65)$$

This relation is known as *Kirchhoff's law*. A major implication is that, since $\alpha \leq 1$, $E(T_s) \leq E_b(T_s)$. Hence *no real surface can have an emissive power exceeding that of a black surface at the same temperature,* and the notion of the blackbody as an ideal emitter is confirmed.

From the definition of the total, hemispherical emissivity, Equation 12.36, an alternative form of Kirchhoff's law, is

$$\frac{\varepsilon_1}{\alpha_1} = \frac{\varepsilon_2}{\alpha_2} = \cdots = 1$$

Hence, for any surface in the enclosure,

$$\varepsilon = \alpha \qquad (12.66)$$

That is, the total, hemispherical emissivity of the surface is equal to its total, hemispherical absorptivity *if* isothermal conditions exist *and* no net radiation heat transfer occurs at any of the surfaces.

We will later find that calculations of radiative exchange between surfaces are greatly simplified *if* Equation 12.66 may be applied to each of the surfaces. However, the restrictive conditions inherent in its derivation should be remembered. In particular, the surface irradiation has been assumed to correspond to emission from a blackbody at the same temperature as the surface. In Section 12.8 we consider other, less restrictive, conditions for which Equation 12.66 is applicable.

The preceding derivation may be repeated for spectral conditions, and for any surface in the enclosure it follows that

$$\varepsilon_\lambda = \alpha_\lambda \qquad (12.67)$$

Conditions associated with the use of Equation 12.67 are less restrictive than those associated with Equation 12.66. In particular, it will be shown that Equation 12.67 is applicable if the irradiation is diffuse *or* if the surface is diffuse. A form of Kirchhoff's law for which there are *no restrictions* involves the spectral, directional properties.

$$\varepsilon_{\lambda,\theta} = \alpha_{\lambda,\theta} \qquad (12.68)$$

This equality is *always* applicable because $\varepsilon_{\lambda,\theta}$ and $\alpha_{\lambda,\theta}$ are *inherent* surface properties. That is, respectively, they are independent of the spectral and directional distributions of the emitted and incident radiation.

More detailed developments of Kirchhoff's law are provided by Planck [1] and Howell et al. [7].

12.8   *The Gray Surface*

In Chapter 13 we will find that the problem of predicting radiant energy exchange between surfaces is greatly simplified if Equation 12.66 may be assumed to apply for each individual surface. It is therefore important to examine whether this equality may be applied to conditions other than those for which it was derived, namely, irradiation due to emission from a blackbody at the same temperature as the surface.

Accepting the fact that the spectral, directional emissivity and absorptivity are equal under any conditions, Equation 12.68, we begin by considering the conditions associated with using Equation 12.67. According to the definitions of the spectral, hemispherical properties, Equations 12.41 and 12.49, we are really asking under what conditions, if any, the following equality will hold:

$$\varepsilon_\lambda = \frac{\int_0^{2\pi} \int_0^{\pi/2} \varepsilon_{\lambda,\theta} \cos\theta \sin\theta \, d\theta \, d\phi}{\int_0^{2\pi} \int_0^{\pi/2} \cos\theta \sin\theta \, d\theta \, d\phi} \stackrel{?}{=} \frac{\int_0^{2\pi} \int_0^{\pi/2} \alpha_{\lambda,\theta} I_{\lambda,i} \cos\theta \sin\theta \, d\theta \, d\phi}{\int_0^{2\pi} \int_0^{\pi/2} I_{\lambda,i} \cos\theta \sin\theta \, d\theta \, d\phi} = \alpha_\lambda \quad (12.69)$$

Since $\varepsilon_{\lambda,\theta} = \alpha_{\lambda,\theta}$, it follows by inspection of Equation 12.69 that Equation 12.67 is applicable if *either* of the following conditions is satisfied:

1. The *irradiation* is *diffuse* ($I_{\lambda,i}$ is independent of θ and ϕ).
2. The *surface* is *diffuse* ($\varepsilon_{\lambda,\theta}$ and $\alpha_{\lambda,\theta}$ are independent of θ and ϕ).

The first condition is a reasonable approximation for many engineering calculations; the second condition is reasonable for many surfaces, particularly for electrically nonconducting materials (Figure 12.16).

Assuming the existence of either diffuse irradiation or a diffuse surface, we now consider what *additional* conditions must be satisfied for Equation 12.66 to be valid. From Equations 12.43 and 12.52, the equality applies if

$$\varepsilon = \frac{\int_0^\infty \varepsilon_\lambda E_{\lambda,b}(\lambda, T) \, d\lambda}{E_b(T)} \stackrel{?}{=} \frac{\int_0^\infty \alpha_\lambda G_\lambda(\lambda) \, d\lambda}{G} = \alpha \quad (12.70)$$

Since $\varepsilon_\lambda = \alpha_\lambda$, it follows by inspection of Equation 12.70 that Equation 12.66 applies if *either* of the following conditions is satisfied:

1. The irradiation corresponds to emission from a blackbody at the surface temperature T, in which case $G_\lambda(\lambda) = E_{\lambda,b}(\lambda, T)$ and $G = E_b(T)$.
2. The *surface* is *gray* (α_λ and ε_λ are independent of λ).

Note that the first condition corresponds to the major assumption required for the derivation of Kirchhoff's law (Section 12.7).

Because the total absorptivity of a surface depends on the spectral distribution of the irradiation, it cannot be stated unequivocally that $\alpha = \varepsilon$. For example, a particular surface may be highly absorbing to radiation in one spectral region and virtually nonabsorbing in another region (Figure 12.25a). Accordingly, for the two possible irradiation fields $G_{\lambda,1}(\lambda)$

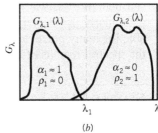

FIGURE 12.25 Spectral distribution of (*a*) the spectral absorptivity of a surface and (*b*) the spectral irradiation at the surface.

and $G_{\lambda,2}(\lambda)$ of Figure 12.25*b*, the values of α will differ drastically. In contrast, the value of ε is independent of the irradiation. Hence there is *no* basis for stating that α is *always* equal to ε.

To assume gray surface behavior, hence the validity of Equation 12.66, it is not necessary for α_λ and ε_λ to be independent of λ over the entire spectrum. Practically speaking, a *gray surface* may be defined as *one for which α_λ and ε_λ are independent of λ over the spectral regions of the irradiation and the surface emission*. For example, from Equation 12.70 it is readily shown that gray surface behavior may be assumed for the conditions of Figure 12.26. That is, the irradiation and surface emission are concentrated in a region for which the spectral properties of the surface are approximately constant. Accordingly,

$$\varepsilon = \frac{\varepsilon_{\lambda,o}\int_{\lambda_1}^{\lambda_2}E_{\lambda,b}(\lambda,T)\,d\lambda}{E_b(T)} = \varepsilon_{\lambda,o} \qquad \text{and} \qquad \alpha = \frac{\alpha_{\lambda,o}\int_{\lambda_3}^{\lambda_4}G_\lambda(\lambda)\,d\lambda}{G} = \alpha_{\lambda,o}$$

in which case $\alpha = \varepsilon = \varepsilon_{\lambda,o}$. However, if the irradiation were in a spectral region corresponding to $\lambda < \lambda_1$ or $\lambda > \lambda_4$, gray surface behavior could not be assumed.

A surface for which $\alpha_{\lambda,\theta}$ and $\varepsilon_{\lambda,\theta}$ are independent of θ and λ is termed a *diffuse, gray surface* (diffuse because of the directional independence and gray because of the wavelength independence). It is a surface for which both Equations 12.66 and 12.67 are

FIGURE 12.26 A set of conditions for which gray surface behavior may be assumed.

satisfied. We *assume* such surface conditions in many of our subsequent considerations, particularly in Chapter 13. However, although the assumption of a gray surface is reasonable for many practical applications, caution should be exercised in its use, particularly if the spectral regions of the irradiation and emission are widely separated.

EXAMPLE 12.10

A diffuse, fire brick wall of temperature $T_s = 500$ K has the spectral emissivity shown and is exposed to a bed of coals at 2000 K.

Determine the total, hemispherical emissivity and emissive power of the fire brick wall. What is the total absorptivity of the wall to irradiation resulting from emission by the coals?

SOLUTION

Known: Brick wall of surface temperature $T_s = 500$ K and prescribed $\varepsilon_\lambda(\lambda)$ is exposed to coals at $T_c = 2000$ K.

Find:

1. Total, hemispherical emissivity of the fire brick wall.
2. Total emissive power of the brick wall.
3. Absorptivity of the wall to irradiation from the coals.

Schematic:

Assumptions:

1. Brick wall is opaque and diffuse.
2. Spectral distribution of irradiation at the brick wall approximates that due to emission from a blackbody at 2000 K.

Analysis:

1. The total, hemispherical emissivity follows from Equation 12.43.

$$\varepsilon(T_s) = \frac{\int_0^\infty \varepsilon_\lambda(\lambda) E_{\lambda,b}(\lambda, T_s)\, d\lambda}{E_b(T_s)}$$

Breaking the integral into parts,

$$\varepsilon(T_s) = \varepsilon_{\lambda,1} \frac{\int_0^{\lambda_1} E_{\lambda,b}\, d\lambda}{E_b} + \varepsilon_{\lambda,2} \frac{\int_{\lambda_1}^{\lambda_2} E_{\lambda,b}\, d\lambda}{E_b} + \varepsilon_{\lambda,3} \frac{\int_{\lambda_2}^\infty E_{\lambda,b}\, d\lambda}{E_b}$$

and introducing the blackbody functions, it follows that

$$\varepsilon(T_s) = \varepsilon_{\lambda,1} F_{(0 \to \lambda_1)} + \varepsilon_{\lambda,2}[F_{(0 \to \lambda_2)} - F_{(0 \to \lambda_1)}] + \varepsilon_{\lambda,3}[1 - F_{(0 \to \lambda_2)}]$$

From Table 12.2

$$\lambda_1 T_s = 1.5\ \mu\text{m} \times 500\ \text{K} = 750\ \mu\text{m} \cdot \text{K}: \qquad F_{(0 \to \lambda_1)} = 0.000$$
$$\lambda_2 T_s = 10\ \mu\text{m} \times 500\ \text{K} = 5000\ \mu\text{m} \cdot \text{K}: \qquad F_{(0 \to \lambda_2)} = 0.634$$

Hence

$$\varepsilon(T_s) = 0.1 \times 0 + 0.5 \times 0.634 + 0.8(1 - 0.634) = 0.610 \qquad \triangleleft$$

2. From Equations 12.32 and 12.36, the total emissive power is

$$E(T_s) = \varepsilon(T_s) E_b(T_s) = \varepsilon(T_s)\sigma T_s^4$$
$$E(T_s) = 0.61 \times 5.67 \times 10^{-8}\ \text{W/m}^2 \cdot \text{K}^4 (500\ \text{K})^4 = 2162\ \text{W/m}^2 \qquad \triangleleft$$

3. From Equation 12.52, the total absorptivity of the wall to radiation from the coals is

$$\alpha = \frac{\int_0^\infty \alpha_\lambda(\lambda)\, G_\lambda(\lambda)\, d\lambda}{\int_0^\infty G_\lambda(\lambda)\, d\lambda}$$

Since the surface is diffuse, $\alpha_\lambda(\lambda) = \varepsilon_\lambda(\lambda)$. Moreover, since the spectral distribution of the irradiation approximates that due to emission from a blackbody at 2000 K, $G_\lambda(\lambda) \propto E_{\lambda,b}(\lambda, T_c)$. It follows that

$$\alpha = \frac{\int_0^\infty \varepsilon_\lambda(\lambda) E_{\lambda,b}(\lambda, T_c)\, d\lambda}{\int_0^\infty E_{\lambda,b}(\lambda, T_c)\, d\lambda}$$

Breaking the integral into parts and introducing the blackbody functions, we then obtain

$$\alpha = \varepsilon_{\lambda,1} F_{(0 \to \lambda_1)} + \varepsilon_{\lambda,2}[F_{(0 \to \lambda_2)} - F_{(0 \to \lambda_1)}] + \varepsilon_{\lambda,3}[1 - F_{(0 \to \lambda_2)}]$$

From Table 12.2

$$\lambda_1 T_c = 1.5\ \mu m \times 2000\ K = 3000\ \mu m \cdot K: \qquad F_{(0 \to \lambda_1)} = 0.273$$

$$\lambda_2 T_c = 10\ \mu m \times 2000\ K = 20,000\ \mu m \cdot K: \qquad F_{(0 \to \lambda_2)} = 0.986$$

Hence

$$\alpha = 0.1 \times 0.273 + 0.5(0.986 - 0.273) + 0.8(1 - 0.986) = 0.395 \qquad \triangleleft$$

Comments:

1. The emissivity depends on the surface temperature T_s, while the absorptivity depends on the spectral distribution of the irradiation, which depends on the temperature of the source T_c.

2. The surface is not gray, $\alpha \neq \varepsilon$. This result is to be expected. Since emission is associated with $T_s = 500$ K, its spectral maximum occurs at $\lambda_{max} \approx 6\ \mu m$. In contrast, since irradiation is associated with emission from a source at $T_c = 2000$ K, its spectral maximum occurs at $\lambda_{max} \approx 1.5\ \mu m$. Even though ε_λ and α_λ are equal, because they are not constant over the spectral ranges of both emission and irradiation, $\alpha \neq \varepsilon$. For the prescribed spectral distribution of $\varepsilon_\lambda = \alpha_\lambda$, ε and α decrease with increasing T_s and T_c, respectively, and it is only for $T_s = T_c$ that $\varepsilon = \alpha$. The foregoing expressions for ε and α may be used to determine their equivalent variation with T_s and T_c, and the following result is obtained:

EXAMPLE 12.11

A *small*, solid metallic sphere has an opaque, diffuse coating for which $\alpha_\lambda = 0.8$ for $\lambda \leq 5\ \mu m$ and $\alpha_\lambda = 0.1$ for $\lambda > 5\ \mu m$. The sphere, which is initially at a uniform temperature of 300 K, is inserted into a *large* furnace whose walls are at 1200 K. Determine the total, hemispherical absorptivity and emissivity of the coating for the initial condition and for the final, steady-state condition.

SOLUTION

Known: Small metallic sphere with spectrally selective absorptivity, initially at $T_s = 300$ K, is inserted into a large furnace at $T_f = 1200$ K.

Find:

1. Total, hemispherical absorptivity and emissivity of sphere coating for the initial condition.
2. Values of α and ε after sphere has been in furnace a long time.

Schematic:

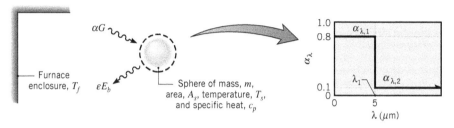

Assumptions:

1. Coating is opaque and diffuse.
2. Since furnace surface is much larger than that of sphere, irradiation approximates emission from a blackbody at T_f.

Analysis:

1. From Equation 12.52 the total, hemispherical absorptivity is

$$\alpha = \frac{\int_0^\infty \alpha_\lambda(\lambda) G_\lambda(\lambda)\, d\lambda}{\int_0^\infty G_\lambda(\lambda)\, d\lambda}$$

or, with $G_\lambda = E_{\lambda,b}(T_f) = E_{\lambda,b}(\lambda, 1200\ \text{K})$,

$$\alpha = \frac{\int_0^\infty \alpha_\lambda(\lambda) E_{\lambda,b}(\lambda, 1200\ \text{K})\, d\lambda}{E_b(1200\ \text{K})}$$

Hence

$$\alpha = \alpha_{\lambda,1} \frac{\int_0^{\lambda_1} E_{\lambda,b}(\lambda, 1200\ \text{K})\, d\lambda}{E_b(1200\ \text{K})} + \alpha_{\lambda,2} \frac{\int_{\lambda_1}^\infty E_{\lambda,b}(\lambda, 1200\ \text{K})\, d\lambda}{E_b(1200\ \text{K})}$$

or

$$\alpha = \alpha_{\lambda,1} F_{(0 \to \lambda_1)} + \alpha_{\lambda,2}[1 - F_{(0 \to \lambda_1)}]$$

From Table 12.2,

$$\lambda_1 T_f = 5\ \mu\text{m} \times 1200\ \text{K} = 6000\ \mu\text{m} \cdot \text{K}: \qquad F_{(0 \to \lambda_1)} = 0.738$$

Hence

$$\alpha = 0.8 \times 0.738 + 0.1(1 - 0.738) = 0.62$$

The total, hemispherical emissivity follows from Equation 12.43.

$$\varepsilon(T_s) = \frac{\int_0^\infty \varepsilon_\lambda E_{\lambda,b}(\lambda, T_s)\, d\lambda}{E_b(T_s)}$$

Since the surface is diffuse, $\varepsilon_\lambda = \alpha_\lambda$ and it follows that

$$\varepsilon = \alpha_{\lambda,1} \frac{\int_0^{\lambda_1} E_{\lambda,b}(\lambda, 300\ \text{K})\, d\lambda}{E_b(300\ \text{K})} + \alpha_{\lambda,2} \frac{\int_{\lambda_1}^\infty E_{\lambda,b}(\lambda, 300\ \text{K})\, d\lambda}{E_b(300\ \text{K})}$$

or,

$$\varepsilon = \alpha_{\lambda,1}\, F_{(0\to\lambda_1)} + \alpha_{\lambda,2}[1 - F_{(0\to\lambda_1)}]$$

From Table 12.2,

$$\lambda_1 T_s = 5\ \mu\text{m} \times 300\ \text{K} = 1500\ \mu\text{m} \cdot \text{K}: \qquad F_{(0\to\lambda_1)} = 0.014$$

Hence

$$\varepsilon = 0.8 \times 0.014 + 0.1(1 - 0.014) = 0.11 \qquad \triangleleft$$

2. Because the spectral characteristics of the coating and the furnace temperature remain fixed, there is no change in the value of α with increasing time. However, as T_s increases with time, the value of ε will change. After a sufficiently long time, $T_s = T_f$, and $\varepsilon = \alpha$ ($\varepsilon = 0.62$).

Comments:

1. The equilibrium condition that eventually exists ($T_s = T_f$) corresponds precisely to the condition for which Kirchhoff's law was derived. Hence α must equal ε.

2. Approximating the sphere as a lumped capacitance and neglecting convection, an energy balance for a control volume about the sphere yields

$$\dot{E}_{\text{in}} - \dot{E}_{\text{out}} = \dot{E}_{\text{st}}$$

$$(\alpha G)A_s - (\varepsilon \sigma T_s^4)A_s = mc_p \frac{dT_s}{dt}$$

The differential equation could be solved to determine $T(t)$ for $t > 0$, and the variation in ε that occurs with increasing time would have to be included in the solution.

12.9   *Environmental Radiation*

Solar radiation is essential to all life on earth. Through the process of photosynthesis, solar radiation satisfies the human need for food, fiber, and fuel. Utilizing thermal and photovoltaic processes, it also has the potential to satisfy considerable demand for heat and electricity.

Together, solar radiation and radiation emitted by surfaces of the earth's land and oceans comprise what is commonly termed *environmental radiation*. It is the interaction of environmental radiation with the earth's atmosphere that determines the temperature of our planet.

12.9.1 Solar Radiation

The sun is a nearly spherical radiation source that is 1.39×10^9 m in diameter and is located 1.50×10^{11} m from the earth. As noted previously, the sun emits approximately as a blackbody at 5800 K. As the radiation emitted by the sun travels through space, the radiation flux decreases because of the greater (spherical) area through which it passes. At the outer edge of the earth's atmosphere, the flux of solar energy has decreased by a factor of $(r_s/r_d)^2$, where r_s is the radius of the sun and r_d is the distance from the sun to the earth. The *solar constant*,[3] S_c, is defined as the flux of solar energy incident on a surface oriented normal to the sun's rays, at the outer edge of the earth's atmosphere, when the earth is at its mean distance from the sun (Figure 12.27). It has a value of 1368 ± 0.65 W/m^2. For a *horizontal* surface (that is, parallel to the earth's surface), solar radiation appears as a beam of nearly *parallel rays* that form an angle θ, the *zenith angle*, relative to the surface normal. The *extraterrestrial solar irradiation*, $G_{S,o}$, defined for a horizontal surface, depends on the geographic latitude, as well as the time of day and year. It may be determined from an expression of the form

$$G_{S,o} = S_c \cdot f \cdot \cos \theta \qquad (12.71)$$

The quantity f is a correction factor to account for the eccentricity of the earth's orbit about the sun ($0.97 \lesssim f \lesssim 1.03$). On a time- and surface area-averaged basis, the earth receives $S_c \times \pi r_e^2 / 4\pi r_e^2 = S_c/4 = 342$ W/m^2 of solar irradiation. The diameter of the earth is $d_e = 2r_e = 1.27 \times 10^7$ m.

As shown in Figure 12.28a, the spectral distribution of extraterrestrial solar irradiation *approximates* that of a blackbody at 5800 K. The radiation is concentrated in the

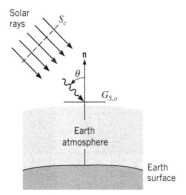

FIGURE 12.27 Directional nature of solar radiation outside the earth's atmosphere.

---

[3]The term *solar constant* is a misnomer in that its value varies with time in a predictable manner. Radiation emitted by the sun undergoes an 11-year cycle with peak emission (+0.65 W/m^2) corresponding to periods of high sunspot activity [8].

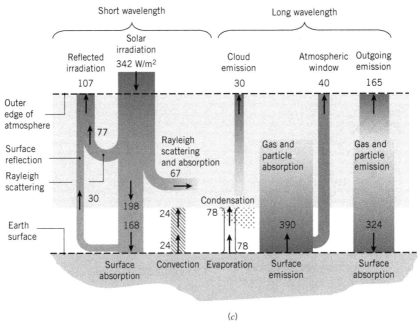

FIGURE 12.28   Solar and environmental radiation. (*a*) Spectral distribution of downward-propagating short wavelength solar radiation. (*b*) Spectral distribution of upward-propagating long wavelength environmental radiation. (*c*) Energy balance on the atmosphere for moderate temperature and cloudy conditions [9].

short wavelength region ($0.2 \lesssim \lambda \lesssim 3$ μm) of the spectrum with the peak occurring at approximately 0.50 μm. However, as the solar radiation propagates through the earth's atmosphere, its magnitude and both its spectral and directional distributions experience substantial modification. This change is due to *absorption* and *scattering* of the radiation by atmospheric constituents. The effect of absorption by the atmospheric gases O_3 (ozone), H_2O, O_2, and CO_2 is shown on the lower curve of Figure 12.28*a* corresponding to solar irradiation at the earth's surface, after it has passed through the atmosphere. Absorption by ozone is strong in the UV region, providing considerable attenuation below 0.4 μm and complete attenuation below 0.3 μm. In the visible region there is some absorption by O_3 and O_2, and in the near and far IR regions absorption is dominated by water vapor. Throughout the solar spectrum, there is also continuous absorption of short wavelength radiation by the dust and aerosol content of the atmosphere, including the products of fossil fuel combustion such as soot.

Atmospheric scattering provides for *redirection* of the sun's rays and therefore also affects solar radiation reaching the earth's surface. Two types of scattering are shown in Figure 12.29. *Rayleigh* (or *molecular*) *scattering* is caused by very small gas molecules. It occurs when the ratio of the effective molecule diameter to the wavelength of the radiation, $\pi D/\lambda$, is much less than unity and provides for nearly uniform scattering of the radiation in all directions. In contrast, *Mie scattering* by larger dust or soot particles occurs when $\pi D/\lambda$ is approximately unity and is concentrated in the direction of the incident rays. Hence virtually all Mie-scattered radiation strikes the earth's surface in directions close to that of the sun's rays.

12.9.2 The Atmospheric Radiation Balance

In addition to solar irradiation from above, the atmosphere is irradiated from below by the earth's surface. Since the average temperature of the earth is approximately 290 K, this upward-propagating radiation is concentrated at longer wavelengths, as shown in Figure 12.28*b*. The spectral distribution of the earth's emission varies smoothly relative to the distribution of the extraterrestrial solar irradiation of Figure 12.28*a*; the smooth variation is also characteristic of many engineered surfaces. However, like the downward-propagating solar irradiation of Figure 12.28*a*, the terrestrial emission is modified by absorption and

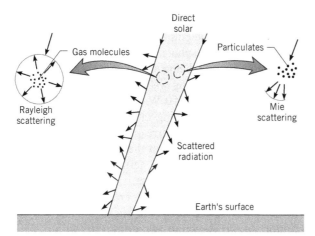

FIGURE 12.29   Scattering of solar radiation in the earth's atmosphere.

scattering as it propagates upward through the atmosphere. Absorption by water vapor occurs throughout the spectrum. Strong ozone absorption is noted in the wavelength region around 9 μm and robust CO_2 absorption spans the wavelength region $13 \lesssim \lambda \lesssim 16$ μm. The majority of terrestrial emission in the $8 \lesssim \lambda \lesssim 13$ μm *atmospheric window* propagates to the outer edge of the atmosphere, except in the spectral range associated with strong ozone absorption. Rayleigh and Mie scattering of the long wavelength terrestrial emission is triggered by the presence of various particles and aerosols in the atmosphere.

The modification of both the downward-propagating extraterrestrial solar irradiation (Figure 12.28a) and the upward-propagating terrestrial emission (Figure 12.28b) due to absorption and scattering have a profound influence on the energy balance of the atmosphere. For *either* the upward- or downward-propagating radiation, the net effect is the heating of the atmosphere since the energy content of the radiation leaving the atmosphere is less than that of the corresponding incoming radiation. However, this heating is balanced by cooling due to radiation emitted by atmospheric constituents.

A representative *equilibrium* energy balance (Figure 12.28c) shows the *partitioning* of the short wavelength solar irradiation and the long wavelength terrestrial emission [9]. Of the surface- and time-averaged value of 342 W/m^2 for the solar irradiation at the outer edge of earth's atmosphere, 77 W/m^2 is reflected back to space, primarily by Rayleigh scattering, while 67 W/m^2 heats the atmosphere through the effects of absorption, including absorption by soot, dust, and clouds. The remaining portion of the solar irradiation (198 W/m^2) reaches the earth's surface, where 30 W/m^2 reflects back to space and 168 W/m^2 is absorbed.

The partitioning of the long wavelength radiation associated with emission from the earth's surface is more complex. The time-averaged surface emission (390 W/m^2) is mainly absorbed by the atmosphere, except for 40 W/m^2 corresponding to the atmospheric window. The remaining 350 W/m^2 component of the surface emission combines with the short wavelength radiation absorption (67 W/m^2), convection (24 W/m^2) from the earth's surface, and condensation in the form of precipitation in the lowest regions of the atmosphere (78 W/m^2) to heat the bulk of the atmosphere. In turn, the heated atmospheric gases emit radiation at long wavelengths resulting in a radiation flux of 165 W/m^2 at the outer edge of the atmosphere, and a corresponding downward radiation flux of 324 W/m^2 at the earth's surface. Emission from clouds accounts for a radiation flux of 30 W/m^2. Since conditions are *assumed* to be in equilibrium, the *net* heat transfer at the outer edge of the atmosphere and at the earth's surface are both zero.

In reality, conditions are not in equilibrium, since the absorption and scattering of both short- and long-wavelength radiation evolve in response to the changing chemical and particulate content of our atmosphere. Anthropogenic activity that influences the makeup of the atmosphere is primarily related to the combustion of fossil fuels, leading to increases in the CO_2 and aerosol content of the atmosphere. Hence gas absorption and aerosol-induced scattering (and absorption) are continually being affected by human activity. In general, as the CO_2 content of the atmosphere increases, long wavelength radiation emitted by the surface of the earth and absorbed by the atmosphere (350 W/m^2) will be absorbed closer to the earth's surface, resulting in a decrease in the outgoing emission (165 W/m^2) of long-wavelength radiation at the top of the atmosphere and a corresponding increase in radiative flux to the earth's surface (324 W/m^2). With a reduction in the outgoing long wavelength emission, the *net* heat transfer at the top of the atmosphere, termed *radiative forcing*, is to the atmosphere, and atmospheric temperatures must increase.

The reduction in the outgoing long-wavelength emission could be offset by increases in other components at the top of the atmosphere such as enhanced reflection of short-wavelength solar irradiation due to Rayleigh scattering (107 W/m^2) [10] or reflection of short-wavelength radiation from the earth's surface (30 W/m²) [11]. However, as engineers improve the efficiency and cleanliness of combustion processes, for example, the concentrations of aerosols and particles that are responsible for scattering are correspondingly reduced. Decreasing the production of pollutants by improving combustion technology has important health benefits for humankind, but ironically may lead to a reduction in the beneficial short-wavelength reflection (107 W/m²), increasing the temperature of the atmosphere and also potentially changing the convection and precipitation that occurs in the lower atmosphere, influencing and modifying weather patterns [12–14]. Clearly, fossil fuel combustion and the associated environmental radiation heat transfer effects are complex, and, while they are not completely understood, they may have a profound impact at the global scale.

12.9.3 Terrestrial Solar Irradiation

Solar irradiation at the earth's surface may be utilized in a wide variety of engineering applications, including but not limited to heating and electricity generation. Increased usage of solar irradiation for these purposes reduces our dependence on fossil fuels and, in turn, can mitigate the potential for atmospheric warming. As such, knowledge of the nature of the solar irradiation at the earth's surface is crucial. Detailed treatment of solar energy technologies is left to the literature [15–19].

The cumulative effect of the scattering processes on the *directional distribution* of solar radiation striking the earth's surface is shown in Figure 12.30a. That portion of the radiation that has penetrated the atmosphere without having been scattered (or absorbed) is in the direction of the zenith angle and is termed the *direct radiation*. The scattered radiation is incident from all directions, although its intensity is largest for directions close to that of the direct radiation. The scattered radiation may vary from approximately 10% of the total solar radiation on a clear day to nearly 100% on a very overcast day. The scattered component of the solar radiation is often *approximated* as being independent of direction (Figure 12.30b), or *diffuse*.

As evident in Figure 12.28c, the long-wavelength forms of the environmental radiation include emission from the earth's surface, as well as gas emission from certain atmospheric constituents. The emissive power from the earth's surface may be computed in the conventional manner. That is,

$$E = \varepsilon \sigma T^4 \tag{12.72}$$

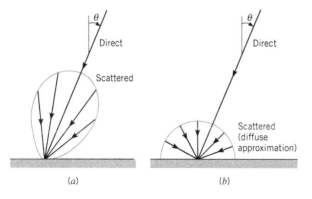

(a)   (b)

Figure 12.30 Directional distribution of solar radiation at the earth's surface. (*a*) Actual distribution. (*b*) Diffuse approximation.

where ε and T are the surface emissivity and temperature, respectively. As implied in Figure 12.28b, emissivities are generally close to unity; that of water, for example, is approximately 0.97. Using $\varepsilon = 0.97$ and $\bar{E} = 390$ W/m² from Figure 12.28c, the effective radiation temperature of the earth is $\bar{T} = 291$ K. Emission is concentrated in the spectral region from approximately 4 to 40 μm, with the peak occurring at approximately 10 μm, as evident in Figure 12.28b.

Downward-propagating atmospheric emission that is incident upon the earth's surface is largely due to the CO_2 and H_2O content of the atmosphere and is concentrated in the spectral regions from 5 to 8 μm and above 13 μm. Although the spectral distribution of atmospheric emission does not correspond to that of a blackbody, its contribution to irradiation of the earth's surface can be estimated using Equation 12.32. In particular, irradiation at the earth's surface due to atmospheric emission may be expressed in the form

$$G_{atm} = \sigma T_{sky}^4 \tag{12.73}$$

where T_{sky} is termed the *effective sky temperature*. Its value depends on atmospheric conditions, and for the moderate temperature, cloudy conditions of Figure 12.28c, $G_{atm} = 324$ W/m² and $T_{sky} = 275$ K. Actual values range from a low of 230 K under a cold, clear sky to a high of approximately 285 K under warm, cloudy conditions. When its value is small, as on a cool, clear night, an exposed pool of water may freeze even though the air temperature exceeds 273 K.

We close by recalling that values of the spectral properties of a surface at short wavelengths may be appreciably different from values at long wavelengths (Figures 12.17 and 12.22). Since solar radiation is concentrated in the short wavelength region of the spectrum and surface emission is at much longer wavelengths, it follows that many surfaces may not be approximated as gray in their response to solar irradiation. In other words, the solar absorptivity of a surface α_S may differ from its emissivity ε. Values of α_S and the emissivity at moderate temperatures are presented in Table 12.3 for representative surfaces. Note that the ratio α_S/ε is an important engineering parameter. Small values are desired if the surface is intended to reject heat; large values are required if the surface is intended to collect solar energy.

TABLE 12.3 Solar absorptivity α_S and emissivity ε of surfaces having spectral absorptivity given in Figure 12.22

Surface	α_S	ε (300 K)	α_S/ε
Evaporated aluminum film	0.09	0.03	3.0
Fused quartz on aluminum film	0.19	0.81	0.24
White paint on metallic substrate	0.21	0.96	0.22
Black paint on metallic substrate	0.97	0.97	1.0
Stainless steel, as received, dull	0.50	0.21	2.4
Red brick	0.63	0.93	0.68
Human skin (Caucasian)	0.62	0.97	0.64
Snow	0.28	0.97	0.29
Corn leaf	0.76	0.97	0.78

EXAMPLE 12.12

A flat-plate solar collector with no cover plate has a selective absorber surface of emissivity 0.1 and solar absorptivity 0.95. At a given time of day the absorber surface temperature T_s is 120°C when the solar irradiation is 750 W/m², the effective sky temperature is −10°C, and the ambient air temperature T_∞ is 30°C. Assume that the convection heat transfer coefficient for the calm day conditions can be estimated from

$$\bar{h} = 0.22(T_s - T_\infty)^{1/3} \text{ W/m}^2 \cdot \text{K}$$

Calculate the useful heat removal rate (W/m²) from the collector for these conditions. What is the corresponding efficiency of the collector?

SOLUTION

Known: Operating conditions for a flat-plate solar collector.

Find:
1. Useful heat removal rate per unit area, q_u'' (W/m²).
2. Efficiency η of the collector.

Schematic:

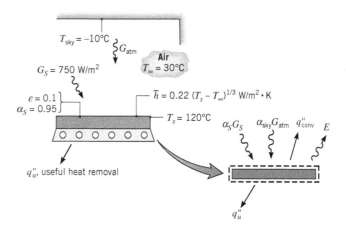

Assumptions:
1. Steady-state conditions.
2. Bottom of collector well insulated.
3. Absorber surface diffuse.

Analysis:
1. Performing an energy balance on the absorber,

$$\dot{E}_{in} - \dot{E}_{out} = 0$$

or, per unit surface area,

$$\alpha_S G_S + \alpha_{\text{sky}} G_{\text{atm}} - q''_{\text{conv}} - E - q''_u = 0$$

From Equation 12.73,

$$G_{\text{atm}} = \sigma T_{\text{sky}}^4$$

Since the atmospheric irradiation is concentrated in approximately the same spectral region as that of surface emission, it is reasonable to assume that

$$\alpha_{\text{sky}} \approx \varepsilon = 0.1$$

With

$$q''_{\text{conv}} = \overline{h}(T_s - T_\infty) = 0.22(T_s - T_\infty)^{4/3} \qquad \text{and} \qquad E = \varepsilon \sigma T_s^4$$

it follows that

$$q''_u = \alpha_S G_S + \varepsilon \sigma T_{\text{sky}}^4 - 0.22(T_s - T_\infty)^{4/3} - \varepsilon \sigma T_s^4$$

$$q''_u = \alpha_S G_S - 0.22(T_s - T_\infty)^{4/3} - \varepsilon \sigma (T_s^4 - T_{\text{sky}}^4)$$

$$q''_u = 0.95 \times 750 \text{ W/m}^2 - 0.22(120 - 30)^{4/3} \text{ W/m}^2$$
$$\qquad - 0.1 \times 5.67 \times 10^{-8} \text{ W/m}^2 \cdot \text{K}^4 (393^4 - 263^4) \text{ K}^4$$

$$q''_u = (712.5 - 88.7 - 108.1) \text{ W/m}^2 = 516 \text{ W/m}^2 \qquad \qquad \triangleleft$$

2. The collector efficiency, defined as the fraction of the solar irradiation extracted as useful energy, is then

$$\eta = \frac{q''_u}{G_S} = \frac{516 \text{ W/m}^2}{750 \text{ W/m}^2} = 0.69 \qquad \qquad \triangleleft$$

Comments:

1. Since the spectral range of G_{atm} is entirely different from that of G_S, it would be *incorrect* to assume that $\alpha_{\text{sky}} = \alpha_S$.

2. The convection heat transfer coefficient is extremely low ($\overline{h} \approx 1 \text{ W/m}^2 \cdot \text{K}$). With a modest increase to $\overline{h} = 5 \text{ W/m}^2 \cdot \text{K}$, the useful heat flux and the efficiency are reduced to $q''_u = 154 \text{ W/m}^2$ and $\eta = 0.21$. A cover plate can contribute significantly to reducing convection (and radiation) heat loss from the absorber plate.

12.10 *Summary*

Many important ideas have been introduced in this chapter, along with a considerable amount of new terminology. However, the subject matter has been developed in a systematic fashion, and careful rereading of the material should make you more comfortable with its application. A glossary has been provided in Table 12.4 to assist you in assimilating the terminology.

TABLE 12.4 Glossary of radiative terms

Term	Definition
Absorption	The process of converting radiation intercepted by matter to internal thermal energy.
Absorptivity	Fraction of the incident radiation absorbed by matter. Equations 12.47, 12.48, and 12.51. Modifiers: *directional, hemispherical, spectral, total.*
Blackbody	The ideal emitter and absorber. Modifier referring to ideal behavior. Denoted by the subscript b.
Diffuse	Modifier referring to the directional independence of the intensity associated with emitted, reflected, or incident radiation.
Directional	Modifier referring to a particular direction. Denoted by the subscript θ.
Directional distribution	Refers to variation with direction.
Emission	The process of radiation production by matter at a non-zero temperature. Modifiers: *diffuse, blackbody, spectral.*
Emissive power	Rate of radiant energy emitted by a surface in all directions per unit area of the surface, E (W/m^2). Modifiers: *spectral, total, blackbody.*
Emissivity	Ratio of the radiation emitted by a surface to the radiation emitted by a blackbody at the same temperature. Equations 12.36, 12.38, 12.39, and 12.40. Modifiers: *directional, hemispherical, spectral, total.*
Gray surface	A surface for which the spectral absorptivity and the emissivity are independent of wavelength over the spectral regions of surface irradiation and emission.
Hemispherical	Modifier referring to all directions in the space above a surface.
Intensity	Rate of radiant energy propagation in a particular direction, per unit area normal to the direction, per unit solid angle about the direction, I ($W/m^2 \cdot sr$). Modifier: *spectral.*
Irradiation	Rate at which radiation is incident on a surface from all directions per unit area of the surface, G (W/m^2). Modifiers: *spectral, total, diffuse.*
Kirchhoff's law	Relation between emission and absorption properties for surfaces irradiated by a blackbody at the same temperature. Equations 12.65, 12.66, 12.67, and 12.68.
Planck's law	Spectral distribution of emission from a blackbody. Equation 12.30.
Radiosity	Rate at which radiation leaves a surface due to emission and reflection in all directions per unit area of the surface, J (W/m^2). Modifiers: *spectral, total.*
Reflection	The process of redirection of radiation incident on a surface. Modifiers: *diffuse, specular.*
Reflectivity	Fraction of the incident radiation reflected by matter. Equations 12.54, 12.55, and 12.57. Modifiers: *directional, hemispherical, spectral, total.*
Semitransparent	Refers to a medium in which radiation absorption is a volumetric process.
Solid angle	Region subtended by an element of area on the surface of a sphere with respect to the center of the sphere, ω (sr). Equations 12.7 and 12.8.

TABLE 12.4 *Continued*

Term	Definition
Spectral	Modifier referring to a single-wavelength (monochromatic) component. Denoted by the subscript λ.
Spectral distribution	Refers to variation with wavelength.
Specular	Refers to a surface for which the angle of reflected radiation is equal to the angle of incident radiation.
Stefan–Boltzmann law	Emissive power of a blackbody. Equation 12.32.
Thermal radiation	Electromagnetic energy emitted by matter at a non-zero temperature and concentrated in the spectral region from approximately 0.1 to 100 μm.
Total	Modifier referring to all wavelengths.
Transmission	The process of thermal radiation passing through matter.
Transmissivity	Fraction of the incident radiation transmitted by matter. Equations 12.59 and 12.60. Modifiers: *hemispherical, spectral, total.*
Wien's displacement law	Locus of the wavelength corresponding to peak emission by a blackbody. Equation 12.31.

Test your understanding of the terms and concepts introduced in this chapter by addressing the following questions.

- What is the nature of radiation? What two important features characterize radiation?
- What is the physical origin of radiation *emission* from a surface? How does emission affect the thermal energy of a material?
- In what region of the electromagnetic spectrum is *thermal radiation* concentrated?
- What is the *spectral intensity* of radiation emitted by a surface? On what variables does it depend? How may knowledge of this dependence be used to determine the rate at which matter loses thermal energy due to emission from its surface?
- What is a *steradian*? How many steradians are associated with a hemisphere?
- What is the distinction between *spectral* and *total* radiation? Between *directional* and *hemispherical* radiation?
- What is *total emissive power*? What role does it play in a surface energy balance?
- What is a *diffuse emitter*? For such an emitter, how is the intensity related to the total emissive power?
- What is *irradiation*? How is it related to the intensity of incident radiation, if the radiation is diffuse?
- What is *radiosity*? What role do the total radiosity and the total irradiation play in a surface energy balance?
- What are the characteristics of a *blackbody*? Does such a thing actually exist in nature? What is the principal role of blackbody behavior in radiation analysis?
- What is the *Planck distribution*? What is *Wien's displacement law*?
- From memory, sketch the spectral distribution of radiation emission from a blackbody at three temperatures, $T_1 < T_2 < T_3$. Identify salient features of the distributions.

- In what region of the electromagnetic spectrum is radiation emission from a surface at room temperature concentrated? What is the spectral region of concentration for a surface at 1000°C? For the surface of the sun?
- What is the *Stefan–Boltzmann law*? How would you determine the total intensity of radiation emitted by a blackbody at a prescribed temperature?
- How would you approximate the total irradiation of a small surface in a large isothermal enclosure?
- In the term *total*, *hemispherical emissivity*, to what do the adjectives *total* and *hemispherical* refer?
- How does the directional emissivity of a material change as the zenith angle associated with emission approaches 90°?
- If the spectral emissivity of a material increases with increasing wavelength, how does its total emissivity vary with temperature?
- What is larger, the emissivity of a polished metal or an oxidized metal? A refractory brick or ice?
- What processes accompany irradiation of a *semitransparent* material? An *opaque* material?
- Are glass and water semitransparent or opaque materials?
- How is the perceived color of a material determined by its response to irradiation in the infrared portion of the spectrum? How is its color affected by its temperature?
- Can snow be thought of as a good absorber or reflector of incident infrared radiation?
- How is the thermal energy of a material affected by the absorption of incident radiation? By the reflection of incident radiation?
- Can the total absorptivity of an opaque surface at a fixed temperature differ according to whether irradiation is from a source at room temperature or from the sun? Can its reflectivity differ? Its emissivity?
- What is a *diffuse reflector*? A *specular reflector*? How does the roughness of a surface affect the nature of reflection for the surface?
- Under what conditions is there equivalence between the spectral, directional emissivity of a surface and the spectral, directional absorptivity? The spectral, hemispherical emissivity and the spectral, hemispherical absorptivity? The total, hemispherical emissivity and the total, hemispherical absorptivity?
- What is a *gray surface*?
- How does the gas and aerosol content of the atmosphere modify the spectral variation of downward-propagating solar radiation? How does the content of the atmosphere modify the spectral variation of upward-propagating terrestrial emission?
- What is meant by radiative forcing, and what is the impact of such forcing on the temperature of earth's atmosphere?
- How can the effective radiation temperature of the earth be calculated from the atmospheric radiation balance? What is the effective sky temperature, and how can it be determined from the atmospheric radiation balance?
- What is the directional nature of solar radiation outside the earth's atmosphere? At the surface of the earth?
- What is the primary difference between Rayleigh and Mie scattering? In the context of solar and environmental radiation, how do these scattering phenomena affect the temperature of earth's atmosphere? How might anthropogenic activity affect radiation scattering, absorption, and emission in the atmosphere?

References

1. Planck, M., *The Theory of Heat Radiation,* Dover Publications, New York, 1959.

2. Zetteli, N., *Quantum Mechanics Concepts and Applications*, Wiley, Chichester, 2001.

3. Gubareff, G. G., J. E. Janssen, and R. H. Torberg, *Thermal Radiation Properties Survey,* 2nd ed., Honeywell Research Center, Minneapolis, 1960.

4. Wood, W. D., H. W. Deem, and C. F. Lucks, *Thermal Radiative Properties,* Plenum Press, New York, 1964.

5. Touloukian, Y. S., *Thermophysical Properties of High Temperature Solid Materials,* Macmillan, New York, 1967.

6. Touloukian, Y. S., and D. P. DeWitt, *Thermal Radiative Properties,* Vols. 7, 8, and 9, from *Thermophysical Properties of Matter,* TPRC Data Series, Y.S. Touloukian and C. Y. Ho, Eds., IFI Plenum, New York, 1970–1972.

7. Howell, J. R., R. Siegel, and M. P. Menguc, *Thermal Radiation Heat Transfer*, 5th ed., Taylor & Francis, New York, 2010.

8. National Academy of Sciences, *Solar Influences on Global Change*, National Academy Press, Washington, D.C. 2004.

9. Kiehl, J. T., and K. E. Trenberth, *Bull. Am. Met. Soc.* **78**, 197, 1997.

10. Myhre, G., *Science*, **325**, 187, 2009.

11. Akbari, H., S. Menon, and A. Rosenfeld, *Climate Change*, **94**, 275, 2009.

12. Arneth, A., N. Unger, M. Kulmala, and M. O. Andreae, *Science*, **326**, 672, 2009.

13. Shindell, D. T., G. Faluvegi, D. M. Koch, G. A. Schmidt, N. Unger, and S. E. Bauer, *Science*, **326**, 716, 2009.

14. Ramanathan, V., P. J. Crutzen, J. T. Kiehl, and D. Rosenfeld, *Science*, **294**, 2119, 2001.

15. Duffie, J. A., and W. A. Beckman, *Solar Engineering of Thermal Processes*, 3rd ed., Wiley, Hoboken, NJ, 2006.

16. Goswami, D. Y., F. Kreith, and J. F. Kreider, *Principles of Solar Energy*, 2nd ed., Taylor & Francis, New York, 2002.

17. Howell, J. R., R. B. Bannerot, and G. C. Vliet, *Solar-Thermal Energy Systems, Analysis and Design,* McGraw-Hill, New York, 1982.

18. Kalogirou, S. A., *Prog. Energy Comb. Sci.*, **30**, 231, 2004.

19. Kalogirou, S. A., *Solar Energy Engineering: Processes and Systems*, Elsevier, Oxford, 2009.

Problems

Radiation Fluxes: Definitions

12.1 Consider an opaque horizontal plate that is well insulated on its back side. The irradiation on the plate is 2500 W/m^2, of which 500 W/m^2 is reflected. The plate is at 227°C and has an emissive power of 1300 W/m^2. Air at 127°C flows over the plate with a convection heat transfer coefficient of 16 W/m$^2 \cdot$ K. Determine the emissivity, absorptivity, and radiosity of the plate. What is the net heat transfer rate per unit area?

12.2 A horizontal, opaque surface at a steady-state temperature of 77°C is exposed to an airflow having a free stream temperature of 27°C with a convection heat transfer coefficient of 28 W/m$^2 \cdot$ K. The emissive power of the surface is 628 W/m^2, the irradiation is 1380 W/m^2, and the reflectivity is 0.40. Determine the absorptivity of the surface. Determine the net radiation heat transfer rate for this surface. Is this heat transfer to the surface or from the surface? Determine the combined heat transfer rate for the surface. Is this heat transfer to the surface or from the surface?

12.3 The top surface of an $L = 5$-mm-thick anodized aluminum plate is irradiated with $G = 1000$ W/m^2 while being simultaneously exposed to convection conditions characterized by $h = 45$ W/m$^2 \cdot$ K and $T_\infty = 30$°C. The bottom surface of the plate is insulated. For a plate temperature of 400 K as well as $\alpha = 0.14$ and $\varepsilon = 0.76$, determine the radiosity at the top plate surface, the net radiation heat flux at the top surface, and the rate at which the temperature of the plate is changing with time.

12.4 A horizontal semitransparent plate is uniformly irradiated from above and below, while air at $T_\infty = 300$ K flows over the top and bottom surfaces, providing a uniform convection heat transfer coefficient of $h = 40$ W/m$^2 \cdot$ K. The absorptivity of the plate to the irradiation is 0.40. Under steady-state conditions measurements made with a radiation detector above the top surface indicate a radiosity (which includes transmission, as well as reflection and emission) of $J = 5000$ W/m^2, while the plate is at a uniform temperature of $T = 350$ K.

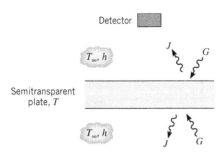

Determine the irradiation G and the emissivity of the plate. Is the plate gray ($\varepsilon = \alpha$) for the prescribed conditions?

Intensity, Emissive Power, and Irradiation

12.5 What is the irradiation at surfaces A_2, A_3, and A_4 of Example 12.1 due to emission from A_1?

12.6 A furnace with an aperture of 20-mm diameter and emissive power of 3.72×10^5 W/m² is used to calibrate a heat flux gage having a sensitive area of 1.6×10^{-5} m².

(a) At what distance, measured along a normal from the aperture, should the gage be positioned to receive irradiation of 1000 W/m²?

(b) If the gage is tilted off normal by 20°, what will be its irradiation?

(c) For tilt angles of 0, 20, and 60°, plot the gage irradiation as a function of the separation distance for values ranging from 100 to 300 mm.

12.7 A small radiant source A_1 emits diffusely with an intensity $I_1 = 1.2 \times 10^5$ W/m² · sr. The radiation detector A_2 is aligned normal to the source at a distance of $L_o = 0.2$ m. An opaque screen is positioned midway between A_1 and A_2 to prevent radiation from the source reaching the detector. The small surface A_m is a perfectly diffuse mirror that permits radiation emitted from the source to be reflected into the detector.

(a) Calculate the radiant power incident on A_m due to emission from the source A_1, $q_{1 \to m}$(W).

(b) Assuming that the radiant power, $q_{1 \to m}$, is perfectly and diffusely reflected, calculate the intensity leaving A_m, I_m (W/m² · sr).

(c) Calculate the radiant power incident on A_2 due to the reflected radiation leaving A_m, $q_{m \to 2}(\mu W)$.

(d) Plot the radiant power $q_{m \to 2}$ as a function of the lateral separation distance y_o for the range $0 \leq y_o \leq 0.2$ m. Explain the features of the resulting curve.

12.8 A radiation detector has an aperture of area $A_d = 10^{-6}$ m² and is positioned at a distance of $r = 2$ m from a surface of area $A_s = 10^{-4}$ m². The angle formed by the normal to the detector and the surface normal is $\theta = 25°$.

The surface is at 500 K and is opaque, diffuse, and gray with an emissivity of 0.8. If the surface irradiation is 2500 W/m², what is the rate at which the detector intercepts radiation from the surface?

12.9 Solar radiation incident on the earth's surface may be divided into the direct (parallel rays incident at a fixed zenith angle θ) and diffuse components (radiation that may be approximated as being diffusely distributed with θ). Consider conditions for a day in which the intensity of the direct solar radiation is $I_{dir} = 210 \times 10^7$ W/m² · sr in the solid angle subtended by the sun with respect to the earth, $\Delta\omega_s = 6.74 \times 10^{-5}$ sr. The intensity of the diffuse radiation is $I_{dif} = 70$ W/m² · sr.

(a) What is the total solar irradiation at the earth's surface when the direct radiation is incident at $\theta = 30°$?

(b) Verify the prescribed value for $\Delta\omega_s$, recognizing that the diameter of the sun is 1.39×10^9 m and the distance between the sun and the earth is 1.496×10^{11} m (1 astronomical unit).

12.10 To initiate a process operation, an infrared motion sensor (radiation detector) is employed to determine the approach of a hot part on a conveyor system. To set the sensor's amplifier discriminator, the engineer needs a relationship between the sensor output signal, S, and the position of the part on the conveyor. The sensor output signal is proportional to the rate at which radiation is incident on the sensor.

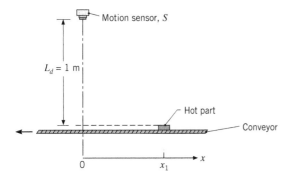

(a) For $L_d = 1$ m, at what location x_1 will the sensor signal S_1 be 75% of the signal corresponding to the position directly beneath the sensor, S_o ($x = 0$)?

(b) For values of $L_d = 0.8$, 1.0, and 1.2 m, plot the signal ratio, S/S_o, versus part position, x, for signal ratios in the range from 0.2 to 1.0. Compare the x-locations for which $S/S_o = 0.75$.

12.11 During radiant heat treatment of a thin-film material, its shape, which may be hemispherical (a) or spherical (b), is maintained by a relatively low air pressure (as in the case of a rubber balloon). Irradiation on the film is due to emission from a radiant heater of area $A_h = 0.0052$ m², which emits diffusely with an intensity of $I_{e,h} = 169,000$ W/m² · sr.

(a) Obtain an expression for the irradiation on the film as a function of the zenith angle θ.

(b) Based on the expressions derived in part (a), which shape provides the more uniform irradiation G and hence provides better quality control for the treatment process?

12.12 Determine the fraction of the total, hemispherical emissive power that leaves a diffuse surface in the directions $\pi/4 \leq \theta \leq \pi/2$ and $0 \leq \phi \leq \pi$.

12.13 Consider a 5-mm-square, diffuse surface ΔA_o having a total emissive power of $E_o = 4000$ W/m². The radiation field due to emission into the hemispherical space above the surface is diffuse, thereby providing a uniform intensity $I(\theta, \phi)$. Moreover, if the space is

a nonparticipating medium (nonabsorbing, nonscattering, and nonemitting), the intensity is independent of radius for any (θ, ϕ) direction. Hence intensities at any points P_1 and P_2 would be equal.

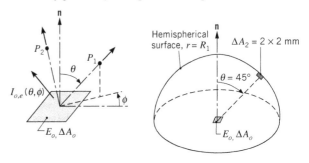

(a) What is the rate at which radiant energy is emitted by ΔA_o, q_{emit}?

(b) What is the intensity $I_{o,e}$ of the radiation field emitted from the surface ΔA_o?

(c) Beginning with Equation 12.13 and presuming knowledge of the intensity $I_{o,e}$, obtain an expression for q_{emit}.

(d) Consider the hemispherical surface located at $r = R_1 = 0.5$ m. Using the conservation of energy requirement, determine the rate at which radiant energy is incident on this surface due to emission from ΔA_o.

(e) Using Equation 12.10, determine the rate at which radiant energy leaving ΔA_o is intercepted by the small area ΔA_2 located in the direction (45°, ϕ) on the hemispherical surface. What is the irradiation on ΔA_2?

(f) Repeat part (e) for the location (0°, ϕ). Are the irradiations at the two locations equal?

(g) Using Equation 12.18, determine the irradiation G_1 on the hemispherical surface at $r = R_1$.

Blackbody Radiation

12.14 The dark surface of a ceramic stove top may be approximated as a blackbody. The "burners," which are integral with the stove top, are heated from below by electric resistance heaters.

(a) Consider a burner of diameter $D = 200$ mm operating at a uniform surface temperature of $T_s = 250°C$ in ambient air at $T_\infty = 20°C$. Without a pot or pan on the burner, what are the rates of heat loss by radiation and convection from the burner? If the efficiency associated with energy transfer from the heaters to the burners is 90%, what is the electric power requirement? At what wavelength is the spectral emission a maximum?

(b) Compute and plot the effect of the burner temperature on the heat rates for $100 \leq T_s \leq 350°C$.

12.15 A small flat plate is positioned just beyond the earth's atmosphere and is oriented such that the normal to the plate passes through the center of the sun. The diameters of the sun and earth are 1.39×10^9 and 1.27×10^7 m, respectively, and the distance between the sun and the earth is 1.5×10^{11} m.

(a) What is the solid angle subtended by the sun about a point on the surface of the plate?

(b) Determine the incident intensity, I_i, on the plate using the known value of the solar irradiation above the earth's atmosphere ($G_S = 1368$ W/m²).

(c) Sketch the incident intensity I_i as a function of the zenith angle θ, where θ is measured from the normal to the plate.

12.16 A spherical aluminum shell of inside diameter $D = 2$ m is evacuated and is used as a radiation test chamber. If the inner surface is coated with carbon black and maintained at 700 K, what is the irradiation on a small test surface placed in the chamber? If the inner surface were not coated and maintained at 700 K, what would the irradiation be?

12.17 The extremely high temperatures needed to trigger nuclear fusion are proposed to be generated by laser-irradiating a spherical pellet of deuterium and tritium fuel of diameter $D_p = 1.8$ mm.

(a) *(b)*

(a) Determine the maximum fuel temperature that can be achieved by irradiating the pellet with 200 lasers, each producing a power of $P = 500$ W. The pellet has an absorptivity $\alpha = 0.3$ and emissivity $\varepsilon = 0.8$.

(b) The pellet is placed inside a cylindrical enclosure. Two laser entrance holes are located at either end of the enclosure and have a diameter of $D_{LEH} = 3$ mm. Determine the maximum temperature that can be generated within the enclosure.

12.18 An enclosure has an inside area of 100 m², and its inside surface is black and is maintained at a constant temperature. A small opening in the enclosure has an area of 0.02 m². The radiant power emitted from this opening is 70 W. What is the temperature of the interior enclosure wall? If the interior surface is maintained at this temperature, but is now polished, what will be the value of the radiant power emitted from the opening?

12.19 Assuming the earth's surface is black, estimate its temperature if the sun has an equivalent blackbody temperature of 5800 K. The diameters of the sun and earth are 1.39×10^9 and 1.27×10^7 m, respectively, and the distance between the sun and earth is 1.5×10^{11} m.

12.20 A proposed method for generating electricity from solar irradiation is to concentrate the irradiation into a cavity that is placed within a large container of a salt with a high melting temperature. If all heat losses are neglected, part of the solar irradiation entering the cavity is used to melt the salt while the remainder is used to power a Rankine cycle. (The salt is melted during the day and is resolidified at night in order to generate electricity around the clock.)

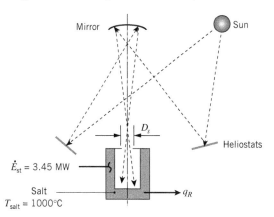

Consider conditions for which the solar power entering the cavity is $q_{sol} = 7.50$ MW and the time rate of change of energy stored in the salt is $\dot{E}_{st} = 3.45$ MW. For a cavity opening of diameter $D_s = 1$ m, determine the rate of heat transfer to the Rankine cycle, q_R. The temperature of the salt is maintained at its melting point, $T_{salt} = T_m = 1000°C$. Neglect heat loss by convection and irradiation from the surroundings.

12.21 Approximations to Planck's law for the spectral emissive power are the Wien and Rayleigh-Jeans spectral distributions, which are useful for the extreme low and high limits of the product λT, respectively.

(a) Show that the Planck distribution will have the form

$$E_{\lambda,b}(\lambda, T) \approx \frac{C_1}{\lambda^5} \exp\left(-\frac{C_2}{\lambda T}\right)$$

when $C_2/\lambda T \gg 1$ and determine the error (compared to the exact distribution) for the condition $\lambda T = 2898 \ \mu\text{m} \cdot \text{K}$. This form is known as Wien's law.

(b) Show that the Planck distribution will have the form

$$E_{\lambda,b}(\lambda, T) \approx \frac{C_1}{C_2} \frac{T}{\lambda^4}$$

when $C_2/\lambda T \ll 1$ and determine the error (compared to the exact distribution) for the condition $\lambda T = 100,000 \ \mu\text{m} \cdot \text{K}$. This form is known as the Rayleigh-Jeans law.

12.22 A furnace with a long, isothermal, graphite tube of diameter $D = 12.5 \ \text{mm}$ is maintained at $T_f = 2000 \ \text{K}$ and is used as a blackbody source to calibrate heat flux gages. Traditional heat flux gages are constructed as blackened thin films with thermopiles to indicate the temperature change caused by absorption of the incident radiant power over the entire spectrum. The traditional gage of interest has a sensitive area of 5 mm² and is mounted coaxial with the furnace centerline, but positioned at a distance of $L = 60 \ \text{mm}$ from the beginning of the heated section. The cool extension tube serves to shield the gage from extraneous radiation sources and to contain the inert gas required to prevent rapid oxidation of the graphite tube.

(a) Calculate the heat flux (W/m²) on the traditional gage for this condition, assuming that the extension tube is cold relative to the furnace.

(b) The traditional gage is replaced by a solid-state (photoconductive) heat flux gage of the same area, but sensitive only to the spectral region between 0.4 and 2.5 μm. Calculate the radiant heat flux incident on the solid-state gage within the prescribed spectral region.

(c) Calculate and plot the total heat flux and the heat flux in the prescribed spectral region for the solid-state gage as a function of furnace temperature for the range $2000 \leq T_f \leq 3000 \text{K}$. Which gage will have an output signal that is more sensitive to changes in the furnace temperature?

12.23 The table below includes the four brightest single stars visible from Earth. The star temperature, star radius (one solar radius is equal to the radius of the sun), and distance between the star and Earth are also provided. Determine the order of decreasing star brightness, as observed by an astronaut in orbit about Earth.

Star	$T_s(\text{K})$	R_s (solar radius)	$D_{s\text{-}E}$ (light years)
Arcturus	4286	25.4	36.7
Canopus	7350	71.4	310
Sirius A	9940	1.71	8.60
Vega	9602	2.50	25.0

12.24 An electrically powered, ring-shaped radiant heating element is maintained at a temperature of $T_h = 3000 \ \text{K}$ and is used in a manufacturing process to heat a small part having a surface area of $A_p = 0.007 \ \text{m}^2$. The surface of the heating element may be assumed to be black.

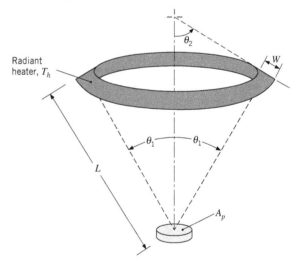

For $\theta_1 = 30°$, $\theta_2 = 60°$, $L = 3 \ \text{m}$, and $W = 30 \ \text{mm}$, what is the rate at which radiant energy emitted by the heater is incident on the part?

12.25 Isothermal furnaces with small apertures approximating a blackbody are frequently used to calibrate heat flux gages, radiation thermometers, and other radiometric devices. In such applications, it is necessary to control power to the furnace such that the variation of temperature and the spectral intensity of the aperture are within desired limits.

(a) By considering the Planck spectral distribution, Equation 12.30, show that the ratio of the fractional change in the spectral intensity to the fractional change in the temperature of the furnace has the form

$$\frac{dI_\lambda / I_\lambda}{dT / T} = \frac{C_2}{\lambda T} \frac{1}{1 - \exp(-C_2/\lambda T)}$$

(b) Using this relation, determine the allowable variation in temperature of the furnace operating at 2000 K to ensure that the spectral intensity at 0.65 μm will not vary by more than 0.5%. What is the allowable variation at 10 μm?

Emissivity

12.26 A small metal object, initially at $T_i = 1000$ K, is cooled by radiation in a low-temperature vacuum chamber. One of two thin coatings can be applied to the object so that spectral hemispherical emissivities vary with wavelength as shown. For which coating will the object most rapidly reach a temperature of $T_f = 500$ K?

12.27 Consider the metallic surface of Example 12.7. Additional measurements of the spectral, hemispherical emissivity yield a spectral distribution which may be approximated as follows:

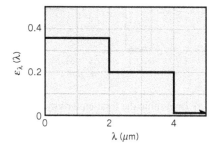

(a) Determine corresponding values of the total, hemispherical emissivity ε and the total emissive power E at 2000 K.

(b) Plot the emissivity as a function of temperature for $500 \leq T \leq 3000$ K. Explain the variation.

12.28 The directional total emissivity of nonmetallic materials may be approximated as $\varepsilon_\theta \approx \varepsilon_n \cos \theta$, where ε_n is the normal emissivity. Show that the total hemispherical emissivity for such materials is 2/3 of the normal emissivity.

12.29 The spectral, directional emissivity of a diffuse material at 2000 K has the following distribution:

Determine the total, hemispherical emissivity at 2000 K. Determine the emissive power over the spectral range 0.8 to 2.5 μm *and* for the directions $0 \leq \theta \leq 30°$.

12.30 A diffuse surface is characterized by the spectral hemispherical emissivity distribution shown. Considering surface temperatures over the range $300 \leq T_s \leq 1000$ K, at what temperature will the emissive power be minimized? Determine the surface hemispherical emissivity and emissive power at $T = 300$, 500, and 700 K. At what wavelength does the peak spectral emission occur when the surface is at these temperatures?

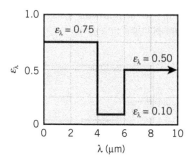

12.31 A sphere is suspended in air in a dark room and maintained at a uniform incandescent temperature. When first viewed with the naked eye, the sphere appears to be brighter around the rim. After several hours, however, it appears to be brighter in the center. Of what type material would you reason the sphere is made? Give plausible reasons for the nonuniformity of brightness of the sphere and for the changing appearance with time.

12.32 A radiation thermometer is a device that responds to a radiant flux within a prescribed spectral interval and is calibrated to indicate the temperature of a blackbody that produces the same flux.

(a) When viewing a surface at an elevated temperature T_s and emissivity less than unity, the thermometer will indicate an apparent temperature referred to as the brightness or spectral radiance temperature T_λ. Will T_λ be greater than, less than, or equal to T_s?

(b) Write an expression for the spectral emissive power of the surface in terms of Wien's spectral distribution (see Problem 12.21) and the spectral emissivity of the surface. Write the equivalent expression using the spectral radiance temperature of the surface and show that

$$\frac{1}{T_s} = \frac{1}{T_\lambda} + \frac{\lambda}{C_2} \ln \varepsilon_\lambda$$

where λ represents the wavelength at which the thermometer operates.

(c) Consider a radiation thermometer that responds to a spectral flux centered about the wavelength $0.65 \, \mu m$. What temperature will the thermometer indicate when viewing a surface with $\varepsilon_\lambda(0.65 \, \mu m) = 0.9$ and $T_s = 1000 \, K$? Verify that Wien's spectral distribution is a reasonable approximation to Planck's law for this situation.

12.33 Sheet steel emerging from the hot roll section of a steel mill has a temperature of 1200 K, a thickness of $\delta = 3 \, mm$, and the following distribution for the spectral, hemispherical emissivity.

The density and specific heat of the steel are $7900 \, kg/m^3$ and $640 \, J/kg \cdot K$, respectively. What is the total, hemispherical emissivity? Accounting for emission from both sides of the sheet and neglecting conduction, convection, and radiation from the surroundings, determine the initial time rate of change of the sheet temperature $(dT/dt)_i$. As the steel cools, it oxidizes and its total, hemispherical emissivity increases. If this increase may be correlated by an expression of

the form $\varepsilon = \varepsilon_{1200}[1200 \, K/T \, (K)]$, how long will it take for the steel to cool from 1200 to 600 K?

12.34 A large body of nonluminous gas at a temperature of 1200 K has emission bands between 2.5 and 3.5 μm and between 5 and 8 μm. The effective emissivity in the first band is 0.8 and in the second 0.6. Determine the emissive power of this gas.

Absorptivity, Reflectivity, and Transmissivity

12.35 An opaque surface with the prescribed spectral, hemispherical reflectivity distribution is subjected to the spectral irradiation shown.

(a) Sketch the spectral, hemispherical absorptivity distribution.

(b) Determine the total irradiation on the surface.

(c) Determine the radiant flux that is absorbed by the surface.

(d) What is the total, hemispherical absorptivity of this surface?

12.36 A small, opaque, diffuse object at $T_s = 400 \, K$ is suspended in a large furnace whose interior walls are at $T_f = 2000 \, K$. The walls are diffuse and gray and have an emissivity of 0.20. The spectral, hemispherical emissivity for the surface of the small object is given below.

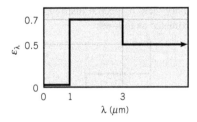

(a) Determine the total emissivity and absorptivity of the surface.

(b) Evaluate the reflected radiant flux and the net radiative flux *to* the surface.

(c) What is the spectral emissive power at $\lambda = 2 \, \mu m$?

(d) What is the wavelength $\lambda_{1/2}$ for which one-half of the total radiation emitted by the surface is in the spectral region $\lambda \geq \lambda_{1/2}$?

12.37 One of two thin coatings may be applied to a metal plate, yielding the following spectral emissivities above and below a *cutoff wavelength* of $\lambda_c = 8\ \mu m$.

Coating	$\varepsilon_\lambda\ (\lambda_c \le 8\ \mu m)$	$\varepsilon_\lambda\ (\lambda_c > 8\ \mu m)$
A	0.75	0.25
B	0.25	0.75
None	0.25	0.25

If the plate temperature is $T_s = 600$ K and the surroundings is at 400 K, determine the radiosity and net radiation heat flux at the surface of the plate for no coating, coating A, and coating B. Which coating provides the largest net radiation flux? If the cutoff wavelength were $\lambda_c = 6\ \mu m$ rather than 8 μm, which coating would provide the largest net radiation flux?

12.38 The spectral reflectivity distribution for white paint (Figure 12.22) can be approximated by the following stair-step function:

$\lambda\ (\mu m)$	<0.4	0.4–3.0	>3.0
α_λ	0.75	0.15	0.96

A small flat plate coated with this paint is suspended inside a large enclosure, and its temperature is maintained at 400 K. The surface of the enclosure is maintained at 3000 K and the spectral distribution of its emissivity has the following characteristics:

$\lambda\ (\mu m)$	<2.0	>2.0
ε_λ	0.2	0.9

(a) Determine the total emissivity, ε, of the enclosure surface.

(b) Determine the total emissivity, ε, and absorptivity, α, of the plate.

12.39 A diffuse, opaque surface at 700 K has spectral emissivities of $\varepsilon_\lambda = 0$ for $0 \le \lambda \le 3\ \mu m$, $\varepsilon_\lambda = 0.5$ for $3\ \mu m < \lambda \le 10\ \mu m$, and $\varepsilon_\lambda = 0.9$ for $10\ \mu m < \lambda < \infty$. A radiant flux of 1000 W/m², which is uniformly distributed between 1 and 6 μm, is incident on the surface at an angle of 30° relative to the surface normal.

Calculate the total radiant power from a 10^{-4} m² area of the surface that reaches a radiation detector positioned along the normal to the area. The aperture of the detector is 10^{-5} m², and its distance from the surface is 1 m.

12.40 The spectral, hemispherical absorptivity of an opaque surface is as shown.

What is the solar absorptivity, α_S? If it is assumed that $\varepsilon_\lambda = \alpha_\lambda$ and that the surface is at a temperature of 340 K, what is its total, hemispherical emissivity?

12.41 The spectral, hemispherical absorptivity of an opaque surface and the spectral distribution of radiation incident on the surface are as shown.

What is the total, hemispherical absorptivity of the surface? If it is assumed that $\varepsilon_\lambda = \alpha_\lambda$ and that the surface is at 1000 K, what is its total, hemispherical emissivity? What is the net radiant heat flux to the surface?

12.42 The spectral absorptivity α_λ and spectral reflectivity ρ_λ for a spectrally selective, diffuse material are as shown.

(a) Sketch the spectral transmissivity τ_λ.

(b) If solar irradiation with $G_S = 750$ W/m² and the spectral distribution of a blackbody at 5800 K is incident on this material, determine the fractions of the irradiation that are transmitted, reflected, and absorbed by the material.

(c) If the temperature of this material is 350 K, determine the emissivity ε.

(d) Determine the net heat flux by radiation to the material.

12.43 The spectral transmissivity of a 1-mm-thick layer of liquid water can be approximated as follows:

$$\tau_{\lambda 1} = 0.99 \qquad 0 \le \lambda \le 1.2 \ \mu m$$
$$\tau_{\lambda 2} = 0.54 \qquad 1.2 \ \mu m < \lambda \le 1.8 \ \mu m$$
$$\tau_{\lambda 3} = 0 \qquad 1.8 \ \mu m < \lambda$$

(a) Liquid water can exist only below its critical temperature, $T_c = 647.3$ K. Determine the maximum possible total transmissivity of a 1-mm-thick layer of liquid water when the water is housed in an opaque container and boiling does not occur. Assume the irradiation is that of a blackbody.

(b) Determine the transmissivity of a 1-mm-thick layer of liquid water associated with melting the platinum wire used in Nukiyama's boiling experiment, as described in Section 10.3.1.

(c) Determine the total transmissivity of a 1-mm-thick layer of liquid water exposed to solar irradiation. Assume the sun emits as a blackbody at $T_s = 5800$ K.

12.44 The spectral transmissivity of plain and tinted glass can be approximated as follows:

Plain glass: $\tau_\lambda = 0.9$ $0.3 \le \lambda \le 2.5 \ \mu m$
Tinted glass: $\tau_\lambda = 0.9$ $0.5 \le \lambda \le 1.5 \ \mu m$

Outside the specified wavelength ranges, the spectral transmissivity is zero for both glasses. Compare the solar energy that could be transmitted through the glasses. With solar irradiation on the glasses, compare the visible radiant energy that could be transmitted.

12.45 Consider a large furnace with opaque, diffuse, gray walls at 3000 K having an emissivity of 0.85. A small, diffuse, spectrally selective object in the furnace is maintained at 300 K.

For the specified points on the furnace wall (A) and the object (B), indicate values for ε, α, E, G, and J.

12.46 The spectral transmissivity of a 50-μm-thick polymer film is measured over the wavelength range 2.5 μm $\le \lambda \le 15 \ \mu$m. The spectral distribution may be approximated as $\tau_\lambda = 0.80$ for 2.5 μm $\le \lambda \le 7 \ \mu$m, $\tau_\lambda = 0.05$ for 7 μm $< \lambda \le 13 \ \mu$m, and $\tau_\lambda = 0.55$ for 13 μm $< \lambda \le 15 \ \mu$m. Transmissivity data outside the range cannot be acquired due to limitations associated with the instrumentation. An engineer wishes to determine the total transmissivity of the film.

(a) Estimate the *maximum possible* total transmissivity of the film associated with irradiation from a blackbody at $T = 30°$C.

(b) Estimate the *minimum possible* total transmissivity of the film associated with irradiation from a blackbody at $T = 30°$C.

(c) Repeat parts (a) and (b) for a blackbody at $T = 600°$C.

Radiation Fluxes and Rates

12.47 An opaque, horizontal plate has a thickness of $L = 21$ mm and thermal conductivity $k = 25$ W/m · K. Water flows adjacent to the bottom of the plate and is at a temperature of $T_{\infty,w} = 25°$C. Air flows above the plate at $T_{\infty,a} = 260°$C with $h_a = 40$ W/m^2 · K. The top of the plate is diffuse and is irradiated with $G = 1450$ W/m^2, of which 435 W/m^2 is reflected. The steady-state top and bottom plate temperatures are $T_t = 43°$C and $T_b = 35°$C, respectively. Determine the transmissivity, reflectivity, absorptivity, and emissivity of the plate. Is the plate gray? What is the radiosity associated with the top of the plate? What is the convection heat transfer coefficient associated with the water flow?

12.48 Solar flux of 1000 W/m^2 is incident on the top side of a plate whose surface has a solar absorptivity of 0.9 and an emissivity of 0.2. The air and surroundings are at 25°C and the convection heat transfer coefficient between the plate and air is 20 W/m^2 · K. Assuming that the bottom side of the plate is insulated, determine the surface convection and radiation heat fluxes.

12.49 Two small surfaces, A and B, are placed inside an isothermal enclosure at a uniform temperature. The enclosure provides an irradiation of 6300 W/m^2 to each of the surfaces, and surfaces A and B absorb incident radiation at rates of 5600 and 630 W/m^2, respectively. Consider conditions after a long time has elapsed.

(a) What are the net heat fluxes for each surface? What are their temperatures?

(b) Determine the absorptivity of each surface.

(c) What are the emissive powers of each surface?

(d) Determine the emissivity of each surface.

12.50 Consider parallel surfaces that are opaque to thermal radiation. The horizontal surfaces are separated by a gap of thickness $L = 0.12$ m. The top surface is at temperature T_2 and is coated with a high emissivity paint, $\varepsilon \approx 1$. The bottom surface is at temperature T_1 and is characterized by:

$$\varepsilon_\lambda = 0.1 \qquad 0 \leq \lambda \leq 10\ \mu m$$
$$\varepsilon_\lambda = 0.8 \qquad \lambda > 10\ \mu m$$

(a) The enclosure formed by the surfaces is filled with air at atmospheric pressure with $T_1 = 500$ K and $T_2 = 300$ K. Determine the heat flux across the enclosure.

(b) The enclosure is filled with liquid water at an average temperature of $\overline{T} = 300$ K. There is a 5 K temperature difference across the enclosure with the bottom surface being hot. Determine the heat flux across the enclosure.

12.51 A special diffuse glass with prescribed spectral radiative properties is heated in a large oven. The walls of the oven are lined with a diffuse, gray refractory brick having an emissivity of 0.75 and are maintained at $T_w = 1800$ K. Consider conditions for which the glass temperature is $T_g = 750$ K.

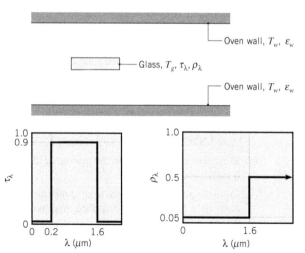

(a) What are the total transmissivity τ, the total reflectivity ρ, and the total emissivity ε of the glass?

(b) What is the net radiative heat flux, $q''_{net,in}(W/m^2)$, to the glass?

(c) For oven wall temperatures of 1500, 1800, and 2000 K, plot $q''_{net,in}$ as a function of glass temperature for $500 \leq T_g \leq 800$ K.

12.52 A small sample of an opaque surface is initially at 1200 K and has the spectral, hemispherical absorptivity shown.

The sample is placed inside a large enclosure whose walls have an emissivity of 0.2 and are maintained at 2400 K.

(a) What is the total, hemispherical absorptivity of the sample surface?

(b) What is its total, hemispherical emissivity?

(c) What are the values of the absorptivity and emissivity after the sample has been in the enclosure a long time?

(d) For a 10-mm-diameter spherical sample in an evacuated enclosure, compute and plot the variation of the sample temperature with time, as it is heated from its initial temperature of 1200 K.

12.53 The 50-mm peephole of a large furnace operating at 450°C is covered with a material having $\tau = 0.8$ and $\rho = 0$ for irradiation originating from the furnace. The material has an emissivity of 0.8 and is opaque to irradiation from a source at room temperature. The outer surface of the cover is exposed to surroundings and ambient air at 27°C with a convection heat transfer coefficient of 50 W/m² · K. Assuming that convection effects on the inner surface of the cover are negligible, calculate the heat loss by the furnace and the temperature of the cover.

12.54 A diffuse surface having the following spectral characteristics is maintained at 500 K when situated in a large furnace enclosure whose walls are maintained at 1500 K:

(a) Sketch the spectral distribution of the surface emissive power E_λ and the emissive power $E_{\lambda,b}$ that the surface would have if it were a blackbody.

(b) Neglecting convection effects, what is the net heat flux to the surface for the prescribed conditions?

(c) Plot the net heat flux as a function of the surface temperature for $500 \leq T \leq 1000$ K. On the same coordinates, plot the heat flux for a diffuse, gray surface with total emissivities of 0.4 and 0.8.

(d) For the prescribed spectral distribution of ε_λ, how do the total emissivity and absorptivity of the surface vary with temperature in the range $500 \leq T \leq 1000$ K?

12.55 Two special coatings are available for application to an absorber plate installed below the cover glass described in Example 12.9. Each coating is diffuse and is characterized by the spectral distributions shown.

Which coating would you select for the absorber plate? Explain briefly. For the selected coating, what is the rate at which radiation is absorbed per unit area of the absorber plate if the total solar irradiation at the cover glass is $G_S = 1000$ W/m^2?

12.56 Sheet steel emerging from the hot roll section of a steel mill has a temperature of 1000 K, a thickness of $\delta = 2.5$ mm, and the following distribution for the spectral, hemispherical emissivity.

The density and specific heat of the steel are 7900 kg/m^3 and 640 J/kg · K, respectively. What is the total, hemispherical emissivity? Accounting for emission from both sides of the sheet and neglecting conduction, convection, and radiation from the surroundings, determine the initial time rate of change of the sheet

temperature $(dT/dt)_i$. As the steel cools, it oxidizes and its total, hemispherical emissivity increases. If this increase may be correlated by an expression of the form $\varepsilon = \varepsilon_{1000}[1000$ K/T (K)], how long will it take for the steel to cool from 1000 to 500 K?

12.57 The window of a large vacuum chamber is fabricated from a material of prescribed spectral characteristics. A collimated beam of radiant energy from a solar simulator is incident on the window and has a flux of 3000 W/m^2. The inside walls of the chamber, which are large compared to the window area, are maintained at 77 K. The outer surface of the window is subjected to surroundings and room air at 25°C, with a convection heat transfer coefficient of 15 W/m^2 · K.

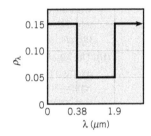

(a) Determine the transmissivity of the window material to radiation from the solar simulator, which approximates the solar spectral distribution.

(b) Assuming that the window is insulated from its chamber mounting arrangement, what steady-state temperature does the window reach?

(c) Calculate the net radiation transfer per unit area of the window to the vacuum chamber wall, excluding the transmitted simulated solar flux.

12.58 A thermocouple inserted in a 4-mm-diameter stainless steel tube having a diffuse, gray surface with an emissivity of 0.4 is positioned horizontally in a large air-conditioned room whose walls and air temperature are 30 and 20°C, respectively.

(a) What temperature will the thermocouple indicate if the air is quiescent?

(b) Compute and plot the thermocouple *measurement error* as a function of the surface emissivity for $0.1 \leq \varepsilon \leq 1.0$.

12.59 A small disk 5 mm in diameter is positioned at the center of an isothermal, hemispherical enclosure. The disk is diffuse and gray with an emissivity of 0.7 and is maintained at 900 K. The hemispherical enclosure, maintained at 300 K, has a radius of 100 mm and an emissivity of 0.85.

Calculate the radiant power leaving an aperture of diameter 2 mm located on the enclosure as shown.

12.60 Radiation leaves a furnace of inside surface temperature 1500 K through an aperture 20 mm in diameter. A portion of the radiation is intercepted by a detector that is 1 m from the aperture, has a surface area of 10^{-5} m^2, and is oriented as shown.

If the aperture is open, what is the rate at which radiation leaving the furnace is intercepted by the detector? If the aperture is covered with a diffuse, semitransparent material of spectral transmissivity $\tau_\lambda = 0.8$ for $\lambda \leq 2$ μm and $\tau_\lambda = 0$ for $\lambda > 2$ μm, what is the rate at which radiation leaving the furnace is intercepted by the detector?

12.61 Referring to the distribution of the spectral transmissivity of low iron glass (Figure 12.23), describe briefly what is meant by the "greenhouse effect." That is, how does the glass influence energy transfer to and from the contents of a greenhouse?

12.62 A cylinder of 30-mm diameter and 150-mm length is heated in a large furnace having walls at 1000 K, while air at 400 K is circulating at 3 m/s. Estimate the steady-state cylinder temperature under the following specified conditions.

(a) The cylinder is in cross flow, and its surface is diffuse and gray with an emissivity of 0.5.

(b) The cylinder is in cross flow, but its surface is spectrally selective with $\alpha_\lambda = 0.1$ for $\lambda \leq 3$ μm and $\alpha_\lambda = 0.5$ for $\lambda > 3$ μm.

(c) The cylinder surface is positioned such that the airflow is longitudinal and its surface is diffuse and gray.

(d) For the conditions of part (a), compute and plot the cylinder temperature as a function of the air velocity for $1 \leq V \leq 20$ m/s.

12.63 A temperature sensor embedded in the tip of a small tube having a diffuse, gray surface with an emissivity of 0.8 is centrally positioned within a large air-conditioned room whose walls and air temperature are 30 and 20°C, respectively.

(a) What temperature will the sensor indicate if the convection coefficient between the sensor tube and the air is 5 W/m² · K?

(b) What would be the effect of using a fan to induce airflow over the tube? Plot the sensor temperature as a function of the convection coefficient for $2 \leq h \leq 25$ W/m² · K and values of $\varepsilon = 0.2, 0.5$, and 0.8.

12.64 The directional absorptivity of a gray surface varies with θ as follows.

(a) What is the ratio of the normal absorptivity α_n to the hemispherical emissivity of the surface?

(b) Consider a plate with these surface characteristics on both sides in earth orbit. If the solar flux incident on one side of the plate is $q_S'' = 1368$ W/m², what equilibrium temperature will the plate assume if it is oriented normal to the sun's rays? What temperature will it assume if it is oriented at 75° to the sun's rays?

12.65 The spectral absorptivity of aluminum coated with a thin layer of silicon dioxide may be approximated as $\alpha_{\lambda,1} = 0.98$ for $\lambda < \lambda_c$ and $\alpha_{\lambda,2} = 0.05$ for $\lambda \geq \lambda_c$ where the *cutoff wavelength* is $\lambda_c = 0.15$ μm under normal circumstances.

(a) Determine the equilibrium temperature of a flat piece of the coated aluminum that is exposed to solar irradiation, $G_S = 1368$ W/m², on its upper surface. The opposite surface is insulated.

(b) The cutoff wavelength can be modified by varying the coating thickness. Determine the value of λ_c that will maximize the equilibrium temperature of the surface.

12.66 A sphere ($k = 185$ W/m · K, $\alpha = 7.25 \times 10^{-5}$ m^2/s) of 30-mm diameter whose surface is diffuse and gray with an emissivity of 0.8 is placed in a large oven whose walls are of uniform temperature at 600 K. The temperature of the air in the oven is 400 K, and the convection heat transfer coefficient between the sphere and the oven air is 15 W/m² · K.

(a) Determine the net heat transfer to the sphere when its temperature is 300 K.

(b) What will be the steady-state temperature of the sphere?

(c) How long will it take for the sphere, initially at 300 K, to come within 20 K of the steady-state temperature?

(d) For emissivities of 0.2, 0.4, and 0.8, plot the elapsed time of part (c) as a function of the convection coefficient for $10 \leq h \leq 25$ W/m² · K.

Radiation Detection

12.67 A thermograph is a device responding to the radiant power from the scene, which reaches its radiation detector within the spectral region 9–12 μm. The thermograph provides an image of the scene, such as the side of a furnace, from which the surface temperature can be determined.

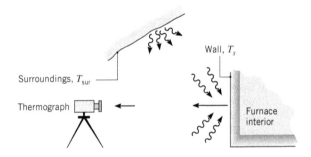

(a) For a black surface at 60°C, determine the emissive power for the spectral region 9–12 μm.

(b) Calculate the radiant power (W) received by the thermograph in the same range (9–12 μm) when viewing, in a normal direction, a small black wall area, 200 mm², at $T_s = 60$°C. The solid angle ω subtended by the aperture of the thermograph when viewed from the target is 0.001 sr.

(c) Determine the radiant power (W) received by the thermograph for the same wall area (200 mm²) and solid angle (0.001 sr) when the wall is a gray, opaque, diffuse material at $T_s = 60$°C with emissivity 0.7 and the surroundings are black at $T_{sur} = 23$°C.

12.68 A radiation detector has an aperture of area $A_d = 10^{-6}$ m² and is positioned at a distance of $r = 1$ m from a surface of area $A_s = 10^{-4}$ m². The angle formed by the normal to the detector and the surface normal is $\theta = 30$°.

The surface is at 500 K and is opaque, diffuse, and gray with an emissivity of 0.7. If the surface irradiation is 1500 W/m², what is the rate at which the detector intercepts radiation from the surface?

12.69 Consider the diffuse, gray opaque disk A_1, which has a diameter of 10 mm, an emissivity of 0.3, and is at a temperature of 400 K. Coaxial to the disk A_1, there is a black, ring-shaped disk A_2 at 1000 K having the dimensions shown in the sketch. The backside of A_2 is insulated and does not directly irradiate the cryogenically cooled detector disk A_3, which is of diameter 10 mm and is located 2 m from A_1.

Calculate the rate at which radiation is incident on A_3 due to emission and reflection from A_1.

12.70 A radiation thermometer measures irradiation reaching its sensor and is calibrated to indicate the temperature of a blackbody that produces the same flux. A steel billet having a diffuse, gray surface of emissivity 0.8 is heated in a furnace whose walls are at 1500 K. A radiation thermometer views the billet through a small hole in the furnace. Estimate the actual temperature of the billet when the radiation thermometer indicates 1160 K. Assume the thermometer is sensitive to total irradiation.

12.71 A charge-coupled device (CCD) infrared imaging system operates in a manner similar to a digital video camera. Instead of being sensitive to irradiation in the visible part of the spectrum, however, each small sensor in the infrared system's CCD array is sensitive in the spectral region 9–12 μm. Note that the system is designed to only view radiation coming from directly in front of it. An experimenter wishes to use the infrared imaging system to map the surface temperature distribution of a heated object in a wind tunnel experiment. The air temperature in the wind tunnel, as well as the surroundings temperature in the laboratory, is 23°C.

(a) In a preliminary test of the concept, the experimenter views a small aluminum billet located in the wind tunnel that is at a billet temperature of 50°C. The aluminum is coated with a high-emissivity paint, $\varepsilon = 0.96$. If the infrared imaging system is calibrated to indicate the temperature of a blackbody, what temperature will be indicated by the infrared imaging system as it is used to view the aluminum billet through a 6-mm-thick fused quartz window?

(b) In a subsequent experiment, the experimenter replaces the quartz window with a thin (130 μm-thick) household polyethylene film with $\tau \approx 0.78$ within the spectral range of the imaging system. What temperature will be indicated by the infrared imaging system when it is used to view the aluminum billet through the polyethylene film?

12.72 A diffuse, spherical object of diameter and temperature 9 mm and 600 K, respectively, has an emissivity of 0.95. Two sensitive radiation detectors, each with an aperture area of 300×10^{-6} m^2, detect the object as it passes over at high velocity from left to right as shown in the schematic. The detectors capture hemispherical irradiation and are equipped with filters characterized by $\tau_\lambda = 0.9$ for $\lambda <$ 2.5 μm and $\tau_\lambda = 0$ for $\lambda \geq 2.5$ μm. At time $t_1 = 0$, detectors A and B indicate irradiations of $G_{A,1} = 5.060$ mW/m^2 and $G_{B,1} = 5.000$ mW/m^2, respectively. At time $t_2 = 4$ ms, detectors A and B indicate irradiations of $G_{A,2} = 5.010$ mW/m^2 and $G_{B,2} = 5.050$ mW/m^2, respectively. The environment is at 300 K. Determine the velocity components of the object, v_x and v_y. Determine when and where the object will strike a horizontal plane located at $y = 0$. *Hint:* The object is located at an elevation above $y = 2$ m when it is detected. Assume the object's trajectory is a straight line in the plane of the page. Recall that the projected area of a sphere is a circle.

12.73 A radiation detector having a sensitive area of $A_d = 4 \times 10^{-6}$ m^2 is configured to receive radiation from a target area of diameter $D_t = 40$ mm when located a distance of $L_t = 1$ m from the target. For the experimental apparatus shown in the sketch, we wish to determine the emitted radiation from a hot sample of diameter $D_s = 20$ mm. The temperature of the aluminum sample is $T_s = 700$ K and its emissivity is $\varepsilon_s = 0.1$. A ring-shaped cold shield is provided to minimize the effect of radiation from outside the sample area, but within the target area. The sample and the shield are diffuse emitters.

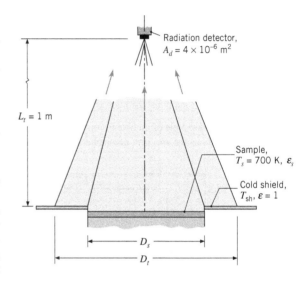

(a) Assuming the shield is black, at what temperature, T_{sh}, should the shield be maintained so that its emitted radiation is 1% of the total radiant power received by the detector?

(b) Subject to the parametric constraint that radiation emitted from the cold shield is 0.05, 1, or 1.5% of the total radiation received by the detector, plot the required cold shield temperature, T_{sh}, as a function of the sample emissivity for $0.05 \leq \varepsilon_s \leq 0.35$.

Applications

12.74 Square plates freshly sprayed with an epoxy paint must be cured at 140°C for an extended period of time. The plates are located in a large enclosure and heated by a bank of infrared lamps. The top surface of each plate has an emissivity of $\varepsilon = 0.8$ and experiences convection with a ventilation airstream that is at $T_\infty = 27$°C and provides a convection coefficient of $h = 20$ W/m$^2 \cdot$ K. The irradiation from the enclosure walls is estimated to be $G_{wall} = 450$ W/m^2, for which the plate absorptivity is $\alpha_{wall} = 0.7$.

(a) Determine the irradiation that must be provided by the lamps, G_{lamp}. The absorptivity of the plate surface for this irradiation is $\alpha_{lamp} = 0.6$.

(b) For convection coefficients of $h = 15$, 20, and 30 $W/m^2 \cdot K$, plot the lamp irradiation, G_{lamp}, as a function of the plate temperature, T_s, for $100 \leq T_s \leq 300°C$.

(c) For convection coefficients in the range from 10 to 30 $W/m^2 \cdot K$ and a lamp irradiation of $G_{lamp} = 3000$ W/m^2, plot the airstream temperature T_∞ required to maintain the plate at $T_s = 140°C$.

12.75 A very small sample of an opaque surface is initially at 1200 K and has the spectral, hemispherical absorptivity shown.

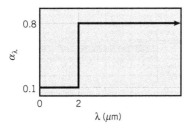

The sample is placed inside a very large enclosure whose walls have an emissivity of 0.2 and are maintained at 2400 K.

(a) What is the total, hemispherical absorptivity of the sample surface?

(b) What is its total, hemispherical emissivity?

(c) What are the values of the absorptivity and emissivity after the sample has been in the enclosure a long time?

(d) For a 10-mm-diameter spherical sample in an evacuated enclosure, compute and plot the variation of the sample temperature with time, as it is heated from its initial temperature of 1200 K.

12.76 A procedure for measuring the thermal conductivity of solids at elevated temperatures involves placement of a sample at the bottom of a large furnace. The sample is of thickness L and is placed in a square container of width W on a side. The sides are well insulated. The walls of the cavity are maintained at T_w, while the bottom

surface of the sample is maintained at a much lower temperature T_c by circulating coolant through the sample container. The sample surface is diffuse and gray with an emissivity ε_s. Its temperature T_s is measured optically.

(a) Neglecting convection effects, obtain an expression from which the sample thermal conductivity may be evaluated in terms of measured and known quantities (T_w, T_s, T_c, ε_s, L). The measurements are made under steady-state conditions. If $T_w = 1400$ K, $T_s = 1000$ K, $\varepsilon_s = 0.85$, $L = 0.015$ m, and $T_c = 300$ K, what is the sample thermal conductivity?

(b) If $W = 0.10$ m and the coolant is water with a flow rate of $\dot{m}_c = 0.1$ kg/s, is it reasonable to assume a uniform bottom surface temperature T_c?

12.77 Photovoltaic materials convert sunlight directly to electric power. Some of the photons that are incident upon the material displace electrons that are in turn collected to create an electric current. The overall efficiency of a photovoltaic panel, η, is the ratio of electrical energy produced to the energy content of the incident radiation. The efficiency depends primarily on two properties of the photovoltaic material, (i) the *band gap*, which identifies the energy states of photons having the potential to be converted to electric current, and (ii) the interband gap conversion efficiency, η_{bg}, which is the fraction of the total energy of photons within the band gap that is converted to electricity. Therefore, $\eta = \eta_{bg}F_{bg}$ where F_{bg} is the fraction of the photon energy incident on the surface within the band gap. Photons that are either outside the material's band gap or within the band gap but not converted to electrical energy are either reflected from the panel or absorbed and converted to thermal energy.

Consider a photovoltaic material with a band gap of $1.1 \leq B \leq 1.8$ eV, where B is the energy state of a photon. The wavelength is related to the energy state of a photon by the relationship $\lambda = 1240$ eV $\cdot$ nm$/B$.

The incident solar irradiation approximates that of a blackbody at 5800 K and $G_S = 1000$ W/m^2.

(a) Determine the wavelength range of solar irradiation corresponding to the band gap.

(b) Determine the overall efficiency of the photovoltaic material if the interband gap efficiency is $\eta_{bg} = 0.50$.

(c) If half of the incident photons that are not converted to electricity are absorbed and converted to thermal energy, determine the rate of heat absorption per unit surface area of the panel.

12.78 One scheme for extending the operation of gas turbine blades to higher temperatures involves applying a ceramic coating to the surfaces of blades fabricated from a superalloy such as inconel. To assess the reliability of such coatings, an apparatus has been developed for testing samples under laboratory conditions. The sample is placed at the bottom of a large vacuum chamber whose walls are cryogenically cooled and which is equipped with a radiation detector at the top surface. The detector has a surface area of $A_d = 10^{-5}$ m^2, is located at a distance of $L_{s-d} = 1$ m from the sample, and views radiation originating from a portion of the ceramic surface having an area of $\Delta A_c = 10^{-4}$ m^2. An electric heater attached to the bottom of the sample dissipates a uniform heat flux, q_h'', which is transferred upward through the sample. The bottom of the heater and sides of the sample are well insulated.

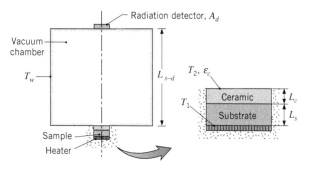

Consider conditions for which a ceramic coating of thickness $L_c = 0.5$ mm and thermal conductivity $k_c = 6$ W/m · K has been sprayed on a metal substrate of thickness $L_s = 8$ mm and thermal conductivity $k_s = 25$ W/m · K. The opaque surface of the ceramic may be approximated as diffuse and gray, with a total, hemispherical emissivity of $\varepsilon_c = 0.8$.

(a) Consider steady-state conditions for which the bottom surface of the substrate is maintained at $T_1 = 1500$ K, while the chamber walls (including the surface of the radiation detector) are maintained at $T_w = 90$ K. Assuming negligible thermal

contact resistance at the ceramic–substrate interface, determine the ceramic top surface temperature T_2 and the heat flux q_h''.

(b) For the prescribed conditions, what is the rate at which radiation emitted by the ceramic is intercepted by the detector?

(c) After repeated experiments, numerous cracks develop at the ceramic–substrate interface, creating an interfacial thermal contact resistance. If T_w and q_h'' are maintained at the conditions associated with part (a), will T_1 increase, decrease, or remain the same? Similarly, will T_2 increase, decrease, or remain the same? In each case, justify your answer.

12.79 A laser-materials-processing apparatus encloses a sample in the form of a disk of diameter $D = 25$ mm and thickness $w = 1$ mm. The sample has a diffuse surface for which the spectral distribution of the emissivity, $\varepsilon_\lambda(\lambda)$, is prescribed. To reduce oxidation, an inert gas stream of temperature $T_\infty = 500$ K and convection coefficient $h = 50$ W/m^2 · K flows over the sample upper and lower surfaces. The apparatus enclosure is large, with isothermal walls at $T_{enc} = 300$ K. To maintain the sample at a suitable operating temperature of $T_s = 2000$ K, a collimated laser beam with an operating wavelength of $\lambda = 0.5$ μm irradiates its upper surface.

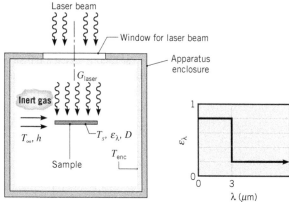

(a) Determine the total emissivity ε of the sample.

(b) Determine the total absorptivity α of the sample for irradiation from the enclosure walls.

(c) Perform an energy balance on the sample and determine the laser irradiation, G_{laser}, required to maintain the sample at $T_s = 2000$ K.

(d) Consider a *cool-down* process, when the laser and the inert gas flow are deactivated. Sketch the total emissivity as a function of the sample temperature, $T_s(t)$, during the process. Identify key features, including the emissivity for the final condition ($t \rightarrow \infty$).

(e) Estimate the time to cool a sample from its operating condition at $T_s(0) = 2000$ K to a *safe-to-touch* temperature of $T_s(t) = 40°C$. Use the lumped capacitance method and include the effect of convection to the inert gas with $h = 50$ W/m² · K and $T_\infty = T_{enc} = 300$ K. The thermophysical properties of the sample material are $\rho = 3900$ kg/m³, $c_p = 760$ J/kg · K, and $k = 45$ W/m · K.

12.80 A cylinder of 30-mm diameter and 150-mm length is heated in a large furnace having walls at 1000 K, while air at 400 K is circulating at 3 m/s. Estimate the steady-state cylinder temperature under the following specified conditions.

(a) The cylinder is in cross flow, and its surface is diffuse and gray with an emissivity of 0.5.

(b) The cylinder is in cross flow, but its surface is spectrally selective with $\alpha_\lambda = 0.1$ for $\lambda \leq 3$ μm and $\alpha_\lambda = 0.5$ for $\lambda > 3$ μm.

(c) The cylinder surface is positioned such that the airflow is longitudinal and its surface is diffuse and gray.

(d) For the conditions of part (a), compute and plot the cylinder temperature as a function of the air velocity for $1 \leq V \leq 20$ m/s.

12.81 A thin coating, which is applied to long, cylindrical copper rods of 10-mm diameter, is cured by placing the rods horizontally in a large furnace whose walls are maintained at 1300 K. The furnace is filled with nitrogen gas, which is also at 1300 K and at a pressure of 1 atm. The coating is diffuse, and its spectral emissivity has the distribution shown.

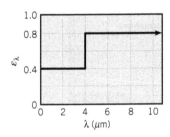

(a) What are the emissivity and absorptivity of the coated rods when their temperature is 300 K?

(b) What is the initial rate of change of their temperature?

(c) What are the emissivity and absorptivity of the coated rods when they reach a steady-state temperature?

(d) Estimate the time required for the rods to reach 1000 K.

12.82 A large combination convection–radiation oven is used to heat-treat a small cylindrical product of diameter 25 mm and length 0.2 m. The oven walls are at a uniform temperature of 1000 K, and hot air at 750 K is in cross flow over the cylinder with a velocity of 5 m/s. The cylinder surface is opaque and diffuse with the spectral emissivity shown.

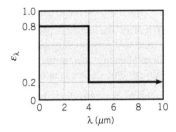

(a) Determine the rate of heat transfer to the cylinder when it is first placed in the oven at 300 K.

(b) What is the steady-state temperature of the cylinder?

(c) How long will it take for the cylinder to reach a temperature that is within 50°C of its steady-state value?

12.83 A 10-mm-thick workpiece, initially at 25°C, is to be annealed at a temperature above 725°C for a period of at least 5 minutes and then cooled. The workpiece is opaque and diffuse, and the spectral distribution of its emissivity is shown schematically. Heating is effected in a large furnace with walls and circulating air at 750°C and a convection coefficient of 100 W/m² · K. The thermophysical properties of the workpiece are $\rho = 2700$ kg/m³, $c = 885$ J/kg · K, and $k = 165$ W/m · K.

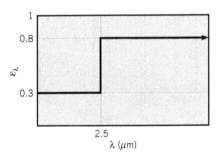

(a) Calculate the emissivity and the absorptivity of the workpiece when it is placed in the furnace at its initial temperature of 25°C.

(b) Determine the net heat flux into the workpiece for this initial condition. What is the corresponding rate of change in temperature, dT/dt, for the workpiece?

(c) Calculate the time for the workpiece to cool from 750°C to a safe-to-touch temperature of 40°C, if the surroundings and cooling air temperature are 25°C and the convection coefficient is 100 W/m² · K.

12.84 In the central receiver concept of solar energy collection, a large number of heliostats (reflectors) provide a concentrated solar flux of $q_s'' = 80,000$ W/m^2 to the receiver, which is positioned at the top of a tower.

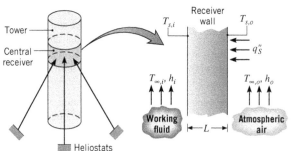

The receiver wall is exposed to the solar flux at its outer surface and to atmospheric air for which $T_{\infty,o} = 300$ K and $h_o = 25$ W/m$^2 \cdot$ K. The outer surface is opaque and diffuse, with a spectral absorptivity of $\alpha_\lambda = 0.9$ for $\lambda < 3$ μm and $\alpha_\lambda = 0.2$ for $\lambda > 3$ μm. The inner surface is exposed to a working fluid (a pressurized liquid) for which $T_{\infty,i} = 700$ K and $h_i = 1000$ W/m$^2 \cdot$ K. The outer surface is also exposed to surroundings for which $T_{\text{sur}} = 300$ K. If the wall is fabricated from a high-temperature material for which $k = 15$ W/m $\cdot$ K, what is the minimum thickness L needed to ensure that the outer surface temperature does not exceed $T_{s,o} = 1000$ K? What is the collection efficiency associated with this thickness?

Space Radiation

12.85 Two plates, one with a black painted surface and the other with a special coating (chemically oxidized copper) are in earth orbit and are exposed to solar radiation. The solar rays make an angle of 30° with the normal to the plate. Estimate the equilibrium temperature of each plate assuming they are diffuse and that the solar flux is 1368 W/m^2. The spectral absorptivity of the black painted surface can be approximated by $\alpha_\lambda = 0.95$ for $0 \le \lambda \le \infty$ and that of the special coating by $\alpha_\lambda = 0.95$ for $0 \le \lambda < 3$ μm and $\alpha_\lambda = 0.05$ for $\lambda \ge 3$ μm.

12.86 A radiator on a proposed satellite solar power station must dissipate heat being generated within the satellite by radiating it into space. The radiator surface has a solar absorptivity of 0.5 and an emissivity of 0.95. What is the equilibrium surface temperature when the solar irradiation is 1000 W/m^2 and the required heat dissipation is 1500 W/m^2?

12.87 An annular fin of thickness t is used as a radiator to dissipate heat for a space power system. The fin is insulated on the bottom and may be exposed to solar irradiation G_S. The fin is coated with a diffuse, spectrally selective material whose spectral reflectivity is specified.

Heat is conducted to the fin through a solid rod of radius r_i, and the exposed upper surface of the fin radiates to free space, which is essentially at absolute zero temperature.

(a) If conduction through the rod maintains a fin base temperature of $T(r_i) = T_b = 400$ K and the fin efficiency is 100%, what is the rate of heat dissipation for a fin of radius $r_o = 0.5$ m? Consider two cases, one for which the radiator is exposed to the sun with $G_S = 1000$ W/m^2 and the other with no exposure ($G_S = 0$).

(b) In practice, the fin efficiency will be less than 100% and its temperature will decrease with increasing radius. Beginning with an appropriate control volume, derive the differential equation that determines the steady-state, radial temperature distribution in the fin. Specify appropriate boundary conditions.

12.88 A rectangular plate of thickness t, length L, and width W is proposed for use as a radiator in a spacecraft application. The plate material has a thermal conductivity of 300 W/m $\cdot$ K, a solar absorptivity of 0.45, and an emissivity of 0.9. The radiator is exposed to solar radiation only on its top surface, while both surfaces are exposed to deep space at a temperature of 4 K.

(a) If the base of the radiator is maintained at $T_b = 80°C$, what is its tip temperature and the rate of heat rejection? Use a computer-based, finite-difference method with a space increment of 0.1 m to obtain your solution.

(b) Repeat the calculation of part (a) for the case when the space ship is on the dark side of the earth and is not exposed to the sun.

(c) Use your computer code to calculate the heat rate and tip temperature for $G_S = 0$ and an extremely large value of the thermal conductivity. Compare your results to those obtained from a hand calculation that assumes the radiator to be at a uniform temperature T_b. What other approach might you use to validate your code?

12.89 A spherical satellite of diameter D is in orbit about the earth and is coated with a diffuse material for which the spectral absorptivity is $\alpha_\lambda = 0.6$ for $\lambda \leq 3$ μm and $\alpha_\lambda = 0.3$ for $\lambda > 3$ μm. Instead of the entire satellite being coated with a material that is spectrally selective, half of the satellite is covered with a diffuse gray coating characterized by $\alpha_1 = 0.6$. The other half of the satellite is coated with a diffuse gray material with $\alpha_2 = 0.3$.

(a) Determine the steady-state satellite temperature when the satellite is on the bright side of the earth with the high-absorptivity coating facing the sun. Determine the steady-state satellite temperature when the low-absorptivity coating faces the sun. *Hint:* Assume one hemisphere of the satellite is irradiated by the sun and the opposite hemisphere is irradiated by the earth.

(b) Determine the steady-state satellite temperature when the satellite is on the dark side of the earth with the high-absorptivity coating facing the earth. Determine the steady-state satellite temperature when the low-absorptivity coating faces the earth.

(c) Identify a scheme to minimize the temperature variations of the satellite as it travels between the bright and dark sides of the earth.

12.90 A spherical capsule of 3-m radius is fired from a space platform in earth orbit, such that it travels toward the center of the sun at 16,000 km/s. Assume that the capsule is a lumped capacitance body with a density–specific heat product of 4×10^6 J/m^3 · K and that its surface is black.

(a) Derive a differential equation for predicting the capsule temperature as a function of time. Solve this equation to obtain the temperature as a function of time in terms of capsule parameters and its initial temperature T_i.

(b) If the capsule begins its journey at 20°C, predict the position of the capsule relative to the sun at which its destruction temperature, 150°C, is reached.

12.91 A solar panel mounted on a spacecraft has an area of 1 m^2 and a solar-to-electrical power conversion efficiency of 12%. The side of the panel with the photovoltaic array has an emissivity of 0.8 and a solar absorptivity of 0.8. The back side of the panel has an emissivity of 0.7. The array is oriented normal to solar irradiation of 1500 W/m^2.

(a) Determine the steady-state temperature of the apanel and the electrical power (W) produced for the prescribed conditions.

(b) If the panel were a thin plate without the solar cells, but with the same radiative properties, determine the temperature of the plate for the prescribed conditions. Compare this result with that from part (a). Are they the same or different? Explain why.

(c) Determine the temperature of the solar panel 1500 s after the spacecraft is eclipsed by a planet. The thermal capacity of the panel per unit area is 9000 J/m^2 · K.

Environmental Radiation

12.92 Solar irradiation of 1100 W/m^2 is incident on a large, flat, horizontal metal roof on a day when the wind blowing over the roof causes a convection heat transfer coefficient of 25 W/m^2 · K. The outside air temperature is 27°C, the metal surface absorptivity for incident solar radiation is 0.60, the metal surface emissivity is 0.20, and the roof is well insulated from below.

(a) Estimate the roof temperature under steady-state conditions.

(b) Explore the effect of changes in the absorptivity, emissivity, and convection coefficient on the steady-state temperature.

12.93 Consider clear sky conditions for which the direct radiation is incident at $\theta = 30°$, with a total flux (based on an area that is normal to the rays) of $q''_{\text{dir}} = 1000$ W/m^2, and the total intensity of the diffuse radiation is $I_{\text{dif}} = 70$ W/m^2 · sr. What is the total solar irradiation at the earth's surface?

12.94 On an overcast day the directional distribution of the solar radiation incident on the earth's surface may be approximated by an expression of the form $I_i = I_n \cos \theta$, where I_n is the total intensity of radiation directed normal to the surface and θ is the zenith angle. What is the value of I_n when the incident solar irradiation is $G = 200 \text{ W/m}^2$?

12.95 A deep cavity of 50-mm diameter approximates a blackbody and is maintained at 250°C while exposed to solar irradiation of 800 W/m² and surroundings and ambient air at 25°C. A thin window of spectral transmissivity and reflectivity 0.9 and 0, respectively, for the spectral range 0.2 to 4 μm is placed over the cavity opening. In the spectral range beyond 4 μm, the window behaves as an opaque, diffuse, gray body of emissivity 0.95. Assuming that the convection coefficient on the upper surface of the window is 10 W/m² · K, determine the temperature of the window and the power required to maintain the cavity at 250°C.

12.96 Solar flux of 900 W/m² is incident on the top side of a plate whose surface has a solar absorptivity of 0.9 and an emissivity of 0.1. The air and surroundings are at 17°C and the convection heat transfer coefficient between the plate and air is 20 W/m² · K. Assuming that the bottom side of the plate is insulated, determine the steady-state temperature of the plate.

12.97 Consider an opaque, gray surface whose directional absorptivity is 0.8 for $0 \le \theta \le 60°$ and 0.1 for $\theta > 60°$. The surface is horizontal and exposed to solar irradiation comprised of direct and diffuse components.

(a) What is the surface absorptivity to direct solar radiation that is incident at an angle of 45° from the normal? What is the absorptivity to diffuse irradiation?

(b) Neglecting convection heat transfer between the surface and the surrounding air, what would be the equilibrium temperature of the surface if the direct

and diffuse components of the irradiation were 600 and 100 W/m², respectively? The back side of the surface is insulated.

12.98 The absorber plate of a solar collector may be coated with an opaque material for which the spectral, directional absorptivity is characterized by relations of the form

$$\alpha_{\lambda,\theta}(\lambda, \theta) = \alpha_1 \cos \theta \qquad \lambda < \lambda_c$$
$$\alpha_{\lambda,\theta}(\lambda, \theta) = \alpha_2 \qquad \lambda > \lambda_c$$

The zenith angle θ is formed by the sun's rays and the plate normal, and α_1 and α_2 are constants.

(a) Obtain an expression for the total, hemispherical absorptivity, α_S, of the plate to solar radiation incident at $\theta = 45°$. Evaluate α_S for $\alpha_1 = 0.93$, $\alpha_2 = 0.25$, and a cut-off wavelength of $\lambda_c = 2 \text{ μm}$.

(b) Obtain an expression for the total, hemispherical emissivity ε of the plate. Evaluate ε for a plate temperature of $T_p = 60°C$ and the prescribed values of α_1, α_2, and λ_c.

(c) For a solar flux of $q_S'' = 1000 \text{ W/m}^2$ incident at $\theta = 45°$ and the prescribed values of α_1, α_2, λ_c, and T_p, what is the net radiant heat flux, q_{net}'', to the plate?

(d) Using the prescribed conditions and the *Radiation/Band Emission Factor* option in the *Tools* section of *IHT* to evaluate $F_{(0 \to \lambda_c)}$, explore the effect of λ_c on α_S, ε, and q_{net}'' for the wavelength range $0.7 \le \lambda_c \le 5 \text{ μm}$.

12.99 It is not uncommon for the night sky temperature in desert regions to drop to −40°C. If the ambient air temperature is 20°C and the convection coefficient for still air conditions is approximately 5 W/m² · K, can a shallow pan of water freeze?

12.100 Irradiation of the earth's surface from the atmosphere exhibits strong spectral variation, mainly associated with the atmospheric window described in Section 12.9.2.

(a) For the range $0 \le \lambda < 8 \text{ μm}$ the spectral emissivity of the sky under normal conditions can be approximated as $\varepsilon_{\lambda,1} = 0.90$, while $\varepsilon_{\lambda,3} = 0.85$ for $\lambda > 13 \text{ μm}$. Within the atmospheric window, 8 μm $\le \lambda \le 13 \text{ μm}$, the spectral emissivity is approximately $\varepsilon_{\lambda,2} = 0.05$, 0.8, and 0.9 for clear, moderately cloudy, and cloudy skies. Calculate the effective sky temperature corresponding to an actual atmosphere temperature of 280 K under the three sky conditions.

(b) Little water exists in the atmosphere under the cold, cloudless conditions of Antarctica where spectral emissivities have been measured to be approximately $\varepsilon_{\lambda,1} = 0.75$, $\varepsilon_{\lambda,2} = 0.03$, and $\varepsilon_{\lambda,3} = 0.75$

corresponding to an actual atmosphere temperature of 220 K. Determine the sky temperature under these extreme conditions. Explain how the coldest surface temperatures recorded on Earth have been found in Antarctica with values less than 200 K.

12.101 Plant leaves possess small channels that connect the interior moist region of the leaf to the environment. The channels, called *stomata*, pose the primary resistance to moisture transport through the entire plant, and the diameter of an individual stoma is sensitive to the level of CO_2 in the atmosphere. Consider a leaf of corn (maize) whose top surface is exposed to solar irradiation of $G_S = 600$ W/m^2 and an effective sky temperature of $T_{sky} = 0°C$. The bottom side of the leaf is irradiated from the ground which is at a temperature of $T_g = 20°C$. Both the top and bottom of the leaf are subjected to convective conditions characterized by $h = 35$ W/m$^2 \cdot$K, $T_\infty = 25°C$ and also experience evaporation through the stomata. Assuming the evaporative flux of water vapor is 50×10^{-6} kg/m$^2 \cdot$s under rural atmospheric CO_2 concentrations and is reduced to 5×10^{-6} kg/m$^2 \cdot$s when ambient CO_2 concentrations are doubled near an urban area, calculate the leaf temperature in the rural and urban locations. The heat of vaporization of water is $h_{fg} = 2400$ kJ/kg and assume $\alpha = \varepsilon = 0.97$ for radiation exchange with the sky and the ground, and $\alpha_S = 0.76$ for solar irradiation.

12.102 A thin sheet of glass is used on the roof of a greenhouse and is irradiated as shown.

The irradiation comprises the total solar flux G_S, the flux G_{atm} due to atmospheric emission (sky radiation), and the flux G_i due to emission from interior surfaces. The fluxes G_{atm} and G_i are concentrated in the far IR region ($\lambda \gtrsim 8$ μm). The glass may also exchange energy by convection with the outside and inside atmospheres. The glass may be assumed to be totally transparent for $\lambda < 1$ μm ($\tau_\lambda = 1.0$ for $\lambda < 1$ μm) and opaque, with $\alpha_\lambda = 1.0$ for $\lambda \geq 1$ μm.

(a) Assuming steady-state conditions, with all radiative fluxes uniformly distributed over the surfaces and the glass characterized by a uniform tempera-

ture T_g, write an appropriate energy balance for a unit area of the glass.

(b) For $T_g = 27°C$, $h_i = 10$ W/m$^2 \cdot$K, $G_S = 1100$ W/m^2, $T_{\infty,o} = 24°C$, $h_o = 55$ W/m$^2 \cdot$K, $G_{atm} = 250$ W/m^2, and $G_i = 440$ W/m^2, calculate the temperature of the greenhouse ambient air, $T_{\infty,i}$.

12.103 A solar furnace consists of an evacuated chamber with transparent windows, through which concentrated solar radiation is passed. Concentration may be achieved by mounting the furnace at the focal point of a large curved reflector that tracks radiation incident directly from the sun. The furnace may be used to evaluate the behavior of materials at elevated temperatures, and we wish to design an experiment to assess the durability of a diffuse, spectrally selective coating for which $\alpha_\lambda = 0.95$ in the range $\lambda \leq 4.5$ μm and $\alpha_\lambda = 0.03$ for $\lambda > 4.5$ μm. The coating is applied to a plate that is suspended in the furnace.

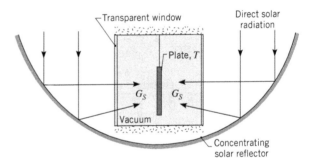

(a) If the experiment is to be operated at a steady-state plate temperature of $T = 2000$ K, how much solar irradiation G_S must be provided? The irradiation may be assumed to be uniformly distributed over the plate surface, and other sources of incident radiation may be neglected.

(b) The solar irradiation may be *tuned* to allow operation over a range of plate temperatures. Compute and plot G_S as a function of temperature for $500 \leq T \leq 3000$ K. Plot the corresponding values of α and ε as a function of T for the designated range.

12.104 Growers use giant fans to prevent grapes from freezing when the effective sky temperature is low. The grape, which may be viewed as a thin skin of negligible thermal resistance enclosing a volume of sugar water, is exposed to ambient air and is irradiated from the sky above and ground below. Assume the grape to be an isothermal sphere of 15-mm diameter, and assume uniform blackbody irradiation over its top and bottom hemispheres due to emission from the sky and the earth, respectively.

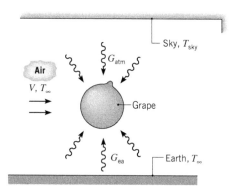

(a) Derive an expression for the rate of change of the grape temperature. Express your result in terms of a convection coefficient and appropriate temperatures and radiative quantities.

(b) Under conditions for which $T_{sky} = 235$ K, $T_\infty = 273$ K, and the fan is off ($V = 0$), determine whether the grapes will freeze. To a good approximation, the skin emissivity is 1 and the grape thermophysical properties are those of sugarless water. However, because of the sugar content, the grape freezes at $-5°C$.

(c) With all conditions remaining the same, except that the fans are now operating with $V = 1$ m/s, will the grapes freeze?

12.105 A circular metal disk having a diameter of 0.4 m is placed firmly against the ground in a barren horizontal region where the earth is at a temperature of 280 K. The effective sky temperature is also 280 K. The disk is exposed to quiescent ambient air at 300 K and direct solar irradiation of 745 W/m². The surface of the disk is diffuse with $\varepsilon_\lambda = 0.9$ for $0 < \lambda < 1$ μm and $\varepsilon_\lambda = 0.2$ for $\lambda > 1$ μm. After some time has elapsed, the disk achieves a uniform, steady-state temperature. The thermal conductivity of the soil is 0.52 W/m · K.

(a) Determine the fraction of the incident solar irradiation that is absorbed.

(b) What is the emissivity of the disk surface?

(c) For a steady-state disk temperature of 340 K, employ a suitable correlation to determine the average free convection heat transfer coefficient at the upper surface of the disk.

(d) Show that a disk temperature of 340 K does indeed yield a steady-state condition for the disk.

12.106 Consider a thin, opaque, horizontal plate with an electrical heater on its back side. The front side is exposed to ambient air that is at 20°C and provides a convection heat transfer coefficient of 20 W/m² · K, solar irradiation of 800 W/m², and an effective sky temperature of −40°C.

What is the electrical power (W/m²) required to maintain the plate surface temperature at $T_s = 60°C$ if the plate is diffuse and has the designated spectral, hemispherical reflectivity?

12.107 The exposed surface of a power amplifier for an earth satellite receiver of area 130 mm × 130 mm has a diffuse, gray, opaque coating with an emissivity of 0.5. For typical amplifier operating conditions, the surface temperature is 58°C under the following environmental conditions: air temperature, $T_\infty = 27°C$; sky temperature, $T_{sky} = -20°C$; convection coefficient, $h = 15$ W/m² · K; and solar irradiation, $G_S = 800$ W/m².

(a) For the above conditions, determine the electrical power being generated within the amplifier.

(b) It is desired to reduce the surface temperature by applying one of the diffuse coatings (A, B, C) shown as follows.

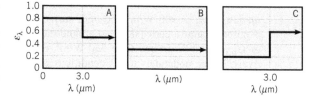

Which coating will result in the coolest surface temperature for the same amplifier operating and environmental conditions?

Heat and Mass Transfer

12.108 It is known that on clear nights a thin layer of water on the ground will freeze before the air temperature drops below 0°C. Consider such a layer of water on a clear night for which the effective sky temperature is −30°C and the convection heat transfer coefficient due to wind motion is $h = 25 \text{ W/m}^2 \cdot \text{K}$. The water may be assumed to have an emissivity of 1.0 and to be insulated from the ground as far as conduction is concerned. Neglecting evaporation, determine the lowest temperature that the air can have without the water freezing. Accounting now for the effect of evaporation, what is the lowest temperature that the air can have without the water freezing? Assume the air to be dry.

12.109 A shallow layer of water is exposed to the natural environment as shown.

Consider conditions for which the solar and atmospheric irradiations are $G_S = 600 \text{ W/m}^2$ and $G_{atm} = 300 \text{ W/m}^2$, respectively, and the air temperature and relative humidity are $T_\infty = 27°C$ and $\phi_\infty = 0.50$, respectively. The reflectivities of the water surface to the solar and atmospheric irradiation are $\rho_S = 0.3$ and $\rho_{atm} = 0$, respectively, while the surface emissivity is $\varepsilon = 0.97$. The convection heat transfer coefficient at the air–water interface is $h = 25 \text{ W/m}^2 \cdot \text{K}$. If the water is at 27°C, will this temperature increase or decrease with time?

12.110 A wet towel hangs on a clothes line under conditions for which one surface receives solar irradiation of $G_S = 900 \text{ W/m}^2$ and both surfaces are exposed to atmospheric (sky) and ground radiation of $G_{atm} = 200 \text{ W/m}^2$ and $G_g = 250 \text{ W/m}^2$, respectively. Under moderately windy conditions, airflow at a temperature of 27°C and a relative humidity of 60% maintains a convection heat transfer coefficient of 20 W/m² · K at both surfaces. The wet towel has an emissivity of 0.96 and a solar absorptivity of 0.65. As a first approximation the properties of the atmospheric air may be evaluated at a temperature of 300 K.

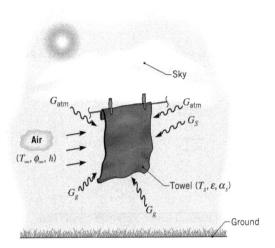

Determine the temperature T_s of the towel. What is the corresponding evaporation rate for a towel that is 0.75 m wide by 1.50 m long?

12.111 Our students perform a laboratory experiment to determine mass transfer from a wet paper towel experiencing forced convection and irradiation from radiant lamps. For the values of T_∞ and T_{wb} prescribed on the sketch, the towel temperature was found to be $T_s = 310 \text{ K}$. In addition, flat-plate correlations yielded average heat and mass transfer convection coefficients of $\bar{h} = 28.7 \text{ W/m}^2 \cdot \text{K}$ and $\bar{h}_m = 0.027 \text{ m/s}$, respectively. The towel has dimensions of $92.5 \text{ mm} \times 92.5 \text{ mm}$ and is diffuse and gray with an emissivity of 0.96.

(a) From the foregoing results, determine the vapor densities, $\rho_{A,s}$ and $\rho_{A,\infty}$, the evaporation rate, n_A (kg/s), and the net rate of radiation transfer to the towel, q_{rad} (W).

(b) Using results from part (a) and assuming that the irradiation G is uniform over the towel, determine the emissive power E, the irradiation G, and the radiosity J.

Radiation Exchange Between Surfaces

13

Having thus far restricted our attention to radiative processes that occur at a *single surface*, we now consider the problem of radiative exchange between two or more surfaces. This exchange depends strongly on the surface geometries and orientations, as well as on their radiative properties and temperatures. Initially, we assume that the surfaces are separated by a *nonparticipating medium*. Since such a medium neither emits, absorbs, nor scatters, it has no effect on the transfer of radiation between surfaces. A vacuum meets these requirements exactly, and most gases meet them to an excellent approximation.

Our first objective is to establish geometrical features of the radiation exchange problem by developing the notion of a *view factor*. Our second objective is to develop procedures for predicting radiative exchange between surfaces that form an *enclosure*. We will limit our attention to surfaces that are assumed to be opaque, diffuse, and gray. We conclude our consideration of radiation exchange between surfaces by considering the effects of a *participating medium*, namely, an intervening gas that emits and absorbs radiation.

13.1 *The View Factor*

To compute radiation exchange between any two surfaces, we must first introduce the concept of a *view factor* (also called a *configuration* or *shape factor*).

13.1.1 The View Factor Integral

The view factor F_{ij} is defined as the *fraction of the radiation leaving surface i that is intercepted by surface j*. To develop a general expression for F_{ij}, we consider the arbitrarily oriented surfaces A_i and A_j of Figure 13.1. Elemental areas on each surface, dA_i and dA_j, are connected by a line of length R, which forms the polar angles θ_i and θ_j, respectively, with the surface normals $\mathbf{n}_i$ and $\mathbf{n}_j$. The values of R, θ_i, and θ_j vary with the position of the elemental areas on A_i and A_j.

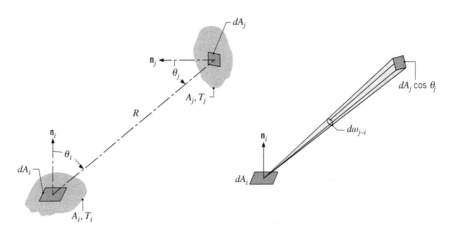

FIGURE 13.1   View factor associated with radiation exchange between elemental surfaces of area dA_i and dA_j.

From the definition of the radiation intensity, Section 12.3.2, and Equation 12.11, the rate at which radiation *leaves* dA_i and is *intercepted* by dA_j may be expressed as

$$dq_{i \to j} = I_{e+r,i} \cos \theta_i \, dA_i \, d\omega_{j-i}$$

where $I_{e+r,i}$ is the intensity of radiation leaving surface i by emission and reflection and $d\omega_{j-i}$ is the solid angle subtended by dA_j when viewed from dA_i. With $d\omega_{j-i} = (\cos \theta_j \, dA_j)/R^2$ from Equation 12.7, it follows that

$$dq_{i \to j} = I_{e+r,i} \frac{\cos \theta_i \cos \theta_j}{R^2} dA_i \, dA_j$$

Assuming that surface i *emits* and *reflects diffusely* and substituting from Equation 12.27, we then obtain

$$dq_{i \to j} = J_i \frac{\cos \theta_i \cos \theta_j}{\pi R^2} dA_i \, dA_j$$

The total rate at which radiation leaves surface i and is intercepted by j may then be obtained by integrating over the two surfaces. That is,

$$q_{i \to j} = J_i \int_{A_i} \int_{A_j} \frac{\cos \theta_i \cos \theta_j}{\pi R^2} dA_j \, dA_i$$

where it is assumed that the radiosity J_i is uniform over the surface A_i. From the definition of the view factor as the fraction of the radiation that leaves A_i and is intercepted by A_j,

$$F_{ij} = \frac{q_{i \to j}}{A_i J_i}$$

it follows that

$$F_{ij} = \frac{1}{A_i} \int_{A_i} \int_{A_j} \frac{\cos \theta_i \cos \theta_j}{\pi R^2} dA_j \, dA_i \tag{13.1}$$

Similarly, the view factor F_{ji} is defined as the fraction of the radiation that leaves A_j and is intercepted by A_i. The same development then yields

$$F_{ji} = \frac{1}{A_j} \int_{A_i} \int_{A_j} \frac{\cos \theta_i \cos \theta_j}{\pi R^2} dA_j \, dA_i \tag{13.2}$$

Either Equation 13.1 or 13.2 may be used to determine the view factor associated with any two surfaces that are *diffuse emitters* and *reflectors* and have *uniform radiosity*.

13.1.2 View Factor Relations

An important view factor relation is suggested by Equations 13.1 and 13.2. In particular, equating the integrals appearing in these equations, it follows that

$$A_i F_{ij} = A_j F_{ji} \tag{13.3}$$

This expression, termed the *reciprocity relation*, is useful in determining one view factor from knowledge of the other.

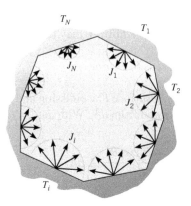

FIGURE 13.2   Radiation exchange in an enclosure.

Another important view factor relation pertains to the surfaces of an *enclosure* (Figure 13.2). From the definition of the view factor, the *summation rule*

$$\sum_{j=1}^{N} F_{ij} = 1 \tag{13.4}$$

may be applied to each of the N surfaces in the enclosure. This rule follows from the conservation requirement that all radiation leaving surface i must be intercepted by the enclosure surfaces. The term F_{ii} appearing in this summation represents the fraction of the radiation that leaves surface i and is directly intercepted by i. If the surface is concave, it *sees itself* and F_{ii} is nonzero. However, for a plane or convex surface, $F_{ii} = 0$.

To calculate radiation exchange in an enclosure of N surfaces, a total of N^2 view factors is needed. This requirement becomes evident when the view factors are arranged in the matrix form:

$$\begin{bmatrix} F_{11} & F_{12} & \cdots & F_{1N} \\ F_{21} & F_{22} & \cdots & F_{2N} \\ \vdots & \vdots & & \vdots \\ F_{N1} & F_{N2} & \cdots & F_{NN} \end{bmatrix}$$

However, all the view factors need not be calculated *directly*. A total of N view factors may be obtained from the N equations associated with application of the summation rule, Equation 13.4, to each of the surfaces in the enclosure. In addition, $N(N-1)/2$ view factors may be obtained from the $N(N-1)/2$ applications of the reciprocity relation, Equation 13.3, which are possible for the enclosure. Accordingly, only $[N^2 - N - N(N-1)/2] = N(N-1)/2$ view factors need be determined directly. For example, in a three-surface enclosure this requirement corresponds to only $3(3-1)/2 = 3$ view factors. The remaining six view factors may be obtained by solving the six equations that result from use of Equations 13.3 and 13.4.

To illustrate the foregoing procedure, consider a simple, two-surface enclosure involving the spherical surfaces of Figure 13.3. Although the enclosure is characterized by $N^2 = 4$ view factors (F_{11}, F_{12}, F_{21}, F_{22}), only $N(N-1)/2 = 1$ view factor need be determined directly. In this case such a determination may be made by *inspection*. In particular, since all radiation leaving the inner surface must reach the outer surface, it follows that $F_{12} = 1$. The same may not be said of radiation leaving the outer surface, since this surface sees itself. However, from the reciprocity relation, Equation 13.3, we obtain

$$F_{21} = \left(\frac{A_1}{A_2} \right) F_{12} = \left(\frac{A_1}{A_2} \right)$$

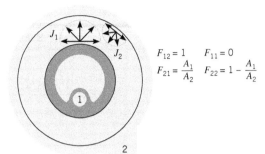

$$F_{12} = 1 \qquad F_{11} = 0$$
$$F_{21} = \frac{A_1}{A_2} \qquad F_{22} = 1 - \frac{A_1}{A_2}$$

FIGURE 13.3   View factors for the enclosure formed by two spheres.

From the summation rule, we also obtain

$$F_{11} + F_{12} = 1$$

in which case $F_{11} = 0$, and

$$F_{21} + F_{22} = 1$$

in which case

$$F_{22} = 1 - \left(\frac{A_1}{A_2} \right)$$

For more complicated geometries, the view factor may be determined by solving the double integral of Equation 13.1. Such solutions have been obtained for many different surface arrangements and are available in equation, graphical, and tabular form [1–4]. Results for several common geometries are presented in Tables 13.1 and 13.2 and Figures 13.4 through 13.6. The configurations of Table 13.1 are assumed to be infinitely long

TABLE 13.1   View factors for two-dimensional geometries.

Geometry	Relation
Parallel Plates with Midlines Connected by Perpendicular 	$F_{ij} = \dfrac{[(W_i + W_j)^2 + 4]^{1/2} - [(W_j - W_i)^2 + 4]^{1/2}}{2W_i}$ $W_i = w_i/L, \ W_j = w_j/L$
Inclined Parallel Plates of Equal Width and a Common Edge	$F_{ij} = 1 - \sin\left(\dfrac{\alpha}{2} \right)$

(continues)

TABLE 13.1 *Continued*

Geometry	Relation

Perpendicular Plates with a Common Edge

$$F_{ij} = \frac{1 + (w_j/w_i) - [1 + (w_j/w_i)^2]^{1/2}}{2}$$

Three-Sided Enclosure

$$F_{ij} = \frac{w_i + w_j - w_k}{2w_i}$$

Parallel Cylinders of Different Radii

$$F_{ij} = \frac{1}{2\pi}\left\{ \pi + [C^2 - (R+1)^2]^{1/2} \right.$$
$$- [C^2 - (R-1)^2]^{1/2}$$
$$+ (R-1)\cos^{-1}\left[\left(\frac{R}{C}\right) - \left(\frac{1}{C}\right)\right]$$
$$\left. - (R+1)\cos^{-1}\left[\left(\frac{R}{C}\right) + \left(\frac{1}{C}\right)\right]\right\}$$

$$R = r_j/r_i, \ S = s/r_i$$
$$C = 1 + R + S$$

Cylinder and Parallel Rectangle

$$F_{ij} = \frac{r}{s_1 - s_2}\left[\tan^{-1}\frac{s_1}{L} - \tan^{-1}\frac{s_2}{L}\right]$$

Infinite Plane and Row of Cylinders

$$F_{ij} = 1 - \left[1 - \left(\frac{D}{s}\right)^2\right]^{1/2}$$
$$+ \left(\frac{D}{s}\right)\tan^{-1}\left[\left(\frac{s^2 - D^2}{D^2}\right)^{1/2}\right]$$

Data from Howell, J. R., *A Catalog of Radiation Configuration Factors*, McGraw-Hill, New York, 1982.

TABLE 13.2 View factors for three-dimensional geometries.

Geometry	Relation
Aligned Parallel Rectangles **(Figure 13.4)** 	$\bar{X} = X/L,\ \bar{Y} = Y/L$ $$F_{ij} = \frac{2}{\pi \bar{X}\bar{Y}} \left\{ \ln\left[\frac{(1+\bar{X}^2)(1+\bar{Y}^2)}{1+\bar{X}^2+\bar{Y}^2}\right]^{1/2} \right.$$ $$+ \bar{X}(1+\bar{Y}^2)^{1/2} \tan^{-1}\frac{\bar{X}}{(1+\bar{Y}^2)^{1/2}}$$ $$\left. + \bar{Y}(1+\bar{X}^2)^{1/2} \tan^{-1}\frac{\bar{Y}}{(1+\bar{X}^2)^{1/2}} - \bar{X}\tan^{-1}\bar{X} - \bar{Y}\tan^{-1}\bar{Y} \right\}$$
Coaxial Parallel Disks **(Figure 13.5)** 	$R_i = r_i/L,\ R_j = r_j/L$ $$S = 1 + \frac{1+R_j^2}{R_i^2}$$ $$F_{ij} = \frac{1}{2}\{S - [S^2 - 4(r_j/r_i)^2]^{1/2}\}$$
Perpendicular Rectangles with a Common Edge **(Figure 13.6)** 	$H = Z/X,\ W = Y/X$ $$F_{ij} = \frac{1}{\pi W}\left(W\tan^{-1}\frac{1}{W} + H\tan^{-1}\frac{1}{H} \right.$$ $$- (H^2+W^2)^{1/2}\tan^{-1}\frac{1}{(H^2+W^2)^{1/2}}$$ $$+ \frac{1}{4}\ln\left\{ \frac{(1+W^2)(1+H^2)}{1+W^2+H^2}\left[\frac{W^2(1+W^2+H^2)}{(1+W^2)(W^2+H^2)}\right]^{W^2} \right.$$ $$\left.\left. \times \left[\frac{H^2(1+H^2+W^2)}{(1+H^2)(H^2+W^2)}\right]^{H^2} \right\} \right)$$

Data from Howell, J. R., *A Catalog of Radiation Configuration Factors*, McGraw-Hill, New York, 1982.

(in a direction perpendicular to the page) and are hence two-dimensional. The configurations of Table 13.2 and Figures 13.4 through 13.6 are three-dimensional.

It is useful to note that the results of Figures 13.4 through 13.6 may be used to determine other view factors. For example, the view factor for an end surface of a cylinder (or a truncated cone) relative to the lateral surface may be obtained by using the results of Figure 13.5 with the summation rule, Equation 13.4. Moreover, Figures 13.4 and 13.6 may be used to obtain other useful results if two additional view factor relations are developed.

The first relation concerns the additive nature of the view factor for a subdivided surface and may be inferred from Figure 13.7. Considering radiation from surface i to surface j, which is divided into n components, it is evident that

$$F_{i(j)} = \sum_{k=1}^{n} F_{ik} \tag{13.5}$$

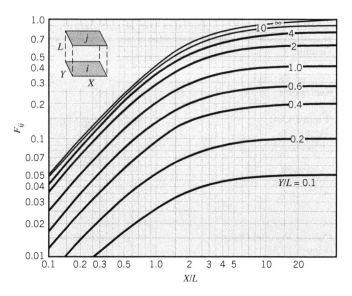

FIGURE 13.4   View factor for aligned parallel rectangles.

where the parentheses around a subscript indicate that it is a composite surface, in which case (*j*) is equivalent to (1, 2, . . . , *k*, . . . , *n*). This expression simply states that radiation reaching a composite surface is the sum of the radiation reaching its parts. Although it pertains to subdivision of the receiving surface, it may also be used to obtain the second view factor relation, which pertains to subdivision of the originating surface. Multiplying

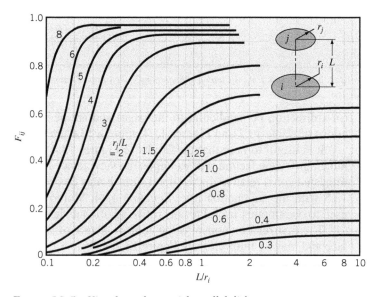

FIGURE 13.5   View factor for coaxial parallel disks.

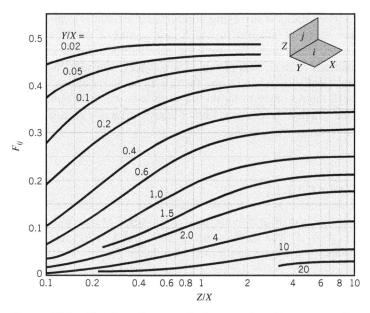

FIGURE 13.6 View factor for perpendicular rectangles with a common edge.

Equation 13.5 by A_i and applying the reciprocity relation, Equation 13.3, to each of the resulting terms, it follows that

$$A_j F_{(j)i} = \sum_{k=1}^{n} A_k F_{ki} \tag{13.6}$$

or

$$F_{(j)i} = \frac{\sum_{k=1}^{n} A_k F_{ki}}{\sum_{k=1}^{n} A_k} \tag{13.7}$$

Equations 13.6 and 13.7 may be applied when the originating surface is composed of several parts.

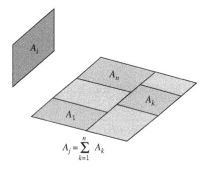

FIGURE 13.7 Areas used to illustrate view factor relations.

For problems involving complicated geometries, analytical solutions to Equation 13.1 may not be obtainable, in which case values of the view factors must be estimated using numerical methods. In situations involving extremely complex structures that may have hundreds or thousands of radiative surfaces, considerable error may be associated with the numerically calculated view factors. In such situations, Equation 13.3 should be used to check the accuracy of individual view factors, and Equation 13.4 should be used to determine whether the conservation of energy principle is satisfied [5].

EXAMPLE 13.1

Consider a diffuse circular disk of diameter D and area A_j and a plane diffuse surface of area $A_i \ll A_j$. The surfaces are parallel, and A_i is located at a distance L from the center of A_j. Obtain an expression for the view factor F_{ij}.

SOLUTION

Known: Orientation of small surface relative to large circular disk.

Find: View factor of small surface with respect to disk, F_{ij}.

Schematic:

Assumptions:
1. Diffuse surfaces.
2. $A_i \ll A_j$.
3. Uniform radiosity on surface A_i.

Analysis: The desired view factor may be obtained from Equation 13.1.

$$F_{ij} = \frac{1}{A_i} \int_{A_i} \int_{A_j} \frac{\cos \theta_i \cos \theta_j}{\pi R^2} dA_j dA_i$$

Recognizing that θ_i, θ_j, and R are approximately independent of position on A_i, this expression reduces to

$$F_{ij} = \int_{A_j} \frac{\cos \theta_i \cos \theta_j}{\pi R^2} dA_j$$

or, with $\theta_i = \theta_j \equiv \theta$,

$$F_{ij} = \int_{A_j} \frac{\cos^2 \theta}{\pi R^2} dA_j$$

With $R^2 = r^2 + L^2$, $\cos \theta = (L/R)$, and $dA_j = 2\pi r \, dr$, it follows that

$$F_{ij} = 2L^2 \int_0^{D/2} \frac{r \, dr}{(r^2 + L^2)^2} = \frac{D^2}{D^2 + 4L^2} \qquad \triangleleft \quad (13.8)$$

Comments:

1. Equation 13.8 may be used to quantify the asymptotic behavior of the curves in Figure 13.5 as the radius of the lower circle, r_i, approaches zero.

2. The preceding geometry is one of the simplest cases for which the view factor may be obtained from Equation 13.1. Geometries involving more detailed integrations are considered in the literature [1, 3].

EXAMPLE 13.2

Determine the view factors F_{12} and F_{21} for the following geometries:

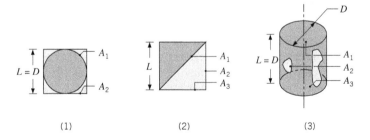

(1) (2) (3)

1. Sphere of diameter D inside a cubical box of length $L = D$.
2. One side of a diagonal partition within a long square duct.
3. End and side of a circular tube of equal length and diameter.

SOLUTION

Known: Surface geometries.

Find: View factors.

Assumptions: Diffuse surfaces with uniform radiosities.

Analysis: The desired view factors may be obtained from inspection, the reciprocity rule, the summation rule, and/or use of the charts.

1. Sphere within a cube:

   By inspection, $F_{12} = 1$ ◁

   By reciprocity, $F_{21} = \dfrac{A_1}{A_2} F_{12} = \dfrac{\pi D^2}{6L^2} \times 1 = \dfrac{\pi}{6}$ ◁

2. Partition within a square duct:

   From summation rule, $F_{11} + F_{12} + F_{13} = 1$ where $F_{11} = 0$

   By symmetry, $F_{12} = F_{13}$

   Hence $F_{12} = 0.50$ ◁

   By reciprocity, $F_{21} = \dfrac{A_1}{A_2} F_{12} = \dfrac{\sqrt{2}L}{L} \times 0.5 = 0.71$ ◁

3. Circular tube:

   From Table 13.2 or Figure 13.5, with $(r_3/L) = 0.5$ and $(L/r_1) = 2$, $F_{13} = 0.172$

   From summation rule, $F_{11} + F_{12} + F_{13} = 1$

   or, with $F_{11} = 0$, $F_{12} = 1 - F_{13} = 0.828$ ◁

   From reciprocity, $F_{21} = \dfrac{A_1}{A_2} F_{12} = \dfrac{\pi D^2/4}{\pi DL} \times 0.828 = 0.207$ ◁

Comment: The geometric surfaces may, in reality, not be characterized by uniform radiosities. The consequences of nonuniform radiosity are discussed in Example 13.3.

13.2 *Blackbody Radiation Exchange*

In general, radiation may leave a surface due to both reflection and emission, and on reaching a second surface, experience reflection as well as absorption. However, matters are simplified for surfaces that may be approximated as blackbodies, since there is no reflection. Hence energy leaves only as a result of emission, and all incident radiation is absorbed.

Consider radiation exchange between two black surfaces of arbitrary shape (Figure 13.8). Defining $q_{i \rightarrow j}$ as the rate at which radiation *leaves* surface i and is *intercepted* by surface j, it follows that

$$q_{i \rightarrow j} = (A_i J_i)F_{ij} \tag{13.9}$$

or, since radiosity equals emissive power for a black surface ($J_i = E_{bi}$),

$$q_{i \rightarrow j} = A_i F_{ij} E_{bi} \tag{13.10}$$

Similarly,

$$q_{j \rightarrow i} = A_j F_{ji} E_{bj} \tag{13.11}$$

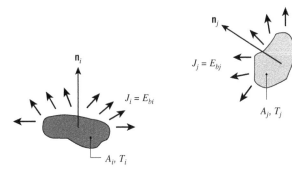

Figure 13.8 Radiation transfer between two surfaces that may be approximated as blackbodies.

The *net radiative exchange* between the two surfaces may then be defined as

$$q_{ij} = q_{i \to j} - q_{j \to i} \qquad (13.12)$$

from which it follows that

$$q_{ij} = A_i F_{ij} E_{bi} - A_j F_{ji} E_{bj}$$

or, from Equations 12.32 and 13.3

$$q_{ij} = A_i F_{ij} \sigma (T_i^4 - T_j^4) \qquad (13.13)$$

Equation 13.13 provides the *net* rate at which radiation *leaves* surface i as a result of its interaction with j, which is equal to the *net* rate at which j *gains* radiation due to its interaction with i.

The foregoing result may also be used to evaluate the net radiation transfer from any surface in an *enclosure* of black surfaces. With N surfaces maintained at different temperatures, the net transfer of radiation from surface i is due to exchange with the remaining surfaces and may be expressed as

$$q_i = \sum_{j=1}^{N} A_i F_{ij} \sigma (T_i^4 - T_j^4) \qquad (13.14)$$

The net radiative heat flux, $q_i'' = q_i / A_i$, was denoted as q_{rad}'' in Chapters 1 and 12. The subscript *rad* has been dropped here for convenience.

EXAMPLE 13.3

A furnace cavity, which is in the form of a cylinder of 50-mm diameter and 150-mm length, is open at one end to large surroundings that are at 27°C. The bottom of the cavity is heated independently, as are three annular sections that form the sides of the cavity. All interior surfaces of the cavity may be approximated as blackbodies and are maintained at 1650°C. What is the required electrical power input to the bottom surface of the cavity? What is the electrical power to the top, middle, and bottom sections of the cavity sides? The backs of the electrically heated surfaces are well insulated.

SOLUTION

Known: Temperature of furnace surfaces and surroundings.

Find: Electrical power required to maintain four sections of the furnace at the prescribed temperature.

Schematic:

Assumptions:

1. Interior surfaces behave as blackbodies with uniform radiosity and irradiation.
2. Heat transfer by convection is negligible.
3. Backs of electrically heated surfaces are adiabatic.

Analysis: Subject to the foregoing assumptions, the only heat loss from the furnace is by radiation through the opening. Because the surroundings are large, the irradiation from the surroundings is equal to emission from a blackbody at T_{sur}, as discussed in Section 12.7. Furthermore, since radiation heat transfer between the furnace and the surroundings must pass through the opening, the radiation exchange may be analyzed as if it were between the furnace and a hypothetical black surface 5 at the opening, with $T_5 = T_{sur}$.

This approach is discussed in detail in Example 13.4. The electrical power delivered to each surface balances the corresponding radiation loss, which may be obtained from Equation 13.14. After employing Equation 13.3, we may write the following equations for surfaces 1 through 4.

Surface 1:
$$q_1 = A_1 F_{15} \sigma (T_1^4 - T_5^4) = A_5 F_{51} \sigma (T_1^4 - T_5^4) \tag{1}$$

Surface 2:
$$q_2 = A_2 F_{25} \sigma (T_2^4 - T_5^4) = A_5 F_{52} \sigma (T_2^4 - T_5^4) \tag{2}$$

Surface 3:
$$q_3 = A_3 F_{35} \sigma (T_3^4 - T_5^4) = A_5 F_{53} \sigma (T_3^4 - T_5^4) \tag{3}$$

Surface 4:
$$q_4 = A_4 F_{45} \sigma (T_4^4 - T_5^4) = A_5 F_{54} \sigma (T_4^4 - T_5^4) \tag{4}$$

We will determine the view factors by first defining two hypothetical surfaces A' and A'' as shown in the schematic. From Table 13.2 with $(r_i/L) = (r_j/L) = (0.025 \text{ m}/0.150 \text{ m}) = 0.167$, $F_{51} = 0.0263$. With $(r_i/L) = (r_j/L) = (0.025 \text{ m}/0.100 \text{ m}) = 0.25$, $F_{5A''} = 0.0557$ so that $F_{52} = F_{5A''} - F_{51} = 0.0557 - 0.0263 = 0.0294$. Likewise, with $(r_i/L) = (r_j/L) = (0.025 \text{ m}/0.050 \text{ m}) = 0.5$, $F_{5A'} = 0.172$ so that $F_{53} = F_{5A'} - F_{5A''} = 0.172 - 0.0557 = 0.1163$. Finally, $F_{54} = 1 - F_{5A'} = 1 - 0.172 = 0.828$. The electrical power delivered to each of the four furnace surfaces can now be determined by solving Equations 1 through 4 for the radiation loss from each surface with $A_5 = \pi D^2/4 = \pi \times (0.05 \text{ m})^2/4 = 0.00196 \text{ m}^2$.

$$q_1 = 0.00196 \text{ m}^2 \times 0.0263 \times 5.67 \times 10^{-8} \text{ W/m}^2 \cdot \text{K}^4 \times (1923 \text{ K}^4 - 300 \text{ K}^4) = 39.9 \text{ W} \quad \lhd$$

$$q_2 = 0.00196 \text{ m}^2 \times 0.0294 \times 5.67 \times 10^{-8} \text{ W/m}^2 \cdot \text{K}^4 \times (1923 \text{ K}^4 - 300 \text{ K}^4) = 44.7 \text{ W} \quad \lhd$$

$$q_3 = 0.00196 \text{ m}^2 \times 0.1163 \times 5.67 \times 10^{-8} \text{ W/m}^2 \cdot \text{K}^4 \times (1923 \text{ K}^4 - 300 \text{ K}^4) = 177 \text{ W} \quad \lhd$$

$$q_4 = 0.00196 \text{ m}^2 \times 0.828 \times 5.67 \times 10^{-8} \text{ W/m}^2 \cdot \text{K}^4 \times (1923 \text{ K}^4 - 300 \text{ K}^4) = 1260 \text{ W} \quad \lhd$$

Comments:

1. Adding the view factors corresponding to surface 5 yields

   $$F_{51} + F_{52} + F_{53} + F_{54} + F_{55} = 0.0263 + 0.0294 + 0.1163 + 0.828 + 0 = 1$$

   Hence the enclosure rule, Equation 13.4, is satisfied, indicating that the view factors have been calculated correctly. Alternatively, the enclosure rule could have been utilized to determine one of the view factors used in the problem solution.

2. The rate of radiation heat loss from the furnace is $q_{\text{tot}} = q_1 + q_2 + q_3 + q_4 = 1522 \text{ W} = 1.522 \text{ kW}$. If the furnace were to have been treated as a single surface f, the rate of heat loss could be quickly calculated as $q_{\text{tot}} = A_5 F_{5f} \sigma (T_f^4 - T_{\text{sur}}^4) = 0.00196 \text{ m}^2 \times 1 \times 5.67 \times 10^{-8} \text{ W/m}^2 \cdot \text{K} \times (1923 \text{ K}^4 - 300 \text{ K}^4) = 1.522 \text{ kW}$. The answer is the same as determined in the problem solution since $F_{5f} = 1 = F_{51} + F_{52} + F_{53} + F_{54} + F_{55}$.

3. We have assumed that each surface i is *isothermal* and characterized by *uniform radiosity*, J_i, as well as *uniform irradiation*, G_i. Because the isothermal furnace walls are treated as blackbodies, $J_i = E_{bi}$ and the assumption of uniform radiosity is valid. However, the irradiation distribution of the furnace surfaces is *not* uniform since, for example, irradiation from the cool surroundings influences the upper region of an annular surface more than the lower region. To quantify this effect, the *local* radiation heat flux along the vertical furnace wall may be determined by considering a ring element of differential area $dA = \pi D dx$ as shown. Both sides of Equation 13.13 may

be divided by dA yielding $q''(x) = F_{dA-A_5}\sigma(T_f^4 - T_{sur}^4)$ where the view factor from the differential ring element to the opening, area A_5, is [4]

$$F_{dA-A_5} = \frac{(x/D)^2 + 1/2}{\sqrt{1 + (x/D)^2}} - x/D$$

Substituting the expression for F_{dA-A_5} into the equation for the heat flux $q''(x)$ yields the heat flux distribution shown below. Also shown are the *average* heat fluxes associated with the three furnace segments, $\overline{q_2''} = q_2/(\pi DL/3) = 44.7$ W/$(\pi \times 0.05$ m $\times 0.15$ m/3) = 5690 W/m^2 = 5.69 kW/m^2, $\overline{q_3''} = q_3/(\pi DL/3) = 22.5$ kW/m^2, and $\overline{q_4''} = q_4/(\pi DL/3) = 160$ kW/m^2.

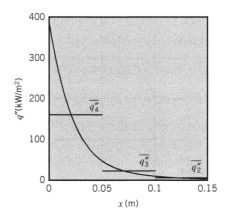

Because of the nonuniformity of the irradiation along the sidewalls of the cavity, the local heat flux is highly nonuniform with the largest value occurring adjacent to the furnace opening. If local temperatures or heat fluxes are of interest, it is necessary to subdivide the various *geometric surfaces* into smaller *radiative surfaces*. This may be done either analytically, as demonstrated here, or numerically. Computational determination of local temperatures or heat fluxes may involve hundreds or perhaps thousands of radiative surfaces, even for simple geometries such as in this example.

13.3 Radiation Exchange Between Opaque, Diffuse, Gray Surfaces in an Enclosure

In general, radiation may leave an opaque surface due to both reflection and emission, and on reaching a second opaque surface, experience reflection as well as absorption. In an enclosure, such as that of Figure 13.9a, radiation may experience multiple reflections off all surfaces, with partial absorption occurring at each.

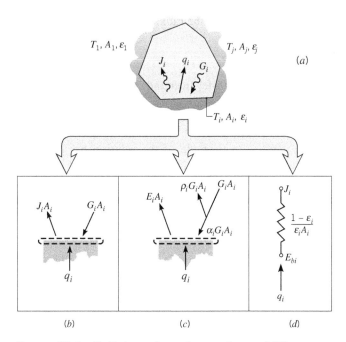

FIGURE 13.9 Radiation exchange in an enclosure of diffuse, gray surfaces with a nonparticipating medium. (*a*) Schematic of the enclosure. (*b*) Radiative balance according to Equation 13.15. (*c*) Radiative balance according to Equation 13.17. (*d*) Resistance representing net radiation transfer from a surface, Equation 13.19.

Analyzing radiation exchange in an enclosure may be simplified by making certain assumptions. Each surface of the enclosure is assumed to be *isothermal* and to be characterized by a *uniform radiosity* and a *uniform irradiation*. The surfaces are also assumed to be *opaque* ($\tau = 0$) and to have emissivities, absorptivities, and reflectivities that are independent of direction (the surfaces are *diffuse*) *and* independent of wavelength (the surfaces are *gray*). It was shown in Section 12.8 that under these conditions the emissivity is equal to the absorptivity, $\varepsilon = \alpha$ (a form of Kirchhoff's law). Finally, the medium within the enclosure is taken to be *nonparticipating*. The problem is generally one in which either the temperature T_i or the net radiative heat flux q_i'' associated with each of the surfaces is known. The objective is to use this information to determine the unknown radiative heat fluxes and temperatures associated with each of the surfaces.

13.3.1 Net Radiation Exchange at a Surface

The term q_i, which is the *net* rate at which radiation *leaves* surface i, represents the net effect of radiative interactions occurring at the surface (Figure 13.9*b*). It is the rate at which energy would have to be transferred to the surface by other means to maintain it at a constant temperature. It is equal to the difference between the surface radiosity and irradiation and from Equation 12.5 may be expressed as

$$q_i = A_i(J_i - G_i) \tag{13.15}$$

Using the definition of the radiosity, Equation 12.4,

$$J_i \equiv E_i + \rho_i G_i \tag{13.16}$$

the net radiative transfer from the surface may be expressed as

$$q_i = A_i(E_i - \alpha_i G_i) \tag{13.17}$$

where use has been made of the relationship $\alpha_i = 1 - \rho_i$ for an opaque surface. This relationship corresponds to Equation 12.6 and is illustrated in Figure 13.9c. Noting that $E_i = \varepsilon_i E_{bi}$ and recognizing that $\rho_i = 1 - \alpha_i = 1 - \varepsilon_i$ for an opaque, diffuse, gray surface, the radiosity may also be expressed as

$$J_i = \varepsilon_i E_{bi} + (1 - \varepsilon_i)G_i \tag{13.18}$$

Solving for G_i and substituting into Equation 13.15, it follows that

$$q_i = A_i\left(J_i - \frac{J_i - \varepsilon_i E_{bi}}{1 - \varepsilon_i}\right)$$

or

$$q_i = \frac{E_{bi} - J_i}{(1 - \varepsilon_i)/\varepsilon_i A_i} \tag{13.19}$$

Equation 13.19 provides a convenient representation for the net radiative heat transfer rate from a surface. This transfer, which is represented in Figure 13.9d, is associated with the driving potential $(E_{bi} - J_i)$ and a *surface radiative resistance* of the form $(1 - \varepsilon_i)/\varepsilon_i A_i$. Hence, if the emissive power that the surface would have if it were black exceeds its radiosity, there is net radiation heat transfer from the surface; if the inverse is true, the net transfer is to the surface.

It is sometimes the case that one of the surfaces is very large relative to the other surfaces under consideration. For example, the system might consist of multiple small surfaces in a large room. In this case, the area of the large surface is effectively infinite $(A_i \rightarrow \infty)$, and we see that its surface radiative resistance, $(1 - \varepsilon_i)/\varepsilon_i A_i$, is effectively zero, just as it would be for a black surface $(\varepsilon_i = 1)$. Hence, $J_i = E_{bi}$, and *a surface which is large relative to all other surfaces under consideration can be treated as if it were a blackbody.* This important conclusion was reached in Section 12.7 and utilized in Example 13.3, where it was based on a physical argument and has now been confirmed from our treatment of gray surface radiation exchange. Again, the physical explanation is that, even though the large surface may reflect some of the irradiation incident upon it, it is so big that there is a high probability that the reflected radiation reaches another point on the same large surface. After many such reflections, all the radiation that was originally incident on the large surface is absorbed by the large surface, and none ever reaches any of the smaller surfaces.

13.3.2 Radiation Exchange Between Surfaces

To use Equation 13.19, the surface radiosity J_i must be known. To determine this quantity, it is necessary to consider radiation exchange between the surfaces of the enclosure.

The irradiation of surface i can be evaluated from the radiosities of all the surfaces in the enclosure. In particular, from the definition of the view factor, it follows that the total rate at which radiation reaches surface i from all surfaces, including i, is

$$A_i G_i = \sum_{j=1}^{N} F_{ji} A_j J_j$$

or from the reciprocity relation, Equation 13.3,

$$A_i G_i = \sum_{j=1}^{N} A_i F_{ij} J_j$$

Canceling the area A_i and substituting into Equation 13.15 for G_i,

$$q_i = A_i \left(J_i - \sum_{j=1}^{N} F_{ij} J_j \right)$$

or, from the summation rule, Equation 13.4,

$$q_i = A_i \left(\sum_{j=1}^{N} F_{ij} J_i - \sum_{j=1}^{N} F_{ij} J_j \right)$$

Hence

$$q_i = \sum_{j=1}^{N} A_i F_{ij} (J_i - J_j) = \sum_{j=1}^{N} q_{ij} \tag{13.20}$$

Radiation Network Approach   Equation 13.20 equates the net rate of radiation transfer from surface i, q_i, to the sum of components q_{ij} related to radiative exchange with the other surfaces. Each component may be represented by a *network element* for which $(J_i - J_j)$ is the driving potential and $(A_i F_{ij})^{-1}$ is a *space* or *geometrical resistance* (Figure 13.10).

Combining Equations 13.19 and 13.20, we then obtain

$$\frac{E_{bi} - J_i}{(1 - \varepsilon_i)/\varepsilon_i A_i} = \sum_{j=1}^{N} \frac{J_i - J_j}{(A_i F_{ij})^{-1}} \tag{13.21}$$

As shown in Figure 13.10, this expression represents a radiation balance for the radiosity *node* associated with surface i. The rate of radiation transfer (current flow) to i through its surface resistance must equal the net rate of radiation transfer (current flows) from i to all other surfaces through the corresponding geometrical resistances.

Note that Equation 13.21 is especially useful when the surface temperature T_i (hence E_{bi}) is known. Although this situation is typical, it does not always apply. In particular, situations may arise for which the net radiation transfer rate at the surface q_i, rather than the temperature T_i, is known. In such cases the preferred form of the radiation balance is Equation 13.20, rearranged as

$$q_i = \sum_{j=1}^{N} \frac{J_i - J_j}{(A_i F_{ij})^{-1}} \tag{13.22}$$

Use of network representations was first suggested by Oppenheim [6]. The network is built by first identifying nodes associated with the radiosities of each of the N surfaces of the enclosure. The method provides a useful tool for visualizing radiation exchange in the

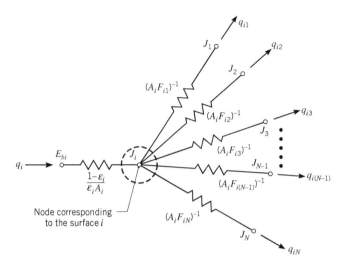

FIGURE 13.10 Network representation of radiative exchange between surface i and the remaining surfaces of an enclosure.

enclosure and, at least for simple enclosures, may be used as the basis for predicting this exchange.

Direct Approach An alternative *direct approach* to solving radiation enclosure problems involves writing Equation 13.21 for each surface at which T_i is known, and writing Equation 13.22 for each surface at which q_i is known. The resulting set of N linear, algebraic equations is solved for $J_1, J_2, \ldots, J_N$. With knowledge of the J_i, Equation 13.19 may then be used to determine the net radiation heat transfer rate q_i at each surface of known T_i or the value of T_i at each surface of known q_i. For any number N of surfaces in the enclosure, the foregoing problem may readily be solved by the iteration or matrix inversion methods of Chapter 4 and Appendix D.

EXAMPLE 13.4

In manufacturing, the special coating on a curved solar absorber surface of area $A_2 = 15$ m^2 is cured by exposing it to an infrared heater of width $W = 1$ m. The absorber and heater are each of length $L = 10$ m and are separated by a distance of $H = 1$ m. The upper surface of the absorber and the lower surface of the heater are insulated.

The heater is at $T_1 = 1000$ K and has an emissivity of $\varepsilon_1 = 0.9$, while the absorber is at $T_2 = 600$ K and has an emissivity of $\varepsilon_2 = 0.5$. The system is in a large room whose walls are at 300 K. What is the net rate of heat transfer to the absorber surface?

SOLUTION

Known: A curved, solar absorber surface with a special coating is being cured by use of an infrared heater in a large room.

Find: Net rate of heat transfer to the absorber surface.

Schematic:

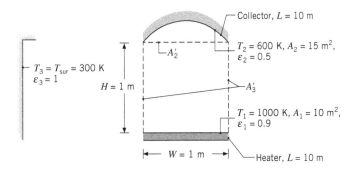

Collector, $L = 10$ m

$T_2 = 600$ K, $A_2 = 15$ m^2, $\varepsilon_2 = 0.5$

A_2'

$T_3 = T_{sur} = 300$ K
$\varepsilon_3 = 1$

$H = 1$ m

A_3'

$T_1 = 1000$ K, $A_1 = 10$ m^2, $\varepsilon_1 = 0.9$

$W = 1$ m

Heater, $L = 10$ m

Assumptions:

1. Steady-state conditions exist.
2. Convection effects are negligible.
3. Absorber and heater surfaces are diffuse and gray and are characterized by uniform irradiation and radiosity.
4. The surrounding room is large and therefore behaves as a blackbody.

Analysis: The system may be viewed as a three-surface enclosure, with the third surface being the large surrounding room, which behaves as a blackbody. We are interested in obtaining the net rate of radiation transfer to surface 2. We solve the problem using both the radiation network and direct approaches.

Radiation Network Approach The radiation network is constructed by first identifying nodes associated with the radiosities of each surface, as shown in step 1 in the following schematic. Then each radiosity node is connected to each of the other radiosity nodes through the appropriate space resistance, as shown in step 2. We will treat the surroundings as having a large but unspecified area, which introduces difficulty in expressing the space resistances $(A_3F_{31})^{-1}$ and $(A_3F_{32})^{-1}$. Fortunately, from the reciprocity relation (Equation 13.3), we can replace A_3F_{31} with A_1F_{13} and A_3F_{32} with A_2F_{23}, which are more readily obtained. The final step is to connect the blackbody emissive powers associated with the temperature of each surface to the radiosity nodes, using the appropriate form of the surface resistance.

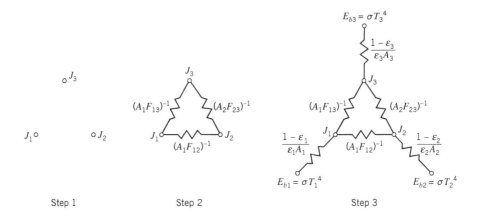

Step 1   Step 2   Step 3

In this problem, the surface resistance associated with surface 3 is zero according to assumption 4; therefore, $J_3 = E_{b3} = \sigma T_3^4 = 459$ W/m^2.

Summing currents at the J_1 node yields

$$\frac{\sigma T_1^4 - J_1}{(1 - \varepsilon_1)/\varepsilon_1 A_1} = \frac{J_1 - J_2}{1/A_1 F_{12}} + \frac{J_1 - \sigma T_3^4}{1/A_1 F_{13}} \qquad (1)$$

while summing the currents at the J_2 node results in

$$\frac{\sigma T_2^4 - J_2}{(1 - \varepsilon_2)/\varepsilon_2 A_2} = \frac{J_2 - J_1}{1/A_1 F_{12}} + \frac{J_2 - \sigma T_3^4}{1/A_2 F_{23}} \qquad (2)$$

The view factor F_{12} may be obtained by recognizing that $F_{12} = F_{12'}$, where A_2' is shown in the schematic as the rectangular base of the absorber surface. Then, from Figure 13.4 or Table 13.2, with $Y/L = 10/1 = 10$ and $X/L = 1/1 = 1$,

$$F_{12} = 0.39$$

From the summation rule, and recognizing that $F_{11} = 0$, it also follows that

$$F_{13} = 1 - F_{12} = 1 - 0.39 = 0.61$$

The last needed view factor is F_{23}. We recognize that, since radiation propagating from surface 2 to surface 3 must pass through the hypothetical surface A_2',

$$A_2 F_{23} = A_2' F_{2'3}$$

and from symmetry $F_{2'3} = F_{13}$. Thus

$$F_{23} = \frac{A_2'}{A_2} F_{13} = \frac{10 \text{ m}^2}{15 \text{ m}^2} \times 0.61 = 0.41$$

We may now solve Equations 1 and 2 for J_1 and J_2. Recognizing that $E_{b1} = \sigma T_1^4 = 56{,}700$ W/m^2 and canceling the area A_1, we can express Equation 1 as

$$\frac{56{,}700 - J_1}{(1 - 0.9)/0.9} = \frac{J_1 - J_2}{1/0.39} + \frac{J_1 - 459}{1/0.61}$$

or

$$-10 J_1 + 0.39 J_2 = -510{,}582 \qquad (3)$$

Noting that $E_{b2} = \sigma T_2^4 = 7348 \text{ W/m}^2$ and dividing by the area A_2, we can express Equation 2 as

$$\frac{7348 - J_2}{(1 - 0.5)/0.5} = \frac{J_2 - J_1}{15 \text{ m}^2/(10 \text{ m}^2 \times 0.39)} + \frac{J_2 - 459}{1/0.41}$$

or

$$0.26J_1 - 1.67J_2 = -7536 \tag{4}$$

Solving Equations 3 and 4 simultaneously yields $J_2 = 12{,}487 \text{ W/m}^2$.

An expression for the net rate of heat transfer *from* the absorber surface, q_2, may be written upon inspection of the radiation network and is

$$q_2 = \frac{\sigma T_2^4 - J_2}{(1 - \varepsilon_2)/\varepsilon_2 A_2}$$

resulting in

$$q_2 = \frac{(7348 - 12{,}487)\,\text{W/m}^2}{(1 - 0.5)/(0.5 \times 15 \text{ m}^2)} = -77.1 \text{ kW}$$

Hence, the net heat transfer rate *to* the absorber is $q_{\text{net}} = -q_2 = 77.1 \text{ kW}$. ◁

Direct Approach Using the direct approach, we write Equation 13.21 for each of the three surfaces. We use reciprocity to rewrite the space resistances in terms of the known view factors from above and to eliminate A_3.

Surface 1

$$\frac{\sigma T_1^4 - J_1}{(1 - \varepsilon_1)/\varepsilon_1 A_1} = \frac{J_1 - J_2}{1/A_1 F_{12}} + \frac{J_1 - J_3}{1/A_1 F_{13}} \tag{5}$$

Surface 2

$$\frac{\sigma T_2^4 - J_2}{(1 - \varepsilon_2)/\varepsilon_2 A_2} = \frac{J_2 - J_1}{1/A_2 F_{21}} + \frac{J_2 - J_3}{1/A_2 F_{23}} = \frac{J_2 - J_1}{1/A_1 F_{12}} + \frac{J_2 - J_3}{1/A_2 F_{23}} \tag{6}$$

Surface 3

$$\frac{\sigma T_3^4 - J_3}{(1 - \varepsilon_3)/\varepsilon_3 A_3} = \frac{J_3 - J_1}{1/A_3 F_{31}} + \frac{J_3 - J_2}{1/A_3 F_{32}} = \frac{J_3 - J_1}{1/A_1 F_{13}} + \frac{J_3 - J_2}{1/A_2 F_{23}} \tag{7}$$

Substituting values of the areas, temperatures, emissivities, and view factors into Equations 5 through 7 and solving them simultaneously, we obtain $J_1 = 51{,}541 \text{ W/m}^2$, $J_2 = 12{,}487 \text{ W/m}^2$, and $J_3 = 459 \text{ W/m}^2$. Equation 13.19 may then be written for surface 2 as

$$q_2 = \frac{\sigma T_2^4 - J_2}{(1 - \varepsilon_2)/\varepsilon_2 A_2}$$

This expression is identical to the expression that was developed using the radiation network. Hence, $q_2 = -77.1 \text{ kW}$. ◁

Comments:

1. In order to solve Equations 5 through 7 simultaneously, we must first multiply both sides of Equation 7 by $(1 - \varepsilon_3)/\varepsilon_3 A_3 = 0$ to avoid division by zero, resulting in the simplified form of Equation 7, which is $J_3 = \sigma T_3^4$.

2. If we substitute $J_3 = \sigma T_3^4$ into Equations 5 and 6, it is evident that Equations 5 and 6 are identical to Equations 1 and 2, respectively.

3. The direct approach is recommended for problems involving $N \geq 4$ surfaces, since radiation networks become quite complex as the number of surfaces increases.

4. As will be seen in Section 13.4, the radiation network approach is particularly useful when thermal energy is transferred to or from surfaces by additional means, that is, by conduction and/or convection. In these *multimode* heat transfer situations, the additional energy delivered to or taken from the surface can be represented by additional current into or out of a node.

5. Recognize the utility of using a hypothetical surface (A_2') to simplify the evaluation of view factors.

6. We could have approached the solution in a slightly different manner. Radiation leaving surface 1 must pass through the openings (hypothetical surface 3′) in order to reach the surroundings. Thus, we can write

$$F_{13} = F_{13'}$$

$$A_1 F_{13} = A_1 F_{13'} = A_3' F_{3'1}$$

A similar relationship can be written for exchange between surface 2 and the surroundings, that is, $A_2 F_{23} = A_3' F_{3'2}$. Thus, the space resistances which connect to radiosity node 3 in the radiation network above can be replaced by space resistances pertaining to surface 3′. The resistance network would be unchanged, and the space resistances would have the same values as those determined in the foregoing solution. However, it may be more convenient to calculate the view factors by utilizing the hypothetical surfaces 3′. With the surface resistance for surface 3 equal to zero, we see that *openings of enclosures that exchange radiation with large surroundings may be treated as hypothetical, nonreflecting black surfaces ($\varepsilon_3 = 1$) whose temperature is equal to that of the surroundings ($T_3 = T_{\text{sur}}$).*

7. The heater and absorber surfaces would not be characterized by uniform irradiation or radiosity. The calculated heat rate could be checked by dividing the heater and absorber into subsurfaces, and repeating the analysis.

13.3.3 The Two-Surface Enclosure

The simplest example of an enclosure is one involving two surfaces that exchange radiation only with each other. Such a two-surface enclosure is shown schematically in Figure 13.11a. Since there are only two surfaces, the net rate of radiation transfer *from* surface 1, q_1, must equal the net rate of radiation transfer *to* surface 2, $-q_2$, and both quantities must equal the net rate at which radiation is exchanged between 1 and 2. Accordingly,

$$q_1 = -q_2 = q_{12}$$

The radiation transfer rate may be determined by applying Equation 13.21 to surfaces 1 and 2 and solving the resulting two equations for J_1 and J_2. The results could then be used with Equation 13.19 to determine q_1 (or q_2). However, in this case the desired result is more readily obtained by working with the network representation of the enclosure shown in Figure 13.11b.

FIGURE 13.11   The two-surface enclosure. (*a*) Schematic. (*b*) Network representation.

From Figure 13.11*b* we see that the total resistance to radiation exchange between surfaces 1 and 2 is comprised of the two surface resistances and the geometrical resistance. Hence, substituting from Equation 12.32, the net radiation exchange between surfaces may be expressed as

$$q_{12} = q_1 = -q_2 = \frac{\sigma(T_1^4 - T_2^4)}{\dfrac{1-\varepsilon_1}{\varepsilon_1 A_1} + \dfrac{1}{A_1 F_{12}} + \dfrac{1-\varepsilon_2}{\varepsilon_2 A_2}} \tag{13.23}$$

The foregoing result may be used for any two isothermal diffuse, gray surfaces that *form an enclosure* and *are each characterized by uniform radiosity and irradiation.* Important special cases are summarized in Table 13.3.

TABLE 13.3   Special diffuse, gray, two-surface enclosures

Large (Infinite) Parallel Planes

$A_1 = A_2 = A$

$F_{12} = 1$

$$q_{12} = \frac{A\sigma(T_1^4 - T_2^4)}{\dfrac{1}{\varepsilon_1} + \dfrac{1}{\varepsilon_2} - 1} \tag{13.24}$$

Long (Infinite) Concentric Cylinders

$\dfrac{A_1}{A_2} = \dfrac{r_1}{r_2}$

$F_{12} = 1$

$$q_{12} = \frac{\sigma A_1(T_1^4 - T_2^4)}{\dfrac{1}{\varepsilon_1} + \dfrac{1-\varepsilon_2}{\varepsilon_2}\left(\dfrac{r_1}{r_2}\right)} \tag{13.25}$$

Concentric Spheres

$\dfrac{A_1}{A_2} = \dfrac{r_1^2}{r_2^2}$

$F_{12} = 1$

$$q_{12} = \frac{\sigma A_1(T_1^4 - T_2^4)}{\dfrac{1}{\varepsilon_1} + \dfrac{1-\varepsilon_2}{\varepsilon_2}\left(\dfrac{r_1}{r_2}\right)^2} \tag{13.26}$$

Small Convex Object in a Large Cavity

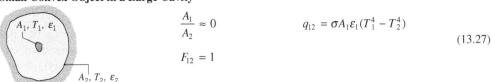

$\dfrac{A_1}{A_2} \approx 0$

$F_{12} = 1$

$$q_{12} = \sigma A_1 \varepsilon_1 (T_1^4 - T_2^4) \tag{13.27}$$

13.3.4   Two-Surface Enclosures in Series and Radiation Shields

The preceding two-surface enclosure analysis can be readily extended to systems having multiple two-surface enclosures arranged in series. An important application of such a scenario is the use of *radiation shields*, typically constructed from low emissivity (high reflectivity) materials, to reduce the net radiation transfer between two surfaces.

An example is shown in Figure 13.12a, where a thin radiation shield is placed between surfaces 1 and 2. If all surfaces exchange energy only by radiation, and are not otherwise heated or cooled, the rate of radiation transfer must be the same at every surface, that is, $q = q_1 = q_{1s} = q_{s2} = -q_2$. This scenario can be represented by a series radiation network (Figure 13.12b) that includes a surface resistance for each of the four surfaces (including both sides of the shield) and a space resistance between each pair of adjacent surfaces. Note that the emissivity of one side of the shield (ε_{s1}) may differ from that of the opposite side (ε_{s2}) and the radiosities will always differ. The resulting heat transfer rate through the system is given by

$$q = \frac{\sigma(T_1^4 - T_2^4)}{\dfrac{1-\varepsilon_1}{\varepsilon_1 A_1} + \dfrac{1}{A_1 F_{1s}} + \dfrac{1-\varepsilon_{s1}}{\varepsilon_{s1} A_s} + \dfrac{1-\varepsilon_{s2}}{\varepsilon_{s2} A_s} + \dfrac{1}{A_s F_{s2}} + \dfrac{1-\varepsilon_2}{\varepsilon_2 A_2}} \tag{13.28}$$

This result is not specific to the configuration of Figure 13.12a, but applies to any geometry in which the shield completely blocks the two other surfaces from directly exchanging radiation ($F_{12} = 0$).

Without the radiation shield, the net rate of radiation transfer between surfaces 1 and 2 is given by Equation 13.23. However, with the radiation shield, additional resistances are present, as evident in Equation 13.28, and the heat transfer rate is reduced. Note that the resistances associated with the radiation shield become large when the emissivities ε_{s1} and ε_{s2} are small.

Equation 13.28 may be used to determine the net heat transfer rate if T_1 and T_2 are known. From knowledge of q, the value of the shield temperature, T_s, can then be determined by expressing Equation 13.23 between surfaces 1 and s1 or between surfaces 2s and 2.

The foregoing procedure may be readily extended to problems involving multiple radiation shields. In the special case of surfaces that are infinite parallel planes (that is, all

FIGURE 13.12 Radiation exchange between surfaces with a radiation shield in place. (*a*) Schematic. (*b*) Network representation.

view factors are unity), and for which all the emissivities are equal, it may be shown that, with N shields,

$$(q_{12})_N = \frac{1}{N+1}(q_{12})_0 \tag{13.29}$$

where $(q_{12})_0$ is the radiation transfer rate with no shields ($N = 0$).

EXAMPLE 13.5

A cryogenic fluid flows through a long tube of 20-mm diameter, the outer surface of which is diffuse and gray with $\varepsilon_1 = 0.02$ and $T_1 = 77$ K. This tube is concentric with a larger tube of 50-mm diameter, the inner surface of which is diffuse and gray with $\varepsilon_2 = 0.05$ and $T_2 = 300$ K. The space between the surfaces is evacuated. Calculate the rate of heat gain by the cryogenic fluid per unit length of tubes. If a thin radiation shield of 35-mm diameter and $\varepsilon_s = 0.02$ (both sides) is inserted midway between the inner and outer surfaces, calculate the change (percentage) in heat gain per unit length of the tubes.

SOLUTION

Known: Concentric tube arrangement with diffuse, gray surfaces of different emissivities and temperatures.

Find:

1. Rate of heat gain by the cryogenic fluid passing through the inner tube.
2. Percentage change in heat gain with radiation shield inserted midway between inner and outer tubes.

Schematic:

Without shield (*wo*)   With shield (*w*)

Assumptions:

1. Surfaces are diffuse and gray and characterized by uniform irradiation and radiosity.
2. Space between tubes is evacuated.
3. Conduction resistance for radiation shield is negligible.
4. Concentric tubes form a two-surface enclosure (end effects are negligible).

Analysis:

1. The network representation of the system without the shield is shown in Figure 13.11, and the heat rate may be obtained from Equation 13.25, where

$$q = \frac{\sigma(\pi D_1 L)(T_1^4 - T_2^4)}{\dfrac{1}{\varepsilon_1} + \dfrac{1-\varepsilon_2}{\varepsilon_2}\left(\dfrac{D_1}{D_2}\right)}$$

Hence

$$q' = \frac{q}{L} = \frac{5.67 \times 10^{-8} \ \text{W/m}^2 \cdot \text{K}^4 \ (\pi \times 0.02 \ \text{m})[(77 \ \text{K})^4 - (300 \ \text{K})^4]}{\dfrac{1}{0.02} + \dfrac{1 - 0.05}{0.05}\left(\dfrac{0.02 \ \text{m}}{0.05 \ \text{m}}\right)}$$

$$q' = -0.50 \ \text{W/m} \qquad \triangleleft$$

2. The network representation of the system with the shield is shown in Figure 13.12, and the heat rate is now

$$q = \frac{E_{b1} - E_{b2}}{R_{\text{tot}}} = \frac{\sigma(T_1^4 - T_2^4)}{R_{\text{tot}}}$$

where

$$R_{\text{tot}} = \frac{1 - \varepsilon_1}{\varepsilon_1 (\pi D_1 L)} + \frac{1}{(\pi D_1 L) F_{1s}} + 2\left[\frac{1 - \varepsilon_s}{\varepsilon_s (\pi D_s L)}\right] + \frac{1}{(\pi D_s L) F_{s2}} + \frac{1 - \varepsilon_2}{\varepsilon_2 (\pi D_2 L)}$$

or

$$R_{\text{tot}} = \frac{1}{L}\left\{\frac{1 - 0.02}{0.02(\pi \times 0.02 \ \text{m})} + \frac{1}{(\pi \times 0.02 \ \text{m})1}\right.$$

$$\left. + 2\left[\frac{1 - 0.02}{0.02(\pi \times 0.035 \ \text{m})}\right] + \frac{1}{(\pi \times 0.035 \ \text{m})1} + \frac{1 - 0.05}{0.05(\pi \times 0.05 \ \text{m})}\right\}$$

$$R_{\text{tot}} = \frac{1}{L}(779.9 + 15.9 + 891.3 + 9.1 + 121.0) = \frac{1817}{L}\left(\frac{1}{\text{m}^2}\right)$$

Hence

$$q' = \frac{q}{L} = \frac{5.67 \times 10^{-8} \ \text{W/m}^2 \cdot \text{K}^4 \ [(77 \ \text{K})^4 - (300 \ \text{K})^4]}{1817(1/\text{m})} = -0.25 \ \text{W/m} \qquad \triangleleft$$

The percentage change in the heat gain is then

$$\frac{q_w' - q_{wo}'}{q_{wo}'} \times 100 = \frac{(-0.25 \ \text{W/m}) - (-0.50 \ \text{W/m})}{-0.50 \ \text{W/m}} \times 100 = -50\% \qquad \triangleleft$$

Comment: Because the geometries are concentric and the specified emissivities and prescribed surface temperatures are spatially uniform, each surface is characterized by uniform irradiation and radiosity distributions. Hence the calculated heat transfer rates would not change if the cylindrical surfaces were to be subdivided into smaller radiative surfaces.

13.3.5   The Reradiating Surface

The assumption of a *reradiating surface* is common to many industrial applications. This idealized surface is characterized by *zero* net radiation transfer ($q_i = 0$). It is closely approached by real surfaces that are well insulated on one side *and* for which convection effects may be neglected on the opposite (radiating) side. With $q_i = 0$, it follows from Equations 13.15 and 13.19 that $G_i = J_i = E_{bi}$. Hence, if the radiosity of a reradiating surface is

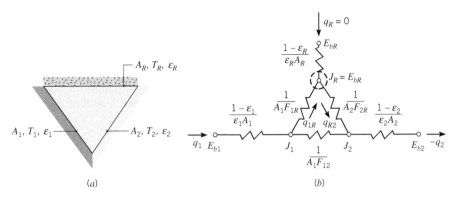

Figure 13.13 A three-surface enclosure with one surface reradiating. (*a*) Schematic. (*b*) Network representation.

known, its temperature is readily determined. In an enclosure, the equilibrium temperature of a reradiating surface is determined by its interaction with the other surfaces, and it is *independent of the emissivity of the reradiating surface.*

A three-surface enclosure, for which the third surface, surface R, is reradiating, is shown in Figure 13.13*a*, and the corresponding network is shown in Figure 13.13*b*. Surface R is presumed to be well insulated, and convection effects are assumed to be negligible. Hence, with $q_R = 0$, the net radiation transfer from surface 1 must equal the net radiation transfer to surface 2. The network is a simple series–parallel arrangement, and from its analysis it is readily shown that

$$q_1 = -q_2 = \frac{E_{b1} - E_{b2}}{\dfrac{1 - \varepsilon_1}{\varepsilon_1 A_1} + \dfrac{1}{A_1 F_{12} + [(1/A_1 F_{1R}) + (1/A_2 F_{2R})]^{-1}} + \dfrac{1 - \varepsilon_2}{\varepsilon_2 A_2}} \qquad (13.30)$$

Knowing $q_1 = -q_2$, Equation 13.19 may be applied to surfaces 1 and 2 to determine their radiosities J_1 and J_2. Knowing J_1, J_2, and the geometrical resistances, the radiosity of the reradiating surface J_R may be determined from the radiation balance

$$\frac{J_1 - J_R}{(1/A_1 F_{1R})} - \frac{J_R - J_2}{(1/A_2 F_{2R})} = 0 \qquad (13.31)$$

The temperature of the reradiating surface may then be determined from the requirement that $\sigma T_R^4 = J_R$.

Note that the general procedure described in Section 13.3.2 may be applied to enclosures with reradiating surfaces. For each such surface, it is appropriate to use Equation 13.22 with $q_i = 0$.

EXAMPLE 13.6

A paint baking oven consists of a long, triangular duct in which a heated surface is maintained at 1200 K and another surface is insulated. Painted panels, which are maintained at 500 K, occupy the third surface. The triangle is of width $W = 1$ m on a side, and the heated and insulated surfaces have an emissivity of 0.8. The emissivity of the panels is 0.4. During steady-state operation, at what rate must energy be supplied to the heated side per unit length of the duct to maintain its temperature at 1200 K? What is the temperature of the insulated surface?

SOLUTION

Known: Surface properties of a long triangular duct that is insulated on one side and heated and cooled on the other sides.

Find:

1. Rate at which heat must be supplied per unit length of duct.
2. Temperature of the insulated surface.

Schematic:

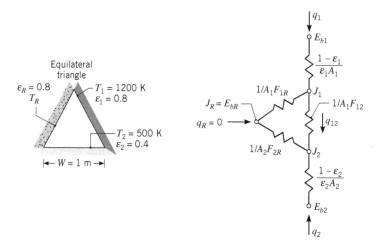

Assumptions:

1. Steady-state conditions exist.
2. All surfaces are opaque, diffuse, gray, and of uniform radiosity and irradiation.
3. Convection effects are negligible.
4. Surface R is reradiating.
5. End effects are negligible.

Analysis:

1. The system may be modeled as a three-surface enclosure with one surface reradiating. The rate at which energy must be supplied to the heated surface may then be obtained from Equation 13.30:

$$q_1 = \frac{E_{b1} - E_{b2}}{\dfrac{1 - \varepsilon_1}{\varepsilon_1 A_1} + \dfrac{1}{A_1 F_{12} + [(1/A_1 F_{1R}) + (1/A_2 F_{2R})]^{-1}} + \dfrac{1 - \varepsilon_2}{\varepsilon_2 A_2}}$$

From symmetry, $F_{12} = F_{1R} = F_{2R} = 0.5$. Also, $A_1 = A_2 = W \cdot L$, where L is the duct length. Hence

$$q_1' = \frac{q_1}{L} = \frac{5.67 \times 10^{-8}\,\text{W/m}^2 \cdot \text{K}^4\,(1200^4 - 500^4)\,\text{K}^4}{\dfrac{1 - 0.8}{0.8 \times 1\,\text{m}} + \dfrac{1}{1\,\text{m} \times 0.5 + (2 + 2)^{-1}\,\text{m}} + \dfrac{1 - 0.4}{0.4 \times 1\,\text{m}}}$$

or

$$q_1' = 37 \text{ kW/m} = -q_2' \qquad \triangleleft$$

2. The temperature of the insulated surface may be obtained from the requirement that $J_R = E_{bR}$, where J_R may be obtained from Equation 13.31. However, to use this expression J_1 and J_2 must be known. Applying the surface energy balance, Equation 13.19, to surfaces 1 and 2, it follows that

$$J_1 = E_{b1} - \frac{1 - \varepsilon_1}{\varepsilon_1 W} q_1' = 5.67 \times 10^{-8} \text{ W/m}^2 \cdot \text{K}^4 (1200 \text{ K})^4$$

$$- \frac{1 - 0.8}{0.8 \times 1 \text{ m}} \times 37{,}000 \text{ W/m} = 108{,}323 \text{ W/m}^2$$

$$J_2 = E_{b2} - \frac{1 - \varepsilon_2}{\varepsilon_2 W} q_2' = 5.67 \times 10^{-8} \text{ W/m}^2 \cdot \text{K}^4 (500 \text{ K})^4$$

$$- \frac{1 - 0.4}{0.4 \times 1 \text{ m}} (-37{,}000 \text{ W/m}) = 59{,}043 \text{ W/m}^2$$

From the energy balance for the reradiating surface, Equation 13.31, it follows that

$$\frac{108{,}323 - J_R}{\dfrac{1}{W \times L \times 0.5}} - \frac{J_R - 59{,}043}{\dfrac{1}{W \times L \times 0.5}} = 0$$

Hence

$$J_R = 83{,}683 \text{ W/m}^2 = E_{bR} = \sigma T_R^4$$

$$T_R = \left(\frac{83{,}683 \text{ W/m}^2}{5.67 \times 10^{-8} \text{ W/m}^2 \cdot \text{K}^4} \right)^{1/4} = 1102 \text{ K} \qquad \triangleleft$$

Comments:

1. We would expect the temperature of the reradiating surface to be higher in regions adjacent to surface 1 and lower in regions closer to surface 2. Our intuition corresponds to the fact that the surface irradiation and radiosity distributions are not uniform, calling into question the validity of Assumption 2. The temperature distribution of the reradiating surface could be determined by use of an analytical or numerical approach, as described in Comment 3 of Example 13.3. If each geometric surface were to be subdivided into 10 smaller elements, however, we would need $(3 \times 10)^2 = 900$ view factors. Precise prediction of radiation heat transfer rates in enclosures whose geometric surfaces are not characterized by uniform radiosity or irradiation distributions involves a trade-off between accuracy and computational effort.

2. The results are independent of the value of ε_R.

3. This problem may also be solved using the direct approach. The solution involves first determining the three unknown radiosities J_1, J_2, and J_R. The governing equations are obtained by writing Equation 13.21 for the two surfaces of known temperature, 1 and 2, and Equation 13.22 for surface R. The three equations are

$$\frac{E_{b1} - J_1}{(1 - \varepsilon_1)/\varepsilon_1 A_1} = \frac{J_1 - J_2}{(A_1 F_{12})^{-1}} + \frac{J_1 - J_R}{(A_1 F_{1R})^{-1}}$$

$$\frac{E_{b2} - J_2}{(1 - \varepsilon_2)/\varepsilon_2 A_2} = \frac{J_2 - J_1}{(A_2 F_{21})^{-1}} + \frac{J_2 - J_R}{(A_2 F_{2R})^{-1}}$$

$$0 = \frac{J_R - J_1}{(A_R F_{R1})^{-1}} + \frac{J_R - J_2}{(A_R F_{R2})^{-1}}$$

Canceling the area A_1, the first equation reduces to

$$\frac{117{,}573 - J_1}{0.25} = \frac{J_1 - J_2}{2} + \frac{J_1 - J_R}{2}$$

or

$$10J_1 - J_2 - J_R = 940{,}584 \qquad (1)$$

Similarly, for surface 2,

$$\frac{3544 - J_2}{1.50} = \frac{J_2 - J_1}{2} + \frac{J_2 - J_R}{2}$$

or

$$-J_1 + 3.33J_2 - J_R = 4725 \qquad (2)$$

and for the reradiating surface

$$0 = \frac{J_R - J_1}{2} + \frac{J_R - J_2}{2}$$

or

$$-J_1 - J_2 + 2J_R = 0 \qquad (3)$$

Solving Equations 1, 2, and 3 simultaneously yields

$$J_1 = 108{,}328 \text{ W/m}^2 \quad J_2 = 59{,}018 \text{ W/m}^2 \quad \text{and} \quad J_R = 83{,}673 \text{ W/m}^3$$

Recognizing that $J_R = \sigma T_R^4$, it follows that

$$T_R = \left(\frac{J_R}{\sigma}\right)^{1/4} = \left(\frac{83{,}673 \text{ W/m}^2}{5.67 \times 10^{-8} \text{ W/m}^2 \cdot \text{K}^4}\right)^{1/4} = 1102 \text{ K}$$

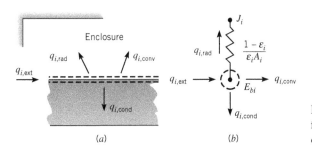

FIGURE 13.14 Multimode heat transfer from a surface in an enclosure. (*a*) Surface energy balance. (*b*) Circuit representation.

13.4 *Multimode Heat Transfer*

Thus far, radiation exchange in an enclosure has been considered under conditions for which conduction and convection could be neglected. However, in many applications, convection and/or conduction are comparable to radiation and must be considered in the heat transfer analysis.

Consider the general surface condition of Figure 13.14*a*. In addition to exchanging energy by radiation with other surfaces of the enclosure, there may be external heat addition to the surface, as, for example, by electric heating, and heat transfer from the surface by convection and conduction. From a surface energy balance, it follows that

$$q_{i,\text{ext}} = q_{i,\text{rad}} + q_{i,\text{conv}} + q_{i,\text{cond}} \tag{13.32}$$

where $q_{i,\text{rad}}$, the net rate of radiation transfer from the surface, is determined by standard procedures for an enclosure. Hence, in general, $q_{i,\text{rad}}$ may be determined from Equation 13.19 or 13.20, while for special cases such as a two-surface enclosure and a three-surface enclosure with one reradiating surface, it may be determined from Equations 13.23 and 13.30, respectively. The surface network element of the radiation circuit is modified according to Figure 13.14*b*, where $q_{i,\text{ext}}$, $q_{i,\text{cond}}$, and $q_{i,\text{conv}}$ represent current flows to or from the surface node. Note, however, that while $q_{i,\text{cond}}$ and $q_{i,\text{conv}}$ are proportional to temperature differences, $q_{i,\text{rad}}$ is proportional to the difference between temperatures raised to the fourth power. Conditions are simplified if the back of the surface is insulated, in which case $q_{i,\text{cond}} = 0$. Moreover, if there is no external heating and convection is negligible, the surface is reradiating.

EXAMPLE 13.7

Consider an air heater consisting of a semicircular tube for which the plane surface is maintained at 1000 K and the other surface is well insulated. The tube radius is 20 mm, and both surfaces have an emissivity of 0.8. If atmospheric air flows through the tube at 0.01 kg/s and $T_m = 400$ K, what is the rate at which heat must be supplied per unit length to maintain the plane surface at 1000 K? What is the temperature of the insulated surface?

SOLUTION

Known: Airflow conditions in tubular heater and heater surface conditions.

Find: Rate at which heat must be supplied and temperature of insulated surface.

Schematic:

Assumptions:

1. Steady-state conditions.
2. Diffuse, gray surfaces experiencing uniform irradiation and radiosity.
3. Negligible tube end effects and axial variations in gas temperature.
4. Fully developed flow.

Properties: Table A.4, air (1 atm, 400 K): $k = 0.0338$ W/m · K, $\mu = 230 \times 10^{-7}$ kg/s · m, $c_p = 1014$ J/kg · K, $Pr = 0.69$.

Analysis: Since the semicircular surface is well insulated and there is no external heat addition, a surface energy balance yields

$$-q_{2,\text{rad}} = q_{2,\text{conv}}$$

Since the tube constitutes a two-surface enclosure, the net radiation transfer to surface 2 may be evaluated from Equation 13.23. Hence

$$\frac{\sigma(T_1^4 - T_2^4)}{\dfrac{1 - \varepsilon_1}{\varepsilon_1 A_1} + \dfrac{1}{A_1 F_{12}} + \dfrac{1 - \varepsilon_2}{\varepsilon_2 A_2}} = hA_2(T_2 - T_m)$$

where the view factor is $F_{12} = 1$ and, per unit length, the surface areas are $A_1 = 2r_o$ and $A_2 = \pi r_o$. With

$$Re_D = \frac{\rho u_m D_h}{\mu} = \frac{\dot{m} D_h}{A_c \mu} = \frac{\dot{m} D_h}{(\pi r_o^2/2)\mu}$$

the hydraulic diameter is

$$D_h = \frac{4 A_c}{P} = \frac{2\pi r_o}{\pi + 2} = \frac{0.04\pi \, \text{m}}{\pi + 2} = 0.0244 \text{ m}$$

Hence

$$Re_D = \frac{0.01 \text{ kg/s} \times 0.0244 \text{ m}}{(\pi/2)(0.02 \text{ m})^2 \times 230 \times 10^{-7} \text{ kg/s} \cdot \text{m}} = 16{,}900$$

From the Dittus–Boelter equation,

$$Nu_D = 0.023 \, Re_D^{4/5} Pr^{0.4}$$

$$Nu_D = 0.023(16,900)^{4/5}(0.69)^{0.4} = 47.8$$

$$h = \frac{k}{D_h} Nu_D = \frac{0.0338 \text{ W/m} \cdot \text{K}}{0.0244 \text{ m}} 47.8 = 66.2 \text{ W/m}^2 \cdot \text{K}$$

Dividing both sides of the energy balance by A_1, it follows that

$$\frac{5.67 \times 10^{-8} \text{ W/m}^2 \cdot \text{K}^4 [(1000)^4 - T_2^4] \text{K}^4}{\dfrac{1-0.8}{0.8} + 1 + \dfrac{1-0.8}{0.8} \dfrac{2}{\pi}} = 66.2 \frac{\pi}{2}(T_2 - 400) \text{ W/m}^2$$

or

$$5.67 \times 10^{-8} T_2^4 + 146.5 T_2 - 115,313 = 0$$

which yields

$$T_2 = 696 \text{ K} \qquad \qquad \triangleleft$$

From an energy balance at the hot surface,

$$q_{1,\text{ext}} = q_{1,\text{rad}} + q_{1,\text{conv}} = q_{2,\text{conv}} + q_{1,\text{conv}}$$

Hence, on a unit length basis,

$$q'_{1,\text{ext}} = h\pi r_o (T_2 - T_m) + h2r_o (T_1 - T_m)$$

$$q'_{1,\text{ext}} = 66.2 \times 0.02[\pi(696 - 400) + 2(1000 - 400)] \text{ W/m}$$

$$q'_{1,\text{ext}} = (1231 + 1589) \text{ W/m} = 2820 \text{ W/m} \qquad \qquad \triangleleft$$

Comments:

1. The irradiation, radiosity, and convection heat flux distributions along the surfaces would not be uniform. As a consequence, we would expect the temperature of the insulated surface to be higher in the corner regions adjacent to surface 1 and lower in the crown of the enclosure. Determination of the irradiation, radiosity, temperature, and convection heat flux distributions along the various surfaces would require a more complex analysis incorporating many radiative surfaces.

2. Applying an energy balance to a differential control volume about the air, it follows that

$$\frac{dT_m}{dx} = \frac{q'_1}{\dot{m}c_p} = \frac{2820 \text{ W/m}}{0.01 \text{ kg/s} (1014 \text{ J/kg} \cdot \text{K})} = 278 \text{ K/m}$$

Hence the air temperature change is significant, and a more representative analysis would subdivide the tube into axial zones and would allow for variations in air and insulated surface temperatures between zones. Moreover, a two-surface analysis of radiation exchange would no longer be appropriate.

13.5   *Implications of the Simplifying Assumptions*

Although we have developed a means for predicting radiation exchange between surfaces, it is important to be cognizant of the inherent limitations of our analyses. Recall that we have analyzed *idealized surfaces* that are isothermal, opaque, and gray that emit and reflect diffusely and are characterized by uniform radiosity and irradiation distributions. Moreover, the enclosures that we have analyzed have been assumed to house media that are nonparticipating; that is, they neither absorb nor scatter the surface radiation, and emit no radiation.

The analysis techniques presented in this chapter may often be used to obtain first estimates and, in many cases, sufficiently accurate results for radiation heat transfer involving multiple surfaces that form an enclosure. In some cases, however, the assumptions are inappropriate, and more refined prediction methods are needed. Although beyond the scope of this text, the methods are discussed in more advanced treatments of radiation transfer [3, 7–12].

13.6   *Radiation Exchange with Participating Media*

Except for our discussion of environmental radiation (Section 12.9), we have said little about gaseous radiation, having confined our attention to radiation exchange at the surface of an opaque solid or liquid. For *nonpolar* gases, such as O_2 or N_2, such neglect is justified, since the gases do not emit radiation and are essentially transparent to incident thermal radiation. However, the same may not be said for polar molecules, such as CO_2, H_2O (vapor), NH_3, and hydrocarbon gases, which emit and absorb over a wide temperature range. For such gases matters are complicated by the fact that, unlike radiation from a solid or a liquid, which is distributed continuously with wavelength, gaseous radiation is concentrated in specific *wavelength intervals* (called bands). Moreover, gaseous radiation is not a surface phenomenon, but is instead a *volumetric* phenomenon.

13.6.1   Volumetric Absorption

Spectral radiation absorption in a gas (or in a semitransparent liquid or solid) is a function of the absorption coefficient κ_λ (1/m) and the thickness L of the medium (Figure 13.15). If a monochromatic beam of intensity $I_{\lambda,0}$ is incident on the medium, the intensity is reduced due to absorption, and the reduction occurring in an infinitesimal layer of thickness dx may be expressed as

$$dI_\lambda(x) = -\kappa_\lambda I_\lambda(x)\, dx \tag{13.33}$$

FIGURE 13.15   Absorption in a gas or liquid layer.

Separating variables and integrating over the entire layer, we obtain

$$\int_{I_{\lambda,0}}^{I_{\lambda,L}} \frac{dI_\lambda(x)}{I_\lambda(x)} = -\kappa_\lambda \int_0^L dx$$

where κ_λ is assumed to be independent of x. It follows that

$$\frac{I_{\lambda,L}}{I_{\lambda,0}} = e^{-\kappa_\lambda L} \tag{13.34}$$

This exponential decay, termed *Beer's law*, is a useful tool in approximate radiation analysis. It may, for example, be used to infer the overall spectral absorptivity of the medium. In particular, with the transmissivity defined as

$$\tau_\lambda = \frac{I_{\lambda,L}}{I_{\lambda,0}} = e^{-\kappa_\lambda L} \tag{13.35}$$

the absorptivity is

$$\alpha_\lambda = 1 - \tau_\lambda = 1 - e^{-\kappa_\lambda L} \tag{13.36}$$

If Kirchhoff's law is assumed to be valid, $\alpha_\lambda = \varepsilon_\lambda$, Equation 13.36 also provides the spectral emissivity of the medium.

13.6.2 Gaseous Emission and Absorption

A common engineering calculation is one that requires determination of the radiant heat flux from a gas to an adjoining surface. Despite the complicated spectral and directional effects inherent in such calculations, a simplified procedure may be used. The method was developed by Hottel [13] and involves determining radiation emission from a hemispherical gas mass of temperature T_g to a surface element dA_1, which is located at the center of the hemisphere's base. Emission from the gas per unit area of the surface is expressed as

$$E_g = \varepsilon_g \sigma T_g^4 \tag{13.37}$$

where the gas emissivity ε_g was determined by correlating available data. In particular, ε_g was correlated in terms of the temperature T_g and total pressure p of the gas, the partial pressure p_g of the radiating species, and the radius L of the hemisphere.

Results for the emissivity of water vapor are plotted in Figure 13.16 as a function of the gas temperature, for a total pressure of 1 atm, and for different values of the product of the vapor partial pressure and the hemisphere radius. To evaluate the emissivity for total pressures other than 1 atm, the emissivity from Figure 13.16 must be multiplied by the correction factor C_w from Figure 13.17. Similar results were obtained for carbon dioxide and are presented in Figures 13.18 and 13.19.

The foregoing results apply when water vapor or carbon dioxide appear *separately* in a mixture with other species that are nonradiating. However, the results may readily be extended to situations in which water vapor and carbon dioxide appear *together* in a mixture with other nonradiating gases. In particular, the total gas emissivity may be expressed as

$$\varepsilon_g = \varepsilon_w + \varepsilon_c - \Delta\varepsilon \tag{13.38}$$

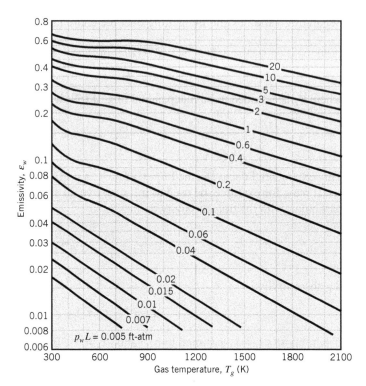

FIGURE 13.16   Emissivity of water vapor in a mixture with nonradiating gases at 1-atm total pressure and of hemispherical shape [13]. Used with permission.

FIGURE 13.17   Correction factor for obtaining water vapor emissivities at pressures other than $(\varepsilon_{w,p\neq1\,\text{atm}} = C_w\varepsilon_{w,p=1\,\text{atm}})$ [13]. Used with permission.

FIGURE 13.18 Emissivity of carbon dioxide in a mixture with nonradiating gases at 1-atm total pressure and of hemispherical shape [13]. Used with permission.

FIGURE 13.19 Correction factor for obtaining carbon dioxide emissivities at pressures other than 1 atm $(\varepsilon_{c,p\neq1\,\mathrm{atm}} = C_c \varepsilon_{c,p=1\,\mathrm{atm}})$ [13]. Used with permission.

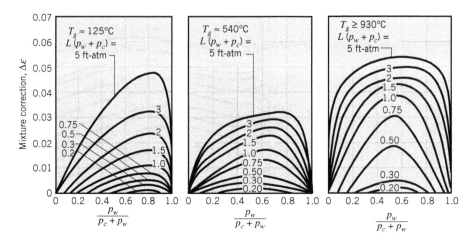

FIGURE 13.20 Correction factor associated with mixtures of water vapor and carbon dioxide [13]. Used with permission.

where the correction factor $\Delta\varepsilon$ is presented in Figure 13.20 for different values of the gas temperature. This factor accounts for the reduction in emission associated with the mutual absorption of radiation between the two species.

Recall that the foregoing results provide the emissivity of a hemispherical gas mass of radius L radiating to an element of area at the center of its base. However, the results may be extended to other gas geometries by introducing the concept of a *mean beam length, L_e*. The quantity was introduced to correlate, in terms of a single parameter, the dependence of gas emissivity on both the size and the shape of the gas geometry. It may be interpreted as the radius of a hemispherical gas mass whose emissivity is equivalent to that for the geometry of interest. Its value has been determined for numerous gas shapes [13], and representative results are listed in Table 13.4. Replacing L by L_e in Figures 13.16 through 13.20, the emissivity associated with the geometry of interest may then be determined.

Using the results of Table 13.4 with Figures 13.16 through 13.20, it is possible to determine the rate of radiant heat transfer to a surface due to emission from an adjoining gas. This heat rate may be expressed as

$$q = \varepsilon_g A_s \sigma T_g^4 \tag{13.39}$$

TABLE 13.4 **Mean beam lengths L_e for various gas geometries**

Geometry	Characteristic Length	L_e
Sphere (radiation to surface)	Diameter (D)	$0.65D$
Infinite circular cylinder (radiation to curved surface)	Diameter (D)	$0.95D$
Semi-infinite circular cylinder (radiation to base)	Diameter (D)	$0.65D$
Circular cylinder of equal height and diameter (radiation to entire surface)	Diameter (D)	$0.60D$
Infinite parallel planes (radiation to planes)	Spacing between planes (L)	$1.80L$
Cube (radiation to any surface)	Side (L)	$0.66L$
Arbitrary shape of volume V (radiation to surface of area A)	Volume to area ratio (V/A)	$3.6V/A$

where A_s is the surface area. If the surface is black, it will, of course, absorb all this radiation. A black surface will also emit radiation, and the net rate at which radiation is exchanged between the surface at T_s and the gas at T_g is

$$q_{\text{net}} = A_s \sigma (\varepsilon_g T_g^4 - \alpha_g T_s^4) \tag{13.40}$$

For water vapor and carbon dioxide the required gas absorptivity α_g may be evaluated from the emissivity by expressions of the form [13]

Water:

$$\alpha_w = C_w \left(\frac{T_g}{T_s} \right)^{0.45} \times \varepsilon_w \left(T_s, \, p_w L_e \frac{T_s}{T_g} \right) \tag{13.41}$$

Carbon dioxide:

$$\alpha_c = C_c \left(\frac{T_g}{T_s} \right)^{0.65} \times \varepsilon_c \left(T_s, \, p_c L_e \frac{T_s}{T_g} \right) \tag{13.42}$$

where ε_w and ε_c are evaluated from Figures 13.16 and 13.18, respectively, and C_w and C_c are evaluated from Figures 13.17 and 13.19, respectively. Note, however, that in using Figures 13.16 and 13.18, T_g is replaced by T_s and $p_w L_e$ or $p_c L_e$ is replaced by $p_w L_e (T_s/T_g)$ or $p_c L_e (T_s/T_g)$, respectively. Note also that, in the presence of both water vapor and carbon dioxide, the total gas absorptivity may be expressed as

$$\alpha_g = \alpha_w + \alpha_c - \Delta\alpha \tag{13.43}$$

where $\Delta\alpha = \Delta\varepsilon$ is obtained from Figure 13.20.

13.7 *Summary*

In this chapter we focused on the analysis of radiation exchange between the surfaces of an enclosure, and in treating this exchange we introduced the concept of a *view factor*. Because knowledge of this geometrical quantity is essential to determining radiation exchange between any two diffuse surfaces, you should be familiar with the means by which it may be determined. You should also be adept at performing radiation calculations for an enclosure of *isothermal*, *opaque*, *diffuse*, and *gray* surfaces of *uniform radiosity* and *irradiation.* Moreover, you should be familiar with the results that apply to simple cases such as an enclosure of black surfaces, a two-surface enclosure, or a three-surface enclosure with a reradiating surface. Finally, be cognizant that in situations where the radiosity and irradiation distributions are nonuniform, it may be necessary to perform a radiation exchange analysis involving many surfaces in order to determine local surface temperatures or local heat fluxes.

Test your understanding of pertinent concepts by addressing the following questions.

- What is a *view factor*? What assumptions are typically associated with computing the view factor between two surfaces?
- What is the *reciprocity relation* for view factors? What is the *summation rule*?

- Can the view factor of a surface with respect to itself be nonzero? If so, what kind of surface exhibits such behavior?
- What is a *nonparticipating medium*?
- What assumptions are inherent in treating radiation exchange between surfaces of an enclosure that may not be approximated as blackbodies? Comment on the validity of the assumptions and when or where they are most likely to break down.
- How is the radiative resistance of a surface in an enclosure defined? What is the driving potential that relates this resistance to the net rate of radiation transfer from the surface? What is the resistance if the surface may be approximated as a blackbody?
- How is the geometrical resistance associated with radiative exchange between two surfaces of an enclosure defined? What is the driving potential that relates this resistance to the net rate of radiation transfer between the surfaces?
- What is a *radiation shield* and how is net radiation transfer between two surfaces affected by an intervening shield? Is it advantageous for a shield to have a large surface absorptivity or reflectivity?
- What is a *reradiating surface*? Under what conditions may a surface be approximated as reradiating? What is the relation between radiosity, blackbody emissive power, and irradiation for a reradiating surface? Does the temperature of such a surface depend on its radiative properties?
- What can be said about a surface of an enclosure for which net radiation heat transfer to the surface is balanced by convection heat transfer from the surface to a gas in the enclosure? Is the surface reradiating? Is the back side of the surface adiabatic?
- Consider a surface in an enclosure for which net radiation transfer from the surface exceeds convection heat transfer from a gas in the enclosure. What other process or processes must occur at the surface, either independently or collectively?
- What molecular features render a gas nonemitting and nonabsorbing? What features allow for emission and absorption of radiation by a gas?
- What features distinguish emission and absorption of radiation by a gas from that of an opaque solid?
- How does the intensity of radiation propagating through a semitransparent medium vary with distance in the medium? What can be said about this variation if the absorption coefficient is very large? If the coefficient is very small?

References

1. Hamilton, D. C., and W. R. Morgan, "Radiant Interchange Configuration Factors," National Advisory Committee for Aeronautics, Technical Note 2836, 1952.

2. Eckert, E. R. G., "Radiation: Relations and Properties," in W. M. Rohsenow and J. P. Hartnett, Eds., *Handbook of Heat Transfer*, 2nd ed., McGraw-Hill, New York, 1973.

3. Howell, J. R., R. Siegel, and M. P. Menguc, *Thermal Radiation Heat Transfer*, 5th ed., Taylor & Francis, New York, 2010.

4. Howell, J. R., *A Catalog of Radiation Configuration Factors*, McGraw-Hill, New York, 1982.

5. Emery, A. F., O. Johansson, M. Lobo, and A. Abrous, *J. Heat Transfer*, **113**, 413, 1991.

6. Oppenheim, A. K., *Trans. ASME*, **65**, 725, 1956.

7. Hottel, H. C., and A. F. Sarofim, *Radiative Transfer*, McGraw-Hill, New York, 1967.

8. Tien, C. L., "Thermal Radiation Properties of Gases," in J. P. Hartnett and T. F. Irvine, Eds., *Advances in Heat Transfer*, Vol. 5, Academic Press, New York, 1968.

9. Sparrow, E. M., "Radiant Interchange Between Surfaces Separated by Nonabsorbing and Nonemitting Media," in W. M. Rohsenow and J. P. Hartnett, Eds., *Handbook of Heat Transfer*, McGraw-Hill, New York, 1973.

10. Dunkle, R. V., "Radiation Exchange in an Enclosure with a Participating Gas," in W. M. Rohsenow and J. P. Hartnett, Eds., *Handbook of Heat Transfer*, McGraw-Hill, New York, 1973.

11. Sparrow, E. M., and R. D. Cess, *Radiation Heat Transfer*, Hemisphere Publishing, New York, 1978.

12. Edwards, D. K., *Radiation Heat Transfer Notes*, Hemisphere Publishing, New York, 1981.

13. Hottel, H. C., and R. B. Egbert, *AIChE J.*, **38**, 531, 1942.

Problems

View Factors

13.1 Determine F_{12} and F_{21} for the following configurations using the reciprocity theorem and other basic shape factor relations. Do not use tables or charts.

(a) Small sphere of area A_1 under a concentric hemisphere of area $A_2 = 2A_1$

(a)

(b) Long duct

(b)

(c) Long inclined plates (point *B* is directly above the center of A_1)

(c)

(d) Long cylinder lying on infinite plane

(d)

(e) Hemisphere–disk arrangement

(e)

(f) Long, open channel

(f)

(g) Long concentric cylinders

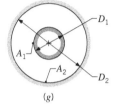

(g)

(h) Long, square rod in a long cylinder. Also, what is F_{22}?

(h)

13.2 A spherical particle of radius $r_1 = 5$ mm is suspended by a thin wire within a cylindrical tube of inner radius $r_2 = 15$ mm and length $L = 0.5$ m. Determine the view factors F_{11}, F_{12}, F_{13}, and F_{14} for $S = 5$ and 250 mm. The view factor between the sphere and the bottom end of the tube (a disk of radius r_3) is $F_{13} = 0.5\left[1 - \left(1 + R_3^2\right)^{-1/2}\right]$ where $R_3 = r_3/S$.

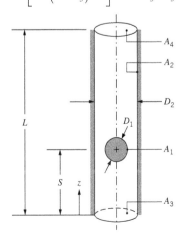

13.3 Consider the following grooves, each of width W, that have been machined from a solid block of material.

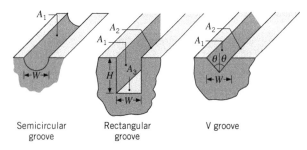

Semicircular groove   Rectangular groove   V groove

(a) For each case obtain an expression for the view factor of the groove with respect to the surroundings outside the groove.

(b) For the V groove, obtain an expression for the view factor F_{12}, where A_1 and A_2 are opposite surfaces.

(c) If $H = 2W$ in the rectangular groove, what is the view factor F_{12}?

13.4 A right-circular cone and a right-circular cylinder of the same diameter and length (A_2) are positioned coaxially at a distance L_o from the circular disk (A_1) shown schematically. The inner base and lateral surfaces of the cylinder may be treated as a single surface, A_2. The hypothetical area corresponding to the opening of the cone and cylinder is identified as A_3.

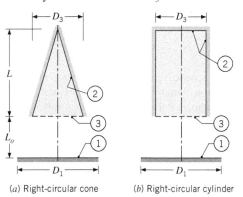

(a) Right-circular cone   (b) Right-circular cylinder

(a) Show that, for both arrangements, $F_{21} = (A_1/A_2)F_{13}$ and $F_{22} = 1 - (A_3/A_2)$, where F_{13} is the view factor between two coaxial, parallel disks (Table 13.2).

(b) For $L = L_o = 50$ mm and $D_1 = D_3 = 50$ mm, calculate F_{21} and F_{22} for the conical and cylindrical configurations and compare their relative magnitudes. Explain any similarities and differences.

(c) Do the relative magnitudes of F_{21} and F_{22} change for the conical and cylindrical configurations as L increases and all other parameters remain fixed? In the limit of very large L, what do you expect will happen? Sketch the variations of F_{21} and F_{22} with L, and explain the key features.

13.5 Consider the two parallel, coaxial, ring-shaped disks shown schematically.

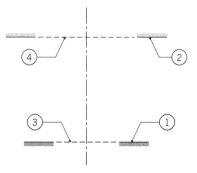

Show that F_{12} can be expressed as

$$F_{12} = \frac{1}{A_1}[A_{(1,3)}F_{(1,3)(2,4)} - A_3F_{3(2,4)} - A_4(F_{4(1,3)} - F_{43})]$$

where all view factors on the right-hand side of the equation can be evaluated from Figure 13.5 or Table 13.2 for coaxial parallel disks.

13.6 The "crossed-strings" method of Hottel [7] provides a simple means to calculate view factors between surfaces that are of infinite extent in one direction. For two such surfaces (a) with unobstructed views of one another, the view factor is of the form

$$F_{12} = \frac{1}{2w_1}[(ac + bd) - (ad + bc)]$$

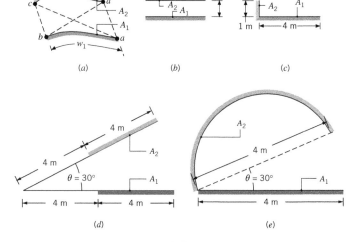

Use this method to evaluate the view factors F_{12} for sketches (b) through (e). Compare your results with those from using the appropriate graphs, analytical expressions, and view factor relations.

13.7 Consider the parallel rectangles shown schematically.

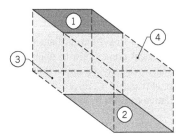

Show that the view factor F_{12} can be expressed as

$$F_{12} = \frac{1}{2A_1} [A_{(1,4)} F_{(1,4)(2,3)} - A_1 F_{13} - A_4 F_{42}]$$

where all view factors on the right-hand side of the equation can be evaluated from Figure 13.4 (see Table 13.2) for aligned parallel rectangles.

13.8 Consider the perpendicular rectangles shown schematically.

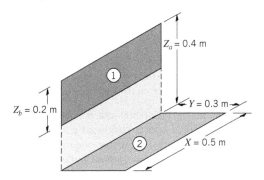

(a) Determine the shape factor F_{12}.

(b) For rectangle widths of $X = 0.5$, 1.5, and 5 m, plot F_{12} as a function of Z_b for $0.05 \leq Z_b \leq 0.4$m. Compare your results with the view factor obtained from the two-dimensional relation for perpendicular plates with a common edge (Table 13.1).

13.9 Determine the shape factor, F_{12}, for the rectangles shown.

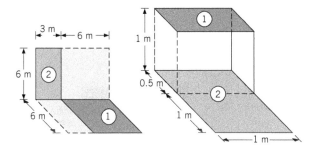

(a) Perpendicular rectangles without a common edge.

(b) Parallel rectangles of unequal areas.

13.10 Consider parallel planes of infinite extent normal to the page as shown in the sketch.

(a) Determine F_{12} using the results of Figure 13.4.

(b) Determine F_{12} using the first case of Table 13.1.

(c) Determine F_{12} using Hottel's crossed-string method described in Problem 13.6.

(d) Determine F_{12} using the second case of Table 13.1.

(e) Determine F_{12} using the results of Figure 13.4 if the dimensions are increased to $w = L = 2$ m.

13.11 Consider two diffuse surfaces A_1 and A_2 on the inside of a spherical enclosure of radius R. Using the following methods, derive an expression for the view factor F_{12} in terms of A_2 and R.

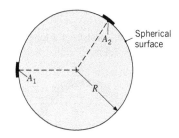

(a) Find F_{12} by beginning with the expression $F_{ij} = q_{i \rightarrow j}/A_i J_i$.

(b) Find F_{12} using the view factor integral, Equation 13.1.

13.12 As shown in the sketch, consider the disk A_1 located coaxially 1 m distant, but tilted $30°$ off the normal, from the ring-shaped disk A_2.

What is the irradiation on A_1 due to radiation from A_2, which is a diffuse, gray surface with an emissivity of 0.7?

13.13 A heat flux gage of 4-mm diameter is positioned normal to and 1 m from the 5-mm-diameter aperture of a blackbody furnace at 1000 K. The diffuse, gray cover shield ($\varepsilon = 0.2$) of the furnace has an outer diameter of 100 mm and its temperature is 350 K. The furnace and gage are located in a large room whose walls have an emissivity of 0.8 and are at 300 K.

- Furnace, $T_f = 1000$ K
- Cover, $T_c = 350$ K, $\varepsilon = 0.2$
- Gage, 4-mm diameter
- $D = 100$ mm
- Aperture, 5-mm diameter
- 1 m

(a) What is the irradiation on the gage, G_g (W/m^2), considering only emission from the aperture of the furnace?

(b) What is the irradiation on the gage due to radiation from the cover and aperture?

Blackbody Radiation Exchange

13.14 A drying oven consists of a long semicircular duct of diameter $D = 1.5$ m.

- $T_1 = 1200$ K
- $T_2 = 350$ K
- $D = 1.5$ m

Materials to be dried cover the base of the oven while the wall is maintained at 1200 K. What is the drying rate per unit length of the oven (kg/s · m) if a water-coated layer of material is maintained at 350 K during the drying process? Blackbody behavior may be assumed for the water surface and the oven wall.

13.15 Consider the arrangement of the three black surfaces shown, where A_1 is small compared to A_2 or A_3.

- A_2
- 2 m
- A_3
- A_1
- 3 m

Determine the value of F_{13}. Calculate the net radiation heat transfer rate from A_1 to A_3 if $A_1 = 0.05$ m^2, $T_1 = 1000$ K, and $T_3 = 600$ K.

13.16 A circular ice rink 25 m in diameter is enclosed by a hemispherical dome 35 m in diameter. If the ice and dome surfaces may be approximated as blackbodies and are at 0 and 15°C, respectively, what is the net rate of radiative transfer from the dome to the rink?

13.17 Consider coaxial, parallel, black disks separated a distance of 0.20 m. The lower disk of diameter 0.40 m is maintained at 500 K and the surroundings are at 300 K.

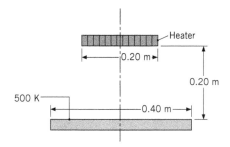

- Heater
- 0.20 m
- 0.20 m
- 500 K
- 0.40 m

What temperature will the upper disk of diameter 0.20 m achieve if electrical power of 17.5 W is supplied to the heater on the back side of the disk?

13.18 A circular plate of 500-mm diameter is maintained at $T_1 = 600$ K and is positioned coaxial to a conical shape. The back side of the cone is well insulated. The plate and the cone, whose surfaces are black, are located in a large, evacuated enclosure whose walls are at 300 K.

- Cone with insulated back side
- T_2
- Surroundings $T_{sur} = 300$ K
- $L = 500$ mm
- $D = 500$ mm
- $T_1 = 600$ K
- Plate with electrical heater

(a) What is the temperature of the conical surface, T_2?

(b) What is the electrical power that would be required to maintain the circular plate at 600 K?

13.19 Two furnace designs, as illustrated and dimensioned in Problem 13.4, are used to heat a disk-shaped work piece (A_1). The power supplied to the conical and cylindrical furnaces (A_2) is 50 W. The work piece is located in a large room at a temperature of 300 K, and its back side is well insulated. Assuming all surfaces are black, determine the temperature of the work piece, T_1, and the temperature of the furnace inner surface, T_2, for each of the designs. In your analysis, use the expressions provided in Problem 13.4 for the view factors, F_{21} and F_{22}.

13.20 A furnace is constructed in three sections, which include insulated circular (2) and cylindrical (3) sections, as well as an intermediate cylindrical section (1) with embedded electrical resistance heaters. The overall length and diameter are 200 mm and 100 mm, respectively, and the cylindrical sections are of equal length. The surroundings are at 0 K.

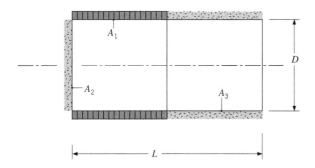

(a) If all the surfaces are black, determine the electrical power, q_1, required to maintain the heated section at 1000 K.

(b) What are the temperatures of the insulated sections, T_2 and T_3?

(c) For $D = 100$ mm, generate a plot of q_1, T_2, and T_3 as functions of the length-to-diameter ratio, with $1 \le L/D \le 5$.

13.21 A cylindrical cavity of diameter D and depth L is machined in a metal block, and conditions are such that the base and side surfaces of the cavity are maintained at $T_1 = 1000$ K and $T_2 = 700$ K, respectively. Approximating the surfaces as black, determine the emissive power of the cavity if $L = 20$ mm and $D = 10$ mm.

13.22 In the arrangement shown, the lower disk has a diameter of 40 mm and a temperature of 500 K. The upper surface, which is at 1000 K, is a ring-shaped disk with inner and outer diameters of 0.15 m and 0.2 m. This upper surface is aligned with and parallel to the lower disk and is separated by a distance of 1 m.

Assuming both surfaces to be blackbodies, calculate their net rate of radiative heat exchange.

13.23 Two plane coaxial disks are separated by a distance $L = 0.20$ m. The lower disk (A_1) is solid with a diameter $D_o = 0.80$ m and a temperature $T_1 = 300$ K. The upper disk (A_2), at temperature $T_2 = 1000$ K, has the same outer diameter but is ring-shaped with an inner diameter $D_i = 0.40$ m.

Assuming the disks to be blackbodies, calculate the net radiative heat exchange between them.

13.24 A radiometer views a small target (1) that is being heated by a ring-shaped disk heater (2). The target has an area of $A_1 = 0.0004$ m², a temperature of $T_1 = 500$ K, and a diffuse, gray emissivity of $\varepsilon_1 = 0.8$. The heater operates at $T_2 = 1000$ K and has a black surface. The radiometer views the entire sample area with a solid angle of $\omega = 0.0008$ sr.

(a) Write an expression for the radiant power leaving the target which is collected by the radiometer, in terms of the target radiosity J_1 and relevant geometric parameters. Leave in symbolic form.

(b) Write an expression for the target radiosity J_1 in terms of its irradiation, emissive power, and appropriate radiative properties. Leave in symbolic form.

(c) Write an expression for the irradiation on the target, G_1, due to emission from the heater in terms of the heater emissive power, the heater area, and an appropriate view factor. Use this expression to numerically evaluate G_1.

(d) Use the foregoing expressions and results to determine the radiant power collected by the radiometer.

13.25 The arrangement shown is to be used to calibrate a heat flux gage. The gage has a black surface that is 10 mm in diameter and is maintained at 17°C by means of a water-cooled backing plate. The heater, 200 mm in diameter, has a black surface that is maintained at 800 K and is located 0.5 m from the gage. The surroundings and the air are at 27°C and the convection heat transfer coefficient between the gage and the air is 15 W/m² · K.

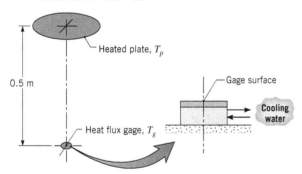

(a) Determine the net radiation exchange between the heater and the gage.

(b) Determine the net transfer of radiation to the gage per unit area of the gage.

(c) What is the net heat transfer rate to the gage per unit area of the gage?

13.26 An electrically heated sample is maintained at a surface temperature of $T_s = 500$ K. The sample coating is diffuse but spectrally selective, with the spectral emissivity distribution shown schematically. The sample is irradiated by a furnace located coaxially at a distance of $L_{sf} = 750$ mm. The furnace has isothermal walls with an emissivity of $\varepsilon_f = 0.7$ and a uniform temperature of $T_f = 3000$ K. A radiation detector of area $A_d = 8 \times 10^{-5}$ m² is positioned at a distance of $L_{sd} = 1.0$ m

from the sample along a direction that is 45° from the sample normal. The detector is sensitive to spectral radiant power only in the spectral region from 3 to 5 μm. The sample surface experiences convection with a gas for which $T_\infty = 300$ K and $h = 20$ W/m² · K. The surroundings of the sample mount are large and at a uniform temperature of $T_{sur} = 300$ K.

(a) Determine the electrical power, P_e, required to maintain the sample at $T_s = 500$ K.

(b) Considering both emission and reflected irradiation from the sample, determine the radiant power that is incident on the detector within the spectral region from 3 to 5 μm.

13.27 Water flowing through a large number of long, circular, thin-walled tubes is heated by means of hot parallel plates above and below the tube array. The space between the plates is evacuated, and the plate and tube surfaces may be approximated as blackbodies.

(a) Neglecting axial variations, determine the tube surface temperature, T_s, if water flows through each tube at a mass rate of $\dot{m} = 0.20$ kg/s and a mean temperature of $T_m = 300$ K.

(b) Compute and plot the surface temperature as a function of flow rate for $0.05 \leq \dot{m} \leq 0.25$ kg/s.

13.28 A row of regularly spaced, cylindrical heating elements is used to maintain an insulated furnace wall at

500 K. The opposite wall is at a uniform temperature of 300 K.

The insulated wall experiences convection with air at 450 K and a convection coefficient of 250 W/m² · K. Assuming the walls and elements are black, estimate the required operating temperature for the elements.

13.29 A manufacturing process calls for heating long copper rods, which are coated with a thin film having $\varepsilon = 1$, by placing them in a large evacuated oven whose surface is maintained at 1650 K. The rods are of 10-mm diameter and are placed in the oven with an initial temperature of 300 K.

(a) What is the initial rate of change of the rod temperature?

(b) How long must the rods remain in the oven to achieve a temperature of 1000 K?

(c) The heating process may be accelerated by routing combustion gases, also at 1650 K, through the oven. For convection coefficients of 10, 100, and 500 W/m² · K, determine the time required for the rods to reach 1000 K.

13.30 Many products are processed in a manner that requires a specified product temperature as a function of time. Consider a product in the shape of long, 10-mm-diameter cylinders that are conveyed slowly through a processing oven as shown in the schematic. The product exhibits near-black behavior and is attached to the conveying apparatus at the product's ends. The surroundings are at 300 K, while the radiation panel heaters are each isothermal at 500 K and have surfaces that exhibit near-black behavior. An engineer proposes a novel oven design with a tilting top surface so as to be able to quickly change the thermal response of the product.

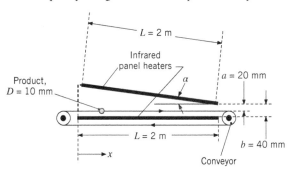

(a) Determine the radiation per unit length incident upon the product at $x = 0.5$ m and $x = 1$ m for $\alpha = 0$.

(b) Determine the radiation per unit length incident upon the product at $x = 0.5$ m and $x = 1$ m for $\alpha = \pi/15$. *Hint:* The view factor from the cylinder to the left-hand side surroundings can be found by summing the view factors from the cylinder to the two surfaces shown as bold dashed lines in the schematic.

Two-Surface Enclosures and Radiation Shields

13.31 Consider two very large parallel plates with diffuse, gray surfaces.

Determine the irradiation and radiosity for the upper plate. What is the radiosity for the lower plate? What is the net radiation exchange between the plates per unit area of the plates?

13.32 Equation 1.7 is subject to several restrictions, such as the surroundings being isothermal and large relative to the smaller surface. Consider the small and large surfaces to be isothermal concentric spheres of radii r_1 and r_2, respectively.

(a) Calculate the radius ratio, r_2/r_1 for which Equation 1.7 yields a radiation heat rate that is within 1% of the heat rate given by Equation 13.26 for the following cases. The opaque surfaces are diffuse and gray.

$$
\begin{aligned}
\text{Case 1: } & \varepsilon_1 = 0.9, \ \varepsilon_2 = 0.9 \\
\text{Case 2: } & \varepsilon_1 = 0.9, \ \varepsilon_2 = 0.1 \\
\text{Case 3: } & \varepsilon_1 = 0.1, \ \varepsilon_2 = 0.9 \\
\text{Case 4: } & \varepsilon_1 = 0.1, \ \varepsilon_2 = 0.1
\end{aligned}
$$

(b) Plot the required radius ratio versus the emissivity of the larger sphere for the two emissivities of the smaller sphere.

13.33 A flat-bottomed hole 6 mm in diameter is bored to a depth of 24 mm in a diffuse, gray material having an emissivity of 0.8 and a uniform temperature of 1000 K.

(a) Determine the radiant power leaving the opening of the cavity.

(b) The effective emissivity ε_e of a cavity is defined as the ratio of the radiant power leaving the cavity to that from a blackbody having the area of the cavity opening and a temperature of the inner surfaces of the cavity. Calculate the effective emissivity of the cavity described above.

(c) If the depth of the hole were increased, would ε_e increase or decrease? What is the limit of ε_e as the depth increases?

13.34 Consider a double-pane window. The glass surface may be treated with a low-emissivity coating to reduce its emissivity from $\varepsilon = 0.95$ to $\varepsilon = 0.05$. Determine the radiation heat flux between two glass sheets for case 1: $\varepsilon_1 = \varepsilon_2 = 0.95$, case 2: $\varepsilon_1 = \varepsilon_2 = 0.05$, and case 3: $\varepsilon_1 = 0.05$, $\varepsilon_2 = 0.95$. The glass temperatures are $T_1 = 20°C$ and $T_2 = -20°C$, respectively.

13.35 Two large parallel plates are maintained at $T_1 = 500$ and $T_2 = 300$ K, respectively. The hot plate has an emissivity of 0.9 while that of the cold plate is 0.7.

(a) Determine the radiation heat flux without and with a radiation shield formed of a flat sheet of foil placed midway between the two plates. Both sides of the shield have an emissivity of 0.05.

(b) A scheme is proposed to further reduce the radiation heat flux between the plates by adding more shield material, achieved by corrugating the flat shield material. Calculate the radiation heat flux associated with the corrugated shield of the sketch with a crease angle of $\alpha = \pi/4$. Is the scheme successful in reducing the heat flux?

(c) Graph the radiation heat flux between the plates with no shield, with a flat shield in place, and with a corrugated shield in place over the range $0 \leq \alpha \leq \pi$. Explain the behavior evident in your graph.

13.36 Consider a long V groove 10 mm deep machined on a block that is maintained at 1000 K.

If the groove surfaces are diffuse and gray with an emissivity of 0.6, determine the radiant flux leaving the groove to its surroundings. Also determine the effective emissivity of the cavity, as defined in Problem 13.33.

13.37 Consider the cavities formed by a cone, cylinder, and sphere having the same opening size (d) and major dimension (L), as shown in the diagram.

(a) Find the view factor between the inner surface of each cavity and the opening of the cavity.

(b) Find the effective emissivity of each cavity, ε_e, as defined in Problem 13.33, assuming the inner walls are diffuse and gray with an emissivity of ε_w.

(c) For each cavity and wall emissivities of $\varepsilon_w = 0.5$, 0.7, and 0.9, plot ε_e as a function of the major dimension-to-opening size ratio, L/d, over a range from 1 to 10.

13.38 Consider the attic of a home located in a hot climate. The floor of the attic is characterized by a width of $L_1 = 10$ m while the roof makes an angle of $\theta = 30°$ from the horizontal direction, as shown in the schematic. The homeowner wishes to reduce the heat load to the home by adhering bright aluminum foil ($\varepsilon_f = 0.07$) onto the surfaces of the attic space. Prior to installation of the foil, the surfaces are of emissivity $\varepsilon_o = 0.85$.

(a) Consider installation on the bottom of the attic roof only. Determine the ratio of the radiation heat transfer after to before the installation of the foil.

(b) Determine the ratio of the radiation heat transfer after to before installation if the foil is installed only on the top of the attic floor.

(c) Determine the ratio of the radiation heat transfer if the foil is installed on both the roof bottom and the floor top.

13.39 A long, thin-walled horizontal tube 100 mm in diameter is maintained at 120°C by the passage of steam through its interior. A radiation shield is installed around the tube, providing an air gap of 10 mm between the tube and the shield, and reaches a surface temperature of 35°C. The tube and shield are diffuse, gray surfaces with emissivities of 0.80 and 0.10, respectively. What is the radiant heat transfer from the tube per unit length?

13.40 A t = 5-mm-thick sheet of anodized aluminum is used to reject heat in a space power application. The edge of the sheet is attached to a hot source, and the sheet is maintained at nearly isothermal conditions at T = 300 K. The sheet is not subjected to irradiation.

(a) Determine the net radiation heat transfer rate from both sides of the 200 mm × 200 mm sheet to deep space.

(b) An engineer suggests boring 3-mm-diameter holes through the sheet. The holes are spaced 5 mm apart. The interior surfaces of the holes are anodized after they are bored. Determine the net radiation heat transfer rate from both sides of the sheet to deep space.

(c) As an alternative design, the 3-mm-diameter flat-bottomed holes are not bored completely through the sheet but are bored to depths of 2 mm on each side, leaving a 1-mm-thick web of aluminum separating the bottoms of the holes located on opposite sides of the sheet. Determine the net radiation heat transfer rate from both sides of the sheet to deep space.

(d) Compare the ratio of the net radiation heat transfer rate to the mass of the sheet for the three designs.

13.41 A very long electrical conductor 10 mm in diameter is concentric with a cooled cylindrical tube 50 mm in diameter whose surface is diffuse and gray with an emissivity of 0.9 and temperature of 27°C. The electrical conductor has a diffuse, gray surface with an emissivity of 0.6 and is dissipating 6.0 W per meter of length. Assuming that the space between the two surfaces is evacuated, calculate the surface temperature of the conductor.

13.42 Liquid oxygen is stored in a thin-walled, spherical container 0.75 m in diameter, which is enclosed within a second thin-walled, spherical container 1.1 m in diameter. The opaque, diffuse, gray container surfaces have an emissivity of 0.05 and are separated by an evacuated space. If the outer surface is at 280 K and the inner surface is at 90 K, what is the mass rate of oxygen lost due to evaporation? (The latent heat of vaporization of oxygen is 2.13×10^5 J/kg.)

13.43 Two concentric spheres of diameter D_1 = 0.8 m and D_2 = 1.2 m are separated by an air space and have surface temperatures of T_1 = 400 K and T_2 = 300 K.

(a) If the surfaces are black, what is the net rate of radiation exchange between the spheres?

(b) What is the net rate of radiation exchange between the surfaces if they are diffuse and gray with ε_1 = 0.5 and ε_2 = 0.05?

(c) What is the net rate of radiation exchange if D_2 is increased to 20 m, with ε_2 = 0.05, ε_1 = 0.5, and D_1 = 0.8 m? What error would be introduced by assuming blackbody behavior for the outer surface (ε_2 = 1), with all other conditions remaining the same?

(d) For D_2 = 1.2 m and emissivities of ε_1 = 0.1, 0.5, and 1.0, compute and plot the net rate of radiation exchange as a function of ε_2 for $0.05 \le \varepsilon_2 \le 1.0$.

13.44 Heat transfer by radiation occurs between two large parallel plates, which are maintained at temperatures T_1 and T_2, with $T_1 > T_2$. To reduce the rate of heat transfer between the plates, it is proposed that they be separated by a thin shield that has different emissivities on opposite surfaces. In particular, one surface has the emissivity $\varepsilon_s < 0.5$, while the opposite surface has an emissivity of $2\varepsilon_s$.

(a) How should the shield be oriented to provide the larger reduction in heat transfer between the plates? That is, should the surface of emissivity ε_s or that of emissivity $2\varepsilon_s$ be oriented toward the plate at T_1?

(b) What orientation will result in the larger value of the shield temperature T_s?

13.45 Consider two large, diffuse, gray, parallel surfaces separated by a small distance. If the surface emissivities are 0.8, what emissivity should a thin radiation shield have to reduce the radiation heat transfer rate between the two surfaces by a factor of 10?

13.46 At the bottom of a large vacuum chamber whose walls are at 300 K, a black panel 0.1 m in diameter is maintained at 77 K. To reduce the heat gain to this panel, a radiation shield of the same diameter D and an emissivity of 0.05 is placed close to the panel. Calculate the net rate of heat gain to the panel.

13.47 A furnace is located next to a dense array of cryogenic fluid piping. The ice-covered piping approximates a plane surface with an average temperature of T_p = 0°C and an emissivity of ε_p = 0.6. The furnace wall has a temperature of T_f = 200°C and an emissivity of ε_f = 0.9. To protect the refrigeration equipment and piping from excessive heat loading, reflective aluminum radiation shielding with an emissivity of $\varepsilon_s = 0.1$ is placed between the piping and the furnace wall, as shown in the schematic. Assume all surfaces are diffuse-gray.

Dense piping array, $T_p = 0°C$, $\varepsilon_p = 0.6$ — Furnace wall $T_f = 200°C$, $\varepsilon_f = 0.9$

Installation criterion $T_{s,N} \leq 30°C$ — Radiation shields, N $\varepsilon_s = 0.1$

If the temperature of the shield closest to the piping $T_{s,N}$ must be less than 30°C, how many radiation shields, N, must be installed between the piping and the furnace wall?

13.48 A diffuse, gray radiation shield of 60-mm diameter and emissivities of $\varepsilon_{2,i} = 0.01$ and $\varepsilon_{2,o} = 0.1$ on the inner and outer surfaces, respectively, is concentric with a long tube transporting a hot process fluid. The tube surface is black with a diameter of 20 mm. The region interior to the shield is evacuated. The exterior surface of the shield is exposed to a large room whose walls are at 17°C and experiences convection with air at 27°C and a convection heat transfer coefficient of 10 W/m² · K.

Shield, $D_2 = 60$ mm
$\varepsilon_{2,o}$
$\varepsilon_{2,i}$
Heated tube, $D_1 = 20$ mm
Evacuated

Determine the operating temperature for the inner tube if the shield temperature is maintained at 42°C.

Enclosures with a Reradiating Surface

13.49 A long conduit is constructed with diffuse, gray walls 0.5 m wide. The top and bottom of the conduit are insulated. The emissivities of the walls are $\varepsilon_1 = 0.45$, $\varepsilon_2 = 0.65$, and $\varepsilon_3 = 0.15$, respectively, while the temperatures of walls 1 and 2 are 500 K and 700 K, respectively.

(a) Determine the temperature of the insulated walls.

(b) Determine the net radiation heat rate from surface 2 per unit conduit length.

A_1, T_1, ε_1 — A_3, T_3, ε_3 — A_2, T_2, ε_2

13.50 Consider the three-surface enclosure shown. The lower plate (A_1) is a black disk of 200-mm diameter and is supplied with a heat rate of 10,000 W. The upper plate (A_2), a disk coaxial to A_1, is a diffuse, gray surface with $\varepsilon_2 = 0.8$ and is maintained at $T_2 = 473$ K. The diffuse, gray sides between the plates are perfectly insulated. Assume convection heat transfer is negligible.

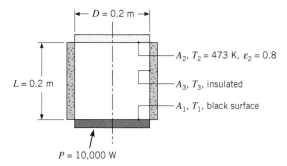

$D = 0.2$ m
$L = 0.2$ m
$A_2, T_2 = 473$ K, $\varepsilon_2 = 0.8$
A_3, T_3, insulated
A_1, T_1, black surface
$P = 10,000$ W

Determine the operating temperature of the lower plate T_1 and the temperature of the insulated side T_3.

13.51 A furnace has the form of a truncated conical section, as shown in the schematic. The floor of the furnace has an emissivity of $\varepsilon_1 = 0.7$ and is maintained at 1000 K with a heat flux of 2200 W/m². The lateral wall is perfectly insulated with an emissivity of $\varepsilon_3 = 0.3$. Assume that all the surfaces are diffuse-gray.

$D_2 = 20$ mm
A_2
A_3
A_1
$L = 50$ mm
$D_1 = 40$ mm

(a) Determine the temperature of the upper surface, T_2, if its emissivity is $\varepsilon_2 = 0.5$. What is the temperature of the lateral wall, T_3?

(b) If conditions for the furnace floor remain unchanged, but all surfaces are black instead of diffuse-gray, determine T_2 and T_3. From the diffuse and blackbody enclosure analyses, what can you say about the influence of ε_2 on your results?

13.52 Two parallel, aligned disks, 0.4 m in diameter and separated by 0.1 m, are located in a large room whose walls are maintained at 300 K. One of the disks is maintained at a uniform temperature of 500 K with an emissivity of 0.6, while the back side of the second disk is well insulated. If the disks are diffuse, gray surfaces, determine the temperature of the insulated disk.

13.53 Coatings applied to long metallic strips are cured by installing the strips along the walls of a long oven of square cross section.

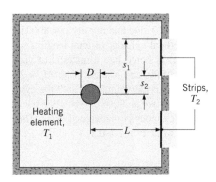

Thermal conditions in the oven are maintained by a long silicon carbide rod (heating element), which is of diameter $D = 20$ mm and is operated at $T_1 = 1700$ K. Each of two strips on one side wall has the same orientation relative to the rod ($s_1 = 60$ mm, $s_2 = 25$ mm, $L = 80$ mm) and is operated at $T_2 = 600$ K. All surfaces are diffuse and gray with $\varepsilon_1 = 0.9$ and $\varepsilon_2 = 0.3$. Assuming the oven to be well insulated at all but the strip surfaces and neglecting convection effects, determine the heater power requirement per unit length (W/m).

13.54 The bottom of a steam-producing still of 200-mm diameter is heated by radiation. The heater, maintained at 1000°C and separated 100 mm from the still, has the same diameter as the still bottom. The still bottom and heater surfaces are black.

(a) By what factor could the vapor production rate be increased if the cylindrical sides (dashed surface) were insulated rather than open to the surroundings maintained at 27°C?

(b) For heater temperatures of 600, 800, and 1000°C, plot the net rate of radiation heat transfer to the still as a function of the separation distance over the range from 25 to 100 mm. Consider the cylindrical sides to be insulated and all other conditions to remain the same.

13.55 A molten aluminum alloy at 950 K is poured into a cylindrical container that is well insulated from large surroundings at 300 K. The inner diameter of the container is 500 mm, and the distance from the surface of the melt to the top of the container is 200 mm.

If the oxidized aluminum at the surface of the melt has an emissivity of 0.55, what is the net rate of radiation heat transfer from the melt?

13.56 The coating on a 1 m × 2 m metallic surface is cured by placing it 0.5 m below an electrically heated surface of equivalent dimensions, and the assembly is exposed to large surroundings at 300 K. The heater is well insulated on its top side and is aligned with the coated surface. Both the heater and coated surfaces may be approximated as blackbodies. During the curing procedure, the heater and coated surfaces are maintained at 700 and 400 K, respectively. Convection effects may be neglected.

(a) What is the electrical power, q_{elec}, required to operate the heater?

(b) What is q_{elec} if the surfaces are connected with reradiating sidewalls?

(c) The facility may be used to cure different surface coatings. With the heater still approximated as a blackbody, compute and plot q_{elec} as a function of the coating's emissivity for values ranging from 0.1 to 1.0. Perform the calculations for both open and reradiating sides.

13.57 A cubical furnace 2 m on a side is used for heat treating steel plate. The top surface of the furnace consists of electrical radiant heaters that have an emissivity of 0.8 and a power input of 1.5×10^5 W. The sidewalls consist of a well-insulated refractory material, while the bottom consists of the steel plate, which has an emissivity of 0.4. Assume diffuse, gray surface behavior for the heater and the plate, and consider conditions for which the plate is at 300 K. What are the corresponding temperatures of the heater surface and the sidewalls?

13.58 Consider a long duct constructed with diffuse, gray walls 1 m wide.

(a) Determine the net radiation transfer from surface A_1 per unit length of the duct.

(b) Determine the temperature of the insulated surface A_3.

(c) What effect would changing the value of ε_3 have on your result? After considering your assumptions,

comment on whether you expect your results to be exact.

13.59 A solar collector consists of a long duct through which air is blown; its cross section forms an equilateral triangle 1 m on a side. One side consists of a glass cover of emissivity $\varepsilon_1 = 0.9$, while the other two sides are absorber plates with $\varepsilon_2 = \varepsilon_3 = 1.0$.

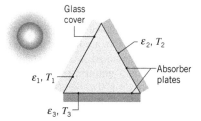

During operation the surface temperatures are known to be $T_1 = 25°C$, $T_2 = 60°C$, and $T_3 = 70°C$. What is the net rate at which radiation is transferred to the cover due to exchange with the absorber plates?

13.60 A frictionless piston of diameter $D_p = 50$ mm, thickness $\delta = 10$ mm, and density $\rho = 8000$ kg/m³ is placed in a vertical cylinder of total length $L = 100$ mm. The cylinder is bored into a material of very low thermal conductivity, and a mass of air $M_a = 75 \times 10^{-6}$ kg is enclosed in the volume beneath the cylinder. The surroundings are at $T_{sur} = 300$ K while the bottom surface of the cylinder is maintained at T_1. The emissivity of all the surfaces is $\varepsilon = 0.3$.

What is the distance between the bottom of the piston and the bottom of the cylinder, L_b, for $T_1 = 300, 450,$ and 600 K? What is the piston temperature? *Hint:* Neglect convection heat transfer and assume that the average air temperature is $\bar{T}_a = (T_p + T_1)/2$.

13.61 A long cylindrical heater element of diameter $D = 10$ mm, temperature $T_1 = 1500$ K, and emissivity $\varepsilon_1 = 1$ is used in a furnace. The bottom area A_2 is a diffuse, gray surface with $\varepsilon_2 = 0.6$ and is maintained at $T_2 = 500$ K. The side and top walls are fabricated from an insulating, refractory brick that is diffuse and gray with $\varepsilon = 0.9$. The length of the furnace normal to the page is very large compared to the width w and height h.

Furnace walls ─── Heating element, A_1
T_1, ε_1
$h = 0.87$ m ── 60°
Bottom, A_2 T_2, ε_2
$w = 1.0$ m

Neglecting convection and treating the furnace walls as isothermal, determine the power per unit length that must be provided to the heating element to maintain steady-state conditions. Calculate the temperature of the furnace wall.

13.62 A composite wall consists of two large plates separated by sheets of refractory insulation, as shown in the schematic. In the installation process, the sheets of thickness $L = 50$ mm and thermal conductivity $k = 0.05$ W/m · K are separated at 1-m intervals by gaps of width $w = 10$ mm. The hot and cold plates have temperatures and emissivities of $T_1 = 400°C$, $\varepsilon_1 = 0.85$ and $T_2 = 35°C$, $\varepsilon_2 = 0.5$, respectively. Assume that the plates and insulation are diffuse-gray surfaces.

Hot side ── Gap ── $w = 10$ mm
$T_1 = 400°C$
$L = 50$ mm
$T_2 = 35°C$
Cold side ── 1 m ── Insulation, $k = 0.05$ W/m · K

(a) Determine the rate of heat loss by radiation through the gap per unit length of the composite wall (normal to the page).

(b) Recognizing that the gaps are located on a 1-m spacing, determine what fraction of the total heat loss through the composite wall is due to transfer by radiation through the insulation gap.

13.63 A small disk of diameter $D_1 = 50$ mm and emissivity $\varepsilon_1 = 0.6$ is maintained at a temperature of $T_1 = 900$ K. The disk is covered with a hemispherical

radiation shield of the same diameter and an emissivity of $\varepsilon_2 = 0.02$ (both sides). The disk and cap are located at the bottom of a large evacuated refractory container ($\varepsilon_4 = 0.85$), facing another disk of diameter $D_3 = D_1$, emissivity $\varepsilon_3 = 0.4$, and temperature $T_3 = 400$ K. The view factor F_{23} of the shield with respect to the upper disk is 0.3.

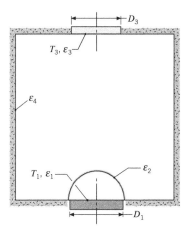

D_3
T_3, ε_3
ε_4
T_1, ε_1 ── ε_2
D_1

(a) Construct an equivalent thermal circuit for the above system. Label all nodes, resistances, and currents.

(b) Find the net rate of heat transfer between the hot disk and the rest of the system.

Enclosures: Three or More Surfaces

13.64 Consider a cylindrical cavity of diameter $D = 100$ mm and depth $L = 50$ mm whose sidewall and bottom are diffuse and gray with an emissivity of 0.6 and are at a uniform temperature of 1500 K. The top of the cavity is open and exposed to surroundings that are large and at 300 K.

(a) Calculate the net radiation heat transfer rate from the cavity, treating the bottom and sidewall of the cavity as one surface (q_A).

(b) Calculate the net radiation heat transfer rate from the cavity, treating the bottom and sidewall of the cavity as two separate surfaces (q_B).

(c) Plot the percentage difference between q_A and q_B as a function of L over the range 5 mm $\leq L \leq$ 100 mm.

13.65 Consider a circular furnace that is 0.3 m long and 0.3 m in diameter. The two ends have diffuse, gray surfaces that are maintained at 400 and 500 K with emissivities of 0.4 and 0.5, respectively. The lateral surface is also diffuse and gray with an emissivity of 0.8 and a temperature of 800 K. Determine the net radiative heat transfer from each of the surfaces.

13.66 Three long, 20-mm-diameter cylindrical bars are to be heat-treated in an oven. The bars are initially at a temperature of 300 K, while the oven walls are char-coated and at 450 K. The bar and oven wall emissivities are 0.85 and 1, respectively. Determine the initial radiation heat rate per unit bar length to each of the three bars if they are all placed in parallel on a horizontal, thin-wire rack within the oven. Consider surface-to-surface separation distances of $s = 0$, 10, and 50 mm. At what separation distance and for which bar(s) is the radiation heat rate maximized? At what separation distance and for which bar(s) is radiation heating minimized? Plot the initial heat rate per unit length to the middle bar and to an exterior bar for separation distances $0 \leq s \leq 1$ m.

13.67 Two convex objects are inside a large vacuum enclosure whose walls are maintained at $T_3 = 300$ K. The objects have the same area, 0.2 m², and the same emissivity, 0.2. The view factor from object 1 to object 2 is $F_{12} = 0.3$. Embedded in object 2 is a heater that generates 400 W. The temperature of object 1 is maintained at $T_1 = 200$ K by circulating a fluid through channels machined into it. At what rate must heat be supplied (or removed) by the fluid to maintain the desired temperature of object 1? What is the temperature of object 2?

13.68 A room is represented by the following enclosure, where the ceiling (1) has an emissivity of 0.8 and is maintained at 40°C by embedded electrical heating elements. Heaters are also used to maintain the floor (2) of emissivity 0.9 at 50°C. The right wall (3) of emissivity 0.7 reaches a temperature of 15°C on a cold winter day. The left wall (4) and end walls (5A) and (5B) are very well insulated. To simplify the analysis, treat the two end walls as a single surface (5).

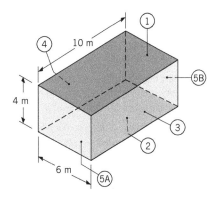

Assuming the surfaces are diffuse and gray, find the net radiation heat transfer from each surface.

13.69 A cylindrical furnace for heat-treating materials in a spacecraft environment has a 90-mm diameter and an overall length of 180 mm. Heating elements in the 135-mm-long section (1) maintain a refractory lining of $\varepsilon_1 = 0.8$ at 800°C. The linings for the bottom (2) and upper (3) sections are made of the same refractory material, but are insulated.

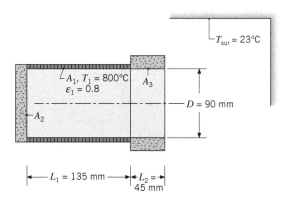

Determine the power required to maintain the furnace operating conditions with the surroundings at 23°C.

13.70 An observation cabin is located in a hot-strip mill directly over the line, as shown in the schematic. The cabin floor is exposed to a portion of the hot strip that is at a temperature of $T_{ss} = 920$°C and has an emissivity of $\varepsilon_{ss} = 0.85$, as well as to *surroundings* within the mill (not shown) at a temperature of $T_{sur} = 80$°C. To protect the operating personnel in the cabin, the floor must be maintained at $T_f = 50$°C.

Determine the coolant system heat removal rate required to maintain the cabin floor at 50°C for the following conditions: (a) when the floor, with an emissivity of $\varepsilon_f = 0.6$, is directly exposed to the strip; and (b) when a radiation shield of emissivity $\varepsilon_s = 0.1$ is installed between the floor and the strip.

13.71 A small oven consists of a cubical box of dimension $L = 0.1$ m, as shown. The floor of the box consists of a heater that supplies $P = 400$ W. The remaining walls lose heat to the surroundings outside the oven, which maintains their temperatures at $T_3 = 400$ K. A spherical object of diameter $D = 30$ mm is placed at the center of the oven. Sometime after the sphere is placed in the oven, its temperature is $T_1 = 420$ K. All surfaces have emissivities of 0.4.

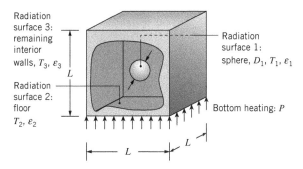

Radiation surface 3: remaining interior walls, T_3, ε_3

Radiation surface 2: floor T_2, ε_2

Radiation surface 1: sphere, D_1, T_1, ε_1

Bottom heating: P

(a) Find the following view factors: F_{12}, F_{13}, F_{21}, F_{31}, F_{23}, F_{32}, F_{33}.

(b) Determine the temperature of the floor and the net rate of heat transfer leaving the sphere due to radiation. Is the sphere under steady-state conditions?

Multimode Heat Transfer: Introductory

13.72 An opaque, diffuse, gray (200 mm × 200 mm) plate with an emissivity of 0.8 is placed over the opening of a furnace and is known to be at 400 K at a certain instant. The bottom of the furnace, having the same dimensions as the plate, is black and operates at 1000 K. The sidewalls of the furnace are well insulated. The top of the plate is exposed to ambient air with a convection coefficient of 25 W/m$^2 \cdot$ K and to large surroundings. The air and surroundings are each at 300 K.

Air $\qquad T_\infty = 300$ K
$\qquad h = 25$ W/m$^2 \cdot$ K

$T_{sur} = 300$ K

Plate, 400 K, $\varepsilon = 0.8$

Insulated sidewalls

200 mm

Furnace bottom, 1000 K

(a) Evaluate the net rate of radiative heat transfer to the bottom surface of the plate.

(b) If the plate has mass and specific heat of 2 kg and 900 J/kg $\cdot$ K, respectively, what will be the change in temperature of the plate with time, dT_p/dt? Assume convection to the bottom surface of the plate to be negligible.

(c) Extending the analysis of part (b), generate a plot of the change in temperature of the plate with time, dT_p/dt, as a function of the plate temperature for $350 \leq T_p \leq 900$ K and all other conditions remaining the same. What is the steady-state temperature of the plate?

13.73 A tool for processing silicon wafers is housed within a vacuum chamber whose walls are black and maintained by a coolant at $T_{vc} = 300$ K. The thin silicon wafer is mounted close to, but not touching, a chuck, which is electrically heated and maintained at the temperature T_c. The surface of the chuck facing the wafer is black. The wafer temperature is $T_w = 700$ K, and its surface is diffuse and gray with an emissivity of $\varepsilon_w = 0.6$. The function of the grid, a thin metallic foil positioned coaxial with the wafer and of the same diameter, is to control the power of the ion beam reaching the wafer. The grid surface is black with a temperature of $T_g = 500$ K. The effect of the ion beam striking the wafer is to apply a uniform heat flux of $q''_{ib} = 600$ W/m^2. The top surface of the wafer is subjected to the flow of a process gas for which $T_\infty = 500$ K and $h = 10$ W/m$^2 \cdot$ K. Since the gap between the wafer and chuck, δ, is very small, flow of the process gas in this region may be neglected.

Ion beam generator

Grid, T_g

Vacuum chamber, $T_{vc} = 300$ K

Gas T_∞, h

q''_{ib}

T_w, ε_w

$L = 200$ mm

Gas flow

Wafer

Chuck, T_c

T_c $\qquad \delta \ll D$

$D = 200$ mm

(a) Represent the wafer schematically, showing a control surface and all relevant thermal processes.

(b) Perform an energy balance on the wafer and determine the chuck temperature T_c.

13.74 Most architects know that the ceiling of an ice-skating rink must have a high reflectivity. Otherwise, condensation may occur on the ceiling, and water may drip onto the ice, causing bumps on the skating surface. Condensation will occur on the ceiling when its surface temperature drops below the dew point of the rink air. Your assignment is to perform an analysis to determine the effect of the ceiling emissivity on the ceiling temperature, and hence the propensity for condensation.

The rink has a diameter of $D = 50$ m and a height of $L = 10$ m, and the temperatures of the ice and walls are $-5°C$ and $15°C$, respectively. The rink air temperature is $15°C$, and a convection coefficient of 5 W/m²·K characterizes conditions on the ceiling surface. The thickness and thermal conductivity of the ceiling insulation are 0.3 m and 0.035 W/m·K, respectively, and the temperature of the outdoor air is $-5°C$. Assume that the ceiling is a diffuse-gray surface and that the walls and ice may be approximated as blackbodies.

(a) Consider a flat ceiling having an emissivity of 0.05 (highly reflective panels) or 0.94 (painted panels). Perform an energy balance on the ceiling to calculate the corresponding values of the ceiling temperature. If the relative humidity of the rink air is 70%, will condensation occur for either or both of the emissivities?

(b) For each of the emissivities, calculate and plot the ceiling temperature as a function of the insulation thickness for $0.1 \leq t \leq 1$ m. Identify conditions for which condensation will occur on the ceiling.

13.75 Coated metallic disks are cured by placing them at the top of a cylindrical furnace whose bottom surface is electrically heated and whose sidewall may be approximated as a reradiating surface. Curing is accomplished by maintaining a disk at $T_2 = 400$ K for an extended period. The electrically heated surface is maintained at $T_1 = 800$ K and is mounted on a ceramic base material of thermal conductivity $k = 20$ W/m · K. The bottom of the base material, as well as the ambient air and large surroundings above the disk, are maintained at a temperature of 300 K. Emissivities of the heater and the disk inner and outer surfaces are $\varepsilon_1 = 0.9$, $\varepsilon_{2,i} = 0.5$, and $\varepsilon_{2,o} = 0.9$, respectively.

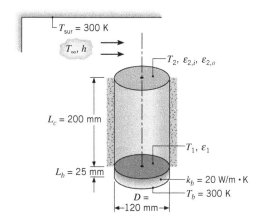

Assuming steady-state operation and neglecting convection within the cylindrical cavity, determine the electrical power that must be supplied to the heater and the convection coefficient h that must be maintained at the outer surface of the disk in order to satisfy the prescribed conditions.

13.76 Boiler tubes exposed to the products of coal combustion in a power plant are subject to fouling by the ash (mineral) content of the combustion gas. The ash forms a solid deposit on the tube outer surface, which reduces heat transfer to a pressurized water/steam mixture flowing through the tubes. Consider a thin-walled boiler tube ($D_t = 0.05$ m) whose surface is maintained at $T_t = 600$ K by the boiling process. Combustion gases flowing over the tube at $T_\infty = 1800$ K provide a convection coefficient of $\bar{h} = 100$ W/m² · K, while radiation from the gas and boiler walls to the tube may be approximated as that originating from large surroundings at $T_{sur} = 1500$ K.

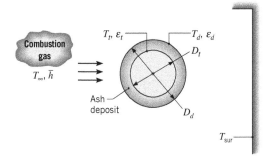

(a) If the tube surface is diffuse and gray, with $\varepsilon_t = 0.8$, and there is *no* ash deposit layer, what is the rate of heat transfer per unit length, q', to the boiler tube?

(b) If a deposit layer of diameter $D_d = 0.055$ m and thermal conductivity $k = 1$ W/m · K forms on the tube, what is the deposit surface temperature, T_d? The deposit is diffuse and gray, with $\varepsilon_d = 0.9$, and T_t, T_∞, $\bar{h}$, and T_{sur} remain unchanged. What is the net rate of heat transfer per unit length, q', to the boiler tube?

(c) Explore the effect of variations in D_d on T_d, q', as well as on relative contributions of convection and radiation to the net heat transfer rate. Represent your results graphically.

13.77 Consider two very large parallel plates. The bottom plate is warmer than the top plate, which is held at a constant temperature of $T_1 = 330$ K. The plates are separated by $L = 0.1$ m, and the gap between the two surfaces is filled with air at atmospheric pressure. The heat flux from the bottom plate is $q'' = 250$ W/m².

(a) Determine the temperature of the bottom plate and the ratio of the convective to radiative heat fluxes for $\varepsilon_1 = \varepsilon_2 = 0.5$. Evaluate air properties at $T = 350$ K.

(b) Repeat part (a) for $\varepsilon_1 = \varepsilon_2 = 0.25$ and 0.75.

13.78 Electrical conductors, in the form of parallel plates of length $L = 40$ mm, have one edge mounted to a ceramic insulated base and are spaced a distance $w = 10$ mm apart. The plates are exposed to large isothermal surroundings at $T_{sur} = 300$ K. The conductor (1) and ceramic (2) surfaces are diffuse and gray with emissivities of $\varepsilon_1 = 0.8$ and $\varepsilon_2 = 0.6$, respectively. For a prescribed operating current in the conductors, their temperature is $T_1 = 500$ K.

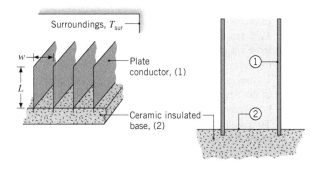

(a) Determine the electrical power dissipated in a conductor plate per unit length, q'_1, considering only radiation exchange. What is the temperature of the insulated base, T_2?

(b) Determine q'_1 and T_2 when the surfaces experience convection with an airstream at 300 K and a convection coefficient of $h = 25$ W/m² · K.

13.79 The spectral absorptivity of a large diffuse surface is $\alpha_\lambda = 0.9$ for $\lambda < 1$ μm and $\alpha_\lambda = 0.3$ for $\lambda \geq 1$ μm. The bottom of the surface is well insulated, while the top may be exposed to one of two different conditions.

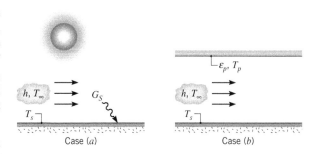

Case (a)   Case (b)

(a) In case (a) the surface is exposed to the sun, which provides an irradiation of $G_S = 1200$ W/m², and to an airflow for which $T_\infty = 300$ K. If the surface temperature is $T_s = 320$ K, what is the convection coefficient associated with the airflow?

(b) In case (b) the surface is shielded from the sun by a large plate and an airflow is maintained between the plate and the surface. The plate is diffuse and gray with an emissivity of $\varepsilon_p = 0.8$. If $T_\infty = 300$ K and the convection coefficient is equivalent to the result obtained in part (a), what is the plate temperature T_p that is needed to maintain the surface at $T_s = 320$ K?

13.80 A long uniform rod of 50-mm diameter with a thermal conductivity of 15 W/m · K is heated internally by volumetric energy generation of 20 kW/m³. The rod is positioned coaxially within a larger circular tube of 60-mm diameter whose surface is maintained at 500°C. The annular region between the rod and the tube is evacuated, and their surfaces are diffuse and gray with an emissivity of 0.2.

(a) Determine the center and surface temperatures of the rod.

(b) Determine the center and surface temperatures of the rod if atmospheric air occupies the annular space.

(c) For tube diameters of 60, 100, and 1000 mm and for both the evacuated and atmospheric conditions, compute and plot the center and surface temperatures as a function of equivalent surface emissivities in the range from 0.1 to 1.0.

13.81 The cross section of a long circular tube, which is divided into two semicylindrical ducts by a thin wall, is shown.

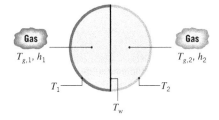

Sides 1 and 2 are maintained at temperatures of $T_1 = 600$ K and $T_2 = 400$ K, respectively, while the mean temperatures of gas flows through ducts 1 and 2 are $T_{g,1} = 571$ K and $T_{g,2} = 449$ K, respectively. The foregoing temperatures are invariant in the axial direction. The gases provide surface convection coefficients of $h_1 = h_2 = 5$ W/m² · K, while all duct surfaces may be approximated as blackbodies ($\varepsilon_1 = \varepsilon_2 = \varepsilon_w = 1$). What is the duct wall temperature, T_w? By performing an energy balance on the gas in side 1, verify that $T_{g,1}$ is, in fact, equal to 571 K.

13.82 Applying high-emissivity paints to radiating surfaces is a common technique used to enhance heat transfer by radiation.

   (a) For large parallel plates, determine the radiation heat flux across the gap when the surfaces are at $T_1 = 350$ K, $T_2 = 300$ K, $\varepsilon_1 = \varepsilon_2 = \varepsilon_s = 0.85$.

   (b) Determine the radiation heat flux when a very thin layer of high-emissivity paint, $\varepsilon_p = 0.98$, is applied to both surfaces.

   (c) Determine the radiation heat flux when the paint layers are each $L = 2$ mm thick and the thermal conductivity of the paint is $k = 0.21$ W/m · K.

   (d) Plot the heat flux across the gap for the bare surface as a function of ε_s, with $0.05 \leq \varepsilon_s \leq 0.95$. Show on the same plot the heat flux for the painted surface with very thin paint layers and the painted surface with $L = 2$-mm-thick paint layers.

Multimode Heat Transfer: Advanced

13.83 Options for thermally shielding the top ceiling of a large furnace include the use of an insulating material of thickness L and thermal conductivity k, case (a), or an air space of equivalent thickness formed by installing a steel sheet above the ceiling, case (b).

(a)   (b)

   (a) Develop mathematical models that could be used to assess which of the two approaches is better.

In both cases the interior surface is maintained at the same temperature, $T_{s,i}$, and the ambient air and surroundings are at equivalent temperatures ($T_\infty = T_{sur}$).

   (b) If $k = 0.090$ W/m · K, $L = 25$ mm, $h_o = 25$ W/m² · K, the surfaces are diffuse and gray with $\varepsilon_i = \varepsilon_o = 0.50$, $T_{s,i} = 900$ K, and $T_\infty = T_{sur} = 300$ K, what is the outer surface temperature $T_{s,o}$ and the rate of heat loss per unit surface area associated with each option?

   (c) For each case, assess the effect of surface radiative properties on the outer surface temperature and the rate of heat loss per unit area for values of $\varepsilon_i = \varepsilon_o$ ranging from 0.1 to 0.9. Plot your results.

13.84 The composite insulation shown, which was described in Chapter 1 (Problem 1.64d), is being considered as a ceiling material.

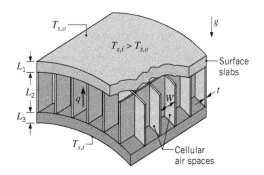

It is proposed that the outer and inner slabs be made from low-density particle board of thicknesses $L_1 = L_3 = 12.5$ mm and that the honeycomb core be constructed from a high-density particle board. The square cells of the core are to have length $L_2 = 50$ mm, width $W = 10$ mm, and wall thickness $t = 2$ mm. The emissivity of both particle boards is approximately 0.85, and the honeycomb cells are filled with air at 1-atm pressure. To assess the effectiveness of the insulation, its total thermal resistance must be evaluated under representative operating conditions for which the bottom (inner) surface temperature is $T_{s,i} = 25°C$ and the top (outer) surface temperature is $T_{s,o} = -10°C$. To assess the effect of free convection in the air space, assume a cell temperature difference of 20°C and evaluate air properties at 7.5°C. To assess the effect of radiation across the air space, assume inner surface temperatures of the outer and inner slabs to be −5 and 15°C, respectively.

13.85 Hot coffee is contained in a cylindrical thermos bottle that is of length $L = 0.3$ m and is lying on its side (horizontally).

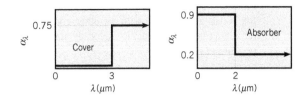

The coffee container consists of a glass flask of diameter $D_1 = 0.07$ m, separated from an aluminum housing of diameter $D_2 = 0.08$ m by air at atmospheric pressure. The outer surface of the flask and the inner surface of the housing are silver coated to provide emissivities of $\varepsilon_1 = \varepsilon_2 = 0.25$. If these surface temperatures are $T_1 = 75°C$ and $T_2 = 35°C$, what is the heat loss from the coffee?

13.86 Consider a vertical, double-pane window for the conditions prescribed in Problem 9.76. That is, vertical panes at temperatures of $T_1 = 22°C$ and $T_2 = -20°C$ are separated by atmospheric air, and the critical Rayleigh number for the onset of convection is $Ra_{L,c} = 2000$.

(a) What is the conduction heat flux across the air gap for the optimal spacing L_{op} between the panes?

(b) If the glass has an emissivity of $\varepsilon_g = 0.90$, what is the total heat flux across the gap?

(c) What is the total heat flux if a special, low-emissivity coating ($\varepsilon_c = 0.10$) is applied to one of the panes at its air-glass interface? What is the total heat flux if both panes are coated?

13.87 A flat-plate solar collector, consisting of an absorber plate and single cover plate, is inclined at an angle of $\tau = 60°$ relative to the horizontal.

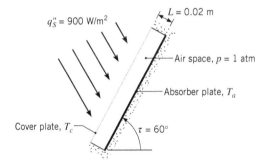

Consider conditions for which the incident solar radiation is collimated at an angle of $60°$ relative to the horizontal and the solar flux is 900 W/m². The cover plate is perfectly transparent to solar radiation ($\lambda \leq 3\ \mu$m) and is opaque to radiation of larger wavelengths. The cover and absorber plates are diffuse surfaces having the spectral absorptivities shown.

The length and width of the absorber and cover plates are much larger than the plate spacing L. What is the rate at which solar radiation is absorbed per unit area of the absorber plate? With the absorber plate well insulated from below and absorber and cover plate temperatures T_a and T_c of 70°C and 27°C, respectively, what is the rate of heat loss per unit area of the absorber plate?

13.88 Consider the tube and radiation shield of Problem 13.39, but now account for free convection in the gap between the tube and the shield.

(a) What is the total rate of heat transfer per unit length between the tube and the shield?

(b) Explore the effect of variations in the shield diameter on the total heat rate, as well as on the contributions due to convection and radiation.

13.89 The lower side of a 400-mm-diameter disk is heated by an electric furnace, while the upper side is exposed to quiescent, ambient air and surroundings at 300 K. The radiant furnace (negligible convection) is of circular construction with the bottom surface ($\varepsilon_1 = 0.6$) and cylindrical side surface ($\varepsilon_2 = 1.0$) maintained at $T_1 = T_2 = 500$ K. The surface of the disk facing the radiant furnace is black ($\varepsilon_{d,1} = 1.0$), while the upper surface has an emissivity of $\varepsilon_{d,2} = 0.8$. Assume the plate and furnace surfaces to be diffuse and gray.

(a) Determine the net heat transfer rate to the disk, $q_{net,d}$, when $T_d = 400$ K.

(b) Plot $q_{net,d}$ as a function of the disk temperature for $300 \le T_d \le 500$ K, with all other conditions remaining the same. What is the steady-state temperature of the disk?

13.90 The fire tube of a hot water heater consists of a long circular duct of diameter $D = 0.07$ m and temperature $T_s = 385$ K, through which combustion gases flow at a temperature of $T_{m,g} = 900$ K. To enhance heat transfer from the gas to the tube, a thin partition is inserted along the midplane of the tube. The gases may be assumed to have the thermophysical properties of air and to be radiatively nonparticipating.

(a) With no partition and a gas flow rate of $\dot{m}_g = 0.05$ kg/s, what is the rate of heat transfer per unit length, q', to the tube?

(b) For a gas flow rate of $\dot{m}_g = 0.05$ kg/s and emissivities of $\varepsilon_s = \varepsilon_p = 0.5$, determine the partition temperature T_p and the total rate of heat transfer q' to the tube.

(c) For $\dot{m}_g = 0.02$, 0.05, and 0.08 kg/s and equivalent emissivities $\varepsilon_p = \varepsilon_s \equiv \varepsilon$, compute and plot T_p and q' as a function of ε for $0.1 \le \varepsilon \le 1.0$. For $\dot{m}_g = 0.05$ kg/s and equivalent emissivities, plot the convective and radiative contributions to q' as a function of ε.

13.91 Consider the conditions of Problem 9.85. Accounting for radiation, as well as convection, across the helium-filled cavity, determine the mass rate at which gaseous nitrogen is vented from the system. The cavity surfaces are diffuse and gray with emissivities of $\varepsilon_i = \varepsilon_o = 0.3$. If the cavity is evacuated, how may surface conditions be altered to further reduce the evaporation? Support your recommendation with appropriate calculations.

13.92 A special surface coating on a square panel that is 5 m×5 m on a side is cured by placing the panel directly under a radiant heat source having the same dimensions. The heat source is diffuse and gray and operates with a power input of 75 kW. The top surface of the heater, as well as the bottom surface of the panel, may be assumed to be well insulated, and the arrangement exists in a large room with air and wall temperatures of 25°C. The

surface coating is diffuse and gray, with an emissivity of 0.30 and an upper temperature limit of 400 K. Neglecting convection effects, what is the minimum spacing that may be maintained between the heater and the panel to ensure that the panel temperature will not exceed 400 K? Allowing for convection effects at the coated surface of the panel, what is the minimum spacing?

13.93 A radiant heater, which is used for surface treatment processes, consists of a long cylindrical heating element of diameter $D_1 = 0.005$ m and emissivity $\varepsilon_1 = 0.80$. The heater is partially enveloped by a long, thin parabolic reflector whose inner and outer surface emissivities are $\varepsilon_{2i} = 0.10$ and $\varepsilon_{2o} = 0.80$, respectively. Inner and outer surface areas per unit length of the reflector are each $A'_{2i} = A'_{2o} = 0.20$ m, and the average convection coefficient for the combined inner and outer surfaces is $\bar{h}_{2(i,o)} = 2$ W/m² · K. The system may be assumed to be in an infinite, quiescent medium of atmospheric air at $T_\infty = 300$ K and to be exposed to large surroundings at $T_{sur} = 300$ K.

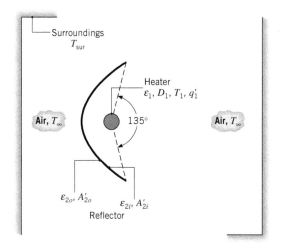

(a) Sketch the appropriate radiation circuit, and write expressions for each of the network resistances.

(b) If, under steady-state conditions, electrical power is dissipated in the heater at $P'_1 = 1500$ W/m and the heater surface temperature is $T_1 = 1200$ K, what is the *net* rate at which *radiant* energy is transferred from the heater?

(c) What is the net rate at which radiant energy is transferred from the heater to the surroundings?

(d) What is the temperature, T_2, of the reflector?

13.94 A wall-mounted natural gas heater uses combustion on a porous catalytic pad to maintain a ceramic plate of emissivity $\varepsilon_c = 0.95$ at a uniform temperature of $T_c = 1000$ K. The ceramic plate is separated from a glass plate by an air gap of thickness $L = 50$ mm. The surface of the glass is diffuse, and its spectral transmissivity and absorptivity may be approximated as $\tau_\lambda = 0$ and $\alpha_\lambda = 1$ for $0 \le \lambda \le 0.4$ μm, $\tau_\lambda = 1$ and $\alpha_\lambda = 0$ for $0.4 < \lambda \le 1.6$ μm, and $\tau_\lambda = 0$ and $\alpha_\lambda = 0.9$ for $\lambda > 1.6$ μm. The exterior surface of the glass is exposed to quiescent ambient air and large surroundings for which $T_\infty = T_{sur} = 300$ K. The height and width of the heater are $H = W = 2$ m.

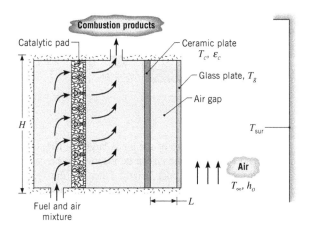

(a) What is the total transmissivity of the glass to irradiation from the ceramic plate? Can the glass be approximated as opaque and gray?

(b) For the prescribed conditions, evaluate the glass temperature, T_g, and the rate of heat transfer from the heater, q_h.

(c) A fan may be used to control the convection coefficient h_o at the exterior surface of the glass. Compute and plot T_g and q_h as a function of h_o for $10 \le h_o \le 100$ W/m$^2 \cdot$ K.

Participating Media

13.95 Pure, solid silicon is produced from a melt, as, for example, in Problem 1.39. Solid silicon is semitransparent, and the spectral absorption coefficient distribution for pure silicon may be approximated as

$\kappa_{\lambda,1} = 10^8 \, \text{m}^{-1}$ ($\lambda \le 0.4 \, \mu$m)

$\kappa_{\lambda,2} = 0 \, \text{m}^{-1}$ ($0.4 \, \mu$m $< \lambda \le 8 \, \mu$m)

$\kappa_{\lambda,3} = 10^2 \, \text{m}^{-1}$ ($8 \, \mu$m $< \lambda \le 25 \, \mu$m)

$\kappa_{\lambda,4} = 0 \, \text{m}^{-1}$ (25μm < 1)

(a) Determine the total absorption coefficient defined as

$$\kappa = \frac{\int_0^\infty \kappa_\lambda G_\lambda(\lambda) d\lambda}{G}$$

for irradiation from surroundings at a temperature equal to the melting point of silicon.

(b) Estimate the total transmissivity, total absorptivity, and total emissivity of an $L = 150$-μm-thick sheet of solid silicon at its melting point.

13.96 A furnace having a spherical cavity of 0.5-m diameter contains a gas mixture at 1 atm and 1400 K. The mixture consists of CO_2 with a partial pressure of 0.25 atm and nitrogen with a partial pressure of 0.75 atm. If the cavity wall is black, what is the cooling rate needed to maintain its temperature at 500 K?

13.97 A gas turbine combustion chamber may be approximated as a long tube of 0.4-m diameter. The combustion gas is at a pressure and temperature of 1 atm and 1000°C, respectively, while the chamber surface temperature is 500°C. If the combustion gas contains CO_2 and water vapor, each with a mole fraction of 0.15, what is the net radiative heat flux between the gas and the chamber surface, which may be approximated as a blackbody?

13.98 A flue gas at 1-atm total pressure and a temperature of 1400 K contains CO_2 and water vapor at partial pressures of 0.05 and 0.10 atm, respectively. If the gas flows through a long flue of 1-m diameter and 400 K surface temperature, determine the net radiative heat flux from the gas to the surface. Blackbody behavior may be assumed for the surface.

13.99 A furnace consists of two large parallel plates separated by 0.75 m. A gas mixture consisting of O_2, N_2, CO_2, and water vapor, with mole fractions of 0.20, 0.50, 0.15, and 0.15, respectively, flows between the plates at a total pressure of 2 atm and a temperature of 1300 K. If the plates may be approximated as blackbodies and are maintained at 500 K, what is the net radiative heat flux to the plates?

13.100 In an industrial process, products of combustion at a temperature and pressure of 2000 K and 1 atm, respectively, flow through a long, 0.25-m-diameter pipe whose inner surface is black. The combustion gas contains CO_2 and water vapor, each at a partial pressure of 0.10 atm. The gas may be assumed to have the thermophysical properties of atmospheric air and to be in fully developed flow with $\dot{m} = 0.25$ kg/s. The pipe is cooled by passing water in cross flow over its outer surface. The upstream velocity and temperature of the water are 0.30 m/s and 300 K, respectively. Determine the pipe wall temperature and heat flux. *Hint:* Emission from the pipe wall may be neglected.

13.101 Waste heat recovery from the exhaust (flue) gas of a melting furnace is accomplished by passing the gas through a vertical metallic tube and introducing saturated water (liquid) at the bottom of an annular region around the tube.

The tube length and inside diameter are 7 and 1 m, respectively, and the tube inner surface is black. The gas in the tube is at atmospheric pressure, with CO_2 and $H_2O(v)$ partial pressures of 0.1 and 0.2 atm, respectively, and its mean temperature may be approximated as $T_g = 1400$ K. The gas flow rate is $\dot{m} = 2$ kg/s. If saturated water is introduced at a pressure of 2.455 bars, estimate the water flow rate $\dot{m}_s$ for which there is complete conversion from saturated liquid at the inlet to saturated vapor at the outlet. Thermophysical properties of the gas may be approximated as $\mu = 530 \times 10^{-7}$ kg/s · m, $k = 0.091$ W/m · K, and $Pr = 0.70$.

Heat and Mass Transfer

13.102 A radiant oven for drying newsprint consists of a long duct ($L = 20$ m) of semicircular cross section. The newsprint moves through the oven on a conveyor belt at a velocity of $V = 0.2$ m/s. The newsprint has a water content of 0.02 kg/m^2 as it enters the oven and is completely dry as it exits. To assure quality, the newsprint must be maintained at room temperature (300 K) during drying. To aid in maintaining this condition, all system components and the

air flowing through the oven have a temperature of 300 K. The inner surface of the semicircular duct, which is of emissivity 0.8 and temperature T_1, provides the radiant heat required to accomplish the drying. The wet surface of the newsprint can be considered to be black. Air entering the oven has a temperature of 300 K and a relative humidity of 20%.

Since the velocity of the air is large, its temperature and relative humidity can be assumed to be constant over the entire duct length. Calculate the required evaporation rate, air velocity u_∞, and temperature T_1 that will ensure steady-state conditions for the process.

13.103 A grain dryer consists of a long semicircular duct of radius $R = 1$ m. One-half of the base surface consists of an electrically heated plate of emissivity $\varepsilon_p = 0.8$, while the other half supports the grain to be dried, which has an emissivity of $\varepsilon_g = 0.9$. In a batch drying process for which the temperature of the grain is $T_g = 330$ K, 2.50 kg of water are to be removed per meter of duct length over a 1-h period.

(a) Neglecting convection heat transfer, determine the required temperature T_p of the heater plate.

(b) If the water vapor is swept from the duct by the flow of dry air, what convection mass transfer coefficient h_m must be maintained by the flow?

(c) If the air is at 300 K, is the assumption of negligible convection justified?

13.104 Wet pharmaceutical tablets are processed in an infrared dryer consisting of a cylindrical drum ($D_d = 0.5$ m) that rotates slowly about a stationary cylindrical infrared heater ($D_h = 0.1$ m). The outside of the drum is well-insulated. Moisture from the tablets is transferred to nitrogen that flows axially through the dryer. As the drum rotates, the tablets tumble due to the combined effects of rotational and gravitational acceleration, forming a thin layer of pharmaceutical product that covers half of the drum's lower interior surface. Consider tablets of temperature $T_p = 320$ K and an emissivity of $\varepsilon_p = 0.95$. If the flow of nitrogen maintains a convection mass transfer coefficient of 0.025 m/s on the surface of the tablets, what is the evaporation rate per unit drum length? Neglecting convection heat transfer, determine the heater temperature, T_h, that must be maintained to drive the evaporation. What is the temperature of the inner drum surface not covered with tablets? The infrared heater has an emissivity $\varepsilon_h = 0.85$. The view factor of a long half-cylinder of diameter D_d to itself, in the presence of a concentric coaxial cylinder of diameter D_h, may be expressed as

$$F_{ii} = 1 - \frac{2}{\pi}\{[1 - (D_h/D_d)^2]^{1/2}$$
$$+ (D_h/D_d)\sin^{-1}(D_h/D_d)\}$$

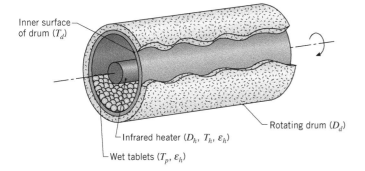

Inner surface of drum (T_d)

Infrared heater (D_h, T_h, ε_h)

Wet tablets (T_p, ε_h)

Rotating drum (D_d)

13.105 A novel *infrared recycler* has been proposed for reclaiming the millions of kilograms of waste plastics produced by the dismantling and shredding of automotive vehicles following their retirement. To address the problem of sorting mixed plastics into components such as polypropylene and polycarbonate, a washed stream of the mixed plastics is routed to an infrared heating system, where it is dried and subsequently heated to a temperature for which one of the components begins to soften, while the others remain rigid. The mixed stream is then routed through steel rollers, to which the softened plastic sticks and is removed from the stream. Heating of the stream is then continued to facilitate removal of a second component, and the heating/removal process is repeated until all of the components are separated.

Consider the initial drying stage for a system comprised of a cylindrical heater aligned coaxially with a rotating drum of diameter $D_d = 1$ m. Shortly after entering the drum, wet plastic pellets may be assumed to fully cover the bottom semicylindrical section and to remain at a temperature of $T_p = 325$ K during the drying process. The surface area of the pellets may be assumed to correspond to that of the semicylinder and to have an emissivity of $\varepsilon_p = 0.95$.

(a) If the flow of dry air through the drum maintains a convection mass transfer coefficient of 0.024 m/s on the surface of the pellets, what is the evaporation rate per unit length of the drum?

(b) Neglecting convection heat transfer, determine the temperature T_h that must be maintained by a heater of diameter $D_h = 0.10$ m and emissivity $\varepsilon_h = 0.8$ to sustain the foregoing evaporation rate. What is the corresponding value of the temperature T_d for the top surface of the drum? The outer surface of the drum is well insulated, and its length-to-diameter ratio is large. As applied to the top *(d)* or bottom *(p)* surface of the drum, the view factor of an infinitely long semicylinder to itself, in the presence of a concentric, coaxial cylinder, may be expressed as

$$F_{ii} = 1 - \frac{2}{\pi}\{[1 - (D_h/D_d)^2]^{1/2}$$
$$+ (D_h/D_d)\sin^{-1}(D_h/D_d)\}$$

Diffusion Mass Transfer

We have learned that heat is transferred if there is a temperature difference in a medium. Similarly, if there is a difference in the concentration of some chemical species in a mixture, mass transfer *must* occur.[1]

Mass transfer is mass in transit as the result of a species concentration difference in a mixture.

Just as a *temperature gradient* constitutes the *driving potential* for heat transfer, a species *concentration gradient* in a mixture provides the *driving potential* for transport of that species.

It is important to clearly understand the context in which the term *mass transfer* is used. Although mass is certainly transferred whenever there is bulk fluid motion, this is not what we have in mind. For example, we do *not* use the term *mass transfer* to describe the motion of air that is induced by a fan or the motion of water being forced through a tube. In both cases, there is gross or bulk fluid motion due to mechanical work. We do, however, use the term to describe the relative motion of species in a mixture due to the presence of concentration gradients. One example is the dispersion of oxides of sulfur released from a power plant smoke stack into the environment. Another example is the transfer of water vapor into dry air, as in a home humidifier.

There are *modes* of *mass transfer* that are similar to the conduction and convection modes of heat transfer. In Chapters 6 through 8 we considered mass transfer by convection, which is *analogous* to convection heat transfer; in this chapter we consider mass transfer by diffusion, which can be *analogous* to conduction heat transfer.

14.1 *Physical Origins and Rate Equations*

From the standpoint of physical origins and the governing rate equations, strong analogies exist between heat and mass transfer by diffusion.

14.1.1 Physical Origins

Consider a chamber in which two different gas species at the same temperature and pressure are initially separated by a partition. If the partition is removed without disturbing the fluid, both species will be transported by diffusion. Figure 14.1 shows the situation as it might exist shortly after removal of the partition. A higher concentration means more molecules per unit volume, and the concentration of species A (light dots) decreases with increasing x, while the concentration of B increases with x. Since mass diffusion is in the direction of decreasing concentration, there is net transport of species A to the right and of species B to the left. The physical mechanism may be explained by considering the imaginary plane shown as a dashed line at x_o. Since molecular motion is random, there is equal probability of any molecule moving to the left or the right. Accordingly, more molecules of species A

---

[1]A species is an identifiable molecule, such as carbon dioxide, CO_2, that can be transported by diffusion and advection and/or converted to some other form by a chemical reaction. A species may be a single atom or a complex polyatomic molecule. It can also be appropriate to identify a mixture (such as air) as a species.

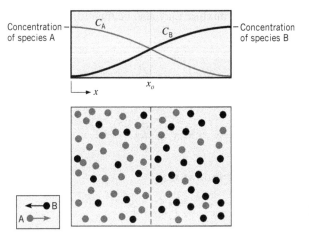

Concentration of species A

C_A

C_B

Concentration of species B

x_o

x

B

A

FIGURE 14.1   Mass transfer by diffusion in a binary gas mixture.

cross the plane from the left (since this is the side of higher A concentration) than from the right. Similarly, the concentration of B molecules is higher to the right of the plane than to the left, and random motion provides for *net* transfer of species B to the left. Of course, after a sufficient time, uniform concentrations of A and B are achieved, and there is no *net* transport of species A or B across the imaginary plane.

Mass diffusion occurs in liquids and solids, as well as in gases. However, since mass transfer is strongly influenced by molecular spacing, diffusion occurs more readily in gases than in liquids and more readily in liquids than in solids. Examples of diffusion in gases, liquids, and solids, respectively, include nitrous oxide from an automobile exhaust in air, dissolved oxygen in water, and helium in Pyrex.

14.1.2   Mixture Composition

Throughout this chapter we will be concerned with the transfer of mass in mixtures. We first review various concepts from thermodynamics. A mixture consists of two or more chemical constituents (*species*), and the amount of any species i may be quantified in terms of its *mass density* ρ_i (kg/m^3) or its *molar concentration* C_i (kmol/m^3). The mass density and molar concentration are related through the species molecular weight, $\mathcal{M}_i$ (kg/kmol), such that

$$\rho_i = \mathcal{M}_i C_i \tag{14.1}$$

With ρ_i representing the mass of species i per unit volume of the mixture, the mixture mass density is

$$\rho = \sum_i \rho_i \tag{14.2}$$

Similarly, the total number of moles per unit volume of the mixture is

$$C = \sum_i C_i \tag{14.3}$$

The amount of species i in a mixture may also be quantified in terms of its *mass fraction*

$$m_i = \frac{\rho_i}{\rho} \tag{14.4}$$

or its *mole fraction*

$$x_i = \frac{C_i}{C} \tag{14.5}[2]$$

From Equations 14.2 and 14.3, it follows that

$$\sum_i m_i = 1 \tag{14.6}$$

and

$$\sum_i x_i = 1 \tag{14.7}$$

For a mixture of ideal gases, the mass density and molar concentration of any constituent are related to the partial pressure of the constituent through the ideal gas law. That is,

$$\rho_i = \frac{p_i}{R_i T} \tag{14.8}$$

and

$$C_i = \frac{p_i}{\mathcal{R} T} \tag{14.9}$$

where R_i is the gas constant for species i and $\mathcal{R}$ is the universal gas constant. Using Equations 14.5 and 14.9 with *Dalton's law* of partial pressures,

$$p = \sum_i p_i \tag{14.10}$$

it follows that

$$x_i = \frac{C_i}{C} = \frac{p_i}{p} \tag{14.11}$$

14.1.3 Fick's Law of Diffusion

Since similar physical mechanisms are associated with heat and mass transfer by diffusion, it is not surprising that the corresponding rate equations are of the same form. The rate

---

[2]Do not confuse x_i, the mole fraction of species i, with the spatial coordinate x. The former variable will always be subscripted with the species designation.

equation for mass diffusion is known as *Fick's law,* and for the transfer of species A in a *binary mixture* of A and B, it may be expressed in vector form as

$$\mathbf{j}_A = -\rho D_{AB} \nabla m_A \qquad (14.12)^3$$

or

$$\mathbf{J}_A^* = -C D_{AB} \nabla x_A \qquad (14.13)^3$$

The form of these expressions is similar to that of Fourier's law, Equation 2.3. Moreover, just as Fourier's law serves to define one important transport property, the thermal conductivity, Fick's law defines a second important transport property, namely, the *binary diffusion coefficient* or *mass diffusivity,* D_{AB}.

The quantity $\mathbf{j}_A$ (kg/s·m^2) is defined as the diffusive mass flux of species A. It is the amount of A that is transferred by diffusion per unit time and per unit area perpendicular to the direction of transfer, and it is proportional to the mixture mass density, $\rho = \rho_A + \rho_B$ (kg/m^3), and to the gradient in the species mass fraction, $m_A = \rho_A/\rho$. The species flux may also be evaluated on a molar basis, where $\mathbf{J}_A^*$ (kmol/s·m^2) is the diffusive molar flux of species A. It is proportional to the total molar concentration of the mixture, $C = C_A + C_B$ (kmol/m^3), and to the gradient in the species mole fraction, $x_A = C_A/C$. The foregoing forms of Fick's law may be simplified when the total mass density ρ or the total molar concentration C is a constant.

14.1.4 Mass Diffusivity

Considerable attention has been given to predicting the mass diffusivity D_{AB} for the binary mixture of two gases, A and B. Assuming ideal gas behavior, kinetic theory may be used to show that

$$D_{AB} \approx \frac{1}{3}\bar{c}\lambda_{mfp} \sim p^{-1}T^{3/2} \qquad (14.14)$$

where T is expressed in kelvins. As noted in Section 2.2.1, $\bar{c}$ increases with increasing temperature and decreasing molecular weight, and therefore the mass diffusivity increases with increasing temperature and decreasing molecular weight. Because λ_{mfp} is inversely proportional to gas pressure, the mass diffusivity decreases with increasing pressure. This relation applies for restricted pressure and temperature ranges and is useful for estimating values of the mass diffusivity at conditions other than those for which data are available. Bird et al. [1–3] provide detailed discussions of available theoretical treatments and comparisons with experiment.

For binary liquid solutions, it is necessary to rely exclusively on experimental measurements. For small concentrations of A (the solute) in B (the solvent), D_{AB} is known to increase with increasing temperature. The mechanism of diffusion of gases, liquids, and solids in solids is complicated and generalized theories are not available. Furthermore, only limited experimental results are available in the literature.

Data for binary diffusion in selected mixtures are presented in Table A.8. Skelland [4] and Poling et al. [5] provide more detailed treatments of this subject.

---

[3]Inherent in Equations 14.12 and 14.13 is the assumption that mass diffusion results only from a concentration gradient. In fact, mass diffusion may also result from a temperature gradient, a pressure gradient, or an external force. In most problems, these effects are negligible and the dominant driving potential is the species concentration gradient. This condition is referred to as *ordinary diffusion.* Treatment of the other (higher-order) effects is presented by Bird et al. [1–3].

EXAMPLE 14.1

Consider the diffusion of hydrogen (species A) in air, liquid water, or iron (species B) at $T = 293$ K. Calculate the species flux on both molar and mass bases if the concentration gradient at a particular location is $dC_A/dx = -1$ kmol/m$^3 \cdot$m. Compare the values of the mass diffusivities to the values of the thermal diffusivities. The mole fraction of the hydrogen, x_A, is much less than unity.

SOLUTION

Known: Concentration gradient of hydrogen in air, liquid water, or iron at $T = 293$ K.

Find: Molar and mass fluxes of hydrogen and the relative values of the mass and thermal diffusivities for the three cases.

Schematic:

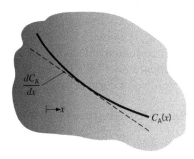

Assumptions: Steady-state conditions.

Properties: Table A.8, hydrogen–air (298 K): $D_{AB} = 0.41 \times 10^{-4}$ m^2/s, hydrogen–water (298 K): $D_{AB} = 0.63 \times 10^{-8}$ m^2/s, hydrogen–iron (293 K): $D_{AB} = 0.26 \times 10^{-12}$ m^2/s. Table A.4, air (293 K): $\alpha = 21.6 \times 10^{-6}$ m^2/s; Table A.6, water (293 K): $k_f = 0.603$ W/m$\cdot$K, $\rho = 998$ kg/m^3, $c_p = 4182$ J/kg$\cdot$K. Table A.1, iron (300 K): $\alpha = 23.1 \times 10^{-6}$ m^2/s.

Analysis:

1. Using Equation 14.14, we find that the mass diffusivity of hydrogen in air at $T = 293$ K is

$$D_{AB,T} = D_{AB,298\,K} \times \left(\frac{T}{298\,K}\right)^{3/2} = 0.41 \times 10^{-4}\ \text{m}^2/\text{s} \times \left(\frac{293\,K}{298\,K}\right)^{3/2} = 0.40 \times 10^{-4}\ \text{m}^2/\text{s}$$

For the case where hydrogen is a *dilute species*, that is, $x_A \ll 1$, the thermal properties of the medium can be taken to be those of the *host medium*, species B. The thermal diffusivity of water is

$$\alpha = \frac{k}{\rho c_p} = \frac{0.603\ \text{W/m} \cdot \text{K}}{998\ \text{kg/m}^3 \cdot 4182\ \text{J/kg} \cdot \text{K}} = 0.144 \times 10^{-6}\ \text{m}^2/\text{s}$$

The ratio of the thermal diffusivity to the mass diffusivity is the Lewis number, *Le*, defined in Equation 6.57.

The molar flux of hydrogen is described by Fick's law, Equation 14.13,

$$J^*_{A,x} = -CD_{AB}\frac{dx_A}{dx}$$

The total molar concentration, C, is approximately constant since A is a dilute species; therefore

$$J^*_{A,x} = -D_{AB}\frac{dC_A}{dx}$$

Hence, for the hydrogen–air mixture,

$$J^*_{A,x} = -0.40 \times 10^{-4}\,\text{m}^2/\text{s} \times \left(-1\frac{\text{kmol}}{\text{m}^3 \cdot \text{m}}\right) = 4 \times 10^{-5}\frac{\text{kmol}}{\text{s} \cdot \text{m}^2} \qquad \triangleleft$$

The mass flux of hydrogen in air is found from the expression

$$j_{A,x} = \mathcal{M}_A J^*_{A,x} = 2\frac{\text{kg}}{\text{kmol}} \times \left(4 \times 10^{-5}\frac{\text{kmol}}{\text{s} \cdot \text{m}^2}\right) = 8 \times 10^{-5}\frac{\text{kg}}{\text{s} \cdot \text{m}^2} \qquad \triangleleft$$

2. The results for the three different mixtures are summarized in the following table. $\triangleleft$

Species B	$\alpha \times 10^6$ (m²/s)	$D_{AB} \times 10^6$ (m²/s)	Le	$j_{A,x} \times 10^6$ (kg/s · m²)
Air	21.6	40	0.54	80
Water	0.144	6.3×10^{-3}	23	13×10^{-3}
Iron	23.1	260×10^{-9}	89×10^6	0.52×10^{-6}

Comments:

1. The thermal diffusivities of the three media vary by two orders of magnitude. We saw in Chapter 5 that this relatively broad range of thermal diffusivities is responsible for the different rates at which objects respond thermally during transient conduction processes. Mass diffusivities can vary by 8 or more orders of magnitude, with the highest diffusivities associated with diffusion in gases and the lowest diffusivities associated with diffusion in solids. Different materials respond to mass transfer at very different rates, depending on whether the host medium is a gas, liquid, or solid.

2. The ratio of the thermal diffusivity to the mass diffusivity, the Lewis number, is typically of order unity for gases. This implies that changes in the thermal and species distributions progress at similar rates in gases that undergo simultaneous heat and mass transfer by diffusion. In solids or liquids, thermal energy is conducted much more readily than chemical species can be transferred by diffusion.

14.2 *Mass Transfer in Nonstationary Media*[4]

14.2.1 Absolute and Diffusive Species Fluxes

We have seen that diffusion mass transfer is similar to conduction heat transfer and that the diffusive fluxes, as given by Equations 14.12 and 14.13, are analogous to the heat flux as expressed by Fourier's law. If there is bulk motion, then, like heat transfer, mass transfer

---

[4]If only problems involving stationary media are of interest, the reader may proceed directly to Section 14.3.

can also occur by advection. Unlike conduction heat transfer, however, the diffusion of a species always involves the movement of molecules or atoms from one location to another. In many cases, this molecular scale motion results in bulk motion. In this section we define the total or *absolute flux* of a species, which includes both diffusive and advective components.

The absolute mass (or molar) flux of a species is defined as the total flux relative to a fixed coordinate system. To obtain an expression for the absolute mass flux, consider species A in a binary mixture of A and B. The absolute mass flux $\mathbf{n}''_A$ is related to the species absolute velocity $\mathbf{v}_A$ by

$$\mathbf{n}''_A = \rho_A \mathbf{v}_A \tag{14.15}$$

A value of $\mathbf{v}_A$ may be associated with any point in the mixture, and it is interpreted as the average velocity of all the A particles in a small volume element about the point. An average, or aggregate, velocity may also be associated with the particles of species B, in which case

$$\mathbf{n}''_B = \rho_B \mathbf{v}_B \tag{14.16}$$

A *mass-average velocity for the mixture* may then be obtained from the requirement that

$$\rho \mathbf{v} = \mathbf{n}'' = \mathbf{n}''_A + \mathbf{n}''_B = \rho_A \mathbf{v}_A + \rho_B \mathbf{v}_B \tag{14.17}$$

giving

$$\mathbf{v} = m_A \mathbf{v}_A + m_B \mathbf{v}_B \tag{14.18}$$

It is important to note that we have defined the velocities ($\mathbf{v}_A$, $\mathbf{v}_B$, $\mathbf{v}$) and the fluxes ($\mathbf{n}''_A, \mathbf{n}''_B, \mathbf{n}''$) as *absolute* quantities. That is, they are referred to axes that are fixed in space. The mass-average velocity $\mathbf{v}$ is a useful parameter of the binary mixture, for two reasons. First, it need only be multiplied by the total mass density to obtain the total mass flux with respect to fixed axes. Second, it is the mass-average velocity which is required in the equations expressing conservation of mass, momentum, and energy such as those presented and discussed in Chapter 6.

We may now define the *mass flux of species* A *relative to the mixture mass-average velocity* as

$$\mathbf{j}_A \equiv \rho_A (\mathbf{v}_A - \mathbf{v}) \tag{14.19}$$

Whereas $\mathbf{n}''_A$ is the *absolute flux* of species A, $\mathbf{j}_A$ is the *relative* or *diffusive flux* of the species and is the quantity previously given by Fick's law, Equation 14.12. It represents the motion of the species relative to the average motion of the mixture. It follows from Equations 14.15 and 14.19 that

$$\mathbf{n}''_A = \mathbf{j}_A + \rho_A \mathbf{v} \tag{14.20}$$

This expression delineates the two contributions to the absolute flux of species A: a contribution due to *diffusion* (i.e., due to the motion of A *relative* to the mass-average motion of the mixture) and a contribution due to *advection* (i.e., due to motion of A *with* the mass-average motion of the mixture). Substituting from Equations 14.12 and 14.17, we obtain

$$\mathbf{n}''_A = -\rho D_{AB} \nabla m_A + m_A (\mathbf{n}''_A + \mathbf{n}''_B) \tag{14.21}$$

If the second term on the right-hand side of Equation 14.21 is zero, mass transfer of species A occurs purely by diffusion, and the situation is analogous to heat transfer purely by conduction. We will later identify special situations for which this occurs.

The foregoing considerations may be extended to species B. The *mass flux of* B *relative to the mixture mass-average velocity* (the *diffusive flux*) is

$$\mathbf{j}_B \equiv \rho_B(\mathbf{v}_B - \mathbf{v}) \tag{14.22}$$

where

$$\mathbf{j}_B = -\rho D_{BA}\nabla m_B \tag{14.23}$$

It follows from Equations 14.17, 14.19, and 14.22 that the diffusive fluxes in a binary mixture are related by

$$\mathbf{j}_A + \mathbf{j}_B = 0 \tag{14.24}$$

If Equations 14.12 and 14.23 are substituted into Equation 14.24, and it is recognized that $\nabla m_A = -\nabla m_B$, since $m_A + m_B = 1$ for a binary mixture, it follows that

$$D_{BA} = D_{AB} \tag{14.25}$$

Hence, as in Equation 14.21, the *absolute* flux of species B may be expressed as

$$\mathbf{n}_B'' = -\rho D_{AB}\,\nabla m_B + m_B(\mathbf{n}_A'' + \mathbf{n}_B'') \tag{14.26}$$

Although the foregoing expressions pertain to *mass* fluxes, the same procedures can be used to obtain results on a *molar* basis. The absolute molar fluxes of species A and B may be expressed as

$$\mathbf{N}_A'' \equiv C_A\mathbf{v}_A \quad \text{and} \quad \mathbf{N}_B'' \equiv C_B\mathbf{v}_B \tag{14.27}$$

and a *molar-average velocity for the mixture,* $\mathbf{v}^*$, is obtained from the requirement that

$$\mathbf{N}'' = \mathbf{N}_A'' + \mathbf{N}_B'' = C\mathbf{v}^* = C_A\mathbf{v}_A + C_B\mathbf{v}_B \tag{14.28}$$

giving

$$\mathbf{v}^* = x_A\mathbf{v}_A + x_B\mathbf{v}_B \tag{14.29}$$

Note that the molar-average velocity is not the same as the mass-average velocity and is therefore not appropriate for use in the conservation equations of Chapter 6.

The significance of the molar-average velocity is that, when multiplied by the total molar concentration C, it provides the total molar flux $\mathbf{N}''$ with respect to a fixed coordinate system. Equation 14.27 provides the *absolute molar flux* of species A and B. In contrast, the molar flux of A relative to the mixture molar average velocity $\mathbf{J}_A^*$, termed the *diffusive flux,* may be obtained from Equation 14.13 or from the expression

$$\mathbf{J}_A^* \equiv C_A(\mathbf{v}_A - \mathbf{v}^*) \tag{14.30}$$

To determine an expression similar in form to Equation 14.21, we combine Equations 14.27 and 14.30 to obtain

$$\mathbf{N}_A'' = \mathbf{J}_A^* + C_A\mathbf{v}^* \tag{14.31}$$

or, from Equations 14.13 and 14.28,

$$\mathbf{N}_A'' = -CD_{AB}\nabla x_A + x_A(\mathbf{N}_A'' + \mathbf{N}_B'') \tag{14.32}$$

Note that Equation 14.32 represents the absolute molar flux as the sum of a diffusive flux and an advective flux. Again, if the second term on the right-hand side is zero, mass transfer is purely by diffusion and is analogous to heat conduction, when formulated in molar quantities instead of mass quantities. For the binary mixture, it also follows that

$$\mathbf{J}_A^* + \mathbf{J}_B^* = 0 \tag{14.33}$$

14.2.2 Evaporation in a Column

Let us now consider diffusion in the binary gas mixture of Figure 14.2. Fixed species concentrations $C_{A,L}$ and $C_{B,L}$ are maintained at the top of a tube containing a liquid layer of species A, and the system is at constant pressure and temperature. Since equilibrium exists between the vapor and liquid phases at the liquid interface, the vapor concentration corresponds to saturated conditions. Since $C_{A,0} > C_{A,L}$, species A *evaporates* from the liquid interface and is transferred upward by diffusion. For steady, one-dimensional conditions with no chemical reactions, species A cannot accumulate in the control volume of Figure 14.2, and the absolute molar flux of A must be constant throughout the column. Hence

$$\frac{dN_{A,x}''}{dx} = 0 \tag{14.34}$$

From the definition of the total molar concentration, $C = C_A + C_B$, $x_A + x_B = 1$ throughout the column. Knowing that $C_{A,0} > C_{A,L}$, we conclude that $x_{A,0} > x_{A,L}$ and $x_{B,0} < x_{B,L}$. Therefore, dx_B/dx is positive, and gas B must diffuse from the top of the column to the liquid interface. However, if species B cannot be absorbed into liquid A, steady-state conditions can be maintained only if $N_{B,x}'' = 0$ everywhere within the control volume of Figure 14.2. The only way this can be possible is if the downward diffusion of gas B is

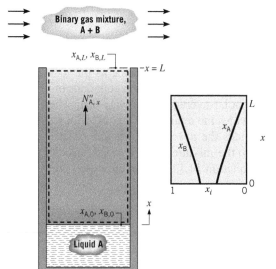

FIGURE 14.2   Evaporation of liquid A into a binary gas mixture, A + B.

exactly balanced by upward advection of gas B. The implication of this important conclusion is that we must account for the advection of the gases in the column in order to successfully predict the mole fraction distributions $x_A(x)$ and $x_B(x)$, and, in turn, the species concentration distributions $C_A(x) = x_A(x)C$ and $C_B(x) = x_B(x)C$, as well as the evaporation rate of liquid A. An appropriate expression for $N''_{A,x}$ may be obtained by substituting the requirement that $N''_{B,x} = 0$ into Equation 14.32, giving

$$N''_{A,x} = -CD_{AB}\frac{dx_A}{dx} + x_A N''_{A,x} \tag{14.35}$$

or, from Equation 14.28,

$$N''_{A,x} = -CD_{AB}\frac{dx_A}{dx} + C_A v^*_x \tag{14.36}$$

From this equation it is evident that the diffusive transport of species A $[-CD_{AB}(dx_A/dx)]$ is augmented by bulk motion $(C_A v^*_x)$. Rearranging Equation 14.35, we obtain

$$N''_{A,x} = -\frac{CD_{AB}}{1 - x_A}\frac{dx_A}{dx} \tag{14.37}$$

For constant p and T, C and D_{AB} are also constant. Substituting Equation 14.37 into Equation 14.34, we then obtain

$$\frac{d}{dx}\left(\frac{1}{1 - x_A}\frac{dx_A}{dx}\right) = 0$$

Integrating twice, we have

$$-\ln(1 - x_A) = C_1 x + C_2$$

Applying the conditions $x_A(0) = x_{A,0}$ and $x_A(L) = x_{A,L}$, the constants of integration may be evaluated and the mole fraction distribution becomes

$$\frac{1 - x_A}{1 - x_{A,0}} = \left(\frac{1 - x_{A,L}}{1 - x_{A,0}}\right)^{x/L} \tag{14.38}$$

Since $1 - x_A = x_B$, we also obtain

$$\frac{x_B}{x_{B,0}} = \left(\frac{x_{B,L}}{x_{B,0}}\right)^{x/L} \tag{14.39}$$

To determine the evaporation rate of species A, Equation 14.38 is first used to evaluate the mole fraction gradient (dx_A/dx). Substituting the result into Equation 14.37, it follows that

$$N''_{A,x} = \frac{CD_{AB}}{L}\ln\left(\frac{1 - x_{A,L}}{1 - x_{A,0}}\right) \tag{14.40}$$

EXAMPLE 14.2

A water-resistant fabric is formed from an impermeable polymer material. To allow water vapor to pass through the fabric, its microstructure consists of open pores of diameter $D = 10 \ \mu m$ that penetrate the entire thickness of the $L = 100$-μm-thick sheet. The small pore diameter prevents liquid water from passing through the fabric. Determine the rate at which water vapor is transmitted through a single pore when saturated liquid exists on the top of the sheet and moist air at $\phi_\infty = 50\%$ relative humidity exists on the bottom side of the sheet. Evaluate the transfer rate at a temperature of $T = 298$ K and a pressure of $p = 1$ atm. Investigate the sensitivity of the transfer rate to the temperature, and compare the transfer rates to rates that are predicted by neglecting the molar-average motion of the mixture in the pore.

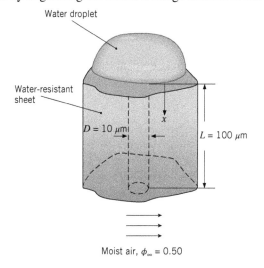

SOLUTION

Known: Thickness and pore diameter of a porous sheet, thermal conditions, and humidity.

Find: Evaporation rate through a single pore, including and neglecting the molar average motion. Determine sensitivity of evaporation rate to temperature.

Assumptions:
1. Steady-state, isothermal, one-dimensional conditions.
2. No chemical reactions.
3. Pore penetrates the thickness of the sheet in a perpendicular manner and is of circular cross section.
4. Ideal gas consisting of water vapor (A) and air (B).

Properties: Table A.6, saturated water, vapor (298 K): $p_{sat} = 0.03165$ bar. Table A.8, water vapor–air (298 K): $D_{AB} = 0.26 \times 10^{-4}$ m²/s.

Analysis:
1. Equation 14.40 may be used to determine the water vapor transferred through a single pore, accounting for the effects of the nonzero molar-averaged velocity. Hence,

$$N_{A,x} = A_{pore} N''_{A,x} = \frac{\pi D^2 C D_{AB}}{4L} \ln\left(\frac{1 - x_{A,L}}{1 - x_{A,0}}\right) \qquad (1)$$

where the total concentration is

$$C = \frac{p}{\mathcal{R}T} = \frac{1.0133 \text{ bar}}{8.314 \times 10^{-2} \frac{\text{m}^3 \cdot \text{bar}}{\text{kmol} \cdot \text{K}} \times 298 \text{ K}} = 40.9 \times 10^{-3} \text{ kmol/m}^3$$

From Equation 14.11, and assuming saturation conditions for the water vapor adjacent to the water droplet, the mole fraction at $x = 0$ is

$$x_{A,0} = \frac{p_{A,sat}}{p} = \frac{0.03165 \text{ bar}}{1.0133 \text{ bar}} = 31.23 \times 10^{-3}$$

while at $x = L$ the mole fraction is

$$x_{A,L} = \frac{\phi_\infty p_{A,sat}}{p} = \frac{0.5 \times 0.03165 \text{ bar}}{1.0133 \text{ bar}} = 15.62 \times 10^{-3}$$

Therefore, the evaporation rate per pore may be evaluated using Equation 1 as

$$N_{A,x} = \frac{\pi \times (10 \times 10^{-6} \text{ m})^2 \times 40.9 \times 10^{-3} \text{ kmol/m}^3 \times 0.26 \times 10^{-4} \text{ m}^2/\text{s}}{4 \times 100 \times 10^{-6} \text{ m}}$$

$$\times \ln\left(\frac{1 - 15.62 \times 10^{-3}}{1 - 31.23 \times 10^{-3}}\right)$$

$$= 13.4 \times 10^{-15} \text{ kmol/s} \qquad \triangleleft$$

2. Neglecting the effects of the molar-averaged velocity, we see that Equation 14.32 reduces to

$$N''_{A,x} = -CD_{AB} \frac{dx_A}{dx}$$

where the total concentration, C, is constant, and the flux of water vapor is independent of x. Hence, the species transfer rate per pore may be expressed as

$$N_{A,x} = A_{pore} N''_{A,x} = \frac{\pi D^2 C D_{AB}}{4L} (x_{A,0} - x_{A,L}) \qquad (2)$$

$$= \frac{\pi \times (10 \times 10^{-6} \text{ m})^2 \times 40.9 \times 10^{-3} \text{ kmol/m}^3 \times 0.26 \times 10^{-4} \text{ m}^2/\text{s}}{4 \times 100 \times 10^{-6} \text{ m}}$$

$$\times (31.23 \times 10^{-3} - 15.62 \times 10^{-3})$$

$$= 13.0 \times 10^{-15} \text{ kmol/s} \qquad \triangleleft$$

A slightly increased evaporation rate is predicted to occur when advection is considered.

3. The temperature dependence of the evaporation rate is determined by accounting for the sensitivity of the binary diffusion coefficient to temperature (Equation 14.14), $D_{AB} \propto T^{3/2}$, and repeating the calculations over the range $300 < T < 360$ K. The results, shown below, indicate a significant dependence of the evaporation rate on the temperature. This strong dependence is due primarily to the large variation of the saturation pressure of water vapor with temperature, as evident in Table A.6.

Comments:

1. The total evaporation rate per area of the fabric could be determined by multiplying the number of pores per unit area by the evaporation rate per pore.

2. The pressure was assumed constant in developing this solution. Since the molar-average velocity is nonzero, there must be a pressure gradient to overcome friction at the pore walls. If the pore is oriented vertically, there is also a static pressure gradient. The pressure gradient to overcome friction can be estimated by first determining the mass-average velocity. From Equation 14.17

$$v_x = \frac{n''_{A,x} + n''_{B,x}}{\rho} = \frac{n''_{A,x}}{\rho} = \frac{N''_{A,x}\mathcal{M}_A}{\rho} = \frac{N_{A,x}\mathcal{M}_A}{A_{\text{pore}}\,\rho}$$

Then the pressure gradient can be estimated by treating the flow as if it were a fully developed flow in a circular tube. From Equation 8.14

$$\left|\frac{dp}{dx}\right| = \frac{32\mu v_x}{D^2} = \frac{32\nu N_{A,x}\mathcal{M}_A}{A_{\text{pore}}D^2}$$

Since the mixture is mostly air, we use the kinematic viscosity of air. Then taking the highest mean velocity case corresponding to $T = 360$ K, $N_{A,x} = 530 \times 10^{-15}$ kmol/s,

$$\left|\frac{dp}{dx}\right| = \frac{32 \times 22.0 \times 10^{-6}\ \text{m}^2/\text{s} \times 530 \times 10^{-15}\ \text{kmol/s} \times 18\ \text{kg/kmol}}{\pi(10 \times 10^{-6}\ \text{m})^4/4}$$

$$= 860 \times 10^3\ \text{Pa/m}$$

Thus the pressure drop to overcome friction is

$$\Delta p_{\text{friction}} = 860 \times 10^3\ \text{Pa/m} \times 100 \times 10^{-6}\ \text{m} = 86\ \text{Pa}$$

If the pore is vertical, the static pressure change is

$$\Delta p_{\text{static}} = \rho_{\text{air}}\,gL = 0.970\ \text{kg/m}^3 \times 9.8\ \text{m/s}^2 \times 100 \times 10^{-6}\ \text{m} = 0.001\ \text{Pa}$$

Both of these pressure differences are negligible relative to atmospheric pressure, so the assumption of constant pressure was appropriate.

3. As the temperature and, in turn, the saturation pressure and mole fraction of the water vapor increase, the molar-average velocity becomes large, and advection effects become important. Alternatively, the molar-average velocity may be neglected when the concentration of water vapor is small. The existence of the two conditions $N''_B \approx 0$ and $x_A \ll 1$ is termed the *stationary medium approximation*.

14.3 *The Stationary Medium Approximation*

Fick's law for the diffusive flux of a species was introduced in Equations 14.12 and 14.13. In Section 14.2, we saw that the molecular motion associated with mass transfer can induce bulk motion within an otherwise stagnant fluid. In this case the absolute or total species flux (given by Equation 14.21 or 14.32) includes both a diffusive component (given by Equation 14.12 or 14.13) and an advective component associated with the bulk motion. In this section we consider a scenario for which it is appropriate to neglect the advective contribution to mass transfer.

When the diffusion of a very small amount of species A occurs within a stagnant species B, the molecular motion associated with the mass transfer will not induce significant bulk motion of the medium. This situation is common when one considers the diffusion of a *dilute* gas or liquid within a stagnant liquid or a solid host medium, such as when water vapor is transferred through the solid wall of a room. In these cases, the medium can be assumed to be *stationary*, and advection can be neglected. For situations where this *stationary medium approximation* is appropriate, the diffusive mass and molar fluxes of Equations 14.12 and 14.13 are identical to the absolute mass and molar fluxes.[5] That is,

$$\mathbf{n}''_A = \mathbf{j}_A = -\rho D_{AB} \nabla m_A \tag{14.41}$$

$$\mathbf{N}''_A = \mathbf{J}^*_A = -C D_{AB} \nabla x_A \tag{14.42}$$

Furthermore, since the concentration of species A is small, the total density (ρ) or concentration (C) is approximately that of the host medium, species B. The important conclusion is that the stationary medium approximation allows one to utilize many of the results of Chapters 2 through 5 by employing an *analogy* between conduction heat transfer and diffusion mass transfer. In the remainder of the chapter, we restrict our attention to cases where the stationary medium approximation is appropriate.

14.4 *Conservation of Species for a Stationary Medium*

Just as the first law of thermodynamics (the law of *conservation of energy*) plays an important role in heat transfer analyses, the law of *conservation of species* plays an important role

[5]For readers who skipped Section 14.2 on mass transfer in *nonstationary* media, the absolute flux of species A measured relative to fixed coordinates is denoted by $\mathbf{n}''_A$ (mass flux) or $\mathbf{N}''_A$ (molar flux). For those who read Section 14.2, note that the stationary medium approximation is equivalent to saying that the host medium (B) is stationary, $\mathbf{n}''_B = 0$ and $\mathbf{N}''_B = 0$, and species A is dilute, $m_A \ll 1$ and $x_A \ll 1$. Thus, in Equations 14.21 and 14.32, the advective component is negligible, resulting in Equations 14.41 and 14.42.

in the analysis of mass transfer problems. In this section we consider a general statement of this law, as well as application of the law to species diffusion in a stationary medium.

14.4.1   Conservation of Species for a Control Volume

A general formulation of the energy conservation requirement, Equation 1.12c was expressed for the control volume of Figure 1.8b. We may now express an analogous species mass conservation requirement for the control volume of Figure 14.3.

> *The rate at which the mass of some species enters a control volume, plus the rate at which the species mass is generated within the control volume, minus the rate at which this species mass leaves the control volume must equal the rate of increase of the species mass stored within the control volume.*

For example, any species A may enter and leave the control volume due to both fluid motion and diffusion across the control surface; these processes are *surface* phenomena represented by $\dot{M}_{A,in}$ and $\dot{M}_{A,out}$.[6] The same species A may also be generated, $\dot{M}_{A,g}$, and accumulated or stored, $\dot{M}_{A,st}$, *within* the control volume. The conservation equation may then be expressed on a rate basis as

$$\dot{M}_{A,in} + \dot{M}_{A,g} - \dot{M}_{A,out} = \frac{dM_A}{dt} \equiv \dot{M}_{A,st} \tag{14.43}$$

Species generation exists when chemical reactions occur in the system. For example, for a dissociation reaction of the form $AB \rightarrow A + B$, there would be net production of species A and B, as well as net reduction of the species AB.

14.4.2   The Mass Diffusion Equation

The foregoing result may be used to obtain a mass, or species, diffusion equation that is analogous to the heat equation of Chapter 2. We will consider a medium that is a binary mixture of species A and B for which the stationary medium approximation holds. That is, mass transfer may be approximated as occurring only by diffusion because advection is negligible. The resulting equation could be solved for the species concentration distribution, which could, in turn, be used with Fick's law to determine the species diffusion rate at any point in the medium.

Allowing for concentration gradients in each of the x, y, and z coordinate directions, we first define a differential control volume, $dx\,dy\,dz$, within the medium (Figure 14.4) and consider the processes that influence the distribution of species A. With the concentration gradients, diffusion must result in the transport of species A through the control surfaces. Moreover, relative to stationary coordinates, the species transport rates at opposite surfaces must be related by

$$n''_{A,x+dx}\,dy\,dz = n''_{A,x}\,dy\,dz + \frac{\partial(n''_{A,x}\,dy\,dz)}{\partial x}\,dx \tag{14.44a}$$

FIGURE 14.3   Conservation of species for a control volume.

---

[6]The mass of species i (kg) is designated as M_i, to distinguish it from the mass fraction of species i, m_i.

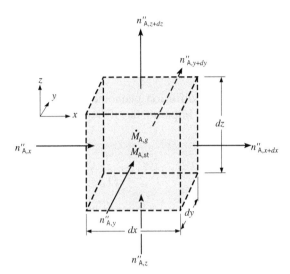

FIGURE 14.4 Differential control volume, $dx \, dy \, dz$, for species diffusion analysis in Cartesian coordinates.

$$n''_{A,y+dy} \, dx \, dz = n''_{A,y} \, dx \, dz + \frac{\partial(n''_{A,y} \, dx \, dz)}{\partial y} \, dy \qquad (14.44b)$$

$$n''_{A,z+dz} \, dx \, dy = n''_{A,z} \, dx \, dy + \frac{\partial(n''_{A,z} \, dx \, dy)}{\partial z} \, dz \qquad (14.44c)$$

In addition, there may be volumetric (also referred to as *homogeneous*) chemical reactions occurring throughout the medium, perhaps nonuniformly. The rate at which species A is generated within the control volume due to such reactions may be expressed as

$$\dot{M}_{A,g} = \dot{n}_A \, dx \, dy \, dz \qquad (14.45)$$

where $\dot{n}_A$ is the rate of increase of the mass of species A per unit volume of the mixture (kg/s · m^3). Finally, these processes may change the mass of species A stored within the control volume, and the rate of change is

$$\dot{M}_{A,st} = \frac{\partial \rho_A}{\partial t} \, dx \, dy \, dz \qquad (14.46)$$

With mass inflow rates determined by $n''_{A,x}$, $n''_{A,y}$, and $n''_{A,z}$ and the outflow rates determined by Equations 14.44, Equations 14.44 through 14.46 may be substituted into Equation 14.43 to obtain

$$-\frac{\partial n''_A}{\partial x} - \frac{\partial n''_A}{\partial y} - \frac{\partial n''_A}{\partial z} + \dot{n}_A = \frac{\partial \rho_A}{\partial t}$$

Then, substituting the x, y, and z components of Equation 14.41, we obtain

$$\frac{\partial}{\partial x}\left(\rho D_{AB} \frac{\partial m_A}{\partial x}\right) + \frac{\partial}{\partial y}\left(\rho D_{AB} \frac{\partial m_A}{\partial y}\right) + \frac{\partial}{\partial z}\left(\rho D_{AB} \frac{\partial m_A}{\partial z}\right) + \dot{n}_A = \frac{\partial \rho_A}{\partial t} \qquad (14.47a)$$

In terms of the molar concentration, a similar derivation yields

$$\frac{\partial}{\partial x}\left(CD_{AB}\frac{\partial x_A}{\partial x}\right) + \frac{\partial}{\partial y}\left(CD_{AB}\frac{\partial x_A}{\partial y}\right) + \frac{\partial}{\partial z}\left(CD_{AB}\frac{\partial x_A}{\partial z}\right) + \dot{N}_A = \frac{\partial C_A}{\partial t} \qquad (14.48a)$$

In subsequent treatments of species diffusion phenomena, we shall work with simplified versions of the foregoing equations. In particular, if D_{AB} and ρ are constant, Equation 14.47a may be expressed as

$$\frac{\partial^2 \rho_A}{\partial x^2} + \frac{\partial^2 \rho_A}{\partial y^2} + \frac{\partial^2 \rho_A}{\partial z^2} + \frac{\dot{n}_A}{D_{AB}} = \frac{1}{D_{AB}}\frac{\partial \rho_A}{\partial t} \qquad (14.47b)$$

Similarly, if D_{AB} and C are constant, Equation 14.48a may be expressed as

$$\frac{\partial^2 C_A}{\partial x^2} + \frac{\partial^2 C_A}{\partial y^2} + \frac{\partial^2 C_A}{\partial z^2} + \frac{\dot{N}_A}{D_{AB}} = \frac{1}{D_{AB}}\frac{\partial C_A}{\partial t} \qquad (14.48b)$$

Equations 14.47b and 14.48b are analogous to the heat equation, Equation 2.21. As for the heat equation, two boundary conditions must be specified for each coordinate needed to describe the system. Conditions are also needed at an *initial time* if the problem of interest is transient. Hence it follows that, for analogous boundary and initial conditions, the solution to Equation 14.47b for $\rho_A(x, y, z, t)$ or to Equation 14.48b for $C_A(x, y, z, t)$ is of the same form as the solution to Equation 2.21 for $T(x, y, z, t)$.

The species diffusion equations may also be expressed in cylindrical and spherical coordinates. These alternative forms can be inferred from the analogous expressions for heat transfer, Equations 2.26 and 2.29, and in terms of the molar concentration are:

Cylindrical Coordinates:

$$\frac{1}{r}\frac{\partial}{\partial r}\left(CD_{AB}r\frac{\partial x_A}{\partial r}\right) + \frac{1}{r^2}\frac{\partial}{\partial \phi}\left(CD_{AB}\frac{\partial x_A}{\partial \phi}\right) + \frac{\partial}{\partial z}\left(CD_{AB}\frac{\partial x_A}{\partial z}\right) + \dot{N}_A = \frac{\partial C_A}{\partial t} \qquad (14.49)$$

Spherical Coordinates:

$$\frac{1}{r^2}\frac{\partial}{\partial r}\left(CD_{AB}r^2\frac{\partial x_A}{\partial r}\right) + \frac{1}{r^2\sin^2\theta}\frac{\partial}{\partial \phi}\left(CD_{AB}\frac{\partial x_A}{\partial \phi}\right)$$
$$+ \frac{1}{r^2\sin\theta}\frac{\partial}{\partial \theta}\left(CD_{AB}\sin\theta\frac{\partial x_A}{\partial \theta}\right) + \dot{N}_A = \frac{\partial C_A}{\partial t} \qquad (14.50)$$

Simpler forms are, of course, associated with the absence of chemical reactions ($\dot{n}_A = \dot{N}_A = 0$) and with one-dimensional, steady-state conditions.

14.4.3 Stationary Media with Specified Surface Concentrations

Consider, for example, one-dimensional diffusion of species A through a planar medium of A and B, as shown in Figure 14.5. For steady-state conditions with no homogeneous chemical reactions, the molar form of the species diffusion equation (14.48a) reduces to

$$\frac{d}{dx}\left(CD_{AB}\frac{dx_A}{dx}\right) = 0 \qquad (14.51)$$

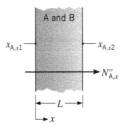

FIGURE 14.5 Mass transfer in a stationary planar medium.

Assuming the total molar concentration and the diffusion coefficient to be constant, Equation 14.51 may be solved and the boundary conditions illustrated in Figure 14.5 may be applied to yield

$$x_A(x) = (x_{A,s2} - x_{A,s1})\frac{x}{L} + x_{A,s1} \tag{14.52}$$

From Equation 14.42, it follows that

$$N''_{A,x} = -CD_{AB}\frac{x_{A,s2} - x_{A,s1}}{L} \tag{14.53}$$

Multiplying by the surface area A and substituting for $x_A \equiv C_A/C$, the molar rate is then

$$N_{A,x} = \frac{D_{AB}A}{L}(C_{A,s1} - C_{A,s2}) \tag{14.54}$$

From this expression we can define a resistance to species transfer by diffusion in a planar medium as

$$R_{m,\text{dif}} \equiv \frac{C_{A,s1} - C_{A,s2}}{N_{A,x}} = \frac{L}{D_{AB}A} \tag{14.55}$$

Comparing the foregoing results with those obtained for one-dimensional, steady-state conduction in a plane wall with no generation (Section 3.1), it is evident that a direct analogy exists between heat and mass transfer by diffusion.

The analogy also applies to cylindrical and spherical systems. For one-dimensional, steady diffusion in a cylindrical, nonreacting medium, Equation 14.49 reduces to

$$\frac{d}{dr}\left(rCD_{AB}\frac{dx_A}{dr}\right) = 0 \tag{14.56}$$

Similarly, for a spherical medium,

$$\frac{d}{dr}\left(r^2CD_{AB}\frac{dx_A}{dr}\right) = 0 \tag{14.57}$$

Equations 14.56 and 14.57, as well as Equation 14.51, dictate that the molar transfer rate, $N_{A,r}$ or $N_{A,x}$, is constant in the direction of transfer (r or x). Assuming C and D_{AB} to be constant, it is a simple matter to obtain general solutions to Equations 14.56 and 14.57. For prescribed surface species concentrations, the corresponding solutions and diffusion

TABLE 14.1   Summary of species diffusion solutions for stationary media with specified surface concentrations[a]

Geometry	Species Mole Fraction Distribution, $x_A(x)$ or $x_A(r)$	Species Diffusion Resistance, $R_{m,\text{dif}}$
	$x_A(x) = (x_{A,s2} - x_{A,s1})\dfrac{x}{L} + x_{A,s1}$	$R_{m,\text{dif}} = \dfrac{L}{D_{AB}A}$ [b]
	$x_A(r) = \dfrac{x_{A,s1} - x_{A,s2}}{\ln(r_1/r_2)}\ln\left(\dfrac{r}{r_2}\right) + x_{A,s2}$	$R_{m,\text{dif}} = \dfrac{\ln(r_2/r_1)}{2\pi L D_{AB}}$ [c]
	$x_A(r) = \dfrac{x_{A,s1} - x_{A,s2}}{1/r_1 - 1/r_2}\left(\dfrac{1}{r} - \dfrac{1}{r_2}\right) + x_{A,s2}$	$R_{m,\text{dif}} = \dfrac{1}{4\pi D_{AB}}\left(\dfrac{1}{r_1} - \dfrac{1}{r_2}\right)$ [c]

[a]Assuming C and D_{AB} are constant.

[b]$N_{A,x} = (C_{A,s1} - C_{A,s2})/R_{m,\text{dif}} = C(x_{A,s1} - x_{A,s2})/R_{m,\text{dif}}$.

[c]$N_{A,r} = (C_{A,s1} - C_{A,s2})/R_{m,\text{dif}} = C(x_{A,s1} - x_{A,s2})/R_{m,\text{dif}}$.

resistances are summarized in Table 14.1. Although species diffusion resistances are exactly analogous to heat conduction resistances, it is not generally valid to combine diffusion resistances in series for multiple layers as was done in Chapter 3 with conduction resistances. Mass transfer is complicated by the existence of discontinuous species concentrations across interfaces between different materials, as will be addressed later in this chapter.

EXAMPLE 14.3

The efficacy of pharmaceutical products is reduced by prolonged exposure to high temperature, light, and humidity. For water vapor–sensitive consumer products that are in tablet or capsule form, and might be stored in humid environments such as bathroom medicine cabinets, *blister packaging* is used to limit the direct exposure of the medicine to humid conditions until immediately before its ingestion.

Consider tablets that are contained in a blister package composed of a flat *lidding sheet* and a second, *formed sheet* that includes troughs to hold each tablet. The formed sheet is

$L = 50$ μm thick and is fabricated of a polymer material. Each trough is of diameter $D = 5$ mm and depth $h = 3$ mm. The lidding sheet is fabricated of aluminum foil. The binary diffusion coefficient for water vapor in the polymer is $D_{AB} = 6 \times 10^{-14}$ m^2/s while the aluminum may be assumed to be impermeable to water vapor. For molar concentrations of water vapor in the polymer at the outer and inner surfaces of $C_{A,s1} = 4.5 \times 10^{-3}$ kmol/m^3 and $C_{A,s2} = 0.5 \times 10^{-3}$ kmol/m^3, respectively, determine the rate at which water vapor is transferred through the trough wall to the tablet.

Blister package

SOLUTION

Known: Molar concentrations of water vapor at the inner and outer surfaces of a polymer sheet and trough geometry.

Find: Rate of water vapor molar diffusive transfer through the trough wall.

Schematic:

Section A-A

Assumptions:

1. Steady-state, one-dimensional conditions.
2. Stationary medium.
3. No chemical reactions.
4. Polymer sheet is thin relative to the dimensions of the trough, and diffusion may be analyzed as though it occurs through a plane wall.

Analysis: The total water vapor transfer rate is the summation of the transfer rate through the cylindrical walls of the trough and the bottom, circular surface of the trough. From Equation 14.54 we may write

$$N_{A,x} = \frac{D_{AB}A}{L}(C_{A,s1} - C_{A,s2}) = \frac{D_{AB}}{L}\left(\frac{\pi D^2}{4} + \pi D h\right)(C_{A,s1} - C_{A,s2})$$

Hence

$$
N_{A,x} = \frac{6 \times 10^{-14}\,\text{m}^2/\text{s}}{50 \times 10^{-6}\,\text{m}} \left[\frac{\pi (5 \times 10^{-3}\,\text{m})^2}{4} + \pi (5 \times 10^{-3}\,\text{m})(3 \times 10^{-3}\,\text{m}) \right]
$$

$$
\times (4.5 \times 10^{-3} - 0.5 \times 10^{-3})\,\text{kmol/m}^3
$$

$$
= 0.32 \times 10^{-15}\,\text{kmol/s} \qquad\qquad \triangleleft
$$

Comments:

1. The mass diffusion rate of water vapor is $n_{A,x} = \mathcal{M}_A N_{A,x} = 18\,\text{kg/kmol} \times 0.32 \times 10^{-15}$ kmol/s $= 5.8 \times 10^{-15}$ kg/s.

2. The *shelf life* of the medicine is inversely proportional to the rate at which water vapor is transferred through the polymer sheet. Shelf life may be extended by increasing the thickness of the sheet, resulting in increased cost of the package. Specification of materials for use in blister packaging involves tradeoffs between shelf life, cost, formability, and recyclability of the polymer material.

14.5 *Boundary Conditions and Discontinuous Concentrations at Interfaces*

In the previous section, expressions for mass transfer resistances were developed by applying constant surface concentration boundary conditions. For a surface at $x = 0$, the constant surface species concentration boundary condition is expressed as

$$
C_A(0,t) = C_{A,s} \tag{14.58a}
$$

or

$$
x_A(0,t) = x_{A,s} \tag{14.58b}
$$

We have already used Equations 14.58a and 14.58b in Examples 14.2 and 14.3. A constant surface concentration boundary condition can equivalently be expressed in terms of a mass fraction or density. While the forms of Equations 14.58a and 14.58b are simple, determination of the appropriate value of $x_{A,s}$ (or $C_{A,s}$) can be complicated, as discussed below.

The second boundary condition that is analogous to the conduction heat transfer conditions of Table 2.2 is that of a constant species flux, $J_{A,s}^*$, at a surface. Using Fick's law, Equation 14.13, the condition is expressed for a surface at $x = 0$ as

$$
-CD_{AB} \frac{\partial x_A}{\partial x} \bigg|_{x=0} = J_{A,s}^* \tag{14.59}
$$

A special case of this condition corresponds to the *impermeable surface* for which $\partial x_A/\partial x \big|_{x=0} = 0$ when a stationary medium is considered. A constant species flux boundary condition can also be expressed on a mass basis.

One phenomenon that makes mass transfer more complex than heat transfer is that species concentrations are typically *discontinuous* at the interface between two materials, whereas temperature is continuous. To take a familiar example, consider a pool of water that is exposed to air. If we are interested in determining the rate at which water vapor is transferred into the air, we would need to specify the water vapor concentration in the air at the air–water interface. We know that the mole fraction of water in the pool is essentially unity (neglecting the small amount of dissolved oxygen or nitrogen in the water). But it would be incorrect to specify $x_{A,s} = 1$ for the water vapor mole fraction *in the air* at the interface. Clearly, the concentration of water is discontinuous across the interface. In general, concentration boundary conditions at the interface that separates two materials describe a *relationship* between the concentrations on either side of the interface. The relationships are either based on theory or are deduced from experiments. They can be expressed in a variety of forms, a few of which are now introduced.

14.5.1 Evaporation and Sublimation

A common mass transfer scenario is the transfer of a species A into a gas stream due to evaporation or sublimation from a liquid or solid surface, respectively (Figure 14.6*a*). Conditions *within the gas phase* are of interest, and the concentration (or partial pressure) of species A in the gas phase at the interface (located at $x = 0$) may readily be determined from *Raoult's law*,

$$p_A(0) = x_A(0)p_{A,\text{sat}} \tag{14.60}$$

where p_A is the partial pressure of A *in the gas phase*, x_A is the mole fraction of species A *in the liquid or solid*, and $p_{A,\text{sat}}$ is the saturation pressure of species A at the surface temperature. Raoult's law applies if the gas phase can be approximated as ideal and the liquid or solid phase has a high concentration of species A. If the liquid or solid is a pure species A, that is, $x_A = 1$, Equation 14.60 simplifies to $p_A(0) = p_{A,\text{sat}}$. That is, the partial pressure of the vapor at the interface corresponds to saturated conditions at the temperature of the interface and may be determined from standard thermodynamic tables. This boundary condition was utilized in the solution of Example 14.2 and in Section 6.7.2.

14.5.2 Solubility of Gases in Liquids and Solids

Another common scenario is mass transfer of species A *from a gas phase into a liquid or solid*, species B (Figure 14.6*b*). Mass transfer within the liquid or solid phase is of interest, and the concentration of species A at the interface is required as a boundary condition.

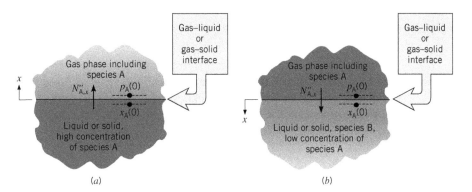

FIGURE 14.6 Species concentration at a gas–liquid or gas–solid interface. (*a*) Evaporation or sublimation of species A from a liquid or solid into a gas. (*b*) Transfer of weakly soluble species A from the gas to a liquid or solid.

If species A is only weakly soluble (x_A is small) in a *liquid, Henry's law* may be used to relate the mole fraction of A in the liquid to the partial pressure of A in the gas phase outside the liquid:

$$x_A(0) = \frac{p_A(0)}{H} \tag{14.61}$$

The coefficient H is known as *Henry's constant*, and values for selected aqueous solutions are listed in Table A.9. Although H depends on temperature, its pressure dependence may generally be neglected for values of p up to 5 bar.

Conditions at a *gas–solid* interface may also be determined if the gas, species A, dissolves in a solid, species B, and a solution is formed. In such cases mass transfer in the solid is independent of the structure of the solid and may be treated as a diffusion process. In contrast, there are many situations for which the porosity of the solid strongly influences gas transport through the solid. Such situations may sometimes be treated in a fashion analogous to heat transfer in a porous medium as discussed in Section 3.1.5 but often must be dealt with using methods described in more advanced texts [2, 4].

Treating the gas and solid as a solution, we can obtain the concentration of the gas in the solid at the interface through use of a property known as the *solubility, S*. It is defined by the expression

$$C_A(0) = S p_A(0) \tag{14.62}$$

where $p_A(0)$ is once again the partial pressure (bar) of the gas adjoining the interface. The molar concentration of A in the solid at the interface, $C_A(0)$, is in units of kilomoles of A per cubic meter of solid, in which case the units of S must be *kilomoles of A per cubic meter of solid per bar (or atm) partial pressure of A*. Values of S for several gas–solid combinations are given in Table A.10. Values of solubility are often presented in units of cubic meters of species A (at STP, Standard Temperature and Pressure, of 0°C and 1 atm) per cubic meter of solid per atm partial pressure of A. Denoting this solubility value as $\tilde{S}$, and recognizing that at STP 1 kmol occupies 22.414 m³, we find that the unit conversion is given by $S = \tilde{S}/(22.414 \text{ m}^3/\text{kmol})$. (Additional conversion between bar and atm may be needed.)

EXAMPLE 14.4

Helium gas is stored at 20°C in a spherical container of fused silica (SiO_2), which has a diameter of 0.20 m and a wall thickness of 2 mm. If the container is charged to an initial pressure of 4 bar, what is the initial rate at which this pressure decreases with time?

SOLUTION

Known: Initial pressure of helium in a spherical, fused silica container of prescribed diameter D and wall thickness L.

Find: The rate of change of the helium pressure, dp_A/dt.

Schematic:

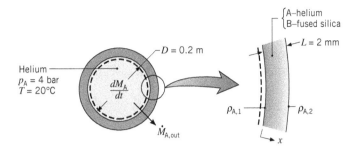

Assumptions:

1. Since $D \gg L$, diffusion may be approximated as being one-dimensional through a plane wall.
2. Quasi-steady diffusion (pressure variation is sufficiently slow to permit assuming steady-state conditions for diffusion through the fused silica at any instant).
3. Stationary medium has uniform density ρ.
4. Pressure of helium in outside air is negligible.
5. Helium exhibits ideal gas behavior.

Properties: Table A.8, helium-fused silica (293 K): $D_{AB} = 0.4 \times 10^{-13}$ m^2/s. Table A.10, helium-fused silica (293 K): $S = 0.45 \times 10^{-3}$ kmol/m$^3 \cdot$ bar.

Analysis: The rate of change of the helium pressure may be obtained by applying the species conservation requirement, Equation 14.43, to a control volume about the helium. It follows that

$$-\dot{M}_{A,out} = \dot{M}_{A,st}$$

or, since the helium outflow is due to diffusion through the fused silica,

$$\dot{M}_{A,out} = n''_{A,x} A$$

and the change in mass storage is

$$\dot{M}_{A,st} = \frac{dM_A}{dt} = \frac{d(\rho_A V)}{dt}$$

the species balance reduces to

$$-n''_{A,x} A = \frac{d(\rho_A V)}{dt}$$

Recognizing that $\rho_A = \mathcal{M}_A C_A$ and applying the ideal gas law

$$C_A = \frac{p_A}{\mathcal{R} T}$$

the species balance becomes

$$\frac{dp_A}{dt} = -\frac{\mathscr{R}T}{\mathcal{M}_A V} A n''_{A,x}$$

For a stationary medium the absolute flux of species A through the fused silica is equal to the diffusion flux, $n''_{A,x} = j_{A,x}$, in which case, from Equation 14.41,

$$n''_{A,x} = -\rho D_{AB} \frac{dm_A}{dx} = -D_{AB} \frac{d\rho_A}{dx}$$

or, for the assumed conditions,

$$n''_{A,x} = D_{AB} \frac{\rho_{A,1} - \rho_{A,2}}{L}$$

The species densities $\rho_{A,1}$ and $\rho_{A,2}$ pertain to conditions *within* the fused silica at its inner and outer surfaces, respectively, and may be evaluated from knowledge of the solubility through Equation 14.62. Hence with $\rho_A = \mathcal{M}_A C_A$,

$$\rho_{A,1} = \mathcal{M}_A S p_{A,i} = \mathcal{M}_A S p_A \qquad \text{and} \qquad \rho_{A,2} = \mathcal{M}_A S p_{A,o} = 0$$

where $p_{A,i}$ and $p_{A,o}$ are helium pressures at the inner and outer surfaces, respectively. Hence

$$n''_{A,x} = \frac{D_{AB} \mathcal{M}_A S p_A}{L}$$

and substituting into the species balance it follows that

$$\frac{dp_A}{dt} = -\frac{\mathscr{R}T A D_{AB} S}{LV} p_A$$

or with $A = \pi D^2$ and $V = \pi D^3/6$

$$\frac{dp_A}{dt} = -\frac{6\mathscr{R}T D_{AB} S}{LD} p_A$$

Substituting numerical values, the rate of change of the pressure is

$$\frac{dp_A}{dt} = [-6(0.08314 \text{ m}^3 \cdot \text{bar/kmol} \cdot \text{K}) 293 \text{ K} (0.4 \times 10^{-13} \text{ m}^2/\text{s})$$

$$\times 0.45 \times 10^{-3} \text{ kmol/m}^3 \cdot \text{bar} \times 4 \text{ bar}] \div [0.002 \text{ m} (0.2 \text{ m})]$$

$$\frac{dp_A}{dt} = -2.63 \times 10^{-11} \text{ bar/s} \qquad \lhd$$

Comments: The foregoing result provides the initial (maximum) leakage rate for the system. The leakage rate decreases as the inside pressure decreases due to the dependence of $\rho_{A,1}$ on p_A.

EXAMPLE 14.5

Hydrogen gas is maintained at 3 bar and 1 bar on opposite sides of a plastic membrane, which is 0.3 mm thick. The temperature is 25°C, and the binary diffusion coefficient of hydrogen in the plastic is 8.7×10^{-8} m²/s. The solubility of hydrogen in the membrane is 1.5×10^{-3} kmol/m³ · bar. What is the diffusive mass flux of hydrogen through the membrane?

SOLUTION

Known: Pressure of hydrogen on opposite sides of a membrane.

Find: The hydrogen diffusive mass flux $n''_{A,x}$ (kg/s · m²).

Schematic:

Assumptions:

1. Steady-state, one-dimensional conditions exist.
2. Membrane is a stationary, nonreacting medium of uniform total molar concentration.

Analysis: For the prescribed conditions, Equation 14.42 reduces to Equation 14.53, which may be expressed as

$$N''_{A,x} = CD_{AB} \frac{x_{A,s1} - x_{A,s2}}{L} = \frac{D_{AB}}{L}(C_{A,s1} - C_{A,s2})$$

The surface molar concentrations of hydrogen may be obtained from Equation 14.62, where

$$C_{A,s1} = 1.5 \times 10^{-3} \text{ kmol/m}^3 \cdot \text{bar} \times 3 \text{ bar} = 4.5 \times 10^{-3} \text{ kmol/m}^3$$
$$C_{A,s2} = 1.5 \times 10^{-3} \text{ kmol/m}^3 \cdot \text{bar} \times 1 \text{ bar} = 1.5 \times 10^{-3} \text{ kmol/m}^3$$

Hence

$$N''_{A,x} = \frac{8.7 \times 10^{-8} \text{ m}^2/\text{s}}{0.3 \times 10^{-3} \text{ m}}(4.5 \times 10^{-3} - 1.5 \times 10^{-3}) \text{ kmol/m}^3$$
$$N''_{A,x} = 8.7 \times 10^{-7} \text{ kmol/s} \cdot \text{m}^2$$

On a mass basis,

$$n''_{A,x} = N''_{A,x} \mathcal{M}_A$$

where the molecular weight of hydrogen is 2 kg/kmol. Hence

$$n''_{A,x} = 8.7 \times 10^{-7} \text{ kmol/s} \cdot \text{m}^2 \times 2 \text{ kg/kmol} = 1.74 \times 10^{-6} \text{ kg/s} \cdot \text{m}^2 \qquad \lhd$$

Comments: The molar concentrations of hydrogen in the gas phase, $C_{A,1}$ and $C_{A,2}$, differ from the surface concentrations in the membrane and may be calculated from the ideal gas equation of state

$$C_A = \frac{p_A}{\mathscr{R}T}$$

where $\mathscr{R} = 8.314 \times 10^{-2} \text{ m}^3 \cdot \text{bar/kmol} \cdot \text{K}$. It follows that $C_{A,1} = 0.121 \text{ kmol/m}^3$ and $C_{A,2} = 0.040 \text{ kmol/m}^3$. Even though $C_{A,s2} < C_{A,2}$, hydrogen transport will occur from the membrane to the gas at $p_{A,2} = 1$ bar. This seemingly anomalous result may be explained by recognizing that the two concentrations are based on *different* volumes; in one case the concentration is per unit volume of the membrane and in the other case it is per unit volume of the adjoining gas phase. For this reason it is not possible to infer the direction of hydrogen transport from a simple comparison of the numerical values of $C_{A,s2}$ and $C_{A,2}$.

14.5.3   Catalytic Surface Reactions

Many mass transfer problems involve specification of the species flux, rather than the species concentration, at a surface. One such problem relates to the process of catalysis, which involves the use of special surfaces to promote *heterogeneous chemical reactions*. Such a reaction occurs at the surface of a material, can be viewed as a *surface phenomenon*, and can be treated as a boundary condition.[7] Often, a one-dimensional diffusion analysis may be used to *approximate* the performance of a catalytic reactor.

Consider the system of Figure 14.7. A catalytic surface is placed in a gas stream to promote a heterogeneous chemical reaction involving species A. Assume that the reaction produces species A at a rate $\dot{N}''_A$, which is defined as the molar rate of production per unit surface area of the catalyst. Once steady-state conditions are reached, the rate of species transfer from the surface, $N''_{A,x}$, must equal the surface reaction rate:

$$N''_{A,x}(0) = \dot{N}''_A \qquad (14.63)$$

It is also assumed that species A leaves the surface as a result of one-dimensional transfer through a thin film of thickness L and that no reactions occur within the film itself. The mole fraction of A at $x = L$, $x_{A,L}$, corresponds to conditions in the mainstream of the mixture and is presumed to be known. Representing the remaining species of the mixture as a single species B and assuming the medium to be stationary, Equation 14.48a reduces to

$$\frac{d}{dx}\left(CD_{AB} \frac{dx_A}{dx} \right) = 0 \qquad (14.64)$$

---

[7]The generation terms that appear in Equations 14.47 through 14.50 are due to chemical reactions that occur *volumetrically*. These volumetric reactions are referred to as *homogeneous chemical reactions* and will be treated in Section 14.6.

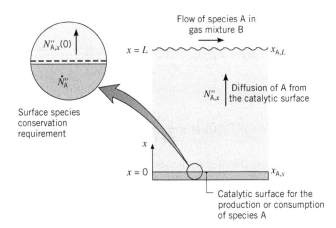

FIGURE 14.7 One-dimensional diffusion with heterogeneous catalysis.

where D_{AB} is the binary diffusion coefficient for A in B, and B may be a multicomponent mixture. Assuming C and D_{AB} to be constant, Equation 14.64 may be solved subject to the conditions that

$$x_A(L) = x_{A,L}$$

and

$$N''_{A,x}(0) = -CD_{AB} \frac{dx_A}{dx}\bigg|_{x=0} = \dot{N}''_A \qquad (14.65)$$

This expression follows from Equation 14.63 and the substitution of Fick's law, Equation 14.42.

For a catalytic surface, the surface reaction rate $\dot{N}''_A$ generally depends on the surface concentration $C_A(0)$. For a *first-order reaction* that results in species consumption at the surface, the reaction rate is of the form

$$\dot{N}''_A = -k''_1 C_A(0) \qquad (14.66)$$

where k''_1 (m/s) is the reaction rate constant. Accordingly, the surface boundary condition, Equation 14.65, reduces to

$$-D_{AB} \frac{dx_A}{dx}\bigg|_{x=0} = -k''_1 x_A(0) \qquad (14.67)$$

Solving Equation 14.64 subject to the above conditions, it is readily verified that the concentration distribution is linear and of the form

$$\frac{x_A(x)}{x_{A,L}} = \frac{1 + (xk''_1/D_{AB})}{1 + (Lk''_1/D_{AB})} \qquad (14.68)$$

At the catalytic surface this result reduces to

$$\frac{x_A(0)}{x_{A,L}} = \frac{1}{1 + (Lk''_1/D_{AB})} \qquad (14.69)$$

and the molar flux is

$$N''_{A,x}(0) = -CD_{AB} \left. \frac{dx_A}{dx} \right|_{x=0} = -k''_1 C x_A(0)$$

or

$$N''_{A,x}(0) = -\frac{k''_1 C x_{A,L}}{1 + (L k''_1 / D_{AB})} \tag{14.70}$$

The negative sign implies mass transfer *to* the surface.

Two limiting cases of the foregoing results are of special interest. For the limit $k''_1 \to 0$, $(L k''_1 / D_{AB}) \ll 1$ and Equations 14.69 and 14.70 reduce to

$$\frac{x_{A,s}}{x_{A,L}} \approx 1 \quad \text{and} \quad N''_{A,x}(0) \approx -k''_1 C x_{A,L}$$

In such cases the rate of reaction is controlled by the reaction rate constant, and the limitation due to diffusion is negligible. The process is said to be *reaction limited*. Conversely, for the limit $k''_1 \to \infty$, $(L k''_1 / D_{AB}) \gg 1$ and Equations 14.69 and 14.70 reduce to

$$x_{A,s} \approx 0 \quad \text{and} \quad N''_{A,x}(0) \approx -\frac{CD_{AB} x_{A,L}}{L}$$

In this case the reaction is controlled by the rate of diffusion to the surface, and the process is said to be *diffusion limited*.

14.6 *Mass Diffusion with Homogeneous Chemical Reactions*

Just as heat diffusion may be influenced by internal sources of energy, species transfer by diffusion may be influenced by homogeneous chemical reactions. We restrict our attention to stationary media, in which case Equation 14.41 or 14.42 determines the absolute species flux. If we also assume steady, one-dimensional transfer in the x-direction and that D_{AB} and C are constant, Equation 14.48b reduces to

$$D_{AB} \frac{d^2 C_A}{dx^2} + \dot{N}_A = 0 \tag{14.71}$$

The volumetric production rate, $\dot{N}_A$, is often described using one of the following forms.

Zero-Order Reaction:

$$\dot{N}_A = k_0$$

First-Order Reaction:

$$\dot{N}_A = k_1 C_A$$

That is, the reaction may occur at a constant rate (zero order) or at a rate that is proportional to the local concentration (first order). The units of k_0 and k_1 are kmol/s $\cdot$ m^3 and s^{-1}, respectively.

If $\dot{N}_A$ is positive, the reaction results in the generation of species A; if it is negative, it results in the consumption of A.

In many applications the species of interest is converted to another form through a first-order chemical reaction, and Equation 14.71 becomes

$$D_{AB} \frac{d^2 C_A}{dx^2} - k_1 C_A = 0 \qquad (14.72)$$

This linear, homogeneous differential equation has the general solution

$$C_A(x) = C_1 e^{mx} + C_2 e^{-mx} \qquad (14.73)$$

where $m = (k_1/D_{AB})^{1/2}$ and the constants C_1 and C_2 depend on the prescribed boundary conditions. The form of this equation is identical to that characterizing heat conduction in an extended surface, Equation 3.71.

Consider the situation illustrated in Figure 14.8. Gas A is soluble in liquid B, where it is transferred by diffusion and experiences a first-order chemical reaction. The solution is dilute, and the concentration of A in the liquid at the interface is a known constant $C_{A,0}$. If the bottom of the container is impermeable to A, the boundary conditions are

$$C_A(0) = C_{A,0} \qquad \text{and} \qquad \frac{dC_A}{dx}\bigg|_{x=L} = 0$$

These species boundary conditions are analogous to the thermal boundary conditions of case B in Table 3.4. Since Equation 14.73 is of the same form as Equation 3.71, it follows that

$$C_A(x) = C_{A,0} \frac{\cosh m(L-x)}{\cosh mL} \qquad (14.74)$$

Quantities of special interest are the concentration of A at the bottom and the flux of A across the gas–liquid interface. Applying Equation 14.74 at $x = L$, we obtain

$$C_A(L) = \frac{C_{A,0}}{\cosh mL} \qquad (14.75)$$

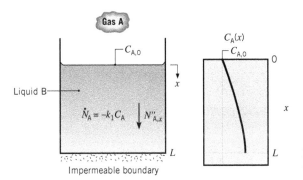

FIGURE 14.8 Diffusion and homogeneous reaction of gas A in liquid B.

Moreover,

$$N''_{A,x}(0) = -D_{AB} \left. \frac{dC_A}{dx} \right|_{x=0}$$

$$= D_{AB}C_{A,0}\, m \left. \frac{\sinh m(L-x)}{\cosh mL} \right|_{x=0}$$

or

$$N''_{A,x}(0) = D_{AB}C_{A,0}\, m \tanh mL \qquad (14.76)$$

Results for a container with its bottom held at a fixed concentration or an infinitely deep container may be obtained by analogy to cases C and D, respectively, of Table 3.4.

EXAMPLE 14.6

Biofilms, which are colonies of bacteria that can cling to living or inert surfaces, can cause a wide array of human infections. Infections caused by bacteria living within biofilms are often chronic because antibiotics that are applied to the surface of a biofilm have difficulty penetrating through the film thickness. Consider a biofilm that is associated with a skin infection. An antibiotic (species A) is applied to the top layer of a biofilm (species B) so that a fixed concentration of medication, $C_{A,0} = 4 \times 10^{-3}$ kmol/m^3, exists at the upper surface of the biofilm. The diffusion coefficient of the medication within the biofilm is $D_{AB} = 2 \times 10^{-12}$ m^2/s. The antibiotic is consumed by biochemical reactions within the film, and the consumption rate depends on the local concentration of medication expressed as $\dot{N}_A = -k_1 C_A$ where $k_1 = 0.1$ s^{-1}. To eradicate the bacteria, the antibiotic must be consumed at a rate of at least 0.2×10^{-3} kmol/s · m^3 ($\dot{N}_A \le -0.2 \times 10^{-3}$ kmol/s · m^3) since, at smaller absolute consumption rates, the bacteria will be able to grow back faster than it is destroyed. Determine the maximum thickness of a biofilm, L, that may be treated successfully by the antibiotic.

SOLUTION

Known: Topical antibiotic and biofilm properties, surface concentration of the medication, and required minimum consumption rate of antibiotic.

Find: Maximum thickness of a bacteria-laden biofilm, L, that may be successfully treated.

Schematic:

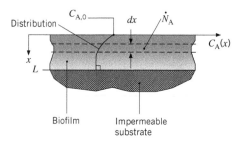

Assumptions:

1. Steady-state, one-dimensional conditions.

2. Stationary, homogeneous medium with constant properties.

3. Impermeable bottom of the biofilm.

Analysis: The absolute antibiotic consumption rate will be smallest at $x = L$, where the antibiotic concentration is smallest. Thus, we require $\dot{N}_A(L) = -0.2 \times 10^{-3}\,\text{kmol/s} \cdot \text{m}^3$. The expression for the first-order reaction may be combined with Equation 14.74 to write

$$\dot{N}_A(L) = -k_1 C_A(L) = -k_1 \frac{C_{A,0}}{\cosh mL} \tag{1}$$

where

$$m = (k_1/D_{AB})^{1/2} = \left(\frac{0.1\,\text{s}^{-1}}{2 \times 10^{-12}\,\text{m}^2/\text{s}} \right)^{1/2} = 2.24 \times 10^5\,\text{m}^{-1}$$

Equation 1 may be solved for the maximum allowable thickness:

$$L = m^{-1} \cosh^{-1} \left(\frac{-k_1 C_{A,0}}{\dot{N}_A(L)} \right) \tag{2}$$

Substituting values into Equation 2 yields

$$L = (2.24 \times 10^5\,\text{m}^{-1})^{-1} \cosh^{-1} \left(\frac{-0.1\,\text{s}^{-1} \times 4 \times 10^{-3}\,\text{kmol/m}^3}{-0.2 \times 10^{-3}\,\text{kmol/s} \cdot \text{m}^3} \right)$$

$$= 5.9 \times 10^{-6}\,\text{m} = 5.9\,\mu\text{m} \qquad \triangleleft$$

Comment: The ability of the antibiotic agent to kill bacteria in thicker biofilms is hampered by the high rate at which the agent is consumed and the slow rate at which it can be diffused through the complex, polymeric matrix of the biofilm [6]. The diffusion rate can be increased by decreasing the size of the molecule of the antibiotic, allowing it to penetrate through the biofilm with less resistance. This is an active area of research in *nanomedicine*.

14.7 *Transient Diffusion*

Results analogous to those of Chapter 5 may be obtained for the transient diffusion of a dilute species A in a stationary medium. Assuming no homogeneous reactions, constant D_{AB} and C, and one-dimensional transfer in the x-direction, Equation 14.48b reduces to

$$\frac{\partial^2 C_A}{\partial x^2} = \frac{1}{D_{AB}} \frac{\partial C_A}{\partial t} \tag{14.77}$$

Assuming an initial uniform concentration,

$$C_A(x, 0) = C_{A,i} \tag{14.78}$$

Equation 14.77 may be solved for boundary conditions that depend on the particular geometry and surface conditions. If, for example, the geometry is a plane wall of thickness $2L$ with the same convection conditions imposed at each exposed surface, the boundary conditions are

$$\left.\frac{\partial C_A}{\partial x}\right|_{x=0} = 0 \tag{14.79}$$

$$C_A(L, t) = C_{A,s} \tag{14.80}$$

Equation 14.79 describes the symmetry requirement at the midplane. Equation 14.80 represents the surface convection condition if the *mass transfer Biot number, $Bi_m = h_m L / D_{AB}$,* is much larger than unity. In this case the resistance to species transfer by diffusion in the medium is much larger than the resistance to species transfer by convection at the surface. If this situation is taken to the limit of $Bi_m \to \infty$, or $Bi_m^{-1} \to 0$, it follows that the species concentration in the fluid is essentially uniform, equal to the free stream species concentration. Then $C_{A,s}$, the species concentration in the medium at its surface, can be found from Equation 14.60 or 14.61, where p_A is the free stream partial pressure of species A.

The analogy between heat and mass transfer may conveniently be applied if we nondimensionalize the above equations. Introducing a dimensionless concentration and time, as follows,

$$\gamma^* \equiv \frac{\gamma}{\gamma_i} = \frac{C_A - C_{A,s}}{C_{A,i} - C_{A,s}} \tag{14.81}$$

$$t_m^* \equiv \frac{D_{AB} t}{L^2} \equiv Fo_m \tag{14.82}$$

and substituting into Equation 14.77, we obtain

$$\frac{\partial^2 \gamma^*}{\partial x^{*2}} = \frac{\partial \gamma^*}{\partial Fo_m} \tag{14.83}$$

where $x^* = x/L$. Similarly, the initial and boundary conditions are

$$\gamma^*(x^*, 0) = 1 \tag{14.84}$$

$$\left.\frac{\partial \gamma^*}{\partial x^*}\right|_{x^*=0} = 0 \tag{14.85}$$

and

$$\gamma^*(1, t_m^*) = 0 \tag{14.86}$$

One need only compare Equations 14.83 through 14.86 with Equations 5.37 through 5.39 and Equation 5.40 for the case of $Bi \to \infty$ to confirm the existence of the analogy. Note that for $Bi \to \infty$, Equation 5.40 reduces to $\theta^*(1, t^*) = 0$, which is analogous to Equation 14.86. Hence the two systems of equations must have equivalent solutions.

The correspondence between variables for transient heat and mass diffusion is summarized in Table 14.2. From this correspondence it is possible to use many of the heat transfer results of Chapter 5 to solve transient mass diffusion problems. For example, if the restriction

TABLE 14.2   Correspondence between
heat and mass transfer variables
for transient diffusion

Heat Transfer	Mass Transfer
$\theta^* = \dfrac{T - T_\infty}{T_i - T_\infty}$	$\gamma^* = \dfrac{C_A - C_{A,s}}{C_{A,i} - C_{A,s}}$
$1 - \theta^* = \dfrac{T - T_i}{T_\infty - T_i}$	$1 - \gamma^* = \dfrac{C_A - C_{A,i}}{C_{A,s} - C_{A,i}}$
$Fo = \dfrac{\alpha t}{L^2}$	$Fo_m = \dfrac{D_{AB} t}{L^2}$
$Bi = \dfrac{hL}{k}$	$Bi_m = \dfrac{h_m L}{D_{AB}}$
$\dfrac{x}{2\sqrt{\alpha t}}$	$\dfrac{x}{2\sqrt{D_{AB} t}}$

$Fo_m > 0.2$ is met, Equation 5.44 could be used by replacing θ_o^* and Fo by γ_o^* and Fo_m, with $Bi \to \infty$ ($\zeta_1 = 1.5708$; $C_1 = 1.2733$), to determine the midplane concentration $C_{A,o}$. The remaining equations may be applied in a similar fashion, along with results obtained for the semi-infinite solid or for objects with constant surface temperatures or surface heat fluxes.

The results of Sections 5.5 and 5.6 must be used with care because, as illustrated in Example 14.1 and in Table A.8, the mass diffusivities associated with many liquid and solid host media are extremely small relative to their thermal diffusivities. Therefore, values of the mass transfer Fourier number, Fo_m, are usually many orders of magnitude smaller than the values of the heat transfer Fourier number, Fo. Hence, the *approximate solutions* of Sections 5.5.2 and 5.6.2 are often of little value in analyzing mass transfer within stationary media, since the approximate solutions are valid only for $Fo_m > 0.2$. Recourse to solving the *exact expressions* of Sections 5.5.1 and 5.6.1 may be required. The infinite series solutions of Equations 5.42, 5.50, and 5.51 may necessitate evaluation of many terms before convergence is obtained. The semi-infinite solid solutions of Section 5.7, on the other hand, are often applicable in mass transfer problems because of the slow diffusion rates of boundary condition information into the host medium. The approximate results of Section 5.8 are also applicable, since results are presented over the entire Fourier number range.

EXAMPLE 14.7

Transdermal drug delivery involves the controlled time-release of medication through the skin to the bloodstream, usually from a *patch* that is adhered to the body. Advantages include steady and mild drug delivery rates that reduce shock to the system as might occur with intravenous infusions, the ability to deliver medication to nauseated or unconscious patients that would be otherwise delivered in oral form, and ease of use.

Consider a square patch of length and width $L = 50$ mm that consists of a host medium containing an initial, uniform density of medication, $\rho_{A,p,i} = 100$ kg/m^3. The patch is applied to the skin, which contains an initial drug concentration of $\rho_{A,s,i} = 0$. At the patch–skin interface located at $x = 0$, the ratio of the medication density on the patch side to the medication density on the patient side is described by a *partition coefficient* of $K = 0.5$.

1. Determine the total amount of medication (dosage) delivered to the patient over a treatment period of one week. Nominal values of the diffusion coefficients of the medication within the patch and skin are $D_{Ap} = 0.1 \times 10^{-12}$ m^2/s and $D_{As} = 0.2 \times 10^{-12}$ m^2/s, respectively.

2. Investigate the sensitivity of the total dosage delivered to the patient to the mass diffusivity of the patch, D_{Ap}, and the mass diffusivity of the patient's skin, D_{As}.

SOLUTION

Known: Initial density of a drug within a transdermal patch, size of the patch, partition coefficient, and mass diffusivities.

Find: Total dosage of medicine delivered to the patient over a one-week time period, sensitivity of the dosage to the mass diffusivity of the patch and skin.

Schematic:

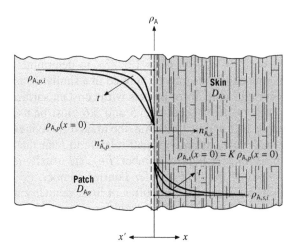

Assumptions:

1. One-dimensional conditions with constant properties.
2. Semi-infinite patch and skin.
3. No chemical reactions.
4. Stationary media.

Analysis:

1. The conservation of species equation applies to both the patch and skin. With the foregoing assumptions, Equation 14.47b becomes

$$\frac{\partial^2 \rho_A}{\partial x^2} = \frac{1}{D_{AB}} \frac{\partial \rho_A}{\partial t}$$

which is analogous to Equation 5.29. Moreover, the initial conditions

$$\rho_A(x < 0, t = 0) = \rho_{A,p,i}; \quad \rho_A(x > 0, t = 0) = \rho_{A,s,i}$$

boundary conditions

$$\rho_{A,p}(x \to -\infty) = \rho_{A,p,i}; \quad \rho_{A,s}(x \to +\infty) = \rho_{A,s,i} \tag{1a}$$

and interface condition

$$n''_{A,x,p}(x = 0) = n''_{A,x,s}(x = 0) \tag{1b}$$

are analogous to the situation in Figure 5.9 and Equation 5.64 where two semi-infinite solids are brought into thermal contact. In this problem, the partitioning of the species must also be accounted for. That is,

$$\rho_{A,s}(x = 0) = K\rho_{A,p}(x = 0) \tag{1c}$$

By analogy to Equation 5.65, we have

$$\frac{-D_{Ap}[\rho_{A,p}(x = 0) - \rho_{A,p,i}]}{\sqrt{\pi D_{Ap} t}} = \frac{D_{As}[\rho_{A,s}(x = 0) - \rho_{A,s,i}]}{\sqrt{\pi D_{As} t}}$$

which, after substituting Equation 1c and noting that $\rho_{A,s,i} = 0$, may be solved to yield

$$\rho_{A,s}(x = 0) = \rho_{A,p,i}\left(\frac{\sqrt{D_{Ap}}}{\sqrt{D_{As}} + \sqrt{D_{Ap}}/K}\right) \tag{2}$$

The instantaneous flux of medication to the patient may be determined by noting the analogy with Equation 5.61:

$$n''_A(x = 0, t) = \frac{D_{As}\rho_{A,s}(x = 0)}{\sqrt{\pi D_{As} t}} \tag{3}$$

Substituting Equation 2 into Equation 3 yields

$$n''_A(x = 0, t) = \frac{\rho_{A,p,i}}{\sqrt{\pi t}}\frac{\sqrt{D_{As} D_{Ap}}}{\sqrt{D_{As}} + \sqrt{D_{Ap}}/K}$$

The dosage, D, delivered to the patient from $t = 0$ to a treatment time of t_t may be expressed as

$$D = L^2 \int_{t=0}^{t_t} n''_A(x = 0, t)dt$$

$$D = \frac{\rho_{A,p,i}L^2}{\sqrt{\pi}}\frac{\sqrt{D_{As} D_{Ap}}}{\sqrt{D_{As}} + \sqrt{D_{Ap}}/K}\int_{t=0}^{t_t} t^{-1/2}\,dt$$

$$= \frac{2\rho_{A,p,i}L^2}{\sqrt{\pi}}\frac{\sqrt{D_{As} D_{Ap}}}{\sqrt{D_{As}} + \sqrt{D_{Ap}}/K}\sqrt{t_t} \tag{4}$$

For a total treatment time of $t_t = 7$ days $\times$ 24 h/day $\times$ 3600 s/h $= 605 \times 10^3$ s, the dosage is

$$D = \frac{2 \times 100 \text{ kg/m}^3 \times (50 \times 10^{-3} \text{ m})^2}{\sqrt{\pi}}$$

$$\times \frac{\sqrt{0.2 \times 10^{-12} \text{ m}^2/\text{s} \times 0.1 \times 10^{-12} \text{ m}^2/\text{s}}}{\sqrt{0.2 \times 10^{-12} \text{ m}^2/\text{s}} + \sqrt{0.1 \times 10^{-12} \text{ m}^2/\text{s}/0.5}} \sqrt{605 \times 10^3 \text{ s}}$$

$$= 29 \times 10^{-6} \text{ kg} = 29 \text{ mg} \qquad \lhd$$

2. The sensitivity of the dosage to the patch and skin mass diffusivities may be evaluated by solving Equation 4 for different combinations of D_{Ap} and D_{As}. Results are shown in the following graphs for $D_{Ap} = 0.1 \times 10^{-12}$ m^2/s and 0.01×10^{-12} m^2/s and for $D_{As} = 0.1 \times 10^{-12}$ m^2/s, 0.2×10^{-12} m^2/s, and 0.4×10^{-12} m^2/s. Note that as either mass diffusivity is increased, the dosage increases.

Comments:

1. The function of the outer layer of the skin, the *epidermis,* is to protect the body from external contamination. Transdermal drug delivery is feasible only for medications characterized by extremely small molecules that can diffuse through the relatively impenetrable epidermis. Mass transfer considerations restrict the number of drugs that can be delivered transdermally.

2. It is desirable to desensitize the dosage to variations in the skin diffusivity, since this parameter varies from patient to patient. Therefore, the host medium of the patch (called the *vehicle*) is designed so that it is the rate-limiting factor in controlling the dosage. The sensitivity of the dosage to the patient's skin diffusivity is reduced by decreasing the mass diffusivity in the vehicle, as evident by comparing the results of part 2 of the problem. In general, it is desirable to design the vehicle so that $D_{Ap}/D_{As} \ll 1$.

3. The patch is designed to deliver medication as though it is a semi-infinite medium. The required patch thickness for this assumption to be valid may be estimated by calculating the location where the density of the medication in the vehicle is reduced to 95% of the difference between $\rho_{A,p,i}$ and $\rho_{A,p}(x = 0)$ over the treatment time. By analogy to Equation 5.60, the *concentration penetration depth,* $\delta_{p,c}$, associated with 5% depletion of the drug may be determined from

$$\frac{\rho_{A,p}(x') - \rho_{A,p}(x' = 0)}{\rho_{A,p,i} - \rho_{A,p}(x' = 0)} = 0.95 = \text{erf} \frac{\delta_{p,c}}{\sqrt{4 D_{Ap} t_t}}$$

Evaluating this expression yields a species penetration depth of 690×10^{-6} m = 0.69 mm. The assumption of a semi-infinite patch is valid for a vehicle thickness greater than 0.69 mm.

4. If the vehicle is much thinner than the species penetration depth calculated in Comment 3, the actual dosage will fall below the desired dosage prior to the end of the one-week treatment period. Therefore, a tradeoff in the design of the patch involves (a) loading enough medication to ensure semi-infinite behavior and (b) minimizing the cost of the patch. In practice, more than 95% of the medication remains in the vehicle after the treatment period, necessitating careful disposal of the patch after its use.

14.8 *Summary*

In this chapter we focused on the analysis of species transfer by diffusion. Although transfer rates are generally small, particularly in liquids and solids, the process is germane to many technologies, as well as to the environmental and life sciences. You should test your understanding of fundamental issues by addressing the following questions.

- If a cube of sugar is placed in a cup of coffee, what is the driving potential for dispersion of the sugar in the coffee? What is the physical mechanism responsible for dispersion if the coffee is stagnant? What is the physical mechanism if the coffee is stirred?
- What is the relationship between the molar concentration and the mass density of a species in a mixture?
- How is the mass density of a mixture defined? The molar concentration of the mixture?
- In using *Fick's law* to determine the mass or molar flux of a species in a mixture, what specifically is being determined?
- Under what conditions does the species diffusion flux equal the absolute flux associated with transport of the species?
- What is the stationary medium approximation?
- Is the species flux N_A'' independent of location for a stationary medium within which there is species transfer by diffusion and production (or consumption) by a homogeneous chemical reaction?
- Under what conditions may a diffusion resistance be used to determine the species flux from knowledge of the species concentrations at the inner and outer surfaces of a medium?
- In heat transfer, equilibrium dictates equivalent temperatures on the gas and liquid (or solid) sides of an interface between the two phases. Can the same be said about the concentration of a chemical species present in the gas and liquid (or solid) phases?
- What is *Raoult's law*? How is the partial pressure of a gas-phase species related to the mole fraction of the same species in an adjacent liquid or solid?
- What is *Henry's constant*? How does the concentration of a chemical species in a liquid vary with the partial pressure of the species in an adjoining gas? How does the concentration vary with the temperature of the liquid?
- What is the difference between a *homogeneous* and a *heterogeneous chemical reaction*?
- In heterogeneous catalysis, what is implied if the process is said to be *reaction limited*? What is implied if the process is said to be *diffusion limited*?

- What is a *zero-order reaction*? A *first-order reaction*?
- If convection mass transfer is associated with gas flow over a liquid or solid in which the gas is transferred by diffusion, what can be said about the ratio of convection to diffusion resistances? How is the *mass transfer Biot number* defined?
- In a transient diffusion process, what can be said about the mass transfer Biot number?

References

1. Bird, R. B., *Adv. Chem. Eng.,* **1**, 170, 1956.

2. Bird, R. B., W. E. Stewart, and E. N. Lightfoot, *Transport Phenomena,* revised 2nd ed. Wiley, Hoboken, NJ, 2007.

3. Hirschfelder, J. O., C. F. Curtiss, and R. B. Bird, *Molecular Theory of Gases and Liquids,* Wiley, Hoboken, NJ, 1964.

4. Skelland, A. H. P., *Diffusional Mass Transfer,* Krieger, Malabar, FL, 1985.

5. Poling, B. E., J. M. Prausnitz, and J. O'Connell, *The Properties of Gases and Liquids*, 5th ed., McGraw-Hill, New York, 2001.

6. Costerton, J. W., P. S. Stewart, and E. P. Greenberg, *Science*, **284**, 1318, 1999.

Problems

Mixture Composition

14.1 Assuming air to be composed exclusively of O_2 and N_2, with their partial pressures in the ratio 0.21:0.79, what are their mass fractions?

14.2 Consider an ideal gas mixture of *n* species.

(a) Derive an equation for determining the mass fraction of species *i* from knowledge of the mole fraction and the molecular weight of each of the *n* species. Derive an equation for determining the mole fraction of species *i* from knowledge of the mass fraction and the molecular weight of each of the *n* species.

(b) In a mixture containing equal mole fractions of O_2, N_2, and CO_2, what is the mass fraction of each species? In a mixture containing equal mass fractions of O_2, N_2, and CO_2, what is the mole fraction of each species?

14.3 A mixture of CO_2 and N_2 is in a container at 25°C, with each species having a partial pressure of 1 bar. Calculate the molar concentration, the mass density, the mole fraction, and the mass fraction of each species.

14.4 A container holds a mixture of O_2 and H_2 at a total pressure of 1 bar. The mass fraction of O_2 is 0.75. If the temperature of the mixture is 45°C, calculate the mass density, mole fraction, and partial pressure of each species.

14.5 A He-Xe mixture containing 0.75 mole fraction of helium is used for cooling of electronics in an avionics application. At a temperature of 300 K and atmospheric pressure, calculate the mass fraction of helium and the mass density, molar concentration, and molecular weight of the mixture. If the cooling system capacity is 20 L, what is the mass of the coolant?

Fick's Law and Mass Diffusivity

14.6 Estimate values of the mass diffusivity D_{AB} for binary mixtures of the following gases at 350 K and 1 atm: ammonia–air and hydrogen–air.

14.7 Plot the mass diffusivity, D_{AB}, versus the molecular weight of Substance A for Substance B being air at $p = 1.5$ atm, $T = 320$ K. Substance A is each of the first 8 entries of Table A.8. Is your plot consistent with kinetic theory? Consult various sources, including Table A.4 and Example 6.2 for molecular weight values.

14.8 A 100-mm-long, hollow iron cylinder is exposed to a 1000°C carburizing gas (a mixture of CO and CO_2) at its inner and outer surfaces of radii 4.30 and 5.70 mm, respectively. Consider steady-state conditions for which carbon diffuses from the inner surface of the iron wall to the outer surface and the total transport amounts to 3.6×10^{-3} kg of carbon over 100 h. The variation of the carbon composition (weight % carbon) with radius is tabulated for selected radii.

r (mm)	4.49	4.66	4.79	4.91	5.16	5.27	5.40	5.53
Wt.C (%)	1.42	1.32	1.20	1.09	0.82	0.65	0.46	0.28

(a) Beginning with Fick's law and the assumption of a constant diffusion coefficient, D_{C-Fe}, show that $d\rho_C/d(\ln r)$ is a constant. Sketch the carbon mass density, $\rho_C(r)$, as a function of $\ln r$ for such a diffusion process.

(b) The foregoing table corresponds to measured distributions of the carbon mass density. Is D_{C-Fe} constant for this diffusion process? If not, does D_{C-Fe} increase or decrease with an increasing carbon concentration?

(c) Using the experimental data, calculate and tabulate D_{C-Fe} for selected carbon compositions.

14.9 Consider air in a closed, cylindrical container with its axis vertical and with opposite ends maintained at different temperatures. Assume that the total pressure of the air is uniform throughout the container.

(a) If the bottom surface is colder than the top surface, what is the nature of conditions within the container? For example, will there be vertical gradients of the species (O_2 and N_2) concentrations? Is there any motion of the air? Does mass transfer occur?

(b) What is the nature of conditions within the container if it is inverted (i.e., the warm surface is now at the bottom)?

14.10 An old-fashioned glass apothecary jar contains a patent medicine. The neck is closed with a rubber stopper that is 20 mm tall, with a diameter of 10 mm at the bottom end, widening to 20 mm at the top end. The molar concentration of medicine vapor in the stopper is 2×10^{-3} kmol/m^3 at the bottom surface and is negligible at the top surface. If the mass diffusivity of medicine vapor in rubber is 0.2×10^{-9} m^2/s, find the rate (kmol/s) at which vapor exits through the stopper.

Nonstationary Media: Column Evaporation

14.11 Consider the evaporation of liquid A into a column containing a binary gas mixture of A and B. Species B cannot be absorbed in liquid A and the boundary conditions are the same as in Section 14.2.2. Show how the ratio of the molar-average velocity to the species velocity of A, $v_x^*/v_{A,x}$, varies with the mole fraction of species A.

14.12 An open vertical tube of diameter 5 mm and height 100 mm (above water at 27°C) is exposed to ambient air at 27°C and 25% relative humidity. Determine the evaporation rate, assuming that only mass diffusion occurs. Determine the evaporation rate, considering bulk motion.

14.13 A spherical droplet of liquid A and radius r_o evaporates into a stagnant layer of gas B. Derive an expression for the evaporation rate of species A in terms of the saturation pressure of species A, $p_A(r_o) = p_{A,sat}$, the partial pressure of species A at an arbitrary radius r, $p_A(r)$, the total pressure p, and other pertinent quantities. Assume

the droplet and the mixture are at a uniform pressure p and temperature T.

14.14 The presence of a small amount of air may cause a significant reduction in the heat rate to a water-cooled steam condenser surface. For a clean surface with pure steam and the prescribed conditions, the condensate rate is 0.020 kg/m$^2 \cdot$ s. With the presence of stagnant air in the steam, the condensate surface temperature drops from 28 to 24°C and the condensate rate is reduced by a factor of 2.

For the air–steam mixture, determine the partial pressure of air as a function of distance from the condensate film.

14.15 A laboratory apparatus to measure the diffusion coefficient of vapor–gas mixtures consists of a vertical, small-diameter column containing the liquid phase that evaporates into the gas flowing over the mouth of the column. The gas flow rate is sufficient to maintain a negligible vapor concentration at the exit plane. The column is 150 mm high, and the pressure and temperature in the chamber are maintained at 0.25 atm and 320 K, respectively.

For calibration purposes, you've been asked to calculate the expected evaporation rate (kg/h $\cdot$ m^2) for a test with water and air under the foregoing conditions, using the known value of D_{AB} for the vapor–air mixture.

Stationary Media: Conservation of Species and Mass Diffusion Equation

14.16 A thin plastic membrane is used to separate helium from a gas stream. Under steady-state conditions the concentration of helium in the membrane is known to be 0.02 and 0.005 kmol/m^3 at the inner and outer surfaces, respectively. If the membrane is 1 mm thick and the binary diffusion coefficient of helium with respect to the plastic is 10^{-9} m^2/s, what is the diffusive flux?

14.17 Beginning with a differential control volume, derive the diffusion equation, on a molar basis, for species A in a three-dimensional (Cartesian coordinates), stationary medium, considering species generation with constant properties. Compare your result with Equation 14.48b.

14.18 Consider the radial diffusion of a gaseous species (A) through the wall of a plastic tube (B), and allow for chemical reactions that provide for the depletion of A at a rate $\dot{N}_A$ (kmol/s · m^3). Derive a differential equation that governs the molar concentration of species A in the plastic. Compare your result with Equation 14.49.

14.19 Beginning with a differential control volume, derive the diffusion equation, on a molar basis, for species A in a one-dimensional, spherical, stationary medium, considering species generation. Compare your result with Equation 14.50.

14.20 Gaseous hydrogen at 10 bar and 27°C is stored in a 110-mm inside diameter spherical tank having a steel wall 2 mm thick. The molar concentration of hydrogen in the steel is 1.50 kmol/m^3 at the inner surface and negligible at the outer surface, while the diffusion coefficient of hydrogen in steel is approximately 0.3×10^{-12} m^2/s. What is the initial rate of mass loss of hydrogen by diffusion through the tank wall? What is the initial rate of pressure drop within the tank?

Discontinuous Concentrations at Interfaces

14.21 A spherical rubber container of diameter 200 mm and wall thickness 1 mm contains a gas at a pressure of 5 bar. The container is placed in air of pressure 1 bar and temperature 20°C. Determine the initial rate at which the pressure in the container changes if the gas is pure nitrogen. Determine the time rate of pressure change if the gas is pure oxygen. Assume air is composed exclusively of O_2 and N_2, with their partial pressures in the ratio 0.21:0.79.

14.22 Hydrogen at a pressure of 2 atm flows within a tube of diameter 40 mm and wall thickness 0.5 mm. The outer surface is exposed to a gas stream for which the hydrogen partial pressure is 0.1 atm. The mass diffusivity and solubility of hydrogen in the tube material are 1.8×10^{-11} m^2/s and 160 kmol/m^3 · atm, respectively.

When the system is at 500 K, what is the rate of hydrogen transfer through the tube per unit length (kg/s · m)?

14.23 Oxygen gas is maintained at pressures of 1.5 bar and 0.5 bar on opposite sides of a rubber membrane that is 0.75 mm thick, and the entire system is at 25°C. What is the molar diffusive flux of O_2 through the membrane? What are the molar concentrations of O_2 on both sides of the membrane (outside the rubber)?

14.24 Insulation degrades (experiences an increase in thermal conductivity) if it is subjected to water vapor condensation. The problem may occur in home insulation during cold periods, when vapor in a humidified room diffuses through the drywall (plaster board) and condenses in the adjoining insulation. Estimate the mass diffusion rate for a 3 m × 5 m wall, under conditions for which the vapor pressure is 0.03 bar in the room air and 0.0 bar in the insulation. The drywall is 10 mm thick, and the solubility of water vapor in the wall material is approximately 5×10^{-3} kmol/m^3 · bar. The binary diffusion coefficient for water vapor in the drywall is approximately 10^{-9} m^2/s.

14.25 Consider Example 14.4. Find the time required for the pressure in the container to drop by 1% and by 10%, taking into account that as the interior pressure drops so does the leakage rate. Compare your result with the time estimated using the initial rate of pressure decrease found in Example 14.4.

14.26 Helium gas at 25°C and 4 bars is stored in a spherical Pyrex container of 200-mm inside diameter and 10-mm thickness. What is the rate of mass loss from the container?

14.27 Consider the blister packaging material of Example 14.3.

(a) Under the same conditions as in the example, determine the solubility of the polymer material (kmol/m^3 · bar) if the temperature is 295 K and the relative humidity inside and outside the package is $\phi_2 = 0.1$ and $\phi_1 = 0.9$, respectively.

(b) By selecting a different packaging material, the diffusivity of the film may be changed. Determine the total water vapor transfer rate associated with reducing the diffusivity to 10% of its original value.

(c) The solubility of the material adjacent to its exposed surface may be modified by coating it with various thin films. Determine the water vapor transfer rate after coating both sides of the original polymer sheet and, in turn, reducing the solubility near both surfaces to 10% of the original value.

(d) Determine the water vapor transfer rate after coating the exterior of the original polymer sheet and reducing its solubility by a factor of 9, while leaving the interior surface untreated.

14.28 An experiment is designed to measure the partition coefficient, K, associated with the transfer of a pharmaceutical product through a polymer material. The partition coefficient is defined as the ratio of the densities of the species of interest (the pharmaceutical) on either side of an interface. In the experiment, liquid pharmaceutical ($\rho_p = 1250 \text{ kg/m}^3$) is injected into a hollow polymer sphere of inner and outer diameters $D_i = 5$ mm and $D_o = 5.1$ mm, respectively. The sphere is exposed to convective conditions for which the density of the pharmaceutical at the outer surface is zero. After one week, the sphere's mass is reduced by $\Delta M = 8.2$ mg. What is the value of the partition coefficient if the mass diffusivity is $D_{AB} = 0.2 \times 10^{-11} \text{ m}^2/\text{s}$?

14.29 Ultra-pure hydrogen is required in applications ranging from the manufacturing of semiconductors to powering fuel cells. The crystalline structure of palladium allows only the transfer of atomic hydrogen (H) through its thickness, and therefore palladium membranes are used to filter hydrogen from contaminated streams containing mixtures of hydrogen and other gases. Hydrogen molecules (H_2) are first adsorbed onto the palladium's surface and are then dissociated into atoms (H), which subsequently diffuse through the metal. The H atoms recombine on the opposite side of the membrane, forming pure H_2. The surface concentration of H takes the form $C_H = K_s p_{H_2}^{0.5}$, where $K_s \approx 1.4 \text{ kmol/m}^3 \cdot \text{bar}^{0.5}$ is known as *Sievert's constant*. Consider an industrial hydrogen purifier consisting of an array of palladium tubes with one tube end connected to a collector plenum and the other end closed. The tube bank is inserted into a shell. Impure H_2 at $T = 600$ K, $p = 15$ bar, $x_{H_2} = 0.85$ is introduced into the shell while pure H_2 at $p = 6$ bar, $T = 600$ K is extracted through the tubes. Determine the production rate of pure hydrogen (kg/h) for $N = 100$ tubes which are of inside diameter $D_i = 1.6$ mm, wall thickness $t = 75$ μm, and length $L = 80$ mm. The mass diffusivity of hydrogen (H) in palladium at 600 K is approximately $D_{AB} = 7 \times 10^{-9} \text{ m}^2/\text{s}$.

Catalytic Surface Reactions

14.30 Nitric oxide (NO) emissions from automobile exhaust can be reduced by using a catalytic converter, and the following reaction occurs at the catalytic surface:

$$NO + CO \rightarrow \tfrac{1}{2} N_2 + CO_2$$

The concentration of NO is reduced by passing the exhaust gases over the surface, and the rate of reduction at the catalyst is governed by a first-order reaction of the form given by Equation 14.66. As a first approximation it may be assumed that NO reaches the surface by one-dimensional diffusion through a thin gas film of thickness L that adjoins the surface. Referring to Figure 14.7, consider a situation for which the exhaust gas is at 500°C and 1.2 bar and the mole fraction of NO is $x_{A,L} = 0.10$. If $D_{AB} = 10^{-4} \text{ m}^2/\text{s}$, $k'' = 0.05$ m/s, and the film thickness is $L = 1$ mm, what is the mole fraction of NO at the catalytic surface and what is the NO removal rate for a surface of area $A = 250 \text{ cm}^2$?

14.31 Pulverized coal pellets, which may be approximated as carbon spheres of radius $r_o = 1$ mm, are burned in a pure oxygen atmosphere at 1450 K and 1 atm. Oxygen is transferred to the particle surface by diffusion, where it is consumed in the reaction $C + O_2 \rightarrow CO_2$. The reaction rate is first order and of the form $N''_{O_2} = -k''_1 C_{O_2}(r_o)$, where $k''_1 = 0.1$ m/s. Neglecting changes in r_o, determine the steady-state O_2 molar consumption rate in kmol/s. At 1450 K, the binary diffusion coefficient for O_2 and CO_2 is $1.71 \times 10^{-4} \text{ m}^2/\text{s}$.

14.32 To enhance the effective surface, and hence the chemical reaction rate, catalytic surfaces often take the form of porous solids. One such solid may be visualized as consisting of a large number of cylindrical pores, each of diameter D and length L.

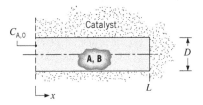

Consider conditions involving a gaseous mixture of A and B for which species A is chemically consumed at the catalytic surface. The reaction is known to be first order, and the rate at which it occurs per unit area of the surface may be expressed as $k''_1 C_A$, where k''_1 (m/s) is the reaction rate constant and C_A (kmol/m³) is the local molar concentration of species A. Under steady-state conditions, flow over the porous solid is known to maintain a fixed value of the molar concentration $C_{A,0}$ at the pore mouth. Beginning from fundamentals, obtain the differential equation that governs the variation of C_A with distance x along the pore. Applying appropriate boundary conditions, solve the equation to obtain an expression for $C_A(x)$.

14.33 A platinum catalytic reactor in an automobile is used to convert carbon monoxide to carbon dioxide in an oxidation reaction of the form $2CO + O_2 \rightarrow 2CO_2$. Species transfer between the catalytic surface and the exhaust gases may be assumed to occur by diffusion in a film of thickness $L = 10$ mm. Consider an exhaust gas that has a pressure of 1.2 bars, a temperature of 500°C, and a CO mole fraction of 0.0012. If the reaction rate constant of the catalyst is $k_1'' = 0.005$ m/s and the diffusion coefficient of CO in the mixture is 10^{-4} m²/s, what is the molar concentration of CO at the catalytic surface? What is the rate of removal of CO per unit area of the catalyst? What is the removal rate if k_1'' is adjusted to render the process diffusion limited?

14.34 A novel process has been proposed to create a composite palladium tube for use as a hydrogen separation membrane in order to produce high-purity hydrogen. To fabricate the composite palladium tube, a gas containing palladium (species A) flows through a porous-walled tube, and the palladium deposits into the pores of the tube wall. The gas mass flow rate is $\dot{m}$ and the inlet mass concentration of palladium is $\rho_{A,m,i}$. The mass transfer coefficient for transfer of palladium between the gas and the surface is h_m, and the deposition rate is proportional to the mass concentration of palladium at the tube surface, that is, $-n_{A,s}'' = k_1\rho_{A,s}$. The palladium is a dilute species, so the total mass flow rate is approximately constant down the length of the tube.

(a) Building upon the convection mass transfer analysis presented in Section 8.9, derive an expression for the variation of the mean species density of palladium with distance from the tube entrance, and determine an expression for the local deposition rate (kg/m² · s) for a tube of diameter D. Neglect any leakage of gas through the porous tube walls.

(b) If the tube is too long, the variation in deposit thickness will be unacceptably large. What is the ratio of the deposition rates at $x = L$ and $x = 0$?

Homogeneous Chemical Reactions: Steady State

14.35 Consider a spherical organism of radius r_o within which respiration occurs at a uniform volumetric rate of $\dot{N}_A = -k_0$. That is, oxygen (species A) consumption is governed by a zero-order, homogeneous chemical reaction.

(a) If a molar concentration of $C_A(r_o) = C_{A,o}$ is maintained at the surface of the organism, obtain an expression for the radial distribution of oxygen, $C_A(r)$, within the organism. From your solution, can you discern any limits on applicability of the result?

(b) Obtain an expression for the rate of oxygen consumption within the organism.

(c) Consider an organism of radius $r_o = 0.10$ mm and a diffusion coefficient for oxygen transfer of $D_{AB} = 10^{-8}$ m²/s. If $C_{A,o} = 5 \times 10^{-5}$ kmol/m³ and $k_0 = 1.2 \times 10^{-4}$ kmol/s · m³, what is the molar concentration of O_2 at the center of the organism?

14.36 Referring to Problem 14.35, a more representative model of respiration in a spherical organism is one for which oxygen consumption is governed by a first-order reaction of the form $\dot{N}_A = -k_1C_A$.

(a) If a molar concentration of $C_A(r_o) = C_{A,o}$ is maintained at the surface of the organism, obtain an expression for the radial distribution of oxygen, $C_A(r)$, within the organism. *Hint:* To simplify solution of the species diffusion equation, invoke the transformation $y \equiv rC_A$.

(b) Obtain an expression for the rate of oxygen consumption within the organism.

(c) Consider an organism of radius $r_o = 0.10$ mm and a diffusion coefficient of $D_{AB} = 10^{-8}$ m²/s. If $C_{A,o} = 5 \times 10^{-5}$ kmol/m³ and $k_1 = 20$ s⁻¹, estimate the corresponding value of the molar concentration at the center of the organism. What is the rate of oxygen consumption by the organism?

14.37 Consider combustion of hydrogen gas in a mixture of hydrogen and oxygen adjacent to the metal wall of a combustion chamber. Combustion occurs at constant temperature and pressure according to the chemical reaction $2H_2 + O_2 \rightarrow 2H_2O$. Measurements under steady-state conditions at a distance of 10 mm from the wall indicate that the molar concentrations of hydrogen, oxygen, and water vapor are 0.10, 0.10, and 0.20 kmol/m³, respectively. The generation rate of water vapor is 0.96×10^{-2} kmol/m³ · s throughout the region of interest. The binary diffusion coefficient for each of the species (H_2, O_2, and H_2O) in the remaining species is 0.6×10^{-5} m²/s.

(a) Determine an expression for and make a qualitative plot of C_{H_2} as a function of distance from the wall.

(b) Determine the value of C_{H_2} at the wall.

(c) On the same coordinates used in part (a), sketch curves for the concentrations of oxygen and water vapor.

(d) What is the molar flux of water vapor at $x = 10$ mm?

14.38 Consider the problem of oxygen transfer from the interior lung cavity, across the lung tissue, to the network of blood vessels on the opposite side. The lung tissue (species B) may be approximated as a plane wall of thickness L. The inhalation process may be assumed to maintain a constant molar concentration $C_A(0)$ of oxygen (species A) in the tissue at its inner surface ($x = 0$), and assimilation of oxygen by the blood may be assumed to maintain a constant molar concentration $C_A(L)$ of oxygen in the tissue at its outer surface ($x = L$). There is oxygen consumption in the tissue due to metabolic processes, and the reaction is zero order, with $\dot{N}_A = -k_0$. Obtain expressions for the distribution of the oxygen concentration in the tissue and for the rate of assimilation of oxygen by the blood per unit tissue surface area.

14.39 As an employee of the Los Angeles Air Quality Commission, you have been asked to develop a model for computing the distribution of NO_2 in the atmosphere. The molar flux of NO_2 at ground level, $N''_{A,0}$, is presumed known. This flux is attributed to automobile and smoke stack emissions. It is also known that the concentration of NO_2 at a distance well above ground level is zero and that NO_2 reacts chemically in the atmosphere. In particular, NO_2 reacts with unburned hydrocarbons (in a process that is activated by sunlight) to produce PAN (peroxyacetylnitrate), the final product of photochemical smog. The reaction is first order, and the local rate at which it occurs may be expressed as $\dot{N}_A = -k_1 C_A$.

(a) Assuming steady-state conditions and a stagnant atmosphere, obtain an expression for the vertical distribution $C_A(x)$ of the molar concentration of NO_2 in the atmosphere.

(b) If an NO_2 partial pressure of $p_A = 2 \times 10^{-6}$ bar is sufficient to cause pulmonary damage, what is the value of the ground level molar flux for which you would issue a smog alert? You may assume an isothermal atmosphere at $T = 300$ K, a reaction coefficient of $k_1 = 0.03$ s^{-1}, and an NO_2–air diffusion coefficient of $D_{AB} = 0.15 \times 10^{-4}$ m^2/s.

Transient Diffusion: Introductory

14.40 In Problem 14.39, NO_2 transport by diffusion in a stagnant atmosphere was considered for steady-state conditions. However, the problem is actually time dependent, and a more realistic approach would account for transient effects. Consider the ground level emission of NO_2 to begin in the early morning (at $t = 0$), when the NO_2 concentration in the atmosphere is everywhere zero. Emission occurs throughout the day at a constant flux $N''_{A,0}$, and the NO_2 again experiences a first-order photochemical reaction in the atmosphere ($\dot{N}_A = -k_1 C_A$).

(a) For a differential element in the atmosphere, derive a differential equation that could be used to determine the molar concentration $C_A(x, t)$. State appropriate initial and boundary conditions.

(b) Obtain an expression for $C_A(x, t)$ under the special condition for which photochemical reactions may be neglected. For this condition what are the molar concentrations of NO_2 at ground level and at 100-m elevation 3 h after the start of the emissions, if $N''_{A,0} = 3 \times 10^{-11}$ kmol/s · m^2 and $D_{AB} = 0.15 \times 10^{-4}$ m^2/s?

14.41 A large sheet of material 40 mm thick contains dissolved hydrogen (H_2) having a uniform concentration of 3 kmol/m^3. The sheet is exposed to a fluid stream that causes the concentration of the dissolved hydrogen to be reduced suddenly to zero at both surfaces. This surface condition is maintained constant thereafter. If the mass diffusivity of hydrogen is 9×10^{-7} m^2/s, how much time is required to bring the density of dissolved hydrogen to a value of 1.2 kg/m^3 at the center of the sheet?

14.42 A common procedure for increasing the moisture content of air is to bubble it through a column of water. Assume the air bubbles to be spheres of radius $r_o = 1$ mm and to be in thermal equilibrium with the water at 25°C. How long should the bubbles remain in the water to achieve a vapor concentration at the center that is 99% of the maximum possible (saturated) concentration? The air is dry when it enters the water.

14.43 Consider Problem 14.42. How long should the bubbles remain in the water to achieve an average vapor concentration that is 50% of the maximum value?

14.44 Steel is carburized in a high-temperature process that depends on the transfer of carbon by diffusion. The value of the diffusion coefficient is strongly temperature dependent and may be approximated as D_{C-S} (m^2/s) $\approx 2 \times 10^{-5}$ exp $[-17,000/T$ (K)$]$. If the process is effected at 1000°C and a carbon mole fraction of 0.02 is maintained at the surface of the steel, how much time is required to elevate the carbon content of the steel from an initial value of 0.1% to a value of 1.0% at a depth of 1 mm?

14.45 A thick plate of pure iron at 1000°C is subjected to a carburization process in which the surface of the plate is suddenly exposed to a gas that induces a carbon concentration $C_{C,s}$ at one surface. The average diffusion coefficient for carbon and iron at this temperature is $D_{C-Fe} = 3 \times 10^{-11}$ m^2/s. Use the correspondence between heat and mass transfer variables in addressing the following questions.

(a) Consider the heat transfer analog to the carburization problem. Sketch the mass and heat transfer systems. Show and explain the correspondence between variables. Provide the analytical solutions to the heat and mass transfer problems.

(b) Determine the carbon concentration ratio, $C_C(x, t)/C_{C,s}$, at a depth of 1 mm after 1 h of carburization.

(c) From the analogy, show that the time dependence of the mass flux of carbon into the plate may be expressed as $n_C'' = \rho_{C,s}(D_{C-Fe}/\pi t)^{1/2}$. Also, obtain an expression for the mass of carbon per unit area entering the iron plate over the time period t.

14.46 A pharmaceutical product is designed to be absorbed in the gastrointestinal tract. The active ingredient is pressed into a tablet with a density of $\rho_A = 15$ kg/m^3 while the remainder of the tablet is composed of inactive ingredients. The partition coefficient is $K = 3 \times 10^{-2}$, and the diffusion coefficient of the active ingredient in the gastrointestinal fluid is $D_{AB} = 0.4 \times 10^{-10}$ m^2/s.

(a) Estimate the dosage delivered over a time period of 5 h for a spherical tablet of diameter $D = 6$ mm. *Hint:* Assume the change in the tablet radius over the dosage period is small.

(b) Estimate the dosage delivered over 5 h for $N = 200$ small, spherical tablets contained in a gelatin capsule that quickly dissolves after ingestion, releasing the medication. The initial mass of the medication is the same as in part (a).

14.47 If an amount of energy Q_o''(J/m^2) is released instantaneously, as, for example, from a pulsed laser, and it is absorbed by the surface of a semi-infinite medium, with no attendant losses to the surroundings, the subsequent temperature distribution in the medium is

$$T(x,t) - T_i = \frac{Q_o''}{\rho c (\pi \alpha t)^{1/2}} \exp(-x^2/4\alpha t)$$

where T_i is the initial, uniform temperature of the medium. Consider an analogous mass transfer process involving deposition of a thin layer of phosphorous (P) on a silicon (Si) wafer at room temperature. If the wafer is placed in a furnace, the diffusion of P into Si is significantly enhanced by the high-temperature environment.

A Si wafer with 1-μm-thick P film is suddenly placed in a furnace at 1000°C, and the resulting distribution of P is characterized by an expression of the form

$$C_P(x,t) = \frac{M_{P,o}''}{(\pi D_{P-Si}t)^{1/2}} \exp(-x^2/4D_{P-Si}t)$$

where $M_{P,o}''$ is the molar area density (kmol/m^2) of P associated with the film of concentration C_P and thickness d_o.

(a) Explain the correspondence between variables in the analogous temperature and concentration distributions.

(b) Determine the mole fraction of P at a depth of 0.1 μm in the Si after 30 s. The diffusion coefficient is $D_{P-Si} = 1.2 \times 10^{-17}$ m^2/s. The mass densities of P and Si are 2000 and 2300 kg/m^3, respectively, and their molecular weights are 30.97 and 28.09 kg/kmol.

Transient Diffusion: Advanced

14.48 The presence of CO_2 in solution is essential to the growth of aquatic plant life, with CO_2 used as a reactant in the photosynthesis. Consider a stagnant body of water in which the concentration of $CO_2(\rho_A)$ is everywhere zero. At time $t = 0$, the water is exposed to a source of CO_2, which maintains the surface ($x = 0$) concentration at a fixed value $\rho_{A,0}$. For time $t > 0$, CO_2 will begin to accumulate in the water, but the accumulation is inhibited by CO_2 consumption due to photosynthesis. The time rate at which this consumption occurs per unit volume is equal to the product of a reaction rate constant k_1 and the local CO_2 concentration $\rho_A(x, t)$.

(a) Write (do not derive) a differential equation that could be used to determine $\rho_A(x, t)$ in the water. What does each term in the equation represent physically?

(b) Write appropriate boundary conditions that could be used to obtain a particular solution, assuming a "deep" body of water. What would be the form of this solution for the special case of negligible CO_2 consumption ($k_1 \approx 0$)?

14.49 Hydrogen gas is used in a process to manufacture a sheet material of 6-mm thickness. At the end of the process, H_2 remains in solution in the material with a uniform concentration of 320 kmol/m^3. To remove H_2 from the material, both surfaces of the sheet are exposed to an airstream at 500 K and a total pressure of 3 atm. Due to contamination, the hydrogen partial pressure is 0.1 atm in the airstream, which provides a convection mass transfer coefficient of 1.5 m/h. The mass diffusivity and solubility of hydrogen (A) in the sheet material (B) are $D_{AB} = 2.6 \times 10^{-8}$ m²/s and $S_{AB} = 160$ kmol/m³ · atm, respectively.

(a) If the sheet material is left exposed to the airstream for a long time, determine the final content of hydrogen in the material (kg/m³).

(b) Identify and evaluate the parameter that can be used to determine whether the transient mass diffusion process in the sheet can be assumed to be characterized by a uniform concentration at any time during the process. *Hint:* This situation is analogous to that used to determine the validity of the lumped-capacitance method for a transient heat transfer analysis.

(c) Determine the time required to reduce the hydrogen mass density at the center of the sheet to twice the limiting value calculated in part (a).

14.50 Consider the hydrogen-removal process described in Problem 14.49, but under conditions for which the mass diffusivity of the hydrogen gas (A) in the sheet material (B) is $D_{AB} = 1.8 \times 10^{-11}$ m²/s (instead of 2.6×10^{-8} m²/s). With the smaller value of D_{AB}, a *uniform concentration* may no longer be assumed to exist in the material during the removal process.

(a) If the sheet material is left exposed to the airstream for a very long time, what is the final content of hydrogen in the material (kg/m³)?

(b) Identify and evaluate the parameters that describe the transient mass diffusion process in the sheet. *Hint:* The situation is analogous to that of transient heat conduction in a plane wall.

(c) Determine the time required to reduce the hydrogen mass density at the center of the sheet to twice the limiting value calculated in part (a).

(d) Assuming a uniform concentration at any time during the removal process, calculate the time required to reach twice the limiting density calculated in part (a). Compare the result with that obtained from part (c), and explain the differences.

14.51 A 1-mm-thick square (100 mm × 100 mm) sheet of polymer is suspended from a precision scale in a chamber characterized by a temperature and relative humidity of $T = 300$ K and $\phi = 0$, respectively. Suddenly, at time $t = 0$, the chamber's relative humidity is raised to $\phi = 0.95$. The measured mass of the sheet increases by 0.012 mg over 24 h and by 0.016 mg over 48 h. Determine the solubility and mass diffusivity of water vapor in the polymer. Preliminary experiments have indicated that the mass diffusivity is greater than 7×10^{-13} m²/s.

14.52 The surface of glass *quickly* develops very small *microcracks* when exposed to high humidity. Although microcracks can be safely ignored in most applications, they can significantly decrease the mechanical strength of very small glass structures such as optical fibers. Consider a glass optical fiber of diameter $D_i = 125$ μm that is coated with an acrylate polymer to form a coated fiber of outer diameter $D_o = 250$ μm. A telecommunications engineer insists that the optical fiber be stored in a low-humidity environment prior to installation so that it is sufficiently strong to withstand rough treatment by technicians in the field. If installation of a roll of fiber requires several hot and humid days to complete, will careful storage beforehand prevent microcracking? The mass diffusivity of water vapor in the acrylate is $D_{AB} = 5.5 \times 10^{-13}$ m²/s while the glass can be considered impermeable.

14.53 A person applies an insect repellent onto an exposed area of $A = 0.5$ m² of their body. The mass of spray used is $M = 10$ grams, and the spray contains 25% (by mass) active ingredient. The inactive ingredient quickly evaporates from the skin surface.

(a) If the spray is applied uniformly and the density of the dried active ingredient is $\rho = 2000$ kg/m³, determine the initial thickness of the film of active ingredient on the skin surface. The temperature, molecular weight, and saturation pressure of the active ingredient are 32°C, 152 kg/kmol, and 1.2×10^{-5} bars, respectively.

(b) If the convection mass transfer coefficient associated with sublimation of the active ingredient to the air is $\bar{h}_m = 5 \times 10^{-3}$ m/s, the partition coefficient associated with the ingredient–skin interface is $K = 0.05$, and the mass diffusivity of the active ingredient in the skin is $D_{AB} = 1 \times 10^{-13}$ m²/s, determine how long the insect repellent remains effective. The partition coefficient is the ratio of the ingredient density in the skin to the ingredient density outside the skin.

(c) If the spray is reformulated so that the partition coefficient becomes very small, how long does the insect repellent remain effective?

14.54 As seen in Section 2.2.1, Fourier's law cannot be used to predict heat transfer phenomena associated with individual molecules. Similarly, Fick's law describes the aggregate behavior of a large number of molecules, and cannot be used to predict the behavior of a small number of molecules.

The human nose can be sensitive to only a few molecules of an aromatic substance. In a classroom demonstration of volumetric heating, your professor microwaves a bag of popcorn and measures its change in temperature. One second after opening the hot bag to insert a thermocouple, students sitting in the last row of the classroom, located 30 m from the professor, smell the hot popcorn. Is it plausible that the popcorn aroma reached the students by advection? Using Fick's law, estimate the time needed for the students to be exposed to a trace amount of the aromatic, $C/C_s = 0.0001$, using the penetration depth concept of Section 5.7 and typical values of the binary diffusion coefficient for gases in air. The concentration C_s corresponds to the maximum aromatic concentration located at the opening of the bag. Can Fick's law be used to explain the speed at which the aromatic substance moved from the front to the back of the classroom?

APPENDIX A

Thermophysical Properties of Matter[1]

---

[1]The convention used to present numerical values of the properties is illustrated by this example:

T (K)	$v \cdot 10^7$ (m²/s)	$k \cdot 10^3$ (W/m · K)
300	0.349	521

where $v = 0.349 \times 10^{-7}$ m²/s and $k = 521 \times 10^{-3} = 0.521$ W/m · K at 300 K.

TABLE A.1   Thermophysical Properties of Selected Metallic Solids[a]

	Melting Point (K)	Properties at 300 K				Properties at Various Temperatures (K) k (W/m · K)/c_p (J/kg · K)									
Composition		ρ (kg/m³)	c_p (J/kg · K)	k (W/m · K)	$\alpha \cdot 10^6$ (m²/s)	100	200	400	600	800	1000	1200	1500	2000	2500
Aluminum Pure	933	2702	903	237	97.1	302 / 482	237 / 798	240 / 949	231 / 1033	218 / 1146					
Alloy 2024-T6 (4.5% Cu, 1.5% Mg, 0.6% Mn)	775	2770	875	177	73.0	65 / 473	163 / 787	186 / 925	186 / 1042						
Alloy 195, Cast (4.5% Cu)		2790	883	168	68.2			174 / —	185 / —						
Beryllium	1550	1850	1825	200	59.2	990 / 203	301 / 1114	161 / 2191	126 / 2604	106 / 2823	90.8 / 3018	78.7 / 3227	— / 3519		
Bismuth	545	9780	122	7.86	6.59	16.5 / 112	9.69 / 120	7.04 / 127							
Boron	2573	2500	1107	27.0	9.76	190 / 128	55.5 / 600	16.8 / 1463	10.6 / 1892	9.60 / 2160	9.85 / 2338				
Cadmium	594	8650	231	96.8	48.4	203 / 198	99.3 / 222	94.7 / 242							
Chromium	2118	7160	449	93.7	29.1	159 / 192	111 / 384	90.9 / 484	80.7 / 542	71.3 / 581	65.4 / 616	61.9 / 682	57.2 / 779	49.4 / 937	
Cobalt	1769	8862	421	99.2	26.6	167 / 236	122 / 379	85.4 / 450	67.4 / 503	58.2 / 550	52.1 / 628	49.3 / 733	42.5 / 674		
Copper Pure	1358	8933	385	401	117	482 / 252	413 / 356	393 / 397	379 / 417	366 / 433	352 / 451	339 / 480			
Commercial bronze (90% Cu, 10% Al)	1293	8800	420	52	14		42 / 785	52 / 460	59 / 545						
Phosphor gear bronze (89% Cu, 11% Sn)	1104	8780	355	54	17		41 / —	65 / —	74 / —						
Cartridge brass (70% Cu, 30% Zn)	1188	8530	380	110	33.9	75 / —	95 / 360	137 / 395	149 / 425						
Constantan (55% Cu, 45% Ni)	1493	8920	384	23	6.71	17 / 237	19 / 362								
Germanium	1211	5360	322	59.9	34.7	232 / 190	96.8 / 290	43.2 / 337	27.3 / 348	19.8 / 357	17.4 / 375	17.4 / 395			

TABLE A.1 *Continued*

| Composition | Melting Point (K) | Properties at 300 K | | | | Properties at Various Temperatures (K) k (W/m · K)/c_p (J/kg · K) | | | | | | | | | |
		ρ (kg/m³)	c_p (J/kg · K)	k (W/m · K)	$\alpha \cdot 10^6$ (m²/s)	100	200	400	600	800	1000	1200	1500	2000	2500
Gold	1336	19300	129	317	127	327 109	323 124	311 131	298 135	284 140	270 145	255 155			
Iridium	2720	22500	130	147	50.3	172 90	153 122	144 133	138 138	132 144	126 153	120 161	111 172		
Iron Pure	1810	7870	447	80.2	23.1	134 216	94.0 384	69.5 490	54.7 574	43.3 680	32.8 975	28.3 609	32.1 654		
Armco (99.75% pure)		7870	447	72.7	20.7	95.6 215	80.6 384	65.7 490	53.1 574	42.2 680	32.3 975	28.7 609	31.4 654		
Carbon steels Plain carbon (Mn ≤ 1%, Si ≤ 0.1%)		7854	434	60.5	17.7			56.7 487	48.0 559	39.2 685	30.0 1169				
AISI 1010		7832	434	63.9	18.8			58.7 487	48.8 559	39.2 685	31.3 1168				
Carbon–silicon (Mn ≤ 1%, 0.1% < Si ≤ 0.6%)		7817	446	51.9	14.9			49.8 501	44.0 582	37.4 699	29.3 971				
Carbon–manganese–silicon (1% < Mn ≤ 1.65%, 0.1% < Si ≤ 0.6%)		8131	434	41.0	11.6			42.2 487	39.7 559	35.0 685	27.6 1090				
Chromium (low) steels $\frac{1}{2}$Cr–$\frac{1}{4}$Mo–Si (0.18% C, 0.65% Cr, 0.23% Mo, 0.6% Si)		7822	444	37.7	10.9			38.2 492	36.7 575	33.3 688	26.9 969				
1 Cr–$\frac{1}{2}$Mo (0.16% C, 1% Cr, 0.54% Mo, 0.39% Si)		7858	442	42.3	12.2			42.0 492	39.1 575	34.5 688	27.4 969				
1 Cr–V (0.2% C, 1.02% Cr, 0.15% V)		7836	443	48.9	14.1			46.8 492	42.1 575	36.3 688	28.2 969				

Properties at various temperatures (K) are given as k (W/m·K) / c_p (J/kg·K).

Composition	Melting Point (K)	ρ (kg/m³)	c_p (J/kg·K)	k (W/m·K)	α·10⁶ (m²/s)	100 K	200 K	400 K	600 K	800 K	1000 K	1200 K	1500 K	2000 K	2500 K
Stainless steels															
AISI 302	1670	8055	480	15.1	3.91			17.3 / 512	20.0 / 559	22.8 / 585	25.4 / 606	28.0 / 640	31.7 / 682		
AISI 304	1670	7900	477	14.9	3.95	9.2 / 272	12.6 / 402	16.6 / 515	19.8 / 557	22.6 / 582	25.4 / 611				
AISI 316		8238	468	13.4	3.48			15.2 / 504	18.3 / 550	21.3 / 576	24.2 / 602				
AISI 347		7978	480	14.2	3.71			15.8 / 513	18.9 / 559	21.9 / 585	24.7 / 606				
Lead	601	11340	129	35.3	24.1	39.7 / 118	36.7 / 125	34.0 / 132	31.4 / 142						
Magnesium	923	1740	1024	156	87.6	169 / 649	159 / 934	153 / 1074	149 / 1170	146 / 1267					
Molybdenum	2894	10240	251	138	53.7	179 / 141	143 / 224	134 / 261	126 / 275	118 / 285	112 / 295	105 / 308	98 / 330	90 / 380	86 / 459
Nickel Pure	1728	8900	444	90.7	23.0	164 / 232	107 / 383	80.2 / 485	65.6 / 592	67.6 / 530	71.8 / 562	76.2 / 594	82.6 / 616		
Nichrome (80% Ni, 20% Cr)	1672	8400	420	12	3.4			14 / 480	16 / 525	21 / 545					
Inconel X-750 (73% Ni, 15% Cr, 6.7% Fe)	1665	8510	439	11.7	3.1	8.7 / —	10.3 / 372	13.5 / 473	17.0 / 510	20.5 / 546	24.0 / 626	27.6 / —	33.0 / —		
Niobium	2741	8570	265	53.7	23.6	55.2 / 188	52.6 / 249	55.2 / 274	58.2 / 283	61.3 / 292	64.4 / 301	67.5 / 310	72.1 / 324	79.1 / 347	
Palladium	1827	12020	244	71.8	24.5	76.5 / 168	71.6 / 227	73.6 / 251	79.7 / 261	86.9 / 271	94.2 / 281	102 / 291	110 / 307		
Platinum Pure	2045	21450	133	71.6	25.1	77.5 / 100	72.6 / 125	71.8 / 136	73.2 / 141	75.6 / 146	78.7 / 152	82.6 / 157	89.5 / 165	99.4 / 179	
Alloy 60Pt–40Rh (60% Pt, 40% Rh)	1800	16630	162	47	17.4			52 / —	59 / —	65 / —	69 / —	73 / —	76 / —		
Rhenium	3453	21100	136	47.9	16.7	58.9 / 97	51.0 / 127	46.1 / 139	44.2 / 145	44.1 / 151	44.6 / 156	45.7 / 162	47.8 / 171	51.9 / 186	
Rhodium	2236	12450	243	150	49.6	186 / 147	154 / 220	146 / 253	136 / 274	127 / 293	121 / 311	116 / 327	110 / 349	112 / 376	
Silicon	1685	2330	712	148	89.2	884 / 259	264 / 556	98.9 / 790	61.9 / 867	42.2 / 913	31.2 / 946	25.7 / 967	22.7 / 992		
Silver	1235	10500	235	429	174	444 / 187	430 / 225	425 / 239	412 / 250	396 / 262	379 / 277	361 / 292			
Tantalum	3269	16600	140	57.5	24.7	59.2 / 110	57.5 / 133	57.8 / 144	58.6 / 146	59.4 / 149	60.2 / 152	61.0 / 155	62.2 / 160	64.1 / 172	65.6 / 189
Thorium	2023	11700	118	54.0	39.1	59.8 / 99	54.6 / 112	54.5 / 124	55.8 / 134	56.9 / 145	56.9 / 156	58.7 / 167			
Tin	505	7310	227	66.6	40.1	85.2 / 188	73.3 / 215	62.2 / 243							

TABLE A.1 *Continued*

Composition	Melting Point (K)	Properties at 300 K				Properties at Various Temperatures (K) k (W/m · K)/c_p (J/kg · K)									
		ρ (kg/m^3)	c_p (J/kg · K)	k (W/m · K)	$\alpha \cdot 10^6$ (m^2/s)	100	200	400	600	800	1000	1200	1500	2000	2500
Titanium	1953	4500	522	21.9	9.32	30.5 / 300	24.5 / 465	20.4 / 551	19.4 / 591	19.7 / 633	20.7 / 675	22.0 / 620	24.5 / 686		
Tungsten	3660	19300	132	174	68.3	208 / 87	186 / 122	159 / 137	137 / 142	125 / 145	118 / 148	113 / 152	107 / 157	100 / 167	95 / 176
Uranium	1406	19070	116	27.6	12.5	21.7 / 94	25.1 / 108	29.6 / 125	34.0 / 146	38.8 / 176	43.9 / 180	49.0 / 161			
Vanadium	2192	6100	489	30.7	10.3	35.8 / 258	31.3 / 430	31.3 / 515	33.3 / 540	35.7 / 563	38.2 / 597	40.8 / 645	44.6 / 714	50.9 / 867	
Zinc	693	7140	389	116	41.8	117 / 297	118 / 367	111 / 402	103 / 436						
Zirconium	2125	6570	278	22.7	12.4	33.2 / 205	25.2 / 264	21.6 / 300	20.7 / 322	21.6 / 342	23.7 / 362	26.0 / 344	28.8 / 344	33.0 / 344	

[a]Adapted from References 1–7.

TABLE A.2 Thermophysical Properties of Selected Nonmetallic Solids[a]

Composition	Melting Point (K)	ρ (kg/m³)	c_p (J/kg·K)	k (W/m·K)	α·10⁶ (m²/s)		100	200	400	600	800	1000	1200	1500	2000	2500
		Properties at 300 K					Properties at Various Temperatures (K) — k (W/m·K) / c_p (J/kg·K)									
Aluminum oxide, sapphire	2323	3970	765	46	15.1	k	450	82	32.4	18.9	13.0	10.5				
						c_p			940	1110	1180	1225				
Aluminum oxide, polycrystalline	2323	3970	765	36.0	11.9	k	133	55	26.4	15.8	10.4	7.85	6.55	5.66	6.00	
						c_p			940	1110	1180	1225				
Beryllium oxide	2725	3000	1030	272	88.0	k			196	111	70	47	33	21.5	15	
						c_p			1350	1690	1865	1975	2055	2145	2750	
Boron	2573	2500	1105	27.6	9.99	k	190	52.5	18.7	11.3	8.1	6.3	5.2			
						c_p			1490	1880	2135	2350	2555			
Boron fiber epoxy (30% vol) composite	590	2080														
k, ‖ to fibers				2.29		k	2.10	2.23	2.28							
k, ⊥ to fibers				0.59		k	0.37	0.49	0.60							
c_p			1122			c_p	364	757	1431							
Carbon Amorphous	1500	1950	—	1.60	—	k	0.67	1.18	1.89	2.19	2.37	2.53	2.84	3.48		
Diamond, type IIa insulator	—	3500	509	2300	—	k	10,000	4000	1540							
						c_p	21	194	853							
Graphite, pyrolytic	2273	2210		1950												
k, ‖ to layers				1950		k	4970	3230	1390	892	667	534	448	357	262	
k, ⊥ to layers				5.70		k	16.8	9.23	4.09	2.68	2.01	1.60	1.34	1.08	0.81	
c_p			709			c_p	136	411	992	1406	1650	1793	1890	1974	2043	
Graphite fiber epoxy (25% vol) composite	450	1400														
k, heat flow ‖ to fibers				11.1		k	5.7	8.7	13.0							
k, heat flow ⊥ to fibers				0.87		k	0.46	0.68	1.1							
c_p			935			c_p	337	642	1216							
Pyroceram, Corning 9606	1623	2600	808	3.98	1.89	k	5.25	4.78	3.64	3.28	3.08	2.96	2.87	2.79		
						c_p			908	1038	1122	1197	1264	1498		

Table A.2 *Continued*

Composition	Melting Point (K)	Properties at 300 K				Properties at Various Temperatures (K) k (W/m · K)/c_p (J/kg · K)									
		ρ (kg/m³)	c_p (J/kg · K)	k (W/m · K)	$\alpha \cdot 10^6$ (m²/s)	100	200	400	600	800	1000	1200	1500	2000	2500
Silicon carbide	3100	3160	675	490	230	—	—	— / 880	— / 1050	— / 1135	87 / 1195	58 / 1243	30 / 1310		
Silicon dioxide, crystalline (quartz)	1883	2650													
k, ∥ to c axis				10.4		39	16.4	7.6	5.0	4.2					
k, ⊥ to c axis				6.21		20.8	9.5	4.70	3.4	3.1					
c_p			745			—	—	885	1075	1250					
Silicon dioxide, polycrystalline (fused silica)	1883	2220	745	1.38	0.834	0.69 / —	1.14 / —	1.51 / 905	1.75 / 1040	2.17 / 1105	2.87 / 1155	4.00 / 1195			
Silicon nitride	2173	2400	691	16.0	9.65	— / —	— / 578	13.9 / 778	11.3 / 937	9.88 / 1063	8.76 / 1155	8.00 / 1226	7.16 / 1306	6.20 / 1377	
Sulfur	392	2070	708	0.206	0.141	0.165 / 403	0.185 / 606								
Thorium dioxide	3573	9110	235	13	6.1			10.2 / 255	6.6 / 274	4.7 / 285	3.68 / 295	3.12 / 303	2.73 / 315	2.5 / 330	
Titanium dioxide, polycrystalline	2133	4157	710	8.4	2.8			7.01 / 805	5.02 / 880	3.94 / 910	3.46 / 930	3.28 / 945			

[a]Adapted from References 1, 2, 3 and 6.

TABLE A.3   Thermophysical Properties of Common Materials[a]

Structural Building Materials

Description/Composition	Typical Properties at 300 K		
	Density, ρ (kg/m^3)	Thermal Conductivity, k (W/m · K)	Specific Heat, c_p (J/kg · K)
Building Boards			
Asbestos–cement board	1920	0.58	—
Gypsum or plaster board	800	0.17	—
Plywood	545	0.12	1215
Sheathing, regular density	290	0.055	1300
Acoustic tile	290	0.058	1340
Hardboard, siding	640	0.094	1170
Hardboard, high density	1010	0.15	1380
Particle board, low density	590	0.078	1300
Particle board, high density	1000	0.170	1300
Woods			
Hardwoods (oak, maple)	720	0.16	1255
Softwoods (fir, pine)	510	0.12	1380
Masonry Materials			
Cement mortar	1860	0.72	780
Brick, common	1920	0.72	835
Brick, face	2083	1.3	—
Clay tile, hollow			
1 cell deep, 10 cm thick	—	0.52	—
3 cells deep, 30 cm thick	—	0.69	—
Concrete block, 3 oval cores			
Sand/gravel, 20 cm thick	—	1.0	—
Cinder aggregate, 20 cm thick	—	0.67	—
Concrete block, rectangular core			
2 cores, 20 cm thick, 16 kg	—	1.1	—
Same with filled cores	—	0.60	—
Plastering Materials			
Cement plaster, sand aggregate	1860	0.72	—
Gypsum plaster, sand aggregate	1680	0.22	1085
Gypsum plaster, vermiculite aggregate	720	0.25	—

TABLE A.3 *Continued*

Insulating Materials and Systems

Description/Composition	Typical Properties at 300 K		
	Density, ρ (kg/m^3)	Thermal Conductivity, k (W/m · K)	Specific Heat, c_p (J/kg · K)
Blanket and Batt			
Glass fiber, paper faced	16	0.046	—
	28	0.038	—
	40	0.035	—
Glass fiber, coated; duct liner	32	0.038	835
Board and Slab			
Cellular glass	145	0.058	1000
Glass fiber, organic bonded	105	0.036	795
Polystyrene, expanded			
Extruded (R-12)	55	0.027	1210
Molded beads	16	0.040	1210
Mineral fiberboard; roofing material	265	0.049	—
Wood, shredded/cemented	350	0.087	1590
Cork	120	0.039	1800
Loose Fill			
Cork, granulated	160	0.045	—
Diatomaceous silica, coarse	350	0.069	—
Powder	400	0.091	—
Diatomaceous silica, fine powder	200	0.052	—
	275	0.061	—
Glass fiber, poured or blown	16	0.043	835
Vermiculite, flakes	80	0.068	835
	160	0.063	1000
Formed/Foamed-in-Place			
Mineral wool granules with asbestos/inorganic binders, sprayed	190	0.046	—
Polyvinyl acetate cork mastic; sprayed or troweled	—	0.100	—
Urethane, two-part mixture; rigid foam	70	0.026	1045
Reflective			
Aluminum foil separating fluffy glass mats; 10–12 layers, evacuated; for cryogenic applications (150 K)	40	0.00016	—
Aluminum foil and glass paper laminate; 75–150 layers; evacuated; for cryogenic application (150 K)	120	0.000017	—
Typical silica powder, evacuated	160	0.0017	—

TABLE A.3 *Continued*

Industrial Insulation

Description/ Composition	Maximum Service Temperature (K)	Typical Density (kg/m³)	Typical Thermal Conductivity, k (W/m · K), at Various Temperatures (K)													
			200	215	230	240	255	270	285	300	310	365	420	530	645	750
Blankets																
Blanket, mineral fiber, metal reinforced	920	96–192									0.038	0.046	0.056	0.078		
	815	40–96									0.035	0.045	0.058	0.088		
Blanket, mineral fiber, glass; fine fiber, organic bonded	450	10				0.036	0.038	0.040	0.043	0.048	0.052	0.076				
		12				0.035	0.036	0.039	0.042	0.046	0.049	0.069				
		16				0.033	0.035	0.036	0.039	0.042	0.046	0.062				
		24				0.030	0.032	0.033	0.036	0.039	0.040	0.053				
		32				0.029	0.030	0.032	0.033	0.036	0.038	0.048				
		48				0.027	0.029	0.030	0.032	0.033	0.035	0.045				
Blanket, alumina–silica fiber	1530	48												0.071	0.105	0.150
		64												0.059	0.087	0.125
		96												0.052	0.076	0.100
		128												0.049	0.068	0.091
Felt, semirigid; organic bonded	480	50–125						0.035	0.036	0.038	0.039	0.051	0.063			
	730	50						0.030	0.032	0.033	0.035	0.051	0.079			
Felt, laminated; no binder	920	120	0.023	0.025	0.026	0.027	0.029						0.051	0.065	0.087	
Blocks, Boards, and Pipe Insulations																
Asbestos paper, laminated and corrugated																
4-ply	420	190								0.078	0.082	0.098				
6-ply	420	255								0.071	0.074	0.085				
8-ply	420	300								0.068	0.071	0.082				
Magnesia, 85%	590	185									0.051	0.055	0.061			
Calcium silicate	920	190									0.055	0.059	0.063	0.075	0.089	0.104

TABLE A.3 *Continued*

Industrial Insulation (Continued)

Description/ Composition	Maximum Service Temperature (K)	Typical Density (kg/m³)	Typical Thermal Conductivity, k (W/m · K), at Various Temperatures (K)													
			200	215	230	240	255	270	285	300	310	365	420	530	645	750
Cellular glass	700	145			0.046	0.048	0.051	0.052	0.055	0.058	0.062	0.069	0.079			
Diatomaceous silica	1145	345												0.092	0.098	0.104
	1310	385												0.101	0.100	0.115
Polystyrene, rigid																
Extruded (R-12)	350	56	0.023	0.023	0.022	0.023	0.023	0.025	0.026	0.027	0.029					
Extruded (R-12)	350	35	0.023	0.023	0.023	0.025	0.025	0.026	0.027	0.029						
Molded beads	350	16	0.026	0.029	0.030	0.033	0.035	0.036	0.038	0.040						
Rubber, rigid foamed	340	70						0.029	0.030	0.032	0.033					
Insulating Cement																
Mineral fiber (rock, slag or glass)																
With clay binder	1255	430									0.071	0.079	0.088	0.105	0.123	
With hydraulic setting binder	922	560									0.108	0.115	0.123	0.137		
Loose Fill																
Cellulose, wood or paper pulp	—	45							0.038	0.039	0.042					
Perlite, expanded	—	105	0.036	0.039	0.042	0.043	0.046	0.049	0.051	0.053	0.056					
Vermiculite, expanded	—	122			0.056	0.058	0.061	0.063	0.065	0.068	0.071					
expanded		80			0.049	0.051	0.055	0.058	0.061	0.063	0.066					

TABLE A.3 *Continued*

Other Materials

Description/Composition	Temperature (K)	Density, ρ (kg/m³)	Thermal Conductivity, k (W/m · K)	Specific Heat, c_p (J/kg · K)
Asphalt	300	2115	0.062	920
Bakelite	300	1300	1.4	1465
Brick, refractory				
Carborundum	872	—	18.5	—
	1672	—	11.0	—
Chrome brick	473	3010	2.3	835
	823		2.5	
	1173		2.0	
Diatomaceous	478	—	0.25	—
silica, fired	1145	—	0.30	
Fireclay, burnt 1600 K	773	2050	1.0	960
	1073	—	1.1	
	1373	—	1.1	
Fireclay, burnt 1725 K	773	2325	1.3	960
	1073		1.4	
	1373		1.4	
Fireclay brick	478	2645	1.0	960
	922		1.5	
	1478		1.8	
Magnesite	478	—	3.8	1130
	922	—	2.8	
	1478		1.9	
Clay	300	1460	1.3	880
Coal, anthracite	300	1350	0.26	1260
Concrete (stone mix)	300	2300	1.4	880
Cotton	300	80	0.06	1300
Foodstuffs				
Banana (75.7% water content)	300	980	0.481	3350
Apple, red (75% water content)	300	840	0.513	3600
Cake, batter	300	720	0.223	—
Cake, fully baked	300	280	0.121	—
Chicken meat, white	198	—	1.60	—
(74.4% water content)	233	—	1.49	
	253		1.35	
	263		1.20	
	273		0.476	
	283		0.480	
	293		0.489	
Glass				
Plate (soda lime)	300	2500	1.4	750
Pyrex	300	2225	1.4	835

TABLE A.3 *Continued*

Other Materials (Continued)

Description/ Composition	Temperature (K)	Density, ρ (kg/m³)	Thermal Conductivity, k (W/m · K)	Specific Heat, c_p (J/kg · K)
Ice	273	920	1.88	2040
	253	—	2.03	1945
Leather (sole)	300	998	0.159	—
Paper	300	930	0.180	1340
Paraffin	300	900	0.240	2890
Rock				
Granite, Barre	300	2630	2.79	775
Limestone, Salem	300	2320	2.15	810
Marble, Halston	300	2680	2.80	830
Quartzite, Sioux	300	2640	5.38	1105
Sandstone, Berea	300	2150	2.90	745
Rubber, vulcanized				
Soft	300	1100	0.13	2010
Hard	300	1190	0.16	—
Sand	300	1515	0.27	800
Soil	300	2050	0.52	1840
Snow	273	110	0.049	—
		500	0.190	—
Teflon	300	2200	0.35	—
	400		0.45	—
Tissue, human				
Skin	300	—	0.37	—
Fat layer (adipose)	300	—	0.2	—
Muscle	300	—	0.5	—
Wood, cross grain				
Balsa	300	140	0.055	—
Cypress	300	465	0.097	—
Fir	300	415	0.11	2720
Oak	300	545	0.17	2385
Yellow pine	300	640	0.15	2805
White pine	300	435	0.11	—
Wood, radial				
Oak	300	545	0.19	2385
Fir	300	420	0.14	2720

[a]Adapted from References 1 and 8–13.

TABLE A.4 Thermophysical Properties
of Gases at Atmospheric Pressure[a]

T (K)	ρ (kg/m^3)	c_p (kJ/kg $\cdot$ K)	$\mu \cdot 10^7$ (N $\cdot$ s/m^2)	$\nu \cdot 10^6$ (m^2/s)	$k \cdot 10^3$ (W/m $\cdot$ K)	$\alpha \cdot 10^6$ (m^2/s)	Pr
Air, $\mathcal{M} = 28.97$ kg/kmol							
100	3.5562	1.032	71.1	2.00	9.34	2.54	0.786
150	2.3364	1.012	103.4	4.426	13.8	5.84	0.758
200	1.7458	1.007	132.5	7.590	18.1	10.3	0.737
250	1.3947	1.006	159.6	11.44	22.3	15.9	0.720
300	1.1614	1.007	184.6	15.89	26.3	22.5	0.707
350	0.9950	1.009	208.2	20.92	30.0	29.9	0.700
400	0.8711	1.014	230.1	26.41	33.8	38.3	0.690
450	0.7740	1.021	250.7	32.39	37.3	47.2	0.686
500	0.6964	1.030	270.1	38.79	40.7	56.7	0.684
550	0.6329	1.040	288.4	45.57	43.9	66.7	0.683
600	0.5804	1.051	305.8	52.69	46.9	76.9	0.685
650	0.5356	1.063	322.5	60.21	49.7	87.3	0.690
700	0.4975	1.075	338.8	68.10	52.4	98.0	0.695
750	0.4643	1.087	354.6	76.37	54.9	109	0.702
800	0.4354	1.099	369.8	84.93	57.3	120	0.709
850	0.4097	1.110	384.3	93.80	59.6	131	0.716
900	0.3868	1.121	398.1	102.9	62.0	143	0.720
950	0.3666	1.131	411.3	112.2	64.3	155	0.723
1000	0.3482	1.141	424.4	121.9	66.7	168	0.726
1100	0.3166	1.159	449.0	141.8	71.5	195	0.728
1200	0.2902	1.175	473.0	162.9	76.3	224	0.728
1300	0.2679	1.189	496.0	185.1	82	257	0.719
1400	0.2488	1.207	530	213	91	303	0.703
1500	0.2322	1.230	557	240	100	350	0.685
1600	0.2177	1.248	584	268	106	390	0.688
1700	0.2049	1.267	611	298	113	435	0.685
1800	0.1935	1.286	637	329	120	482	0.683
1900	0.1833	1.307	663	362	128	534	0.677
2000	0.1741	1.337	689	396	137	589	0.672
2100	0.1658	1.372	715	431	147	646	0.667
2200	0.1582	1.417	740	468	160	714	0.655
2300	0.1513	1.478	766	506	175	783	0.647
2400	0.1448	1.558	792	547	196	869	0.630
2500	0.1389	1.665	818	589	222	960	0.613
3000	0.1135	2.726	955	841	486	1570	0.536
Ammonia (NH$_3$), $\mathcal{M} = 17.03$ kg/kmol							
300	0.6894	2.158	101.5	14.7	24.7	16.6	0.887
320	0.6448	2.170	109	16.9	27.2	19.4	0.870
340	0.6059	2.192	116.5	19.2	29.3	22.1	0.872
360	0.5716	2.221	124	21.7	31.6	24.9	0.872
380	0.5410	2.254	131	24.2	34.0	27.9	0.869

TABLE A.4 *Continued*

T (K)	ρ (kg/m³)	c_p (kJ/kg · K)	$\mu \cdot 10^7$ (N · s/m²)	$\nu \cdot 10^6$ (m²/s)	$k \cdot 10^3$ (W/m · K)	$\alpha \cdot 10^6$ (m²/s)	Pr
Ammonia (NH₃) *(continued)*							
400	0.5136	2.287	138	26.9	37.0	31.5	0.853
420	0.4888	2.322	145	29.7	40.4	35.6	0.833
440	0.4664	2.357	152.5	32.7	43.5	39.6	0.826
460	0.4460	2.393	159	35.7	46.3	43.4	0.822
480	0.4273	2.430	166.5	39.0	49.2	47.4	0.822
500	0.4101	2.467	173	42.2	52.5	51.9	0.813
520	0.3942	2.504	180	45.7	54.5	55.2	0.827
540	0.3795	2.540	186.5	49.1	57.5	59.7	0.824
560	0.3708	2.577	193	52.0	60.6	63.4	0.827
580	0.3533	2.613	199.5	56.5	63.8	69.1	0.817
Carbon Dioxide (CO₂), $\mathscr{M}$ **= 44.01 kg/kmol**							
280	1.9022	0.830	140	7.36	15.20	9.63	0.765
300	1.7730	0.851	149	8.40	16.55	11.0	0.766
320	1.6609	0.872	156	9.39	18.05	12.5	0.754
340	1.5618	0.891	165	10.6	19.70	14.2	0.746
360	1.4743	0.908	173	11.7	21.2	15.8	0.741
380	1.3961	0.926	181	13.0	22.75	17.6	0.737
400	1.3257	0.942	190	14.3	24.3	19.5	0.737
450	1.1782	0.981	210	17.8	28.3	24.5	0.728
500	1.0594	1.02	231	21.8	32.5	30.1	0.725
550	0.9625	1.05	251	26.1	36.6	36.2	0.721
600	0.8826	1.08	270	30.6	40.7	42.7	0.717
650	0.8143	1.10	288	35.4	44.5	49.7	0.712
700	0.7564	1.13	305	40.3	48.1	56.3	0.717
750	0.7057	1.15	321	45.5	51.7	63.7	0.714
800	0.6614	1.17	337	51.0	55.1	71.2	0.716
Carbon Monoxide (CO), $\mathscr{M}$ **= 28.01 kg/kmol**							
200	1.6888	1.045	127	7.52	17.0	9.63	0.781
220	1.5341	1.044	137	8.93	19.0	11.9	0.753
240	1.4055	1.043	147	10.5	20.6	14.1	0.744
260	1.2967	1.043	157	12.1	22.1	16.3	0.741
280	1.2038	1.042	166	13.8	23.6	18.8	0.733
300	1.1233	1.043	175	15.6	25.0	21.3	0.730
320	1.0529	1.043	184	17.5	26.3	23.9	0.730
340	0.9909	1.044	193	19.5	27.8	26.9	0.725
360	0.9357	1.045	202	21.6	29.1	29.8	0.725
380	0.8864	1.047	210	23.7	30.5	32.9	0.729
400	0.8421	1.049	218	25.9	31.8	36.0	0.719
450	0.7483	1.055	237	31.7	35.0	44.3	0.714
500	0.67352	1.065	254	37.7	38.1	53.1	0.710
550	0.61226	1.076	271	44.3	41.1	62.4	0.710
600	0.56126	1.088	286	51.0	44.0	72.1	0.707

TABLE A.4 *Continued*

T (K)	ρ (kg/m³)	c_p (kJ/kg · K)	$\mu \cdot 10^7$ (N · s/m²)	$\nu \cdot 10^6$ (m²/s)	$k \cdot 10^3$ (W/m · K)	$\alpha \cdot 10^6$ (m²/s)	Pr
Carbon Monoxide (CO) *(continued)*							
650	0.51806	1.101	301	58.1	47.0	82.4	0.705
700	0.48102	1.114	315	65.5	50.0	93.3	0.702
750	0.44899	1.127	329	73.3	52.8	104	0.702
800	0.42095	1.140	343	81.5	55.5	116	0.705
Helium (He), $\mathcal{M} = $ **4.003 kg/kmol**							
100	0.4871	5.193	96.3	19.8	73.0	28.9	0.686
120	0.4060	5.193	107	26.4	81.9	38.8	0.679
140	0.3481	5.193	118	33.9	90.7	50.2	0.676
160	—	5.193	129	—	99.2	—	—
180	0.2708	5.193	139	51.3	107.2	76.2	0.673
200	—	5.193	150	—	115.1	—	—
220	0.2216	5.193	160	72.2	123.1	107	0.675
240	—	5.193	170	—	130	—	—
260	0.1875	5.193	180	96.0	137	141	0.682
280	—	5.193	190	—	145	—	—
300	0.1625	5.193	199	122	152	180	0.680
350	—	5.193	221	—	170	—	—
400	0.1219	5.193	243	199	187	295	0.675
450	—	5.193	263	—	204	—	—
500	0.09754	5.193	283	290	220	434	0.668
550	—	5.193	—	—	—	—	—
600	—	5.193	320	—	252	—	—
650	—	5.193	332	—	264	—	—
700	0.06969	5.193	350	502	278	768	0.654
750	—	5.193	364	—	291	—	—
800	—	5.193	382	—	304	—	—
900	—	5.193	414	—	330	—	—
1000	0.04879	5.193	446	914	354	1400	0.654
Hydrogen (H₂), $\mathcal{M} = $ **2.016 kg/kmol**							
100	0.24255	11.23	42.1	17.4	67.0	24.6	0.707
150	0.16156	12.60	56.0	34.7	101	49.6	0.699
200	0.12115	13.54	68.1	56.2	131	79.9	0.704
250	0.09693	14.06	78.9	81.4	157	115	0.707
300	0.08078	14.31	89.6	111	183	158	0.701
350	0.06924	14.43	98.8	143	204	204	0.700
400	0.06059	14.48	108.2	179	226	258	0.695
450	0.05386	14.50	117.2	218	247	316	0.689
500	0.04848	14.52	126.4	261	266	378	0.691
550	0.04407	14.53	134.3	305	285	445	0.685

TABLE A.4 *Continued*

T (K)	ρ (kg/m³)	c_p (kJ/kg · K)	$\mu \cdot 10^7$ (N · s/m²)	$\nu \cdot 10^6$ (m²/s)	$k \cdot 10^3$ (W/m · K)	$\alpha \cdot 10^6$ (m²/s)	Pr
Hydrogen (H₂) *(continued)*							
600	0.04040	14.55	142.4	352	305	519	0.678
700	0.03463	14.61	157.8	456	342	676	0.675
800	0.03030	14.70	172.4	569	378	849	0.670
900	0.02694	14.83	186.5	692	412	1030	0.671
1000	0.02424	14.99	201.3	830	448	1230	0.673
1100	0.02204	15.17	213.0	966	488	1460	0.662
1200	0.02020	15.37	226.2	1120	528	1700	0.659
1300	0.01865	15.59	238.5	1279	568	1955	0.655
1400	0.01732	15.81	250.7	1447	610	2230	0.650
1500	0.01616	16.02	262.7	1626	655	2530	0.643
1600	0.0152	16.28	273.7	1801	697	2815	0.639
1700	0.0143	16.58	284.9	1992	742	3130	0.637
1800	0.0135	16.96	296.1	2193	786	3435	0.639
1900	0.0128	17.49	307.2	2400	835	3730	0.643
2000	0.0121	18.25	318.2	2630	878	3975	0.661
Nitrogen (N₂), $\mathcal{M} = 28.01$ kg/kmol							
100	3.4388	1.070	68.8	2.00	9.58	2.60	0.768
150	2.2594	1.050	100.6	4.45	13.9	5.86	0.759
200	1.6883	1.043	129.2	7.65	18.3	10.4	0.736
250	1.3488	1.042	154.9	11.48	22.2	15.8	0.727
300	1.1233	1.041	178.2	15.86	25.9	22.1	0.716
350	0.9625	1.042	200.0	20.78	29.3	29.2	0.711
400	0.8425	1.045	220.4	26.16	32.7	37.1	0.704
450	0.7485	1.050	239.6	32.01	35.8	45.6	0.703
500	0.6739	1.056	257.7	38.24	38.9	54.7	0.700
550	0.6124	1.065	274.7	44.86	41.7	63.9	0.702
600	0.5615	1.075	290.8	51.79	44.6	73.9	0.701
700	0.4812	1.098	321.0	66.71	49.9	94.4	0.706
800	0.4211	1.122	349.1	82.90	54.8	116	0.715
900	0.3743	1.146	375.3	100.3	59.7	139	0.721
1000	0.3368	1.167	399.9	118.7	64.7	165	0.721
1100	0.3062	1.187	423.2	138.2	70.0	193	0.718
1200	0.2807	1.204	445.3	158.6	75.8	224	0.707
1300	0.2591	1.219	466.2	179.9	81.0	256	0.701
Oxygen (O₂), $\mathcal{M} = 32.00$ kg/kmol							
100	3.945	0.962	76.4	1.94	9.25	2.44	0.796
150	2.585	0.921	114.8	4.44	13.8	5.80	0.766
200	1.930	0.915	147.5	7.64	18.3	10.4	0.737
250	1.542	0.915	178.6	11.58	22.6	16.0	0.723
300	1.284	0.920	207.2	16.14	26.8	22.7	0.711

TABLE A.4 *Continued*

T (K)	ρ (kg/m^3)	c_p (kJ/kg · K)	$\mu \cdot 10^7$ (N · s/m^2)	$\nu \cdot 10^6$ (m^2/s)	$k \cdot 10^3$ (W/m · K)	$\alpha \cdot 10^6$ (m^2/s)	Pr
Oxygen (O$_2$) *(continued)*							
350	1.100	0.929	233.5	21.23	29.6	29.0	0.733
400	0.9620	0.942	258.2	26.84	33.0	36.4	0.737
450	0.8554	0.956	281.4	32.90	36.3	44.4	0.741
500	0.7698	0.972	303.3	39.40	41.2	55.1	0.716
550	0.6998	0.988	324.0	46.30	44.1	63.8	0.726
600	0.6414	1.003	343.7	53.59	47.3	73.5	0.729
700	0.5498	1.031	380.8	69.26	52.8	93.1	0.744
800	0.4810	1.054	415.2	86.32	58.9	116	0.743
900	0.4275	1.074	447.2	104.6	64.9	141	0.740
1000	0.3848	1.090	477.0	124.0	71.0	169	0.733
1100	0.3498	1.103	505.5	144.5	75.8	196	0.736
1200	0.3206	1.115	532.5	166.1	81.9	229	0.725
1300	0.2960	1.125	588.4	188.6	87.1	262	0.721
Water Vapor (Steam), $\mathcal{M} = 18.02$ kg/kmol							
380	0.5863	2.060	127.1	21.68	24.6	20.4	1.06
400	0.5542	2.014	134.4	24.25	26.1	23.4	1.04
450	0.4902	1.980	152.5	31.11	29.9	30.8	1.01
500	0.4405	1.985	170.4	38.68	33.9	38.8	0.998
550	0.4005	1.997	188.4	47.04	37.9	47.4	0.993
600	0.3652	2.026	206.7	56.60	42.2	57.0	0.993
650	0.3380	2.056	224.7	66.48	46.4	66.8	0.996
700	0.3140	2.085	242.6	77.26	50.5	77.1	1.00
750	0.2931	2.119	260.4	88.84	54.9	88.4	1.00
800	0.2739	2.152	278.6	101.7	59.2	100	1.01
850	0.2579	2.186	296.9	115.1	63.7	113	1.02

[a]Adapted from References 8, 14, and 15.

TABLE A.5 Thermophysical Properties of Saturated Fluids[a]

Saturated Liquids

T (K)	ρ (kg/m³)	c_p (kJ/kg · K)	$\mu \cdot 10^2$ (N · s/m²)	$\nu \cdot 10^6$ (m²/s)	$k \cdot 10^3$ (W/m · K)	$\alpha \cdot 10^7$ (m²/s)	Pr	$\beta \cdot 10^3$ (K⁻¹)
Engine Oil (Unused)								
273	899.1	1.796	385	4280	147	0.910	47,000	0.70
280	895.3	1.827	217	2430	144	0.880	27,500	0.70
290	890.0	1.868	99.9	1120	145	0.872	12,900	0.70
300	884.1	1.909	48.6	550	145	0.859	6400	0.70
310	877.9	1.951	25.3	288	145	0.847	3400	0.70
320	871.8	1.993	14.1	161	143	0.823	1965	0.70
330	865.8	2.035	8.36	96.6	141	0.800	1205	0.70
340	859.9	2.076	5.31	61.7	139	0.779	793	0.70
350	853.9	2.118	3.56	41.7	138	0.763	546	0.70
360	847.8	2.161	2.52	29.7	138	0.753	395	0.70
370	841.8	2.206	1.86	22.0	137	0.738	300	0.70
380	836.0	2.250	1.41	16.9	136	0.723	233	0.70
390	830.6	2.294	1.10	13.3	135	0.709	187	0.70
400	825.1	2.337	0.874	10.6	134	0.695	152	0.70
410	818.9	2.381	0.698	8.52	133	0.682	125	0.70
420	812.1	2.427	0.564	6.94	133	0.675	103	0.70
430	806.5	2.471	0.470	5.83	132	0.662	88	0.70
Ethylene Glycol [$C_2H_4(OH)_2$]								
273	1130.8	2.294	6.51	57.6	242	0.933	617	0.65
280	1125.8	2.323	4.20	37.3	244	0.933	400	0.65
290	1118.8	2.368	2.47	22.1	248	0.936	236	0.65
300	1114.4	2.415	1.57	14.1	252	0.939	151	0.65
310	1103.7	2.460	1.07	9.65	255	0.939	103	0.65
320	1096.2	2.505	0.757	6.91	258	0.940	73.5	0.65
330	1089.5	2.549	0.561	5.15	260	0.936	55.0	0.65
340	1083.8	2.592	0.431	3.98	261	0.929	42.8	0.65
350	1079.0	2.637	0.342	3.17	261	0.917	34.6	0.65
360	1074.0	2.682	0.278	2.59	261	0.906	28.6	0.65
370	1066.7	2.728	0.228	2.14	262	0.900	23.7	0.65
373	1058.5	2.742	0.215	2.03	263	0.906	22.4	0.65
Glycerin [$C_3H_5(OH)_3$]								
273	1276.0	2.261	1060	8310	282	0.977	85,000	0.47
280	1271.9	2.298	534	4200	284	0.972	43,200	0.47
290	1265.8	2.367	185	1460	286	0.955	15,300	0.48
300	1259.9	2.427	79.9	634	286	0.935	6780	0.48
310	1253.9	2.490	35.2	281	286	0.916	3060	0.49
320	1247.2	2.564	21.0	168	287	0.897	1870	0.50

TABLE A.5 *Continued*

Saturated Liquids (Continued)

T (K)	ρ (kg/m^3)	c_p (kJ/kg · K)	$\mu \cdot 10^2$ (N · s/m^2)	$\nu \cdot 10^6$ (m^2/s)	$k \cdot 10^3$ (W/m · K)	$\alpha \cdot 10^7$ (m^2/s)	Pr	$\beta \cdot 10^3$ (K^{-1})
Refrigerant-134a (C$_2$H$_2$F$_4$)								
230	1426.8	1.249	0.04912	0.3443	112.1	0.629	5.5	2.02
240	1397.7	1.267	0.04202	0.3006	107.3	0.606	5.0	2.11
250	1367.9	1.287	0.03633	0.2656	102.5	0.583	4.6	2.23
260	1337.1	1.308	0.03166	0.2368	97.9	0.560	4.2	2.36
270	1305.1	1.333	0.02775	0.2127	93.4	0.537	4.0	2.53
280	1271.8	1.361	0.02443	0.1921	89.0	0.514	3.7	2.73
290	1236.8	1.393	0.02156	0.1744	84.6	0.491	3.5	2.98
300	1199.7	1.432	0.01905	0.1588	80.3	0.468	3.4	3.30
310	1159.9	1.481	0.01680	0.1449	76.1	0.443	3.3	3.73
320	1116.8	1.543	0.01478	0.1323	71.8	0.417	3.2	4.33
330	1069.1	1.627	0.01292	0.1209	67.5	0.388	3.1	5.19
340	1015.0	1.751	0.01118	0.1102	63.1	0.355	3.1	6.57
350	951.3	1.961	0.00951	0.1000	58.6	0.314	3.2	9.10
360	870.1	2.437	0.00781	0.0898	54.1	0.255	3.5	15.39
370	740.3	5.105	0.00580	0.0783	51.8	0.137	5.7	55.24
Refrigerant-22 (CHClF$_2$)								
230	1416.0	1.087	0.03558	0.2513	114.5	0.744	3.4	2.05
240	1386.6	1.100	0.03145	0.2268	109.8	0.720	3.2	2.16
250	1356.3	1.117	0.02796	0.2062	105.2	0.695	3.0	2.29
260	1324.9	1.137	0.02497	0.1884	100.7	0.668	2.8	2.45
270	1292.1	1.161	0.02235	0.1730	96.2	0.641	2.7	2.63
280	1257.9	1.189	0.02005	0.1594	91.7	0.613	2.6	2.86
290	1221.7	1.223	0.01798	0.1472	87.2	0.583	2.5	3.15
300	1183.4	1.265	0.01610	0.1361	82.6	0.552	2.5	3.51
310	1142.2	1.319	0.01438	0.1259	78.1	0.518	2.4	4.00
320	1097.4	1.391	0.01278	0.1165	73.4	0.481	2.4	4.69
330	1047.5	1.495	0.01127	0.1075	68.6	0.438	2.5	5.75
340	990.1	1.665	0.00980	0.0989	63.6	0.386	2.6	7.56
350	920.1	1.997	0.00831	0.0904	58.3	0.317	2.8	11.35
360	823.4	3.001	0.00668	0.0811	53.1	0.215	3.8	23.88
Mercury (Hg)								
273	13,595	0.1404	0.1688	0.1240	8180	42.85	0.0290	0.181
300	13,529	0.1393	0.1523	0.1125	8540	45.30	0.0248	0.181
350	13,407	0.1377	0.1309	0.0976	9180	49.75	0.0196	0.181
400	13,287	0.1365	0.1171	0.0882	9800	54.05	0.0163	0.181
450	13,167	0.1357	0.1075	0.0816	10,400	58.10	0.0140	0.181
500	13,048	0.1353	0.1007	0.0771	10,950	61.90	0.0125	0.182
550	12,929	0.1352	0.0953	0.0737	11,450	65.55	0.0112	0.184
600	12,809	0.1355	0.0911	0.0711	11,950	68.80	0.0103	0.187

TABLE A.5 *Continued*

Saturated Liquid–Vapor, 1 atm[b]

Fluid	T_{sat} (K)	h_{fg} (kJ/kg)	ρ_f (kg/m³)	ρ_g (kg/m³)	$\sigma \cdot 10^3$ (N/m)
Ethanol	351	846	757	1.44	17.7
Ethylene glycol	470	812	1111[c]	—	32.7
Glycerin	563	974	1260[c]	—	63.0[c]
Mercury	630	301	12,740	3.90	417
Refrigerant R-134a	247	217	1377	5.26	15.4
Refrigerant R-22	232	234	1409	4.70	18.1

[a]Adapted from References 15–19.
[b]Adapted from References 8, 20, and 21.
[c]Property value corresponding to 300 K.

TABLE A.6 Thermophysical Properties of Saturated Water[a]

Temperature, T (K)	Pressure, p (bar)[b]	Specific Volume (m³/kg)		Heat of Vaporization, h_{fg} (kJ/kg)	Specific Heat (kJ/kg·K)		Viscosity (N·s/m²)		Thermal Conductivity (W/m·K)		Prandtl Number		Surface Tension, $\sigma_f \cdot 10^3$ (N/m)	Expansion Coefficient, $\beta_f \cdot 10^6$ (K⁻¹)	Temperature, T (K)
		$v_f \cdot 10^3$	v_g		$c_{p,f}$	$c_{p,g}$	$\mu_f \cdot 10^6$	$\mu_g \cdot 10^6$	$k_f \cdot 10^3$	$k_g \cdot 10^3$	Pr_f	Pr_g			
273.15	0.00611	1.000	206.3	2502	4.217	1.854	1750	8.02	569	18.2	12.99	0.815	75.5	−68.05	273.15
275	0.00697	1.000	181.7	2497	4.211	1.855	1652	8.09	574	18.3	12.22	0.817	75.3	−32.74	275
280	0.00990	1.000	130.4	2485	4.198	1.858	1422	8.29	582	18.6	10.26	0.825	74.8	46.04	280
285	0.01387	1.000	99.4	2473	4.189	1.861	1225	8.49	590	18.9	8.81	0.833	74.3	114.1	285
290	0.01917	1.001	69.7	2461	4.184	1.864	1080	8.69	598	19.3	7.56	0.841	73.7	174.0	290
295	0.02617	1.002	51.94	2449	4.181	1.868	959	8.89	606	19.5	6.62	0.849	72.7	227.5	295
300	0.03531	1.003	39.13	2438	4.179	1.872	855	9.09	613	19.6	5.83	0.857	71.7	276.1	300
305	0.04712	1.005	29.74	2426	4.178	1.877	769	9.29	620	20.1	5.20	0.865	70.9	320.6	305
310	0.06221	1.007	22.93	2414	4.178	1.882	695	9.49	628	20.4	4.62	0.873	70.0	361.9	310
315	0.08132	1.009	17.82	2402	4.179	1.888	631	9.69	634	20.7	4.16	0.883	69.2	400.4	315
320	0.1053	1.011	13.98	2390	4.180	1.895	577	9.89	640	21.0	3.77	0.894	68.3	436.7	320
325	0.1351	1.013	11.06	2378	4.182	1.903	528	10.09	645	21.3	3.42	0.901	67.5	471.2	325
330	0.1719	1.016	8.82	2366	4.184	1.911	489	10.29	650	21.7	3.15	0.908	66.6	504.0	330
335	0.2167	1.018	7.09	2354	4.186	1.920	453	10.49	656	22.0	2.88	0.916	65.8	535.5	335
340	0.2713	1.021	5.74	2342	4.188	1.930	420	10.69	660	22.3	2.66	0.925	64.9	566.0	340
345	0.3372	1.024	4.683	2329	4.191	1.941	389	10.89	664	22.6	2.45	0.933	64.1	595.4	345
350	0.4163	1.027	3.846	2317	4.195	1.954	365	11.09	668	23.0	2.29	0.942	63.2	624.2	350
355	0.5100	1.030	3.180	2304	4.199	1.968	343	11.29	671	23.3	2.14	0.951	62.3	652.3	355
360	0.6209	1.034	2.645	2291	4.203	1.983	324	11.49	674	23.7	2.02	0.960	61.4	697.9	360
365	0.7514	1.038	2.212	2278	4.209	1.999	306	11.69	677	24.1	1.91	0.969	60.5	707.1	365
370	0.9040	1.041	1.861	2265	4.214	2.017	289	11.89	679	24.5	1.80	0.978	59.5	728.7	370
373.15	1.0133	1.044	1.679	2257	4.217	2.029	279	12.02	680	24.8	1.76	0.984	58.9	750.1	373.15
375	1.0815	1.045	1.574	2252	4.220	2.036	274	12.09	681	24.9	1.70	0.987	58.6	761	375
380	1.2869	1.049	1.337	2239	4.226	2.057	260	12.29	683	25.4	1.61	0.999	57.6	788	380
385	1.5233	1.053	1.142	2225	4.232	2.080	248	12.49	685	25.8	1.53	1.004	56.6	814	385
390	1.794	1.058	0.980	2212	4.239	2.104	237	12.69	686	26.3	1.47	1.013	55.6	841	390
400	2.455	1.067	0.731	2183	4.256	2.158	217	13.05	688	27.2	1.34	1.033	53.6	896	400
410	3.302	1.077	0.553	2153	4.278	2.221	200	13.42	688	28.2	1.24	1.054	51.5	952	410
420	4.370	1.088	0.425	2123	4.302	2.291	185	13.79	688	29.8	1.16	1.075	49.4	1010	420
430	5.699	1.099	0.331	2091	4.331	2.369	173	14.14	685	30.4	1.09	1.10	47.2		430

TABLE A.6 *Continued*

Temperature, T (K)	Pressure, p (bar)[b]	Specific Volume (m³/kg)		Heat of Vaporization, h_{fg} (kJ/kg)	Specific Heat (kJ/kg·K)		Viscosity (N·s/m²)		Thermal Conductivity (W/m·K)		Prandtl Number		Surface Tension, $\sigma_f \cdot 10^3$ (N/m)	Expansion Coefficient, $\beta_f \cdot 10^6$ (K⁻¹)	Temperature, T (K)
		$v_f \cdot 10^3$	v_g		$c_{p,f}$	$c_{p,g}$	$\mu_f \cdot 10^6$	$\mu_g \cdot 10^6$	$k_f \cdot 10^3$	$k_g \cdot 10^3$	Pr_f	Pr_g			
440	7.333	1.110	0.261	2059	4.36	2.46	162	14.50	682	31.7	1.04	1.12	45.1	—	440
450	9.319	1.123	0.208	2024	4.40	2.56	152	14.85	678	33.1	0.99	1.14	42.9	—	450
460	11.71	1.137	0.167	1989	4.44	2.68	143	15.19	673	34.6	0.95	1.17	40.7	—	460
470	14.55	1.152	0.136	1951	4.48	2.79	136	15.54	667	36.3	0.92	1.20	38.5	—	470
480	17.90	1.167	0.111	1912	4.53	2.94	129	15.88	660	38.1	0.89	1.23	36.2	—	480
490	21.83	1.184	0.0922	1870	4.59	3.10	124	16.23	651	40.1	0.87	1.25	33.9	—	490
500	26.40	1.203	0.0766	1825	4.66	3.27	118	16.59	642	42.3	0.86	1.28	31.6	—	500
510	31.66	1.222	0.0631	1779	4.74	3.47	113	16.95	631	44.7	0.85	1.31	29.3	—	510
520	37.70	1.244	0.0525	1730	4.84	3.70	108	17.33	621	47.5	0.84	1.35	26.9	—	520
530	44.58	1.268	0.0445	1679	4.95	3.96	104	17.72	608	50.6	0.85	1.39	24.5	—	530
540	52.38	1.294	0.0375	1622	5.08	4.27	101	18.1	594	54.0	0.86	1.43	22.1	—	540
550	61.19	1.323	0.0317	1564	5.24	4.64	97	18.6	580	58.3	0.87	1.47	19.7	—	550
560	71.08	1.355	0.0269	1499	5.43	5.09	94	19.1	563	63.7	0.90	1.52	17.3	—	560
570	82.16	1.392	0.0228	1429	5.68	5.67	91	19.7	548	76.7	0.94	1.59	15.0	—	570
580	94.51	1.433	0.0193	1353	6.00	6.40	88	20.4	528	76.7	0.99	1.68	12.8	—	580
590	108.3	1.482	0.0163	1274	6.41	7.35	84	21.5	513	84.1	1.05	1.84	10.5	—	590
600	123.5	1.541	0.0137	1176	7.00	8.75	81	22.7	497	92.9	1.14	2.15	8.4	—	600
610	137.3	1.612	0.0115	1068	7.85	11.1	77	24.1	467	103	1.30	2.60	6.3	—	610
620	159.1	1.705	0.0094	941	9.35	15.4	72	25.9	444	114	1.52	3.46	4.5	—	620
625	169.1	1.778	0.0085	858	10.6	18.3	70	27.0	430	121	1.65	4.20	3.5	—	625
630	179.7	1.856	0.0075	781	12.6	22.1	67	28.0	412	130	2.0	4.8	2.6	—	630
635	190.9	1.935	0.0066	683	16.4	27.6	64	30.0	392	141	2.7	6.0	1.5	—	635
640	202.7	2.075	0.0057	560	26	42	59	32.0	367	155	4.2	9.6	0.8	—	640
645	215.2	2.351	0.0045	361	90	—	54	37.0	331	178	12	26	0.1	—	645
647.3[c]	221.2	3.170	0.0032	0	∞	∞	45	45.0	238	238	∞	∞	0.0	—	647.3[c]

[a]Adapted from Reference 22.
[b]1 bar = 10⁵ N/m².
[c]Critical temperature.

TABLE A.7 Thermophysical Properties of Liquid Metals[a]

Composition	Melting Point (K)	T (K)	ρ (kg/m³)	c_p (kJ/kg · K)	$\nu \cdot 10^7$ (m²/s)	k (W/m · K)	$\alpha \cdot 10^5$ (m²/s)	Pr
Bismuth	544	589	10,010	0.1444	1.617	16.4	1.138	0.0142
		811	9738	0.1545	1.133	15.6	1.035	0.0110
		1033	9454	0.1645	0.8343	15.6	1.001	0.0083
Lead	600	644	10,540	0.159	2.276	16.1	1.084	0.024
		755	10,412	0.155	1.849	15.6	1.223	0.017
		977	10,140	—	1.347	14.9	—	—
Potassium	337	422	807.3	0.80	4.608	45.0	6.99	0.0066
		700	741.7	0.75	2.397	39.5	7.07	0.0034
		977	674.4	0.75	1.905	33.1	6.55	0.0029
Sodium	371	373	928.0	1.38	7.532	86.0	6.69	0.011
		644	860.2	1.30	3.270	72.3	6.48	0.0051
		977	778.5	1.26	2.285	59.7	6.12	0.0037
NaK, (56%/44%)	292	366	887.4	1.130	6.522	25.6	2.552	0.026
		644	821.7	1.055	2.871	27.5	3.17	0.0091
		977	740.1	1.043	2.174	28.9	3.74	0.0058
NaK, (22%/78%)	262	366	849.0	0.946	5.797	24.4	3.05	0.019
		672	775.3	0.879	2.666	26.7	3.92	0.0068
		1033	690.4	0.883	2.118	—	—	—
PbBi, (44.5%/55.5%)	398	422	10,524	0.147	—	9.05	0.586	—
		644	10,236	0.147	1.496	11.86	0.790	0.189
		922	9835	—	1.171	—	—	—
Mercury	234			See Table A.5				

[a]Adapted from Reference 23.

TABLE A.8   Binary Diffusion Coefficients at One Atmosphere[a,b]

Substance A	Substance B	T (K)	D_{AB} (m^2/s)
Gases			
NH_3	Air	298	0.28×10^{-4}
H_2O	Air	298	0.26×10^{-4}
CO_2	Air	298	0.16×10^{-4}
H_2	Air	298	0.41×10^{-4}
O_2	Air	298	0.21×10^{-4}
Acetone	Air	273	0.11×10^{-4}
Benzene	Air	298	0.88×10^{-5}
Naphthalene	Air	300	0.62×10^{-5}
Ar	N_2	293	0.19×10^{-4}
H_2	O_2	273	0.70×10^{-4}
H_2	N_2	273	0.68×10^{-4}
H_2	CO_2	273	0.55×10^{-4}
CO_2	N_2	293	0.16×10^{-4}
CO_2	O_2	273	0.14×10^{-4}
O_2	N_2	273	0.18×10^{-4}
Dilute Solutions			
Caffeine	H_2O	298	0.63×10^{-9}
Ethanol	H_2O	298	0.12×10^{-8}
Glucose	H_2O	298	0.69×10^{-9}
Glycerol	H_2O	298	0.94×10^{-9}
Acetone	H_2O	298	0.13×10^{-8}
CO_2	H_2O	298	0.20×10^{-8}
O_2	H_2O	298	0.24×10^{-8}
H_2	H_2O	298	0.63×10^{-8}
N_2	H_2O	298	0.26×10^{-8}
Solids			
O_2	Rubber	298	0.21×10^{-9}
N_2	Rubber	298	0.15×10^{-9}
CO_2	Rubber	298	0.11×10^{-9}
He	SiO_2	293	0.4×10^{-13}
H_2	Fe	293	0.26×10^{-12}
Cd	Cu	293	0.27×10^{-18}
Al	Cu	293	0.13×10^{-33}

[a]Adapted with permission from References 24, 25, and 26.
[b]Assuming ideal gas behavior, the pressure and temperature dependence of the diffusion coefficient for a binary mixture of gases may be estimated from the relation

$$D_{AB} \propto p^{-1}T^{3/2}$$

TABLE A.9   Henry's Constant for Selected Gases in Water at Moderate Pressure[a]

T (K)	$H = p_{A,i}/x_{A,i}$ (bar)							
	NH$_3$	Cl$_2$	H$_2$S	SO$_2$	CO$_2$	CH$_4$	O$_2$	H$_2$
273	21	265	260	165	710	22,880	25,500	58,000
280	23	365	335	210	960	27,800	30,500	61,500
290	26	480	450	315	1300	35,200	37,600	66,500
300	30	615	570	440	1730	42,800	45,700	71,600
310	—	755	700	600	2175	50,000	52,500	76,000
320	—	860	835	800	2650	56,300	56,800	78,600
323	—	890	870	850	2870	58,000	58,000	79,000

[a]Adapted with permission from Reference 27.

TABLE A.10   The Solubility of Selected Gases and Solids[a]

Gas	Solid	T (K)	$S = C_{A,i}/p_{A,i}$ (kmol/m^3 · bar)
O$_2$	Rubber	298	3.12×10^{-3}
N$_2$	Rubber	298	1.56×10^{-3}
CO$_2$	Rubber	298	40.15×10^{-3}
He	SiO$_2$	293	0.45×10^{-3}
H$_2$	Ni	358	9.01×10^{-3}

[a]Data from Reference 26.

TABLE A.11 Total, Normal (*n*) or Hemispherical (*h*) Emissivity of Selected Surfaces

Metallic Solids and Their Oxides[a]

Description/Composition		Emissivity, ε_n or ε_h, at Various Temperatures (K)										
		100	200	300	400	600	800	1000	1200	1500	2000	2500
Aluminum												
Highly polished, film	(h)	0.02	0.03	0.04	0.05	0.06						
Foil, bright	(h)	0.06	0.06	0.07								
Anodized	(h)			0.82	0.76							
Chromium												
Polished or plated	(n)	0.05	0.07	0.10	0.12	0.14						
Copper												
Highly polished	(h)			0.03	0.03	0.04	0.04	0.04				
Stably oxidized	(h)					0.50	0.58	0.80				
Gold												
Highly polished or film	(h)	0.01	0.02	0.03	0.03	0.04	0.05	0.06				
Foil, bright	(h)	0.06	0.07	0.07								
Molybdenum												
Polished	(h)					0.06	0.08	0.10	0.12	0.15	0.21	0.26
Shot-blasted, rough	(h)					0.25	0.28	0.31	0.35	0.42		
Stably oxidized	(h)					0.80	0.82					
Nickel												
Polished	(h)					0.09	0.11	0.14	0.17			
Stably oxidized	(h)					0.40	0.49	0.57				
Platinum												
Polished	(h)						0.10	0.13	0.15	0.18		
Silver												
Polished	(h)			0.02	0.02	0.03	0.05	0.08				
Stainless steels												
Typical, polished	(n)			0.17	0.17	0.19	0.23	0.30				
Typical, cleaned	(n)			0.22	0.22	0.24	0.28	0.35				
Typical, lightly oxidized	(n)						0.33	0.40				
Typical, highly oxidized	(n)						0.67	0.70	0.76			
AISI 347, stably oxidized	(n)					0.87	0.88	0.89	0.90			
Tantalum												
Polished	(h)							0.11	0.11	0.17	0.23	0.28
Tungsten												
Polished	(h)							0.10	0.13	0.18	0.25	0.29

TABLE A.11 *Continued*

Nonmetallic Substances[b]

Description/Composition		Temperature (K)	Emissivity ε
Aluminum oxide	(n)	600	0.69
		1000	0.55
		1500	0.41
Asphalt pavement	(h)	300	0.85–0.93
Building materials			
Asbestos sheet	(h)	300	0.93–0.96
Brick, red	(h)	300	0.93–0.96
Gypsum or plaster board	(h)	300	0.90–0.92
Wood	(h)	300	0.82–0.92
Cloth	(h)	300	0.75–0.90
Concrete	(h)	300	0.88–0.93
Glass, window	(h)	300	0.90–0.95
Ice	(h)	273	0.95–0.98
Paints			
Black (Parsons)	(h)	300	0.98
White, acrylic	(h)	300	0.90
White, zinc oxide	(h)	300	0.92
Paper, white	(h)	300	0.92–0.97
Pyrex	(n)	300	0.82
		600	0.80
		1000	0.71
		1200	0.62
Pyroceram	(n)	300	0.85
		600	0.78
		1000	0.69
		1500	0.57
Refractories (furnace liners)			
Alumina brick	(n)	800	0.40
		1000	0.33
		1400	0.28
		1600	0.33
Magnesia brick	(n)	800	0.45
		1000	0.36
		1400	0.31
		1600	0.40
Kaolin insulating brick	(n)	800	0.70
		1200	0.57
		1400	0.47
		1600	0.53
Sand	(h)	300	0.90
Silicon carbide	(n)	600	0.87
		1000	0.87
		1500	0.85
Skin	(h)	300	0.95
Snow	(h)	273	0.82–0.90

TABLE A.11   *Continued*

Nonmetallic Substances[b]

Description/Composition		Temperature (K)	Emissivity ε
Soil	(h)	300	0.93–0.96
Rocks	(h)	300	0.88–0.95
Teflon	(h)	300	0.85
		400	0.87
		500	0.92
Vegetation	(h)	300	0.92–0.96
Water	(h)	300	0.96

[a]Data from Reference 1.
[b]Data from References 1, 9, 28, and 29.

TABLE A.12   Solar Radiative Properties for Selected Materials[a]

Description/Composition	α_S	ε^b	α_S/ε	τ_S
Aluminum				
Polished	0.09	0.03	3.0	
Anodized	0.14	0.84	0.17	
Quartz overcoated	0.11	0.37	0.30	
Foil	0.15	0.05	3.0	
Brick, red (Purdue)	0.63	0.93	0.68	
Concrete	0.60	0.88	0.68	
Galvanized sheet metal				
Clean, new	0.65	0.13	5.0	
Oxidized, weathered	0.80	0.28	2.9	
Glass, 3.2-mm thickness				
Float or tempered				0.79
Low iron oxide type				0.88
Metal, plated				
Black sulfide	0.92	0.10	9.2	
Black cobalt oxide	0.93	0.30	3.1	
Black nickel oxide	0.92	0.08	11	
Black chrome	0.87	0.09	9.7	
Mylar, 0.13-mm thickness				0.87
Paints				
Black (Parsons)	0.98	0.98	1.0	
White, acrylic	0.26	0.90	0.29	
White, zinc oxide	0.16	0.93	0.17	
Plexiglas, 3.2-mm thickness				0.90
Snow				
Fine particles, fresh	0.13	0.82	0.16	
Ice granules	0.33	0.89	0.37	
Tedlar, 0.10-mm thickness				0.92
Teflon, 0.13-mm thickness				0.92

[a]Based on tables from Reference 29.
[b]The emissivity values in this table correspond to a surface temperature of approximately 300 K.

References

1. Touloukian, Y. S., and C. Y. Ho, Eds., *Thermophysical Properties of Matter,* Vol. 1, *Thermal Conductivity of Metallic Solids;* Vol. 2, *Thermal Conductivity of Nonmetallic Solids;* Vol. 4, *Specific Heat of Metallic Solids;* Vol. 5, *Specific Heat of Nonmetallic Solids;* Vol. 7, *Thermal Radiative Properties of Metallic Solids;* Vol. 8, *Thermal Radiative Properties of Nonmetallic Solids;* Vol. 9, *Thermal Radiative Properties of Coatings,* Plenum Press, New York, 1972.

2. Touloukian, Y. S., and C. Y. Ho, Eds., *Thermophysical Properties of Selected Aerospace Materials,* Part I: Thermal Radiative Properties; Part II: Thermophysical Properties of Seven Materials. Thermophysical and Electronic Properties Information Analysis Center, CINDAS, Purdue University, West Lafayette, IN, 1976.

3. Ho, C. Y., R. W. Powell, and P. E. Liley, *J. Phys. Chem. Ref. Data,* **3,** Supplement 1, 1974.

4. Desai, P. D., T. K. Chu, R. H. Bogaard, M. W. Ackermann, and C. Y. Ho, Part I: Thermophysical Properties of Carbon Steels; Part II: Thermophysical Properties of Low Chromium Steels; Part III: Thermophysical Properties of Nickel Steels; Part IV: Thermophysical Properties of Stainless Steels. CINDAS Special Report, Purdue University, West Lafayette, IN, September 1976.

5. American Society for Metals, *Metals Handbook,* Vol. 1, *Properties and Selection of Metals,* 8th ed., ASM, Metals Park, OH, 1961.

6. Hultgren, R., P. D. Desai, D. T. Hawkins, M. Gleiser, K. K. Kelley, and D. D. Wagman, *Selected Values of the Thermodynamic Properties of the Elements,* American Society of Metals, Metals Park, OH, 1973.

7. Hultgren, R., P. D. Desai, D. T. Hawkins, M. Gleiser, and K. K. Kelley, *Selected Values of the Thermodynamic Properties of Binary Alloys,* American Society of Metals, Metals Park, OH, 1973.

8. American Society of Heating, Refrigerating and Air Conditioning Engineers, *ASHRAE Handbook of Fundamentals,* ASHRAE, New York, 1981.

9. Mallory, J. F., *Thermal Insulation,* Van Nostrand Reinhold, New York, 1969.

10. Hanley, E. J., D. P. DeWitt, and R. E. Taylor, "The Thermal Transport Properties at Normal and Elevated Temperature of Eight Representative Rocks," *Proceedings of the Seventh Symposium on Thermophysical Properties,* American Society of Mechanical Engineers, New York, 1977.

11. Sweat, V. E., "A Miniature Thermal Conductivity Probe for Foods," American Society of Mechanical Engineers, Paper 76-HT-60, August 1976.

12. Kothandaraman, C. P., and S. Subramanyan, *Heat and Mass Transfer Data Book,* Halsted Press/Wiley, Hoboken, NJ, 1975.

13. Chapman, A. J., *Heat Transfer,* 4th ed., Macmillan, New York, 1984.

14. Vargaftik, N. B., *Tables of Thermophysical Properties of Liquids and Gases,* 2nd ed., Hemisphere Publishing, New York, 1975.

15. Eckert, E. R. G., and R. M. Drake, *Analysis of Heat and Mass Transfer,* McGraw-Hill, New York, 1972.

16. Vukalovich, M. P., A. I. Ivanov, L. R. Fokin, and A. T. Yakovelev, *Thermophysical Properties of Mercury,* State Committee on Standards, State Service for Standards and Handbook Data, Monograph Series No. 9, Izd. Standartov, Moscow, 1971.

17. Tillner-Roth, R., and H. D. Baehr, *J. Phys. Chem. Ref. Data,* **23,** 657, 1994.

18. Kamei, A., S. W. Beyerlein, and R. T. Jacobsen, *Int. J. Thermophysics,* **16,** 1155, 1995.

19. Lemmon, E. W., M. O. McLinden, and M. L. Huber, *NIST Standard Reference Database* 23: Reference Fluid Thermodynamic and Transport Properties-REFPROP, Version 7.0 National Institute of Standards and Technology, Standard Reference Data Program, Gaithersburg, 2002.

20. Bolz, R. E., and G. L. Tuve, Eds., *CRC Handbook of Tables for Applied Engineering Science,* 2nd ed., CRC Press, Boca Raton, FL, 1979.

21. Liley, P. E., private communication, School of Mechanical Engineering, Purdue University, West Lafayette, IN, May 1984.

22. Liley, P. E., Steam Tables in SI Units, private communication, School of Mechanical Engineering, Purdue University, West Lafayette, IN, March 1984.

23. *Liquid-Metals Handbook,* 2nd ed., The Atomic Energy Commission, Department of the Navy, Washington, DC, 1952.

24. Perry, J. H., Ed., *Chemical Engineer's Handbook,* 4th ed., McGraw-Hill, New York, 1963.

25. Geankoplis, C. J., *Mass Transport Phenomena,* Holt, Rinehart & Winston, New York, 1972.

26. Barrer, R. M., *Diffusion in and Through Solids,* Macmillan, New York, 1941.

27. Spalding, D. B., *Convective Mass Transfer,* McGraw-Hill, New York, 1963.

28. Gubareff, G. G., J. E. Janssen, and R. H. Torborg, *Thermal Radiation Properties Survey,* Minneapolis-Honeywell Regulator Company, Minneapolis, MN, 1960.

29. Kreith, F., and J. F. Kreider, *Principles of Solar Energy,* Hemisphere Publishing, New York, 1978.

APPENDIX **B**

Mathematical Relations and Functions

B.1 *Hyperbolic Functions*[1]

x	$\sinh x$	$\cosh x$	$\tanh x$	x	$\sinh x$	$\cosh x$	$\tanh x$
0.00	0.0000	1.0000	0.00000	2.00	3.6269	3.7622	0.96403
0.10	0.1002	1.0050	0.09967	2.10	4.0219	4.1443	0.97045
0.20	0.2013	1.0201	0.19738	2.20	4.4571	4.5679	0.97574
0.30	0.3045	1.0453	0.29131	2.30	4.9370	5.0372	0.98010
0.40	0.4108	1.0811	0.37995	2.40	5.4662	5.5569	0.98367
0.50	0.5211	1.1276	0.46212	2.50	6.0502	6.1323	0.98661
0.60	0.6367	1.1855	0.53705	2.60	6.6947	6.7690	0.98903
0.70	0.7586	1.2552	0.60437	2.70	7.4063	7.4735	0.99101
0.80	0.8881	1.3374	0.66404	2.80	8.1919	8.2527	0.99263
0.90	1.0265	1.4331	0.71630	2.90	9.0596	9.1146	0.99396
1.00	1.1752	1.5431	0.76159	3.00	10.018	10.068	0.99505
1.10	1.3356	1.6685	0.80050	3.50	16.543	16.573	0.99818
1.20	1.5095	1.8107	0.83365	4.00	27.290	27.308	0.99933
1.30	1.6984	1.9709	0.86172	4.50	45.003	45.014	0.99975
1.40	1.9043	2.1509	0.88535	5.00	74.203	74.210	0.99991
1.50	2.1293	2.3524	0.90515	6.00	201.71	201.72	0.99999
1.60	2.3756	2.5775	0.92167	7.00	548.32	548.32	1.0000
1.70	2.6456	2.8283	0.93541	8.00	1490.5	1490.5	1.0000
1.80	2.9422	3.1075	0.94681	9.00	4051.5	4051.5	1.0000
1.90	3.2682	3.4177	0.95624	10.000	11013	11013	1.0000

[1]The hyperbolic functions are defined as

$$\sinh x = \tfrac{1}{2}(e^{x} - e^{-x}) \qquad \cosh x = \tfrac{1}{2}(e^{x} + e^{-x}) \qquad \tanh x = \frac{e^{x} - e^{-x}}{e^{x} + e^{-x}} = \frac{\sinh x}{\cosh x}$$

The derivatives of the hyperbolic functions of the variable u are given as

$$\frac{d}{dx}(\sinh u) = (\cosh u)\frac{du}{dx} \qquad \frac{d}{dx}(\cosh u) = (\sinh u)\frac{du}{dx} \qquad \frac{d}{dx}(\tanh u) = \left(\frac{1}{\cosh^{2} u}\right)\frac{du}{dx}$$

B.2 *Gaussian Error Function*[1]

w	erf w	w	erf w	w	erf w
0.00	0.00000	0.36	0.38933	1.04	0.85865
0.02	0.02256	0.38	0.40901	1.08	0.87333
0.04	0.04511	0.40	0.42839	1.12	0.88679
0.06	0.06762	0.44	0.46622	1.16	0.89910
0.08	0.09008	0.48	0.50275	1.20	0.91031
0.10	0.11246	0.52	0.53790	1.30	0.93401
0.12	0.13476	0.56	0.57162	1.40	0.95228
0.14	0.15695	0.60	0.60386	1.50	0.96611
0.16	0.17901	0.64	0.63459	1.60	0.97635
0.18	0.20094	0.68	0.66378	1.70	0.98379
0.20	0.22270	0.72	0.69143	1.80	0.98909
0.22	0.24430	0.76	0.71754	1.90	0.99279
0.24	0.26570	0.80	0.74210	2.00	0.99532
0.26	0.28690	0.84	0.76514	2.20	0.99814
0.28	0.30788	0.88	0.78669	2.40	0.99931
0.30	0.32863	0.92	0.80677	2.60	0.99976
0.32	0.34913	0.96	0.82542	2.80	0.99992
0.34	0.36936	1.00	0.84270	3.00	0.99998

[1]The Gaussian error function is defined as

$$\text{erf } w = \frac{2}{\sqrt{\pi}} \int_0^w e^{-v^2} dv$$

The complementary error function is defined as

$$\text{erfc } w \equiv 1 - \text{erf } w$$

B.3   *The First Four Roots of the Transcendental Equation, $\xi_n \tan \xi_n = Bi$, for Transient Conduction in a Plane Wall*

$Bi = \dfrac{hL}{k}$	ξ_1	ξ_2	ξ_3	ξ_4
0	0	3.1416	6.2832	9.4248
0.001	0.0316	3.1419	6.2833	9.4249
0.002	0.0447	3.1422	6.2835	9.4250
0.004	0.0632	3.1429	6.2838	9.4252
0.006	0.0774	3.1435	6.2841	9.4254
0.008	0.0893	3.1441	6.2845	9.4256
0.01	0.0998	3.1448	6.2848	9.4258
0.02	0.1410	3.1479	6.2864	9.4269
0.04	0.1987	3.1543	6.2895	9.4290
0.06	0.2425	3.1606	6.2927	9.4311
0.08	0.2791	3.1668	6.2959	9.4333
0.1	0.3111	3.1731	6.2991	9.4354
0.2	0.4328	3.2039	6.3148	9.4459
0.3	0.5218	3.2341	6.3305	9.4565
0.4	0.5932	3.2636	6.3461	9.4670
0.5	0.6533	3.2923	6.3616	9.4775
0.6	0.7051	3.3204	6.3770	9.4879
0.7	0.7506	3.3477	6.3923	9.4983
0.8	0.7910	3.3744	6.4074	9.5087
0.9	0.8274	3.4003	6.4224	9.5190
1.0	0.8603	3.4256	6.4373	9.5293
1.5	0.9882	3.5422	6.5097	9.5801
2.0	1.0769	3.6436	6.5783	9.6296
3.0	1.1925	3.8088	6.7040	9.7240
4.0	1.2646	3.9352	6.8140	9.8119
5.0	1.3138	4.0336	6.9096	9.8928
6.0	1.3496	4.1116	6.9924	9.9667
7.0	1.3766	4.1746	7.0640	10.0339
8.0	1.3978	4.2264	7.1263	10.0949
9.0	1.4149	4.2694	7.1806	10.1502
10.0	1.4289	4.3058	7.2281	10.2003
15.0	1.4729	4.4255	7.3959	10.3898
20.0	1.4961	4.4915	7.4954	10.5117
30.0	1.5202	4.5615	7.6057	10.6543
40.0	1.5325	4.5979	7.6647	10.7334
50.0	1.5400	4.6202	7.7012	10.7832
60.0	1.5451	4.6353	7.7259	10.8172
80.0	1.5514	4.6543	7.7573	10.8606
100.0	1.5552	4.6658	7.7764	10.8871
∞	1.5708	4.7124	7.8540	10.9956

B.4 *Bessel Functions of the First Kind*

x	$J_0(x)$	$J_1(x)$
0.0	1.0000	0.0000
0.1	0.9975	0.0499
0.2	0.9900	0.0995
0.3	0.9776	0.1483
0.4	0.9604	0.1960
0.5	0.9385	0.2423
0.6	0.9120	0.2867
0.7	0.8812	0.3290
0.8	0.8463	0.3688
0.9	0.8075	0.4059
1.0	0.7652	0.4400
1.1	0.7196	0.4709
1.2	0.6711	0.4983
1.3	0.6201	0.5220
1.4	0.5669	0.5419
1.5	0.5118	0.5579
1.6	0.4554	0.5699
1.7	0.3980	0.5778
1.8	0.3400	0.5815
1.9	0.2818	0.5812
2.0	0.2239	0.5767
2.1	0.1666	0.5683
2.2	0.1104	0.5560
2.3	0.0555	0.5399
2.4	0.0025	0.5202

B.5 *Modified Bessel Functions[1] of the First and Second Kinds*

x	$e^{-x}I_0(x)$	$e^{-x}I_1(x)$	$e^{x}K_0(x)$	$e^{x}K_1(x)$
0.0	1.0000	0.0000	∞	∞
0.2	0.8269	0.0823	2.1407	5.8334
0.4	0.6974	0.1368	1.6627	3.2587
0.6	0.5993	0.1722	1.4167	2.3739
0.8	0.5241	0.1945	1.2582	1.9179
1.0	0.4657	0.2079	1.1445	1.6361
1.2	0.4198	0.2152	1.0575	1.4429
1.4	0.3831	0.2185	0.9881	1.3010
1.6	0.3533	0.2190	0.9309	1.1919
1.8	0.3289	0.2177	0.8828	1.1048
2.0	0.3085	0.2153	0.8416	1.0335
2.2	0.2913	0.2121	0.8056	0.9738
2.4	0.2766	0.2085	0.7740	0.9229
2.6	0.2639	0.2046	0.7459	0.8790
2.8	0.2528	0.2007	0.7206	0.8405
3.0	0.2430	0.1968	0.6978	0.8066
3.2	0.2343	0.1930	0.6770	0.7763
3.4	0.2264	0.1892	0.6579	0.7491
3.6	0.2193	0.1856	0.6404	0.7245
3.8	0.2129	0.1821	0.6243	0.7021
4.0	0.2070	0.1787	0.6093	0.6816
4.2	0.2016	0.1755	0.5953	0.6627
4.4	0.1966	0.1724	0.5823	0.6453
4.6	0.1919	0.1695	0.5701	0.6292
4.8	0.1876	0.1667	0.5586	0.6142
5.0	0.1835	0.1640	0.5478	0.6003
5.2	0.1797	0.1614	0.5376	0.5872
5.4	0.1762	0.1589	0.5279	0.5749
5.6	0.1728	0.1565	0.5188	0.5633
5.8	0.1696	0.1542	0.5101	0.5525
6.0	0.1666	0.1520	0.5019	0.5422
6.4	0.1611	0.1479	0.4865	0.5232
6.8	0.1561	0.1441	0.4724	0.5060
7.2	0.1515	0.1405	0.4595	0.4905
7.6	0.1473	0.1372	0.4476	0.4762
8.0	0.1434	0.1341	0.4366	0.4631
8.4	0.1398	0.1312	0.4264	0.4511
8.8	0.1365	0.1285	0.4168	0.4399
9.2	0.1334	0.1260	0.4079	0.4295
9.6	0.1305	0.1235	0.3995	0.4198
10.0	0.1278	0.1213	0.3916	0.4108

[1]$I_{n+1}(x) = I_{n-1}(x) - (2n/x)I_n(x)$

APPENDIX C

Thermal Conditions Associated with Uniform Energy Generation in One-Dimensional, Steady-State Systems

In Section 3.5 the problem of conduction with thermal energy generation is considered for one-dimensional, steady-state conditions. The form of the heat equation differs, according to whether the system is a plane wall, a cylindrical shell, or a spherical shell (Figure C.1). In each case, there are several options for the boundary condition at each surface, and hence a greater number of possibilities for specific forms of the temperature distribution and heat rate (or heat flux).

An alternative to solving the heat equation for each possible combination of boundary conditions involves obtaining a solution by prescribing *boundary conditions of the first kind,* Equation 2.31, at both surfaces and then applying an energy balance to each surface at which the temperature is unknown. For the geometries of Figure C.1, with uniform temperatures $T_{s,1}$ and $T_{s,2}$ prescribed at each surface, solutions to appropriate forms of the heat equation are readily obtained and are summarized in Table C.1. The temperature distributions may be used with Fourier's law to obtain corresponding distributions for the heat flux and heat rate. If $T_{s,1}$ and $T_{s,2}$ are both known for a particular problem, the expressions of Table C.1 provide all that is needed to completely determine related thermal conditions. If $T_{s,1}$ and/or $T_{s,2}$ are not known, the results may still be used with surface energy balances to determine the desired thermal conditions.

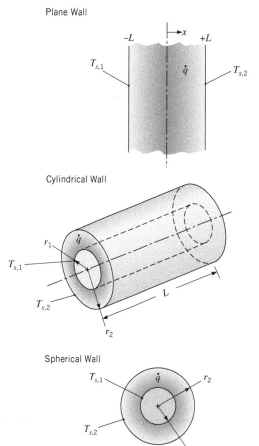

FIGURE C.1   One-dimensional conduction systems with uniform thermal energy generation: a plane wall with asymmetric surface conditions, a cylindrical shell, and a spherical shell.

TABLE C.1 One-Dimensional, Steady-State Solutions to the Heat Equation for Plane, Cylindrical, and Spherical Walls with Uniform Generation and Asymmetrical Surface Conditions

Temperature Distribution

Plane Wall

$$T(x) = \frac{\dot{q}L^2}{2k}\left(1 - \frac{x^2}{L^2}\right) + \frac{T_{s,2} - T_{s,1}}{2}\frac{x}{L} + \frac{T_{s,1} + T_{s,2}}{2} \tag{C.1}$$

Cylindrical Wall

$$T(r) = T_{s,2} + \frac{\dot{q}r_2^2}{4k}\left(1 - \frac{r^2}{r_2^2}\right) - \left[\frac{\dot{q}r_2^2}{4k}\left(1 - \frac{r_1^2}{r_2^2}\right) + (T_{s,2} - T_{s,1})\right]\frac{\ln(r_2/r)}{\ln(r_2/r_1)} \tag{C.2}$$

Spherical Wall

$$T(r) = T_{s,2} + \frac{\dot{q}r_2^2}{6k}\left(1 - \frac{r^2}{r_2^2}\right) - \left[\frac{\dot{q}r_2^2}{6k}\left(1 - \frac{r_1^2}{r_2^2}\right) + (T_{s,2} - T_{s,1})\right]\frac{(1/r) - (1/r_2)}{(1/r_1) - (1/r_2)} \tag{C.3}$$

Heat Flux

Plane Wall

$$q''(x) = \dot{q}x - \frac{k}{2L}(T_{s,2} - T_{s,1}) \tag{C.4}$$

Cylindrical Wall

$$q''(r) = \frac{\dot{q}r}{2} - \frac{k\left[\frac{\dot{q}r_2^2}{4k}\left(1 - \frac{r_1^2}{r_2^2}\right) + (T_{s,2} - T_{s,1})\right]}{r\ln(r_2/r_1)} \tag{C.5}$$

Spherical Wall

$$q''(r) = \frac{\dot{q}r}{3} - \frac{k\left[\frac{\dot{q}r_2^2}{6k}\left(1 - \frac{r_1^2}{r_2^2}\right) + (T_{s,2} - T_{s,1})\right]}{r^2[(1/r_1) - (1/r_2)]} \tag{C.6}$$

Heat Rate

Plane Wall

$$q(x) = \left[\dot{q}x - \frac{k}{2L}(T_{s,2} - T_{s,1})\right]A_x \tag{C.7}$$

Cylindrical Wall

$$q(r) = \dot{q}\pi L r^2 - \frac{2\pi L k}{\ln(r_2/r_1)} \cdot \left[\frac{\dot{q}r_2^2}{4k}\left(1 - \frac{r_1^2}{r_2^2}\right) + (T_{s,2} - T_{s,1})\right] \tag{C.8}$$

Spherical Wall

$$q(r) = \frac{\dot{q}4\pi r^3}{3} - \frac{4\pi k\left[\frac{\dot{q}r_2^2}{6k}\left(1 - \frac{r_1^2}{r_2^2}\right) + (T_{s,2} - T_{s,1})\right]}{(1/r_1) - (1/r_2)} \tag{C.9}$$

Alternative surface conditions could involve specification of a uniform surface heat flux (*boundary condition of the second kind,* Equation 2.32 or 2.33) or a convection condition (*boundary condition of the third kind,* Equation 2.34). In each case, the surface temperature would not be known but could be determined by applying a surface energy balance. The forms that such balances may take are summarized in Table C.2. Note that, to accommodate situations for which a surface of interest may adjoin a composite wall in which there is no generation, the boundary condition of the third kind has been applied by using the overall heat transfer coefficient U in lieu of the convection coefficient h.

TABLE C.2   Alternative Surface Conditions and Energy Balances for
One-Dimensional, Steady-State Solutions to the Heat Equation for
Plane, Cylindrical, and Spherical Walls with Uniform Generation

Plane Wall

Uniform Surface Heat Flux

$$x = -L: \qquad q''_{s,1} = -\dot{q}L - \frac{k}{2L}(T_{s,2} - T_{s,1}) \tag{C.10}$$

$$x = +L: \qquad q''_{s,2} = \dot{q}L - \frac{k}{2L}(T_{s,2} - T_{s,1}) \tag{C.11}$$

Prescribed Transport Coefficient and Ambient Temperature

$$x = -L: \qquad U_1(T_{\infty,1} - T_{s,1}) = -\dot{q}L - \frac{k}{2L}(T_{s,2} - T_{s,1}) \tag{C.12}$$

$$x = +L: \qquad U_2(T_{s,2} - T_{\infty,2}) = \dot{q}L - \frac{k}{2L}(T_{s,2} - T_{s,1}) \tag{C.13}$$

Cylindrical Wall

Uniform Surface Heat Flux

$$r = r_1: \qquad q''_{s,1} = \frac{\dot{q}r_1}{2} - \frac{k\left[\frac{\dot{q}r_2^2}{4k}\left(1 - \frac{r_1^2}{r_2^2}\right) + (T_{s,2} - T_{s,1})\right]}{r_1 \ln(r_2/r_1)} \tag{C.14}$$

$$r = r_2: \qquad q''_{s,2} = \frac{\dot{q}r_2}{2} - \frac{k\left[\frac{\dot{q}r_2^2}{4k}\left(1 - \frac{r_1^2}{r_2^2}\right) + (T_{s,2} - T_{s,1})\right]}{r_2 \ln(r_2/r_1)} \tag{C.15}$$

Prescribed Transport Coefficient and Ambient Temperature

$$r = r_1: \qquad U_1(T_{\infty,1} - T_{s,1}) = \frac{\dot{q}r_1}{2} - \frac{k\left[\frac{\dot{q}r_2^2}{4k}\left(1 - \frac{r_1^2}{r_2^2}\right) + (T_{s,2} - T_{s,1})\right]}{r_1 \ln(r_2/r_1)} \tag{C.16}$$

$$r = r_2: \qquad U_2(T_{s,2} - T_{\infty,2}) = \frac{\dot{q}r_2}{2} - \frac{k\left[\frac{\dot{q}r_2^2}{4k}\left(1 - \frac{r_1^2}{r_2^2}\right) + (T_{s,2} - T_{s,1})\right]}{r_2 \ln(r_2/r_1)} \tag{C.17}$$

Spherical Wall

Uniform Surface Heat Flux

$$r = r_1: \qquad q''_{s,1} = \frac{\dot{q}r_1}{3} - \frac{k\left[\frac{\dot{q}r_2^2}{6k}\left(1 - \frac{r_1^2}{r_2^2}\right) + (T_{s,2} - T_{s,1})\right]}{r_1^2[(1/r_1) - (1/r_2)]} \tag{C.18}$$

$$r = r_2: \qquad q''_{s,2} = \frac{\dot{q}r_2}{3} - \frac{k\left[\frac{\dot{q}r_2^2}{6k}\left(1 - \frac{r_1^2}{r_2^2}\right) + (T_{s,2} - T_{s,1})\right]}{r_2^2[(1/r_1) - (1/r_2)]} \tag{C.19}$$

TABLE C.2   *Continued*

Prescribed Transport Coefficient and Ambient Temperature

$$r = r_1: \qquad U_1(T_{\infty,1} - T_{s,1}) = \frac{\dot{q}r_1}{3} - \frac{k\left[\dfrac{\dot{q}r_2^2}{6k}\left(1 - \dfrac{r_1^2}{r_2^2}\right) + (T_{s,2} - T_{s,1})\right]}{r_1^2[(1/r_1) - (1/r_2)]} \tag{C.20}$$

$$r = r_2: \qquad U_2(T_{s,2} - T_{\infty,2}) = \frac{\dot{q}r_2}{3} - \frac{k\left[\dfrac{\dot{q}r_2^2}{6k}\left(1 - \dfrac{r_1^2}{r_2^2}\right) + (T_{s,2} - T_{s,1})\right]}{r_2^2[(1/r_1) - (1/r_2)]} \tag{C.21}$$

As an example, consider a plane wall for which a uniform (known) surface temperature $T_{s,1}$ is prescribed at $x = -L$ and a uniform heat flux $q''_{s,2}$ is prescribed at $x = +L$. Equation C.11 may be used to evaluate $T_{s,2}$, and Equations C.1, C.4, and C.7 may then be used to determine the temperature, heat flux, and heat rate distributions, respectively.

Special cases of the foregoing configurations involve a plane wall with one adiabatic surface, a solid cylinder (a circular rod), and a sphere (Figure C.2). Subject to the requirements that $dT/dx|_{x=0} = 0$ and $dT/dr|_{r=0} = 0$, the corresponding forms of the heat equation may be solved to obtain Equations C.22 through C.24 of Table C.3. The solutions are based on prescribing

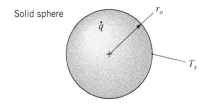

FIGURE C.2 One-dimensional conduction systems with uniform thermal energy generation: a plane wall with one adiabatic surface, a cylindrical rod, and a sphere.

TABLE C.3   One-Dimensional, Steady-State Solutions to the Heat
Equation for Uniform Generation in a Plane Wall with One
Adiabatic Surface, a Solid Cylinder, and a Solid Sphere

Temperature Distribution

Plane Wall

$$T(x) = \frac{\dot{q}L^2}{2k}\left(1 - \frac{x^2}{L^2}\right) + T_s \qquad\qquad (C.22)$$

Circular Rod

$$T(r) = \frac{\dot{q}r_o^2}{4k}\left(1 - \frac{r^2}{r_o^2}\right) + T_s \qquad\qquad (C.23)$$

Sphere

$$T(r) = \frac{\dot{q}r_o^2}{6k}\left(1 - \frac{r^2}{r_o^2}\right) + T_s \qquad\qquad (C.24)$$

Heat Flux

Plane Wall $\qquad q''(x) = \dot{q}x \qquad\qquad (C.25)$

Circular Rod $\qquad q''(r) = \dfrac{\dot{q}r}{2} \qquad\qquad (C.26)$

Sphere $\qquad q''(r) = \dfrac{\dot{q}r}{3} \qquad\qquad (C.27)$

Heat Rate

Plane Wall $\qquad q(x) = \dot{q}x A_x \qquad\qquad (C.28)$

Circular Rod $\qquad q(r) = \dot{q}\pi L r^2 \qquad\qquad (C.29)$

Sphere $\qquad q(r) = \dfrac{\dot{q}4\pi r^3}{3} \qquad\qquad (C.30)$

a uniform temperature T_s at $x = L$ and $r = r_o$. Using Fourier's law with the temperature distributions, the heat flux (Equations C.25 through C.27) and heat rate (Equations C.28 through C.30) distributions may also be obtained. If T_s is not known, it may be determined by applying a surface energy balance, appropriate forms of which are summarized in Table C.4.

TABLE C.4   Alternative Surface Conditions and
Energy Balances for One-Dimensional, Steady-
State Solutions to the Heat Equation for Uniform
Generation in a Plane Wall with One Adiabatic
Surface, a Solid Cylinder, and a Solid Sphere

Prescribed Transport Coefficient and Ambient Temperature
Plane Wall

$x = L$: $\dot{q}L = U(T_s - T_\infty) \qquad\qquad (C.31)$

Circular Rod

$r = r_o$: $\dfrac{\dot{q}r_o}{2} = U(T_s - T_\infty) \qquad\qquad (C.32)$

Sphere

$r = r_o$: $\dfrac{\dot{q}r_o}{3} = U(T_s - T_\infty) \qquad\qquad (C.33)$

APPENDIX D

The Gauss–Seidel Method

The Gauss–Seidel method is an example of an iterative approach for solving systems of linear algebraic equations, such as that represented by Equation 4.47, reproduced below.

$$
\begin{aligned}
a_{11}T_1 + a_{12}T_2 + a_{13}T_3 + \cdots + a_{1N}T_N &= C_1 \\
a_{21}T_1 + a_{22}T_2 + a_{23}T_3 + \cdots + a_{2N}T_N &= C_2 \\
\vdots \qquad \vdots \qquad \vdots \qquad \vdots \qquad \vdots \qquad \vdots& \\
a_{N1}T_1 + a_{N2}T_2 + a_{N3}T_3 + \cdots + a_{NN}T_N &= C_N
\end{aligned}
\tag{4.47}
$$

For small numbers of equations, Gauss–Seidel iteration can be performed by hand. Application of the Gauss–Seidel method to the system of equations represented by Equation 4.47 is facilitated by the following procedure.

1. To whatever extent possible, the equations should be reordered to provide diagonal elements whose magnitudes are larger than those of other elements in the same row. That is, it is desirable to sequence the equations such that $|a_{11}| > |a_{12}|$, $|a_{13}|, \ldots, |a_{1N}|$; $|a_{22}| > |a_{21}|$, $|a_{23}|, \ldots, |a_{2N}|$; and so on.

2. After reordering, each of the N equations should be written in explicit form for the temperature associated with its diagonal element. Each temperature in the solution vector would then be of the form

$$
T_i^{(k)} = \frac{C_i}{a_{ii}} - \sum_{j=1}^{i-1} \frac{a_{ij}}{a_{ii}} T_j^{(k)} - \sum_{j=i+1}^{N} \frac{a_{ij}}{a_{ii}} T_j^{(k-1)}
\tag{D.1}
$$

where $i = 1, 2, \ldots, N$. The superscript k refers to the level of the iteration.

3. An initial ($k = 0$) value is assumed for each temperature T_i. Subsequent computations may be reduced by selecting values based on rational estimates of the actual solution.

4. Setting $k = 1$ in Equation D.1, values of $T_i^{(1)}$ are then calculated by substituting assumed (second summation, $k - 1 = 0$) or new (first summation, $k = 1$) values of T_j into the right-hand side. This step is the first ($k = 1$) iteration.

5. Using Equation D.1, the iteration procedure is continued by calculating new values of $T_i^{(k)}$ from the $T_j^{(k)}$ values of the current iteration, where $1 \le j \le i - 1$, and the $T_j^{(k-1)}$ values of the previous iteration, where $i + 1 \le j \le N$.

6. The iteration is terminated when a prescribed *convergence criterion* is satisfied. The criterion may be expressed as

$$
\left| T_i^{(k)} - T_i^{(k-1)} \right| \le \varepsilon
\tag{D.2}
$$

where ε represents an error in the temperature that is considered to be acceptable.

   If step 1 can be accomplished for each equation, the resulting system is said to be *diagonally dominant,* and the rate of convergence is maximized (the number of required iterations is minimized). However, convergence may also be achieved in many situations for which diagonal dominance cannot be obtained, although the rate of convergence is slowed. The manner in which new values of T_i are computed (steps 4 and 5) should also be noted. Because the T_i for a particular iteration are calculated sequentially, each value can be computed by using the *most recent estimates* of the other T_i. This feature is implicit in Equation D.1, where the value of each unknown is updated as soon as possible, that is, for $1 \le j \le i - 1$.

———————————

 An example problem that utilizes the Gauss–Seidel method is included in Section 4S.2.

APPENDIX **E**

The Convection Transfer Equations

In Chapter 2 we considered a stationary substance in which heat is transferred by conduction and developed means for determining the temperature distribution within the substance. We did so by applying *conservation of energy* to a differential control volume (Figure 2.11) and deriving a differential equation that was termed the *heat equation*. For a prescribed geometry and boundary conditions, the equation may be solved to determine the corresponding temperature distribution.

If the substance is not stationary, conditions become more complex. For example, if conservation of energy is applied to a stationary differential control volume in a moving fluid, the effects of fluid motion (*advection*) on energy transfer across the surfaces of the control volume must be considered, along with those of conduction. The resulting differential equation, which provides the basis for predicting the temperature distribution, now requires knowledge of the velocity equations derived by applying *conservation of mass* and *Newton's second law of motion* to a differential control volume.

In this appendix we consider conditions involving flow of a *viscous fluid* in which there is concurrent *heat* and *mass transfer*. We restrict our attention to the *steady, two-dimensional flow* of an *incompressible fluid* with *constant properties* in the x- and y-directions of a Cartesian coordinate system, and present the differential equations that may be used to predict velocity, temperature, and species concentration fields within the fluid. These equations can be derived by applying Newton's second law of motion and conservation of mass, energy, and species to a differential control volume in the fluid.

E.1 Conservation of Mass

One conservation law that is pertinent to the flow of a viscous fluid is that matter can be neither created nor destroyed. For steady flow, this law requires that *the net rate at which mass enters a control volume* (inflow − outflow) *must equal zero*. Applying this law to a differential control volume in the flow yields

$$\frac{\partial u}{\partial x} + \frac{\partial v}{\partial y} = 0 \tag{E.1}$$

where u and v are the x- and y-components of the *mass average velocity*.

Equation E.1, the *continuity equation*, is a general expression of the *overall mass conservation requirement*, and it must be satisfied at every point in the fluid. The equation applies for a single species fluid, as well as for mixtures in which species diffusion and chemical reactions may be occurring, provided that the fluid can be approximated as *incompressible*, that is, constant density.

E.2 Newton's Second Law of Motion

The second fundamental law that is pertinent to the flow of a viscous fluid is *Newton's second law of motion*. For a differential control volume in the fluid, under steady conditions, this requirement states that *the sum of all forces acting on the control volume must equal the net rate at which momentum leaves the control volume* (outflow − inflow).

These equations are derived in Section 6S.1.

Two kinds of forces may act on the fluid: *body forces*, which are proportional to the volume, and *surface forces*, which are proportional to area. Gravitational, centrifugal, magnetic, and/or electric fields may contribute to the total body force, and we designate the *x*- and *y*-components of this force per unit volume of fluid as X and Y, respectively. The surface forces are due to the fluid static pressure as well as to *viscous stresses*.

Applying Newton's second law of motion (in the *x*- and *y*-directions) to a differential control volume in the fluid, accounting for body and surface forces, yields

$$\rho\left(u\frac{\partial u}{\partial x} + v\frac{\partial u}{\partial y}\right) = -\frac{\partial p}{\partial x} + \mu\left(\frac{\partial^2 u}{\partial x^2} + \frac{\partial^2 u}{\partial y^2}\right) + X \tag{E.2}$$

$$\rho\left(u\frac{\partial v}{\partial x} + v\frac{\partial v}{\partial y}\right) = -\frac{\partial p}{\partial y} + \mu\left(\frac{\partial^2 v}{\partial x^2} + \frac{\partial^2 v}{\partial y^2}\right) + Y \tag{E.3}$$

where p is the pressure and μ is the fluid viscosity.

We should not lose sight of the physics represented by Equations E.2 and E.3. The two terms on the left-hand side of each equation represent the *net* rate of momentum flow from the control volume. The terms on the right-hand side, taken in order, account for the net pressure force, the net viscous forces, and the body force. These equations must be satisfied at each point in the fluid, and with Equation E.1 they may be solved for the velocity field.

E.3 *Conservation of Energy*

As mentioned at the beginning of this Appendix, in Chapter 2 we considered a stationary substance in which heat is transferred by conduction and applied conservation of energy to a differential control volume (Figure 2.11) to derive the heat equation. When conservation of energy is applied to a differential control volume *in a moving fluid* under steady conditions, it expresses that the net rate at which energy enters the control volume, plus the rate at which heat is added, minus the rate at which work is done by the fluid in the control volume, is equal to zero. After much manipulation, the result can be rewritten as a *thermal energy equation*. For steady, two-dimensional flow of an incompressible fluid with constant properties, the resulting differential equation is

$$\rho c_p\left(u\frac{\partial T}{\partial x} + v\frac{\partial T}{\partial y}\right) = k\left(\frac{\partial^2 T}{\partial x^2} + \frac{\partial^2 T}{\partial y^2}\right) + \mu\Phi + \dot{q} \tag{E.4}$$

where T is the temperature, c_p is the specific heat at constant pressure, k is the thermal conductivity, $\dot{q}$ is the volumetric rate of thermal energy generation, and $\mu\Phi$, the *viscous dissipation*, is defined as

$$\mu\Phi \equiv \mu\left\{\left(\frac{\partial u}{\partial y} + \frac{\partial v}{\partial x}\right)^2 + 2\left[\left(\frac{\partial u}{\partial x}\right)^2 + \left(\frac{\partial v}{\partial y}\right)^2\right]\right\} \tag{E.5}$$

The same form of the thermal energy equation, Equation E.4, also applies to an ideal gas with negligible pressure variation.

In Equation E.4, the terms on the left-hand side account for the net rate at which thermal energy leaves the control volume due to bulk fluid motion (advection), while the terms on the right-hand side account for net inflow of energy due to conduction, viscous dissipation, and generation. Viscous dissipation represents the net rate at which mechanical work is irreversibly converted to thermal energy due to viscous effects in the fluid. The generation term characterizes conversion from other forms of energy (such as chemical, electrical, electromagnetic, or nuclear) to thermal energy.

E.4 *Conservation of Species*

If the viscous fluid consists of a binary mixture in which there are species concentration gradients, there will be *relative* transport of the species, and *species conservation* must be satisfied at each point in the fluid. For steady flow, this law requires that *the net rate at which species* A *enters a control volume* (inflow − outflow) *plus the rate at which species* A *is produced in the control volume (by chemical reactions) must equal zero.* Applying this law to a differential control volume in the flow yields the following differential equation, which has been expressed on a molar basis:

$$u\frac{\partial C_A}{\partial x} + v\frac{\partial C_A}{\partial y} = D_{AB}\left(\frac{\partial^2 C_A}{\partial x^2} + \frac{\partial^2 C_A}{\partial y^2}\right) + \dot{N}_A \tag{E.6}$$

where C_A is the molar concentration of species A, D_{AB} is the binary diffusion coefficient, and $\dot{N}_A$ is the molar rate of production of species A per unit volume. Again, this equation has been derived assuming steady, two-dimensional flow of an incompressible fluid with constant properties. Terms on the left-hand side account for net transport of species A due to bulk fluid motion (advection), while terms on the right-hand side account for net inflow due to diffusion and production due to chemical reactions.

An example problem involving the solution of the convection transfer equations is included in Section 6S.1.

Boundary Layer Equations for Turbulent Flow

It has been noted in Section 6.3 that turbulent flow is inherently *unsteady*. This behavior is shown in Figure F.1, where the variation of an arbitrary flow property P is plotted as a function of time at some location in a turbulent boundary layer. The property P could be a velocity component, the fluid temperature, or a species concentration, and at any instant it may be represented as the sum of a *time-mean* value $\bar{P}$ and a fluctuating component P'. The average is taken over a time that is large compared with the period of a typical fluctuation, and if $\bar{P}$ is independent of time, the time-mean flow is said to be *steady*.

Since engineers are typically concerned with the time-mean properties, $\bar{P}$, the difficulty of solving the time-dependent governing equations is often eliminated by averaging the equations over time. For steady (in the mean), incompressible, constant property, boundary layer flow with negligible viscous dissipation, using well-established time-averaging procedures [1], the following forms of the continuity, x-momentum, energy, and species conservation equations may be obtained:

$$\frac{\partial \bar{u}}{\partial x} + \frac{\partial \bar{v}}{\partial y} = 0 \tag{F.1}$$

$$\bar{u}\frac{\partial \bar{u}}{\partial x} + \bar{v}\frac{\partial \bar{u}}{\partial y} = -\frac{1}{\rho}\frac{d\bar{p}_\infty}{dx} + \frac{1}{\rho}\frac{\partial}{\partial y}\left(\mu\frac{\partial \bar{u}}{\partial y} - \rho\overline{u'v'}\right) \tag{F.2}$$

$$\bar{u}\frac{\partial \bar{T}}{\partial x} + \bar{v}\frac{\partial \bar{T}}{\partial y} = \frac{1}{\rho c_p}\frac{\partial}{\partial y}\left(k\frac{\partial \bar{T}}{\partial y} - \rho c_p\overline{v'T'}\right) \tag{F.3}$$

$$\bar{u}\frac{\partial \bar{C}_A}{\partial x} + \bar{v}\frac{\partial \bar{C}_A}{\partial y} = \frac{\partial}{\partial y}\left(D_{AB}\frac{\partial \bar{C}_A}{\partial y} - \overline{v'C_A'}\right) \tag{F.4}$$

The equations are like those for the laminar boundary layer, Equations 6.27 through 6.30 (after neglecting viscous dissipation), except for the presence of additional terms of the form $\overline{a'b'}$. These terms account for the effect of the turbulent fluctuations on momentum, energy, and species transport.

On the basis of the foregoing results, it is customary to speak of a *total* shear stress and *total* heat and species fluxes, which are defined as

$$\tau_{\text{tot}} = \left(\mu\frac{\partial \bar{u}}{\partial y} - \rho\overline{u'v'}\right) \tag{F.5}$$

$$q''_{\text{tot}} = -\left(k\frac{\partial \bar{T}}{\partial y} - \rho c_p\overline{v'T'}\right) \tag{F.6}$$

$$N''_{A,\text{tot}} = -\left(D_{AB}\frac{\partial \bar{C}_A}{\partial y} - \overline{v'C_A'}\right) \tag{F.7}$$

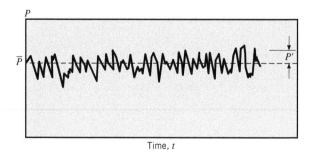

FIGURE F.1   Property variation with time at some point in a turbulent boundary layer.

and consist of contributions due to molecular diffusion and turbulent mixing. From the form of these equations we see how momentum, energy, and species transfer rates are enhanced by the existence of turbulence. The term $-\rho\overline{u'v'}$ appearing in Equation F.5 represents the momentum flux due to the turbulent fluctuations, and it is often termed the *Reynolds stress*. The terms $\rho c_p\overline{v'T'}$ and $\overline{v'C_A'}$ in Equations F.6 and F.7, respectively, represent the heat and species fluxes due to the turbulent fluctuations. Unfortunately, these new terms introduced by the time-averaging process are additional unknowns, so that the number of unknowns exceeds the number of equations. Resolving this problem is the subject of the field of *turbulence modeling* [2].

References

1. Kays, W. M., M. E. Crawford, and B. Weigand, *Convective Heat and Mass Transfer*, 4th ed., McGraw-Hill Higher Education, Boston, 2005.

2. Wilcox, D. C., *Turbulence Modeling for CFD*, 2nd ed., DCW Industries, La Cañada, 1998.

APPENDIX **G**

An Integral Laminar Boundary Layer Solution for Parallel Flow over a Flat Plate

An alternative approach to solving the boundary layer equations involves the use of an approximate *integral* method. The approach was originally proposed by von Kárman [1] in 1921 and first applied by Pohlhausen [2]. It is without the mathematical complications inherent in the *exact* (*similarity*) method of Section 7.2.1; yet it can be used to obtain reasonably accurate results for the key boundary layer parameters (δ, δ_t, δ_c, C_f, h, and h_m). Although the method has been used with some success for a variety of flow conditions, we restrict our attention to parallel flow over a flat plate, subject to the same restrictions enumerated in Section 7.2.1, that is, *incompressible laminar flow* with *constant fluid properties* and *negligible viscous dissipation.*

To use the method, the boundary layer equations, Equations 7.4 through 7.7, must be cast in integral form. These forms are obtained by integrating the equations in the *y*-direction across the boundary layer. For example, integrating Equation 7.4, we obtain

$$\int_0^\delta \frac{\partial u}{\partial x} dy + \int_0^\delta \frac{\partial v}{\partial y} dy = 0 \tag{G.1}$$

or, since $v = 0$ at $y = 0$,

$$v(y = \delta) = -\int_0^\delta \frac{\partial u}{\partial x} dy \tag{G.2}$$

Similarly, from Equation 7.5, we obtain

$$\int_0^\delta u \frac{\partial u}{\partial x} dy + \int_0^\delta v \frac{\partial u}{\partial y} dy = \nu \int_0^\delta \frac{\partial}{\partial y} \left(\frac{\partial u}{\partial y} \right) dy$$

or, integrating the second term on the left-hand side by parts,

$$\int_0^\delta u \frac{\partial u}{\partial x} dy + uv \Big|_0^\delta - \int_0^\delta u \frac{\partial v}{\partial y} dy = \nu \frac{\partial u}{\partial y} \Big|_0^\delta$$

Substituting from Equations 7.4 and G.2, we obtain

$$\int_0^\delta u \frac{\partial u}{\partial x} dy - u_\infty \int_0^\delta \frac{\partial u}{\partial x} dy + \int_0^\delta u \frac{\partial u}{\partial x} dy = -\nu \frac{\partial u}{\partial y} \Big|_{y=0}$$

or

$$u_\infty \int_0^\delta \frac{\partial u}{\partial x} dy - \int_0^\delta 2u \frac{\partial u}{\partial x} dy = \nu \frac{\partial u}{\partial y} \Big|_{y=0}$$

Therefore

$$\int_0^\delta \frac{\partial}{\partial x} (u_\infty \cdot u - u \cdot u) dy = \nu \frac{\partial u}{\partial y} \Big|_{y=0}$$

Rearranging, we then obtain

$$\frac{d}{dx} \left[\int_0^\delta (u_\infty - u) u \, dy \right] = \nu \frac{\partial u}{\partial y} \Big|_{y=0} \tag{G.3}$$

Equation G.3 is the integral form of the boundary layer momentum equation. In a similar fashion, the following integral forms of the boundary layer energy and species continuity equations may be obtained:

$$\frac{d}{dx} \left[\int_0^{\delta_t} (T_\infty - T) u \, dy \right] = \alpha \frac{\partial T}{\partial y} \Big|_{y=0} \tag{G.4}$$

$$\frac{d}{dx}\left[\int_0^{\delta_c} (\rho_{A,\infty} - \rho_A)u\, dy\right] = D_{AB}\left.\frac{\partial \rho_A}{\partial y}\right|_{y=0} \tag{G.5}$$

Equations G.3 through G.5 satisfy the *x*-momentum, the energy, and the species conservation requirements in an *integral* (or *average*) fashion over the entire boundary layer. In contrast, the original conservation equations, (7.5) through (7.7), satisfy the conservation requirements *locally,* that is, at each point in the boundary layer.

The integral equations can be used to obtain *approximate* boundary layer solutions. The procedure involves first *assuming* reasonable functional forms for the unknowns u, T, and ρ_A in terms of the corresponding (*unknown*) boundary layer thicknesses. The assumed forms must satisfy appropriate boundary conditions. Substituting these forms into the integral equations, expressions for the boundary layer thicknesses may be determined and the assumed functional forms may then be completely specified. Although this method is approximate, it frequently leads to accurate results for the surface parameters.

Consider the hydrodynamic boundary layer, for which appropriate boundary conditions are

$$u(y = 0) = \left.\frac{\partial u}{\partial y}\right|_{y=\delta} = 0 \quad \text{and} \quad u(y = \delta) = u_\infty$$

From Equation 7.5 it also follows that, since $u = v = 0$ at $y = 0$,

$$\left.\frac{\partial^2 u}{\partial y^2}\right|_{y=0} = 0$$

With the foregoing conditions, we could, for example, approximate the velocity profile as a third-degree polynomial of the form

$$\frac{u}{u_\infty} = a_1 + a_2\left(\frac{y}{\delta}\right) + a_3\left(\frac{y}{\delta}\right)^2 + a_4\left(\frac{y}{\delta}\right)^3$$

and apply the conditions to determine the coefficients a_1 through a_4. It is easily verified that $a_1 = a_3 = 0$, $a_2 = \frac{3}{2}$ and $a_4 = -\frac{1}{2}$, in which case

$$\frac{u}{u_\infty} = \frac{3}{2}\frac{y}{\delta} - \frac{1}{2}\left(\frac{y}{\delta}\right)^3 \tag{G.6}$$

The velocity profile is then specified in terms of the unknown boundary layer thickness δ. This unknown may be determined by substituting Equation G.6 into G.3 and integrating over y to obtain

$$\frac{d}{dx}\left(\frac{39}{280}u_\infty^2\delta\right) = \frac{3}{2}\frac{\nu u_\infty}{\delta}$$

Separating variables and integrating over x, we obtain

$$\frac{\delta^2}{2} = \frac{140}{13}\frac{\nu x}{u_\infty} + \text{constant}$$

However, since $\delta = 0$ at the leading edge of the plate ($x = 0$), the integration constant must be zero and

$$\delta = 4.64\left(\frac{\nu x}{u_\infty}\right)^{1/2} = \frac{4.64x}{Re_x^{1/2}} \tag{G.7}$$

Substituting Equation G.7 into Equation G.6 and evaluating $\tau_s = \mu(\partial u/\partial y)_s$, we also obtain

$$C_{f,x} = \frac{\tau_s}{\rho u_\infty^2/2} = \frac{0.646}{Re_x^{1/2}} \tag{G.8}$$

Despite the approximate nature of the foregoing procedure, Equations G.7 and G.8 compare quite well with results obtained from the exact solution, Equations 7.19 and 7.20.

In a similar fashion one could assume a temperature profile of the form

$$T^* = \frac{T - T_s}{T_\infty - T_s} = b_1 + b_2\left(\frac{y}{\delta_t}\right) + b_3\left(\frac{y}{\delta_t}\right)^2 + b_4\left(\frac{y}{\delta_t}\right)^3$$

and determine the coefficients from the conditions

$$T^*(y = 0) = \left.\frac{\partial T^*}{\partial y}\right|_{y=\delta_t} = 0$$

$$T^*(y = \delta_t) = 1$$

as well as

$$\left.\frac{\partial^2 T^*}{\partial y^2}\right|_{y=0} = 0$$

which is inferred from the energy equation (7.6). We then obtain

$$T^* = \frac{3}{2}\frac{y}{\delta_t} - \frac{1}{2}\left(\frac{y}{\delta_t}\right)^3 \tag{G.9}$$

Substituting Equations G.6 and G.9 into Equation G.4, we obtain, after some manipulation and assuming $Pr \gtrsim 1$,

$$\frac{\delta_t}{\delta} = \frac{Pr^{-1/3}}{1.026} \tag{G.10}$$

This result is in good agreement with that obtained from the exact solution, Equation 7.24. Moreover, the heat transfer coefficient may then be computed from

$$h = \frac{-k\,\partial T/\partial y\,\big|_{y=0}}{T_s - T_\infty} = \frac{3}{2}\frac{k}{\delta_t}$$

Substituting from Equations G.7 and G.10, we obtain

$$Nu_x = \frac{hx}{k} = 0.332\,Re_x^{1/2}\,Pr^{1/3} \tag{G.11}$$

This result agrees precisely with that obtained from the exact solution, Equation 7.23. Using the same procedures, analogous results may be obtained for the concentration boundary layer.

References

1. von Kárman, T., *Z. Angew. Math. Mech.*, **1**, 232, 1921.   2. Pohlhausen, K., *Z. Angew. Math. Mech.*, **1**, 252, 1921.

Index